IB	IIB	IIIA	IVA	VA	VIA	VIIA	VIIIA
							2 He
		5 B	6 C	7 N	8 O	9 F	10 Ne
		13 Al	14 Si	15 P	16 S	17 Cl	18 Ar
29 Cu	30 Zn	31 Ga	32 Ge	33 As	34 Se	35 Br	36 Kr
47 Ag	48 Cd	49 In	50 Sn	51 Sb	52 Te	53 I	54 Xe
79 Au	80 Hg	81 Tl	82 Pb	83 Bi	84 Po	85 At	86 Rn

65 Tb	66 Dy	67 Ho	68 Er	69 Tm	70 Yb	71 Lu
97 Bk	98 Cf	99 Es	100 Fm	101 Md	102 No	103 Lr

CHEMISTRY

Edward L. Kemp
UNIVERSITY OF COLORADO, BOULDER

CHEMISTRY

PAINTER HOPKINS PUBLISHERS
Sausalito California

The symbol on the cover is the ancient Gnostic sign representing the four elements—fire, air, water, and earth.

Printed in the United States of America. Published simultaneously in Canada by Clarke, Irwin & Company, Limited, Toronto and Vancouver.

For information contact Painter Hopkins Publishers, P.O. Box 1829, Sausalito, California 94965.

Library of Congress Cataloging in Publication Data

King, Edward Louis, 1920–
Chemistry.
Includes bibliographical references.
1. Chemistry. I. Title.
QD31.2.K54 540 78-25576

ISBN 0-525-07926-2

First Edition 1 2 3 4 5 6 7 8 9 10

Preface

This book is intended for students in a beginning, year course in chemistry. I assume that students using this book will have had an introduction to chemistry in high school and that they will have developed some facility with algebraic operations. Thus the text should appeal particularly to students with interests in science or engineering.

Because the students with this background have at least a minimal knowledge of atomic and molecular structure, I have deferred these topics until the third part of the book, Chapters 8–11. The first two parts, Chapters 1–7, provide background for introductory laboratory work, which can deal with simple stoichiometric relationships in chemical reactions, calorimetric study of enthalpy changes in reactions, studies involving gases (e.g., the molecular weight from vapor density), and studies of the properties of solutions, including phase diagrams and molecular-weight determinations from freezing point depression. The last half of the book, Chapters 12–22, provides the basis for much laboratory work, which can deal with acid–base reactions, ionic equilibria (including pH measurements on solutions of weak acids and bases, solubility studies, and distribution equilibria), oxidation–reduction chemistry, reaction kinetics, the chemistry of selected elements, and studies of light absorption by chemical systems. The basis for experiments involving organic compounds is provided in Chapters 11 and 20.

I have developed quantitative aspects of each subject to the point where students can master the practical calculations involved. In the presentation of these calculations, units are included in each step. Although this procedure leads to cumbersome expressions in some instances, it may reveal mistakes in faulty calculations. I have used units also with the values of equilibrium constants; this procedure has many advantages. The results of calculations come automatically with the correct units. There is no uncertainty in what units were employed in the derivation of the equilibrium-constant value if these units are included. If the equilibrium constants are given with units, students will not be tempted to compare values that shouldn't be compared, for example, K_s ($PbSO_4$) and K_s [$Mg(OH)_2$], which at 25°C are 1.59×10^{-8} mol^2 L^{-2} and 1.26×10^{-11} mol^3 L^{-3}. The question of standard states is a different matter, and a calculation of ΔG^0, as derived in Chapter 12, involves a factor that eliminates the units before the logarithm is taken.

In Chapters 1, 3, and 4, SI units are defined, and the relationships of these to units conventionally used in recent American chemical literature are given. Both SI and non-SI units are used in the body of the text; a practicing engineer, scientist, or medical doctor will encounter both types of units in the last decades of this century.

Although there is no chapter entitled "Thermodynamics," I have stressed the relationship between chemistry and energy from Chapter 3 onward. Thermodynamics never is learned in one exposure, and students respond favorably to the gradual introduction of this important subject. The presentation of thermodynamics, started in Chapter 3, continues in Chapter 4 (with material on the expansion of gases), Chapter 5 (with heat capacities of solids), Chapter 6 (with changes of state), Chapter 7 (with material on solutions), Chapter 9 (with many quantitatively drawn diagrams for Born–Haber cycles), Chapter 10 (with bond energies), Chapter 12 (with chemical equilibrium in the gas phase), Chapters 13 and 14 (with chemical equilibrium in reactions in solution), and Chapter 16 (with thermodynamics of galvanic cells). Potential diagrams, of the type introduced by W. M. Latimer, are introduced in Chapter 16, and then are used in Chapters 17, 18, and 19 to correlate oxidation–reduction chemistry of many elements.

The valence-shell electron-pair repulsion (VSEPR) model is so successful in predicting the geometry of covalently bonded molecules and ions that I have emphasized this approach. All students of beginning chemistry can understand this simple model, but many find the concept of orbital hybridization difficult. I also have stressed the structural relationships of molecules and ions that are isoelectronic, an emphasis which can be correlated nicely with the VSEPR model.

The chapter on kinetics and reaction mechanisms comes early enough (Chapter 15) to allow discussion of this aspect of chemistry in the chapters dealing with selected metals and nonmetals. Although the chapter on kinetics makes some use of calculus notation, a student who has not had calculus can handle most topics in this chapter, and can learn from the chapter the principles that relate rate laws to reaction mechanisms.

Advances in chemical research have an impact on courses in beginning chemistry. The areas of recent advances that, because of their conceptual simplicity, I have found particularly tractable in a beginning course are:

1. gas-phase proton association, the presentation of which enriches the discussion of acidity and basicity in solution.
2. gas-phase negative ion stability, which has expanded greatly the list of elements for which accurate values of electron affinity are known. This has been tied in nicely with our discussion of the recently prepared compounds of alkali-metal anions.
3. photochemistry, which has developed greatly in recent decades; this allows discussion of the simple chemistry of excited atomic and molecular species.

Although individual teachers may choose to deviate from the order of presentation given in the book, I believe there is merit in an arrangement that has the more mathematical topics spread throughout the entire year of the course. Thus the nonmathematical chapter introducing organic chemistry, Chapter 11, can be an end to a first-semester course that has been two thirds quantitative in nature. This chapter emphasizes structural principles that are introduced in Chapter 9. (Additional organic chemistry along with biochemistry is presented in Chapter 20.) In the framework of this schedule of topics, the second semester's work also is approximately two thirds quantitative. Many worked-out examples are used throughout the text to illustrate the solution of problems, and problems and questions are given at the end of each chapter. Answers to representative problems are given in Appendix 4.

The periodic tables presented on the front endpaper of the book and in Chapter 8 label the A and B subgroups as they are given in most American textbooks, that is, the A subgroups consist of the elements in which *s* and *p* orbitals are being filled. However, this designation of the A and B columns of the periodic table is not identical with that recommended in 1970 by IUPAC in *Nomenclature of Inorganic Chemistry, Definitive Rules,* in which the IIIA–VIIA columns are headed by Sc, Ti, V, Cr, and Mn. Many chemistry teachers do not use the A and B designations, so this inconsistency need not be troublesome.

I have presented at the end of each chapter a brief biographical sketch of the scientists whose names are mentioned in the chapter. Some of you may wish to pursue the historical aspects in more detail, but others may simply note the century in which a particular scientist worked.

The development of this book has been influenced by teachers, colleagues, and students during the past four decades. The freshman chemistry course of today is very different from the course I took or the first courses I taught as a teaching assistant and instructor. The actual writing of this book (which has not extended over the entire four decades, even though it seems like a project with its inception in the indistinct past) has benefited from reviews by many teachers. My students at the University of Colorado have used successive partial versions in duplicated form, and their response has helped in refining the presentation. I am grateful to both of these groups. But I am particularly grateful to Dr. James L. Hall, editor, who has given invaluable help in preparing the final revision of the manuscript and in its transformation into a published book.

January 1979

EDWARD L. KING
Chemistry Department
University of Colorado
Boulder, Colorado 80309

Contents

CHEMISTRY

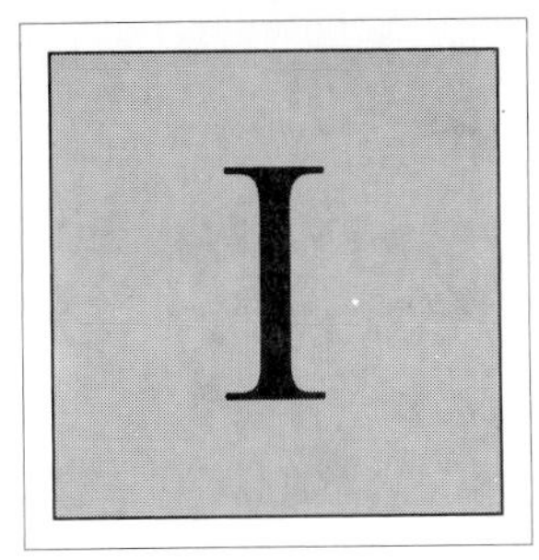

INTRODUCTION TO CHEMISTRY

Modern chemistry is a vast subject. We are part of a material world, and we are surrounded by a material universe; chemistry is the science of these materials. Chemistry deals with both naturally occurring and synthetic substances and materials, and its range is enormous (Figure I). At one extreme are the complex substances upon which life is based: chemists, working with biologists and geneticists, attempt to understand the molecular basis of living organisms. At another extreme are simple molecules, the detailed understanding of which is the goal of chemists working on the borderline between chemistry and astrophysics (ammonia, NH_3, formic acid, HCOOH, and more than twenty other small molecules have been detected in interstellar space). Simple molecules also interest chemists concerned about the pollution of our atmosphere. The simple molecules carbon monoxide, CO, nitrogen oxides, NO and NO_2, and sulfur dioxide, SO_2, are major contributors to air pollution in our urban centers. Between these extremes of the work on complex molecules and simple molecules, there are many areas of intense research activity.

The solid state is the concern of both chemists and physicists. Their development of semiconductor materials has revolutionized the electronics industry. Structural materials are rarely pure substances, and with metallurgists and engineers, chemists work to understand factors that determine the strength of materials. Geochemistry, on the border between geology and chemistry, is not simply the analysis of minerals. The chemical basis for the

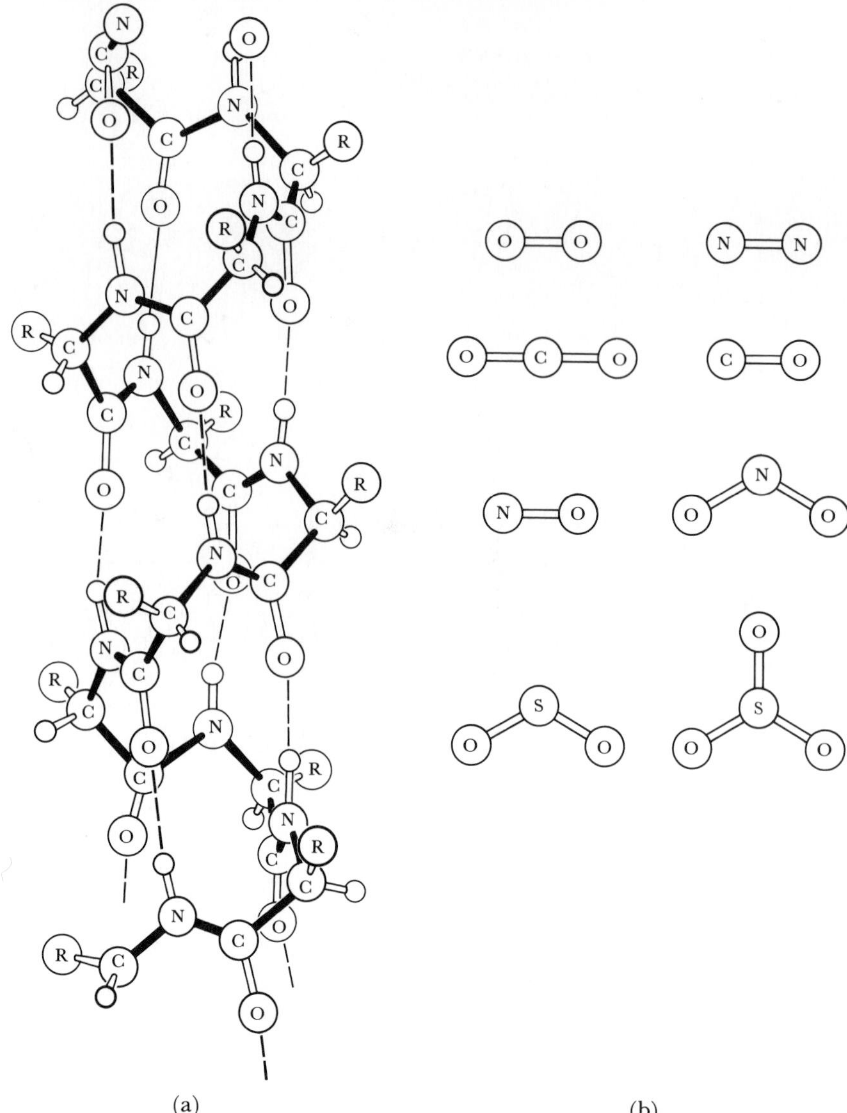

FIGURE I

Extremes of chemical complexity. (a) The helical arrangement of atoms in a protein chain. The protein backbone consists of the sequence of atoms –C–C–N–C–C–N–, as shown with solid lines connecting atoms. Hydrogen atoms not labeled; R designates group of carbon and hydrogen atoms. [Adapted from L. Pauling, *Nature of the Chemical Bond*, Cornell University Press, Ithaca, N.Y., 3rd ed. (1960) p. 500.] (b) Simple molecules in the air we breathe. Oxygen, O_2, nitrogen, N_2, carbon dioxide, CO_2, carbon monoxide, CO, nitric oxide, NO, nitrogen dioxide, NO_2, sulfur dioxide, SO_2, and sulfur trioxide, SO_3. These ball-and-stick molecular models show small atoms to allow clearer representation of atomic arrangement in molecules.

distribution of the elements on Earth and the other planets is one interest of the modern geochemist. The oceans are potential sources of many elements, and geochemists and oceanographers work together to satisfy some of our needs from this vast source.

A major activity of chemists is the preparation of new substances (e.g., synthetic drugs) and materials (e.g., synthetic fibers). New synthetic drugs have helped eliminate some of the sicknesses (including mental illness) which have plagued man for ages. Man-made fibers, the use of which increases each year, are examples of synthetic macromolecules, molecules that are made of a very large number ($>10^3$) of atoms. A significant fraction of all research carried out in the chemical industry is in the area of macromolecular chemistry. Materials that withstand extremely high temperatures are needed in many industries, and their preparation is a goal involving the chemist. The demonstrated existence of some new compounds expands our understanding of chemical bonding: compounds of the noble gas xenon and some compounds of boron fall into this category. The economical production of chemical fertilizers in underdeveloped parts of the world and the desalinization of sea water or brackish water are chemical and engineering problems of the first importance. The field of electrochemistry is an old one, founded in work by MICHAEL FARADAY[1] in the early part of the nineteenth century. As in most fields, so in electrochemistry there are now new areas that attract the attention of chemists. The fuel cell is a battery in which the combustion of a fuel occurs with the direct production of electrical energy. Fuel cells aboard space ships, in which the reaction $2H_2 + O_2 = 2H_2O$ occurs, provide drinking water for the astronauts in addition to electrical energy. Theoretical calculations about chemical systems can be made with varying degrees of rigor; today, computers of greater speed and information storage capacity allow calculations regarding more complicated molecules. Chemists now look beyond understanding of the diatomic hydrogen molecule to such achievable goals as calculation of the structure of liquid water and of the arrangement of atoms in coiled macromolecules.

The approaches used in solving chemical problems are varied. For some problems, rigorous theory helps the planning of new experiments. In others, rigorous theory is not of much help because the systems are too complicated. Progress in many areas of chemistry is based on approximations and intuition. This intuition develops in an individual over a long time, often starting with his earliest exposure to science in grade school. The development of your chemical intuition will continue as you study these chapters.

In the first two chapters, the concepts of elements and compounds will be presented in the framework of the development of atomic theory. The important concept of the mole will be introduced, and its use in a variety of quantitative calculations will be illustrated. Chemical reactions are the processes by which chemical substances are transformed into one another; quantitative calculations regarding chemical reactions are considered in Chapter 2. There are important relationships between chemistry and energy, and this subject is introduced in the third of these introductory chapters. That chapter also presents the first of many applications of the laws of thermodynamics, which we will consider in the course of our study.

---

[1] MICHAEL FARADAY (1791–1867) first studied science while he was an apprentice bookbinder in London. He became associated with Sir Humphrey Davy in the Royal Institution in London in 1813; in 1825 Faraday became director of this laboratory. He later held a special research professorship in the institution. In addition to pioneering the study of electrochemistry, he made notable contributions in the liquefaction of gases and in metallurgy.

1

Elements and Compounds

1–1 Introduction

From the point of view of a chemist, the elements are the ultimate constituents of matter. A material object, of whatever composition, can be converted into its constituent elements by appropriate mechanical, physical, and chemical means. But such ordinary means cannot convert an element to simpler varieties of matter.

Consider a complex heterogeneous material, consisting of several phases; this material can be separated by mechanical means into the individual homogeneous phases. If a single homogeneous phase is a mixture of substances, these substances can be separated from one another by appropriate changes of state. The resulting pure substances, if compounds, can be decomposed to simpler substances, *the elements*. The elements cannot be decomposed further by ordinary chemical means (although transmutation of the elements is possible, and this will be discussed in Chapter 22).

This introductory paragraph has used a number of words that may be unfamiliar. These words and their definitions are:

Matter That which has mass, mass being the property of a material object that accounts for its inertia. A force is necessary to accelerate a mass; NEWTON's[1] second law relates these quantities:

$$\text{force} = \text{mass} \times \text{acceleration}$$

[1] In the course of our study, we refer to the work of many scientists. For most of these scientists, a brief biographical sketch will be given at the end of the chapter in which the scientist is first mentioned.

Heterogeneous A material sample is said to be heterogeneous if it has regions with different properties (e.g., density or color), these regions being separated from one another by physical boundaries.

Phase The sum of all of the regions of a material sample having the same properties is a phase. (Thus a heterogeneous sample consists of more than one phase.)

Homogeneous A material sample is said to be homogeneous if it consists of a single phase.

State The term "state," as used in the paragraph in question, refers to the gaseous, liquid, and solid states. (These terms are familiar ones, and later chapters deal with each.)

Substance Matter with a particular chemical composition that is stable enough to allow its characterization is called a substance.

Compound A substance that can be decomposed by chemical means into simpler substances is a compound.

Element A substance that cannot be decomposed by chemical means into simpler substances is an element.

To clarify those terms further, consider an example: the discharge from an oil well whose flow has been stimulated by injection of water. The discharge from such a well is a heterogeneous mixture of two liquid phases, an aqueous phase and a hydrocarbon phase. These phases have different densities, and they can be separated simply by draining the more dense phase from the bottom of the containing vessel. To simplify this example, we will consider the aqueous phase to be a pure substance, the compound water. This compound can be decomposed by electrolysis (the passage of an electrical current) to give the elements, hydrogen and oxygen. The hydrocarbon phase is not a pure substance but is a liquid solution of many different substances. Simplified, this phase can be considered to be a solution of two compounds, benzene and normal octane. The tendencies for these two substances to vaporize are different, and an appropriate distillation procedure can separate the solution of benzene and octane into the individual compounds, each of which can be decomposed to give its constituent elements, carbon and hydrogen. Figure 1–1 gives a diagram of the separation of the complex heterogeneous mixture being considered into its constituent elements.

In the classification of matter that we have just considered and in the specific example used to illustrate this classification, the constitution of elements and compounds has not been mentioned. From your previous exposure to chemistry or physics in high school, you know something about atoms and molecules. Later in this chapter the basis for the atomic theory will be presented, but for the present no justification will be offered for our accepting the theory that each element is made up of atoms of a particular kind and that each compound is made up of molecules that consist of particular combinations of atoms. Although a detailed discussion of atomic structure will be deferred to Chapter 8 and a corresponding discussion of molecular structure to Chapter 10, we now will review the simplest picture of atoms and molecules that you have learned in your previous study.

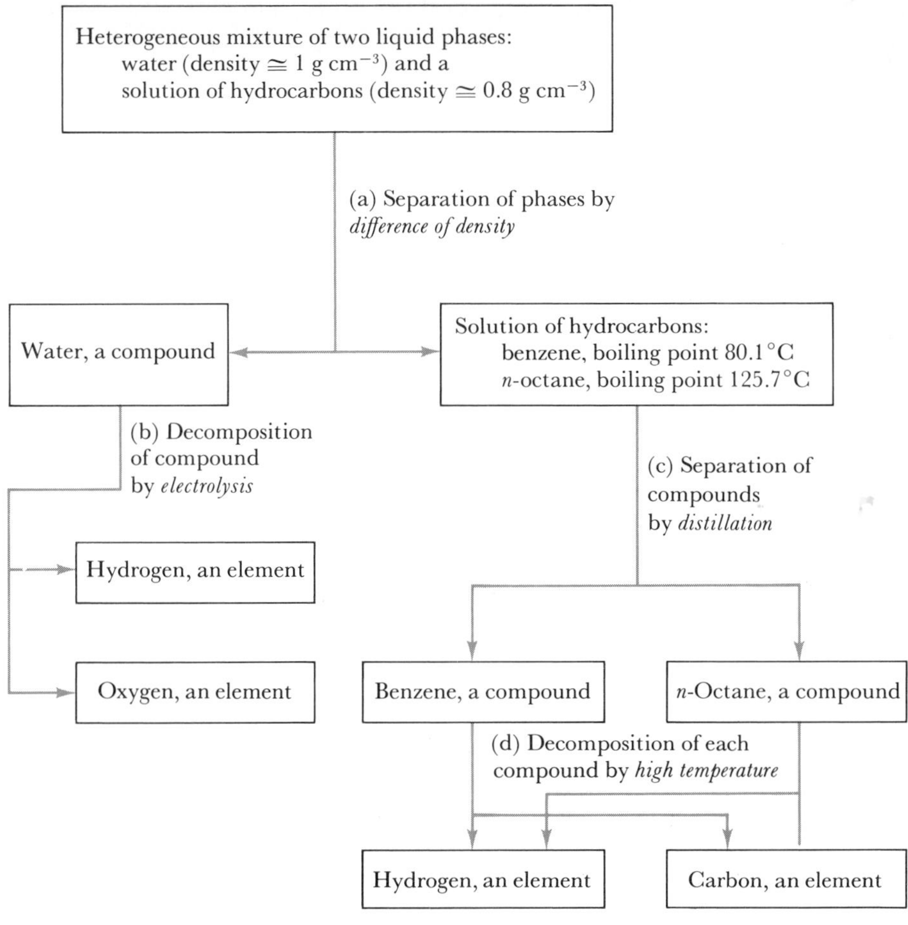

FIGURE 1–1
The separation of a heterogeneous mixture into the constituent elements. Processes (a) and (c) do not involve chemical change; processes (b) and (d) involve chemical change.

Each element is made up of atoms, which consist of small positively charged nuclei surrounded by negatively charged electrons. The nucleus, which has most of the atom's mass, contains two kinds of fundamental particles, protons and neutrons. Protons are positively charged, and neutrons have no charge. The absolute magnitude of the charge of the electron is the same as the charge of the proton. The net charge on a neutral atom is zero; this is the consequence of the number of electrons around the nucleus being equal to the number of protons in the nucleus. The point is illustrated with some examples in Table 1–1. The 106 elements that are now known consist of atoms with 1 to 106 protons in their nuclei; this number, which identifies the element, is called the *atomic number* of the element and is denoted by Z. Table 1–2 lists the atomic numbers, names, and symbols for the 105 elements[2] with $Z = 1$ to $Z = 105$. The symbols are

[2]The names and symbols for elements 104 and 105 given in the table have not been accepted officially. Element 106 has not been named. In 1976, there was an unconfirmed report that Element 107 had been produced. The existence of some superheavy elements ($Z =$ 116, 124, and 127) in nature is being investigated.

TABLE 1–1
The Composition of Atoms of Some Elements

	Nucleus		
	Protons	Neutrons	Electrons
Hydrogen[a]	1	0	1
Carbon[a]	6	6	6
Oxygen[a]	8	8	8
Phosphorus	15	16	15
Chlorine[b]	17	18	17
Chlorine[b]	17	20	17
Manganese	25	30	25
Gold	79	118	79

[a]The composition of atoms given for these elements is that for the predominant naturally occurring isotope. (Isotopes are atoms of a particular element that have different numbers of neutrons in their nuclei.)
[b]Chlorine exists in nature as a mixture of these two isotopes.

used also to represent the chemical composition of molecules of the elements and compounds.

Some elemental substances exist as molecules containing a single atom (e.g., the noble gases, He, Ne, Ar, Kr, Xe, and Rn), and others exist as molecules containing two or more atoms (e.g., H_2, N_2, O_2, O_3, Cl_2, P_4, and S_8). (We see in this tabulation that oxygen can exist in two different molecular forms, ordinary oxygen, O_2, and ozone, O_3.)

Chemical compounds are composed of two or more elements. Atoms of the elements making up a compound are combined in particular proportions; these proportions are given by the chemical formulas, some examples of which are:

water	H_2O	*n*-octane	C_8H_{18}
ammonia	NH_3	sulfuric acid	H_2SO_4 or $(HO)_2SO_2$
methane	CH_4	tetraethyllead	PbC_8H_{20} or $Pb(C_2H_5)_4$
common salt	NaCl	limestone	$CaCO_3$
benzene	C_6H_6		

Each of these chemical formulas involves the symbols of the constituent elements. The numerical subscripts, one being understood if there is no subscript, give the relative numbers of atoms of each element in the compound. The two versions of the chemical formulas given for sulfuric acid and for tetraethyllead show how a line formula can convey some structural information. The formulas H_2SO_4, which is the common representation of sulfuric acid, and PbC_8H_{20}, which is not the common representation of tetraethyllead, convey information about the composition of these compounds but do not convey information about structures of the molecules. However, formulas $(HO)_2SO_2$ and $Pb(C_2H_5)_4$ convey the information

TABLE 1–2
The Elements

Atomic number, Z	Name	Symbol	Atomic number, Z	Name	Symbol
1	Hydrogen	H	54	Xenon	Xe
2	Helium	He	55	Cesium	Cs
3	Lithium	Li	56	Barium	Ba
4	Beryllium	Be	57	Lanthanum	La
5	Boron	B	58	Cerium	Ce
6	Carbon	C	59	Praseodymium	Pr
7	Nitrogen	N	60	Neodymium	Nd
8	Oxygen	O	61	Promethium	Pm
9	Fluorine	F	62	Samarium	Sm
10	Neon	Ne	63	Europium	Eu
11	Sodium	Na	64	Gadolinium	Gd
12	Magnesium	Mg	65	Terbium	Tb
13	Aluminum	Al	66	Dysprosium	Dy
14	Silicon	Si	67	Holmium	Ho
15	Phosphorus	P	68	Erbium	Er
16	Sulfur	S	69	Thulium	Tm
17	Chlorine	Cl	70	Ytterbium	Yb
18	Argon	Ar	71	Lutetium	Lu
19	Potassium	K	72	Hafnium	Hf
20	Calcium	Ca	73	Tantalum	Ta
21	Scandium	Sc	74	Tungsten	W
22	Titanium	Ti	75	Rhenium	Re
23	Vanadium	V	76	Osmium	Os
24	Chromium	Cr	77	Iridium	Ir
25	Manganese	Mn	78	Platinum	Pt
26	Iron	Fe	79	Gold	Au
27	Cobalt	Co	80	Mercury	Hg
28	Nickel	Ni	81	Thallium	Tl
29	Copper	Cu	82	Lead	Pb
30	Zinc	Zn	83	Bismuth	Bi
31	Gallium	Ga	84	Polonium	Po
32	Germanium	Ge	85	Astatine	At
33	Arsenic	As	86	Radon	Rn
34	Selenium	Se	87	Francium	Fr
35	Bromine	Br	88	Radium	Ra
36	Krypton	Kr	89	Actinium	Ac
37	Rubidium	Rb	90	Thorium	Th
38	Strontium	Sr	91	Protactinium	Pa
39	Yttrium	Y	92	Uranium	U
40	Zirconium	Zr	93	Neptunium	Np
41	Niobium	Nb	94	Plutonium	Pu
42	Molybdenum	Mo	95	Americium	Am
43	Technetium	Tc	96	Curium	Cm
44	Ruthenium	Ru	97	Berkelium	Bk
45	Rhodium	Rh	98	Californium	Cf
46	Palladium	Pd	99	Einsteinium	Es
47	Silver	Ag	100	Fermium	Fm
48	Cadmium	Cd	101	Mendelevium	Md
49	Indium	In	102	Nobelium	No
50	Tin	Sn	103	Lawrencium	Lr
51	Antimony	Sb	104	Rutherfordium	Rf
52	Tellurium	Te	105	Hahnium	Ha
53	Iodine	I			

that the hydrogen in sulfuric acid is bonded to oxygen, not sulfur, and that the hydrocarbon part of tetraethyllead is contained in four ethyl groups (C_2H_5). In general, you should write chemical formulas to convey as much structural information as is relevant to the problem at hand. If a line formula that conveys structural information is no more cumbersome than one that conveys only composition, you should use the more informative formula.

In the chemical formula for ammonia, NH_3, for instance, the subscripts disclose that this compound has three times as many hydrogen atoms as nitrogen atoms. In the chemical formula for common salt, NaCl, the understood subscripts disclose that this compound has equal numbers of sodium atoms and chlorine atoms. The two substances, ammonia and common salt, are very different in nature (Figure 1–2). Ammonia is a gas under ordinary conditions, and the chemical formula NH_3 represents not only the composition (the relative number of atoms of each kind) but also the actual number of atoms of each kind in the ammonia molecule. The concept of molecule is well defined when it is applied to a gaseous substance; a molecule in a gas is the combination of atoms that makes up the discrete material entity. In ammonia gas these entities are NH_3, not N_2H_6 or N_3H_9 (each of which has the same relative numbers of nitrogen atoms and hydrogen atoms as NH_3). However, common salt, sodium chloride, is a crystalline solid under ordinary conditions. Even the tiniest crystal of this substance contains an enormous number of atoms. The chemical formula NaCl denotes the correct relative number of atoms of each of the constituent elements, but it does not denote a molecule of salt under ordinary conditions. (At high temperatures sodium chloride can exist at an appreciable

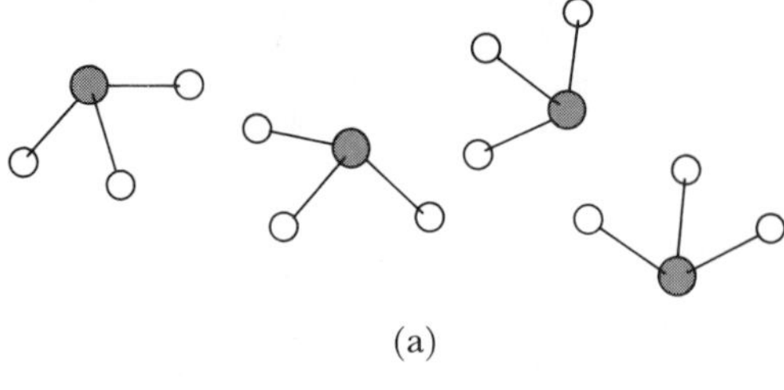

(a)

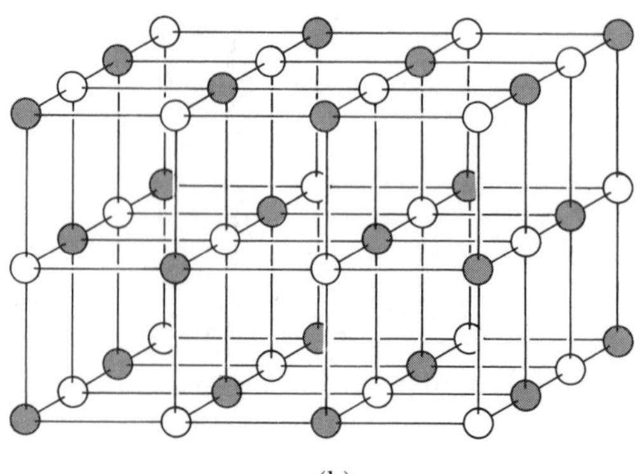

(b)

FIGURE 1–2
Ammonia and sodium chloride. (a) Discrete molecules of ammonia, each containing 1 nitrogen atom and 3 hydrogen atoms. (b) A portion of a crystal of sodium chloride, containing equal numbers of sodium atoms and chlorine atoms (each is present in ionized form, Na^+ and Cl^-).

concentration in the gas phase as discrete NaCl molecules, but this point is not relevant in the present context.) If a molecule is considered to be the grouping of atoms that gives the substance its characteristic properties, a molecule of sodium chloride under ordinary conditions is a crystal of sodium chloride with the composition $(NaCl)_n$, in which n is an extremely large number. (A typical crystal of salt from a salt shaker, a cube with an edge 0.2 mm long, contains 2×10^{17} sodium atoms and 2×10^{17} chlorine atoms, each in an ionized from, Na^+ and Cl^-.)

There are a number of ways in which the elements can be classified. An element can be classified according to the state in which it exists under ordinary conditions (i.e., a temperature of about 25°C and a pressure of about one atmosphere). There are eleven elements that are gases under such conditions; they are argon, chlorine, fluorine, helium, hydrogen, krypton, neon, nitrogen, oxygen, radon, and xenon. There are two elements that are liquids under ordinary conditions; they are bromine and mercury. (Gallium, with a melting point of 29.8°C, is close to being a liquid at 25°C.) The remaining elements are solids.

A more useful classification of the elements is as metals or nonmetals. Metals in the uncombined state are characterised by:

1. a lustrous appearance,
2. a high conductance for heat and electricity, and
3. a high malleability and ductility; that is, most metals can be hammered into thin sheets and drawn into fine wires.

Most metals have relatively high melting points and boiling points. None of the eleven elements that are gaseous under ordinary conditions are metals. Mercury, a liquid under ordinary conditions, is lustrous and a good conductor; it is a metal. The elements classified as metals also have certain types of chemical properties. We will learn these properties as the subject is developed further, rather than use them as criteria for classification. The elements which do not exhibit the properties of metals are classified as non-metals. Because the traits distinguishing metals and nonmetals are qualitative in nature, some elements do not fall unambiguously in either category. These elements, which include silicon, germanium, arsenic, antimony, and tellurium, are called *metalloids* or *semimetals*. One objective of the study of chemistry is to find an explanation for this classification of the elements as metals, metalloids, and nonmetals. The explanation will follow from the electronic structure of atoms of the elements.

Among the elements, carbon holds a special position. It is a constituent of most compounds of which living matter (plants and animals) is composed. The enormous variety in organic chemistry (the chemistry of carbon compounds) is attributable to the electronic structure of the carbon atom.

Elements other than carbon also are found in living matter. At the present time, twenty-four elements have been shown to be essential to mammalian life. These elements are listed in Table 1–3. Highly refined analytical methods as well as carefully controlled growth conditions for

TABLE 1–3
Elements Essential to Life[a]

Atomic number	Element	Composition of human body (atom %)[b]
1	H	63
8	O	25.5
6	C	9.5
7	N	1.4
20	Ca	0.31
15	P	0.22
19	K	0.06
16	S	0.05
11	Na	0.03
17	Cl	0.03
12	Mg	0.01

Atomic number	Element[c]	Atomic number	Element[c]
9	F	29	Cu
14	Si	30	Zn
23	V	34	Se
24	Cr	42	Mo
25	Mn	50	Sn
26	Fe	53	I
27	Co		

[a]E. Frieden, "The Chemical Elements of Life," *Sci. Am.*, **227** (July 1972), p. 52.
[b]Atom % means the percentage of the total number of atoms.
[c]The occurrence of these elements in the human body is less than 0.01 atom %.

test animals are needed to show that a particular element is essential to life. We can expect that the list of essential elements will be expanded in the future as analytical methods become more sensitive.

1–2 Some Quantitative Properties of Matter

Because chemistry is a quantitative science, we must have definite unambiguous units to measure the properties of matter. For the present, the quantities mass (and weight), volume, density, and temperature will be considered.

The concepts of mass and weight are related, and in the past you may have equated them. Televised views of weightless astronauts hovering in space should have dispelled confusion about the difference between mass and weight. The mass of an object is determined by the actual amount of matter it contains. The weight of an object on the Earth's surface is determined by the attractive force between the object and the Earth. On other

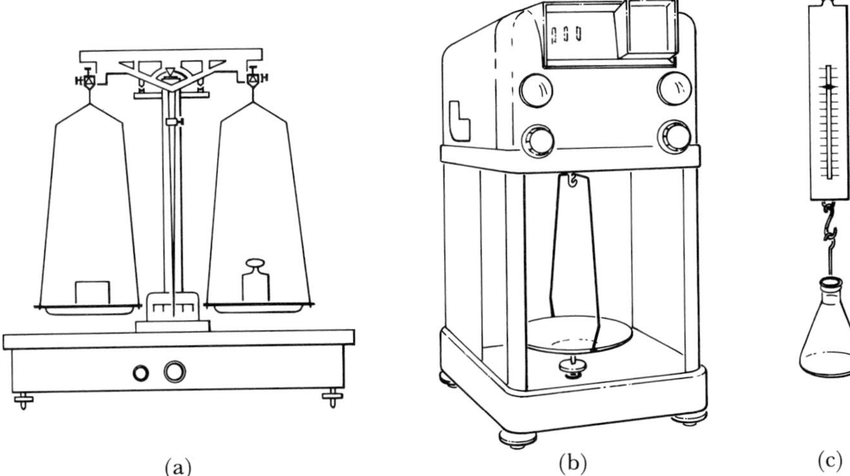

FIGURE 1–3
The measurement of mass and weight. (a) The double-pan balance, which compares the mass of an object with the mass of standard "weights." The modern automatic single-pan balance (b) is based upon the same principle. (c) A spring scale, which measures weight.

planets or on the moon the weight of an object is determined by an analogous attractive force. Newton's law of universal gravitation,

$$F = G\frac{Mm}{r^2}$$

in which G is the proportionality constant, M is the mass of the Earth, planet, or moon, m is the mass of the object in question, and r is the distance between the centers of the object and the Earth (or planet or the moon), gives the attractive force, F. The force of attraction also can be expressed in terms of Newton's second law of motion,

$$F = ma$$

in which a is acceleration. For the attraction between a planetary body and an object, this equation is written in terms of the acceleration of gravity, g,

$$F = mg = \text{weight}$$

The weight of an object is this force of attraction. Comparison of the two preceding equations for the force of attraction shows that $g = GM/r^2$. The standard value of g for the Earth is 9.80665 meters per second per second (9.80665 m s^{-2}).[3]

The mass of a sample can be determined by comparison with known masses. This is done with a double-pan balance (Figure 1–3a), on which

[3]The acceleration of gravity varies over the surface of the Earth. The variation is not large, but it is large enough (9.783 m s^{-2} measured at a site in India, and 9.826 m s^{-2} measured at a site in Norway) to be significant in certain types of work.

We will commonly express the dimensions of physical quantities by using the symbols for the units with the appropriate power shown as an exponent to each symbol. This usage is described fully in Appendix 1.

the mass of the object is balanced with a known mass. (If the known masses, the "weights," and the object being weighed have different densities, a bouyancy correction is necessary. The nature of this correction will be described in Chapter 4.) In measurements with a spring balance, Figure 1–3b, the extension of the spring is determined by the weight of the sample, not the mass. (A bouyancy correction is necessary in this situation also.) The dimensions of weight and mass clearly are not the same; the correct dimensions of weight are those of force, mass times acceleration. Nonetheless, it is customary, to speak of weight as though it had the dimensions of mass. A sample with a mass of one kilogram is commonly said to weigh one kilogram, it being understood that this sample is being weighed at a place where the acceleration of gravity is the standard value. The weight of a one kilogram mass is not the same on the surface of the moon, where the acceleration of gravity is 1.613 $m\ s^{-2}$.

Units in which mass (and weight) are commonly given are:

Kilogram (kg) The basic SI[4] unit of mass. It is the mass of a particular piece of platinum–iridium alloy kept in the International Bureau of Weights and Measures at Sevres, France.

Gram (g) $1\ g = 10^{-3}\ kg$

Pound (lb)[5] $1\ lb \equiv 0.45359237\ kg$

Ounce (oz)[5] $1\ oz = \frac{1}{16}\ lb \equiv 0.028349523\ kg$

In most laboratory operations chemists deal with amounts of material that have masses in the range 10^{-3} to 10^3 grams (10^{-6} to 1 kilogram), and these metric units are used to express the masses. However, situations may arise in which you will find it convenient to use common English units, and facility in making conversions between units is necessary.

EXAMPLE: If the price of gold is \$179.45 per ounce, what is the price per gram?

In solving this problem, you multiply the cost in dollars per ounce by unit factors which convert the unit of mass to the gram:

$$\text{cost} = \$179.45\ \text{oz}^{-1}$$

$$= \$179.45\ \cancel{\text{oz}^{-1}} \times \frac{1\ \cancel{\text{oz}}}{0.02835\ \cancel{\text{kg}}} \times \frac{1\ \cancel{\text{kg}}}{1000\ \text{g}}$$

$$= \$6.33\ \text{g}^{-1}$$

Each of the factors (1 oz/0.02835 kg) and (1 kg/1000 g) is a quotient

[4]The International System of Units (designated SI) is a metric system of units proposed in 1960 by the General Conference on Weights and Measures. These units are summarized in Appendix 1.

[5]These non-SI units are now defined in terms of the basic SI unit of mass, the kilogram, as given here.

in which the numerator is equal to the denominator. Such factors are called *unit factors*. Notice that cancellation of units shows that the answer has the appropriate dimensions, cost per gram. Pay attention to the units in all numerical problems!

Quantities related to length are involved in many contexts in chemistry. For instance, the first power of length (l) is involved in the distances between the centers of atoms in a molecule or in a crystal, or in the distances that gas molecules travel between collisions. The second power of length (l^2) arises in the cross-sectional area of an opening in a vessel through which gas molecules escape or the cross-sectional area of the gas molecules, which is a factor in determining the probability that gas molecules will collide with one another. The third power of length (l^3) arises both in the volume of the vessel containing a gas and in the volume of a sample of liquid or solid. Units in which length are commonly given are:

Meter (m) The basic SI unit of length. The meter is defined as the length equal to 1,650,763.73 wavelengths (in vacuum) of the radiation corresponding to a transition between two particular energy levels of atomic krypton. (Some terms in this definition may be unfamiliar.)
Centimeter (cm) $1 \text{ cm} = 10^{-2} \text{ m}$
Millimeter (mm) $1 \text{ mm} = 10^{-3} \text{ m}$
Angstrom unit (Å) $1 \text{ Å} = 10^{-10} \text{ m}$ (this unit was named for A. J. ÅNGSTRÖM)
Inch (in.) $1 \text{ in.} \equiv 0.0254 \text{ m}$ (definition)

Units in which volume are commonly expressed are:

Cubic meter (m^3) Because the meter is the basic unit of length in SI, the cubic meter is the basic unit of volume.
Cubic decimeter (dm^3) $1 \text{ dm}^3 = (0.1 \text{ m})^3 = 10^{-3} \text{ m}^3$
The cubic decimeter is the volume of a cube with a side of length 1 dm (0.1 m). This unit plays a special role in volume measurement since 1 liter, a commonly used non-SI unit is now defined as 1 dm^3.
Liter (L) $1 \text{ L} \equiv 1 \text{ dm}^3$ (definition)
Cubic centimeter (cm^3) $1 \text{ cm}^3 = 10^{-6} \text{ m}^3$
[This unit also is abbreviated cc, and it is also equal to the milliliter (ml), a non-SI unit.]

The mass, m, (or weight) and the volume, V, of a sample of matter are each directly proportional to the amount of material in the sample. The mass and volume are *extensive* properties of the sample. (This is the term used to describe a property whose value depends upon the size of the sample.) The ratio of these quantities, the density, d,

$$d = \frac{m}{V}$$

does not depend upon the amount of material in the sample. A property of matter that does not depend upon the sample size is an *intensive* property of the material. The dimension of density is the dimension of mass divided by the dimension of volume. The commonly used dimension for density is g cm^{-3}, but in SI the appropriate dimension is kg m^{-3}.

EXAMPLE: The density of liquid mercury at 0°C is 13.595 g cm^{-3}. What is this density in kg m^{-3}?

$$d = 13.595\ \cancel{\text{g}}\ \cancel{\text{cm}^{-3}} \times \left(\frac{10^2\ \cancel{\text{cm}}}{1\ \text{m}}\right)^3 \times \frac{1\ \text{kg}}{10^3\ \cancel{\text{g}}}$$

$$= 1.3595 \times 10^4\ \text{kg m}^{-3}$$

EXAMPLE: At 25°C, water has a density of 0.9971 g cm^{-3}. What is the density in pounds per cubic foot?

$$d = 0.9971\ \text{g cm}^{-3} \times \left(\frac{1\ \text{lb}}{453.6\ \text{g}}\right) \times \left(\frac{2.54\ \text{cm}}{1\ \cancel{\text{in}}} \times \frac{12\ \cancel{\text{in}}}{1\ \text{ft}}\right)^3$$

$$= 0.9971\ \cancel{\text{g}}\ \cancel{\text{cm}^{-3}} \times \left(\frac{1\ \text{lb}}{453.6\ \cancel{\text{g}}}\right) \times \frac{2.832 \times 10^4\ \cancel{\text{cm}^3}}{1\ \text{ft}^3}$$

$$= 62.25\ \text{lb ft}^{-3}$$

In each of these two examples, as in the previous example, the unit-factor method was used.

The density of a substance may be used to identify it. "Fool's gold," iron pyrites (FeS_2), with a metallic gold luster, has a density of 5.0 g cm^{-3}, much smaller than the density of elemental gold, 19.3 g cm^{-3}. A measurement of density serves to distinguish these substances.

An extremely important characteristic of matter is its temperature. The concept of temperature is much less easy to visualize than the concepts of mass, length, area, volume, and density, just described. But the term will be used in this introductory chapter, and some preliminary discussion is needed. Temperature is a measure of the hotness of a sample of matter; it determines the direction of heat flow between a sample of matter and its surroundings. Heat flows from a hot object to a less hot object. If two samples of matter at different temperatures are brought into contact, and heat flows from the first to the second, the first sample is said to have a higher temperature than the second. Such qualitative comparisons may be adequate for some purposes, but a quantitative measure of temperature is necessary.

A simple quantitative measure of temperature is provided by the common mercury-sealed-in-glass thermometer. The density of mercury depends upon temperature more strongly than does the density of glass; therefore mercury in a glass tube expands to fill more of the tube when its temperature is raised. Convenient fixed points in calibrating a thermometer of this type are the freezing point of water (the temperature at which pure ice is in equilibrium with liquid water saturated with air at one atmosphere pressure) and the normal boiling point of water (the temperature at which pure liquid water is in equilibrium with water vapor at one atmosphere pressure). In the past, these temperatures were defined as 0 and 100 on the

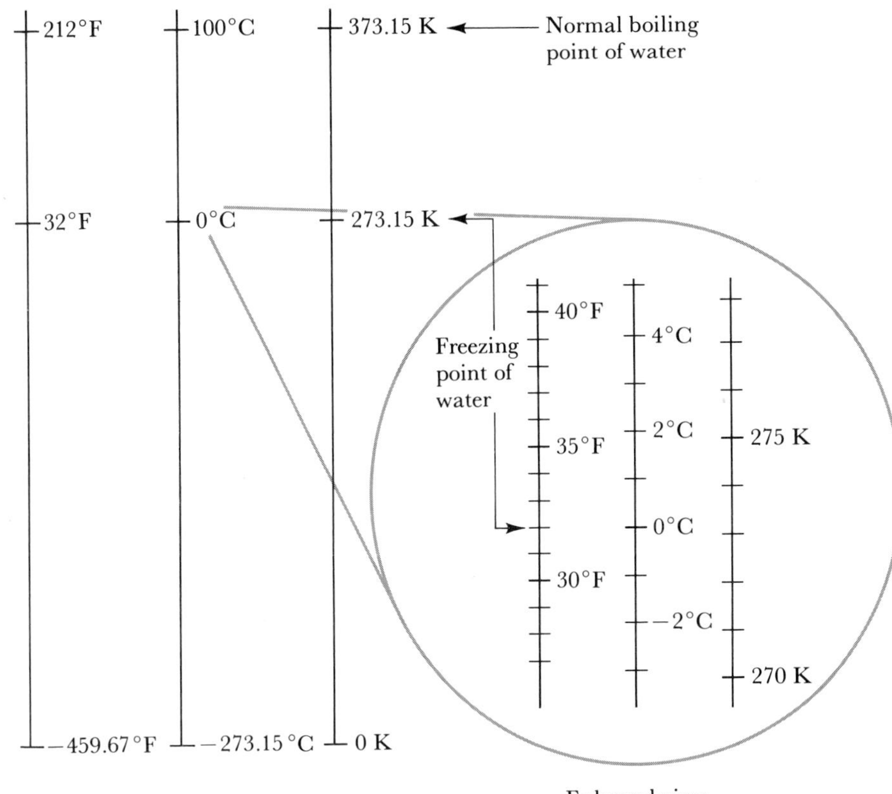

FIGURE 1–4
The Fahrenheit, Celsius, and Kelvin temperature scales.

CELSIUS temperature scale and 32 and 212 on the FAHRENHEIT temperature scale. These temperatures are depicted in Figure 1–4. Because the temperature differences between the two fixed points are 100 Celsius degrees and 180 Fahrenheit degrees, the size of these degrees are related:

$$100 \text{ Celsius degrees} = 180 \text{ Fahrenheit degrees}$$

$$1 \text{ Celsius degree} = \frac{180 \text{ Fahrenheit degrees}}{100}$$

$$= \frac{9}{5} \text{ Fahrenheit degrees}$$

$$1 \text{ Fahrenheit degree} = \frac{100 \text{ Celsius degrees}}{180}$$

$$= \frac{5}{9} \text{ Celsius degrees}$$

An equation for converting a temperature from the Fahrenheit scale, t_f, to the corresponding temperature on the Celsius scale, t_c, can be derived:

$$t_c = (t_f - 32°\text{F}) \times \frac{5°\text{C}}{9°\text{F}}$$

The first factor on the right side of the equation is the number of Fahrenheit degrees above the freezing point of water, defined as 0°C. The second factor takes account of the relative size of the temperature units on the two temperature scales. This equation can be rearranged to allow conversion of a temperature on the Fahrenheit scale to a temperature on the Celsius scale:

$$\frac{9°\text{F}}{5°\text{C}} \times t_c = t_f - 32°\text{F}$$

$$t_f = 32°\text{F} + \frac{9°\text{F}}{5°\text{C}} \times t_c$$

EXAMPLE: Normal body temperature is 98.6°F. What is this temperature on the Celsius scale?

$$t_c = (98.6°\text{F} - 32°\text{F}) \times \frac{5°\text{C}}{9°\text{F}}$$

$$= 66.6\cancel{°\text{F}} \times \frac{5°\text{C}}{9\cancel{°\text{F}}}$$

$$= 37.0°\text{C}$$

The concept of an absolute temperature scale will be developed in our discussion of gases in Chapter 4, but this development is not necessary to introduce the SI definition of the kelvin unit of temperature:

Kelvin unit (K) The kelvin unit (named for LORD KELVIN) of thermodynamic temperature is the fraction 1/273.16 of the thermodynamic temperature of the triple point of water. That is, this unit of temperature is 1/273.16 of the difference of temperature between absolute zero (0 K) and the triple point of water (273.16 K). (The triple point of water is shown in Figure 1–5.)

On this temperature scale the freezing point and boiling point of water are 273.15 K, and 373.15 K, respectively. Thus to the accuracy of any current measurements, the sizes of the Celsius degree and the kelvin unit are the same, and a temperature difference has the same numerical value expressed in Celsius degrees or in kelvin units: the range of temperature over which water is a liquid at 1 atm pressure is 100.00°C or 100.00 K. The symbols °C and K are used both for temperatures and temperature intervals. Conversion of temperatures on the kelvin and Celsius scale involves only addition and subtraction:[6]

$$\frac{T_K}{\text{K}} = \frac{t_c}{°\text{C}} + 273.15$$

[6]In these equations, the quotients $t_c/°\text{C}$ and T_K/K are numbers. If $t_c = 37°\text{C}$, $t_c/°\text{C} = 37°\text{C}/°\text{C} = 37$. (You will find additional discussion of this point in Chapter 3 and in Appendix 1.)

$$\frac{t_c}{°C} = \frac{T_K}{K} - 273.15$$

Thus normal body temperature on the kelvin scale is

$$\frac{T_K}{K} = \frac{37.0°C}{°C} + 273.15$$

$$= 310.15$$

$$T_K = 310.15 \text{ K}$$

(Notice that there is no degree sign with the symbol K in the SI unit of temperature.)

EXAMPLE: The absolute zero of temperature on the Celsius scale is −273.15°C. What is this temperature on the Fahrenheit scale?

$$t_f = 32°F + \frac{9°F}{5°C}(-273.15°C)$$

$$= 32°F - 491.67°F$$

$$= -459.67°F$$

1–3 The Masses of Atoms

The work of the modern chemist makes almost constant use of the relative masses of atoms of the elements. And the concept that matter consists of atoms with particular relative masses played a very important role in the development of the atomic theory.

DALTON'S ATOMS AND THEIR RELATIVE MASSES

Although atomism was conceived by ancient philosophers, it was JOHN DALTON, JOSEPH GAY–LUSSAC, AMADEO AVOGADRO, and STANISLAO CANNIZARO whose experimental observations in the first half of the nineteenth century firmly established the atomic theory of matter as we now know it. These experimental observations dealt with the weight percentage composition of different compounds of the same elements and the relative volumes of gases taking part in chemical reactions.

Dalton's atomic theory can be summarized in a series of postulates:

1. Matter is composed of indivisible and indestructible atoms.
2. Atoms of each element have a particular characteristic mass and a particular characteristic set of properties.
3. Molecules of chemical compounds consist of combinations of atoms of the different elements, and in chemical reactions these combinations are altered, but the individual atoms are not destroyed.

In arriving at these postulates, Dalton placed emphasis upon the weights of the atoms. Consideration of the oxides of nitrogen shows how weight percentage composition data bear on the postulates of Dalton's atomic theory. Analysis of three oxides of nitrogen, designated here as compounds I, II, and III, allows expression of the composition on a weight percentage basis. For instance, a 5.00-g sample of Compound I contains 1.82 g of oxygen and 3.18 g of nitrogen, corresponding to the weight percentages:

$$\text{O:} \quad \frac{1.82\ \cancel{g}}{5.00\ \cancel{g}} \times 100 = 36.4\%$$

$$\text{N:} \quad \frac{3.18\ \cancel{g}}{5.00\ \cancel{g}} \times 100 = 63.6\%$$

The weight percentage data for the three oxides of nitrogen under consideration are:

	wt % O	wt % N
Compound I	36.4	63.6
Compound II	53.3	46.7
Compound III	69.6	30.4

A cursory examination of these data does not disclose a simple composition relationship among the three different compounds. A more informative way of expressing the compositions of these compounds is the ratio of the weight of oxygen per unit weight of nitrogen, w_O/w_N. Expressed in this way, these data become:

	w_O/w_N
Compound I	$36.4/63.6 = 0.572$
Compound II	$53.3/46.7 = 1.14$
Compound III	$69.6/30.4 = 2.29$

These ratios, which have the approximate relative values 1:2:4, illustrate the law of multiple proportions, which can be stated: *The relative masses of two elements in different compounds of these elements are in the ratios of small whole numbers.*

Although the law of multiple proportions is consistent with Dalton's atomic theory, data that illustrate only this law (e.g., the weight percentage data for the nitrogen oxides) do not settle either of the questions:

1. What are the simplest chemical formulas of compounds I, II, and III?
2. What are the relative masses of oxygen atoms and nitrogen atoms?

If one of the compounds is assumed to have a particular simplest chemical formula, the simplest chemical formulas of the other compounds and the relative masses of the two kinds of atoms can be derived from the data. Suppose that Compound I has the simplest formula NO. Since this compound contains 0.572 g of oxygen per 1.000 g of nitrogen, the relative masses of atoms of these two elements, m_O and m_N, would be:

$$\frac{m_O}{m_N} = \frac{0.572\ \cancel{g}}{1.000\ \cancel{g}} = 0.572$$

With this value for the relative masses, the simplest formulas of the other two compounds can be derived:

Compound II, N_aO_b

$$\frac{w_O}{w_N} = \frac{b \times m_O}{a \times m_N} = \frac{b}{a} \times \left(\frac{m_O}{m_N}\right)$$

$$1.14 = \frac{b}{a} \times 0.572$$

$$\frac{b}{a} = \frac{1.14}{0.572} \cong 2$$

and the simplest formula is NO_2.

Compound III, N_cO_d

$$\frac{w_O}{w_N} = \frac{d \times m_O}{c \times m_N} = \frac{d}{c} \times \left(\frac{m_O}{m_N}\right)$$

$$2.29 = \frac{d}{c} \times 0.572$$

$$\frac{d}{c} = \frac{2.29}{0.572} \cong 4$$

and the simplest formula is NO_4.

Other possible simplest chemical formulas can be assumed for Compound I, and for each of the different assumptions the relative atomic masses (m_O/m_N) and simplest chemical formulas of the other compounds can be calculated, just as we have done for the assumption that Compound I is NO. This is summarized:

Assumed simplest formula for Compound I:	NO	N_2O	N_3O_2
m_O/m_N:	0.57	1.14	0.86
Simplest formula for Compound II:	NO_2	NO	N_3O_4
Simplest formula for Compound III:	NO_4	NO_2	N_3O_8

To determine which of these assumptions is correct, we will take account of the relative volumes of nitrogen gas and oxygen gas produced in the decomposition of these compounds, and also the formulas of nitrogen gas and oxygen gas. We can do so thanks to the discoveries of Gay–Lussac and Avogadro.

In 1809, Gay–Lussac reported that at the same conditions of temperature and pressure gaseous substances react with one another in volume ratios that are simple (the ratios of small whole numbers). For example, he reported that one volume of oxygen gas reacted with two volumes of hydrogen gas to give two volumes of water vapor:

1 volume oxygen gas + 2 volumes hydrogen gas → 2 volumes water vapor

To explain this observation (and other similar observations), Avogadro, in 1811, proposed the hypothesis that at a particular temperature and pressure equal volumes of different gases contain the same number of molecules. The observation that one volume of oxygen gas reacted with two volumes of hydrogen gas to give two volumes of water vapor becomes, with Avogadro's hypothesis, equivalent to the statement:

$$1 \text{ molecule oxygen} + 2 \text{ molecules hydrogen} \rightarrow 2 \text{ molecules water}$$

Each molecule of water contains a particular number of oxygen atoms, so from the above statement each molecule of elemental oxygen must contain two times this number. If the water molecule contains one atom of oxygen, the oxygen molecule contains two atoms; if the water molecule contains two atoms of oxygen, the oxygen molecule contains four atoms; and so on. Therefore the oxygen molecule must contain an even number of atoms, for example, O_2, or O_4, or O_6. The uncertainty about the oxygen molecule (among other uncertainties, including analytical data of uncertain quality) led to an uncertain interval in the development of the atomic theory in the period 1827–1857. With many more data about gases, Cannizaro, in 1858, showed that it was not necessary to assume a formula for oxygen gas more complicated than O_2. (If the oxygen molecule were O_4, it would seem reasonable that one volume of oxygen gas would give four volumes of some oxygen-containing compound. No such compound has been found.) Analogous observations on other reactions have demonstrated that nitrogen gas is N_2.

With the molecular formula of oxygen and nitrogen established (O_2 and N_2), we now can reconsider the nitrogen oxides. Experiments show that one volume of Compound I decomposes to yield one volume of nitrogen gas (N_2) and one half volume of oxygen gas (O_2). With Avogadro's hypothesis, this observation is equivalent to the statement:

$$1 \text{ molecule Compound I} \rightarrow 1 \text{ molecule } N_2 + \tfrac{1}{2} \text{molecule } O_2$$

Therefore Compound I must be N_2O; this is the molecular formula, not merely the simplest formula. The volume data for decomposition of the other compounds are:

	$\frac{\text{Vol } N_2}{\text{Vol Cmpd}}$	$\frac{\text{Vol } O_2}{\text{Vol Cmpd}}$	Molecular formula
Compound II	0.50	0.50	NO
Compound III	0.50	1.00	NO_2

These volume data are equivalent to the statements:

$$1 \text{ molecule Compound II} \rightarrow \tfrac{1}{2} \text{molecule } N_2 + \tfrac{1}{2} \text{molecule } O_2$$
$$1 \text{ molecule Compound III} \rightarrow \tfrac{1}{2} \text{molecule } N_2 + 1 \text{ molecule } O_2$$

which establish the molecular formulas for these compounds given in the tabulation. The common names for the compounds are:

Compound I, N_2O	nitrous oxide
Compound II, NO	nitric oxide
Compound III, NO_2	nitrogen dioxide

The relative masses of atoms of oxygen, atoms of nitrogen, and molecules of these oxides of nitrogen also are established. Because 1.14 g of oxygen is combined with 1.00 g of nitrogen in Compound II, nitric oxide (NO), the relative masses of atoms of oxygen and atoms of nitrogen are

$$\frac{m_{O}}{m_{N}} = \frac{1.14}{1}$$

The masses of molecules of three oxides of nitrogen relative to the mass of the nitrogen atom can be calculated from the relative atomic masses. With the relative masses of oxygen atoms and nitrogen atoms being 1.14 and 1.00, respectively, the relative masses of these molecules are:

$$\frac{m_{N_2O}}{m_{N}} = \frac{(2 \times 1) + 1.14}{1} = 3.14$$

$$\frac{m_{NO}}{m_{N}} = \frac{1 + 1.14}{1} = 2.14$$

$$\frac{m_{NO_2}}{m_{N}} = \frac{1 + (2 \times 1.14)}{1} = 3.28$$

More will be said about molecular weights after the modern atomic weight scale is presented.

THE MODERN ATOMIC WEIGHT SCALE; THE MOLE

As just illustrated, only the relative masses of atoms of different elements and the relative masses of molecules of different compounds are derivable from data on combining proportions of elements. Therefore a scale of relative masses of atoms can be based upon an arbitrary assignment of a particular definite mass to a particular nuclide,[7] which is an atom with a particular nuclear composition. The present atomic weight scale, adopted internationally in 1961 by both chemists and physicists, is based upon assignment of exactly 12 as the relative mass of the carbon-12 nuclide [a carbon atom with a nucleus containing 12 nucleons, i.e., 12 (neutrons + protons)]. The number of nucleons in the nucleus of an atom is called the *mass number*; it is given as a superscript left of the symbol, for example, ^{12}C. Therefore the unit of mass on the present atomic weight scale is one twelfth of the mass of an atom of carbon-12. (This unit also is called an atomic mass unit (amu) or a dalton; neither of these terms is used in SI.)

[7]The word "nuclide" is not a synonym for isotope. Nuclides of a particular element with different numbers of neutrons are called *isotopes*.

A complete set of atomic weights based upon the $^{12}C \equiv 12$ standard is given in Table 1–4. These relative atomic masses, or atomic weights (symbol AW) as they are commonly called, are given in Table 1–4 for the elements as they exist in materials of terrestrial origin. The number of significant figures[8] that can be justified for the atomic weight of an element is a consequence of the accuracy of the chemical analyses upon which the quantity is based and the constancy of the isotopic composition of the element as it is found on Earth. In Table 1–4, each atomic weight is given to four significant figures or to the nearest 0.01 unit. These values are adequate for most of our purposes, but a table with the best values known at this time is given in the back end papers.

Let us consider the atomic weights for three elements, hydrogen (AW = 1.008), fluorine (AW = 19.00), and mercury (AW = 200.6). Certain statements can be made to illustrate the meaning of these numbers:

1. The average mass of hydrogen atoms is 1.008/12 times the mass of atoms of ^{12}C.
2. The average mass of fluorine atoms is 19.00/12 times the mass of atoms of ^{12}C.
3. The average mass of mercury atoms is 200.6/12 times the mass of atoms of ^{12}C.
4. Comparisons need not involve the standard; the atomic weights given also allow this statement: The average mass of fluorine atoms is 19.00/200.6 times the average mass of mercury atoms.

These statements reflect the relative nature of atomic weights as they are defined; atomic weights are relative atomic masses.

In chemistry, it is convenient to designate a particular number of atoms, molecules, ions (atoms with a net charge), electrons, or other particles with a special name. This particular unit of amount is the *mole*.[9] In SI, the definition of mole is:

> The mole is the amount of a substance that contains as many entities as there are atoms in exactly 12 g (0.012 kg) of carbon-12. (The entities must be specified; they may be atoms, molecules, ions, electrons, other particles, or specified groups of such particles.) The symbol for mole is mol.

---

[8]A discussion of significant figures is presented in Appendix 2.

[9]Other names that are commonly used synonyms for mole are:

1. gram molecular weight
2. gram formula weight (used in connection with substances, such as sodium chloride, which commonly do not occur as discrete molecules; see Figure 1–2b)
3. gram atomic weight (used for the number of atoms in 12 grams of carbon-12)

We will not use these names.

TABLE 1–4
Atomic Masses

Name	Symbol	Atomic number	Atomic mass[a]	Name	Symbol	Atomic number	Atomic mass[a]
Actinium	Ac	89	[227]	Mercury	Hg	80	200.6
Aluminum	Al	13	26.98	Molybdenum	Mo	42	95.94
Americium	Am	95	[243]	Neodymium	Nd	60	144.2
Antimony	Sb	51	121.8	Neon	Ne	10	20.18
Argon	Ar	18	39.95	Neptunium	Np	93	237.0
Arsenic	As	33	74.92	Nickel	Ni	28	58.70
Astatine	At	85	[210]	Niobium	Nb	41	92.91
Barium	Ba	56	137.3	Nitrogen	N	7	14.01
Berkelium	Bk	97	[247]	Nobelium	No	102	[259]
Beryllium	Be	4	9.012	Osmium	Os	76	190.2
Bismuth	Bi	83	209.0	Oxygen	O	8	16.00
Boron	B	5	10.81	Palladium	Pd	46	106.4
Bromine	Br	35	79.90	Phosphorus	P	15	30.97
Cadmium	Cd	48	112.4	Platinum	Pt	78	195.1
Calcium	Ca	20	40.08	Plutonium	Pu	94	[244]
Californium	Cf	98	[251]	Polonium	Po	84	[209]
Carbon	C	6	12.01	Potassium	K	19	39.10
Cerium	Ce	58	140.1	Praseodymium	Pr	59	140.9
Cesium	Cs	55	132.9	Promethium	Pm	61	[145]
Chlorine	Cl	17	35.45	Protactinium	Pa	91	231.0
Chromium	Cr	24	52.00	Radium	Ra	88	226.0
Cobalt	Co	27	58.93	Radon	Rn	86	[222]
Copper	Cu	29	63.55	Rhenium	Re	75	186.2
Curium	Cm	96	[247]	Rhodium	Rh	45	102.9
Dysprosium	Dy	66	162.5	Rubidium	Rb	37	85.47
Einsteinium	Es	99	[254]	Ruthenium	Ru	44	101.1
Erbium	Er	68	167.3	Samarium	Sm	62	150.4
Europium	Eu	63	152.0	Scandium	Sc	21	44.96
Fermium	Fm	100	[257]	Selenium	Se	34	78.96
Fluorine	F	9	19.00	Silicon	Si	14	28.09
Francium	Fr	87	[223]	Silver	Ag	47	107.9
Gadolinium	Gd	64	157.3	Sodium	Na	11	22.99
Gallium	Ga	31	69.72	Strontium	Sr	38	87.62
Germanium	Ge	32	72.59	Sulfur	S	16	32.06
Gold	Au	79	197.0	Tantalum	Ta	73	180.9
Hafnium	Hf	72	178.5	Technetium	Tc	43	[97]
Helium	He	2	4.003	Tellurium	Te	52	127.6
Holmium	Ho	67	164.9	Terbium	Tb	65	158.9
Hydrogen	H	1	1.008	Thallium	Tl	81	204.4
Indium	In	49	114.8	Thorium	Th	90	232.0
Iodine	I	53	126.9	Thulium	Tm	69	168.9
Iridium	Ir	77	192.2	Tin	Sn	50	118.7
Iron	Fe	26	55.85	Titanium	Ti	22	47.90
Krypton	Kr	36	83.80	Tungsten	W	74	183.9
Lanthanum	La	57	138.9	Uranium	U	92	238.0
Lawrencium	Lr	103	[260]	Vanadium	V	23	50.94
Lead	Pb	82	207.2	Xenon	Xe	54	131.3
Lithium	Li	3	6.941	Ytterbium	Yb	70	173.0
Lutetium	Lu	71	175.0	Yttrium	Y	39	88.91
Magnesium	Mg	12	24.31	Zinc	Zn	30	65.38
Manganese	Mn	25	54.94	Zirconium	Zr	40	91.22
Mendelevium	Md	101	[258]				

[a] Expressed to 4 significant figures based on the atomic mass of $^{12}C = 12$ exactly; number in brackets denotes isotope of longest known half-life.

Therefore a mole of each element has a mass equal to the atomic weight in grams:

$$1 \text{ mol H} = 1.008 \text{ g H}$$
$$1 \text{ mol F} = 19.00 \text{ g F}$$
$$1 \text{ mol Hg} = 200.6 \text{ g Hg}$$

Thus 1.008 g of hydrogen, 19.00 g of fluorine, and 200.6 g of mercury each contain the same number of atoms as are contained in exactly 12 g of carbon-12. The number of moles of an element in a particular sample is obtained by multiplying the mass of the sample by a unit factor. For instance, the number of moles in 4.53 g of hydrogen is

$$4.53 \text{ g H} = 4.53 \cancel{\text{g H}} \times \frac{1 \text{ mol H}}{1.008 \cancel{\text{g H}}} = 4.49 \text{ mol H}$$

Example: How many moles of mercury are present in the following masses of mercury: 1.00 g, 1.75 mg, 3.73 kg?

$$1.00 \text{ g Hg} = 1.00 \cancel{\text{g Hg}} \times \frac{1 \text{ mol Hg}}{200.6 \cancel{\text{g Hg}}}$$
$$= 4.99 \times 10^{-3} \text{ mol Hg}$$

$$1.75 \times 10^{-3} \text{ g Hg} = 1.75 \times 10^{-3} \cancel{\text{g Hg}} \times \frac{1 \text{ mol Hg}}{200.6 \cancel{\text{g Hg}}}$$
$$= 8.72 \times 10^{-6} \text{ mol Hg}$$

$$3.73 \times 10^{3} \text{ g Hg} = 3.73 \times 10^{3} \cancel{\text{g Hg}} \times \frac{1 \text{ mol Hg}}{200.6 \cancel{\text{g Hg}}}$$
$$= 18.6 \text{ mol Hg}$$

The atomic weights are relative atomic masses, and as such are dimensionless. However, in many calculations it is appropriate to use the mass of one mole of an element. This quantity does have dimensions:

H	1.008 g mol^{-1}
F	19.00 g mol^{-1}
Hg	200.6 g mol^{-1}

Example: The density of liquid mercury at 0°C is 13.595 g cm^{-3}. What is the volume of one mole of this element under these conditions?

The volume of one mole is obtained by multiplication of the weight of one mole in grams by the volume of one gram (which is the reciprocal of the density):

$$\frac{V}{n} = \underbrace{200.6 \cancel{\text{g}} \text{ mol}^{-1}}_{\text{weight of one mole}} \times \underbrace{\frac{1}{13.595 \cancel{\text{g}} \text{ cm}^{-3}}}_{\text{volume of one gram}} = 14.76 \text{ cm}^{3} \text{ mol}^{-1}$$

For many calculations in chemistry, it is not necessary to know the number of entities in a mole. For some purposes, however, the value of this quantity is necessary; it can be determined by a number of experimental methods. The currently accepted "best value" of Avogadro's constant, $\mathcal{N}_A$, as this quantity is known, is

$$\mathcal{N}_A = \text{number of atoms in exactly 12 g of } {}^{12}\text{C}$$
$$= 6.0220943 \times 10^{23}\ \text{mol}^{-1}$$

In most of our uses, the value for this quantity will be taken as $6.02 \times 10^{23}\ \text{mol}^{-1}$. Notice two points in connection with this expression of Avogadro's constant. We will call this Avogadro's constant, not Avogadro's number, because it is a quantity with dimensions, not a pure number. The dimensions are expressed:

$$\mathcal{N}_A = 6.02 \times 10^{23}\ \text{mol}^{-1}$$
$$\text{not} \quad \mathcal{N}_A = 6.02 \times 10^{23}/\text{mol}$$

even though in speaking you may wish to say 6.02×10^{23} per mole. Giving the exponents for dimensions is unambiguous, but slash marks may be ambiguous for complicated dimensions (see Appendix 1).

Example: How many atoms are there in samples of mercury and fluorine each weighing 10.00 g?

$$\text{Hg} \quad 10.00\ \text{g} = 10.00\ \cancel{\text{g}} \times \frac{1\ \cancel{\text{mol}}}{200.6\ \cancel{\text{g}}} \times 6.02 \times 10^{23}\ \cancel{\text{mol}^{-1}}$$
$$= 3.00 \times 10^{22}$$

$$\text{F} \quad 10.00\ \text{g} = 10.00\ \cancel{\text{g}} \times \frac{1\ \cancel{\text{mol}}}{19.00\ \cancel{\text{g}}} \times 6.02 \times 10^{23}\ \cancel{\text{mol}^{-1}}$$
$$= 3.17 \times 10^{23}$$

In these calculations, and in many which follow, the chemical symbol is omitted from the dimensions of the numerical quantity. There is a risk of error in this omission, but it is cumbersome to include the chemical symbols in dimensions.

Example: How many atoms are there in a spherical droplet of mercury with a diameter of 1.00 mm ($r = 5.00 \times 10^{-2}$ cm)?

$$V = \frac{4}{3}\pi r^3$$
$$= \frac{4}{3}\pi(5.00 \times 10^{-2}\ \text{cm})^3$$
$$= 5.24 \times 10^{-4}\ \text{cm}^3$$

$$5.24 \times 10^{-4}\ \text{cm}^3 = 5.24 \times 10^{-4}\ \cancel{\text{cm}^3} \times \frac{1\ \cancel{\text{mol}}}{14.76\ \cancel{\text{cm}^3}} \times 6.02 \times 10^{23}\ \cancel{\text{mol}^{-1}}$$
$$= 2.14 \times 10^{19}$$

(The molar volume of mercury, 14.76 $\text{cm}^3\ \text{mol}^{-1}$, was calculated in a previous example.)

1–4 Compounds

Most of our discussion of chemistry will deal with the properties and reactions of compounds, not the uncombined elements. The relative masses of compounds are described on the same scale as are the relative masses of atoms, and this old subject has some modern facets, as you will learn. But there are many qualities of compounds other than relative mass, and we will introduce you to some of these by considering the familiar compound, water.

THE COMPOSITION OF CHEMICAL COMPOUNDS; MOLECULAR WEIGHTS

The concept of a chemical compound consisting of molecules that are combinations of atoms has already been introduced in connection with discussion of Dalton's atomic theory. The compounds nitrous oxide (N_2O), nitric oxide (NO), and nitrogen dioxide (NO_2) were used to illustrate the law of multiple proportions and the determination of the relative masses of atoms of nitrogen and oxygen. The molecular weight (symbol MW) corresponding to a chemical formula is the sum of the atomic weights of the constituent atoms. It is, therefore, the relative mass of a formula unit (a molecule) of the compound on the same scale as the atomic weight scale. With nitrogen and oxygen having the atomic weights

$$\mathrm{AW(N)} = 14.01$$
$$\mathrm{AW(O)} = 16.00$$

the molecular weights of the oxides of nitrogen are:

$$\mathrm{MW(N_2O)} \qquad 2 \times 14.01 + 16.00 = 44.02$$
$$\mathrm{MW(NO)} \qquad 14.01 + 16.00 = 30.01$$
$$\mathrm{MW(NO_2)} \qquad 14.01 + 2 \times 16.00 = 46.01$$

Thus we can make statements regarding these compounds that are analogous to those made already for elements:

1. The average mass of nitrous oxide molecules is 44.02/12 times the mass of atoms of ^{12}C.
2. The average mass of nitric oxide molecules is 30.01/12 times the mass of atoms of ^{12}C.
3. The average mass of nitrogen dioxide molecules is 46.01/12 times the mass of atoms of ^{12}C.
4. One mole of nitrous oxide, which contains 6.02×10^{23} molecules, weighs 44.02 g.
5. One mole of nitric oxide, which contains 6.02×10^{23} molecules, weighs 30.01 g.
6. One mole of nitrogen dioxide, which contains 6.02×10^{23} molecules, weighs 46.01 g.

Calculations of molecular weights of other compounds are:

Sodium chloride, NaCl

$$1 \times 22.99 = 22.99$$
$$1 \times 35.45 = 35.45$$
$$\mathrm{MW(NaCl)} = 58.44$$

Methane, CH_4

$$1 \times 12.01 = 12.01$$
$$4 \times 1.008 = 4.032$$
$$\mathrm{MW(CH_4)} = 16.04$$

Calcium carbonate, $CaCO_3$

$$1 \times 40.08 = 40.08$$
$$1 \times 12.01 = 12.01$$
$$3 \times 16.00 = 48.00$$
$$\mathrm{MW(CaCO_3)} = 100.09$$

Sucrose, $C_{12}H_{22}O_{11}$

$$12 \times 12.01 = 144.12$$
$$22 \times 1.008 = 22.176$$
$$11 \times 16.00 = 176.00$$
$$\mathrm{MW(C_{12}H_{22}O_{11})} = 342.30$$

Just as for atomic weights and molecular weights of the nitrogen oxides, these molecular weights are relative masses (relative to the mass of atoms of $^{12}C \equiv 12$ exactly) and are dimensionless. But we also will be interested in the mass of one mole of a compound; for these compounds,

$$1 \text{ mol NaCl} = 58.44 \text{ g}$$
$$1 \text{ mol } CH_4 = 16.04 \text{ g}$$
$$1 \text{ mol } CaCO_3 = 100.09 \text{ g}$$
$$1 \text{ mol } C_{12}H_{22}O_{11} = 342.30 \text{ g}$$

The weight percentage composition of a compound is easily calculated from the chemical formula. Consider sucrose, for which the molecular weight has been calculated to be 342.30. One mole of sucrose (342.30 g) contains:

144.12 g C
22.176 g H
176.00 g O

The weight percentage of each element is obtained by dividing these weights by the weight of one mole:

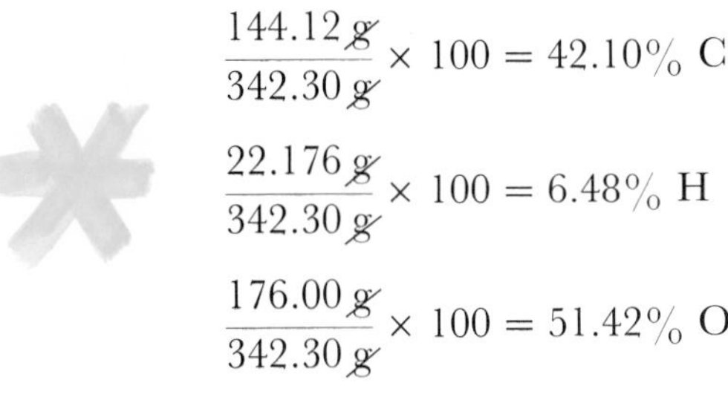

$$\frac{144.12 \cancel{g}}{342.30 \cancel{g}} \times 100 = 42.10\% \text{ C}$$

$$\frac{22.176 \cancel{g}}{342.30 \cancel{g}} \times 100 = 6.48\% \text{ H}$$

$$\frac{176.00 \cancel{g}}{342.30 \cancel{g}} \times 100 = 51.42\% \text{ O}$$

A common calculation chemists must make is the reverse of that just outlined. That is, appropriate analysis gives the weight percentage of each element in a material, believed to be a pure substance; what is the simplest formula of the compound? As an example, consider a compound of carbon, hydrogen, and chlorine with the weight percentage composition: 49.02% C, 2.74% H, and 48.24% Cl. The relative number of atoms of each element in this compound is the same as the relative number of moles of atoms of each element in any particular weight of the substance, for example, in a 100.00-g sample. In a 100.00-g sample, there are:

$$\begin{array}{r} 49.02 \text{ g C} \\ 2.74 \text{ g H} \\ 48.24 \text{ g Cl} \end{array}$$

The numbers of moles of atoms of each element in this 100.00-g sample are obtained by dividing each of these weights by the weight of one mole of that element:

$$\text{C} \quad \frac{49.02 \text{ g}}{12.01 \text{ g mol}^{-1}} = 4.08 \text{ mol}$$

$$\text{H} \quad \frac{2.74 \text{ g}}{1.008 \text{ g mol}^{-1}} = 2.72 \text{ mol}$$

$$\text{Cl} \quad \frac{48.24 \text{ g}}{35.45 \text{ g mol}^{-1}} = 1.36 \text{ mol}$$

These numbers of moles of atoms of each kind in a 100.00-g sample are the relative numbers of atoms of each kind in the substance. Since chemical formulas generally are written with integers as subscripts ($C_{4.08}H_{2.72}Cl_{1.36}$ won't do),[10] these quantities are divided by an appropriate common divisor, 1.36:

$$\text{C} \quad \frac{4.08 \text{ mol}}{1.36} = 3 \text{ mol}$$

$$\text{H} \quad \frac{2.72 \text{ mol}}{1.36} = 2 \text{ mol}$$

$$\text{Cl} \quad \frac{1.36 \text{ mol}}{1.36} = 1 \text{ mol}$$

This operation does not change the relative numbers of moles of atoms of each element in this compound. The simplest integers that are calculated in this way give the simplest formula, or empirical formula, of the compound. The empirical formula of this compound is C_3H_2Cl. The molecular formula can be any whole-number multiple of this; the weight percentage compositions of C_3H_2Cl, $C_6H_4Cl_2$, $C_9H_6Cl_3$ are the same. However, the molecular weights corresponding to these chemical formulas differ:

$$\begin{aligned} \text{MW}(C_3H_2Cl) &= 73.50 \\ \text{MW}(C_6H_4Cl_2) &= 147.00 \\ \text{MW}(C_9H_6Cl_3) &= 220.50 \end{aligned}$$

[10] Solid compounds with formulas having nonintegral subscripts are described in Chapter 9.

An experimental determination of molecular weight (e.g., by mass spectrometry, described in this chapter, or by measurement of the vapor density, described in Chapter 4) is needed to determine which of these formulas is the correct molecular formula.

WATER, A FAMILIAR CHEMICAL COMPOUND

You have been introduced to some features of molecular composition in previous sections of this chapter. Other features now will be discussed with reference to a familiar substance, water. Many properties of bulk portions of this compound can be described in terms that do not mention atoms or molecules; however, the description may involve a detailed discussion of the molecular structure. In this introductory discussion of water, many new concepts (e.g., phase diagrams, triple point, critical point, equilibrium, Le Chatelier's principle, molecular geometry, entropy, and hydrogen bonding) will be mentioned. These concepts will recur throughout the book, and you are not expected to gain a complete understanding of them in this introduction.

In common experience, water is known as a solid, a liquid, and a gas. Whether a bulk sample of water is solid, liquid, or gas depends upon the temperature and pressure,[11] as is shown in Figure 1–5. The combinations of temperature and pressure defined by lines separating two regions of the figure are conditions under which two of the phases (solid and liquid, solid and gas, or liquid and gas) exist together at equilibrium. The point at which the three lines come together, the triple point[12] (0.0100°C and a pressure of 4.588 torr) specifies the only set of conditions under which these three phases (gas, liquid, and ordinary ice) coexist at equilibrium in a system containing water as the only component. The extremity of the line defining the conditions of temperature and pressure under which the liquid and gaseous states exist together at equilibrium is the critical point (374.2°C and 217.7 atm); the critical temperature 374.2°C is the highest temperature at which water can exist as separate gas and liquid phases.

The figure shows that the temperature at which liquid and solid water coexist at equilibrium decreases as the pressure increases. At the triple

[11]You probably already are familiar with the concept of pressure. It will be discussed more fully in Chapter 4. For the present, pressure will be expressed in two commonly used units, the torr and the atmosphere. A pressure of 1 torr is the pressure exerted by a column of mercury 1 mm high at 25°C, at a place where the acceleration of gravity is 9.80665 m s^{-2}. The torr is named after Evangelista Torricelli (1608–1647), an Italian mathematician and scientist, who invented the barometer. One atmosphere is equivalent to 760 torr. Neither the torr nor the atmosphere is an SI unit. The SI unit of pressure will be given in Chapter 4.

[12]The triple point for water is not identical to the normal melting point, which is the temperature at which the solid and liquid are in equilibrium in the presence of air at 1 atm pressure. The normal melting point of water is lower than its triple point because of the pressure difference (an increase of pressure causes the melting point of water to decrease because liquid water is denser than solid water) and because of the solubility of air in liquid water.

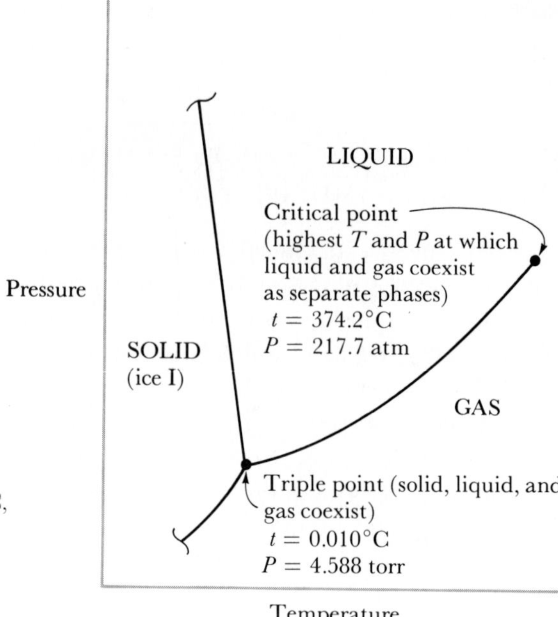

FIGURE 1–5
The phases of water existing in the range $-10°C < T < 400°C$, $P < 300$ atm. (Schematic; ordinate and abscissa are not linear.)

point ($P = 4.588$ torr), the liquid and solid coexist at a temperature of 0.0100°C. At a pressure of 10 atm these two phases coexist at −0.065°C. At 2040 atm, these two phases coexist at −22°C.[13] This trend in freezing temperature with pressure is a consequence of the greater density of liquid water compared to ice; at 0°C and 1 atm pressure, the densities of liquid and solid water are: liquid, 0.9998 g cm^{-3}; solid, 0.9168 g cm^{-3}. Water is unusual in this respect; the solid phase for most substances is more dense than the liquid phase. An increase in pressure stabilizes the phase with the greater density. This fact illustrates LE CHATELIER's principle, which can be stated: *When a system at equilibrium is subjected to a stress, a change occurs that partially offsets this stress.* In this example the increase in pressure (the stress) needed to reduce the volume of the system by a particular amount is lower than it otherwise would be because the stress has caused the formation of the more dense phase. Other illustrations of Le Chatelier's principle will be given throughout the book.

A description of water in each of the three states (illustrated in Figure 1–6) can be made in the following parallel ways:

Solid	The external shape of a sample is fixed, and it does not depend upon the shape of the container.	Molecules are in fixed positions relative to one another. Although there is atomic motion in the solid, the molecules do not move about freely.

[13]This is the lowest temperature at which ordinary liquid water can exist under equilibrium conditions. This combination of temperature and pressure is the triple point at which liquid water, ordinary ice (ice I), and a solid phase called ice III coexist.

Liquid	The volume of a sample is fixed (depending upon temperature and pressure), but the shape of the sample is not. The shape is dictated by the shape of the container.	Molecules are in contact with one another, but there is motion of molecules relative to one another. The individual molecules attract one another strongly.
Gas	The sample fills the container completely.	Molecules are separated by relatively large distances. The interaction of molecules with one another is minimal.

You will understand the correlation of properties given in the left column with the molecular description in the right column if you remember that molecular motion increases with an increase in temperature and molecular attraction increases with a decrease in average intermolecular separation. At low temperature and high pressure, water exists as a solid in which the mutual attraction of the individual molecules holds them in fixed, regularly repeating positions; although there is some motion of atoms, this motion is not sufficient to overcome the attractive forces. At higher temperatures the balance between attractive forces and molecular motion holds the molecules in contact with one another, but not in fixed, regularly repeating positions. The molecules can flow over one another, and the substance flows to fill the lowest accessible regions of the container. At still higher temperatures and low pressures, the average distance between molecules is large, and molecular attraction is not appreciable. The individual, independently moving molecules fill the container completely.

The attraction of water molecules for one another is high. One consequence of this attraction is the high critical temperature for water; it is much higher than that for other substances of comparable molecular weight (CH_4, $T_c = -82.5°C$; Ne, $T_c = -228.7°C$). To understand this (as well as the high melting point and high boiling point), we must consider the structure of the water molecule.

Enroute to consideration of this structure, we will go through several steps, which involve answering these questions:

1. What is the composition in percentage by weight?

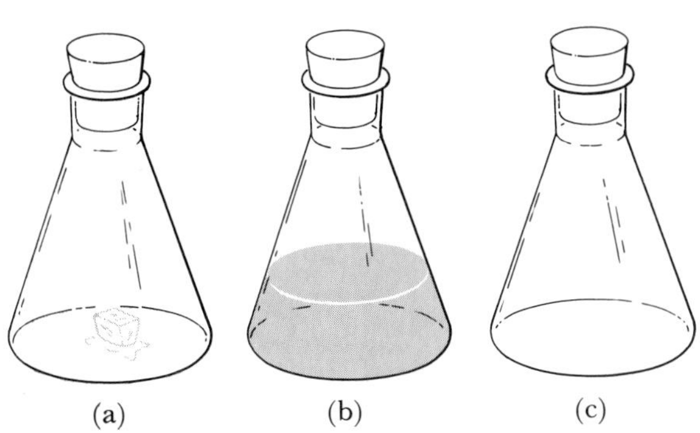

FIGURE 1–6
Water in each of the three states of matter. (a) Solid: density at triple point = 0.9168 g cm^{-3} (a 30.0-g ice cube under these conditions occupies 32.7 cm^3). (b) Liquid: density at triple point = 0.9998 g cm^{-3} (200.00 g of liquid water under these conditions occupies 200.04 cm^3). (c) Gas: density at triple point = 4.84×10^{-6} g cm^{-3} (500 cm^3 contains 2.42 mg water vapor).

2. What is the simplest formula? This is obtained from the composition by a simple calculation.
3. What is the molecular formula? This may be the same as the simplest formula, or it may be a multiple of the simplest formula. A new experiment is needed to establish this value.
4. What is the spatial arrangement of atoms in the molecule? Methods of determining molecular geometry are described in Chapters 5, 9, and 21.
5. What is the distribution of electrons in the molecule? Indirect information on this complicated matter is obtained from various chemical and physical properties, and theoretical calculations also shed light on it. However, this last aspect of molecular structure is the most difficult to ascertain.

For water, the answers to these questions are:

1. Analysis gives the weight percentages:

   11.2% hydrogen

   88.8% oxygen

2. Calculation using the weight percentage composition given in (1) and the atomic weights gives: A 100.0-g sample of water contains 11.2 g of hydrogen, and 88.8 g of oxygen.

$$\text{H} \qquad 11.2\text{ g} = 11.2\cancel{\text{g}} \times \frac{1\text{ mol}}{1.008\cancel{\text{g}}} = 11.1\text{ mol}$$

$$\text{O} \qquad 88.8\text{ g} = 88.8\cancel{\text{g}} \times \frac{1\text{ mol}}{16.00\cancel{\text{g}}} = 5.55\text{ mol}$$

   If each of these results is divided by the smaller number (5.55), we have

$$\frac{11.1\text{ mol H atoms}}{5.55} = 2\text{ mol H atoms}$$

$$\frac{5.55\text{ mol O atoms}}{5.55} = 1\text{ mol O atoms}$$

   which give the relative numbers of atoms of hydrogen and oxygen in this compound. Thus the simplest formula (the empirical formula) for water is H_2O.

3. Measurement of the density of water vapor shows that 18.02 g occupy the same volume as 32.00 g of oxygen gas (at the same temperature and pressure). Because it has been shown that oxygen gas is diatomic (O_2) with molecular weight 32.00, application of Avogadro's principle to these data gives 18.02 as the molecular weight for water; therefore H_2O is the molecular formula.
4. Experimental measurements of a type not easy to explain in an introductory chapter allow chemists to determine the distances between the centers of the atoms in gas molecules. For the gaseous water molecule these distances are given in Figure 1–7a. The two hydrogen–oxygen

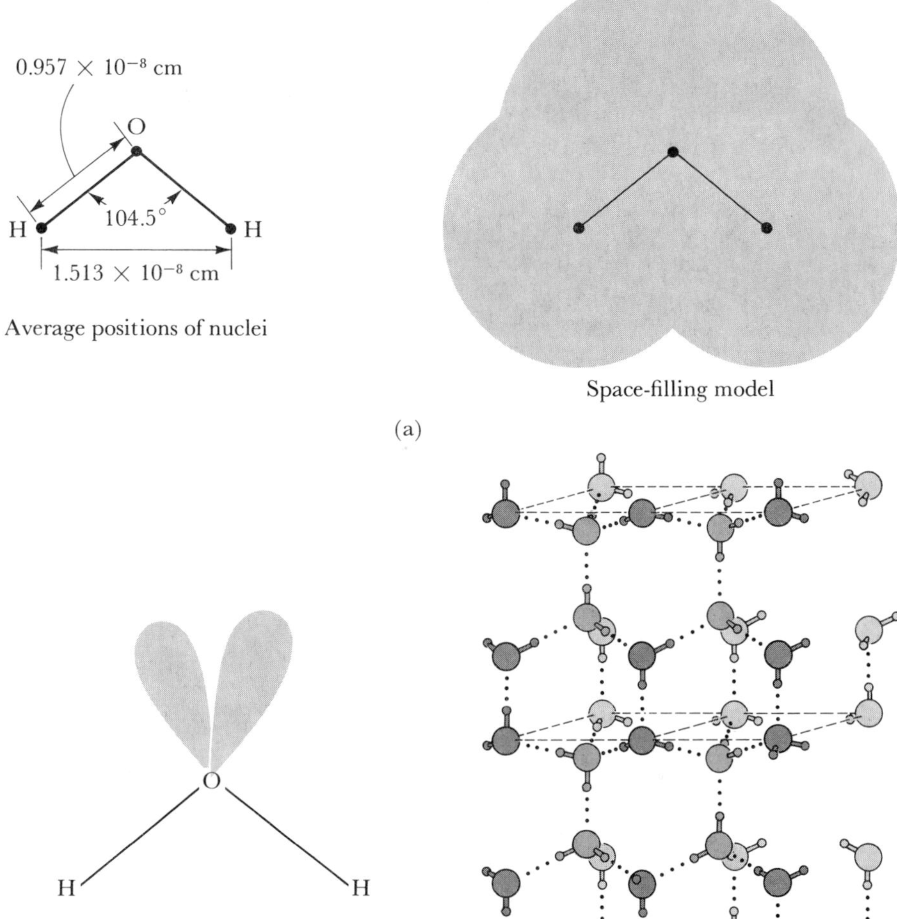

FIGURE 1–7
Water. (a) The geometry of the gaseous molecule. (b) The electronic structure; a shared electron pair shown with a line, and an unshared pair shown as shaded region. (c) The structure of ice.

distances, d(H–O), are equal, and the molecule is not linear. The hydrogen–oxygen distance is short enough to imply O–H chemical bonding; the longer hydrogen–hydrogen distance does not imply chemical bonding between these atoms.

5. This nonlinear geometry is consistent with a simplified view of the electronic structure of the molecule. The total number of electrons in the water molecule is ten; each atom of hydrogen ($Z = 1$) has one electron and the atom of oxygen ($Z = 8$) has eight electrons. Two of these electrons are in the inner electron shell of the oxygen atom and the other eight electrons are grouped in four pairs, with a spatial arrangement in which each pair is directed toward the corner of a tetrahedron with the oxygen nucleus at its center. Each hydrogen atom shares one of these pairs of electrons with the oxygen atom, and the other two pairs (the

unshared or lone pairs) are directed toward the other two corners of the tetrahedron. Each shared pair of electrons constitutes a covalent chemical bond, as conceived by G. N. LEWIS in 1916. This picture of the water molecule is shown in Figure 1–7b.

This structure (meaning both the geometry defined by the position of the centers of the atoms and the distribution of electrons) helps explain the extraordinary attraction of water molecules for one another. In ice, water molecules are arranged with each oxygen atom surrounded by four hydrogen atoms, two of which "belong" to it and two of which "belong" to other water molecules. The hydrogen atoms belonging to other water molecules are attracted to the unshared pairs of electrons of the water molecule in question. The atomic arrangement in ice is shown in Figure 1–7c. The key features of this pattern are repeated over and over; each oxygen atom in a perfect ice crystal is surrounded in this way by four hydrogen atoms. Although this very regular arrangement is broken down partially in liquid water, the O—H---O attraction persists, and over small regions in liquid water there are atomic arrangements similar to that in ice. (The ice structure is one with much open space, and melting causes this open structure to collapse partially; this explains why liquid water has a greater density than solid water at the melting point.) It is only when the liquid is converted to the vapor that the strong attractive forces are overcome.

In the atomic arrangement of a hydrogen atom between two nonmetal atoms, X—H---Y, attraction results from interaction of the covalently bonded relatively positive hydrogen atom and an unshared pair of electrons on the adjacent atom; this arrangement is called a *hydrogen bond*. The idea that a hydrogen atom could bind two other atoms together was considered very novel when it was proposed in 1920 by LATIMER and RODEBUSH, but this type of attraction is now known to be common. Particularly interesting are the roles of hydrogen bonding in determining the properties of aqueous solutions, the coiling of the long-chain protein molecules, and the pairing of DNA molecules in operation of the genetic code.

We have not discussed one important difference among the solid, liquid, and gaseous states that also is implied in Figure 1–6. The order in the positions of the water molecules is greatest in the solid, least in the gas, and intermediate between these extremes in the liquid. This order in the positions of the molecules in a sample of matter is related to an extensive property called *entropy*, which we shall meet again and again in the course of our study.

1–5 Isotopes and Masses of Isotopic Molecular Species

For calculations used in most operations in the chemical laboratory and in chemical industries, the atomic weights of the elements as they occur in nature, the values tabulated in Table 1–4, are the numerical quantities needed. Recent developments, however, have made the weights of particular isotopic molecular species of interest to the chemist.

In this connection, our principal concern is to describe the chemical analysis of substances by high-resolution mass spectrometry, but first we will consider certain points about isotopes and their masses. Most elements occur in nature as a mixture of atoms of different definite masses. For instance, on the $^{12}C = 12$ scale, the atomic weights of the three stable oxygen isotopes are 15.994915, 16.999133, and 17.999160. The atomic weight of naturally occurring oxygen (99.759% ^{16}O, 0.037% ^{17}O, and 0.204% ^{18}O) is the weighted average of these relative masses; in calculating a weighted average, appropriate account is taken of the relative abundances. With only six significant figures, a calculation of the weighted average of these atomic weights is

$$\begin{aligned} 0.99759 \times 15.9949 &= 15.9564 \\ 0.00037 \times 16.9991 &= 0.0063 \\ 0.00204 \times 17.9992 &= 0.0367 \\ \hline \mathrm{AW(O)} &= 15.9994 \end{aligned}$$

which is the atomic weight of oxygen. Atomic weights for the particular isotopes of oxygen given above as well as the conventional atomic weights given in Table 1–4 are for *neutral atoms*.

Notice that the atomic weight of oxygen on the carbon-12 scale is within 0.004% of 16 (exactly). A scale of atomic weights based on naturally occurring oxygen having an atomic weight of exactly 16 was used by chemists for many decades prior to adoption of the present scale. Many numerical data and derived numerical quantities found in chemical handbooks are based upon the old chemical atomic weight scale. The O = 16 and $^{12}C = 12$ atomic weight scales are so close that the errors in numerical quantities based upon the O = 16 scale can be ignored.

Of the stable elements occurring in the Earth's crust, twenty occur as a single nuclide: these are ^{9}Be, ^{19}F, ^{23}Na, ^{27}Al, ^{31}P, ^{45}Sc, ^{55}Mn, ^{59}Co, ^{75}As, ^{89}Y, ^{93}Nb, ^{103}Rh, ^{127}I, ^{133}Cs, ^{141}Pr, ^{159}Tb, ^{165}Ho, ^{169}Tm, ^{197}Au, and ^{209}Bi. Therefore the atomic weights given for these elements in Table 1–4 are the atomic weights for these particular nuclides. At the other extreme, the element with the largest number of stable isotopes is tin ($Z = 50$), with ten isotopes; with their natural abundance in parentheses, these are: ^{112}Sn (1.0%), ^{114}Sn (0.7%), ^{115}Sn (0.4%), ^{116}Sn (14.3%), ^{117}Sn (7.6%), ^{118}Sn (24.1%), ^{119}Sn (8.5%), ^{120}Sn (32.5%), ^{122}Sn (4.8%), and ^{124}Sn (6.1%). The atomic weight for tin given in Table 1–4 is the weighted average of the atomic weights of these ten nuclides.

There are no stable nuclides for any element with an atomic number equal to or greater than 84 (polonium); thus bismuth ($Z = 83$) is the element with largest atomic number that exists as a stable nuclide. However, there are elements with larger atomic numbers occurring in nature as unstable nuclides. Some of these nuclides (e.g., ^{238}U with a half-life of 4.5×10^9 years) have a half-life similar to or longer than the age of the Earth (about 4.6×10^9 years); others (e.g., ^{226}Ra with a half-life of 1620 years) exist in nature despite a relatively short half-life because they are produced continuously in the disintegration of long-lived unstable parent nuclei. (The rate of decay of an unstable nuclide can be characterized by the time interval during which one half of the atoms in a sample undergo decay; this is the half-life of the nuclide. It is discussed thoroughly in Chapter 22.)

Two elements of atomic number less than 84 do not exist on Earth as stable nuclides; one is technetium ($Z = 43$), the longest-lived isotope of which is ^{97}Tc, with a half-life of 2.6×10^6 years, and the other is promethium ($Z = 61$), the longest-lived isotope of which is ^{145}Pm, with a half-life of 25 years.

The chemical properties of different isotopes of most elements are the same. For compounds of the lightest elements, where the percentage differences between the masses of the isotopes are appreciable, this generalization is not strictly true. Chemical isotope effects are particularly striking for compounds of hydrogen, for which the two stable isotopes differ in mass by a factor of nearly two: ^1H, atomic weight 1.0078; ^2H (called deuterium), atomic weight 2.0141. A reaction that breaks a covalent chemical bond to hydrogen may be as much as eighteen-fold faster for the hydrogen-1 compound than for the hydrogen-2 compound.

In early studies of electrical discharge in gases at low pressure, scientists observed positive rays with a much lower ratio of charge to mass than was characteristic of cathode rays (electrons). At the time of their discovery, these were called canal rays. The canal rays were positive ions, atoms or molecules from which an electron had been removed, and the existence of isotopes was discovered in these early studies, which involved ancestors of the modern mass spectrograph.

The operation of a mass spectrograph is based on two properties of charged particles: a charged particle is accelerated in an electric field, and a

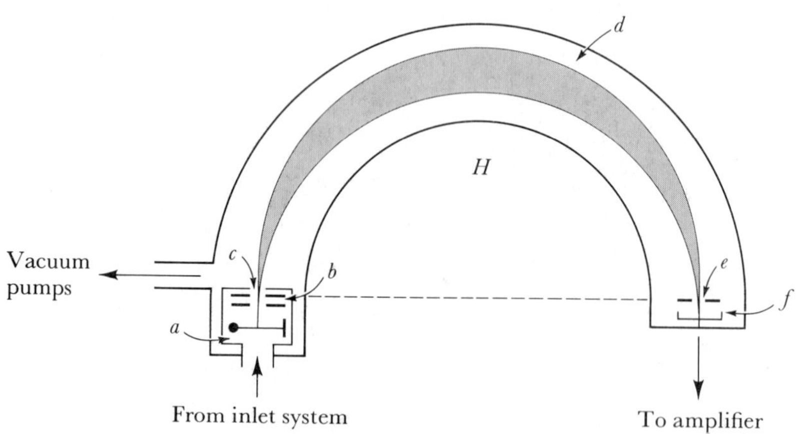

FIGURE 1–8

Dempster's apparatus for studying positive rays: (*a*) ion source, (*b*) plates between which there is a voltage drop of a few thousand volts, (*c*) source slit, (*d*) deflected beam in magnetic field perpendicular to plane of diagram, (*e*) exit slit, where beam is refocused after deflection of 180°, and (*f*) detector. *H* denotes the magnetic field. [From K. Biemann, *Mass Spectrometry*, McGraw-Hill Book Co., New York (1962) p. 6.]

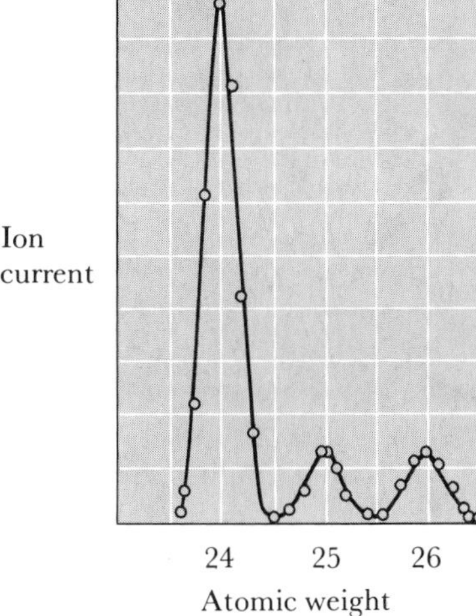

FIGURE 1–9
Data obtained in an analysis of magnesium by Dempster in 1920. Modern measurements give the abundance of magnesium isotopes as ^{24}Mg (78.8%), ^{25}Mg (10.1%), and ^{26}Mg (11.1%). [From J. D. Stranathan, *The Particles of Modern Physics*, Blakiston, New York (1942) p. 155.]

moving charged particle is deflected in a magnetic field. Mass spectrographs have been made with various designs using electric and magnetic fields to accelerate and deflect charged particles in such a way that particles with a particular mass-to-charge ratio have a particular trajectory. A simple design is that which was employed by DEMPSTER in his early studies on the masses of atoms. In Dempster's equipment (Figure 1–8), the charged particles that reach the detector have trajectories with a particular radius of curvature. The ratio of mass to charge corresponding to this radius of curvature is determined by the accelerating voltage and the magnetic field. When the accelerating voltage is changed, particles of different mass-to-charge ratio have a trajectory that allows them to go through the exit slit to be detected. Figure 1–9 shows a plot of data obtained by Dempster in 1920. The resolution is adequate to show the three isotopes of magnesium, but is far inferior to that obtained using a modern high-resolution instrument (see Figure 1–10).

Modern mass spectrometers give highly accurate values for the relative masses of atomic and molecular species. (Because the ion source produces primarily ions of charge +1, the mass-to-charge ratio gives us the mass.) Table 1–5 gives atomic weights for some particular nuclides.

It may surprise you that a highly accurate molecular weight for a particular isotopic species identifies the chemical composition of the species, but this is indeed the case. Consider the simple example of the three stable molecules with a nominal relative mass of 28: $^{12}C^{16}O$, $^{14}N^{14}N$, and $^{12}C_2$-$^{1}H_4$. The molecular weights of these species are (to six significant figures): 27.9949, 28.0062, and 28.0313, respectively.[14] If we consider only molecular

[14]These are molecular weights of the particular isotopic species, not conventional molecular weights for the substances containing the naturally occurring mixture of isotopes. The conventional molecular weights for these substances are CO, 28.0104; N_2, 28.0134; and C_2H_4, 28.0536.

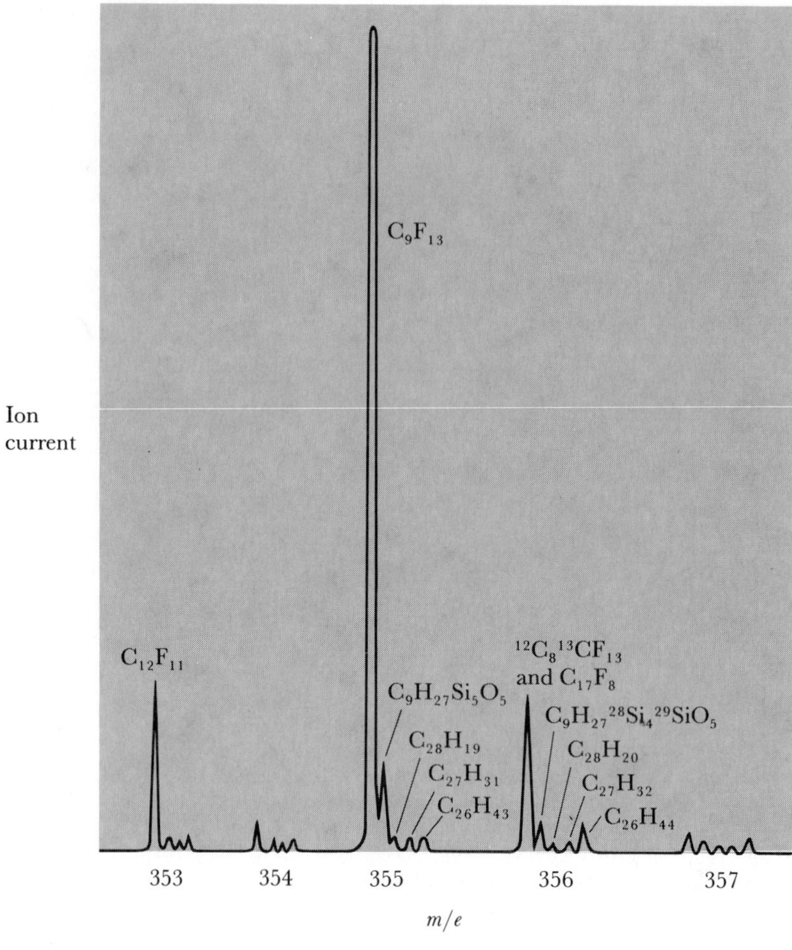

FIGURE 1–10
High-resolution mass spectrum of hydrocarbon and fluorocarbon species in the vicinity of mass number 355. Notice that at each mass number (355 or 356), there are hydrocarbon species that differ from one another by one ^{12}C and twelve ^{1}H (e.g., $C_{28}H_{20}$, $C_{27}H_{32}$, and $C_{26}H_{44}$). These species differ in *m/e* by $12 \times 1.00782 - 12 = 0.09384$. The silicon-containing species arise from silicone grease in the equipment. [From J. H. Beynon, in R. M. Elliott, *Advances in Mass Spectrometry*, Vol. II, Pergamon Press, London (1962).]

species containing isotopes that are present in the element in amounts greater than 2%, these are the only stable species of this nominal mass.[15] These species have molecular weights that can be distinguished using modern high-resolution mass spectrometers. If the molecular weight of a particular isotopic species was known to be 28.01 ± 0.02, the substance would not be identified by this information, but knowing the substance to have a molecular weight of 28.006 ± 0.002 would identify the substance to

[15]The isotopic species $^{13}C_2{}^{1}H_2$ (MW 28.0224), $^{1}H^{13}C^{14}N$ (MW 28.0143), $^{2}H^{12}C^{14}N$ (MW 28.0172), or $^{1}H^{12}C^{15}N$ (MW 28.0079) are not being considered because these molecular species involve isotopes that are present in nature in very small amounts. The relative abundance of the rarer isotopes of elements in these species are: ^{13}C, 1.11% of carbon; ^{2}H, 0.015% of hydrogen; and ^{15}N, 0.37% of nitrogen.

TABLE 1–5
Atomic Weights for Certain Nuclides

Element	Nuclide	Atomic weight
Hydrogen	^{1}H	1.00782
	^{2}H	2.01410
Carbon	^{12}C	12 (definition)
	^{13}C	13.0034
	^{14}C	14.0032
Nitrogen	^{14}N	14.0031
	^{15}N	15.0001
Oxygen	^{16}O	15.9949
	^{17}O	16.9991
	^{18}O	17.9992
Fluorine	^{19}F	18.9984
Uranium	^{235}U	235.0439
	^{238}U	238.0508

be molecular nitrogen, N_2. The resolution ability of the best instruments now available exceeds that expressed by the error limits given above, and the elemental composition of compounds now is being determined routinely by high-resolution mass spectrometry.

Chlorine exists in nature as a mixture of two isotopes present in the amounts:

$$^{35}Cl \quad 75.53 \text{ atom } \%$$
$$^{37}Cl \quad 24.47 \text{ atom } \%$$

Therefore the presence of a single chlorine atom in a molecule is suggested by a mass spectrum in which there is a pair of peaks separated by two units of mass, the peak at the higher mass having one third of the intensity of the peak at the lower mass. Such a spectrum is shown in Figure 1–11.

Calculating the relative abundances of different isotopic molecules containing two or more chlorine atoms is a more complicated problem. The laws of probability dictate the relative abundances. Some feeling for the problem can be obtained by considering the relative abundances of the three kinds of diatomic chlorine molecules: $^{35}Cl^{35}Cl$, $^{35}Cl^{37}Cl$, and ^{37}Cl-^{37}Cl. We can simplify the problem by assuming that the two isotopes are present in the proportions three atoms of ^{35}Cl to one atom of ^{37}Cl (i.e., 75% ^{35}Cl and 25% ^{37}Cl). The relative amounts of the three kinds of diatomic chlorine molecules are the same as the relative numbers of different combinations of pairs of balls drawn from a large supply in which there are three times as many white balls as black balls. The probability is 3/4 for drawing a white ball from such a supply, and it is 1/4 for drawing a black ball. The probability of drawing balls of two particular colors in succession is the

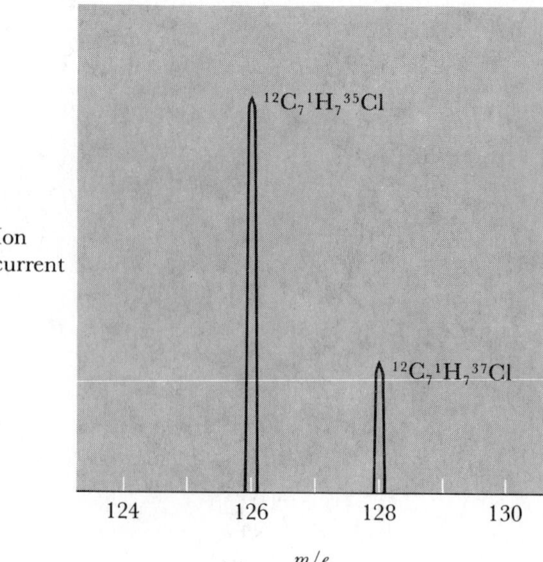

FIGURE 1–11
The mass spectrum of a substance containing one chlorine atom per molecule (C_7H_7Cl, benzyl chloride).

product of the individual probabilities. In the example we are considering, the proportions of the three different combinations are:

$$\text{white, white } (^{35}\text{Cl}^{35}\text{Cl}) \qquad \frac{3}{4} \times \frac{3}{4} = \frac{9}{16}$$

$$\text{white, black } (^{35}\text{Cl}^{37}\text{Cl}) \qquad 2 \times \frac{3}{4} \times \frac{1}{4} = \frac{6}{16}$$

$$\text{black, black } (^{37}\text{Cl}^{37}\text{Cl}) \qquad \frac{1}{4} \times \frac{1}{4} = \frac{1}{16}$$

The proportions are 9:6:1. The factor two is included in calculation of the relative number of pairs of unlike balls (white, black or $^{35}Cl^{37}Cl$) because there are two ways of drawing these combinations of balls: the white one may be drawn first, or the black one may be drawn first.

In addition to revealing the elemental composition of a substance, the mass spectrum also may shed light on its molecular structure. Processes in addition to removal of a single electron from a molecule may occur in the ionization chamber of a mass spectrometer. The molecule may be fragmented, and the masses of the fragment ions may reveal the structure of the parent molecule. Let us reconsider the three molecular species of nominal molecular weight 28, carbon monoxide (CO), nitrogen (N_2), and ethylene (C_2H_4), which cannot be distinguished by the molecular weights of the predominant molecular species if these relative masses are known only to the nearest 0.1 unit.

Even with such low resolving power, however, the substances can be distinguished if fragment ions as well as the parent ions are detected. The possible fragment ions of these molecules are:

parent:	$^{12}C^{16}O^+$	mass 28.0	parent:	$^{12}C_2{}^1H_4$	mass 28.0
fragments:	$^{12}C^+$	12.0	fragments:	$^{12}C_2{}^1H_3^+$	27.0
	$^{16}O^+$	16.0		$^{12}C_2{}^1H_2^+$	26.0
				$^{12}C_2{}^1H^+$	25.0
parent:	$^{14}N_2^+$	mass 28.0		$^{12}C_2^+$	24.0
fragment:	$^{14}N^+$	14.0		$^{12}C^1H_2^+$	14.0
				$^{12}C^1H^+$	13.0
				$^{12}C^+$	12.0

The fragmentation pattern is different for each of these substances, and it is clear that they could be distinguished by measurement of the mass spectrum under low resolution.

Notice that the fragment ion CH_3^+ is not reported in the mass spectrum of C_2H_4. Ethylene with its molecular structure H_2CCH_2 cannot give CH_3 in the absence of rearrangement. This simple example shows an additional use of mass spectrometry. The fragmentation pattern may give information about the molecular structure as well as about the elemental composition of a substance.

Biographical Notes

ANDERS J. ÅNGSTRÖM (1814–1874), Swedish physicist and astronomer, who was a founder of the field of spectroscopy; the angstrom unit was named in his honor. He was a pioneer in making a spectral analysis of the sun and the first to observe hydrogen in the sun.

AMADEO AVOGADRO (1776–1856) was Professor of Physics in his native town of Turin, Italy. He also practiced law. The hypothesis mentioned in the text, now known as Avogadro's hypothesis, was largely ignored during his lifetime.

STANISLAO CANNIZARO (1826–1910), an Italian chemist, was Professor of Chemistry at several universities (Alessandria, Genoa, Palermo, and Rome). He made important contributions to organic chemistry, as well as atomic theory. Earlier he participated in the Sicilian revolt.

ANDERS CELSIUS (1701–1744), a Swedish astronomer, devised the temperature scale that bears his name. (This scale was formerly called the centigrade scale.) He also verified Newton's theory that the Earth was flattened at the poles.

JOHN DALTON (1766–1844) was a school teacher with broad interests. He helped lay the foundations of meteorology as a science, and in 1794 gave the first description of color blindness, from which he suffered. He formulated his atomic theory in the period 1800–1803. The first volume of Dalton's book, *New System of Chemical Philosophy*, was published in two parts in the period 1808–1810; this book helped call attention to his atomic theory.

ARTHUR J. DEMPSTER (1886–1950), an American physicist, was born in Canada and received his undergraduate training there (University of Toronto). After advanced study at Göttingen, Munich, and Wurzburg, he went to the University of Chicago, where he received his Ph.D. in 1916. He was a faculty member at the University of Chicago for many years. In his research with the mass spectrometer, he discovered uranium-235, the isotope of uranium that is fissionable with slow neutrons.

GABRIEL D. FAHRENHEIT (1686–1736), a physicist, was born in Danzig, then part of Poland, and died in the Netherlands. He invented the alcohol-in-glass and mercury-in-glass thermometers and he devised the temperature scale that bears his name.

JOSEPH L. GAY–LUSSAC (1778–1850) was a French physicist and chemist. In addition to studies on the combining volumes of gases, his work included electrochemical studies and the development of manufacturing processes for sulfuric acid.

LORD KELVIN (see WILLIAM THOMSON).

WENDELL M. LATIMER (1893–1955) was Professor of Chemistry in the University of California at Berkeley. He is known principally for his book *Oxidation Potentials*, which did much to systematize oxidation–reduction chemistry. The diagrams used in our discussion of the electromotive force for half-reactions (Chapter 16) were introduced in his book, and they sometimes are called Latimer diagrams.

HENRI L. LE CHATELIER (1850–1936), a French chemist, developed his ideas about the effect of pressure upon equilibrium in connection with the chemistry of cements. He was Professor of Chemistry in the École des Mines.

GILBERT N. LEWIS (1875–1946), an American chemist, was educated at Nebraska and Harvard (A.B., 1896; Ph.D., 1899); he also studied in Europe. He was a faculty member at the Massachusetts Institute of Technology from 1907 to 1912, when he moved to Berkeley where for the rest of his life he did outstanding work on atomic and molecular structure and in chemical thermodynamics.

ISAAC NEWTON (1643–1726), English mathematician, physicist, and astronomer, was a true genius. Largely self-taught in science, he discovered calculus, which was developed simultaneously by Leibniz (1646–1716), developed the law of universal gravitation and the laws of motion upon which classical mechanics is based, and made other notable contributions to science.

WORTH H. RODEBUSH (1887–1959) was a Professor of Chemistry in the University of Illinois for many years after receiving his doctor's degree at Berkeley.

WILLIAM THOMSON, the first Lord Kelvin (1824–1907), was a British mathematician and physicist, who was educated at Cambridge; he was professor of natural philosophy at the University of Glasgow, Scotland, for 53 years (1846–1899). He did important work in thermodynamics, and also in electricity and magnetism. The kelvin unit was named after Thomson.

Problems and Questions

1–1 When a certain liquid is boiled, it boils away completely at a constant temperature. However, when this liquid is cooled and slow solidification occurs, the temperature of freezing decreases as solidification continues. Is this liquid a pure substance or a solution?

1–2 For each of the following categories list six examples that you encounter in everyday life:

heterogeneous mixtures	compounds
liquid solutions	elements

1–3 Convert the melting points (given on the Celsius scale) of each of these elements to a temperature on the Fahrenheit scale:

N_2	$-209.9°C$	Na	$97.8°C$
Cl_2	$-101.0°C$	Fe	$1535°C$

1–4 The density of aluminum metal is $2.70\ g\ cm^{-3}$. What is the mass of a block of aluminum $1.00\ ft \times 1.00\ ft \times 6.00\ in.$?

1–5 Strict adherence to SI units would require you to express density in the units $kg\ m^{-3}$. What is the density of aluminum expressed in this set of units?

1–6 What is the molecular weight of each of these common chemical compounds?

sodium bicarbonate, $NaHCO_3$	laughing gas, N_2O
silica, SiO_2	limestone, $CaCO_3$
epsom salts, $MgSO_4 \cdot 7H_2O$	ozone, O_3

1–7 What is the mass of each of the following samples of matter?

1.00×10^{20} molecules of C_2H_6	3.40 mol CO_2
1.50×10^{-4} mol CH_3CO_2H	1.50 mmol $C_6H_{12}O_6$

What fraction of the atoms in each of these compounds are carbon atoms? What fraction of the mass of each of these compounds is carbon?

1–8 How many atoms are present in each of the following samples of matter?

1.00 mg	H_2O	1.00 mmol	$Ca(OH)_2$
1.00 kg	O_3	3.00 mol	CO_2
1.00 kg	O_2	40.0 mg	CO
40.0 g	SO_3		

1–9 How many moles are present in each of the following samples?

100 g NaCl	1.00 kg O_3
15.0 kg Al_2O_3	1.00 kg O_2
$10.0\ dm^3$ liquid *n*-octane (C_8H_{18}) (density = $0.703\ g\ cm^{-3}$)	1.00 oz gold
10.0 lb sugar (sucrose, $C_{12}H_{22}O_{11}$)	

1–10 Arrange the following in order of increasing number of atoms:

4.0 g hydrogen	4.0 g helium
1.0 mol O_2	32 g SO_2
30 g water	32 g SO_3
300 g uranium	1.0 mol O_3

1–11 The densities at 20°C of several elements and compounds are given below:

Al(*s*)	$2.70\ g\ cm^{-3}$	NaCl(*s*)	$2.17\ g\ cm^{-3}$
Be(*s*)	$1.85\ g\ cm^{-3}$	$SiO_2(s)$	$2.66\ g\ cm^{-3}$
$Br_2(l)$	$3.12\ g\ cm^{-3}$	*n*-$C_8H_{18}(l)$	$0.703\ g\ cm^{-3}$
Ir(*s*)	$22.42\ g\ cm^{-3}$	$C_{12}H_{22}O_{11}(s)$	$1.58\ g\ cm^{-3}$

Fe(*s*)	7.87 g cm^{-3}	$CO(NH_2)_2(s)$	1.32 g cm^{-3}
U(*s*)	18.95 g cm^{-3}	$C_6H_4Cl_2(s)$	1.29 g cm^{-3}

What is the volume of one mole of each of these substances? How many atoms are in 1.00 cm^3 of each of these substances?

1–12 The principal uranium ore found in the United States is U_3O_8. Rich deposits contain 0.50% U_3O_8 by weight. What weight of such an ore would have to be processed to obtain 1000 kg of uranium metal, if no loss occurred?

1–13 Sulfuric acid (H_2SO_4) is the chemical that is produced in the largest amount in the United States. In 1977, 6.88×10^{10} lbs were produced. How many lbs of sulfur were used to produce this amount of sulfuric acid? How many moles of S_8 is this? The density of elemental sulfur (rhombic modification) is 2.07 g cm^{-3}. What volume of elemental sulfur was used to produce this amount of sulfuric acid?

1–14 What is the weight percentage of iron in each of the oxides of iron: FeO, Fe_3O_4, and Fe_2O_3? Show that these values are consistent with the law of multiple proportions.

1–15 The percentage composition by weight of three compounds of nitrogen and hydrogen and the volumes of the constituent element gases combined in one volume of each of these compounds are:

	Weight percentage: H	Weight percentage: N	Vol H_2 / Vol Cmpd	Vol N_2 / Vol Cmpd
Compound I	17.75	82.25	1.50	0.50
Compound II	12.58	87.42	2.00	1.00
Compound III	2.34	97.66	0.50	1.50

Using the same steps as employed in the chapter (for the three compounds of nitrogen and oxygen), derive the molecular formulas of these compounds and the relative atomic weights of hydrogen and nitrogen.

1–16 Known oxides of uranium have the following weight percentages of oxygen: (a) 11.85%, (b) 15.20%, and (c) 16.78. Show that the compositions of these compounds conform to the law of multiple proportions. With the atomic weight of oxygen given [AW(O) = 16.00], give two different possible values of the atomic weight for uranium; for each possible atomic weight of uranium, give the simplest chemical formula for each of these oxides.

1–17 Given below are the weights of each of several gaseous chlorine-containing compounds that occupy the same volume as 32.0 g of oxygen gas. (These volumes were measured at the same temperature and pressure.) Also given are the percentages of chlorine in each compound.

Substance	Weight occupying same volume as 32.0 g oxygen	Weight % Cl
chlorine	70.90 g	100
methyl chloride	50.49 g	70.2
chloroform	119.38 g	89.1
chlorine oxide (1)	67.45 g	52.6
chlorine oxide (2)	86.91 g	81.6

Calculate the weight of chlorine in one mole of each compound. What is the largest value of the atomic weight for chlorine that is consistent with these data? Given this atomic weight, how many atoms of chlorine are there in a molecule of each of these substances?

1–18 An iron carbide, cementite, present in cast iron, has the composition (weight percentage) 93.31% Fe, 6.69% C. What is its simplest formula?

1–19 In the elemental state under ordinary conditions, bromine is a liquid with density 3.12 g cm^{-3}; metallic lead, a solid, has a density of 11.34 g cm^{-3}. Equal volumes of solid lead and liquid bromine are mixed; quantitative formation of lead dibromide, $PbBr_2$ (density = 6.66 g cm^{-3}), occurs. Which reagent was in excess? What weight of lead dibromide forms if 10.00 g of metallic lead is present initially? What is the volume of the mixture before and after reaction if the initial mixture contains 10.00 g of lead?

1–20 A substance A_2B has the composition by weight 40% A, 60% B. What is the weight percentage composition of the substance AB_2?

1–21 A pure substance has the composition by weight 53.3% C, 15.7% H, 31.0% N. What is the simplest formula of this compound?

1–22 A pure substances has the composition by weight 74.03% C, 8.70% H, and 17.27% N. Show that the compositions of this compound and the compound of Problem 1–21 are consistent with the law of multiple proportions. What is the simplest formula of this compound?

1–23 A substance containing carbon, hydrogen, and oxygen has the following percentage composition by weight: 48.6% C, 8.2% H, and 43.2% O. What is the simplest formula for this compound?

1–24 In the analysis of organic compounds containing C, H, and O, the sample is burned in excess oxygen; this converts all of the carbon to CO_2 and all of the hydrogen to H_2O. These products then are separated and weighed. Analysis of a sample weighing 0.1542 g gives 0.2211 g CO_2 and 0.1207 g H_2O. What is the simplest formula of this compound? (Assume that all of the original compound not accounted for by carbon and hydrogen is oxygen.)

1–25 Upon combustion, a 0.524-g sample of another organic compound containing C, H, and O produces 1.190 g CO_2 and 0.488 g H_2O. What is the simplest formula of this compound?

1–26 A pure substance containing chlorine, hydrogen, carbon, and oxygen has the percentage composition by weight 49.6% Cl, 2.80% H, 25.2% C, and 22.4% O. What is the simplest formula for this compound?

1–27 What fraction of the weight of the hydrated salt $CaCl_2 \cdot 6H_2O$ is lost when it is heated strongly enough to drive off all of the water?

1–28 The two chlorofluorocarbons used as propellants in aerosol dispensers, and implicated in ozone destruction in the atmosphere, have weight percentage data:

I 10.0% C, 58.6% Cl, 31.4% F

II 11.5% C, 33.9% Cl, 54.6% F

Show that these data are consistent with the law of multiple proportions. Using the atomic weights of these elements, derive the empirical formulas of these compounds.

1–29 What is the percentage composition of each of the compounds $C_{10}H_{20}O_2$ and $C_{11}H_{22}O_2$? What accuracy in analysis for each of the constituent elements is necessary to allow these substances to be distinguished by percentage composition data?

1–30 Chlorine occurs in nature as a mixture of two isotopes of mass number 35 and 37; the atom percentages are 75.53% ^{35}Cl and 24.47% ^{37}Cl. The atomic weights of these two isotopes are 34.969 and 36.966, respectively. What is the conventional atomic weight of chlorine? How many grams of ^{35}Cl are present in 100.0 g of natural chlorine?

1–31 Rubidium (AW = 85.4678) exists on Earth as a mixture, a stable isotope ^{85}Rb (AW = 84.9117) and a very long-lived unstable isotope ^{87}Rb. If the ratio of the number of atoms of the two isotopes $n_{85}/n_{87} = 2.591$, what is the atomic weight of rubidium-87?

1–32 The conventional atomic weight of copper is 63.546. This element in nature is made up of two isotopes: ^{63}Cu (AW = 62.930) and ^{65}Cu (AW = 64.928). What fraction of the atoms in natural copper is ^{65}Cu?

1–33 Silicon exists in nature as a mixture of three stable isotopes 92.2% ^{28}Si (AW = 27.977), 4.7% ^{29}Si (AW = 28.977), and 3.1% ^{30}Si (AW = 29.974). What is the conventional atomic weight of silicon?

1–34 Two molecules with nominal values of the molecular weight equal to 32 are molecular oxygen ($^{16}O^{16}O$) and methyl alcohol ($^{1}H_3{}^{12}C^{16}O^{1}H$). High-precision mass spectrometry of an unknown substance gives a parent peak at mass = 32.0262. Which of these substances is the unknown?

1–35 Three molecules with nominal molecular weight 30 are $^{12}C_2{}^{1}H_6$, $^{14}N^{16}O$, and $^{1}H_2{}^{12}C^{16}O$. Study of a substance in a high-resolution mass spectrometer gives the molecular weight of the parent peak to be 30.0105. Which of the above molecules is responsible for this peak?

2

Chemical Reactions

2–1 Introduction

In the preceding chapter, we learned that chemical compounds have compositions determined by the particular combinations of elements that they contain. The processes by which chemical substances (compounds or elements) are transformed into other chemical substances are called *chemical reactions*. Some chemical reactions occur in nature without our intervention, but man causes others to occur. Some reactions involve simple substances (elements or compounds consisting of molecules containing, at most, a few atoms), and other reactions involve complex molecules. The reaction of nitrogen (N_2) with oxygen (O_2) to form nitric oxide (NO) is a reaction of simple substances that occurs in nature and also is caused by man. This reaction, which occurs in the atmosphere during an electrical storm, is part of the nitrogen cycle, a complex sequence of reactions that converts atmospheric nitrogen into chemical compounds and living matter (Figure 2–1). The cycle is completed when these substances, in turn, decompose to yield nitrogen, which escapes to the atmosphere. This same reaction occurs in the internal combustion engines of automobiles and in stationary power plants, where air is heated to high temperatures by the combustion of a fuel. This reaction is both a benefit and a detriment. Life depends upon the nitrogen cycle, but excessive production of nitric oxide in urban centers is the first step in the production of photochemical smog.

(a) THE NITROGEN CYCLE

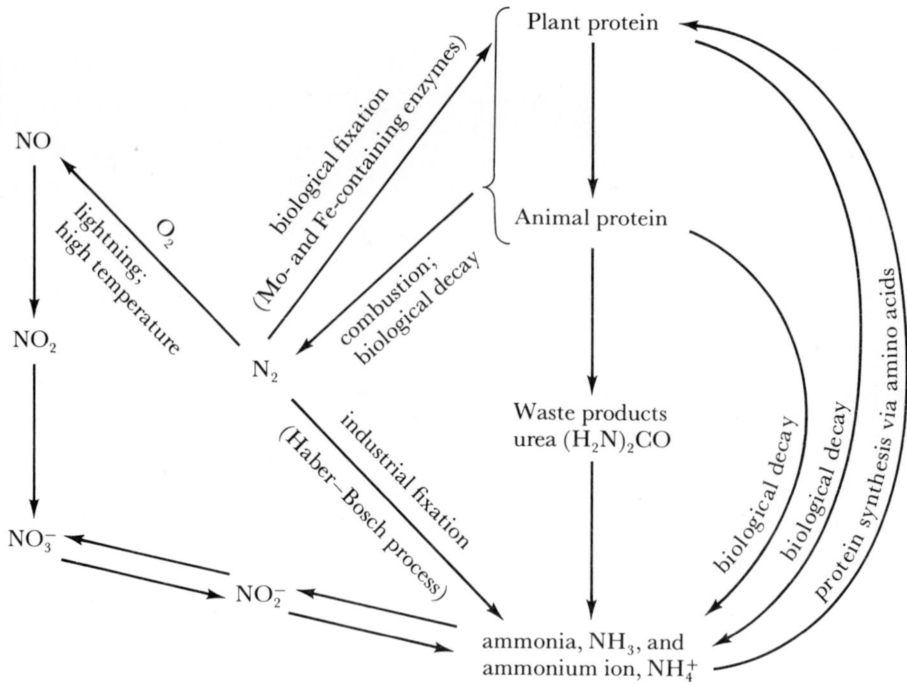

(b) THE CHEMICAL REACTIONS

$$N_2(g) + O_2(g) = 2NO(g)$$
$$2NO(g) + O_2(g) = 2NO_2(g)$$
$$3NO_2(g) + H_2O = NO(g) + 2H^+ + 2NO_3^-$$

FIGURE 2–1
The reaction of nitrogen and oxygen to give nitric oxide, NO, nitrogen dioxide, NO_2, and nitrate ion, NO_3^-: (a) as part of the nitrogen cycle; (b) the chemical reactions.

The statement *nitrogen plus oxygen gives nitric oxide* defines this reaction, but it does not reveal quantitative features of the reaction. Such features are contained in a balanced chemical equation for the reaction:

$$N_2(g) + O_2(g) = 2NO(g)$$

This chemical equation, which includes designation of the physical state (*g* for gas) of each reactant and the product, is balanced in the sense that the same number of atoms of each kind appears on the product side and the reactant side of the equation. The chemical equation can be read to mean that the reaction of one molecule of nitrogen (N_2) with one molecule of oxygen (O_2) produces two molecules of nitric oxide (NO). The coefficients of formulas in the chemical equation also give the relative numbers of moles of reactants and products involved in the chemical reaction. Using the atomic weights of nitrogen and oxygen, we can give the chemical equation meaning also in terms of relative masses of reactants and products.

For reactions involving gases, as in the present example, we can use Avogadro's principle and interpret the coefficients in terms of the relative volumes of the reactants and products, the volumes being measured at a particular temperature and pressure. All of this can be summarized:

	$N_2(g)$	+ $O_2(g)$	= $2NO(g)$
Relative numbers of molecules:	1 molecule	1 molecule	2 molecules
Relative numbers of moles:	1 mole	1 mole	2 moles
Relative masses:	1 mol × (2 × 14.01 g mol^{-1}) = 28.02 g	1 mol × (2 × 16.00 g mol^{-1}) = 32.00 g	2 mol × (30.01 g mol^{-1}) = 60.02 g
Relative volumes:	1 volume	1 volume	2 volumes

The quantitative relationships expressed in this chemical equation are called the stoichiometric[1] relationships of the reaction. (This term is used in other contexts with a similar meaning. The fixed proportions of nitrogen and oxygen in nitric oxide are called stoichiometric proportions.) The chemical equation gives us no information on certain points:

1. the conditions of temperature and pressure of reactant and product gases under which the reaction occurs at a measurable rate,
2. the extent to which the reaction occurs under particular conditions of temperature and pressure,
3. the possibility of occurrence of other reactions of the reactants (and/or products) under the conditions in question; for instance, the chemical equation for nitric oxide synthesis gives no information regarding the possible formation of nitrous oxide (N_2O) from the same reactants:

$$2N_2(g) + O_2(g) = 2N_2O(g)$$

2–2 Balanced Chemical Equations

In a balanced chemical equation, atoms of each kind (i.e., atoms of each participating element) are conserved. For a reaction involving ions (atomic and molecular species with a net charge) the sum of the charges on the reactants must be equal to the sum of the charges on the products. To write a correctly balanced equation, you first must know the correct formulas of the reactants and products. Such information is obtained from direct experimental observation of the chemical reaction in question, or perhaps from knowledge of the general properties of the class of substances being considered. The products of a reaction *should not* be predicted simply on the basis that the resulting equation is easily balanced. This point can be illustrated with an example. When potassium chlorate ($KClO_3$) is heated,

[1]From the Greek word *stoicheio*, meaning component.

a gas evolves and the solid residue is potassium chloride (KCl). Not knowing the identity of the gaseous product, you might be tempted to complete the equation as follows:

$$KClO_3(s) = KCl(s) + O_3(g)$$

but *this is not the reaction that occurs*. The gaseous product is molecular oxygen (O_2), not ozone (O_3). The correctly balanced equation,

$$2KClO_3(s) = 2KCl(s) + 3O_2(g)$$

can be obtained by starting with a statement that involves correct chemical formulas of reactant and products:

$$KClO_3 \text{ gives } KCl \text{ plus } O_2$$

Inspection shows that a coefficient of 2 for $KClO_3$ and 3 for O_2 will balance oxygen atoms, and a coefficient of 2 for KCl will balance potassium and chlorine atoms.

A more complicated example is the burning of heptane (C_7H_{16}) with excess oxygen, which gives carbon dioxide plus water:

$$C_7H_{16} \text{ plus } O_2 \text{ gives } CO_2 \text{ plus } H_2O$$

To balance the equation for a complicated reaction, you begin by choosing the coefficient for some one reactant or product. A good rule of thumb is to set the coefficient of the most complex substance as one. In the present example, the most complex reactant or product is heptane; the steps are:

1. select the coefficient 1 for C_7H_{16};
2. this establishes the coefficient 7 for CO_2 and 8 for H_2O to balance carbon atoms and hydrogen atoms, respectively;
3. then (2) establishes the coefficient 11 for O_2 to balance oxygen atoms; this gives the equation

$$1C_7H_{16} + 11O_2 = 7CO_2 + 8H_2O$$

It is customary to omit one as the coefficient in a chemical equation. Therefore this chemical equation is written

$$C_7H_{16} + 11O_2 = 7CO_2 + 8H_2O$$

The coefficient one is understood where no coefficient is given explicitly.

In the presence of insufficient oxygen, carbon monoxide is produced when heptane is burned; the balanced equation for this reaction, obtained by the same steps as just used, is

$$C_7H_{16} + \tfrac{15}{2}O_2 = 7CO + 8H_2O$$

This equation gives the correct relative proportions of reactants and products, and the equation is acceptable. However, it is preferable to have an equation balanced with whole-number coefficients; this is accomplished by doubling each coefficient to give

$$2C_7H_{16} + 15O_2 = 14CO + 16H_2O$$

Suppose that under some particular set of conditions a sample of heptane is burned to give one half carbon monoxide and one half carbon dioxide. This chemical change is expressed by the above two independent chemical equations plus a qualifying statement regarding the relative amounts of heptane reacting in each of the reactions. This chemical change *should not* be expressed in terms of a single chemical equation obtained by adding the equations for the *two independent* reactions. Such a procedure (the addition of chemical equations for independent chemical reactions) is not generally useful, because the resulting chemical equation implies stoichiometric relationships that may hold over only a very limited range of conditions.

The balanced equations for the two oxidation reactions of heptane,

$$2C_7H_{16} + 15O_2 = 14CO + 16H_2O$$
$$C_7H_{16} + 11O_2 = 7CO_2 + 8H_2O$$

represent overall chemical changes. It is certain that each of these reactions occurs in a sequence of steps, with the partially oxidized product of one step being oxidized further in a following step. Although the identity of these intermediates is a subject of interest, the balanced equation for the *overall reaction* contains no information about these details.

Many substances dissociate to give ions in solvents with appropriate properties (high polarity and appropriate solvating ability, properties to be discussed later). In the balanced equations for reactions of ionic species in such solvents, account is taken of this dissociation. The equation for neutralization of an aqueous (water) solution of sodium hydroxide (NaOH) with an aqueous solution of hydrochloric acid (HCl) can be written

$$HCl + NaOH = H_2O + NaCl$$

but this chemical equation takes no account of the essentially complete dissociation of hydrochloric acid, sodium hydroxide, and sodium chloride in aqueous solution. If these solutes are represented by the dissociated ions, the equation

$$H^+ + Cl^- + Na^+ + OH^- = H_2O + Na^+ + Cl^-$$

results. But sodium ion and chloride ion appear on each side of this equation. The neutralization reaction has not altered their amounts. Deletion of these species from each side of the equation gives the *balanced equation for the net reaction*:

$$H^+ + OH^- = H_2O \qquad \text{or} \qquad H_3O^+ + OH^- = 2H_2O$$

if account is taken of the solvation of the proton by one water molecule. The balanced equation for the net reaction can be derived in the manner just outlined or by considering the predominant species in the solution before and after reaction. The predominant solute species in the reactant solutions and in the product solution for the present example are:

Reactant solutions:	hydrochloric acid	H_3O^+ and Cl^-
	sodium hydroxide	Na^+ and OH^-
Product solution: sodium chloride		Na^+ and Cl^-

Therefore it is clear that the neutralization reaction has consumed hydronium ion (H_3O^+) and hydroxide ion (OH^-), but has not consumed sodium ion or chloride ion. The balanced equation for the net reaction shows this; it does not involve either sodium ion or chloride ion.

The neutralization of an aqueous solution of acetic acid (a slightly dissociated acid) with sodium hydroxide is different. The predominant solute species in the reactant and product solutions are:

Reactant solutions:	acetic acid	HOAc (Ac stands for C_2H_3O)
	sodium hydroxide	Na^+ and OH^-
Product solution: sodium acetate		Na^+ and OAc^-

Therefore the equation for the net reaction is

$$HOAc + OH^- = H_2O + OAc^-$$

The precipitation of mercury(II)[2] sulfide (HgS) from a solution of mercury(II) in hydrochloric acid solution by the action of hydrogen sulfide (a very slightly dissociated acid) involves reactant and product solutions containing the following predominant species:

Reactant solutions:	hydrogen sulfide	H_2S
	mercury(II) in HCl solution	$HgCl_4^{2-}$
Product solution:	precipitated mercury(II) sulfide	HgS
	chloride ion	Cl^-

The net reaction is

$$HgCl_4^{2-} + H_2S = HgS + 2H^+ + 4Cl^-$$

or

$$HgCl_4^{2-} + H_2S + 2H_2O = HgS + 2H_3O^+ + 4Cl^-$$

if the hydration of hydrogen ion is shown.

An important part of the formula of a chemical species is its net charge. A balanced chemical equation has the same net charge on each side. For some reactions, this property of a balanced chemical equation may provide you with a useful step in the process of balancing the equation. Sodium metavanadate ($NaVO_3$) is soluble in aqueous solutions of sodium hydroxide. In such solutions the vanadium is present as orthovanadate ion (VO_4^{3-}). To balance the equation for the reaction VO_3^- gives VO_4^{3-}, you recognize that the charge on the two sides of the equation is balanced by hydroxide ion (OH^-) on the reactant side:

$$VO_3^- + 2OH^- = VO_4^{3-}$$

[2]The Roman numeral II in this name designates the oxidation state of mercury in this compound. The concept of oxidation state is described fully in Chapters 11 and 16.

You then balance the hydrogen by adding water on the product side:

$$VO_3^- + 2OH^- = VO_4^{3-} + H_2O$$

and the result is a balanced equation. It is clear that to balance chemical equations for reactions in solution, we must know the correct chemical formulas for the species present in reactant and product solutions.

The chemical equations for some reactions involving oxidation and reduction are so complex that special balancing procedures have been developed. These will be described fully in Chapter 16.

2–3 Calculations Involving Chemical Reactions

The quantitative aspects of a balanced chemical equation already have been introduced. The relative masses and relative numbers of moles of reactants and products are given by the chemical equation. Using this information, we can do many types of practical quantitative calculations, which are called *stoichiometric calculations*.

Here are some stoichiometric calculations pertaining to the synthesis of ammonia (NH_3) from its constituent elements. The balanced equation for the reaction of nitrogen (N_2) with hydrogen (H_2) to form ammonia (NH_3),

$$N_2(g) + 3H_2(g) = 2NH_3(g)$$

conveys certain information:

1. In this reaction of one mole of nitrogen gas with three moles of hydrogen gas, two moles of ammonia gas are produced.
2. Use of atomic weights [AW(H) = 1.008 and AW(N) = 14.01] allows the statement under (1) to be converted to: In this reaction, for every 28.02 g of nitrogen gas that react with 6.048 g of hydrogen gas, 34.07 g of ammonia gas are produced.
3. Use of Avogadro's principle (equal volumes of any gas at a particular temperature and pressure contain the same number of molecules) allows the statement under (1) to be converted to: In a reaction of one volume of nitrogen gas with three volumes of hydrogen gas, two volumes of ammonia gas are produced (each volume being measured at the same temperature and pressure).
4. As we will learn in Chapter 4, one mole of an ideal gas occupies 22.41 liters at standard conditions (1 atm pressure and 0°C). This fact allows the statements under (1) and (2) to be converted to: In reaction of 22.41 liters of nitrogen gas with 67.23 liters of hydrogen gas, 44.82 liters of ammonia gas are produced (with the volume of each gas measured at standard conditions). (This statement is based upon the assumption that these gases are ideal. The meaning and correctness of this assumption about gas-ideality will be clarified in Chapter 4.)

The information conveyed by the balanced equation can be summarized:

	$N_2(g)$ +	$3H_2(g)$ =	$2NH_3(g)$
Relative numbers of moles:	1 mole	3 moles	2 moles
Relative masses:	1 mol × 28.02 g mol^{-1} = 28.02 g	3 mol × 2.016 g mol^{-1} = 6.05 g	2 mol × 17.034 g mol^{-1} = 34.07 g
Relative volumes:	1	3	2
Relative volumes (at std. cond.):	1 mol × 22.41 L mol^{-1} = 22.41 L	3 mol × 22.41 L mol^{-1} = 67.23 L	2 mol × 22.41 L mol^{-1} = 44.82 L

We can answer questions regarding relative amounts of reactants and products participating in the reaction by considering the meaning of the balanced chemical equation as expressed above.

EXAMPLE: What weight of ammonia is produced in the reaction of 10.50 g of hydrogen gas with an excess of nitrogen gas?

The answer to this question follows simply from the number of moles of hydrogen:

$$10.50\,\cancel{g} \times \frac{1\text{ mol H}_2}{2.016\,\cancel{g}} = 5.208\text{ mol H}_2$$

The number of moles of ammonia produced is obtained by considering the relative coefficients in the chemical equation

$$3H_2 \rightarrow 2NH_3$$

$$5.208\,\cancel{\text{mol H}_2} \times \frac{2\text{ mol NH}_3}{3\,\cancel{\text{mol H}_2}} = 3.472\text{ mol NH}_3$$

The weight of ammonia is then obtained:

$$3.472\,\cancel{\text{mol}} \times 17.03\text{ g}\,\cancel{\text{mol}^{-1}} = 59.13\text{ g}$$

These steps can be combined:

$$\text{wt NH}_3 = \underbrace{\underbrace{\underbrace{\frac{10.50\text{ g}}{2.016\text{ g mol}^{-1}}}_{\text{moles H}_2} \times \frac{2\text{ mol}}{3\text{ mol}}}_{\text{moles NH}_3} \times 17.03\text{ g mol}^{-1}}_{\text{weight NH}_3}$$

$$= 59.13\text{ g}$$

Separating the steps in this calculation serves the purpose of clarifying the approach to stoichiometric calculations such as this. The combined setup is convenient for numerical accuracy when using a pocket calculator. (If in a multistep calculation the answer in each step is rounded off to an appropriate number of significant figures, the final calculation may give an answer that is less exact than is justified by the data.[3]) Tying stoichiometric calculations to the mole concept is systematic, and it helps avoid errors. However, this particular problem can be solved without involving the numbers of moles or the balanced equation. Using this alternate approach, you simply calculate the weight of ammonia that contains 10.50 g of hydrogen; that is,

$$\underbrace{\underbrace{10.50\text{ g}}_{\text{weight H}} \times \underbrace{\frac{17.03\text{ g}}{3 \times 1.008\text{ g}}}_{\text{weight NH}_3\text{ per 1.000 g H}}}_{\text{weight NH}_3\text{ per 10.50 g hydrogen}} = 59.13\text{ g}$$

There are situations in which the reactants are mixed in proportions that are not stoichiometric, and there are conditions under which a reaction does not go to completion. Stoichiometric calculations can be made for such situations. These calculations may involve the balanced chemical equation or they may not; an example will illustrate this.

Example: In a reaction mixture initially containing 6.00 mol of hydrogen gas and 1.60 mol of nitrogen gas, incomplete reaction produces 2.50 mol of ammonia gas. How many moles of nitrogen and hydrogen are present in the final mixture?

To help you solve problems such as this one, indicate the initial and final amounts of reactants and products under the chemical equation:

	$3H_2(g)$	+	$N_2(g)$	=	$2NH_3(g)$
Amount (initial):	6.00 mol		1.60 mol		0 mol
Amount (final):	x		y		2.50 mol

The chemical equation shows that production of 1 mol of ammonia consumes 0.5 mol of nitrogen and 1.5 mol of hydrogen. Therefore

$$x = 6.00\text{ mol} - 2.50\text{ mol} \times 1.5 = 2.25\text{ mol}$$

$$y = 1.60\text{ mol} - 2.50\text{ mol} \times 0.5 = 0.35\text{ mol}$$

To do this calculation without the chemical equation, you simply recognize that 2.50 mol of NH_3 contains

7.50 mol H

2.50 mol N

The initial amounts of nitrogen atoms and hydrogen atoms were

$$6.00\text{ mol } H_2 = 12.00\text{ mol H}$$

$$1.60\text{ mol } N_2 = 3.20\text{ mol N}$$

[3]This point is illustrated in Appendix 2.

Therefore the moles of hydrogen atoms and nitrogen atoms remaining after reaction are

$$\text{H:} \quad 12.00 \text{ mol} - 7.50 \text{ mol} = 4.50 \text{ mol H} = 2.25 \text{ mol } H_2$$

$$\text{N:} \quad 3.20 \text{ mol} - 2.50 \text{ mol} = 0.70 \text{ mol N} = 0.35 \text{ mol } N_2$$

EXAMPLE: What is the percentage yield of ammonia in the preceding example?

To calculate the percentage yield that the 2.50 mol NH_3 represents, we must determine what the yield would have been if the reaction went to completion. First, let us determine which reagent was in limited supply; the amount of this reagent establishes the maximum yield. In the original mixture the ratio of nitrogen to hydrogen was

$$\frac{n_{N_2}}{n_{H_2}} = \frac{1.60 \text{ mol}}{6.00 \text{ mol}} = 0.267$$

which is less than the stoichiometric ratio, 0.333. Thus nitrogen is the limiting reagent. The balanced equation shows that if 1 mol N_2 reacts, 2 mol NH_3 is formed. If all of the nitrogen (1.60 mol) was converted to ammonia,

$$2 \times 1.60 \text{ mol} = 3.20 \text{ mol}$$

would have been produced. Thus the percentage yield is

$$\% \text{ yield} = \frac{\text{yield}}{\text{maximum possible yield}} \times 100$$

$$= \frac{2.50 \text{ mol}}{3.20 \text{ mol}} \times 100 = 78.1$$

You also can calculate the volumes of gaseous reactants and products. The ammonia-synthesis reaction provides examples.

EXAMPLE: What volume of ammonia gas (measured at standard temperature and pressure, STP) is produced from 10.50 g of hydrogen?

The first two steps in solution of this problem are the same as in the initial solution of an earlier problem. The final step is

$$\text{Vol of } NH_3 \text{ (at STP)} = 3.472 \cancel{\text{mol}} \times 22.41 \text{ L } \cancel{\text{mol}^{-1}}$$
$$= 77.81 \text{ L}$$

As in the earlier problem, these steps can be combined:

$$\text{Vol } NH_3 = \underbrace{\underbrace{\underbrace{\frac{10.50 \cancel{\text{g}}}{2.016 \cancel{\text{g mol}^{-1}}}}_{\text{moles } H_2} \times \frac{2 \cancel{\text{mol}}}{3 \cancel{\text{mol}}}}_{\text{moles } NH_3} \times \frac{22.41 \text{ L}}{1 \cancel{\text{mol}}}}_{\text{volume } NH_3}$$

$$= 77.81 \text{ L}$$

EXAMPLE: A sample of N_2 (1.0 L) is mixed with 6.5 L of H_2; incomplete reaction gives 6.0 L of gas (which contains the product NH_3 and the unreacted N_2 and H_2). What fraction of the molecules in the final mixture are ammonia molecules? (Each volume is measured at the same temperature and pressure.)

Let x = volume N_2 that reacted. The final volume will be made up of:

N_2: 1.0 L − x
H_2: 6.5 L − $3x$ (since 1 vol N_2 reacts with 3 vol H_2)
NH_3: $2x$ (since 1 vol N_2 gives 2 vol NH_3)

The sum of these volumes is

$$1.0\text{ L} - x + 6.5\text{ L} - 3x + 2x = 6.0\text{ L}$$

which gives

$$-2x = -1.5\text{ L}$$
$$x = 0.75\text{ L}$$

The fraction of the molecules in the final mixture that is NH_3 is the same as the fraction of the volume that is NH_3. Therefore the fraction of ammonia in the final mixture is

$$\frac{2 \times 0.75\,\cancel{\text{L}}}{6.0\,\cancel{\text{L}}} = 0.25$$

In this example the actual numbers of moles of reactant and product gases cannot be calculated because the temperature and pressure were not specified.

In the examples we have considered to this point, no more than a single chemical equation was needed for each system considered. However, the point has already been made that two or more chemical equations are required to express the changes that occur in some systems. Appropriate algebraic manipulation allows stoichiometric calculations to be made for reactions in such systems.

EXAMPLE: At elevated temperatures, a metal carbonate decomposes to give carbon dioxide and the corresponding metal oxide. For magnesium carbonate and barium carbonate the reactions are

$$MgCO_3(s) = MgO(s) + CO_2(g)$$
$$BaCO_3(s) = BaO(s) + CO_2(g)$$

A mixture of $MgCO_3$ and $BaCO_3$ weighing 3.45 g is heated and converted to 1.93 g of metal oxides. What weight of magnesium carbonate was present in the initial sample?

First, the molecular weights of these substances must be calculated from the atomic weights: Mg, 24.31; Ba, 137.3; C, 12.01; O, 16.00.

They are

$$MW(MgCO_3) = 84.32 \qquad MW(BaCO_3) = 197.3$$
$$MW(MgO) = 40.31 \qquad MW(BaO) = 153.3$$

Because the weight of $MgCO_3$ is the quantity sought, it will be designated as x; then the weight of $BaCO_3$ must be $(3.45 \text{ g} - x)$. The balanced equations show that one mole of each metal oxide is formed from one mole of the metal carbonate. The total weight of metal oxide is the sum of the weights of MgO and BaO. This leads to

$$1.93 \text{ g} = \underbrace{\underbrace{\frac{x}{84.32 \text{ g mol}^{-1}}}_{\substack{\text{moles } MgCO_3 = \\ \text{moles MgO}}} \times 40.31 \text{ g mol}^{-1}}_{\text{weight of MgO}}$$

$$+ \underbrace{\underbrace{\frac{3.45 \text{ g} - x}{197.3 \text{ g mol}^{-1}}}_{\substack{\text{moles } BaCO_3 = \\ \text{moles BaO}}} \times 153.3 \text{ g mol}^{-1}}_{\text{weight BaO}}$$

$$1.93 \text{ g} = 0.4781x + (3.45 \text{ g} - x)(0.7770)$$
$$0.299x = 0.751 \text{ g}$$
$$x = 2.51 \text{ g}$$

Example: A sample of liquid heptane weighing 11.500 g is mixed with 1.3000 mol of oxygen gas. The heptane is burned completely to a mixture of carbon monoxide and carbon dioxide. The amount of gas present after reaction (carbon dioxide, carbon monoxide, and unreacted oxygen) is 1.0500 mol. (These gases are measured with the water produced present as liquid.) How many moles of carbon monoxide are produced and how many moles of oxygen are unreacted?

The number of moles of heptane present initially is calculated from the weight and the molecular weight:

$$\frac{11.500 \text{ g}}{100.20 \text{ g mol}^{-1}} = 0.1148 \text{ mol}$$

Examination of the balanced chemical equations,

$$2C_7H_{16}(l) + 15O_2(g) = 14CO(g) + 16H_2O(l)$$
$$C_7H_{16}(l) + 11O_2(g) = 7CO_2(g) + 8H_2O(l)$$

shows that the total number of moles of CO (x) and CO_2 (y) produced is seven times the number of moles of heptane:

$$x + y = 7 \times 0.1148 \text{ mol} = 0.8036 \text{ mol}$$

The equations also show that for every 14 moles of CO produced, 1 mole of gas disappears (15 moles $O_2 \rightarrow$ 14 moles CO), and for every 7 moles of CO_2 produced, 4 moles of gas disappear (11 moles $O_2 \rightarrow$ 7 moles CO_2). In this example (1.3000 mol − 1.0500 mol) = 0.2500 mol of gas disappears. Therefore the amount of carbon dioxide is

$$y = 0.8036 \text{ mol} - x$$

and the equation for the amount of gas which disappears is

$$\begin{aligned} 0.2500 \text{ mol} &= \tfrac{1}{14}x + \tfrac{4}{7}(0.8036 \text{ mol} - x) \\ &= \tfrac{1}{14}x - \tfrac{8}{14}x + 0.4592 \text{ mol} \\ 0.500x &= 0.2092 \text{ mol} \\ x &= 0.4184 \text{ mol} \\ (0.8036 \text{ mol} - x) &= 0.3852 \text{ mol} \end{aligned}$$

The amount of oxygen remaining unreacted is the initial amount (1.3000 mol) minus the amount used in the reactions, which is

$$\tfrac{15}{14}(\text{amt of CO}) + \tfrac{11}{7}(\text{amt of CO}_2) = \tfrac{15}{14}x + \tfrac{11}{7}y$$

Therefore the amount of oxygen remaining is

$$1.3000 \text{ mol} - \tfrac{15}{14}(0.4184 \text{ mol}) - \tfrac{11}{7}(0.3852 \text{ mol}) = 0.2464 \text{ mol}$$

2–4 Chemical Reactions of Unstable Compounds

The terms *stable* and *unstable* applied to chemical compounds lack rigor. Whether a substance is stable or unstable depends upon the conditions of temperature and pressure to which it is subjected, and also upon the other chemical substances that are present. These factors determine the chemical reactions which may occur. The stability of a chemical substance should be described with respect to the possible chemical reactions it may undergo. A time scale of the observations also must be specified.

The igneous silicate minerals that make up some mountains are compounds which apparently have extraordinary stability. But these substances, with the exception of quartz, are subject to the weathering action of oxygen, water, and carbon dioxide. Sedimentary rocks—shale, sandstone, and limestone—are the result. These transformations, of which the degradation of anorthite, $CaAl_2Si_2O_8$, a feldspar, by water and carbon dioxide is an example,

$$CaAl_2Si_2O_8 + CO_2 + xH_2O = CaCO_3 + Al_2O_3 \cdot xH_2O + 2SiO_2$$

occur over very long periods of time at ordinary temperatures.

Kept away from oxygen and water, there are many, many chemical substances that persist for an indefinite time. These same substances may react with either or both water and oxygen, thereby limiting their stability in air. An example is sodium hydride, NaH, which has a negligible tendency to decompose at ordinary temperatures to give the constituent elements. In the presence of water or oxygen, this substance is quite unstable; each of the reactions

$$NaH + H_2O = H_2 + NaOH$$
$$2NaH + O_2 = 2NaOH$$

occurs readily. Therefore it is appropriate to describe sodium hydride as being: (a) stable with respect to the constituent elements, (b) unstable with respect to reaction with water, and (c) unstable with respect to reaction with oxygen. Consideration of hydrogen peroxide (H_2O_2) further illustrates these points. The decomposition to give the constituent elements:

$$H_2O_2(l) = H_2(g) + O_2(g)$$

does not occur, but decomposition to give water and oxygen:

$$2H_2O_2(l) = 2H_2O(l) + O_2(g)$$

goes essentially to completion, *if enough time is allowed.* The rate of this decomposition reaction is influenced by many different chemical substances. Some increase the rate; these substances, called *catalysts*, include simple substances [e.g., iron(III) ion, Fe^{3+}] and complex substances [e.g., the enzyme, catalase, which has a molecular weight of 250,000 and which contains iron(III)]. The mechanism of decomposition of hydrogen peroxide is complicated, and some added substances interfere with one or more of the reactive intermediate species which are present during decomposition; such added substances inhibit the decomposition. (The label of a bottle of dilute hydrogen peroxide in a drug store reveals the presence of acetanilide as an inhibitor.)

Many useful chemical substances, like hydrogen peroxide, are unstable, but decompose so slowly that they can be used. Other substances decompose slowly enough to be used, but too quickly to be packaged and stored. Such compounds must be prepared at the time of use. Ozone is an example, the decomposition reaction

$$2O_3(g) = 3O_2(g)$$

going essentially to completion over a period of hours. However, ozone can be generated in air or pure oxygen by an electric discharge, and the flowing gas mixture can be used where a powerful oxidizing agent is needed. One such use is to purify municipal water supplies, ozone replacing chlorine in the initial purification. (A trace of chlorine would then be added to protect the water from contamination during distribution.) The electric discharge used in preparation of ozone dissociates some oxygen molecules to give atoms:

$$O_2 \rightarrow 2O$$

These atoms can combine with molecular oxygen to produce ozone by the reaction

$$O_2(g) + O(g) = O_3(g)$$

Ozone is stable with respect to atomic oxygen plus molecular oxygen, but is unstable with respect to molecular oxygen:

$$2O_3(g) = 3O_2(g)$$

The energetics of these transformations will be discussed in the next chapter.

2–5 Other Aspects of Chemical Reactions

There are important aspects of chemical reactions not yet mentioned in this preliminary discussion. These will be presented in due course, but an introduction will be given here. These aspects are:

1. the energetics of chemical reactions,
2. the reversibility of chemical reactions and chemical equilibrium, and
3. the rates at which chemical reactions occur.

In some chemical reactions the energy involved is the most important feature of the reaction. Heptane can be used as a fuel because of the heat evolved in the combustion reaction. The balanced chemical equation, already given, can be expanded to include a statement of the heat evolved per mole of heptane burned:

$$C_7H_{16}(l) + 11O_2(g) = 7CO_2(g) + 8H_2O(l) + 1{,}150 \text{ kcal mol}^{-1}$$

It takes energy to make some chemical reactions occur. One important argument for recycling metallic aluminum is that an enormous amount of electrical energy is needed to produce the metal from the oxide, which is the important ore. The reaction

$$2Al_2O_3(s) = 4Al(s) + 3O_2(g)$$

is one in which the energy content of the products is greater than that of the reactant. This energy must be added to the system to bring about the reaction. The important subject of energetics of chemical reactions will be introduced in the next chapter and will recur in many later chapters.

The reaction of hydrogen with nitrogen to give ammonia, a reaction used in the examples of stoichiometric calculations, illustrates the remaining aspects of reactions to be mentioned here. The chemical equation

$$N_2(g) + 3H_2(g) = 2NH_3(g)$$

is balanced, but it discloses no information about the conditions under which the reaction goes forward at an appreciable rate or about the extent to which the reaction take place. At ordinary temperatures, ammonia gas is stable with respect to nitrogen and hydrogen, but the rate of its formation from its constituent elements is very low. At high pressures (>200 atm) and a moderate temperature (500°C) in the presence of an appropriate catalyst, the reaction is the basis for large-scale production of ammonia (the HABER–BOSCH process). This reaction is readily reversible, and at 1000°C the decomposition of ammonia to give its constituent elements is essentially complete. This reverse reaction is useful for production of pure hydrogen. Under conditions intermediate between these extremes, mixtures of hydrogen, nitrogen, and ammonia are formed whether the starting material is ammonia or a mixture of hydrogen and nitrogen. This situation is an example of *chemical equilibrium*. The final state with all three substances present (both reactants and products) is one of maximum stability, which will persist forever at the particular temperature and pressure in question. Why the extent of reaction at equilibrium varies with temperature and pressure will be discussed in Chapter 12.

Most chemical reactions occur in a sequence of steps, the product of one step being the reactant of a subsequent step. These individual steps are called elementary reactions, and one objective of the study of chemical reactions is to trace the sequence of elementary reactions that make up an overall reaction. Even reactions that appear to be simple may occur in a sequence of steps. A possible example is the gas-phase reaction of hydrogen and iodine to give hydrogen iodide:

$$H_2(g) + I_2(g) = 2HI(g)$$

This reaction was long thought to be an elementary reaction, that is, to occur in a single step, but experimental measurements (to be described in Chapter 21) suggest that the reaction occurs in a sequence of two steps:

$$I_2 \rightleftarrows 2I$$
$$2I + H_2 \rightarrow 2HI$$

If the stoichiometry of a reaction is complex, it is certain that the reaction occurs in a sequence of elementary reactions. A reaction important in analytical chemistry, the oxidation of iron(II) ion by dichromate ion,

$$6Fe^{2+} + Cr_2O_7^{2-} + 14H_3O^+ = 6Fe^{3+} + 2Cr^{3+} + 21H_2O$$

is such a reaction. The chapter on transition-metal chemistry, Chapter 19, includes the multistep mechanism for this reaction. However, the point being made in this introduction is a simple one: Reactions with complex stoichiometry (anything more complicated than two or three molecules coming together) almost certainly occur by mechanisms consisting of two or more steps, and even reactions with simple stoichiometry may occur by mechanisms more complex than a single elementary reaction.

Biographical Notes

CARL BOSCH (1874–1940), a German chemist, was a leader in the German chemical industry. He was responsible for the development of large-scale industrial production of ammonia. He received the Nobel Prize in Chemistry in 1931.

FRITZ HABER (1868–1934), a German chemist, was educated at Berlin and Heidelberg. He was Professor of Physical Chemistry in the University of Berlin and also Director of the Kaiser Wilhelm Institute for Physical Chemistry. For his work in developing the method for fixing atmospheric nitrogen, he was awarded the Nobel Prize in Chemistry in 1918. (His name also will come up in Chapter 9 in connection with the Born–Haber cycle for the formation of ionic crystals.)

Problems and Questions

2–1 Balance the following chemical equations (all reactants and products are given):

(a) $C_3H_8 + O_2 = CO_2 + H_2O$
(b) $Al_2O_3 + HCl = AlCl_3 + H_2O$
(c) $F_2 + H_2O = HF + O_2$
(d) $Fe_2O_3 + CO = Fe_3O_4 + CO_2$
(e) $PH_3 + O_2 = P_4O_{10} + H_2O$
(f) $CO_2 + Al = Al_2O_3 + CO$
(g) $H_3PO_4 + CaO = Ca_3(PO_4)_2 + H_2O$
(h) $F_2 + C_3H_8O = HF + CF_4 + O_2$
(i) $CaCN_2 + H_2O = NH_3 + CaCO_3$
(j) $MnO_2 + HCl = MnCl_2 + Cl_2 + H_2O$
(k) $NH_3 + O_2 = NO + H_2O$
(l) $Al_4C_3 + H_2O = Al(OH)_3 + CH_4$
(m) $FeS + O_2 = Fe_2O_3 + SO_2$
(n) $Fe_2O_3 + H_2O = Fe(OH)_3$
(o) $H_2S + SO_2 = S + H_2O$
(p) $CuFeS_2 + O_2 = Cu + FeO + SO_2$
(q) $ZnS + O_2 = ZnO + SO_2$
(r) $NaHCO_3 + SO_3 = Na_2SO_4 + CO_2 + H_2O$

2–2 Balance the following chemical equations for reactions occurring in solution (all reactants and products are given):

(a) $Al + H^+ = Al^{3+} + H_2$
(b) $Al + H_2O + OH^- = AlO_2^- + H_2$
(c) $I_2 + S_2O_3^{2-} = I^- + S_4O_6^{2-}$
(d) $H_3PO_4 + OH^- = PO_4^{3-} + H_2O$
(e) $BaCO_3 + H^+ = Ba^{2+} + CO_2 + H_2O$
(f) $Zn(OH)_2 + OH^- = Zn(OH)_4^{2-}$
(g) $La(OH)_3 + HCO_3^- = La_2(CO_3)_3 + OH^- + H_2O$
(h) $BH_4^- + H_2O = H_2 + B(OH)_4^-$
(i) $CaCO_3 + CO_2 + H_2O = Ca^{2+} + HCO_3^-$
(j) $Ag_2S + H^+ = H_2S + Ag^+$
(k) $Ag_2O + H^+ + Cl^- = AgCl + H_2O$
(l) $Eu^{2+} + H^+ = Eu^{3+} + H_2$
(m) $SO_3^{2-} + S_8 = S_2O_3^{2-}$
(n) $HCO_3^- = CO_2 + CO_3^{2-} + H_2O$
(o) $Ag_2S + CN^- + H_2O = Ag(CN)_2^- + HS^- + OH^-$
(p) $PCl_3 + H_2O = H_3PO_3 + H^+ + Cl^-$

2–3 What weight of hydrogen gas is needed to react with 10.00 g of each of the compounds AgCl, Ag_2Se, Ag_3PO_4? The reaction with each compound produces silver metal plus an acid, HCl, H_2Se, or H_3PO_4.

2–4 Hydrogen gas is produced when a very active metal (e.g., Ca) is treated with water:

$$Ca + 2H_2O = Ca(OH)_2 + H_2$$

What weight of hydrogen gas is produced when 1.00 g of metallic calcium reacts with excess water?

2–5 Hydrogen gas is produced when an ionic hydride (e.g., CaH_2) is treated with water:

$$CaH_2 + 2H_2O = Ca(OH)_2 + 2H_2$$

What weight of hydrogen gas is produced when 1.00 g of calcium hydride reacts with excess water? Compare this answer with that for Problem 2–4; explain the relative values obtained in the calculations.

2–6 What weight of hydrogen gas is produced in the complete decomposition of 1.00 kg of ammonia (NH_3)?

2–7 It was shown in this chapter (Section 2–3) that certain problems of chemical stoichiometry can be solved easily without using a balanced chemical equation; for others, a balanced chemical equation is necessary. Contrast Problems 2–5 and 2–6 in this respect.

2–8 By a sequence of chemical reactions $CoCl_2$ can be converted to $Co(NH_3)_6Cl_3$. A student obtained 9.70 g of $Co(NH_3)_6Cl_3$ from 5.43 g of $CoCl_2$. What was the percentage yield in this experiment?

2–9 Consider the series of chemical reactions, given in Figure 2–1, that are responsible for the conversion of atmospheric nitrogen, $N_2(g)$, to nitrate ion, NO_3^-. Write a single balanced equation for the overall reaction with excess oxygen. What is the maximum weight of nitric acid that can be formed from 1.00 kg N_2?

2–10 The sequence of reactions by which nitric acid is produced commercially includes the reaction of water with nitrogen dioxide:

$$3NO_2 + H_2O = NO + 2HNO_3$$

How many moles of nitric acid are produced when 3.50 kg of NO_2 react in this reaction?

2–11 The exothermic gas-forming reaction

$$(CH_3)_2NNH_2(s) + 2N_2O_4(l) = 2CO_2(g) + 3N_2(g) + 4H_2O(g)$$

is a source of energy in some space vehicle operations. What weight of dinitrogen tetraoxide is needed to react with 10.00 kg of dimethylhydrazine $[(CH_3)_2NNH_2]$?

2–12 Mercury and sulfur react to form a stable sulfide, HgS. A 3.00-g sample of mercury is mixed with 1.00 g of sulfur. If the reaction occurs to completion, what weight of what substance remains unreacted?

2–13 An equimolar gaseous mixture of nitrogen and hydrogen is heated to produce ammonia. It is found that the final volume of the mixture of gases (N_2, H_2, and NH_3) measured at the same pressure and temperature is 83.5% of the initial volume. What fraction of the hydrogen gas reacted? What fraction of the nitrogen gas reacted?

2–14 A reaction believed to produce dilute acid in the water from mines is the air oxidation of iron pyrites (FeS_2):

$$4FeS_2 + 15O_2 + 8H_2O = 2Fe_2O_3 + 8SO_4^{2-} + 16H^+$$

What weight of FeS_2 reacts in the production of each mole of hydrogen ion (H^+)?

2–15 Acetylene (C_2H_2) is produced when calcium carbide (CaC_2) is treated with water. Calcium carbide, in turn, is produced by the reaction of calcium oxide (CaO) and carbon at elevated temperatures. The chemical equations for these reactions are:

$$CaC_2(s) + 2H_2O(l) = Ca(OH)_2(s) + C_2H_2(g)$$
$$CaO(s) + 3C(s) = CaC_2(s) + CO(g)$$

(a) What weight of calcium carbide is needed to produce 1.00 kg of acetylene?
(b) What volume of acetylene, measured at standard conditions, is produced when 535 g of calcium carbide react with excess water?
(c) How many moles of calcium oxide are needed to produce 535 g of calcium carbide?

2–16 Magnesium metal reacts quantitatively with oxygen to give magnesium oxide, MgO. If 5.00 g of Mg and 5.00 g of O_2 are allowed to react, what weight of MgO is formed, and what weight of which reactant is left in excess?

2–17 The common gaseous hydrocarbons containing two carbon atoms per molecule are C_2H_2 (acetylene), C_2H_4 (ethylene), and C_2H_6 (ethane). For which of these does complete combustion to give gaseous CO_2 and H_2O result in no change of volume (all gases being measured at the same temperature and pressure)?

2–18 The complete combustion of a sample of octane (C_8H_{18}) with 50.00 g O_2 gives a mixture of products that contains 13.43 g of unreacted oxygen. What weight of octane was present initially, and how many moles of water were produced in the reaction?

2–19 If heated to a high temperature, metal carbonates decompose to give carbon dioxide and the metal oxide. For calcium carbonate, the reaction is

$$CaCO_3(s) = CaO(s) + CO_2(g)$$

(a) What weight of residue (calcium oxide) remains after a 15.5-g sample of calcium carbonate is decomposed completely?
(b) Magnesium carbonate decomposes in an analogous reaction. Complete decomposition of a mixture of $CaCO_3$ and $MgCO_3$ weighing 5.40 g yields 2.81 g of oxide residue. What weight of calcium carbonate was present in the original sample? What is the ratio (atoms Ca)/(atoms Mg) in the sample?

2–20 A sample of silver weighing 1.00 g is heated in the absence of oxygen with 1.00 g of sulfur. Silver sulfide (Ag_2S) is formed. After the reaction is complete, the excess sulfur is dissolved in carbon disulfide (which does not dissolve silver sulfide). This solution of sulfur is evaporated to remove the solvent, and the residual sulfur is then converted to sulfur dioxide (SO_2). What weight of sulfur dioxide is formed? What volume will the sulfur dioxide gas occupy at standard conditions (0°C, 1 atm)?

2–21 A mixture of copper(I) oxide (Cu_2O), and copper(II) oxide (CuO) is reduced to copper metal with hydrogen. A 5.350-g sample of the mixture gives 4.372 g of metallic copper. What percentage by weight of the original sample was copper(I) oxide?

2–22 A mixture of FeO and Fe_3O_4 weighing 5.430 g is heated in pure oxygen, which converts each of these compounds to Fe_2O_3. The final oxidized sample weighs 5.779 g. What weight of FeO was in the original sample? What weight of metallic iron would be produced if this original sample had been converted to the metal by reaction with gaseous hydrogen?

2–23 A mixture of silver oxide, Ag_2O, and copper oxide, CuO, weighing 1.450 g is heated with excess hydrogen gas. (Each reaction produces the pure metal plus water vapor, H_2O.) The weight of water vapor produced is 0.252 g. (a) How many moles of water were produced? (b) What weight of copper oxide was present in the initial sample? (c) What weight of copper is present in the final mixture of metals?

2–24 A gaseous mixture of 1.00 mol of butane (C_4H_{10}) and 12.00 mol of oxygen reacts to produce carbon dioxide and carbon monoxide (both gaseous) plus liquid water. The amount of oxygen in the final mixture is 6.80 mol. How many moles of carbon monoxide are in the final mixture?

3

Chemistry and Energy

3–1 Introduction

Energy plays a central role in your life. Your body performs work, and the energy expended in doing this work is provided by the food you eat. Energy is required to heat your home and to run your means of transportation. The production of farm crops and manufactured goods requires energy. Interest now is high in the discovery and use of energy resources and in energy conservation. Chemistry is involved in one way or another in most aspects of energy production and energy consumption. This chapter will give you an introduction to some of the relationships between chemistry and energy.

The reactions of oxygen are fundamental in any discussion of chemistry and energy. Because oxygen is so highly reactive with most elements, both metals and nonmetals, the mass of oxygen atoms was used in defining the atomic weight scale that served chemists for a half century. In geologic history, most metals in the Earth's crust have been converted to oxides in reactions that evolve heat. To produce from these oxide ores the free metals, which are important in modern technology, one must reverse the oxidation reactions. The reactions by which oxides of aluminum and iron decompose to give the free metals,

$$2Al_2O_3 = 4Al + 3O_2$$
$$2Fe_2O_3 = 4Fe + 3O_2$$

require energy. To produce aluminum from its ore, bauxite (Al_2O_3), electrical energy is used, and carbon is used to produce iron from hematite (Fe_2O_3), a principal ore of iron:

$$3C + Fe_2O_3 = 3CO + 2Fe$$

The energy needed for this reaction, which is provided by the absorption of heat at elevated temperatures, is less than that of the simple decomposition reaction, because this reaction involves also the conversion of carbon to carbon monoxide. The energy to create the high temperatures needed to produce metallic iron is provided by burning coal (represented here as carbon),

$$C + O_2 = CO_2$$

The high stability of carbon dioxide and water is responsible for the production of heat in the reaction of oxygen with substances we use as fuels. Heat is evolved in each of these reactions:

1. the metabolism of stearic acid, $C_{18}H_{36}O_2$, a component of animal fats: $C_{18}H_{36}O_2 + 26O_2 = 18CO_2 + 18H_2O$
2. the oxidation of glucose, $C_6H_{12}O_6$, during respiration: $C_6H_{12}O_6 + 6O_2 = 6CO_2 + 6H_2O$
3. the burning of methane, CH_4, the principal component of natural gas: $CH_4 + 2O_2 = CO_2 + 2H_2O$
4. the burning of heptane, C_7H_{16}, a constituent of gasoline: $C_7H_{16} + 11O_2 = 7CO_2 + 8H_2O$

In the burning of methane in a gas stove or a furnace, the heat given off in the reaction is our primary interest. In the burning of heptane in the engine of an automobile, the mechanical energy derived from the heat is of interest. The metabolism of carbohydrates (e.g., glucose) and fats provides the heat that keeps our bodies warm, but these reactions do much more. Complex mechanisms, to be introduced in Chapter 20, link these oxidation reactions to other processes of life.

The relationships between chemistry and energy are much deeper than has been implied by our discussion to this point. Using appropriate thermal data for chemical substances, data pertaining to the heat evolved or absorbed in chemical reactions, and the heat required to raise the temperature of chemical substances, it is possible to calculate the extent to which a chemical reaction can occur. This type of calculation is part of the subject of *chemical thermodynamics*, which is introduced in this chapter. The development of this subject, some parts of which are abstract and difficult for the beginning student, will be spread over a number of chapters.

3–2 Energy, Heat, and Work

Before we study the application of thermodynamics to chemical problems, we must consider some simple concepts of energy; this will allow us also to extend our list of SI units by introducing the units for *time*, *force*, *energy*, and *electrical current*. When you lift a book from the table, you recognize that energy is required. Thus it will be easy for you to recognize that the raising of weights on a piston by the expansion of a gas in a cylinder also requires energy. Much of our introduction to thermodynamics will involve gases in cylinders, but our goal, already stated, is the understanding of its application to chemistry.

POTENTIAL AND KINETIC ENERGY; SOME DEFINITIONS AND UNITS

The evolution of heat in the reactions mentioned in the introduction is a result of differences of the potential and kinetic energies of atoms and molecules in the reactants and in the products of the reactions. To introduce the concepts of potential energy and kinetic energy, we will start with an equation defining mechanical work, for energy is the ability to do work. Work, w, is a force, F, acting through a distance, l,

$$w = F \times l$$

and force is defined by Newton's law,

$$F = m \times a$$

in which m is mass and a is acceleration. To use these equations, we can employ the SI units for mass and length introduced in Chapter 1, and the SI unit for time, the second:

Second (s) The basic SI unit of time. This is defined as the duration of a particular number of periods of particular radiation from the cesium-133 atom.[1]

The SI unit for force follows directly from Newton's law and the units for mass, distance, and time; to accelerate a one-kilogram mass one meter per second per second requires a force of one newton (N):

$$\begin{aligned} F &= 1\ \text{kg} \times 1\ \text{m s}^{-2} \\ &= 1\ \text{kg m s}^{-2} = 1\ \text{N} \end{aligned}$$

That is, the SI unit of force, the newton, is

$$1\ \text{N} = 1\ \text{kg m s}^{-2}$$

The energy equivalent to the work done when a force of one newton acts

[1]The meaning of this definition probably is not very clear to you at this time. For the present, simply recognize that this contemporary definition of a unit of time is based upon an atomic process, not as a definite fraction (1/86400) of a mean solar day, as a second was defined in the past.

through a distance of one meter is the joule (symbol J; named for JAMES JOULE):

$$\begin{aligned} w &= F \times l \\ &= (1\ \text{kg m s}^{-2}) \times (1\ \text{m}) \\ &= 1\ \text{kg m}^2\ \text{s}^{-2} = 1\ \text{J} \end{aligned}$$

That is, the SI unit of energy (and work) is the joule:

$$\begin{aligned} 1\ \text{J} &= 1\ \text{N m} \\ &= 1\ \text{kg m}^2\ \text{s}^{-2} \end{aligned}$$

In the cgs (centimeter-gram-second) system of units, the unit of force is the dyne:

$$1\ \text{dyn} = 1\ \text{g cm s}^{-2}$$

and the unit of energy is the erg:

$$1\ \text{erg} = 1\ \text{g cm}^2\ \text{s}^{-2}$$

The relationships between the gram and kilogram and between the centimeter and meter allow calculation of the relationships of the newton and dyne:

$$1\ \text{N} = 1\ \text{kg m s}^{-2} \times \frac{10^3\ \text{g}}{1\ \text{kg}} \times \frac{10^2\ \text{cm}}{1\ \text{m}} = 10^5\ \text{dyn}$$

and the joule and erg:

$$1\ \text{J} = 1\ \text{kg m}^2\ \text{s}^{-2} \times \frac{10^3\ \text{g}}{1\ \text{kg}} \times \left(\frac{10^2\ \text{cm}}{1\ \text{m}}\right)^2 = 10^7\ \text{erg}$$

A unit of energy commonly used by many chemists, the calorie, once defined as the heat required to raise the temperature of one gram of water by one centigrade degree (from 14.5°C to 15.5°C) is now defined as 4.184 joules:

$$1\ \text{cal} \equiv 4.184\ \text{J}$$

(The calorie used in nutrition is actually the kilocalorie.) The chemical literature is full of data and derived quantities expressed in terms of calories, and you should remember this definition. The pace with which SI units will be adopted is uncertain, and a contemporary student of chemistry or a person working in chemistry or related fields must be ready to cope both with SI units and with other commonly used non-SI units. We will use both, but SI units will be used predominately in later chapters. If all physical quantities include the correct dimensions, no ambiguity will arise.

EXAMPLE: What work is done in raising a 500-g mass 15.0 cm?

The work (w) is the product of force (F) and distance (l):

$$w = F \times l$$

The force is given by Newton's law,

$$F = m \times a = m \times g$$

in which a is the acceleration of gravity, $g = 9.81\ \text{m s}^{-2}$:

$$\begin{aligned} w &= 0.500\ \text{kg} \times 9.81\ \text{m s}^{-2} \times 0.150\ \text{m} \\ &= 0.736\ \text{kg m}^2\ \text{s}^{-2} = 0.736\ \text{J} \end{aligned}$$

[Notice that when the mass and distance are changed to SI units (15.0 cm = 0.150 m, and 500 g = 0.500 kg), the equation gives the result directly in the SI unit for energy.]

EXAMPLE: What force must be applied to a 3.0-kg mass to bring about an acceleration of 50 cm s^{-2}?

$$\begin{aligned} F &= m \times a \\ &= 3.0\ \text{kg} \times 0.50\ \text{m s}^{-2} \\ &= 1.5\ \text{kg m s}^{-2} = 1.5\ \text{N} \end{aligned}$$

At this point, we are ready to consider two forms of energy, kinetic energy and potential energy. *Kinetic energy* is the energy that an object has because it is moving. The kinetic energy, E_k, is given by the equation

$$E_k = \tfrac{1}{2}mu^2$$

in which u is the velocity of the object of mass m. Numerical calculations can be made simply with this equation using SI units. Thus the energy of a one-kilogram mass moving with a velocity of one meter per second is one half joule:

$$\begin{aligned} E_k &= \tfrac{1}{2} \times 1\ \text{kg} \times (1\ \text{m s}^{-1})^2 \\ &= \tfrac{1}{2}\ \text{kg m}^2\ \text{s}^{-2} = \tfrac{1}{2}\ \text{J} \end{aligned}$$

Potential energy is the energy an object has because of its position, if there is a force acting to change that position. A familiar instance of potential energy is the energy of an object subject to the force of gravity. The potential energy of an object on the table top relative to its potential energy on the floor is equal to the work required to raise the object from the floor to the table top. This is illustrated in Figure 3–1. In State 1, with the object on the floor, the potential energy is defined as zero. To raise the object to the table top, at height h above the floor, the work needed is

$$\begin{aligned} w &= F \times h \\ &= (m \times g) \times h \end{aligned}$$

In State 2, with the object on the table top, the potential energy, E_p, is, therefore, equal to the product of the mass, the height of the table top, and the acceleration of gravity:

$$E_p = mgh$$

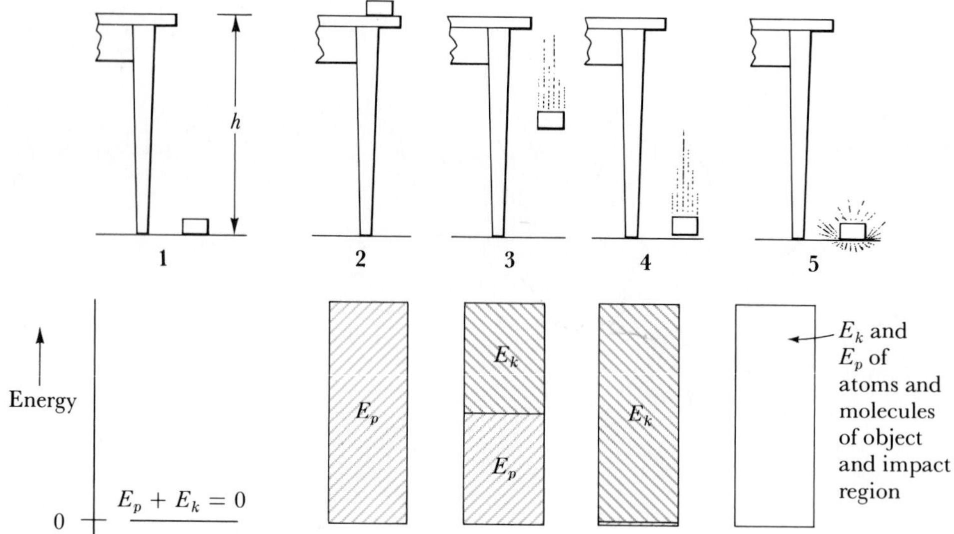

FIGURE 3–1
The conversion of potential energy (E_p) into kinetic energy (E_k) as an object falls (States 2–5). Upon impact (State 5), the kinetic energy of the moving object is converted to kinetic and potential energy of the atoms and molecules of the object and the impact area of the floor.

As the object falls to the floor, this potential energy is converted to kinetic energy (State 3). Neglecting friction of the air, the kinetic energy just before impact of the object on the floor (State 4) is equal to the loss of potential energy:

$$\tfrac{1}{2}mu^2 = mgh$$

which gives an equation for the velocity just before impact:

$$u = \sqrt{2gh}$$

During its fall, the object conserves its energy; the sum of its kinetic energy and potential energy is a constant. After impact (State 5), the kinetic energy of the object is zero, and its potential energy with respect to the defined origin is zero. What has been the fate of this energy? It has been converted to additional kinetic energy of the atoms and molecules making up the object and the floor at the place of impact and also to additional potential energy of interaction of these atoms and molecules.

EXAMPLE: What is the velocity and what is the kinetic energy of a 1.50-kg mass that has fallen 0.800 m where the acceleration of gravity is 9.81 m s^{-2}? (Assume no friction due to air.)

$$\begin{aligned} E_k = E_p(\text{initial}) &= mgh \\ &= 1.50 \text{ kg} \times 9.81 \text{ m s}^{-2} \times 0.800 \text{ m} \\ &= 11.8 \text{ kg m}^2 \text{ s}^{-2} = 11.8 \text{ J} \end{aligned}$$

The velocity of the object after falling 0.800 m does not depend upon its mass; this is shown in the equation derived when the kinetic energy and potential energy are equated:

$$\begin{aligned}\tfrac{1}{2}mu^2 &= mgh\\ u &= \sqrt{2gh}\\ &= \sqrt{2 \times 9.80 \text{ m s}^{-2} \times 0.800 \text{ m}}\\ &= 3.96 \text{ m s}^{-1}\end{aligned}$$

The situation depicted in Figure 3–1 for States 2–5 (the interconversion of potential energy and kinetic energy) is analogous to a chemical system of two iodine atoms colliding to form a diatomic iodine molecule in the presence of another atom, an atom of neon. Because of their attraction for one another, the separated iodine atoms have a positive potential energy relative to that of the atoms combined in an iodine molecule. As the iodine atoms approach one another, this potential energy is converted to kinetic energy; the atoms approaching one another move more rapidly as they come closer. When the iodine atoms are separated by a distance similar to their separation in the iodine molecule, the atom of neon joins the collision and takes away some of the energy of the colliding iodine atoms. The increased kinetic energy of the neon atom is analogous to the increased kinetic energy of atoms of the floor in the example shown in Figure 3–1. Figure 3–2 shows the stages of the collision just discussed in the terms used in Figure 3–1 (States 3–5). The potential energy lost in forming the chemical bond in the iodine molecule has been transformed partially into kinetic energy of the neon atom. (As we will learn in Chapter 15, the foreign gas molecule in the reaction

$$2\text{I} + \text{Ne} \rightarrow \text{I}_2 + \text{Ne}$$

is needed to allow the colliding iodine atoms to stick together as a molecule.)

ELECTRICAL ENERGY; SOME DEFINITIONS AND UNITS

Both electrical units and electrical measurements play a large role in many types of quantitative chemical studies. Much of chemistry involves particles (ions and electrons) having a net electrical charge, and some chemical reactions can be harnessed to produce electrical energy. Conversely, electrical energy can be used to bring about some chemical reactions, as in the production of aluminum metal from bauxite (Al_2O_3), already mentioned. Electrical calibration can be used in calorimetric studies, thereby tying such studies to SI units. The SI unit of electrical quantities is the ampere (symbol A; named for ANDRÉ AMPERE) a defined unit of electrical current:

Ampere (A) The ampere is that constant current which, if maintained in two straight parallel conductors of infinite length, of negligible circular cross section, and placed one meter apart in a vacuum, would produce between these conductors a force equal to 2×10^{-7} newtons per meter of length.

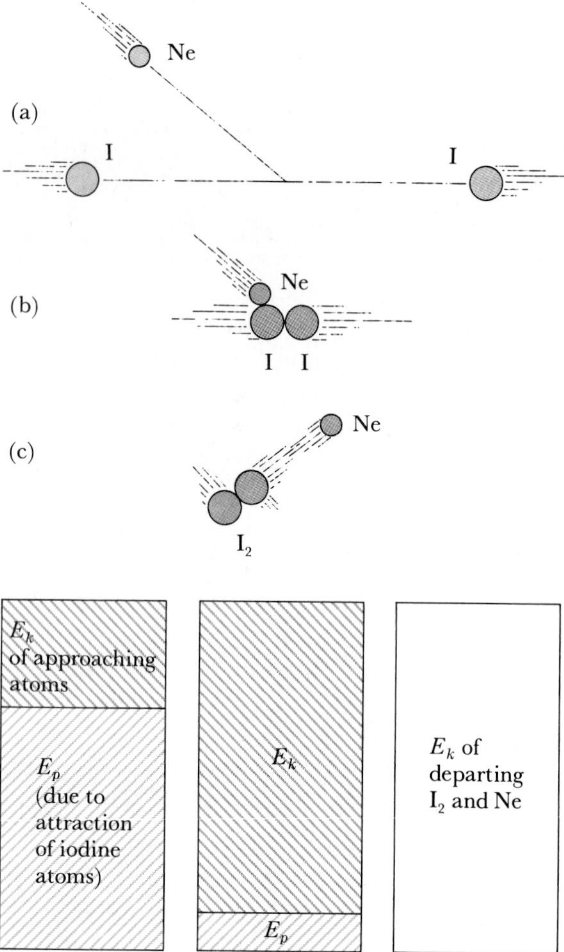

FIGURE 3–2
The interconversion of potential and kinetic energy as two iodine atoms collide in the presence of a neon atom. Zero energy is defined here as the states of the neon atom and the iodine molecule at rest. States a, b, and c are analogous to States 3, 4, and 5 of Figure 3–1.

From this definition of the ampere, and the defined unit of time, the second, the defined unit of electrical charge is obtained. The *coulomb* (symbol C; named for C. A. DE COULOMB) is the quantity of electrical charge corresponding to the flow of one ampere of current for one second:

$$1\ \text{C} = 1\ \text{A s}$$

Electrically charged particles attract or repel one another, depending upon whether the charges have the opposite sign or the same sign. Because of attraction and repulsion of electrical charges, a charged particle has a different potential energy at different positions in a space between electrically charged bodies, as depicted in Figure 3–3. Work would be required to move an electron from the position shown in Figure 3–3d to the position shown in Figure 3–3a. The energy corresponding to this work is determined by the magnitude of the charge and the potential difference (measured in

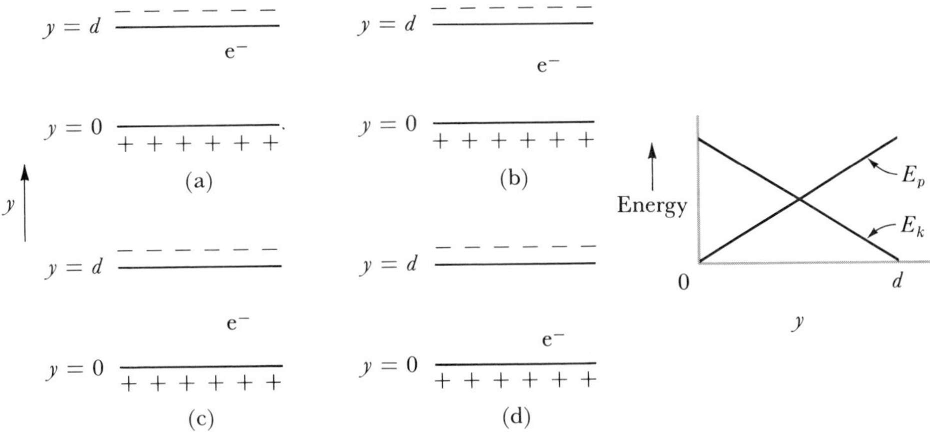

FIGURE 3–3
An electrically charged particle (an electron) at different positions in an electrical field. The potential energy of the electron is greatest in (a) and least in (d). In moving from (a) to (d), the electron loses potential energy and gains kinetic energy, just as the object in Figure 3–1 does as it falls in the gravitational field.

volts). The volt is defined as the potential difference through which the motion of one coulomb gives (or requires) one joule:

$$1\ \text{J} = 1\ \text{V C}$$
$$1\ \text{V} = 1\ \text{J C}^{-1}$$

In terms of basic SI units,

$$\begin{aligned} 1\ \text{V} &= 1\ (\text{kg m}^2\ \text{s}^{-2})(\text{A s})^{-1} \\ &= 1\ \text{kg m}^2\ \text{s}^{-3}\ \text{A}^{-1} \end{aligned}$$

because

$$1\ \text{J} = 1\ \text{kg m}^2\ \text{s}^{-2} \qquad \text{and} \qquad 1\ \text{C} = 1\ \text{A s}$$

Example: A current of 0.50 A flows for 15 min through a resistor in which the voltage drop is 10 V. What is the energy dissipated during this period?

$$\begin{aligned} \text{energy} &= \text{voltage drop} \times \text{quantity of electricity} \\ &= 10\ \text{V} \times 0.50\ \text{A} \times 15\ \text{min} \times \frac{60\ \text{s}}{1\ \text{min}} \\ &= 4.5 \times 10^3\ \text{V A s} = 4.5 \times 10^3\ \text{J} \end{aligned}$$

In everyday life you probably are familiar with the wattage rating of electrical appliances and light bulbs. The watt (symbol W) is a unit of power, which is the rate of energy consumption or production:

$$1\ \text{watt} = 1\ \text{joule per second}$$
$$1\ \text{W} = 1\ \text{J s}^{-1}$$

If one electronic charge flows through a potential difference of one volt, the energy is one electronvolt (symbol eV); to relate this quantity of energy to the joule, we must know the charge on the electron. The currently accepted best value for this quantity is:

$$\begin{aligned} 1 \text{ electron charge} &= 1.6022 \times 10^{-19}\ \text{C} \\ 1\ \text{eV} &= 1.6022 \times 10^{-19}\ \text{C V} \\ &= 1.6022 \times 10^{-19}\ (\text{A s})(\text{J A}^{-1}\ \text{s}^{-1}) \\ &= 1.6022 \times 10^{-19}\ \text{J} \end{aligned}$$

We may measure the energy change in chemical reactions in terms of the energy that corresponds to the flow of one mole of electrons through a particular potential difference of one volt. To calculate this, we must calculate the charge on one mole of electrons, which is called the *Faraday constant*, represented by $\mathscr{F}$,

$$\begin{aligned} \mathscr{F} &= 1.6022 \times 10^{-19}\ \text{C} \times 6.0221 \times 10^{23}\ \text{mol}^{-1} \\ &= 96{,}486\ \text{C mol}^{-1} \end{aligned}$$

For most of our purposes, we will round this to

$$\mathscr{F} = 9.65 \times 10^{4}\ \text{C mol}^{-1}$$

If this amount of electrical charge flows through a potential difference of one volt, the energy is:

$$\begin{aligned} 1\ \mathscr{F}\ \text{V} &= 9.65 \times 10^{4}\ \text{C V mol}^{-1} \\ &= 9.65 \times 10^{4}\ \text{J mol}^{-1} \\ &= 9.65 \times 10^{4}\ \text{J mol}^{-1} \times \frac{1\ \text{cal}}{4.184\ \text{J}} \\ &= 2.31 \times 10^{4}\ \text{cal mol}^{-1} \end{aligned}$$

There are other kinds of energy, and some of these are important in chemistry. Radiant energy is one; it will be considered in several chapters, where we shall introduce the concepts and units needed to discuss it. Although not especially relevant in an introduction to chemistry, relativistic energy is central to analyses of nuclear stability, radioactivity and nuclear reactions, and this subject will be considered in detail in Chapter 22.

THE FIRST LAW OF THERMODYNAMICS

One fundamental principle of thermodynamics is the conservation of energy, developed by a number of investigators and stated in 1841 by J. R. MAYER. The conversion of potential energy into kinetic energy depicted in Figures 3–1 and 3–2 conforms to this principle. There is a conversion of forms of energy in each of these examples, but the total energy of the system (the falling object in Figure 3–1) and its surroundings (the floor in Figure 3–1) is constant. The *first law of thermodynamics* is the law of conservation of energy with heat taken into account. Like the other laws of thermodynamics to be introduced later, it is simply a generalization derived from experience.

In thermodynamic calculations we will consider the exchange of heat and work between a chemical system and its surroundings. There is no exchange of work between a system and its surroundings if the chemical system is contained in a vessel of constant volume. In such situations the exchange of heat simply changes the property of the system that is called its *internal energy* (symbol U),

$$\Delta U = U_{\text{final}} - U_{\text{initial}} = q_v$$

This equation defines internal energy. The symbol Δ used in this equation and in similar equations means the change of a property from its initial state to its final state, final value *minus* initial value. The heat *added to the system* at constant volume is designated q_v. Thus the internal energy of the constant-volume system increases if q_v is positive and decreases if q_v is negative.

The internal energy of a system can be changed also by the exchange of work with the surroundings. A simple example is the experiment that Joule did to show the equivalence of heat and mechanical energy. The work done by a paddle wheel rotating in a fluid increases the internal energy of the fluid; this work could be measured by the potential energy change of the falling weight that drives the paddle wheel. If $q = 0$,

$$\Delta U = w$$

Most current books on chemical thermodynamics define work, w, as the work done *on the system* by the surroundings. (In many older books, w is the work done *on the surroundings*.) The two situations that we have discussed are depicted in Figure 3–4. If both heat and work are exchanged between the system and its surroundings, the change in internal energy of the system is given by

$$\Delta U = q + w$$

This equation is a statement of the *first law of thermodynamics*.

We can illustrate this law by comparing the amounts of heat required to raise the temperature of a sample of gas in a constant-volume container and in a container that allows the pressure to be kept constant (e.g., a cylinder with a movable piston on which there is a set of weights). The experiments could be performed by monitoring the time required for a heater with a particular wattage to raise the temperature of the gas by a particular amount. Although a rigorous definition of heat capacity involves differential calculus, it will be sufficient at this point to define the *molar heat capacity* as the heat required to raise the temperature of one mole of the substance by one kelvin. (The heat required to raise the temperature of one gram by one kelvin is called the *specific heat*.) The experimentally

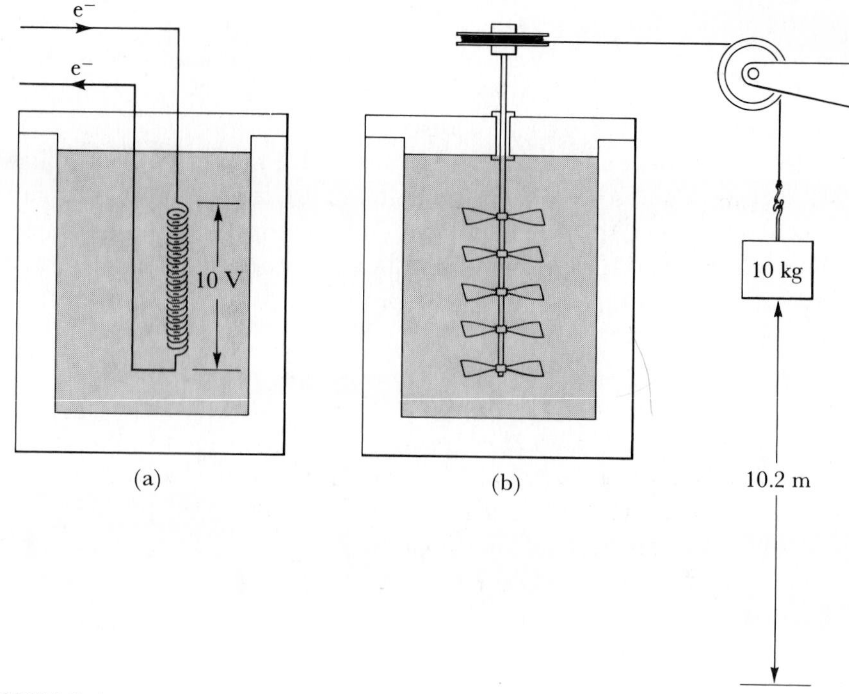

FIGURE 3–4
The increase of internal energy of a fluid at constant volume; $\Delta U = 1.000$ kJ. (a) Heat added to system by electrical heater. 100 C flows through a voltage drop of 10 V; $100\ \text{C} \times 10\ \text{V} = 1000\ \text{C V} = 1000\ \text{J}$. (b) Mechanical work done on system by paddle wheel run by falling weight. A 10.00-kg mass falls through 10.20 m; $10.0\ \text{kg} \times 9.81\ \text{m s}^{-2} \times 10.2\ \text{m} = 1000\ \text{kg m}^2\ \text{s}^{-2} = 1000\ \text{J}$.

determined molar heat capacity at constant volume, C_v, for helium is

$$C_v = 2.98\ \text{cal K}^{-1}\ \text{mol}^{-1}$$
$$= 12.5\ \text{J K}^{-1}\ \text{mol}^{-1}$$

The amount of heat required to raise the temperature of a sample of helium in a constant-volume container from a temperature t_i to a temperature t_f is the product of the amount of helium in moles (n), its molar heat capacity (C_v), and the temperature increase:

$$q_v = n \times C_v \times (t_f - t_i)$$

For example, to raise the temperature of 5.30 mol of helium in a constant-volume container by 15.0 K would require

$$q_v = 5.30\ \text{mol} \times 12.5\ \text{J K}^{-1}\ \text{mol}^{-1} \times 15.0\ \text{K}$$
$$= 994\ \text{J}$$

In our comparison of the constant-volume and constant-pressure conditions, we will calculate the amount of heat required to increase the temperature of 1.00 mol He from 20.0°C to 90.0°C ($\Delta t = 70.0°\text{C} = 70.0$ K, because the magnitudes of the Celsius degree and the kelvin are the same.)

For constant volume, the heat added, q_v, is given by

$$q_v = 1.00 \text{ mol} \times 12.5 \text{ J K}^{-1} \text{ mol}^{-1} \times 70.0 \text{ K}$$
$$= 875 \text{ J}$$

To calculate the heat required under constant pressure, we use the molar heat capacity at constant pressure, C_p; for helium, the experimentally determined value is

$$C_p = 20.8 \text{ J K}^{-1} \text{ mol}^{-1}$$

The quantity of heat added is

$$q_p = = 1.00 \text{ mol} \times 20.8 \text{ J K}^{-1} \text{ mol}^{-1} \times 70.0 \text{ K}$$
$$= 1456 \text{ J}$$

More heat is required to raise the temperature of the gas under constant pressure because the gas expands, and work is done on the surroundings by the system. (The weights on the piston are raised.) For the system being considered, one mole of helium at low pressure, the internal energy depends only upon the temperature. Thus the change in internal energy, ΔU, is the same under both sets of conditions. This fact allows us to calculate the amount of work done on the surroundings $(-w)$ under constant pressure:

$$-w = q_p - \Delta U$$
$$= 1456 \text{ J} - 875 \text{ J} = 581 \text{ J}$$

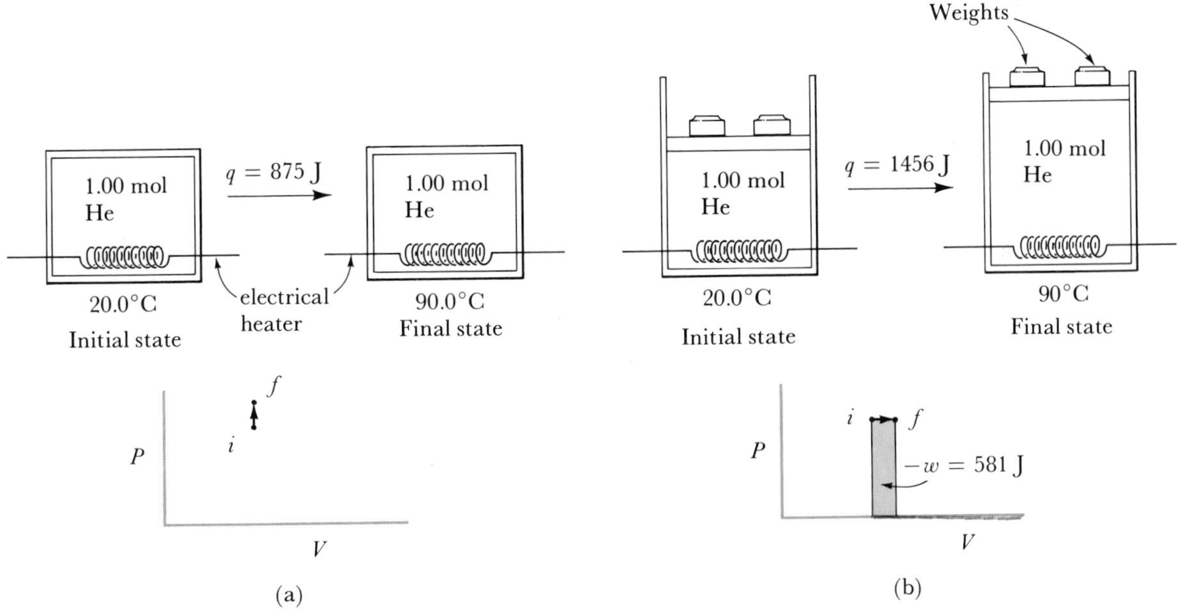

FIGURE 3–5
Experiments in which the temperature of one mole of helium gas is raised from 20.0°C to 90.0°C ($\Delta T = 70.0$ K). (a) Volume is constant: $\Delta U = 875$ J, $q = 875$ J, $w = 0$. (b) Pressure is constant: $\Delta U = 875$ J, $q = 1456$ J, $w = -581$ J.

These experiments are depicted in Figure 3–5. No work is done in the experiment in Figure 3–5a; only the internal energy of the gas in the container is changed. Work is done on the surroundings in the experiment in Figure 3–5b; the expanding gas raises the weights on the movable piston. Because the weights on the piston have a particular mass, they exert a particular pressure, that is, a particular force over the area of the piston. The situation depicted in Figure 3–5b is an expansion of ΔV at constant pressure P. The work done on the surroundings, $-w$, is the area under the curve in a plot of pressure versus volume:

$$-w = P(V_f - V_i) = P\Delta V$$

We see that this area is zero when $\Delta V = 0$, and is greater than zero when the pressure is constant. That the product $P\Delta V$ is work is shown by the equivalence of the dimensions for a pressure acting through a volume and for a force acting through a distance:

$$(F \times l^{-2}) \times l^3 \equiv F \times l$$

$$\text{pressure} \times \text{volume} \equiv \text{force} \times \text{distance}$$

In this example the only type of work that is possible is $P\Delta V$ work; there are other types of work, for example, electrical work, and these would be included in an application of the first law of thermodynamics to situations where other types of work are done.

In using the preceding example to illustrate the first law of thermodynamics, it was not necessary to specify the nature of the increase in internal energy of the system with the absorption of heat. But our understanding of the subject may be enhanced by such knowledge. The increase of internal energy in this example is simply the increase in the kinetic energy of the helium atoms' translational motion accompanying the increase of temperature. This correspondence between temperature and kinetic energy will be discussed in the next chapter.

INTERNAL ENERGY AND ENTHALPY

In calculating the heat required to raise the temperature of one mole of helium gas, we saw that it was important to specify whether the volume or the pressure was held constant. The heat added at constant volume is the more interesting quantity if we are concerned primarily with the internal changes in the chemical system caused by the temperature increase (the changes in potential energy and kinetic energy of the atoms). But conditions of constant pressure are common in chemistry; reactions are carried out in vessels open to the atmosphere. For this reason, an energy function related to the internal energy that also contains a contribution from the volume of the system is useful. This function, called *enthalpy* (symbol H), is defined

$$H \equiv U + PV$$

(We already have seen that the product PV has the same dimensions as

U.) For a process carried out from an initial state i to a final state f at *constant pressure* ($P_f = P_i = P$):

$$\begin{aligned}\Delta H &= H_f - H_i \\ &= (U_f - U_i) + (P_f V_f - P_i V_i) \\ &= (U_f - U_i) + P(V_f - V_i) \\ &= \Delta U + P\Delta V\end{aligned}$$

In the experiment of raising the temperature of one mole of helium gas at constant pressure,

$$q = \Delta U - w$$

but the increase of temperature resulted in an expansion of the gas, and work, $P\Delta V$, was done on the surroundings by the system:

$$-w = P\Delta V$$

which gives

$$q = \Delta U + P\Delta V$$

which is ΔH. That is,

$$q_p = \Delta H$$

The heat added to the system in an experiment at constant pressure (and in which only $P\Delta V$ work is done) is the change of enthalpy of the system.

The internal energy, U, and enthalpy, H, are *state functions*. This means that the values of ΔU and ΔH depend only upon the initial and final states and not upon the route between the states. The work done on the system, w, and the heat added to the system, q, are *not* state functions. The concept of state function is an important one; its meaning will be clarified by considering further the change:

$$\text{He}(g, 20°\text{C}, P = P_i) = \text{He}(g, 90°\text{C}, P = P_i)$$

Two different routes for this change are depicted in Figure 3–6. The first route (a) is that already discussed: heat is added to raise the temperature of the system (one mole of helium) kept at constant pressure. The changes of internal energy and enthalpy are directly obtained from the temperature change and the molar heat capacities, C_v and C_p:

$$\Delta U = q_v = C_v \Delta T = 12.5\ \text{J K}^{-1}\ \text{mol}^{-1} \times 70.0\ \text{K} = 875\ \text{J mol}^{-1}$$
$$\Delta H = q_p = C_p \Delta T = 20.8\ \text{J K}^{-1}\ \text{mol}^{-1} \times 70.0\ \text{K} = 1456\ \text{J mol}^{-1}$$

Because internal energy and enthalpy are state functions, these are also the values of ΔU and ΔH for the three-step process shown in Figure 3–6b. In this sequence of steps, the first step, State 1 → State 3, involves addition of heat to the system at constant volume (this increases the pressure and the temperature). The second step, State 3 → State 4, involves the addition of heat to the system at constant pressure (this increases the volume and the temperature). The third step, State 4 → State 2, involves the loss of

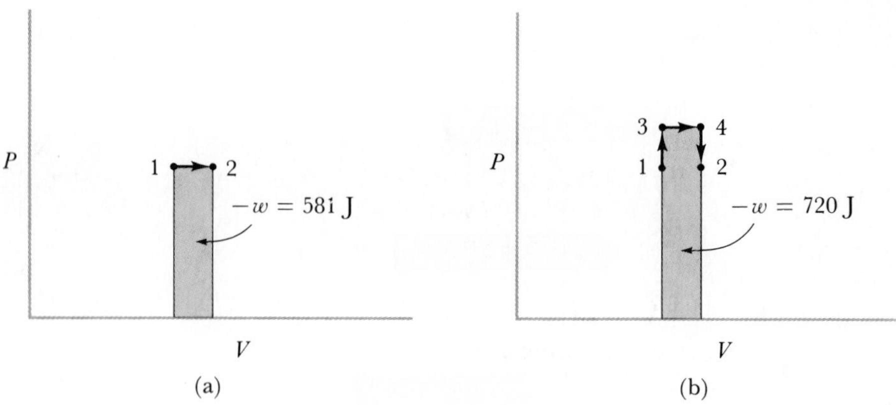

FIGURE 3–6
Pressure–volume diagrams for the change

$$\mathrm{He}(g, 20.0°\mathrm{C}, P = P_i) = \mathrm{He}(g, 90.0°\mathrm{C}, P = P_i)$$

occurring by two routes:

(a) Constant-pressure process $(1 \rightarrow 2)$

$$\Delta U = 875 \text{ J mol}^{-1} \qquad \Delta H = 1456 \text{ J mol}^{-1}$$
$$w = -581 \text{ J mol}^{-1} \qquad q = 1456 \text{ J mol}^{-1}$$

(b) Pathway $(1 \rightarrow 3 \rightarrow 4 \rightarrow 2)$ $(T_3 > T_1;\ T_4 > T_2;\ T_4 > T_3)$

$$\Delta U = 875 \text{ J mol}^{-1} \qquad \Delta H = 1456 \text{ J mol}^{-1}$$
$$w = -720 \text{ J mol}^{-1} \qquad q = 1595 \text{ J mol}^{-1}$$

heat by the system (which decreases the pressure and the temperature). The expansion of the gas in going from State 1 to State 2 involves work done by the system upon the surroundings, and therefore w is negative in each case. The value of $-w$ is the area $P\Delta V$ under the line giving pressure versus volume in each case. It is larger for the three-step process. The first law allows us to obtain q, the net heat added to the system in each case:

$$q = \Delta U - w$$

Since $-w$ is larger for the three-step process, q is also larger for this process.

A final word about heat and work: It is incorrect to say that a system contains heat or contains work. A system contains energy; it contains enthalpy; but it does not contain heat or work. Heat is energy being transferred from a hot body to a cold body. Work, similarly, is a form of energy transfer. In the present example, work was done by the expansion of the gas to raise the weights on the movable piston.

3–3 Energy Changes in Chemical Reactions

Now that we have learned about the first law of thermodynamics and the state functions, internal energy and enthalpy, we can go on to consider processes with chemical aspects, chemical reactions and changes of state of chemical substances. In the course of this, we will learn an additional law of thermodynamics that guides us in determining the direction in which a process occurs.

In general, the reactants and products of chemical reactions have different internal energies and different enthalpies. Therefore a chemical reaction is accompanied by the evolution or the absorption of energy. If the reaction evolves energy, the reaction is said to be *exothermic*; if the reaction absorbs energy, the reaction is said to be *endothermic*. In some cases, qualitative observations may disclose whether a reaction or process is exothermic or endothermic. The burning of a fuel obviously is an exothermic reaction; the evolved energy raises the temperature of the system, so heat will be transferred to the surroundings. The melting of ice, $H_2O(s) = H_2O(l)$, is an endothermic process; the energy required to melt the ice cubes immersed in a beverage lowers the temperature of the mixture.

The energy evolved or absorbed during a chemical reaction can be measured by a calorimeter. In the simplest calorimeter, shown in Figure 3–7, the reaction is carried out in an insulated vessel (e.g., a vacuum bottle), which transmits no heat to or from its surroundings. Under these conditions, which are called *adiabatic*, the energy change accompanying the reaction raises or lowers the temperature of the system (the reaction mixture and its container). The total heat capacity of the calorimeter and its contents can be measured independently with a previously characterized reaction or can be calibrated with an electrical heater. Electrical calibration has the advantage of tying the energy of reaction directly to SI energy units.

EXAMPLE: A solution containing 1.043×10^{-3} mol of hydrochloric acid (HCl) in 10.0 cm^3 of water is mixed with 90.0 cm^3 of sodium

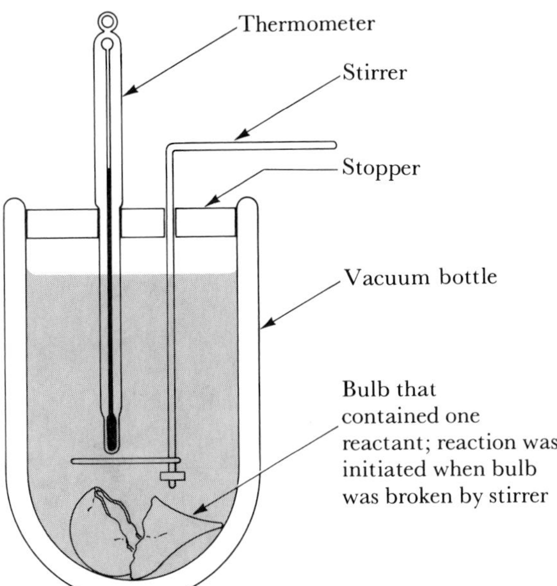

FIGURE 3–7
An adiabatic calorimeter. Because heat cannot be transferred to or from the system, the enthalpy change in the reaction raises or lowers the temperature of the system (the reaction mixture and inner walls of the container).

hydroxide solution (containing an excess of sodium hydroxide) in an insulated vessel. The net neutralization reaction,

$$H^+ + OH^- = H_2O$$

occurs, and the temperature increases from 25.051°C to 25.142°C. When reaction is complete, an electrical calibration, with the product solution in the calorimeter, shows that a current of 0.500 A flowing for 3.00 min, with a voltage drop across the heater of 12.03 V, has increased the temperature to 26.832°C. What is the enthalpy change for the reaction?

The total heat capacity of the reaction mixture and the calorimeter is obtained from the electrical calibration; the quantity of energy added in the calibration is

$$12.03\ \text{V} \times 0.500\ \text{A} \times 180\ \text{s} = 1080\ \text{V A s} = 1080\ \text{J}$$

This energy raised the temperature from 25.142°C to 26.832°C. The quantity of enthalpy in the exothermic chemical reaction ($-\Delta H \times n$) is proportional to the temperature rise that it has caused, and the porportionality constant is determined by the electrical calibration:

$$-\Delta H \times n = \underbrace{\underbrace{(25.142°\text{C} - 25.051°\text{C})}_{\Delta t \text{ in experiment}} \times \underbrace{\frac{1080\ \text{J}}{(26.832°\text{C} - 25.142°\text{C})}}_{\text{J per K in calibration}}}_{\text{Enthalpy decrease in reaction}}$$

$$= 0.091\ \text{K} \times 639\ \text{J K}^{-1}$$

$$-\Delta H \times n = 58.2\ \text{J}$$

This enthalpy decrease resulted from a reaction involving 1.043×10^{-3} mol of acid, and for one mole the value of ΔH is

$$\Delta H = -58.2\ \text{J} \times \frac{1}{1.043 \times 10^{-3}\ \text{mol}}$$

$$= -55{,}800\ \text{J mol}^{-1} = -55.8\ \text{kJ mol}^{-1}$$

There also are types of calorimeters in which the temperature is maintained constant. That condition is called *isothermal*. One way to maintain the temperature of the reaction mixture constant during a slow endothermic reaction is concurrently to add electrical energy, which can be monitored. Another way to maintain constant temperature is that of the BUNSEN ice calorimeter.[2] In this apparatus the reaction vessel at 0°C is surrounded by a vessel containing an ice–water mixture at 0°C. (See Figure 3–8.) The flow of heat between the reaction vessel and the mixture

[2] A calorimetric experiment appropriate for student use involving a Bunsen ice calorimeter is described in the article by B. H. Mahan, *J. Chem. Educ.* **37**, 634 (1960).

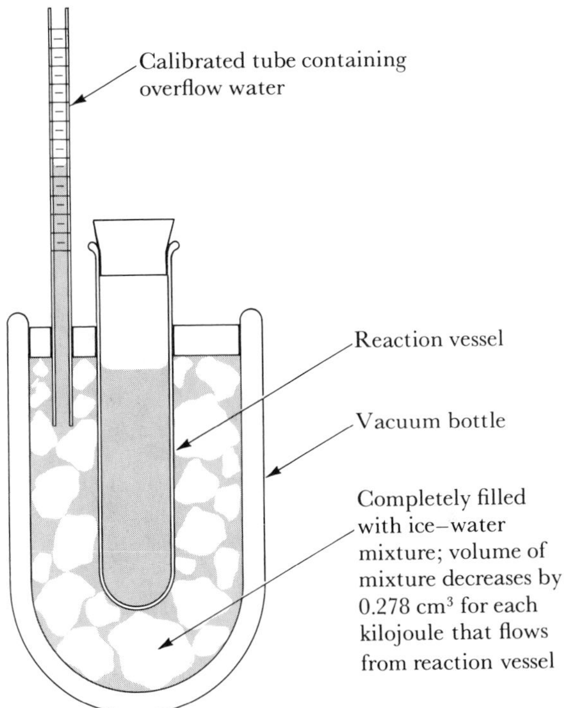

FIGURE 3–8
A Bunsen ice calorimeter.

of ice and liquid water results in the occurrence of the process

$$H_2O(s, 0°C) = H_2O(l, 0°C) \qquad \Delta H = +6.01 \text{ kJ mol}^{-1}$$
$$\Delta V = -1.67 \text{ cm}^3 \text{ mol}^{-1}$$

to the right if the reaction evolves energy and to the left if it absorbs energy. From the quantities given we see that the proportionality between volume change, which is the measured quantity, and heat evolved in the reaction vessel is

$$\frac{\Delta V}{\Delta H} = \frac{-1.67 \text{ cm}^3 \text{ mol}^{-1}}{6.01 \text{ kJ mol}^{-1}} = -0.278 \text{ cm}^3 \text{ kJ}^{-1}$$

Some calorimeters work with fixed volumes and others with fixed pressures. The observed amount of heat absorbed by the chemical system during reaction (defined as q) depends upon the type of calorimeter employed. If a constant-volume calorimeter is used, no work is done by the system on the surroundings nor is work done by the surroundings on the system. The observed heat transferred between the system and its surroundings is due solely to the change in the internal energy of the chemical system, ΔU. It is equal to the heat absorbed by the system in the experiment at constant volume, q_v:

$$q_v = \Delta U$$

If the internal energy of the chemical system decreases with the occurrence of reaction, $\Delta U < 0$, and q_v is negative.

If a constant-pressure calorimeter is used, work is done on the surroundings by the chemical system or vice versa, depending upon whether the volume of the chemical system increases or decreases in the reaction. The amount of work is appreciable if the reaction involves the net production or consumption of gases, but it is very small if only condensed phases are involved in the reaction. The heat absorbed by the system in a reaction at constant pressure, q_p, is different from ΔU by this amount of work done on the system (the product of pressure and the change of volume):

$$\begin{aligned} w &= -P(V_f - V_i) \\ &= -P\Delta V \\ \Delta U &= q_p + w \\ &= q_p - P\Delta V \\ q_p &= \Delta U + P\Delta V = \Delta H \end{aligned}$$

The effect of a net production of gas upon the relative values of ΔU and ΔH can be shown with two examples involving changes of state for water. A change of state that does not involve gas is the melting process,

$$H_2O(s,\ 0°C) = H_2O(l,\ 0°C)$$

At constant pressure (1.00 atm) this process involves only a small change of volume, $\Delta V = -1.67\ cm^3\ mol^{-1}$ and $P\Delta V = -0.169\ J\ mol^{-1}$. (You will learn in the next chapter how to convert $P\Delta V = -1.67\ cm^3\ atm\ mol^{-1}$ to $P\Delta V = -0.169\ J\ mol^{-1}$.) For this process $\Delta H = +6.01\ kJ\ mol^{-1}$ and, therefore, to this number of significant figures, the value of ΔU is the same as the value of ΔH,

$$\begin{aligned} \Delta U &= \Delta H - P\Delta V \\ &= 6.01\ kJ\ mol^{-1} - 1.69 \times 10^{-4}\ kJ\ mol^{-1} \\ &= 6.01\ kJ\ mol^{-1} \end{aligned}$$

A change of state that involves production of gas is the vaporization process,

$$H_2O(l,\ 100°C) = H_2O(g,\ 100°C)$$

and for this process at constant pressure (1.00 atm) the change of volume is appreciable, $\Delta V = +30.6\ L\ mol^{-1}$; thus $P\Delta V = +30.6\ L\ atm\ mol^{-1}$ or $+3.10\ kJ\ mol^{-1}$. For this process $\Delta H = 40.66\ kJ\ mol^{-1}$, and

$$\begin{aligned} \Delta U &= \Delta H - P\Delta V \\ &= 40.66\ kJ\ mol^{-1} - 3.10\ kJ\ mol^{-1} \\ &= 37.56\ kJ\ mol^{-1} \end{aligned}$$

For most purposes, we will consider $\Delta U \cong \Delta H$ if only condensed states are involved in the process, but the term $P\Delta V$ cannot be neglected if the process involves the net consumption or production of gas.

STANDARD ENTHALPIES OF FORMATION OF SUBSTANCES; THERMOCHEMICAL CALCULATIONS

Now we will consider the relative enthalpy content of gaseous molecules containing one carbon atom combined with hydrogen and oxygen. These substances are methane (CH_4), methyl alcohol (CH_3OH), formaldehyde (H_2CO), formic acid ($HCOOH$), carbon dioxide (CO_2), and carbon monoxide (CO). We can write a chemical equation for the formation of each of these carbon compounds from the constituent elements, even though the reaction may not occur cleanly under realizable conditions. These reactions and the reactions forming water in the liquid and gaseous states are given in Table 3–1. Given with each of these chemical equations for the formation of one mole of product is the change of enthalpy associated with the reaction at 25°C. These changes of enthalpy are designated ΔH_f^0, the subscript f meaning that each reaction involves formation of one mole of product from the constituent elements, and the superscript 0 meaning that each reactant is in its standard state. The standard state of

TABLE 3–1
Enthalpy Changes for the Formation of Some Gaseous Molecules Containing One Carbon Atom

Reaction	$\dfrac{\Delta H_f^{0}}{\text{kJ mol}^{-1}}$ [a]	$\dfrac{\Delta H_f^{0}}{\text{kcal mol}^{-1}}$ [a]
$C(s) + 2H_2(g) = CH_4(g)$	−74.85	−17.89
$C(s) + \frac{1}{2}O_2(g) + 2H_2(g) = CH_3OH(g)$	−201.25	−48.10
$C(s) + \frac{1}{2}O_2(g) + H_2(g) = H_2CO(g)$	−115.9	−27.70
$C(s) + O_2(g) + H_2(g) = HCOOH(g)$	−362.6	−86.67
$C(s) + O_2(g) = CO_2(g)$	−393.5	−94.05
$C(s) + \frac{1}{2}O_2(g) = CO(g)$	−110.5	−26.42
$H_2(g) + \frac{1}{2}O_2(g) = H_2O(g)$	−241.8	−57.80
$H_2(g) + \frac{1}{2}O_2(g) = H_2O(l)$	−285.9	−68.32

[a] Notice the form in which the column headings are written. This form allows the entries to be recorded as numbers. If

$$\Delta H_f^0 = -74.85 \text{ kJ mol}^{-1}$$

the quotient

$$\frac{\Delta H_f^0}{\text{kJ mol}^{-1}} = \frac{-74.85 \cancel{\text{kJ mol}^{-1}}}{\cancel{\text{kJ mol}^{-1}}} = -74.85$$

is a number. This type of heading will be used in many column headings, as used here, and in the designation of dimensions in coordinates for many graphs. The use of this notation also will make it possible to give the dimensions of terms in some equations more concisely than otherwise might be possible.

an element is the form in which it exists at ordinary conditions of temperature and pressure. For carbon, the standard state is graphite, which is more stable than diamond.[3] For oxygen and hydrogen, the standard states are the gaseous diatomic molecules, $O_2(g)$ and $H_2(g)$. (For gases, the standard state implies also a standard pressure, generally one atmosphere, and gas ideality. This latter point will be made clearer in the next chapter.)

The balanced equations in Table 3–1 can be combined by addition and subtraction to give balanced equations for other reactions. For example, the equation for the oxidation of formaldehyde, H_2CO, to formic acid, HCOOH,

$$H_2CO(g) + \tfrac{1}{2}O_2(g) = HCOOH(g)$$

is the combination

$$C(s) + H_2(g) + O_2(g) = HCOOH(g)$$
$$\Delta H^0 \equiv \Delta H_f^0[HCOOH(g)]$$

minus

$$C(s) + H_2(g) + \tfrac{1}{2}O_2(g) = H_2CO(g)$$
$$\Delta H^0 \equiv \Delta H_f^0[H_2CO(g)]$$

The value of ΔH^0 for the oxidation of formaldehyde to give formic acid is the corresponding algebraic combination of values of ΔH_f^0; it is the standard enthalpy of formation of formic acid (the product) minus the standard enthalpy of formation of formaldehyde (the reactant):

$$\Delta H^0 = \Delta H_f^0[HCOOH(g)] - \Delta H_f^0[H_2CO(g)]$$

$$\frac{\Delta H^0}{\text{kcal mol}^{-1}} = -86.67 - (-27.70)$$

$$\Delta H^0 = -58.97 \text{ kcal mol}^{-1}$$

The basis for this calculation can be shown graphically, as in Figure 3–9, which gives the relative enthalpy content of each of the molecules involved. The enthalpy content of one mole of formic acid is 58.97 kcal more negative than that of one mole of formaldehyde plus one half mole of oxygen. Also shown in this graph is the enthalpy content of one mole of carbon dioxide and one mole of water, the completely oxidized forms of carbon and hydrogen. The enthalpy content of formic acid and formaldehyde also can be reckoned relative to carbon dioxide and water; these values are given along the right side of the figure. The reaction under consideration also is an algebraic combination of the combustion reactions for formaldehyde and formic acid and the corresponding changes of enthalpy, ΔH_c^0:

[3]The structures of diamond and graphite are discussed in Chapter 11.

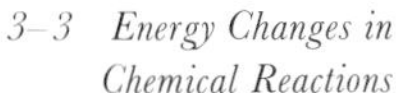

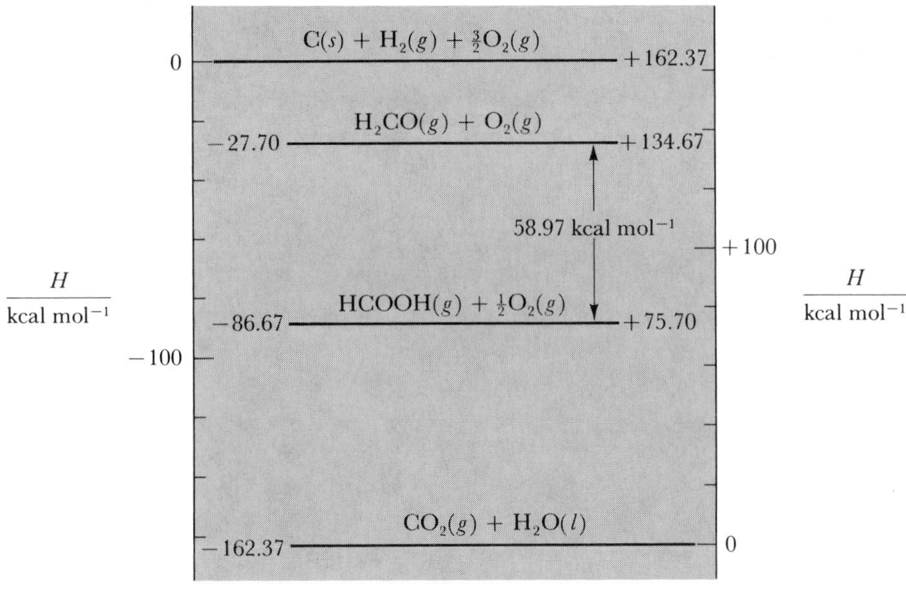

FIGURE 3–9
The relative enthalpy content (H) of some compounds containing carbon, hydrogen, and oxygen. The enthalpies are for amounts of material containing 1 mol C, 2 mol H, and 3 mol O. The enthalpy values on the left side are values relative to the constituent elements in their standard states; the values on the right side are values relative to carbon dioxide plus water.

$$H_2CO(g) + O_2(g) = CO_2(g) + H_2O(l) \qquad \Delta H^0 = \Delta H_c^0[H_2CO(g)]$$

minus

$$\frac{HCOOH(g) + \tfrac{1}{2}O_2(g) = CO_2(g) + H_2O(l)}{H_2CO(g) + \tfrac{1}{2}O_2(g) = HCOOH(g)} \qquad \begin{array}{l} \Delta H^0 = \Delta H_c^0[HCOOH(g)] \\ \Delta H^0 = \Delta H_c^0[H_2CO(g)] \\ \qquad - \Delta H_c^0[HCOOH(g)] \end{array}$$

$$\frac{\Delta H^0}{\text{kcal mol}^{-1}} = -134.67 - (-75.70) \qquad \Delta H^0 = -58.97 \text{ kcal mol}^{-1}$$

These arithmetic manipulations are examples of the HESS law of additivity of reaction heats. A particular relative enthalpy content can be assigned to each set of compounds containing a particular amount of each element (one mole of carbon atoms, two moles of hydrogen atoms, and three moles of oxygen atoms in the present example) because enthalpy is a state function. The sequence of reactions by which a compound is formed does not influence its enthalpy content.

The calculations we have performed illustrate that the value of ΔH^0 for a reaction can be calculated using:

1. the values of ΔH_f^0 for the reactants and products of the reaction, $\Delta H^0 = \sum \Delta H_f^0(\text{products}) - \sum \Delta H_f^0(\text{reactants})$,

2. the values of ΔH^0 for the combustion reactions of the reactants and products, $\Delta H^0 = \sum \Delta H_c^0(\text{reactants}) - \sum \Delta H_c^0(\text{products})$.

The calorimetric study of combustion reactions for carbon-containing compounds has been developed into a highly accurate technique, which has been used to determine the enthalpy content for most such compounds. However, even though the enthalpy content of a substance has been established by combustion calorimetry,[4] the data generally are summarized in terms of the value of the standard enthalpy of formation, ΔH_f^0. This will be illustrated with data for benzoic acid, $C_7H_6O_2(s)$. For the combustion reaction

$$C_7H_6O_2(s) + 7\tfrac{1}{2}O_2(g) = 7CO_2(g) + 3H_2O(l)$$

the value of ΔH^0 for 25°C derived from the calorimetric measurements is -771.2 kcal mol^{-1}. This value of ΔH^0 is related to the standard enthalpies of formation by the following equation, which takes into account the coefficients in the chemical equation:

$$\Delta H = 3\,\Delta H_f^0[H_2O(l)] + 7\,\Delta H_f^0[CO_2(g)] - \Delta H_f^0[C_7H_6O_2(s)]$$

which can be rearranged to give:

$$\Delta H_f^0[C_7H_6O_2(s)] = 3\,\Delta H_f^0[H_2O(l)] + 7\,\Delta H_f^0[CO_2(g)] - \Delta H$$

The standard enthalpies of formation of $H_2O(l)$ and $CO_2(g)$ are known, -68.32 kcal mol^{-1} and -94.05 kcal mol^{-1}, respectively. Therefore the calculation is

$$\frac{\Delta H_f^0[C_7H_6O_2(s)]}{\text{kcal mol}^{-1}} = 3(-68.32) + 7(-94.05) - (-771.2) = -92.1$$

$$\Delta H_f^0[C_7H_6O_2(s)] = -92.1 \text{ kcal mol}^{-1}$$

Thus the value of ΔH_f^0 for benzoic acid can be obtained from heat of combustion data.

Table 3–2 is a tabulation of values of the standard enthalpy of formation for a selection of chemical compounds. With these values of ΔH_f^0, you can calculate the value of ΔH^0 for any reaction involving as reactants and products substances included in the compilation. For instance, consider the reaction of calcium oxide with carbon dioxide:

$$CaO(s) + CO_2(g) = CaCO_3(s)$$

This reaction is the following algebraic combination of reactions for forming the products and reactants from the constituent elements:

[4]Combustion calorimetry generally involves a constant-volume calorimeter operating in a bath of liquid water at ~ 25°C. The reactants are initially at this temperature, and in the final state the temperature is slightly higher than this. It is the rise in temperature that allows calculation of the heat evolved. In the final state the water produced in the reaction is in the liquid state. Because the system has a constant volume, the measured temperature change allows calculation of ΔU; the relationship of ΔU to the value of ΔH will be presented in Chapter 4.

$$\begin{aligned} Ca(s) + C(s) + \tfrac{3}{2}O_2(g) &= CaCO_3(s) \\ -[C(s) + O_2(g) &= CO_2(g)] \\ -[Ca(s) + \tfrac{1}{2}O_2(g) &= CaO(s)] \end{aligned}$$

The enthalpy change for the reaction is the corresponding algebraic combination of the heats of formation of the product and reactants:

$$\Delta H^0 = \Delta H_f^0[CaCO_3(s)] - \Delta H_f^0[CO_2(g)] - \Delta H_f^0[CaO(s)]$$

$$\frac{\Delta H^0}{\text{kcal mol}^{-1}} = -288.45 - (-94.05) - (-151.8) = -42.6$$

$$\Delta H^0 = -42.6 \text{ kcal mol}^{-1}$$

EXAMPLE: Calculate the standard enthalpy change in the reaction $4NH_3(g) + 5O_2(g) = 4NO(g) + 6H_2O(g)$.

$$\Delta H^0 = 6\,\Delta H_f^0[H_2O(g)] + 4\,\Delta H_f[NO(g)] - 4\,\Delta H_f[NH_3(g)] - 5\,\Delta H_f[O_2(g)]$$

The term $\Delta H_f(O_2)$ is zero because $O_2(g)$ is the standard form of oxygen. This gives

$$\frac{\Delta H^0}{\text{kcal mol}^{-1}} = 6(-57.80) + 4(21.57) - 4(-11.02) - 5(0) = -216.44$$

$$\Delta H^0 = -216.44 \text{ kcal mol}^{-1}$$

TABLE 3–2

Standard Enthalpies of Formation of Some Common Chemical Substances (at 298.2 K)[a]

Substance	ΔH_f^0 / kJ mol^{-1}	ΔH_f^0 / kcal mol^{-1}	Substance	ΔH_f^0 / kJ mol^{-1}	ΔH_f^0 / kcal mol^{-1}
$BaO(s)$	−582	−139.1	$C_8H_{18}(g)$[c]	−208	−49.7
$Ba(OH)_2(s)$	−946	−226.1	$Fe_2O_3(s)$	−822.2	−196.5
$BaCO_3(s)$	−1219	−291.3	$Fe_3O_4(s)$	−1117.1	−267.0
$BaSO_4(s)$	−1465	−350.1	$H_2(g)$	0	0
$CaO(s)$	−635	−151.8	$H_2O(g)$	−241.8	−57.80
$CaCO_3(s)$	−1207	−288.45	$H_2O(l)$	−285.9	−68.32
C(s, graphite)	0	0	$MgO(s)$	−602	−143.9
C(s, diamond)	+1.88	+0.45	$NH_3(g)$	−46.11	−11.02
$CO(g)$	−110.5	−26.42	$NO(g)$	+90.25	+21.57
$CO_2(g)$	−393.5	−94.05	$O_2(g)$	0	0
$HCOOH(l)$	−425	−101.6	$O_3(g)$	+143	+34.1
$H_2CO(l)$	−139	−33.2	$O(g)$	+249.2	+59.55
$CH_3OH(l)$	−239	−57.1	$SO_2(g)$	−296.8	−70.94
$C_2H_6(g)$	−84.67	−20.236	$SO_3(g)$	−395.7	−94.58
$C_7H_6O_2(s)$[b]	−385.3	−92.1			

[a] A more detailed table of values of ΔH_f^0 is given in Appendix 3.
[b] $C_7H_6O_2(s)$ is benzoic acid.
[c] $C_8H_{18}(g)$ is octane.

The dimensions given for this answer include (mol^{-1}), although it is the enthalpy change for the reaction of 4 moles of ammonia. This is conventional; a value of ΔH^0 pertains to the chemical equation with the particular set of coefficients given. For example, for the reaction

$$2NH_3(g) + 2\tfrac{1}{2}O_2(g) = 2NO(g) + 3H_2O(g)$$

which is one half the previous balanced equation, the value of ΔH^0 is one half of the value given above:

$$\begin{aligned}\Delta H^0 &= \tfrac{1}{2} \times (-216.44 \text{ kcal mol}^{-1}) \\ &= -108.22 \text{ kcal mol}^{-1}\end{aligned}$$

In the preceding chapter, it was stated that ozone (O_3) was unstable with respect to diatomic oxygen (O_2), but was stable with respect to diatomic oxygen plus atomic oxygen. The entries in Table 3–2 allow us to calculate the enthalpy change in each of these reactions. For the reaction

$$2O_3(g) = 3O_2(g)$$

the standard enthalpy change is

$$\Delta H^0 = 3\Delta H_f^0(O_2) - 2\,\Delta H_f^0(O_3)$$

$$\frac{\Delta H^0}{\text{kcal mol}^{-1}} = 0 - 2(34.1) = -68.2$$

$$\Delta H^0 = -68.2 \text{ kcal mol}^{-1}$$

For the reaction

$$O_3(g) = O_2(g) + O(g)$$

the standard enthalpy change is

$$\Delta H^0 = \Delta H_f^0(O_2) + \Delta H_f^0(O) - \Delta H_f^0(O_3)$$

$$\frac{\Delta H^0}{\text{kcal mol}^{-1}} = 0 + 59.55 - 34.1 = +25.5$$

$$\Delta H^0 = +25.5 \text{ kcal mol}^{-1}$$

It is the decrease of enthalpy that is primarily responsible for the instability of ozone with respect to diatomic oxygen, and also for the formation of ozone in a mixture of atomic oxygen and diatomic oxygen. This latter reaction, which occurs in the upper atmosphere, is the reverse of that just considered. This reaction is exothermic:

$$O(g) + O_2(g) = O_3(g) \qquad \Delta H^0 = -25.5 \text{ kcal mol}^{-1}$$

because the changes of any state function (e.g., enthalpy) for two processes that are the forward and reverse of one another are equal in magnitude but opposite in sign. Figure 3–10 shows the enthalpy content of various forms of oxygen.

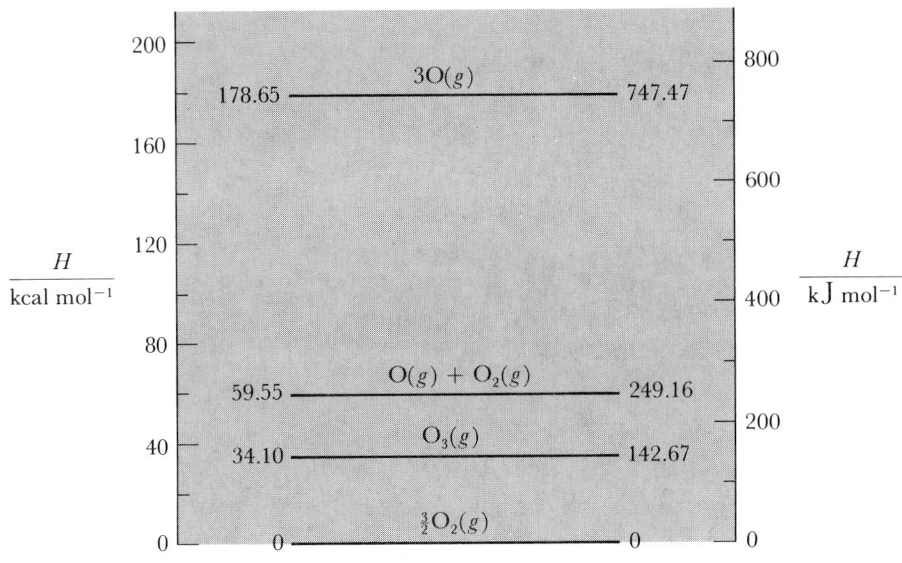

FIGURE 3–10
The enthalpy contents of various combinations of three moles of oxygen atoms. Scale at left, kcal mol^{-1}; at right, kJ mol^{-1}.

THE DEPENDENCE OF ΔH ON TEMPERATURE

Most tabulations of thermochemical data are for a particular temperature, generally 25°C. The values of ΔH_f^0 in Table 3–2 are for this temperature. A question of some importance is whether values of ΔH_f^0 for chemical compounds and ΔH^0 for chemical reactions depend upon temperature. The answer is provided by the thermochemical cycle shown in Figure 3–11, which considers a specific example, the reaction

$$C(s) + 2H_2(g) = CH_4(g)$$

at 25°C and 75°C. The value of ΔH^0 for this reaction at 25°C is -17.89 kcal mol^{-1}; this is the value of ΔH_f^0 for methane. This magnitude is shown as the vertical distance between the reactant and product levels for 25°C. The levels for both reactants and product at 75°C are higher than those for 25°C because the enthalpy increases with an increase in temperature. For the product, one mole of methane, with $C_p = 8.60$ cal K^{-1} mol^{-1}, the increase of the enthalpy is

$$\begin{aligned} H_{75} - H_{25} &= C_p(75°\text{C} - 25°\text{C}) \\ &= 8.60 \text{ cal K}^{-1} \text{ mol}^{-1} \times 50 \text{ K} \\ &= 430 \text{ cal mol}^{-1} \end{aligned}$$

For the reactants, one mole of graphite ($C_p = 2.00$ cal K^{-1} mol^{-1}) and two moles of hydrogen gas ($C_p = 6.90$ cal K^{-1} mol^{-1}), the increase of the enthalpy is

$$\begin{aligned} H_{75} - H_{25} &= (2.00 \text{ cal K}^{-1} \text{ mol}^{-1} + 2 \times 6.90 \text{ cal K}^{-1} \text{ mol}^{-1}) \times 50 \text{ K} \\ &= 790 \text{ cal K}^{-1} \text{ mol}^{-1} \end{aligned}$$

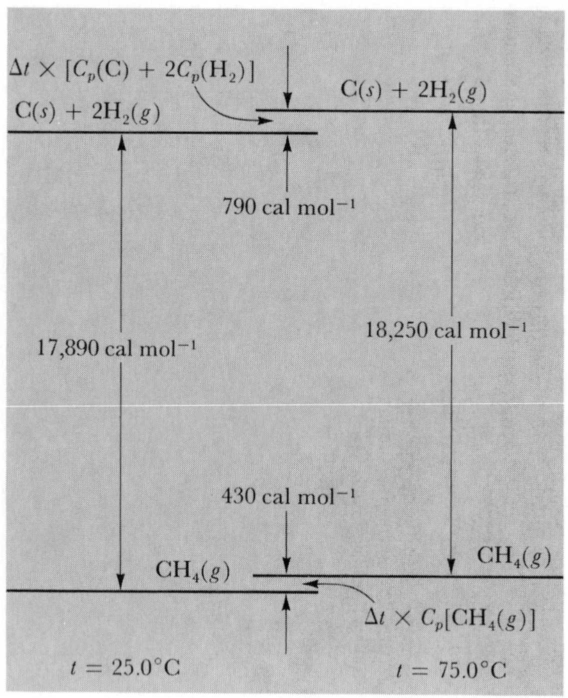

FIGURE 3–11
The relative enthalpy content of one mole of methane compared to one mole of carbon plus two moles of hydrogen at 25.0°C and 75.0°C.

Therefore, for the reaction $C(s) + 2H_2(g) = CH_4(g)$,

$$\Delta H_{75} - \Delta H_{25} = (H_{75} - H_{25})_{\text{products}} - (H_{75} - H_{25})_{\text{reactants}}$$

$$\frac{\Delta H_{75} - \Delta H_{25}}{\text{kcal mol}^{-1}} = 430 - 790 = -360$$

$$\frac{\Delta H_{75}}{\text{kcal mol}^{-1}} = \frac{\Delta H_{25}}{\text{kcal mol}^{-1}} - 360$$

$$= -17{,}890 - 360 = 18{,}250$$

$$\Delta H_{75} = -18{,}250 \text{ cal mol}^{-1}$$

Now look back at the way in which the terms 430 cal mol^{-1} and 790 cal mol^{-1} arose in the calculation; it is clear that

$$\Delta H_{75} = \Delta H_{25} + \Delta C_p \times 50 \text{ K}$$

in which ΔC_p is the heat capacity of the product minus the heat capacity of the reactants. In general terms,

$$\Delta H_{t_2} = \Delta H_{t_1} + (t_2 - t_1)\Delta C_p$$

The enthalpy change in a chemical reaction varies with temperature if the heat capacity of the products differs from the heat capacity of the reactants. In this calculation it was assumed that the molar heat capacities of the several substances [$C(s)$, $H_2(g)$, and $CH_4(g)$] were constant over the temperature range;

molar heat capacities vary with temperature, and in accurate calculations in which the temperature change is large this must be taken into account.

THE SECOND LAW OF THERMODYNAMICS; THE CONCEPT OF ENTROPY

One of the most important problems in chemistry is to determine the extent to which a chemical reaction can occur under a particular set of conditions (e.g., temperature). This problem is in the domain of thermodynamics. But it is a problem that cannot be solved by using the first law of thermodynamics and the associated thermochemical procedures just outlined. Another principle is needed. This principle is the second law of thermodynamics, which involves the concept of entropy. One statement of the *second law of thermodynamics* is:

> The occurrence of a spontaneous, irreversible process is accompanied by an increase of the entropy of the universe, consisting of the system in which the process occurs and its surroundings.

The meaning of this statement will be made clear in what follows.

We will consider first an isolated system of a mole of gas (e.g., helium) in an insulated container partitioned into two compartments, as shown in Figure 3–12. The gas is present initially in one of the compartments (Figure 3–12a); removal of the partition allows the gas to fill the entire container.

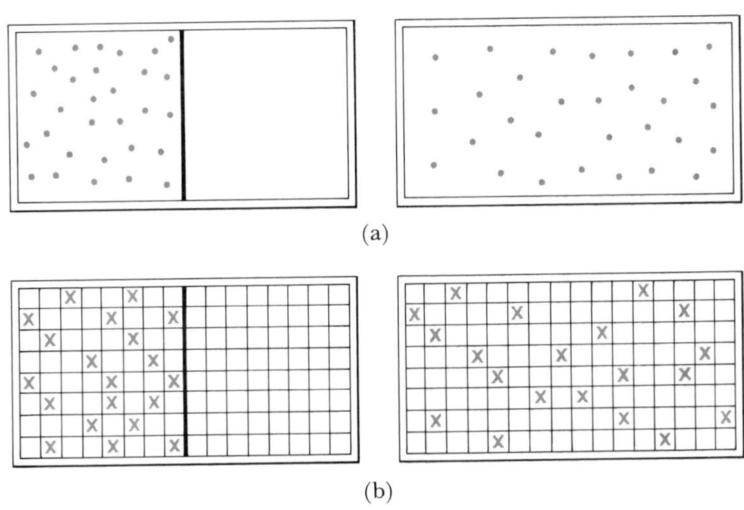

FIGURE 3–12
The spontaneous expansion of a gas. (a) Removal of the partition allows the gas to occupy the entire volume, and this process occurs spontaneously even though the internal energy of the system is unchanged by the expansion. (b) A model of the spontaneous process. Initially there are 20 molecules in 64 cells, and finally there are 20 molecules in 128 cells. The ratio of the number of possible arrangements of molecules in cells in these two states is $\Omega_2/\Omega_1 = 6.1 \times 10^6$.

The process

$$He(V = 50 \text{ L mol}^{-1}) = He(V = 100 \text{ L mol}^{-1})$$

occurs spontaneously. (The reverse process, in which the mole of helium gas contracts into the 50-L space at one end of the larger container, does not occur spontaneously.) No work is exchanged between the system (the mole of helium) and the surroundings in this free expansion. Since the container is insulated, no exchange of heat occurs; thus

$$w = 0 \qquad q = 0 \qquad \Delta U = 0$$

Because the internal energy of the system does not change in the process, no tendency toward a state of lower internal energy could account for the process. The process occurs because the entropy of a mole of helium occupying a volume of 100 L is greater than the entropy of a mole of helium occupying a volume of 50 L (at the same temperature).

The relationship at a particular temperature between the molar entropy of a gas and the volume it occupies will be derived in Chapter 4 without reference to the molecular nature of the gas; here we will introduce this subject by considering the statistical approach that relates the entropy of a gas to the randomness of the distribution of gas molecules in space. We can think of the molecules of a gas as being distributed among an enormous number of tiny cells of equal volume. There are many more of these cells than there are gas molecules, so most cells are empty and essentially no cell contains two molecules. In the sense in which it is used here, *randomness* means the number of different ways of arranging the molecules in the cells; this quantity is represented by Ω, that is,

$$\Omega = \text{number of arrangements}$$

A derivation of the value of Ω involves simple probability theory.[5] For a sample of N molecules, the ratio of the values of Ω for the gas in two different volumes is derived to be:

$$\frac{\Omega_2}{\Omega_1} = \left(\frac{V_2}{V_1}\right)^N$$

For the case of doubling the volume of a mole of helium gas,

$$\frac{\Omega_2}{\Omega_1} = 2^{6.02 \times 10^{23}}$$

which is a very large number indeed ($2^{6.02 \times 10^{23}} \cong 10^{1.81 \times 10^{23}}$). Perhaps

[5] The number of arrangements of N molecules in Z cells is the number of arrangements of Z objects, the cells, in two groups, one with N objects (the occupied cells) and the other with $(Z - N)$ objects (the unoccupied cells); this is given by $Z!/(N!(Z - N)!)$, where $X!$ (X is a number) denotes the factorial of X, $X! = X \times (X - 1) \times (X - 2) \times \cdots 1$. The equation $\Omega_2/\Omega_1 = (V_2/V_1)^N$ given in the text is the result if Z and N are very large numbers with $Z \gg N$ and $Z \propto V$. We will develop the relationship of entropy and randomness (as measured by the value of Ω) in Chapter 5.

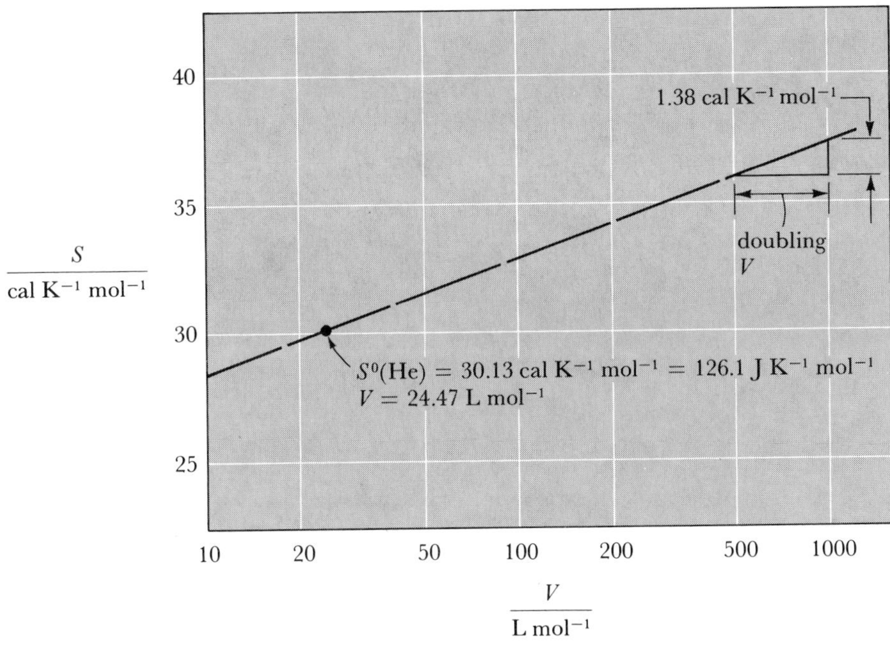

FIGURE 3–13
The molar entropy of helium gas at 25.0°C as a function of the molar volume. (Notice the logarithmic abscissa.) For each doubling of molar volume, S increases by 1.38 cal K^{-1} mol^{-1}.

you will get a qualitative sense of this relationship by considering the lower part of Figure 3–12. In this picture, there are twenty molecules shown, and the number of cells increases from 64 to 128. The ratio of the number of possible arrangements in the two cases is

$$\frac{\Omega_2}{\Omega_1} = 6.1 \times 10^6$$

which is a measure of the increase of randomness when the 20 molecules occupy the two-fold larger number of cells.

There is a logarithmic relationship between the molar entropy of a gas, designated as S, and the molar volume of the gas:

$$S = a + b \ln V$$

For the present, we need not be concerned with the values a and b, which depend upon the nature of the gas molecule and the temperature. Figure 3–13 shows the dependence of the molar entropy of helium upon the molar volume; for each increase of volume by a factor of 2, the entropy increases by 1.38 cal K^{-1} mol^{-1} or 5.76 J K^{-1} mol^{-1}.[6]

[6]Notice that entropy has the same dimension as heat capacity. The reason for this will be made clear. (In the chemical literature, one finds some usage of the "entropy unit" (symbol e.u.) as the dimension for entropy. This generally means cal K^{-1} mol^{-1} but the term "entropy unit" is inherently ambiguous, and it is preferable to avoid it.)

In considering the dependence of the molar entropy of a gas at a particular temperature upon the volume it occupies, we have focused attention on randomness of distribution of molecules in space. In Chapter 5 we will consider another type of randomness related to the distribution of atoms among the energy states that they occupy. The statistical approach to entropy will be considered in greater detail at that point.

The relationship between the entropy of a gas and its molar volume also is pertinent in the vaporization of liquid water,

$$H_2O(l) = H_2O(g)$$

and the reverse process, the condensation of water vapor,

$$H_2O(g) = H_2O(l)$$

You have observed each of these processes occurring spontaneously. Liquid water in an open dish evaporates. This process is endothermic ($\Delta H = +10.52$ kcal mol^{-1}), and heat is absorbed from the surroundings to bring about the change. The evaporation of perspiration from your body has a cooling effect because heat is required for the vaporization process. The reverse process, the condensation of water vapor, an exothermic process ($\Delta H = -10.52$ kcal mol^{-1}), occurs spontaneously when dew appears on grass after a cool night. It also occurs when you exhale on a cold morning.

Whether the process $H_2O(l) = H_2O(g)$ occurs with the absorption of 10.52 kcal mol^{-1} or the process $H_2O(g) = H_2O(l)$ occurs with the evolution of 10.52 kcal mol^{-1}, the change conforms to the first law of thermodynamics. What then determines that the process can go spontaneously in one direction or the other, depending upon the conditions—the temperature and the concentration of the water vapor? We will explore part of this question by considering the vaporization process at 298.2 K:

$$H_2O(l, 298.2\text{ K}) = H_2O(g, 298.2\text{ K}) \qquad \Delta H = +10.52\text{ kcal mol}^{-1}$$

In spite of the endothermic nature of this process, the process is spontaneous if the pressure of the water vapor is appropriately low. Three sets of conditions will be considered:

1. $H_2O(l, 760\text{ torr}, 298.2\text{ K}) = H_2O(g, 760\text{ torr}, 298.2\text{ K})$
2. $H_2O(l, 760\text{ torr}, 298.2\text{ K}) = H_2O(g, 23.76\text{ torr}, 298.2\text{ K})$
3. $H_2O(l, 760\text{ torr}, 298.2\text{ K}) = H_2O(g, 10.0\text{ torr}, 298.2\text{ K})$

Each of these changes is depicted in Figure 3–14, which shows the volume change accompanying the vaporization process. The pressure of the gaseous water determines the volume that one mole occupies. The value of ΔH for each of these processes is the same,[7] $+10.52$ kcal mol^{-1}. The first process does not occur: liquid water confined in a cylinder at 298.2 K and a pressure

[7]This statement is only approximately correct if nonideality of water vapor is taken into account.

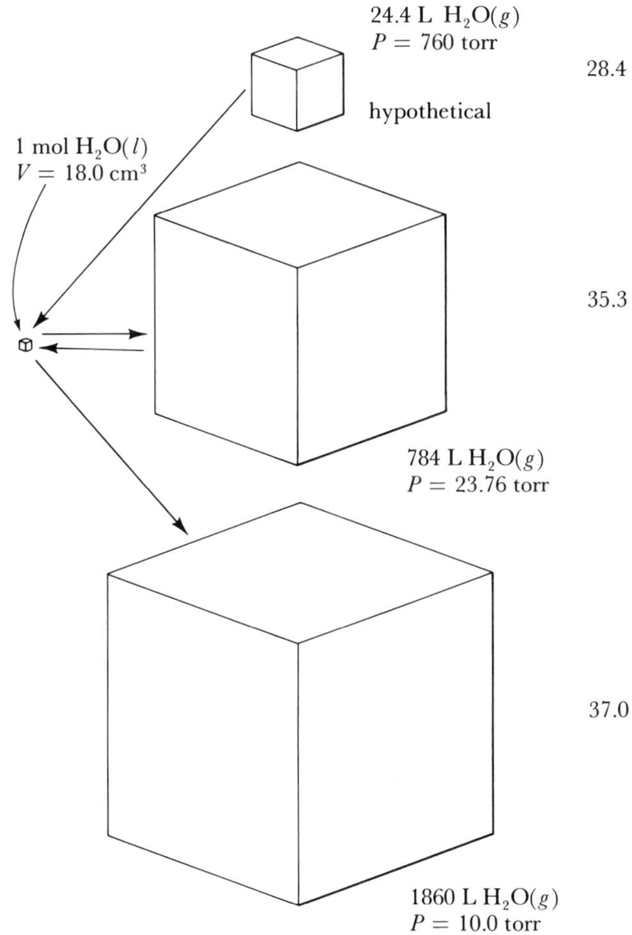

FIGURE 3–14
The transformation of one mole of liquid water to water vapor at 298.2 K and various pressures. (a) 760 torr (hypothetical) (b) 23.76 torr, and (c) 10.0 torr. The volume of liquid water is 18.0 cm^3. The volumes of water vapor at these pressures and temperature are 24.4 L (hypothetical), 784 L, and 1860 L, respectively.

of 760 torr *will not* spontaneously become water vapor. If the process did occur, 24.4 L of vapor would be produced. The third process does occur: reduction of the pressure on the sample from 760 torr to 10 torr would result in irreversible vaporization of the entire sample (of one mole) to occupy a volume of 1860 L. The second process is a special one. Depending upon the amount of heat transferred to the sample from the surroundings, varying amounts of water vaporize; the volume occupied by this water vapor is 784 L mol^{-1}. At this pressure, 23.76 torr, a stable situation exists for all relative proportions of liquid and vapor phases. At 298.2 K, water vapor at this pressure is in equilibrium with liquid water. The vaporization of water under this set of conditions of temperature and pressure is *reversible.* A process is said to be reversible if an infinitesimal change of conditions causes the process to occur in the reverse direction. This is the case with Process 2. If the pressure were raised slightly above 23.76 torr, condensation of water vapor would occur; if the pressure were lowered slightly below

23.76 torr, vaporization of liquid water would occur. The arrows going in both directions for Process 2 in Figure 3–14 denote reversibility. The tendency for the vaporization process to occur depends both upon the change in enthalpy, ΔH, which is the same for each set of conditions depicted in Figure 3–14, and the change of entropy, ΔS, which is different for each set of conditions.

Chemical substances have an extensive property called entropy, a state function, which is a measure of the randomness of spatial arrangement of atoms and molecules in the material and also the randomness of distribution of the total potential and kinetic energy among the various atoms and molecules. These are not simple concepts, but their meaning will become clearer with discussion. For the present, the point can be illustrated with values at 298.2 K of the molar entropy (S) of water in various states being considered:

	$\dfrac{S}{\text{cal K}^{-1}\text{ mol}^{-1}}$	$\dfrac{S}{\text{J K}^{-1}\text{ mol}^{-1}}$
$H_2O(l)$	16.7	70.0
$H_2O(g$, 760 torr$)$	45.1	188.7
$H_2O(g$, 23.76 torr$)$	52.0	217.5
$H_2O(g$, 10.0 torr$)$	53.7	224.7

This variation in the molar entropy for water is consistent with the statement that entropy is a measure of randomness of spatial arrangement of the molecules. This randomness is greatest if the one mole of water is a gas occupying 1860 L and is least if it is a liquid occupying 18 cm^3. The entropy of the water increases in each of the vaporization processes under consideration, but it increases more the larger the volume occupied by the gaseous water. The values of ΔS for the three processes can be calculated from the values of entropy just given, $\Delta S = S[H_2O(g)] - S[H_2O(l)]$; they are:

Pressure $[H_2O(g)]$	$\dfrac{\Delta S}{\text{cal K}^{-1}\text{ mol}^{-1}}$	$\dfrac{\Delta S}{\text{J K}^{-1}\text{ mol}^{-1}}$
1. 760 torr	28.4	118.7
2. 23.76 torr	35.3	147.5
3. 10.0 torr	37.0	154.7

Since the value of ΔS for the vaporization of water depends upon the pressure of the gas, it is convenient to have a special designation for the change of entropy under some particular set of conditions, the *standard conditions*. For gases, standard conditions generally are taken to be the gas at 1 atmosphere pressure (1 atm ≡ 760 torr). Thus the first of the values of ΔS given above can be designated ΔS^0:

$$H_2O(l, 298.2\text{ K}) = H_2O(g, 298.2\text{ K}) \qquad \Delta S^0 = 28.4\text{ cal K}^{-1}\text{ mol}^{-1}$$

The enthalpy change and the entropy change both play roles in determining whether a process occurs spontaneously. The way in which the changes of enthalpy and entropy are combined to determine the tendency for reaction to occur at a particular temperature is

$$\Delta H - T\,\Delta S$$

The quantity $(H - TS)$ is given the name "free energy" or GIBBS free energy, and it is represented by the symbol G,[8]

$$G = H - TS$$

In this equation, and in all related equations, the temperature must be an absolute temperature. In chemistry the Kelvin scale is used uniformly, but in some engineering work, the Rankine scale is used: this is an absolute temperature scale with a degree the size of a fahrenheit degree. With the dimensions cal K^{-1} mol^{-1} or J K^{-1} mol^{-1} for molar entropy, the product of temperature and molar entropy has the dimensions of energy per mole, cal mol^{-1}, or J mol^{-1}, just as enthalpy does. For the processes under consideration, each at 298.2 K, the changes in free energy can be calculated:

1. $H_2O(g$, 760 torr) is produced

$$\Delta G = +10{,}520 \text{ cal mol}^{-1} - 298.2 \text{ K} \times 28.4 \text{ cal K}^{-1} \text{ mol}^{-1}$$
$$= +2050 \text{ cal mol}^{-1}$$

2. $H_2O(g$, 23.76 torr) is produced

$$\Delta G = +10{,}520 \text{ cal mol}^{-1} - 298.2 \text{ K} \times 35.3 \text{ cal K}^{-1} \text{ mol}^{-1}$$
$$= 0 \text{ cal mol}^{-1}$$

3. $H_2O(g$, 10.0 torr) is produced

$$\Delta G = +10{,}520 \text{ cal mol}^{-1} - 298.2 \text{ K} \times 37.0 \text{ cal K}^{-1} \text{ mol}^{-1}$$
$$= -510 \text{ cal mol}^{-1}$$

The important points emerging from these calculations are:

$\Delta G > 0$	transformation does not occur
$\Delta G = 0$	transformation is at equilibrium
$\Delta G < 0$	transformation does occur

Figure 3–15 shows a plot of ΔH, $T\Delta S$, and ΔG for the process as a function of the volume occupied by the water vapor. Clearly, for all pressures of water vapor lower than 23.76 torr (all molar volumes of water vapor greater than 784 L), the value of ΔG is negative, and the vaporization of water under these conditions occurs spontaneously. At this temperature, it is only for $V \geqq 784$ L mol^{-1} that the change in entropy of the system is positive enough to offset the positive change of enthalpy in the vaporization process.

[8] The symbol F also is used for free energy in some books.

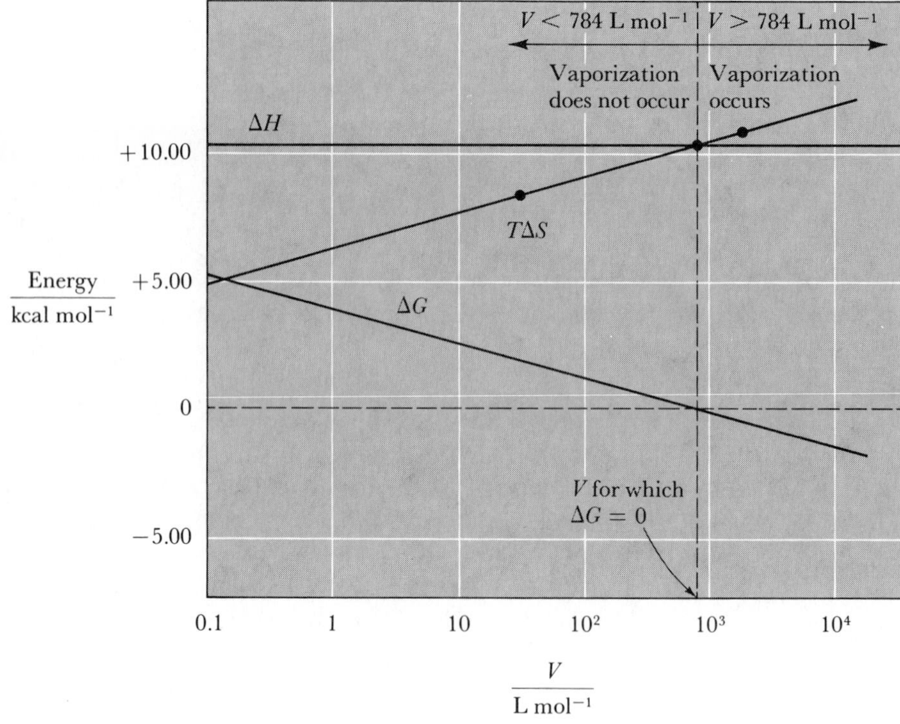

FIGURE 3–15
The dependence of ΔH, ΔS, and ΔG for the process

$$H_2O(l,\ 298.2\ K) = H_2O(g,\ 298.2\ K)$$

upon the volume occupied by the vapor. The points shown on $T\Delta S$ line are those considered in text ($V = 24.4$, 784, and 1860 L mol^{-1}). (The abscissa is a logarithmic scale.)

The relationship $\Delta G = \Delta H - T\Delta S$ shows the following important points:

1. The more negative the value of ΔH, the more negative the value of ΔG.
2. The more positive the value of ΔS, the more negative the value of ΔG.
3. The contribution to ΔG from the change in entropy, ΔS, is greater the higher the temperature.
4. Each of these quantities, ΔH, ΔS, and T, plays a role in determining the magnitude of ΔG.

The criterion for a spontaneous process given previously in the statement of the second law of thermodynamics involved the change of entropy of the system plus the surroundings,

$$\Delta S_{\text{system}} + \Delta S_{\text{surroundings}} > 0$$

for a spontaneous process. The criterion just presented,

$$\Delta G_{\text{system}} < 0$$

may appear to be a different criterion. In fact, these criteria are equivalent.

In the example just discussed, the vaporization of water was considered at a particular temperature, and emphasis was placed upon the molar volume of the gaseous product. In the next examples, a particular pressure of gases will be considered, and emphasis will be placed upon the role of temperature.

The Water-Gas Reaction. The "water-gas" reaction is the conversion of carbon to carbon monoxide by the action of steam:

$$C(s) + H_2O(g) = CO(g) + H_2(g)$$

The value of ΔH^0 at 25°C for this reaction can be calculated from standard heats of formation of carbon monoxide and water vapor at 25°C:

$$\Delta H^0 = \Delta H_f^0[CO(g)] + \Delta H_f^0[H_2(g)] - \Delta H_f^0[C(s)] - \Delta H_f^0[H_2O(g)]$$

$$\frac{\Delta H^0}{\text{kcal mol}^{-1}} = -26.42 + 0 - 0 - (-57.80) = +31.38$$

$$\Delta H^0 = +31.38 \text{ kcal mol}^{-1}$$

This calculation also is represented in Figure 3–16, which shows the enthalpy change of the reaction as the difference of the enthalpies of formation of the product, carbon monoxide, and the reactant, water. (The other participants in this reaction are elements, with $\Delta H_f^0 = 0$.) In this reaction two moles of gas are produced from one mole of gas and one mole of solid, and the value of ΔS^0 is positive as a consequence:

$$\Delta S^0 = +32.0 \text{ cal K}^{-1} \text{ mol}^{-1} \text{ at } 298.2 \text{ K}$$

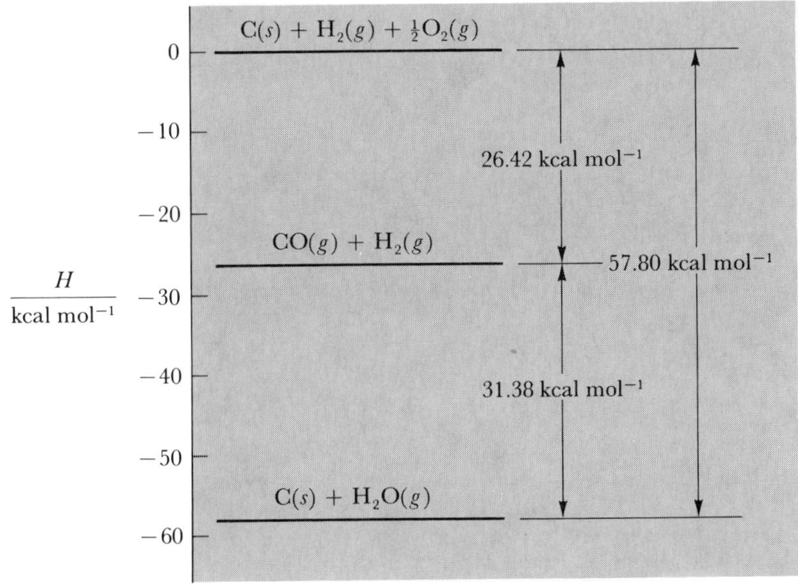

FIGURE 3–16
The relative enthalpy content of reactants and products for the water-gas reaction (and the constituent elements).

FIGURE 3–17
The water-gas reaction,

$$C(s) + H_2O(g) = CO(g) + H_2(g)$$

in which one mole of gas gives two moles of gas. At a constant pressure (1.00 atm) of each gaseous substance, the randomness of the system increases with occurrence of reaction.

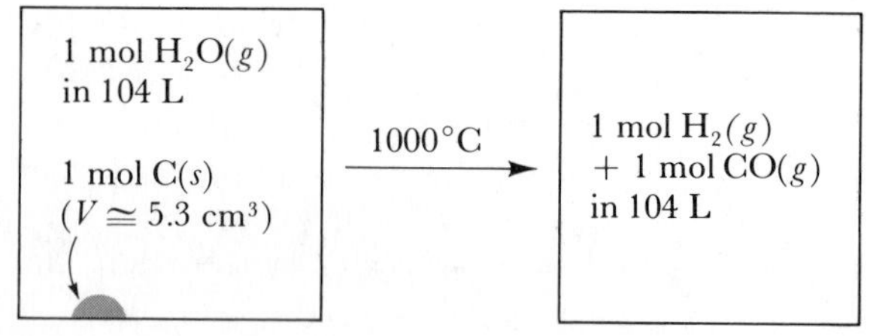

Figure 3–17 shows the situation in which a mole of carbon (graphite), with a molar volume of 5.3 cm^3, is converted into a mole of carbon monoxide gas that occupies a volume of 104 L at 100°C and 1 atm. The randomness of position of the carbon atoms certainly has increased in this process. With ΔS^0 positive, the contribution of $T\Delta S^0$ toward making ΔG^0 negative is greater the higher the temperature.

Figure 3–18 shows the way in which ΔG^0 ($\Delta H^0 - T\Delta S^0$) changes with temperature. At high temperatures, where the water-gas reaction occurs at a convenient rate, the yield of products is high. This reaction is an efficient producer of hydrogen for use as a chemical reagent or as a fuel. The product mixture can be used as a fuel because both carbon monoxide and hydrogen burn exothermically:

$$2CO(g) + O_2(g) = 2CO_2(g) \qquad \Delta H^0 = -135.3 \text{ kcal mol}^{-1}$$
$$2H_2(g) + O_2(g) = 2H_2O(g) \qquad \Delta H^0 = -115.6 \text{ kcal mol}^{-1}$$

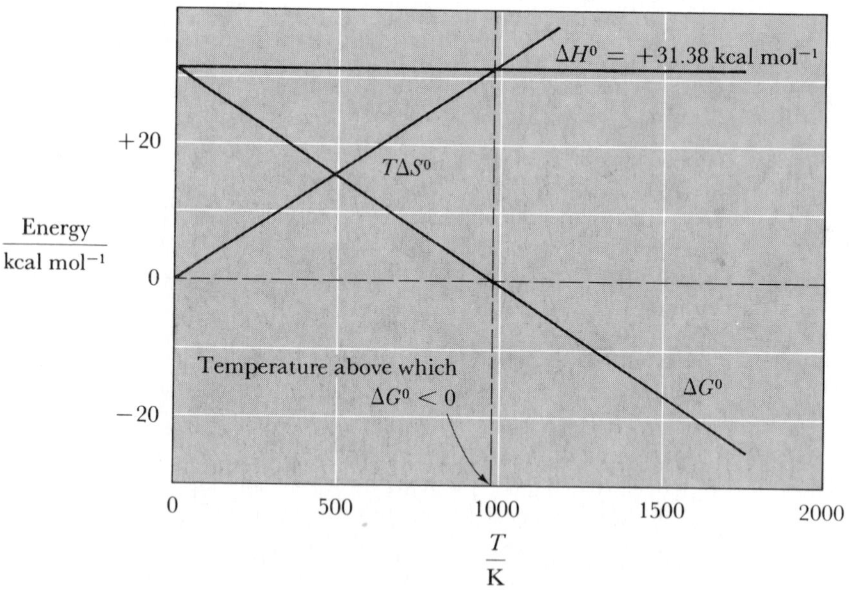

FIGURE 3–18
The dependence of ΔG^0 upon temperature for the water-gas reaction, $C(s) + H_2O(g) = CO(g) + H_2(g)$, (a reaction for which ΔS^0 is *positive*.) The straight-line plots are based upon the assumption that ΔH^0 and ΔS^0 do not depend upon temperature. Thus, as drawn here, ΔH^0 line is horizontal, and $T\Delta S^0$ line is straight; the slope of this line is ΔS^0.

Therefore the molar fuel value of water gas is

$$\tfrac{1}{4}(135.3 \text{ kcal mol}^{-1} + 115.6 \text{ kcal mol}^{-1}) = 62.7 \text{ kcal mol}^{-1}$$

which is not as high as that for methane (191.8 kcal mol^{-1}), the principal constituent of natural gas, but is high enough to make the water-gas reaction a practical way of gasifying coal. Other related reactions, to be discussed later, also are used in coal gasification.

The Dissociation of Molecular Iodine. The dissociation of a gaseous diatomic or polyatomic molecule to its constituent atoms is an endothermic process, but at a high enough temperature a reaction of this type may occur to some extent because it is accompanied by an increase of entropy. Let us consider iodine as an example. The energy required to dissociate the iodine molecule is 36.1 kcal mol^{-1}:

$$I_2(g) = 2I(g) \qquad \Delta H^0 = 36.1 \text{ kcal mol}^{-1}$$

but this dissociation, producing 2 gas particles from 1 gas particle (with each gaseous species at a pressure of 1 atm), is accompanied by an increase in entropy:

$$I_2(g) = 2I(g) \qquad \Delta S^0 = 24.1 \text{ cal K}^{-1} \text{ mol}^{-1}$$

The magnitude of the change of free energy in this process depends upon the temperature:

$$\Delta G^0 = \Delta H^0 - T\Delta S^0$$

$$\frac{\Delta G^0}{\text{cal mol}^{-1}} = +36{,}100 - 24.1 \text{ K}^{-1} \times T$$

Figure 3–19 shows a plot of ΔG^0 versus T; clearly, with increasing temperature the value of ΔG^0 becomes less positive, and at temperatures above ~1500 K it becomes negative.[9]

Two factors compete in establishing the extent to which a chemical reaction occurs: to achieve greater stability, a reaction occurs to allow

1. the enthalpy of the system to decrease, and/or
2. the entropy of the system to increase.

For the reaction under consideration, the enthalpy factor favors molecular iodine, but the entropy factor favors atomic iodine. At chemical equilibrium, the two factors balance. The higher the temperature, the less important the enthalpy factor, and the more important the entropy factor becomes. However, keep in mind that ΔG^0 is the change of free energy for the reaction occurring with each gaseous species in its standard state:

$$I_2(g, 1 \text{ atm}) = 2I\ (g, 1 \text{ atm})$$

[9] This calculation assumes that the values of ΔH^0 and ΔS^0 do not depend upon temperature. The assumption is not strictly correct, but for the purpose of this example, it introduces no conceptual error. The actual temperature at which $\Delta G^0 = 0$ is 1454 K.

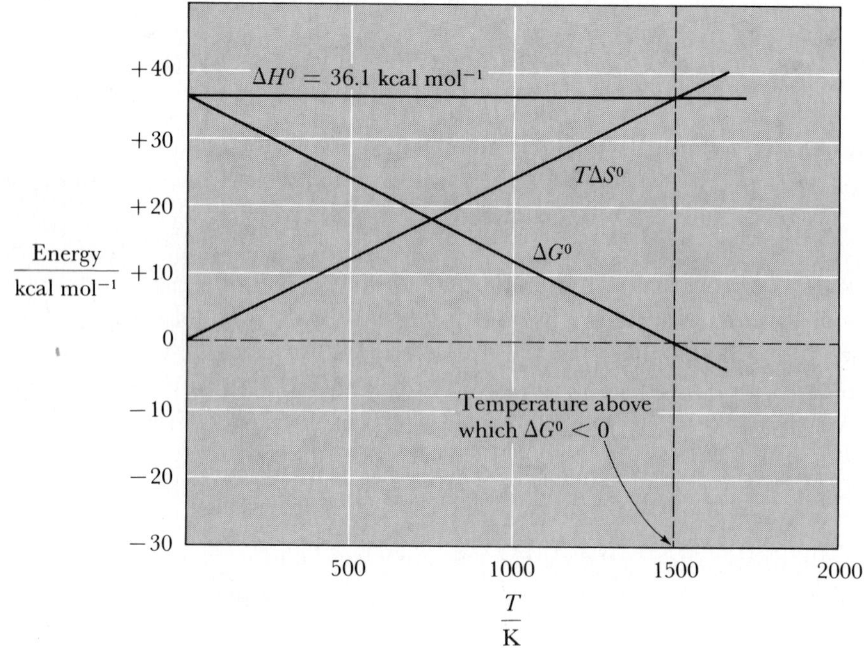

FIGURE 3–19
The dependence of ΔG^0 upon temperature for the dissociation of molecular iodine, $I_2(g) = 2I(g)$, (a reaction for which ΔS^0 is positive).

With the reactant and product species at other pressures, the value of ΔG for the reaction will become negative at other temperatures.

The Formation of Ammonia from Its Constituent Elements. So far, we have considered only endothermic reactions accompanied by an increase in the number of gas molecules. Now we will consider an exothermic reaction accompanied by a decrease in the number of gas molecules, the formation of ammonia from its constituent elements:

$$N_2(g) + 3H_2(g) = 2NH_3(g) \qquad \Delta H^0 = -22.04 \text{ kcal mol}^{-1}$$
$$\Delta S^0 = -47.4 \text{ cal K}^{-1} \text{ mol}^{-1}$$

If, as in the preceding example, we assume that ΔH^0 and ΔS^0 are independent of temperature, we have for ΔG^0:

$$\Delta G^0 = \Delta H^0 - T\Delta S^0$$

$$\frac{\Delta G^0}{\text{cal mol}^{-1}} = -22{,}040 + 47.4 \text{ K}^{-1} \times T$$

This equation is plotted in Figure 3–20. We see that $\Delta G^0 < 0$ for temperatures less than 465 K, and $\Delta G^0 > 0$ for temperatures greater than 465 K. For this reaction, in which 4 moles of gas give 2 moles of gas, the change of entropy is a factor that works against the occurrence of reaction. At high temperatures ($T > 465$ K), this decrease of entropy prevents the exothermic formation of ammonia from its elements in their standard states (i.e., $P_{N_2} = P_{H_2} = P_{NH_3} = 1$ atm).

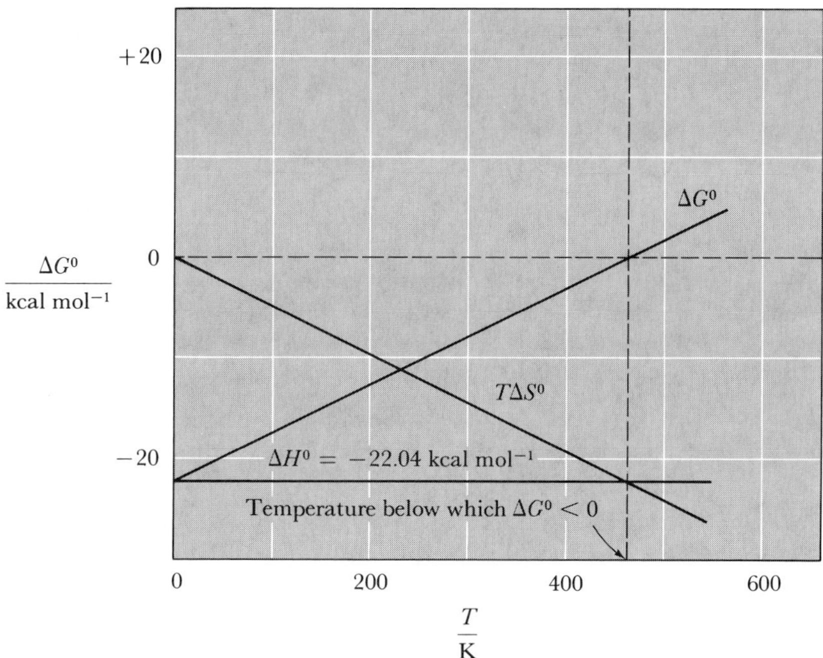

FIGURE 3–20
The dependence of ΔG^0 upon temperature for the production of ammonia from its constituent elements, $N_2(g) + 3H_2(g) = 2NH_3(g)$ (a reaction for which ΔS^0 is negative).

3–4 A Summary of Thermodynamic Functions

In this chapter several thermodynamic functions have been presented. These are:

internal energy	U
enthalpy	$H = U + PV$
entropy	S
Gibbs free energy	$G = H - TS$

Each of the functions U, H, S, and G is an *extensive* property; that is, the value of each is proportional to the amount of material. For instance, the entropy of 1 mol of liquid water at 298.2 K and 1 atm is 16.72 cal K^{-1} and the entropy of 2 mol is 33.44 cal K^{-1}. These functions are *state functions*, which means that the value of the property is a function of the state (defined by the temperature and pressure) but not a function of the path by which that state was achieved. This was illustrated in the discussion of the temperature dependence of ΔH. From the quantities summarized in Figure 3–11, it can be seen that the enthalpy change for the process

$$C(s, 298.2\text{ K}) + 2H_2(g, 298.2\text{ K}) = CH_4(g, 348.2\text{ K})$$

is -17.46 kcal mol^{-1}, whether the path is

$$C(s,\ 298.2\ K) + 2H_2(g,\ 298.2\ K) = CH_4(g,\ 298.2\ K)$$

followed by

$$CH_4(g,\ 298.2\ K) = CH_4(g,\ 348.2\ K)$$

or an alternate path,

$$C(s,\ 298.2\ K) + 2H_2(g,\ 298.2\ K) = C(s,\ 348.2\ K) + 2H_2(g,\ 348.2\ K)$$

followed by

$$C(s,\ 348.2\ K) + 2H_2(g,\ 348.2\ K) = CH_4(g,\ 348.2\ K)$$

The value of ΔH depends only upon the initial and final states!

Two other quantities have been introduced in this chapter, *work* and *heat*. These quantities do not pertain to states of chemical substances. These quantities arise in processes, and the magnitude of each depends upon the path between the initial and final states of the process. This important point will be illustrated further in examples dealing with the expansion of gases, in Chapter 4.

3–5 Chemistry and Energy: Some Practical Aspects

One can hardly discuss chemistry without considering energy—the connection has been made in the earlier parts of this chapter. It is equally true that one can hardly discuss energy without considering chemistry. Solar radiation is the ultimate source of most of the energy available to us. The net chemical change in photosynthesis involves carbon dioxide and water being converted to oxygen and carbohydrate, for example, cellulose, a high-molecular-weight substance with the simplest formula CH_2O:

$$nCO_2 + nH_2O = (CH_2O)_n + nO_2$$

This transformation is not one that occurs spontaneously; the radiant energy is the agent that brings about the change. The products of the reaction, cellulose and oxygen, have stored this energy. When wood (cellulose) is burned, this stored solar energy is released. The chemical reaction occurring in the burning of cellulose is the spontaneous exothermic change

$$(CH_2O)_n + nO_2 = nCO_2 + nH_2O$$
$$\Delta H \cong -110 \text{ kcal mol}^{-1} \text{ per unit of } n$$

which is the reverse of the photosynthesis reaction.

Over the ages, decomposition of carbohydrate (formed in photosynthesis) has given hydrocarbons (oil) and carbon (a component of coal):

$$(CH_2O)_n = (CH_2)_n + \frac{n}{2}O_2$$

$$(CH_2O)_n = C + nH_2O$$

When hydrocarbon fuels or coal burn, the stored solar energy is released.

The fossil fuels that have stored the solar energy of the past are being depleted rapidly; this is particularly true of liquid hydrocarbons. In the immediate future we must replace naturally occurring hydrocarbon fuels with substitutes, possibly some derived from the more abundant fossil fuel, coal. In the longer range, we must replace all fossil fuels. The direct use and storage of solar energy is a promising possibility.

THE GASIFICATION OF COAL

We already have considered the water-gas reaction and the reaction forming methane from its constituent elements; these reactions are used to gasify coal. The synthetic gas produced in gasification of coal can be made into a cleaner fuel than the coal itself, because the sulfur in coal can be removed in the gasification process.

The complete gasification process varies in its complexity, depending upon the use intended for the product. If we seek a substitute for natural gas intended for pipeline distribution, we need almost pure methane (CH_4). If we want a gas for heat in industrial operations, we may find on-the-site production of hydrogen and carbon monoxide mixed with nitrogen (from air) the most economical. If a raw material for synthesis of ammonia (NH_3) is our objective, the gas, called synthesis gas, should be rich in hydrogen.

Because the water-gas reaction of steam and carbon (the principal constituent of coal[10]),

$$C(s) + H_2O(g) = CO(g) + H_2(g) \qquad \Delta H = +31.38 \text{ kcal mol}^{-1}$$

goes to an appreciable extent only at high temperatures (see Figure 3–18), the reaction needs heat. This can be provided by an exothermic chemical reaction; in practice, by one of the following:

1. The burning in the system of some of the coal with oxygen (either pure oxygen or air). The reaction

   $$C(s) + O_2(g) = CO_2(g) \qquad \Delta H^0 = -94.05 \text{ kcal mol}^{-1}$$

   occurs where the oxygen is in excess at its place of admission, followed by reaction of the carbon dioxide with carbon:

   $$C(s) + CO_2(g) = 2CO(g) \qquad \Delta H^0 = +41.21 \text{ kcal mol}^{-1}$$

   The net result of these two reactions in the exothermic oxidation of carbon to carbon monoxide,

   $$2C(s) + O_2(g) = 2CO(g) \qquad \Delta H^0 = -52.84 \text{ kcal mol}^{-1}$$

   (Because carbon monoxide has fuel value, this procedure in itself is a gasification of coal.)
2. The burning, externally, of coal in excess air.

[10]Coal contains appreciable hydrogen; a typical coal can be represented by $CH_{0.8}$.

The water-gas itself (an equimolar mixture of hydrogen and carbon monoxide) can be used as a fuel, or the ratio of hydrogen to carbon monoxide can be altered by the reversible reaction,

$$CO(g) + H_2O(g) = CO_2(g) + H_2(g)$$

The production of methane involves the reaction of hydrogen either with carbon,

$$C(s) + 2H_2(g) = CH_4(g)$$

or with carbon monoxide,

$$CO(g) + 3H_2(g) = CH_4(g) + H_2O(g)$$

Each of these reactions is exothermic. Carbon monoxide and hydrogen also are the reactants in the principal reaction for large-scale commercial production of methyl alcohol,

$$CO(g) + 2H_2(g) = CH_3OH(g)$$

Methyl alcohol burns exothermically:

$$3O_2(g) + 2CH_3OH(l) = 2CO_2(g) + 4H_2O(g)$$

and its use as a fuel for automobiles is being studied.

HYDROGEN AS A FUEL

Current discussion of the nation's future energy needs includes consideration of hydrogen as a medium for storing and transporting energy and as a fuel. Because hydrogen does not occur in nature in the uncombined state, it cannot be considered a primary fuel (like coal or oil), but once produced it has many characteristics of an ideal fuel; in addition, its transportation through a pipeline system is economically feasible, and stored in tanks or as easily decomposed metal hydrides (e.g., $TiH_{1.7}$, a nonstoichiometric solid[11]), it is a convenient means of storing energy.

The fuel value of hydrogen is less than that of methane on a molar basis (which is equivalent to a volume basis) but greater than that of methane by weight:

	$-\Delta H_c^0$	
	kcal mol^{-1}	kcal g^{-1}
H_2	68.32	33.9
CH_4	212.80	13.3

The greater fuel value of hydrogen by weight is important in transportation uses where the fuel is carried along. Its lower fuel value compared to methane on a molar basis, which is the same as a volume basis, would be a disadvantage in sending it through pipelines were it not for the fact that

[11]Nonstoichiometric compounds are discussed in Chapter 9.

hydrogen gas has a lower viscosity, thereby reducing the energy required for pumping it.

The use of hydrogen as a fuel is not complicated by some of the pollution problems associated with the burning of a hydrocarbon fuel. If hydrogen is burned with oxygen, only water is produced; if it burned with air, nitrogen oxides form, but apparently to a lesser extent than in the internal combustion engine using gasoline.

Some who look far into the future see large-scale production of hydrogen gas by electrolysis of water (the passage of an electric current through water) using nuclear power:

$$2H_2O \xrightarrow{\text{electrolysis}} 2H_2(g) + O_2(g)$$

The oxygen produced also is a useful commodity. Such nuclear power stations could be located on platforms at sea, and the gases could be piped to shore.

Although the thermal decomposition of water, $2H_2O = 2H_2 + O_2$, occurs only at very high temperatures ($\sim$2800 K), some sequences of reactions, which include both endothermic and exothermic reactions occurring at lower temperatures, have as a net result the decomposition of water. One such sequence is the following series of three reactions:

$$2[C(s) + H_2O(g) = CO(g) + H_2(g)] \qquad \Delta H > 0$$
$$2[CO(g) + 2Fe_3O_4(s) = C(s) + 3Fe_2O_3(s)] \qquad \Delta H < 0$$
$$6Fe_2O_3(s) = 4Fe_3O_4(s) + O_2(g) \qquad \Delta H > 0$$

The sum of these reactions is the decomposition of water, $2H_2O(g) = 2H_2(g) + O_2(g)$. Figure 3–18 shows that the first of these reactions, an endothermic reaction, can go at temperatures greater than $\sim$900 K. The second reaction is exothermic, and ΔG^0 is negative for this reaction at temperatures below $\sim$500 K. The third reaction, like the first, is an endothermic reaction with a positive value of the standard entropy change. (Notice that this reaction produces gas.) This reaction, the decomposition of hematite[12] (Fe_2O_3) to magnetite[12] (Fe_3O_4) and oxygen gas, goes at $\sim$1800 K and above. Each of the endothermic reactions requires a high temperature and the absorption of heat from the surroundings. Proposals for large-scale production of hydrogen using these reactions include use of heat generated in nuclear reactors, heat that might otherwise go to waste. Although the temperature required to make each of these reactions go is very much less than that required for the "unassisted" thermal decomposition of water, the temperature required for the last step of this sequence ($\sim$1800 K) exceeds that for which there is now economically feasible technology. A number of other, more complex sequences of reactions have been proposed. Although none of these proposals has been developed into a large-scale operation, scientists are working to realize this goal.

[12] These are the names applied to these oxides of iron as they occur in nature; each is an important ore.

Biographical Notes

ANDRÉ M. AMPERE (1775–1836), a French physicist and mathematician, was the first person to show that two parallel conductors carrying currents in the same direction attract one another. The ampere unit was named after him.

ROBERT W. BUNSEN (1811–1899), a German chemist, made many contributions to 19th-century chemistry. He did much to develop spectral analysis and used it to discover cesium (in 1860) and rubidium (in 1861). Among his inventions was the gas burner that bears his name.

CHARLES A. DE COULOMB (1736–1806), a French physicist, studied electricity and magnetism. He first demonstrated that the force of electrical attraction or repulsion between charged spheres is proportional to the product of the charges and inversely proportional to the square of the distance between the centers of the spheres. This relationship is called Coulomb's law.

JOSIAH WILLARD GIBBS (1839–1903) was Professor of Mathematical Physics in Yale University from 1871 to 1903. He was one of the most outstanding American scientists of all time. His work laid the foundation of many areas of physics, physical chemistry, and mathematics.

GERMAIN H. HESS (1802–1850), born in Switzerland and a student of Berzelius in Stockholm, was professor of chemistry at St. Petersburg, Russia, at the time of his death. He is considered to be the founder of thermochemistry. He formulated the Hess law in 1840.

JAMES P. JOULE (1818–1889), initially a Manchester brewer, was an English physicist who, in his twenties, discovered that mechanical work could be converted to heat; he also was the first to show that the heat (q) produced by an electrical current (i) is proportional to the resistance (r), the square of the current, and the time (t) ($q = i^2rt$). He made many other contributions in the fields of thermodynamics and electricity.

JULIUS R. VON MAYER (1814–1878) was a German physicist and physician. He applied the law of conservation of energy to living phenomena as well as to simpler examples.

Problems and Questions

3–1 Derive the following relationships presented in this chapter: 1 N = 10^5 dyn, and 1 J = 10^7 erg.

3–2 A 1.00-kg mass is 3.00 m above the floor. The acceleration of gravity at this point is 9.81 m s^{-2}. What is the potential energy of this configuration relative to the mass resting on the floor? If the mass falls to the floor, what is its velocity immediately before striking the floor? (Assume no friction.) Instead of striking the floor, the object falls into an insulated vessel containing 5 kg of water. Initially the water and the object have a temperature of 20.000°C. What is their temperature in the final state? The specific heat of water is 1.00 cal K^{-1} g^{-1} and that of the object is 0.59 cal K^{-1} g^{-1}. (Assume the container to have a negligible heat capacity.)

3–3 The muzzle velocity of a bullet shot vertically from a gun is 2000 ft s^{-1}. To what height would the bullet rise if there were no friction from the air? (Assume g = 9.81 m s^{-2}.)

3–4 A unit of energy commonly employed in engineering and in describing the fuel value of various fuels is the Btu, the British thermal unit. It is the energy required to raise the temperature of one pound of water one Fahrenheit degree. How many calories and how many joules are there in one Btu?

3–5 The calorie (denoted "Cal" in this problem) used in describing energy values of foods is the kilocalorie. Express the energy values of each of the following foods in joules:

bacon: 50 Cal/strip

frankfurters: 150 Cal/frankfurter

pork chops: 130 Cal/chop ($\sim\frac{1}{3}$ lb raw)

chocolate cake: 400 Cal/2 in. wedge (2-layer cake with frosting)

raw apple (medium): 70 Cal/apple

waffle: 240 Cal/waffle

If your diet specified food with a total of 2500 "Cal" per day, what would this be in joules per day?

3–6 A 160-lb man has an office on the third floor, 28 ft above street level. How many times must be climb the stairs to his office to use the energy provided by a piece of chocolate cake? (See Problem 3–5.)

3–7 In Chapter 22, we will learn that the formation of one atom of oxygen from eight atoms of hydrogen plus eight neutrons, $8^1\text{H} + 8\text{n} = {}^{16}\text{O}$, is accompanied by an energy change $\Delta U = -2.05 \times 10^{-11}$ J. What is this energy change expressed in electronvolts (eV), the energy unit commonly used in nuclear physics?

3–8 How much heat must be added at constant volume to 15.0 g He(g) to raise its temperature from 23.0°C to 79.5°C? If this heat is furnished by an electrical heater operating with a voltage drop of 13.0 V, how many coulombs of electricity must flow?

3–9 An electrical heater in a system consisting of a cylinder with a movable piston delivers 4.53 kJ of heat to the system. The piston moves against an external pressure such that $P\Delta V = 2.74$ kJ. What is the value of ΔU for the system?

3–10 What is the temperature increase of 1.50 kg of water in an insulated vessel upon passage of 5.00 kC through a voltage drop of 25.0 V? (Assume the heat capacity of the vessel is zero.)

3–11 A sample of metallic aluminum weighing 50.0 g is heated to 85.0°C. It is then added to an insulated vessel containing 250.0 g of H_2O at 25.0°C. The final temperature of the mixture is 27.5°C. What is the approximate value for the specific heat of aluminum?

3–12 Calculate the change of enthalpy for the reactions (at 298.2 K):

$$2\text{Mg}(s) + \text{O}_2(g) = 2\text{MgO}(s)$$
$$\text{Mg}(s) + \text{H}_2\text{O}(l) = \text{MgO}(s) + \text{H}_2(g)$$
$$\text{Mg}(s) + \text{CO}_2(g) = \text{MgO}(s) + \text{CO}(g)$$

Explain how these values show that fire extinguishers involving water or carbon dioxide may not be effective in magnesium fires.

3–13 Calculate the values of ΔH^0 (298.2 K) for each of these reactions in which

solved in 400 g of water? The specific heat of the solution is 3.47 J K^{-1} g^{-1} and the initial temperature of salt and water is 23°C.

3–23 The average power of the solar energy in the southwestern deserts of the United States is ~260 W m^{-2}. This average illumination on one acre for an hour is equivalent to how many gallons of octane in fuel value? (The density of octane is 0.698 g cm^{-3}. The area of an acre is 43,560 ft^2.)

3–24 The enthalpy changes for combustion of certain organic substances to give $CO_2(g)$ and $H_2O(l)$ are

benzene, $C_6H_6(l)$	−782.3 kcal mol^{-1}
acetylene, $C_2H_2(g)$	−312.0 kcal mol^{-1}

What is the standard enthalpy of formation of each of these substances, and what is the value of ΔH^0 for the reaction $3C_2H_2(g) = C_6H_6(l)$?

3–25 The heat of fusion of water is 1436 cal mol^{-1}. This is the enthalpy change for the process $H_2O(s) = H_2O(l)$. To a portion of liquid water weighing 570 g at 23.0°C is added a portion of ice weighing 110 g at 0.0°C. What is the final temperature?

3–26 A typical "ice cube" is 2″ × 1″ × 1″. The density of ice is 0.918 g cm^{-3}. If the temperature of iced tea is to be <5°C, what is minimum number of ice cubes that must be added to 100 cm^3 of concentrated tea at 23°C? (Assume the temperature of the ice cubes to be 0°C, and the specific heat of the tea to be the same as that of pure water.)

3–27 For nitrogen gas the values of C_v and C_p at 25°C are 20.8 J K^{-1} mol^{-1} and 29.1 J K^{-1} mol^{-1}, respectively. When a sample of nitrogen is heated at constant pressure, what fraction of the heat is used to increase the internal energy of the gas? How is the remainder of the energy used? How much heat is required to raise the temperature of 100.0 g N_2 from 25.0°C to 85.0°C in a vessel having a constant volume?

3–28 The heat of vaporization of water at the normal boiling point, 373.2 K, is 9717 cal mol^{-1}. The heat capacity of liquid water is 1.00 cal K^{-1} g^{-1} and of gaseous water is ~~0.50 cal K^{-1} g^{-1}~~. Assume these values are independent of temperature. What is the heat of vaporization of water at 340.2 K? Consider the vaporization process to be the reaction $H_2O(l) = H_2O(g)$. 8.0 cal K^{-1} mol^{-1}

3–29 Consider the following reactions at the same temperature:

$$N_2(g) + 3H_2(g) = 2NH_3(g)$$
$$2SO_2(g) + O_2(g) = 2SO_3(g)$$
$$2H_2O(g) = 2H_2(g) + O_2(g)$$
$$Ca(OH)_2(s) = CaO(s) + H_2O(g)$$

For which of these reactions is ΔH^0 more positive than ΔU^0? Predict the sign of ΔS^0 for each reaction. (The gases in each reaction are at a particular pressure, 1 atm.)

3–30 The enthalpy changes in the combustion of each of the following compounds containing two carbon atoms are (in kcal mol^{-1}): ethane, $C_2H_6(g)$, −368.4; ethyl alcohol, $C_2H_5OH(l)$, −327.6; acetaldehyde, $CH_3CHO(l)$, −279.0; and acetic acid, $CH_3CO_2H(l)$, −209.4. In each reaction the value of ΔH is for the reaction in which the water produced is liquid. From these data and the values of $\Delta H_f^0[H_2O(l)]$ and $\Delta H^0[CO_2(g)]$ given in Table 3–2, calculate the value of ΔH_f^0 for each of these compounds.

3–31 For the reaction at 298.2 K:

$$2HgO(s) = 2Hg(l) + O_2(g)$$

the value of ΔS^0 is $+59.4$ cal K^{-1} mol^{-1}. What is the value of ΔS^0 in J K^{-1} mol^{-1}?

3–32 Using relationships derived in this chapter, calculate the values of ΔS and ΔG for each of the processes:

$$H_2O(l, 298.2\ K) = H_2O(g, V = 1000\ L\ mol^{-1})$$
$$H_2O(l, 298.2\ K) = H_2O(g, V = 100\ L\ mol^{-1})$$

Does either of these processes occur spontaneously?

3–33 If the molar entropy of helium gas at 25.0°C is 30.13 cal K^{-1} mol^{-1} at $V = 24.47$ L mol^{-1}, calculate the entropy of:

$$0.10\ mol\ He\ (g, V = 50.0\ L\ mol^{-1})$$
$$3.0\ mol\ He\ (g, V = 1000.0\ L\ mol^{-1})$$

(Give the answer in J K^{-1}.)

3–34 At 1500 K, the process

$$I_2(g, 10\ atm) = 2I(g, 10\ atm)$$

does not occur, but the process

$$I_2(g, 0.10\ atm) = 2I(g, 0.10\ atm)$$

does occur. Explain.

3–35 For the reaction at 298.2 K,

$$2NO_2(g) = N_2O_4(g)$$

the values of ΔH^0 and ΔS^0 are -13.87 kcal mol^{-1} and -42.21 cal K^{-1} mol^{-1}, respectively. What is the value of ΔG^0 at 298.2 K? Assume that ΔH^0 and ΔS^0 do not depend upon temperature. At what temperature is $\Delta G^0 = 0$? Is ΔG^0 negative above, or below, this temperature?

3–36 Write chemical equations for the complete combustion of methane (CH_4), ethane (C_2H_6), and octane (C_8H_{18}). What is the value of ΔH^0 for each of these reactions at 298.2 K? Compare the fuel values of equal weights of each of these hydrocarbons.

3–37 The heat of combustion when gaseous hydrogen (H_2) produces liquid water is -68.32 kcal mol^{-1}. For gaseous methane (CH_4), the heat of combustion is -212.80 kcal mol^{-1}. What weight of hydrogen is present in a 5.00-g mixture of hydrogen and methane that burns in excess oxygen in a constant-pressure calorimeter to liberate 121.2 kcal?

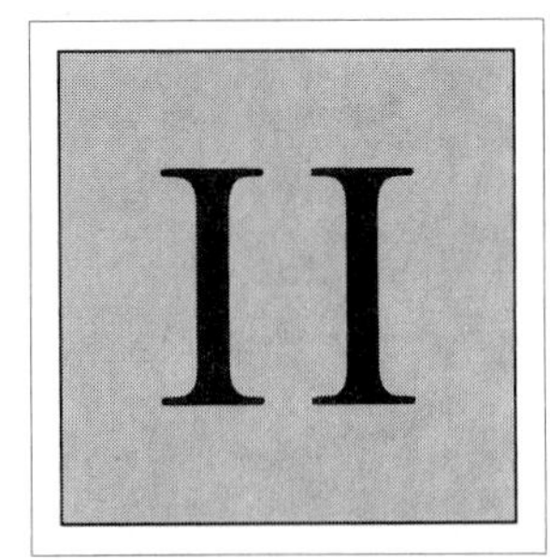

STATES OF MATTER; SOLUTIONS

The next three chapters discuss the gaseous, solid, and liquid states of matter. Most of the discussion in these chapters examines pure substances. Then the fourth chapter of this section deals with solutions, homogeneous mixtures of two or more substances.

The states of matter differ from one another in ways which were outlined in Chapter 1. To the differences already cited there can be added two extensive thermodynamic quantities that you met in Chapter 3, enthalpy and entropy. The conversions of solid to liquid and liquid to vapor are endothermic processes. Therefore the relative enthalpy contents of these states of matter are

$$H(\text{solid}) < H(\text{liquid}) < H(\text{vapor})$$

The conversions

$$\text{solid} = \text{liquid}$$
$$\text{liquid} = \text{vapor}$$

can be carried out reversibly, and the entropy for the system increases in each process; the entropy change for a system undergoing a reversible process at a temperature T is q_{rev}/T. We will consider the increase of entropy calculated in this way for each of these processes and correlate these calculations with the relationship between entropy and disorder

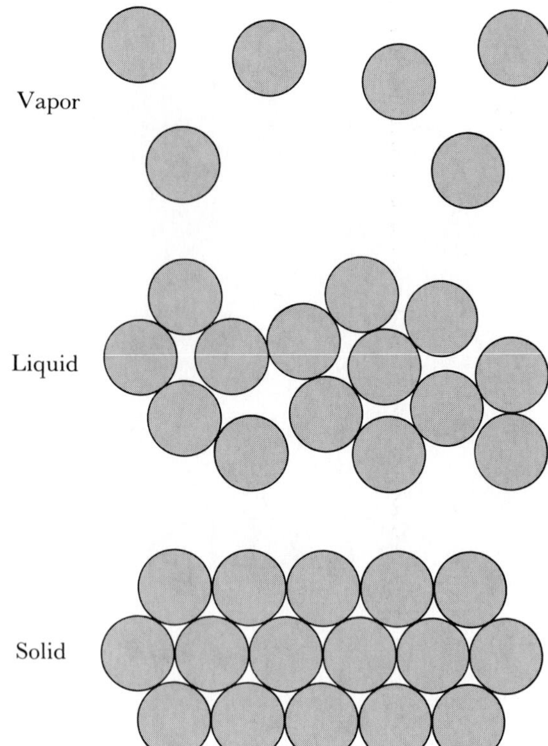

FIGURE II
The increase of disorder in atomic positions in the transitions

Solid → Liquid → Vapor

$S_{solid} < S_{liquid} < S_{gas}$

already discussed in Chapter 3. The increase in disorder of atomic arrangement in going from solid to liquid to gas is depicted in Figure II. Throughout our discussions of the properties of substances in each of these states, we shall ask how the properties are explained by the molecular nature of the substances.

In our discussions of solutions, we will pay particular attention to the way the vapor pressures of the components vary with the concentration. This variance has many practical applications, for example, the separation of substances by distillation, and the determination of molecular weights of solutes by study of freezing points of the solutions. Questions about the attraction of molecules for one another in the liquid phase also are answered by vapor-pressure data.

4

The Gaseous State

4–1 Introduction

In this chapter we shall consider the properties of gases. First we shall examine the experimentally observed relationship between the amount of a gas (n, the number of moles), its temperature (T), pressure (P), and volume (V), then we shall consider the absolute temperature scale. We shall use a model for the properties of gases, the kinetic molecular theory, to derive an equation relating these same quantities, n, T, P, and V. Comparison of the equation based on this model with the empirical equation will allow us to identify the temperature of a gas with the average kinetic energy of the molecules. This relationship will be very important in our study of the dependence of chemical properties upon temperature. Certain properties of molecules, their size and attraction for one another, make the behavior of real gases somewhat more complicated than predicted by the idealized model of the kinetic molecular theory, and we shall consider some of these complications. We also shall develop our knowledge of thermodynamics through the study of gases.

The concept of pressure was mentioned in the preceding chapters. You are familiar with this concept from everyday experience and also from other courses in science; consequently we have been able to do without a full discussion of the units in which pressure can be expressed. However, now it is important to go into some detail.

Since pressure is force divided by area, the SI unit of pressure is the pressure exerted by the SI unit of force, one newton ($1 \text{ N} = 1 \text{ kg m s}^{-2}$), over the SI unit of area, one square meter; this unit of pressure, 1 N m^{-2}, is called the *pascal* (symbol Pa; named for BLAISE PASCAL):

$$1 \text{ Pa} = 1 \text{ N m}^{-2} = \frac{1 \text{ kg m s}^{-2}}{1 \text{ m}^2} = 1 \text{ kg m}^{-1} \text{ s}^{-2}$$

The pascal is being adopted only very slowly as the unit of pressure, but meanwhile you will often be using the more common units, the atmosphere (atm), the torr (torr), and its practical equivalent, the conventional millimeter of mercury (mm Hg). These quantities are most easily introduced in terms of atmospheric pressure, although the atmosphere and torr are now defined in terms of the pascal.

A standard method for measuring atmospheric pressure is the torricellian mercury barometer (named after EVANGELISTA TORRICELLI), shown in Figure 4–1. This barometer balances the pressure of the atmosphere against the pressure of a column of mercury. It is the height of the column that determines the pressure; although an increase in the cross-section area of the barometer tube increases the mass of mercury, it does not change the

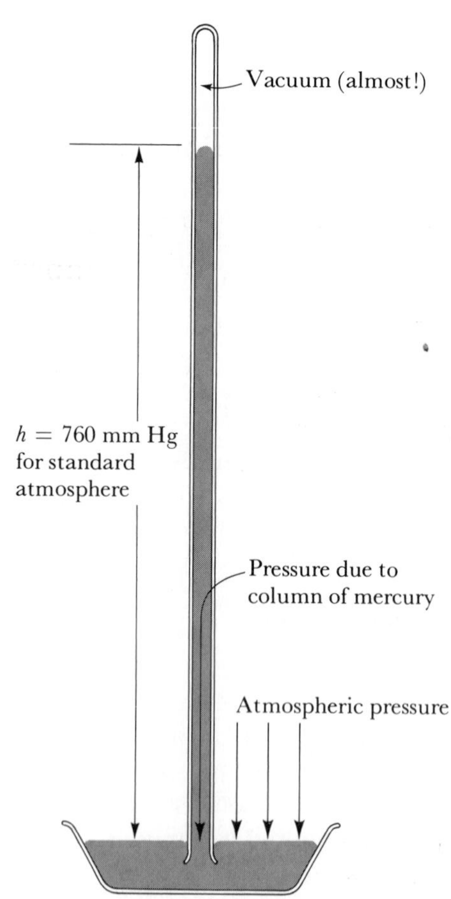

FIGURE 4–1
A torricellian barometer. The tube, completely filled with mercury, is inverted in a dish of mercury. The mercury flows out of the tube until the pressure in the tube at the level of the mercury–air interface is equal to the pressure of the atmosphere. For the standard atmosphere $h = 760$ mm Hg.

force per unit area. The standard atmosphere is the pressure equal to that of a column of mercury (a liquid of density 13.5951 g cm^{-3} at 0°C) exactly 0.76 m high at a place where the acceleration of gravity is 9.80665 m s^{-2}:

$$P = F \times l^{-2}$$
$$= m \times g \times l^{-2}$$

We will consider a column of mercury of one square meter cross section. To express the mass in kilograms, the density of mercury and volume of the column of mercury are expressed in the appropriate units:

$$\text{mass} = \text{density} \times \text{volume}$$
$$= 13.5951 \text{ g cm}^{-3} \times \frac{1 \text{ kg}}{10^3 \text{ g}} \times \frac{10^6 \text{ cm}^3}{1 \text{ m}^3} \times 0.76 \text{ m}^3$$
$$= 10{,}332.3 \text{ kg}$$

Therefore the standard atmosphere is:

$$1 \text{ atm} = 10{,}332.3 \text{ kg} \times 9.806\,65 \text{ m s}^{-2} \times (1 \text{ m})^{-2}$$
$$= 101{,}325 \text{ kg m}^{-1} \text{ s}^{-2}$$
$$= 101{,}325 \text{ N m}^{-2} = 101{,}325 \text{ Pa}$$

This is now the definition of the standard atmosphere. The torr is defined as (1/760) times this pressure:

$$1 \text{ torr} = \frac{1}{760} \text{ atm} = \frac{101{,}325 \text{ Pa}}{760} = 133.322 \text{ Pa}$$

4–2 The Ideal Gas Law; The Absolute Temperature Scale

Now that presentation of the quantitative units for expressing pressure is complete, we can go on with a discussion of experimental observations on the properties of gases. Molecules in the gaseous state are separated from one another by relatively large distances, which makes the specific nature of molecules of a particular gas unimportant relative to some properties. It is these properties on which we focus attention in developing the concept of the ideal gas.

BOYLE'S LAW

Over three hundred years ago ROBERT BOYLE used a simple apparatus (Figure 4–2) to demonstrate an approximately inverse relationship between the volume and pressure of a sample of gas held at a particular temperature. Some of his data are given in Table 4–1. Figure 4–3a gives Boyle's data in a plot of P versus V. This nonlinear plot does not disclose the exact functional relationship between pressure and volume, but it does show that the volume

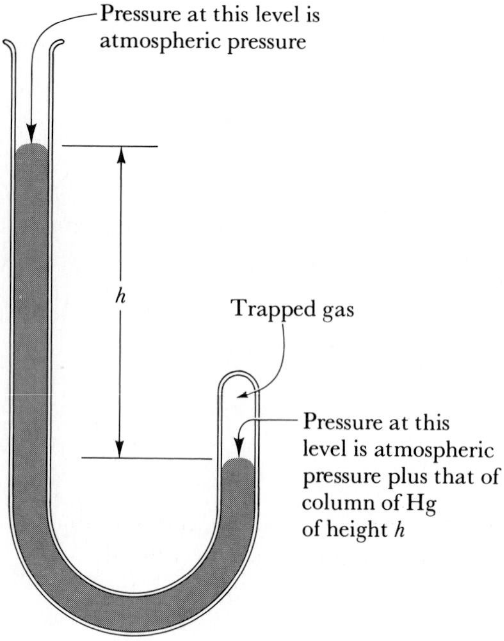

FIGURE 4–2
Apparatus for study of relationship between volume and pressure of gas. The pressure of the trapped gas is atmospheric pressure plus that due to the indicated column of mercury. The mercury flows until the pressure is the same at each particular height within the body of the mercury.

TABLE 4–1
Boyle's Data on the Compressibility of Air[a]

$\frac{V}{\text{in.}^3}$	P / in. Hg: P_{Hg}	P_{air}	P_{total}	$\frac{10^{-2}\,PV}{\text{in.}^3\,(\text{in. Hg})}$
48.0	0	29.1	29.1	14.0
40.0	6.2	29.1	35.3	14.1
32.0	15.1	29.1	44.2	14.1
24.0	29.7	29.1	58.8	14.1
20.0	41.6	29.1	70.7	14.1
16.0	58.1	29.1	87.2	14.0
12.0	88.4	29.1	117.5	14.1

[a]This table presents every 5th entry from Boyle's table of data. Taken from J. B. Conant, *On Understanding Science*, Yale University Press, New Haven, (1947) p. 54. The units of volume are not presented in this source. For the present purpose, it is acceptable to assume the unit is cubic inches. In this calculation of PV it is assumed that V is known to three significant figures, and the pressure is given to that number of significant figures also.

decreases as the pressure increases. We can find the mathematical relationship of volume and pressure by plotting the data in other ways. If the volume is exactly inversely proportional to the pressure,

$$V \propto \frac{1}{P}$$

a plot of V versus $1/P$ will be a straight line through the origin:

$$V = \text{constant} \times \frac{1}{P}$$

Figure 4–3b is such a plot, and we see that it does represent the data. The relationship can be rearranged to

$$PV = \text{constant}$$

and the horizontal line in Figure 4–3c is a third way to plot the data. This relationship of pressure and volume for a particular amount of gas at a particular temperature commonly is called *Boyle's law*.

Boyle's law is not, however, a highly accurate description of the relationship between pressure and volume. Accurate data such as those in

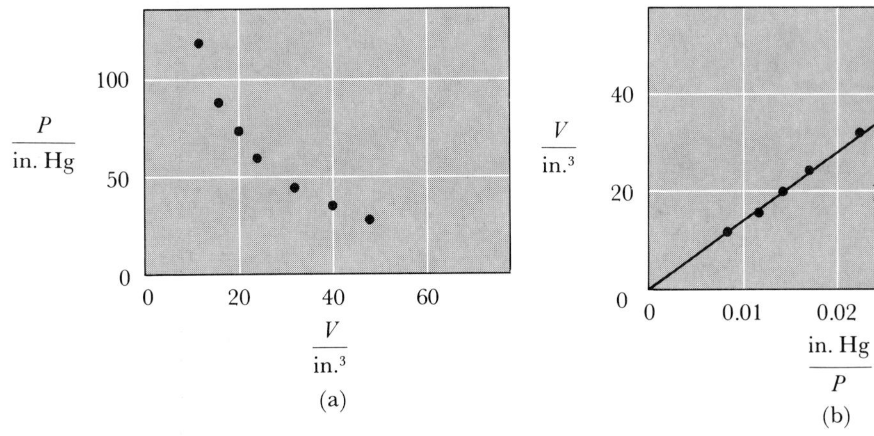

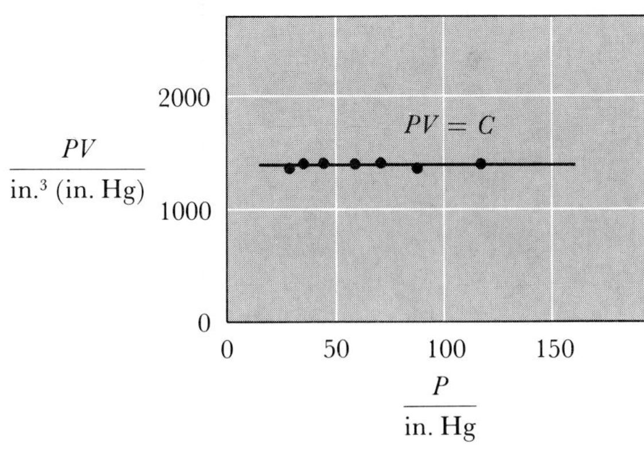

FIGURE 4–3
Various plots of Boyle's data. (a) P versus V; (b) V versus $1/P$; (c) PV versus P.

Figure 4–4 show that the product of pressure and volume of a sample of gas at constant temperature depends upon the pressure; at low pressures ($P < 1$ atm), the dependence of the product PV upon P is essentially linear:

$$PV = \alpha + \beta P$$

This is the equation for a straight line ($y = b + mx$) in a plot of PV versus P, with α the intercept and β the slope. (An equation of this type, with the product PV expressed as a series of terms in increasing powers of P (i.e., $\alpha P^0 + \beta P^1$), is called a *virial equation.* As we shall explain later, virial equations involving higher powers of P are used to correlate the PV product at higher pressures.) The equation involving only the first term on the right side:

$$PV = \alpha$$

is, of course, Boyle's law. With the term βP added, the equation can handle mild deviations from Boyle's law.

If different amounts of gas are studied, one finds the product PV is directly proportional to the number of moles of gas. The linear virial equation applicable to different amounts of gas is written

$$\frac{PV}{n} = \alpha + \beta P$$

Figure 4–4 shows us that the value of α is the same for each of the gases (22.415 L atm mol^{-1} at 273.15 K), but the value of β is different for each gas. (We will see later that the value of β gives information about the attraction of gas molecules for one another.) If an ideal gas is defined as one which obeys Boyle's law at each particular temperature, the linear relationship $PV/n = \alpha + \beta P$ obeyed by real gases at low pressures shows that a real gas approaches ideal gas behavior if

$$\beta P \ll \alpha \qquad \text{or} \qquad P \ll \frac{\alpha}{\beta}$$

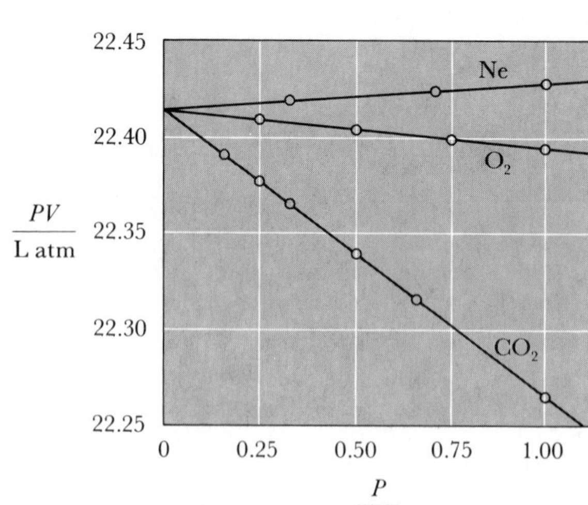

FIGURE 4–4
The dependence of the product of pressure and volume for one mole of real gases upon the pressure at 0.00°C. [From L. P. Hammett, *Introduction to the Study of Physical Chemistry*, McGraw-Hill Book Co., N.Y. (1952) p. 6.]

Ideal gas behavior (*exact* conformity to Boyle's law) is not achieved exactly at any pressure, but it is approached if the pressure is very small compared to the quotient of the parameters α/β. Much of our discussion of gases will pertain to the *ideal gas*. The preceding treatment of pressure–volume data for real gases will have shown you that an ideal gas is a fictitious entity. Nonetheless, the behavior of real gases at low pressures approaches that of an ideal gas, and even for moderate pressures, calculations based upon the ideal gas approximation (i.e., Boyle's law, PV/n = constant) for a gas at a particular temperature are useful. Considering only the gases shown in Figure 4–4, the values of PV/n at 1 atm differ from 22.415 L atm mol^{-1} by +0.08% (Ne), −0.07% (O_2), and −0.65% (CO_2).

CHARLES' LAW; THE ABSOLUTE TEMPERATURE SCALE

Accurate data on the PV product for a gas can be obtained with the gas sample maintained at various particular temperatures. The data for each temperature are analogous to those for 0.00°C, illustrated in Figure 4–4, and they can be fit to the virial equation; in this way, a value of α is obtained for each temperature studied. Studies at 50.00°C and 100.00°C yield the following values of α:

$$50.00°\text{C} \qquad \alpha = 26.518 \text{ L atm mol}^{-1}$$

$$100.00°\text{C} \qquad \alpha = 30.621 \text{ L atm mol}^{-1}$$

These values of α and the value for 0.00°C are a linear function of the temperature:

$$\alpha = \gamma + \delta t_c$$

in which t_c is the temperature on the Celsius scale. Figure 4–5a shows this linear plot, in which the intercept at t_c = 0°C is equal to 22.415 L atm mol^{-1} and the slope δ is 0.08206 L atm mol^{-1} (°C)$^{-1}$. This line can be extrapolated to the temperature at which $\alpha = 0$, as shown in Figure 4–5b. For $\alpha = 0$, $t_c = -\gamma/\delta$; this temperature is −273.15°C. This equation can be transformed into one in which the intercept is zero by shifting the origin of the temperature scale. The algebraic manipulation is:

$$\alpha = \gamma + \delta t_c$$

$$= \delta\left(\frac{\gamma}{\delta} + t_c\right)$$

$$= \delta T$$

in which

$$T = t_c + \frac{\gamma}{\delta}$$

$$= t_c + \frac{22.415 \text{ L atm mol}^{-1}}{0.08206 \text{ L atm mol}^{-1}\ (°\text{C})^{-1}}$$

$$T = t_c + 273.15°\text{C}$$

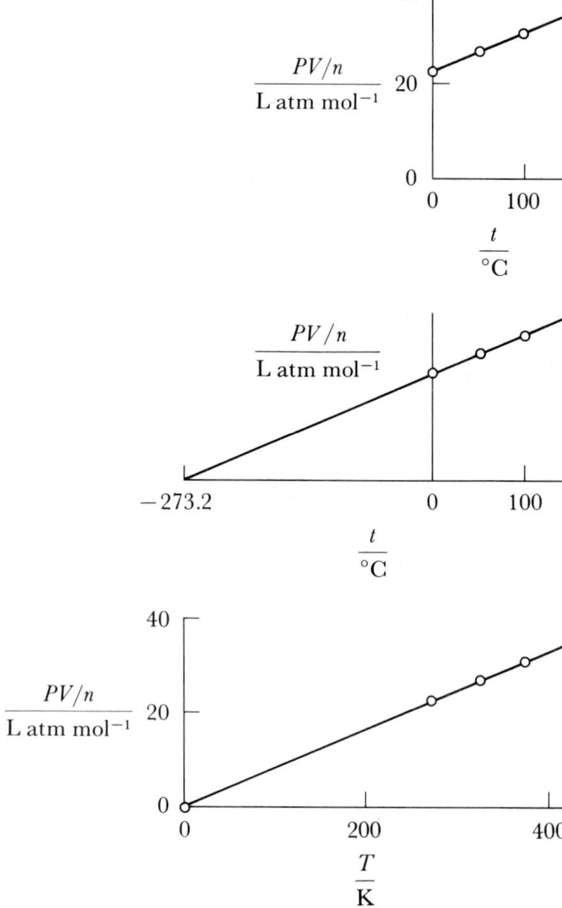

FIGURE 4–5
The dependence of PV/n (extrapolated to $P = 0$) upon temperature. (Experimental points are for 0.00, 50.00, and 100.00°C.) (a) α versus t_C; (b) α versus t_C, with extrapolation to $\alpha = 0$ at $t_c = -273.2$°C; (c) α versus T ($T/K = (t_C + 273.2°C)/°C$).

This expression for T defines the absolute temperature scale on the basis of the pressure–volume properties of ideal gases (real gases, the properties of which are extrapolated to zero pressure).[1] Figure 4–5c shows these same data in a plot of α versus T; in this plot the intercept is zero, $\alpha = 0$ at $T =$ 0 K. The linear dependence of the volume of a gas (at constant pressure) upon temperature was first observed by JACQUES CHARLES in 1787, and the relationship of volume and temperature is generally called *Charles' law*.

Because the relationship between the temperature of a gas and its product PV is linear, as shown in Figure 4–5, temperature-measuring devices can be based upon this proportionality. Gas thermometers are used in studies demanding high accuracy, and also to calibrate mercury-in-glass thermometers. These gas thermometers are of several kinds; a schematic diagram of a constant-volume gas thermometer is given in Figure 4–6. The pressure required to maintain the volume of the gas constant is a measure of the temperature. A gas thermometer can be calibrated with

[1]Recall that the temperature of the triple point of water now is defined as 273.16 K in SI; this also defines the size of the kelvin unit of temperature.

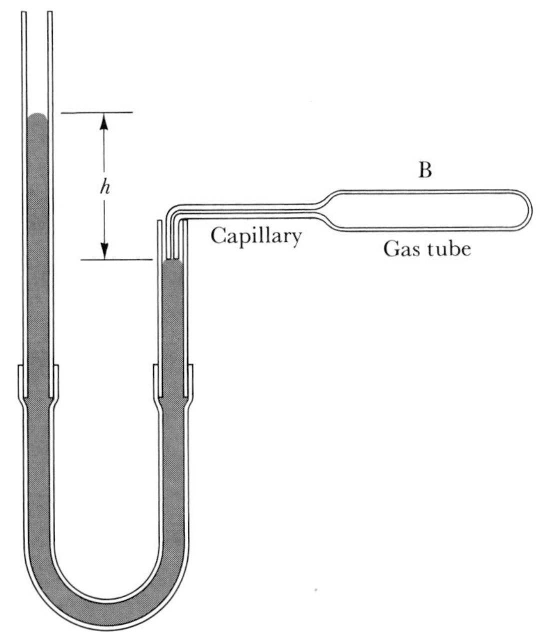

FIGURE 4–6
A gas themometer with a constant volume. The gas is contained in tube B, which is immersed in the medium whose temperature is being measured. The flexible tubing allows adjustment of the pressure of the mercury to make the volume of the gas exactly the same in each use. [From M. W. Zemansky, *Heat and Thermodynamics*, 2nd ed., McGraw-Hill Book Co., N.Y. (1943).]

known temperatures (e.g., the freezing point and boiling point of water). That is, the pressures needed to maintain the volume at these temperatures can be determined. Linear interpolation and extrapolation allows each particular pressure reading to be interpreted as a temperature. Since the nonideality of the gas complicates the reliability of a gas thermometer, low pressures are used in the most accurate work. For practical purposes it is necessary to have accurately measured reference temperatures for calibration of thermometers. Some useful reference temperatures are given in Table 4–2.

TABLE 4–2
Reference Temperatures[a] Used in Calibration of Thermometers

References		$\frac{T}{\text{K}}$
H_2 normal boiling point		20.28
O_2 triple point		54.361
O_2 normal boiling point		90.188
Hg triple point		234.308
H_2O normal melting point		273.15
H_2O triple point	exactly	273.16
H_2O normal boiling point		373.15
Zn normal melting point		692.73
Ag normal melting point		1235.08
Au normal melting point		1337.58

[a] The definitions of normal melting point and normal boiling point are given in Chapter 6. The triple point was defined in Chapter 1.

THE IDEAL GAS LAW

It is possible to combine Boyle's law,

$$\frac{PV}{n} = \text{constant (at constant } T)$$

and Charles' law,

$$\frac{V}{n} = \text{constant} \times T \text{ (at constant } P)$$

into a single equation,

$$\frac{PV}{n} = \text{constant} \times T \quad \text{or} \quad PV = nRT$$

in which R, called the *gas constant*, is the proportionality constant. The gas constant, an important parameter in physical chemistry, is the slope of the plot of PV/n versus T given in Figure 4–5c; to four significant figures, the value is

$$R = 0.08206 \text{ L atm K}^{-1} \text{ mol}^{-1}$$

To remember an approximate value of R in these units, which is useful to know if you are solving P–V–T problems (using the units pressure in atmospheres and volume in liters), remember that one mole of an ideal gas occupies 22.4 liters at one atmosphere pressure at 273 K:

$$R = \frac{PV}{nT}$$

$$= \frac{1 \text{ atm} \times 22.4 \text{ L}}{1 \text{ mol} \times 273 \text{ K}} = 0.0821 \text{ L atm K}^{-1} \text{ mol}^{-1}$$

From the standard atmosphere,

$$1 \text{ atm} \equiv 101{,}325 \text{ kg m}^{-1} \text{ s}^{-2}$$

and the liter,

$$1 \text{ L} \equiv 1 \text{ dm}^3 = 10^{-3} \text{ m}^3$$

we can calculate the value R in SI units:

$$R = 0.08206 \text{ L atm K}^{-1} \text{ mol}^{-1} \times \frac{10^{-3} \text{ m}^3}{\text{L}} \times \frac{101{,}325 \text{ kg m}^{-1} \text{ s}^{-2}}{\text{atm}}$$

$$= 8.315 \text{ kg m}^2 \text{ s}^{-2} \text{ K}^{-1} \text{ mol}^{-1} = 8.315 \text{ J K}^{-1} \text{ mol}^{-1}$$

Conversion of this to dimensions involving calories gives

$$R = 8.315 \text{ J K}^{-1} \text{ mol}^{-1} \times \frac{1 \text{ cal}}{4.184 \text{ J}} = 1.987 \text{ cal K}^{-1} \text{ mol}^{-1}$$

For internal consistency we have used four significant figures in these calculations. For many problems three significant figures are adequate:

$$0^\circ\mathrm{C} = 273\ \mathrm{K}$$
$$R = 8.31\ \mathrm{J\ K^{-1}\ mol^{-1}}\ \text{(see footnote 2)}$$
$$R = 1.99\ \mathrm{cal\ K^{-1}\ mol^{-1}}$$
$$R = 0.0821\ \mathrm{L\ atm\ K^{-1}\ mol^{-1}}$$

You can solve many practical problems using the ideal gas law, but you should recognize the uncertainties introduced by use of this limiting law.

The equation $PV = nRT$ involves four variables: P, V, n, and T. Any three of these determine the fourth, and problems arise in which any one of the four variables is the unknown.

EXAMPLE: A 1.000-g sample of water is allowed to vaporize completely into a 10.00-L container. What is the pressure of the water vapor at 200°C?

The gas law can be rearranged to give pressure as the unknown:

$$P = \frac{nRT}{V}$$

The number of moles of water vapor is

$$n = \frac{1.000\ \mathrm{g}}{18.02\ \mathrm{g\ mol^{-1}}} = 0.0555\ \mathrm{mol}$$

and the absolute temperature is

$$200^\circ\mathrm{C} \times \frac{1\ \mathrm{K}}{1^\circ\mathrm{C}} + 273\ \mathrm{K} = \underline{473\ \mathrm{K}}$$

Therefore the pressure is

$$P = \frac{0.0555\ \mathrm{mol} \times 0.0821\ \mathrm{L\ atm\ K^{-1}\ mol^{-1}} \times 473\ \mathrm{K}}{10.00\ \mathrm{L}}$$
$$= 0.216\ \mathrm{atm}$$

EXAMPLE: A balloon is to be filled with 30.0 kg of helium gas. What volume can be filled to a pressure of 1.15 atm if the temperature is 20.0°C?

The ideal gas law gives

$$V = \frac{nRT}{P}$$

[2]The value of $R = 8.31\ \mathrm{J\ K^{-1}\ mol^{-1}}$ results from rounding the value $R = 8.3148\ \mathrm{J\ K^{-1}\ mol^{-1}}$, which is the value of R consistent with the molar volume $22.415\ \mathrm{L\ mol^{-1}}$ at 273.15 K.

To calculate the volume, the number of moles of helium must be known:

$$n = \frac{30.0 \text{ kg}}{0.00400 \text{ kg mol}^{-1}} = 7.50 \times 10^3 \text{ mol}$$

and the temperature on the Celsius scale must be converted to kelvin units:

$$20.0°\text{C} \times \frac{1 \text{ K}}{1°\text{C}} + 273 \text{ K} = 293 \text{ K}$$

The calculation of volume is

$$V = \frac{7.50 \times 10^3 \text{ mol} \times 0.0821 \text{ L atm K}^{-1} \text{ mol}^{-1} \times 293 \text{ K}}{1.15 \text{ atm}}$$

$$= 1.57 \times 10^5 \text{ L}$$

EXAMPLE: An unknown gas is admitted to an evacuated vessel having a volume of 0.500 L. The vessel is closed with the measured pressure equal to 700 torr at 25°C. The vessel is weighed, and the mass of the contents is found to be 0.982 g. What is the molecular weight of this gaseous substance?

Because the number of moles is the mass of the sample, w, divided by the molecular weight, MW, the gas law can be written

$$PV = \frac{w}{\text{MW}} RT$$

which gives for the molecular weight

$$\text{MW} = \frac{w}{V} \times \frac{RT}{P}$$

Since the value of R which we have been using in these problems is expressed in the units

$$\text{L atm K}^{-1} \text{ mol}^{-1}$$

the pressure must be expressed in atmospheres:

$$P = 700 \text{ torr} = 700 \text{ torr} \times \frac{1 \text{ atm}}{760 \text{ torr}} = 0.921 \text{ atm}$$

The calculation of molecular weight is:

$$\text{MW} = \frac{0.982 \text{ g}}{0.500 \text{ L}} \times \frac{0.0821 \text{ L atm K}^{-1} \text{ mol}^{-1} \times 298 \text{ K}}{0.921 \text{ atm}}$$

$$= 52.2 \text{ g mol}^{-1}$$

(Notice that the quotient w/V, which appears in the equation for the molecular weight, is the density of the gas.)

EXAMPLE: This example illustrates the use of SI units in calculations. An unknown gas has a density of 1.275 kg m^{-3} at a pressure of 5.00 × 10^4 Pa (= 5.00 × 10^4 kg m^{-1} s^{-2}) at a temperature of 350.0 K. What is its molecular weight?

In SI $R = 8.31$ J K^{-1} mol^{-1} = 8.31 kg m^2 s^{-2} K^{-1} mol^{-1}, and the calculation is:

$$\text{MW} = \frac{w}{V} \times \frac{RT}{P}$$

$$= \frac{1.275 \text{ kg m}^{-3} \times 8.31 \text{ kg m}^2 \text{ s}^{-2} \text{ K}^{-1} \text{ mol}^{-1} \times 350.0 \text{ K}}{5.00 \times 10^4 \text{ kg m}^{-1} \text{ s}^{-2}}$$

$$= 0.0742 \text{ kg mol}^{-1}$$

Because the pressure is given only to three significant figures, this is the number of significant figures used for R and retained in the answer. Accurate measurement of the molecular weight should involve measurement of the gas density at different pressures. Extrapolation of the apparent value of the molecular weight to zero pressure would give the correct value uninfluenced by gas nonideality.

EXAMPLE: In a gas thermometer, the pressure needed to fix the volume of 0.2000 g of helium at 0.5000 L is 850.0 torr. What is the temperature?

$$T = \frac{PV}{nR}$$

$$P = 850.0 \text{ torr} \times \frac{1 \text{ atm}}{760 \text{ torr}} = 1.118 \text{ atm}$$

$$n = \frac{0.2000 \text{ g}}{4.003 \text{ g mol}^{-1}} = 0.04996 \text{ mol}$$

$$T = \frac{1.118 \text{ atm} \times 0.5000 \text{ L}}{0.04996 \text{ mol} \times 0.08206 \text{ L atm K}^{-1} \text{ mol}^{-1}} = 136.4 \text{ K}$$

It was stated in Chapter 1 that buoyancy corrections may be necessary in measurements of the mass or weight of an object. This is a consequence of ARCHIMEDES' principle, which states that a body completely or partially immersed in a fluid is subject to a buoyant force equal to the weight of the displaced fluid. In calculating the buoyancy correction, the necessary items of data are the densities of the object, the weights, and the air (in which the masses of the object and the weights are compared).

EXAMPLE: An aluminum object weighed on a double-pan balance appears to have the same mass as brass weights with a correct mass of 10.5432 g. The densities of aluminum and brass are 2.702 g cm^{-3} and 8.40 g cm^{-3}, respectively. Atmospheric pressure is 753 torr and the temperature is 21°C. What is the correct mass of the object?

The approximate density of air can be calculated from the ideal gas law; the molecular weight of air is 28.97 (which is 0.78 × MW (N_2) + 0.21 × MW(O_2) + 0.01 × MW(Ar):

$$\frac{w}{V} = \frac{PMW}{RT}$$

$$= \frac{(753 \text{ torr}/760 \text{ torr atm}^{-1}) \times 28.97 \text{ g mol}^{-1}}{0.0821 \text{ L atm K}^{-1} \text{ mol}^{-1} \times 294 \text{ K}}$$

$$= 1.19 \text{ g L}^{-1}$$

$$= 1.19 \text{ g L}^{-1} \times \frac{1 \text{ L}}{1000 \text{ cm}^3} = 1.19 \times 10^{-3} \text{ g cm}^{-3}$$

The volume of the weights, V_1, is

$$V_1 = \frac{10.5432 \text{ g}}{8.40 \text{ g cm}^{-3}} = 1.26 \text{ cm}^3$$

and the approximate volume[3] of the sample, V_2, is

$$V_2 = \frac{10.5432 \text{ g}}{2.702 \text{ g cm}^{-3}} = 3.90 \text{ cm}^3$$

Because the volume of the object is larger than the volume of the weights, there is a net buoyant force acting on the object. It is the difference of these volumes multiplied by the density of air:

$$(3.90 \text{ cm}^3 - 1.26 \text{ cm}^3) \times 1.19 \times 10^{-3} \text{ g cm}^{-3} = 0.0031 \text{ g}$$

Thus the correct mass of the aluminum object is

$$10.5432 \text{ g} + 0.0031 \text{ g} = 10.5463 \text{ g}$$

The concept of the volume concentration of chemical species, measured in moles per liter, mol L^{-1}, (or in moles per cubic decimeter, mol dm^{-3}, if SI units are used), is commonly employed in chemical studies of liquids and liquid solutions. Volume concentration also is an important concept in certain gas-phase studies. The volume concentration of a gas, n/V, is given by the ideal gas law,

$$\frac{n}{V} = \frac{P}{RT}$$

In particular, notice that the concentration of a gas is proportional to its pressure, and that the proportionality constant involves the temperature.

EXAMPLE: What is the concentration of a gas at a pressure of 150 torr at 300 K?

[3]This volume is approximate because this mass is the uncorrected mass of the object.

$$\frac{n}{V} = \frac{150 \text{ torr} \times \dfrac{1 \text{ atm}}{760 \text{ torr}}}{0.0821 \text{ L atm K}^{-1} \text{ mol}^{-1} \times 300 \text{ K}}$$

$$= 8.01 \times 10^{-3} \text{ mol L}^{-1}$$

What is the concentration of this gas in molecules per cubic centimeter?

$$\frac{N_A n}{V} = 6.02 \times 10^{23} \text{ mol}^{-1} \times 8.01 \times 10^{-3} \text{ mol L}^{-1} \times \frac{1 \text{ L}}{10^3 \text{ cm}^3}$$

$$= 4.82 \times 10^{18} \text{ cm}^{-3}$$

EXAMPLE: The density of common ether [diethyl ether, $(C_2H_5)_2O$, MW = 74.12] as a liquid at 25°C is 0.708 g cm^{-3}. At this temperature, the pressure of the vapor in equilibrium with the liquid is 233 torr. What is the concentration (in moles per liter) for ether in each of these two states?

For the liquid state:

$$\text{Conc} = \frac{n}{V} = \frac{d}{\text{MW}} = \frac{0.708 \text{ g cm}^{-3}}{74.12 \text{ g mol}^{-1}} \times \frac{10^3 \text{ cm}^3}{1 \text{ L}}$$

$$= 9.55 \text{ mol L}^{-1}$$

For the gaseous state:

$$\text{Conc} = \frac{n}{V} = \frac{P}{RT}$$

$$= \frac{233 \text{ torr} \times \dfrac{1 \text{ atm}}{760 \text{ torr}}}{0.0821 \text{ L atm K}^{-1} \text{ mol}^{-1} \times 298 \text{ K}}$$

$$= 0.0125 \text{ mol L}^{-1}$$

This last calculation shows how much less concentrated a typical substance is in its vapor as compared to its liquid.

Many problems dealing with gases involve a change of conditions from an initial state to a final state, with the amount, the pressure, the volume, *or* the temperature remaining unchanged. These problems can be solved without explicit use of the gas constant, R. For each of these states,

$$\frac{PV}{nT} = R$$

therefore the relationship between these variables for an initial state (designated 1) and a final state (designated 2) is:

$$\frac{P_2V_2}{n_2T_2} = \frac{P_1V_1}{n_1T_1}$$

Appropriately rearranged with constant factors canceled, this equation can be used to solve many problems.

EXAMPLE: A sample of gas occupies 773 cm^3 at 300 K and a pressure of 187 torr. What will the volume be if the temperature is raised to 380 K and the pressure is raised to 200 torr?

Rearrangement of the equation just given to yield V_2, the new volume, gives

$$V_2 = V_1 \times \frac{P_1}{P_2} \times \frac{T_2}{T_1}$$

$$= 773 \text{ cm}^3 \times \frac{187 \text{ torr}}{200 \text{ torr}} \times \frac{380 \text{ K}}{300 \text{ K}} = 915 \text{ cm}^3$$

This expression also can be derived simply by recognizing that V_2 is equal to V_1 multiplied by a ratio of pressures and a ratio of absolute temperatures. Each of these two ratios is chosen to increase or decrease the calculated volume, V_2, depending upon what common sense suggests the consequences of the change in each variable, pressure and temperature, to be. In this example, an increase of pressure decreases the volume, and an increase of temperature increases the volume. Thus the ratio of pressures is less than 1 $\left(\frac{187 \text{ torr}}{200 \text{ torr}}\right)$, and the ratio of temperatures is greater than 1 $\left(\frac{380 \text{ K}}{300 \text{ K}}\right)$.

MIXTURES OF GASES

Each component of a mixture of gases completely fills the vessel in which the mixture is contained. The total pressure P is the sum of the pressures p_1, p_2, p_3, and so on, due to each of the components:

$$P = p_1 + p_2 + p_3 + \cdots$$

If each of the partial pressures is expressed in terms of the amount of each gas, $n_1, n_2, n_3, \cdots$, and the ideal gas law, for example, $p_1 = n_1RT/V$, this equation becomes:

$$P = \frac{n_1RT}{V} + \frac{n_2RT}{V} + \frac{n_3RT}{V} + \cdots$$

$$= (n_1 + n_2 + n_3 + \cdots)\frac{RT}{V} = \frac{nRT}{V}$$

The ideal gas law is as applicable to a mixture of gases, with n the sum of the number of moles of all of the gases, $n = n_1 + n_2 + n_3 + \cdots$, as it is to a single gas. That is, at total pressures that are low enough for the ideal gas law to be approximately correct, the law may be used for mixtures as well as for single gases. Under these conditions it also is possible to calculate the partial pressure due to each gas in the mixture; for the ith component,

$$p_i = \frac{n_i RT}{V} = \frac{n_i}{n} P$$

because $RT/V = P/n$. That is, the partial pressure of each gas in a mixture is the product of the total pressure (which is measurable) and the mole fraction of that gas, n_i/n (which can be determined by analysis).

One familiar mixture of gases is air; the principal components of a dried sample of air are nitrogen (78.08 mole %), oxygen (20.95 mole %), argon (0.934 mole %), and carbon dioxide (0.033 mole %). If the total pressure of this mixture of gases is 743.0 torr, the pressure of each component can be obtained as the product of total pressure and the individual mole fractions (n_i/n):

N_2	0.7808×743.0 torr $= 580.1$ torr
O_2	0.2095×743.0 torr $= 155.7$ torr
Ar	0.00934×743.0 torr $= 6.94$ torr
CO_2	0.00033×743.0 torr $= 0.25$ torr

In certain common laboratory operations, gases are collected over liquids in which they do not dissolve appreciably. In the space above the liquid, both the gas in question and vapor of the liquid are present.

An experiment which you may have done involves the decomposition of potassium chlorate to give oxygen gas:

$$2KClO_3(s) = 2KCl(s) + 3O_2(g)$$

the gas being collected over water as shown in Figure 4–7a. This experimental arrangement with a calibrated collection vessel allows us to measure the volume occupied by the mixture of gases (oxygen and water vapor). We measure this volume after adjusting the depth of immersion of the collection vessel with the trapped gas so the total pressure of the gas is the same as atmospheric pressure, which can be measured. We then have for the total amount of gas, n, and the amount of water vapor, n_{H_2O},

$$n = \frac{PV}{RT} \qquad n_{H_2O} = \frac{P_{H_2O}V}{RT}$$

Because $n = n_{O_2} + n_{H_2O}$,

$$\begin{aligned} n_{O_2} &= n - n_{H_2O} \\ &= \frac{PV}{RT} - \frac{P_{H_2O}V}{RT} \\ &= \frac{(P - P_{H_2O})V}{RT} \end{aligned}$$

Example: The decomposition of a mixture of potassium chloride and potassium chlorate weighing 1.43 g gives 327 cm^3 of oxygen gas.

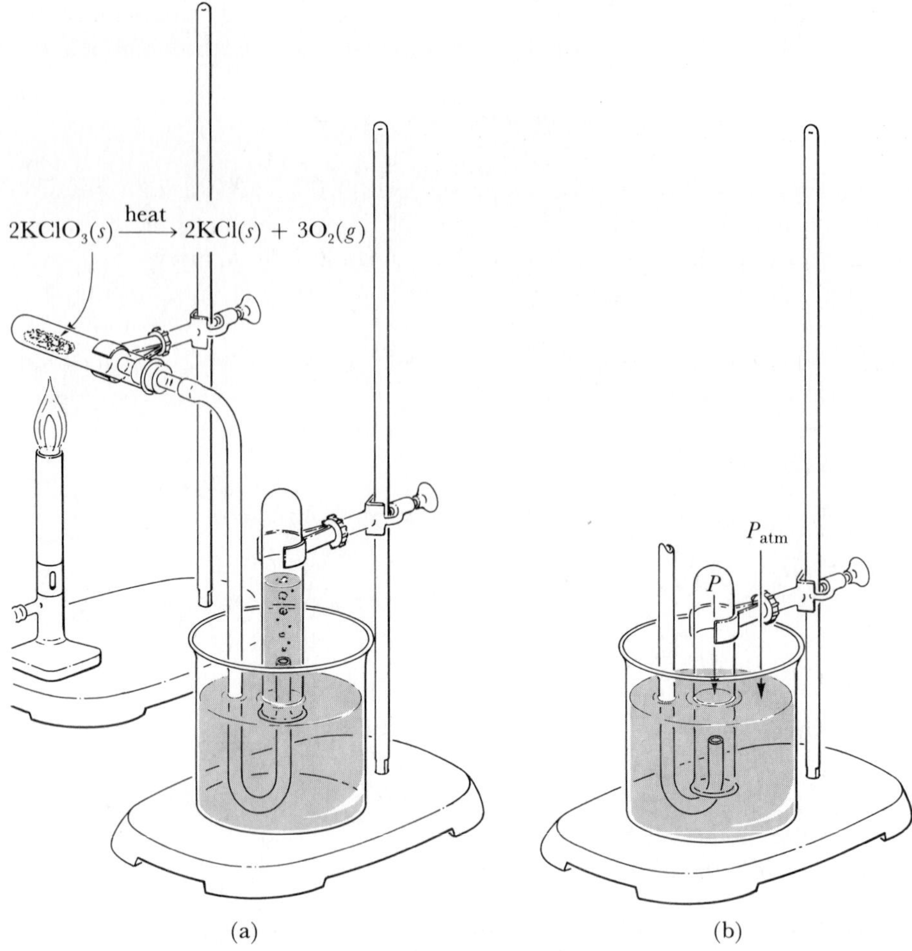

FIGURE 4–7
Collection of oxygen over water. (a) Collection of gas. (b) Adjustment of collection vessel to have the water level the same inside (where $P = P_{O_2} + P_{H_2O}$) and outside (where $P = P_{atm}$, which is measurable); when levels are the same, the volume is read.

The oxygen is collected over water at 22.0°C, and the total pressure of the gas is 743.0 torr. What fraction of the original sample is potassium chlorate? (The vapor pressure of water at 22.0°C is 21.07 torr.)

Use of the preceding equation allows us to calculate the number of moles of oxygen:

$$n_{O_2} = \frac{(743.0\text{ torr} - 21.07\text{ torr}) \times \dfrac{1\text{ atm}}{760\text{ torr}} \times 0.327\text{ L}}{0.0821\text{ L atm K}^{-1}\text{ mol}^{-1} \times 295.2\text{ K}}$$

$$= 0.0128\text{ mol}$$

Because 1 mol of potassium chlorate ($KClO_3$) gives $1\frac{1}{2}$ mol of oxygen gas, this amount of oxygen is produced in the decomposition of

$$\tfrac{2}{3} \times 0.0128\text{ mol} = 0.00853\text{ mol}$$

of potassium chlorate. Therefore the mass of potassium chlorate is

$$8.53 \times 10^{-3}\ \text{mol} \times 122.55\ \text{g mol}^{-1} = 1.046\ \text{g KClO}_3$$

The fraction of $KClO_3$ in the sample is

$$\frac{1.046\ \text{g}}{1.43\ \text{g}} = 0.73$$

4–3 The Kinetic Molecular Theory

Most developments in science involve a mixture of empirical observations and theoretical concepts. This was true for the correlation of the pressure–volume–temperature behavior of gases. We have discussed some of the empirical observations, now let us consider some of the theoretical concepts.

KINETIC ENERGY OF GAS MOLECULES AND ABSOLUTE TEMPERATURE

The ideal gas law, $PV = nRT$, is an empirical equation; at no stage in the development of this equation from experimental observations was a theoretical model involved. Now we will derive an equation for the quantity (PV/n) in terms of the mass and velocity of the gas molecules. The basis for this derivation is the *kinetic molecular theory*, the postulates of which, as originally proposed by BERNOULLI, are:

1. Molecules of a gas are in constant motion. Collisions of these molecules with walls of the vessel are responsible for the pressure of the gas.
2. Molecules of a gas are separated from one another by very large distances relative to the size of the molecules.
3. Molecules of a gas exert no attractive or repulsive forces on one another.

Real gases, of course, have properties that do not conform to these postulates, and we will consider later how the ideal gas law can be modified to take account of the finite size of gas molecules and the attractive forces between the molecules. In the derivation that follows, the molecules are assumed to be infinitesimal in size, which is the limit of Postulate 2. Such molecules will collide only with the walls of the container.[4]

Let us consider a sample of n moles of gas consisting of molecules with a particular mass m in a cubical container at a particular temperature. No transfer of heat occurs to or from the surroundings. Molecules of this gas move randomly in all directions. We wish to derive an equation that relates the pressure and volume of the gas to the mass and velocity of the gas

[4]The probability of two gaseous molecules colliding with one another depends upon the square of the sum of the radii of the two molecules. For infinitesimal molecules (i.e., mathematical points of zero radius), this probability of collision is zero.

molecules. Pressure is force per unit area, and the force exerted on the walls of the container is the rate of transfer of momentum to the wall by collisions of the molecules. We will focus our attention on a representative molecule with velocity u. The kinetic energy of this molecule is

$$\text{kinetic energy} = \tfrac{1}{2}mu^2$$

and the momentum of this molecule is

$$\text{momentum} = mu$$

The collisions of this representative molecule with the walls are perfectly elastic: no energy is lost or gained in such collisions. Consider collisions of this molecule with one of the two walls perpendicular to the x axis (see Figure 4–8). The time between successive collisions of the representative molecule and this wall is determined by the component of the velocity along the x axis. The velocity u can be resolved into its components along directions of each of the axes defined by the three edges of the cubical container of edge length l. Because the molecules are moving randomly in all directions, the representative molecule will be assumed to be moving in a direction such that

$$u_x = \pm u_y = \pm u_z$$

That is, it prefers no one direction. The components of the velocity along the x, y, and z axes are related to the total velocity by the Pythagorean theorem (as indicated in the figure):

$$\begin{aligned} u^2 &= u_x^2 + u_y^2 + u_z^2 \\ &= 3u_x^2 \end{aligned}$$

$$u_x = \frac{1}{\sqrt{3}}u$$

(It is u_x that is relevant in determining the frequency of impacts of the mole-

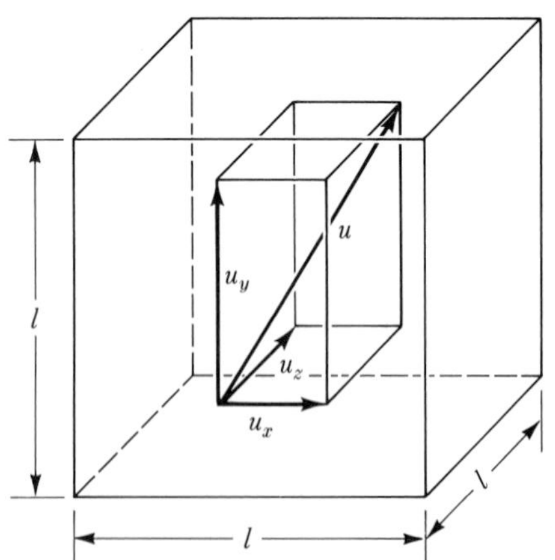

FIGURE 4–8
Resolution of the velocity of a gas molecule in a cubical box into its components along the x, y, and z axes. The x, y, and z components of the velocity are given by the edges of the rectangular parallelopiped whose diagonal is the velocity vector $u^2 = u_x^2 + u_y^2 + u_z^2$.

cule with the wall in question, but it is u that determines the kinetic energy of the molecule.) The time interval between impacts of the molecule and the particular wall is the time required for the molecule to move a distance $2l$ in the x direction:

$$\text{time between impacts} = \frac{2l}{u_x}$$

and the number of impacts per unit time is the reciprocal of this:

$$\text{frequency of impacts} = \frac{u_x}{2l}$$

It is not meaningful to calculate the force on the wall due to a single impact, but it is meaningful to calculate the average force from the impacts of this representative molecule over a period of time:

$$\text{average force} = \text{momentum transfer per impact} \times \text{impact frequency}$$

The x component of momentum of the molecule changes from $-mu_x$ to $+mu_x$ in a collision, as is shown in Figure 4–9. Therefore the change of momentum is

$$\Delta(mu_x) = 2mu_x$$

Momentum of this magnitude is transferred to the wall in each impact; therefore the average force on the wall in question due to the collisions of this representative molecule is:

momentum change per impact

$$\text{average force} = (2mu_x) \times \left(\frac{u_x}{2l}\right)$$

impact frequency

$$= \frac{mu_x^2}{l}$$

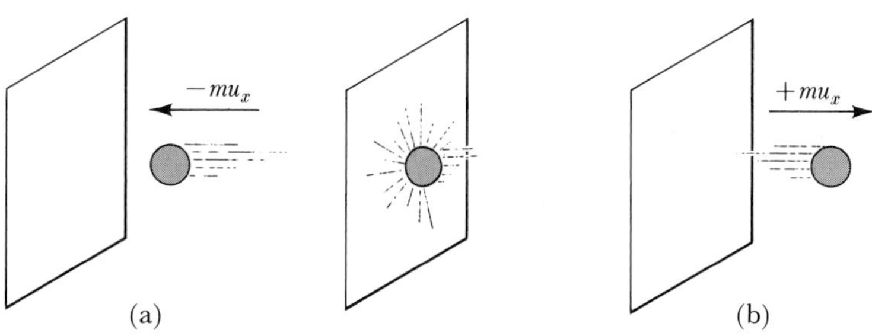

FIGURE 4–9
The transfer of momentum from moving molecule to wall in a collision of a gas molecule with a wall. (a) Before collision, $(\text{momentum})_x = -mu_x$. (b) After collision, $(\text{momentum})_x = +mu_x$. Change of momentum $= mu_x - (-mu_x) = 2mu_x$.

We now wish to express this equation in terms of the total velocity, u, rather than u_x. With $u_x = u/\sqrt{3}$, the average force becomes

$$\text{average force} = \frac{mu^2}{3l}$$

The average pressure on the walls due to collisions of this representative molecule is the average force divided by the area of the square face (l^2) of this cubical container:

$$\text{average pressure} = \frac{mu^2}{3l} \times \frac{1}{l^2} = \frac{mu^2}{3l^3}$$

and because the volume of the cube is l^3,

$$\text{average pressure} = \frac{mu^2}{3V}$$

To obtain the pressure due to all of the gas, nN_A molecules, where N_A is Avogadro's constant, we identify this representative molecule as one with a value of u^2 that is the same as the average of the values of u^2 for all of the molecules, $\overline{u^2}$:

$$P = nN_A \times \frac{m\overline{u^2}}{3V}$$

This equation can be converted to an equation for the quantity (PV/n) involving the average kinetic energy of the molecules (average kinetic energy $= m\overline{u^2}/2$) by multiplying both sides of the equation by (V/n) and multiplying the numerator and denominator of the right side by 2:

$$\begin{aligned}\frac{PV}{n} &= N_A \times \frac{2m\overline{u^2}}{3 \times 2} \\ &= \tfrac{2}{3} \times N_A \times (\tfrac{1}{2}m\overline{u^2})\end{aligned}$$

Therefore we have two equations for the quantity (PV/n):

$$(1) \qquad \frac{PV}{n} = RT$$

an empirical equation in which the gas constant is a constant independent of the nature of the gas, and

$$(2) \qquad \frac{PV}{n} = \tfrac{2}{3} \times N_A \times (\tfrac{1}{2}m\overline{u^2})$$

a theoretical equation based upon the kinetic molecular theory. If these equations for PV/n are equated, we have

$$\tfrac{2}{3} \times N_A \times (\tfrac{1}{2}m\overline{u^2}) = RT \qquad \text{or} \qquad (\tfrac{1}{2}m\overline{u^2}) = \tfrac{3}{2} \times \frac{R}{N_A} \times T$$

The quotient, R/N_A, which can be viewed as the gas constant per molecule, is called the BOLTZMANN constant, k,

$$k = \frac{R}{N_A} = \frac{8.315 \text{ J K}^{-1} \text{ mol}^{-1}}{6.022 \times 10^{23} \text{ mol}^{-1}} = 1.381 \times 10^{-23} \text{ J K}^{-1}$$

The temperature is the only variable on the right side of the equation; therefore we can conclude that the average kinetic energy of molecules is proportional to the temperature of the gas. Put differently, the temperature of the gas as defined by a gas thermometer is simply a manifestation of the average kinetic energy of the gas molecules.

The *absolute temperature* is fundamental to the kinetic molecular theory. We saw in Chapter 3 that it has a corresponding central role in thermodynamics. Keep this in mind when you are confronted in science with quantitative problems that involve temperature. It is the *absolute temperature* that is involved in the calculation. Only if the problem involves taking a difference of two temperatures will use of temperatures on the Celsius scale be correct, and the difference in two temperatures on the Celsius scale is the same as the difference in kelvin units.

The value of the gas constant R is independent of the mass of the gas molecules, and at a particular temperature, there is an inverse relationship between the root-mean-square velocity, u_{rms}, of gas molecules and the square root of the mass of the molecules:

$$u_{rms} = \sqrt{\overline{u^2}} = \left(\frac{3R}{mN_A} \times T\right)^{1/2}$$

The grouping of terms in the quotient $R/(mN_A)$ of this equation can be viewed in either of two ways:

1. R is the gas constant per mole and mN_A is the mass of one mole of molecules, or
2. (R/N_A) is the gas constant per molecule (k, the Boltzmann constant) and m is the mass of one molecule.

In the calculations of root-mean-square velocity, the value of R must be expressed in SI units:

$$\begin{aligned} R &= 8.31 \text{ J K}^{-1} \text{ mol}^{-1} \\ &= 8.31 \text{ kg m}^2 \text{ s}^{-2} \text{ K}^{-1} \text{ mol}^{-1} \end{aligned}$$

EXAMPLE: What is the root-mean-square velocity of hydrogen molecules at 300 K and at 400 K?

At 300 K,

$$\begin{aligned} u_{rms} &= \left(\frac{3 \times 8.31 \text{ kg m}^2 \text{ s}^{-2} \text{ K}^{-1} \text{ mol}^{-1} \times 300 \text{ K}}{0.002016 \text{ kg mol}^{-1}}\right)^{1/2} \\ &= (3.71 \times 10^6 \text{ m}^2 \text{ s}^{-2})^{1/2} \\ &= 1.93 \times 10^3 \text{ m s}^{-1} \end{aligned}$$

This velocity can be expressed in miles per hour to allow comparison with velocities in everyday experience:

$$u_{rms} = 1.93 \times 10^3 \text{ m s}^{-1} \times 3600 \text{ s hr}^{-1} \times \frac{1 \text{ inch}}{2.54 \times 10^{-2} \text{ m}}$$

$$\times \frac{1 \text{ feet}}{12 \text{ inch}} \times \frac{1 \text{ mile}}{5280 \text{ feet}}$$

$$= 4.32 \times 10^3 \text{ miles hr}^{-1}$$

The root-mean-square velocity of hydrogen molecules at 300 K is much greater than the velocities of automobiles or airplanes, but it is not as large as the velocity that a space vehicle must have to escape from the Earth's gravitational field ($\sim$25,000 miles hr^{-1}).

Because u_{rms} is proportional to $T^{1/2}$, the value at 400 K is

$$u_{rms} \text{ (400 K)} = u_{rms} \text{ (300 K)} \times \sqrt{\frac{400 \text{ K}}{300 \text{ K}}}$$

$$= 1.93 \times 10^3 \text{ m s}^{-1} \times 1.15$$

$$= 2.23 \times 10^3 \text{ m s}^{-1}$$

Because the root-mean-square velocity of gas molecules is inversely proportional to the square root of the molecular weight, the values for other gases at 300 K can be calculated from the value for hydrogen just derived:

$$H_2 \text{ (MW} = 2.016) \qquad u_{rms} = 1.93 \times 10^3 \text{ m s}^{-1}$$

For O_2 (MW = 32.00):

$$u_{rms} = 1.93 \times 10^3 \text{ m s}^{-1} \times \sqrt{\frac{2.016}{32.00}} = 4.84 \times 10^2 \text{ m s}^{-1}$$

For I_2 (MW = 253.8):

$$u_{rms} = 1.93 \times 10^3 \text{ m s}^{-1} \times \sqrt{\frac{2.016}{253.8}} = 1.72 \times 10^2 \text{ m s}^{-1}$$

The dependence of the root-mean-square velocity upon mass, shown in these calculations, is the basis for the dependence of the rate of effusion of gases upon mass.

In 1846, GRAHAM observed that the rate of effusion of different gases was inversely proportional to the square root of the molecular weight of the gas; comparing two gases of molecular weights MW_1 and MW_2, he observed effusion rates r_1 and r_2, which were related:

$$\frac{r_1}{r_2} = \sqrt{\frac{MW_2}{MW_1}}$$

This relationship is known as *Graham's law*. Figure 4–10 shows features of the equipment needed to study effusion; an opening small enough so the

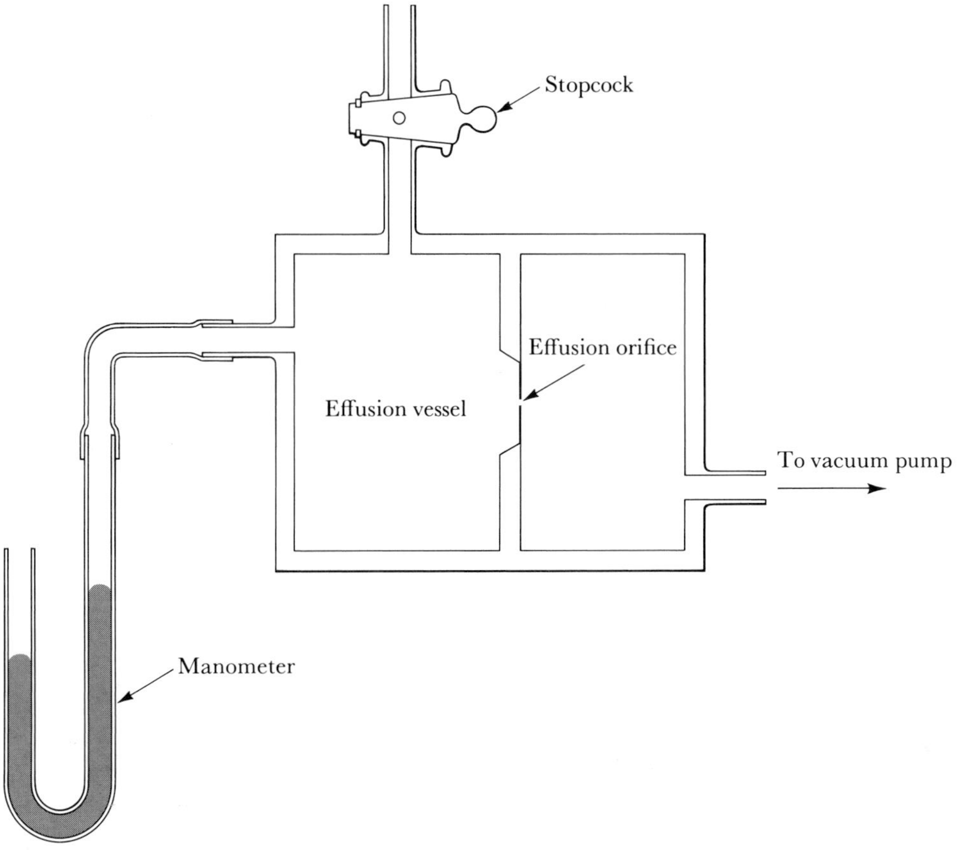

FIGURE 4–10
Apparatus for study of effusion of gases. In this apparatus, molecules leave the left vessel singly, without collisions with one another in the effusion orifice. [From W. Kauzmann, *Kinetic Theory of Gases*, W. A. Benjamin, Inc., Menlo Park, Calif. (1966) p. 62.]

molecules do not collide with one another in passing through the opening, and an evacuated space into which the gas molecules leak.

The dependence of effusion rate upon molecular weight allows effusion to be used in the partial separation of isotopic molecules. If an equimolar mixture of $^1H_2$ (MW = 2.016) and $^2H_2$ (D_2) (MW = 4.029) is present in an effusion cell, the ratio of the rates of effusion would be

$$\frac{r_{H_2}}{r_{D_2}} = \sqrt{\frac{4.029}{2.016}} = 1.414$$

and the first traces of gas in the right side of the apparatus pictured in Figure 4–10 would be enriched in H_2.

The dependence of effusion rate upon molecular weight is the basis for large-scale separation of the uranium isotopes ^{235}U and ^{238}U. Uranium-235 (0.72% abundance in natural uranium) undergoes fission with slow neutrons, and the separation of this isotope from the more abundant isotope was an important goal of the Manhattan Project in World War II. The gaseous species separated by effusion were $^{235}U^{19}F_6$ and $^{238}U^{19}F_6$

(molecular weights 349.33 and 352.34, respectively). Since fluorine exists in nature as only a single nuclide, ^{19}F, these were the only two isotopic species present in uranium hexafluoride. Although the ratio $\sqrt{MW_2/MW_1}$ is only 1.0043, the production of predominantly $^{235}UF_6$ was possible; it required effusion through many, many successive barriers. A plant at Oak Ridge, Tennessee, with acres of such barriers, successfully separated enough uranium-235 to construct the atomic bomb that destroyed Hiroshima, Japan, on August 6, 1945. This separation procedure is one of those now used to prepare uranium enriched in uranium-235 for use in nuclear power plants.

Another phenomenon involving the flow of gases that depends upon their masses is diffusion. This term refers to the net flow that makes the concentration of a substance uniform throughout a given space. The basis for the relationship between diffusion rate and molecular weight is more complex than simply the velocity of molecular motion, but the result is the same as that for effusion:

$$\text{diffusion rate} \propto (\text{MW})^{-1/2}$$

In deriving the relationship of the quantity (PV/n) to the average kinetic energy of molecules of an ideal gas $(\frac{1}{2}m\overline{u^2})$, we considered the average rate of change of momentum of the molecules upon collision with the walls of the container $(\Delta mu/\Delta t)$ to be the product of two factors:

$$\frac{\Delta mu}{\Delta t} = \left(\frac{\Delta mu}{\text{impact}}\right) \times \left(\frac{\text{impacts}}{\Delta t}\right)$$

We can explain the P–V–T–n relationships of the ideal gas law, $PV = nRT$, in a similar way, viewing the pressure of a gas as a consequence of three factors:

$$P \propto \left(\frac{\Delta mu}{\text{impact}}\right) \times \left(\frac{\text{impacts}}{\Delta t}\right) \times \left(\frac{1}{\text{area}}\right)$$

A summary of the changes of each of these factors for a two-fold increase of pressure, volume, amount of gas, and temperature is presented in Table 4–3. The average change of momentum per impact is proportional to the square root of the temperature:

$$\left(\frac{\Delta mu}{\text{impact}}\right) \propto T^{1/2}$$

The total impact frequency is proportional to the number of molecules and their average velocity, and inversely proportional to the average distance the molecules travel between successive impacts (which is proportional to the cube root of the volume):

$$\frac{\Delta\ \text{impacts}}{\Delta t} \propto n \times T^{1/2} \times V^{-1/3}$$

TABLE 4–3
Changes of the Factors[a] that Influence the Pressure of a Gas

$$P \propto \left(\frac{\Delta mu}{\text{impact}}\right) \times \left(\frac{\text{impacts}}{\Delta t}\right) \times \left(\frac{1}{\text{area}}\right)$$

Variables: Held constant	Variables: Changed	$\frac{\Delta mu}{\text{impact}}$	$\frac{\text{impacts}}{\Delta t}$	$\frac{1}{\text{area}}$
n, T	$P_2 = 2P_1$ $V_2 = \frac{1}{2}V_1$	no change	increases by factor of 1.26	increases by factor of 1.59
n, P	$T_2 = 2T_1$ $V_2 = 2V_1$	increases by factor of 1.41	increases by factor of 1.12	decreases by factor of 1.59
n, V	$T_2 = 2T_1$ $P_2 = 2P_1$	increases by factor of 1.41	increases by factor of 1.41	no change
T, P	$n_2 = 2n_1$ $V_2 = 2V_1$	no change	increases by factor of 1.59	decreases by factor of 1.59
T, V	$n_2 = 2n_1$ $P_2 = 2P_1$	no change	increases by factor of 2.00	no change
P, V	$n_2 = 2n_1$ $T_2 = \frac{1}{2}T_1$	decreases by factor of 1.41	increases by factor of 1.41	no change

[a] $2^{1/6} = 1.12$, $2^{1/3} = 1.26$, $2^{1/2} = 1.41$, $2^{2/3} = 1.59$.

The area of the walls of the vessel is proportional to the volume raised to the two thirds power:

$$\frac{1}{\text{area}} \propto \frac{1}{V^{2/3}}$$

Combining these factors, we have

$$P \propto (T^{1/2}) \times (n \times T^{1/2} \times V^{-1/3}) \times V^{-2/3}$$

or

$$P \propto \frac{nT}{V}$$

which is the proportionality of the ideal gas law.

Example: The temperature of a sample of gas in a cylinder with a movable piston is raised from 300 K to 400 K. The pressure is constant.

By what factor does the momentum transfer per impact change, and by what factor does the total impact frequency per unit area change?

Because the momentum transfer per impact is proportional to the momentum (mu), this increase in temperature increases the momentum by the factor:

$$\frac{\left(\dfrac{\Delta mu}{\text{impact}}\right)_{T_2}}{\left(\dfrac{\Delta mu}{\text{impact}}\right)_{T_1}} = \sqrt{\frac{T_2}{T_1}} = \sqrt{\frac{400\text{ K}}{300\text{ K}}} = 1.15$$

Because the pressure is constant, this 15% increase in the momentum transfer per impact is offset by a decrease in the impact frequency per unit area:

$$\frac{\left(\dfrac{\text{impact frequency}}{\text{area}}\right)_{T_2}}{\left(\dfrac{\text{impact frequency}}{\text{area}}\right)_{T_1}} = \sqrt{\frac{300\text{ K}}{400\text{ K}}} = 0.866$$

HEAT CAPACITIES OF GASES

In Chapter 3 the heat capacities of gases at constant volume and at constant pressure were used to introduce the first law of thermodynamics. The numerical calculations involved experimentally determined heat capacity values. Now we will explore the subject further, and we will relate the experimental values of C_v and C_p to the gas constant, R.

First, we will consider a monatomic gas (e.g., helium). When heat is added to a sample of such a gas in a container of constant volume, $w = 0$ and $q = \Delta U$. For helium at ordinary temperatures, changes of the internal energy are simply changes of the kinetic energy of translation of the helium atoms. In the kinetic molecular theory, the kinetic energy of one mole of gas is related to the temperature:

$$N_{\text{A}} \times (\tfrac{1}{2}\, m\overline{u^2}) = \tfrac{3}{2}RT$$

Now let us consider the heat required to raise the temperature from T_1 to T_2. Before addition of heat (State 1):

temperature	T_1
kinetic energy (for one mole)	$\frac{3}{2}RT_1$

After addition of heat (State 2):

temperature	T_2
kinetic energy (for one mole)	$\frac{3}{2}RT_2$

Therefore the heat capacity is

$$C_v = \frac{\text{heat added}}{\text{temperature increase}} = \frac{\text{gain in kinetic energy of molecules}}{\text{temperature increase}}$$

$$= \frac{\frac{3}{2}RT_2 - \frac{3}{2}RT_1}{T_2 - T_1} = \frac{\frac{3}{2}R(T_2 - T_1)}{T_2 - T_1} = \frac{3}{2}R$$

$$= \frac{3}{2} \times 8.31 \text{ J K}^{-1} \text{ mol}^{-1} = 12.47 \text{ J K}^{-1} \text{ mol}^{-1}$$

The observed values of C_v for helium and the other noble gases, all monatomic, agree very well with this calculated value.

The value of C_p is greater than the value of C_v because with an increase of temperature at constant pressure the system expands and work is done against the surroundings. The situation is illustrated in Figure 4-11, which includes the constant-volume situation just described. In raising the temperature at constant pressure, work $(-w)$ is done against the external pressure P; this work done on the surroundings is

$$-w = P\Delta V = PV_2 - PV_1$$

If we substitute for V_2 and V_1 the values given by the ideal gas law ($V = RT/P$ for 1 mole), we obtain:

$$-w = P\left(\frac{RT_2}{P}\right) - P\left(\frac{RT_1}{P}\right) = RT_2 - RT_1 = R(T_2 - T_1)$$

$\bar{E}_k = \frac{3}{2}RT_1$ | Heat added $= \frac{3}{2}R(T_2 - T_1)$ | $\bar{E}_k = \frac{3}{2}RT_2$

(a) Constant V

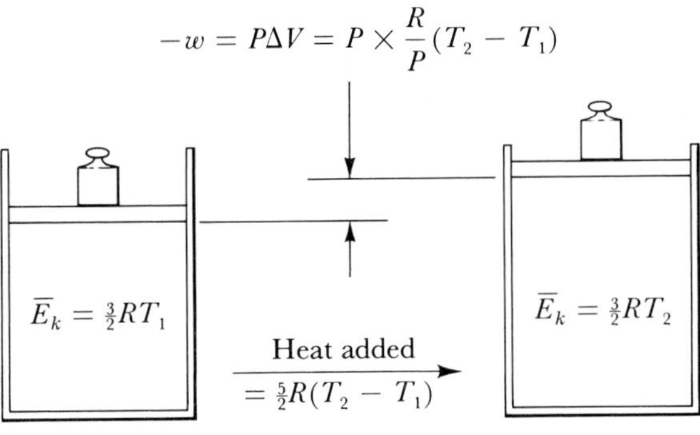

(b) Constant P

FIGURE 4–11
One mole of monatomic gas (e.g., helium) before and after addition of heat. (a) At constant volume: only the translational energy of the gas is increased. (b) At constant pressure: the translational energy of the gas is increased, and work $(-w)$ is done on the surroundings by increasing the volume against the externally applied pressure.

To obtain the heat required to raise the temperature from T_1 to T_2 at constant pressure, we must add to this the heat required to increase the average kinetic energy of the gas molecules (ΔU, already calculated):

$$\begin{aligned} q &= \Delta U - w \\ &= \tfrac{3}{2}R(T_2 - T_1) + R(T_2 - T_1) \\ &= \tfrac{5}{2}R(T_2 - T_1) \end{aligned}$$

Therefore the heat capacity at constant pressure is

$$\begin{aligned} C_p &= \frac{\text{heat added}}{\text{temperature increase}} \\ &= \frac{\tfrac{5}{2}R(T_2 - T_1)}{T_2 - T_1} = \frac{5}{2}R \\ &= \frac{5}{2} \times 8.31\ \text{J K}^{-1}\ \text{mol}^{-1} = 20.8\ \text{J K}^{-1}\ \text{mol}^{-1} \end{aligned}$$

This calculated value also agrees with the experimental value presented in Chapter 3. The monatomic nature of helium was pertinent in our consideration of C_v, but it is not pertinent in consideration of the work done in the expansion accompanying the increase of temperature at constant pressure. Therefore the molar heat capacity of any ideal gas, monatomic or polyatomic, at constant pressure includes a contribution of R for the work done against the external pressure:

$$C_p = C_v + R$$

Values of the molar heat capacities of polyatomic gases are greater than the values for monatomic gases, such as helium, as is shown in Table 4–4, which presents experimentally determined values of C_v and C_p for various gases. The greater values of C_v and C_p for polyatomic gases are explained by taking into account the rotational and vibrational motion of atoms within a polyatomic molecules. To raise the temperature of a polyatomic gas by a given amount requires additional heat, because with an increase of temperature the energies of rotation and vibration increase as well as the translational energy. The more complex a molecule, the greater its molar heat capacity. (The vibration of diatomic molecules is discussed in Chapter 10, and the rotation and vibration of molecules are discussed in Chapter 21.) The ratio C_p/C_v decreases as the value of C_v increases:

$$\frac{C_p}{C_v} = \frac{C_v + R}{C_v} = 1 + \frac{R}{C_v}$$

The molar heat capacities of substances depend upon the temperature, but many calculations may be made with adequate accuracy by assuming that the values of C_p and C_v are constant. In this case, the heat necessary to raise n moles of a substance from T_1 to T_2 is

$$q_v = n \times C_v \times (T_2 - T_1) \quad \text{and} \quad q_p = n \times C_p \times (T_2 - T_1)$$

TABLE 4–4
Molar Heat Capacities of Various Gases at 298 K

Gas	$\frac{C_v}{\text{J K}^{-1}\text{ mol}^{-1}}$	$\frac{C_p}{\text{J K}^{-1}\text{ mol}^{-1}}$	$\frac{C_p}{C_v}$
He, Ne, Ar[a]	12.47	20.80	1.67
H_2	20.54	28.86	1.41
N_2	20.71	29.03	1.40
N_2O	30.38	38.70	1.27
CO_2	28.95	37.27	1.29
C_2H_6	44.60	52.92	1.19

[a]And other monatomic gases.

EXAMPLE: What amount of heat is required to raise the temperature at constant pressure of 1.00 g of hydrogen gas and 1.00 g of carbon dioxide gas from 10.0°C to 40.0°C? What fraction of this energy is required to increase the internal energy of each gas?

The temperature range is small (30.0 K) and includes 298 K; this allows us to assume that C_p is essentially constant, with its value equal to the value tabulated for 298 K for each gas.

For hydrogen:

$$q_p = \frac{1.00\text{ g}}{2.016\text{ g mol}^{-1}} \times 28.86\text{ J K}^{-1}\text{ mol}^{-1} \times (40.0°\text{C} - 10.0°\text{C}) \times \frac{1\text{ K}}{1°\text{C}}$$

$$= 429\text{ J}$$

The fraction of this energy required to increase the internal energy is the ratio C_v/C_p. Thus

$$\frac{\Delta U}{q_p} = \frac{C_v}{C_p} = \frac{20.54\text{ J K}^{-1}\text{ mol}^{-1}}{28.86\text{ J K}^{-1}\text{ mol}^{-1}} = 0.712$$

For carbon dioxide the corresponding calculations are

$$q_p = \frac{1.00\text{ g}}{44.01\text{ g mol}^{-1}} \times 37.27\text{ J K}^{-1}\text{ mol}^{-1} \times (40.0°\text{C} - 10.0°\text{C}) \times \frac{1\text{ K}}{1°\text{C}}$$

$$= 25.4\text{ J}$$

and

$$\frac{\Delta U}{q_p} = \frac{C_v}{C_p} = \frac{28.95\text{ J K}^{-1}\text{ mol}^{-1}}{37.27\text{ J K}^{-1}\text{ mol}^{-1}} = 0.777$$

In Chapter 3, we saw that ΔH^0 and ΔU^0 for a reaction do not differ appreciably if there is neither net production nor net consumption of gas. For other reactions, ΔH^0 and ΔU^0 differ appreciably. We now can calculate this difference. Consider the occurrence of each of the reactions

(a) $H_2(g) + Cl_2(g) = 2HCl(g)$
(b) $N_2(g) + 3H_2(g) = 2NH_3(g)$
(c) $2NO_2(g) = N_2(g) + 2O_2(g)$

in a cylinder with a movable piston. The pressure and temperature are maintained constant. (The pressure is low enough that the gaseous mixtures can be considered ideal.) For each of these reactions,

$$\Delta H^0 = \Delta U^0 + \Delta(PV)$$

which for constant pressure is

$$\Delta H^0 = \Delta U^0 + P\Delta V$$

But the change of volume is given by the ideal gas law and the change of the number of gas molecules in the chemical equation, Δn,

$$\Delta V = \frac{RT\Delta n}{P}$$

or

$$\Delta H^0 = \Delta U^0 + RT\Delta n$$

Therefore at 298.2 K,

$$\begin{aligned} \Delta H^0 - \Delta U^0 &= RT\Delta n \\ &= 8.31 \text{ J K}^{-1} \text{ mol}^{-1} \times 298.2 \text{ K} \times \Delta n \\ &= 2480 \text{ J mol}^{-1} \times \Delta n \end{aligned}$$

For these reactions:

(a) $\Delta n = 2 - 2 = 0$
(b) $\Delta n = 2 - 4 = -2$
(c) $\Delta n = 3 - 2 = 1$

Then:

$$\begin{aligned} \Delta H^0 - \Delta U^0 &= 0 \text{ for Reaction (a)} \\ &= -4.96 \text{ kJ mol}^{-1} \text{ for Reaction (b)} \\ &= +2.48 \text{ kJ mol}^{-1} \text{ for Reaction (c)} \end{aligned}$$

Each of these reactions is exothermic, and the heat transferred to the surroundings in a calorimetric experiment run at constant pressure and constant temperature, $-\Delta H^0$, will be for:

Reaction (a) The same as $(-\Delta U^0)$ because there is no transfer of work
Reaction (b) Greater than $(-\Delta U^0)$ because the surroundings does work on the system during the reaction
Reaction (c) Less than $(-\Delta U^0)$ because the system does work on the surroundings during the reaction

This is summarized in Figure 4–12. You should keep in mind this relationship of the changes of enthalpy and internal energy in gaseous reactions. From time to time you will be confronted with values of ΔH^0 when you wish to know ΔU^0 and vice versa. If a reaction involves only condensed phases, the difference between ΔH^0 and ΔU^0 is small, but it is not exactly zero. (Recall that in Chapter 3 we already have considered the melting of ice, for which $\Delta V = -1.63 \text{ cm}^3 \text{ mol}^{-1}$.)

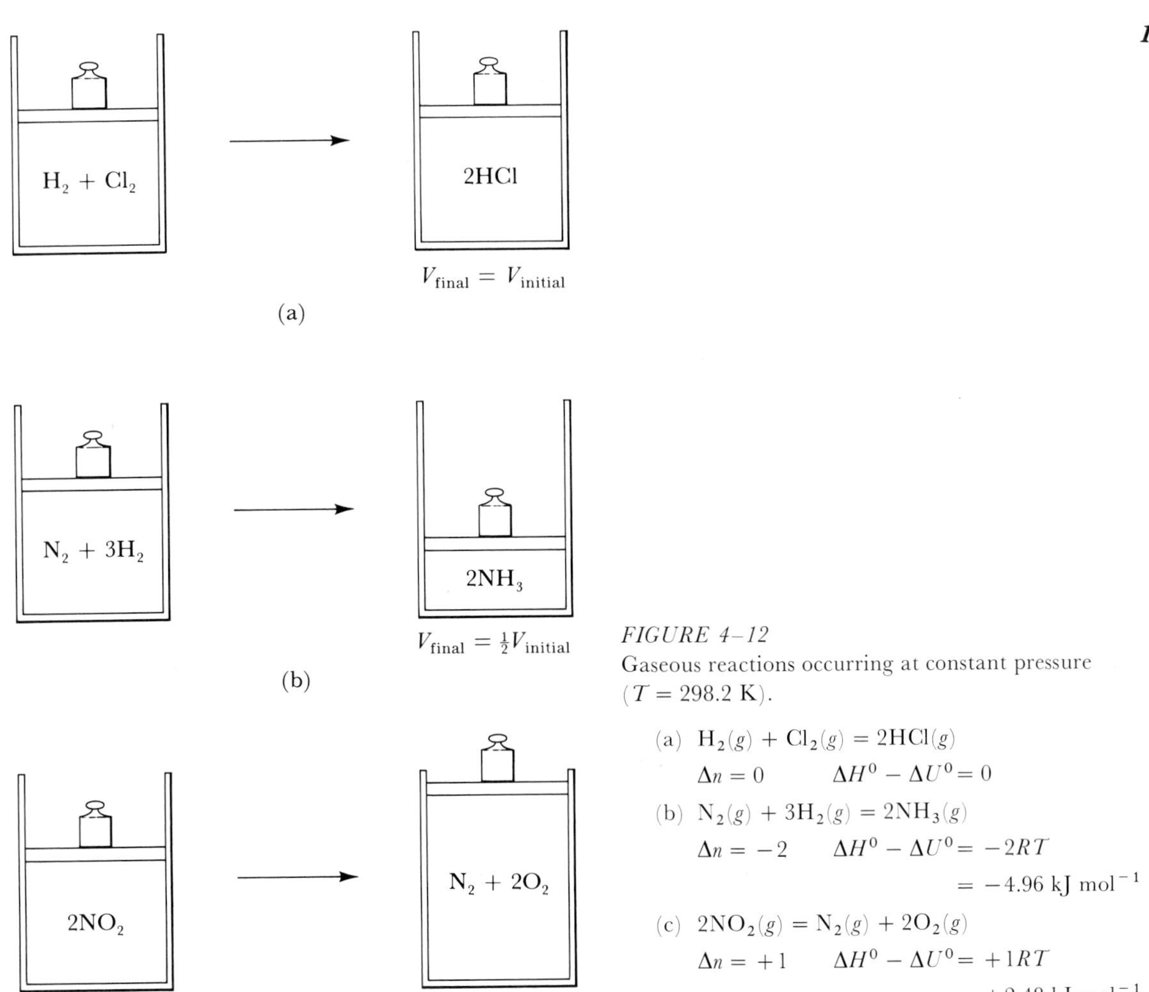

FIGURE 4–12
Gaseous reactions occurring at constant pressure ($T = 298.2$ K).

(a) $H_2(g) + Cl_2(g) = 2HCl(g)$

$\Delta n = 0 \qquad \Delta H^0 - \Delta U^0 = 0$

(b) $N_2(g) + 3H_2(g) = 2NH_3(g)$

$\Delta n = -2 \qquad \Delta H^0 - \Delta U^0 = -2RT$

$= -4.96 \text{ kJ mol}^{-1}$

(c) $2NO_2(g) = N_2(g) + 2O_2(g)$

$\Delta n = +1 \qquad \Delta H^0 - \Delta U^0 = +1RT$

$= +2.48 \text{ kJ mol}^{-1}$

DISTRIBUTION OF KINETIC ENERGIES OF GAS MOLECULES

In our discussion to this point, we have considered the root-mean-square velocity and the average kinetic energy of gas molecules. Many properties of a gas (e.g., its heat capacity) could be explained by the kinetic molecular theory if all gas molecules in a sample had the same kinetic energy. But to explain other properties (e.g., the dependence of the rate of a chemical reaction in the gas phase upon temperature), account must be taken of the range of kinetic energies of gas molecules at each temperature. The collisions of gas molecules are governed by definite laws of physics, those dealing with the conservation of energy and momentum; the exchange of energy and momentum between colliding molecules generally will cause the velocity of a particular molecule to change with each collision. The net result is a range of velocities and kinetic energies of the gas molecules in a sample at

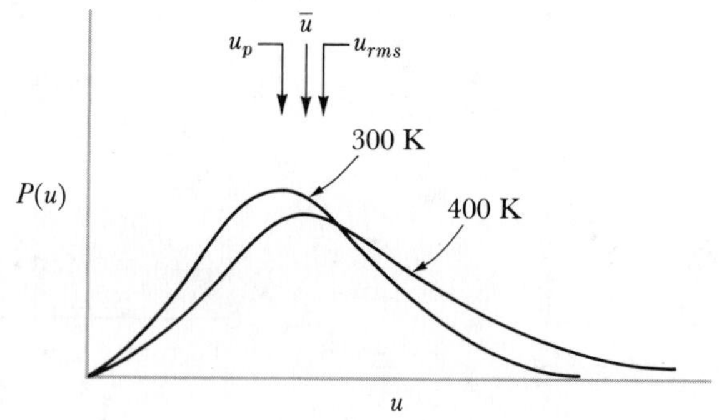

FIGURE 4–13
$P(u)$, the relative probability of gas molecules having velocities in the range u to $(u + \Delta u)$, as a function of u for 300 K and 400 K. The equation for each of these lines,

$$P(u) = \left(\frac{2}{\pi}\right)^{1/2} \left(\frac{m}{kT}\right)^{3/2} u^2 e^{-mu^2/2kT}$$

is known as the Maxwell–Boltzmann distribution law. For the distribution at 300 K, the values of u_p (the most probable velocity), $\bar{u}$ (the average velocity), and u_{rms} (the root-mean-square velocity) are indicated.

a particular temperature. This distribution of velocities can be described in terms of a quantity $P(u)$. The probability that a molecule has a velocity in a small range Δu between u and $(u + \Delta u)$ is $P(u) \times \Delta u$. Figure 4–13 shows a plot of $P(u)$ versus u for two temperatures. Several features of this distribution, called the MAXWELL–BOLTZMANN distribution, deserve our attention. There are relatively few molecules with velocities very much smaller or very much larger than the most probable velocity (u_p, the velocity corresponding to the maximum). The most probable velocity and the average velocity ($\bar{u}$) are proportional to the root-mean-square velocity:[5]

$$u_p = \sqrt{\frac{2}{3}}\, u_{rms} \qquad \bar{u} = \sqrt{\frac{8}{3\pi}}\, u_{rms}$$

The values of these porportionality constants do not depend upon temperature; therefore each of these velocities, u_p and $\bar{u}$, depends upon temperature in the same way as does the root-mean-square velocity:

$$u_p \propto T^{1/2} \qquad \bar{u} \propto T^{1/2}$$

This is a very mild dependence, as shown in a previous example where we calculated that the value of u_{rms} for hydrogen molecules increased by ~15% from 1.93×10^3 m s^{-1} to 2.23×10^3 m s^{-1} when the temperature increased from 300 K to 400 K. But the fraction of hydrogen molecules with a velocity greater than 4.0×10^3 m s^{-1} goes up by a factor of 3.2 for this increase of temperature from 300 K to 400 K. This effect of temperature upon the fraction of molecules with very high velocity is particularly relevant to the temperature dependence of the velocity of chemical reactions, a subject which we will consider in Chapter 15.

4–4 The Nonideality of Gases

If we consider a gas at 273.2 K and $P = 0.10$ atm, the behavior is close to ideal (see Figure 4–4). The average separation of gas molecules under these

[5]The numerical values of the proportionality factors result from appropriate mathematical derivations involving the equation for $P(u)$. We will not consider these derivations. If you have had calculus, you will recognize that u_p is the value of u for which $dP(u)/du = 0$.

conditions is ~72 Å, and they do not interact appreciably with one another. For each factor of ten by which the pressure is increased, the average separation of the molecules is decreased by a factor of ~2.2 ($\sqrt[3]{10} = 2.154$). With this decrease in the average separation of molecules, they interact with one another more strongly, an interaction that involves both attraction and repulsion. This interaction leads to gas nonideality, which we now will consider.

EMPIRICAL REPRESENTATION OF GAS NONIDEALITY

We already know from the data plotted in Figure 4–4 that real gases do not conform exactly to the ideal gas law, which can be written:

$$\frac{PV}{nRT} = 1$$

Deviations from ideality can be incorporated into this equation by adding to the right side of the equation terms that involve successive positive powers of the pressure:

$$\frac{PV}{nRT} = 1 + B'P + C'P^2 + D'P^3 + \cdots$$

This, like the simple linear equation $PV = \alpha + \beta P$ presented earlier, is a virial equation, and B', C', and D' are called the second, third, and fourth virial coefficients, respectively. The virial coefficients depend upon temperature, as is shown in Figure 4–14, which gives data for oxygen gas at 273.2 K, 373.2 K, and 473.2 K.

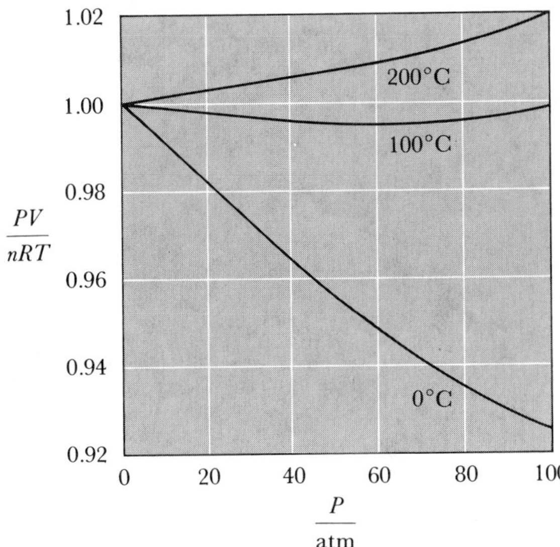

FIGURE 4–14
The dependence of PV/nRT upon P for oxygen at different temperatures. Qualitative examination of these curves shows that the Boyle temperature for O_2 is between 100°C and 200°C. (The observed value is 150°C.) [From L. P. Hammett, *Introduction to the Study of Physical Chemistry*, McGraw-Hill Book Co., N.Y. (1952).]

TABLE 4–5
The Boyle Temperature for Certain Substances

Substance	Boyle temperature K	Critical temperature K	$T_B{}^a/T_c$
H_2	120	33.2	3.6
He	35	5.3	6.6
N_2	323	126.0	2.6
O_2	423	154.3	2.7
CO_2	650	304.2	2.1

[a]The quotient of Boyle temperature, T_B, and critical temperature, T_c.

For oxygen at 273.2 K, the equation for the compressibility factor:[6]

$$\frac{PV}{nRT} = 1 - 9.93 \times 10^{-4}\ \text{atm}^{-1}\ P + 2.19 \times 10^{-6}\ \text{atm}^{-2}\ P^2$$

allows the value of $PV/(nRT)$ to be calculated within 0.1% of the observed value up to a pressure of 100 atm. The coefficients in this equation are the virial coefficients for 273.2 K:

$$B' = -9.93 \times 10^{-4}\ \text{atm}^{-1} \qquad C' = 2.19 \times 10^{-6}\ \text{atm}^{-2}$$

The value of B', which is the slope at low values of P in a plot of (PV/nRT) versus P, becomes more positive the higher the temperature, as is shown in Figure 4–14. For each gas there is a temperature at which $B' = 0$. At this temperature, called the Boyle temperature, the pressure–volume relationship for a real gas is fortuitously ideal at low pressures. Boyle temperatures (and critical temperatures) for a number of gases are given in Table 4–5.

The virial equation for representing the dependence of (PV/nRT) upon P is an empirical equation. No model for the behavior of gases was used in development of the equation. However, we can derive an equation of state for real gases by starting with the ideal gas law and then including allowance for:

1. the finite size of molecules, and
2. the attraction of molecules for one another.

The equation of state that we will derive in this way is the van der Waals' equation.

THE VAN DER WAALS EQUATION OF STATE

VAN DER WAALS modified the ideal gas law to account for the finite size of molecules and the attraction between them in a way that contrasts with the empirical power series approach of the virial equation. To arrive at the van

[6]The quotient $PV/(nRT)$ generally is called the compressibility factor for a gas.

der Waals equation of state for a real gas, we will start with the ideal gas law, solved for the pressure:

1. $P = \dfrac{nRT}{V}$

   which holds for a gas consisting of point molecules with no attraction for one another.
2. The finite size of the molecules makes the volume accessible to the molecules smaller than the total volume of the vessel. To take this into account, an excluded volume nb (with b the excluded volume per mole) is subtracted from the total volume of the container; V in the ideal gas law is replaced by $V - nb$:

$$P = \frac{nRT}{V - nb}$$

3. The attraction of the gas molecules for one another reduces the pressure that the molecules exert on the walls of the container. Because the attractive interaction primarily involves pairs of molecules, the effect is proportional to the square of the concentration, $(n/V)^2$, and the equation for pressure becomes

$$P = \frac{nRT}{V - nb} - a\left(\frac{n}{V}\right)^2$$

   in which a is the proportionality constant between the square of the gas concentration $(n/V)^2$ and the diminution of pressure.

The preceding equation can be rearranged to

$$\left[P + a\left(\frac{n}{V}\right)^2\right](V - nb) = nRT$$

the *van der Waals equation of state*. The van der Waals constants a and b for a number of gases are given in Table 4–6.

For spherical molecules, the significance of the quantity b is seen easily in Figure 4–15. The centers of two spherical molecules of radius r can approach one another no closer than a distance $2r$. The volume of the sphere that this distance defines is $\frac{4}{3}\pi(2r)^3$; this volume is not accessible to occupancy by the center of a second molecule. Such a volume is excluded for each pair of molecules in the sample; therefore in a mole the excluded volume b is given by

$$b = \frac{N_A}{2} \times \frac{4}{3}\pi(2r)^3 = 4 \times N_A \times \frac{4}{3}\pi r^3$$

That is, the excluded volume is expected to be four times the volume actually occupied by the spherical molecules. The values of b given in Table 4–6 are somewhat larger than molar volumes as determined from the densities of liquids. For instance, the molar volume of liquid xenon is

TABLE 4–6
Constants for the van der Waals Equation of State,
$\left[P + a\left(\frac{n}{V}\right)^2\right](V - nb) = nRT$

Gas	$10^6\left(\frac{a}{\text{m}^6 \text{ atm mol}^{-2}}\right)$	$\frac{b}{\text{cm}^3 \text{ mol}^{-1}}$
H_2	0.2444	26.6
N_2	1.390	39.1
O_2	1.360	31.8
Cl_2	6.493	56.2
CO_2	3.592	42.7
CH_4	2.253	42.8
He	0.0341	23.7
Ar	1.345	32.2
Xe	4.194	51.1

37 cm^3 mol^{-1}. If the spherical xenon atoms are closely packed in the liquid, this molar volume is ~ 1.4 times the actual volume of the atoms.[7] Therefore we calculate the actual volume of one mole of xenon atoms to be 26 cm^3 from the density of the liquid and 12.8 cm^3 from the value of b. The trends in the values of b given in the table are reasonable (e.g., in the noble gases $b_{He} < b_{Ar} < b_{Xe}$), but quantitative interpretation of the value of b as being four times the actual volume of the molecule is not justified.

Now we will relate the van der Waals equation of state to the virial equation of state. We can do this most directly for the virial equation

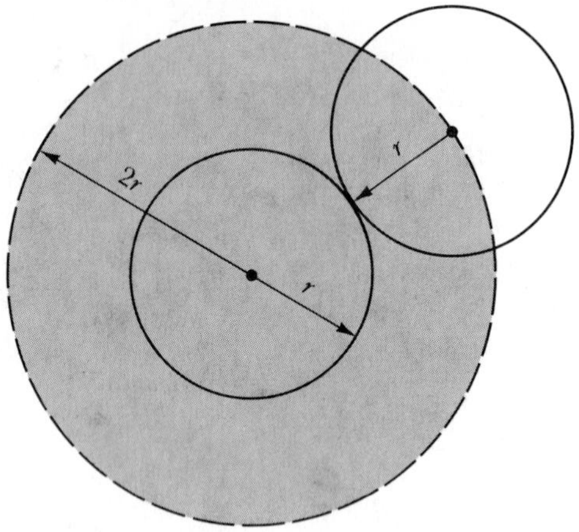

FIGURE 4–15
The excluded volume. The sphere (dotted line) of radius $2r$ around a molecule of radius r cannot be occupied by the center of a second molecule.

[7]The basis for this value (1.4) will be given in Chapter 5; the value is appropriate for the solid. The packing of atoms in the liquid is less compact.

expressed in terms of the reciprocal of the molar volume (n/V) rather than in terms of the pressure, P; this virial equation is

$$\frac{PV}{nRT} = 1 + B \times \frac{n}{V} + C \times \left(\frac{n}{V}\right)^2 + \cdots$$

The van der Waals equation, written in the form

$$P = \frac{nRT}{V - nb} - a\left(\frac{n}{V}\right)^2$$

can be converted to an equation for (PV/nRT) by multiplication of each side of this equation by V/nRT, giving

$$\frac{PV}{nRT} = \frac{V}{V - nb} - a\left(\frac{n}{V}\right)^2\left(\frac{V}{n}\right) \times \frac{1}{RT} = \frac{V}{V - nb} - \frac{a}{RT}\left(\frac{n}{V}\right)$$

The term $V/(V - nb)$ can be expressed as a power series in (n/V) simply by carrying out the indicated division $V/(V - nb)$:

$$\begin{array}{r l}
 & 1 + \dfrac{nb}{V} + \left(\dfrac{nb}{V}\right)^2 + \cdots \\
\hline
V - nb \,) & V \\
 & \underline{V - nb} \\
 & \quad + nb \\
 & \quad \underline{+ nb - \dfrac{(nb)^2}{V}} \\
 & \qquad + \dfrac{(nb)^2}{V} \\
 & \qquad \underline{+ \dfrac{(nb)^2}{V} - \dfrac{(nb)^3}{V^2}}
\end{array}$$

and so on. Because at moderate gas densities nb is very small compared to V, $nb/V < 1$, the cubic and higher power terms in this series are very small compared to unity, and the rearranged van der Waals equation becomes

$$\frac{PV}{nRT} = 1 + \frac{n}{V}\left(b - \frac{a}{RT}\right) + \left(\frac{n}{V}\right)^2 b^2 + \cdots$$

Comparison of this equation with the virial equation identifies the second virial coefficient, B, in terms of a and b, the constants of the van der Waals equation:

$$B = b - \frac{a}{RT}$$

The virial coefficients in the equation involving powers of the gas pressure are related to those in the equation involving powers of the molar volume;

for the second virial coefficient,

$$B' = \frac{B}{RT}$$

therefore, in terms of the van der Waals parameters a and b, the value of B' is

$$B' = \frac{\left(b - \dfrac{a}{RT}\right)}{RT}$$

We see that both b and a appear in the equation for the second virial coefficient. The term expressing the attraction of the molecules for one another $(-a/RT)$, which makes a negative contribution to B or B', becomes less important the higher the temperature. At low temperatures, under which conditions the molecules move more slowly, the effect of intermolecular attraction dominates the second virial coefficient. At high temperatures the value of $-a/RT$ becomes smaller, and the term related to the size of the molecules, b, becomes dominant, making the value of B' positive. The trend in B' values with temperature, shown in Figure 4–14, is consistent with this derivation.

THE POTENTIAL ENERGY OF INTERMOLECULAR INTERACTION

The model that we used in deriving the van der Waals equation included provision for the size of the molecules and for their attraction for one another. Each of these properties of gas molecules is shown in Figure 4–16, which gives the potential energy of a pair of argon atoms as a function of the separation of the centers of the atoms. If two argon atoms (monatomic

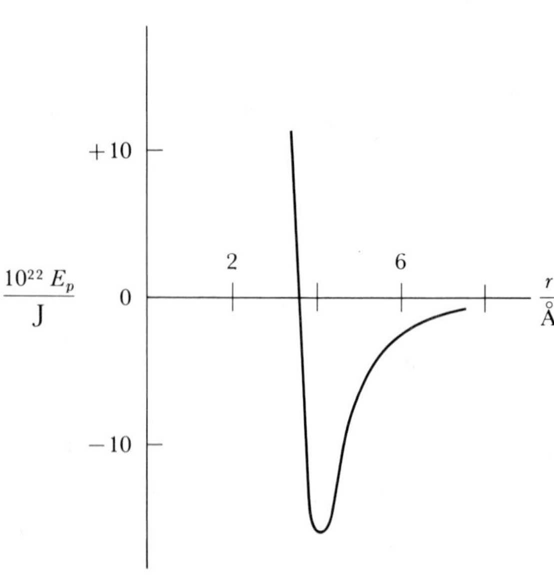

FIGURE 4–16
The potential energy, E_p, of a pair of argon atoms as a function of the separation of their centers.

molecules) are separated by large distances, >50 Å, the attraction of the atoms for one another is negligible. We will define the potential energy of interaction of the atoms in this situation as zero. As the separation of the two atoms becomes smaller their attraction for one another increases, and the potential energy of the system decreases. At still smaller distances, repulsion of the atoms becomes dominant, and the potential energy increases with decreasing separation.

To help you understand the potential-energy curve given in Figure 4–16, let us consider the head-on elastic collision of two argon atoms. Figure 4–17a shows the pair of argon atoms at three different distances of separation. These atoms had a particular amount of kinetic energy when they were separated by a large distance, at which the potential energy is defined as zero. This amount of kinetic energy is the total energy of the pair of argon atoms throughout the collision; Figure 4–17b shows the energy as a horizontal line. The total kinetic energy of the pair of argon atoms increases as the potential energy becomes more negative because the sum of the kinetic energy and potential energy is a constant. This dependence of kinetic energy upon interatomic separation is shown in Figure 4–17c. As the atoms approach, the attraction causes them to speed up, but at smaller distances from one another they slow down. Then the velocity becomes zero at the atomic separation where the potential energy becomes equal to the total energy. The system of two atoms in this configuration can become more stable (i.e., have a lower potential energy) if the atoms move away from each other. The atoms bounce off one another, and the magnitude of the velocity changes as the separation increases just as it did when the

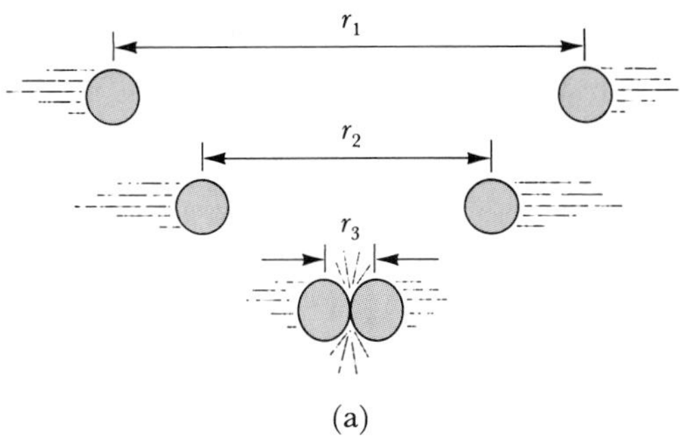

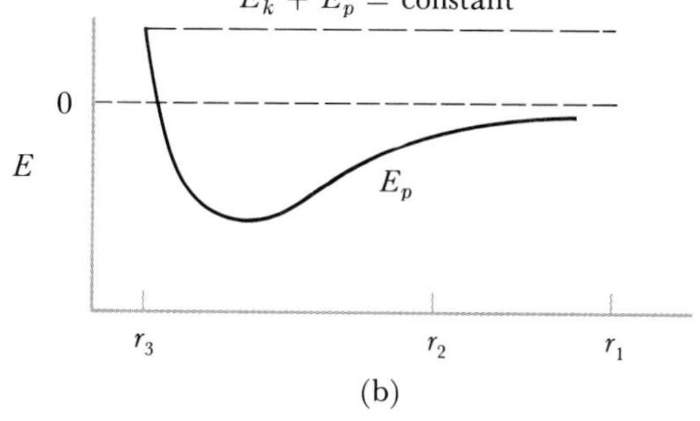

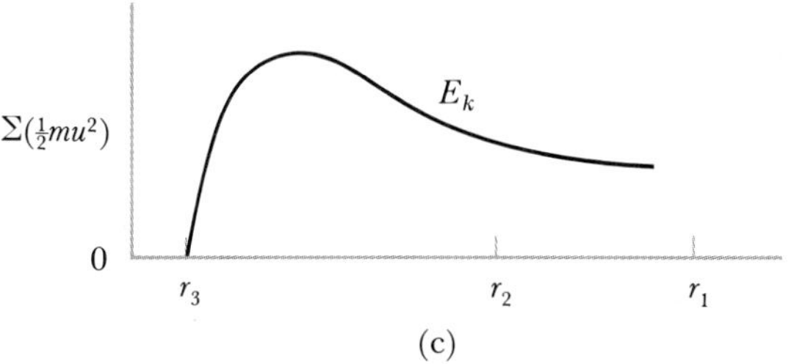

FIGURE 4–17
The head-on collision of two argon atoms. (a) The atoms at various distances from one another. (b) The total energy of the pair of atoms with the potential-energy curve. (c) The kinetic energy of the pair of atoms. This is zero at r_3, where E_p = total energy.

atoms approached one another. This discussion of the potential-energy curve brings out an important point about the size of atoms. Although the potential energy increases steeply at small distances of separation, its line is not vertical at these distances. Thus, if we considered the radius of an atom to be one half the smallest distance of separation in a head-on collision between like atoms, the value of the radius, so-defined, would depend upon the initial velocities of the approaching atoms. An atom is not an infinitely hard sphere; rather it is more like an inflated rubber ball.

An equation for the potential energy of a pair of molecules as a function of their separation must allow both for the decrease as the molecules approach one another and the steep increase as they approach one another closely. Many chemists use the LENNARD-JONES equation,

$$U(r) = \frac{A}{r^{12}} - \frac{B}{r^6}$$

in which A and B are constants and r is the distance between centers of the molecules. The positive term, A/r^{12}, corresponds to repulsion of the molecules for one another, and the negative term, $-B/r^6$, corresponds to their attraction. This is shown in Figure 4–18. With an inverse twelfth power dependence of the positive term, the potential energy rises steeply at small values of r. The significance of the two constants, A and B, in determining the quantitative characteristics of the potential-energy curve is not easy to see. However, we can write the equation for the Lennard-Jones potential in terms of two constants that do have simple significance in the graph. These constants are ϵ, the absolute magnitude of the potential energy at the minimum, and σ, the value of r at which the potential-energy curve crosses the axis, $U(r) = 0$. The Lennard-Jones potential written in terms of

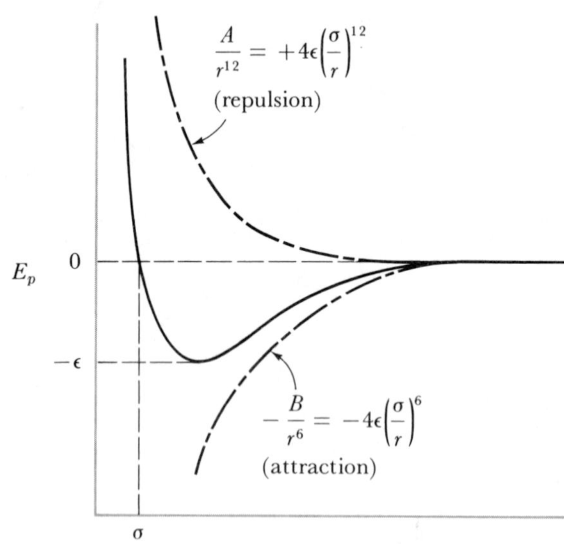

FIGURE 4–18
The potential energy of a pair of atoms as a function of the separation of the atoms:

$$E_p = U(r) = \frac{A}{r^{12}} - \frac{B}{r^6} = 4\epsilon\left[\left(\frac{\sigma}{r}\right)^{12} - \left(\frac{\sigma}{r}\right)^6\right]$$

TABLE 4–7
Lennard-Jones Parameters for Some Gases[a]

Gas	$\frac{\sigma}{Å}$	$\left(\frac{d}{Å}\right)^b$	$10^{22} \times \left(\frac{\epsilon}{J}\right)^c$	$\frac{\Delta H_{vap}}{kJ\ mol^{-1}}$
Ne	2.75	3.14	4.91	1.80
Ar	3.40	3.72	16.53	6.53
Kr	3.60	4.04	23.60	9.04
Xe	4.10	4.38	30.50	12.64
N_2	3.70	—	13.12	5.56
CH_4	3.82	—	20.45	8.20

[a]From W. Kauzmann, *Kinetic Theory of Gases*, W. A. Benjamin, Inc., Menlo Park, Calif. (1966) p. 88. The values of ϵ and σ were obtained from interpretation of the nonideality of these gases.
[b]From R. W. G. Wyckoff, *Crystal Structure*, Wiley-Interscience, New York (1963). (These are distances of separation of centers of adjacent atoms in crystals at low temperature.)
[c]The energy quantity ϵ pertains to the interaction of two atoms.

these parameters is

$$U(r) = 4\epsilon\left[\left(\frac{\sigma}{r}\right)^{12} - \left(\frac{\sigma}{r}\right)^{6}\right]$$

It is possible to relate the experimentally determined second virial coefficient as a function of temperature to the values of ϵ and σ. Table 4–7 gives values of these parameters, so determined. We see that the values of σ for the noble gases are slightly smaller than values of the atomic diameter derived from the separation of atoms in the solid. The values of ϵ can be correlated with the change of internal energy in the vaporization process [i.e., the process $Ar(l) = Ar(g)$]. A nicely linear correlation involving the enthalpy change is shown in Figure 4–19.

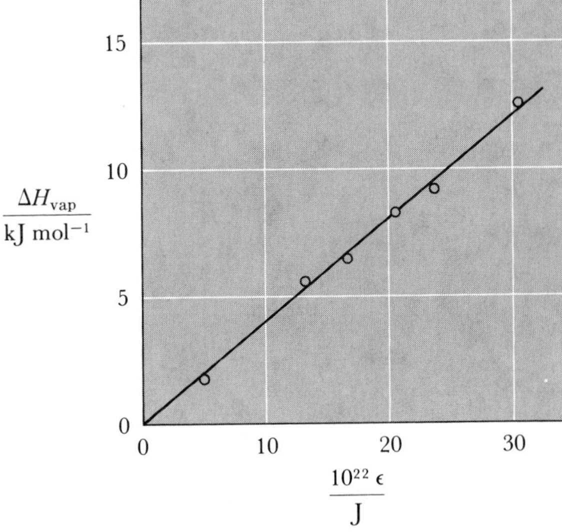

FIGURE 4–19
The change of enthalphy upon vaporization of 1 mol of certain liquids versus the Lennard-Jones parameter, ϵ, which is a measure of the attraction of pairs of molecules in the gas phase.

4–5 Thermal Effects in the Expansion of Gases

We met the first and second laws of thermodynamics in Chapter 3. Now we will increase our knowledge of thermodynamics by considering the expansion of gases under a variety of conditions. In specifying the conditions, we will want to know (a) whether the expansion of the gas is carried out isothermally, that is, at a constant temperature, or adiabatically, that is, in an insulated vessel which does not exchange heat with the surroundings; (b) whether work is done by the expanding gas; and (c) the extent to which the process approaches reversibility. We also shall consider some consequences of gas nonideality.

Let us consider a system of one mole of an ideal gas in a cylinder with an idealized weightless, frictionless piston maintained at a constant temperature, 298.2 K. The initial volume of the gas is 1.00 m^3 and the initial pressure is 2480 Pa.[8] The cylinder is in a constant temperature thermostat bath (the surroundings), which can absorb heat from the system or give heat to the system to allow its temperature to remain constant at 298.2 K. There also are weights, which can be placed on the piston to fix the external pressure at whatever value is desired. The experimental arrangement is pictured in Figure 4–20. Four experiments will be considered. In Experiment 1, expansion of the gas to 10.0 m^3 will occur with no weights on the piston; the final volume in this case is fixed by the position of a stop. Because no weights are on the piston, no work is done on the surrounding. In the other experiments, weights on the piston are removed in stages, and the final volume at each stage is determined by the external pressure at that stage; at the completion of each stage $P_{gas} = P_{ext}$, and the volume is calculated using the gas law:

$$V = \frac{nRT}{P_{gas}} = \frac{nRT}{P_{ext}}$$

The final pressure in each experiment is 248 Pa, and the final volume is 10.0 m^3. Experiments 2, 3, and 4 are:

2. Ninety percent of the weight is removed from the piston at once, making the external pressure one tenth of the initial value. The piston moves to $V = 10.0$ m^3. The weight remaining on the piston is raised in this expansion; work is done on the surroundings ($-w > 0$).
3. Weights are removed in two stages. The external pressure at the intermediate stage is 1364 Pa, and the volume of the gas during this stage increases from 1.00 m^3 to 1.82 m^3. Then in the final stage, additional weights are removed until the external pressure is 248.0 Pa and the volume increases from 1.82 m^3 to 10.0 m^3. In this process more work is done on the surroundings than in Experiment 2.

[8] In this example, SI units will be used exclusively. Recall that the SI unit of pressure is the pascal: 1 Pa = 1 N m^{-2} = 1 kg m^{-1} s^{-2}.

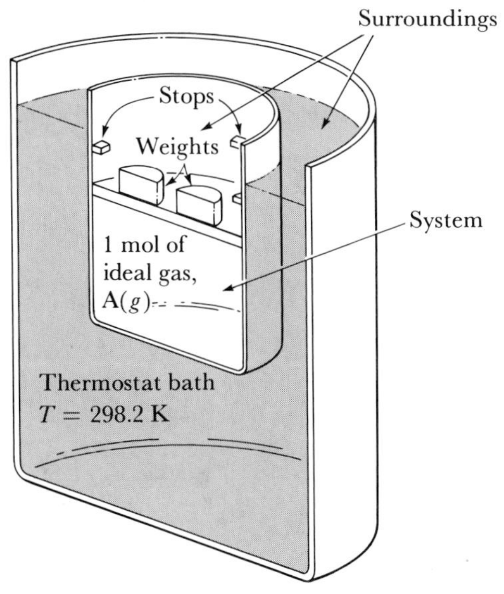

FIGURE 4–20
The idealized experimental arrangement of a cylinder with a frictionless, weightless piston in a thermostat bath at 298.2 K (the surroundings) containing one mole of an ideal gas (the system). There is a vacuum above the piston so the weights alone determine the external pressure.

4. Weights are removed in an infinite number of infintesimal stages. In this process the volume increases infinitesimally in each stage, and the pressure of the gas at any point in the process is only infinitesimally greater than the external pressure due to the weights. This idealized process is reversible; at any point its direction can be reversed by adding an infinitesimal weight to the piston.

We have learned earlier that the work done on the surroundings ($-w$) in the expansion of a gas against an external pressure is the area under the curve in a graph of P_{ext} versus V. In the present example, the external pressure during each stage of expansion is fixed by the mass of the weights on the piston during the expansion and the area of the piston. (In a real experiment carried out in this way, the rate at which heat flowed into the gas from the thermostat during expansion would determine how constant the temperature would remain. The temperature would eventually return to the temperature of the thermostat, however. Also, the rate of motion of the piston would determine whether the gas had a uniform density during expansion. However, neither of these points is really relevant in discussing the exchange of work between the system and the surroundings in the expansion because this is determined by the external pressure.)

The idealized plots of P_{ext} versus V are given for the four experiments in Figure 4–21. We see that in the free expansion (Experiment 1, Figure 4–21a), where the external pressure was zero during the expansion, the work is zero. In Experiments 2, 3, and 4, work is done on the surroundings, and it is larger, the larger the number of steps; it is a maximum for the reversible process. Calculations of the work done on the surroundings for

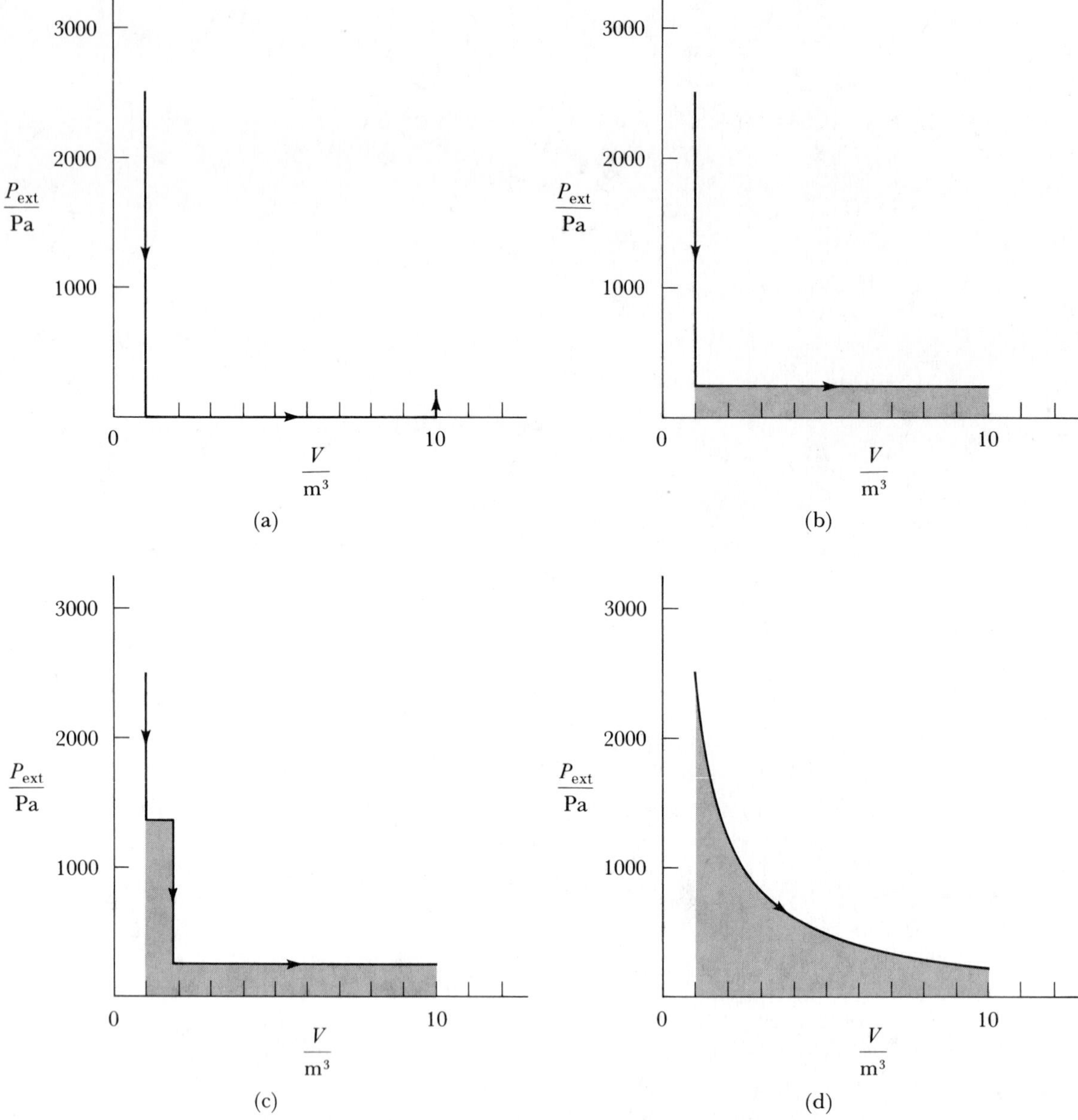

FIGURE 4–21
Pressure versus volume for different pathways for the process (at $T = 298.2$ K):

$$A(g, P = 2480 \text{ Pa}) = A(g, P = 248.0 \text{ Pa})$$

Arrows show direction of change. (a) Experiment 1: free expansion, $P_{ext} = 0$ during expansion. (b) Experiment 2: 1-step expansion, $P_{ext} = 248.0$ Pa during expansion from 1.00 m^3 to 10.0 m^3. (c) Experiment 3: 2-step expansion, $P_{ext} = 1364$ Pa during expansion from 1.00 m^3 to 1.82 m^3, and 248.0 Pa during expansion from 1.82 m^3 to 10.0 m^3. (d) Experiment 4: reversible, infinite-step expansion, $P_{ext} \cong P_{gas} = \dfrac{nRT}{V}$.

these experiments are:

Experiment 2. $P_{ext} = 248$ Pa

$$-w = P_{ext}\,\Delta V$$
$$= 248 \text{ Pa} \times (10.0 \text{ m}^3 - 1.0 \text{ m}^3)$$
$$= 2230 \text{ Pa m}^3 = 2230 \text{ kg m}^{-1}\text{ s}^{-2}\text{ m}^3$$
$$= 2230 \text{ J}$$

Experiment 3. P_{ext} (intermediate stage) = 1364 Pa

$$V \text{(intermediate stage)} = \frac{1 \text{ mol} \times 8.31 \text{ J K}^{-1}\text{ mol}^{-1} \times 298.2 \text{ K}}{1364 \text{ Pa}}$$
$$= \frac{1 \text{ mol} \times 8.31 \text{ kg m}^2\text{ s}^{-2}\text{ K}^{-1}\text{ mol}^{-1} \times 298.2 \text{ K}}{1364 \text{ kg m}^{-1}\text{ s}^{-2}}$$
$$= 1.82 \text{ m}^3$$

P_{ext} (final stage) = 248 Pa

$$-w = 1364 \text{ Pa } (1.82 \text{ m}^3 - 1.00 \text{ m}^3) + 248 \text{ Pa } (10.0 \text{ m}^3 - 1.82 \text{ m}^3)$$
$$= 3150 \text{ Pa m}^3 = 3150 \text{ kg m}^{-1}\text{ s}^{-2}\text{ m}^3$$
$$= 3150 \text{ J}$$

Experiment 4. $P_{ext} = P_{gas} = \dfrac{nRT}{V}$

The calculation of the area under this curve involves calculus, with which you may not be familiar. The result of such a calculation is $-w = 5710$ J.[9]

Because the initial and final temperatures are the same, maintained constant by the thermostat bath, the value of ΔU for the process

$$\text{A}(g, T = 298.2 \text{ K}, P = 2480 \text{ Pa}) = \text{A}(g, T = 298.2 \text{ K}, P = 248.0 \text{ Pa})$$

is zero regardless of the pathway. (Remember that internal energy, U, is a state function.) From the first law of thermodynamics, $\Delta U = q + w$, and because $\Delta U = 0$, we have $q = -w$. Thus, the work done on the surroundings in Experiments 2, 3, and 4 does not use the internal energy of the gas; rather it uses heat that flows from the thermostat. The heat that flows from the thermostat to the system, q, like w, is not a state function; it depends

[9] If you have had integral calculus, you will recognize the solution:

$$-w = \int_{1\text{ m}^3}^{10\text{ m}^3} P\,dV = \int_{1\text{ m}^3}^{10\text{ m}^3} \frac{nRT}{V}\,dV = nRT \ln \frac{10 \text{ m}^3}{1 \text{ m}^3}$$
$$= 1 \text{ mol} \times 8.31 \text{ J K}^{-1}\text{ mol} \times 298.2 \text{ K} \times 2.303 = 5710 \text{ J}$$

upon the pathway:

Experiment 1. $q = -w = 0$
Experiment 2. $q = -w = 2230\ \text{J}$
Experiment 3. $q = -w = 3150\ \text{J}$
Experiment 4. $q_{\text{rev}} = -w_{\text{rev}} = 5710\ \text{J}$

Because the enthalpy is a state function, the value of ΔH for the overall change is the same regardless of the pathway. Its value is

$$\Delta H = \Delta U + \Delta(PV)$$

But $PV = nRT$, which gives

$$\begin{aligned}\Delta H &= \Delta U + \Delta(nRT)\\ &= 0 + 0 = 0\end{aligned}$$

because $T_{\text{final}} = T_{\text{initial}}$.

We can calculate the value of the entropy change for this process by considering Experiment 4, in which the process occurs by a reversible pathway. The value of ΔS for the system undergoing a reversible change is the heat added to the system, q_{rev}, divided by the temperature at which the heat is added; *this is the definition of the change of entropy*:[10]

$$\begin{aligned}\Delta S &= \frac{q_{\text{rev}}}{T}\\ &= \frac{5710\ \text{J mol}^{-1}}{298.2\ \text{K}} = 19.1\ \text{J K}^{-1}\ \text{mol}^{-1}\end{aligned}$$

This value of the change of entropy, ΔS, calculated by considering the reversible pathway, is the value of ΔS for the system undergoing this process by any pathway because entropy is a state function.

Whether a process occurs spontaneously depends on the change of entropy of the system and the surroundings. Let us calculate the change of entropy of the surroundings for the experiments we have just discussed.

---

[10] In Footnote 9, it was shown that the work done on the surroundings in the reversible expansion of a gas from V_1 to V_2 is

$$-w = nRT \ln \frac{V_2}{V_1}$$

Because $q_{\text{rev}} = -w$, and $\Delta S = q_{\text{rev}}/T$, we have for the entropy change accompanying the isothermal expansion of a gas from V_1 to V_2:

$$\Delta S = nR \ln \frac{V_2}{V_1}$$

which is the same as the equation given in Chapter 3. Application of this equation to the present example gives

$$\Delta S = 1\ \text{mol} \times 8.31\ \text{J K}^{-1}\ \text{mol}^{-1} \times \ln \frac{10\ \text{m}^3}{1\ \text{m}^3} = 19.1\ \text{J K mol}^{-1}$$

The value of q, the heat flowing from the surroundings, is different in each case. The large thermostat bath suffers no appreciable temperature change with the flow of this heat, therefore

$$\Delta S_{\text{surroundings}} = \frac{-q}{T}$$

in which $-q$ is the heat added to the thermostat. For each of the pathways, the values of ΔS for the surroundings are:

1. $\Delta S_{\text{surroundings}} = -\dfrac{0}{298.2\ \text{K}} = 0$

2. $\Delta S_{\text{surroundings}} = -\dfrac{2230\ \text{J}}{298.2\ \text{K}} = -7.48\ \text{J K}^{-1}$

3. $\Delta S_{\text{surroundings}} = -\dfrac{3150\ \text{J}}{298.2\ \text{K}} = -10.6\ \text{J K}^{-1}$

4. $\Delta S_{\text{surroundings}} = -\dfrac{5710\ \text{J}}{298.2\ \text{K}} = -19.1\ \text{J K}^{-1}$

Values of the total entropy change for the system and the surroundings are:

$\Delta S = \Delta S_{\text{system}} + \Delta S_{\text{surroundings}}$

1. $\Delta S = 19.1\ \text{J K}^{-1} - 0 = +19.1\ \text{J K}^{-1}$
2. $\Delta S = 19.1\ \text{J K}^{-1} - 7.48\ \text{J K}^{-1} = +11.6\ \text{J K}^{-1}$
3. $\Delta S = 19.1\ \text{J K}^{-1} - 10.6\ \text{J K}^{-1} = +8.5\ \text{J K}^{-1}$
4. $\Delta S = 19.1\ \text{J K}^{-1} - 19.1\ \text{J K}^{-1} = 0$

We see that there is an increase in the entropy of the system plus the surroundings for Experiments 1, 2, and 3. For the reversible process (Experiment 4), the change of entropy of the system and the change of entropy of the surroundings exactly offset one another, making $\Delta S = 0$. You recognize Experiments 1, 2, and 3 as ones that will occur spontaneously; if weights are taken from the piston, the piston moves upward and the gas expands. For each of these cases, the total change of entropy is positive, which confirms your intuition that the expansion occurring by these pathways is spontaneous. You also recognize that one cannot bring about the process in a finite time by the reversible pathway, Experiment 4. The ideal limiting reversible case in this example, and in all others, is a limit which can be approached but not achieved. And for this limiting, reversible, unachievable pathway the change of entropy of the system plus surroundings is zero.

The change of free energy for the system, $\Delta G = \Delta H - T\,\Delta S$, can be calculated from the quantities already found; Table 4–8 summarizes results of the calculations we have made. You should notice three features in particular:

1. The irreversible, isothermal expansion of an ideal gas is a spontaneous process because the total change of entropy of the system and the surroundings is positive ($\Delta S > 0$).

TABLE 4–8
Summary of Changes of Thermodynamic Quantities for the Process:[a]

$$\text{A}(g, P = 2480\ \text{Pa}) = \text{A}(g, P = 248.0\ \text{Pa}) \qquad T = 298.2\ \text{K}$$

Changes of state functions for gas (independent of pathway):

$$\Delta U = 0 \qquad \Delta S = +19.1\ \text{J K}^{-1}$$
$$\Delta H = 0 \qquad \Delta G = -5710\ \text{J}$$

Experiment	$\frac{w}{\text{J}}$	$\frac{q}{\text{J}}$	$\frac{\Delta S_{\text{surroundings}}}{\text{J K}^{-1}}$	$\frac{\Delta S_{\text{total}}}{\text{J K}^{-1}}$
1	0	0	0	+19.1
2	−2230	+2230	−7.48	+11.6
3	−3150	+3150	−10.6	+8.5
4	−5710	+5710	−19.1	0

[a]For one mole of the ideal gas A.

2. The change of entropy for the gas can be calculated from the heat absorbed by the gas in the reversible process, divided by the temperature ($\Delta S_{\text{gas}} = q_{\text{rev}}/T$). But this value of ΔS_{gas} also is the value of ΔS_{gas} for this process by all pathways because entropy is a state function.
3. The work done on the surroundings ($-w$) is a maximum for the reversible isothermal process.

We will learn more about the application of thermodynamics to gases if we now consider the opposite process for the ideal gas A:

$$\text{A}(g, P = 248.0\ \text{Pa}) = \text{A}(g, P = 2480\ \text{Pa}) \qquad T = 298.2\ \text{K}$$

Changes of the state functions ΔU, ΔH, ΔS, and ΔG all are the negative of the values already calculated, because the identities of initial and final states simply are reversed. For one mole of gas, these changes are:

$$\Delta U = 0 \qquad \Delta S = -19.1\ \text{J K}^{-1}$$
$$\Delta H = 0 \qquad \Delta G = +5710\ \text{J}$$

As in the expansion, we can bring about this compression of the gas along different pathways. We can place weights on the movable piston to increase the external pressure from 248.0 Pa to 2480 Pa in one step, in an infinite number of infinitesimal steps, or in some intermediate number of steps. Figure 4–22 depicts only the two extremes. For the compression of gas, work is done on the gas, $w > 0$, and heat is given to the surroundings, $q < 0$. When the compression occurs in one step, the external pressure is abruptly increased from 248.0 Pa to 2480 Pa, and this increased pressure acts through a volume change $V = 10.0\ \text{m}^3$ to $V = 1.0\ \text{m}^3$. The work done

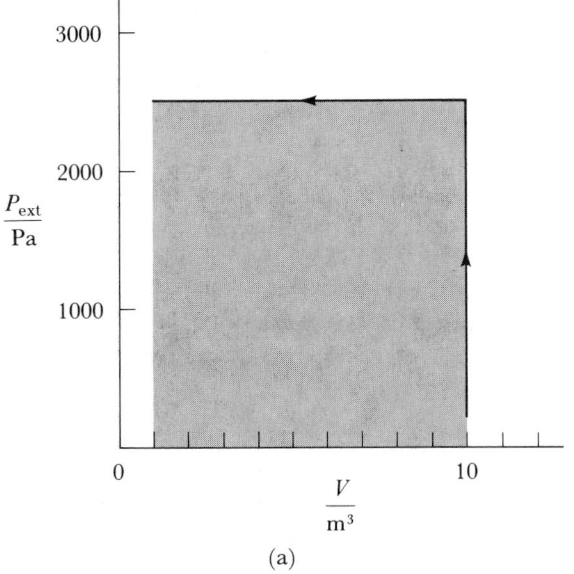

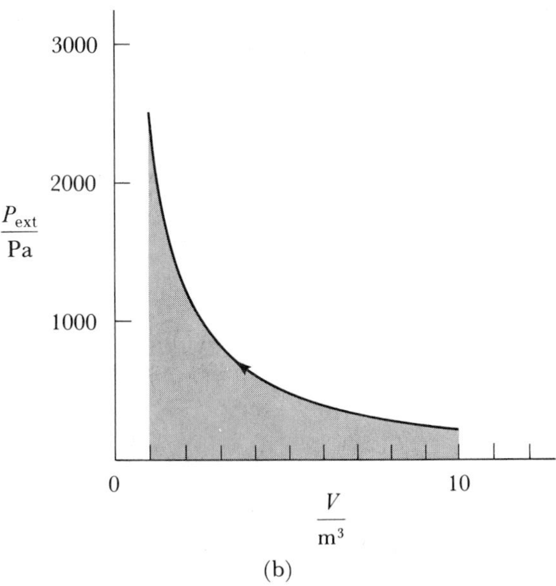

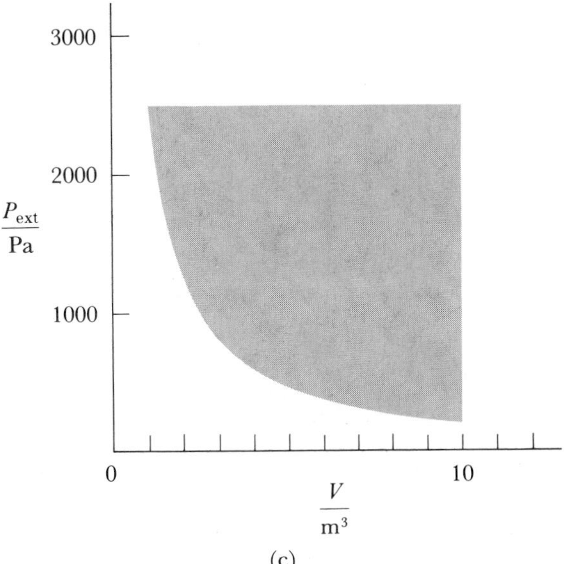

FIGURE 4–22
Pressure versus volume for different pathways for the process (at $T = 298.2$ K):

$$A(g, P = 248.0 \text{ Pa}) = A(g, P = 2480 \text{ Pa})$$

Arrows show the direction of change. (a) 1-step compression, $P_{ext} = 2480$ Pa during contraction from 10.0 m^3 to 1.00 m^3. (b) reversible, infinite-step compression, $P_{ext} \cong P_{gas} = \dfrac{nRT}{V}$. (c) the work wasted in using pathway (a).

on the gas, the shaded area in the graph, is

$$\begin{aligned} w &= 2480 \text{ Pa} \times (10.0 \text{ m}^3 - 1.0 \text{ m}^3) \\ &= 2.23 \times 10^4 \text{ Pa m}^3 = 2.23 \text{ kg m}^{-1} \text{ s}^{-2} \times \text{m}^3 \\ &= 2.23 \times 10^4 \text{ J} \end{aligned}$$

Therefore, for this experiment,

$$q = -2.23 \times 10^4 \text{ J}$$

and this amount of heat flows into the surroundings—the thermostat. The entropy change for the surroundings is calculated in the same way as before; we divide the heat added to the surroundings, $-q$, by the temperature:

$$\Delta S_{\text{surroundings}} = \frac{-q}{T} = \frac{2.23 \times 10^4 \text{ J}}{298.2 \text{ K}} = 74.8 \text{ J K}^{-1}$$

and the total change of entropy of the system plus surroundings is positive:

$$\begin{aligned} \Delta S &= \Delta S_{\text{system}} + \Delta S_{\text{surroundings}} \\ &= -19.1 \text{ J K}^{-1} + 74.8 \text{ J K}^{-1} \\ &= +55.7 \text{ J K}^{-1} \end{aligned}$$

The process, which occurs spontaneously when the extra weights are placed upon the piston, is accompanied by an increase of the total entropy of the system plus the surroundings. For the reversible contraction of the gas, the work done on the gas is exactly the same as the work done on the surroundings in the reversible expansion:

$$w = 5710 \text{ J} \qquad \text{and} \qquad q = -5710 \text{ J}$$

The change of entropy of the surroundings, obtained by dividing the heat added to the surroundings by the temperature, is

$$\begin{aligned} \Delta S_{\text{surroundings}} &= \frac{5710 \text{ J}}{298.2 \text{ K}} \\ &= +19.1 \text{ J K}^{-1} \end{aligned}$$

and the total entropy change for the reversible contraction of this gas is zero:

$$\begin{aligned} \Delta S &= \Delta S_{\text{system}} + \Delta S_{\text{surroundings}} \\ &= -19.1 \text{ J K}^{-1} + 19.1 \text{ J K}^{-1} = 0 \end{aligned}$$

Comparison of the irreversible compression and reversible compression of the gas illustrates some important points. It took more work to compress the gas irreversibly (22.3 kJ) than to compress it reversibly (5.71 kJ); the difference of these quantities (16.6 kJ) can be viewed as *wasted work.* All irreversible processes waste work, and in such processes the total entropy increases.

We see that the reversible pathways for expansion and contraction are the exact reverse of one another. The work done on the surroundings in the reversible expansion is exactly equal to the work done on the system in the reversible contraction. In all irreversible cycles in which the gas is brought back to its initial state by compression after the expansion, more work is done on the gas in the compression than is done on the surroundings in the expansion. This net expenditure of work on the system in the irreversible cycle (wasted work) adds heat to the surroundings and thereby causes the total entropy to increase.

There is an additional feature to note about the contraction of a gas. If we ask whether the process could occur without action of weights on the piston, that is, whether the molecules simply could move to occupy only one tenth of the volume of the vessel, we are confronted with a situation where $\Delta S_{\text{gas}} < 0$, $\Delta S_{\text{surroundings}} = 0$, and the total change of entropy is negative; the process is not spontaneous.

In the plots of P versus V in Figures 4–21d and 4–22b, the reversible pathway for the isothermal change of volume is a hyperbolic plot because $PV = nRT$.[11] The corresponding plot for a reversible adiabatic expansion is not hyperbolic because the temperature changes during the expansion. Application of the first law of thermodynamics,

$$\Delta U = q + w$$

to the adiabatic expansion of a gas gives

$$\Delta U = +w$$

because no heat is transferred to or from the gas, that is, $q = 0$. If work is done on the surroundings ($-w > 0$), as in an expansion, the change in the gas's internal energy is negative ($\Delta U < 0$). The average kinetic energy of the molecules decreases in the expansion. The energy required for the work done against the surroundings subtracts from the internal energy of the molecules. The amount by which the temperature decreases and the volume increases when the pressure is reduced depends upon the molar heat capacity of the gas. It seems reasonable that the temperature decrease in the adiabatic expansion of a gas would be smaller the larger the heat capacity of the gas. The larger the heat capacity of the gas, the greater the amount of energy it will give up per kelvin unit decrease of temperature. This is the observed behavior.

Conditions approaching adiabatic conditions are realized momentarily when a compressed gas escapes rapidly from a cylinder into the atmosphere. The escaping gas quickly pushes back the atmosphere, and heat is transferred from the surroundings slowly. To the extent that $q = 0$ during the process, the work is done on the surroundings at the expense of the internal energy of the gas, which is cooled as a result. This phenomenon is used to liquefy gases. In a multistage process, the cooling that results from the expansion lowers the temperature sufficiently to cause part of the gaseous substance to liquefy.

In the thermal effects accompanying expansion of a real gas, there are factors not present in the expansion of an ideal gas. For real gases the internal energy depends upon the pressure as well as the temperature. The adiabatic expansion of a real gas into an evacuated space ($w = 0$) is accompanied by a change of temperature. If in the initial state the value of (PV/nRT) is less than one, the gas nonideality is due mainly to attraction

[11] You may recall from a course in mathematics that the equation for a hyperbola is $xy =$ constant.

of the molecules for one another. When the gas expands to occupy a larger volume, the molecules are farther apart; in the limiting case they are far enough apart for the gas to approach ideality. The energy required to overcome the molecular attraction comes from the internal energy of the gas. The gas molecules slow down in moving away from one another, and the gas cools with the expansion.

However, if the value of (PV/nRT) is greater than one in the initial state, the nonideality is due mainly to repulsion of the molecules for one another. With expansion into an evacuated space, the mutual repulsion of the gas molecules makes them speed up as they move farther away from one another, and the temperature of the gas increases.

Biographical Notes

ARCHIMEDES (~287 B.C.–212 B.C.) was a Greek mathematician and physicist. He concerned himself with many problems of geometry, and he invented the science of hydrostatics.

DANIEL BERNOULLI (1700–1782) came from a family that included many prominent mathematicians and scientists. He was Professor of Mathematics at St. Petersburg and Basel. It was in his most famous book, *Hydrodynamica* (published in 1738), that his contribution to the kinetic theory of gases appeared. This contribution was largely overlooked for 100 years.

LUDWIG E. BOLTZMANN (1844–1906), an Austrian physicist, did much of the fundamental work on the kinetic theory of gases. He was associated with universities in Vienna, Graz, Munich, and Leipzig. His name will come up again in Chapter 5 in connection with the relationship of entropy and randomness.

ROBERT BOYLE (1627–1691) was a leader of British science and a charter member of the Royal Society. He came from a prominent family; he was the seventh son of the first Earl of Cork, who was knighted in 1603 and who later purchased the estates of Sir Walter Raleigh. Robert Boyle's scientific work was carried out in Oxford (until 1668) and later in London.

JACQUES CHARLES (1746–1823), a French scientist, taught physics at the Sorbonne. He was the first to use hydrogen gas for inflating balloons (1783), and he made many balloon ascents.

THOMAS GRAHAM (1805–1869), a Scottish chemist who was educated at Glasgow, taught at Edinburgh and London before becoming master of the mint. In addition to working on effusion of gases, he studied the use of dialysis to separate molecules of varying size and the chemistry of phosphoric acid.

JOHN E. LENNARD-JONES (1894–1954), British physicist and chemist, made important contributions to our knowledge of the electronic structure of molecules, the structure of liquids, and intermolecular forces. He was associated with Cambridge University from 1932 to 1953.

JAMES CLARK MAXWELL (1831–1879), a Scottish physicist, is considered by many to be the greatest theoretical physicist of the last century. After teaching in Aberdeen and London, he held the newly established chair in theoretical physics at Cambridge, where he directed the founding of the Cavendish laboratory. His contributions were enormous, the most important of which were in the areas of electricity and magnetism.

BLAISE PASCAL (1623–1662), a great French mathematician and physicist, invented a calculating machine, helped develop the modern theory of probability, made many observations on hydrostatics, and is generally considered to be the founder of hydrodynamics.

EVANGELISTA TORRICELLI (1608–1647), an Italian mathematician, invented the mercury barometer to measure atmospheric pressure. He also invented a primitive microscope, and he did pioneering work in fluid mechanics.

JOHANNES D. VAN DER WAALS (1837–1923), a Dutch physicist, did pioneering work in several areas of physical chemistry. He was Professor of Physics at the University of Amsterdam from 1877 to 1907. For his work on the equation of state for gases, he received the Nobel Prize in Physics in 1910.

Problems and Questions

4–1 A sample of gas occupies a volume of 4.00 L at 30.0°C with the pressure equal to 500 torr. What is the pressure if this same sample of gas occupies a volume of 3.00 L at 75.0°C?

4–2 What pressure will 5.00 g of oxygen gas (O_2) exert in a cylindrical vessel 1.00 m in length and 10.0 cm in diameter at 400 K?

4–3 A certain sample of gas occupies a volume of 5.00 m^3 at 500 K with the pressure 4.00 Pa. At what temperature will this sample occupy 35.0 m^3 if the pressure is 0.200 Pa?

4–4 A gaseous compound has a density of 1.038 g dm^{-3} at 25.0°C and P = 714.4 torr. What is the approximate molecular weight of the compound?

4–5 The normal boiling point of benzene (C_6H_6) is 80.1°C. What is the density of benzene vapor (in g L^{-1}) at this set of conditions (1.00 atm, 80.1°C)? (Assume benzene vapor is an ideal gas.)

4–6 A gaseous compound has the empirical formula CHCl. At 100°C, its density at 750 torr is 3.12×10^{-3} g cm^{-3}. What is the molecular formula of this compound?

4–7 The volume of a sample of gas held at constant pressure increases by a factor of 1.170 when the temperature is increased by 25.0 K. What is the initial temperature of this gas?

4–8 What percentage error is there in the approximate conversion factor of energy units 100 J = 1 L atm?

4–9 A hot-air balloon with a volume of 50,000 ft^3 lifts a load of 0.5 ton in air at 740 torr at 24°C. What is the temperature of the air in the balloon?

4–10 A mixture of helium (He) and nitrogen (N_2) has a density of 0.403 g L^{-1} at 25°C and 740 torr. What fraction of the gas molecules is helium? What fraction of the weight of the gas is helium?

4–11 Early evidence for the existence of noble gases was the discrepancy between the apparent molecular weight of nitrogen obtained from the atmosphere (by removal of the oxygen, carbon dioxide, and water vapor) and that obtained from the decomposition of pure ammonium nitrite ($NH_4NO_2 = N_2 + 2H_2O$). The values so obtained were:

28.01	(pure N_2)
28.15	(N_2 contaminated with Ar)

If the discrepancy is due solely to the presence of argon [AW(Ar) = 39.95], what is the ratio n_{Ar}/n_{N_2} in the atmosphere?

4–12 A sample of oxygen is collected over water. With the total pressure of 735 torr, the volume is 550 cm^3 and the temperature is 30°C. How many moles of oxygen are present? (The pressure of the water vapor is 31.82 torr; this value is given in Table 6–1.)

4–13 A weatherman on television announces that the temperature during the next day is expected to double, to go from 25°F to 50°F. Criticize the wording of this statement.

4–14 A mixture of CaO(*s*) and BaO(*s*) weighing 5.14 g is added to a 1.50-L vessel containing $CO_2(g)$ at a pressure of 750 torr at a temperature of 30.0°C. After the reaction $MO(s) + CO_2(g) = MCO_3(s)$ occurs to completion, the pressure of $CO_2(g)$ is 230 torr. What weight of CaO was in the sample of oxides?

4–15 The values of the product PV for a sample of CO_2 at two pressures at 0°C are:

$\frac{P}{\text{atm}}$	0.250	0.500
$\frac{PV}{\text{L atm}}$	8.9510	8.9359

What is the value of PV at $P = 1$ atm, and how many moles of CO_2 are there in the sample?

4–16 A gaseous compound at 273.2 K has a density as a function of pressure:

$\frac{P}{\text{atm}}$	0.2500	0.5000	0.7500	1.0000
$\frac{d}{\text{g L}^{-1}}$	0.17893	0.35808	0.53745	0.71707

What is the accurate value of the molecular weight for this compound? Is 273.2 K above, or below, the Boyle temperature for this compound? What is the value of the second virial coefficient B' for this gas at 273.2 K?

4–17 For nitrogen at 0°C, the values of PV/n at 1 atm, 10 atm, and 50 atm are 22.4043 L atm mol^{-1}, 22.3386 L atm mol^{-1}, and 22.0829 L atm mol^{-1}. Do these data conform to the equation $PV/n = \alpha + \beta P$? From the data at $P = 1$ atm and $P = 10$ atm, calculate the value of β. Use this value of β to calculate the value of PV/n at $P = 50$ atm and compare it with the experimental value.

4–18 What is the root-mean-square velocity of molecules of each of the following gases at 273.15 K: $^1H^2H$, $^2H_2$, $^{235}UF_6$, $^{238}UF_6$, CO_2 (predominant isotopic species)?

4–19 The average velocity of gaseous N_2 molecules at 350 K is 5.14×10^4 cm s^{-1}. What is the average velocity of N_2 molecules at 500 K? What is the average velocity of H_2 molecules at 350 K?

4–20 An aquanaut is living in a chamber on the bottom of the ocean. The atmosphere is a mixture of helium and oxygen, and the temperature is 70°C. Despite the fact that the ambient temperature is normal, the aquanaut feels cold. Explain.

4–21 Given the virial equation for oxygen at 273.15 K

$$\frac{PV}{nRT} = 1 - 9.93 \times 10^{-4}\ \text{atm}^{-1} \times P + 2.19 \times 10^{-6}\ \text{atm}^{-1} \times P^2$$

calculate the value of the compressibility factor (PV/nRT) at the pressures 1 atm, 5 atm, 10 atm, and 50 atm.

4–22 A sample of N_2O gas weighing 10.0 g is heated at constant pressure to raise the temperature from 25.5°C to 61.6°C. The value of C_p is 38.7 J K^{-1} mol^{-1}. How much heat is required, and what is the increase in internal energy of the sample?

4–23 Consider two vessels having the same volume at the same temperature. One contains helium (He) and the other contains oxygen (O_2); the pressure is the same in each vessel. Compare the quantities (Δmu/impact) and (impacts/Δt) for the two gases.

4–24 Consider a gas in a vessel of constant volume. The temperature is raised from 300 K to 350 K. By what factor do each of quantites (Δmu/impact), (impacts/Δt), and (1/area) change?

4–25 In a 90 cm^3 nickel vessel at 25°C fluorine gas (F_2) at 9.60 atm was mixed with xenon (1.06 atm). The vessel was sealed and heated at 400°C for one hour and then cooled to a temperature at which xenon and xenon fluoride are nonvolatile solids. The fluorine then was transferred to another 90 cm^3 vessel where its pressure at 25°C was found to be 7.48 atm. What is the formula of the xenon fluoride formed? (Assume that all of the xenon reacted.)

4–26 The equation $PV/n = 0.08206$ L atm K^{-1} $mol^{-1} \times T$ is used to calculate the volume of a sample of carbon dioxide gas at 273.2 K at $P = 0.500$ atm. What error arises in this calculation because of nonideality of carbon dioxide? (See Figure 4–4.)

4–27 Consider a mole of CO_2 at 310 K, ~6 K above the critical temperature. What is the pressure if the volume is 1 L (10^{-3} m^3)? Calculate this quantity using (a) the ideal gas law, and (b) the van der Waals equation of state.

4–28 (*This problem is for students who have had differential calculus.*) The Lennard-Jones equation for the potential energy of interaction of two gaseous molecules is

$$U(r) = \frac{A}{r^{12}} - \frac{B}{r^6}$$

The quantities ϵ and σ, defined in Figure 4–18, are related to A and B. Derive these relationships by recognizing the following: (a) $U(r) = 0$ for $r = \sigma$, and (b) $dU(r)/dr = 0$ for $U(r) = -\epsilon$.

4–29 Examine the equation for the Maxwell–Boltzmann distribution given in the caption to Figure 4–13. Identify the factor responsible for the increase of $P(u)$ with increasing u at low values of u, and the factor responsible for the decrease in $P(u)$ with increasing u at high values of u.

4–30 (*This problem is for students who have had differential calculus.*) The most probable velocity of a gaseous molecule is the velocity for which $dP(u)/du = 0$. (See Figure 4–13.) Derive an equation for the most probable velocity (u_p) as a function of m and T, starting with the equation given in the caption to this figure.

√4–31 The molar heat capacities for carbon dioxide at 298.2 K are:

$$C_v = 28.95 \text{ J K}^{-1} \text{ mol}^{-1}$$
$$C_p = 37.27 \text{ J K}^{-1} \text{ mol}^{-1}$$

The molar entropy of carbon dioxide gas at 298.2 K and $P = 1$ atm is:

$$S^0 = 213.64 \text{ J K}^{-1} \text{ mol}^{-1}$$

Calculate the heat required to raise the temperature of one mole of carbon dioxide gas to 350 K for each situation: V = constant, and P = constant. Each of these processes is reversible, and the entropy change can be calculated:

$$\Delta S = \int_{T_1}^{T_2} \frac{C dT}{T} = C \ln \frac{T_2}{T_1}$$

Using this equation, calculate the entropy of carbon dioxide gas for these two situations:

$$CO_2(g, 350 \text{ K}, P = 1.000 \text{ atm}, V = 28.72 \text{ L mol}^{-1})$$
$$CO_2(g, 350 \text{ K}, P = 1.174 \text{ atm}, V = 24.47 \text{ L mol}^{-1})$$

Is the difference of S for these two states consistent with the relationship $\Delta S = R \ln (V_2/V_1)$?

4–32 Consider the irreversible cycle involving an ideal gas at 298.2 K:
(1) $A(g, P = 2480 \text{ Pa}) = A(g, P = 248 \text{ Pa})$
(2) $A(g, P = 248 \text{ Pa}) = A(g, P = 2480 \text{ Pa})$
The pathway for (1) is Step (b), Figure 4–21, and that for (2) is Step (a), Figure 4–22a. What is the total wasted work done on the gas in this cycle? What is the total gain in entropy of the surroundings in this cycle?

4–33 The concept of *wasted work* was developed in connection with the irreversible compression of a gas. Discuss this concept in connection with the irreversible expansion of a gas.

√4–34 An ideal gas is compressed adiabatically. Does the temperature increase, or decrease? How does the process of making the volume decrease cause the molecular velocities to change in the way demanded by your answer to the first part of this question?

√4–35 Consider the process discussed in the chapter:

$$A(g, P = 2480 \text{ Pa}) = A(g, P = 248 \text{ Pa})$$

at 298.2 K. If the pressure is reduced in the stages $P_i = 2480$ Pa, $P_1 = 1000$ Pa, $P_2 = 500$ Pa, and $P_f = 248$ Pa, what are the values of w and q? What is the change of entropy for the system and surroundings if the process is carried out in this way?

4–36 One mole of an ideal gas is contained in a cylinder with a movable piston. The temperature is constant at 350 K. Weights are removed abruptly from the piston to give in turn the sequence of three pressures:
(1) $P_1 = 5.00$ atm (initial state)
(2) $P_2 = 2.24$ atm
(3) $P_3 = 1.00$ atm (final state)
What is the volume of the gas at each step after the system is at 350 K?

What is the total work done on the surroundings in going from the initial to the final conditions? Give your answer in joules. What would be the total work done on the surroundings if the process were carried out reversibly?

4–37 Consider the isothermal expansion of one mole of ideal gas at 300 K. The volume increases from 30 L to 40 L. Calculate the values of q, w, ΔU, ΔH, ΔS, and ΔG for two situations: (a) a free expansion, and (b) a reversible expansion.

5

The Solid State

5–1 Introduction

Most substances and materials of everyday life are solids. There are naturally occurring materials: minerals, which give the Earth its structure, cellulose, which gives plants their structure, and bone, which gives animals their structure. There are solid articles fabricated from metals, from wood, and from synthetic plastics. Although most biological molecules occur naturally in liquid solution, much of our knowledge of these substances has come from structural studies of solids isolated from the solutions.

In the gaseous state, molecules are separated by relatively large distances. In the limiting case of very low pressure (therefore of low concentration of molecules), the molecules of a gas exert essentially no influence on one another. The molecules of a gas are randomly arranged relative to one another, and the fraction of the total volume occupied by molecules is very small. For this reason the exact sizes and shapes of molecules do not influence the properties of a dilute gas. In the solid state, the situation is very different. The atomic or molecular units that compose the solid are separated by relatively small distances; nearest-neighbor atoms or molecules are in contact with one another. The sizes and shapes of the atomic or molecular units do play an important role in determining the way in which these units are arranged in a crystalline solid.

A regular arrangement of atomic or molecular units in close proximity gives the system its lowest potential energy, and a tendency toward this energy is responsible for the crystalline nature of solids. Figure 5–1 shows the regular arrangement of atoms or molecules in the crystal lattices of several substances. The attraction within crystal lattices may result from the formation of chemical bonds, as in diamond (a solid form of carbon), the attraction of oppositely charged particles for one another in an ionic solid, as in common salt, NaCl (containing Na^+ and Cl^-), the attraction of neutral molecules for one another due to hydrogen bonding, as in ice, and the weaker attraction of all types of molecular units for one another due to van der Waals forces, as in solid noble gases and solid methane. In this chapter we shall consider the experimental methods by which atomic arrangements in solids are determined. Each substance adopts its particular atomic arrangement because of the electronic structures of its constituent atoms; we shall explain these arrangements only when we have established the appropriate background.

The regular arrangement of atoms or molecules is responsible for the visible characteristics of crystalline solids. Crystals have regular symmetrical

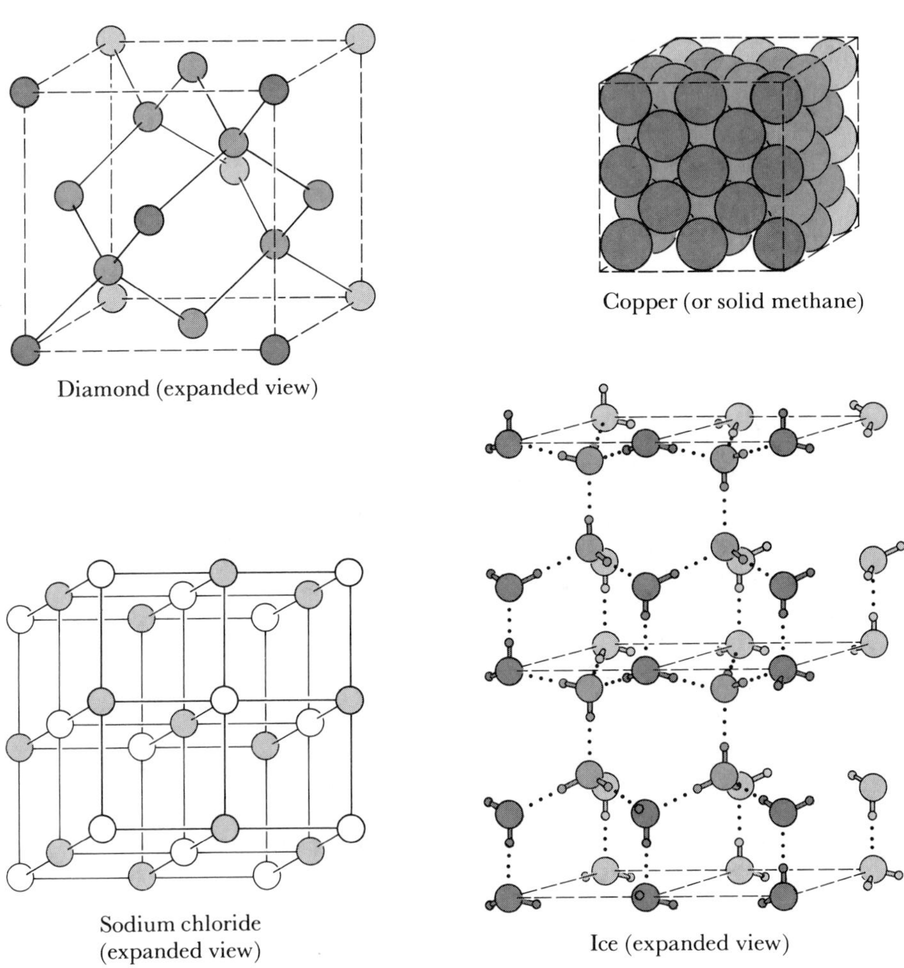

FIGURE 5–1
The arrangement of atoms and molecules in some crystalline solids. (Molecules of methane, CH_4, are represented as spheres.)

geometrical forms (called habits) with planar faces, and the angles between these faces, defining the shape of the crystal, have characteristic values. Figure 5–2 presents photographs of some crystals showing their regular geometrical forms. If cleaved, a crystalline material has the property of splitting along certain planes; these planes are parallel to the plane faces of the original crystal. (If wear has obscured the faces of the crystal, cleavage will reexpose the habits that the material exhibits.)

There are noncrystalline materials that have some characteristics of solids, namely a particular volume and shape independent of the volume and shape of the container. These materials are called *amorphous*; common glass is an example. Unlike a crystalline solid, which melts at a particular temperature, an amorphous solid softens over a range of temperature as it is heated. Cleavage of an amorphous solid does not give fragments with planar faces as cleavage of a crystalline solid does. In amorphous solids the arrangement of atoms and molecules is not as regular as in a crystalline material. Some substances, such as silica, SiO_2, can exist either as a crystalline solid (e.g., quartz, a naturally occurring mineral) or as an amorphous solid. The conditions under which each is formed reveal something about the contrasting nature of these solids. If liquid silica at a temperature of 2000°C is cooled very, very slowly, the thermal motion of the atoms decreases with a decrease of temperature, but there is time for the atoms to find the positions of greatest stability when the thermal motion becomes too slow to maintain the liquid state. The resulting solid is crystalline because

FIGURE 5–2
Photographs of crystalline materials: (a) halite, NaCl, (b) quartz, SiO_2, and (c) pyromorphite, $Pb_5Cl(PO_4)_3$.

the atoms are arranged in a regular array. With rapid cooling of the liquid, the atoms do not find the positions of greatest stability, and the amorphous material produced has some of the disorder of atomic arrangement associated with the liquid state.

To learn about the solid state we will focus primarily on the crystal structures of the elements. This limitation simplifies the discussion. The atoms of each solid are all the same size, making it easier to calculate the packing of these atoms. However, these geometrical calculations still give us the background we will need in considering ionic crystals in Chapter 9 and other solids throughout the book. We also will enlarge our knowledge of thermodynamics by considering the thermal properties of solids, a discussion that will introduce the third law of thermodynamics.

5–2 Some Polyhedra and Their Geometrical Properties

In studying the arrangements of atoms in crystalline solids or in individual polyatomic molecules, we will be considering the geometry of various regular polyhedra. To help in this study, we now will review certain relationships in these structures. The regular polyhedra with four, six, eight, and twelve corners are summarized in Table 5–1. Figure 5–3 shows these polyhedra, and Figure 5–4 shows how the tetrahedron and octahedron can be inscribed in a cube. The central role that the cube plays in the arrangement of atoms in many crystalline substances makes it appropriate for us to review its characteristics. The Pythagorean theorem (named for PYTHAGORAS of Samos) relating the length of sides, a and b, and hypotenuse, h, of a right-angled triangle is all we need to know:

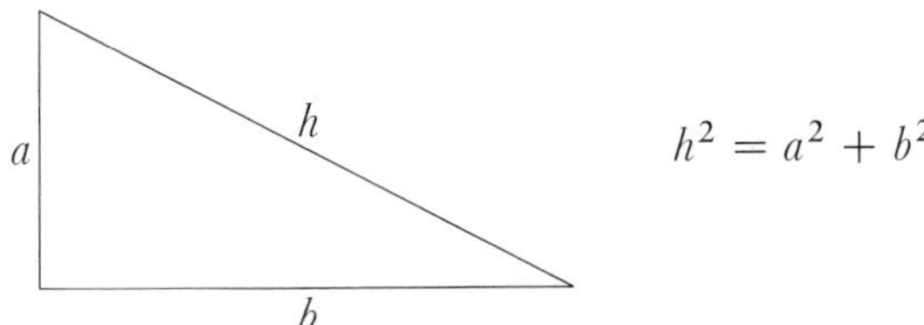

$$h^2 = a^2 + b^2$$

The calculations are summarized in Figure 5–5.

TABLE 5–1
Regular Polyhedra

Polyhedron	Number of: Corners	Edges	Faces
Tetrahedron	4	6	4
Octahedron	6	12	8
Cube	8	12	6
Icosahedron	12	30	20

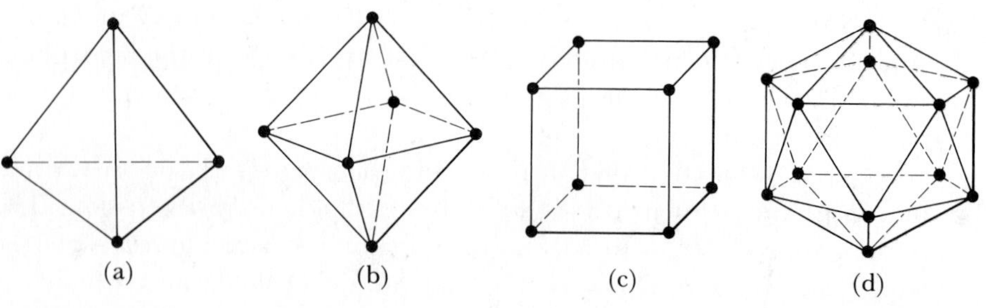

FIGURE 5–3
Regular polyhedra: (a) the tetrahedron (the molecule P_4), (b) the octahedron (the ion B_6^{2-} in CaB_6), (c) the cube (the arrangement of nearest ions in NaCl), (d) the icosahedron (the atomic arrangement of boron atoms in B_{12} units in elemental boron and in $B_{12}H_{12}^{2-}$).

FIGURE 5–4
Relationship of the tetrahedron and octahedron to the cube. (a) A tetrahedron is defined by alternate corners of a cube. (b) An octahedron is defined by the centers of the six faces of a cube.

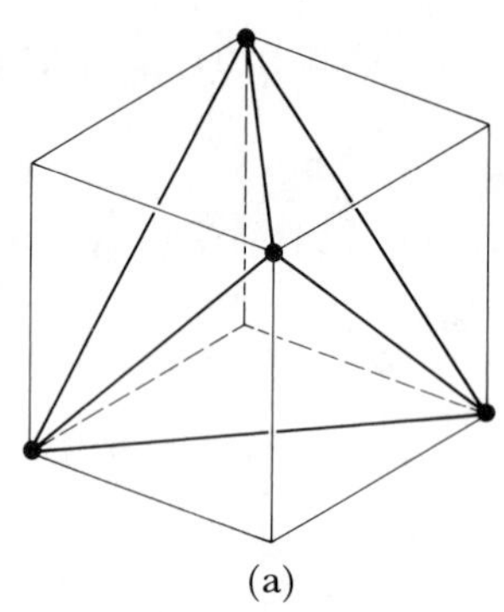

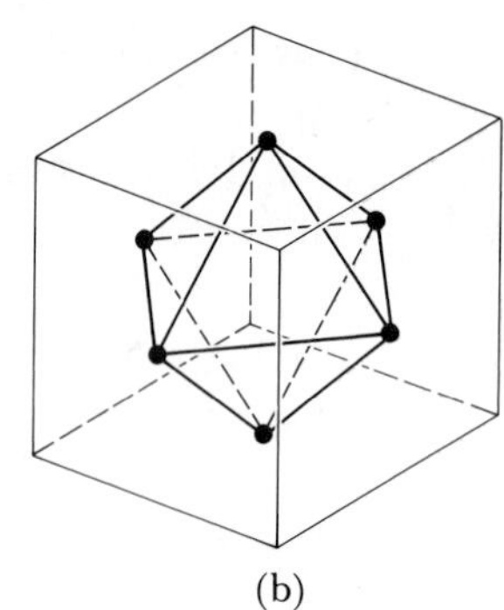

FIGURE 5–5
Relationship of the lengths of the edge, *e*, the diagonal of a face, *d*, and the diagonal of the body of a cube, *D*:

$$e:d:D = 1:\sqrt{2}:\sqrt{3}$$

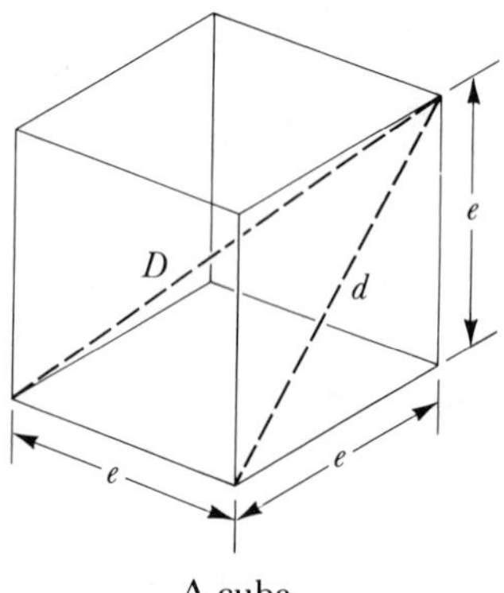

length of edge $= e$
length of diagonal of face $= d$

$$d^2 = e^2 + e^2 = 2e^2$$
$$d = \sqrt{2}\,e$$

length of diagonal of body $= D$

$$D^2 = d^2 + e^2$$
$$= 2e^2 + e^2 = 3e^2$$
$$D = \sqrt{3}\,e$$

Now let us apply the results of this calculation to the sodium chloride structure shown in Figure 5–1. The atoms in this structure that are closest to one another are the sodium ions and chloride ions that define the corners of each smallest cube. The distance between the centers of these adjacent ions, as measured by a technique to be described, is

$$0.2184 \text{ nm} \qquad \text{or} \qquad 2.184 \text{ Å}$$

The separation of the centers of the closest like ions (sodium ion–sodium ion or chloride ion–chloride ion) is defined by the diagonal of a face of the cube; this distance is $\sqrt{2}$ times the length of the cube's edge:

$$\sqrt{2} \times 2.814 \text{ Å} = 3.980 \text{ Å}$$

The next-to-closest separation of unlike ions is defined by a diagonal of the body of the cube; this is $\sqrt{3}$ times the length of an edge of the cube:

$$\sqrt{3} \times 2.814 \text{ Å} = 4.874 \text{ Å}$$

We will need to make other calculations like these later on.

5–3 X-ray Diffraction

The principal method for determining the arrangement of atoms in crystalline solids is x-ray diffraction. You are familiar with x rays and some of their properties from visits to the dentist where your teeth may have been "x rayed" or to the doctor for treatment of a broken bone. This use of x radiation is possible in medicine because it is able to penetrate matter to varying extents, depending upon the nature of the matter, and because its transmitted radiation can expose photographic film just as visible light does. A very different property of x rays is responsible for their use in the x-ray diffraction study of crystal structures. X radiation, like visible light, is electromagnetic radiation, which is an oscillating electric and magnetic field that travels at a characteristic speed, the speed of light, c, which is

$$c = 3.00 \times 10^8 \text{ m s}^{-1} \qquad \text{(see footnote 1)}$$

Some properties of electromagnetic radiation are consistent with its being a wave. In particular, light that has passed through a series of appropriately spaced slits forms a diffraction pattern, and x rays in passing through a crystalline sample form a diffraction pattern. Here "diffraction pattern" means an alternating reinforcement and diminution of intensity of the radiation scattered from the slits or from atoms of the crystal as shown in Figure 5–6. A series of slits can diffract visible light because the separation of the slits and the wavelength of the visible light have a similar magnitude. Analogously, a crystal can diffract x rays due to the similar magnitude of the wavelength of x rays useful in crystallography (typically 0.5–2.3 Å) and typical distances between the centers of atoms in solids. For example, the smallest oxygen atom–oxygen atom distance in ice is 2.75 Å, and the smallest copper atom–copper atom distance in metallic copper is 2.556 Å (see Figure 5–1).

[1]The currently accepted most accurate measurement of this quantity is $c = 2.99792458 \times 10^8 \text{ m s}^{-1}$. As will be explained in Chapter 21, this value is the speed of light in a vacuum. The speed of light through matter is slightly smaller than this.

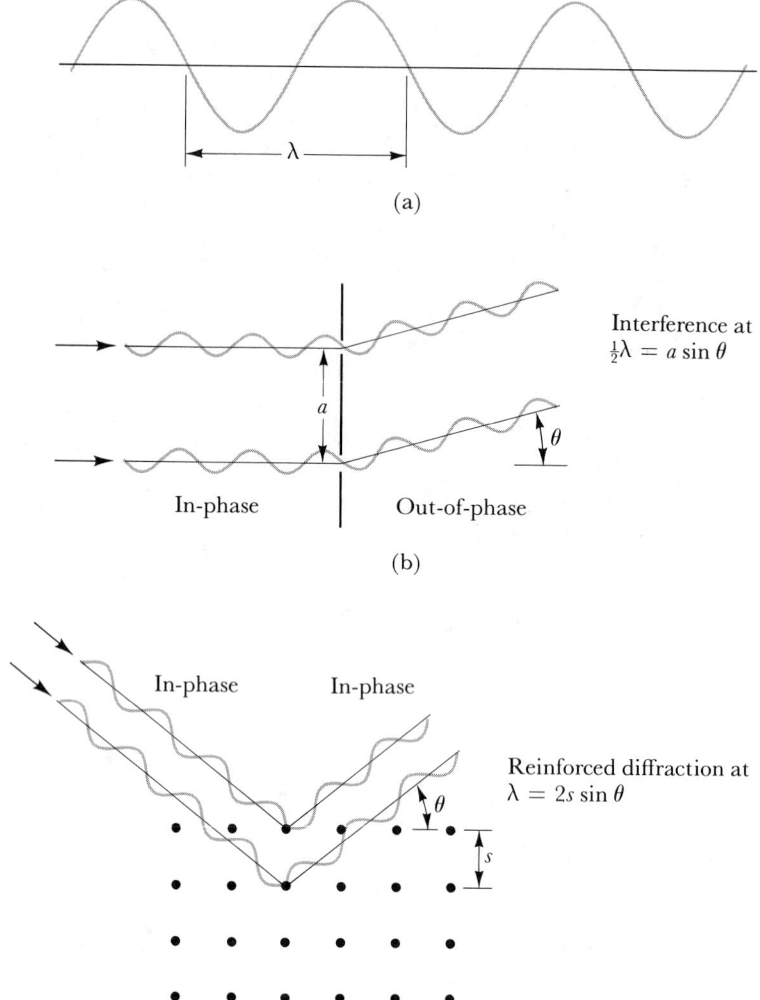

FIGURE 5–6
Diffraction of electromagnetic radiation. (a) The wavelength; λ. (b) Diffraction of visible light ($\lambda \cong (4.0 - 7.0) \times 10^{-7}$ m) by a pair of slits. (c) Diffraction of x rays [$\lambda \cong (0.5 - 2.3) \times 10^{-10}$ m; range of wavelengths useful in x-ray diffraction] by atoms in a crystal.

Because x radiation is an oscillating electric field, it is scattered when interacting with the electrons of atoms. In the scattering from a sample of many atoms, the intensity of the scattered radiation is enhanced or diminished as a function of the angle between the scattered radiation and the incident beam of x rays. If the atoms in a sample are arranged completely randomly, the pattern of the scattered radiation tells us nothing about the separation of the atoms. However, the regular arrangement of atoms in crystalline solids involves many recurring interatomic distances. Also, the various interatomic distances are related to one another by numerical factors which are determined by the structure. (This we have already seen in calculations presented in Figure 5–5.) The pattern of scattered radiation from such an arrangement of atoms tells much about this arrangement.

The diffraction of x rays by the atoms of a crystalline solid can be thought of as reinforced reflection of the radiation from various equally spaced planes of atoms. This is the point of view that W. H. BRAGG and

W. L. BRAGG used in deriving the relationship known as Bragg's law,

$$s = \frac{n\lambda}{2 \sin \theta}$$

where s is the separation of the planes of atoms, n is an integer (1, 2, 3, · · ·), λ is the wavelength of the radiation, and θ is one half the angle between the incident beam and the diffracted beam. Measurement of the angle at which reinforced diffraction occurs gives information about the separation of the planes of atoms, which in turn tells us about the interatomic distances, as indicated in Figure 5–6c. (Here and elsewhere, "interatomic distance" means distance separating centers of atoms.)

We will learn something about the relationship of the separation of planes of atoms and interatomic distances in a crystal by considering the sodium chloride crystal, as shown in Figure 5–7. In this figure we see a set of planes that contains equal numbers of sodium ions and chloride ions separated by the closest sodium ion–chloride ion distance of 2.814 Å, another set of planes that contains equal numbers of sodium ions and chloride ions separated by 1.990 Å (which is 2.814 Å/$\sqrt{2}$), a set of planes that contains either all sodium ions or all chloride ions separated by 1.625 Å (which is 2.814 Å/$\sqrt{3}$), and many other sets of parallel planes. We see in this example that the separations of the planes of atoms (s in the Bragg equation), which were derived from values of the angle of diffraction, are related to the closest separation of sodium ions and chloride ions in this crystal. The various sets of parallel planes of atoms have particular orientations relative to one another. These orientations and the relative magnitudes of the derived values of s tell us the atomic arrangement in the solid. If sodium chloride had a different structure, the relationship of the observed values of s would not be $1:1/\sqrt{2}:1/\sqrt{3}$. (Of course, there are many more than three values of s determined in a diffraction study.) The power of an atom to scatter x radiation depends upon the number of electrons it has, therefore intensities of the individual diffracted beams contribute additional information about the structure of crystalline materials.

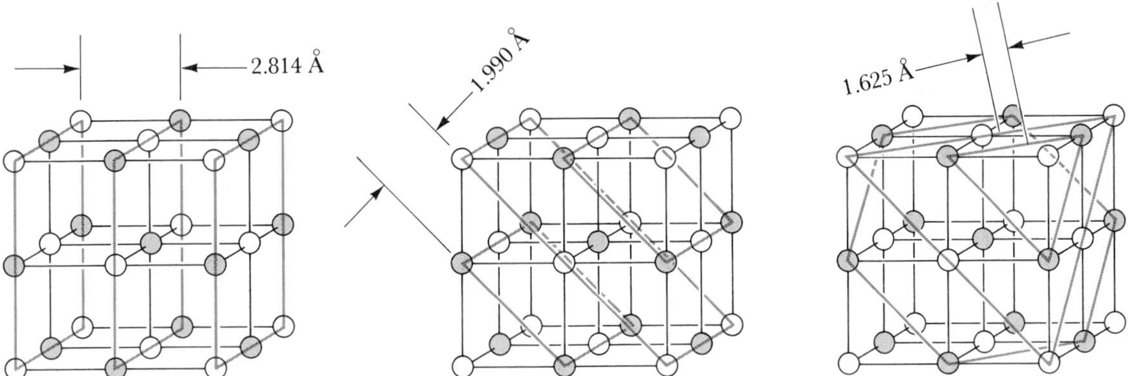

FIGURE 5–7
Planes of atoms in the sodium chloride crystal. The planes that are shown are separated by d, $d/\sqrt{2}$, and $d/\sqrt{3}$; d = 2.814 Å, the closest Na^+–Cl^- distance.

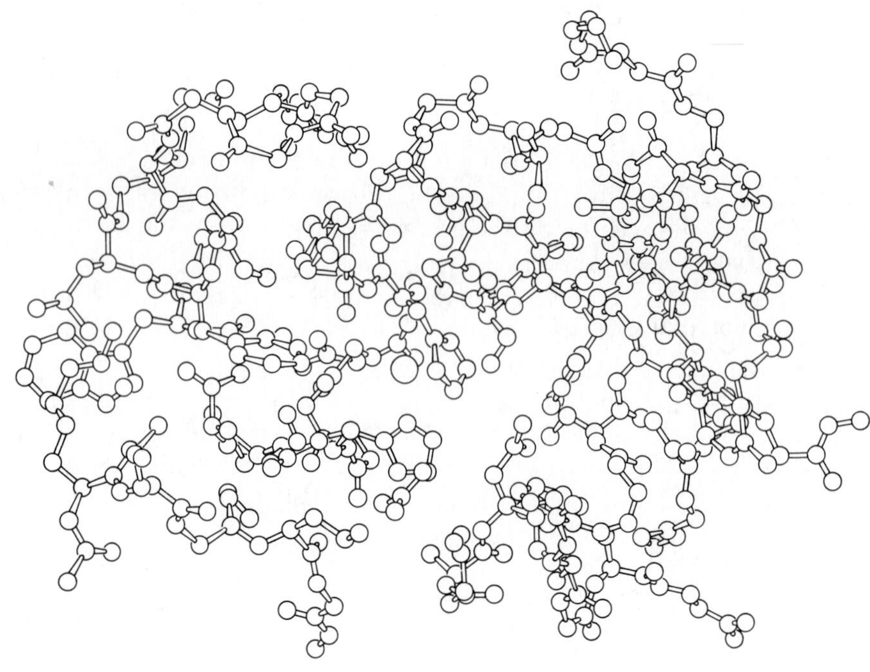

FIGURE 5–8
The structure of the enzyme carboxypeptidase-A, which contains 307 linked amino acids. (The linking of amino acids in proteins is discussed in Chapter 20.) [From W. N. Lipscomb, *Proceedings of Robert A. Welch Foundation Conference on Chemical Research*, Vol. 15 (1971) p. 140.]

X-ray diffraction studies have progressed greatly since the technique was developed early in this century by the Braggs and others. The current success of x-ray diffraction in determining the structures of complex materials, including enzymes (see Figure 5–8), is possible only because high-speed computers can be used to program the x-ray diffractometer to take the data and also can help analyze the data.

5–4 Solids Containing Identical Atoms

Some of the points that we have discussed about simple geometrical relationships in a cube are illustrated by structures of the solid elements. In most metals and in the solid noble-gas elements, the atoms are packed together as closely as possible. In other solid elements the arrangement of the atoms is less compact. Carbon in the form of diamond is an example. Most of the nonmetallic elements form discrete molecules containing two to eight atoms (e.g., O_2 and S_8). In the solid state these molecules are packed together in a manner that minimizes the potential energy, but the energy of attraction between the individual molecules is much less than that between the bonded atoms within a molecule.

Among the atomic arrangements shown in Figure 5–1 is that for metallic copper. The crystal structures of this element, of many other metals, and of the solid noble-gas elements are easy to understand in terms of the packing of identical spheres. If the attractive forces between atoms are not strongly directional, the potential energy of a large number of atoms is minimized if the atoms are packed together as closely as possible, that is, with a minimum amount of empty space. There are two equally compact ways of packing identical spheres, and each of these arrangements is a known arrangement of the atoms in crystalline metals. Before considering these structures, we will discuss a two-dimensional example, the packing in a plane of identical circles of radius r. Two patterns seem reasonable. In one, there is a triangular arrangement of circles, and in the other, there is a square arrangement. In the triangular arrangement, each circle is in contact with six neighboring circles (Figure 5–9a). In the square arrangement, each circle is in contact with four neighboring circles (Figure 5–9b). In each case we will consider a unit of area representative of the entire array, such that the entire array can be generated by translation of this unit along the directions of each of its sides. (In three dimensional arrangements, such a unit of volume is called a *unit cell.*) Each of these two-dimensional unit cells, with sides equal in length to the diameter of the circle ($2r$), contains exactly one circle. The fraction of the area of the square arrangement which is occupied by the circles is

$$\frac{\text{Area of circle}}{\text{Area of square}} = \frac{\pi r^2}{(2r)^2} = \frac{\pi}{4} = 0.785$$

For the triangular arrangement the two-dimensional unit cell is a parallelogram in which the fraction of the area occupied by the circles is

$$\frac{\text{Area of circle}}{\text{Area of parallelogram}} = \frac{\pi r^2}{2r \times \sqrt{3}r} = \frac{\pi}{2\sqrt{3}} = 0.907$$

Therefore less area is empty in the triangular arrangement of circles. No

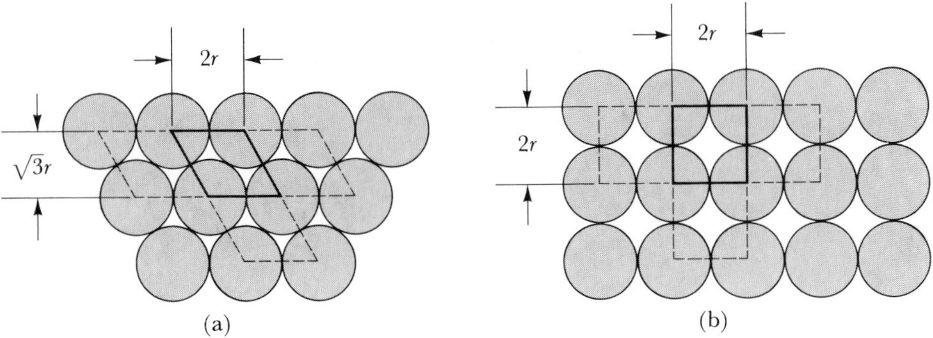

FIGURE 5–9
The packing of identical circles in a plane. (a) Triangular packing; 0.907 of area is occupied. (b) Square packing; 0.785 of area is occupied. In each arrangement the two-dimensional unit cell is shown.

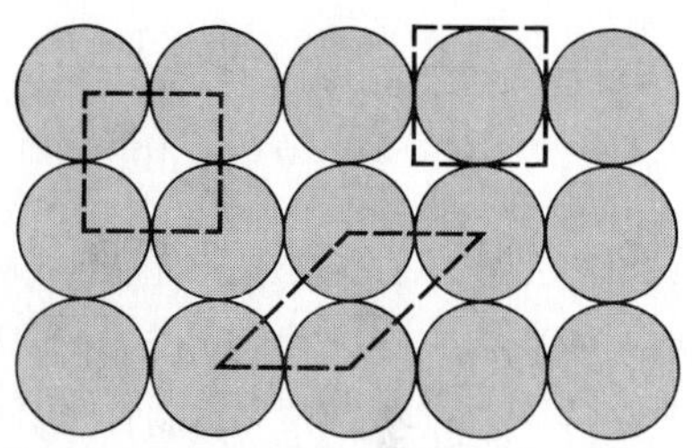

FIGURE 5–10
Alternative two-dimensional unit-cell arrangements for the square packing of circles in a plane. Each unit cell contains one circle.

other arrangement of identical circles fills an area as completely as this. We can call this a closest-packed arrangement.

The two-dimensional unit cells for these arrangements have been drawn with corners at the centers of the circles. This is not necessary, and the size of a unit cell and its contents are unchanged if alternate ways of placing the cells are selected. This is shown in Figure 5–10. Placement of the unit cell for the square arrangement with a single circle completely contained in the unit cell is particularly appropriate for illustrating the content of a unit cell. The point that there is more than one way of superimposing the unit cell upon an orderly two-dimensional arrangement of circles is equally true for orderly three-dimensional arrangements of spheres.

Our consideration of the closest packing of equal-sized spheres in three dimensions will start with the possible arrangements in one layer of spheres. This is equivalent to the possible arrangements of circles in a plane, which we have just considered. Therefore the triangular packing shown in Figure 5–9 is a closest-packed layer of identical spheres. To build a three-dimensional lattice of spheres, another layer of spheres will be added to each side of this first layer. Layers of spheres identical to the first layer fit in the dimples (the spaces between the circles) of the first layer. If we examine the parallelogram unit cell of the closest-packed layer, we see that it contains one complete circle and two dimples. This tells us the composition of a closest-packed layer of spheres: there are two times as many dimples in a layer as there are closest-packed spheres. Viewed from above, these triangular dimples can be divided into two equivalent sets, as is shown in Figure 5–11. For one set, the triangles point in one direction and for the other set they point in the other direction. Another layer of spheres can sit in one or the other of these sets of dimples on each side of the

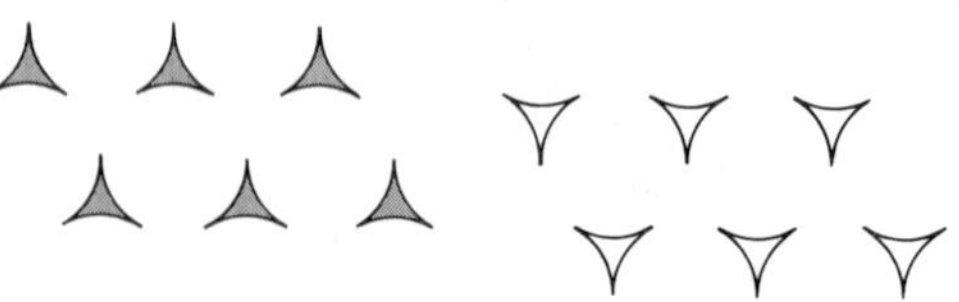

FIGURE 5–11
A layer of closest-packed spheres, showing the two sets of dimples (one set is darkened; the other is not). Either set of dimples, but not both sets, on each side of this layer can be occupied by spheres of another closest-packed layer.

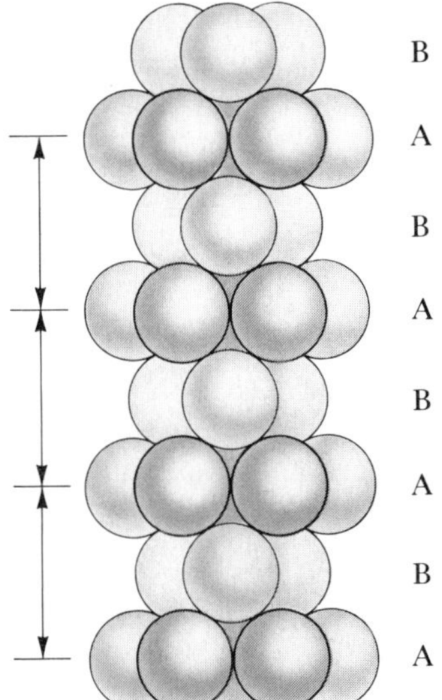

FIGURE 5–12
The arrangement of closest-packed layers in hexagonal closest-packing of identical spheres. The height of a unit cell is shown with arrows.

original layer. These two additional layers, one on each side of the middle layer, can either sit in the same set of dimples of the middle layer or in different sets of dimples of the middle layer. In the first of these arrangements, spheres in the top and bottom layers of the three-layer array are directly over one another; this arrangement, if continued indefinitely, can be designated ABABAB · · · , and a crystal made up of atoms with this arrangement has hexagonal symmetry. This arrangement is called *hexagonal closest-packed*, and one view of it is shown in Figure 5–12. In the second of these arrangements, spheres in the top and bottom layers of the three-layer array are not directly over one another; the atoms of the top and bottom layers are in different sets of dimples of the middle layer. This arrangement, if continued indefinitely, can be designated ABCABCA · · · , and a crystal made up of atoms with this arrangement has cubic symmetry. This arrangement is called *cubic closest-packed*, and one view of it is shown in Figure 5–13.

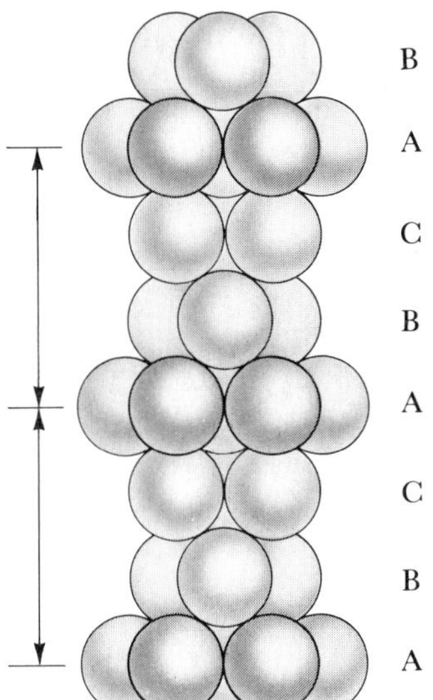

FIGURE 5–13
The arrangement of closest-packed layers in cubic closest-packing of identical spheres. The repeat distance for these layers is shown with arrows. The unit cell for this structure is seen clearly in Figure 5–14.

In these closest-packed arrangements, each sphere touches twelve others. Figure 5–11 shows a sphere in contact with six others in the same layer, and it also is in contact with three spheres in each of the two adjacent layers. The number of nearest-neighbor atoms to a particular atom generally is called its *coordination number*; therefore we can say that the coordination number of the identical atoms in a closest-packed lattice, either hexagonal or cubic, is twelve.

The cubic symmetry of the cubic closest-packed lattice is more apparent when the lattice is oriented appropriately. Figure 5–14 shows an array of fourteen spheres in layers containing one, six, six, and one spheres, with the ABCA pattern oriented to show a cubic arrangement in which eight of the spheres are at the eight corners of the cube and one is in the middle of each of the six faces of the cube. This cubic closest-packed arrange-

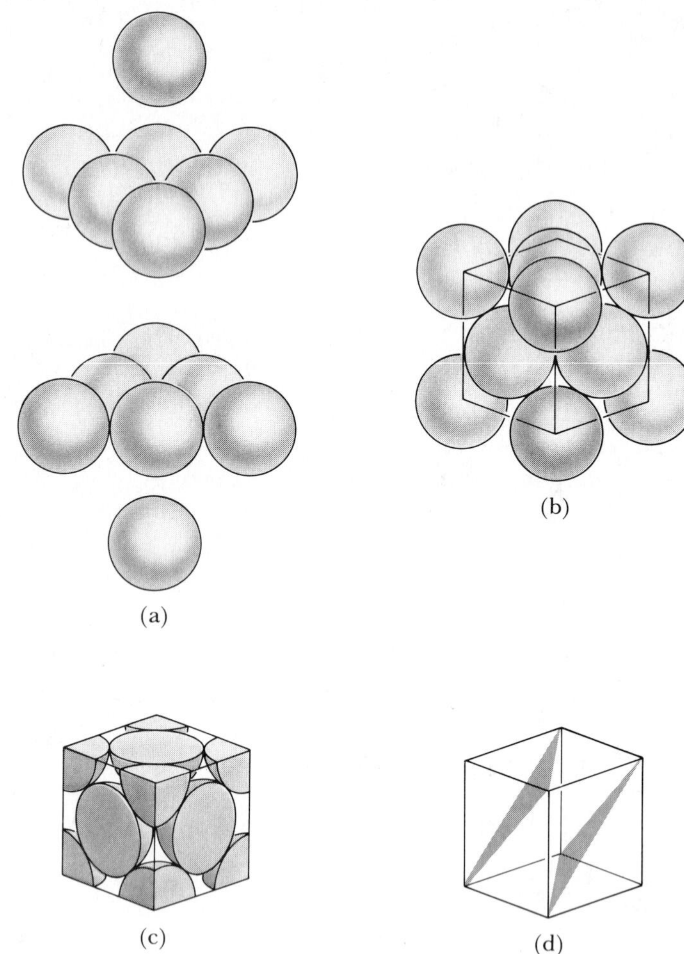

FIGURE 5–14
The cubic closest-packed arrangement of identical spheres. (a) Four layers of closest-packed spheres. (*Note*: Repeat distance in Figure 5–13 involves all or part of the four layers.) (b) The fourteen spheres from (a) oriented to show face-centered cubic arrangement. (c) A unit cell (contains four spheres). (d) The orientation of the closest-packed layers.

ment also is called a *face-centered cubic* arrangement. A unit cell for this arrangement is defined by the centers of the eight spheres at the corners of the cube. We can calculate the content of the unit cell by recognizing that in the three-dimensional array of identical unit cells, each corner of the cubic unit cell is common to eight unit cells and each face of a unit cell is common to two unit cells. Thus the content of a unit cell of the face-centered cubic arrangement of identical spheres is:

$$8 \text{ corners} \times \frac{\frac{1}{8}\text{ sphere}}{\text{corner}} = 1 \text{ sphere}$$

$$6 \text{ faces} + \frac{\frac{1}{2}\text{ sphere}}{\text{face}} = \underline{3 \text{ spheres}}$$

$$4 \text{ spheres}$$

To determine the size of the cubic unit cell, we must decide which spheres actually are in contact. For this arrangement the spheres along the diagonal of each face are in contact. (Figure 5–14 shows that the planes defined by the intersecting face diagonals are closest-packed planes.) Therefore the diagonal of each face has a length of $4r$. Because the face diagonal of a cube is $\sqrt{2}$ times as long as its edge, the edge of the cube is

$$e = \frac{d}{\sqrt{2}} = \frac{4r}{\sqrt{2}} = 2.828r$$

The fraction of the unit cell that is occupied by spheres is given by the quotient:

$$\frac{\text{Volume of 4 spheres of radius } r}{\text{Volume of cube with edge } \dfrac{4r}{\sqrt{2}}} = \frac{4 \times \left(\dfrac{4}{3}\pi r^3\right)}{\left(\dfrac{4r}{\sqrt{2}}\right)^3} = \frac{\pi}{3\sqrt{2}} = 0.740$$

Although this calculation is for the cubic closest-packed arrangement, the result (0.740 of the volume occupied) also is valid for hexagonal closest-packing of identical spheres.

Among the metals with a cubic closest-packed arrangement of atoms are copper, aluminum, iron, cobalt, and nickel. Magnesium and zinc are common metals with a hexagonal closest-packed arrangement of atoms. There are some metals, calcium being one, that crystallize in either of these closest-packed arrangements. The factors determining the arrangement adopted by a particular metal are not well understood.

The model of a metallic structure that we have been discussing, a cubic closest-packed lattice of spherical atoms, allows us to calculate the size of the atoms from the measured density of the metal. For instance, consider metallic aluminum, with a density of 2.702 g cm^{-3}, which has this structure. We can calculate the length of the edge of the cubic unit cell from the density; the radius of an atom of aluminum is simply $\sqrt{2}/4$ times this edge length. (The diagonal of the face is $\sqrt{2}$ times the length of an edge of the cubic unit cell, and this diagonal is four times the atomic radius.)

The molar volume of aluminum is

$$V = \frac{26.98 \text{ g mol}^{-1}}{2.702 \text{ g cm}^{-3}} = 9.985 \text{ cm}^3 \text{ mol}^{-1}$$

The volume of the cubic unit cell, which contains four atoms, is

$$v = \frac{\text{volume of one mole}}{\text{number of unit cells in one mole}}$$

$$v = \frac{9.985 \text{ cm}^3 \text{ mol}^{-1}}{6.02 \times 10^{23} \text{ mol}^{-1} \times \frac{1}{4}} = 6.635 \times 10^{-23} \text{ cm}^3$$

The length of the edge of a cube is the cube root of the volume:

$$l = (6.635 \times 10^{-23} \text{ cm}^3)^{1/3} = 4.048 \times 10^{-8} \text{ cm}$$

and the value of the radius of the aluminum atom is

$$r = \frac{\sqrt{2}}{4} \times 4.048 \times 10^{-8} \text{ cm} = 1.43 \times 10^{-8} \text{ cm} = 1.43 \text{ Å}$$

Alternatively, we can multiply the molar volume of aluminum by the fraction of this volume that is occupied and divide by Avogadro's constant to obtain the volume of one spherical aluminum atom:

$$v = \frac{9.985 \text{ cm}^3 \text{ mol}^{-1} \times 0.740}{6.02 \times 10^{23} \text{ mol}^{-1}} = 1.227 \times 10^{-23} \text{ cm}^3$$

From this the radius of the aluminum atom can be calculated:

$$r = \left(\frac{3}{4\pi} V\right)^{1/3}$$

$$= \left(\frac{3}{4\pi} \times 1.227 \times 10^{-23} \text{ cm}^3\right)^{1/3} = 1.43 \times 10^{-8} \text{ cm}$$

The closest packing of identical spheres is important not only in connection with the crystal structures of metals and solid noble-gas elements, but also in connection with ionic compounds in which the ions of one charge sign (generally the negative ions) have a cubic or hexagonal closest-packed arrangement and the ions of the opposite charge occupy empty regions of the closest-packed lattices. These structures will be considered in detail in Chapter 9.

THE BODY-CENTERED CUBIC STRUCTURE

The alkali metals (Li, Na, K, Rb, and Cs) and some of the transition metals (e.g. Cr and Fe) crystallize with a structure which is different from the closest-packed arrangements just discussed. Figure 5–15 shows this structure, called the *body-centered cubic* structure, in which each atom has

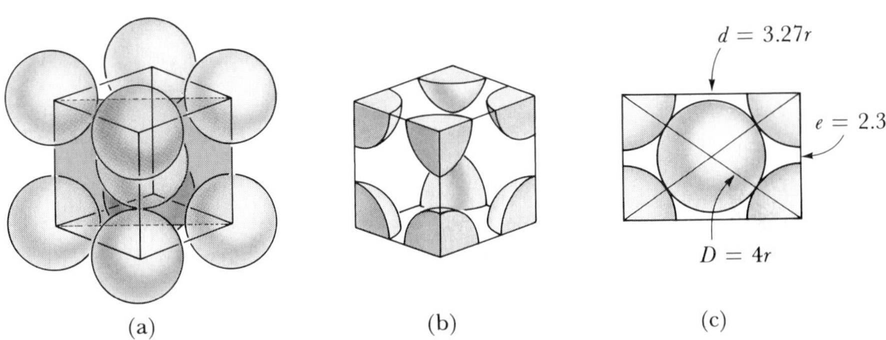

FIGURE 5–15
The body-centered cubic arrangement of identical spheres. (a) The arrangement of spheres (with spheres not in contact). The plane shown in (c) is indicated. (b) The unit cell (omitting central sphere). (c) A cross section showing spheres in contact along body diagonal.

eight atoms as nearest neighbors. It is easy to calculate the fraction of the volume occupied by the spherical atoms in this structure. The result can then be compared with the corresponding figure for the closest-packed structures. The unit cell, shown in Figure 5–15b, contains two atoms, represented as spheres. There is one eighth of a sphere at each of the eight corners of the cube and one sphere in the center of the cube:

$$8 \text{ corners} \times \frac{\frac{1}{8}\text{ sphere}}{\text{corner}} = 1 \text{ sphere}$$

$$1 \text{ center} \times \frac{1 \text{ sphere}}{\text{center}} = \frac{1 \text{ sphere}}{2 \text{ spheres}}$$

In this structure the spheres are in contact along the body diagonal of the cubic unit cell, which therefore has a length equal to $4r$, $D = 4r$. The length of an edge of the cubic unit cell $e = D/\sqrt{3} = 4r/\sqrt{3}$. The fraction of the space occupied is given by the quotient:

$$\frac{\text{Volume of 2 spheres of radius } r}{\text{Volume of cube with edge } \dfrac{4r}{\sqrt{3}}} = \frac{2 \times \dfrac{4}{3}\pi r^3}{\left(\dfrac{4r}{\sqrt{3}}\right)^3} = \frac{\sqrt{3}\,\pi}{8} = 0.680$$

The packing of spheres in this structure is 0.680/0.740, or 92%, as compact as in the closest-packed structures. The percentage may seem surprisingly large, since in the body-centered cubic structure each sphere has only eight nearest neighbors, rather than the twelve characteristic of the closest-packed arrangements. But in the body-centered cubic structure there are six relatively close next-to-nearest neighbors. If you consider the neighbors of a sphere in the center of one cubic unit cell of this structure, you will see that its next-to-the nearest neighbors are in the centers of each of the six cubes that share faces with the cube in question. The distance, s_2, between next-to-nearest neighbor atoms therefore is the same as the length of an edge of the cube, already shown to be $4r/\sqrt{3}$:

$$s_2 = \frac{4r}{\sqrt{3}} = 2.31r$$

Thus these six next-to-nearest neighbor spheres are separated by only 15.5% more than the separation of nearest neighbors ($s_1 = 2.00r$).

In Chapter 6 we shall compare the structure of solid sodium with the structure of the corresponding liquid; for that purpose we shall calculate the relative separations of more distantly separated atoms in the body-centered cubic arrangement. Figure 5–16 shows a cubic unit cell with one each of the adjacent cubic unit cells that share a face, an edge, and a corner with the cube in question. After the six next-to-nearest neighbors is a set of neighboring atoms at the centers of cubes that share an edge with the central cube. Their separation is equal to the length of a face diagonal of the cube:

$$s_3 = \sqrt{2} \times e$$

$$= \sqrt{2} \times \frac{4r}{\sqrt{3}} = 3.27r$$

Because a cube has twelve edges and there is one atom in the center of these cubes, there are twelve neighbors at this distance. Next comes a set of twenty-four neighbors, which are at the four distant corners of each of the six cubes which share faces with the central cube. Examination of Figure 5–16 will show you that the separation of these neighbors is equal to the length of a hypotenuse of a right triangle with sides equal to one and one half edges of

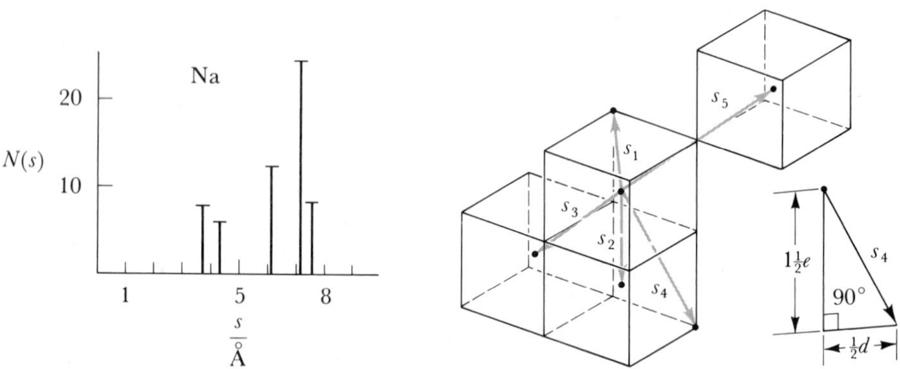

FIGURE 5–16
The relative separation of atoms in a body-centered cubic lattice, showing a central cubic unit cell plus one each of unit cells that share a face, an edge, and a corner with the central cube. If r = radius of spherical atom, body diagonal (D) = $4r$, edge (e) = $4r/\sqrt{3}$, and face diagonal (d) = $\sqrt{2} \times 4r/\sqrt{3}$. Separations ($s$) of sets of neighbors are (with number in each set): $s_1 = \frac{1}{2}D = 2r(8)$, $s_2 = e = 4r/\sqrt{3} = 2.31r(6)$, $s_3 = d = \sqrt{2} \times 4r/\sqrt{3} = 3.27r(12)$, $s_4 = 3.83r(24)$ (see detailed insert), $s_5 = D = 4r(8)$. At the left is the radial distribution curve for this structure with distances for sodium, which has this structure.

the cube and one half of a diagonal of the face of the cube:

$$s_4 = \sqrt{\left(\frac{3e}{2}\right)^2 + \left(\frac{d}{2}\right)^2}$$

$$= r\sqrt{\left(\frac{3}{2} \times \frac{4}{\sqrt{3}}\right)^2 + \left(\frac{1}{2} \times \frac{4\sqrt{2}}{\sqrt{3}}\right)^2} = 3.83r$$

The next neighbors are at the centers of the eight cubes that share corners with the central cube. These neighbors are separated by a distance that is equal to the length of the body diagonal of a cube:

$$s_5 = D = 4r$$

This distribution is represented in Figure 5–16, which shows the number of neighbors as a function of the distance. This graph, with its specific separations for each set of neighbors, implies a static crystal. However, the atoms in a crystal vibrate about their equilibrium positions, and this makes the distribution curve somewhat diffuse, a feature not shown in this figure. In the following chapter on the liquid state, you will see an analogous distribution curve for liquid sodium, in which the distances separating atoms are even more smeared out.

SOME SOLID ELEMENTS WITH DIRECTIONAL BONDS

The discussion in the preceding two sections assumed the stability of the closest-packed arrangements of atoms and of an arrangement that was almost as compact (the body-centered cubic lattice). Closest-packed arrangements would be the rule for all solid elements if their atoms were approximately spherical and if the attraction and repulsion between atoms had no directional characteristics. However, there are solid elements in whose structures the atoms have coordination numbers smaller than twelve, the value for the closest-packed structures, or eight, the value for the body-centered cubic structure. Carbon and boron are such elements. Features of the electronic structure of the carbon atom and the boron atom are responsible for stability of the observed structures, and these electronic structures will be presented in due course. Now we simply will contrast these structures with those of the metals.

The structure of diamond is shown in Figure 5–17. In diamond each carbon atom is surrounded by four other carbon atoms, which define a regular tetrahedron. Figure 5–17a shows a unit cell for the diamond structure. In this structure each atom has four nearest neighbors at a distance of 1.54 Å; to be more stable than a closest-packed structure, in which each atom has twelve nearest neighbors, the carbon atom must form four directional bonds of extraordinary strength.

The atomic arrangement in one of the crystalline forms of elemental boron is shown in Figure 5–18. This structure has a unit cell containing fifty atoms. Most of the boron atoms in this structure are arranged in

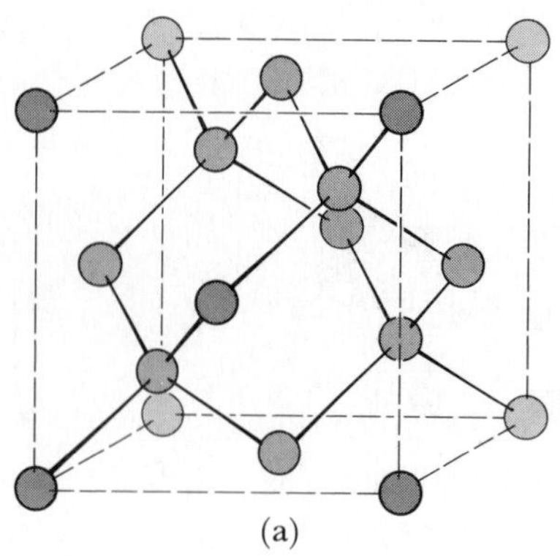

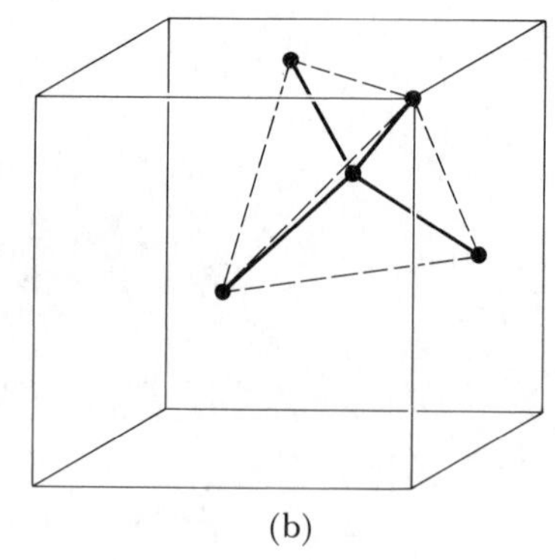

FIGURE 5–17
The arrangement of atoms in diamond. (a) A unit cell (showing, however, full spheres on faces of cube). (b) Part of a unit cell showing four carbon atoms tetrahedrally arranged around a carbon atom.

regular icosahedral units of twelve boron atoms; there are four icosahedral groupings per unit cell. The boron atoms of a particular icosahedron are 1.80 Å from each of five other boron atoms in the icosahedron, and approximately the same distance from a boron atom of another icosahedron; the distance from a bridging boron atom is shorter, ~1.62 Å. The icosahedral arrangement of boron atoms in this crystalline form of the element is found also in certain compounds of boron (see Chapter 12). Although the bonding of boron atoms with one another in this structure is not our principal concern at this stage, it is clear that each line in the figure *does not* stand for a pair of electrons. (The boron atom has only three electrons in its outer shell.)

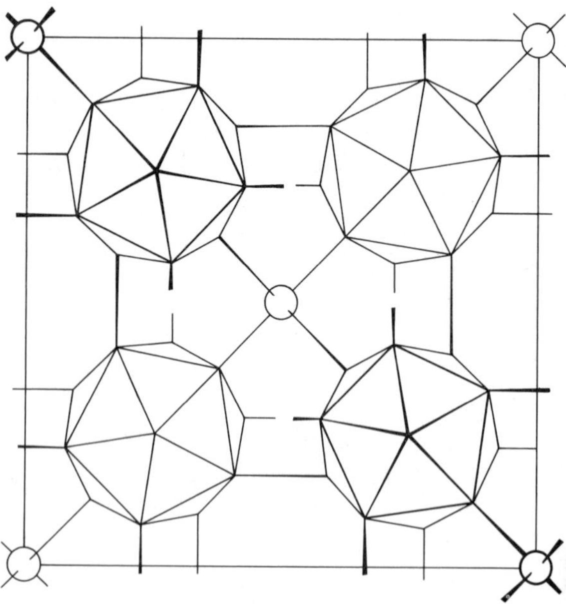

FIGURE 5–18
The atomic arrangement in tetragonal boron. A boron atom is at each intersection of lines. Each unit cell contains 50 boron atoms: 48 of which are contained in 4 icosahedra of 12 atoms each, and the remaining two are interstitial, helping to join different icosahedra. Each boron atom that is part of an icosahedron has coordination number 6; interstitial boron atoms have coordination number 4.

The coordination numbers of boron atoms in tetragonal boron, and of carbon atoms in diamond (6 and 4, respectively) are smaller than those of the metals lithium (8, body-centered cubic) and beryllium (12, hexagonal closest-packed), which precede boron and carbon in order of atomic number. However, the molar volumes of these elements,

$$
\begin{array}{llll}
{}_{3}\mathrm{Li} & 13.09\ \mathrm{cm^3\ mol^{-1}} & {}_{5}\mathrm{B} & 4.62\ \mathrm{cm^3\ mol^{-1}} \\
{}_{4}\mathrm{Be} & 4.87\ \mathrm{cm^3\ mol^{-1}} & {}_{6}\mathrm{C} & 3.42\ \mathrm{cm^3\ mol^{-1}}
\end{array}
$$

do not reflect the smaller coordination numbers of the elements boron and carbon because atoms of these elements are closer to one another in the solid than are the atoms in the metals. The interatomic distances in these metals are: lithium, 3.04 Å, and beryllium, 2.26 Å.

SOME SOLID ELEMENTS CONSISTING OF DISCRETE MOLECULES

Molecules of most nonmetals consist of a few atoms, for example, two atoms (as in Cl_2), four atoms (as in P_4), and eight atoms (as in S_8). In the solid state, atoms of these elements have two kinds of neighboring atoms: there are nearest-neighbor atoms to which each atom is bonded, and there are neighboring atoms belonging to other molecules. The interatomic separations of the bonded atoms are much smaller than the closest separations of atoms in different molecules. However, the latter type of interatomic separation is small enough to account for the attractive interactions between molecules in molecular crystals. These attractive forces are the same as those that are partially responsible for nonideality of the noble gases, namely, van der Waals attractions. Table 5–2 summarizes the interatomic distances within the molecules and between the molecules Cl_2, P_4, and S_8. There is a clear distinction between the distance separating bonded atoms in the same molecule and the distance separating adjacent atoms in different molecules.

TABLE 5–2
Interatomic Distances in Certain Solid Nonmetals

Element	Molecule, X_n	X—X distance in molecule	Closest X-----X distance between molecules
${}_{15}P$	P_4	2.20 Å	3.8 Å
${}_{16}S$	S_8	2.06 Å	3.7 Å
${}_{17}Cl$	Cl_2	1.99 Å	3.6 Å

5–5 The Thermal Properties of Solids

In two previous chapters, you were introduced to two views of entropy: its relationship to randomness of atomic position was discussed in Chapter 3, and its relationship to thermal properties of gases was presented in Chapter 4. Now consideration of the thermal properties of solids will expand your understanding of entropy. In this section we will extend the concept of randomness, as it is related to entropy, to include randomness of the distribution of energies among atoms (or molecules), in addition to the randomness of atomic position.

MOLAR HEAT CAPACITIES OF SOLIDS

When heat is added to a crystalline solid, its internal energy and its temperature increase. What molecular motion in the solid increases when the temperature increases? As in the analogous discussion for the gases (Section 4–3), it will be simpler to introduce this subject by considering a solid that does not contain discrete polyatomic units. The metals and carbon conform to this requirement. The molar heat capacities of some of these elements are plotted as a function of temperature in Figure 5–19. Key features of each of these plots are:

1. the molar heat capacity approaches zero as the temperature approaches 0 K,
2. the molar heat capacity becomes $\sim 3R$ (6.0 cal K^{-1} mol^{-1} or 25 J K^{-1} mol^{-1}) at high temperatures, and
3. the molar heat capacity approaches the limiting high temperature value at different temperatures for different elements.

We will not concern ourselves with quantitative theories to explain the dependence of C_v upon temperature, but rather will seek only a qualitative explanation. Most theories of the thermal properties of solids assume that atoms of the solid vibrate to and fro in three dimensions about their most stable positions in the crystal lattice and that only certain vibrational

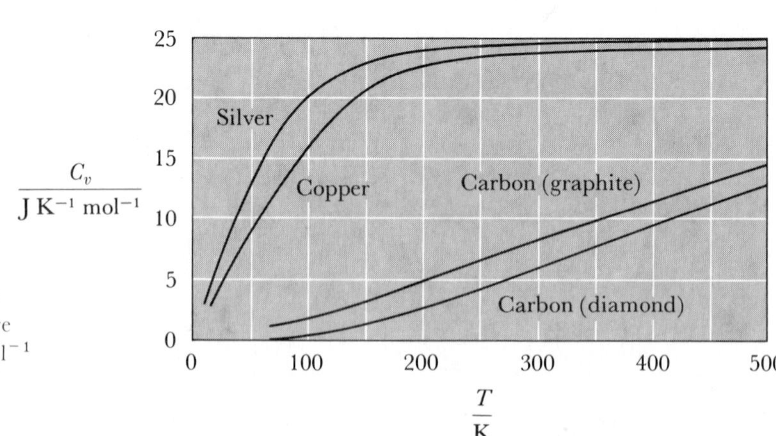

FIGURE 5–19
The heat capacities of solid elements that do not contain discrete polyatomic molecules. The limiting high temperature value of C_v is $\sim 3R$: 3×8.31 J K^{-1} mol^{-1} $\cong 24.9$ J K^{-1} mol^{-1}.

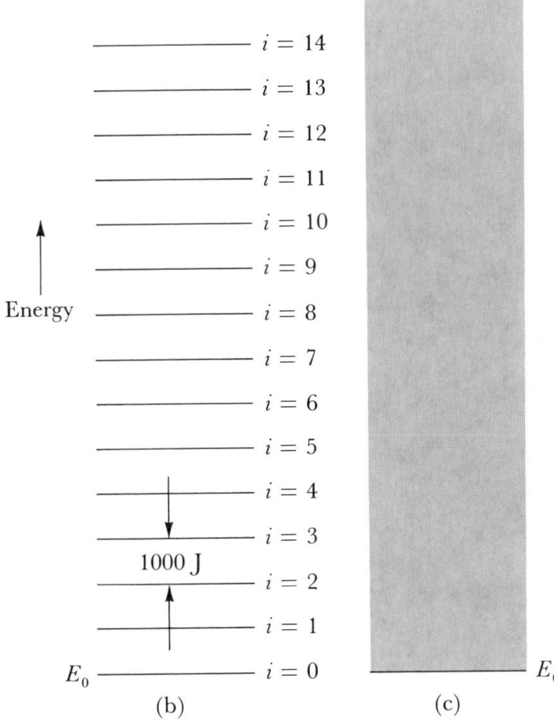

FIGURE 5–20
The contrast between quantized energy and continuous energy. (a) A one-dimensional lattice; each atom can vibrate about its most stable position. (b) A set of equally spaced energy levels separated by 1000 J mol^{-1}. Each horizontal line represents an allowed energy. (c) A continuous range of energy.

energies are possible. If only certain vibrational energies are possible, the vibrational energy is said to be *quantized.* (We will learn about quantization of electronic energy in atoms in Chapter 8 and quantization of vibrational energy of molecules in Chapter 10. The quantization of electronic, vibrational, and rotational energy of molecules also is discussed in Chapter 21.) In the simple one-dimensional model of a crystalline solid, atoms are equally spaced along a line; each atom can vibrate to and fro along this line about its most stable position. The allowed energies of vibration of each atom are shown by the series of equally spaced energy levels in Figure 5–20b. In this figure, energy is plotted vertically, and each horizontal line represents an allowed vibrational energy. This figure is to be contrasted with Figure 5–20c, in which the shaded region is intended to show that all energies above zero are allowed. This continuous range of allowed energies is an appropriate representation of the translational energy of gas molecules.[2]

Although the model with equally spaced energy levels is too simple to explain the thermal properties of metals quantitatively, it is adequate to

[2]The translational energy of gas molecules is quantized, but the spacing of the allowed energies is very close ($\sim 10^{-11}$ J mol^{-1}). Because the average kinetic energy of gas molecules at 300 K is 3.7×10^3 J mol^{-1}, this quantization does not play any role in ordinary chemistry.

illustrate the Boltzmann distribution, to show why the molar heat capacity approaches zero at very low temperatures, and to provide a basis for discussing the entropies of solids. In using this model, it is essential to know the relative numbers of atoms that vibrate with each allowed energy. These are given by the Boltzmann distribution, which describes a system in thermal equilibrium,

$$\frac{n_i}{n_0} = e^{-E_i/RT}$$

in which n_i/n_0 is the ratio of the number of atoms in the ith vibrational state to the number of atoms in the lowest vibrational state, e is the base of the natural logarithms, $e = 2.718\cdots$,[3] and E_i is the energy of the ith vibrational state relative to the energy of the lowest vibrational state. (You will recognize the exponential term $e^{-E_i/RT}$ as analogous to one factor in the Maxwell–Boltzmann distribution law for the velocities of gas molecules, presented in Figure 4–13.) Thus for the situation depicted in Figure 5–20b, $E_1 = 1000\ \text{J mol}^{-1}$, $E_2 = 2000\ \text{J mol}^{-1}$, $E_3 = 3000\ \text{J mol}^{-1}$, and so on.

The consequences of this distribution law are:

1. As the temperature approaches 0 K, the value of $e^{-E_i/RT}$ for each E_i approaches zero:

$$e^{-E_i/R\times 0} = e^{-\infty} = 0$$

That is, practically all of the atoms have the lowest allowed vibrational energy at very low temperatures.

2. As the temperature approaches infinity, the limiting value of $e^{-E_i/RT}$ is one:

$$e^{-E_i/R\times\infty} = e^0 = 1$$

That is, at very high temperatures, there are almost the same number of atoms in each allowed vibrational energy state.

3. The temperature at which appreciable numbers of atoms have vibrational energy greater than E_0 depends upon the magnitude of E_1, which varies from substance to substance. The quotient E_1/R has the units of temperature ($\text{J mol}^{-1}/\text{J K}^{-1}\ \text{mol}^{-1} \equiv \text{K}$), and the relative value of this quotient and the temperature determine the value of n_1/n_0. For the case shown in Figure 5–20b, $E_1 = 1000\ \text{J mol}^{-1}$, the value of E_1/R is $1000\ \text{J mol}^{-1}/8.31\ \text{J K}^{-1}\ \text{mol}^{-1} = 120.3$ K. At 120.3 K the value of n_1/n_0 is $e^{-1.00}$:

$$\frac{n_1}{n_0} = e^{-1000\ \text{J mol}^{-1}/(8.31\ \text{J K}^{-1}\ \text{mol}^{-1} \times 120.3\ \text{K})}$$

$$= e^{-1.00} = 0.368$$

[3]The value of e is given by the sum of the infinite series $e = 1 + \frac{1}{1} + \frac{1}{2!} + \frac{1}{3!} + \frac{1}{4!} + \cdots$.

and the value of n_2/n_0 is $e^{-2.00}$:

$$\frac{n_2}{n_0} = e^{-2000\,\text{J mol}^{-1}/(8.31\,\text{J K}^{-1}\,\text{mol}^{-1} \times 120.3\,\text{K})}$$

$$= e^{-2.00} = (0.368)^2 = 0.135$$

We see that for a Boltzmann distribution among equally spaced levels, the occupancy of each level is a particular fraction of the occupancy of the previous level. In this example the fraction is 0.368:

$$\frac{n_1}{n_0} = 0.368 \qquad \frac{n_2}{n_1} = 0.368 \qquad \frac{n_3}{n_2} = 0.368$$

Therefore

$$\frac{n_2}{n_0} = (0.368)^2 = 0.135$$

$$\frac{n_3}{n_0} = (0.368)^3 = 0.0498$$

Certain points about this model can be made in calculations for a sample of 100 atoms. These atoms will be distributed according to the Boltzmann distribution among two series of equally spaced energy levels ($E_1 = 1000$ J mol^{-1} or 1.66×10^{-21} J, and $E_1 = 2500$ J mol^{-1} or 4.15×10^{-21} J) at several different temperatures: 0 K, 10 K, 100 K, and 300 K. For $E_1 = 1000$ J mol^{-1} ($E_i = i \times 1000$ J mol^{-1}), the values of n_i/n_0 are equal to

$$e^{-1000 \times i/8.31\,\text{K}^{-1} \times T}$$

At these temperatures the values of n_i/n_0 are:

0 K	0	100 K	$(0.300)^i$
10 K	$(5.94 \times 10^{-6})^i$	300 K	$(0.670)^i$

For $E_1 = 2500$ J mol^{-1} ($E_i = i \times 2500$ J mol^{-1}), the values of n_i/n_0 are equal to

$$e^{-2500 \times i/8.31\,\text{K}^{-1} \times T}$$

At these temperatures the values of n_i/n_0 are:

0 K	0	100 K	$(0.0494)^i$
10 K	$(8.60 \times 10^{-14})^i$	300 K	$(0.367)^i$

The results of these calculations, given in Figure 5–21, show how more and more atoms acquire vibrational energy as the temperature increases. The calculations also show that the wider the spacing of the energy levels the greater the number of atoms in the lowest vibrational state at a particular temperature. Although a sample of one hundred atoms is not an adequate model for a macroscopic sample, we see in this figure that for $T = 0$–10 K for either $E_1 = 1000$ J mol^{-1} or 2500 J mol^{-1}, all atoms are in the $i = 0$

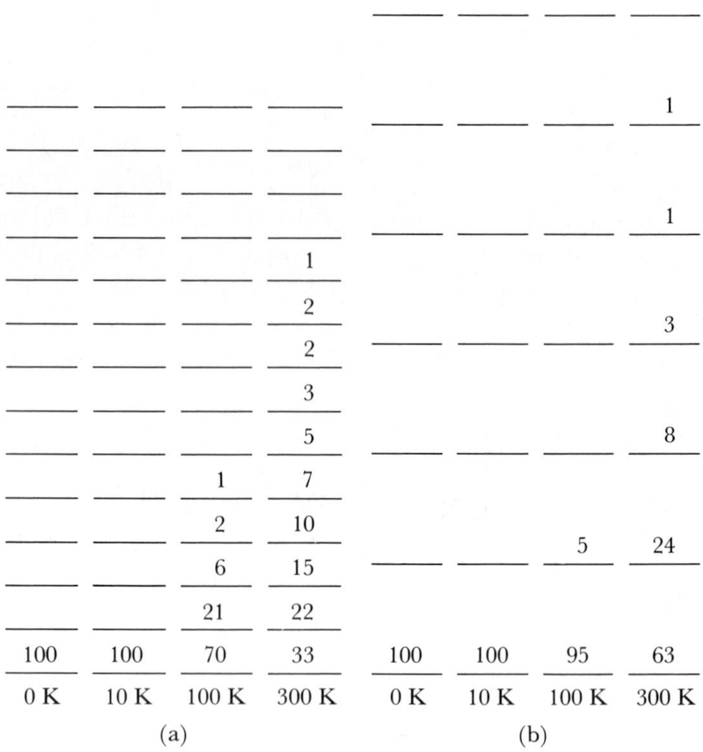

Energy level	0 K	10 K	100 K	300 K
9				1
8				2
7				2
6				3
5				5
4			1	7
3			2	10
2			6	15
1			21	22
0	100	100	70	33

(a)

Energy level	0 K	10 K	100 K	300 K
5				1
4				1
3				3
2				8
1			5	24
0	100	100	95	63

(b)

FIGURE 5–21
The Boltzmann distribution of one hundred atoms among equally spaced vibrational energy levels. The number on each line is the number of atoms with this vibrational energy. (a) Spacing is 1000 J mol^{-1} (1.66×10^{-21} J). (b) Spacing is 2500 J mol^{-1} (4.15×10^{-21} J). (Calculations made with equation $n_i/n_0 = e^{-E_i/RT}$ with rounded integers given.)

state. Therefore, over the temperature range 0–10 K, the value of C_v would be zero, because in this range of temperatures the atoms have no energy in excess of that corresponding to the energy of the $i = 0$ state. At higher temperatures, however, there are atoms in the $i = 1, 2, 3, \cdots$ states, and to increase the temperature, more energy is needed. Thus the value of C_v would be larger.

If the model involved a very large number of atoms that could vibrate in three directions (along the x, y, and z axis), the calculation would show that at high temperatures the value of C_v should be $3R$:

$$\begin{aligned} C_v &= 3R \\ &= 3 \times 8.31 \text{ J K}^{-1} \text{ mol}^{-1} \\ &= 24.9 \text{ J K}^{-1} \text{ mol}^{-1} \end{aligned}$$

which is the observed limiting value shown in Figure 5–19.

The Law of Dulong and Petit. Our previous discussion of the composition and molecular formulas of nitrogen oxides, and of the associated problem of assigning correct relative atomic masses (atomic weights) to nitrogen and oxygen, showed several facts:

1. The weight percentage composition data for compounds do not establish the correct atomic weights of the elements, although such data do establish possibilities that include the correct values.

2. Volumetric relationships for gaseous reactions interpreted in the light of Avogadro's principle allow choice of the correct atomic weights from those consistent with weight percentage composition data. However, the volumetric relationships for gaseous reactions could not be used to determine the atomic weights of elements which did not form gaseous compounds. For this reason the atomic weights for many metals proposed by BERZELIUS in 1814 were multiples of the correct values. For instance, the atomic weights he assigned at that time to iron and silver were 110.98 and 430.11, respectively.[4] He based these values upon weight percentage composition data for oxides of these metals, assumed to be FeO_2 and AgO_2. The oxides he studied were, in fact, FeO and Ag_2O, and therefore his proposed atomic weights for these metals were in error by factors of approximately two and four, respectively.

A relationship between the heat capacity of solid elements and their atomic weights, proposed in 1819, helped correct these (and other) errors. Having recorded the heat capacities of many solid elements, DULONG and PETIT proposed that the product of the specific heat (measured at constant pressure) of a solid element (in cal g^{-1} K^{-1}) and its atomic weight (in g mol^{-1}) is ~6.0 cal K^{-1} mol^{-1} (or ~25 J K^{-1} mol^{-1}). This relationship, which became known as the *law of Dulong and Petit*, was potentially very useful in settling uncertainties about values of atomic weights. Guidance from this law was not accepted uniformly, because some solid elements seemed not to conform. For instance, the solid nonmetals carbon (diamond), boron, and silicon have molar heat capacities at room temperature of 1.49 cal K^{-1} mol^{-1}, 2.65 cal K^{-1} mol^{-1}, and 4.71 cal K^{-1} mol^{-1}, respectively.

The preceding discussion provides a basis for the law of Dulong and Petit and also explains the exceptions just noted. For most metals, the spacing of the vibrational energy states is small compared to the magnitude of RT at 298 K. Under these circumstances there is no preponderance of atoms in the lowest vibrational state, and theory predicts the molar heat capacity $C_v = 3R = 5.96$ cal K^{-1} mol^{-1} (24.9 J K^{-1} mol^{-1}). Because of the small $P\Delta V$ work done on the surroundings when a solid is heated at constant pressure, the expected value of C_p is slightly larger. Theory confirms the law of Dulong and Petit.

For the solids cited previously that do not obey the Dulong and Petit law (carbon, silicon, and boron), the spacing of the vibrational energy states is not small compared to RT at 298 K. As a result, the molar heat capacities of these elements at 298 K are lower than the limiting values.

EXAMPLE: Use of the law of Dulong and Petit to establish an atomic weight can be illustrated with data for silver, an example already cited. A sample of silver oxide weighing 0.7580 g decomposes to give

[4] Berzelius assigned atomic weights on a scale in which oxygen = 100; the values given above are recalculated to the present standard.

0.7057 g of silver and 0.0523 g of oxygen. Therefore the weight of silver per 16 g of oxygen (1 mol of oxygen atoms) is

$$\frac{16 \text{ g O}}{1 \text{ mol O atoms}} \times \frac{0.7057 \text{ g Ag}}{0.0523 \text{ g O}} = 215.9 \text{ g Ag (mol O atoms)}^{-1}$$

The atomic weight of silver is given by this datum if the formula of the oxide is known. For different assumed formulas, the data give different atomic weights:

$$\begin{array}{rl} AgO_2 & AW(Ag) = 431.8 \\ AgO & AW(Ag) = 215.9 \\ Ag_2O & AW(Ag) = 107.9 \end{array}$$

The specific heat of silver is 0.24 J K^{-1} g^{-1}, which allows an approximate calculation of the atomic weight. The product of the specific heat and the atomic weight, designated x, will give the molar heat capacity:

$$0.24 \text{ J K}^{-1} \text{ g}^{-1} (x) \cong 25 \text{ J K}^{-1} \text{ mol}^{-1}$$

$$x \cong 100 \text{ g mol}^{-1}$$

Although this value is not accurate, it clearly distinguishes between the possibilities which were consistent with the more accurate analytical data:

$$AW(Ag) = 107.9$$

ENTROPY OF CRYSTALLINE SOLIDS; THE THIRD LAW OF THERMODYNAMICS

In Chapter 4 we learned that the change of entropy of a system undergoing a reversible change at some particular temperature is the heat transferred to the system, q, divided by the temperature:

$$\Delta S = \frac{q_{\text{rev}}}{T}$$

An analogous relationship holds for the change of entropy of the system if its temperature changes when heat is added. To analyze this relationship correctly, we should use integral calculus,[5] but we can make do with a graphical solution.

First, let us suppose that the heat capacity of the substance is essentially constant over the temperature range in question. From Figure 5–19 we

[5]In calculus notation, $dq = C_p\,dT$, and $dS = (C_p\,dT/T)$. Integration of this equation between the limits T_1 to T_2 gives

$$\Delta S = \int_{T_1}^{T_2} \frac{C_p\,dT}{T} = \int_{T_1}^{T_2} C_p\,d(\ln T)$$

learn that this is the case for silver at high temperatures ($T > 250$ K) where $C_p = 25.5$ J K^{-1} mol^{-1}. We will calculate the approximate increase in entropy of silver accompanying its increase of temperature from 288.2 K to 298.2 K by dividing the heat added by the average temperature over this temperature range:

$$S_{298.2\,\mathrm{K}} - S_{288.2\,\mathrm{K}} = \frac{C_p \times (T_2 - T_1)}{T_{\mathrm{average}}}$$

$$= \frac{25.5\ \mathrm{J\ K^{-1}\ mol^{-1}} \times (298.2\ \mathrm{K} - 288.2\ \mathrm{K})}{293.2\ \mathrm{K}}$$

$$= 0.870\ \mathrm{J\ K^{-1}\ mol^{-1}}$$

But this calculation is that of the area enclosed in a plot of (C_p/T) versus T. Figure 5–22a shows the rectangular region which corresponds to the calculation just given. Figure 5–22b shows the correct plot in which the ~3% variation in C_p/T over this temperature range is shown. The areas enclosed in 5–22a and 5–22b are approximately the same. The complete plot of the C_p/T versus T for silver from very low temperatures is given in Figure 5–22c. We see in this graph that the value of C_p/T approaches zero as the temperature goes to zero. The area under the curve from $T = 0$ K to $T = 298.2$ K gives $S_{298.2} - S_0$. This is the value of $S_{298.2}$ because *the entropy of every perfect crystalline material at 0 K is zero*. This is a statement of the *third law of thermodynamics*, which evolved during the first twenty years of this century in the work of T. W. RICHARDS, W. NERNST, G. N. LEWIS, and G. E. GIBSON. With the third law of thermodynamics, it is possible to determine the entropy of a substance from measurements of its heat capacity from very low temperatures to the temperature under consideration. Table 5–3 gives values of the entropy of certain elements and compounds determined in this way.

The assignment of zero as the entropy of a perfect crystalline compound at 0 K is consistent with what has been said about entropy and randomness or disorder. In a perfect crystal at 0 K, there is no disorder. Each lattice site is occupied by the appropriate atom, and each atom has the lowest possible vibrational energy. This is perfect order. With an increase of temperature, the additional vibrational energy can be distributed among the atoms of the crystal in a number of ways. This is a type of randomness or disorder, and it contributes to the entropy of a substance just as did the randomness of spatial distribution of molecules discussed in Chapter 3. However, we will probe the relationship between randomness and entropy more thoroughly here than we did in discussing the dependence of molar entropy of a gas upon its volume.

Before applying simple probability theory to the distribution of 100 atoms among the various energy states shown in Figure 5–21b, we will consider a system of three atoms with a total of one unit of energy. There is only *one macroscopic state* for this system. This state has one atom with one unit of energy and two atoms with zero units of energy. There are *three*

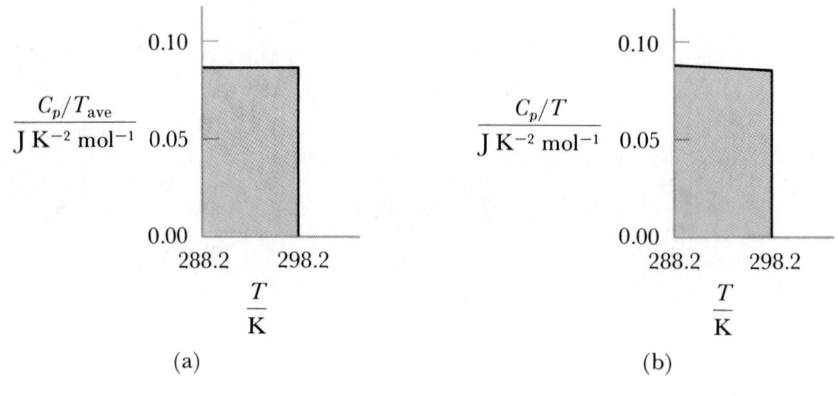

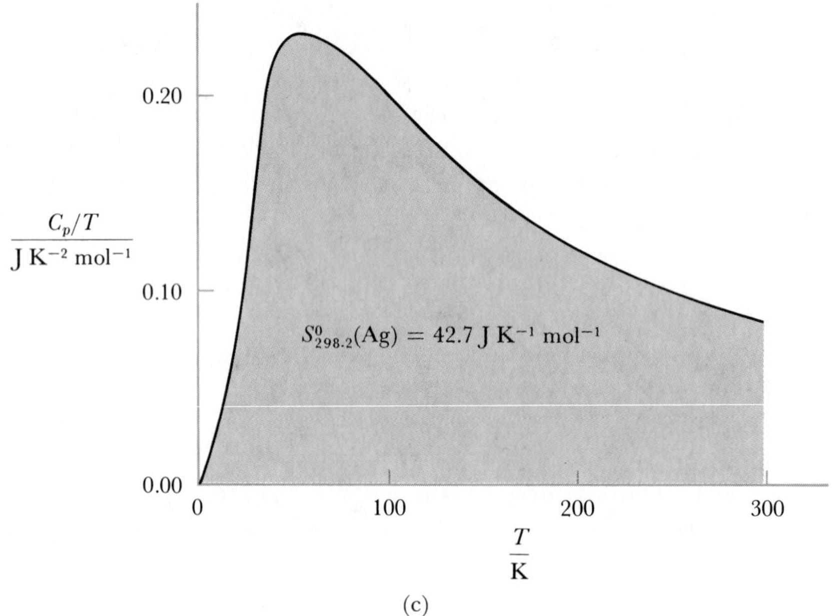

FIGURE 5–22
The change of entropy of silver metal undergoing a change of temperature.

$$\Delta T = (298.2\ \text{K} - 288.2\ \text{K})$$

(a) A rectangular plot C_p/T_{ave} versus T.

$$\text{Area} = \frac{25.5\ \text{J K}^{-1}\ \text{mol}^{-1}}{293.2\ \text{K}} \times 10.0\ \text{K} = 0.87\ \text{J K}^{-1}\ \text{mol}^{-1}$$

(b) A continuous plot of C_p/T versus T. Under the curve from 288.2 K to 298.2 K, the area is 0.87 J K^{-1} mol^{-1}.

$$\Delta T = (298.2\ \text{K} - 0\ \text{K})$$

(c) A continuous plot of C_p/T versus T. Under the curve from 0 K to 298.2 K, the area is 42.7 J K^{-1} mol^{-1}. This is $S^0_{298.2}$ for silver.

TABLE 5–3
The Entropies of Solid Elements and Solid Compounds at 298.2 K

Element	$\frac{S}{\text{cal K}^{-1}\text{ mol}^{-1}}$	$\frac{S}{\text{J K}^{-1}\text{ mol}^{-1}}$
C(diamond)	0.59	2.5
Si	4.47	18.7
Ge	10.14	42.43
Li	6.70	28.0
Na	12.20	51.0
K	15.20	63.6
Ag	10.21	42.7

Compound	$\frac{S}{\text{cal K}^{-1}\text{ mol}^{-1}}$	$\frac{S}{\text{J K}^{-1}\text{ mol}^{-1}}$
NaF	14.0	58.6
KCl	19.76	82.7
RbBr	25.88	108.3
CsI	31.0	129.7
MgO	6.40	26.8

microscopically different arrangements corresponding to this *one macroscopic state*. They can be depicted

$i = 1$	$E = E_1$	a	b	c
$i = 0$	$E = E_0$	b, c	a, c	a, b

in which the three atoms are designated a, b, and c. This can be derived by inspection as we have done, but it also can be calculated as the number of arrangements of three objects in two groups, one group with two objects and the other group with one object. This is given by

$$\Omega = \frac{3!}{2!\,1!} = \frac{3 \times 2 \times 1}{2 \times 1 \times 1} = 3$$

This is simply the equation derived from probability theory for the number of ways of arranging N objects in groups of $n_0, n_1, n_2, \cdots,$

$$\Omega = \frac{N!}{n_0!\,n_1!\,n_2!}$$

where $n!$ is the factorial of the number n:

$$n! = n \times (n-1) \times (n-2) \times \cdots \times 2 \times 1$$

This number of microscopically different arrangements, denoted by the symbol Ω, is a measure of disorder in the sense in which we are using the term. The relationship between entropy and disorder is given by Boltzmann's equation,

$$S = k \ln \Omega$$

in which $k = 1.38 \times 10^{-23}\ \mathrm{J\ K^{-1}}$ (Boltzmann's constant).[6]

Now let us calculate Ω and S for the system of 100 atoms distributed among the states separated by 4.15×10^{-21} J or 2500 J $\mathrm{mol^{-1}}$ (see Figure 5–21b):

0 K, and 10 K: $\Omega = \dfrac{100!}{100!} = 1$ $\qquad S = k \ln 1 = 0$

100 K: $\Omega = \dfrac{100!}{95!\,5!} = 7.53 \times 10^{7}$ $\qquad S = k \ln(7.53 \times 10^{7}) = 2.50 \times 10^{-22}\ \mathrm{J\ K^{-1}}$

300 K: $\Omega = \dfrac{100!}{63!\,24!\,8!\,3!} = 3.14 \times 10^{41}$ $\qquad S = k \ln(3.14 \times 10^{41}) = 1.32 \times 10^{-21}\ \mathrm{J\ K^{-1}}$

We see that the disorder, as measured by the value of Ω, increases dramatically as the temperature increases, with the concomitant increase in the number of states among which the atoms are distributed. As a result of the increase of Ω with an increase of temperature, the entropy increases also. Although the entropies calculated here are very small compared to those tabulated in Table 5–3, these calculated values are for 100 atoms, not a mole.

We can find the entropy of a sample of matter both by applying the third law of thermodynamics to heat capacity data and by using the equation $S = k \ln \Omega$, but only for substances for which we know the energies of all allowed vibrational and rotational states. The two independent approaches to evaluating the molar entropy agree, and this agreement supports the validity of the third law of thermodynamics.

Now we can consider why the Boltzmann distribution is more probable than other macroscopic states with the same energy content. For simplicity, we will consider a smaller sample, a sample of 40 atoms with a total of 18 units of energy distributed among equally spaced energy levels. Figure 5–23 shows five macroscopically different states corresponding to this energy content. The first of these, in which each state has one third as many atoms as the preceding state, is a Boltzmann distribution; the others are not. The value of Ω calculated in the way already outlined is the largest for the Boltzmann distribution. (If a larger number of atoms were considered, the Boltzmann distribution would be even more favored than in this simple calculation.)

---

[6]The equation $S = k \ln \Omega$ is carved on Boltzmann's tombstone in Vienna, Austria.

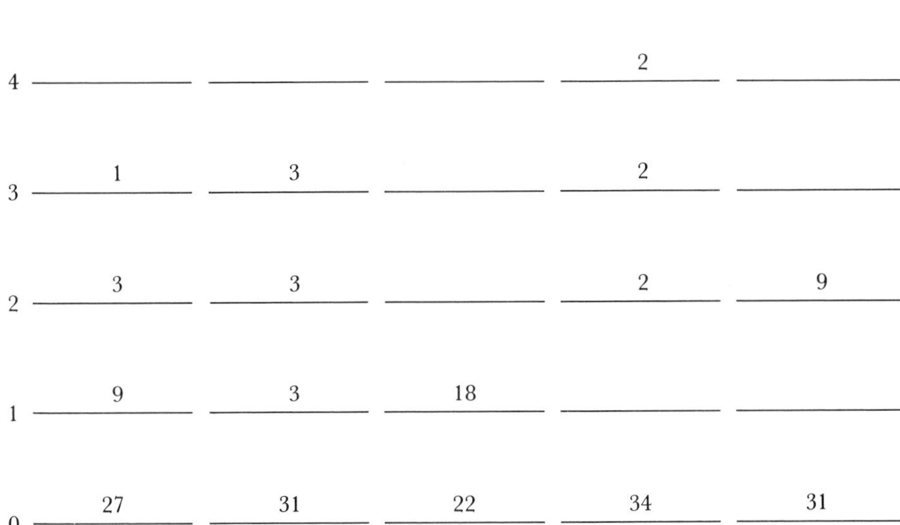

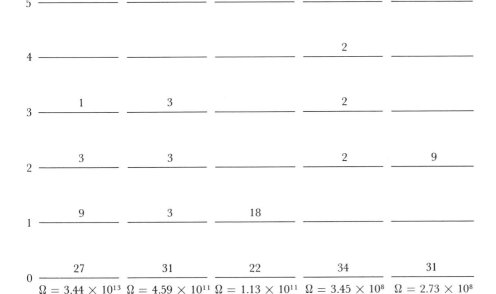

FIGURE 5–23
Macroscopically different distributions of 40 atoms with a total of 18 units of energy. The value of Ω is given for each distribution. It is largest for the Boltzmann distribution, in which each state has a particular fraction of the number of atoms in the next lower state (here $n_{j+1}/n_j = 1/3$).

The discussion of this section has given you an introduction to some important topics: the third law of thermodynamics and a quantitative relationship between disorder and entropy. Although these topics have been introduced in connection with the vibration of atoms in solids, the concepts have broader applications.

Biographical Notes

JÖNS J. BERZELIUS (1779–1848), a Swedish chemist, is considered to be one of the founders of modern chemistry. Educated as an M.D., he started his studies in chemistry in his spare time. His early work dealt with analyses and atomic-weight determinations, but he also made important contributions to electrochemistry, structures of compounds, catalysis, and isomerism.

WILLIAM HENRY BRAGG (1862–1942) and his son WILLIAM LAWRENCE BRAGG (1890–1971), British scientists, received the Nobel Prize in Physics in 1915 for their pioneering work in developing the field of x-ray diffraction.

PIERRE L. DULONG (1785–1838) became Professor of Physics at the École Polytechnique in 1820, although he was trained initially as a physician. In addition to studying the heat capacities of solids, he discovered the unstable compound nitrogen trichloride. He lost an eye and two fingers in an explosion of this substance.

GEORGE E. GIBSON (1884–1954) was Professor of Chemistry in the University of California from 1913 until his retirement. His first work in thermodynamics was carried out in Germany. In addition to collaborating with Lewis, he directed the research of many graduate students who went on to do important work in physical chemistry; among his students were H. Eyring and W. M. Latimer.

WALTHER H. NERNST (1864–1941), a German physicist and chemist, was Professor of Physical Chemistry at Göttingen (1891) and Berlin (1905); in 1925 he became Director of the Institute of Physics of the University of Berlin. In addition to his work in areas related to the third law of thermodynamics, he did pioneering work in electrochemistry (see Chapter 16). He received the Nobel Prize in Chemistry in 1920.

ALEXIS T. PETIT (1791–1820) was a French physicist. He was trained at the École Polytechnique and was Professor of Physics there before his death. With Dulong, he studied the heat capacities of solids.

PYTHAGORAS of Samos (an island in the Aegean Sea) lived in the 6th century B.C. A philosopher and mathematician, he is credited with many discoveries in simple geometry in addition to the theorem by which he is known. He left no writings, however, and it is possible that some of these discoveries were made by his followers.

THEODORE W. RICHARDS (1868–1928), Professor of Chemistry at Harvard University, was the first American to receive the Nobel Prize in Chemistry (in 1914). This award recognized his accurate determinations of atomic weights. He did important work in many areas of physical chemistry. Among his students was G. N. Lewis.

Problems and Questions

5–1 Is the portion of a sodium chloride lattice shown in Figure 5–3c a unit cell?

5–2 What are the lengths of the edge and face diagonal of a cube that has a body diagonal of 2.30 Å?

5–3 The volume of a cube is 7.42×10^{-30} m^3. What are the lengths of the edge, the face diagonal, and the body diagonal of this cube?

5–4 The volume of a pyramid is given by the equation $V = \frac{1}{3}b \times h$ (b is the area of the base and h is the height of the pyramid). What fraction of the volume of each cube shown in Figure 5–4 is occupied by the inscribed tetrahedron and octahedron?

5–5 Examine Figure 5–6c and derive Bragg's law $s = n\lambda/(2 \sin\theta)$. (The meaning of these symbols is given in the figure and the text.)

5–6 Draw a two-dimensional lattice of equal circles in which the coordination number of each circle is three. Indicate the two-dimensional unit cell, and calculate the fraction of the space occupied by circles in this lattice. What relationship does this lattice have with the triangular packing of circles shown in Figure 5–9?

5–7 Consider the calculations associated with the packing of circles in a plane (Figure 5–9). If canned beverages were packed in hexagonal seven-packs instead of rectangular six-packs, what would be the maximum realizable percentage increase in storage capacity of a truck or a refrigerator?

5–8 Consider an arrangement of identical spheres in a simple cubic lattice. (The element polonium crystallizes in this atomic arrangement.) Draw two possible unit cells for this structure. What fraction of the volume is occupied by spheres? If the closest center-to-center distance between spheres is $2r$, give the four next closest center-to-center distances and the number of neighbors at each of these distances.

5–9 Metallic silver crystallizes in the cubic closest-packed lattice. Silver atoms, in contact along the diagonal of a face of the cubic unit cell, are separated by 2.889 Å (center of atom to center of atom distance). What is the length of a side of the unit cell and what is its volume? What is the density of metallic silver?

5–10 Metallic cesium crystallizes in the body-centered cubic lattice. Cesium atoms, in contact along the diagonal of the body of the cubic unit cell, are separated by 5.324 Å. What is the length of a side of the cubic unit cell and what is its volume? What is the density of metallic cesium?

5–11 Study the sodium chloride structure shown in Figure 1-2. Identify the planes of atoms that are separated by 2.814 Å$/\sqrt{5}$. (The closest sodium ion–chloride ion distance is 2.814 Å.)

5–12 The density of metallic potassium is 0.86 g cm^{-3}. The structure of this metal is body-centered cubic. What is the length of a side of the cubic unit cell, and what is the atomic radius of potassium in potassium metal?

5–13 The density of metallic copper is 8.92 g cm^{-3}. The structure of this metal is cubic closest-packed. What is the length of a side of the cubic unit cell, and what is the atomic radius of copper in copper metal? Calculate the latter quantity r(Cu) in another way, starting with the fact that 74.0% of the space is occupied.

5–14 How many atoms are there in a unit cell for a hexagonal closest-packed arrangement of identical atoms? (Study Figures 5–9, 5–11, and 5–12 to help answer the question.)

5–15 Study Figures 5–15 and 5–16, and calculate in terms of r the values of s_6 and s_7, the sixth and seventh smallest separations of atoms in the body-centered cubic arrangement. (Values of s_1–s_5 have been calculated in the text.) How many atoms are at each of these separations?

5–16 How many atoms are in the diamond unit cell, shown in Figure 5–17?

5–17 Which do you expect to have the larger entropy, 1 mol crystalline quartz (SiO_2) or 1 mol silica glass (SiO_2)?

5–18 What are the values of E_i/RT for which the exponential factor $e^{-E_i/RT}$ has the values 0.200, 0.040, 0.010, and 0.002?

5–19 What is the Boltzmann factor $e^{-E_1/RT}$ for $E_1 = 1000$ J mol^{-1} and $E_1 = 2500$ J mol^{-1} at 200 K? What fraction of the atoms in a one-dimensional crystalline solid would be in the lowest vibrational state for each of these cases? (*Hint*: Show by long division that $1/(1 - x) = 1 + x + x^2 + x^3 + \cdots$ and, therefore, $1/(1 + x + x^2 + x^3 + \cdots) = 1 - x$. You can use this relationship to simplify this calculation.)

5–20 An oxide of yttrium is 78.74% (by weight) yttrium. The specific heat of yttrium metal is 0.071 cal g^{-1}. What is the atomic weight of yttrium?

5–21 How many microscopic states are there that correspond to a macroscopic state with 9 atoms having zero units of energy, 3 atoms with one unit of energy, and 1 atom with two units of energy? This group of 13 atoms has a total of five units of energy. Another macroscopic state with this amount of energy has 8 atoms with zero units of energy and 5 atoms with one unit of energy. How many microscopic states correspond to this macroscopic state?

5–22 Consider a collection of 7 atoms that has a total of four units of energy. The allowed energy levels are equally spaced one unit apart. The macroscopic state with the Boltzmann distribution is one with 4 atoms having zero energy, 2 atoms having one unit of energy and 1 atom having two units of energy. Calculate Ω for this arrangement. What other macroscopic states are possible for this number of atoms with this amount of energy, and what is the value of Ω for each?

6

The Liquid State; Changes of State

6–1 Introduction

In many of its properties, the liquid state is intermediate between the gaseous and solid states. Like solids but unlike gases, liquids are relatively incompressible; the density of a liquid is not much affected by moderate variation of pressure. At 25°C the density of water is 0.99707 g cm^{-3} at 1 atm, and it is 1.046 g cm^{-3} at 1065 atm, a change by a factor of ~1.05. At 400°C, where water is a gas at all pressures, water has a density of 3.26×10^{-4} g cm^{-3} at 1 atm and a density of 0.157 g cm^{-3} at 242 atm, a change by a factor of ~482. Like gases but unlike solids, liquids are free to flow. To attain the lowest possible potential energy in the Earth's gravitational field, a liquid flows to fill the bottom of its container.

Liquids are intermediate between solids and gases in energy or enthalpy content. For water,

$$H_2O(s, 0°C) = H_2O(l, 0°C) \qquad \Delta H^0 = +1{,}440 \text{ cal mol}^{-1} = +6{,}020 \text{ J mol}^{-1}$$

$$H_2O(l, 0°C) = H_2O(g, 0°C) \qquad \Delta H^0 = +10{,}750 \text{ cal mol}^{-1} = +44{,}980 \text{ J mol}^{-1}$$

Thus in enthalpy content at 0°C, liquid water is 11.8% of the way from solid water to water vapor:

$$100 \times \frac{6{,}020 \text{ J mol}^{-1}}{6{,}020 \text{ J mol}^{-1} + 44{,}980 \text{ J mol}^{-1}} = 11.8\%$$

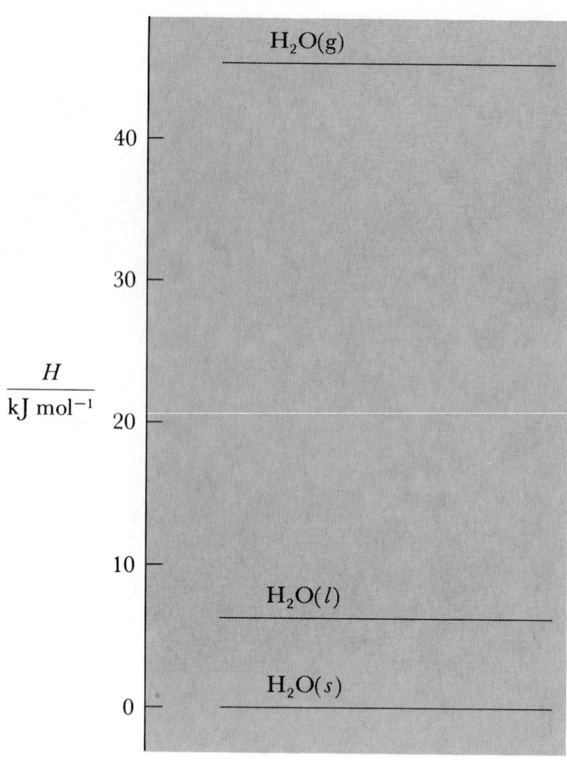

FIGURE 6–1
The relative enthalpies of solid water, liquid water, and gaseous water at 0°C.

This is shown in Figure 6–1. In each of the endothermic processes,

$$\text{solid} \rightarrow \text{liquid} \qquad \text{liquid} \rightarrow \text{gas}$$

there is an increase in the entropy of the substance being considered. The increase in randomness in going from solid to liquid to gas already has been mentioned (see Figure II), and in Chapter 3 we related the tendency for liquid water to vaporize (an endothermic process) to the entropy increase for the water in the process. Now we will concern ourselves further with the energy and entropy changes associated with changes of state.

6–2 The Structure of Liquids

We learned in Chapter 5 that atoms in a crystalline solid vibrate about their positions of maximum stability. In the liquid state there also is molecular motion, some of which resembles the vibration of the atoms in a solid and some of which involves motion over greater distances. Because liquids can flow and because molecules can diffuse within a liquid, these molecules can move relative to one another by appreciable distances. Nonetheless, there are preferred intermolecular separations in a liquid. The closest distance between molecules in a liquid would be expected to be similar to that observed for the solid. We can use x-ray diffraction to learn about the average positions of molecules in a liquid, as we did in studying crystalline

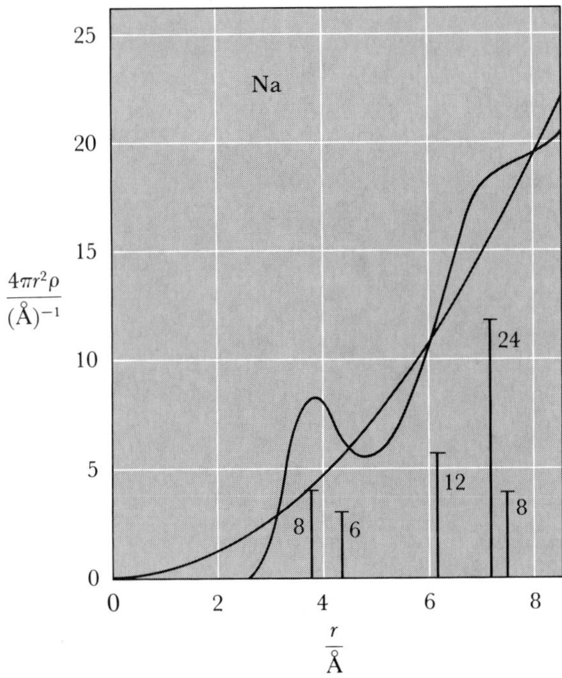

FIGURE 6–2
The structure of liquid sodium; $4\pi r^2\rho(r)$ versus r, in which r is the distance from the center of any one atom. The parabolic line is that for $\rho(r) = \text{constant}$, which corresponds to the average density. The vertical lines are from Figure 5–16 for solid sodium. [Based upon G. W. Castellan, *Physical Chemistry*, Addison-Wesley, Reading, Mass. (1964).]

solids. But a liquid has no parallel planes of atoms, as those shown in Figure 5–7 for crystalline sodium chloride.

Figure 5–16 gave the interatomic separations in solid sodium, a body-centered cubic lattice. Figure 6–2 is an analogous plot for liquid sodium. In this figure the ordinate and abscissa have the same significance as in the earlier figure; the number of atoms at a distance between r and $r + \Delta r$ from any one atom is plotted against this distance, r. The function plotted is $4\pi r^2\rho(r)$, in which $\rho(r)$ is the radial distribution function. This is the concentration of atoms as a function of r averaged over the time of observation in a diffraction experiment. Features that you should notice in Figure 6–2 are:

1. The shortest distance between sodium atoms in the liquid is comparable to this quantity for the solid. The vertical line at 3.72 Å for the shortest distance in the crystalline solid is the average separation of the vibrating atoms.
2. The function $4\pi r^2\rho(r)$ shows that separations between atoms in the liquid are similar to those in the solid. The thermal motion of atoms in the liquid and the lack of long-range order in the liquid are shown by the continuous nature of this curve. Only for $r < \sim 3$ Å is the value of $4\pi r^2\rho(r)$ equal to zero; at all other distances from an atom, there is a finite probability of finding another atom.
3. At large separations, the function $4\pi r^2\rho(r)$ has essentially the same value as the concentration calculated under the assumption that the liquid is a structureless medium of constant density [$\rho(r) = \text{constant}$].

The qualitative features of the radial distribution function for liquid sodium are shown by other liquids also. But when we consider liquids made up of complex molecules, we discover many additional features of liquid structure. Liquid water is special in many ways, and intermolecular hydrogen bonding is responsible for many of its unusual properties.

Some mention also must be made of a type of matter commonly known as *liquid crystals*. The heating of certain solid organic compounds (compounds of carbon) gives not the expected transition from an opaque solid to a transparent liquid at a sharply defined temperature, but rather transitions to liquidlike phases at several temperatures. The appearance and other characteristics of the substance change at each transition temperature. The situation can be summarized:

$$\text{Crystalline solid} \underset{T_1}{\rightleftarrows} \text{Mesophase I} \underset{T_2}{\rightleftarrows} \text{Mesophase II} \underset{T_3}{\rightleftarrows} \text{Isotropic liquid}$$

Here, Mesophase I and Mesophase II are the names that are used for the two liquid-crystal phases. Each of these phases has some of the properties that identify liquids: the ability to flow when poured and to take the shape of the bottom of their container. The adjective "isotropic," used to describe the liquid produced at the highest transition temperature, means that the properties of the liquid are the same in all directions, a characteristic of most liquids (e.g., liquid sodium, the structure of which has just been described). To oversimplify a very complex subject, the structural differences among the four phases can be regarded as:

Crystalline solid	Order in 3 dimensions
Mesophase I	Order in 2 dimensions; disorder in 1 dimension
Mesophase II	Order in 1 dimension; disorder in 2 dimensions
Isotropic liquid	Disorder in 3 dimensions

To form liquid-crystal phases, a compound must have certain features of molecular structure, generally an elongated rodlike shape. The substance 5-chloro-6-*n*-heptyloxy-2-napthoic acid,[1]

```
              H       Cl
              |       |            H  H    H  H    H  H
              C       C             \/      \/      \/
       H    //  \   /   \\     O     C       C       C      H
         \ C      C       C  /   \  /  \    /  \    /  \   /
           |      ||      |       C      C        C      C
           C      C       C      / \    / \      / \      \
   O      / \\   /  \   // \    H   H  H   H    H   H  H   H
    \\  C        C       C    H
 H—O  /          |       |
                 H       H
```

[1]At this time, don't worry about the meaning of this name and the structural formula. Simply recognize that this is a rodlike molecule.

exhibits the behavior outlined, with $T_1 = 165.5°C$, $T_2 = 176.5°C$, and $T_3 = 201°C$. Unlike isotropic liquid phases, liquid-crystal phases are opaque, and they reflect light. This is the property of liquid crystals that is the basis for their use in display devices, for example, digital watches.[2]

6–3 Thermodynamic Relationships for Liquid–Vapor Interconversion

Now we will consider in detail the conversion of liquid to vapor, a subject which we introduced previously. We will learn how to do a number of quantitative calculations. In some calculations, we will use a strictly empirical equation, that is, one which is not derived from theory, but is based on its success in correlating experimental data. In other calculations, we will use a closely related equation derived from thermodynamic theory. In the course of this, you will see a relationship of experiment and theory:

experimental data → empirical equation

thermodynamic theory → theoretical equation

This theoretical equation involves the changes of enthalpy and entropy in the vaporization process. We also will learn the scientific basis for familiar phenomena, the boiling and freezing of liquids.

EQUILIBRIUM BETWEEN A LIQUID AND ITS VAPOR

Liquid water left in an open container at room temperature (e.g., 25°C, a warm room) slowly disappears. It evaporates; the process

$$H_2O(l, 25°C) = H_2O(g, 25°C)$$

occurs spontaneously. To define this process more completely, we should specify the pressure of the gaseous water. The system of an open container in a room is not defined well enough for careful study. Consider instead the experiment pictured in Figure 6–3. Initially the liquid is separated from an evacuated space by a partition. The pressure gauge registers zero. Then the partition is removed; the evacuated space now is open to molecules from the liquid. Evaporation of some liquid gives the situation at the bottom of the figure. The gauge shows an increase in pressure when the partition is removed; the pressure then remains steady at some particular value (23.76 torr if the liquid is water at 25.0°C) as long as the temperature is maintained constant. This pressure is called the *vapor pressure* of the liquid, and a vapor with this pressure is called the *saturated vapor*. A state of equilibrium exists between liquid and saturated vapor. We discussed this transformation in Chapter 3:

$$H_2O(l, 25°C) = H_2O(g, 23.76 \text{ torr}, 25°C)$$

[2] G. H. Heilmeier, "Liquid–Crystal Display Devices," *Sci. Am.*, April 1970, pp. 100–106.

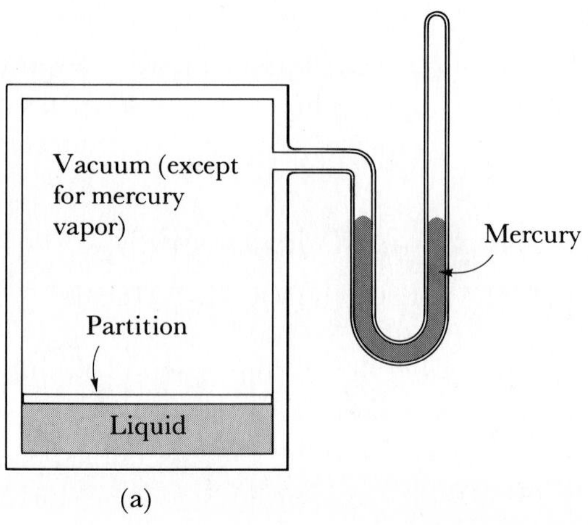

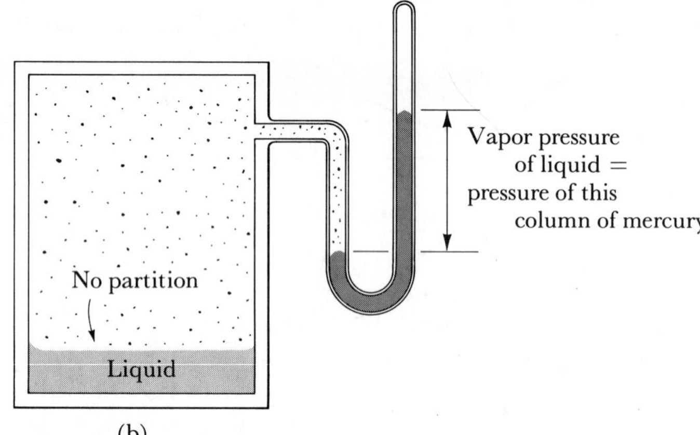

FIGURE 6–3
The pressure of the vapor in equilibrium with a liquid. (a) Liquid is separated from open space by partition. (b) Partition removed, and liquid vaporizes.

It is a process for which $\Delta G = 0$. In this state of equilibrium, the rate of the process

$$H_2O(l) \rightarrow H_2O(g)$$

is the same as the rate of the process

$$H_2O(g) \rightarrow H_2O(l)$$

The pressure on the walls of the container above the liquid and on the liquid's surface results from collisions between molecules in the vapor space and these surfaces. This pressure is related to the temperature and the concentration of gas by the ideal gas law,

$$P = \left(\frac{n}{V}\right)RT$$

or by an equation of state more appropriate for the nonideality of the gas phase. It is not particularly fruitful, however, to consider the pressure of the gas *in equilibrium with the liquid* in terms of the gas laws. In the situations discussed in Chapter 4, the concentration of the gas and the temperature

TABLE 6–1
The Vapor Pressure of Water as a Function of Temperature[a]

$\frac{t}{°C}$	$\frac{P}{torr}$	$\frac{V^b/n}{L\ mol^{-1}}$	$\frac{t}{°C}$	$\frac{P}{torr}$	$\frac{V^b/n}{L\ mol^{-1}}$
−10.0	2.149	7639	50.0	92.51	218
0.0	4.579	3721	60.0	149.38	139
10.0	9.209	1918	70.0	233.7	91.6
20.0	17.535	1043	80.0	355.1	62.0
25.0	23.756	784	90.0	525.76	43.1
30.0	31.824	594	100.0	760.0	30.6
40.0	55.324	353	110.0	1074.6	22.2

[a]1 torr ≡ 1/760 atm
1 atm ≡ 101,325 kg m^{-1} s^{-1} = 101,325 Pa
[b]These values of the molar volume of the vapor have been calculated from the ideal gas law:

$$\frac{V}{n} = \frac{RT}{P}$$

These values agree with experimental values within 1% for all temperatures below ~90°C.

were varied independently by allowing the pressure to vary. The temperature and the concentration of a saturated vapor cannot be varied independently. Experiment shows that the vapor pressure of a liquid depends upon temperature, but does not depend upon the size of the container or upon the amount of excess liquid present.[3]

The vapor pressure of a liquid increases with an increase in temperature; the data for water are presented in Table 6–1, which also gives the molar volume of the saturated vapor at each temperature. This dependence of vapor pressure upon temperature can be discussed either from the point of view of thermodynamics or from the point of view of the rates of the vaporization and condensation processes.

The thermodynamics of the vaporization process were introduced in Chapter 3, and should be reviewed at this point. The process $H_2O(l) = H_2O(g)$ is endothermic; the enthalpy change at 25°C is

$$\Delta H = 44{,}020 \text{ J mol}^{-1}$$

and the change in internal energy is slightly smaller than this value:

$$\Delta U = \Delta H - \Delta(PV) = \Delta H - [(PV)_{\text{gas}} - (PV)_{\text{liquid}}]$$
$$\cong \Delta H - nRT$$

[3]The vapor pressure of a pure liquid from tiny droplets is greater than that from a bulk liquid sample, and the vapor pressure of a pure liquid depends also upon the total pressure in the system. This latter point will be discussed further in Chapter 7.

The change of the quantity PV is approximately equal to the value of PV for the gas, because the liquid occupies a much smaller volume than the gas. Therefore the change in internal energy is

$$\Delta U = 44{,}020 \text{ J mol}^{-1} - 1 \text{ mol} \times 8.31 \text{ J K}^{-1} \text{ mol}^{-1} \times 298.2 \text{ K}$$
$$= 41{,}540 \text{ J mol}^{-1}$$

This increase of internal energy upon vaporization is an increase of the potential energy of the water molecules in the vapor state, as compared to the liquid state. The water molecules attract one another, and their potential energy is more positive when they are separated by large distances (in the gas phase) than when they are close together (in the liquid phase). The change is qualitatively analogous to the change of potential energy with the separation of two gas molecules, depicted in Figure 4–16. The discussion of the vaporization process is Chapter 3 helped introduce the concept of entropy, and to discuss the dependence of vapor pressure upon temperature we must consider the change of entropy in the vaporization process, as well as the change of enthalpy. For each temperature, the equilibrium process

$$H_2O(l) = H_2O(g, P = P_{\text{satd}}, V/n = RT/P_{\text{satd}})$$

is a process for which $\Delta G = 0$. The change of entropy can be calculated from the enthalpy change:

$$\Delta S = \frac{\Delta H}{T} - \frac{\Delta G}{T} = \frac{\Delta H}{T} \qquad \text{if } \Delta G = 0$$

In using this equation to calculate ΔS, we must use the value of ΔH appropriate for the temperature in question. Because the molar heat capacities for liquid water and water vapor are different:[4]

$$C_p[H_2O(l)] = 75.3 \text{ J K}^{-1} \text{ mol}^{-1}$$
$$C_p[H_2O(g)] = 37.2 \text{ J K}^{-1} \text{ mol}^{-1}$$

the value of ΔH depends upon temperature. Using the equation developed in Section 3–3 we can calculate the value of ΔH for the vaporization of water at a temperature T_2 from the value at 298.2 K ($+44{,}020 \text{ J mol}^{-1}$):

$$\frac{\Delta H(T_2)}{\text{J mol}^{-1}} = +44{,}020 + (T_2 - 298.2 \text{ K}) \times (37.2 - 75.3) \text{ K}^{-1}$$

Thus at 373.2 K,

$$\frac{\Delta H}{\text{J mol}^{-1}} = +44{,}020 + (373.2 \text{ K} - 298.2 \text{ K})(-38.1 \text{ K}^{-1}) = 41{,}160$$

The values of ΔS for the vaporization process to give saturated vapor at

[4]Each of these values depends mildly upon temperature; this will not be taken into account.

298.2 K and at 373.2 K are:

	$T = 298.2$ K	$T = 373.2$ K
P_{satd}/torr	23.76	760
$(V_{satd} - V_l)$/L mol^{-1}	784	30.6
ΔH/J mol^{-1}	44,020	41,160
ΔS/J K^{-1} mol^{-1}	147.6	110.3

The increase of entropy in conversion of one mole of liquid to one mole of saturated vapor is smaller at the higher temperature. The principal factor making the change of entropy smaller at 373.2 K is the smaller molar volume of the saturated vapor at this temperature.

When liquid–vapor equilibrium is established, as depicted in the lower part of Figure 6–3, the concentration of molecules in the vapor space remains constant not because vaporization has stopped, but because the rates of vaporization and condensation are equal. The temperature dependence of the vapor pressure is a consequence of the relationship between temperature and the rates of these two opposing processes. Experiments show that the vaporization rate of liquids increases dramatically with an increase of temperature.[5] An increase in temperature increases the fraction of the molecules that have enough energy to overcome their mutual attraction in the liquid phase. (Recall that in discussing Figure 4–13 we showed that the fraction of gas molecules with high velocities, and therefore high energies, increases with an increase of temperature.) The rate of evaporation can be expected also to depend upon the concentration of molecules in the surface of the liquid; this decreases very slightly with an increase of temperature. The rate of the condensation process also can be viewed as a product of two factors, a concentration factor and a specific rate factor. The specific rate factor for the condensation process depends only mildly upon the temperature, being proportional to the average velocity of the gas molecules ($\bar{u} \propto T^{1/2}$). Most gas molecules need no energy to condense into a liquid; if a gas molecule hits the surface, it sticks. Therefore, to maintain the vapor pressure equilibrium with an increase in temperature, the concentration in the vapor phase must increase sharply so that the rate of the condensation process will equal the greatly increased rate of the vaporization process. This relationship is summarized in Table 6–2.

The vapor pressures of water and ethyl alcohol are plotted as a function of temperature in Figure 6–4a. The dependence is not linear. For clarity in presentation of data, scientists generally seek ways of plotting data to give straight lines. Then too, a linear correlation may provide guidance in establishing a theoretical framework for interpreting the data. These vapor pressure data give lines that are *approximately* straight in plots of the logarithm of the vapor pressure versus the reciprocal of the temperature, as is shown

[5]A simple experimental study of the vaporization rate of liquids is given in J. F. Brennan, J. S. Shapiro, and E. C. Walton, *J. Chem. Educ.* **51**, 276 (1974).

TABLE 6–2
The Factors that Determine Rates of the Processes[a]:

Vaporization $H_2O(l) \rightarrow H_2O(g)$
Condensation $H_2O(g) \rightarrow H_2O(l)$

	298.2 K	373.2 K
Vaporization		
Specific rate	1	29
Surface concentration	1	0.97
Condensation		
Specific rate	1	1.1
Concentration of vapor	1	25.6

[a]For purposes of this comparison, each factor is defined as unity at 298.2 K.

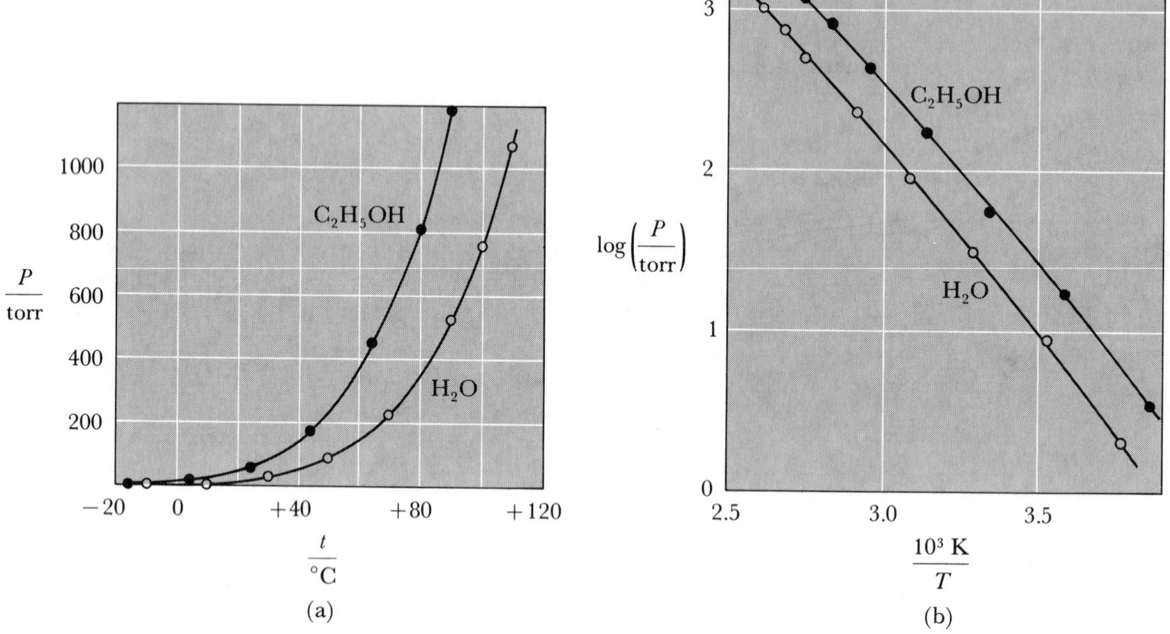

FIGURE 6–4
The vapor pressure of water and ethyl alcohol as a function of temperature. (a) P versus T. (b) log (P/torr) versus ($1/T$).

in Figure 6–4b. The equation for a straight line in this set of coordinates is:

$$\log\left(\frac{P}{\text{torr}}\right) = \frac{A}{T} + \text{B} \qquad \text{(see footnote 6)}$$

[6]Since one can take the logarithm of a number, but not the logarithm of a physical quantity, the ordinate in this graph is log (P/torr), not log P.

The slope of the line is A and the intercept at $1/T = 0$ is B. Since plots of this type are not exactly linear, the values of A and B that correlate the data for a particular compound depend upon the range of temperatures being considered. The parameters A and B that correlate data for water and ethyl alcohol over a large range of temperatures are:

	Temp. range / K	A / K	B
Water (H_2O)	273–373	−2264.5	8.9672
Ethyl alcohol (C_2H_5OH)	218–363	−2276.9	9.3847

These parameters can be used to calculate approximate values of the vapor pressure at a particular temperature in this range, or the temperature at which a liquid has a particular vapor pressure.

EXAMPLE: What is the vapor pressure of ethyl alcohol at 20.0°C?

$$\log\left(\frac{P}{\text{torr}}\right) = -\frac{2276.9\text{ K}}{293.2\text{ K}} + 9.3847 = 1.619$$

$$P = 41.6\text{ torr}$$

(The experimental value is 43.9 torr.)

EXAMPLE: At what temperature is the vapor pressure of ethyl alcohol equal to 760 torr?

The equation can be rearranged to:

$$T = \frac{A}{\log\left(\frac{P}{\text{torr}}\right) - B} = \frac{-2276.9\text{ K}}{\log 760 - 9.3847}$$

$$= 350.1\text{ K } (76.9°\text{C})$$

(The experimental value is 78.3°C.)

In these calculations we see the error in assuming a strictly linear relationship between $\log(P/\text{torr})$ and T^{-1}.

At this point we have not learned the significance of the parameters A and B. Thermodynamic theory provides the answer under the assumptions:[7]

1. the molar volume of the liquid is very much smaller than the molar volume of the vapor,
2. the vapor behaves as an ideal gas, and
3. the values of ΔH^0 and ΔS^0 do not depend upon temperature.

The resulting equation gives for the dependence of vapor pressure of a liquid upon temperature:

$$\ln\left(\frac{P}{\text{torr}}\right) = -\frac{\Delta H^0_{\text{vap}}}{R} \times \frac{1}{T} + \frac{\Delta S^0_{\text{torr}}}{R}$$

[7]See O. L. I. Brown, *J. Chem. Educ.* **28**, 428 (1951).

The parameters A and B in the empirical equation involving decadic logarithms therefore are identified as

$$A = -\frac{\Delta H^0_{\text{vap}}}{2.303R}$$

$$B = \frac{\Delta S^0_{\text{torr}}}{2.303R}$$

But use of these empirical parameters to obtain ΔH^0_{vap} and ΔS^0_{torr}:

$$\Delta H^0_{\text{vap}} = -2.303R \times A$$

$$\Delta S^0_{\text{torr}} = 2.303R \times B$$

is justified only if the temperature range over which the data are correlated is a range in which the three previously stated assumptions are valid.

These assumptions are valid for water at low temperatures where the molar volume of saturated water vapor is large. The vapor pressure of water at intervals of 5 K between 10°C and 40°C is plotted in Figure 6–5a. The linear equation

$$\log\left(\frac{P}{\text{torr}}\right) = -\frac{2301.6 \text{ K}}{T} + 9.094$$

allows us to calculate values of vapor pressure over a smaller range of temperature with an average difference between the observed and calculated values of 0.06 torr. The parameters correspond to:

$$\begin{aligned}\Delta H^0_{\text{vap}} &= 2301.6 \text{ K} \times 8.315 \text{ J K}^{-1} \text{ mol}^{-1} \times 2.303\\ &= 44{,}070 \text{ J mol}^{-1}\\ \Delta S^0_{\text{torr}} &= 9.094 \times 8.315 \text{ J K}^{-1} \text{ mol}^{-1} \times 2.303\\ &= 174.1 \text{ J K}^{-1} \text{ mol}^{-1}\end{aligned}$$

We will interpret these values as suitable for the midpoint of the temperature range, 298.2 K. This value of ΔH^0_{vap} derived from the vapor pressure data agrees very well with the value presented earlier (44.02 kJ mol^{-1}) derived from calorimetric studies. Figure 6–5b shows these data with the linear plot extending to $T^{-1} = 0$; it is clear that a long extrapolation is needed to obtain the value of ΔS^0_{torr} as the intercept.

The value of ΔS^0 for a vaporization process depends upon the standard state chosen for the vapor. Since the plot in Figure 6–5 involves vapor pressures in torr, the value of ΔS^0 calculated above, designated as ΔS^0_{torr}, is for the process

$$H_2O(l, 298.2 \text{ K}) = H_2O(g, 1 \text{ torr}, 298.2 \text{ K})$$

For vaporization processes in which the vapor has a different standard state, for example,

$$H_2O(l, 298.2 \text{ K}) = H_2O(g, 1 \text{ atm}, 298.2 \text{ K})$$

or

$$H_2O(l, 298.2 \text{ K}) = H_2O(g, 1 \text{ Pa}, 298.2 \text{ K})$$

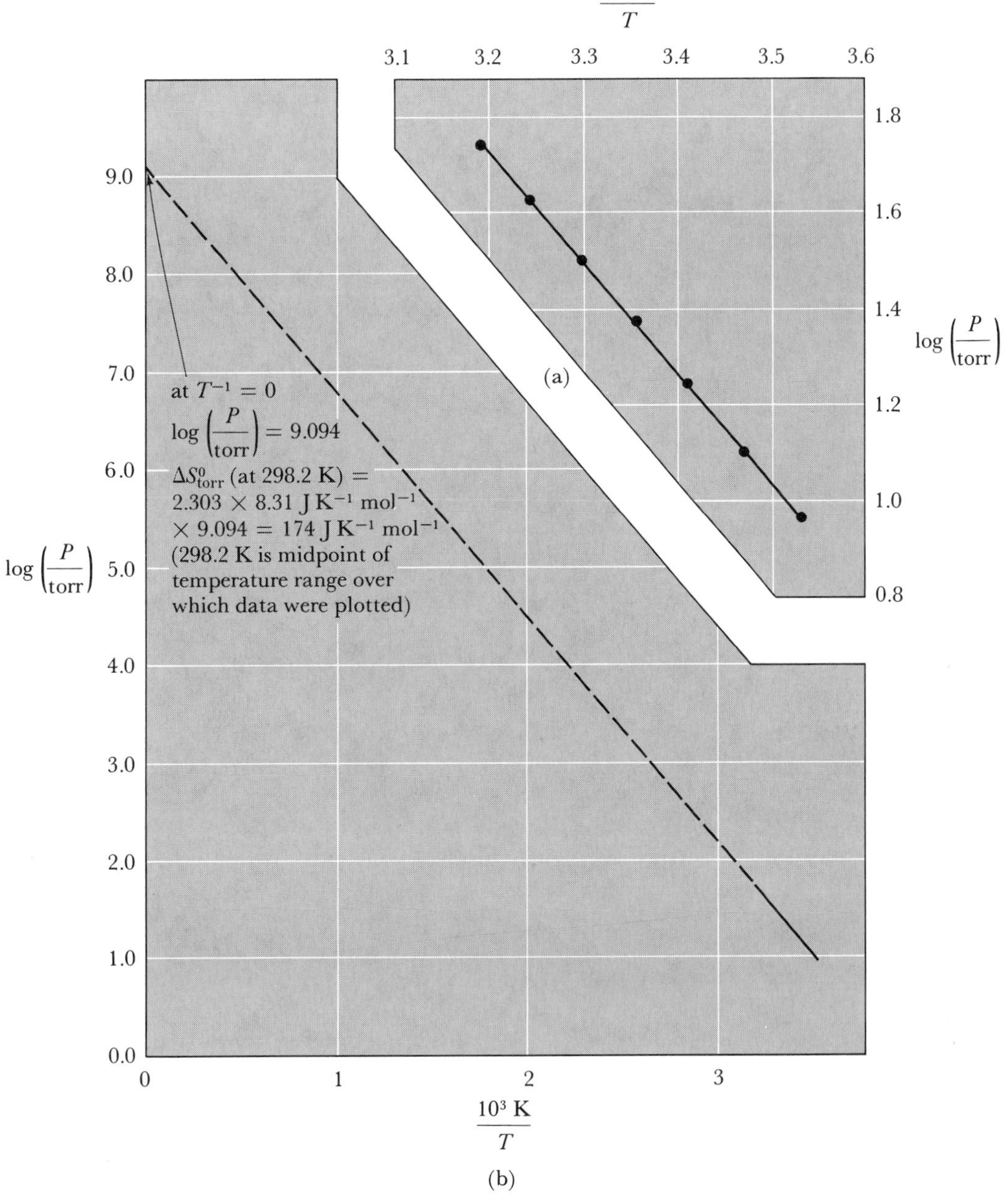

FIGURE 6–5

The vapor pressure of water as a function of temperature. (a) log (P/torr) versus T^{-1} over range T = 283.2 K to 313.2 K (midpoint 298.2 K). The line is

$$\log (P/\text{torr}) = -\frac{2301.6}{T} + 9.094$$

(b) The line given in (a) is extrapolated to $T^{-1} = 0$, where log (P/torr) = 9.094; $\Delta S^0 = 2.303 \times 8.31\ \text{J K}^{-1}\ \text{mol}^{-1} \times 9.094 = 174\ \text{J K}^{-1}\ \text{mol}^{-1}$.

the values of ΔS^0 are different; this point was discussed in Chapter 3, where ΔS^0_{atm} was presented:

$$\Delta S^0_{atm} = +28.39 \text{ cal K}^{-1}\text{ mol}^{-1} = +118.8 \text{ J K}^{-1}\text{ mol}^{-1}$$

The value of ΔS^0_{atm} is less positive than the value of ΔS^0_{torr} because the molar volume of the gas is smaller if the pressure of the gas is 1 atm than if the pressure of the gas is 1 torr. The values of ΔS^0 with the gas in different standard states are related by an equation that involves the ratio of the molar volumes of the gas in the two standard states, designated by subscripts 1 and 2,

$$\Delta S^0_2 = \Delta S^0_1 + R \ln \frac{V_2}{V_1}$$

Since $V_2/V_1 = P_1/P_2$, this becomes

$$\Delta S^0_2 = \Delta S^0_1 + R \ln \frac{P_1}{P_2}$$

Applied to the vaporization of water at 298.2 K, we have for ΔS^0_{Pa} (1 torr = 133.3 Pa):

$$\Delta S^0_{Pa} = \Delta S^0_{torr} + R \ln 133.3$$

$$\frac{\Delta S^0_{Pa}}{\text{J K}^{-1}\text{ mol}^{-1}} = 174.1 + 8.31 \times 4.89 = 214.8$$

$$\Delta S^0_{Pa} = 214.8 \text{ J K}^{-1}\text{ mol}^{-1}$$

Plots of log (P/torr) versus T^{-1} for most substances are approximately linear over a much larger range of temperature than would be expected on the assumptions, already mentioned, which are necessary in deriving the equation

$$\log\left(\frac{P}{\text{torr}}\right) = -\frac{\Delta H^0_{vap}}{2.303RT} + \frac{\Delta S^0_{torr}}{R}$$

Errors cancel one another when all of the assumptions fail at high temperature, where the molar volume of the saturated vapor is relatively small.

We can use this equation to estimate the enthalpy change for a vaporization process if we know the vapor pressure of a liquid at two temperatures. If the equation is written for two temperatures:

$$\log\left(\frac{P_1}{\text{torr}}\right) = -\frac{\Delta H^0_{vap}}{2.303RT_1} + \frac{\Delta S^0_{torr}}{R}$$

$$\log\left(\frac{P_2}{\text{torr}}\right) = -\frac{\Delta H^0_{vap}}{2.303RT_2} + \frac{\Delta S^0_{torr}}{R}$$

and these equations are subtracted from one another, the equation

$$\log\left(\frac{P_1}{P_2}\right) = -\frac{\Delta H_{vap}}{2.303R}\left(\frac{1}{T_1} - \frac{1}{T_2}\right)$$

is obtained.

Example: The vapor pressure of mercury is 0.01267 torr at 50.0°C and is 0.2729 torr at 100.0°C. What is the enthalpy of vaporization in this range of temperature?

$$\Delta H_{\text{vap}} = -2.303R \times \frac{\log\left(\frac{P_1}{P_2}\right)}{\frac{1}{T_1} - \frac{1}{T_2}}$$

$$= \frac{-2.303 \times 8.315 \text{ J K}^{-1} \text{ mol}^{-1} \times \log\left(\frac{0.2729}{0.01267}\right)}{\frac{1}{373.2 \text{ K}} - \frac{1}{323.2 \text{ K}}}$$

$$= 61.59 \text{ kJ mol}^{-1}$$

With a value of ΔH_{vap} known, the vapor pressure at some particular temperature can be estimated if the vapor pressure at another temperature is known; the equation can be rearranged to

$$\log\left(\frac{P_1}{\text{torr}}\right) = \log\left(\frac{P_2}{\text{torr}}\right) - \frac{\Delta H_{\text{vap}}}{2.303R}\left(\frac{1}{T_1} - \frac{1}{T_2}\right)$$

Example: Calculate the vapor pressure of mercury at 70.0°C, given the vapor pressure at 50.0°C and the value of ΔH^0_{vap} from the preceding example:

$$\log\left(\frac{P}{\text{torr}}\right) = \log(0.01267) - \frac{61{,}590 \text{ J mol}^{-1}}{2.303 \times 8.315 \text{ J K}^{-1} \text{ mol}^{-1}} \times \left(\frac{1}{343.2 \text{ K}} - \frac{1}{323.2 \text{ K}}\right)$$

$$= -1.897 + 0.580 = -1.317$$

or

$$P = 0.0482 \text{ torr}$$

Examination of Figure 6–4 suggests that liquid ethyl alcohol has a higher vapor pressure than liquid water at all temperatures. This conclusion is correct. But we could compare substances for which the vapor pressure versus temperature curves cross. Figure 6–6 shows such a comparison of methyl alcohol (CH_3OH), which forms hydrogen bonds, and normal hexane (C_6H_{14}), which does not form hydrogen bonds. The molar enthalpies of vaporization for these substances are in the order

$$\Delta H^0_{\text{vap}}(C_6H_{14}) < \Delta H^0_{\text{vap}}(CH_3OH)$$

and we see that at low temperatures ($T < 322$ K) the vapor pressure of hexane is greater, but that at high temperatures ($T > 322$ K) the vapor pressure of methyl alcohol is greater. We see in this figure that the vapor pressure of a liquid is more sensitive to changes of temperature the larger

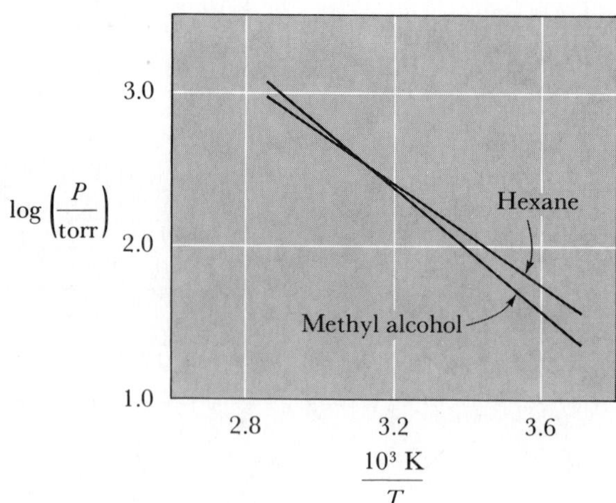

FIGURE 6–6
The vapor pressure as a function of temperature for methyl alcohol (which forms hydrogen bonds) and hexane (which does not form hydrogen bonds). Vapor pressures of these liquids are equal at log $(P/\text{torr}) = 2.587$, $10^3\ \text{K}/T = 3.105$; that is, $P = 386$ torr at $T = 322.0$ K.

the value of ΔH^0_{vap}. In this comparison, the enthalpy change for vaporization of the hydrogen bonded compound is larger, and its vapor pressure changes more with temperature.

The energy required for vaporization of a liquid is supplied by the surroundings if the sample is in thermal contact with its environment. If the sample is insulated from its surroundings, the energy comes from the sample itself. The residual liquid is cooled by evaporation. This can be understood in terms of the kinetic energies of the molecules of the liquid; as already mentioned, it is the faster molecules that escape. The average kinetic energy of the molecules in the liquid decreases if faster ones are lost and if no additional heat flows into the system. The remaining liquid, with its slower molecules, has a lower temperature, because temperature is a measure of the kinetic energy of molecular motion. A fine stream of ethyl chloride (with boiling point 12.3°C) is an effective local anesthetic. Its evaporation will freeze a small region of skin, thereby making it insensitive to minor pain.

BOILING; THE NORMAL BOILING POINT

The phenomenon of boiling differs from vaporization. Simple vaporization occurs from the surface of the liquid, but in boiling, vapor forms within the body of the liquid. This situation, illustrated in Figure 6–7, results if the vapor pressure of the liquid exceeds the applied pressure. The temperature at which the vapor pressure of a liquid is equal to the pressure of the atmosphere is called the *boiling point*, and the temperature at which the vapor pressure is the standard atmosphere (1 atm ≡ 760 torr ≡ 101,325 N m^{-2}) is called the *normal boiling point*. In plots of vapor pressure as a function of temperature, the normal boiling point can be read from the graph as the temperature at which the vapor pressure curve crosses the line $P = 760$ torr.

We can estimate the normal boiling point for a substance from the enthalpy of vaporization and the vapor pressure at some temperature, but

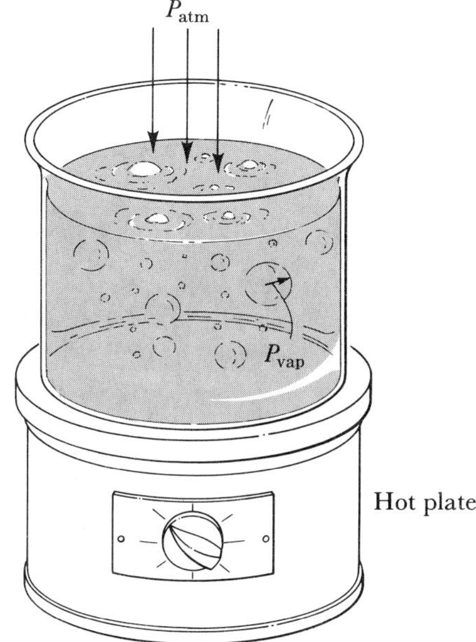

FIGURE 6–7
Boiling of a liquid. For the bubble of vapor to exist in the liquid, the pressure within the bubble (the vapor pressure of the liquid) must be infinitesimally greater than the external pressure (atmospheric pressure plus the pressure due to the liquid over the bubble.)

we may be caught in a long extrapolation if we assume, incorrectly, that log (P/torr) is an accurately linear function of T^{-1}. The equation rearranged to allow calculation of the normal boiling point, T_b, is

$$T_b = \left[\frac{1}{T} + \frac{2.303R}{\Delta H_{\text{vap}}} \log \frac{P}{760 \text{ torr}}\right]^{-1}$$

EXAMPLE: Estimate the normal boiling point of mercury using the values of ΔH_{vap} and $P(100°\text{C})$ given previously.

$$T_b = \left[\frac{1}{373.2 \text{ K}} + \frac{2.303 \times 8.315 \text{ J K}^{-1} \text{ mol}^{-1}}{61{,}590 \text{ J}} \log\left(\frac{0.2729}{760}\right)\right]^{-1}$$

$$= 621.7 \text{ K}$$

(The experimental value is 629.9 K.)

Because atmospheric pressure varies with elevation, the temperature at which water (or any other liquid) boils in an open vessel varies from place to place. Table 6–3 gives the elevations of certain geographical locations and the corresponding temperatures at which water boils. Altitude must be taken into account in cooking at the boiling temperature. For instance, it takes ~7 minutes to soft-boil an egg at Leadville, Colorado.[8]

There are circumstances under which we wish to boil a liquid at temperatures other than the ordinary boiling point. To cook food in water at temperatures greater than the ordinary boiling point we must use a utensil such as a pressure cooker, a closed vessel with an adjustable vent system that maintains a particular pressure greater than atmospheric pressure. Since boiling is much more rapid than simple vaporization, this

---

[8] Information provided by the cook at the Golden Burro Cafe.

TABLE 6–3
The Boiling Point of Water at Various Elevations

Location	Elevation feet	P_{atm} torr	T_b °C
Death Valley (Calif.)	−282	770	100.3
New York City, N.Y.	10	760	100
Madison, Wisc.	900	730	99
Boulder, Colo.	5,430	610	94
Top of Mt. Washington (N.H.)	6,293	590	93
Leadville, Colo.	10,150	510	89
Top of Mt. Whitney (Calif.)	14,494	430	85
Top of Mt. McKinley (Alaska)	20,320	340	79
Top of Mt. Everest (Tibet)	29,028	240	70

way of converting a liquid to its vapor is preferred for many chemical operations (e.g., separating a volatile solvent from nonvolatile solutes, and separating the volatile components of a liquid solution from one another by distillation). Yet the ordinary boiling temperature of certain substances may be so high that the substances decompose in boiling. Under these circumstances, we use a vacuum pump to maintain low pressures, so that boiling occurs at temperatures low enough to prevent decomposition.

When heat is added continuously to a liquid, the liquid's temperature and vapor pressure increase. If boiling starts when the vapor pressure reaches atmospheric pressure, the temperature stays constant at the boiling point; the heat being added gives the system the needed enthalpy of vaporization. If boiling actually does not occur when the vapor pressure reaches atmospheric pressure, the temperature continues to rise and the liquid is said to be *superheated*. A superheated liquid is in a very unstable condition, and when it finally boils, it does so violently. A liquid superheats because many molecules must cooperate to form its vapor bubbles. Small foreign particles may provide nuclei for bubble formation, so glass beads or broken pottery chips are used in some laboratory operations to prevent superheating of liquids.

THE ENTROPY OF VAPORIZATION; TROUTON'S RULE

Discussion in Chapter 3 and previously in this chapter has emphasized the importance of the entropy change in determining the extent of the vaporization process. Values of the entropy change for vaporization for liquids also give us information about the liquid state. This is shown in Table 6–4, which gives values of ΔS_{vap} at the normal boiling point for a number of liquids. F. T. TROUTON called attention to the fact that for many liquids the value of $\Delta H_{vap}/T_b \cong 21$ cal K^{-1} mol^{-1} or 88 J K^{-1} mol^{-1}, and this generalization is known as *Trouton's rule*. Because the liquids being considered

TABLE 6–4
The Entropy of Vaporization of Various Liquids

	At normal boiling point[a]			$V = 49.5$ L mol^{-1}	
Substance	T_b / K	ΔH_{vap} / kJ mol^{-1}	ΔS_{vap} / J K^{-1} mol^{-1}	T / K[b]	ΔS_{vap} / J K^{-1} mol^{-1}
CH_4	112	8.20	73.2	92.6	94.1
Xe	165	12.64	76.6	145	87.0
n-C_6H_{14}	342	28.9	84.5	323.1	92.9
C_6H_6	353	30.7	87.0	334.6	95.0
Hg	630	58.1	92.2	632.5	92.0
H_2O	373.2	41.16	110.3	302	145.6
C_2H_5OH	352	38.6	109.7	286	149.8

[a]At the normal boiling point the volume of the saturated vapor (ignoring nonideality) is given by the gas law:

$$V = 22.41 \text{ L mol}^{-1} \times \frac{T_b}{273.15 \text{ K}}$$

[b]At this temperature the molar volume of each saturated vapor is 49.5 L mol^{-1}.

have a range of boiling points, the molar volumes of the saturated vapors differ. We have seen that the molar volume of a vapor is the principal factor in determining its entropy, and this factor makes a different contribution to the value of ΔS_{vap} for different substances at their normal boiling points. J. H. HILDEBRAND has reasoned that it would be more rational to compare values of the entropy change on vaporization if the values were calculated for temperatures at which the molar volume of each saturated vapor was the same; this point is illustrated in Figure 6–8.

Table 6–4 also presents values of the entropy change on vaporization at temperatures where the molar volume of the saturated vapor of each substance is 49.5 L mol^{-1}. (This volume is an arbitrary selection; it was used by Hildebrand in his correlations.) Let us notice two features of these tabulated quantities. For the hydrocarbons, xenon, and mercury, values for the molar entropy of vaporization are much closer to one another at temperatures where the molar volumes of the saturated vapors are the same, rather than at temperatures where the vapor pressures of each liquid is 760 torr. These data support Hildebrand's rule, which states that the entropy changes for vaporization of different liquids are equal at equal molar volumes of the saturated vapor. Values of the entropy of vaporization of liquids capable of hydrogen bonding (water and ethyl alcohol) are abnormally high compared to those of other liquids, whether the comparison is made at the normal boiling points or at temperatures where the molar volumes of the saturated vapors are equal. The greater order (less randomness) in a hydrogen-bonded liquid makes the increase in randomness (increase in entropy) upon vaporization of such a liquid greater than that for other liquids.

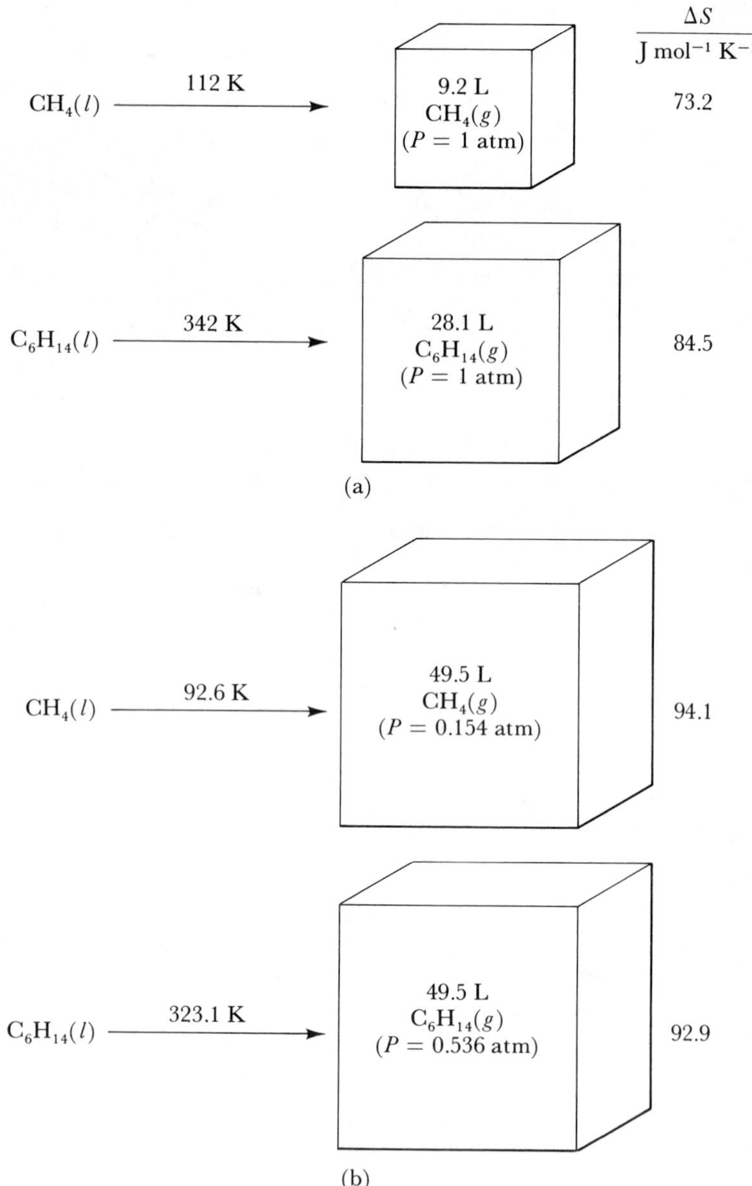

FIGURE 6–8
Comparison of the entropy of vaporization of methane (CH_4) and normal hexane (n-C_6H_{14}) (a) at the normal boiling points (where the molar volumes of the vapors are 9.2 L mol^{-1} and 28.1 L mol^{-1}, respectively), and (b) at temperatures where the molar volume of each vapor is 49.5 L mol^{-1} (92.6 K and 323.1 K, respectively). The more nearly constant values for the process in which the molar volume of the vapor is constant supports Hildebrand's rule.

THE CRITICAL STATE

The critical point was shown in the phase diagram for water given in Figure 1–5. In a graph of pressure versus temperature, the critical point is the end of the line that gives the vapor pressure of the liquid as a function of temperature. For water, this extremity is

$$T_c = 647.4 \text{ K} \qquad P_c = 217.7 \text{ atm}$$

This temperature and pressure are called the *critical temperature* and *critical pressure*, respectively.

Now we will consider the critical point in a graph of pressure versus molar volume. At very high temperatures ($T \gg T_c$), the pressure versus molar volume plots conform closely to the ideal gas law,

$$P\left(\frac{V}{n}\right) = RT$$

This curve is shown in Figure 6–9 at $T = \Theta_1$. At lower temperatures ($T < T_c$), the ideal gas law may be approximately correct for the gas phase if the pressures are not too high. The curved line segment **CD** ($T = \Theta_4$) is calculated using the ideal gas law. But the ideal gas law does not apply to the liquid phase. Curved line segment **AB** ($T = \Theta_4$) shows that the liquid is relatively incompressible. This line is almost vertical; the relative change in volume is very small for a large change of pressure. Point **B** (liquid phase) and Point **C** (gas phase) give the (P, V/n) data for two phases which are in equilibrium with one another. If the graph dealt with water with $\Theta_4 =$ 25.0°C,

Point **B** (liquid)	$P = 23.756$ torr
	$V/n = 0.01807 \text{ L mol}^{-1}$
Point **C** (vapor)	$P = 23.756$ torr
	$V/n = 784.0 \text{ L mol}^{-1}$

At this temperature the molar volumes of these two phases differ enormously.

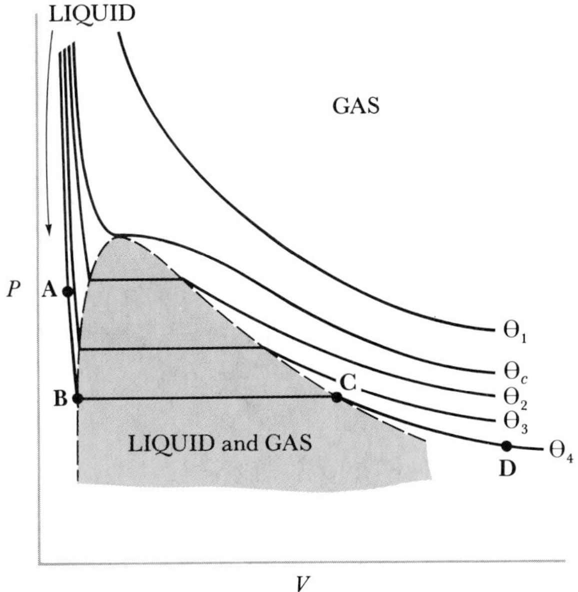

FIGURE 6–9
Pressure–molar volume isotherms:

$T = \Theta_1$ (very much greater than the critical temperature),

$$P\left(\frac{V}{n}\right) = RT$$

$T = \Theta_c$ (the critical temperature). Isotherm drawn to conform to van der Waals equation of state.

$T = \Theta_2$, Θ_3, and Θ_4 (below critical temperature). In the shaded region, both the liquid and gas phases are present.

TABLE 6–5
The Density of Liquid Water and Saturated Water Vapor

$\frac{T}{°C}$	$\frac{V/n}{L\ mol^{-1}}$ Liquid	Vapor	Heat of vaporization kJ mol^{-1}
100	0.0188	30.6	41.16
150	0.0196	7.07	37.0
200	0.0208	2.29	34.0
250	0.0225	0.900	30.0
300	0.0252	0.390	24.6
320	0.0268	0.279	21.7
340	0.0293	0.194	18.0
360	0.0327	0.125	12.6
374.2[a]	0.0563	0.0563	0

[a]Critical temperature.

With an increase of temperature, the molar volume of the liquid increases and that of the saturated vapor decreases; they approach one another, and they become equal at the critical temperature. We can see this in the data for water summarized in Table 6–5, and also by comparing the molar volumes indicated by points at the extremities of the horizontal lines in Figure 6–9. At the critical temperature, Θ_c, the points corresponding to the values of (V/n) for the two phases coalesce to a single point. The critical temperature is the temperature above which a gas cannot be liquified by applying pressure. It is the temperature at which the properties of the two fluid phases, the liquid and its saturated vapor, become identical.

The meaning of the line **ABCD** in Figure 6–9 will become clearer as we consider in detail a liquid under high pressure thermostatted at $T = \Theta_4$ in a cylinder with a movable piston. At Point **A**, only the liquid phase is present. Weights then are removed from the piston until the pressure exerted on the liquid is only infinitesimally greater than the vapor pressure of the liquid at that temperature; this is Point **B**. With a further infinitesimal reduction in the applied pressure, a vapor phase forms; along the horizontal segment **BC** two phases are present, the liquid and the vapor. The relative amount of material in the two phases varies along the line, 100% liquid at **B** and 100% vapor at **C**. With further reduction in pressure on the vapor at **C**, the volume of the vapor increases; this is the line segment **CD**. Along this line segment, the sample is entirely gaseous. This is illustrated in Figure 6–10.

In the isothermal vaporization of a liquid, like that depicted along **ABCD** in Figure 6–9, both liquid and vapor phases are present during part of the overall process (the part **B** → **C**). However, there are other routes from liquid to vapor in which the system consists of a single phase at all stages of the overall process. This is shown as **AGFED** in Figure 6–10

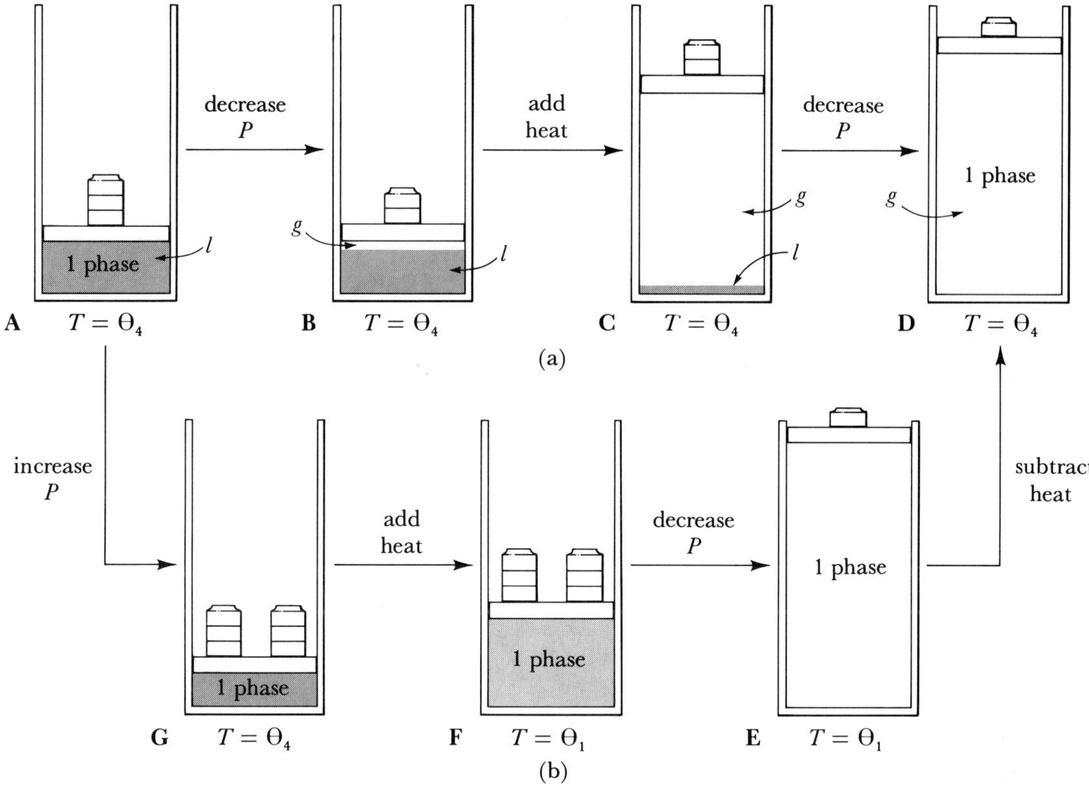

FIGURE 6–10
The transformation of liquid to vapor by two routes: (a) Route **ABCD**; notice that two phases are present in all stages between **B** and **C**. (b) Route **AGFED**; only one phase is present in all stages between **A** and **D**.

and in Figure 6–11 on both pressure–volume and pressure–temperature diagrams. (The isothermal process **ABCD** carried out at temperature Θ_4, shown in Figure 6–9, also is shown here.) In Step **A** → **G**, the liquid is compressed isothermally to a pressure above the critical pressure. Then heat is added at constant pressure, and the temperature is raised to a temperature Θ_1, above the critical temperature; this is Step **G** → **F**. At this new temperature, Θ_1, the pressure is reduced isothermally to its value at Point **D**; this is Step **F** → **E**. The temperature then is reduced to Θ_4 while the pressure is held constant; this is Step **E** → **D**. When the temperature is Θ_4, the state of the system is that specified by Point **D** (all vapor). Along this entire route, **AGEFD**, only a single phase is present at any point despite the fact that at the start, Point **A**, the fluid is a liquid, and at the end, Point **D**, the fluid is a gas.

A liquid is removed by vaporization along a route analogous to **AGFED** in Figure 6–11 in a procedure called *critical-point drying*. The procedure is useful in drying biological materials for examination under an electron

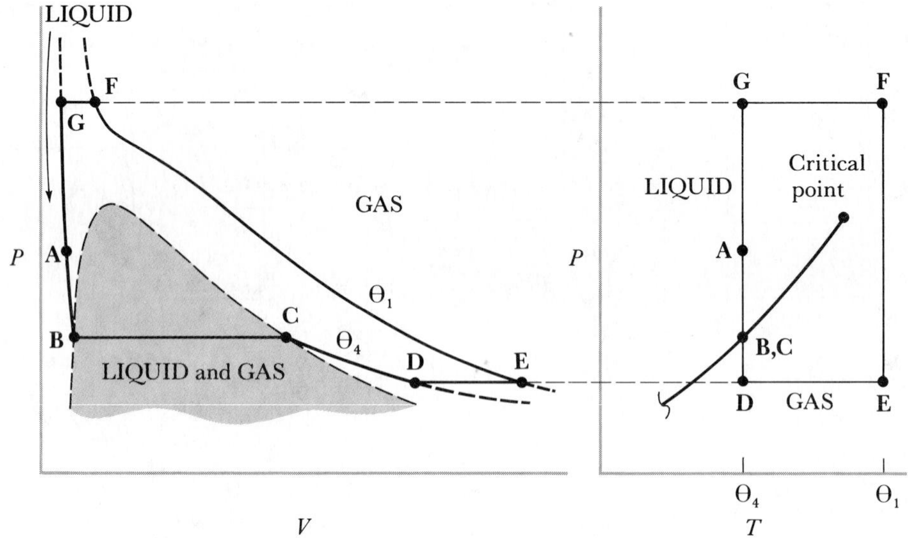

FIGURE 6–11
Conversion of liquid (**A**) at temperature Θ_4 to vapor (**D**) at temperature Θ_4. Isothermal route, **ABCD**; route along which there is only one phase, **AGFED**. At left, pressure versus volume. At right, pressure versus temperature.

microscope. Simply drying such materials in air damages them structurally as the water film–air boundary passes through the sample. No such air–liquid interface ever passes through the sample in critical-point drying. In this procedure a fluid with an appropriately low critical temperature is needed. Carbon dioxide ($T_c = 31.1°C$), nitrous oxide ($T_c = 36.5°C$), and monochlorotrifluoromethane, Freon ($T_c = 28.9°C$) have been used.

6–4 Phase Equilibria Involving Solid, Liquid, and Vapor

Equilibrium between the liquid and vapor phases for pure substances exists at particular combinations of temperature and pressure. This is true also for equilibrium between solid and liquid, and solid and vapor. Conditions for these equilibria were shown in the phase diagram for water in Figure 1–5. Figure 6–12 repeats this figure and also presents the phase diagrams for carbon dioxide and sulfur. These three diagrams have common features and contrasting features. Raising the temperature of water at a constant pressure of 1 atm from 260 K to 380 K results in the successive transformations:

$$\text{solid} \rightarrow \text{liquid at 273 K}$$
$$\text{liquid} \rightarrow \text{vapor at 373 K}$$

But raising the temperature of carbon dioxide at a constant pressure of

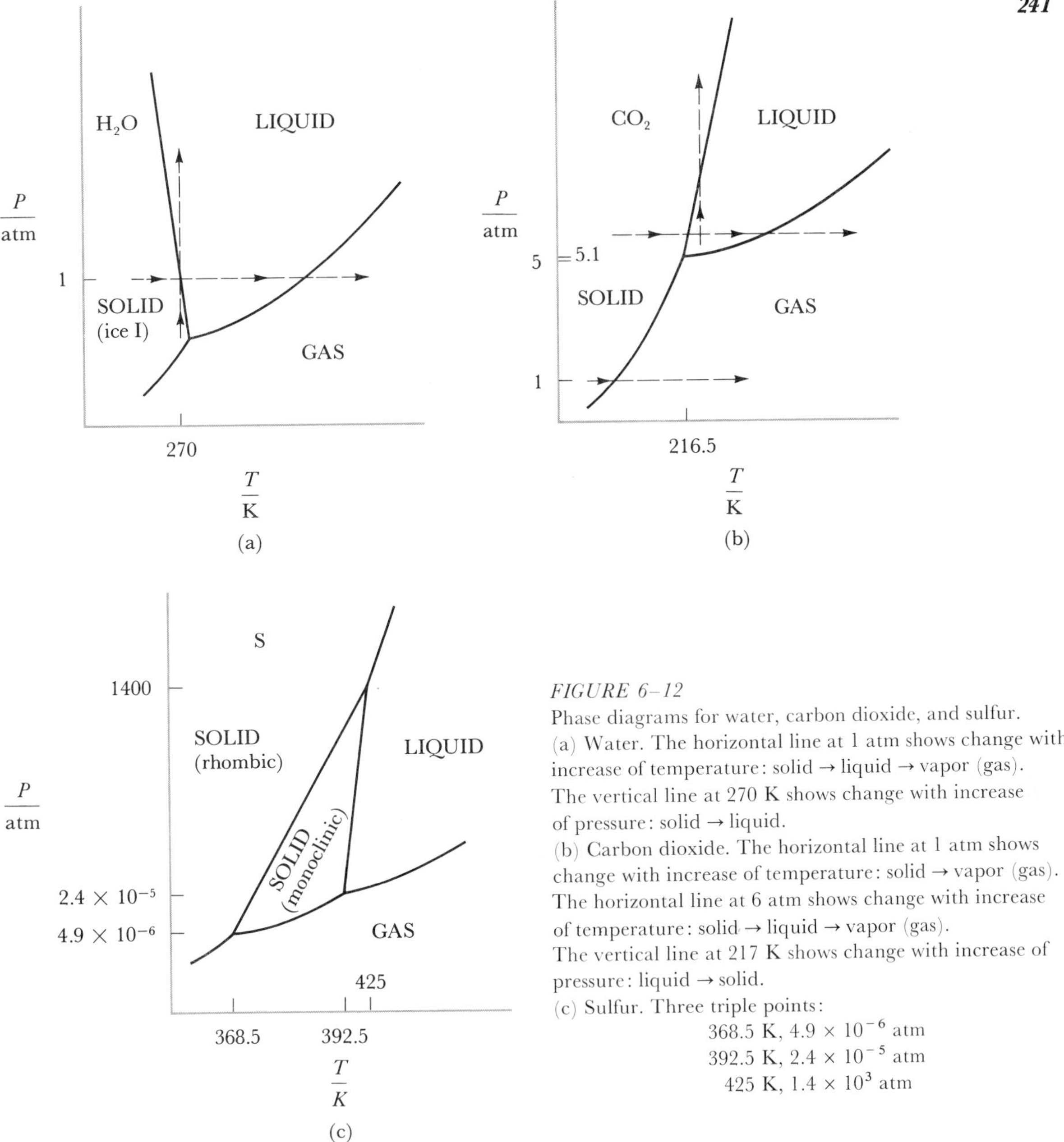

FIGURE 6–12
Phase diagrams for water, carbon dioxide, and sulfur.
(a) Water. The horizontal line at 1 atm shows change with increase of temperature: solid → liquid → vapor (gas). The vertical line at 270 K shows change with increase of pressure: solid → liquid.
(b) Carbon dioxide. The horizontal line at 1 atm shows change with increase of temperature: solid → vapor (gas). The horizontal line at 6 atm shows change with increase of temperature: solid → liquid → vapor (gas). The vertical line at 217 K shows change with increase of pressure: liquid → solid.
(c) Sulfur. Three triple points:
368.5 K, 4.9×10^{-6} atm
392.5 K, 2.4×10^{-5} atm
425 K, 1.4×10^{3} atm

1 atm from 190 K to 310 K results in only one transformation:

solid → vapor at 194.7 K

You probably are familiar with this transformation, which occurs when Dry Ice (solid carbon dioxide) sublimes away. (The change of solid directly to vapor is called *sublimation.*) For carbon dioxide to exist as a liquid, the pressure must be greater than the pressure at the triple point, 5.1 atm. For water, the corresponding pressure is 4.588 torr (the pressure at the triple point). Atmospheric pressure on the surface of Mars at the present time is less than this value, so liquid water cannot exist there. At a pressure of 1 atm and temperatures below $\sim 0°C$, liquid water cannot exist. Snow sublimes directly on cold days, as one would expect from the phase diagram.

The phase diagram for sulfur introduces a new feature: more than one solid phase. This phenomenon, called *polymorphism*, is exhibited by many substances, including water. For sulfur, the two solid phases have different crystalline forms, but each solid contains discrete S_8 molecules. These solids are called rhombic sulfur and monoclinic sulfur, to designate the crystalline forms. For sulfur the sequence of transformations brought about by increase of temperature can be:

$$\text{solid (rhombic)} \rightarrow \text{vapor},$$
$$\text{solid (rhombic)} \rightarrow \text{solid (monoclinic)} \rightarrow \text{vapor}, \quad \text{or}$$
$$\text{solid (rhombic)} \rightarrow \text{solid (monoclinic)} \rightarrow \text{liquid} \rightarrow \text{vapor},$$

depending upon the pressure.[9] Because of the polymorphism of sulfur, this phase diagram has three triple points: 368.5 K and 4.9×10^{-6} atm, 392.5 K and 2.4×10^{-5} atm, and 425 K and 1.4×10^{3} atm.

The water and carbon dioxide phase diagrams contrast with respect to the sign of the slope of the solid–liquid line, which is determined by the relative densities of the two phases. In contrast to water, for which the solid phase (ice I) is less dense than the liquid, solid carbon dioxide ($d = 1.56$ g cm^{-3}) is more dense than the liquid ($d = 1.10$ g cm^{-3}). For the transformation

$$CO_2(s) = CO_2(l)$$

the change of volume is

$$\begin{aligned}\Delta V &= V_l - V_s \\ &= 40.0 \text{ cm}^3 \text{ mol}^{-1} - 28.2 \text{ cm}^3 \text{ mol}^{-1} = 11.8 \text{ cm}^3 \text{ mol}^{-1}\end{aligned}$$

Thus an increase of pressure increases the temperature at which solid and liquid carbon dioxide are in equilibrium. If the pressure on an equilibrium mixture of solid and liquid carbon dioxide is increased, the phase with the larger molar volume (the liquid) is converted to the phase with the smaller molar volume (the solid). This direction of change, liquid → solid, with an increase of pressure is simply an example of Le Chatelier's principle.

In the phase diagram for water shown in Figure 6–13, each line giving the vapor pressure of a condensed phase is extended past the triple point. At temperatures less than that of the triple point (0.01°C), the liquid phase has a larger vapor pressure than the solid phase; at temperatures greater than that of the triple point, the solid phase has a larger vapor pressure.

[9]Remember that these phase diagrams are for equilibria. In the sulfur system, equilibrium between the rhombic and monoclinic phases is not established rapidly. Rapid heating of rhombic sulfur can melt this sulfur at ~ 386 K. The mixture of liquid sulfur and rhombic sulfur so produced is not stable; in time, it will be transformed to solid monoclinic sulfur.

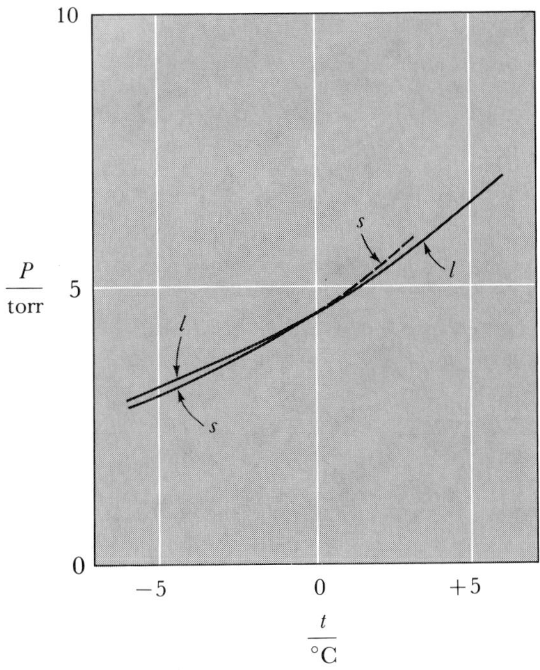

FIGURE 6–13
The phase diagram for water showing the vapor pressure for solid and liquid extrapolated into regions where each of these condensed phases is unstable. The curve for liquid water at $t < 0°C$ is based upon experimental study of supercooled liquid water. The curve for ice at $t > 0°C$ is hypothetical, obtained by extrapolation.

The phase with the larger vapor pressure is unstable with respect to the phase with the smaller vapor pressure:

$$\text{at } T > 0.01°\text{C} \qquad \text{solid} \rightarrow \text{liquid}$$

and

$$\text{at } T < 0.01°\text{C} \qquad \text{liquid} \rightarrow \text{solid}$$

are the transformations that occur spontaneously. If the vapor pressures of the two phases are equal, the phases are in equilibrium.

The vapor pressure curves (P versus T) in Figure 6–13 are drawn to scale, and we see that the plots are curved. In Figure 6–4 the data for liquid water were presented in an approximately linear plot of the logarithm of the vapor pressure versus the reciprocal of the temperature. The same type of plot is approximately linear for the vapor pressure of ice, as is shown in Figure 6–14. Thermodynamic theory identifies the slope in this plot with the enthalpy change for vaporization of the solid; the equation is the same as that already presented:

$$\log \frac{P_1}{P_2} = -\frac{\Delta H_{\text{vap}}}{2.303R}\left(\frac{1}{T_1} - \frac{1}{T_2}\right)$$

in which ΔH_{vap} is the enthalpy change for vaporization of the solid.

EXAMPLE: The vapor pressure of ice is 4.579 torr at 0.00°C and 0.0966 torr at −40.00°C. Estimate the enthalpy change in the process $H_2O(s) = H_2O(g)$.

The equation solved for ΔH_{vap} is:

$$\Delta H_{vap} = -2.303R \times \frac{\log\left(\frac{P_1}{P_2}\right)}{\frac{1}{T_1} - \frac{1}{T_2}}$$

$$= \frac{-2.303 \times 8.315\text{ J K}^{-1}\text{ mol}^{-1} \times \log\left(\frac{4.579}{0.0966}\right)}{\frac{1}{273.2\text{ K}} - \frac{1}{233.2\text{ K}}}$$

$$= 51{,}100\text{ J mol}^{-1}$$

The enthalpy change for melting of ice at 0.0°C and the enthalpy change for vaporization of water at 0.0°C were given in the introduction to the chapter (Section 6–1). The sum of the enthalpy changes for these two processes is equal to the enthalpy change for vaporization of the solid:

$$\Delta H(s \rightarrow g) = \Delta H(s \rightarrow l) + \Delta H(l \rightarrow g)$$
$$= 6{,}020\text{ J mol}^{-1} + 44{,}980\text{ J mol}^{-1} = 51{,}000\text{ J mol}^{-1}$$

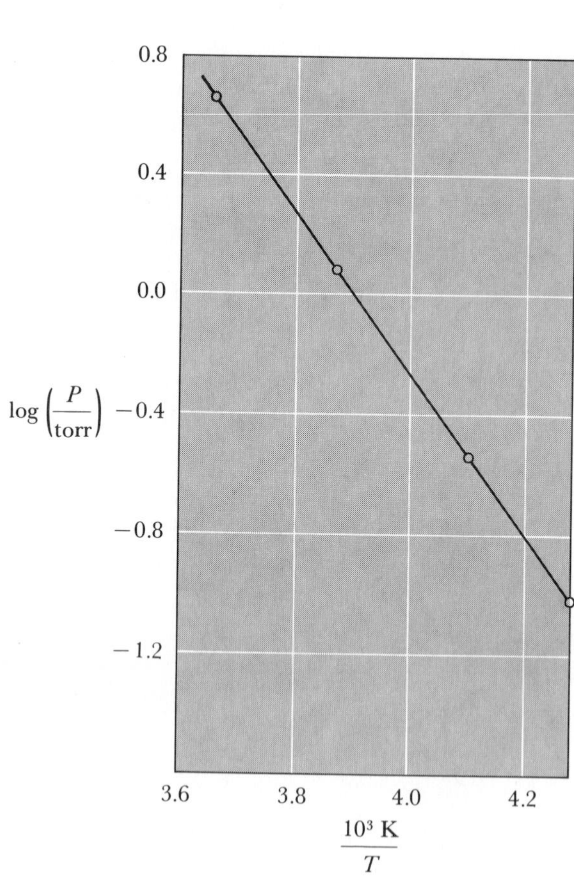

FIGURE 6–14
The vapor pressure of ice. From the slope of the line,

$$\frac{\Delta \log\left(\frac{P}{\text{torr}}\right)}{\Delta\left(\frac{1}{T}\right)} = -2.67 \times 10^3\text{ K}$$

the ΔH_{vap} for ice can be calculated:

$$\Delta H_{vap} = 2.303 \times R \times \frac{-\Delta \log\left(\frac{P}{\text{torr}}\right)}{\Delta\left(\frac{1}{T}\right)}$$

$$= 2.303 \times 8.31\text{ J K mol}^{-1} \times 2.67 \times 10^3\text{ K}$$
$$= 51.1\text{ kJ mol}^{-1}$$

This agrees very well with the value just calculated from the temperature coefficient of the vapor pressure of ice.

We already have considered the fact that liquids can be superheated, that is, heated to a temperature above the boiling point, without boiling. Liquids also can be *supercooled*, that is, cooled to a temperature below the equilibrium crystallization temperature (the true freezing point) without forming a solid phase. Liquids can be supercooled easily because a very large number of atoms or molecules is required to form a crystal of the right size to promote continued growth. It appears that aggregates of atoms or molecules smaller than some critical size do not grow (even in a supercooled liquid). The formation of a crystal of this critical size from the randomly oriented atoms or molecules of the liquid is a highly improbable event. The presence of foreign particles in a liquid can serve as nuclei for the formation of the crystalline solid. In the complete absence of such nuclei, homogeneous nucleation can occur, but the temperature at which this occurs may be very much lower than the true freezing temperature. The amounts of supercooling observed for some liquids are: water, 40 K; gallium metal, 70 K; sodium chloride, 165 K; and copper metal, 240 K.

With the supercooling of liquids such a common phenomenon, how can the true freezing temperature be measured accurately? This can be done by measuring the temperature of the sample as a function of time as heat flows to the colder surroundings. If supercooling does not occur, the temperature remains constant at the true freezing temperature even though heat is flowing from the sample to the surroundings, because the freezing process liquid → solid is exothermic. The energy evolved in this process goes to the surroundings. The cooling curve for a liquid that freezes at the true freezing temperature is shown in Figure 6–15a. If supercooling occurs, the temperature drops below the true freezing point, but then rapidly rises to the true freezing temperature once the crystallization process starts. When the temperature of the sample has risen to the true freezing temperature, it cannot rise further; the exothermic process liquid → solid does

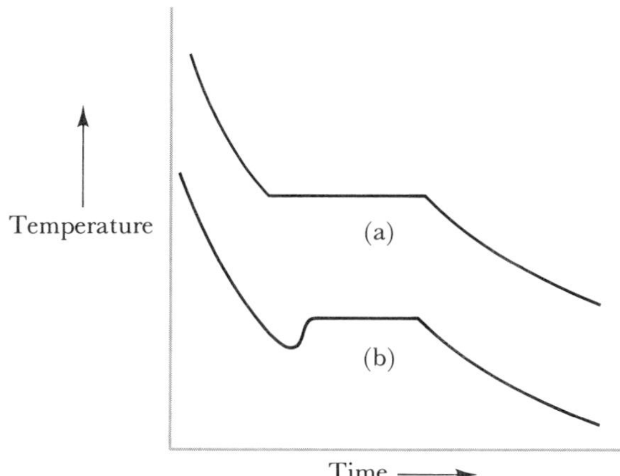

FIGURE 6–15
Cooling curves for a pure liquid that undergoes solidification: (a) without supercooling; (b) with supercooling. The true freezing temperature is the temperature corresponding to the horizontal portion of each graph.

not occur at higher temperatures. Rather, the temperature remains constant, just as it did in the situation where supercooling did not occur. At this temperature the conversion of liquid to solid continues to occur at the rate allowed by the heat flow from the system to the colder surroundings. When the sample has solidified completely, its temperature then can drop; eventually it reaches the temperature of the surroundings. The cooling curve for the sample that exhibits supercooling is given in Figure 6–15b.

Biographical Notes

J. H. HILDEBRAND (1881–), Professor Emeritus of Chemistry in the University of California (Berkeley), has made many important contributions in the study of liquid solutions and solubility phenomena. He has been leader in chemical education. In addition, he was president of the Sierra Club (1937–1940) and manager of the U. S. Olympic Ski Team in 1936.

FREDERICK T. TROUTON (1863–1922) was educated at Dublin, from which he received M.A. and D.Sc. degrees. His paper dealing with the approximate constancy of the ratio $\Delta H_{vap}/T_b$ was published in 1884, when he was an undergraduate. He was Professor of Physics at University College, London.

Problems and Questions

6–1 The vapor pressure of water at 30.0°C is 31.824 torr; at this temperature the density of liquid water is 0.99567 g cm^{-3}. What is the ratio of the average distance between water molecules in the liquid and in the saturated vapor at this temperature?

6–2 From the data given in the introduction to this chapter, what is the enthalpy change for the process $H_2O(s, 0°C) = H_2O(g, 0°C)$?

6–3 If the radial distribution functions for $^1H_2{}^{16}O$, $^1H^2H^{16}O$, and $^2H_2{}^{16}O$ are identical, what do you expect the densities of each of these pure liquids to be at 30.0°C if the density given in Problem 6–1 is for ordinary water, a liquid that is 99.985% $^1H_2{}^{16}O$ and 0.015% $^1H^2H^{16}O$?

6–4 Dry nitrogen gas is bubbled through liquid benzene (C_6H_6) at 20.0°C. From 100.0 L of the gaseous mixture of nitrogen and benzene, 24.7 g of benzene is condensed by passing the mixture through a trap at −70°C (where nitrogen is gaseous and the vapor pressure of benzene is negligible). What is the vapor pressure of benzene at 20.0°C?

6–5 Dry nitrogen gas (N_2) weighing 100.0 g is bubbled through liquid water at 25.0°C. The gaseous mixture of nitrogen and water vapor escapes at a total pressure of 700 torr. What weight of water vaporizes?

6–6 The molar enthalpy of vaporization of water at 373.2 K is 41.16 kJ mol^{-1}. What fraction of this energy goes to change the internal energy of the water, and what fraction goes to do work against the atmosphere? (Assume that water vapor is an ideal gas.)

6–7 A sealed vessel with a volume of 10.00 L contains 1.00 g H_2O. At what temperature is 50.0% of the water in the liquid phase?

6–8 Values of the vapor pressure of liquid ammonia at two temperatures are −50.0°C, 306.6 torr; −40.0°C, 538.3 torr. From these data calculate the value of ΔH for the vaporization process and predict the normal boiling point of ammonia.

6–9 Assuming the molar heat capacities of liquid water and water vapor given in the text are independent of temperature, estimate the molar enthalpy of vaporization of liquid water at 323.2 K from the value given in Table 6–4 for 373.2 K.

6–10 Using vapor pressure data in Table 6–1, calculate the molar enthalpy of vaporization of water for 298.2 K. Do two calculations; in one use the data for 20.0°C and 30.0°C, and in the other use the data for 10.0°C and 40.0°C. (The midpoint of each of these temperature ranges is 25.0°C, or 298.2 K.)

6–11 From the vapor-pressure data for water at 0.0, 20.0, 80.0, and 100.0°C, estimate the value of ΔH_{vap} at 10.0°C and at 90.0°C. Compare the two calculated values, and estimate from this comparison the value of ΔC_p $[C_p(g) - C_p(l)]$ for the process $H_2O(l) = H_2O(g)$. (Compare this value with the value calculated from the individual values of C_p given in the text.)

6–12 The vapor pressure of ice is 4.579 torr at 0.0°C and 0.776 torr at −20.0°C. Estimate the vapor pressure at −15.0°C.

6–13 The vapor pressure of solid carbon dioxide as a function of temperature is (given T, P): −100.0°C, 104.81 torr; −90.0°C, 279.5 torr; −80.0°C, 672.2 torr; −70.0°C, 1486.1 torr. Plot these data in a graph $\log P$ (or $\ln P$) versus $1/T$. Determine the value of the enthalpy change for vaporization.

6–14 Using data given in the text, calculate the value of ΔS for the process

$$H_2O(s, 0°C) = H_2O(g, 0°C, P = 4.579 \text{ torr})$$

6–15 Two vessels maintained at −10°C are connected through the vapor space. One contains ice and the other contains supercooled liquid water. With time, which sample becomes larger?

6–16 In Figure 6–6 we see that the vapor pressure of methyl alcohol and hexane are approximately equal at 48.8°C. At this temperature, the vaporization of which liquid has a larger value of ΔS?

6–17 The normal boiling point of methyl alcohol (CH_3OH) is 64.7°C and that of chloroform ($CHCl_3$) is 61.2°C. Using Trouton's rule, predict the enthalpy change for vaporization of these substances. Which of these predictions do you expect to be more accurate? Why?

6–18 The normal boiling points of water (100.0°C) and normal heptane (C_7H_{16}, 98.4°C) are very close to one another. Heptane, unlike water, does not form hydrogen bonds. Predict which of these liquids has the higher vapor at 50.0°C and at 150.0°C. Give your reasoning.

6–19 The normal boiling point for carbon tetrachloride (CCl_4) is 76.54°C. Assume this substance obeys Trouton's rule, and under this assumption, and the assumption that ΔH_{vap} does not depend upon temperature, estimate the vapor pressure of this compound at 25.0°C.

6–20 It is possible to superheat most liquids, and it is possible to supercool some liquids. It is not possible, however, to superheat a pure crystalline solid, because a solid melts to give the liquid when its temperature reaches the melting point. From what was said about the cause of superheating and supercooling of liquids, suggest a reason for the fact that solids cannot be superheated.

6–21 A sample of Dry Ice left in an open vessel evaporates with no appearance of liquid CO_2, but a sample of Dry Ice sealed in a thick-walled test tube develops a liquid phase. Explain.

6–22 Ethylene glycol ($HOCH_2CH_2OH$) and glycerol [$(HOCH_2)_2CHOH$] are organic molecules containing more than one hydroxyl group. The normal boiling points and molar heats of vaporization are:

	$\frac{T_b}{°C}$	$\frac{\Delta H_{vap}}{kJ\ mol^{-1}}$
Ethylene glycol	198	58.7
Glycerol	284(est)*	76.1

(*Glycerol decomposes at temperatures below the normal boiling point.) What is the entropy change for the vaporization process at the normal boiling point for each of these substances? Do these values conform to Trouton's rule?

6–23 The normal boiling point and enthalpy change upon vaporization for nitric oxide (NO) are 121.4 K and 13.78 kJ mol^{-1}. Calculate the value of ΔS^0 for the vaporization process at the normal boiling point. Is this value normal? Compare this value of ΔS^0_{vap} with that for nitrous oxide, N_2O ($T_b = 184.6$ K, $\Delta H^0_{vap} = 16.56$ kJ mol^{-1}).

6–24 From the phase diagram for sulfur (Figure 6–12), draw a conclusion regarding the relative density of the two forms of solid sulfur. (Check your conclusion by using a chemical handbook.)

6–25 The vapor pressure of a substance at a single temperature (P_1, T_1) cannot be used to calculate ΔH_{vap}. However, use of one datum (P_1, T_1) plus assumption of the validity of Trouton's rule [ΔS_{vap} (gas at 1 atm) = 88 J K^{-1} mol^{-1}] does allow estimation of ΔH_{vap}. Derive the equation for doing this.

6–26 (*This problem is for students who have had differential calculus.*) An isotherm (a plot of P versus V) for the critical temperature has the following properties at the critical point: $dP/dV = 0$ and $d^2P/dV^2 = 0$. Show that these conditions allow evaluation of the van der Waals constants a and b in terms of values of the critical pressure P_c and critical volume V_c.

7

Solutions

7–1 Introduction

A solution is a homogeneous mixture of two or more substances; solutions of particular components can have a range of compositions. In this respect, solutions are unlike stoichiometric compounds, which have fixed definite compositions. Solutions can exist in the solid, liquid, or gaseous state. There are important solutions in each of these states. Many alloys are solid solutions; certain alloys are more useful than pure metals, and chemists and metallurgists are increasing their efforts to develop new alloys. Liquid solutions are familiar to all. Most beverages are solutions: even the water from the tap is an aqueous solution containing carbon dioxide and some salts. If these are calcium, magnesium, or iron salts, the water is said to be hard, and before such water can be used for some purposes the cations of these salts must be removed. The oceans are aqueous solutions containing vast amounts of dissolved material. It is now economically feasible to use the ocean as a source of certain elements, for example, magnesium, bromine, and iodine, and with further research and development the list will expand. Any gaseous mixture is a solution, because it consists of a single phase; pure air is a gaseous solution: it is a homogeneous mixture of oxygen, nitrogen, argon, carbon dioxide, water vapor and other minor constituents. (Impure air containing solid particles is not homogeneous.)

In Chapter 1 we saw that the way a homogeneous material undergoes a change of state tells us whether the material is a solution or a pure substance. In the separation of benzene and octane by distillation, shown in Figure 1–1, the boiling temperature increases as the distillation proceeds. If the nature of this liquid had not been known, this fact would have shown it to be a solution, not a pure substance. It also is generally true that liquid solutions do not solidify at a constant temperature. For instance, a liquid solution of copper and nickel (50% by weight of each metal) solidifies to give an alloy of these metals over the range of temperature ~1520–1590 K. Each of the pure metals freezes at a definite temperature, nickel at 1726 K and copper at 1356 K. The nonconstancy of its freezing temperature or boiling temperature shows a liquid to be a solution, but constancy of either of these properties does not prove the liquid to be a pure substance. We will encounter examples of liquid solutions that have constant freezing temperatures or constant boiling temperatures later in this chapter.

We shall consider many aspects of solutions in this chapter, including the quantitative composition of solutions, the vapor pressures of the components of a solution, and the boiling and freezing temperatures of solutions. We also shall study some thermodynamic aspects of solutions.

7–2 The Composition of Solutions

Because solutions can have a range of compositions, it is important to have quantitative ways of describing the compositions; a number are possible. The nickel–copper alloy mentioned previously was described as 50% nickel–50% copper (by weight). A 25.0-g sample contains 12.5 g of nickel and 12.5 g of copper. A 5% tincture of iodine solution contains 5 g of iodine per 100 g of solution. Although the weight percentage is a simple basis for expressing the composition of a solution, it is not as useful for most chemical purposes as composition units that involve the mole. Some units of concentration are based upon the volume of solution; other composition units do not involve measurements of volume. The latter will be considered first.

THE MOLE-FRACTION UNIT

The mole fraction of each component of a solution is defined as the number of moles of that component divided by the total number of moles of all components in the solution. Let us consider a solution of ethylene glycol ($C_2H_6O_2$, MW = 62.07) and water (H_2O, MW = 18.02) that is 40.00 weight percent glycol. The mole fraction of glycol in the solution x_G can be calculated from the number of moles of glycol, n_G, and the number of moles of water, n_W:

$$x_G = \frac{n_G}{n_G + n_W}$$

$$= \frac{\dfrac{40.00\text{ g}}{62.07\text{ g mol}^{-1}}}{\dfrac{40.00\text{ g}}{62.07\text{ g mol}^{-1}} + \dfrac{60.00\text{ g}}{18.02\text{ g mol}^{-1}}}$$

$$= \frac{0.644\text{ mol}}{0.644\text{ mol} + 3.330\text{ mol}} = 0.162$$

The sum of the mole fractions of all of the components of a solution must be one. The mole fraction of water, x_W, in this solution can be calculated in either of two ways:

$$x_W = 1.000 - x_G$$
$$= 1.000 - 0.162 = 0.838$$

or

$$x_W = \frac{n_W}{n_G + n_W}$$

$$= \frac{3.330\text{ mol}}{3.330\text{ mol} + 0.644\text{ mol}} = 0.838$$

To prepare a solution with a particular mole fraction composition, one must weigh each component of the solution. Weighing can be done accurately, and much high-quality quantitative work on solutions involves mole-fraction units. The mole fraction of a component of a solution is the fraction of all of the molecules in the solution that are molecules of that component. In the glycol–water solution that we have considered, the fractions of all of the molecules that are glycol molecules and water molecules are 0.162 and 0.838, respectively. Because of this simple meaning in terms of relative numbers of molecules, mole fraction units are used to relate many properties of solutions to the appropriate theory.

The term mole percent is used in some descriptions of solution composition; this composition unit is simply the mole fraction multiplied by 100:

$$\text{mole \%} = 100 \times \text{mole fraction}$$

THE MOLAL UNIT

The molality of a component of a solution is defined as the number of moles of that component per kilogram of solvent. This composition unit can be used with solutions for which it is convenient or conventional to consider one of the components as the solvent. Usually the component of a liquid solution that is present in the larger amount is considered to be the solvent, but not necessarily so. The other components are called solutes. To calculate the molality of a solution, consider the following example. If 5.00 g of pure acetic acid ($C_2H_4O_2$, MW = 60.05) are dissolved in 350 g of water, the molality of the solution is:

$$\text{molality} = \underbrace{\underbrace{\underbrace{\frac{5.00\text{ g}}{60.05\text{ g mol}^{-1}}}_{\text{Moles of solute}} \times \frac{1}{350\text{ g}}}_{\text{Moles of solute per gram of solvent}} \times \frac{1000\text{ g}}{1\text{ kg}}}_{\text{Moles of solute per kilogram of solvent}}$$

$$= 0.238\text{ mol kg}^{-1} = 0.238\text{ molal} = 0.238m$$

THE WEIGHT-PERCENTAGE UNIT

Many chemical reagents are sold as aqueous solutions with the weight percentage of the solute specified. Bottles of reagent grade concentrated aqueous ammonia are specified to contain 29.0% NH_3, meaning 29.0% by weight. From this information the molality of ammonia in the solution can be calculated. The calculation starts:

100.0 g of solution contains:

$0.290 \times 100\text{ g} = 29.0\text{ g } NH_3$ and

$0.710 \times 100\text{ g} = 71.0\text{ g } H_2O$

The molality of ammonia [MW(NH_3) = 17.03] in this solution is:

$$\underbrace{\underbrace{\underbrace{\frac{29.0\text{ g}}{17.03\text{ g mol}^{-1}}}_{\text{Moles of NH}_3} \times \frac{1}{71.0\text{ g}}}_{\text{Moles of NH}_3\text{ per one gram of water}} \times \frac{1000\text{ g}}{1\text{ kg}}}_{\text{Moles of NH}_3\text{ per kilogram of water}} = 24.0\text{ mol kg}^{-1} = 24.0\text{ molal}$$

There may be situations in which a composition in molality units must be converted to weight-percentage units. Let us consider an aqueous solution of urea [$CO(NH_2)_2$, MW = 60.06] which has a composition 1.34 molal. The weight percentage of urea in this solution is:

$$\text{wt \% urea} = \frac{\text{wt urea}}{\text{wt urea} + \text{wt water}} \times 100\%$$

$$= \frac{1.34\text{ mol kg}^{-1} \times 60.06\text{ g mol}^{-1}}{1.34\text{ mol kg}^{-1} \times 60.06\text{ g mol}^{-1} + 1000\text{ g kg}^{-1}} \times 100\%$$

$$= 7.45\text{ wt \%}$$

THE PARTS-PER-MILLION UNIT

This unit is used to describe the relative amount of a minor constituent (often an impurity) in a mixture, either heterogeneous or homogeneous (a solution). It generally is understood to be a weight composition unit. If a substance is present in a solution at a level of 1 part per million (1 ppm), 1 million grams of solution contain one gram of the substance. Sea water with a density of 1.024 g cm^{-3} contains 65 ppm of bromide ion.

Mole fraction, molality, weight percentage, and parts per million units, which depend only on the relative masses of the components, express composition in terms that do not vary as the solution expands or contracts with a change of temperature. For a particular amount of solution, changes in volume with changes in temperature do not alter the numerical value of the composition expressed in these units.

THE MOLARITY UNIT

The concentration unit most widely used in chemical studies of liquid solutions is the molarity unit, which is based upon measurements of volume. The molarity of a solute is defined as the moles of solute per liter of solution. (1 L ≡ 1 dm^3.) Because the volume of a liquid solution changes slightly with changes in temperature, the molarity of a solution is a function of temperature. The temperature at which a stated molarity applies should be specified. Wide use of molarity units is due, in part, to the fact that volumes of liquids are measured easily; Figure 7–1 shows some common items of volumetric glassware.

The molarity of a solution of sodium chloride (NaCl) prepared by dissolving 15.0 g of sodium chloride (MW = 58.45) in water and adjusting the final volume to 250.0 cm^3 (by use of a volumetric flask) is:

$$\text{molarity} = \frac{\text{moles of NaCl}}{\text{liters of solution}}$$

$$= \underbrace{\underbrace{\underbrace{\frac{15.00\text{ g}}{58.45\text{ g mol}^{-1}}}_{\text{Moles of NaCl}} \times \frac{1}{250.0\text{ cm}^3}}_{\substack{\text{Moles of NaCl}\\ \text{per cm}^3\text{ solution}}} \times \frac{1000\text{ cm}^3}{1\text{ L}}}_{\substack{\text{Moles of NaCl}\\ \text{per liter of solution}}}$$

$$= 1.027 \text{ mol L}^{-1} = 1.027 \text{ mol dm}^{-3} = 1.027M$$

The units for molarity in SI are mol dm^{-3}. Because the liter is so commonly used as a volume unit in chemistry, we shall use mol L^{-1} in most quantitative descriptions of the concentrations of solutions. The symbol M, which is not an SI unit, will be used rarely. It is a useful symbol for the label on a reagent bottle, for example, 1.027M NaCl, but it is not useful in nu-

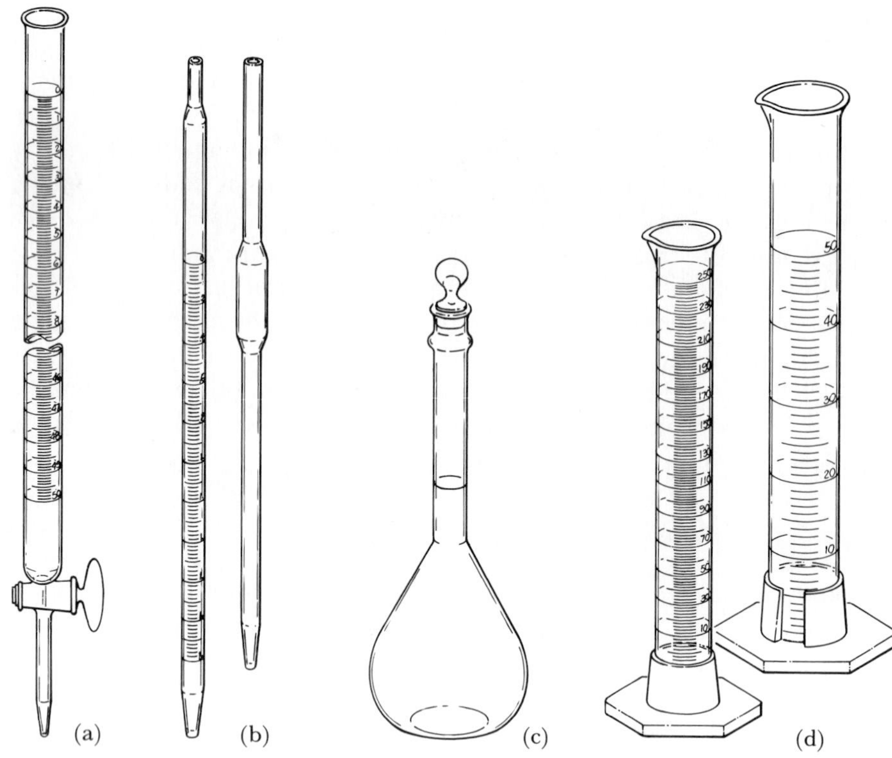

FIGURE 7–1
Volumetric glassware. (a) A buret. The graduations allow measurement of the volume of a liquid which has been *delivered*. (b) Pipets. Used for the *delivery* of definite volumes of liquid (e.g., 5.00 cm^3, 10.00 cm^3, 50.00 cm^3). (c) Volumetric flasks. When filled to the calibration mark, a volumetric flask *contains* a definite volume of liquid. (d) Graduated cylinders. Used for less accurate measurement of volumes of liquids.

merical calculations in which either the unit mol or the unit L^{-1} may be canceled, but not both.

CALCULATIONS INVOLVING CONCENTRATIONS OF SOLUTIONS

We shall use two different types of notation to describe the concentrations of solutions. We shall use the symbol C to stand for the stoichiometric concentration. In cases of possible ambiguity, the chemical formula of the solute will be indicated either as a subscript or in parenthesis. Thus for the sodium chloride solution just considered,

$$C_{\mathrm{NaCl}} = 1.027 \text{ mol L}^{-1} \quad \text{or} \quad C(\mathrm{NaCl}) = 1.027 \text{ mol L}^{-1}$$

For many purposes it is necessary to specify the concentration of each chemical species in solution. As we will learn in Chapter 9, sodium chloride in aqueous solution is completely dissociated to give sodium ions and chloride ions; the reaction

$$NaCl(aq) \rightarrow Na^{+}(aq) + Cl^{-}(aq)$$

goes essentially to completion. The symbols $[Na^{+}]$ and $[Cl^{-}]$ are used to stand for the concentrations of these species, sodium ion and chloride ion. For the solution just considered,

$$[Na^{+}] = 1.027 \text{ mol L}^{-1} \qquad [Cl^{-}] = 1.027 \text{ mol L}^{-1}$$

The formula of a chemical species enclosed in square brackets stands for the concentration of that species.

The equation

$$\text{molarity} = \frac{\text{moles of solute}}{\text{volume of solution (in liters)}} = \frac{n}{V}$$

contains three quantities. If two are known, the third can be calculated. In the example already cited, we calculated the molarity of sodium chloride, given the amount of solute and the volume of solution. For another example, we can calculate the volume of a particular solution that contains a certain amount of solute. What volume of $6.30M$ HCl (hydrochloric acid) contains 0.100 mol HCl?

$$V = \frac{\text{moles of solute}}{\text{molarity}} = \frac{0.100 \text{ mol}}{6.30 \text{ mol L}^{-1}} = 0.0159 \text{ L} = 15.9 \text{ cm}^3$$

Often it is convenient to prepare a dilute solution by diluting a more concentrated solution with pure solvent. This procedure can be quantified by use of a pipet and volumetric flask (pictured in Figure 7–1).

EXAMPLE: What is the concentration of a perchloric acid ($HClO_4$) solution prepared by diluting 50.00 cm^3 of $4.352M$ $HClO_4$ with water to the mark in a 2.000-L volumetric flask?

The amount of perchloric acid taken is:

$$\begin{aligned} \text{amount of } HClO_4 &= \text{molarity} \times \text{volume} \\ &= 4.352 \text{ mol L}^{-1} \times \frac{50.00 \text{ cm}^3}{1000 \text{ cm}^3 \text{ L}^{-1}} \\ &= 0.2176 \text{ mol} \end{aligned}$$

This also is the amount of perchloric acid in the entire portion of diluted solution; the concentration of the diluted solution is, therefore,

$$\begin{aligned} C &= \frac{n}{V} \\ &= \frac{0.2176 \text{ mol}}{2.000 \text{ L}} = 0.1088 \text{ mol L}^{-1} \end{aligned}$$

This calculation can be carried out in one step. Because the amount of solute is the same before and after dilution, we may write the equation involving concentration (C) and volume (V):

$$n = C_1V_1 = C_2V_2$$

relating the concentrations and volumes before dilution (subscript 1) and after dilution (subscript 2). Applying this equation to the present problem gives:

$$\underbrace{0.0500 \text{ L} \times 4.352 \text{ mol L}^{-1}}_{\text{Moles of perchloric acid}} = \underbrace{2.000 \text{ L} \times C_2}_{\text{Moles of perchloric acid}}$$

$$C_2 = \frac{0.0500 \text{ L} \times 4.352 \text{ mol L}^{-1}}{2.000 \text{ L}} = 0.1088 \text{ mol L}^{-1}$$

(Notice that careless mistakes may be avoided by keeping account of the dimensions of each of the quantities in the equation.)

To interconvert values of concentrations in molar units and molal (or mole fraction) units, we must know the density of the solution. For example, the molality of 29.0% aqueous ammonia already has been calculated; the density of this solution at 20°C is 0.90 g cm^{-3}. If we start with these data, the calculation of the molarity is

$$C = \underbrace{\underbrace{\underbrace{\underbrace{\frac{29.0 \text{ g}}{17.03 \text{ g mol}^{-1}}}_{\text{Moles NH}_3\text{ in 100 g solution}} \times \frac{1}{100 \text{ g}}}_{\text{Moles NH}_3\text{ in 1 g solution}} \times 0.90 \text{ g cm}^{-3}}_{\text{Moles NH}_3\text{ in 1 cm}^3\text{ solution}} \times \frac{1000 \text{ cm}^3}{1 \text{ L}}}_{\text{Moles of NH}_3\text{ per liter of solution}}$$

$$C = 15.3 \text{ mol L}^{-1}$$

Starting with the molality (24.0 mol kg^{-1}), we can arrive at molarity by calculating the volume occupied by a kilogram of solvent plus 24.0 mol of ammonia:

$$V = \frac{\text{weight of solution}}{\text{density of solution}}$$

$$V = \frac{1000 \text{ g} + 24.0 \text{ mol} \times 17.03 \text{ g mol}^{-1}}{9.0 \times 10^2 \text{ g L}^{-1}} = \frac{1409 \text{ g}}{9.0 \times 10^2 \text{ g L}^{-1}}$$

$$= 1.57 \text{ L}$$

Therefore the concentration of this solution is

$$C = \frac{24.0 \text{ mol}}{1.57 \text{ L}} = 15.3 \text{ mol L}^{-1}$$

The opposite conversion, molarity to molality, is illustrated with a solution of sulfuric acid $14.1M\ H_2SO_4$ (MW = 98.08), a solution having a density of $1.73\ g\ cm^{-3}$. One liter of this solution, which has a mass of 1.73×10^3 g, contains:

$$1\ L \times 14.1\ mol\ L^{-1} \times 98.08\ g\ mol^{-1} = 1380\ g\ H_2SO_4 \quad \text{and}$$
$$1730\ g - 1380\ g = 350\ g\ H_2O$$

Therefore the molality is:

$$\underbrace{\underbrace{\underbrace{14.1\ mol\ L^{-1}}_{\text{Moles } H_2SO_4 \text{ per liter}} \times \frac{1\ L}{350\ g}}_{\text{Moles } H_2SO_4 \text{ per gram solvent}} \times \frac{1000\ g}{1\ kg}}_{\text{Moles } H_2SO_4 \text{ per kilogram solvent}} = 40\ mol\ kg^{-1}$$

It often is necessary to interconvert molarity and molality units. Many of the most accurate thermodynamic data for solutions of inorganic solutes in water are tabulated as a function of the molality of the solute. You may wish to use these data to interpret studies in which the molar concentration scale is used. To conclude our discussion of these concentration units, a special point must be made. Because one liter of a dilute aqueous solution (with a density close to $1.00\ g\ cm^{-3}$) contains approximately one thousand grams of water, numerical values of the molality and molarity of such dilute aqueous solutions are very similar. For instance, for a dilute solution of methyl alcohol (CH_3OH) in water at 298 K,

$$1.000 \times 10^{-3}\ \text{molar}\ CH_3OH = 1.003 \times 10^{-3}\ \text{molal}\ CH_3OH$$

This is not true, of course, for solutions in other solvents, nor is it true for concentrated aqueous solutions. For a dilute solution of methyl alcohol in carbon tetrachloride (density = $1.594\ g\ cm^{-3}$) at 298 K,

$$1.000 \times 10^{-3}\ \text{molar}\ CH_3OH = 6.27 \times 10^{-4}\ \text{molal}\ CH_3OH$$

Our calculations showed for concentrated aqueous solutions of ammonia and sulfuric acid:

$$15.3\ \text{molar}\ NH_3 = 24.0\ \text{molal}\ NH_3$$
$$14.1\ \text{molar}\ H_2SO_4 = 40\ \text{molal}\ H_2SO_4$$

We mentioned previously that the concentration of bromide ion in sea water ($d = 1.024\ g\ cm^{-3}$) is 65 ppm. To calculate the molarity of bromide ion in this solution, we can assume that we have 1.00 L ($1.00 \times 10^3\ cm^3$) of solution. This weighs

$$1.00 \times 10^3\ cm^3 \times 1.024\ g\ cm^{-3} = 1.024 \times 10^3\ g$$

The amount of bromide in this liter of solution is its molarity:

$$C(\mathrm{Br}^-) = \underbrace{\underbrace{\underbrace{\frac{65\ \mathrm{g}}{1.00 \times 10^6\ \mathrm{g}}}_{\text{Weight Br}^-\text{ per 1 g solution}} \times \frac{1}{79.90\ \mathrm{g\ mol}^{-1}}}_{\text{Moles Br}^-\text{ per 1 g solution}} \times 1.024 \times 10^3\ \mathrm{g\ L}^{-1}}_{\text{Moles of Br}^-\text{ per liter of solution}}$$

$$= 8.3 \times 10^{-4}\ \mathrm{mol\ L}^{-1}$$

THE INTERCONVERSION OF CONCENTRATION UNITS

There are precise equations that convert a concentration in one set of units into a concentration in another set of units. The equation for converting molality to molarity follows directly from the definitions given in Table 7–1:

$$\frac{C_B}{m_B} = \frac{n_B}{V} \div \frac{n_B}{n_A \mathrm{MW}_A/1000}$$

$$= \frac{\text{wt of solvent (in kg)}}{\text{volume of solution (in L)}}$$

$$= \underbrace{d}_{\substack{\text{g solution per cm}^3\\ \text{or}\\ \text{kg solution per L}}} \times \underbrace{\frac{1000\ \mathrm{g}}{1000\ \mathrm{g} + m_B \mathrm{MW(B)}}}_{\substack{\text{Fraction (by}\\ \text{weight) of solution}\\ \text{that is solvent}}}$$

$$= \frac{d \times 1000\ \mathrm{g}}{1000\ \mathrm{g} + m_B \mathrm{MW(B)}}$$

Let us use this equation to calculate the ratio C_{NH_3}/m_{NH_3} for 24.0 molal NH_3 ($d = 0.90\ \mathrm{g\ cm}^{-3} = 0.90\ \mathrm{kg\ L}^{-1}$)

$$\frac{C_{NH_3}}{m_{NH_3}} = \frac{0.90\ \mathrm{kg\ L}^{-1} \times 1000\ \mathrm{g}}{1000\ \mathrm{g} + 24.0\ \mathrm{mol} \times 17.03\ \mathrm{g\ mol}^{-1}}$$

$$= 0.64\ \mathrm{kg\ L}^{-1}$$

The ratio of these concentrations, which we already have calculated, is

$$\frac{C_{NH_3}}{m_{NH_3}} = \frac{15.3\ \mathrm{mol\ L}^{-1}}{24.0\ \mathrm{mol\ kg}^{-1}} = 0.64\ \mathrm{kg\ L}^{-1}$$

If the solution is very dilute, that is, if $1000\ \mathrm{g} \gg m_B \mathrm{MW(B)}$, the conversion equation becomes

$$\frac{C_B}{m_B} = d \cong d_A$$

That is, the ratio of the molarity to the molality of a very dilute solution is equal to the density of the solution, d, which is approximately equal to the density of the pure solvent A, d_A, for very dilute solutions.

TABLE 7–1
A Summary of Concentration Units Commonly Used in Chemistry[a]

		Dimensions
Mole fraction	$x_A = \dfrac{n_A}{n_A + n_B + n_C + \cdots}$	1^b
	$x_B = \dfrac{n_B}{n_A + n_B + n_C + \cdots}$	
Molality	$m_B = \dfrac{n_B}{\text{wt of A (in kg)}}$	mol kg^{-1}
	$= \dfrac{n_B}{(n_A\,MW_A)/1000}$	
Molarity	$C_B = \dfrac{n_B}{V \text{ (in L)}}$	mol L^{-1} or mol dm^{-3}

[a]Meaning of symbols: A is the solvent; B, C, etc., are solutes; n_A = number of moles of A; MW_A = molecular weight of A; V = volume of solution. For some solutions, the roles of solute and solvent need not be defined.
[b]This means the quantity is dimensionless.

dimensionless

The conversion of mole fraction units into molarity units also follows from the definitions:

$$\frac{C_B}{x_B} = \frac{n_B}{V} \div \frac{n_B}{n_B + n_A} = \frac{n_B + n_A}{V} = \frac{n_B}{V} + \frac{n_A}{V}$$

$$= C_B + \underbrace{[1000 \text{ cm}^3 \text{ L}^{-1} \times d - C_B\text{MW(B)}]/\text{MW(A)}}_{\dfrac{\text{Weight of A per liter of solution}}{\text{MW}_A}}$$

which can be rearranged to

$$\frac{C_B}{x_B} = \frac{1000 \text{ cm}^3 \text{ L}^{-1} \times d}{\text{MW(A)}} + C_B\left(1 - \frac{\text{MW(B)}}{\text{MW(A)}}\right)$$

The equation in this form is useful for calculating the ratio C_B/x_B if the molarity of the solution is known. If the mole fraction is known, a rearranged equation can be used:

$$\frac{C_B}{x_B} = \frac{1000 \text{ cm}^3 \text{ L}^{-1} \times d}{\text{MW(A)}\left[1 + x_B\left(\dfrac{\text{MW(B)}}{\text{MW(A)}} - 1\right)\right]}$$

For the ethylene glycol–water solution that we considered earlier (40.00 wt % glycol, $x_G = 0.162$), which has a density of 1.053 g cm^{-3}, the calculation of C_G/x_G is

$$\frac{C_G}{x_G} = \frac{1000 \text{ cm}^3 \text{ L}^{-1} \times 1.053 \text{ g cm}^{-3}}{18.02 \text{ g mol}^{-1} \left[1 + 0.162\left(\dfrac{62.07 \text{ g mol}^{-1}}{18.02 \text{ g mol}^{-1}} - 1\right)\right]}$$

$$= 41.9 \text{ mol L}^{-1}$$

Thus the molarity of the solution is

$$C(\text{ethylene glycol}) = 0.162 \times 41.9 \text{ mol L}^{-1} = 6.79 \text{ mol L}^{-1}$$

The equation for C_B/x_B reduces to a simple form in approaching very low concentrations. For this limiting case,

$$\left[1 \gg x_B\left(\frac{\text{MW(B)}}{\text{MW(A)}} - 1\right)\right]$$

$$\frac{C_B}{x_B} = \frac{1000 \text{ cm}^3 \text{ L}^{-1} \times d}{\text{MW(A)}} \cong \frac{1000 \text{ cm}^3 \text{ L}^{-1} \times d_A}{\text{MW(A)}}$$

but this quotient is simply the molarity of the pure solvent or the reciprocal of the molar volume of the pure solvent. Thus for dilute aqueous solutions,

$$\frac{C_B}{x_B} = \frac{1000 \text{ cm}^3 \text{ L}^{-1} \times 1.00 \text{ g cm}^{-3}}{18.02 \text{ g mol}^{-1}}$$

$$= 55.5 \text{ mol L}^{-1}$$

These derivations of the conversion factors C_B/m_B and C_B/x_B have an important consequence:

For dilute solutions, the concentrations in molarity, molality, and mole fraction units are directly proportional to one another:

$$C_B = d_A \times m_B$$

$$C_B = \frac{1}{V_A^0} \times x_B$$

in which V_A^0 is the molar volume of the solvent in L, $V_A^0 = \text{MW}_A/(1000 \text{ cm}^3 \text{ L}^{-1} \times d_A)$.

7–3 The Thermodynamics of Liquid Solution Formation; Solubility

Although we will consider many types of solutions, including those with solutes that are dissociated to give ions in solution, we will approach the thermodynamics of solution formation through less complex systems. In a liquid solution there are attractive interactions between the solution components. These interactions in the solution account for the energy change as a liquid solution forms from pure gaseous components. The process involving gaseous acetone (A) and gaseous chloroform (C), shown in

Figure 7–2,

$$\text{acetone } (g, P_A = 170 \text{ torr}) + \text{chloroform } (g, P_C = 144 \text{ torr}) \\ = \text{liquid solution}$$

occurs spontaneously at 35.2°C. Each of the initial pressures before mixing is less than the vapor pressure of the pure liquid; thus no liquid phase is present initially.

In the formation from gaseous components of one mole of solution ($x_A = 0.50$, $x_C = 0.50$), 33.6 kJ of energy is evolved. This is a measure of the attractive interaction between molecules in the solution. These attractive interactions involve:

Acetone molecules	A---A
Chloroform molecules	C---C
Acetone molecules and chloroform molecules	A---C

When a liquid solution forms from pure components in the liquid or solid state, the situation is more complex. The attractive interactions of the unlike molecules in the solution, represented as A---C, come at the

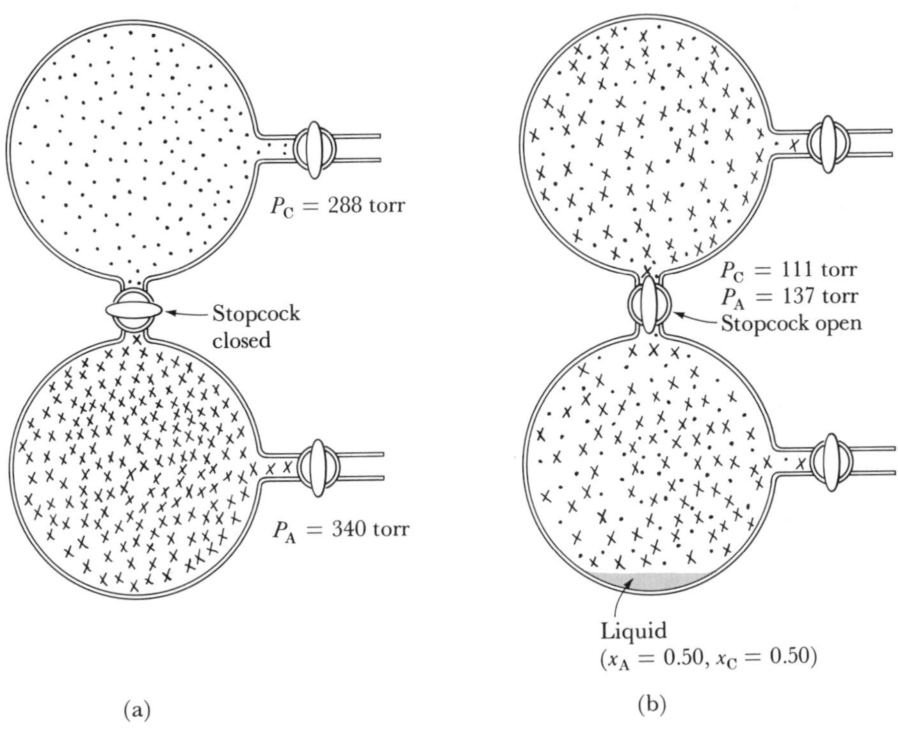

FIGURE 7–2
The formation of a liquid solution of chloroform and acetone from vapor components. (a) Separate vapor phases separated by stopcock (chloroform in upper vessel; acetone in lower vessel). (b) Stopcock opened; the differences of densities and pressures plus diffusion causes the vapors to mix, and a small amount of liquid forms.

expense of interactions of like molecules in pure liquid A and pure liquid C, represented as A---A and C---C. This "trade-off" of attractive interactions can be represented by

$$A{-}{-}{-}A + C{-}{-}{-}C = A{-}{-}{-}C$$

Whether the solution forms with the evolution of energy, the absorption of energy, or with no appreciable energy effect depends upon the energy of the A---C interactions compared to those of the A---A and C---C interactions. From the sign and magnitude of this energy effect, which can be measured calorimetrically, we can determine the relative magnitudes of these various interactions, A---A, C---C, and A---C. For instance, let us consider the energy accompanying the mixing of one half mole of each of the following liquid components to give one mole of solution:

System	ΔU
Benzene–Toluene	67 J
Acetone–Chloroform	−1880 J
Ethyl alcohol–Benzene	1270 J

Compared to the latter two systems, the energy of mixing benzene and toluene is very small. These molecules are very similar: the chemical formula for benzene is C_6H_6, and that for toluene is $C_6H_5CH_3$; toluene has the benzene structure with the grouping of atoms $-CH_3$, a methyl group, replacing one hydrogen atom of benzene. The very small energy effect suggests that like-molecule interactions and unlike-molecule interactions are very similar. Since energy is evolved when liquid acetone and liquid chloroform are mixed, the attractive interaction between the unlike molecules, A---C, must be stronger than between the like molecules, A---A, and C---C. The endothermic mixing in the ethyl alcohol–benzene system suggests the opposite. The attractive interaction between the unlike molecules, A----C, must be weaker than those between the like molecules, A----A and C----C. Further support for this interpretation of the heat-of-mixing data will come from vapor-pressure data, to be presented later in this chapter.

As in other processes, the entropy of the system changes when pure substances are mixed to form a solution. Simple reasoning leads us to expect a positive entropy change upon mixing of unlike molecules. Figure 7–3 shows the greater randomness in the binary mixture relative to the pure

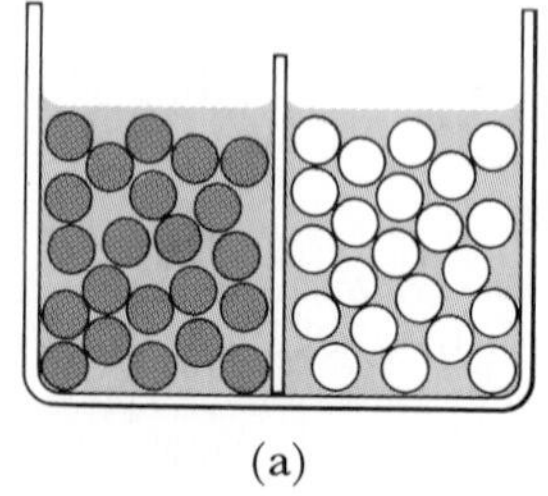

(a)

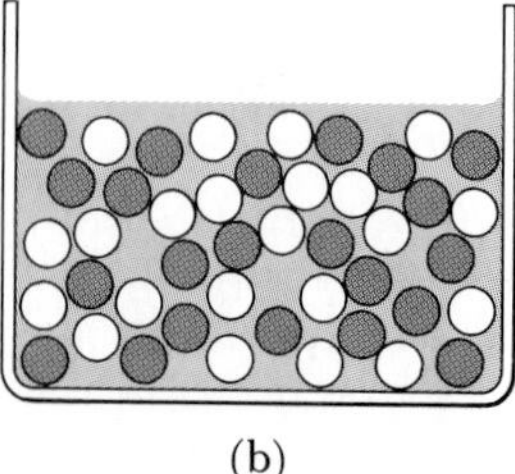

(b)

FIGURE 7–3
The increase of randomness accompanying mixing of two liquids to give a solution. (a) Before mixing: equal amounts of two substances. (b) After mixing: a solution with equal mole fractions of each component ($x_1 = x_2 = 0.50$).

components. Of course, the positive entropy change in forming a solution is the factor making it possible for substances that mix endothermically nonetheless to form a solution. For the solution to form when the pure components are mixed, the change of Gibbs free energy for the process, ΔG, must be negative. If the change of enthalpy is positive, the change of Gibbs free energy will be negative only if the change of entropy is sufficiently positive. We can see this from the following: because $\Delta G = \Delta H - T\,\Delta S$, $\Delta G < 0$, if $\Delta S > \dfrac{\Delta H}{T}$. The entropy change in the process of forming a solution from a liquid solvent and one mole of liquid or solid solute depends upon composition of the solution, being more positive the lower the concentration. For substances with a limited solubility, this solubility limit is the concentration at which the value of ΔG for the process

$$\mathrm{A}(l \text{ or } s) = \mathrm{A}(\text{saturated solution})$$

is zero. Thus it is the concentration for which the entropy change is equal to the enthalpy change divided by the temperature:

$$\Delta S = \frac{\Delta H}{T}$$

We will consider a solution of iodine in perfluoroheptane (C_7F_{16}).[1] Iodine dissolves in this liquid only to a very limited extent at 25°C (8.00×10^{-4} mol L^{-1}), and the solution process is endothermic:

$$I_2(s) = I_2 \text{ (in } C_7F_{16}) \qquad \Delta H = 42.43 \text{ kJ mol}^{-1}$$

The value of ΔH is independent of the concentration of iodine in the dilute range being considered. But the change of entropy that results when one mole of iodine is dissolved in perfluoroheptane depends upon the volume of the solution that contains this amount of solute. (This is analogous to the entropy of vaporization of a mole of liquid water depending upon the molar volume of the water vapor; see Section 3–3.) Consider the processes of dissolving one mole of iodine in 100 L of perfluoroheptane (a hypothetical process because iodine does not dissolve to this extent), in 1250 L of perfluoroheptane (to give a saturated solution at 25°C), and in 1.00×10^4 L

[1] Perfluoroheptane was chosen for the example because of the very low solubility of iodine in this solvent. The analogy of the changes of entropy in the solution process

$$I_2(s) = I_2 \text{ (in solution)}$$

and in the vaporization process

$$I_2(s) = I_2(g)$$

is appropriate only for very dilute solutions. For more concentrated solutions, the entropy change of the solvent also must be considered. To do this, which is beyond our present goal, the mole fraction of each component of the solution is taken into account.

of perfluoroheptane; the changes of entropy are:

Process	$\frac{\Delta S}{\text{J K}^{-1}\text{ mol}^{-1}}$
$I_2(s) = I_2$ (0.0100 mol L^{-1} in C_7F_{16})	121.3
$I_2(s) = I_2$ (8.00 × 10^{-4} mol L^{-1} in C_7F_{16})	142.3
$I_2(s) = I_2$ (1.00 × 10^{-4} mol L^{-1} in C_7F_{16})	159.4

These processes are depicted in Figure 7–4, which is analogous to Figure 3–14. The changes of Gibbs free energy for these three processes calculated from

$$\frac{\Delta G}{\text{J mol}^{-1}} = 42{,}430 - 298.2\text{ K} \times \frac{\Delta S}{\text{J mol}^{-1}}$$

are $+6.26$ kJ mol^{-1}, 0, and -5.10 kJ mol^{-1}, respectively. The way in which the volume of solution influences the value of ΔS, and therefore the value of ΔG, is shown in Figure 7–5. For all volumes less than 1250 liters per mole of iodine, the value of $T\,\Delta S$ is less positive than ΔH, and the value

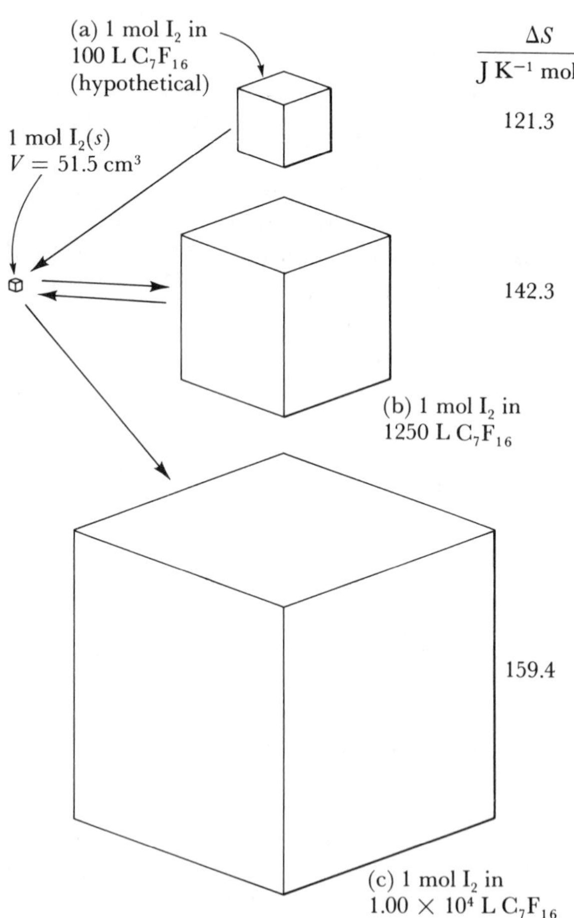

FIGURE 7–4
The dissolving of one mole of solid iodine (volume = 51.5 cm^3) in C_7F_{16} at 25.0°C to give solutions with volume (a) 100 L (hypothetical), (b) 1250 L, and (c) 1.00 × 10^4 L. (Compare this figure with Figure 3–14.)

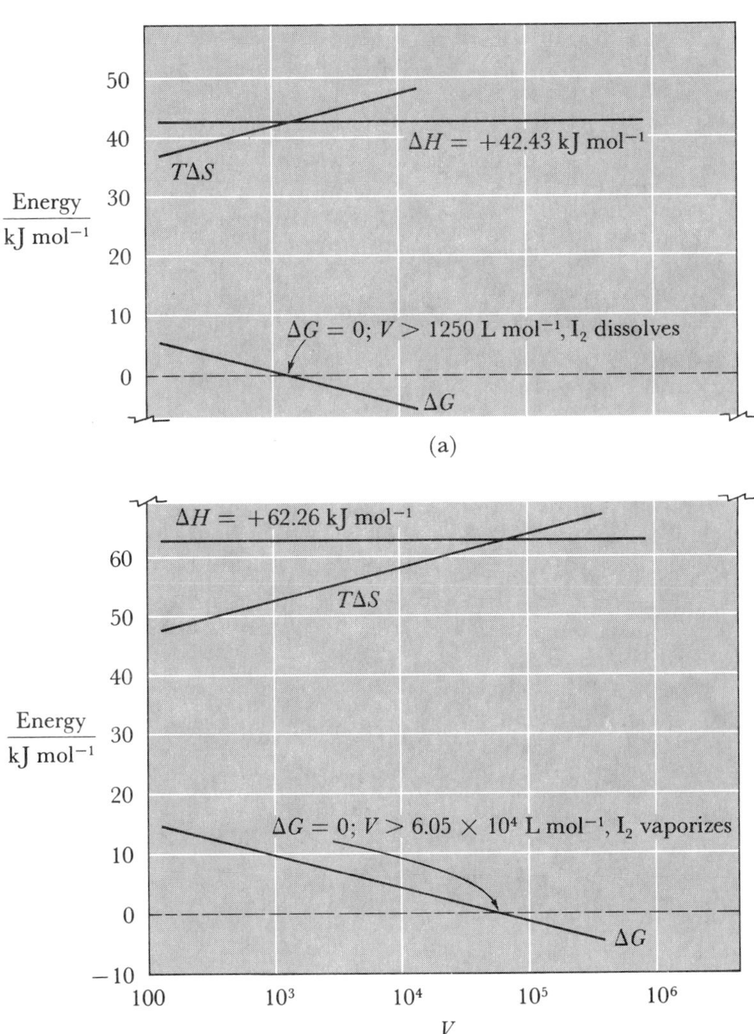

FIGURE 7–5
The dependence of ΔH, $T\Delta S$, and ΔG upon the volume occupied by one mole of iodine in solution or in the gas phase ($T = 298.2$ K). (a) $I_2(s) = I_2$ (in C_7F_{16}). (b) $I_2(s) = I_2(g)$ (The vapor pressure of solid iodine at 298.2 K is 0.307 torr, making the molar volume 6.05×10^4 L mol^{-1}.)

of ΔG is positive. This means that the iodine does not dissolve to give a solution with a concentration greater than 8.00×10^{-4} mol L^{-1} at the temperature in question (298.2 K). For a volume of 1250 L mol^{-1}, the value of ΔG is zero, and this concentration (1 mol/1250 L = 8.00×10^{-4} mol L^{-1}) is the solubility of iodine in perfluoroheptane at 25.0°C.

The analogy between the processes

$$I_2(s) = I_2 \text{ (in } C_7F_{16}) \qquad \Delta H = 42.43 \text{ kJ mol}^{-1}$$
$$I_2(s) = I_2(g) \qquad \Delta H = 62.26 \text{ kJ mol}^{-1}$$

is shown in the plots of $T\,\Delta S$ versus the logarithm of the volume occupied by one mole of iodine. This comparison is not meant to imply that all of the volume of the solution is accessible to the solute iodine molecules. But whether we are considering the liquid solution process or the vaporization process, a ten-fold increase in the volume occupied by one mole of iodine results in an increase of the value of the entropy of the iodine by the same amount ($R \ln 10 = 19.1$ J K^{-1} mol^{-1}). Notice that the plots of $T\Delta S$ versus V (logarithmic abscissa) in the two parts of Figure 7–5 are parallel.

The temperature coefficient of the solubility of a solute in a solvent is determined by the sign of ΔH for the dissolving of an infinitesimal amount of solute in the solution undersaturated by an infinitesimal amount. If this solution process is exothermic, the solubility decreases as the temperature increases; if the process is endothermic, the solubility increases as the temperature increases. Examples of each type of behavior are:

$$I_2(s) = I_2\ (\text{in } C_7F_{16}) \qquad \Delta H = 42.43 \text{ kJ mol}^{-1}$$

$$\text{Solubilities:}\quad 2.3 \times 10^{-4} \text{ mol L}^{-1} \text{ at } 5°\text{C}$$

$$8.0 \times 10^{-4} \text{ mol L}^{-1} \text{ at } 25°\text{C}$$

$$CaCO_3(s) = Ca^{2+}(aq) + CO_3^{2-}(aq) \qquad \Delta H = -20.1 \text{ kJ mol}^{-1}$$

$$\text{Solubilities:}\quad 1.1 \times 10^{-4} \text{ mol L}^{-1} \text{ at } 5°\text{C}$$

$$8.5 \times 10^{-5} \text{ mol L}^{-1} \text{ at } 25°\text{C}$$

These are simple examples of the operation of Le Chatelier's principle. If heat is added to a system in equilibrium consisting of two phases, a pure solute and a saturated solution, the process that absorbs energy occurs, and the temperature increase is lower than it otherwise would be for the addition of a particular amount of heat.

7–4 Dilute Solutions

Now we will consider the equilibrium between a liquid solution and its vapor. We will learn that the partial pressure of each solution component is lower than its vapor pressure as a pure substance at the same temperature. For the solvent component, the *vapor pressure lowering* by a solute is responsible for the *freezing point depression* and the *boiling point elevation*. These related properties of the solution, as well as others (e.g., osmotic pressure, also to be discussed), are called *colligative properties*.

SOLUTIONS OF VOLATILE SOLUTES; HENRY'S LAW

In Chapters 3 and 6 we learned something of the tendency for a substance in the liquid or solid state to vaporize. For a particular substance A, the extent of occurrence of the process

$$A(s \text{ or } l) = A(g)$$

as measured by the pressure of A in the gas phase, is determined primarily by the temperature. If the substance A is a component of a liquid or solid solution, the pressure of A in the gas phase is determined also by the composition of the condensed phase. Consider an undissociated solute in dilute solution. Experimental studies have shown that at a particular temperature the vapor pressure of such a solute over the solution is directly proportional to its concentration in solution. This proportionality is *Henry's law* (named for WILLIAM HENRY), which can be expressed using a proportionality constant k and any concentration unit. For mole fraction, molality, and molarity units, Henry's law for Component 2, the solute, is

$$P_2 = k_x x_2 \qquad P_2 = k_m m_2 \qquad P_2 = k_c C_2$$

(In dilute solution, the concentrations of a solute on these concentration scales are directly proportional to one another; each of these equations is obeyed in dilute solution.)

Henry's law also can be stated as the dependence of the solubility of a gas in a liquid upon the pressure of the gas over the liquid: The solubility of a gas in a liquid is directly proportional to the pressure of the gas over the solution. Table 7–2 gives data showing that dilute aqueous solutions of oxygen and methyl chloride (CH_3Cl) conform to Henry's law. Solutions of ethyl alcohol (C_2H_5OH) in carbon tetrachloride obey Henry's law only at concentrations below $\sim$0.1 mol L^{-1}. The quotient P_2/C_2 (which is equal to the proportionality constant k_c if Henry's law is obeyed) decreases at concentrations above $\sim$0.1 mol L^{-1}. (An ethyl alcohol molecule has a hydroxyl group, and in the concentration range where deviations from Henry's law occur, an appreciable fraction of alcohol molecules in solution associate with one another through the formation of hydrogen bonds. This association occurs to a larger extent the higher the concentration of ethyl alcohol.)

EXAMPLE: What is the concentration at 25°C of oxygen in water which is saturated with air at atmospheric pressure 735 torr?

Since air is 20.9 mole % oxygen, the partial pressure of oxygen is

$$P(O_2) = 0.209 \times 735 \text{ torr} = 154 \text{ torr}$$

and the concentration of oxygen in solution is

$$[O_2] = \frac{P(O_2)}{k_c}$$

$$= \frac{154 \text{ torr}}{6.0 \times 10^5 \text{ torr L mol}^{-1}} = 2.6 \times 10^{-4} \text{ mol L}^{-1}$$

Figure 7–6 shows the distribution of a volatile solute between the liquid and vapor phases at two different concentrations in a system that obeys Henry's law. A two-fold increase in the concentration in one of the phases is accompanied by a two-fold increase in the other phase. This is reasonable if we consider the rates of the vaporization and condensation processes:

$$A(\text{in solution}) \rightarrow A(g) \qquad A(g) \rightarrow A(\text{in solution})$$

Equilibrium exists when these rates are equal, and the rate of each process is proportional to the concentration of the reactant species in that process;

TABLE 7–2
The Pressure of Volatile Solutes Over Liquid Solutions

$[O_2]$ mol L^{-1}	$P(O_2)$ torr	$P(O_2)/[O_2]$ torr L mol^{-1}
Oxygen–Water at 25°C		
1.0×10^{-4}	60	6.0×10^5
3.0×10^{-4}	180	6.0×10^5
1.2×10^{-3}	720	6.0×10^5
		$k_c = 6.0 \times 10^5$ torr L mol^{-1}

$[CH_3Cl]$ mol L^{-1}	$P(CH_3Cl)$ torr	$P(CH_3Cl)/[CH_3Cl]$ torr L mol^{-1}
Methyl chloride–Water at 20°C		
2.4×10^{-3}	25.5	1.06×10^4
1.77×10^{-2}	183	1.03×10^4
4.61×10^{-2}	480	1.04×10^4
		$k_c = 1.04 \times 10^4$ torr L mol^{-1}

$[C_2H_5OH]$ mol L^{-1}	$P(C_2H_5OH)$ torr	$P(C_2H_5OH)/[C_2H_5OH]$ torr L mol^{-1}
Ethyl alcohol–Carbon tetrachloride at 20°C		
1.30×10^{-2}	1.34	1.03×10^2
5.19×10^{-2}	5.25	1.01×10^2
1.04×10^{-1}	10.1	9.7×10
2.09×10^{-1}	15.9	7.6×10
4.21×10^{-1}	20.7	4.9×10
		$k_c = 1.04 \times 10^2$ torr L mol^{-1} (obtained by extrapolation of values of $P(C_2H_5OH)/[C_2H_5OH]$ to $[C_2H_5OH] = 0$)

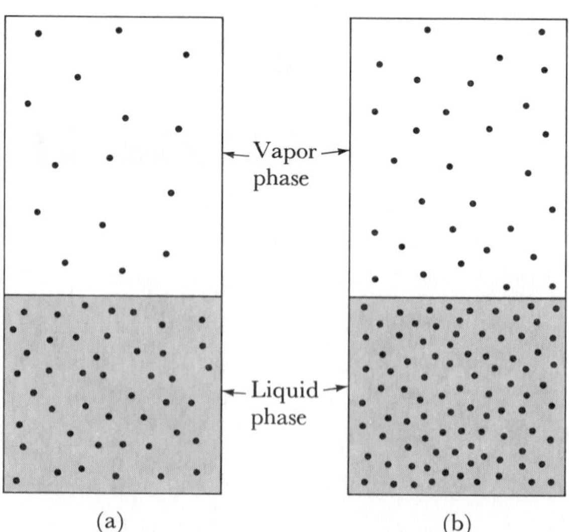

FIGURE 7–6
The distribution of a volatile solute between a liquid solution and the vapor phase. The concentration of the solute in the liquid solution is 2.00-fold higher in (b) than in (a); the pressure of this component in the vapor phase is 2.00-fold higher in (b) than in (a). The system conforms to Henry's law, $P = k_cC$.

for vaporization,

$$\text{Rate of vaporization} = k_v[\text{A}]$$

and for condensation, where instead of concentration the partial pressure will be used,

$$\text{Rate of condensation} = k_c P_\text{A}$$

(The individual constants, k_v and k_c, pertain to rates and should not to be mistaken for Henry's law constants. That the partial pressure of a gas is proportional to its concentration was shown in Chapter 4.) Equating these two rates gives

$$k_c P_\text{A} = k_v[\text{A}]$$

which can be converted to

$$P_\text{A} = \frac{k_v}{k_c}[\text{A}]$$

or with $k_v/k_c = k$,

$$P_\text{A} = k[\text{A}]$$

which is the equation for Henry's law.

For Henry's law to apply, the solute must be nondissociating. If a solute dissociates extensively, the failure of Henry's law is dramatic; if the dissociation is slight, the deviations from Henry's law may be negligible. These two situations can be illustrated with data for hydrogen chloride and carbon dioxide.

First let us consider hydrogen chloride; Table 7–3 gives the partial pressure of hydrogen chloride over its solutions in cyclohexane (C_6H_{12})

TABLE 7–3
The Partial Pressure of Hydrogen Chloride Over Its Solutions in Cyclohexane and Water at 25°C

In cyclohexane (C_6H_{12})

[HCl] mol L^{-1}	P_{HCl} torr	$P_{\text{HCl}}/[\text{HCl}]$ torr L mol^{-1}
3.34×10^{-3}	177	5.30×10^4
7.12×10^{-3}	373	5.24×10^4
1.23×10^{-2}	668	5.43×10^4
		$k_c = 5.3 \times 10^4$ torr L mol^{-1}

In water

C_{HCl} mol L^{-1}	P_{HCl} torr	$P_{\text{HCl}}/C_{\text{HCl}}$ torr L mol^{-1}
0.0050	8.15×10^{-9}	0.163×10^{-5}
0.050	6.5×10^{-7}	1.3×10^{-5}
0.50	5.4×10^{-5}	10.8×10^{-5}

and in water. Within experimental error, hydrogen chloride in cyclohexane solution obeys Henry's law, but the quotient P_{HCl}/C_{HCl} increases by a factor of 66 as the concentration of hydrogen chloride in aqueous solution increases by a factor of one hundred. The contrasting behavior of these two systems is explained by the nature of the solute in the two solvents. In cyclohexane, hydrogen chloride dissolves as a simple molecular solute; in water, hydrogen chloride dissociates into ions. The vaporization of hydrogen chloride from each of these solvents can be represented by the chemical equations:

$$\mathrm{HCl\ (in\ C_6H_{12})} = \mathrm{HCl}(g)$$
$$\mathrm{H^+}(aq) + \mathrm{Cl^-}(aq) = \mathrm{HCl}(g)$$

The stoichiometries of these processes,

$$1 \text{ solute particle} = 1 \text{ gaseous molecule}$$

and

$$2 \text{ solute particles} = 1 \text{ gaseous molecule}$$

make the first system obey Henry's law and the second system fail to obey Henry's law. (Justification for this will be given when we consider equilibrium in more detail.)

Solutions of carbon dioxide in water provide a contrast with those of hydrogen chloride. Although there is some dissociation to give ions, the reaction

$$\mathrm{CO_2}(aq) + \mathrm{H_2O}(l) = \mathrm{H^+}(aq) + \mathrm{HCO_3^-}(aq)$$

occurs at $[CO_2] = 0.010$ mol L^{-1} to an extent of only 0.67%. The predominant form of carbon dioxide in water is molecular carbon dioxide, and the vaporization process can be represented

$$\mathrm{CO_2}(aq) = \mathrm{CO_2}(g)$$
$$1 \text{ solute particle} = 1 \text{ gaseous molecule}$$

Data showing the conformity to Henry's law of aqueous solutions of carbon dioxide (despite its very slight dissociation) are given in Table 7–4a. Table 7–4b presents values of the Henry's law constant ($k_c = P(CO_2)/[CO_2]$) as a function of temperature. The solubility of carbon dioxide in an aqueous solution prepared by bubbling the gas through water increases as the pressure of the gas increases and as the temperature decreases. Commercially bottled carbonated beverages are saturated with carbon dioxide at a temperature of 1–2°C and a gas pressure of about 2 atm.

Carbon dioxide or other volatile solutes can be removed from a solution by boiling. However, this fact does not imply that the solubility of the volatile solute is zero at the boiling temperature. The term "solubility" pertains to equilibrium; a solution in which the solvent and solute are boiling away is not a system at equilibrium. A dissolved gas is removed from solution by boiling because the escaping vapor of the boiling solvent sweeps

TABLE 7–4
Partial Pressure of Carbon Dioxide Over Aqueous Solutions

(*a*) $T = 25.0°C$

$[CO_2]$ / $mol\ L^{-1}$	$P(CO_2)$ / torr	$P(CO_2)/[CO_2]$ / $torr\ L\ mol^{-1}$
0.0050	112	2.24×10^4
0.010	224	2.24×10^4
0.034	760	2.24×10^4
		$k_c = 2.24 \times 10^4$ torr L mol^{-1}

(*b*)

T / °C	$k_c = P(CO_2)/[CO_2]$ / $torr\ L\ mol^{-1}$
0	0.979×10^4
10	1.42×10^4
20	1.95×10^4
25	2.24×10^4
30	2.57×10^4
40	3.32×10^4
50	3.99×10^4

away with it the vaporized solute. With continued escape of solvent, more and more of dissolved solute goes into the vapor phase to be swept away; the concentration of the solute in the liquid solution becomes lower and lower and can be made to approach zero if enough solvent is boiled away.

THE EFFECT OF A SOLUTE ON THE SOLVENT'S VAPOR PRESSURE

We have just considered solutions of volatile solutes and have been introduced to Henry's law, $P_2 = k_x x_2$, which is valid for dilute solutions of undissociated solutes. Now we focus attention on the vapor pressure of the solvent, Component 1, and you will be introduced to *Raoult's law* (named for FRANCOIS RAOULT),

$$P_1 = P_1^0 x_1$$

in which P_1 and P_1^0 are the partial pressures of the solvent over the solution and the pure solvent, respectively, and x_1 is the mole fraction of the solvent. Figure 7–7 shows the pressure of water vapor over aqueous solutions of sucrose. In dilute solution (i.e., high mole fraction of solvent, $x_1 > 0.96$), the vapor pressure of water is given with an error of $<1\%$ by Raoult's law. At higher concentrations the deviations become appreciable; at $x_1 = 0.9$, the error is $\sim 6\%$. For some purposes, it is useful to rearrange this equation to give the fractional lowering of the vapor pressure $(P_1^0 - P_1)/P_1^0$. If each side of the equation

$$P_1 = P_1^0 x_1$$

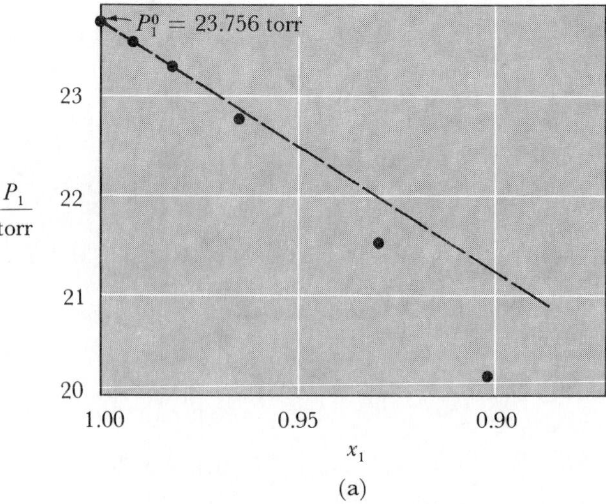

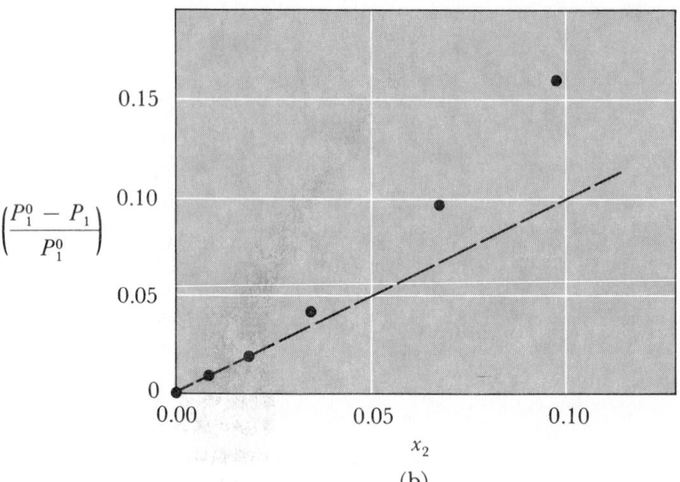

FIGURE 7–7
The vapor pressure of water over aqueous solutions of sucrose (common sugar) at 25.0°C. (a) P_1 versus x_1. (b) $(P_1^0 - P)/P_1^0$ versus x_2. The dashed line is the Raoult law line in each figure.

is subtracted from P_1^0, and if each side of the resulting equation is divided by P_1^0, we obtain an equation for the fractional lowering of vapor pressure:

$$P_1^0 - P_1 = P_1^0 - P_1^0 x_1$$

$$\frac{P_1^0 - P_1}{P_1^0} = \frac{P_1^0(1 - x_1)}{P_1^0} = 1 - x_1$$

But for a binary solution, $(1 - x_1)$ is the mole fraction of the solute, x_2. Thus Raoult's law can be given as

$$\frac{P_1^0 - P_1}{P_1^0} = x_2$$

which relates the fractional vapor pressure lowering to the mole fraction of the solute.

Figure 7–8 shows an experimental arrangement that demonstrates that a solute can lower the vapor pressure of a volatile solvent (e.g., H_2O).

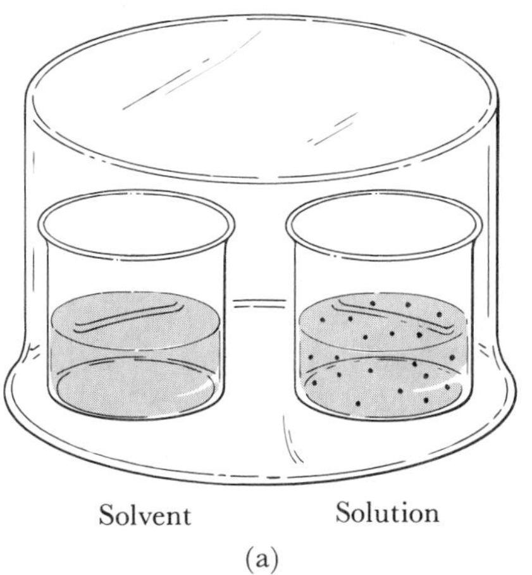

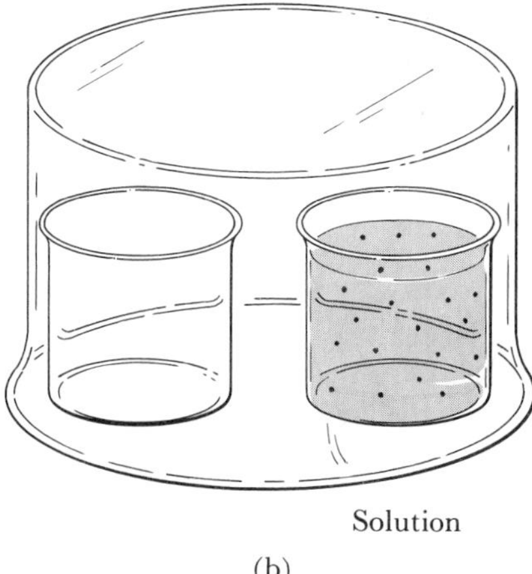

FIGURE 7–8
The transfer of water from the pure liquid to a solution containing a nonvolatile solute (e.g., sugar). (a) Initial condition: Beakers containing pure water and a solution of sugar are enclosed under an evacuated bell jar. (b) After a long time: Water has distilled to the beaker containing the solution. The temperature is maintained constant in this experiment; the transfer occurs more rapidly with air removed from vapor space.

In an enclosed space two beakers are present; one contains a solution of a nonvolatile solute and the other contains the pure solvent. With the passage of time, the volume of solvent decreases in the beaker containing pure solvent, and the volume of the solution increases. The explanation is simple. The pure solvent has a larger vapor pressure than the solution, and the concentration of solvent molecules in the vapor exceeds that which would be present in equilibrium with the solution. The net process with water as the solvent,

$$H_2O(l, x_1 = 1) \rightarrow H_2O(g) \rightarrow H_2O(l, x_1 < 1)$$

occurs in the direction indicated. Given enough time, all of the solvent distills into the beaker containing the nonvolatile solute.

A comparison of Henry's and Raoult's laws is informative. Each is an equation giving a proportionality between the partial pressure (P) of a component of a solution and its mole fraction in solution (x):

$P_2 = k_x x_2$	Henry's law applied to solute, Component 2
$P_1 = P_1^0 x_1$	Raoult's law applied to solvent, Component 1

For Henry's law, the proportionality constant must be evaluated experimentally for the system and temperature in question: its value depends upon the identity of both components of the solution. For Raoult's law, however, the proportionality constant is specified: it is the vapor pressure of the pure component at the temperature in question.

Besides lowering the vapor pressure of a solvent, a solute increases the boiling point of the solution if the solute is nonvolatile, and decreases the freezing point of the solution if the solute does not form solid solutions with

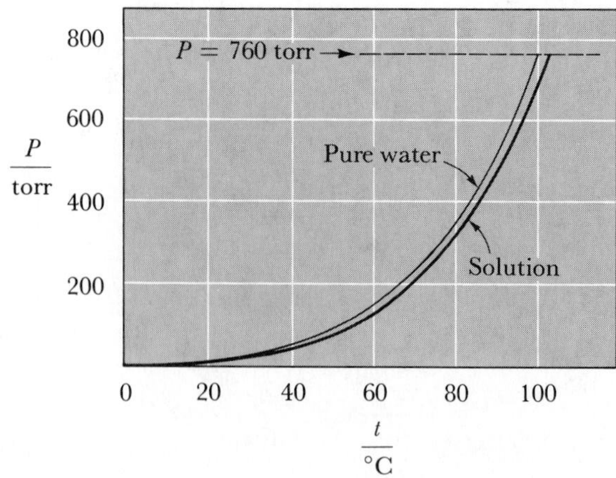

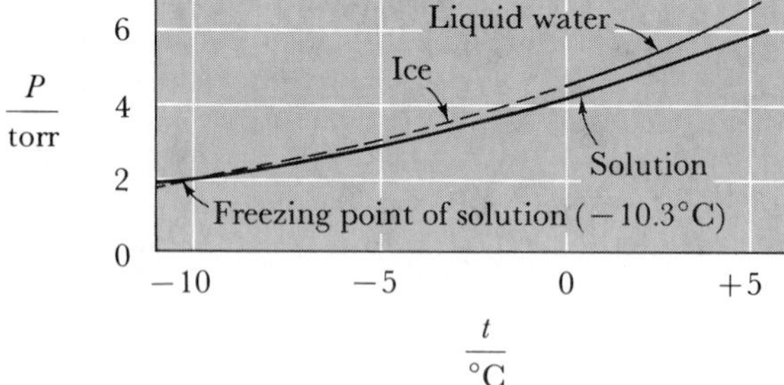

FIGURE 7–9
The vapor pressure of water as a function of temperature. Also shown is the vapor pressure of water over a solution of a nondissociating solute in which the mole fraction of water is 0.90 ($x_{solute} = 0.10$). This solution is one in which water obeys Raoult's law. Enlarged portion shows $t = -11°C$ to $+5°C$.

the solvent. Figure 7–9 shows the vapor pressure of water as a function of temperature. Although related to Figure 1–5, this diagram is different because it describes water in the presence of air, with the total pressure equal to one atmosphere; Figure 1–5 describes a system containing only water (no air). Figure 7–9 also shows the vapor pressure of the solvent component, water, over a solution in which its mole fraction is 0.90. At each temperature the vapor pressure of water is 90% of the value for pure water. The solute is one that does not form solid solutions with ice; therefore the vapor pressure curve for solid water is not influenced by the presence of solute in the liquid phase. (Very few substances form solid solutions with ice.) The normal boiling point of the solution, the temperature at which the vapor pressure is equal to 760 torr, is higher than 100°C. The freezing point, the temperature at which the vapor pressure of water over the solid is equal to the vapor pressure of water over the solution, is lower than 0°C. This is shown also in Figure 7–10, which enlarges the regions around the boiling point and freezing point for a solution with $x_2 = 0.00180$ (0.100 molal).

In solutions that are dilute and obey Raoult's law, the boiling point elevation and the freezing point depression are each proportional to the mole fraction of the solute, x_2. These quantities are given by equations

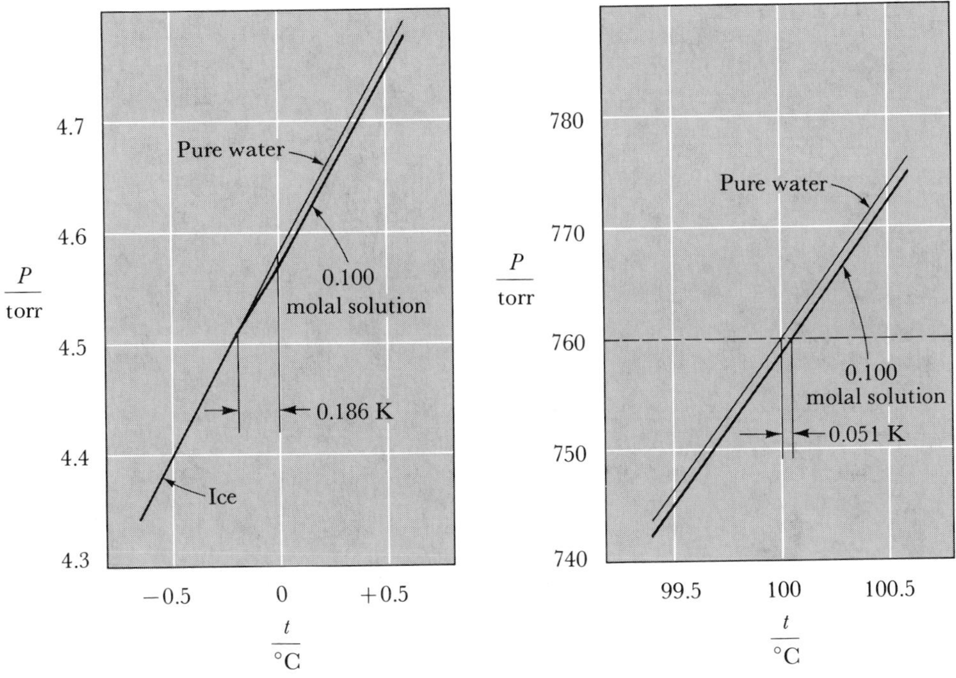

FIGURE 7–10
The vapor pressure of water over pure water and over a solution of a nondissociating solute. The concentration of the solute is 0.100 mol kg^{-1}; the mole fraction of the solute is:

$$x_2 = \frac{n_2}{n_2 + n_1} = \frac{0.100 \text{ mol}}{0.100 \text{ mol} + \dfrac{1000 \text{ g}}{18.02 \text{ g mol}^{-1}}} = 0.00180$$

that can be derived directly.[2] The equations are:

Boiling point elevation $$\Delta T_b = \frac{RT_b^2 x_2}{\Delta H_{vap}}$$

Freezing point depression $$\Delta T_f = \frac{RT_f^2 x_2}{\Delta H_{fus}}$$

For the aqueous solution with $x_2 = 0.00180$, shown in Figure 7–10, the boiling point elevation can be calculated using the molar heat of

---

[2]The derivations will not be given here. You can derive the equations using simple geometrical reasoning in connection with the plot of $\log P$ versus $(1/T)$. Recall from Chapter 6 that the slope, $\Delta \log P/\Delta(1/T)$, is related to the value of ΔH_{vap}. In the derivation, use is made of the relationship $-\ln x \cong 1 - x$. (See Appendix 2 for a discussion of the relationship between natural logarithms and power series.) The derivations are found in textbooks of physical chemistry.

vaporization, $\Delta H_{\text{vap}} = 41{,}160 \text{ J mol}^{-1}$:

$$\Delta T_b = \frac{8.31 \text{ J K}^{-1} \text{ mol}^{-1} \times (373.2 \text{ K})^2}{41{,}160 \text{ J mol}^{-1}} \times 1.80 \times 10^{-3} = 0.051 \text{ K}$$

Calculation of the freezing point depression of this solution uses the molar heat of fusion, $\Delta H_{\text{fus}} = 6020 \text{ J mol}^{-1}$:

$$\Delta T_f = \frac{8.31 \text{ J mol}^{-1} \text{ K}^{-1} \times (273.2 \text{ K})^2}{6020 \text{ J mol}^{-1}} \times 1.80 \times 10^{-3} = 0.186 \text{ K}$$

This particular concentration of solute ($x_2 = 0.00180$) was chosen because it is equal to 0.100 molal, and the molal concentration scale commonly is used in studies of this type. In such studies it is common to express the proportionality between either the boiling point elevation or the freezing point depression and the molality:

$$\Delta T_b = K_b \times \text{molality}$$
$$\Delta T_f = K_f \times \text{molality}$$

From the calculation for the aqueous solution just considered, we have for the proportionality constants:

Boiling point elevation constant K_b

$$K_b = \frac{0.051 \text{ K}}{0.100 \text{ mol kg}^{-1}} = 0.51 \text{ K kg mol}^{-1}$$

Freezing point depression constant K_f

$$K_f = \frac{0.186 \text{ K}}{0.100 \text{ mol kg}^{-1}} = 1.86 \text{ K kg mol}^{-1}$$

These values of K_b and K_f for water and values for some other useful solvents are tabulated in Table 7–5.

TABLE 7–5
Molal Boiling Point Elevation Constants and Molal Freezing Point Depression Constants for Various Substances

Substance	T_b / K	K_b[a]	T_f / K	K_f[a]
Water	373.2	0.51	273.2	1.86
Acetic acid	391.3	2.93	290.2	3.9
Benzene	353.4	2.53	278.65	5.12
Napthalene	—	—	353.2	6.8
Ethylene bromide	—	—	283.2	12.5

[a]Dimensions of K_b and K_f are K kg mol^{-1}.

EXAMPLE: What is the freezing point of a solution in which 0.0150 mol of a nondissociating solute is dissolved in 53.4 g of benzene?

$$\Delta T_f = K_f \times \text{molality}$$

$$= 5.12 \text{ K kg mol}^{-1} \times \frac{0.0150 \text{ mol}}{53.4 \text{ g}} \times \frac{1000 \text{ g}}{1 \text{ kg}} = 1.44 \text{ K}$$

$$T_f = 278.65 \text{ K} - 1.44 \text{ K} = 277.21 \text{ K}$$

In using the tabulated values of the proportionality constants K_f and K_b, you should remember that in the derivation of these equations, it is assumed that the solution is dilute enough so that

1. Raoult's law is obeyed by the solvent, and
2. the molality is proportional to the mole fraction.

The phenomena of boiling point elevation and freezing point depression are used in a number of ways. We commonly add a second component [usually ethylene glycol, $C_2H_4(OH)_2$] to water in the cooling system of an automobile to lower the freezing point of water below 0°C.

As we learned in Chapter 1, the molecular formula of a substance is not given by percentage composition data alone. However, percentage composition data plus the molecular weight do give the molecular formula. Although the uses of mass spectrometry in determining molecular formulas has reduced use of the freezing point depression method, this classical method still is used.

EXAMPLE: A white crystalline solid was analyzed and found to have the empirical formula CH_2O. A sample of this substance weighing 0.135 g was dissolved in 10.00 g of water. The freezing point of this solution was found to be −0.140°C. What is the molecular formula of the white solid?

The molecular weight can be calculated by the following procedure:

$$\text{molality} = \frac{-\Delta T}{K_f} = \frac{0.140 \text{ K}}{1.86 \text{ K kg mol}^{-1}} = 0.0753 \text{ mol kg}^{-1}$$

The molality of the solution is also given by:

$$\text{molality} = \frac{\text{moles of solute}}{\text{1 kilogram of solvent}}$$

$$= \underbrace{\underbrace{\underbrace{\frac{0.135 \text{ g}}{MW_2}}_{\text{Moles of solute}} \times \frac{1}{10.00 \text{ g}}}_{\text{Moles of solute per g solvent}} \times \frac{1000 \text{ g}}{1 \text{ kg}}}_{\text{Moles of solute per kilogram of solvent}} = \frac{13.5 \text{ g kg}^{-1}}{MW_2}$$

in which MW_2 is the molecular weight of the solute.[3] These two expressions of molality can be equated:

$$\frac{13.5 \text{ g kg}^{-1}}{MW_2} = 0.0753 \text{ mol kg}^{-1}$$

This can be solved for MW_2:

$$MW_2 = \frac{13.5 \text{ g kg}^{-1}}{0.0753 \text{ mol kg}^{-1}} = 179 \text{ g mol}^{-1}$$

Now the molecular formula can be determined:

$$(CH_2O)_n \text{ with } (12 + 2 \times 1 + 16)_n = 179$$

$$n = \frac{179}{12 + 2 + 16} = \frac{179}{30} = 6$$

Therefore the formula is $(CH_2O)_6$ or $C_6H_{12}O_6$.

Although either freezing point depression or boiling point elevation methods can be used in evaluating molecular weights, the freezing point depression method is more practical because:

1. freezing points are less sensitive to small variation in atmospheric pressure;
2. the values of K_f generally are larger than the values of K_b (Why is this?);
3. if the compound in question (the solute) decomposes slowly, it is less liable to do so during the measurement at a lower temperature, where the decomposition rate is lower.

Direct study of vapor pressure lowering as a means of determining molecular weight is not as practical as study of the freezing point depression, but commercial instruments are now available that measure the vapor pressure lowering of a solute. They do so by determining the temperature difference between a sample of pure solvent and the solution necessary to make the vapor pressures over the two liquids (the pure solvent and the solution) the same.

OSMOSIS, REVERSE OSMOSIS, AND OSMOTIC PRESSURE

Osmosis occurs when water or another solvent passes through a membrane that stops the flow of solute molecules. Such a membrane is known as a *semipermeable membrane*, and it can be thought of as a very fine sieve, which allows only small molecules to pass through. (Cellophane is a membrane material used in many biochemical laboratories. It is permeable to substances with molecular weight less than $\sim 10^4$, but impermeable to solutes with higher molecular weight. Other membrane materials are available

[3]The dimension g kg^{-1} may seem faulty. This is because it is incomplete; it means g of solute per kg of solvent.

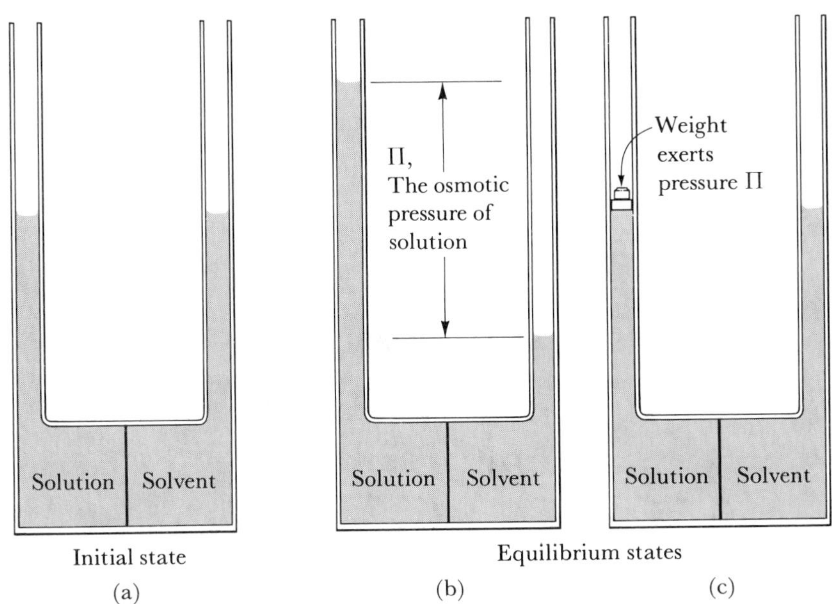

FIGURE 7–11
Osmosis and osmotic pressure.
(a) Initial state: Pure solvent and solution separated by membrane permeable only to the solvent. At membrane surface the vapor pressure of the pure solvent is greater than the vapor pressure of the solution. Thus solvent will flow from pure solvent to solution. This is *osmosis*.
(b) Equilibrium state: Osmosis of pure solvent into the solution has created at the membrane surface a greater static pressure on the solution, thereby raising its vapor pressure; when this pressure is equal to the vapor pressure of the pure solvent, there is no further net flow of solvent.
(c) Equilibrium state: Added pressure on the solution has raised the vapor pressure of the solution at the membrane to that of the solvent. Thus there is no net flow of solvent.

that discriminate at much higher or much lower molecular weights.) If a solution and pure solvent are separated by a semipermeable membrane, as shown in Figure 7–11, there is a net flow of solvent through the membrane to the side of the vessel containing solution. In this respect, the experiment resembles that depicted in Figure 7–8, in which there was a net flow of solvent through the vapor phase from pure solvent to a solution. But in that experiment, the distillation of solvent continued until all of the pure solvent had disappeared, whereas here, the net flow of solvent through the semipermeable membrane stops when there is a particular excess of pressure on the solution side of the membrane. A state of equilibrium then exists, and there is no net tendency for solvent to flow in either direction.

The explanation for this phenomenon lies in the effect the total pressure on a liquid has on its vapor pressure. Theory (thermodynamics), confirmed by experiment, gives the relationship

$$\frac{\ln P_1/P_2}{(p_1 - p_2)} = \frac{V_s^0}{RT}$$

in which P_1 and P_2 are the vapor pressure of the liquid at the externally applied pressures p_1 and p_2, and V_s^0 is the molar volume of the liquid. The change of vapor pressure is very small (negligible) for moderate changes of pressure. For instance, changing the external pressure from 1.00 atm to 0.50 atm (the ordinary atmospheric pressure at an elevation of ~17,000 ft) changes the vapor pressure of water (with $V_s^0 = 0.0180 \text{ L mol}^{-1}$) at 25°C by only ~0.04%:

$$2.303 \log \frac{P_1}{P_2} = \frac{V_s^0(p_1 - p_2)}{RT}$$

$$\log \frac{P_1}{P_2} = \frac{0.0180 \text{ L mol}^{-1} \ (0.50 - 1.00) \text{ atm}}{0.0821 \text{ L atm K}^{-1} \text{ mol}^{-1} \times 298.2 \text{ K} \times 2.303}$$

$$\log \frac{P_1}{P_2} = -1.60 \times 10^{-4}$$

$$P_1 = 0.9996 P_2$$

What is important here is not the fact that very minor changes in vapor pressure are caused by changes of the external pressure; rather, it is the possibility of offsetting the effect that a solute has upon the vapor pressure of the solvent with an increase of the external pressure. The external pressure needed to raise the vapor pressure of a solution to the value for the pure solvent at that temperature is called the osmotic pressure, represented by Π. The excess pressure due to the height of the liquid column shown in Figure 7–11 is the osmotic pressure of the solution; thus in terms of the equation just given,

$$\Pi = p_1 - p_2$$

Combining the equation for the dependence of vapor pressure upon external pressure (just given) and the equation for the dependence of vapor pressure upon solute concentration (Raoult's law), we obtain an equation for osmotic pressure. A solute present at mole fraction x_2 causes the vapor pressure to be lowered from P^0, the vapor pressure of the pure solvent, to P:

$$\frac{P}{P^0} = 1 - x_2$$

$$\ln \frac{P^0}{P} = -\ln (1 - x_2)$$

For small values of x_2, $-\ln (1 - x_2) \cong x_2$; therefore (see footnote 4)

$$\ln \frac{P^0}{P} = x_2$$

[4]Check this with a table of natural logarithms or using a pocket calculator. For instance, for $x_2 = 0.02$, the correct value for $-\ln (1 - 0.02)$ is 0.020203.

The equation giving $\ln (P^0/P)$ in terms of the osmotic pressure is

$$\ln \frac{P^0}{P} = \frac{\Pi V_s^0}{RT}$$

Combination of these two equations gives

$$\Pi = \frac{x_2}{V_s^0} RT$$

If x_2 is small, x_2/V_s^0 is equal to the molar concentration of the solute:

$$\frac{x_2}{V_s^0} = \frac{n_2}{(n_2 + n_1) V_s^0}$$

since $x_2 = n_2/(n_2 + n_1)$; for $n_2 \ll n_1$, this becomes

$$\frac{x_2}{V_s^0} = \frac{n_2}{n_1 V_s^0}$$

But $n_1 V_s^0$ is the total volume of solvent, V, therefore,

$$\frac{x_2}{V_s^0} = \frac{n_2}{V} = C_2$$

Thus the equation for the osmotic pressure is

$$\Pi = C_2 RT$$

This quantity, Π, is easily measured even for very dilute solutions, and the phenomenon is useful in determining the molecular weight of substances with large molecular weights. Such substances are difficult or impossible to characterize by vapor pressure lowering (or freezing point depression) measurements. Consider an aqueous solution containing a dissolved protein of molecular weight 1.0×10^5 at a concentration of 6.00 g per 1000 g of water ($C_2 \cong 6 \times 10^{-5}$ mol L^{-1}). According to Raoult's law, the vapor pressure lowering at 25°C is:

$$P_1^0 - P_1 = x_2 P_1^0 = \frac{n_2}{n_1 + n_2} P_1^0$$

$$= \frac{\dfrac{6.00\ \text{g}}{1 \times 10^5\ \text{g mol}^{-1}}}{\dfrac{1000\ \text{g}}{18.01\ \text{g mol}^{-1}}} \times 23.76\ \text{torr}$$

$$= 2.6 \times 10^{-5}\ \text{torr}$$

Contrast this immeasurably small vapor pressure lowering (a decrease of vapor pressure of $\sim 10^{-4}\%$) with the osmotic pressure of the solution:

$$\begin{aligned}\Pi &= C_2 RT \\ &= 6.00 \times 10^{-5}\ \text{mol L}^{-1} \times 0.0821\ \text{L atm K}^{-1}\ \text{mol}^{-1} \times 298.2\ \text{K} \\ &= 1.47 \times 10^{-3}\ \text{atm} \\ &= 1.12\ \text{torr}\end{aligned}$$

which is an easily measured quantity.

The derivation of the simple equation $\Pi = C_2RT$ is based upon:

1. the assumption of solution ideality (conformity to Raoult's law), and
2. mathematical approximations that are valid only at very low concentrations.

Therefore, we must measure osmotic pressure to find the molecular weight of a solute only in dilute solutions. There are several instruments for determining molecular weight that depend upon measurement of the osmotic pressure of a solution. The most useful of these senses the pressure differential across the membrane between the solvent and solution and counterbalance it with an externally applied pressure before appreciable solvent transport occurs. Such instruments can be used to determine molecular weights in the range 20 thousand to 1 million.

The spontaneous process that occurs during osmosis is:

$$\text{concentrated solution} + \text{solvent} \rightarrow \text{dilute solution}$$

What of the reverse process, which is called *reverse osmosis*? This process does not occur spontaneously in an isolated system, but it can be made to occur if work is done on the system. If an external pressure greater than the osmotic pressure of the solution were applied to the solution in the left side of the apparatus shown in Figure 7–11, the process

$$\text{dilute solution} \rightarrow \text{concentrated solution} + \text{solvent}$$

would occur. The minimum work needed to produce a certain volume of pure solvent, V, is simply the $P\Delta V$ work, which is

$$w = \Pi V$$

Reverse osmosis may be very useful in efforts to clean contaminated water supplies, claim fresh water from brackish water or from sea water, and so on. For these purposes, researchers are developing semipermeable membranes impermeable to simple ionic solutes (e.g., Na^+ and Cl^-) but permeable to water.

Our study of colligative properties has dealt primarily with nondissociating solutes. For solutes that dissociate, it is the total concentration of solute particles that is relevant. That is, the total concentration of solute particles in $0.0300M$ HCl is 0.0600 mol L^{-1}, $[Na^+] = 0.0300$ mol L^{-1} and $[Cl^-] = 0.300$ mol L^{-1}, and the osmotic pressure of this solution at 298.2 K is

$$\begin{aligned}\Pi &\cong 0.0600 \text{ mol L}^{-1} \times 0.0821 \text{ L atm K}^{-1}\text{ mol}^{-1} \times 298.2 \text{ K}\\ &\cong 1.47 \text{ atm}\end{aligned}$$

However, solutions of ionic solutes are very nonideal at moderate-to-high concentrations, and there are deviations for such solutions from the simple relationships of the colligative properties that we have learned here. We will learn something of these deviations in Chapter 9.

7–5 Liquid Solutions and Solid Solutions

Many of the solutions with which all of us, including chemists, deal are dilute solutions; we have just considered such solutions. But there also is much interest in concentrated solutions—solutions in which the two or more components are present in similar amounts. For these solutions, use of the terms "solute" and "solvent" serves no purpose, and the concentration units (molality and molarity) in which the identity of solute and solvent plays roles are not generally used. But the mole-fraction unit is suited for use with such solutions, and it will be used primarily in the material we will be discussing now.

LIQUID SOLUTIONS OF VOLATILE COMPONENTS

We have been considering dilute solutions in which either the solute is volatile (to allow Henry's law to be illustrated) or the solvent is volatile (to allow Raoult's law to be illustrated). Now we will consider solutions in which both components are volatile, and our attention will not be confined to dilute solutions. We will start with an idealized system of two liquid components, A and B, which dissolve in one another in all proportions, each component obeying Raoult's law over the entire composition range. Such a system is said to form *ideal solutions*.

We will consider this system at a particular temperature where the vapor pressures of the pure components are

$$P_A^0 = 400 \text{ torr} \qquad P_B^0 = 600 \text{ torr}$$

Figure 7–12 shows the vapor pressures of each of these components as a function of the composition of the solution. Notice that the vapor pressure

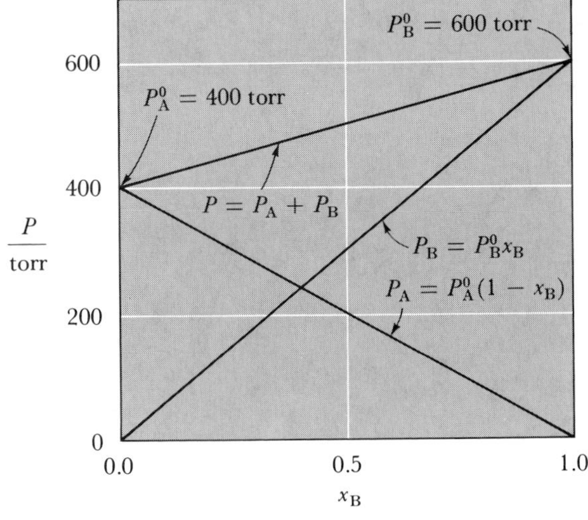

FIGURE 7–12
The vapor pressures of the components of binary liquid solutions of A and B, which obey Raoult's law:

$$P_A^0 = 400 \text{ torr} \qquad P_B^0 = 600 \text{ torr}$$
$$P_A = 400 \text{ torr}(1 - x_B) \qquad P_B = 600 \text{ torr } x_B$$
$$P = P_A + P_B = 400 \text{ torr} + 200 \text{ torr } x_B$$

of each component is given by a straight line connecting the vapor pressure point of the pure liquid, that is, $P_i = P_i^0$ at $x_i = 1$, with the point where $P_i = 0$ at $x_i = 0$; x_i is the mole fraction of the component in question, either A or B. The total vapor pressure (P) for this system is the sum of the partial pressures of the two components. In Figure 7–12, it is given by a straight line connecting P_A^0 and P_B^0:

$$\begin{aligned} P &= P_A + P_B \\ &= P_A^0 x_A + P_B^0 x_B \\ &= P_A^0(1 - x_B) + P_B^0 x_B \\ P &= P_A^0 + x_B(P_B^0 - P_A^0) \end{aligned}$$

A straightforward calculation gives us the composition of the vapor in equilibrium with liquid of a particular composition. If y_A and y_B are the mole fractions of A and B in the vapor, the relationships of these quantities to the partial pressures of A and B are:

$$y_A = \frac{n'_A}{n'_A + n'_B} = \frac{P_A}{P_A + P_B}$$

and

$$y_B = \frac{n'_B}{n'_A + n'_B} = \frac{P_B}{P_A + P_B}$$

in which n'_A and n'_B are the molar amounts of A and B in the vapor phase; each partial pressure is proportional to the amount, n', $P = n'(RT/V)$. Upon substitution, $P_A = P_A^0 x_A$ and $P_B = P_B^0 x_B$, these equations become:

$$y_A = \frac{x_A P_A^0}{x_A P_A^0 + x_B P_B^0}$$

$$y_B = \frac{x_B P_B^0}{x_A P_A^0 + x_B P_B^0}$$

For $x_A = 0.250$, $x_B = 0.750$, the vapor composition is

$$y_A = \frac{0.250 \times 400 \text{ torr}}{0.250 \times 400 \text{ torr} + 0.750 \times 600 \text{ torr}} = \frac{100 \text{ torr}}{100 \text{ torr} + 450 \text{ torr}}$$

$$= \frac{100}{550} = 0.182$$

and

$$y_B = \frac{0.750 \times 600 \text{ torr}}{0.25 \times 400 \text{ torr} + 0.750 \times 600 \text{ torr}} = \frac{450 \text{ torr}}{100 \text{ torr} + 450 \text{ torr}}$$

$$= \frac{450}{550} = 0.818$$

The vapor is richer in the more volatile component than the liquid is ($y_B = 0.818$, $x_B = 0.750$).

EXAMPLE: What composition of liquid solution in the A–B system just described has a vapor in equilibrium with it having $y_A = 0.700$?

Using the equation already presented and the values of P_A^0 and P_B^0, we have

$$y_A = 0.700 = \frac{x_A \times 400 \text{ torr}}{x_A \times 400 \text{ torr} + (1 - x_A) \times 600 \text{ torr}}$$

$$= \frac{400x_A}{600 - 200x_A}$$

This can be rearranged to give

$$420 - 140x_A = 400x_A$$

$$x_A = \frac{420}{540} = 0.778$$

A real system that obeys Raoult's law closely is the benzene–toluene system, for which the vapor pressure data are given in Figure 7–13. The molecular structures of these compounds are very similar:[5]

```
    H       H              H       H
     \     /                \     /
      C═C                    C═C          H
     /    \                 /    \       /
H—C        C—H         H—C        C—C—H
     \\   //                \\   //      \
      C—C                    C—C          H
     /    \                 /    \
    H       H              H       H

     Benzene                    Toluene
```

and we already have learned that the energy change when these liquids are mixed is very small. The close conformity of the system to Raoult's law supports the conclusion already drawn from the energy of mixing, namely that the interaction energies of like and unlike molecules are very similar.

If the components of a binary liquid solution are less similar than benzene and toluene, their vapor pressures over the solution probably will deviate from Raoult's law. For some systems the deviations are positive,

$$P_i > P_i{}^0 x_i \qquad (i = 1 \text{ and } 2)$$

and for others the deviations are negative,

$$P_i < P_i^0 x_i \qquad (i = 1 \text{ and } 2)$$

(There are some systems that deviate positively in one part of the composition range and negatively in another, but we will not consider such systems.) Figure 7–14 presents vapor pressure data for the acetone–chloroform

[5]The lines in these formulas represent a type of chemical bonding; at this point we are not concerned with a description of this bonding. In each of these molecules, all carbon atoms are in the same plane.

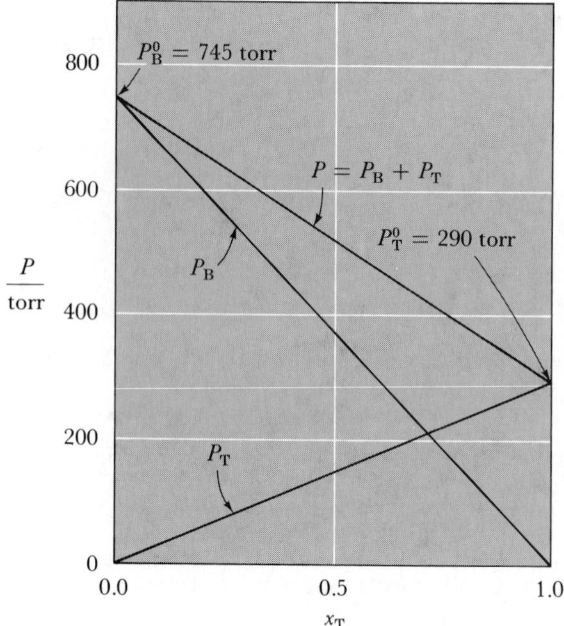

FIGURE 7–13
The vapor pressures of the components of liquid solutions of benzene (B) and toluene (T) at 79.0°C. This system obeys Raoult's law closely.

[$(CH_3)_2CO-HCCl_3$] system, which shows negative deviations from Raoult's law, and Figure 7–15 presents such data for the ethyl alcohol–benzene [$C_2H_5OH-C_6H_6$] system, which shows positive deviations from Raoult's law. For these nonideal systems, we can correlate the type of deviation from Raoult's law with the energy change upon mixing to form a

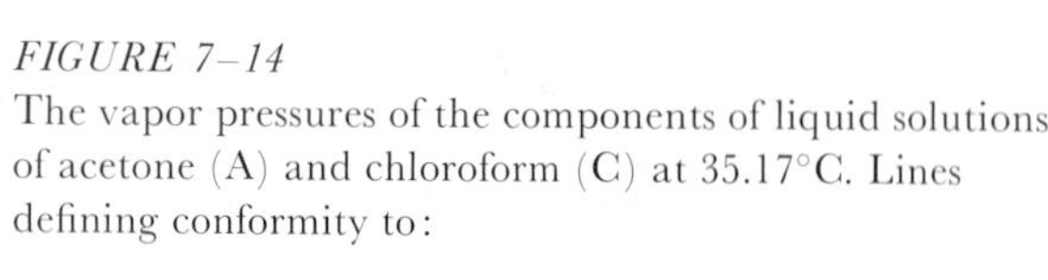
FIGURE 7–14
The vapor pressures of the components of liquid solutions of acetone (A) and chloroform (C) at 35.17°C. Lines defining conformity to:

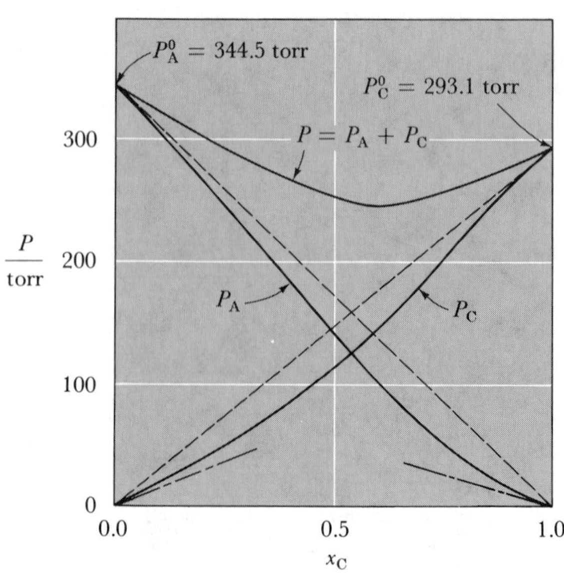

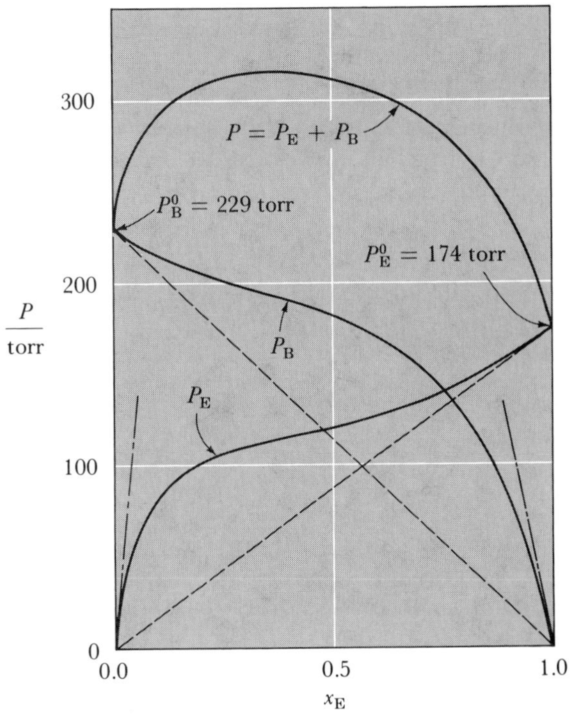

FIGURE 7–15
The vapor pressures of the components of liquid solutions of ethyl alcohol (E) and benzene (B) at 45.0°C. Lines defining conformity to:

Raoult's law _ _ _ _ _ _ _ _ _ _ _

Henry's law —·—·—·—·—·—

solution with each mole fraction = 0.50. The correlation is:

Acetone–Chloroform

$\Delta U = -1880\,\text{J (mol soln)}^{-1}$ negative deviations from Raoult's law

Ethyl alcohol–Benzene

$\Delta U = 1270\,\text{J (mol soln)}^{-1}$ positive deviations from Raoult's law

If A---B attractive interactions are stronger than A---A attractive interactions and/or B---B attractive interactions, energy is evolved when liquid A is mixed with liquid B, and the vapor pressure of each component of the solution is lower than expected from Raoult's law. That is, negative deviations from Raoult's law are observed. Because the vapor, assumed to be ideal, with which the liquid solution is in equilibrium consists of individual molecules, increasing the interaction of unlike molecules in the liquid phase will decrease the number of molecules in the vapor phase. Some systems with this type of nonideality have components that interact to form a new species (e.g., AB from components A and B). The structures of A and B may make the formation of AB likely. Such is the case for the acetone–chloroform system pictured in Figure 7–14. With a hydrogen atom in the one molecule (chloroform, $HCCl_3$) and an oxygen in the other [acetone, $(CH_3)_2CO$], it seems reasonable to expect a one-to-one interaction forming a complex involving hydrogen bonding:

$$(CH_3)_2CO + HCCl_3 \rightleftarrows (CH_3)_2CO{-}{-}{-}HCCl_3$$

This association in the liquid phase causes the true mole fractions of uncombined acetone and chloroform to be lower than the stoichiometric mole fractions. Under the assumption that the solution is an ideal mixture of species—acetone, chloroform, and the acetone–chloroform complex—the extent of the association reaction can be calculated from the vapor pressure data. We will perform such a calculation to illustrate the approach.

In a solution containing 0.360 mol of chloroform and 0.640 mol of acetone, the vapor pressure of chloroform is 72.3 torr and that of acetone is 200.8 torr at 35.2°C. This vapor will be assumed to contain no acetone–chloroform complex. (This assumption is justified by the low concentration of each component in the vapor phase.) At this same temperature the vapor pressures of the pure liquids are: chloroform, 293.1 torr; acetone, 344.5 torr. The true mole fraction of the species chloroform is given by use of Raoult's law:

$$x_{\mathrm{HCCl_3}} = \frac{P_{\mathrm{HCCl_3}}}{P^0_{\mathrm{HCCl_3}}} = \frac{72.3\ \text{torr}}{293.1\ \text{torr}} = 0.247$$

This mole fraction is related to the amount of each species in the liquid. Consider the stoichiometry of the reaction:

	$(CH_3)_2CO$ +	$HCCl_3$	= $(CH_3)_2CO$---$HCCl_3$
Amount if no reaction:	0.640 mol	0.360 mol	0
Amount at equilibrium:	(0.640 mol − z)	(0.360 mol − z)	z

in which z = amount of complex formed. The true mole fraction of chloroform (0.247) is equal to:

$$x_{\mathrm{HCCl_3}} = \frac{\text{moles of HCCl}_3}{(\text{moles of HCCl}_3 + \text{moles of (CH}_3)_2\text{CO} + \text{moles of complex})}$$

$$= \frac{0.360\ \text{mol} - z}{(0.360\ \text{mol} - z) + (0.640\ \text{mol} - z) + z}$$

$$= \frac{0.360\ \text{mol} - z}{1.000\ \text{mol} - z} = 0.247$$

which can be solved for z:

$$0.247\ \text{mol} - 0.247z = 0.360\ \text{mol} - z$$

$$0.753z = 0.113\ \text{mol}$$

$$z = 0.150\ \text{mol}$$

If this interpretation of the system (i.e., specific interaction to form a one-to-one complex) is correct, the vapor pressure of acetone in this solution should be consistent with this value of z. Raoult's law gives:

$$P_{(CH_3)_2CO} = P^0_{(CH_3)_2CO} \times \frac{\text{moles of } (CH_3)_2CO}{(\text{moles of } HCCl_3 + \text{moles of } (CH_3)_2CO + \text{moles of complex})}$$

$$= 344.5 \text{ torr} \times \frac{0.640 \text{ mol} - z}{1.000 \text{ mol} - z}$$

$$= 344.5 \text{ torr} \times \frac{0.640 \text{ mol} - 0.150 \text{ mol}}{1.000 \text{ mol} - 0.150 \text{ mol}}$$

$$= 344.5 \text{ torr} \times \frac{0.490 \text{ mol}}{0.850 \text{ mol}} = 199 \text{ torr}$$

This value agrees with the observed value (200.8 torr), which supports, but does not prove, the hypothesis that the hydrogen-bonded complex, $(CH_3)_2CO$---$HCCl_3$, exists. (Additional support comes from correlation of the data for this composition with the data for other compositions.)

A solution may form from two liquid components, in one or both of which the attractive intermolecular forces are stronger than those in the liquid mixture, that is, A---A and/or B---B interactions are stronger than A---B interactions. In such cases energy is absorbed when the solution forms, and the system has positive deviations from Raoult's law. The ethyl alcohol–benzene system exhibits this type of behavior because of the hydrogen bonding between ethyl alcohol molecules, each with a hydroxyl group. The structure of the ethyl alcohol molecule, including the unshared pairs of electrons on the oxygen atom, is

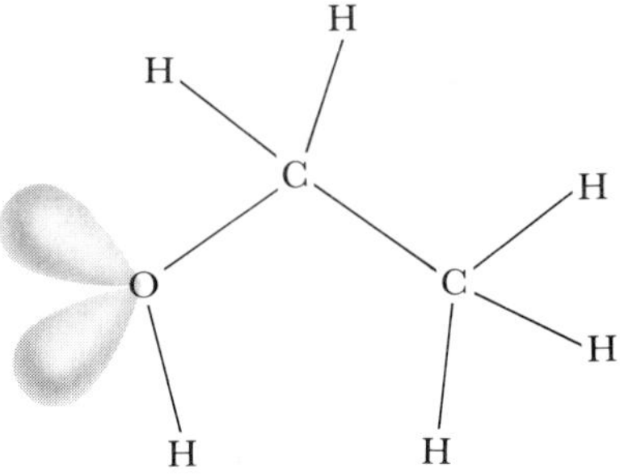

Like the water molecule shown in Figure 1–7, this molecule has a hydroxyl group, —OH, with a hydrogen atom bonded to a nonmetal atom, which also has unshared pairs of electrons. This structure allows for hydrogen-bonded clusters of ethyl alcohol molecules in pure liquid ethyl alcohol and also in solutions of ethyl alcohol in benzene.[6] Endothermic mixing of

[6]Table 6–4 gives data showing that ethyl alcohol has an abnormally high value of $\Delta H_{vap}/T_b$, indicative of hydrogen bonding in the liquid state.

benzene with ethyl alcohol results because energy is needed to break some of the hydrogen-bonded clusters present in the pure liquid ethyl alcohol. Quantitative calculations for this system are more complicated than those for the acetone–chloroform system, because this system probably has several different sized clusters of ethyl alcohol molecules, such as $(C_2H_5OH)_2$, $(C_2H_5OH)_3$, and $(C_2H_5OH)_4$. However, it can be shown simply that if one component of a binary solution self-associates, the other will show a positive deviation from Raoult's law. For instance, let us consider a solution prepared from 0.500 mol each of ethyl alcohol and benzene. If this solution obeyed Raoult's law, the vapor pressure of benzene, P_B, would be:

$$P_B = x_B P_B^0 = 0.500 P_B^0$$

because the stoichiometric mole fraction of benzene is 0.500. However, if the ethyl alcohol is present as clusters of five molecules of ethyl alcohol, $(C_2H_5OH)_5$, the composition of the solution in terms of the species present is

$$0.500 \text{ mol } C_6H_6 \qquad 0.100 \text{ mol } (C_2H_5OH)_5$$

and the true mole fraction of benzene, x'_B, is

$$x'_B = \frac{0.500 \text{ mol}}{0.500 \text{ mol} + 0.100 \text{ mol}} = 0.833$$

The vapor pressure of benzene, if the solution is an ideal mixture of these species, is

$$P_B = x'_B P_B^0 = 0.833 P_B^0$$

Thus the self-association of ethyl alcohol in this solution makes the true mole fraction of benzene larger than the stoichiometric mole fraction, thereby causing positive deviations from Raoult's law.

You should notice certain features of the systems in Figures 7–14 and 7–15. Even though these systems are nonideal, the partial pressure of each component shows approximate obedience to Raoult's law in the composition range where the mole fraction of the component is close to one, and approximate obedience to Henry's law in the composition range where the mole fraction of the component is close to zero. For the system showing positive deviations from Raoult's law, the total vapor pressure may exceed the vapor pressure of either pure component, but the partial pressure of each component of the solution does not exceed the vapor pressure of that component in the pure state (at the same temperature).

SEPARATING A SOLUTION'S VOLATILE COMPONENTS BY DISTILLATION

Distillation is a process that partially or completely separates the components of a liquid solution by converting the liquid to vapor and then condensing the vapor. We will gain an understanding of this process by considering the toluene–benzene system, for which the vapor pressure as a function of solution composition at 79°C was given in Figure 7–13. The same vapor-pressure data can be used to construct Figure 7–16, a type of phase diagram, which gives the phases present as a function of composition and pressure at this temperature. The line **ABC** gives the total vapor pressure as a function of the composition of the liquid phase; this is the same line as that giving the total vapor pressure in Figure 7–13. The line **ADC** gives the composition of the vapor that is responsible for each particular total vapor pressure. Thus we can read from this graph that a liquid with the composition

$$x_{\text{toluene}} = 0.500 \quad (\text{Point } \mathbf{B})$$

has a total vapor pressure of

$$P = 518 \text{ torr}$$

We can read also that the vapor that exerts this pressure has a mole fraction of toluene of 0.274,

$$y_{\text{toluene}} = 0.274 \quad (\text{Point } \mathbf{D})$$

The horizontal line at $P = 518$ torr ties together the compositions of liquid and vapor which are in equilibrium at this pressure. Such a line is commonly called a *tie line*.

The regions in this figure are significant. In the region above the line **ABC**, the pressure is so high that no vapor is present; in the region below the line **ADC**, the pressure is so low that the entire sample is in the vapor state; in the region enclosed between these two lines, both liquid and vapor are present. Consider what happens if the pressure is lowered on a liquid sample at 79°C with a 0.500 mole fraction of toluene. When the pressure

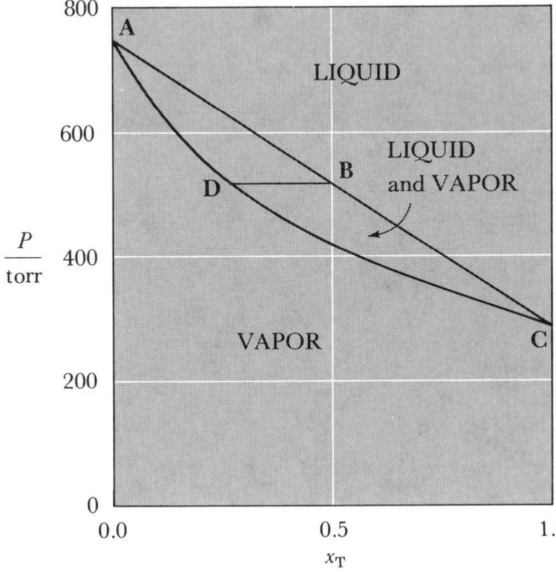

FIGURE 7–16
Phase diagram for the benzene–toluene system at 79.0°C. Each region is labeled with the phase(s) present. In the enclosed region **ABCDA**, both liquid and vapor phases are present in equilibrium with one another. The horizontal tie line **DB** at $P = 518$ torr connects the compositions of liquid and vapor in equilibrium at this pressure: $x_T = 0.500$; $y_T = 0.274$ (y_T = mole fraction of toluene in vapor phase.)

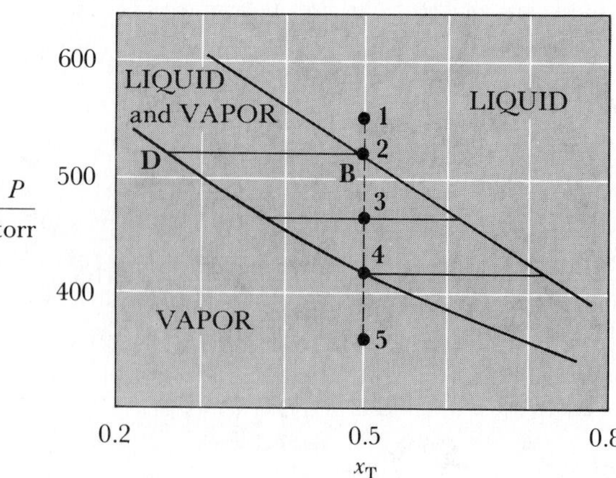

FIGURE 7–17
Stages in the process of lowering the pressure on a sample of toluene and benzene ($x_T = 0.500$, $t = 79.0°C$).
1 $P = 550$ torr; all liquid
2 $P = 518$ torr (**DB** tie line from Figure 7–16); liquid with $x_T = 0.500$, trace vapor with $y_T = 0.274$
3 $P = 465$ torr; liquid with $x_T = 0.613$, vapor with $y_T = 0.375$
4 $P = 412$ torr; trace liquid with $x_T = 0.726$, vapor with $y_T = 0.500$
5 $P = 360$ torr; all vapor

becomes 518 torr, a trace of liquid vaporizes to give vapor with a 0.274 mole fraction of toluene. If this trace of vapor were condensed, the process would be found to have enriched the benzene in the condensate, as compared to the original solution. If instead of allowing the initial trace of vapor to condense, we continue to lower the applied pressure, the system goes through stages depicted in Figure 7–17. At a pressure of 412 torr, the entire sample has become vapor, the last trace of liquid having had the composition given by the right extremity of the tie line at this pressure,

$$x_{\text{toluene}} = 0.726$$

In practice, fractional distillation is performed at a particular pressure and varying temperatures, not at a particular temperature and varying pressures. To consider such a distillation for the benzene–toluene system, we will use a boiling temperature–composition diagram. The diagram for a pressure of 750 torr is given in Figure 7–18. This figure, like Figure 7–16,

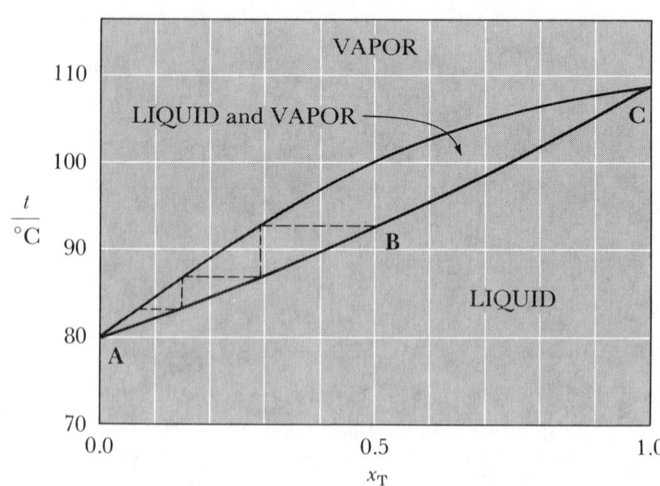

FIGURE 7–18
A boiling temperature–composition diagram for the benzene–toluene system at 750 torr. The line **ABC** gives the boiling temperature as a function of composition. Horizontal tie lines shown for 93°C ($x_T = 0.50, y_T = 0.29$), 86°C ($x_T = 0.29, y_T = 0.15$), and 83°C ($x_T = 0.15$, $y_T = 0.06$).

is a phase diagram in which the regions show the phases that are present. In the two-phase region, the compositions of the two phases in equilibrium with one another are given by the extremes of the horizontal tie lines at each temperature. We can consider an idealized distillation in which the temperature of a liquid with $x_{\text{toluene}} = 0.50$ and $x_{\text{benzene}} = 0.50$ is raised. Boiling at 750 torr starts when the temperature reaches 93°C, and the vapor produced initially has the composition $x_{\text{toluene}} = 0.29$, $x_{\text{benzene}} = 0.71$. If this small amount of vapor were condensed and then heated to boiling, which would start at 86°C, the composition of the vapor produced initially in this stage would be $x_{\text{toluene}} = 0.15$, $x_{\text{benzene}} = 0.85$. The distillation steps depicted in Figure 7–18 show how each successive step produces a condensate richer in the more volatile component, benzene in this example. But in a procedure in which each step involves distillation of only a small fraction of the liquid, only a very small yield of the more volatile component is obtained.

A modern distillation column is constructed to allow successive re-equilibrations of liquid and vapor phases at different heights in the column (see Figure 7–19). As the temperature of the distillation flask reaches the boiling temperature for the original liquid composition, boiling starts and vapor moves up the column. Some vapor condenses in the column, and the liquid so formed flows back down toward the flask. The more thoroughly the rising vapor and falling liquid come in contact in the column and have a chance to equilibrate, the more steps analogous to those considered in Figure 7–18 there will be as the vapor rises through the column. The temperature decreases as the distance up the column from the distillation flask increases; if the distillation proceeds slowly, at each particular height in the column the liquid and vapor compositions are close to the equilibrium compositions corresponding to the temperature at that height. Under such operating conditions, distillate will come over when the temperature at the top of the column reaches 79.5°C, the temperature at which benzene, the more volatile component, has a vapor pressure of 750 torr, and the initial distillate will be pure benzene.

The benzene–toluene system, which we have discussed in detail, closely obeys Raoult's law. Systems that show marked deviations from Raoult's law, such as those whose vapor–pressure diagrams are given in Figures 7–14 and 7–15, have more complex boiling temperature–composition diagrams. An example is the ethyl alcohol–benzene system pictured in Figure 7–20. We see that this system has, for $p = 760$ torr, a minimum boiling temperature, 68.2°C for a solution with a 0.45 mole fraction of ethyl alcohol. A mixture of ethyl alcohol and benzene with this composition is called an *azeotropic mixture*. If a liquid mixture with this composition were boiled at 760 torr, the entire sample would boil away at 68.2°C, the liquid and vapor compositions remaining constant. (Thus constancy of boiling temperature cannot be used as proof that a liquid is a pure compound; the liquid could be a solution with an azeotropic composition.)

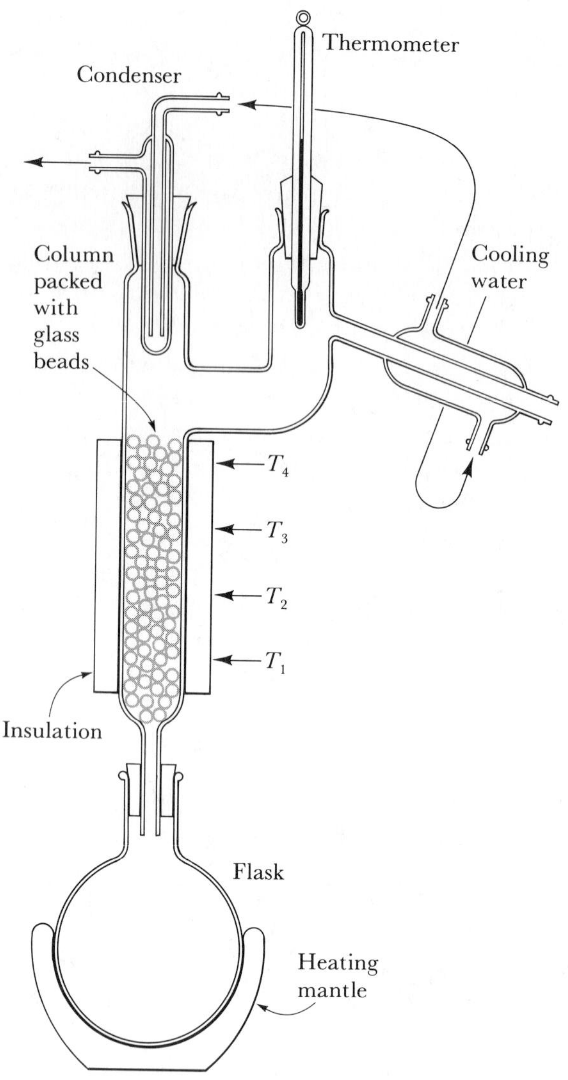

FIGURE 7–19
A distillation apparatus. Heat is added to the flask by electrical heating mantle, and the highest temperature is at this point. The temperature decreases with increasing height in column (i.e., $T_1 > T_2 > T_3 > T_4$). Ideally, at each height in column, liquid and vapor compositions are those given by the tie line for the temperature at that height.

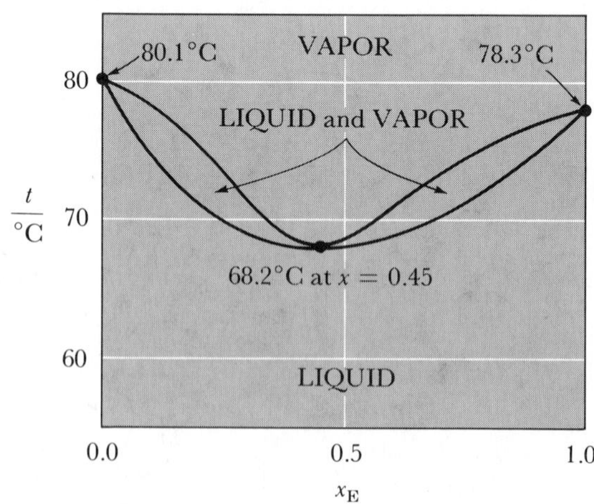

FIGURE 7–20
A boiling temperature–composition diagram for ethyl alcohol–benzene system at 760 torr. This system shows positive deviations from Raoult's law (see Figure 7–15), and an azeotrope exists in this system.

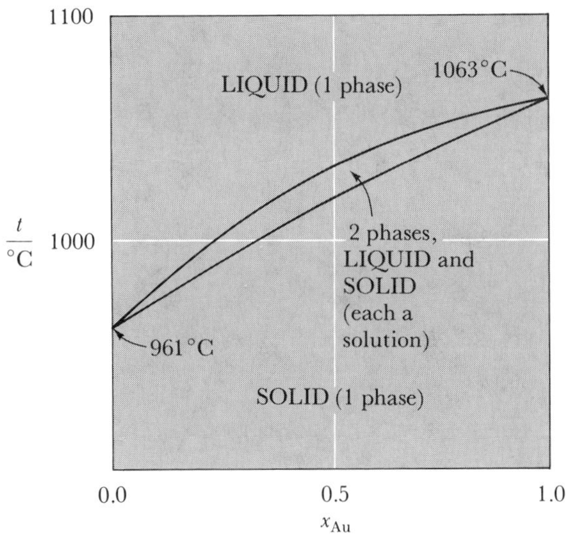

FIGURE 7–21
A freezing temperature–composition diagram for the silver–gold system. Components are completely soluble in one another in each phase. The system is close to ideal.

THE FREEZING OF SOLUTIONS

In the binary mixtures considered in our discussion of liquid–vapor equilibria, there was a single liquid phase for each system. That is, the components of the system were completely soluble in one another in the liquid phase. This property is relevant in determining the form of the phase diagram. In considering solid–liquid equilibria for binary mixtures, we will find that freezing temperature–composition diagrams differ appreciably, depending upon whether there is one solid phase or more. Figure 7–21 is a freezing temperature–composition diagram for the silver–gold system. This diagram resembles the boiling temperature–composition diagram for the benzene–toluene system, a system that obeys Raoult's law. The resemblance is understandable. In each of these systems, the components are completely soluble in both phases. Not only are silver and gold completely soluble in their liquid and solid solutions, these solutions also are close to ideal.[7] That silver and gold should form an approximately ideal solid solution is reasonable. Each pure metal crystallizes in a cubic closest-packed lattice, and the atomic radii of the two elements in the two pure solids are almost identical ($r_{Ag} = 2.889$ Å, $r_{Au} = 2.884$ Å). In the solid solutions, silver and gold atoms are randomly arranged in a cubic closest-packed lattice. The radius of the copper atom is appreciably smaller ($r_{Cu} = 2.556$ Å), and the systems copper–silver and copper–gold have freezing temperature–composition diagrams, shown in Figure 7–22, which contrast with that for the silver–gold system. The diagram for the gold–copper system shows only one liquid phase and one solid phase, but the system is far from ideal. (This freezing temperature–composition diagram resembles the boiling tem-

[7]There are experimental methods of studying the ideality of solid solutions. Some of these methods are electrochemical (see Chapter 16).

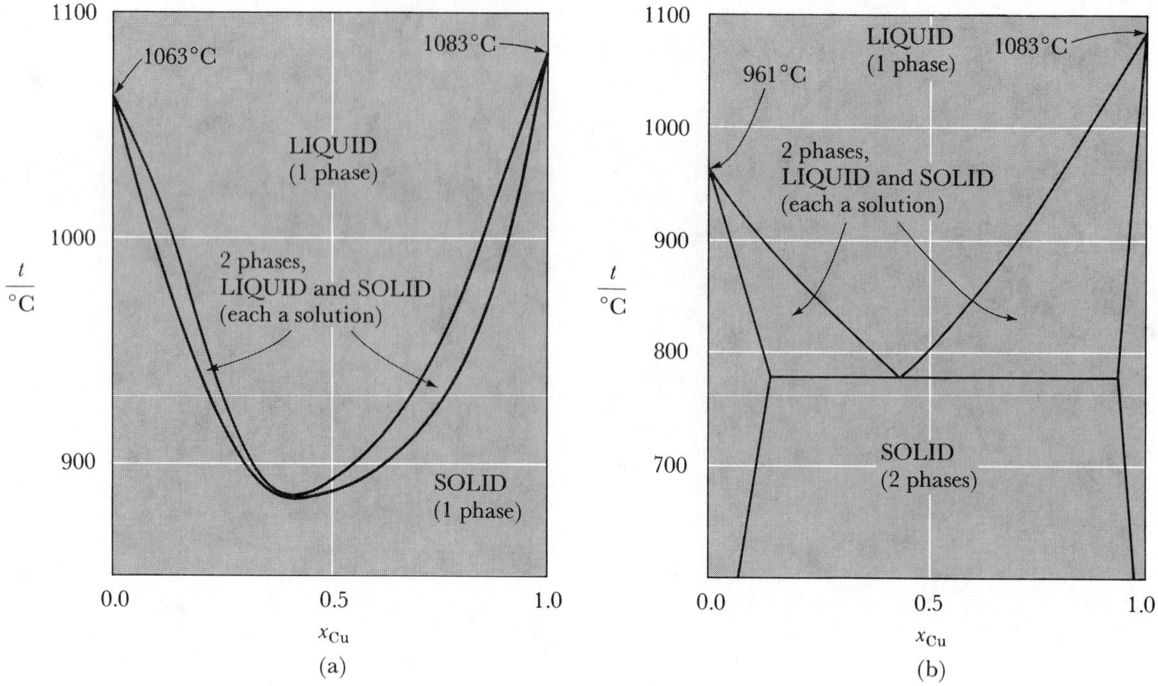

FIGURE 7–22
Freezing temperature–composition diagrams for the systems gold–copper and silver–copper. (a) Gold–copper system. Components are completely soluble in one another in each phase. The system is nonideal. (b) Silver–copper system. Components are completely soluble in the liquid phase but not in the solid; there are two solid phases, each a solution.

perature–composition diagram for the ethyl alcohol–benzene system (Figure 7–15), a nonideal system with positive deviations from Raoult's law.) If this type of nonideality of a phase is greater, a limit is reached and the components do not dissolve completely in one another in that phase at any temperature; a diagram like that for the silver–copper system results. Each of the two solid phases is a solution, but there is limited solubility of copper in silver and limited solubility of silver in copper.

Now we will consider in some detail the freezing temperature–composition diagram for a binary system in which the components are soluble in one another in the liquid phase but are insoluble in one another in the solid phase. This system consists of napthalene ($C_{10}H_8$, the constituent of common moth balls) and phenanthrene ($C_{14}H_{10}$); the freezing temperature–composition diagram is given in Figure 7–23. We can consider this figure to consist of two figures, one showing the lowering of the freezing temperature of naphthalene by phenanthrene (the branch from **A** to **B**) and the other showing the lowering of the freezing temperature of phenanthrene by naphthalene (the branch from **C** to **B**). The intersection of these two branches, Point **B**, is called the *eutectic point*. The eutectic temperature for a system is the temperature below which no liquid phase can exist at equilibrium.

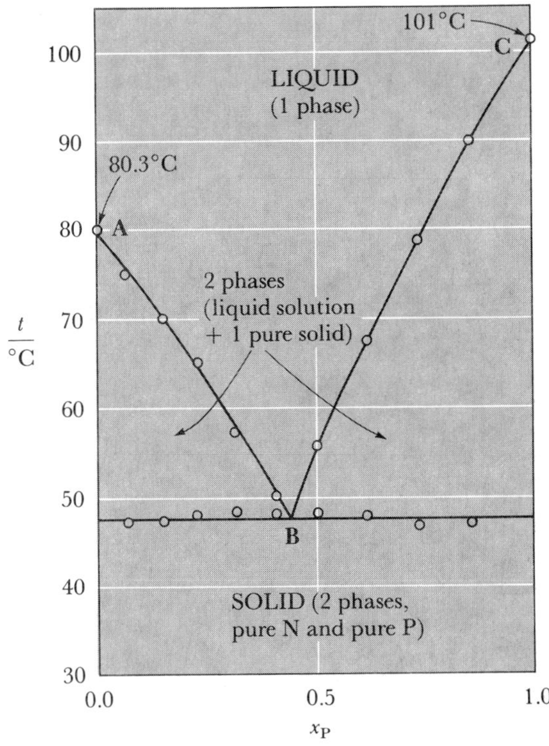

FIGURE 7–23
Freezing temperature–composition diagram for the naphthalene–phenanthrene (N–P) system. Components are completely soluble in the liquid phase, but there is essentially zero mutual solubility in the solid. Each solid phase is a pure compound.

The data upon which figures such as 7–21, 7–22, and 7–23 are based are cooling curves for mixtures of various composition. (A cooling curve for a pure substance was given in Figure 6–15.) The cooling curves (idealized) for naphthalene–phenanthrene mixtures of various composition are shown in Figure 7–24. Consider first the cooling of pure liquid napthalene, starting above its melting temperature (Curve 1); this plot is analogous to Figure 6–15. When the temperature of the sample reaches 80.2°C, solid starts to crystallize from the melt. The transformation liquid to solid, $C_{10}H_8(l) = C_{10}H_8(s)$, is accompanied by evolution of energy. This balances the heat flowing from the system, and the temperature of the sample stays at 80.2°C until solidification is complete. Then the temperature decreases as heat continues to flow from the system to cooler surroundings. The cooling curve for a solution of 20 mole % phenanthrene is shown in Curve 2. When the temperature reaches 80.2°C, there is no break in the curve because the phenanthrene has lowered the freezing point of naphthalene. At 67°C, however, solidification starts, and pure solid naphthalene separates from the melt. (Phenanthrene does not dissolve in solid naphthalene.) The residual liquid becomes richer in phenanthrene, and the freezing temperature of the solution decreases. Because the solidification of naphthalene from this melt liberates energy, the temperature of the system does not decrease as rapidly as it did before the solid started to appear. When the temperature of the melt reaches the eutectic temperature, 47.6°C, the flow of additional heat from the system causes two solid phases, solid naphthalene and solid phenanthrene, to separate from the melt. The pure

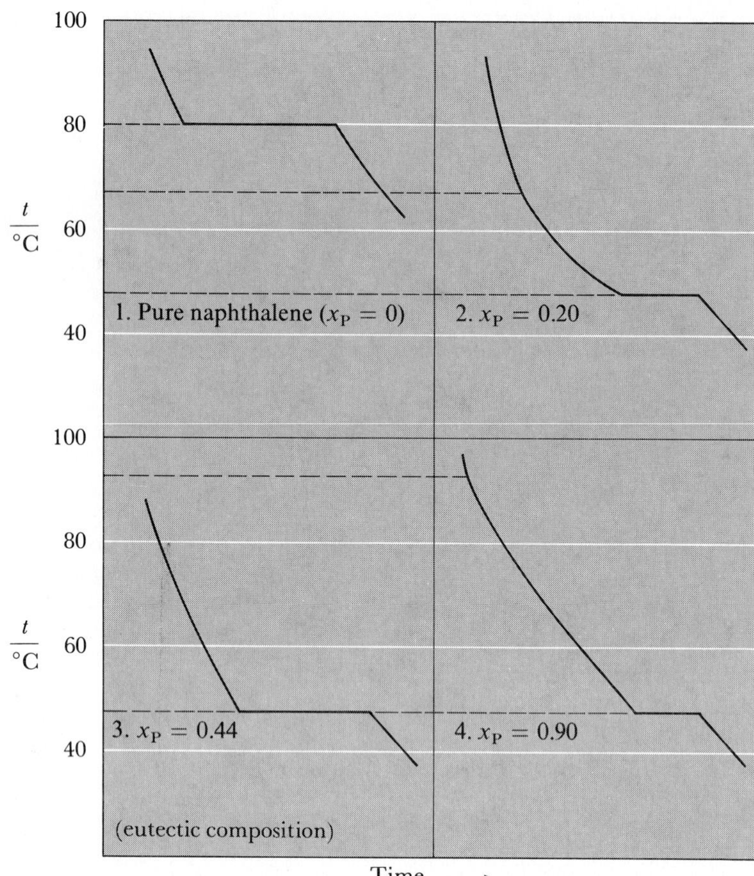

FIGURE 7–24
Cooling curves for various mixtures of naphthalene (N) and phenanthrene (P). Curves are idealized in showing no supercooling.

solid phases separate in the same proportions as they are present in the liquid solution. Therefore the composition of the liquid solution remains constant as solidification occurs, and the temperature also remains constant at the eutectic temperature until the entire sample has solidified. Further removal of heat from the system causes the temperature to drop. The sample at this point is a mixture of two phases, solid naphthalene and solid phenanthrene.

When a liquid solution with the eutectic composition (Curve 3) is cooled, no solid appears until the eutectic temperature is reached. At this temperature the two solid phases separate in the eutectic proportions, and as solidification proceeds the composition of the liquid does not change. The temperature remains at the eutectic temperature until all of the sample has solidified. The cooling curve for a liquid solution with the eutectic composition has the same form as that for a pure liquid, and the reason is the same. In each situation, the liquid composition does not change as solidification proceeds.

The meaning of the cooling curves (Figure 7–24) has been described with reference to the previously presented phase diagram (Figure 7–23). In practice, however, it is the experimental cooling curves that allow the phase diagram to be constructed. The temperatures at which there are breaks in cooling curves for liquids of various compositions are the points that are plotted in Figure 7–23. The exact shape of a cooling curve also discloses information. For instance, a horizontal line shows the kind of change of state in which the composition of the material solidifying is the same as the composition of the liquid. A horizontal cooling curve shows the liquid to be a pure substance (Curve 1 in Figure 7–24) or a liquid with the eutectic composition (Curve 3 in Figure 7–24). Also, a horizontal line at the same temperature for different compositions shows that temperature to be a eutectic temperature; the horizontal portions of Curves 2, 3, and 4 in Figure 7–24 are all at 47.6°C. To interpret cooling curves to yield a phase diagram, we need not observe the sample directly, but only the temperature as a function of time.

The relatively low freezing temperature of a solution with the eutectic composition may make such a liquid useful. Metallic sodium (mp = 97.5°C) and metallic potassium (mp = 63.4°C) form a liquid solution at temperatures as low as −12.3°C if the composition is the eutectic composition (66 mole % K). This liquid has been proposed as a heat-transfer agent in nuclear reactors. Ethylene glycol (mp = −15.6°C) is an antifreeze, which protects automobile engines to temperatures far lower than −15.6°C because the eutectic temperature for the water–ethylene glycol system is −49.3°C.

The process of zone melting, used to purify silicon and other materials, makes use of the differing solubilities in liquid and solid phases (see Figure 7–25). As an example, we will consider the silver–gold system (see Figure 7–21). If a horizontal rod of gold containing silver as an impurity is traversed by a moving heater at a temperature above the melting temperature, the material in the heated zone will be in the liquid state. Freezing then will occur as the heater moves on along the rod. In the freezing process the impurity, silver in this example, being more soluble in the liquid than in the solid, is concentrated in the liquid phase. After one complete pass of the heater, the silver is concentrated in the end of the rod that was heated last. Successive passes of the heater would purify the gold still further. Although this procedure is not used to purify gold that contains silver as an impurity, it is used to produce highly purified germanium and silicon (<1 part impurity in 10^{10}) for use in semiconductor devices. The need for ultrapure germanium and silicon in the communication industry provided the impetus for developing the zone-refining procedure in 1952. Since then the technique has been used to purify many elements and compounds, both inorganic and organic.

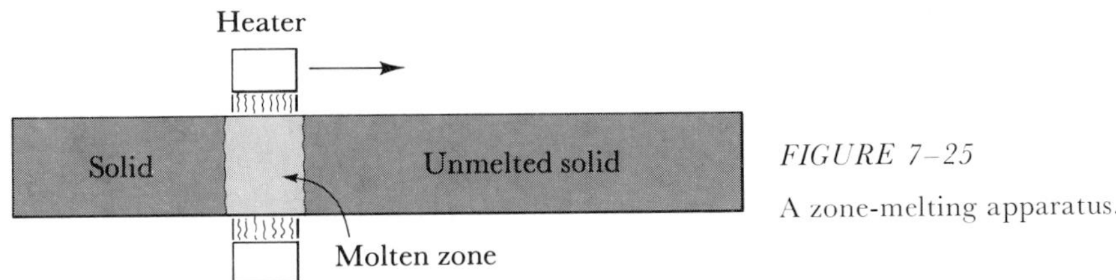

FIGURE 7–25

A zone-melting apparatus.

Biographical Notes

WILLIAM HENRY (ca. 1774–1836), a close friend of John Dalton, formulated the law regarding the solubility of gases in liquids in 1801. He also was the author of a textbook, *Elements of Experimental Chemistry*, which appeared in eleven editions between 1799 and 1830.

FRANCOIS M. RAOULT (1830–1901) was a French Professor of Chemistry in the University of Grenoble. In formulating his law, he studied the way in which a solute depresses the freezing point of a solvent and lowers its vapor pressure.

Problems and Questions

7–1 You wish to prepare solutions with each of the following compositions: 0.250*M* NaCl, 1.3*M* NaOH, and 0.1000*M* KIO_3. If you are to use a 500-cm^3 volumetric flask, what weight of each compound will you need?

7–2 What volume of each of the following aqueous solutions must be taken to obtain 0.0500 mol of solute: 0.150*M* HCl, 0.350*M* H_2SO_4, and 1.43*M* NH_3?

7–3 A solution is prepared by mixing equal weights of water (H_2O) and glycerol ($C_3H_8O_3$). At 20°C this solution has a density of 1.126 g cm^{-3}. What is the mole fraction and the molarity of glycerol in this solution? If water is considered to be the solvent, what is the molality of glycerol?

7–4 A solution is prepared by mixing 10.0-g portions of water (H_2O), methyl alcohol (CH_3OH), and acetone [$(CH_3)_2CO$]. What is the mole fraction of each of these components?

7–5 A solution containing sucrose ($C_{12}H_{22}O_{11}$), 93.0 g L^{-1}, has a density of 1.034 g cm^{-3} at 22°C. What is the molarity and the molality of sucrose in this solution?

7–6 An aqueous solution of acetone [$(CH_3)_2CO$], which is 8.50% (by weight) acetone, has a density of 0.9867 g cm^{-3} at 20.0°C. What are the mole fraction, the molality, and the molarity of acetone in this solution? (Acetone does not dissociate in aqueous solution.)

7–7 An aqueous solution of ethanol (C_2H_5OH) was prepared by adding 69.00 g to a 1-L volumetric flask and diluting to the mark. The resulting solution has a density of 0.9862 g cm^{-3}. What are the mole fraction, the molality, and the molarity of this solution?

7–8 The principal solute in vinegar is acetic acid ($C_2H_3O_2H$). If a sample of vinegar has 5.5 wt % acetic acid and a density of 1.006 g cm^{-3}, what is the molarity of acetic acid?

7–9 A portion of water–methyl alcohol solution ($x_{water} = 0.680$) weighing 150.0 g is mixed with 45.0 g of pure water. What is the mole fraction of water in the resulting solution?

7–10 Values of the weight percentage of the acid and the density (at 20°C) of concentrated aqueous solutions of reagent grade acids commonly provided by chemical manufacturers are:

HCl	36.0%	1.19 g cm^{-3}
HNO_3	69.5%	1.42 g cm^{-3}
H_2SO_4	96.0%	1.84 g cm^{-3}

Calculate the molarity and molality of acid in each of these solutions. What weight of each of these solutions must be taken to prepare 1.00 L of solution with a concentration of 0.100 mol L^{-1}?

7–11 A solution of hydrochloric acid has a concentration of 8.43 mol L^{-1}. What is the concentration of hydrochloric acid in a solution prepared by diluting 25.00 cm^3 of this solution to a volume of 1.00 L?

7–12 What volume of 3.47*M* NaOH (sodium hydroxide) contains 0.145 mol of this compound?

7–13 The solubilities of iodine (I_2) in water and in carbon tetrachloride at 25.0°C are: $S = 1.331 \times 10^{-3}$ mol L^{-1} (in H_2O), and $S = 0.119$ mol L^{-1} in (CCl_4). These two solvents are not soluble in one another. In a mixture of excess solid iodine, water, and carbon tetrachloride, what is the ratio of concentrations $[I_2]_{CCl_4}/[I_2]_{H_2O}$? (This problem introduces you to solvent distribution, which will be discussed further in Chapter 14.)

7–14 When a solution is prepared by mixing chloroform and ether, heat is evolved. Do you expect this solution to show positive, or negative, deviations from Raoult's law?

7–15 For the process at 25.0°C,

$$I_2(s) = I_2(1.00 \times 10^{-4} \text{ mol L}^{-1} \text{ in } C_7F_{16}) \qquad \Delta S = +159.4 \text{ J K}^{-1} \text{ mol}^{-1}$$

What is the value of ΔS for the analogous process in which the concentration of iodine in solution is 2.50×10^{-4} mol L^{-1}? What is the value of ΔG for this latter process?

7–16 What volume of pure oxygen gas at 760 torr dissolves in 1.000 L of water at 25.0°C? Express the Henry's law constant (k) for aqueous solutions of oxygen as a dimensionless quotient, the concentration (in mol L^{-1}) in the gas phase divided by the concentration (in mol L^{-1}) in the aqueous phase.

7–17 A solution of carbon dioxide in water is prepared at 20.0°C by saturating the solution with carbon dioxide gas at 2.00 atm. What is the concentration of carbon dioxide in the solution?

7–18 The enthalpy of vaporization of a pure liquid is related to the slope of a plot of ln (P/torr) versus T^{-1} (as was illustrated in Chapter 6). An analogous relationship holds for the partial pressure of a volatile solute over a solution. Using the data in Table 7–4b, plot ln (k_c/torr L mol^{-1}) versus T^{-1} and from it evaluate the enthalpy change in the process $CO_2(aq) = CO_2(g)$.

7–19 If the Henry's law constant for methyl chloride in aqueous solution at 20°C is 1.04×10^4 torr L mol^{-1} (see Table 7–2), what is the Henry's law constant for this same system with the solution composition expressed in mole-fraction units?

7–20 The vapor pressure of water at 25.0°C is 23.756 torr. Estimate the vapor pressure of water over a solution prepared by mixing 150 g of water and 20 g of urea [$CO(NH_2)_2$], assuming that Raoult's law is obeyed. Predict the boiling point and freezing point of this solution. (Urea is nonvolatile in the temperature range being considered here.)

7–21 An aqueous ammonia solution which is 15.55 wt % NH_3 has a vapor pressure of water of 44.1 torr at 40.0°C. Does water obey Raoult's law in this solution? If not, are the deviations positive, or negative?

7–22 Two liquids, A and B, form solutions that obey Raoult's law closely. The vapor pressures of pure A and B are 700 torr and 475 torr, respectively.

What is the vapor pressure of each component over a solution with $x_B = 0.320$? What composition of solution has a total vapor pressure of 550 torr? What composition of solution has a vapor in equilibrium with it which is 50 mol % B?

7–23 The vapor pressures of pure acetone and pure chloroform at 35.2°C are 344.5 torr and 293.1 torr, respectively. The partial pressure of acetone over a solution with the stoichiometric mole fraction of acetone equal to 0.42 is 109 torr. If for this system deviations of the partial pressures from Raoult's law are to be explained by the association of acetone and chloroform in the liquid phase, what will the vapor pressure of chloroform be over this solution?

7–24 A solution is prepared by dissovling 1.53 g of an unknown solute in 50.0 g of benzene. The freezing temperature of this solution is measured and found to be 1.83 K below the normal freezing temperature of benzene. What is the molecular weight of this solute? What do you expect the boiling point elevation of this solution to be?

7–25 What weight of ethylene glycol ($C_2H_6O_2$) must be added to two gallons of water in a car's radiator to prevent freezing at +5°F?

7–26 The normal boiling temperature of cyclohexane (C_6H_{12}) is 80.7°C and its enthalpy of vaporization at this temperature is 30.14 kJ mol^{-1}. What is the boiling point elevation constant for this solvent?

7–27 A solvent with an unknown enthalpy of fusion has a normal freezing temperature of 12.40°C. A solution prepared by dissolving 0.500 mol of solute in 7.05 mol of the solvent freezes at 10.63°C. What is the enthalpy of fusion of this solvent?

7–28 The boiling point elevation of an aqueous solution of sucrose is 0.048 K. What is the molality of sucrose in this solution? What is the osmotic pressure of this solution at 298.2 K?

7–29 The freezing temperature of an aqueous solution of a nonvolatile solute is −1.05°C. Predict the boiling temperature of this solution ($P_{atm} =$ 760 torr) and its vapor pressure at 25.0°C.

7–30 Some brackish water has an osmotic pressure of 630 torr. This water is purified by reverse osmosis. What is the minimum energy which must be expended to produce 1000 gallons of pure water?

7–31 A solution of a high polymer is prepared by dissolving 0.430 g in 100 cm^3 of water. The osmotic pressure of this solution at 298.2 K was found to be 2.34 torr. What is the molecular weight of this substance? What is the mole fraction of the solute in this solution?

7–32 Two substances, A and B, are miscible in all proportions in the liquid phase. No solid solutions form, but A and B form a solid compound AB. Sketch a freezing temperature–composition diagram (such as Figure 7–13) for this system. (*Hint*: Think of this diagram as two such diagrams, one for the system A + AB and the other for the system AB + B.)

7–33 Sketch the cooling curves expected if a liquid solution of each of the following is cooled until the entire sample solidifies. (Read temperatures from figures in this chapter as accurately as you can.)

80 mole % Cu, 20 mole % Au

80 mole % Cu, 20 mole % Ag

80 mole % Au, 20 mole % Ag

7–34 What interpretation do you place on each of the cooling curves shown below?

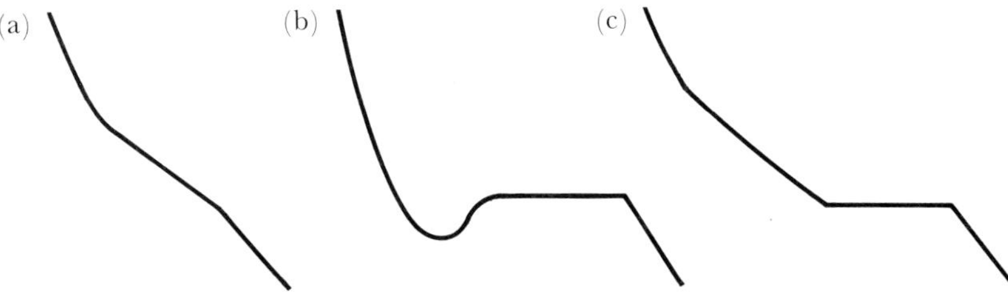

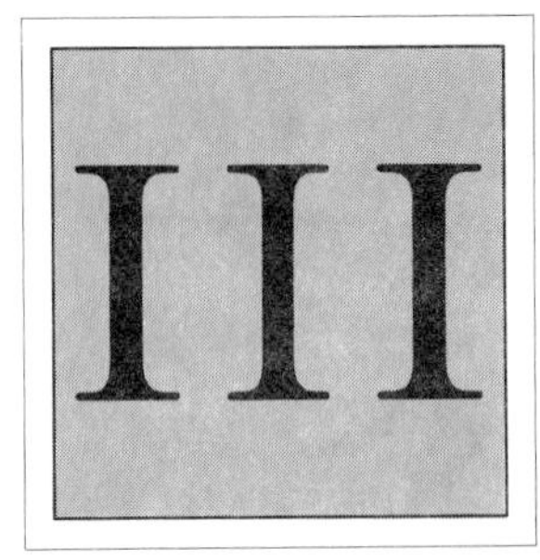

ATOMIC AND MOLECULAR STRUCTURE

Your study of Parts I and II has prepared you to consider many properties of chemical substances. Now you are able to calculate quantities related to chemical stoichiometry, the properties of gases and liquids, the geometrical arrangements of atoms in solids, thermodynamics, the transitions of substances from one state to another, and the colligative properties of solutions. But most of the previous discussion has not involved knowledge of the electronic structure of atoms and molecules. The next four chapters will help fill this gap in your knowledge.

You may know that many attempts to correlate the properties of different elements culminated, in the nineteenth century, in the development of the periodic table. Then, early in this century, the quantum mechanical theory of the electronic structure of atoms gave us a theoretical basis for understanding that table. We shall study the electronic structures of the elements, and then, because the formation of chemical compounds involves the loss, the gain, or the sharing of electrons by atoms, we shall pay particular attention to the energy of processes in which electrons are lost or gained by isolated atoms. We shall consider in detail the formation of simple ionic compounds (e.g., ordinary salt, NaCl). As we will learn in Chapter 9, the factors influencing the stability of such compounds can be analyzed by considering their formation as the sum of a sequence of simple steps. Some of these steps involve the loss and gain of electrons, but others involve the breaking of chemical bonds and the formation of an ionic crystal from gaseous ions. In calculations involving this last step, we use some of the geometrical principles that we learned in Chapter 5. Trends in the energies of these steps will be correlated with positions of the elements in the periodic table.

Consideration of simple covalent compounds in Chapter 10 brings in new concepts of electronic and geometric structures of molecules. We shall learn about the application of quantum mechanics to diatomic molecules and about qualitative theories that allow prediction of the shapes of polyatomic molecules. All of this will be useful in our study of more complex molecules. Chapter 11, an introduction to organic chemistry, will show the application to compounds of carbon of some of the principles that we have learned.

8

Atomic Structure and the Periodic Table

8–1 Introduction

In the last years of the eighteenth century and the first half of the nineteenth, the atomic theory of matter was developed. The ideas proposed by Dalton, Gay–Lussac, Avogadro, and Cannizzaro, described in Chapter 1, laid the foundation for much progress in chemistry. The postulate that matter was made up of atoms with different masses and different properties for each of the elements went far to explain both the chemical properties of substances and the reactions that convert one substance into another. Empirical rules regarding the combining capacities of the various elements made it possible to systematize some of their chemical properties. The empirical rules assigned to each element one or more valences, numbers that described the element's combining capacity. Thus carbon was assigned a valence of 4 and hydrogen a valence of 1, and an increasing understanding of the consequences of these and other valence assignments made the latter part of the nineteenth century a time of great progress in organic chemistry, the chemistry of carbon compounds. During this same period, valence rules for the other elements were part of the basis for developing the periodic table. It was not until the end of the nineteenth century and the start of the present century, however, that the structure of the atom was elucidated. It is the electronic structure of the atom that provides a rational basis for valence rules for the elements.

In this chapter modern atomic theory based upon quantum mechanics will be presented at an elementary level. These are the people and their discoveries that made possible the quantum mechanical theory of the atom:

1898–1903 J. J. THOMSON elucidated the nature of cathode rays (electrons) and measured the ratio of charge to mass for the electron.

1901 M. PLANCK developed a quantum theory to explain the spectrum of radiation from an incandescent solid.

1905–1906 ALBERT EINSTEIN explained the photoelectric effect in terms of the quantum theory of radiant energy in an extension of Planck's proposal.

1909 ROBERT A. MILLIKAN performed the "oil-drop" experiment, which made it possible to measure the charge on the electron.

1911 ERNEST RUTHERFORD developed the description of the nuclear atom based upon experimental observations of the scattering of alpha particles by thin metallic films. Rutherford gave the name "proton" to the simplest positively charged nucleus.

1913 NIELS BOHR developed a quantum theory of the hydrogen atom—the Bohr atom.

1913–1914 HENRY G. J. MOSELEY correlated the frequency of characteristic x radiation from an element with its nuclear charge, and explained the relationship of nuclear charge to an element's position in the periodic table.

1925–26 WERNER HEISENBERG, LOUIS DE BROGLIE, and ERWIN SCHRÖDINGER developed the quantum mechanical theory of the atom.

1932 JAMES CHADWICK discovered the neutron.

If this outline covered all aspects of subatomic phenomena, it would include many post-World War II discoveries and theories. Ordinary chemical phenomena do not, however, involve the particles that are the focus of attention in contemporary high-energy physics, and these particles will not be discussed.[1]

8–2 The Electron, the Proton, and the Neutron

If a potential difference is imposed between electrodes (metallic plates) in a partially evacuated tube, several types of electrical discharge can occur. The study of these phenomena was responsible for our earliest knowledge of the structure of atoms. It was found that a heated cathode (the electrode with excess negative charge) emitted rays of negative electricity. J. J. Thomson showed that these rays could be deflected by electric and magnetic fields. The principles mentioned in Chapter 1 in connection with the mass

[1]The positron, a positively charged electron, will be mentioned in connection with radioactivity in Chapter 22.

spectrograph allowed Thomson to determine the ratio of charge to mass for the particles making up these rays. It was found that this ratio, designated e/m, had the same value regardless of the cathode material or the residual gas in the discharge tube.

In 1911, R. A. Millikan confirmed the particle nature of cathode rays and also evaluated the charge on these particles, the electrons. He studied the motion of tiny oil droplets under the influence of gravity and an electric field in equipment pictured in Figure 8–1. Some droplets of oil acquire a net charge in the atomization process. With a microscope, Millikan observed particular droplets falling under the influence of gravity and changing their velocities as a result of the electric field. Combining the rates of fall in the absence and presence of an electric field, Millikan could determine the charge on a particle. He found that for the oil drops that had a charge, the charge was a certain value or some whole-number multiple of this value. To explain this phenomenon, it was necessary to postulate that the charged droplets of oil had an excess of the smallest units of negative electricity, the electrons. A charged droplet may have an excess of one electron, two electrons, or some other number, but a droplet cannot have an excess of one half electron. The minimum charge which Millikan found was the charge on the electron. With the charge (e) evaluated, the previously determined value of e/m made it possible to calculate the value of m. These values are given in Table 8–1.

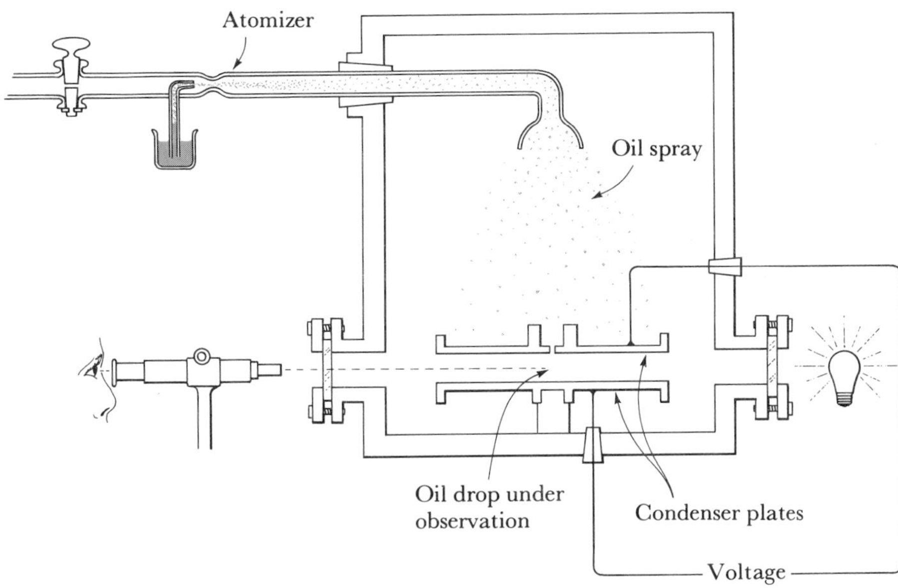

FIGURE 8–1
The Millikan oil-drop experiment. Some of the droplets produced in the upper chamber fall through an opening into the region between plates. A particular droplet can be observed with the electric field on or off. In 1911, Millikan determined the charge on the electron to be 4.89×10^{-10} esu. (Compare this with the current best value in Table 8–1.)

TABLE 8–1
The Properties of Some Fundamental Particles

	Mass	
	$\frac{\text{mass}}{\text{kg}}$	Relative atomic mass[a]
Proton	1.6735×10^{-27} [b]	1.00782[b]
Neutron	1.6749×10^{-27}	1.00866
Electron	9.11×10^{-31}	0.000549

		Charge	
	Relative charge	$\frac{\text{charge}^{c}}{\text{C}}$	$\frac{\text{charge}^{d}}{\text{g}^{1/2}\ \text{cm}^{3/2}\ \text{s}^{-1}}$
Proton	+1	$+1.602 \times 10^{-19}$	$+4.802 \times 10^{-10}$
Neutron	0	0	0
Electron	−1	-1.602×10^{-19}	-4.802×10^{-10}

[a] $^{12}C = 12$ exactly.
[b] The tabulated mass of the proton is the mass of the neutral hydrogen atom.
[c] The basic unit of charge in SI is the coulomb, symbol C.
[d] These charges are expressed in esu, electrostatic units of charge. One esu exerts a force of 1 dyn on an identical charge at a distance of 1 cm. To obtain the dimensions of charge in cgs units, we write Coulomb's law

$$F = 1\ \text{g cm s}^{-2} = \frac{(1\ \text{esu})^2}{(1\ \text{cm})^2}$$

$$1\ \text{esu} = 1\ \text{g}^{1/2}\ \text{cm}^{3/2}\ \text{s}^{-1}$$

Studies of the motion of positive rays in discharge tubes showed that these rays had ratios of charge to mass much smaller than the ratio e/m for the electron, and that the value depended upon the nature of the residual gas. If the gas was hydrogen, H_2, and the conditions were appropriate, the molecule was dissociated and ionized to give the proton, H^+.[2] The ratio of charge to mass for positive rays is the largest for the proton.

With the net charge on a hydrogen atom (a proton plus an electron) shown to be zero, the charge on the ionized hydrogen atom, the proton, had to be equal in magnitude but opposite in sign to the charge on the electron. The mass of the proton could be calculated from the charge and the ratio of mass to charge.

These early studies led to the view that atoms of all elements were made up of positive and negative charges, the positive charge being due to the protons, which also gave the atom most of its mass, and the negative charge being due to the electrons. But more intimate details of atomic structure were not known. In 1911, Rutherford and his associates, Geiger

[2] The convention that allows one to represent ionic species in solution by chemical formulas that do not include solvation results in representing the hydrogen ion in acidic solution as H^+ (i.e., as a proton). As will be shown in later chapters, the hydrogen ion in water and other solvents is strongly solvated.

and Marsden, studied the way thin metallic films scattered a beam of rapidly moving alpha particles (helium ions, He^{2+}, produced by radioactive distegration), and they showed that an atom's positive charge was localized in a very small region, a region much smaller than the size of the atom. This experiment is depicted in Figure 8–2. If the positive charge of an atom were diffusely spread out, alpha particles impinging on a thin metallic film would be deflected only slightly from their paths. In fact, in Rutherford's experiments some alpha particles were scattered directly backward. (Rutherford is reported to have described the discovery of this result as the most incredible event in his life.) For an alpha particle to be deflected through a large angle by interaction with another positive charge, the alpha particle must approach this positive charge closely. (Study of

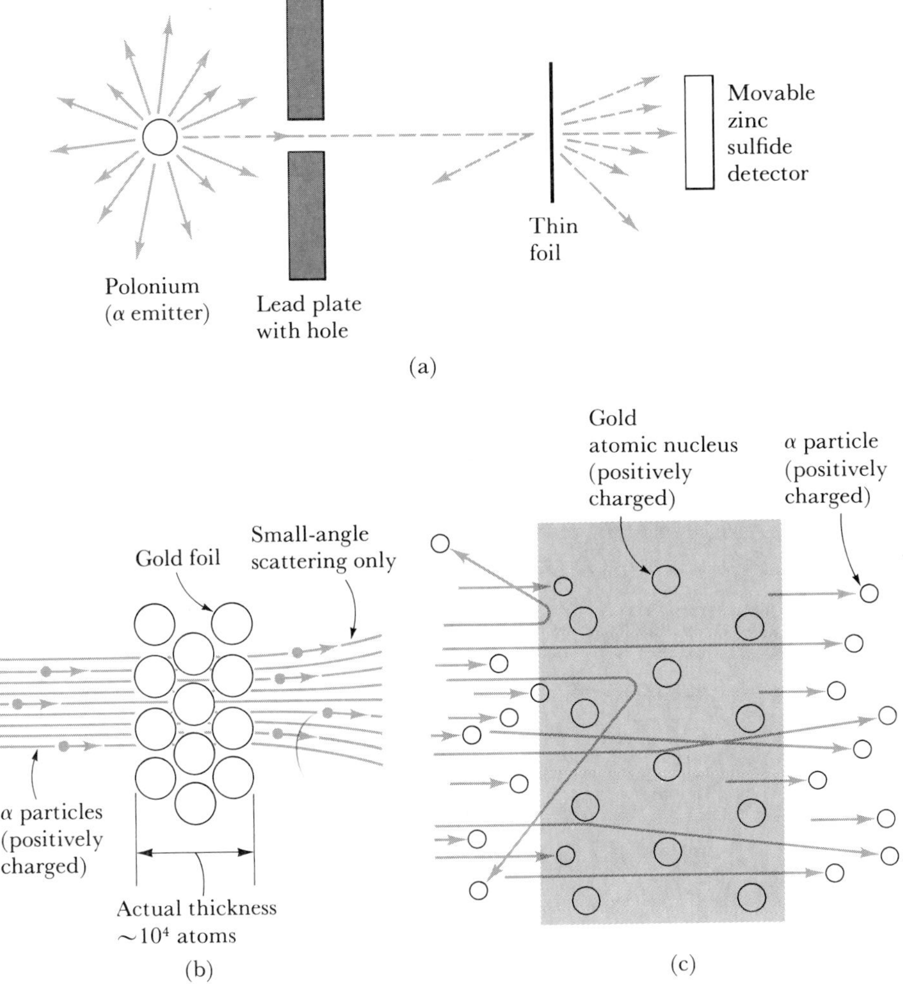

FIGURE 8–2
The scattering of alpha particles (He^{2+}) by atoms in a thin metal foil. (a) The experiment. (b) Small-angle deflection expected for diffuse positive charge (an atomic model proposed by J. J. Thomson). (c) Observed large-angle deflection due to the close approach of an α particle to nuclei of high charge (e.g., +79 for gold atoms).

alpha particle scattering as a function of the thickness of the metallic film showed that the large deflections were not the result of successive small deflections.) Through mathematical analysis of his results, Rutherford could define an upper limit for the size of the positively charged region of the atom, and also could measure the charge. These early results showed the radius of the compact, positively charged region of the atom, the nucleus, to be $\sim 10^{-14}$ m, much smaller than radii deduced from densities of metals ($\sim 10^{-10}$ m). (You will recall that we did calculations in Chapter 5 that related atomic size and density.) The early studies of alpha particle scattering by films of gold, silver, and copper gave values of the nuclear charge which were close to values based upon the atomic numbers of these elements.

Before 1932, it was assumed that the nucleus of an atom contained a sufficient number of protons to account for the atomic weight of that nuclide and a number of electrons sufficient to make the net positive charge on the nucleus equal to the atomic number times the charge on the proton. In 1932, in experiments on the artificial disintegration of nuclei, Chadwick discovered a neutral constituent of atomic nuclei, the neutron, making it unnecessary to assume that atomic nuclei contain electrons.

Rutherford's description of the nuclear atom left much about the structure of the atom unanswered. In particular, it did not explain how the electrons are arranged around the positively charged nucleus, nor why the electrostatic attraction doesn't simply cause the electrons to coalesce into the nucleus. Niels Bohr, who was a twenty-six-year-old postdoctorate in Rutherford's laboratory in 1911, did much to provide an answer to these questions, as we will learn.

8–3 The Old Quantum Theory and the Bohr Atom

Our concepts of the structure of matter underwent drastic changes in the early part of this century. Although the Bohr atom was superceded by the quantum mechanical atom over a half century ago, it is useful for us to consider it and the origin of ideas about quantization before starting our study of the newer theories.

RADIANT ENERGY

You were introduced to electromagnetic radiation in Chapter 5, where x-ray diffraction was discussed. Figure 5–6 showed the wave nature of electromagnetic radiation; it is the wavelength of the radiation that is important in considering the diffraction of x radiation by crystalline solids. There is another view of electromagnetic radiation, however, that developed from a proposal by Max Planck in 1901. To explain the distribution of radiant energy from a heated solid, he proposed that the vibrational energy of atoms in the solid could have only certain values,

$$E_{\text{vib}} = nh\nu$$

in which n is an integer, h is a constant, and ν is the vibrational frequency of the atoms. The vibrational energy levels in this model were equally spaced, as they were in the simple model we used in Chapter 5 to discuss the heat capacity and entropy of solids. Einstein carried the theory further and proposed that radiant energy was corpuscular in nature, with an energy per light quantum given by an analogous equation,

$$E = h\nu$$

in which ν is the vibrational frequency of the light. The proportionality constant, h, in these equations is called *Planck's constant*; it has the value

$$h = 6.626 \times 10^{-34}\ \text{J s}$$

Figure 8–3 shows the relationship between the wavelength of electro-

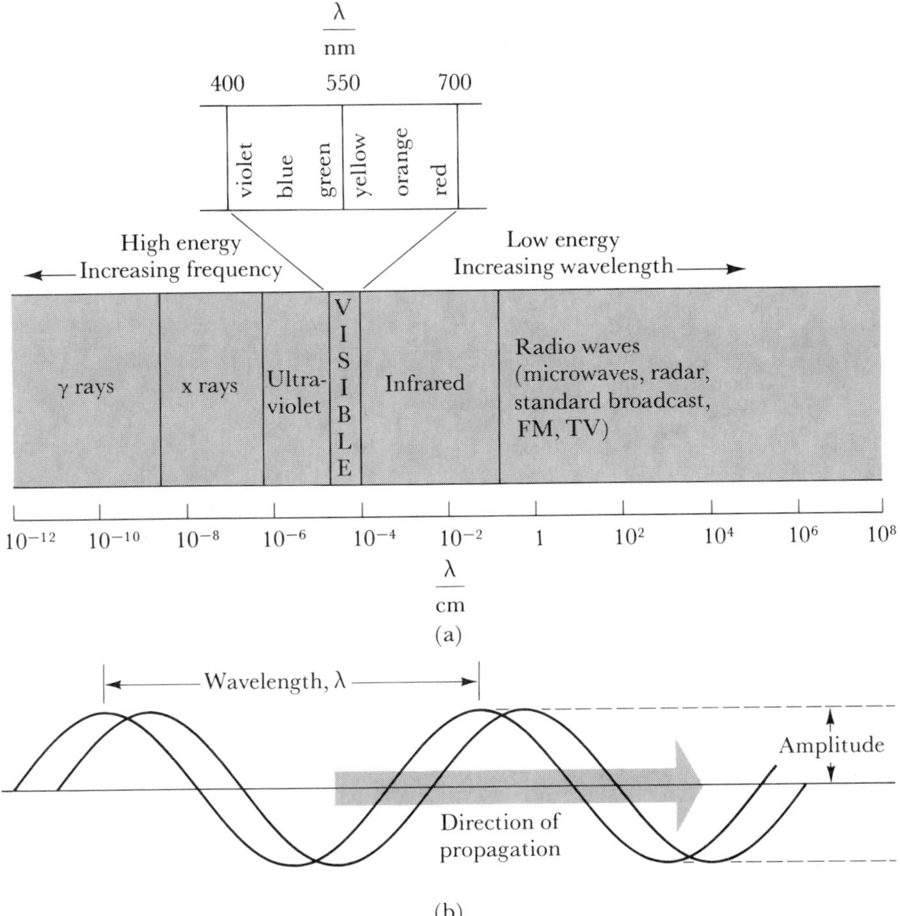

FIGURE 8–3
Electromagnetic radiations. (a) The various regions of the electromagnetic spectrum. (b) The relationship of frequency and wavelength. If the train of waves of length λ passes the viewing point at a velocity c, the number of waves passing the point per unit time, the frequency ν, is

$$\nu = \frac{c}{\lambda}$$

magnetic radiation and its frequency:

$$\nu = \frac{c}{\lambda}$$

Light quanta are called *photons*, and the energy per photon of light of wavelength λ can be calculated:

$$E = 6.626 \times 10^{-34}\ \text{J s} \times \frac{2.998 \times 10^{8}\ \text{m s}^{-1}}{\lambda}$$

$$= \frac{1.986 \times 10^{-25}\ \text{J m}}{\lambda}$$

EXAMPLE: The shortest and longest wavelengths of visible light are ~400 nm (4000 Å) and ~700 nm (7000 Å). The energies of photons of light of these wavelengths are calculated:

$\lambda = 400$ nm

$$E = \frac{1.986 \times 10^{-25}\ \text{J m}}{4.00 \times 10^{-7}\ \text{m}} = 4.97 \times 10^{-19}\ \text{J}$$

$\lambda = 700$ nm

$$E = \frac{1.986 \times 10^{-25}\ \text{J m}}{7.00 \times 10^{-7}\ \text{m}} = 2.84 \times 10^{-19}\ \text{J}$$

For some purposes it is more convenient to consider the energy of one mole of photons. Multiplication of each of these energies by Avogadro's constant gives:

$$E = 4.97 \times 10^{-19}\ \text{J} \times 6.02 \times 10^{23}\ \text{mol}^{-1}$$
$$= 2.99 \times 10^{5}\ \text{J mol}^{-1}\ (\lambda = 400\ \text{nm})$$
$$E = 2.84 \times 10^{-19}\ \text{J} \times 6.02 \times 10^{23}\ \text{mol}^{-1}$$
$$= 1.71 \times 10^{5}\ \text{J mol}^{-1}\ (\lambda = 700\ \text{nm})$$

(Conversion of these values of E to calories per mole gives 71.5 kcal mol^{-1} and 40.9 kcal mol^{-1}, respectively.)

Einstein confirmed his theory that light is corpuscular in nature in his interpretation of the *photoelectric effect*. If light shines on a metal surface, electrons may be ejected from the surface. Experiments show, however, that electrons are not emitted if the light has a wavelength greater than some critical value characteristic for each metal. For sodium metal, the critical wavelength is 544 nm (green light). Light with wavelengths longer than this does not eject electrons from sodium metal no matter how intense the light is. If the wavelength is shorter than this, electrons are emitted, and the maximum kinetic energy of the emitted electrons is greater the shorter the wavelength. This situation can be analyzed in terms of the equation giving the kinetic energy of the ejected electron,

$$\tfrac{1}{2}mv^2 = h\nu - W$$

TABLE 8–2
The Work Functions of Several Metals

	$\frac{W}{\text{eV}}$	$\frac{W}{\text{kJ mol}^{-1}}$		$\frac{W}{\text{eV}}$	$\frac{W}{\text{kJ mol}^{-1}}$
Li	2.49	240	Ag	4.7	450
Na	2.28	220	Hg	4.5	430
K	2.24	216	Au	4.8	460
Ca	2.71	261	Al	4.1	400

in which W is the energy required to cause electron ejection; this is called the *work function*.[3] This equation states simply that the kinetic energy of an ejected electron is equal to the excess energy of the photon relative to the work function. For sodium, the work function can be calculated from the value of the critical wavelength ($\lambda = 544$ nm for $mv^2/2 = 0$):

$$W = E = \frac{hc}{\lambda} = \frac{1.986 \times 10^{-25}\ \text{J m}}{5.44 \times 10^{-7}\ \text{m}} = 3.65 \times 10^{-19}\ \text{J}$$

If light of $\lambda = 400$ nm (blue light) is used, the kinetic energy of the ejected electron is

$$\tfrac{1}{2}mv^2 = 4.97 \times 10^{-19}\ \text{J} - 3.65 \times 10^{-19}\ \text{J} = 1.32 \times 10^{-19}\ \text{J}$$

The velocity of the electron is

$$v = \sqrt{\frac{2 \times (\frac{1}{2}mv^2)}{m}}$$

$$= \sqrt{\frac{2 \times 1.32 \times 10^{-19}\ \text{kg m}^2\ \text{s}^{-2}}{9.11 \times 10^{-31}\ \text{kg}}} = 5.4 \times 10^5\ \text{m s}^{-1}$$

The photoelectric effect is the basis for the common photographic light meter, in which the magnitude of the current is proportional to the intensity of the light. The work function of the instrument's element is small enough to allow it to operate with visible light. Table 8–2 gives the work functions of several metals.

THE SPECTRUM OF ATOMIC HYDROGEN

Data from studies of the spectra of atoms have been particularly important clues in characterizing the electronic structures of atoms. Here we will consider the emission spectrum of the simplest atom, the hydrogen atom. The emission spectrum of atomic hydrogen is obtained by subjecting gaseous hydrogen to an electric discharge. The high-energy electrons of the discharge dissociate the hydrogen molecules and excite the atoms to higher-energy states. The excitation energy is lost by emission of radiant energy. This

[3]The work function should not be mistaken for the ionization energy, which is the energy required to remove an electron from the gaseous atom.

radiant energy, the *emission spectrum*, can be analyzed using an instrument called a spectrograph; studies have found that the spectrum of atomic hydrogen consists of light with certain well-defined wavelengths, as shown in Figure 8–4. These spectral data can be correlated by an equation involving the reciprocal of the wavelength of the radiation,

$$\frac{1}{\lambda} = R\left(\frac{1}{a^2} - \frac{1}{b^2}\right)$$

in which R is called the RYDBERG constant, and a and b are integers. The series of spectral lines shown in Figure 8–4 are those with $a = 1$, the LYMAN series, $a = 2$, the BALMER series, and $a = 3$, the PASCHEN series. We can evaluate the Rydberg constant from spectral data. For instance, the spectral line in the Balmer series corresponding to $a = 2$ and $b = 3$, has a wavelength of 656.3 nm (orange light). From this wavelength we obtain the value of R:

$$R = \frac{1}{\lambda} \times \frac{1}{\dfrac{1}{a^2} - \dfrac{1}{b^2}}$$

$$= \frac{1}{6.563 \times 10^{-7}\ \text{m}} \times \frac{1}{\dfrac{1}{2^2} - \dfrac{1}{3^2}} = 1.097 \times 10^7\ \text{m}^{-1}$$

Spectroscopists often express the reciprocal of the wavelength of a spectral line in cm^{-1}; thus in this unit, the value of R is $1.097 \times 10^5\ \text{cm}^{-1}$. This unit also is called the *wave number*.

The expected positions of other spectral lines in the Balmer series can be calculated. For $a = 2$, $b = 4$:

$$\frac{1}{\lambda} = 1.097 \times 10^7\ \text{m}^{-1} \times \left(\frac{1}{4} - \frac{1}{16}\right) = 2.057 \times 10^6\ \text{m}^{-1}$$

$$\lambda = 486.1\ \text{nm}$$

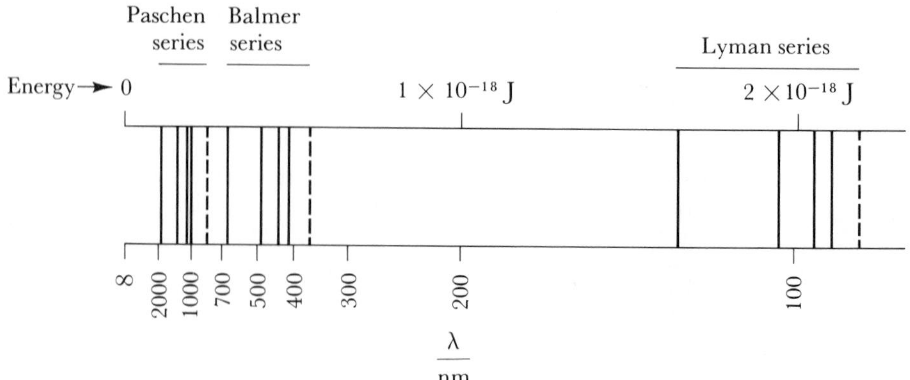

FIGURE 8–4
The emission spectrum of atomic hydrogen: Lyman series, $a = 1$; Balmer series, $a = 2$; Paschen series, $a = 3$. (The scale is linear in λ^{-1}, or linear in energy.)

For $a = 2$, $b = 5$:

$$\frac{1}{\lambda} = 1.097 \times 10^7\ \text{m}^{-1} \times \left(\frac{1}{4} - \frac{1}{25}\right) = 2.304 \times 10^6\ \text{m}^{-1}$$

$$\lambda = 434.0\ \text{nm}$$

Figure 8–4 shows spectral lines at these wavelengths.

The longest wavelength of a spectral line in the Lyman series is that which corresponds to $a = 1$ and $b = 2$:

$$\frac{1}{\lambda} = 1.097 \times 10^7\ \text{m}^{-1} \times \left(\frac{1}{1} - \frac{1}{4}\right) = 8.228 \times 10^6\ \text{m}^{-1}$$

$$\lambda = 121.5\ \text{nm}$$

This spectral line is in the far ultraviolet region of the electromagnetic spectrum.

THE BOHR ATOM

The Rutherford nuclear atom and the empirical Rydberg equation for expressing the wavelengths of lines in the emission spectrum of atomic hydrogen provide the basis for the next great step in explaining the structure of the atom. It was Niels Bohr who took this step. He proposed that the electron in a hydrogen atom (or in other one-electron atomic species, such as He^+) moves around the nucleus only in certain circular orbits. To calculate the radius of each of these orbits and the corresponding energy of the electron, Bohr imposed a quantum condition on the basic equations of classical mechanics and electrostatics. In this system of a positively charged, massive nucleus and one electron, the force of attraction, F, between the electron of charge $-e$ and the nucleus of charge $+Ze$, in which Z is the atomic number (the nuclear charge in multiples of the charge on the proton) separated by a distance r is given by Coulomb's law:[4]

$$F = \frac{Ze^2}{r^2}$$

The electron moving at constant velocity in a circular orbit has angular momentum, p, which is the product of its mass and velocity and the radius of curvature of its orbit:

$$p = mvr$$

---

[4]Here Coulomb's law is written in the form that is correct if the electron charge is expressed in esu and the distance r is expressed in centimeters. If SI units are to be used, Coulomb's law has the form

$$F = \frac{Ze^2}{4\pi\epsilon_0 r^2}$$

Here the electron charge is expressed in coulombs, the distance in meters, and ϵ_0, the permittivity of a vacuum, is $\epsilon_0 = 8.856 \times 10^{-12}\ \text{C}^2\ \text{kg}^{-1}\ \text{m}^{-3}\ \text{s}^2$.

These equations for force and momentum are classical equations. However, the notion that the electron could circulate around the nucleus in an orbit with a particular radius was not classical. In classical theory, the electron would move in a spiral orbit losing potential energy, gaining kinetic energy, and radiating energy enroute to a crash with the nucleus. Bohr proposed the nonclassical postulate that the allowed circular orbits were those in which the angular momentum of the electron had the values

$$p = mvr = n\frac{h}{2\pi}$$

in which n, called the *principal quantum number*, has the allowed values $n = 1, 2, 3, \cdots$, and h is Planck's constant. Another equation relating the velocity of the electron and the radius of its orbit can be obtained by equation, the Coulomb force of attraction with that given by Newton's equation,

$$F = ma$$

For a particle moving at constant velocity in a circular orbit the acceleration is

$$a = \frac{v^2}{r}$$

Thus we have

$$F = m\left(\frac{v^2}{r}\right) = \frac{Ze^2}{r^2}$$

This equation can be rearranged to give an equation for v^2:

$$v^2 = \frac{Ze^2}{mr}$$

Another equation for v^2 can be obtained from the equation for the allowed values of the angular momentum:

$$v^2 = \frac{n^2h^2}{4\pi^2m^2r^2}$$

Equating these two equations for v^2 gives an equation which can be solved for r:

$$\frac{Ze^2}{mr} = \frac{n^2h^2}{4\pi^2m^2r^2}$$

$$r = \left(\frac{h^2}{4\pi^2me^2}\right) \times \frac{n^2}{Z}$$

Substitution of the values of quantities in the parenthesis gives:

$$r = \frac{(6.626 \times 10^{-27}\ \text{g cm}^2\ \text{s}^{-1})^2}{4 \times \pi^2 \times 9.11 \times 10^{-28}\ \text{g} \times (4.802 \times 10^{-10}\ \text{g}^{1/2}\ \text{cm}^{3/2}\ \text{s}^{-1})^2} \times \frac{n^2}{Z}$$

$$= 5.29 \times 10^{-9}\ \text{cm} \times \frac{n^2}{Z}$$

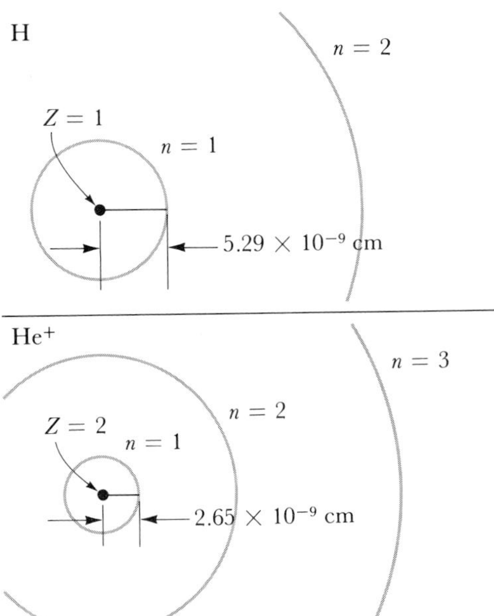

FIGURE 8–5
The circular Bohr orbits for the one-electron species H and He^+:

$$r = \left(\frac{h^2}{4\pi^2 me^2}\right) \times \frac{n^2}{Z}$$
$$= 5.29 \times 10^{-9} \text{ cm} \times n^2 \quad \text{for } Z = 1, \text{H}$$
$$= 2.65 \times 10^{-9} \text{ cm} \times n^2 \quad \text{for } Z = 2, \text{He}^+$$

Figure 8–5 shows the radii of some allowed circular orbits for atomic hydrogen ($Z = 1$) and for a helium ion with the charge $+1$ ($Z = 2$), each being a one-electron atom to which the Bohr theory applies. Despite its flaws in the light of current theory (to be described), this picture contains correct notions that you should remember. One is the inverse relationship between radius and nuclear charge: The more highly charged the nucleus, the closer an electron is attracted to it. The other is the proportionality between the radius of the orbit and the square of the quantum number, n.

The calculated radii for the circular orbits in hydrogen,

$$n = 1 \qquad r = 5.29 \times 10^{-9} \text{ cm} = 0.529 \text{ Å}$$
$$n = 2 \qquad r = 2.12 \times 10^{-8} \text{ cm} = 2.12 \text{ Å}$$

are similar in magnitude to atomic radii encountered in previous chapters, but there is no rigorously derived connection between these calculated radii and observed interatomic distances. Therefore the qualitative reasonableness of the calculated values of r cannot be cited as support for the Bohr theory. In Bohr's time, however, the Rydberg correlation of spectral lines for atomic hydrogen was available. For this reason, the energy of an electron in an allowed Bohr orbit is of more interest to us than the radii of these orbits are.

The energy of an electron in an atom is the sum of its kinetic energy and its potential energy. The kinetic energy is given by

$$E_k = \tfrac{1}{2}mv^2$$

The potential energy is due to the Coulombic attraction between the positively charged nucleus and the negatively charged electron. The force of attraction is zero at $r = \infty$, and this is the defined origin for calculating the potential energy:

$$E_p = 0 \text{ at } r = \infty$$

At closer distances the potential energy is negative:

$$E_p = -\frac{Ze^2}{r} \text{ at } r < \infty$$

The sum of these terms is

$$E = \tfrac{1}{2}mv^2 - \frac{Ze^2}{r}$$

from which v^2 can be eliminated by use of the equation

$$v^2 = \frac{Ze^2}{mr}$$

presented previously. This gives

$$E = \frac{1}{2}m\left(\frac{Ze^2}{mr}\right) - \frac{Ze^2}{r}$$

$$= \frac{1}{2}\frac{Ze^2}{r} - \frac{Ze^2}{r} = -\frac{1}{2}\frac{Ze^2}{r}$$

Thus we see that the kinetic energy of an electron at a distance r from the nucleus is one half the magnitude of the potential energy at this same distance. The total energy is expressed in terms of the value of n by substitution of the allowed values for r, derived previously, into this equation:

$$E = -\frac{1}{2}\frac{Ze^2}{r}$$

$$= -\frac{1}{2}Ze^2 \times \frac{4\pi^2 me^2}{h^2} \times \frac{Z}{n^2}$$

$$= -\left(\frac{2\pi^2 me^4}{h^2}\right) \times \frac{Z^2}{n^2}$$

Substitution of values for the quantities in the parentheses gives:

$$E = -\frac{2\pi^2 \times 9.11 \times 10^{-28}\text{ g} \times (4.802 \times 10^{-10}\text{ g}^{1/2}\text{ cm}^{3/2}\text{ s}^{-1})^4}{(6.626 \times 10^{-27}\text{ g cm}^2\text{ s}^{-1})^2} \times \frac{Z^2}{n^2}$$

$$= -2.178 \times 10^{-11}\text{ g cm}^2\text{ s}^{-2} \times \frac{Z^2}{n^2} = -2.178 \times 10^{-18}\text{ J} \times \frac{Z^2}{n^2}$$

Figure 8–6 shows the energy levels for the allowed states of the hydrogen atom according to the preceding equation. The emission spectrum for atomic hydrogen can be compared to that predicted by this equation. The radiant energy associated with a spectral line is the energy change when the electron goes from an orbit with $n = b$ to an orbit with $n = a$. The energy change,

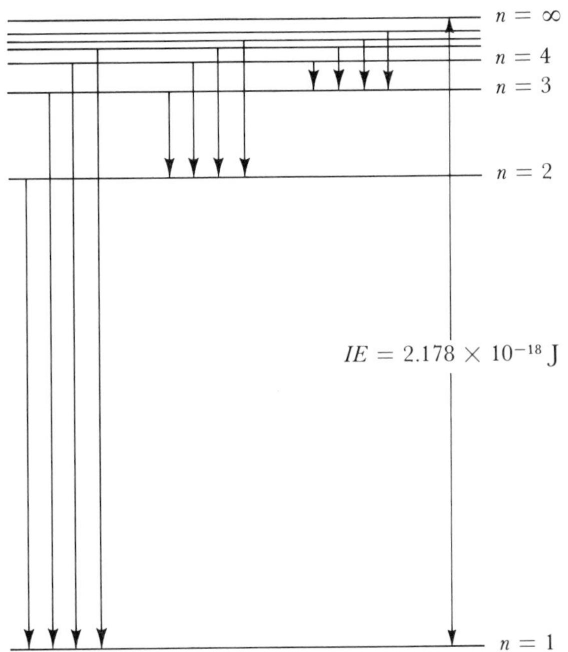

FIGURE 8–6
The allowed electronic energies of the hydrogen atom calculated from the Bohr theory. The vertical lines show transitions to $n = 1$ (Lyman series), $n = 2$ (Balmer series), and $n = 3$ (Paschen series). The four transitions shown for each series are those giving spectral lines of Figure 8–4.

ΔE, is given by the equation

$$\Delta E = E_b - E_a = 2.178 \times 10^{-18} \text{ J} \times \left(-\frac{1}{b^2} + \frac{1}{a^2}\right)$$

which has the same dependence upon the squares of integers as does the empirical Rydberg equation for the reciprocal of the wavelength of the spectral lines. To compare the empirical Rydberg constant, $R = 1.097 \times 10^7 \text{ m}^{-1}$, with the theoretically derived constant, we must convert the theoretically derived coefficient in the equation for ΔE in joules into a coefficient in m^{-1} for the equation for λ^{-1}; the relationship of ΔE and λ^{-1} is obtained from Planck's equation:

$$E = \frac{hc}{\lambda}$$

$$\frac{1}{\lambda} = \frac{E}{hc}$$

Therefore the value for R obtained from the theoretically derived coefficient, 2.178×10^{-18} J, is

$$R = \frac{2.178 \times 10^{-18} \text{ J}}{6.626 \times 10^{-34} \text{ J s} \times 2.998 \times 10^8 \text{ m s}^{-1}}$$

$$= 1.096 \times 10^7 \text{ m}^{-1}$$

which agrees with the empirical value. This essentially quantitative agreement between theory and experiment was a triumph for the Bohr theory.

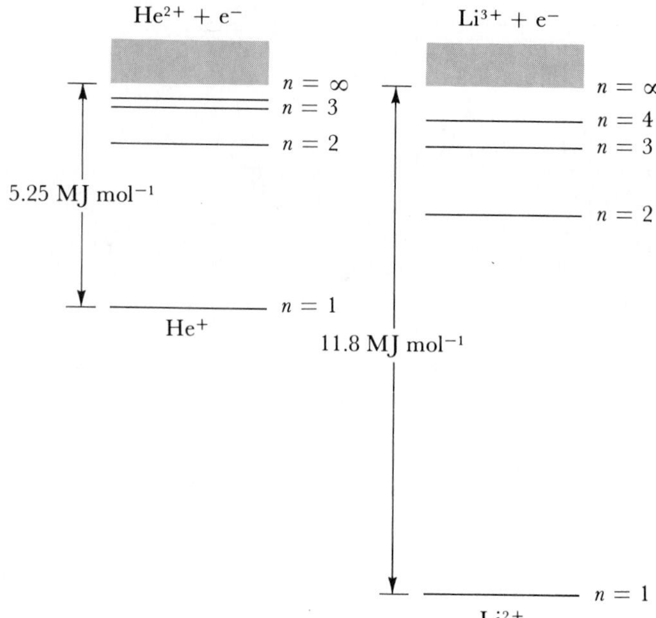

FIGURE 8–7
The allowed electronic energies of the one-electron atomic species He^+ and Li^{2+}.

The distance in Figure 8–6 between the $n = 1$ level and the $n = \infty$ level is the energy for the ionization process:

$$H(g) = H^+(g) + e^- \qquad \Delta E = +2.178 \times 10^{-18}\ \text{J}$$

We often will need to use this energy quantity expressed on a molar basis:

$$\Delta E = +2.178 \times 10^{-18}\ \text{J} \times 6.022 \times 10^{23}\ \text{mol}^{-1}$$
$$= 1312\ \text{kJ mol}^{-1} = 313.6\ \text{kcal mol}^{-1}$$

This energy is called the *ionization energy* for hydrogen.

For other one-electron atomic species (e.g., He^+ and Li^{2+}), the ionization energy and the energies of the states with particular values of n are the values for hydrogen multiplied by the factor Z^2 ($Z = 2$ for He^+ and $Z = 3$ for Li^{2+}). Figure 8–7 is the energy-level diagram for these additional one-electron atoms.

Even though features of the Bohr theory are known now to be incorrect, it does introduce us to the quantization of electronic energies in atoms, and the equation that it provides for the energy of a one-electron atom is the same as that derived in quantum mechanics.

8–4 The Quantum-Mechanical Theory of the Atom

The quantum revolution that started with the work of Planck, Einstein, and Bohr continued in the decade and a half that followed the publication of Bohr's theory of atomic structure. Although much progress still is being made, it was during the period that we now will consider, the 1920s, that our present picture of the atom developed.

We already have learned that electromagnetic radiation has the properties of a wave (a wavelength and the ability to be diffracted) and the properties of a particle (a discrete amount of energy per quantum). If a wave can have the properties of a particle, perhaps a particle can have the properties of a wave. This was proposed by LOUIS DE BROGLIE in 1924. The equation that he put forward for the wavelength of a moving particle,

$$\lambda = \frac{h}{mv}$$

was suggested by the Planck equation, which can be rearranged to relate the wavelength and energy of electromagnetic radiation:

$$\lambda = \frac{hc}{E}$$

If Einstein's relationship between energy and mass, $E = mc^2$, in which c is the velocity of light, is applied to a photon, this equation becomes

$$\lambda = \frac{h}{mc}$$

De Broglie simply proposed that this relationship could be applied to material particles with c replaced by the velocity of the particle, v. The equation $\lambda = h/mv$ also can be written in terms of the kinetic energy, E_k, of the particle:

$$\tfrac{1}{2}mv^2 = E_k$$

Multiplying both sides of the equation by $2m$ gives:

$$2m \times (\tfrac{1}{2}mv^2) = (mv)^2 = 2m \times E_k$$

$$mv = (2mE_k)^{1/2}$$

$$\lambda = \frac{h}{mv} = \frac{h}{\sqrt{2mE_k}}$$

You will obtain some feeling for this relationship if we calculate the wavelength of an electron accelerated by a potential of 200 V and the wavelength of a neutron with a kinetic energy of (3/2) kT at 300 K. (This is a neutron with the average kinetic energy of a thermal neutron at 300 K.)

Electron ($m = 9.11 \times 10^{-31}$ kg) with $E_k = 200$ eV

$$E_k = 200 \text{ eV} \times \frac{1.602 \times 10^{-19}\text{ C}}{1\text{ e}} = 3.20 \times 10^{-17}\text{ V C}$$

$$= 3.20 \times 10^{-17}\text{ kg m}^2\text{ s}^{-2}$$

$$\lambda = \frac{h}{\sqrt{2mE_k}} = \frac{6.626 \times 10^{-34}\text{ kg m}^2\text{ s}^{-1}}{\sqrt{2 \times 9.11 \times 10^{-31}\text{ kg} \times 3.20 \times 10^{-17}\text{ kg m}^2\text{ s}^{-2}}}$$

$$= 8.68 \times 10^{-11}\text{ m} = 0.868\text{ Å}$$

Neutron ($m = 1.675 \times 10^{-27}$ kg) with $E_k = \frac{3}{2}kT$ at 300 K

$$E_k = \frac{3}{2} \times 1.38 \times 10^{-23}\ \text{kg m}^2\ \text{s}^{-2}\ \text{K}^{-1} \times 300\ \text{K}$$

$$= 6.21 \times 10^{-21}\ \text{kg m}^2\ \text{s}^{-2}$$

$$\lambda = \frac{h}{\sqrt{2mE_k}}$$

$$= \frac{6.626 \times 10^{-34}\ \text{kg m}^2\ \text{s}^{-1}}{\sqrt{2 \times 1.675 \times 10^{-27}\ \text{kg} \times 6.21 \times 10^{-21}\ \text{kg m}^2\ \text{s}^{-2}}}$$

$$= 1.45 \times 10^{-10}\ \text{m} = 1.45\ \text{Å}$$

We see that the wavelength associated with a thermal neutron or an electron having an energy of a few hundred electron volts is comparable to the interatomic separation of atoms in molecules (1–3 Å). De Broglie's equation was tested by looking for diffraction of electrons from crystals (analogous to x-ray diffraction, discussed in Chapter 5). Confirmation came in 1927 with experiments performed independently by G. P. THOMSON and by C. J. DAVISSON and L. H. GERMER, who observed diffraction of an electron beam from a crystal of nickel. Just as the diffraction of x rays provides information about the arrangements of atoms in crystals, so also does the diffraction of electrons and of neutrons. Electron diffraction is useful in studying the atomic arrangement in gaseous molecules, and neutron diffraction is useful in locating the positions of hydrogen atoms in crystals. Neither of these tasks is accomplished easily by x-ray diffraction.

THE QUANTUM-MECHANICAL THEORY OF THE ONE-ELECTRON ATOM

A triumph of the Bohr theory of the hydrogen atom was its success in predicting the energies of spectral lines for atomic hydrogen. However, the theory was not successful in treating the energies of allowed states for atoms with two or more electrons. During the period 1920–1925, many physicists carefully studied the old quantum theory (i.e., modifications of the Bohr theory), paving the way for the development of quantum mechanics in 1925–26. A period of intense activity in applying quantum mechanics to physical problems followed.

Quantum mechanics is too complex a subject to be presented satisfactorily at this level, but we shall outline the results of applying quantum mechanics to the electronic structure of the hydrogen atom. You probably will be able to understand these results, and this will help you to understand the electronic structures of atoms of the other elements, the basis for the periodic table and the chemical properties of the elements. The application of quantum mechanics to atomic and molecular systems pervades modern chemistry, and we begin our discussion with the hydrogen atom.

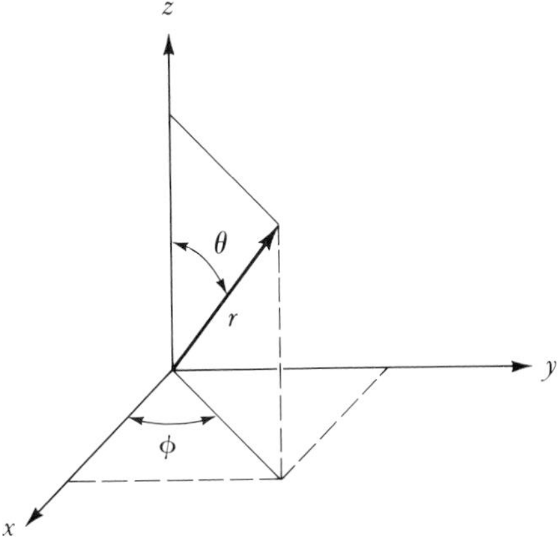

FIGURE 8–8
The system of polar coordinates r, θ, and ϕ used in the Schrödinger equation for the hydrogen atom.

The SCHRÖDINGER wave equation, which is the heart of quantum mechanics, is a differential equation (an equation involving derivatives) for a function ψ, called the *wave function*, in terms of coordinates of the particles in the system. For the hydrogen atom, a polar coordinate system, described in Figure 8–8, is most convenient for defining the position of the electron relative to the nucleus. The calculation starts by expressing the potential energy of the system as a function of the coordinates; this is the same equation that was used in the Bohr treatment of the hydrogen atom,

$$E_p = -\frac{Ze^2}{r}$$

The factor Z is included, as before, so that the results can be applied to other one-electron atoms, for example, He^+ ($Z = 2$), Li^{2+} ($Z = 3$), and so on. This relationship is substituted into the wave equation, which can be solved by straightforward methods.[5] The results are:

1. A solution of the wave equation is possible only for certain definite values of the total energy of the system, E. These allowed values of E are

$$E = -\frac{2\pi^2 me^4}{h^2} \times \frac{Z^2}{n^2}$$

in which n is an integer, $n = 1, 2, 3, \cdots$. This equation for the energy of allowed states is exactly the same as that derived in the Bohr theory. In the Bohr theory, however, the quantum condition is introduced arbitrarily, whereas the quantization of energy in the quantum mechanical

---

[5]With the equation $E_p = -Ze^2/r$ substituted into the wave equation, it is

$$\frac{1}{r^2}\frac{\partial}{\partial r}\left(r^2\frac{\partial\psi}{\partial r}\right) + \frac{1}{r^2\sin^2\theta}\frac{\partial^2\psi}{\partial\phi^2} + \frac{1}{r^2\sin\theta}\frac{\partial}{\partial\theta}\left(\sin\theta\frac{\partial\psi}{\partial\theta}\right) + \frac{8\pi^2 m}{h^2}\left(E + \frac{Ze^2}{r}\right)\psi = 0$$

Fortunately, you do not have to be able to solve this equation to use the solutions.

treatment of the hydrogen atom arises in solving the wave equation, as a mathematical necessity.

2. In solving the equation, two other quantum numbers are introduced:

   a quantum number l, called the *azimuthal* (or *orbital-shape*) *quantum number*, which can have integral values from 0 to $(n-1)$:

$$l = 0, 1, \cdots, n-1$$

   and a quantum number m_l, called the *magnetic quantum number*, which can have integral values from $-l$ to $+l$:

$$m_l = -l, -l+1, \cdots +l-1, +l$$

   These quantum numbers are related to the angular momentum of the electron. The allowed values of the angular momentum (represented by p) are

$$p = \frac{h}{2\pi}\sqrt{l(l+1)}$$

   and the component of angular momentum along the z direction can have the allowed values

$$p_z = m_l \frac{h}{2\pi}$$

Each allowed value of n, l, and m_l defines a wave function $\psi(r, \theta, \phi)$ that is a solution of the wave equation. The wave functions corresponding to $n = 1$ and $n = 2$ are given in Table 8–3.

TABLE 8–3
Solutions of the Schrödinger Wave Equation for a One-Electron Atom[a]

n	l	m_l	Orbital	
1	0	0	$1s$	$\psi_{1s} = \frac{1}{\sqrt{\pi}}\left(\frac{Z}{a_0}\right)^{3/2} e^{-\sigma}$
2	0	0	$2s$	$\psi_{2s} = \frac{1}{4\sqrt{2\pi}}\left(\frac{Z}{a_0}\right)^{3/2} (2-\sigma)e^{-\sigma/2}$
2	1	0	$2p_z$	$\psi_{2p_z} = \frac{1}{4\sqrt{2\pi}}\left(\frac{Z}{a_0}\right)^{3/2} \sigma e^{-\sigma/2}\cos\theta$
2	1	± 1	$2p_x$	$\psi_{2p_x} = \frac{1}{4\sqrt{2\pi}}\left(\frac{Z}{a_0}\right)^{3/2} \sigma e^{-\sigma/2}\sin\theta\cos\phi$
			$2p_y$	$\psi_{2p_y} = \frac{1}{4\sqrt{2\pi}}\left(\frac{Z}{a_0}\right)^{3/2} \sigma e^{-\sigma/2}\sin\theta\sin\phi$

[a] $a_0 = \frac{h^2}{4\pi^2 me^2}$ = radius of Bohr orbit for $n = 1$ for H; $\sigma = Z\left(\frac{r}{a_0}\right)$.

The significance of the function ψ is the following. The probability that the electron is in a volume element dV in size at the coordinates r, θ, and ψ is $\psi^2(r, \theta, \phi)\, dV$. A wave function ψ, defined by a set of allowed values of n, l, and m_l, is called an *orbital*. Since $\psi^2(r, \theta, \phi)$ defines the probability distribution of the electron, it is convenient but superficial to think of an orbital as a region where an electron can be. The different orbitals for an atom are distinguished from one another by the values of n, l, and m_l. The numerical value of n is given, and it is customary to designate the l value with a letter:

$$l = 0, s \qquad l = 1, p \qquad l = 2, d \qquad l = 3, f$$

Thus the orbital associated with the function ψ for $n = 1$, $l = 0$ is called a 1*s* orbital; for $n = 2$, $l = 0$, a 2*s* orbital; for $n = 2$, $l = 1$, a 2*p* orbital; and so on. To differentiate the orbitals associated with the same values of n and l but with different values of m_l, a letter subscript, which refers to the angular orientation of the orbital, is added. The three 2*p* orbitals are designated $2p_x$, $2p_y$, and $2p_z$, with x, y, and z indicating the direction along which the probability is high.

Figure 8–9 shows the dependence of the probabilities for different orbitals upon r. To show this most meaningfully the figure accounts for the dependence of the volume dV upon radius. The volume of a spherical shell of thickness dr at distance r from the nucleus is $dV = 4\pi r^2\, dr$, and in Figure 8–9, $4\pi r^2 \psi^2(r)$ is plotted versus r. Thus if we consider only the part of

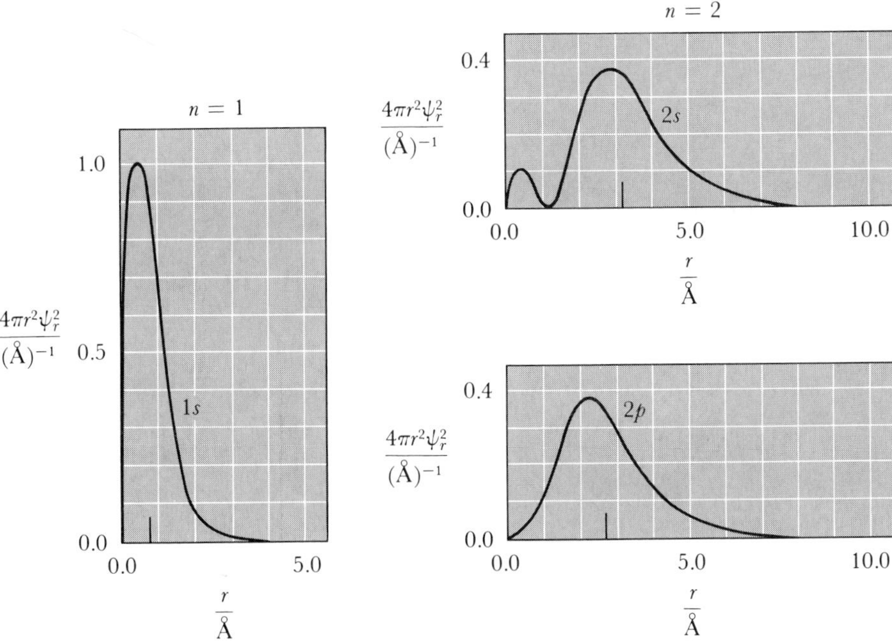

FIGURE 8–9
The radial probability distribution functions $4\pi r^2 \psi_r^2$ for an electron in 1*s*, 2*s*, and 2*p* orbitals. The value of ψ_r^2 at a particular value of r is multiplied by $4\pi r^2$ to take into account the relative volume of a spherical shell at a distance r from the nucleus. The average value of r for each orbital is shown as a vertical line.

the function involving r, the functions for hydrogen ($Z = 1$) plotted in Figure 8–9 are:

$$4\pi r^2\psi_{1s}^2(r) = C_1 \times r^2 \times e^{-2r/a_0}$$

$$4\pi r^2\psi_{2s}^2(r) = C_2 \times r^2 \times \left(2 - \frac{r}{a_0}\right)^2 \times e^{-r/a_0}$$

$$4\pi r^2\psi_{2p}^2(r) = C_2 \times r^2 \times \sigma^2 \times e^{-r/a_0}$$

in which C_1 and C_2 are constants. You can see in these equations the factors that give each radial distribution its form, as shown in the figure:

1. For $n = 1$, $l = 0$, $m_l = 0$, the radial distribution has a maximum[6] at $r = a_0$, which is the radius of the circular Bohr orbit for $n = 1$:

$$a_0 = \frac{h^2}{4\pi^2 me^2} = 5.29 \times 10^{-9} \text{ cm}$$

2. There is an exponential factor e^{-2r/a_0} (for $n = 1$) or e^{-r/a_0} (for $n = 2$), which is responsible for the decrease of $\psi^2(r)$ at large values of r.
3. There is a factor $\left(2 - \frac{r}{a_0}\right)^2$ in the equation for $\psi^2(r)$ for $n = 2$, $l = 0$. This factor is zero at $r = 2a_0$.
4. The radial part of the equation for ψ does not depend upon the value of m_l.

It is possible to derive an equation for the average separation of the electron and the nucleus, $\bar{r}$. This equation is

$$\bar{r} = \underbrace{\left(\frac{n^2 a_0}{Z}\right)}_{\text{The radius of the circular Bohr orbit.}}\left(1 + \frac{1}{2}\left[1 - \frac{l(l+1)}{n^2}\right]\right)$$

We can calculate values of $\bar{r}$ for an electron in the orbitals under consideration (with $Z = 1$):

$$\begin{array}{ll} 1s\ (n = 1, l = 0) & \bar{r} = a_0(1 + \frac{1}{2}) = \frac{3}{2}a_0 \\ 2s\ (n = 2, l = 0) & \bar{r} = 4a_0(1 + \frac{1}{2}) = 6a_0 \\ 2p\ (n = 2, l = 1) & \bar{r} = 4a_0[1 + \frac{1}{2}(1 - \frac{1}{2})] = 5a_0 \end{array}$$

These values of $\bar{r}$ are indicated in Figure 8–9. We see that the average separation of the nucleus and an electron in the $n = 2$ orbitals is 3.33- or 4.00-fold larger than the average separation for an electron in the $n = 1$ orbital. In the Bohr theory the radius of the circular orbits is proportional to n^2.

The angular dependence of the probability is given by the factors in the wave functions which involve θ and ϕ. The dependences for $2s$ and

[6] If you have had differential calculus, perform the operation $d(r^2e^{-2r/a_0})/dr$ and equate this to zero. The result is $r = a_0$, the radius at which $r^2\psi_{1s}^2(r)$ is a maximum.

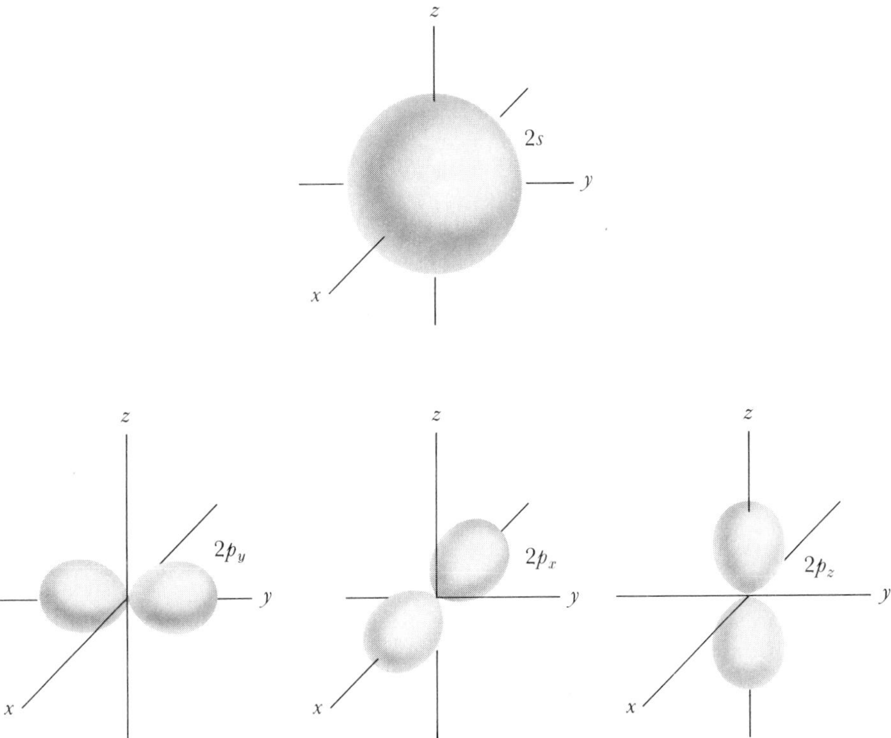

FIGURE 8–10
Pictorial representation of angular probability distribution functions for an electron in the 2s, and the $2p_x$, $2p_y$, and $2p_z$ orbitals. The distance from origin to surface is a measure of the relative probability at that angle. The distance to the spherical surface for the 2s orbital is the same for all angles; the 2s orbital is spherically symmetric. For each of the p orbitals there is a high probability along one of the orthogonal x, y, and z axes.

2p orbitals are depicted in Figure 8–10. For each orbital a surface is shown; the relative probability of finding the electron in a solid angle $d\theta\, d\phi$ at a particular value of θ and ϕ is proportional to the distance from the origin to the surface. For the 1s and 2s orbitals, the surface is a sphere with its center at the origin. It is equally probable that the electron will be found at each value of θ and ϕ. [The wave functions $\psi(1s)$ and $\psi(2s)$ do not contain θ and ϕ.] For each of the three 2p orbitals, the probability is at a maximum along one axis; this is summarized in the relative values of $\psi^2(2p_x)$, $\psi^2(2p_y)$, and $\psi^2(2p_z)$ for the values of θ and ϕ that correspond to the directions of the x, y, and z axes:

		x axis	y axis	z axis
	θ	90°	90°	0°
	ϕ	0°	90°	any value
$\psi^2(2p_x)$		1	0	0
$\psi^2(2p_y)$		0	1	0
$\psi^2(2p_z)$		0	0	1

In connection with later discussion of systems with more than one electron, it is important to recognize that the sum of the angular parts of the probabilities for the $2p_x$, $2p_y$, and $2p_z$ orbitals is spherically symmetric (i.e., does not contain θ or ϕ). We can check this by adding the squares of the angular parts of the equations for $\psi(2p_x)$, $\psi(2p_y)$, and $\psi(2p_z)$ given in Table 8–3:

$$(\sin\theta\cos\phi)^2 + (\sin\theta\sin\phi)^2 + (\cos\theta)^2 = \sin^2\theta\,(\cos^2\phi + \sin^2\phi) + \cos^2\theta$$
$$= \sin^2\theta + \cos^2\theta = 1$$

because the sum of the squares of the sine and cosine of any angle α is unity $(\sin^2\alpha + \cos^2\alpha \equiv 1)$.

Although the terminology may not be completely appropriate, groups of orbitals with the same value of n sometimes are called shells or subshells. (It is primarily n, as already seen, that determines the average distance of an electron from the nucleus.) As we will use the terms, a shell is the complete group of orbitals with the same value of n, and a subshell is the complete group of orbitals with the same values of n and l. The shell corresponding to the largest value of n for which one or more orbitals are occupied in a neutral atom is often called the valence shell. In the literature dealing with atomic structure, the shells may be designated by letters. The $n = 1$ shell is designated K, the $n = 2$ shell is designated L, and so on in alphabetical order.

Certain experimentally observed features of the spectra of gaseous atoms were not explained by the model of atomic structure derived from the Schrödinger equation. To account for these observations, it was necessary to ascribe to the electron an intrinsic angular momentum in addition to the angular momentum due to its orbital motion. This intrinsic angular momentum has the value

$$p = \frac{h}{2\pi}\sqrt{s(s+1)}$$

with $s = \frac{1}{2}$, and this angular momentum can have two orientations in space:

$$p_z = m_s\frac{h}{2\pi}$$

where $m_s = +\frac{1}{2}$ or $-\frac{1}{2}$. This fourth quantum number, m_s, commonly called the *spin quantum number*, was introduced by GOUDSMIT and UHLENBECK in 1926. A simplified view of the origin of this angular momentum ascribes a spinning motion to the electron. A more complete quantum-mechanical treatment of atomic structure by DIRAC, which provided for relativity, introduces the quantum number m_s without the connotation that the electron spins on its axis.

The inherent uncertainty in the position of an electron in an atom implied by the meaning of $\psi^2(r, \theta, \phi)$ is consistent with the *Heisenberg uncertainty principle*. This principle states that the momentum and position of a particle cannot be determined simultaneously without uncertainty in the value of each, such that

$$\Delta p \, \Delta q \cong \frac{h}{4\pi}$$

in which Δp and Δq are the uncertainties in the momentum and the coordinate, respectively. Heisenberg's principle applies to macroscopic moving objects as well, but the uncertainty in momentum or position for such objects is negligible. However, for an electron moving in an atom the uncertainty is appreciable. Calculations show that the uncertainty in electron position implied by the distribution curves given in Figures 8–9 and 8–10 are consistent with the value of Δq calculated according to Heisenberg's uncertainty principle.

We noted the success of the equation

$$E = -\frac{2\pi^2 me^4}{h^2} \times \frac{Z^2}{n^2}$$

in correlating energies of allowed states of the hydrogen atom when we discussed the Bohr theory and the spectrum of atomic hydrogen. The experimentally measured energies for ionization of other one-electron atoms verifies the dependence of energy upon the square of the nuclear charge. For the process $H(g) = H^+(g) + e^-$, $\Delta E = +1312$ kJ mol^{-1}. For other species, He^+ ($Z = 2$), Li^{2+} ($Z = 3$), and Be^{3+} ($Z = 4$), the values of ΔE are:

Process	ΔE kJ mol^{-1}
$He^+(g) = He^{2+}(g) + e^-$	5,248
$Li^{2+}(g) = Li^{3+}(g) + e^-$	11,810
$Be^{3+}(g) = Be^{4+}(g) + e^-$	20,990

and these values are nicely proportional to Z^2; the quotient $\Delta E/Z^2$ is constant:

$$\frac{5{,}248 \text{ kJ mol}^{-1}}{4} = 1312 \text{ kJ mol}^{-1}$$

$$\frac{11{,}810 \text{ kJ mol}^{-1}}{9} = 1312 \text{ kJ mol}^{-1}$$

$$\frac{20{,}990 \text{ kJ mol}^{-1}}{16} = 1312 \text{ kJ mol}^{-1}$$

The energies of spectral transitions for these ionic one-electron species also are related to the corresponding energies for hydrogen by the factor Z^2. Thus corresponding to the Balmer line at 656.3 nm (the transition for $a = 2$, $b = 3$), there is a spectral line at

$$\tfrac{1}{4} \times 656.3 \text{ nm} = 164.1 \text{ nm}$$

in the spectrum of $He^+(g)$. (The energy levels for this system and for $Li^{2+}(g)$ were shown in Figure 8–7.)

In applying the Schrödinger equation to an atom with more than one electron, we must add more terms for the potential energy as a function of the coordinates. For instance, in an atom with two electrons (e.g., He, Li^+, Be^{2+}), there are three terms in the equation for potential energy:

$$E_p(r_{01}, r_{02}, r_{12}) = -\frac{Ze^2}{r_{01}} - \frac{Ze^2}{r_{02}} + \frac{e^2}{r_{12}}$$

in which r_{01} and r_{02} are the distances between the nucleus and each electron and r_{12} is the distance between the two electrons. This addition makes a rigorous solution of the Schrödinger equation for an atom having two electrons virtually impossible; even an approximate solution is difficult. (This is an area of continuing research.) Nonetheless, concepts arising in treatment of the hydrogen atom are used in discussion of atoms with more than one electron. The electronic structures of such atoms can be described in terms of the quantum numbers already introduced in connection with the one-electron atom. A necessary additional specification is provided by the *Pauli exclusion principle* (named for WOLFGANG PAULI), which states that no two electrons in an atom can have the same set of values for the four quantum numbers. For example, in a neutral helium atom ($Z = 2$), the two electrons *cannot* both have the set of quantum numbers: $n = 1$, $l = 0$, $m_l = 0$, and $m_s = +\frac{1}{2}$, but one electron can have this set of quantum numbers if the other electron has the set: $n = 1$, $l = 0$, $m_l = 0$, and $m_s = -\frac{1}{2}$. The most stable electronic configuration of a helium atom is that just described; the configuration can be stated concisely as $1s^2$, in which the prefix gives the value n ($n = 1$ in this case), the letter designates the value of l ($l = 0$, designated by s, in this case), and the superscript designates the total number of electrons in the orbital (or orbitals) in question. That the two electrons indicated by this superscript are in the same orbital with opposite spin is understood because there is only one $1s$ orbital and the two electrons in this orbital must have opposite spin.

The Pauli exclusion principle can be stated in other terms. *Electrons with the same spin cannot occupy the same region of space.* This prohibition is stronger than the Coulombic repulsion of particles of like charge, which tends only to make electrons of either spin repel one another.

In the ground state (the most stable state) of a neutral atom, electrons equal in number to the atomic number occupy the orbitals with the lowest energy (subject to the Pauli exclusion principle). Although orbitals corresponding to different values of l and m_l but the same value of n have the same energy in a one-electron atom, this is not true for atoms with more than one electron. For such atoms, the energies of orbitals with the same value of n but with different values of l are:

Lowest energy $\qquad$ Highest energy

$$s(l = 0) < p(l = 1) < d(l = 2) < f(l = 3)$$

This dependence of the energy upon l in atoms with more than one electron is related to electron–electron repulsion. The electron–electron repulsion reduces the net attraction of the nucleus for an electron; the electron in question is said to be "screened" from the nucleus by the other electrons. The energy required to remove one electron from gaseous atoms of the elements hydrogen ($Z = 1$) through sodium ($Z = 11$), presented in Table 8–4, tells us something more about screening. The process being considered is

$$X(g) = X^+(g) + e^- \qquad \Delta E = IE \text{ (the ionization energy)}$$

The term "ionization energy" (IE) will be used to mean the energy required to detach an electron from the most stable energy state of an atomic or molecular species. Usually, it is a neutral species from which the electron is being detached, and this sometimes is called the *first ionization energy*, with subscript one added to IE, IE_1. The energies required to remove successive electrons are called the *second*, *third*, · · · , ionization energies, and corresponding subscripts can be added to IE, IE_2, IE_3, and so on.

We already have seen that for one-electron atoms, $H(g)$, $He^+(g)$, $Li^{2+}(g)$, and $Be^{3+}(g)$, the energy required for ionization of the ground state of each of these species is given by

$$\Delta E = 1312 \text{ kJ mol}^{-1} \times Z^2$$

If ionization from an excited state ($n > 1$) of these species were being considered, the equation

$$\Delta E = 1312 \text{ kJ mol}^{-1} \times \frac{Z^2}{n^2}$$

would be used. For atoms with more than one electron, the energy required

TABLE 8–4
Electronic Structures and First Ionization Energies of Elements with $Z = 1$ to 11

Element Z	Symbol	Electronic configuration of ground state of atom[a]	IE_1 / kJ mol^{-1}: Experimental value	Calculated, $Z_{eff} = Z$	Calculated, $Z_{eff} = 1$
1	H	$1s$	1311.7	1311.7	1311.7
2	He	$1s^2$	2371.9	5246.8	1311.7
3	Li	$1s^22s^1$	520.1	2951.3	327.9
4	Be	$1s^22s^2$	899.1	5246.8	327.9
5	B	$1s^22s^22p^1$	800.4	8198	327.9
6	C	$1s^22s^22p^2$	1086.2	1.181×10^4	327.9
7	N	$1s^22s^22p^3$	1402.1	1.607×10^4	327.9
8	O	$1s^22s^22p^4$	1313.8	2.10×10^4	327.9
9	F	$1s^22s^22p^5$	1681.1	2.66×10^4	327.9
10	Ne	$1s^22s^22p^6$	2080.3	3.28×10^4	327.9
11	Na	$1s^22s^22p^63s^1$	495.8	1.76×10^4	145.8

[a]The value of m_l is not specified in this description of the electronic configuration.

for ionization cannot be calculated in this way. Because of electron–electron repulsion, removal of an electron requires less energy. For helium,

Process	$\frac{\Delta E}{\text{kJ mol}^{-1}}$
$He(g) = He^{+}(g) + e^{-}$	2372
$He^{+}(g) = He^{2+}(g) + e^{-}$	5248

In each of these processes the electron is being removed from the 1*s* orbital, and the nuclear charge is the same, 2+. But only for the second process is the energy change given by

$$\Delta E = 1312 \text{ kJ mol}^{-1} \times \frac{Z^2}{n^2}$$

$$= 1312 \text{ kJ mol}^{-1} \times \frac{2^2}{1^2} = 5248 \text{ kJ mol}^{-1}$$

However, we can use this equation in a rearranged form to define an effective nuclear charge, Z_{eff}, for the ionization process:

$$Z_{\text{eff}} = n \times \sqrt{\frac{\Delta E}{1312 \text{ kJ mol}^{-1}}}$$

For removal of the first electron from the helium atom, the calculation of Z_{eff} is

$$Z_{\text{eff}} = 1 \times \sqrt{\frac{2372 \text{ kJ mol}^{-1}}{1312 \text{ kJ mol}^{-1}}} = 1.34$$

Thus the electron–electron repulsion has the effect of making the energy required in the ionization process equal to the energy needed to remove the electron from the 1*s* orbital of a one-electron atom with a nucleus having $Z = 1.34$.

Figure 8–11a presents the first ionization energies for elements with $Z = 1$ to 11. Also shown in the figure are the values of *IE* expected under two extreme assumptions:

$Z_{\text{eff}} = Z$	No electron–electron repulsion
$Z_{\text{eff}} = 1$	Perfect screening of the nucleus by all electrons except the most easily removed electron

The experimentally determined values of the ionization energy are between the values calculated under these extreme assumptions. We can calculate a value of Z_{eff} from the experimental values of *IE*, as we already have done for helium. For lithium and sodium, the calculations are:

$_{3}\text{Li}$ ($IE = 520 \text{ kJ mol}^{-1}$, $n = 2$) $\qquad Z_{\text{eff}} = 2 \times \sqrt{\frac{520 \text{ kJ mol}^{-1}}{1312 \text{ kJ mol}^{-1}}} = 1.26$

$_{11}\text{Na}$ ($IE = 496 \text{ kJ mol}^{-1}$, $n = 3$) $\qquad Z_{\text{eff}} = 3 \times \sqrt{\frac{496 \text{ kJ mol}^{-1}}{1312 \text{ kJ mol}^{-1}}} = 1.84$

Figure 8–11b shows the values of Z_{eff} calculated in this way.

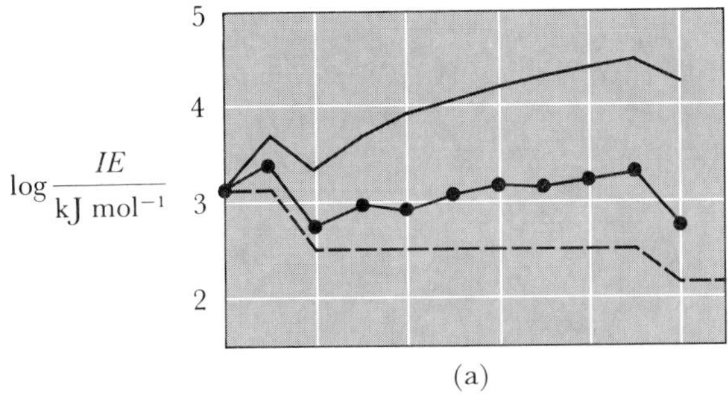

(a)

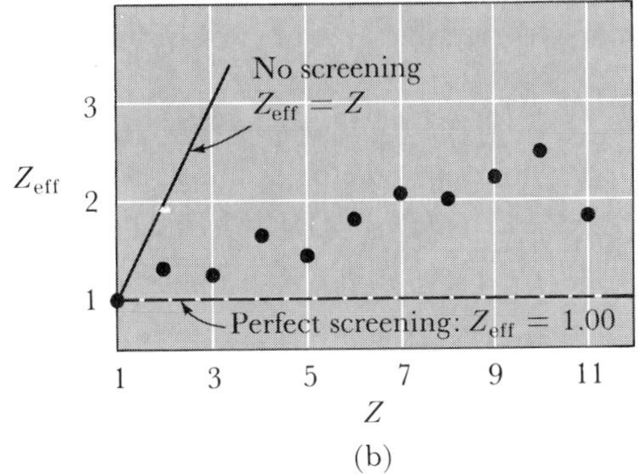

(b)

FIGURE 8–11
The ionization energies (IE) for atoms with $Z = 1$–11 and the effective nuclear charges (Z_{eff}) calculated from IE.

(a) $\frac{IE}{\text{kJ mol}^{-1}}$ versus Z (notice logarithmic ordinate)

—— $IE = 1312 \text{ kJ mol}^{-1} \times \frac{Z^2}{n^2}$ (no screening)

---- $IE = 1312 \text{ kJ mol}^{-1} \times \frac{1}{n^2}$ (perfect screening)

●–● Experimental points

(b) Z_{eff} versus Z.

$$Z_{eff} = n \times \sqrt{\frac{IE}{1312 \text{ kJ mol}^{-1}}}$$

The plot of ionization energy versus atomic number is dramatically discontinuous between each noble gas and the following alkali metal. The principal cause of this phenomenon is the dependence $IE \propto n^{-2}$ and the fact that the n value for a noble gas is smaller than that for the following alkali metal. The electron being removed from the alkali metal atom is, on the average, farther from the nucleus.

Examination of the energy required to ionize an excited sodium atom with its most loosely held electron in an s orbital having a value of n (designated as a) greater than 3 shows the increased effectiveness of the screening as the electron being ionized is farther and farther away from the kernel (the nucleus plus all of the other electrons). The experimentally determined values of the energy required to bring about the process

$$\text{Na}(1s^22s^22p^6 \cdots as^1) \rightarrow \text{Na}^+(1s^22s^22p^6) + e^-$$

for different values of a are:

a	$\frac{\Delta E}{\text{kJ mol}^{-1}}$	$Z_{eff} = a\sqrt{\frac{\Delta E}{1312 \text{ kJ mol}^{-1}}}$
4	187.9	1.51
5	98.7	1.37
6	60.7	1.29
7	41.0	1.24

The effective nuclear charge decreases, approaching 1 as the electron being removed is farther and farther from the kernel, which has the neon structure.

The value of the quantum number l for an electron is relevant in establishing how effectively it will be screened by other electrons in the atom. In particular, the closer an electron approaches the nucleus, the less it is subject to screening by other electrons; this close approach is called *penetration*. The relative penetration of an electron in orbitals with different values of l but the same value of n are:

$$\text{Most penetrating} \quad s > p > d > f \cdots \quad \text{Least penetrating}$$

That an s orbital is more penetrating than a p orbital was shown in Figure 8–9; notice the small peak at $r = 0.404$ Å in the probability distribution for the $2s$ orbital. For electrons with $n = 2$, the s orbital is more stable than the p orbitals because it penetrates closer to the nucleus, and therefore is less effectively screened from the nucleus. The situation for atoms with more than two electrons can be summarized for $n = 2$, 3, and 4:

For $n = 2$,
- $2s$ orbital more stable than $2p$ orbital

For $n = 3$,
- $3s$ orbital more stable than $3p$ orbital
- $3p$ orbital more stable than $3d$ orbital

For $n = 4$,
- $4s$ orbital more stable than $4p$ orbital
- $4p$ orbital more stable than $4d$ orbital
- $4d$ orbital more stable than $4f$ orbital

Thus the electronic configuration of the ground state of the lithium atom is

$$\text{Li} \quad 1s^2 2s^1$$

and not

$$\text{Li} \quad 1s^2 2p^1$$

The electronic configurations given in Table 8–4 do not specify the m_l values. Although the description for lithium,

$$\text{Li} \quad 1s^2 2s^1$$

is unambiguous, that for carbon,

$$\text{C} \quad 1s^2 2s^2 2p^2$$

is ambiguous. It could mean any of the following configurations, each of which is consistent with the Pauli exclusion principle:

	$1s$	$2s$	$2p_x$	$2p_y$	$2p_z$
C	(↑↓)	(↑↓)	(↑)	(↑)	( )
C	(↑↓)	(↑↓)	(↑)	(↓)	( )
C	(↑↓)	(↑↓)	(↑↓)	( )	( )

Each parenthesis represents an orbital, and orientation of an arrow depicts the electron spin:

$$\uparrow \text{ means } m_s = +\tfrac{1}{2}$$
$$\downarrow \text{ means } m_s = -\tfrac{1}{2}$$

The three $2p$ orbitals are designated $2p_x$, $2p_y$, and $2p_z$ as in Table 8–3, but in the absence of an electric or magnetic field to define directions, these labels have no special significance other than to show that there are three different $2p$ orbitals. Thus do not attach any significance to the fact that the p_z orbital is shown to be unoccupied. The first of the possible electronic configurations for carbon given above is more stable than either of the other two. This experimental observation and corresponding observations for other atoms are summarized in *Hund's rule* (named for F. H. HUND), which states that an atom with two or more electrons in orbitals with the same value of n and a particular value of l ($l = 1, 2, \cdots$) is more stable if these electrons have the same value of m_s. Electrons with the same values of n, l, and m_s must have different values of m_l. The first configuration given for carbon, that which follows Hund's rule, is one in which the two $2p$ electrons are farther apart on the average than they are in the last configuration, in which they occupy the same orbital.

The electronic configurations of elements H ($Z = 1$) through Ar ($Z = 18$), including orbital occupancy, are presented in Table 8–5. The configurations are exactly what is expected on the basis of the Pauli exclusion principle and Hund's rule.

The entries in Table 8–5 show that it is the Pauli exclusion principle that determines the capacities of the first and second shells. There are not three different combinations of $n = 1$, $l = 0$, $m_l = 0$, and $m_s = +\frac{1}{2}$ or $-\frac{1}{2}$, and therefore three electrons in the first shell would not be allowed by the Pauli principle. For a three-electron system one electron must have a value of $n \neq 1$. Similarly, for an eleven-electron system, one electron must have a value of n not equal to 1 or 2. By considering the allowed combinations of values of n, l, m_l, and m_s, you can show that the capacity of an electron shell is $2n^2$. Thus

$$\begin{array}{ll} n = 1 & 2n^2 = 2 \\ n = 2 & 2n^2 = 8 \\ n = 3 & 2n^2 = 18 \\ n = 4 & 2n^2 = 32 \\ \text{etc.} & \end{array}$$

Table 8–5 has no surprises, but Table 18–1 (dealing with $Z > 18$) will. We shall consider the electronic structures of the heavier metallic elements briefly now, reserving the details for Chapter 18. At argon, the $3s$ and $3p$ orbitals are filled, but the third shell (i.e., all orbitals with $n = 3$) is not filled completely. For $n = 3$, l can have the value 2 as well as 0 or 1; therefore there are d orbitals for $n = 3$. The five different $3d$ orbitals correspond to the five different possible values of m_l: -2, -1, 0, $+1$, and $+2$.

TABLE 8–5
Electronic Configuration and Orbital Occupancy for the Ground States of Elements H through Ar

Element			Orbital occupancy[a]									IE_1
Z	Symbol	Configuration	$1s$	$2s$	$2p_x$	$2p_y$	$2p_z$	$3s$	$3p_x$	$3p_y$	$3p_z$	kJ mol^{-1}
1	H	$1s^1$	(↑)									1311.7
2	He	$1s^2$	(↑↓)									2371.9
3	Li	$1s^2 2s^1$	(↑↓)	(↑)								520.1
4	Be	$1s^2 2s^2$	(↑↓)	(↑↓)								899.1
5	B	$1s^2 2s^2 2p^1$	(↑↓)	(↑↓)	(↑)							800.4
6	C	$1s^2 2s^2 2p^2$	(↑↓)	(↑↓)	(↑)	(↑)						1086.2
7	N	$1s^2 2s^2 2p^3$	(↑↓)	(↑↓)	(↑)	(↑)	(↑)					1402.1
8	O	$1s^2 2s^2 2p^4$	(↑↓)	(↑↓)	(↑↓)	(↑)	(↑)					1313.8
9	F	$1s^2 2s^2 2p^5$	(↑↓)	(↑↓)	(↑↓)	(↑↓)	(↑)					1681.1
10	Ne	$1s^2 2s^2 2p^6$	(↑↓)	(↑↓)	(↑↓)	(↑↓)	(↑↓)					2080.3
11	Na	$1s^2 2s^2 2p^6 3s^1$	(↑↓)	(↑↓)	(↑↓)	(↑↓)	(↑↓)	(↑)				495.8
12	Mg	$1s^2 2s^2 2p^6 3s^2$	(↑↓)	(↑↓)	(↑↓)	(↑↓)	(↑↓)	(↑↓)				737.6
13	Al	$1s^2 2s^2 2p^6 3s^2 3p^1$	(↑↓)	(↑↓)	(↑↓)	(↑↓)	(↑↓)	(↑↓)	(↑)			577.4
14	Si	$1s^2 2s^2 2p^6 3s^2 3p^2$	(↑↓)	(↑↓)	(↑↓)	(↑↓)	(↑↓)	(↑↓)	(↑)	(↑)		786.2
15	P	$1s^2 2s^2 2p^6 3s^2 3p^3$	(↑↓)	(↑↓)	(↑↓)	(↑↓)	(↑↓)	(↑↓)	(↑)	(↑)	(↑)	1063
16	S	$1s^2 2s^2 2p^6 3s^2 3p^4$	(↑↓)	(↑↓)	(↑↓)	(↑↓)	(↑↓)	(↑↓)	(↑↓)	(↑)	(↑)	1000
17	Cl	$1s^2 2s^2 2p^6 3s^2 3p^5$	(↑↓)	(↑↓)	(↑↓)	(↑↓)	(↑↓)	(↑↓)	(↑↓)	(↑↓)	(↑)	1255
18	Ar	$1s^2 2s^2 2p^6 3s^2 3p^6$	(↑↓)	(↑↓)	(↑↓)	(↑↓)	(↑↓)	(↑↓)	(↑↓)	(↑↓)	(↑↓)	1520

[a]Electron spins (m_s) designated by arrow: ↑ for $+1/2$ and ↓ for $-1/2$.

The next element after argon, potassium ($Z = 19$), might be expected to have the electronic configuration

$$\text{K} \quad 1s^2 2s^2 2p^6 3s^2 3p^6 3d^1$$

but the ground state of the potassium atom has the configuration

$$\text{K} \quad 1s^2 2s^2 2p^6 3s^2 3p^6 4s^1$$

The next element, calcium ($Z = 20$), has the ground-state electronic configuration

$$\text{Ca} \quad 1s^2 2s^2 2p^6 3s^2 3p^6 4s^2$$

It is not until scandium ($Z = 21$) that a $3d$ orbital is occupied; the electronic configuration of the ground state of the scandium atom is

$$\text{Sc} \quad 1s^2 2s^2 2p^6 3s^2 3p^6 3d^1 4s^2$$

From scandium through zinc ($Z = 30$), the five $3d$ orbitals are progressively filled; the ground-state electronic configuration of zinc is

$$\text{Zn} \quad 1s^2 2s^2 2p^6 3s^2 3p^6 3d^{10} 4s^2$$

Later in this chapter we will consider the periodic table and its relationship with the electronic structures of the elements.

Figure 8–12 shows the first ionization energies of most elements; as in Figure 8–11a, which summarizes a smaller range of atomic numbers, this plot reveals maxima at the noble gases ($Z = 2$, 10, 18, 36, 54, and 86) followed by minima at the alkali metals ($Z = 3$, 11, 19, 37, 55, and 87). There are other discontinuities in the region $Z = 1$–18 which deserve our attention; these discontinuities are:

$$IE(_4\text{Be}) > IE(_5\text{B})$$
$$IE(_{12}\text{Mg}) > IE(_{13}\text{Al})$$

which involve comparison of elements with 2 and 3 electrons in their valence shells, and

$$IE(_7\text{N}) > IE(_8\text{O})$$
$$IE(_{15}\text{P}) > IE(_{16}\text{S})$$

which involve comparison of elements with 5 and 6 electrons in their valence shells. For our discussion we will focus attention on the elements with an $n = 2$ valence shell. The electronic structures of the reactant and product species of the ionization processes in question are:

	$2s$	$2p$			$2s$	$2p$
$_4$Be	(↑↓)	( )( )( )	→	$_4\text{Be}^+$	(↑)	( )( )( )
$_5$B	(↑↓)	(↑)( )( )	→	$_5\text{B}^+$	(↑↓)	( )( )( )
$_7$N	(↑↓)	(↑)(↑)(↑)	→	$_7\text{N}^+$	(↑↓)	(↑)(↑)( )
$_8$O	(↑↓)	(↑↓)(↑)(↑)	→	$_8\text{O}^+$	(↑↓)	(↑)(↑)(↑)

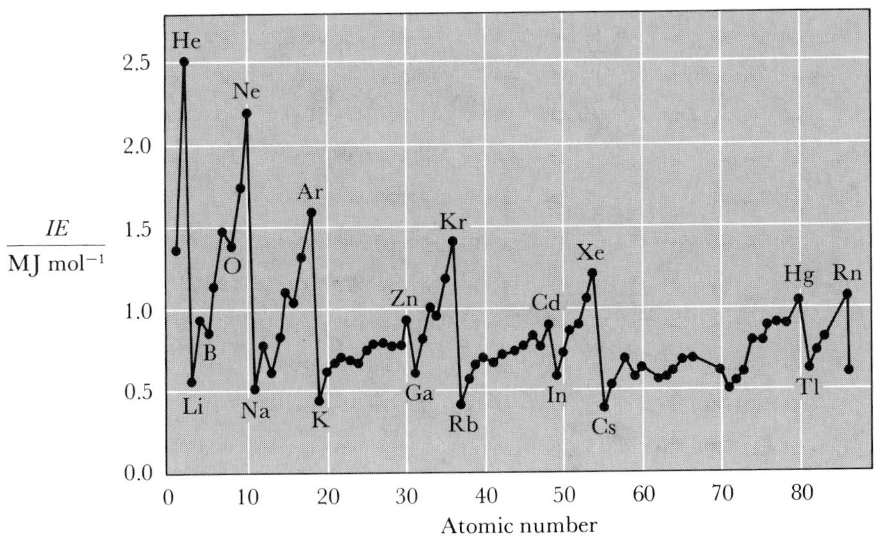

FIGURE 8–12
Ionization energies (IE) of elements with atomic numbers 1–90 (1 MJ = 10^6 J).

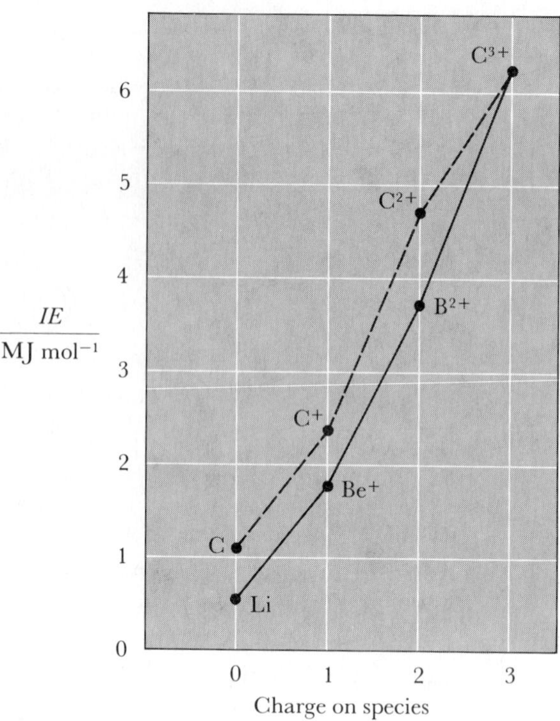

FIGURE 8–13
Ionization energies for removal of one electron from neutral and charged species.

—— Isoelectronic species: Li, Be^+, B^{2+}, C^{3+}

---- Successive ionization of carbon atom: C, C^+, C^{2+}, C^{3+}

The beryllium—boron (or magnesium—aluminum) discontinuity is reasonable when you recognize that the electron being removed from the atom with larger Z (boron in this comparison) is being removed from a less penetrating orbital. (A p orbital is less penetrating than an s orbital with the same value of n.) The nitrogen–oxygen (or phosphorus–sulfur) discontinuity is reasonable when you recognize that the electron being removed from the atom with larger Z (oxygen in this comparison) is being removed from an orbital that contains two electrons. The electron being removed from the atom with smaller Z (nitrogen in this comparison) is being removed from an orbital that contains only one electron.

The energy required to remove an electron from an atom with a net positive charge is greater than the ionization energy of a neutral atom. This is shown in Figure 8–13, which gives experimental values for the energy required to remove an electron from members of two series of atomic species:

1. the isoelectronic species,[7] Li, Be^+, B^{2+}, and C^{3+}

$$IE_1(\text{Li}) < IE_2(\text{Be}) < IE_3(\text{B}) < IE_4(\text{C})$$

[7]Monatomic species are said to be *isoelectronic* if each has the same number of electrons. Each of the species in this series has the electronic structure of a neutral lithium atom. The concept of isoelectronic structures also will be applied to polyatomic species in Chapter 10.

2. the species formed by successive ionization of an atom of a particular element, carbon, C, C^+, C^{2+}, and C^{3+}

$$IE_1(\mathrm{C}) < IE_2(\mathrm{C}) < IE_3(\mathrm{C}) < IE_4(\mathrm{C})$$

The ionization energies for the carbon species are larger because the nuclear charge is larger, but each series shows the very large effect that a net positive charge has upon the energy required to remove an additional electron from an atom.

In the course of removal of successive electrons, there is a striking discontinuity when an electron is removed from the shell with principal quantum number one less than that of the valence shell. This is shown by removal of an electron from species with the helium electronic structure ($1\ \mathrm{MJ} = 10^6\ \mathrm{J}$):

(He	$IE_1 = 2.372\ \mathrm{MJ\ mol^{-1}}$)
Li	$IE_2 = 7.30\ \mathrm{MJ\ mol^{-1}}$
Be	$IE_3 = 14.84\ \mathrm{MJ\ mol^{-1}}$
B	$IE_4 = 25.02\ \mathrm{MJ\ mol^{-1}}$
C	$IE_5 = 37.82\ \mathrm{MJ\ mol^{-1}}$

These energies of ionization are enormously greater than the values plotted in Figure 8–13 for atoms of these same elements but with one more electron. The large energies arise because

1. the value of Z_{eff} is increased by ~1 unit, and, more importantly, because
2. the electron being removed is from the $n = 1$ shell, not the $n = 2$ shell.

ELECTRON AFFINITIES

The addition of an electron to a neutral gaseous atom of some elements evolves energy. This energy is called the *electron affinity* (symbol *EA*), and its sign convention is the opposite of that for ΔE and *IE*. For instance,

$$\mathrm{O}(g) + \mathrm{e}^- = \mathrm{O}^-(g) \qquad \Delta E = -141\ \mathrm{kJ\ mol^{-1}}$$

and

$$EA = -\Delta E = +141\ \mathrm{kJ\ mol^{-1}}$$

Experimentally determined electron affinity values, given in Table 8–6, reveal some unexpected results:

1. The alkali metals, lithium, sodium, potassium, rubidium, and cesium, which lose electrons to form positive ions in most of their compounds, nonetheless have positive electron affinities. Surprisingly, most metals have positive electron affinities. However, ionic compounds with metal anions generally are not stable. The reason for this will be clarified in Chapter 18.
2. The electron affinity decreases as one goes to heavier halogen atoms, that is,

$$EA(\mathrm{Cl}) > EA(\mathrm{Br}) > EA(\mathrm{I})$$

TABLE 8–6
Electron Affinities for Some Gaseous Atoms[a]

	EA/kJ mol^{-1}		EA/kJ mol^{-1}		EA/kJ mol^{-1}
H	72.7	Be	<0	F	327.8
Li	59.8	B	27	Cl	348.7
Na	52.6	C	122.3	Br	324.5
K	48.3	N	≤0	I	295.2
Rb	46.9	O	141.0		
Cs	45.5	Mg	<0		
		Al	44.4		
		Si	133.6		
		P	71.7		
		S	200.3		

[a]For the reaction $X(g) + e^- = X^-(g)$ for which $EA = -\Delta E$.

but it is smaller for fluorine than it is for chlorine,

$$EA(\mathrm{Cl}) > EA(\mathrm{F})$$

There is the same trend in comparison of other first- and second-row elements:

$$EA(\mathrm{S}) > EA(\mathrm{O}) \qquad EA(\mathrm{P}) > EA(\mathrm{N}) \qquad EA(\mathrm{Si}) > EA(\mathrm{C})$$

3. The electron affinities of the elements of the lithium–fluorine row show discontinuities at the same electronic configurations at which there were the discontinuities in ionization energies. This is shown in Figure 8–14. The addition of an electron to an atom of beryllium or an atom of nitrogen is endothermic. The hypothesis that election affinity increases with increasing Z does not predict the comparisons

$$EA(\mathrm{Be}) < EA(\mathrm{Li})$$
$$EA(\mathrm{N}) < EA(\mathrm{C})$$

In the first case, an electron added to a beryllium atom would go into a less-penetrating orbital, and in the second case an electron added to a nitrogen atom would go into an orbital which already contains an electron.

The process by which an electron is added to a neutral atom is the reverse of the process of ionizing the corresponding anion. Thus for the process

$$C^-(g) = C(g) + e^- \qquad \Delta E = EA(\mathrm{C}) = +122.3 \text{ kJ mol}^{-1}$$

With this ionization energy and the stepwise ionization energies for the

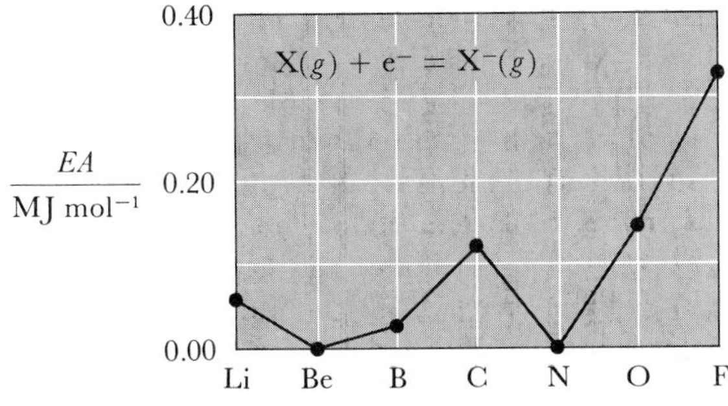

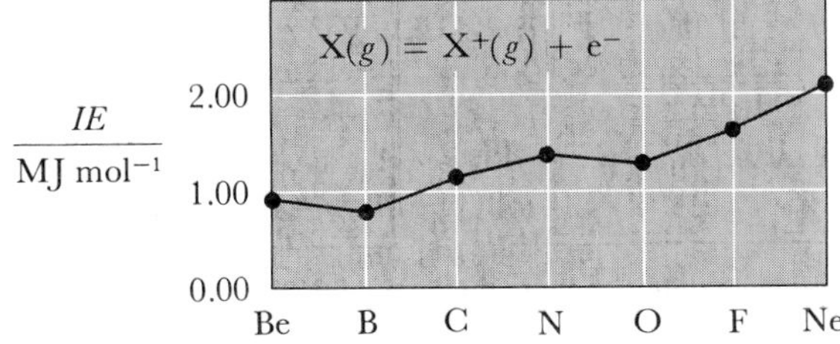

FIGURE 8–14
The electron affinities for elements with $Z = 3$–9 and the ionization energies for elements with $Z = 4$–10. Directly over one another are processes involving species with the same electronic configuration, for example,

$$_7N(g) + e^- = {_7N^-}(g)$$
$$_8O^+(g) + e^- = {_8O}(g)$$

The explanation for the minimum in the EA versus Z plot at $_7N$ is the same as that for the minimum in IE versus Z plot at $_8O$.

carbon atom plotted in Figure 8–13 and given earlier, we can calculate values for Z_{eff} for the successive processes:[8]

$$C^- \xrightarrow{Z_{eff}=0.6} C \xrightarrow{Z_{eff}=1.8} C^+ \xrightarrow{Z_{eff}=2.7} C^{2+} \xrightarrow{Z_{eff}=3.8} C^{3+} \xrightarrow{Z_{eff}=4.4} C^{4+} \xrightarrow{Z_{eff}=5.4} C^{5+}$$

We see that the successive values of Z_{eff} differ from one another by approximately 1 unit.

EXCITED ELECTRONIC STATES OF ATOMS

You were introduced to excited electronic states of atoms earlier in this chapter in our discussion of the emission spectrum of atomic hydrogen. For a one-electron atom such as atomic hydrogen, the energies of the various excited states depend only upon the principal quantum number n. Electron–electron interaction in atoms with more than one electron makes the situation more complex. The energy of an atom with more than one electron depends upon the values of n, l, m_l, and m_s for each of the atom's electrons. This can be illustrated with the simplest atom having more than one electron, the helium atom, with two electrons. For atoms of helium, electronic states with five different energies result from electronic configurations in which one or two electrons are in the $1s$ orbital and one or zero electron is in orbitals with $n = 2$. The energies of these states are shown in Figure 8–15. Each state of the neutral helium atom is more stable than

[8]Calculated as before, $Z_{eff} = n\sqrt{\dfrac{\Delta E}{1312\ \text{kJ mol}^{-1}}}$, with $n = 2$ for all steps except $C^{4+} \rightarrow C^{5+}$, for which $n = 1$.

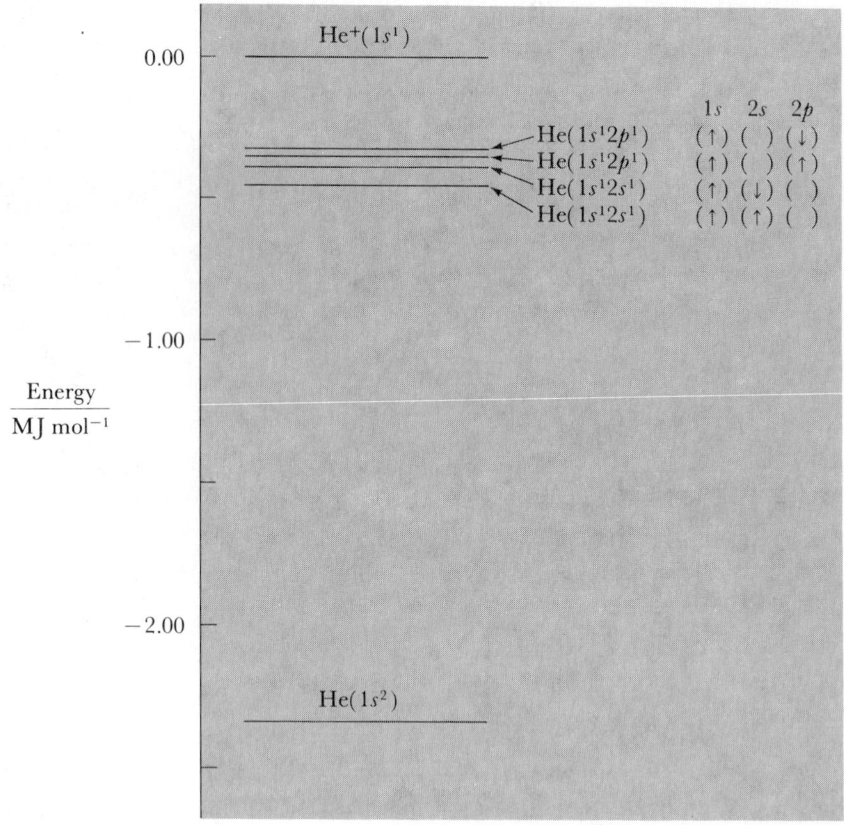

FIGURE 8–15
Energies of electronic states of the helium atom. The values are *relative* to the energy of the ground state of the +1 helium ion, He^+ $(1s^1)$.

$He^+(1s^1)$, and the energies of these states are negative relative to the energy of this species. The energies and these electronic configurations are:

	1s	2s	2p	E / kJ mol^{-1}
He	(↑↓)	( )	( )( )( )	−2372
He	(↑)	(↑)	( )( )( )	−460
He	(↑)	(↓)	( )( )( )	−383
He	(↑)	( )	(↑)( )( )	−350
He	(↑)	( )	(↓)( )( )	−325
He^+	(↑)	( )	( )( )( )	0

Notice that the states with one $2p$ electron are less stable than those with one $2s$ electron; notice also that the states in which both electrons have the same spin are more stable than the corresponding ones in which the spins are opposite. This is consistent with the notion that electrons with the same spin tend to stay farther away from one another than an equivalent pair of electrons with opposite spin. Therefore there is less electron–electron repulsion for a state with the electrons having parallel spin.

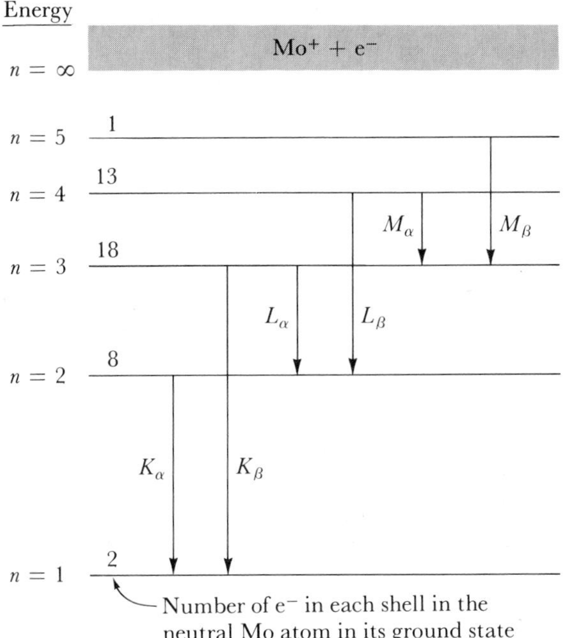

FIGURE 8–16
The transitions that result when an electron in the $n = 1$ orbital of $_{42}Mo$ is ionized from the atom. The transition labeled K_α is that considered in the text. (The energy levels are not drawn to scale.)

THE CHARACTERISTIC X RAYS FROM ELEMENTS

An atom in its most stable electronic state can be promoted to an excited electronic state in a number of ways. One common way is to have it absorb light, and another is to bombard it with high-energy electrons. If a beam of high-energy electrons strikes a target of a metallic element (e.g., molybdenum[9]), the interaction of a bombarding electron with an electron in the $1s$ orbital can ionize this $1s$ electron:

$$\text{Mo} \quad 1s^2 2s^2 2p^6 3s^2 3p^6 3d^{10} 4s^2 4p^6 4d^5 5s^1 \rightarrow \text{Mo}^+ \quad 1s^1 2s^2 2p^6 3s^2 3p^6 3d^{10} 4s^2 4p^6 4d^5 5s^1$$

The vacancy in the $1s$ orbital is filled by an electron from an $n = 2$ or $n = 3$ orbital. The energy the electron loses in this transition is given off as radiant energy. Another vacancy is created in this process, and it is filled by an electron from an orbital with a higher value of n ($n = 3$, or 4, or 5). Radiant energy is given off in this transition also. These transitions are depicted in Figure 8–16. High-energy radiation is given off in electron transitions between orbitals with $n = 2$ and $n = 1$ because these orbitals are not shielded effectively from the large nuclear charge. For molybdenum ($Z = 42$), for instance, the transition of an electron between orbitals with $n = 2$ and $n = 1$ is accompanied by radiation with $\lambda = 7.1 \times 10^{-9}$ cm. (Contrast this with $\lambda = 1.215 \times 10^{-5}$ cm for the spectral line in the Lyman series for

[9]Since much heat is generated in the bombardment of the target with electrons, the target material in modern x-ray tubes generally has a high melting point; for Mo, mp = 2883 K.

hydrogen, which arises from a transition between orbitals with these same values of n.) From our previous discussion, you will expect a relationship between the wavelength of this radiation, which is called x radiation, and the nuclear charge for the element; for the transition between orbitals with $n = 2$ and $n = 1$, the Rydberg-type relationship, with Z replaced by Z'_{eff}, an effective nuclear charge for this process, is

$$\frac{1}{\lambda} = R(Z'_{\text{eff}})^2\left(\frac{1}{1^2} - \frac{1}{2^2}\right)$$

$$= \frac{3}{4}R(Z'_{\text{eff}})^2$$

A study of the characteristic wavelength of x radiation from different elements played a major part in establishing the fundamental role of nuclear charge in atomic phenomena. In 1913 and 1914, MOSELEY determined the wavelength of x radiation from thirty-eight elements and he attempted to correlate the data with their atomic weights. There were exceptions, however, to a smooth correlation between $\lambda^{-1/2}$ and atomic weight. One of these exceptions was the order of cobalt and nickel, with atomic weights 58.93 and 58.71, respectively. The wavelength of the characteristic x rays indicated that cobalt should precede nickel, not follow it, as was suggested by the atomic weights. Moseley found that the correlation of characteristic wavelengths was vastly improved if they were related to integers corresponding to the position of the element in the periodic table. Figure 8–17 shows the correlation between the wavelength of the characteristic x radiation and the integer representing the element's position, the integer we now call the *atomic number*, Z. The linear correlation given in the figure can be written:

$$\sqrt{\frac{1}{\lambda}} = KZ - C$$

which can be rearranged to

$$\sqrt{\frac{1}{\lambda}} = K\left(Z - \frac{C}{K}\right) = KZ'_{\text{eff}}$$

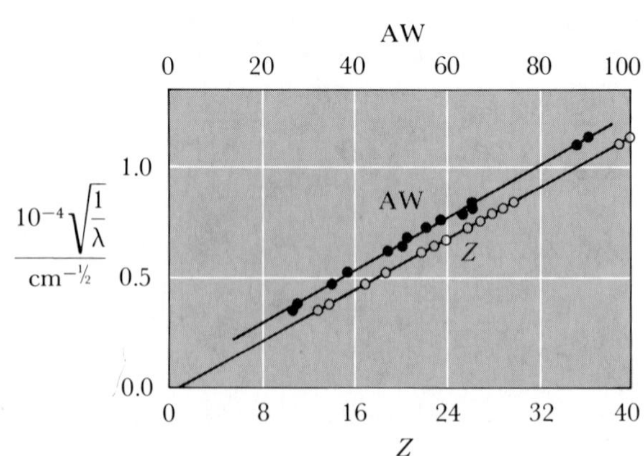

FIGURE 8–17
Plot of $\sqrt{1/\lambda}$ versus Z for K_α x radiation from different elements studied by Moseley. Notice that the correlation with atomic number is superior to that with atomic weight.

This equation has the form that we derived by use of the Rydberg formula and an effective nuclear charge, with $K = \sqrt{3R/4}$.

Let us use this equation to calculate the effective nuclear charge for the $n = 2$ to $n = 1$ transitions in cobalt and nickel:

$_{27}$Co, $\lambda = 1.793 \times 10^{-8}$ cm

$$Z'_{\text{eff}} = \sqrt{\frac{4}{3R\lambda}} = \sqrt{\frac{4}{3 \times 1.097 \times 10^5\ \text{cm}^{-1} \times 1.793 \times 10^{-8}\ \text{cm}}}$$

$$= 26.04$$

$_{28}$Ni, $\lambda = 1.662 \times 10^{-8}$ cm

$$Z'_{\text{eff}} = \sqrt{\frac{4}{3 \times 1.097 \times 10^5\ \text{cm}^{-1} \times 1.662 \times 10^{-8}\ \text{cm}}}$$

$$= 27.04$$

We see that this calculation places cobalt before nickel, and further, that the value of Z'_{eff} for each of these elements is 0.96 units less than its correct atomic number; this difference reflects screening. Moseley's work firmly established a basis for ordering the elements in the periodic table, and we know now that this order, based upon the square root of the reciprocal of the wavelength of characteristic x rays, is the order of increasing nuclear charge.

8–5 The Periodic Table and Trends in the Elements' Properties

A chemist, confronted with a question about the chemistry of a particular element, immediately will visualize where the element is in the periodic table. This knowledge alone makes it possible to answer some questions, and it helps in the rational pursuit of answers to more difficult questions. The reason for an immediate reliance on the periodic table is simple enough: it is the frame on which chemists systematize knowledge of the properties of the elements. This is because, as now we will learn, the position in the periodic table of an element and its electronic structure are directly related. In this section, we will introduce you to the table, but our extensive use of it will come in later chapters where we study the chemistry of selected elements.

THE PERIODIC TABLE

With the background already acquired, we readily can understand the relationship between the arrangement of elements in the periodic table and their electronic structures. However, these relationships were not obvious to chemists of the early nineteenth century. Rational classifications of the elements had to wait until a large number of elements was known (60 elements were identified by the early 1860s). Reasonably accurate atomic weights also were necessary, since early classifications were based upon atomic weights. Early attempts at classifying the known elements succeeded in grouping together elements with similar properties (e.g., between 1817 and 1829 DÖBEREINER developed his triads, which included the series: lithium, sodium, and potassium; calcium, strontium, and barium; chlorine, bromine, and iodine; and sulfur, selenium; and tellurium). However, the precursor to the modern periodic table was developed in 1869 by MENDELEEV in Russia; MEYER in Germany developed a similar table at approximately the same time. In an improved table that Mendeleev published in 1871, he predicted the properties of three missing elements. These elements, scandium, germanium, and gallium, were discovered within the next fifteen years, and the relationship of the properties of these elements (atomic weight and characteristic valence) with the predictions of Mendeleev convinced chemists of that day of the validity of the periodic table. Chemists of the present day do not need to be convinced of the usefulness of the periodic table; they know that an element's electronic structure, which determines its chemical properties, is related to its position in the table.

Elements in a particular column of the long form of the periodic table (Figure 8–18) have analogous electronic structures. The principal quantum number of the valence shell increases by one unit for each successive element in a column, but occupancy of the valence-shell orbitals is the same for elements in each column. The electrons of the occupied valence-shell orbitals as well as the unoccupied valence-shell orbitals are involved in chemical combination. This leads to similarities in the chemistry of elements in each particular column.

In corresponding segments of each row of the periodic table, orbitals belonging to a particular subshell (i.e., a particular value of l) are being filled as follows:

1. in the two-element segment of each row containing the columns headed by hydrogen and beryllium, the s orbital of the valence shell is being filled;
2. in the six-element segment of each row containing the columns headed by the sequence from boron through fluorine and neon, the three p orbitals of the valence shell are being filled;
3. in the ten-element segment of each row containing the columns headed by scandium through zinc, the five d orbitals of the shell beneath the valence shell are being filled; and
4. in the two rows lanthanum through lutecium and actinium through lawrencium there are fourteen-element segments in which the seven f orbitals in the shell two shells beneath the valence shell are being filled.

Thus a glance at the periodic table will tell you which subshell is being filled in each region of the table. This is summarized in Figure 8–19a.

The relationships between electronic structure and position in the periodic table will help you correlate the chemical properties of the elements

IA	IIA	IIIB	IVB	VB	VIB	VIIB	VIIIB			IB	IIB	IIIA	IVA	VA	VIA	VIIA	VIIIA
1 H																	2 He
3 Li	4 Be											5 B	6 C	7 N	8 O	9 F	10 Ne
11 Na	12 Mg											13 Al	14 Si	15 P	16 S	17 Cl	18 Ar
19 K	20 Ca	21 Sc	22 Ti	23 V	24 Cr	25 Mn	26 Fe	27 Co	28 Ni	29 Cu	30 Zn	31 Ga	32 Ge	33 As	34 Se	35 Br	36 Kr
37 Rb	38 Sr	39 Y	40 Zr	41 Nb	42 Mo	43 Tc	44 Ru	45 Rh	46 Pd	47 Ag	48 Cd	49 In	50 Sn	51 Sb	52 Te	53 I	54 Xe
55 Cs	56 Ba	57-71	72 Hf	73 Ta	74 W	75 Re	76 Os	77 Ir	78 Pt	79 Au	80 Hg	81 Tl	82 Pb	83 Bi	84 Po	85 At	86 Rn
87 Fr	88 Ra	89-103	104 Rf	105 Ha													

57 La	58 Ce	59 Pr	60 Nd	61 Pm	62 Sm	63 Eu	64 Gd	65 Tb	66 Dy	67 Ho	68 Er	69 Tm	70 Yb	71 Lu
89 Ac	90 Th	91 Pa	92 U	93 Np	94 Pu	95 Am	96 Cm	97 Bk	98 Cf	99 Es	100 Fm	101 Md	102 No	103 Lr

FIGURE 8–18
The periodic table.

and make predictions regarding their properties. We will try to discover reasons for the successes and failures of simple predictions. You already have seen an instance where a simple prediction based upon position in the periodic table fails. With the order of electron affinities

$$EA(\mathrm{Cl}) > EA(\mathrm{Br}) > EA(\mathrm{I})$$

it would seem reasonable to predict

$$EA(\mathrm{F}) > EA(\mathrm{Cl})$$

Before the electron affinity for fluorine had been measured accurately, it had been predicted that it would be greater than that for chlorine. But this is not correct:

$$EA(\mathrm{F}) = 327.8 \text{ kJ mol}^{-1} \qquad EA(\mathrm{Cl}) = 348.7 \text{ kJ mol}^{-1}$$

However, if knowledge that chlorine, bromine, and iodine form the compounds $MgCl_2$, $MgBr_2$, and MgI_2 leads you to predict the existence of MgF_2, you will be completely correct. We will see in an abundance of predictions based upon the periodic table many which are fulfilled and others which fail.

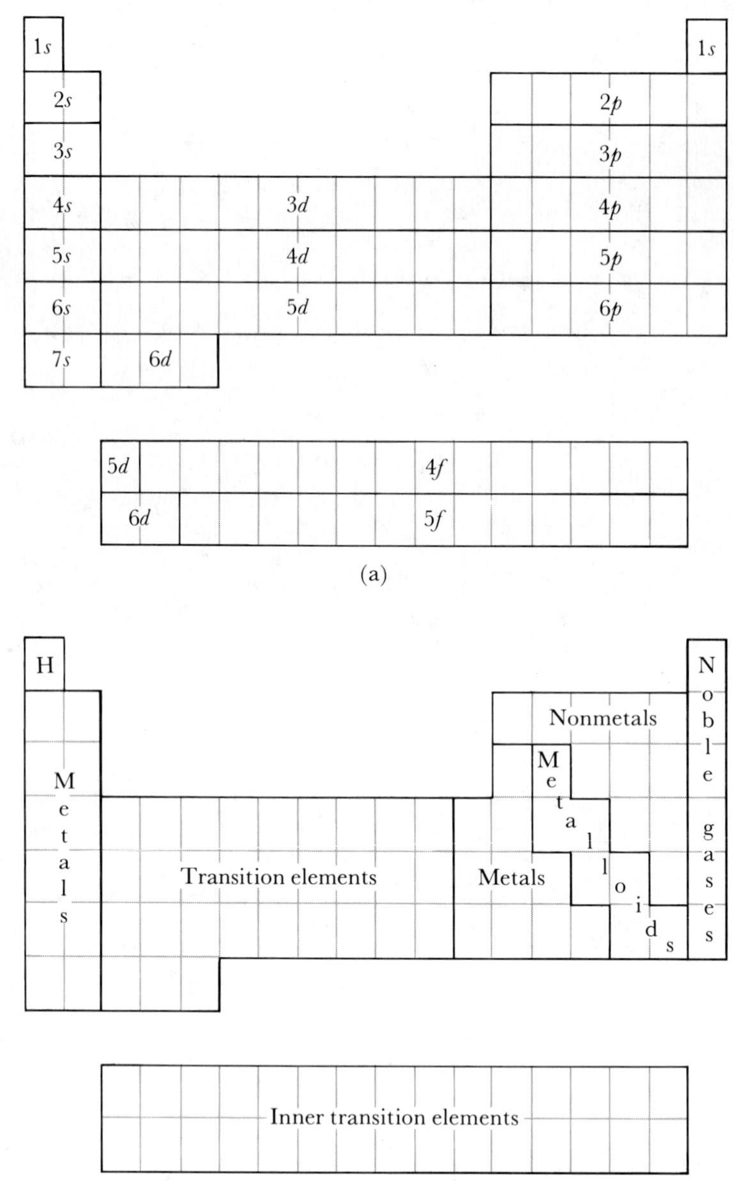

FIGURE 8–19
The periodic table. (a) Table showing what subshell is being filled as the table is developed from $Z = 1$ to $Z = 105$. (b) Classification of the elements:

Hydrogen (by itself, but a nonmetal)

Nontransition elements—metals, metalloids, nonmetals, and noble gases

Transition elements

Inner transition elements

As a guide in thinking about chemical and physical properties of the elements, smooth trends in columns of the periodic table are anticipated. For many properties smooth trends are observed, but for others apparent anomalies exist.

An example of the former type of correlation is the first ionization energy (IE_1) of the alkali metal atoms:

Li	520.1 kJ mol^{-1}
Na	495.8 kJ mol^{-1}
K	418.8 kJ mol^{-1}
Rb	402.9 kJ mol^{-1}
Cs	375.7 kJ mol^{-1}

This is a series in which there is an increase in the principal quantum number of the shell from which the electron is being removed, and an increase in the effective nuclear charge. Calculations of the effective nuclear charge have been presented already for lithium ($Z_{eff} = 1.26$) and sodium ($Z_{eff} = 1.84$). The corresponding calculation for cesium (in which $n = 6$ for the valence shell) is

$$Z_{eff} = 6 \times \sqrt{\frac{375.7 \text{ kJ mol}^{-1}}{1312 \text{ kJ mol}^{-1}}} = 3.21$$

This calculation shows how the less-than-perfect screening causes the effective nuclear charge to increase with increasing value of n for the valence shell. These two trends in going from lithium to cesium—a 9-fold change in n^2 ($2^2 = 4$ to $6^2 = 36$), and a 6.48-fold change in Z_{eff}^2 [$(1.26)^2 = 1.59$ to $(3.21)^2 = 10.30$]—operate in opposing ways to decrease the first ionization energy in going from lithium to cesium by a factor of 1.39 ($6.48/9 = 1/1.39$).

A property that does not change smoothly in going down a column of the periodic table is the enthalpy change for reaction of the alkali metals with acid at 298.2 K:

Reaction	ΔH^0 / kJ mol^{-1}
$Li(s) + H^+(aq) = Li^+(aq) + \frac{1}{2}H_2(g)$	−278.7
$Na(s) + H^+(aq) = Na^+(aq) + \frac{1}{2}H_2(g)$	−239.7
$K(s) + H^+(aq) = K^+(aq) + \frac{1}{2}H_2(g)$	−251.0
$Rb(s) + H^+(aq) = Rb^+(aq) + \frac{1}{2}H_2(g)$	−246.4
$Cs(s) + H^+(aq) = Cs^+(aq) + \frac{1}{2}H_2(g)$	−247.9

Here there is no trend in the values. An important point is illustrated by these examples. There is a smooth trend in values of the first ionization energy of the alkali metals, in spite of the fact that two factors act in opposition (increasing n and increasing Z_{eff}) as one goes down the column of the periodic table. The process of an alkali metal dissolving in acid is enormously more complex, and the net result of various factors is the seemingly chaotic pattern of the values of ΔH. In our discussion of the chemistry of the various elements, we will make valuable use of the periodic table, but this example will prepare you to use caution in making quantitative predictions based simply on the positions in the periodic table of the elements under consideration.

At this point, we also will recognize the placement in the periodic table of the broad classes of elements, the nontransition elements, which include metals, metalloids, nonmetals, and noble gases, and the transition elements, all metals. The transition elements sometimes are divided into transition elements and inner transition elements; this subdivision of the periodic table is shown in Figure 8–19b. The nontransition elements have either completely filled or completely empty subshells of d and f orbitals. The transition elements have partially filled subshells of d or f orbitals in either the free element or in its compounds. Later chapters will show the usefulness of these classifications.

Biographical Notes

JOHANN J. BALMER (1825–1898), a Swiss mathematician and physicist, was one of the first to work on the spectra of atoms. It was his studies which did much to provide experimental support for Bohr's theory.

NIELS BOHR (1885–1962), a Danish physicist, made epic contributions to atomic physics. After receiving his education in Denmark, he spent two years at Manchester and then returned to Denmark to become Professor of Theoretical Physics at the University of Copenhagen. His Institute for Theoretical Physics in Copenhagen (founded in 1920) was a center for research in atomic and nuclear physics. Bohr received the Nobel Prize in Physics in 1922.

JAMES CHADWICK (1891–1974), a British physicist, was associated with Rutherford in the Cavendish Laboratory at Cambridge, where he participated in many pioneering studies of nuclear transmutation. He received the Nobel Prize in Physics in 1935 for his discovery of the neutron.

CLINTON J. DAVISSON (1881–1958) was educated at Chicago (B.S., 1908) and Princeton (Ph.D., 1911). He was associated with Bell Telephone Laboratories from 1917 to 1946. He received the Nobel Prize in Physics in 1937 for his work on electron diffraction.

LOUIS DE BROGLIE (1892–), a French physicist, was associated with the Sorbonne from 1926–1962. He received the Nobel Prize in Physics in 1926 for his proposal that matter has both wave and particle properties.

PAUL A. M. DIRAC (1902–) is a British physicist who formulated the mathematical theory to describe the consequences of relativity in quantum mechanics. For this work he received (with E. Schrödinger) the Nobel Prize in Physics in 1933. Professor Dirac was educated at Bristol and Cambridge, where he became Professor of Mathematics in 1932.

JOHANN W. DÖBEREINER (1780–1849) was Professor of Pharmacy and Chemistry in the University of Jena. In addition to helping classify the elements, he invented a lamp that used the catalytic effect of platinum in the reaction of hydrogen and oxygen (Döbereiner's lamp).

ALBERT EINSTEIN (1879–1955) was a German theoretical physicist who developed the theory of relativity during the first two decades of this century. After many appointments in Germany, Switzerland, and Holland, he came to the United States in 1933, where he was associated with the Institute for Advanced Studies in Princeton. He received the Nobel Prize in Physics in 1921 for his interpretation of the photoelectric effect.

LESTER H. GERMER (1896–1971) was educated at Cornell University (A.B., 1917) and Columbia University (Ph.D., 1927). He was associated with Bell Telephone Laboratories from 1925 to 1953.

SAMUEL GOUDSMIT (1900–1978) was born in the Netherlands. He received his Ph.D. from the University of Leyden in 1927. With Uhlenbeck, he developed the concept of electron spin. In the United States, to which he came in 1927, he was Professor of Physics at the University of Michigan from 1927 to 1946. In 1948 he moved to Brookhaven National Laboratory.

WERNER HEISENBERG (1901–1976), a German physicist, received his D.Phil. degree from the University of Munich in 1923. His paper on the uncertainty principle was published in 1927; for this work he received the Nobel Prize in Physics in 1932. He was Professor of Physics at universities in Leipzig, Berlin, Göttingen, and Munich.

F. H. HUND (1896–), a German physicist, was professor of theoretical physics at Göttingen University. His interests were in areas of quantum theory applied to atoms, molecules, and solids.

THEODORE LYMAN (1874–1954) was a Professor of Physics at Harvard. Between 1906 and 1914 he discovered the lines in the ultraviolet region of the spectrum of atomic hydrogen that arise in transitions to a final state with $n = 1$.

DMITRII I. MENDELEEV (1834–1907) was Professor of Chemistry in the University of St. Petersburg and later the director of the Bureau of Weights and Measures. His genius was in leaving blank spaces in his periodic table in instances where the next known element had properties different from those of the element above it or below it.

LOTHAR J. MEYER (1830–1895) was Professor of Chemistry at Tübingen. In addition to this work on the periodic classification of the elements, he is noted for early studies on the action of carbon monoxide on blood.

ROBERT A. MILLIKAN (1868–1953), an American physicist, was associated with the University of Chicago from 1896 to 1921; from 1921 until his retirement in 1945, he was chief executive officer of the California Institute of Technology. For his measurement of the charge on the electron, he received the Nobel Prize in Physics in 1923. He also verified Einstein's explanation of the photoelectric effect.

HENRY G. J. MOSELEY (1887–1915), a British physicist, was educated at Oxford and studied with Rutherford at Manchester where he started his work on the characteristic x rays of the elements. He returned to Oxford in 1914 to continue this work. He was killed in battle in World War I.

FRIEDRICH PASCHEN (1865–1940), a German physicist, studied the fine structure of atomic spectral lines for evidence of effects due to relativity.

WOLFGANG PAULI (1900–1958), an Austrian physicist, received the Nobel Prize in

Physics in 1945. In addition to proposing the exclusion principle, which bears his name, he postulated the existence of the neutrino (see Chapter 22).

MAX K. E. L. PLANCK (1858–1947) was a German physicist who did much to develop quantum theory. He also was interested in thermodynamics, particularly in the concept of entropy. He received the Nobel Prize in Physics in 1918.

ERNEST RUTHERFORD (1871–1937) was born in New Zealand and received his undergraduate training there. After graduate work at Trinity College, Cambridge, he taught physics at McGill University, Manchester University, and Cambridge University. He was a pioneer in the study of radioactivity; for this he received the Nobel Prize in Chemistry in 1908. His associate, J. H. W. Geiger (1882–1945), was a German physicist who worked in Manchester from 1906 to 1912. He did much to develop instruments to detect and measure alpha particles and beta particles (electrons); the Geiger–Müller counter is well known.

JOHANNES R. RYDBERG (1854–1919), a Professor of Physics in the University of Lund (Sweden), was the first (1890) to recognize the usefulness of expressing the "positions" of spectral lines in wave numbers ($1/\lambda$). The proportionality constant R is named after him.

ERWIN SCHRÖDINGER (1887–1961), an Austrian physicist, was associated with a number of universities in Austria, Switzerland, Germany, and Ireland. He received the Nobel Prize in Physics (jointly with P. A. M. Dirac) in 1933.

GEORGE P. THOMSON (1892–1975), a British physicist, the son of J. J. Thomson, was educated at Trinity College, Cambridge. He was associated with the universities at Cambridge and Aberdeen (Scotland), and with Imperial College. In 1937, he shared the Nobel Prize in Physics with Davisson for experiments demonstrating the wave nature of the electron.

J. J. THOMSON (1856–1940), a British physicist, was associated with Cambridge University from 1876, when he went there as a student, until he died. He was made Professor of Physics in 1884. J. J. Thomson received the Nobel Prize in Physics in 1906.

GEORGE E. UHLENBECK (1900–   ) was born in the Dutch East Indies. He received his Ph.D. from the University of Leyden in 1927. While a student, he developed the concept of electron spin with Goudsmit (also a student) in 1925. He came to the United States in 1927. He has served on the faculties of the Universities of Michigan and Utrech and of Rockefeller University.

Problems and Questions

8–1 What is the ratio of the masses of the electron and proton? (Give your answer to 5 significant figures.)

8–2 What is the magnitude of the esu of charge in SI?

8–3 Calculate the energy of a photon of light having each of the following wavelengths: 100 nm, 500 nm, and 2500 nm. Which of these wavelengths is in the visible region of the spectrum?

8–4 Light of what wavelengths consists of photons with the energies 200 kJ mol^{-1}, 500 kJ mol^{-1}, and 800 kJ mol^{-1}?

8–5 The bright yellow light emitted by a sodium vapor lamp has wavelengths 589.6 nm and 589.0 nm. What are the energies of these transitions in kJ mol^{-1}?

8–6 The range of frequencies shown on the dial of your FM radio is 88 to 108, which means $88 \times 10^6\ s^{-1}$ to $108 \times 10^6\ s^{-1}$. What wavelengths correspond to these two limits?

8–7 The range of frequencies shown on the dial of your AM radio is 540 to 1600, which means $540 \times 10^3\ s^{-1}$ to $1600 \times 10^3\ s^{-1}$. What wavelengths correspond to these two limits?

8–8 In Chapter 1, it was stated that the meter was defined as 1,650,763.73 times the wavelength of radiation of a particular line in the spectrum of Kr (krypton-86; specification of the mass is necessary if the highest accuracy is needed. This is a point that we will not develop in our discussion of allowed energy states of atoms.) What is the frequency of this radiation?

8–9 What are the kinetic energies of electrons that have de Broglie wavelengths of 2.00 Å, 1.00 Å, and 0.20 Å?

8–10 What is the kinetic energy of a helium atom that has a de Broglie wavelength of 1.00 Å?

8–11 What is the maximum wavelength of light that will cause emission of electrons from metallic calcium? If light of wavelength 300 nm impinges on a surface of calcium, what is the energy of an emitted electron?

8–12 Using the equation for the radius of circular orbits in the Bohr theory, calculate the value of r for $n = 1$, 2, and 3 for $Z = 1$. (Keep track of dimensions in this calculation.)

8–13 A line in the Lyman series for atomic hydrogen has the wavelength $\lambda = 9.723 \times 10^{-6}$ cm. What is the value of n for the initial state in this transition?

8–14 The series of spectral lines in the emission spectrum of atomic hydrogen in which the final state has $n = 3$ is called the Paschen series. What wavelengths of radiation are associated with transitions having the four lowest energies in this series?

8–15 The spectral lines in the Lyman series for atomic hydrogen converge at a short wavelength limit. (See Figure 8–4.) Think about the meaning of this limit, and calculate the wavelength of light associated with this limit from another item of data given in this chapter. What are the short wavelength limits for the Balmer and Paschen series of spectral lines?

8–16 What are the ionization energies for the ground states of the one-electron atoms Li^{2+}, Be^{3+}, and B^{4+}?

8–17 The Lyman spectral lines for atomic hydrogen are in the far ultraviolet region of the electromagnetic spectrum, but some lines of the Balmer series are in the visible region. Spectral lines for He^+ have shorter wavelengths than do the corresponding lines in the spectrum of H. What is the smallest value of n for which the transition of an electron from the $n + 1$ shell to the n shell in He^+ corresponds to visible light ($\lambda > 400$ nm)?

8–18 A point has the Cartesian coordinates $x = 1.1$ cm, $y = 1.5$ cm, $z = 3.2$ cm. What are the values of the polar coordinates r, θ, and ϕ for the point? (See Figure 8–8 for the relation of the Cartesian coordinates and the polar coordinates.)

8–19 What is the average value of r (the distance between the nucleus and the electron) for an electron in a hydrogen atom with the sets of quantum numbers: $n = 2$, $l = 0$ and 1; $n = 3$, $l = 0$, 1, and 2? Compare these average values of r with the radii of Bohr orbits with $n = 2$ and 3. (See problem 8–12.)

8–20 The p_x orbital has a maximum probability along the x axis. Using the equations for $\psi(p_x)$ given in Table 8–3, calculate the relative values of

$\psi^2(p_x)$ at $\phi = 0°$, $\theta = 90°$ (along the x axis), at $\phi = 30°$, $\theta = 30°$, and at $\phi = 60°$, $\theta = 5°$ (almost along the z axis).

8–21 The electronic configurations of the five lowest energy states of the helium atom were given in the text. Suggest the electronic configurations for the six lowest energy states of the beryllium atom.

8–22 The wavelength of the K_α line in the x-ray spectrum of zinc is 1.439 Å. What is the frequency of this radiation? Calculate the effective nuclear charge that governs this transition.

8–23 Although we have focused attention on x radiation arising in transitions between the $n = 2$ and $n = 1$ shells, x-ray spectra contain lines caused by other transitions. What are the energies of x-ray photons resulting from $n = 3$ to $n = 2$ transitions in copper, $\lambda = 15.22$ Å; silver, $\lambda = 4.71$ Å; and gold, $\lambda = 1.46$ Å?

8–24 For a one-electron atom, what orbitals have the same energy as that of the 3s orbital? Are the energies of these orbitals the same for atoms with more than one electron? Explain.

8–25 We have seen that the process of adding an electron to the neutral nitrogen atom is analogous to the reverse of the ionization of neutral oxygen. The corresponding process of adding an electron to a beryllium atom involves the ionization of what neutral atom? What can be said about this correspondence?

8–26 Scan the atomic weights of the elements arranged in order of increasing atomic number and find the instances, other than cobalt and nickel, in which an element with higher atomic number has a lower atomic weight.

8–27 The first ionization energy of sulfur is smaller than that of phosphorus (1000 kJ mol^{-1} compared to 1063 kJ mol^{-1}). Explain this.

8–28 Find the maximum wavelength of light that can detach an electron from each of the following gaseous anions: $Na^-(g)$, $O^-(g)$, $Cl^-(g)$. (It is by such measurements on gaseous anions using laser radiation that the most accurate values of electron affinity are determined.)

8–29 The ionization energies of potassium and rubidium are 418.8 kJ mol^{-1} and 402.9 kJ mol^{-1}, respectively. Calculate the effective nuclear charge governing each of these processes. Using the values of this quantity given in the text for the other alkali metals, make plots of Z_{eff}^2 versus Z, n^{-2} versus Z, and Z_{eff}^2/n^2 versus Z for the alkali metals.

8–30 Considering only the alkali metals and halogens, the reaction

$$M(g) + X(g) = M^+(g) + X^-(g)$$

is most endothermic for which M and X? Is the reaction exothermic for any M–X combination? No

8–31 Correlate the discussion of this chapter regarding discontinuities in the plots of IE versus Z with the commonly stated generalizations that atoms with (a) filled electron shells, (b) filled electron subshells, and (c) half-filled electron subshells have particularly stable electronic structures.

8–32 What three elements have the largest ionization energies? What three elements have the smallest ionization energies?

8–33 Examine Figure 8–11 and notice the elements with ionization energies approximately equal to that for hydrogen. What semiquantitive statement can you make regarding the effective nuclear charge governing the ionization process for these elements?

9

Binary Ionic Compounds

9–1 Introduction

The subject of this chapter is the large class of chemical substances known as ionic compounds. A familiar ionic compound is sodium chloride, NaCl, ordinary salt. It is formed from its constituent elements in an exothermic reaction:

$$\mathrm{Na}(s) + \tfrac{1}{2}\mathrm{Cl_2}(g) = \mathrm{NaCl}(s) \qquad \Delta H^0 = -411.0\ \mathrm{kJ\ mol^{-1}} \qquad (298.2\ \mathrm{K})$$

The structure of this compound, established by x-ray diffraction, was shown in Figure 5–1. Its properties, including the standard enthalpy of formation just given, are consistent with this structure, made up of equal numbers of sodium ions (Na^+) and chloride ions (Cl^-). A simple explanation for the formation of a compound with this particular composition (NaCl), rather than others, for example, Na_2Cl or $NaCl_2$, is that in Na^+Cl^- both the sodium ion (Na^+) and chloride ion (Cl^-) have the electronic structure of a noble gas.[1] In Chapter 8, we learned about the energy changes associated with forming these ions, Na^+ and Cl^-, in the gas phase. In this chapter, we will dissect the process of forming crystalline ionic compounds such as $Na^+Cl^-(s)$ into a series of simple steps, including two which were studied in Chapter 8.

[1] In writing the formula of an ionic compound, one usually does not show the charges on the constituent ions. However, if you wish to emphasize the ionic nature of the compound, you may show the charges, as above. And if you are considering the separate ions, you must indicate the charges.

Like sodium ion and chloride ion, the constituent ions in many ionic compounds have the electronic structures of noble gases. Ionic compounds are known that contain:

Group IA metal cations, M^+ (e.g., Na^+, $K^+ \cdots$)
M (noble-gas structure; ns^1) $\rightarrow M^+$ (noble-gas structure) $+ e^-$
Group IIA metal cations, M^{2+} (e.g., Mg^{2+}, $Ba^{2+} \cdots$)
M (noble-gas structure; ns^2) $\rightarrow M^{2+}$ (noble-gas structure) $+ 2e^-$
Group IIIA metal cations, M^{3+} (e.g., Al^{3+})
M (noble-gas structure; ns^2, np^1) $\rightarrow M^{3+}$ (noble-gas structure) $+ 3e^-$
Group VIIA nonmetal anions, X^- (e.g., F^-, Cl^-, $\cdots$)
$X(ns^2, np^5) + e^- \rightarrow X^-$ (noble-gas structure)
Group VIA nonmetal anions, X^{2-} (e.g., O^{2-}, S^{2-})
$X(ns^2, np^4) + 2e^- \rightarrow X^{2-}$ (noble-gas structure)
Group VA nonmetal anions, X^{3-} (e.g., N^{3-}, P^{3-})
$X(ns^2, np^3) + 3e^- \rightarrow X^{3-}$ (noble-gas structure)

The charge on each of these monatomic ions is known as the ionic valence or the oxidation number,[2] and it is clear from these examples that the principal ionic valence that nontransition elements exhibit follows directly from the electronic configuration. In an ionic compound, the molar proportions of cation and anion are such as to give an electrically neutral compound. This is illustrated by the formulas of binary ionic compounds of sodium, magnesium, and aluminum with nitrogen, oxygen, and fluorine:

	F^-	O^{2-}	N^{3-}
Na^+	NaF	Na_2O	Na_3N
Mg^{2+}	MgF_2	MgO	Mg_3N_2
Al^{3+}	AlF_3	Al_2O_3	AlN

In this introduction no mention has been made of ionic compounds containing ions of charge ± 4. Although some compounds are known in which the assignment of these ionic valences is reasonable (e.g., thorium tetrafluoride, ThF_4, contains Th^{4+}, and aluminum carbide, Al_4C_3, contains C^{4-}), monatomic ions with these ionic valences ($4+$ and $4-$) are not common.

The formation of ionic compounds from their constituent elements illustrates the dramatic change in properties that can occur in chemical reactions. Consider the formation of sodium chloride, already mentioned:

$$Na(s) + \tfrac{1}{2}Cl_2(g) = NaCl(s)$$

$Na(s)$: a soft metal ($d = 0.97$ g cm^{-3} at 20°C, 1 atm)

mp 97.8°C bp 892°C

$Cl_2(g)$: a yellowish gas ($d = 2.95 \times 10^{-3}$ g cm^{-3} at 20°C, 1 atm)

mp -101.0°C bp -34.6°C

[2]The term "oxidation number" (or oxidation state) has much broader usage, as we will see.

$NaCl(s)$: a hard, white solid ($d = 2.16\ g\ cm^{-3}$ at 20°C, 1 atm)

mp 801°C bp 1413°C

Metallic sodium conducts electricity in the solid and liquid states; electrons are the current carriers. Chlorine does not conduct electricity in either of these states. Sodium chloride does not conduct electricity in the solid state, but it does in the liquid state, in which the ions Na^+ and Cl^- are the current carriers.

In much of the discussion in this chapter, we will picture the ionic crystal as being made up of spherical ions of particular sizes. We learned in Chapter 8 that the sum of the angular probability distributions for one *s* orbital plus the three *p* orbitals (which are completely filled in an ion with a noble-gas electronic configuration) is a function that does not contain θ or ϕ. Therefore such ionic species are spherical. The radial probability distributions derived from quantum mechanics do not support the idea that isolated ions are hard spheres with a particular size. But in the lattice of an ionic compound, there are particular interatomic separations, as was illustrated for sodium chloride in Figure 5–7. The shortest sodium ion–chloride ion distance in this lattice is 2.814 Å. If we assume that these ions are in contact, this distance is the sum of the radii,

$$(r_{Na^+}) + (r_{Cl^-}) = 2.814\ \text{Å}$$

but this equation does not allow us to evaluate the individual radii. Much effort over a period of more than one half century has not solved this problem satisfactorily. Some approaches have been strictly empirical (e.g., finding the set of radius values that allows accurate prediction of interionic distances in many crystals) and other approaches have used concepts from quantum mechanics and experimental observations of electron distribution in atoms.

Table 9–1 presents a set of ionic radius values for alkali metal ions and halide ions; these values are the product of the semitheoretical approach.

TABLE 9–1
Values of Ionic Radii

		$\frac{r}{\text{Å}}$
	H^-	1.3
Li^+		0.86
	F^-	1.19
Na^+		1.12
	Cl^-	1.70
K^+		1.44
	Br^-	1.87
Rb^+		1.58
	I^-	2.12
Cs^+		1.84

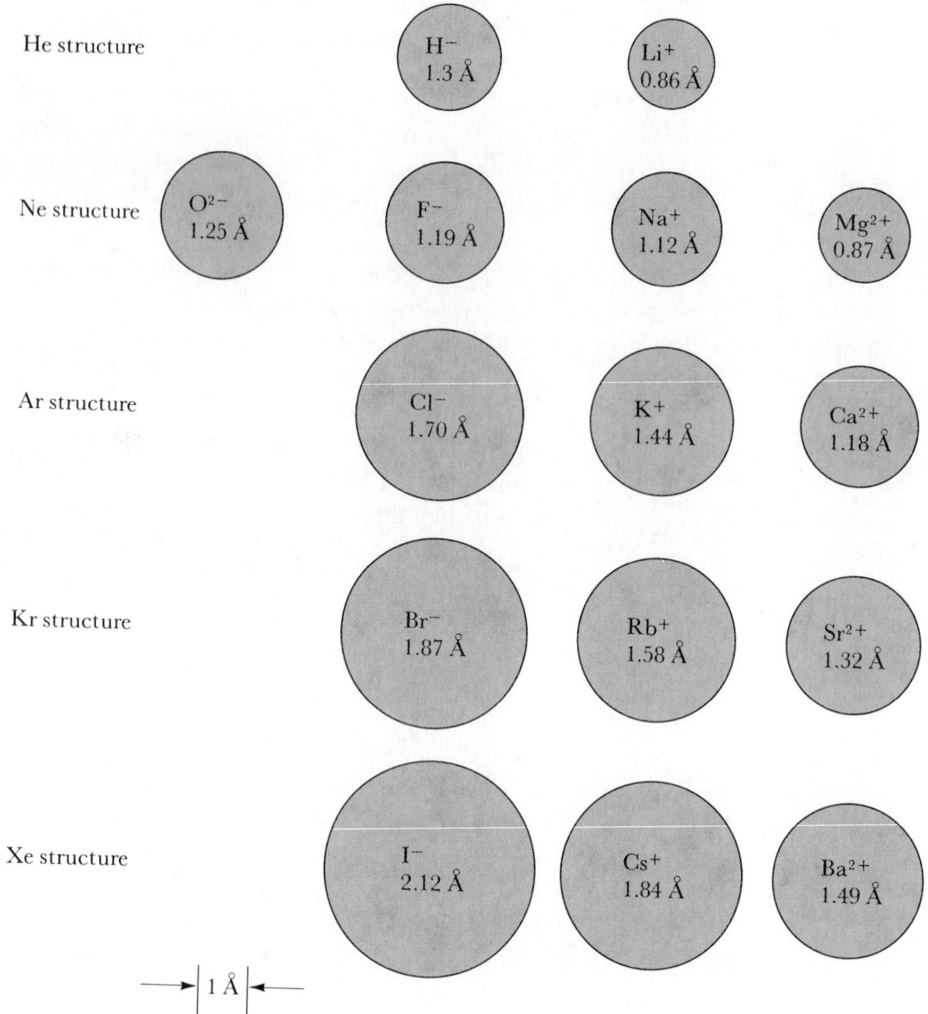

FIGURE 9–1
The size of ions with noble-gas electronic structures. Notice a contraction in size as nuclear charge increases for each electronic structure.

Because of the uncertainties already mentioned and others that will be mentioned later, one must be cautious in using these values in calculations. Nonetheless, the values do show some expected trends. The ionic radius increases in going down a column of the periodic table:

$$r_{F^-} < r_{Cl^-} < r_{Br^-} < r_{I^-}$$

$$r_{Li^+} < r_{Na^+} < r_{K^+} < r_{Rb^+} < r_{Cs^+}$$

and the increase in nuclear charge in going from a halide ion to the alkali metal ion of the same noble-gas structure causes the ionic radius to decrease:

$$r_{F^-} > r_{Na^+} \qquad r_{Cl^-} > r_{K^+}$$

$$r_{Br^-} > r_{Rb^+} \qquad r_{I^-} > r_{Cs^+}$$

For isoelectronic ions, the ion with the larger nuclear charge has the smaller radius because of the greater attraction between the nucleus and the imperfectly screened outermost electrons. These trends in ionic size for ions with noble-gas structures are shown in Figure 9–1.

9–2 The Ionic Crystal Lattice

In the sodium chloride lattice, shown in Figure 5–1, there is a regular arrangement of sodium ions (Na^+) and chloride ions (Cl^-). In this regular arrangement the energy of attraction of ions with unlike charge exceeds the energy of repulsion of ions with like charge, and this structure is very stable. The high melting point already mentioned is one manifestation of this stability. There are other possible arrangements of ions in ionic crystal lattices, and now we will consider some of these and also the factors that determine the most stable structure for a particular compounds.

COORDINATION NUMBERS IN IONIC LATTICES

The coordination number of an atom is the number of atoms with which it is in contact. This concept was introduced in our discussion of structures of solid elements in Chapter 5. We saw that the closest packing of identical spherical atoms gives each atom a coordination number of twelve. We also saw that poorly understood factors in the electronic structure of atoms of some metals (e.g., the alkali metals) cause these elements to crystallize in the body-centered cubic lattice with the coordination number of eight for each atom. In some elements (e.g., carbon), the electronic structure of the atom causes much smaller coordination numbers in the solid (four in diamond), for reasons we understand well. (These well-understood factors are related to the tendency for an atom to form directional covalent bonds, to be discussed in the next chapter.) The concept of coordination number will play an important part in our discussion of the structures of ionic solids.

We will focus attention on binary ionic compounds, ionic compounds containing two elements. The factors that determine the composition and structure of an ionic solid, M_aX_b, containing the ions M^{m+} and X^{x-} are:

1. The magnitude of the charges, $m+$ and $x-$, determine the empirical composition of the compound. Because the compound is electrically neutral,

$$a(m+) + b(x-) = 0 \quad \text{or} \quad \frac{a}{b} = \frac{x}{m}$$

   This point already has been illustrated.
2. The relative sizes of the ions determine the coordination numbers in the solid. This point will be discussed in terms of the ratio of the radii of the smaller ion (r) and the larger ion (R), r/R. We will be able to explain why six is the coordination number of sodium ion and of chloride ion in sodium chloride.

Ions of unlike charge attract one another, and those of like charge repel one another. The potential energy of coulombic interaction of two charges, Z_1 and Z_2, separated by a distance d, is given by the equation

$$U(d) = \frac{Z_1 Z_2 e^2}{d}$$

in which e is the charge on the electron. For real ions the energy of interaction consists of this coulombic energy plus contributions of the same type of attractive and repulsive interactions shown in Figure 4–16 for the potential energy of interaction of gas molecules. The net result is a curve for $U(r)$ as shown in Figure 9–2. In an ionic crystal the coordination number of each kind of ion tends to be a maximum, to make the attractive interaction energy as negative as possible without the repulsive interaction energy being prohibitively positive. If we think of the ions as having particular sizes, the repulsive interaction energy of ions of like charge becomes prohibitive when they touch, that is, when their distance of separation is the sum of their radii.

Using simple trigonometry, we will calculate the critical values of the radius ratio r/R for the coordination numbers that are common in ionic solids. These common coordination numbers are four, six, and eight, and

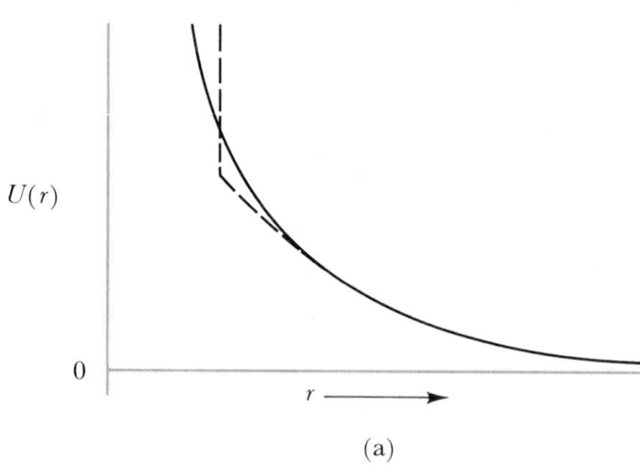

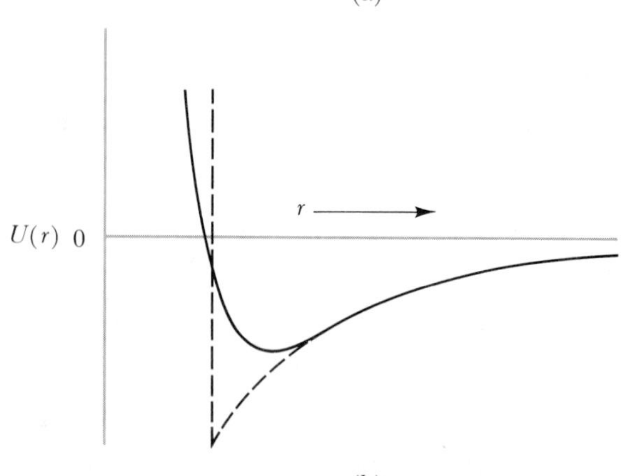

FIGURE 9–2
The potential energy of interaction of ions of charge Z_1 and Z_2 as a function of interatomic separation. (a) $Z_1 = 1+$, $Z_2 = 1+$ or $Z_1 = 1-$, $Z_2 = 1-$. (b) $Z_1 = 1+$, $Z_2 = 1-$. At long distances of separation, the interaction is coulombic; at short distances the repulsive interaction is governed by a term $U(r) \propto r^{-n}$, in which n is large ($n = 5$ to 12). The dashed lines show hypothetical hard-sphere ions.

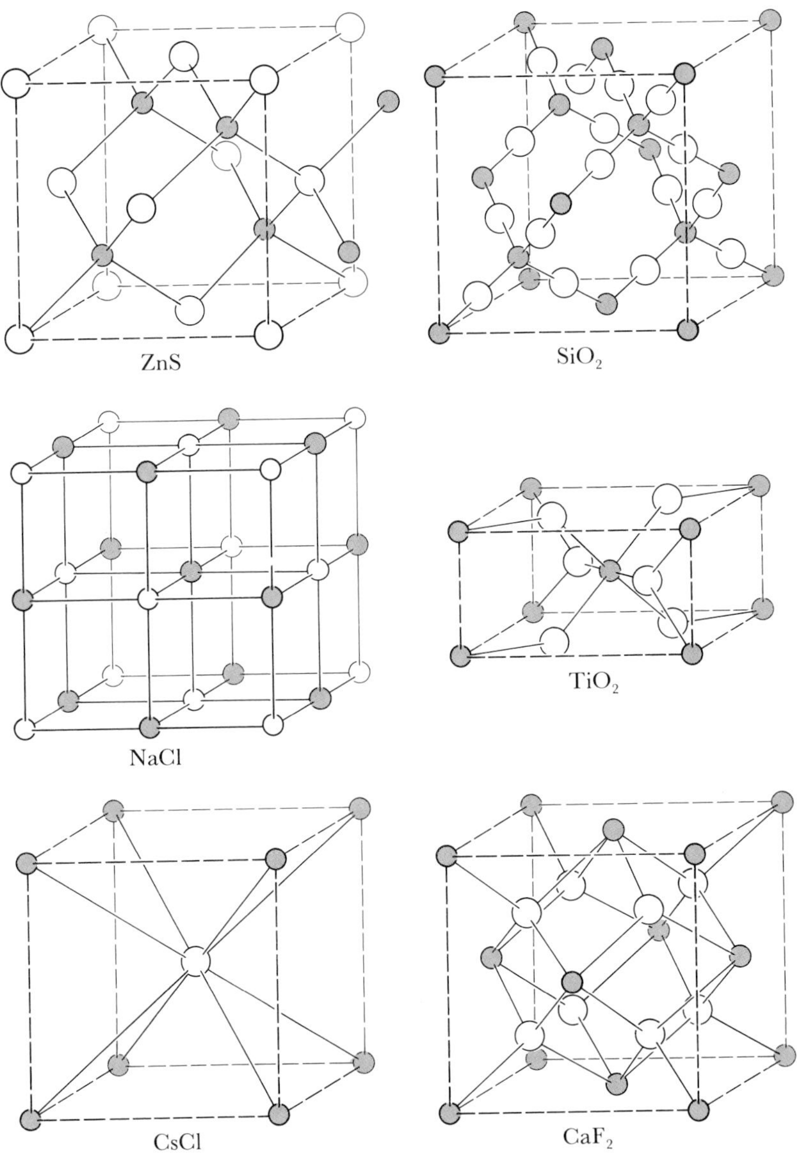

FIGURE 9–3
Ionic lattices. In the left column: MX compounds with coordination numbers of cation (and anion) being 4 (ZnS), 6 (NaCl), and 8 (CsCl). In the right column: MX_2 compounds with coordination numbers of cation being 4 (SiO_2), 6 (TiO_2) and 8 (CaF_2). In these MX_2 compounds, the coordination number of the anion is one half that of the cation.

they are exhibited by the cations in ZnS, NaCl, and CsCl shown in Figure 9–3. (These also are the coordination numbers of the anions in these solids, with one-to-one stoichiometry. More will be said about this.)

A cation with radius r will have the largest possible coordination number that allows contact between the cation and the surrounding anions with radius R *without* allowing the anions to touch. The critical value of

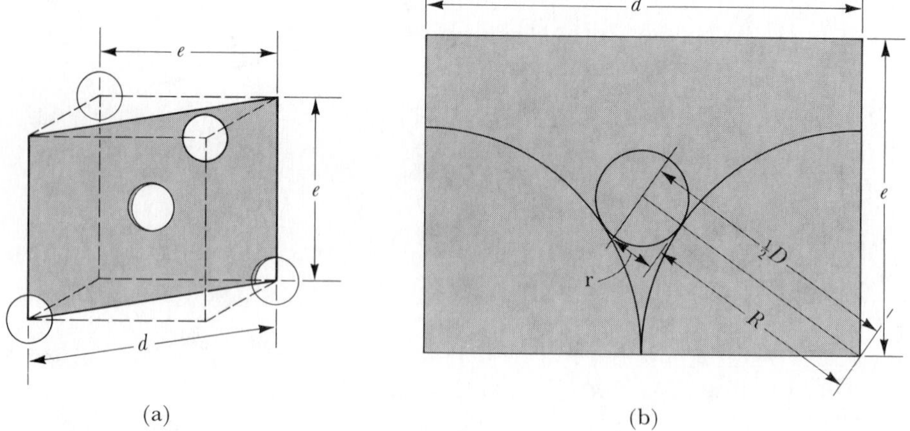

FIGURE 9–4
The coordination of four anions in contact with a cation. (a) tetrahedron of anions (of radius R) around a cation (of radius r). (b) The indicated vertical diagonal plane showing the anion–anion and cation–anion contact. Symbols e, d, and D have the same significance as in Figure 5–5.

r/R is that for which there is both cation–anion contact and anion–anion contact. Anions generally are larger than cations, but there are compounds in which the cation is larger. For such compounds, r is the anion radius and R is the cation radius, and the critical value of r/R is that for which there is both cation–anion contact and cation–cation contact. The common three-dimensional arrangements of anions are ones in which four anions define a tetrahedron, six anions define an octahedron, and eight anions define a cube. A cation occupies the empty space at the center of the polyhedron defined by these anions.

Critical Value of r/R for Coordination Number 4. Figure 9–4 shows an expanded view of a tetrahedral arrangement of anions around a cation. A cross section of this figure shows the ions in contact. The anions are in contact along a diagonal, d, of each face of the cube that is defined by the tetrahedron of anions. Therefore

$$d = 2R$$

and the sum of the cation radius (r) and anion radius (R) is one half the length of a body diagonal, D:

$$r + R = \tfrac{1}{2}D$$

But the body diagonal and face diagonal are related (see Figure 5–5):

$$D = \sqrt{\frac{3}{2}}\,d$$

therefore,

$$r + R = \frac{1}{2}\sqrt{\frac{3}{2}}\,d = \sqrt{\frac{3}{2}}\,R$$

$$\frac{r}{R} = \sqrt{\frac{3}{2}} - 1 = 0.225$$

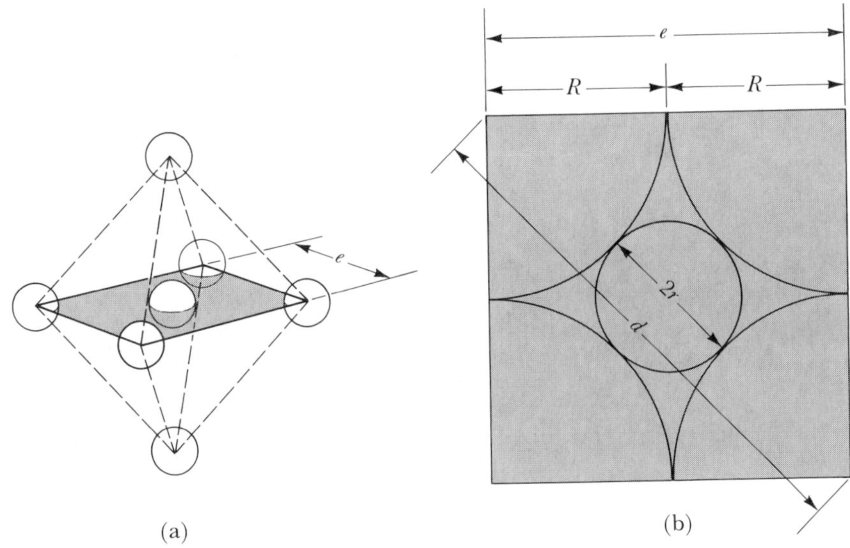

FIGURE 9–5
The coordination of six anions in contact with a cation. (a) An octahedron of anions (of radius R) around a cation (of radius r). (b) The indicated horizontal plane showing the anion–anion and cation–anion contact.

Critical Value of r/R for Coordination Number 6. Figure 9–5 shows an expanded view of an octahedron of anions surrounding a cation. A cross section of this arrangement shows the ions in contact. The anions are in contact along each edge of the octahedron. Four coplanar anions define a square, and the cation is in contact with anions along a diagonal of this square. Therefore we have the edge of the square, e,

$$e = 2R$$

and the diagonal, d,

$$d = 2r + 2R = \sqrt{2}e = 2\sqrt{2}R$$

$$\frac{r}{R} = \sqrt{2} - 1 = 0.414$$

Critical Value of r/R for Coordination Number 8. Figure 9–6 shows an expanded view of a cube of anions surrounding a cation. A cross section shows the ions in contact. In this structure the anions are in contact along each edge of the cube, e, and the cation and anions are in contact along the body diagonal, D,

$$e = 2R$$

$$D = 2r + 2R = \sqrt{3}e = 2\sqrt{3}R$$

$$\frac{r}{R} = \sqrt{3} - 1 = 0.732$$

We see in these calculations that the larger the value of r/R, the larger the possible coordination number. The situation is summarized in Table

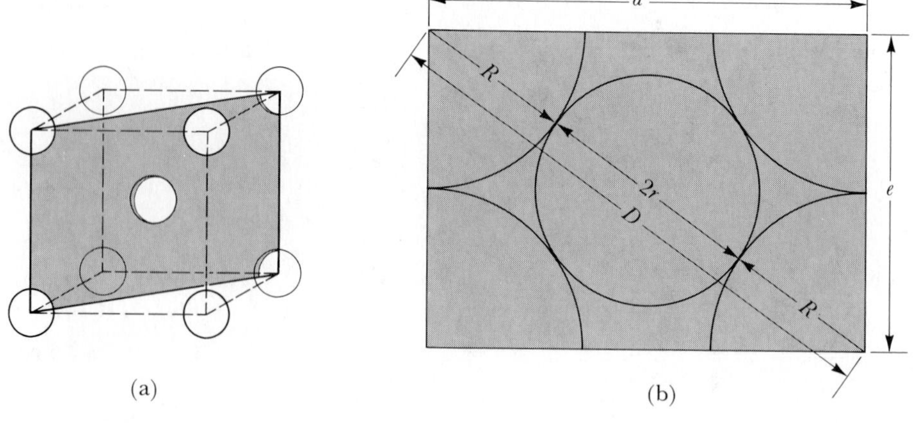

FIGURE 9–6
The coordination of eight anions in contact with a cation. (a) A cube of anions (of radius R) around a cation (of radius r). (b) The indicated vertical plane showing the anion–anion and cation–anion contact.

9–2. Structures of most binary ionic compounds can be explained in terms of the critical radius ratio values just calculated. An ionic compound adopts the lattice with the largest coordination number that does not force ions of the same charge to be in contact. Consider the cases of sodium chloride and cesium chloride, involving ions with the radii:

$$r_{Na^+} = 1.12 \text{ Å} \qquad r_{Cs^+} = 1.84 \text{ Å} \qquad r_{Cl^-} = 1.70 \text{ Å}$$

These values give the ratios:

$$\frac{r}{R} = \frac{r_{Na^+}}{r_{Cl^-}} = 0.66 \quad \text{and} \quad \frac{r}{R} = \frac{r_{Cl^-}}{r_{Cs^+}} = 0.92$$

For sodium chloride the value of the radius ratio r/R is in the range that allows six as the largest possible coordination number. This is the coordination number of sodium ion and chloride ion in sodium chloride (see Figure 9–3). For cesium chloride the cation is larger than the anion and the value of r/R is in the range that allows coordination numbers eight, the experimentally determined coordination number for the ions in cesium chloride (see Figure 9–3). However, the radius ratio values based upon ionic radii given in Table 9–1 do not provide infallible guidance in predicting whether an alkali metal halide has the sodium chloride or cesium chloride structures. For instance, RbI, with $r/R = 0.75$, has the sodium chloride structure under ordinary conditions. However, it does have the cesium chloride structure at high pressures ($P > 5000$ kg cm^{-2}). The approach we have employed is over-simplified, but it discloses one of the factors playing a role in determining the structures of ionic crystals.

In Figure 9–3, the MX compound with coordination number 4 is zinc sulfide, ZnS. The ionic radii for these ions are: Zn^{2+}, $r = 0.70$ Å, and

TABLE 9–2
Radius Ratio Values for Various Coordination Numbers[a]

	Coordination number					
	4		6		8	
$\frac{r}{R}$	Contacts +/−	+/+ or −/−	Contacts +/−	+/+ or −/−	Contacts +/−	+/+ or −/−
>0.732	yes	no	yes	no	yes	no
0.732	yes	no	yes	no	*yes*	*yes*
<0.732 >0.414	yes	no	yes	no	no	yes
0.414	yes	no	*yes*	*yes*	no	yes
<0.414 >0.225	yes	no	no	yes	no	yes
0.225	*yes*	*yes*	no	yes	no	yes

[a]Examples below the heavy line are those for which there is contact of ions with like charge, but no contact of ions with unlike charge; these structures are unstable.

S^{2-}, $r = 1.8$ Å. The ratio r/R is 0.39, which is between 0.414 and 0.225, the range that is consistent with coordination number 4. Figure 9–3 also shows the structures of a series of MX_2 compounds in which the cations have coordination numbers 4, 6, and 8. In each of these compounds the coordination number of the anion is one half the coordination number of the cation, that is, 2, 3, and 4, respectively. This relationship is generally true for ionic compounds with three-dimensional lattices:

> For an ionic compound M_aX_b, the ratio of the coordination numbers is
>
> $$\frac{\text{CN of } M^{m+}}{\text{CN of } X^{x-}} = \frac{b}{a}$$
>
> In an ionic compound M_aX_b there are (b/a) times as many X^{x-} ions as M^{m+} ions, and this stoichiometric relationship requires that the coordination number of M^{m+} must be (b/a) times the coordination number of X^{x-}.

The structures shown in Figure 9–3 of MX compounds in which the coordination number of the cation is 4 or 6 have an arrangement of anions that is like the packing of cubic closest-packed spheres (see Chapter 5). In a lattice of closest-packed spherical anions, there are two types of interstitial sites (empty spaces in the lattice) that may be occupied by cations. One type of interstitial site is surrounded tetrahedrally by four spheres; the other type is surrounded octahedrally by six spheres. These are pictured

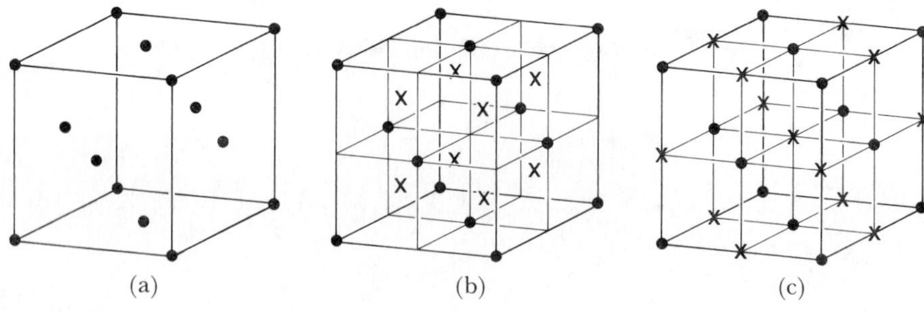

FIGURE 9–7
Interstitial sites in a cubic closest-packed lattice of anions. (a) A unit cell of cubic closest-packed anions; there are 4 complete spheres in the unit cell. (b) The eight tetrahedral sites. Each tetrahedral site is completely in this unit cell. (c) The four octahedral sites. One site is in the center of the unit cell, and one fourth site is at the middle of each edge of the unit cell.

in Figure 9–7, which shows a unit cell of cubic closest-packed anions containing 4 anions. This cubic unit cell can be subdivided into eight smaller cubes. In Figure 9–7b, we see that the center of each of the smaller cubes is a tetrahedral site. Thus there are *eight* tetrahedral sites in a unit cell containing *four* closest-packed anions. In 9–7c, we see one octahedral site at the center of the large cube and one fourth of an octahedral site at the middle of each of the twelve edges of the large cube. Thus this unit cell contains four octahedral sites:

$$1 + 12 \times \tfrac{1}{4} = 4$$

Visualization of this relationship between the number of closest-packed anions and the number of interstitial sites is easier for the cubic closest-packed arrangement than for hexagonal closest packing, but the relationships are the same for both:

> In a closest-packed lattice of ions of one sign (+ or −), there are *two times as many tetrahedral sites* for occupancy by ions of the other sign as there are closest-packed ions, and there also is an *equal number of octahedral sites* and closest-packed ions.

In zinc sulfide, ZnS, zinc ions are present in one half of the tetrahedral sites of the cubic closest-packed lattice of sulfide ions. In sodium chloride, NaCl, sodium ions are present in all of the octahedral sites of the cubic closest-packed lattice of chloride ions. If an ionic compound has a closest-packed lattice of the ions of one sign, the kind of interstitial sites that are occupied by ions of the other sign is determined primarily by radius-ratio considerations, and the fraction of the sites occupied is determined by the stoichiometric proportions of the compound. This latter point is shown in Table 9–3.

TABLE 9–3
Halides, Oxides, and Sulfides with Closest-Packed Arrangements of Anions[a]

Substance	Fraction of interstitial sites occupied	
	Tetrahedral sites	Octahedral sites
Li_2O, Na_2O	all (c)	
CuCl, CuBr, CuI	$\frac{1}{2}$ (c)	
ZnS (sphalerite)[b]	$\frac{1}{2}$ (c)	
ZnS (wurtzite)	$\frac{1}{2}$ (h)	
AgCl, NaCl[b], MnO		all (c)
FeS, CoS, NiS		all (h)
CaF_2[b,c]	all (c)[c]	
TiO_2, FeF_2, $FeCl_2$		$\frac{1}{2}$ (c)
$FeBr_2$, $CoBr_2$, $NiBr_2$		$\frac{1}{2}$ (h)
$ScCl_3$, $FeCl_3$, $FeBr_3$		$\frac{1}{3}$ (h)
$ThCl_4$, ZrI_4, SnI_4	$\frac{1}{8}$ (c)	
Fe_2O_3 (hematite)		$\frac{2}{3}$ (h)
Fe_3O_4 (magnetite)	$\frac{1}{8}$ by Fe(III) (c)	$\frac{1}{4}$ by Fe(II) (c)
		$\frac{1}{4}$ by Fe(III) (c)

[a]The symbols (c) = cubic and (h) = hexagonal.
[b]See Figure 9–3.
[c]Calcium ions in cubic closest-packed arrangement with fluoride ions in all tetrahedral sites.

THE LATTICE ENERGY

An important factor determining the stability of an ionic compound is its lattice energy, U. This positive quantity is defined as the increase in internal energy of the system at 0 K when the constituent ions of an ionic compound are separated to an infinite distance from one another. The lattice energy is not subject to direct experimental evaluation, but it can be calculated with reasonable accuracy, and it can be determined indirectly from experimental data (as we will see). If the ions in a crystalline compound were hard spheres that interacted only by electrostatic attraction and repulsion (as shown with dashed lines in Figure 9–2), the lattice energy would be the sum of the energies due to coulombic interactions of all ions of the crystal.

Let us consider a sodium chloride lattice, and focus attention on a particular sodium ion. This sodium ion attracts the six nearest neighbor chloride ions that are separated from it by a distance, a. The contribution of this attraction to U is

$$-6 \times \frac{Z_1 Z_2 e^2}{a} = +6 \times \frac{e^2}{a}$$

since $Z_1 = +1$ and $Z_2 = -1$. This sodium ion repels the twelve next-to-nearest neighbor sodium ions at a distance $\sqrt{2}\,a$. The contribution of this repulsion to U is

$$-12 \times \frac{Z_1 Z_2 e^2}{\sqrt{2}\,a} = -12 \frac{e^2}{\sqrt{2}\,a}$$

The next-closest ions to the sodium ion under consideration are eight chloride ions at a distance of $\sqrt{3}\,a$. The contribution of this attraction to U is

$$-8 \times \frac{Z_1 Z_2 e^2}{\sqrt{3}\,a} = +8 \frac{e^2}{\sqrt{3}\,a}$$

Figure 9–8 shows these three distances of separation in a sodium chloride lattice. The coulombic contribution to the lattice energy of sodium chloride

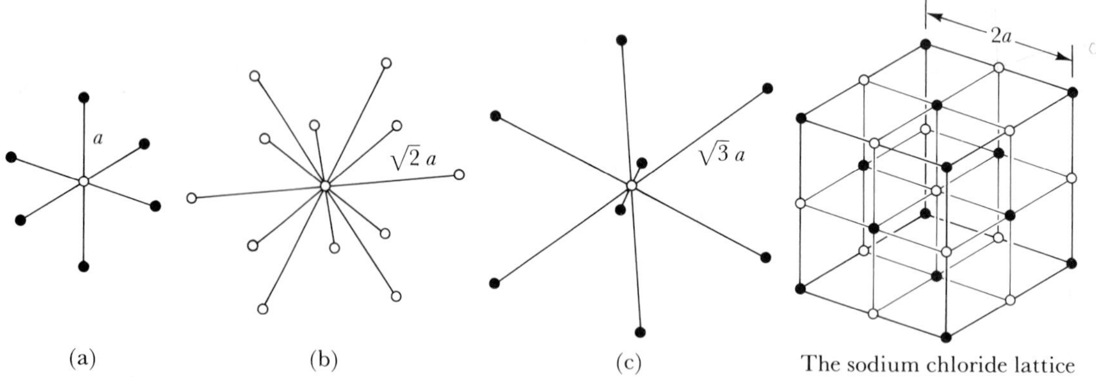

FIGURE 9–8
The distances between a sodium ion in the center of a cube, with a side equal to $2a$, and other ions in the sodium chloride lattice. (a) Distance a to the chloride ion at the center of each of six faces of cube. (b) Distance $\sqrt{2}\,a$ to sodium ions at the middle of each of 12 edges of cube. (c) Distance $\sqrt{3}\,a$ to chloride ions at each of 8 corners of the cube.

is the summation of these terms and the terms arising from the interaction with more distant ions:

$$U = \frac{e^2}{a}\left(\frac{6}{1} - \frac{12}{\sqrt{2}} + \frac{8}{\sqrt{3}} - \frac{6}{2} + \frac{24}{\sqrt{5}} - \cdots\right)$$

The numerator of each term in the parentheses is the number of interactions, and the denominator of each term is the separation of the ions in units of the closest separation, a. The quantity in the parentheses has a limiting value if a large number of terms is included in the series. For the sodium chloride lattice the limiting value is 1.74756. Therefore the contribution to the lattice energy of an ionic solid with the sodium chloride structure due to the coulombic attraction and repulsion of the ions is

$$U = -\frac{1.75 N_A Z_1 Z_2 e^2}{a}$$

which includes Avogadro's constant to take the following into account:

1. A mole of the salt MX contains a mole of cations to which this reasoning applies plus a mole of anions to which the reasoning also applies.
2. In such an accounting, each interaction is counted twice.

Therefore the factor of Avogadro's constant, N_A, comes from $2N_A \times \frac{1}{2} = N_A$. The factor Z_1Z_2 allows us to calculate the lattice energy of solids with ions having charges other than $\pm 1e$, for example, MgO, with the sodium chloride lattice, for which $Z_1 = +2$ and $Z_2 = -2$. For lattices with other arrangements of ions, the constant A, called the MADELUNG constant, in the equation

$$U = -\frac{A N_A Z_1 Z_2 e^2}{a}$$

has values that are given in Table 9–4.

A complete equation for calculating lattice energy takes into account the shape of the curve for the potential energy of ion interaction at close distances; the equation is

$$U = -\frac{A N_A Z_1 Z_2 e^2}{a}\left(1 - \frac{1}{n}\right)$$

in which the value of n depends upon the electronic structure of the ions of the compound. For sodium chloride, $n = 8$; for cesium iodide, $n = 12$.

For sodium chloride (with $a = 0.281$ nm), the lattice energy calculated from this equation using electrostatic units ($e = 4.80 \times 10^{-13}\ g^{1/2}\ m^{3/2}\ s^{-1}$) is:

$$U = -\frac{1.75 \times 6.02 \times 10^{23}\ mol^{-1} \times (+1) \times (-1) \times (4.80 \times 10^{-13}\ g^{1/2}\ m^{3/2}\ s^{-1})^2}{0.281 \times 10^{-9}\ m}\left(1 - \frac{1}{8}\right)$$

$$= 7.56 \times 10^{8}\ g\ m^2\ s^{-2}\ mol^{-1} = 7.56 \times 10^{5}\ kg\ m^2\ s^{-2}\ mol^{-1}$$

$$= 756\ kJ\ mol^{-1}$$

TABLE 9–4
Madelung Constants, *A*, for Various Ionic Lattices[a]

Structure	Coordination numbers		*A*
	Cation	Anion	
MX:			
Sphalerite (cubic ZnS)	4	4	1.63805
Sodium chloride	6	6	1.74756
Cesium chloride	8	8	1.76267
MX_2:			
β-Quartz (SiO_2)	4	2	4.4394
Rutile (TiO_2)	6	3	4.816
Fluorite (CaF_2)	8	4	5.03878

[a]The Madelung constants for MX_2 structures contain the factor 2 for the charge on the cation. They are larger than approximately twice the Madelung constants for MX structures because there are two moles of anions per mole of MX_2. The effect of the additional number of attractive cation–anion interactions is partially compensated for by the additional number of repulsive anion–anion interactions.

The agreement with the value based upon thermochemical data (786 kJ mol^{-1}) is adequate, considering the simplicity of this calculation. The calculation does not take into account minor factors contributing to the lattice energy. Calculated values of the lattice energy of sodium chloride reported in the scientific literature range from 750 kJ mol^{-1} to 780 kJ mol^{-1}. The variation is due to differences in estimates of these minor factors, for example, van der Waals attractive interactions (the factor responsible for the *a* term in van der Waals equation of state for gases).

9–3 The Stability of Ionic Compounds

The lattice energies of ionic compounds are enormous; a value $U = 786$ kJ mol^{-1} for sodium chloride is ~18 times the energy of vaporization of water. Neither sodium chloride nor any other ionic compound can be vaporized to give the separated ions at any easily realizable temperature. But the stability of these compounds relative to other possible products, the constituent elements or the gaseous molecules, is a topic of immediate concern. Now we will explore the energy changes in these processes. In this exploration we will use the concept of lattice energy as well as energy quantities introduced in Chapter 8, ionization energy and electron affinity.

THE BORN–HABER CYCLE

The stabilities of ionic compounds relative to their constituent elements can be discussed very conveniently by resolving the overall reaction into a sequence of simple steps called the BORN–HABER cycle. For example, the formation of sodium chloride,

$$Na(s) + \tfrac{1}{2}Cl_2(g) = NaCl(s)$$

can be resolved into:

(a) $Na(s) = Na(g)$

(b) $Na(g) = Na^+(g) + e^-$

(c) $\tfrac{1}{2}Cl_2(g) = Cl(g)$

(d) $Cl(g) + e^- = Cl^-(g)$

(e) $Na^+(g) + Cl^-(g) = NaCl(s)$

The sum of these steps, (a) + (b) + (c) + (d) + (e), is the overall reaction. The enthalpy difference between the products and the reactants of the overall reaction, $\Delta H_f^0(NaCl)$, is the sum of the enthalpy differences between the products and reactants of the individual steps (a) through (e):

$$\Delta H_f^0(NaCl) = \Delta H_a^0 + \Delta H_b^0 + \Delta H_c^0 + \Delta H_d^0 + \Delta H_e^0$$

The validity of this equation, which is a Hess law relationship, does not depend on this sequence of steps being a pathway for the formation of sodium chloride. Enthalpy is a state function, and the enthalpy change in the reaction

$$Na(s) + \tfrac{1}{2}Cl_2(g) = NaCl(s)$$

does not depend upon the pathway by which the reaction occurs. We resolve the overall reaction into these steps for two reasons. First, if we know the changes of enthalpy associated with the overall reaction and with each of the steps except one, we can calculate the unknown change of enthalpy. Second, we may recognize the factors influencing the relative stabilities of ionic compounds more clearly in trends of the enthalpy changes associated with the simple steps than in the enthalpy changes associated with the overall reactions. Of course, a complete discussion of relative stabilities should include consideration of the changes of entropy as well as the changes of enthalpy. For most of the comparisons we will study, there are only small differences of ΔS^0 for the overall reactions forming the compounds being compared.

The enthalpy changes associated with the steps of the Born–Haber cycle generally are referred to by particular names and are given particular symbols:[3] (a) the sublimation enthalpy (of the metal), $\Delta H_a^0 = S$; (b) the ionization energy (of the gaseous metal atom), $\Delta H_b^0 = IE$; (c) the dissociation enthalpy (of the diatomic nonmetal), $\Delta H_c^0 = \tfrac{1}{2}D$ (one half mole of

[3]The Born–Haber cycle is written in terms of changes of enthalpy, but some of the quantities just referred to—*IE*, ionization energy; *D*, dissociation energy; *EA*, electron affinity; and *U*, lattice energy—pertain to changes of internal energy. Of these, the dissociation energy and lattice energy are associated with steps in which there is a change in the number of gas particles; therefore, for these steps, $\Delta H \neq \Delta U$.

diatomic nonmetal is used); (d) the electron affinity (of the nonmetal), $\Delta H^0_{\text{d}} = -EA$; and (e) the lattice energy of the crystal, $\Delta H^0_{\text{e}} = -U$. Steps (a), (b), and (c) are endothermic; Steps (d) and (e) are exothermic.[4] The Born–Haber cycle for sodium chloride is shown in Figure 9–9, in which a vertical enthalpy scale is used. The steps in the cycle are:

(a) the sublimation of one mole of sodium *requires*:	108.7 kJ
(b) the removal of an electron from one mole of gaseous sodium atoms *requires*:	495.8 kJ
(c) the dissociation of one half mole of chlorine gas *requires*:	119.5 kJ
(d) the addition of an electron to one mole of chlorine atoms *produces*:	348.7 kJ
(e) the formation of one mole of solid sodium chloride from the gaseous ions *produces*:	786.3 kJ

Therefore the net energy *produced* for the reaction

$$\text{Na}(s) + \tfrac{1}{2}\text{Cl}_2(g) = \text{NaCl}(s)$$

is

$$\Delta H^0_f(\text{NaCl}) = S + IE + \tfrac{1}{2}D - EA - U$$
$$\Delta H^0_f(\text{NaCl}) = -411.0 \text{ kJ mol}^{-1}$$

Study of Figure 9–9 allows us to draw certain general conclusions:

1. The larger the sublimation energy of the metal, the larger the dissociation energy of the nonmetal, and the larger the ionization energy of the metal, the more positive will be the enthalpy of the ionic compound relative to the enthalpy of its constituent elements.
2. The larger the electron affinity of the nonmetal, the larger the lattice energy of the ionic compound, the more negative will be the enthalpy of the ionic compound relative to the enthalpy of its constituent elements.

It is informative to consider the sum of Steps (b) and (d),

$$\text{Na}(g) + \text{Cl}(g) = \text{Na}^+(g) + \text{Cl}^-(g)$$

a process in which an electron transfer gives product species each with a noble-gas electronic configuration:

$$\text{Na}^+(g)\ 1s^2 2s^2 2p^6 \qquad \text{(Ne structure)}$$
$$\text{Cl}^-(g)\ 1s^2 2s^2 2p^6 3s^2 3p^6 \qquad \text{(Ar structure)}$$

The value of ΔH^0 for this process is

$$\begin{aligned}\Delta H^0 &= \Delta H^0_{\text{b}} + \Delta H^0_{\text{d}} \\ &= +495.8 \text{ kJ mol}^{-1} - 348.7 \text{ kJ mol}^{-1} \\ &= +147.1 \text{ kJ mol}^{-1}\end{aligned}$$

[4]As we will learn, Step (d) is not exothermic for the anionic component of ionic oxides.

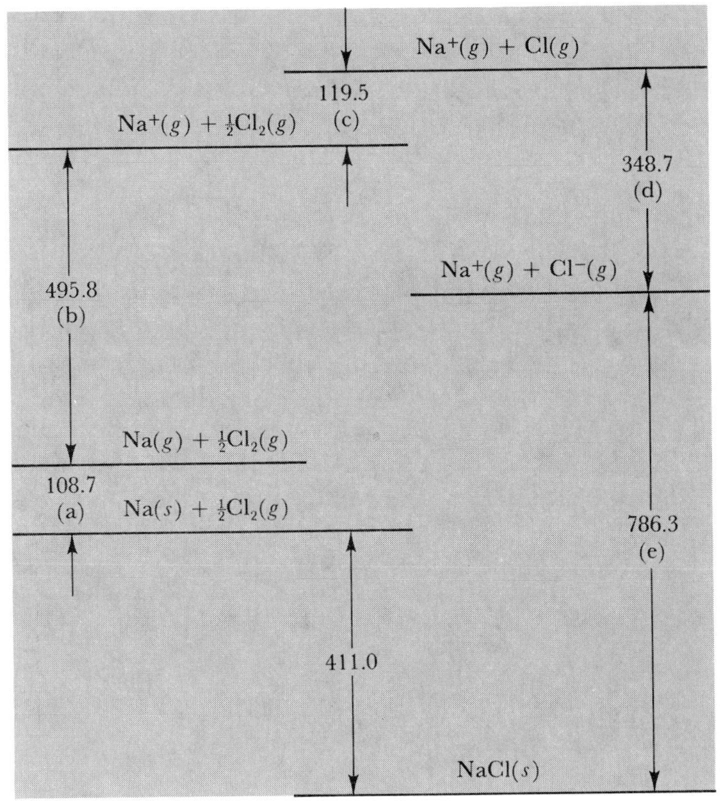

FIGURE 9–9
The Born–Haber cycle for sodium chloride. Energy quantities are in kJ mol^{-1}.

This process, which produces ions with noble-gas configurations, is endothermic. Although simple explanations of chemical compound formation often stress the importance of atoms' acquiring noble-gas electronic configurations, the process under consideration achieves this but is endothermic nonetheless. The electrostatic attraction of the ions so formed makes the electron transfer an exothermic process if the process forms gaseous sodium chloride instead of separated ions:

$$Na^+(g) + Cl^-(g) = Na^+Cl^-(g) \qquad \Delta H^0 = -558.4 \text{ kJ mol}^{-1}$$

Gaseous sodium chloride molecules are known at high temperatures; the condensation of these gaseous molecules to give crystalline sodium chloride is exothermic:

$$Na^+Cl^-(g) = NaCl(s) \qquad \Delta H^0 = -227.9 \text{ kJ mol}^{-1}$$

The lattice energy $(-\Delta H^0_e)$ is the sum of the energy changes of these two steps:

$$-(-558.4 \text{ kJ mol}^{-1} - 227.9 \text{ kJ mol}^{-1}) = +786.3 \text{ kJ mol}^{-1}$$

Using the Born–Haber cycle, we now can make comparisons that illustrate certain points; the compounds we shall compare are:

1. lithium fluoride and sodium hydride
2. silver chloride and potassium chloride
3. sodium fluoride and magnesium oxide

Lithium Fluoride and Sodium Hydride. One reason for comparing the stabilities of these two compounds is to present the subject of saline (i.e., saltlike) hydrides. The bonding of hydrogen in most of its compounds is covalent, but binary compounds of hydrogen with some metals are ionic. The addition of an electron to a hydrogen atom produces hydride ion, H^-, which has the electronic structure of helium. Sodium hydride, which crystallizes with the sodium chloride structure (see Figure 9–3), will be compared with lithium fluoride (which also has the sodium chloride structure), because this comparison involves compounds containing isoelectronic ions. Hydride ion is isoelectronic with lithium ion (each has the helium structure, $1s^2$), and fluoride ion is isoelectronic with sodium ion (each has the neon structure, $1s^2 2s^2 2p^6$). The comparison is presented in Table 9–5 and Figure 9–10. Sodium hydride is stable with respect to its constituent elements, but it is much less stable in this respect than lithium fluoride. The steps in the Born–Haber cycle which tend to make lithium fluoride more stable are:

(b) $\frac{1}{2}D(F_2) \ll \frac{1}{2}D(H_2)$

(d) $EA(F) \gg EA(H)$

(e) $U(LiF) > U(NaH)$

The steps which tend to make sodium hydride more stable are:

(a) $S(Na) < S(Li)$

(c) $IE(Na) < IE(Li)$

We learn certain points from this example. They are:

1. The large dissociation energy of hydrogen is a major factor contributing to the marginal stability of saline hydrides. Only metals for which the

TABLE 9–5
The Born–Haber Cycles for Formation of Lithium Fluoride and Sodium Hydride

		$\frac{\Delta H^0}{\text{kJ mol}^{-1}}$	
		LiF	NaH
(a)	$M(s) = M(g)$	160.7	108.7
(b)	$M(g) = M^+(g) + e^-$	520.1	495.8
(c)	$\frac{1}{2}X_2(g) = X(g)$	77.0	216.0
(d)	$X(g) + e^- = X^-(g)$	−327.8	−72.7
(e)	$M^+(g) + X^-(g) = MX(s)$	−1046.9	−804.2
	$M(s) + \frac{1}{2}X_2(g) = MX(s)$	−616.9	−56.4
	$Li(s) + \frac{1}{2}F_2(g) = LiF(s)$	$\Delta H^0 = \Delta H_f^0(LiF) = -616.9$ kJ mol^{-1}	
	$Na(s) + \frac{1}{2}H_2(g) = NaH(s)$	$\Delta H^0 = \Delta H_f^0(NaH) = -56.4$ kJ mol^{-1}	

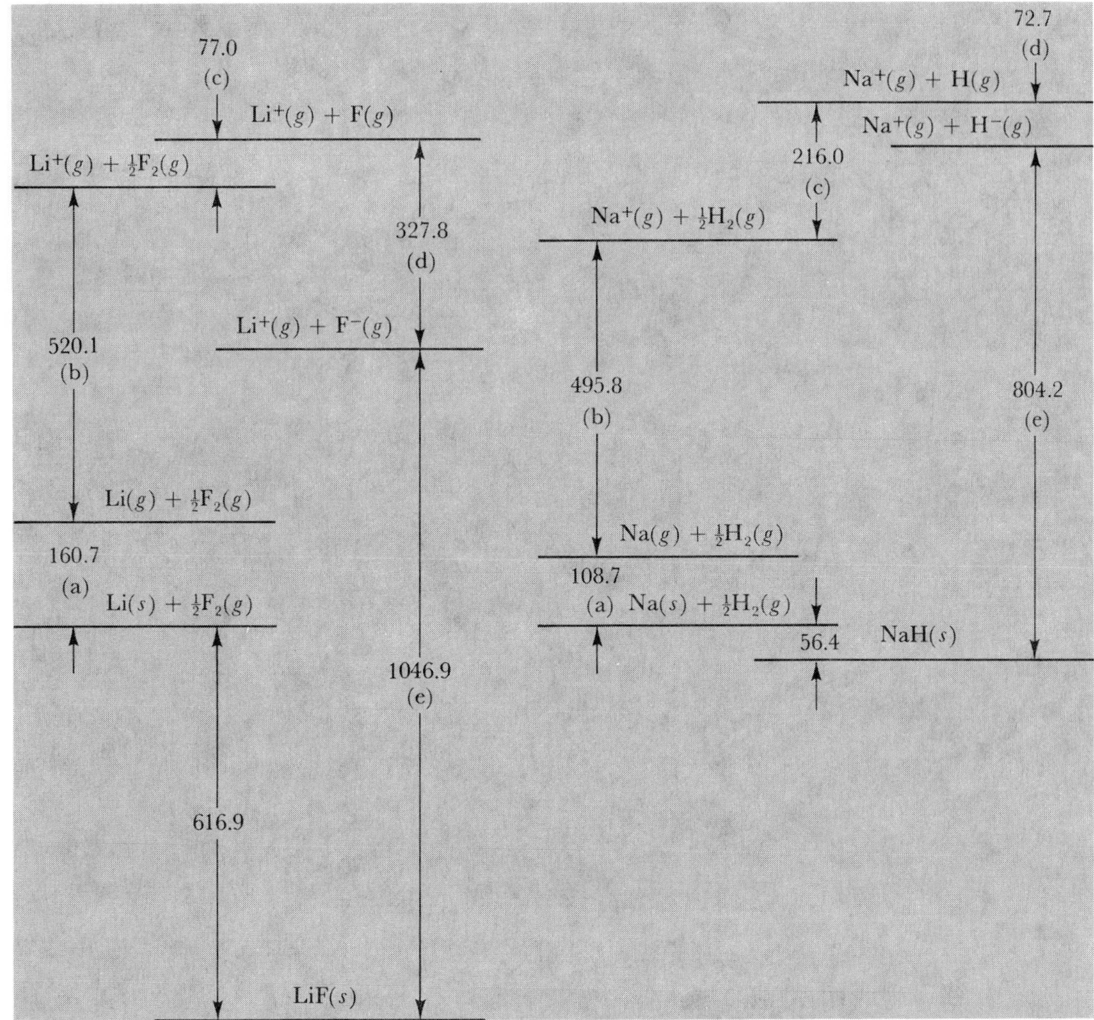

FIGURE 9–10
Born–Haber cycles for lithium fluoride and sodium hydride. The steps are lettered as in the cycle for sodium chloride. Energy quantities are in kJ mol^{-1}.

sublimation and ionization steps [Steps (a) and (b)] are not excessively endothermic will form stable saline hydrides. These metals are called *active metals*.

2. Prediction of the relative changes of enthalpy in the formation of ionic compounds containing different metals and different nonmetals is not necessarily simple. In this example, three of the steps would make the one compound (lithium fluoride) more stable, but two of the steps favor the other compound (sodium hydride).

Silver Chloride and Potassium Chloride. It is interesting to compare these two compounds because of the contrasting electronic structures of silver(I) ion and potassium(I) ion:

$$Ag^+ \text{ (Kr structure), } 4d^{10} \qquad K^+ \text{ (Ar structure)}$$

The ionic radii of these ions are similar (~ 1.3 Å for Ag^+ and ~ 1.4 Å for K^+), but the enthalpies of formation of the two compounds differ greatly:

$$Ag(s) + \tfrac{1}{2}Cl_2(g) = AgCl(s) \qquad \Delta H^0 = \Delta H_f^0(AgCl) = -127.0 \text{ kJ mol}^{-1}$$
$$K(s) + \tfrac{1}{2}Cl_2(g) = KCl(s) \qquad \Delta H^0 = \Delta H_f^0(KCl) = -436.7 \text{ kJ mol}^{-1}$$

Figure 9–11 shows Born–Haber cycles for the formation of these compounds. Each step involving only the metal, sublimation and ionization, tends to make potassium chloride more stable. The difference in the enthalpy changes upon sublimation is striking: 89.1 kJ mol^{-1} for potassium, and 286.6 kJ mol^{-1} for silver. Silver is a hard metal with atoms that attract one another strongly (the boiling point of silver is 2466 K); potassium (and the other alkali metals) are soft metals with atoms that attract one another

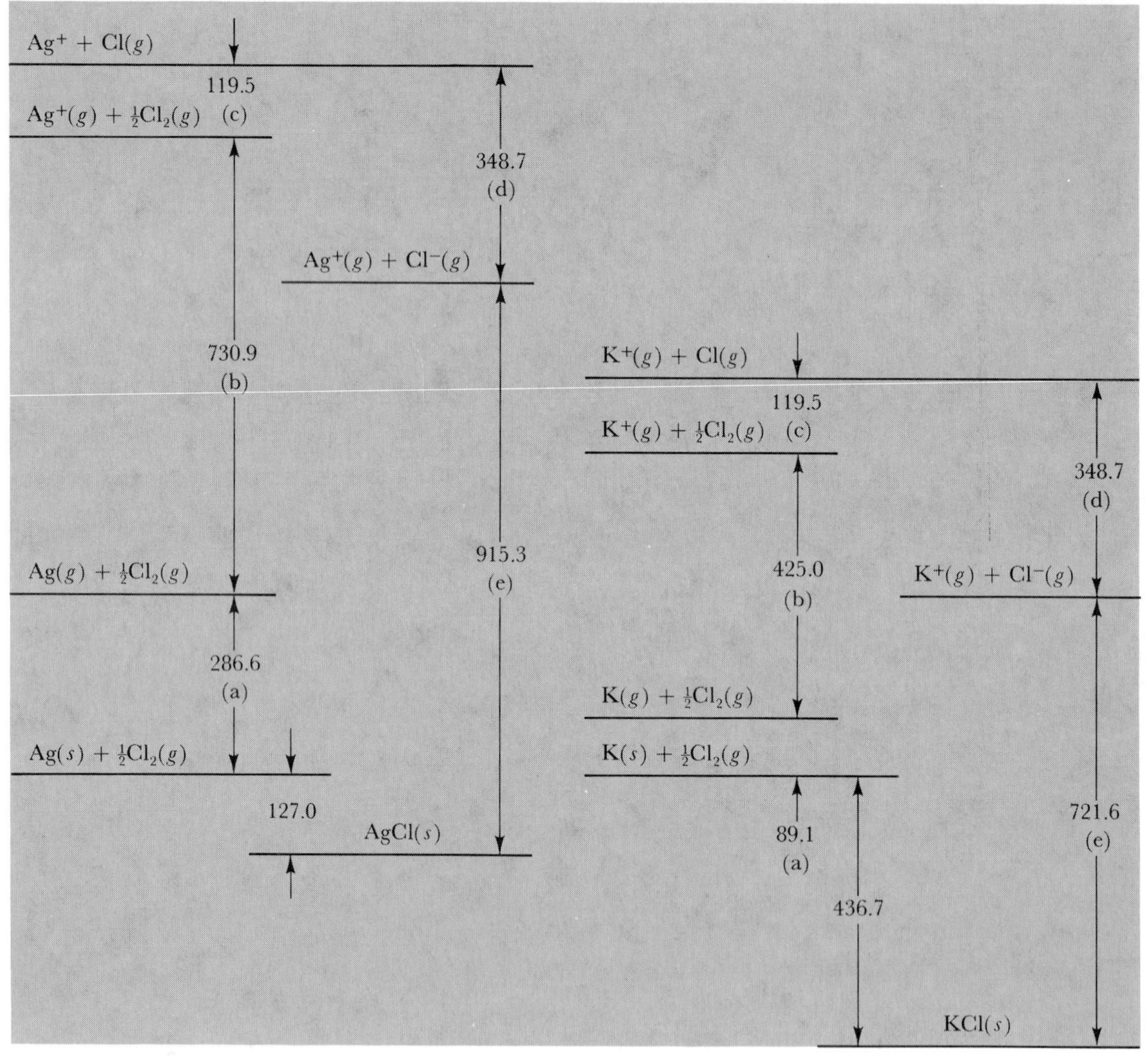

FIGURE 9–11

Born–Haber cycles for silver chloride and potassium chloride. Energy quantities are in kJ mol^{-1}.

less strongly (the boiling point of potassium is 1030 K). The ionization energies differ by even more (425.0 kJ mol^{-1} for potassium, and 730.9 kJ mol^{-1} for silver).

The lattice energy for silver(I) chloride used in this cycle was calculated from other known energy changes in the cycle:

$$U = S + IE + \tfrac{1}{2}D - EA - \Delta H^0$$

$$\frac{U}{\text{kJ mol}^{-1}} = 286.6 + 730.9 + 119.5 - 348.7 + 127.0 = 915.3$$

$$U = 915.3 \text{ kJ mol}^{-1}$$

Values of lattice energy calculated using an ionic model are much smaller than this, being ~820 kJ mol^{-1}. This large discrepancy exists because the bonding in solid silver chloride is not entirely ionic.

Magnesium Oxide and Sodium Fluoride. These two substances, each with the sodium chloride structure, involve isoelectronic ions (Mg^{2+}, Na^+, O^{2-}, and F^- have the neon electronic structure). The Born–Haber cycles are shown in Figure 9–12. The points of particular interest are:

1. The energy required to remove two electrons from the magnesium atom is enormously greater than that required to remove one electron from the sodium atom, 2189 kJ mol^{-1} and 495.8 kJ mol^{-1}, respectively.
2. The lattice energy of an ionic solid containing cations and anions of charge 2 is enormously greater than that of an ionic solid containing ions of charge 1. For this example, the difference of lattice energies is 3002 kJ mol^{-1}, which more than offsets the difference in ionization energies, 1693 kJ mol^{-1}.
3. The addition of two electrons to a gaseous oxygen atom does not evolve energy. Although addition of one electron,

$$O(g) + e^- = O^-(g)$$

is exothermic ($\Delta H^0 = -EA_1 = -141$ kJ mol^{-1}), addition of the second electron,

$$O^-(g) + e^- = O(g)^{2-}$$

is endothermic ($\Delta H^0 = -EA_2 = 878$ kJ mol^{-1}). Oxide ion, O^{2-}, does not exist in the gaseous state; it exists in ionic solids because of the large lattice energy of an ionic solid containing ions of higher charge.

Because oxide ion O^{2-} is not formed in the gas phase, you may wonder how the value $\Delta H^0 = 878$ kJ mol^{-1} for the process $O^-(g) + e^- = O^{2-}(g)$ was estimated. The Born–Haber cycle for magnesium oxide gives the value of oxygen's affinity for the second electron, EA_2 ($-\Delta H$ for the above process), as

$$EA_2 = S + \tfrac{1}{2}D + IE - EA_1 - U - \Delta H^0$$

an equation in which each of the other quantities is known. The value of U

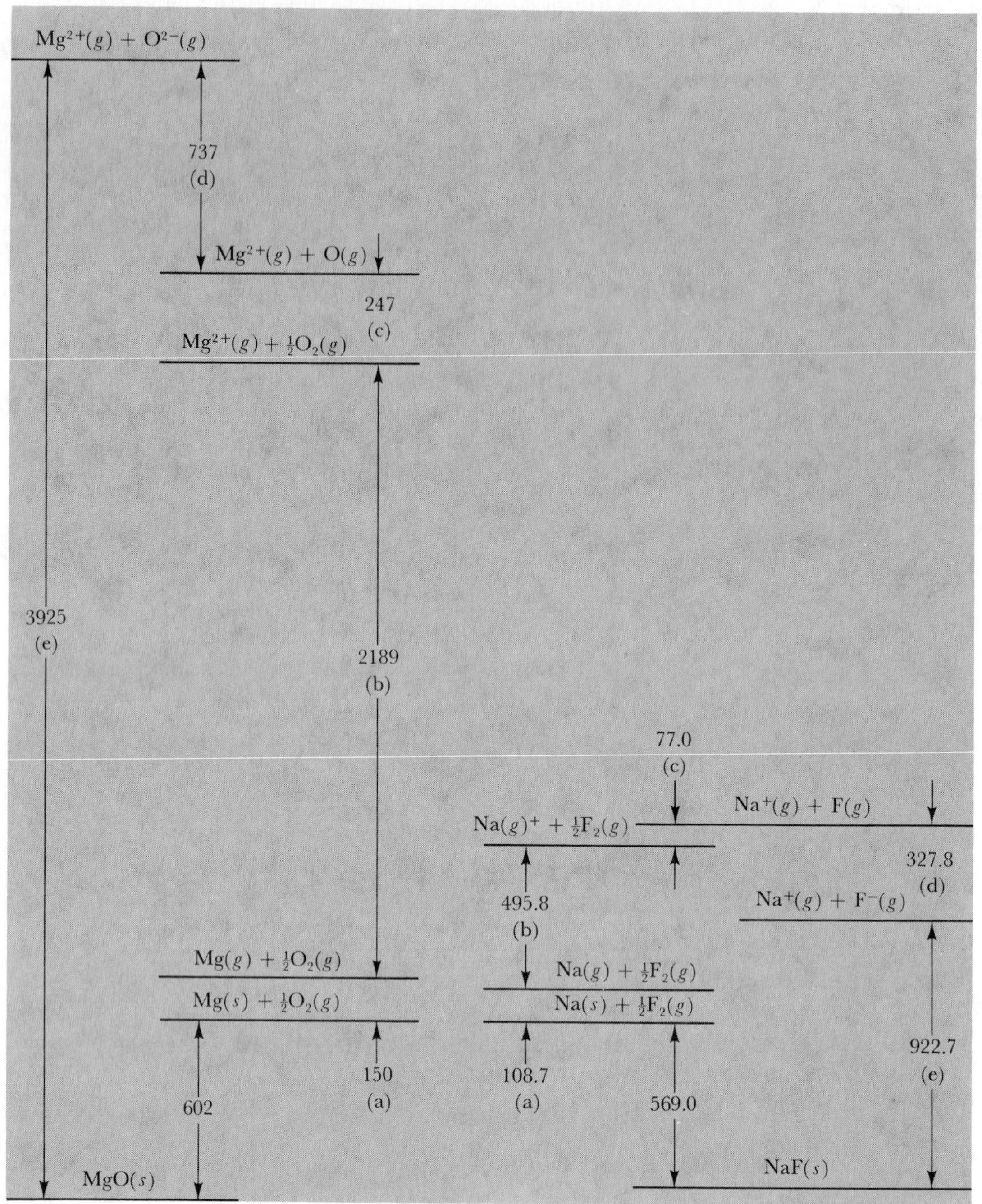

FIGURE 9–12
Born–Haber cycles for magnesium oxide and sodium fluoride. Energy quantities are in kJ mol^{-1}.

comes from a calculation analogous to that outlined previously in this chapter; the remaining quantities are measured experimentally. The values are $\Delta H^0 = -602$ kJ mol^{-1}, $S = 150$ kJ mol^{-1}, $\frac{1}{2}D = 247$ kJ mol^{-1}, $IE =$ 2189 kJ mol^{-1}, $EA_1 = 141$ kJ mol^{-1}, and $U = 3925$ kJ mol^{-1}; these values give $EA_2 = -878$ kJ mol^{-1}.

TRENDS IN CERTAIN PROPERTIES OF METAL HALIDES

We have seen that several factors play roles in determining the enthalpy changes in the formation of ionic compounds. Variation in the importance of these factors for different series of compounds makes it difficult to predict qualitative trends of enthalpy changes. The enthalpy changes in the formation of alkali metal fluorides and iodides are given in Figure 9–13. We see that lithium fluoride is formed most exothermically, and there is a monotonous trend from $\Delta H_f^0(\text{LiF}) = -617$ kJ mol^{-1} to $\Delta H_f(\text{CsF}) = -555$ kJ mol^{-1}. It would be risky, however, to base a prediction of the trend of enthalpy changes in the formation of alkali metal iodides upon this trend for the fluorides. The values of the enthalpy change upon formation of alkali metal iodides range from $\Delta H_f^0(\text{LiI}) = -270$ kJ mol^{-1} to $\Delta H_f^0(\text{CsI}) =$ -337 kJ mol^{-1}. The trends in enthalpies of sublimation of the metals and their ionization energies both operate to make cesium salts more stable. The energy quantity of the Born–Haber cycle which makes lithium fluoride the most stable alkali metal fluoride is the lattice energy, that for lithium fluoride being 290 kJ mol^{-1} greater than that for cesium fluoride. For substances with a particular crystal structure, the cation–anion separation determines the lattice energy. There is a larger percentage change in this

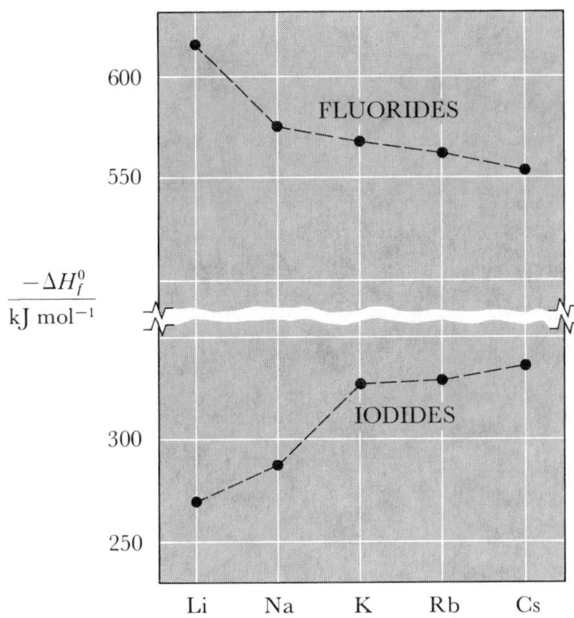

FIGURE 9–13
The standard enthalpy change in formation of alkali metal fluorides and iodides.

quantity in going from $(r_{Li^+} + r_{F^-})$ to $(r_{Cs^+} + r_{F^-})$ than in going from $(r_{Li^+} + r_{I^-})$ to $(r_{Cs^+} + r_{I^-})$:

$$\frac{(r_{Cs^+} + r_{F^-})}{(r_{Li^+} + r_{F^-})} = \frac{1.84 \text{ Å} + 1.19 \text{ Å}}{0.86 \text{ Å} + 1.19 \text{ Å}} = 1.48$$

$$\frac{(r_{Cs^+} + r_{I^-})}{(r_{Li^+} + r_{I^-})} = \frac{1.84 \text{ Å} + 2.12 \text{ Å}}{0.86 \text{ Å} + 2.12 \text{ Å}} = 1.33$$

For alkali metal iodides the lattice energy varies by only 150 kJ mol^{-1} in going from the lithium salt to the cesium salt. This comparison illustrates a general principle: there is a smaller variation of lattice energy in a series of salts with a common large anion than in a corresponding series with a common small anion.

One can learn about the attraction of ions in an ionic solid by considering the normal melting and boiling temperatures. When an ionic solid melts, thermal motion of the ions partially overcomes the attraction of the ions for one another. The stronger this attraction in the solid, the higher the melting temperature. Factors that influence the attraction of the ions for one another are:

1. the charges on the ions,
2. the interatomic separation of the ions,
3. the value of r/R for the ions of the solid, and
4. the relationship of the coordination numbers of the cation and anion in the solid to the stoichiometric ratio in discrete molecules of the compound.

We may see the effect of ionic charge if we compare isoelectronic ionic substances; Table 9–6 gives the melting temperatures of two pairs of such substances, sodium fluoride and magnesium oxide, and potassium fluoride and calcium oxide. In each pair, the ionic compound containing ions of charge 2+ and 2− has a much higher melting temperature than the compound containing ions of charge 1+ and 1−. The table also shows the greater density (and smaller molar volume) of the compounds containing the more highly charged ions.

TABLE 9–6
Comparisons of Properties of Isoelectronic Substances

	mp °C	Density g cm^{-3}	V cm^3 mol^{-1}
NaF	992	2.79	15.0
MgO	2800	3.65	11.0
KF	880	2.48	23.4
CaO	2580	3.31	16.9

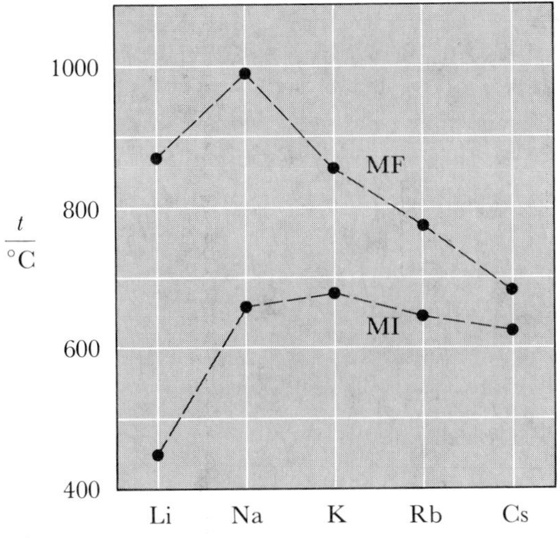

FIGURE 9–14
Melting temperatures of alkali metal fluorides and alkali metal iodides.

We may see the effect of interatomic separation if we compare the melting temperatures of alkali metal halides:

	$\frac{r_{M^+} + r_{X^-}}{Å}$	$\frac{mp}{°C}$
LiF	2.05	870
NaCl	2.82	801
KBr	3.31	734
RbI	3.70	647

Each of the substances being compared has the sodium chloride structure, and the values of r_+/r_- are appreciably greater than the value 0.414, which is the lower limit for this structure (see Table 9–2).

Radius-ratio effects play a role in determining the melting temperatures of ionic compounds. In particular, the melting temperatures of lithium salts are abnormally low. This is shown in Figure 9–14. For lithium iodide, which has the sodium chloride structure, the ratio r/R is 0.41, at the stability limit for a structure with coordination number 6.

The effect of the coordination numbers in the solid relative to the stoichiometric ratio in discrete molecules is shown in a comparison of silicon tetrafluoride (SiF_4) with tin tetrafluoride (SnF_4); the melting and boiling temperatures and coordination numbers of the metal or metalloid component of these compounds are:

	$\frac{mp}{°C}$	$\frac{bp}{°C}$	Coordination number of metal or metalloid: in solid	in molecule
SiF_4	−90	−95	4	4
SnF_4		704	6	4

The melting temperature for SiF_4 is that for the compound under pressure, and the boiling temperatures actually are sublimation temperatures at atmospheric pressure.

The enormous contrast between the sublimation temperatures of these two tetrafluorides follows from the relationship of the coordination numbers in the solid and vapor phases. For silicon tetrafluoride, there are discrete SiF_4 molecules in each phase, and vaporization does not involve breaking any silicon–fluorine bonds. For tin tetrafluoride (SnF_4), the coordination number of tin changes from six in the solid to four in the vapor (which consists of discrete molecules). Vaporization involves breaking the tin–fluorine bonds. Consequently the sublimation temperature is much higher for tin tetrafluoride.

9–4 The Defect Solid State

In previous sections of this chapter we have depicted crystalline ionic solids made up of nicely ordered arrangements of ions. However, a different class of solids, defect crystalline solids, is important also. Structural defects play roles in the reactions of solids and in catalysis by solids. The electrical conductivity and spectral properties of a solid also are dramatically dependent upon the presence of defects. We will introduce the subject with reference to ionic substances, but defect solids are possible for all types of substances.

IMPERFECT CRYSTALS

Even though a crystalline ionic solid has an ideal stoichiometric composition, it may have structural defects. Two types of defects are possible. These are pictured in Figures 9–15; they are called Schottky defects and Frenkel defects. A Schottky defect (named for W. SCHOTTKY) is a vacant lattice site. These occur in lattices of identical atoms (e.g., metals or covalently bonded elemental solids), in lattices of ionic solids, and in lattices of molecular substances (e.g., organic compounds). A Frenkel defect (named for Y. I. FRENKEL) is an atom or ion displaced to an interstitial position not occupied in a perfect crystal.

Relative to the perfect crystal, crystals with either Schottky or Frenkel defects have higher enthalpy content. The process

$$MX(s, \text{perfect crystal}) = MX(s, \text{imperfect crystal})$$

is endothermic. The stability of an imperfect crystal is attributable to the

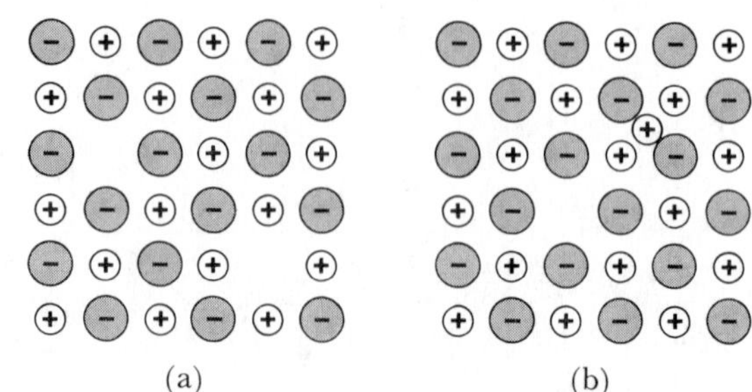

FIGURE 9–15
Defects in crystalline ionic solids (schematic). (a) Schottky defect—a vacant site. For an ionic solid, an equal number of cation and anion sites are vacant, as shown. (b) Frenkel defect—an atom or an ion of either sign is present at an inappropriate site.

positive entropy change in the above process. In a perfect crystal there is complete order of atomic positions; in the imperfect crystal there is disorder. The entropy of the crystal with defects is greater. In determining the tendency for crystalline materials to have imperfections, the entropy factor plays a larger role the higher the temperature. Only at a temperature of absolute zero is a perfect crystal more stable than an imperfect crystal.

There are both experimental and theoretical approaches to studying the number of defects in crystalline solids. For sodium chloride at room temperature, the number of Schottky defects has been estimated to be approximately 10^6 per cm^3. (Because there are 4.5×10^{22} ions present in one cm^3 of sodium chloride, the fraction of the ionic sites that are vacant is minute.)

NONSTOICHIOMETRIC CRYSTALS

In another type of defect solid the constituent elements are present in proportions different from those predicted by simple valence rules. An example is sodium chloride in equilibrium with excess metallic sodium. This material has a deficiency of chlorine; its composition can be represented as $NaCl_{1-x}$ (with $x < 0.01$). In this material, a small fraction of the sites occupied by chloride ions in the perfect crystalline solid are occupied instead by electrons. An electron trapped in a vacant anion site is called an F center (F is an abbreviation for Farbe, German for color), and the trapped electrons are responsible for the optical and electrical properties of this material. (The nonstoichiometric sodium chloride $NaCl_{1-x}$ is blue, and it conducts electricity.) Intense ultraviolet radiation can create F centers in sodium chloride (or another alkali halide salt). In this process with sodium chloride, chlorine gas is lost from such crystals:

$$NaCl(s) + \text{light} = NaCl_{1-x}(s) + \frac{x}{2}Cl_2(g)$$

The range of composition of nonstoichiometric compounds may be enormously greater than that exhibited by sodium chloride. One form of titanium oxide, with the sodium chloride structure, has a range of composition from $TiO_{1.32}$ to $TiO_{0.69}$. Both cation and anion sites are unoccupied in this system; even in the material with the one-to-one stoichiometric composition, TiO, 15% each of the cation and anion sites are vacant. If different fractions of the cation and anion sites are vacant, the solid has a nonstoichiometric composition. A solid with vacant sites has a lower density than expected on the basis of the atomic spacings (as determined by x-ray diffraction). Measurements of density give the information tabulated in Table 9–7.

If stoichiometric nickel(II) oxide, NiO, a pale green ionic compound which is an insulator, is heated (1100–1300 K) with excess oxygen, the reaction

$$(1 - x)NiO + \frac{x}{2}O_2 = Ni_{1-x}O$$

TABLE 9–7
Vacancies from Titanium and Oxygen Sites in Titanium Oxides

Composition	Ti sites vacant, %	O sites vacant, %
$TiO_{1.32}$	26	2
$TiO_{1.12}$	19	9
TiO	15	15
$TiO_{0.69}$	4	34

occurs. The black nonstoichiometric product, a conductor of electricity, has vacant nickel sites in the nickel(II) oxide lattice (sodium chloride type). Since $Ni_{1-x}O$ contains fewer nickel ions than oxide ions of charge 2−, the average charge on the nickel ions must be greater than 2+. If we consider the nickel to be present as Ni^{2+} and Ni^{3+}, we can represent the solid as

$$Ni_y^{3+}Ni_{1-x-y}^{2+}O^{2-}$$

and use the electroneutrality condition

$$3y + 2(1 - x - y) - 2 = 0$$

to calculate y:

$$y = 2x$$

That is, the ratio of nickel(III) to nickel(II) in $Ni_{1-x}O$ is $y/(1-x-y) = 2x/(1-3x)$. The electrical conductivity of this compound (and other nonstoichiometric compounds of this type), a subject of practical interest, results from transfer of electrons between nickel atoms. The transfer of an electron from a nickel(II) ion to a nickel(III) ion gives a nickel(III) ion and a nickel(II) ion:

$$Ni^{2+} \overset{e^-}{\frown} Ni^{3+} \rightarrow Ni^{3+} + Ni^{2+}$$

The net result is the apparent migration of nickel(III) ion in the NiO lattice.

For nonstoichiometric nickel oxide, $Ni_{1-x}O$, the electrical conductivity depends upon the value of x, which is determined by the conditions of preparation, especially the partial pressure of oxygen and the temperature. But the range of values of x that can be introduced in this way is limited, and the value of x is not easily controlled. This problem can be circumvented by introduction of lithium ion, an ion of charge 1+, to occupy cation sites in the nickel(II) oxide lattice of the nonstoichiometric compound. (The ionic radii of Li^+ and Ni^{2+} are similar, 0.86 Å and 0.82 Å, respectively.) The oxidation reaction now can be written

$$(1 - x)NiO + \frac{x}{2}Li_2O + \frac{x}{4}O_2 = Li_xNi_{1-x}O$$

The product of this reaction has no vacant cation sites, that is,

$$x + (1 - x) = 1$$

The relative amounts of nickel(II) and nickel(III) in the product can be calculated using the approach which we employed earlier,

$$Li_xNi_{1-x}O \quad \text{is} \quad Li_x^+Ni_y^{3+}Ni_{1-x-y}^{2+}O^{2-}$$

which gives

$$x + 3y + 2(1 - x - y) - 2 = 0$$
$$y = x$$

If for every Ni^{3+} there is one Li^+, the average of the charges on these two ions is 2+, and no cation sites are vacant. The composition of the compound $Li_xNi_{1-x}O$ is $Li_x^+Ni_x^{3+}Ni_{1-2x}^{2+}O^{2-}$. (Check for yourself that this composition satisfies the electroneutrality condition.) Unlike the nonstoichiometric $Ni_{1-x}O$, the compound $Li_xNi_{1-x}O$ has no defects in the lattice; every metal site is occupied by Li^+, Ni^{3+}, or Ni^{2+}. The composition can be controlled exactly by the relative amounts of lithium oxide and nickel oxide used in the preparation.

9–5 Ionic Compounds in Other States

We have been describing ionic compounds in the solid state, but we also should consider these substances in the pure liquid state, in liquid solution, and in the gaseous state. Liquid ionic compounds (also called molten salts) are capable of dissolving many compounds that do not dissolve appreciably in common liquid solvents, and some ionic compounds are useful as liquid solvents in a range of high temperatures not accessible with other liquid solvents. In practical work with ionic compounds at high temperatures, we must pay attention to their vapor pressures, as the loss of material through vaporization can be appreciable. The study of gaseous ionic compounds also provides information about the deviation of real ions from the hard-sphere model that was the basis for some of the calculations we have considered earlier in this chapter. Despite the strong attraction of oppositely charged ions for one another, many solid ionic compounds dissolve to a large extent in water and other solvents to give solutions in which the ions are dissociated from one another. Each of these topics will be considered in the concluding section of this chapter.

MOLTEN SALTS

Molten salts, first studied systematically by Michael Faraday, have many properties that differ greatly from those of ordinary liquids. The intensity of study in this field has varied since Faraday's time, but interest now is high in the possibility of using molten salts as solvents for high-temperature reactions.

The temperature range over which a molten salt is liquid may be much greater than that for other liquid solvents. For instance, sodium chloride has a liquid range of 613 K (from 1073.6 K to 1686.4 K), much larger than the 100 K range for water (273.2 K to 373.2 K) or carbon tetrachloride (250.2 K to 350.0 K). The electrical conductance of molten salts is enormously greater than that of water or other common solvents. Molten sodium chloride has a conductance 10^8 times that of water. The reactivity of many molten salts is very much lower than that of other solvents, an advantageous feature. For instance, aluminum is produced from its principal ore, bauxite (Al_2O_3), by electrolysis in a molten salt, as was mentioned in Chapter 2. In molten cryolite (Na_3AlF_6, mp 1276 K), electrolysis produces the net reaction $2Al_2O_3 = 4Al + 3O_2$. If an electrical current is passed through an aqueous solution of bauxite dissolved in acid, no aluminum metal is produced; rather the water decomposes: $2H_2O = 2H_2 + O_2$.

Although the high temperatures that can be used with molten salts as solvents have some advantages (e.g., high reaction rates), there are circumstances under which this property of molten salts is not needed. Temperatures lower than the melting point of a pure molten salt can be realized with liquid molten salt mixtures. Figure 9–16 shows the phase diagrams for the sodium chloride–lead chloride system and the sodium chloride–potassium chloride system; we see that in each of these systems there are mixtures that are liquid at temperatures below the melting points of the pure compounds. These phase diagrams can be compared with those given in Chapter 7 for the systems gold–copper (Figure 7–22a) and napthalene–phenanthrene (Figure 7–23). The solubility relationships in the solid phases are the same in systems having diagrams that resemble one another (i.e., solid solutions do not form in the sodium chloride–lead chloride system and the napthalene–phenanthrene system, and solid solutions do form in the sodium chloride–potassium chloride system and the gold–copper system).

GASEOUS IONIC COMPOUNDS

The vapor pressures of ionic compounds are very low at ordinary temperatures, but vapor species for some ionic compounds are stable and are present in appreciable concentrations at high temperatures. Study of these species reveals some important information about the bonding in ionic compounds. Although an ionic compound M^+X^- is electrically neutral, this electroneutrality does not mean that the molecule is electrically symmetrical. In a gaseous ionic compound, one end of the molecule is positive and the other end is negative. Such a molecule is said to be *polar*, and a measure of its polarity is the *dipole moment*.

The simplest model for an electrically neutral but electrically unsymmetrical molecule is a dipole, a positive charge $\delta+$ separated from a negative charge of the same magnitude $\delta-$ by a distance d. For such a

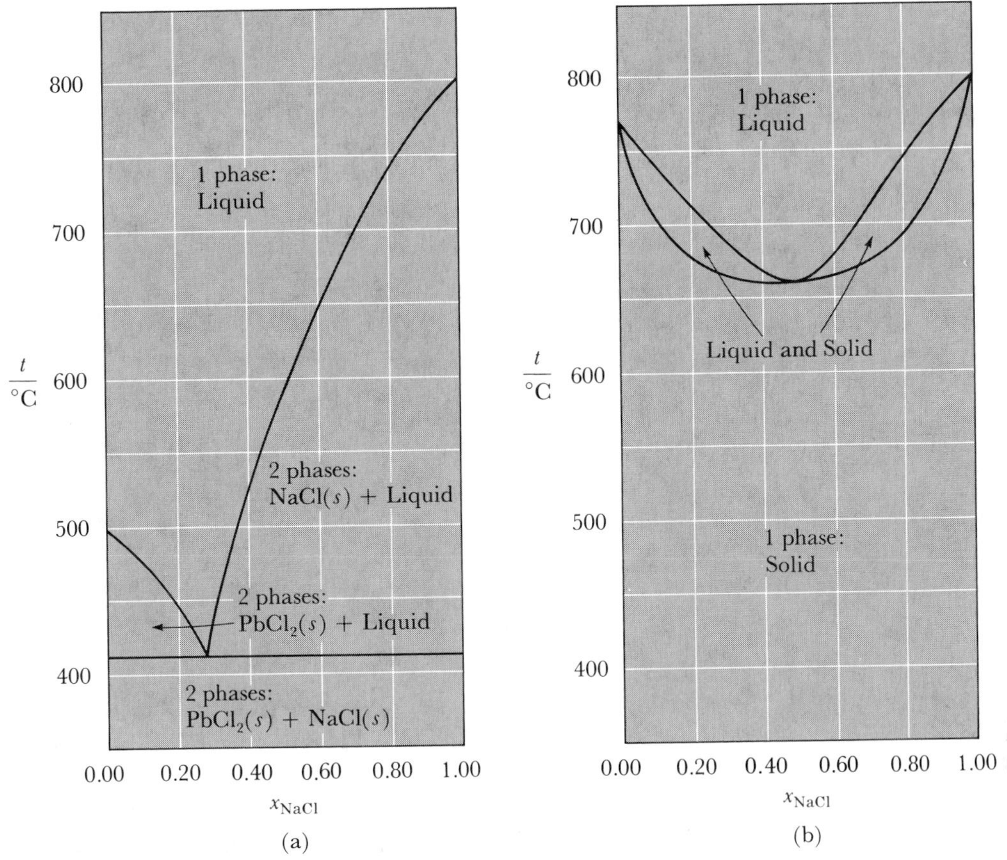

FIGURE 9–16
The phase diagrams for the systems (a) NaCl + $PbCl_2$ (these salts do not form solid solutions) and (b) NaCl + KCl (these salts are completely miscible in the solid state). In the 2-phase regions, each phase is a solution.

model of a polar diatomic molecule, the dipole moment p[5] is the product of the charge $\delta+$ and the distance d

$$p = (\delta +)d = |\delta -|d$$

Dipole moments are commonly expressed in Debye units (symbol D; named for PETER DEBYE), with

$$1 \text{ D} = 10^{-18} \text{ g}^{1/2} \text{ cm}^{5/2} \text{ s}^{-1}$$

The dipole moment resulting if $\delta = 10^{-10}$ esu of charge and $d = 1$ Å (10^{-8} cm) is 1 D. With this definition of the debye, the dipole moments of many simple polar molecules are in the range 0.1 D to 10 D. (Recall that one electrostatic unit of charge is that charge which exerts a force of 1 dyn upon an identical charge separated from it by 1 cm. The dimensions of this unit of charge are $\text{g}^{1/2} \text{ cm}^{3/2} \text{ s}^{-1}$, as explained in a footnote in Table

[5]The symbol μ also is used to denote dipole moment.

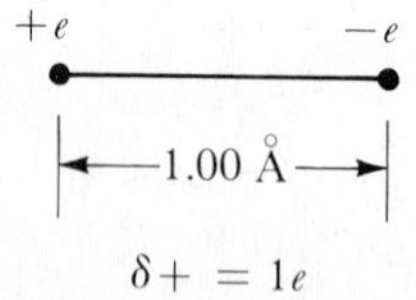

$+e$ $-e$

$\vdash$ 1.00 Å $\dashv$

$\delta + = 1e$

$$p = 4.802 \times 10^{-10}\ g^{1/2}\ cm^{3/2}\ s^{-1} \times 1.00 \times 10^{-8}\ cm$$
$$= 4.80 \times 10^{-18}\ g^{1/2}\ cm^{5/2}\ s^{-1}$$
$$= 4.80\ D$$

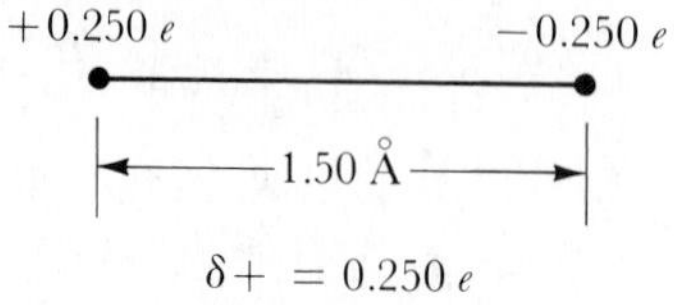

$+0.250\ e$ $-0.250\ e$

$\vdash$ 1.50 Å $\dashv$

$\delta + = 0.250\ e$

$$p = 0.250 \times 4.802 \times 10^{-10}\ g^{1/2}\ cm^{3/2}\ s^{-1} \times 1.50 \times 10^{-8}\ cm$$
$$= 1.80 \times 10^{-18}\ g^{1/2}\ cm^{5/2}\ s^{-1}$$
$$= 1.80\ D$$

FIGURE 9–17
The dipole moments of two idealized polar diatomic molecules.

8–1. In SI, the magnitude of the charge is expressed in coulombs and the distance is expressed in meters. In these units, the debye is 1 D = 3.336×10^{-30} C m.)

Figure 9–17 shows two idealized arrangements of charges with the calculated dipole moments. There are experimental methods for determining the dipole moment of a gaseous molecule, and Table 9–8 gives such values for some compounds. Also given are values calculated on the assumption that these compounds are made of hard-sphere ions, with charges $-e$ and $+e$ at the observed separation of the ions in the gaseous molecule. We see that an observed dipole moment is smaller than the value calculated under this simple assumption. A conclusion to be drawn is that in each of these gaseous molecules the pair of ions are not hard-sphere ions. The deviation from hard-sphere properties is accounted for by the *polarizability* of the ions.

TABLE 9–8
Dipole Moments of Diatomic Alkali Metal Halide Gas Molecules

	p/D Experimental value	p/D Calculated (100% ionic)[a]	% Ionic character[b]
LiF	6.33	7.25	87
LiCl	7.13	9.36	76
LiBr	7.27	10.23	71
LiI	7.43	11.48	65
NaCl	9.00	11.33	79
KCl	10.27	12.82	80
CsCl	10.42	13.97	75

[a]The dipole moment is calculated with the equation $p = e \times d$, in which d is observed separation of ions in the gas phase.
[b]The percent ionic character = $(P_{obs}/P_{calcd}) \times 100$.

TABLE 9–9
The Polarizability of Ions and Molecules[a]

O^{2-}	F^-	Ne	Na^+	Mg^{2+}
~3	0.81	0.39	0.24	0.10
	OH^-	HF		
	1.89	~0.7		
		H_2O		
		1.44		
		NH_3		
		2.2		
		CH_4	NH_4^+	
		2.60	1.65	
S^{2-}	Cl^-	Ar	K^+	Ca^{2+}
~9	2.98	1.62	1.00	0.60
Se^{2-}	Br^-		Rb^+	
~11	4.24		1.50	
	I^-		Cs^+	
	6.45		2.40	

[a]Polarizability values in $cm^3 \times 10^{24}$.

When an atom, a molecule, or an ion is subjected to an electric field, the electron cloud moves relative to the nucleus (or nuclei); idealized hard-sphere ions would not react in this way. The ease with which the electron cloud of an atomic or molecular species can be distorted by an electric field is called its polarizability. The polarizability of an atomic or molecular species, α, is defined as the ratio of the induced dipole to the electric field:

$$\alpha = \frac{\Delta \text{ dipole moment}}{\Delta \text{ electric field}}$$

The dimension of polarizability, so-defined, is $(\text{length})^3$, and values of α are comparable to molecular volumes determined in other ways.[6] The polarizabilities of some molecules and ions are given in Table 9–9. We see that for a series of isoelectronic monatomic species (e.g., O^{2-}, F^-, Ne, Na^+, and Mg^{2+}, all with the neon electronic structure) the polarizability decreases as the nuclear charge increases. This is reasonable because a greater nuclear charge will hold the electron cloud more tightly. We see also that with the total nuclear charge spread among more atoms, there is an easier distortion of the electron cloud (e.g., Ne, HF, H_2O, NH_3, and

[6]The dimension of cubic length follows from use of the electrostatic units in expressing both charge and field strength (field strength is given in potential (volts) per unit length, and potential times charge is energy):

$$\alpha \equiv \frac{\text{charge} \times \text{length}}{(\text{energy} \div \text{charge}) \div \text{length}} \equiv \frac{\text{charge}^2 \times \text{length}^2}{\text{energy}} \equiv \frac{(g^{1/2}\ cm^{3/2}\ s^{-1})^2 \times cm^2}{g\ cm^2\ s^{-2}} \equiv cm^3$$

FIGURE 9–18
A gaseous diatomic molecule consisting of two ions: (a) hard-sphere ions and (b) polarizable ions.

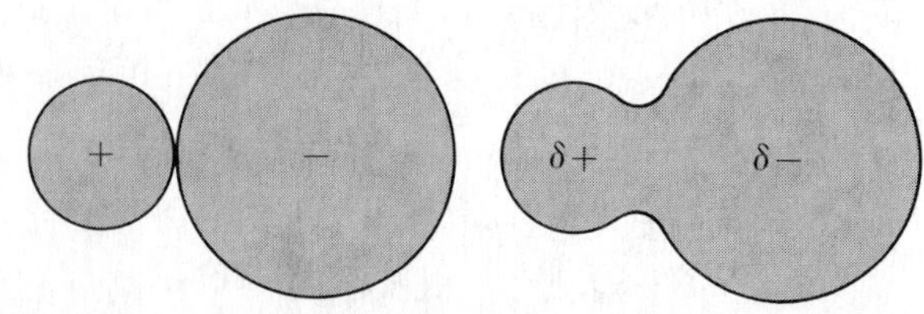

CH_4, each a species with ten electrons and a total nuclear charge of +10.). As you probably would expect, among species with analogous structures but involving atoms in different rows of the periodic table (e.g., the halide ions, F^-, Cl^-, Br^-, and I^-), the polarizability increases with an increase in the number of electrons.

In a gaseous diatomic molecule made up of two oppositely charged ions, the ions exert polarizing effects on each other's electron clouds, as depicted in Figure 9–18. This mutual polarization makes the dipole moment of the molecule smaller than it would be for a molecule consisting of hard-sphere ions. With increasing polarization of the electron clouds, the electron distribution in the gaseous ionic compound approaches that which corresponds to electron sharing by the two atoms. Hence bonding in the gaseous alkali halides may be said to have some small amount of covalent character.

SOLUTIONS OF IONIC COMPOUNDS IN WATER

Many ionic compounds dissolve to an appreciable extent in water and certain other solvents. These solutions have some properties not observed in solutions of the type discussed in Chapter 7. A solution of sodium chloride in water is a good conductor of electricity; sodium chloride and other solutes that give conducting solutions are called *electrolytes*. A solution of sugar in water is an extremely poor conductor of electricity;[7] sugar and other solutes which give nonconducting solutions are called *nonelectrolytes*. The vapor pressure lowering, the boiling point elevation, and the freezing point depression of a dilute solution of sodium chloride in water are about twice as large as these same quantities for a solution of sugar with the same molal concentration. This is shown in Table 9–10, which gives the freezing point of solutions of sodium chloride. Values of the quotient $(-\Delta t/\text{molality})$ are approximately two times larger than 1.86 K kg mol^{-1}, the molal freezing point constant for water presented in Chapter 7. The freezing point depression of electrolytes often is discussed in terms of the quotient $i = (-\Delta t/\text{molality})/K_f$, which is called the van't Hoff factor (named for JACOBUS VAN'T HOFF).

In 1883, ARRHENIUS proposed most clearly that sodium chloride and similar solutes are largely dissociated into ions in aqueous solution. The net

[7]The conductance of a dilute solution of sugar in water is approximately the same as that of pure water.

TABLE 9–10
The Freezing Point Lowering of Aqueous Solutions of Sodium Chloride

$\frac{\text{molality}}{\text{mol kg}^{-1}}$	$\frac{\Delta t}{\text{K}}$	$i = \frac{-\Delta t/\text{molality}}{K_f}$
0.100	−0.347	1.87
0.300	−1.022	1.83
0.500	−1.694	1.82
1.00	−3.391	1.82

reaction occurring when solid sodium chloride dissolves in water is

$$NaCl(s) = Na^+(aq) + Cl^-(aq)$$

Aqueous sodium chloride solution conducts electricity because sodium ions (Na^+) and chloride ions (Cl^-) migrate in an electrical field, the sodium ions in one direction and the chloride ions in the other. The abnormal vapor pressure lowering, boiling point elevation, and freezing point depression for sodium chloride as a solute are caused by its dissociation; one mole of dissolved sodium chloride gives two moles of solute particles: one mole of sodium ion and one mole of chloride ion.

The heat effect associated with dissolving sodium chloride in water is very small ($\Delta H^0 = +3.9$ kJ mol^{-1}). Considering the strong attraction between ions of opposite charge, this experimental result may seem surprising. The reaction

$$NaCl(s) = Na^+(g) + Cl^-(g) \qquad \Delta H^0 = +786.3 \text{ kJ mol}^{-1}$$

in which the oppositely charged ions become separated, is accompanied by an enormous increase of enthalpy; the formally similar reaction,

$$NaCl(s) = Na^+(aq) + Cl^-(aq) \qquad \Delta H^0 = +3.9 \text{ kJ mol}^{-1}$$

which also involves separation of the oppositely charged ions, is accompanied by a very small increase of enthalpy. One conclusion from this comparison is that much heat is evolved in the dissolving of the gaseous sodium and chloride ions in water. Subtraction of the first of the equations from the second gives

$$Na^+(g) + Cl^-(g) = Na^+(aq) + Cl^-(aq)$$

and the enthalpy change for this process is obtained by subtracting the two values of ΔH^0:

$$\frac{\Delta H^0}{\text{kJ mol}^{-1}} = 3.9 - 786.3 = -782.4$$

$$\Delta H^0 = -782.4 \text{ kJ mol}^{-1}$$

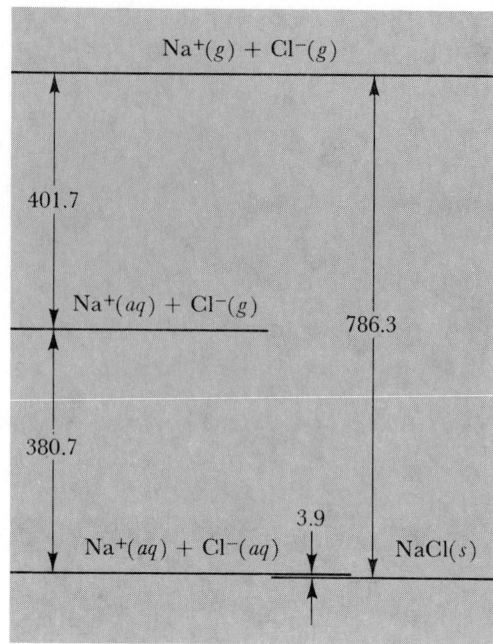

FIGURE 9–19
Thermochemical cycle for the solution of sodium chloride in water. Energy quantities are in kJ mol^{-1}.

The relationship of these several energy quantities is shown in Figure 9–19.

The very small enthalpy effect of dissolving solid sodium chloride in water is a consequence of the very nearly equal magnitudes of the lattice energy of sodium chloride and the sum of the hydration energies of gaseous sodium ion and gaseous chloride ion. The enthalpy change associated with dissolving an ionic solid in water varies, depending upon the substance; it may be positive or it may be negative, but for all substances it is much less positive than the enthalpy change in producing the gaseous ions from the solid. This is simply a consequence of the strong attractive interaction between the dissolved ions and molecules of water.

The sum of the changes of enthalpy associated with the two "reactions,"

$$Na^+(g) = Na^+(aq) \qquad Cl^-(g) = Cl^-(aq)$$

the hydration of one mole of gaseous sodium ion and one mole of gaseous chloride ion, is -782.4 kJ mol^{-1}. These two "reactions" cannot be studied individually, and the heat effect caused by dissolving one mole each of gaseous sodium ion and gaseous chloride ion cannot rigorous be resolved into the heat effects associated with the individual ions. A number of different approaches to this problem give results that agree reasonably well. For sodium ion and chloride ion, the hydration processes are each strongly exothermic:

$$Na^+(g) = Na^+(aq) \qquad \Delta H^0 = -401.7 \text{ kJ mol}^{-1}$$
$$Cl^-(g) = Cl^-(aq) \qquad \Delta H^0 = -380.7 \text{ kJ mol}^{-1}$$

The enthalpies of hydration of common cations and anions are presented

TABLE 9–11
Enthalpies of Hydration of Gaseous Ions[a]

H^+				
−1084				
Li^+	Be^{2+}			F^-
−510	−2473			−506
Na^+	Mg^{2+}	Al^{3+}	S^{2-}	Cl^-
−402	−1908	−4640	−1381	−381
K^+	Ca^{2+}			Br^-
−314	−1577			−343
Rb^+	Sr^{2+}			I^-
−289	−1431			−297
Cs^+	Ba^{2+}			
−255	−1289			

[a]Numerical values are in kJ mol^{-1}.

in Table 9–11, where two important trends can be noted:

1. For ions of a particular charge the enthalpy of hydration is greater the smaller the ion. (Recall the trends in ionic radius values as the principal quantum number increases, shown in Figure 9–1.)
2. The enthalpy of hydration increases with increasing charge; a cursory examination of the table suggests a dependence upon the square of the charge. For instance, lithium(I) ion and magnesium(II) ion have approximately equal radii; the heats of hydration of these two ions are −510 kJ mol^{-1} and −1908 kJ mol^{-1}, respectively. The ratio of these values is 3.74, which is $2^{1.9}$.

The representations $Na^+(aq)$ and $Cl^-(aq)$ do not disclose the nature of the interaction of these ions with water. We will go into these interactions in some detail in later chapters.

The enthalpy data just presented have shown the strong interaction between solute ions and solvent water molecules and the dependence of this interaction upon the magnitude of the ionic charge. These hydration phenomena manifest themselves also in the entropy change in the dissolving of an ionic solute. The effect of ionic charge can be seen in the entropy change for reactions in which isoelectronic salts (KCl and CaS) dissolve in water at 298.2 K:

$$KCl(s) = K^+(aq) + Cl^-(aq) \qquad \Delta S^0 = +75 \text{ J K}^{-1}\text{ mol}^{-1}$$
$$CaS(s) = Ca^{2+}(aq) + S^{2-}(aq) \qquad \Delta S^0 = -138 \text{ J K}^{-1}\text{ mol}^{-1}$$

These are conventional standard entropy changes; therefore they pertain to the reaction in which the ions are produced in the ideal 1-molal solution.

The effect of ionic size can be seen by comparing potassium chloride with lithium fluoride at 298.2 K:

$$\mathrm{LiF}(s) = \mathrm{Li}^+(aq) + \mathrm{F}^-(aq) \qquad \Delta S^0 = -36\ \mathrm{J\ K^{-1}\ mol^{-1}}$$

In these reactions, solid ionic compounds dissolve in water to give solutions in which each ion is present at a concentration of 1.0 molal. The ions gain randomness, but the solvent water *loses randomness*. The higher the charge on the ions and the smaller the size of the ions, the more strongly do the ions interact with the water molecules, thereby reducing their randomness. This effect is so great that the values of ΔS^0 for the solution of lithium fluoride and calcium sulfide are negative.[8]

If the solvent water in a solution of a completely dissociated ionic solute obeyed Raoult's law (appropriate account being taken of the total number of solute particles, ions of both sign), the freezing point depression constants for different types of salts would be simple multiples of 1.86 K kg mol^{-1}. That is, the van't Hoff coefficient would be equal to ν, the number of ions into which the electrolyte dissociates. Table 9–10 shows that i (NaCl) is appreciably less than 2, and decreases with increasing concentration. Figure 9–20 shows the same general behavior for other electrolytes. Although the values of $(-\Delta t/\text{molality})$ approach the ideal value at very low concentrations,

$$-\frac{\Delta t}{\text{molality}} = \nu \times 1.86\ \mathrm{K\ kg\ mol^{-1}}$$

there are wide deviations from ideality at higher concentrations. For some electrolytes the value of i decreases and then increases as the concentration increases. We see in the figure that the nonelectrolytes sucrose and hydrogen peroxide, which form hydrogen bonds with water, give the ideal freezing point depression at concentrations where the electrolytes show wide deviations from ideality. The nonideality of solutions of electrolytes at low concentrations is due to the long-range attractive forces between ions of opposite charge. Each ion is influenced by all other ions, although the effect is greatest for nearby ions; this phenomenon has given rise to the concept of *ion-atmosphere*. The ion-atmosphere for an ion is all other ions in the system. Ion-atmosphere effects are displayed in most phenomena involving electrolytes in solution. In addition to the effects on the freezing point depression shown in Figure 9–20, there are effects on the conductance of solutions of electrolytes.

If the ions of a completely dissociated ionic solute were independent of one another, the conductance of the solution would be directly propor-

[8]The quantities being examined here are standard entropy changes in the reactions dissolving the anhydrous salts. The discussion has focused on the product ions in solution. These entropy changes depend also, of course, upon the entropies of the reactant solid salts. However, variation of the entropies of the solid salts (KCl, 82.7 J K^{-1} mol^{-1}; CaS, 56.5 J K^{-1} mol^{-1}; and LiF, 36 J K^{-1} mol^{-1}) is not large enough to alter the qualitative and semiquantitative comparisons made here involving values of ΔS^0. More is said about entropies of ions in solution in Appendix 3.

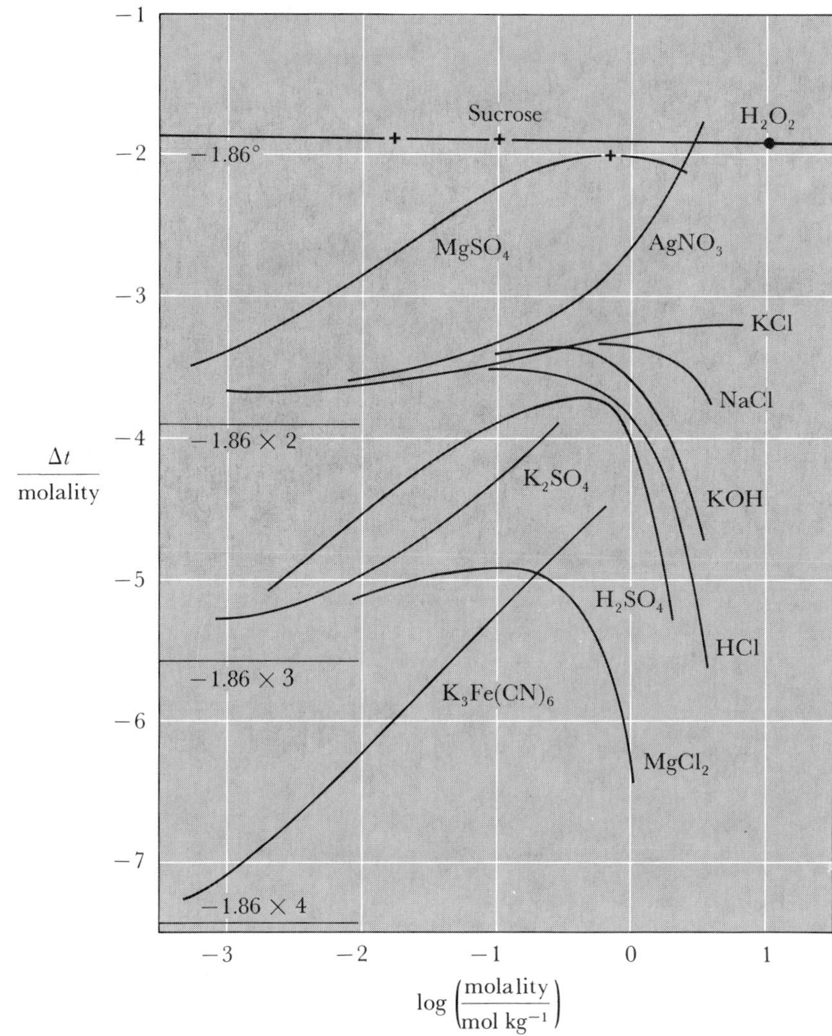

FIGURE 9–20
The freezing point depression of aqueous solutions of various electrolytes. Notice the points for the nonelectrolytes that form hydrogen bonds with H_2O: sucrose +, hydrogen peroxide ●. [From J. H. Hildebrand, *J. Chem. Educ.* **48**, 224 (1971).]

tional to the concentration. However, it is found in practice that the conductance of solutions of electrolytes does not increase with concentration as rapidly as the concentration increases. This is shown in Figure 9–21, in which the conductance (the reciprocal of resistance) of solutions of several chloride salts is presented as a function of concentration. Possible explanations for the decrease in conductance with increasing concentration are:

1. partial association of the ions of opposite charge to form neutral molecules [e.g., $Na^+(aq) + Cl^-(aq) = NaCl(aq)$] or species of lower charge [e.g., $Mg^{2+}(aq) + Cl^-(aq) = MgCl^+(aq)$]. Such association reactions lower the current-carrying ability of the solution, and occur to a greater extent with an increasing concentration of solute. (The dependence of

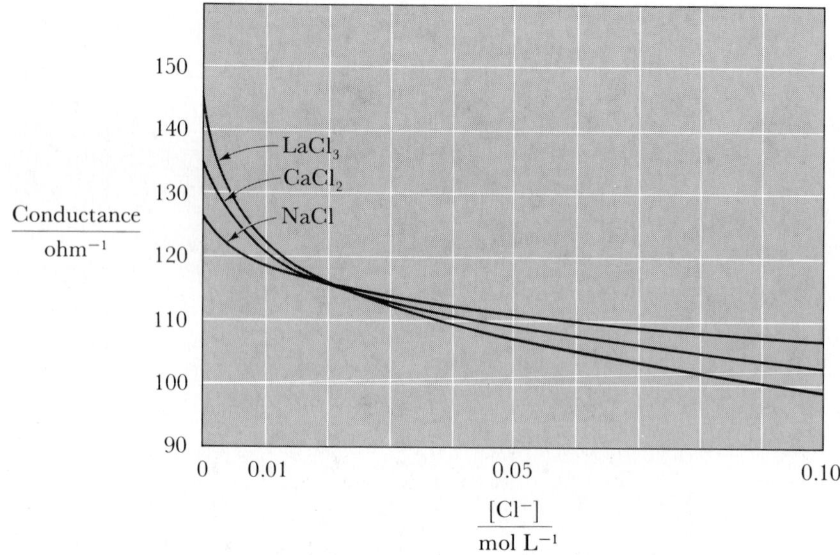

FIGURE 9–21
The total conductance of aqueous solutions containing one mole of chloride ion. The experimental arrangement to which these conductance values correspond is one in which the distance between the parallel plane electrodes is 1 cm. The total amount of solution is such that 1 mol of Cl^- is present. Thus, for instance at $[Cl^-] = 0.10$ mol L^{-1}, $C_{CaCl_2} =$ 0.050 mol L^{-1} and the total volume is 10.0 L. (Of course, the actual measurements are made with smaller cells.)

the extent of ion association upon concentration is discussed thoroughly in Chapter 13.)

2. the influence of the ion-atmosphere on the mobility of ions. In an electrolyte solution, the current is carried by ions of both charges, the positive and negative ions moving in opposite directions. Even though the ions of unlike charge are not associated with one another, an ion-atmosphere exerts a drag on the motion of an ion in solution. This drag increases with an increase of concentration because the ions are closer together on the average when the concentration is increased.

Quantitative correlation of the conductance with concentration shows that the first of these possible explanations *cannot* be the only one for these electrolytes. Other types of data also support the view that sodium chloride, calcium chloride, and lanthanum chloride ($LaCl_3$) are completely dissociated in aqueous solution. Such electrolytes are called *strong electrolytes*; to explain the properties of solutions of such electrolytes, account is taken of ion-atmosphere effects.

There are other ionic solutes (e.g., cadmium halides and certain acids) that are associated to an appreciable extent in solutions of moderate concentrations (0.01–1 mol L^{-1}), and for such solutes, called *weak electrolytes*, the quantitative description of the nonideality must take this association into account.

The conductance curves in Figure 9–21 can be extrapolated to zero concentration to disclose information about the mobility of the individual

TABLE 9–12

The Mobility of Ions, u, in Aqueous Solution at 25°C

	$\frac{10^8\ u}{m^2\ V^{-1}\ s^{-1}}$		$\frac{10^8\ u}{m^2\ V^{-1}\ s^{-1}}$
H^+	36.3	OH^-	20.5
Li^+	4.00	F^-	5.74
Na^+	5.20	Cl^-	7.91
K^+	7.62	Br^-	8.10
Rb^+	8.06	I^-	7.96
Cs^+	8.01	ClO_4^-	6.98
Mg^{2+}	5.50	SO_4^{2-}	8.29
Al^{3+}	6.5	PO_4^{3-}	9.62
Ba^{2+}	6.60		

ions in an electrical field.[9] Limiting ionic mobilities for some common ions are given in Table 9–12. The velocity of an ion is proportional to the electrical field; the mobility of an ion, u, is given as its velocity (in meters per second) in a field of one volt per meter. The dimension of u is $(m\ s^{-1})/(V\ m^{-1}) = m^2\ V^{-1}\ s^{-1}$. These mobility values provide some surprises, in particular the extraordinary mobilities of hydrogen ion and hydroxide ion. These abnormal values are accounted for by the unusual way these ions move. Their migration occurs by a cooperative transfer of protons from hydronium ion to water molecules or from water molecules to hydroxide ion. This can be depicted as involving the proton motion

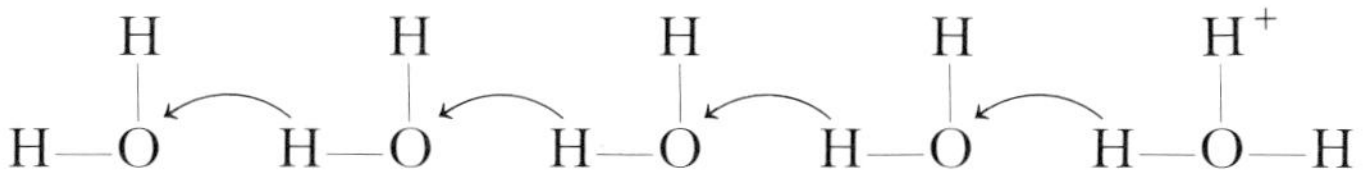

```
     H          H          H          H          H+
     |  ↙⌒      |  ↙⌒      |  ↙⌒      |  ↙⌒      |
  H—O       H—O       H—O       H—O       H—O—H
```

which gives

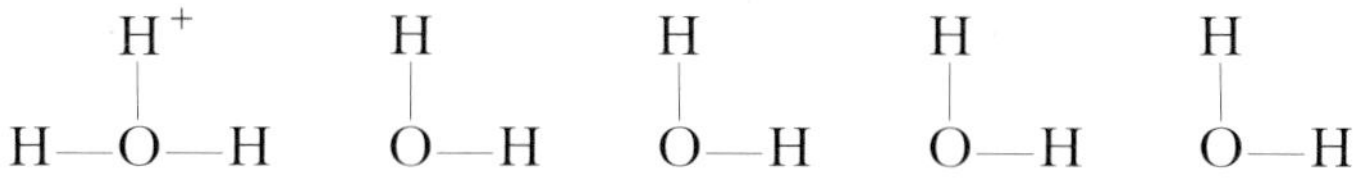

```
     H+         H          H          H          H
     |          |          |          |          |
  H—O—H       O—H        O—H        O—H        O—H
```

or the proton motion

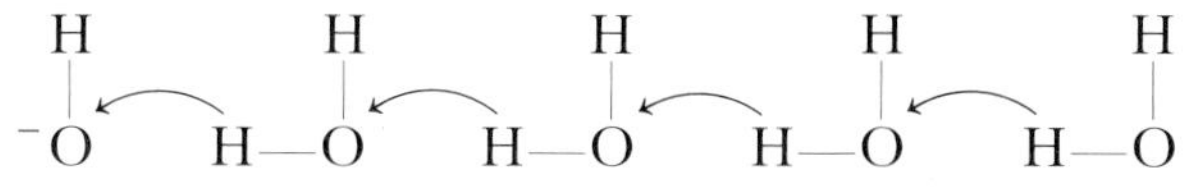

```
     H          H          H          H          H
     |  ↙⌒      |  ↙⌒      |  ↙⌒      |  ↙⌒      |
    ⁻O      H—O       H—O       H—O       H—O
```

which gives

```
     H          H          H          H          H
     |          |          |          |          |
     O—H        O—H        O—H        O—H        O⁻
```

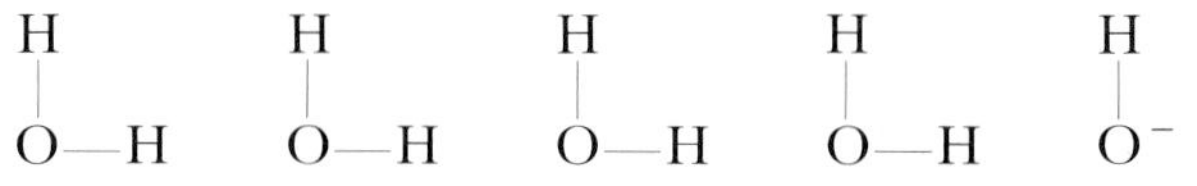

In each case the apparent distance traveled by the hydronium ion (hydrated hydrogen ion) or hydroxide ion is not the distance traveled by a particular

[9]Unlike the hydration enthalpy for individual ions, individual ion mobilities can be evaluated rigorously with appropriate experimental data.

ion; rather, shuffling of protons has made it seem that the ion has moved a greater distance. This mechanism is called a Grotthuss chain (named for C. J. D. VON GROTTHUSS). Table 9–12 also presents data that show unexpected dependences of ion mobility upon ionic size and ionic charge. If the solvent were a structureless medium, the mobility of an ion would be proportional to its charge and inversely proportional to its radius:

$$u \propto \frac{Z}{r}$$

in which Z is the magnitude of the charge of the ion and r is its radius. But the series of alkali metal ions (Li^+, Na^+, K^+, Rb^+, and Cs^+), isoelectronic monatomic cations (Na^+, Mg^{2+}, and Al^{3+}), and isoelectronic polyatomic anions (ClO_4^-, SO_4^{2-}, and PO_4^{3-}) do not show these dependences. The relationship $u \propto (Z/r)$ is based upon a model to which these systems do not conform. The smaller the ion and the larger the charge the greater is the hydration of the ion, and an ion's mobility is decreased if it must drag water molecules with it as it moves under the influence of the electric field. Thus we find a more mild dependence upon charge than first power, and a decrease in the radius of monatomic ions actually decreases the ion mobility.

Biographical Notes

SVANTE A. ARRHENIUS (1859–1927), Swedish physical chemist, is most noted for his work on solutions of electrolytes and his studies of the effect of changes of temperature upon reaction rate (see Chapter 15). He received the Nobel Prize in Chemistry in 1903 for the former work.

MAX BORN (1882–1970) and FRITZ HABER (1868–1934), were a prominent German physicist and chemist, respectively. Born, who became a British subject in 1939, is famous for his work in the theory of relativity, quantum theory, and atomic structure. He received the Nobel Prize in Physics in 1954. Haber, who received the Nobel Prize in Chemistry in 1918, is noted for his work developing the most widely used method for fixing atmospheric nitrogen. Their work on the Born–Haber cycle was published in 1919.

PETER J. W. DEBYE (1884–1966) was born in the Netherlands and received his Ph.D. from the University of Munich. In Europe, he was associated with universities in Zurich, Utrecht, Göttingen, Leipzig, and Berlin. He moved to the United States in 1940, where he was Professor of Chemistry at Cornell University. He did truly outstanding work in many areas of physical chemistry: he studied the properties of polar molecules, solutions of electrolytes, and powder x-ray diffraction methods. Debye received the Nobel Prize in Chemistry in 1936.

Y. I. FRENKEL (1894–1954), a Russian physicist, was educated at St. Petersburg University. Most of his professional life was spent at the Technical Institute in Leningrad. He did outstanding work in many areas of theoretical physics. His work on solids, in which he described the defects now known as Frenkel defects, dates from the middle 1920s.

FRITZ HABER (see Max Born above and Haber's biographical note in Chapter 2).

E. MADELUNG, a German physicist, published the paper that gave his name to the numerical factor in the equation for the lattice energy of an ionic solid in 1918, while he was associated with the University at Freiburg.

W. SCHOTTKY (1886–), a German physicist, did much to elucidate the electrical properties of semiconductor materials. He was Professor of Physics at Rostock and later lecturer in the University of Erlangen.

JACOBUS H. VAN'T HOFF (1852–1911), Dutch chemist, was educated in Holland and began his career there. He was later Professor at Leipzig and Berlin. Van't Hoff is considered one of the founders of physical chemistry. He made pioneering contributions in thermochemistry, in the properties of solutions, and in the theory of optical activity of chemical compounds. In 1901, he received the first Nobel Prize in Chemistry.

C. J. D. VON GROTTHUSS (1785–1822), a physicist, was born in Germany and died in Russia. His early ideas on the electrolysis of water contained features of the mechanism of motion of hydronium ion and hydroxide ion. In 1819, he formulated the law that only absorbed light can cause chemical change in substances.

Problems and Questions

9–1 Give the chemical formulas for all possible ionic compounds involving the alkaline earth ions and the anions S^{2-}, Cl^-, OH^-, and PO_4^{3-}.

9–2 From your knowledge of the crystal structure of sodium chloride, calculate the density of this solid. Compare your calculated value with the experimental value given in the introduction to this chapter.

9–3 The density of cesium chloride is 3.97 g cm^{-3}. From this datum and the structure of cesium chloride shown in Figure 9–3, calculate the closest cesium–chlorine distance in the solid. Compare this with the sum of the ionic radii for these ions tabulated in Table 9–1.

9–4 Study the structures shown in Figure 9–3. In each case, is the cube (or for TiO_2, the rectangular parallelopiped) a unit cell? How many atoms of each kind are present in each unit cell?

9–5 Consider a two-dimensional lattice, M^+X^-, with a square arrangement of ions, each cation (and anion) having a coordination number of four. The closest cation–anion separation is d. Give the first five terms in the equation for the lattice energy of this structure.

9–6 Consider a planar arrangement of a cation surrounded by three anions at the corners of an equilateral traingle. What is the critical value of r/R for which there are both cation–anion and anion–anion contacts?

9–7 Consider the cesium chloride lattice, a body-centered cubic lattice. (This lattice for metals was considered in Chapter 5.) Give the first five terms in the equation for the lattice energy of this structure.

9–8 In each of the compounds K_2O, CuI, and ZrI_4 the anions form a face-centered cubic lattice with cations occupying tetrahedral sites. How many cations are coordinated to each anion?

9–9 Table 9–6 gives the properties of some isoelectronic ionic compounds. What other compounds could be included in the two series listed in this table? From a handbook or other source, learn what you can about the properties of these other compounds.

9–10 Using the ionic model, calculate the lattice energy of potassium bromide, which has the NaCl structure with $a = 3.29$ Å and $n = 9$. Compare this value with the value given in Problem 9–16.

9–11 Which do you expect to have the higher melting point, RbF or SrO, MgO or CaS, LiBr or KBr?

√9–12 The standard enthalpy of formation of magnesium fluoride (MgF_2) and magnesium oxide (MgO) are -1076 kJ mol^{-1} and -602 kJ mol^{-1}. Compare the Born–Haber cycles for these compounds and indicate which steps tend to make magnesium fluoride more stable.

9–13 Some active metals (e.g., calcium and lithium) form nitrides (Ca_3N_2 and Li_3N), which have properties suggesting that they are ionic compounds containing nitride ion N^{3-}. Explain this nitride formation in view of the fact that for the process $N(g) + e^- = N^-(g)$, $EA(N) \cong 0$.

9–14 The Born–Haber cycle for sodium hydride was given in Figure 9–10 (and in Table 9–5). The standard heat of formation of cesium hydride is less negative than that of sodium hydride. To what step(s) in the Born–Haber cycle can this be attributed? (Here not all of the needed data are given. However, qualitative trends in the heats of vaporization of the alkali metals may be obtained by checking values for other alkali metals given in this chapter.)

9–15 Consider the following pairs of substances: CaO and CdS, ZnI_2 and HgI_2, HgO and HgSe. For which compound in each pair will the lattice energy calculated on the basis of an ionic model be more inaccurate? Why?

√9–16 The dissociation energy for bromine (Br_2) is 190.0 kJ mol^{-1} and the heat of vaporization of bromine liquid is 30.9 kJ mol^{-1}. The lattice energy of potassium bromide is 672.4 kJ mol^{-1}, and its standard enthalpy of formation from solid potassium and liquid bromine is -392.2 kJ mol^{-1}. With these data and the appropriate data for potassium (given in Figure 9–11), calculate the electron affinity of bromine. Using hydration enthalpies of potassium ion and bromide ion given in Table 9–11, calculate the enthalpy change for the solution process $KBr(s) = K^+(aq) + Br^-(aq)$.

9–17 The following salts are isoelectronic: $KClO_4$, $CaSO_4$, and $ScPO_4$. The solubilities of these compounds in water decrease in the order listed. Name factors which could be responsible for this trend.

9–18 What is the average charge per nickel atom in a material with composition $Ni_{0.90}O$? What is the ratio of Ni^{2+} to Ni^{3+} in this material? What is the composition of a lithium nickel oxide without lattice defects that has the same relative amounts of nickel(II) and nickel(III) as the nonstoichiometric solid $Ni_{0.90}O$?

9–19 Table 9–7 gives the vacancies of titanium and oxygen sites in solids with composition $TiO_{1.32}$ to $TiO_{0.69}$. If the density of ideal TiO, in which all sites are occupied, is designated as d, what is the density of each of the substances listed in the table? Make a graph in which the fraction of each kind of site that is vacant is plotted versus x, where x is the number of oxygen atoms per titanium atom (TiO_x). From this graph predict the fraction of titanium and oxygen sites vacant in a material with the compositions $TiO_{0.9}$ and $TiO_{1.2}$.

9–20 Barium titanate, $BaTiO_3$, is a nonconductor of electricity, but replacement of some Ba^{2+} with La^{3+} produces $La_xBa_{1-x}TiO_3$, an important semiconductor. What is ratio Ti^{3+}/Ti^{4+} in this compound?

9–21 What is the dipole moment of a gaseous ionic compound in which ions of unit charge ($\pm 1\ e$) are separated by 1.83 Å, assuming hard-sphere ions?

9–22 The dipole moment of gaseous CsF is 7.88 D and the interatomic distance in this molecule is 0.255 nm. What is the percent ionic character in this molecule, as deduced from these data?

9–23 On the basis of data presented in Table 9–9, predict the polarizabilities of NH_2^-, H_3O^+, and BH_4^-. (These species are isoelectronic with other species in the table; you may be helped in making your prediction by placing these species in the table.)

9–24 Phase diagrams (Figure 9–16) reveal that sodium chloride forms solid solutions with potassium chloride but does not form solid solutions with lead chloride. Suggest a reason for this difference.

9–25 Plot the expected cooling curves for liquid mixtures of NaCl with $PbCl_2$ and NaCl with KCl, each with 80 mole percent NaCl.

9–26 If aqueous solutions of the following strong electrolytes behaved ideally, what would you expect for the freezing point of 0.0500-molal solutions: $CaCl_2$, KBr, and $AlBr_3$?

9–27 What are the concentrations of the individual ionic species in each of the following three solutions: 0.054*M* NaCl, 0.0720*M* $CaCl_2$, and 0.142*M* $LaCl_3$?

9–28 Give the concentrations of the individual ionic species in a mixed electrolyte solution prepared by putting each of the following solutions in a 500-cm^3 volumetric flask and adding water to the calibration mark: 0.520*M* NaCl (100.0 cm^3), 0.123*M* $CaCl_2$ (100.0 cm^3), and 0.0540*M* $LaCl_3$ (200.0 cm^3).

9–29 The freezing point of 0.100 molal K_2SO_4 is -0.432°C. What is the van't Hoff factor i for potassium sulfate in this solution?

10

Covalent Bonding

10–1 Introduction

The preceding chapter dealt with chemical substances, such as common salt, in which ionized atoms of metals (positive ions) and nonmetals (negative ions) are held together primarily by the electrostatic attraction of the unlike charges. In contrast, this chapter deals with chemical substances in which atoms are bonded together by the sharing of electrons. The concept of a shared pair of electrons constituting a chemical bond, the covalent bond, was introduced by G. N. Lewis in 1916. Quantum mechanics has allowed a more quantitative development of concepts that Lewis originally proposed in a qualitative way. Many of the original qualitative concepts are still useful, and this chapter will be a mixture of the old and the new.

The description of the composition and structure of a chemical substance, given with reference to water in Chapter 1, concluded with the geometry of the molecule and the electron distribution in the molecule. In this chapter, we shall describe each of these aspects of molecular structure for more complex molecules. We also shall concern ourselves with the energy changes associated with formation of chemical bonds in molecules.

We introduce the subject of covalent bonding by considering two elements, hydrogen and fluorine, and the compound containing these elements, hydrogen fluoride. These elements exist in the free state as diatomic molecules H_2 and F_2, and hydrogen fluoride also is a diatomic molecule, HF. Atoms of hydrogen and fluorine each have a single electron in one valence-shell orbital:

			1s	2s	2p
$_1$H	$1s^1$	H	(↑)		
$_9$F	$1s^2 2s^2 2p^5$	F	(⇅)	(⇅)	(⇅)(⇅)(↑)

In each of the diatomic molecules involving these atoms, a pair of electrons is shared. These molecules can be represented by simple electron-dot structures, called *Lewis structures*,

$$\mathrm{H:H} \qquad \mathrm{:\ddot{F}:\ddot{F}:} \qquad \mathrm{H:\ddot{F}:}$$

in which only valence-shell electrons are shown. These Lewis structures for H_2, F_2, and HF show the two, fourteen, and eight valence-shell electrons belonging to the bonded atoms.

The concepts of atomic orbital and electron spin were not known when electron-dot representations were introduced, but these concepts now are used in specifying the prerequisites for formation of a covalent bond. An atom can form an electron-pair bond for each stable valence-shell orbital, and the electrons of the shared pair must have opposite spin. In the diatomic molecules in question, the hydrogen atom is using its 1*s* orbital, and the fluorine atom is using one of its 2*p* orbitals. These orbitals overlap in the region in which the bonding pair of electrons is located. Certain points that arose in Chapter 8 in connection with the electron distribution in atoms are equally appropriate here. The electrons in a molecule, like the electrons in an atom, are not stationary in space. They have energy, made up of both the kinetic energy of their motion and the potential energy of their electrostatic interaction with positively charged nuclei and with other electrons. The electrons shown as the bonding pair have a probability distribution that favors the region between the bonded atoms, but there is an appreciable probability away from the molecular axis. This point will be presented further in connection with the *molecular-orbital* description of chemical bonding, to be discussed later in this chapter.

In the homonuclear diatomic molecules H_2 and F_2 the bonding electrons are shared equally by the bonded atoms, and these molecules have no net polarity. In hydrogen fluoride, HF, the bonding electrons are not shared equally, and this substance is polar. The dipole moment (p) of hydrogen fluoride is 1.82 D, a value much smaller than the dipole moments of gaseous alkali metal halides [e.g., $p(\mathrm{LiF}) = 6.33$ D]. In Chapter 8 we learned that the electron affinity of fluorine is much greater than the electron affinity of hydrogen [$EA(\mathrm{F}) = 327.8\ \mathrm{kJ\ mol^{-1}}$, $EA(\mathrm{H}) = 72.7\ \mathrm{kJ\ mol^{-1}}$], and on this basis it is reasonable to assume that this molecule has the polarity:

$$\overset{\delta+}{\mathrm{H}}\text{—}\overset{\delta-}{\mathrm{F}}$$

in which $\delta+$ and $\delta-$ represent partial charges, as in Figure 9–17.

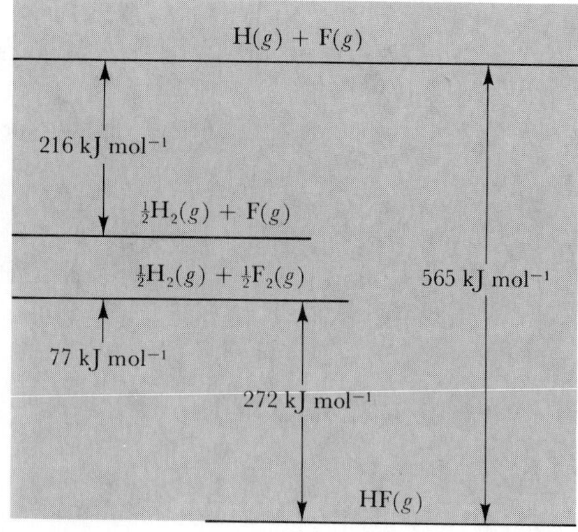

FIGURE 10–1
The energy of different combinations of one mole each of hydrogen atoms and fluorine atoms.

Chapter 9 emphasized the Born–Haber cycles for the formation of crystalline ionic compounds. We can consider the formation of covalently bonded hydrogen fluoride from its constituent elements,

$$H_2(g) + F_2(g) = 2HF(g)$$

also in terms of a cycle of steps:

(a) $H_2(g) = 2H(g)$

(b) $F_2(g) = 2F(g)$

(c) $HF(g) = H(g) + F(g)$

The formation reaction is the combination (a) + (b) − 2 × (c). The change of internal energy in each of these reactions is the bond-dissociation energy of the molecule, which is represented by the symbol D. The bond-dissociation energies are $D(\text{H–H}) = 432\ \text{kJ mol}^{-1}$, $D(\text{F–F}) = 154\ \text{kJ mol}^{-1}$, and $D(\text{H–F}) = 565\ \text{kJ mol}^{-1}$. The change of enthalpy in the reaction forming two moles of hydrogen fluoride can be calculated:

$$\Delta H^0 = 2\Delta H_f^0(\text{HF}) = D(\text{H}_2) + D(\text{F}_2) - 2D(\text{HF}) \qquad \text{(see footnote 1)}$$

$$\frac{\Delta H^0}{\text{kJ mol}^{-1}} = +432 + 154 - 2 \times 565 = -544$$

$$\Delta H^0 = -544\ \text{kJ mol}^{-1}$$

This calculation and Figure 10–1 show that the formation of diatomic hydrogen fluoride from the reactant diatomic molecules is exothermic. This is so because the bond-dissociation energy of hydrogen fluoride is greater than the average of the bond-dissociation energies of hydrogen and fluorine.

---

[1]In the overall reaction $H_2(g) + F_2(g) = 2HF(g)$, there is no change in the number of moles of gas, therefore $\Delta H = \Delta U$. Thus there is no correction necessary in using the bond-dissociation energies (which are values of ΔU) in calculation of ΔH.

We now will explore further the concept of bond energy that we have used already in thermochemical calculations. We can clarify this concept by graphing the total energy of two hydrogen atoms with electrons of opposite spin as a function of the distance of separation of the two nuclei. A graph of this relationship is given in Figure 10–2. The general shape of the curve is similar to that already given for the interaction of two argon atoms (Figure 4–16) or the interaction of two ions of opposite charge (Figure 9–2), but the mathematical equations for these three curves are not the same, and the energy at the minimum is very different for the van der Waals interaction of argon atoms ($E = -0.995$ kJ mol^{-1} at $r = 3.82$ Å) than for the interaction of sodium ion and chloride ion ($E = -517$ kJ mol^{-1} at 2.4 Å) or the interaction of two hydrogen atoms ($E = -457.9$ kJ mol^{-1} at 0.74 Å). The curved line in this figure gives the total energy of two hydrogen atoms with a fixed separation as a function of the interatomic separation. The total energy given by this line includes the potential energy of interaction of the four particles (two protons and two electrons) and the kinetic energy of the electrons. (This is analogous to the energy of the hydrogen atom as calculated in the Bohr derivation, in which the contributions of both kinetic energy and potential energy are taken into account.) We will think of the curve as giving the *net potential energy* of the system relative to that of the separated atoms; this net potential energy includes a positive contribution of the kinetic energy of the electrons. Points in this figure that are not on the curved line correspond to systems in which the two hydrogen nuclei are moving relative to one another. We shall discuss four points shown in the figure, each for a particular separation of the atoms, r_1. The value of the ordinate for each point gives a total energy, E, the sum of the net potential energy, E_p, and kinetic energy of relative nuclear motion, E_k. For Point a, $E < E_p$, which would correspond to a negative amount of kinetic energy;

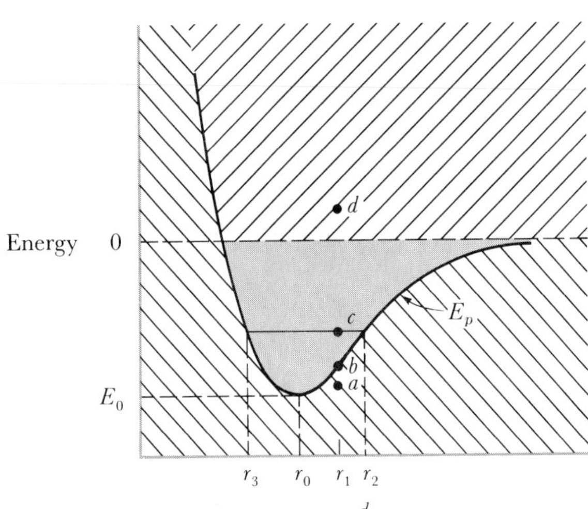

FIGURE 10–2
The potential energy of two hydrogen atoms with electrons having opposite spin. Significance of regions:

(hatched)	$E > 0$	atoms not combined in molecule
(shaded)	$0 > E > E_p$	atoms combined in molecule
(cross-hatched)	$E_p > E$	impossible

this is an impossible situation. For Point b, $E = E_p$; this would correspond to the system of two hydrogen atoms having no kinetic energy of motion relative to one another. For Point c, $E > E_p$; the atoms have positive kinetic energy of motion relative to one another, but the total energy is less than that of two separated hydrogen atoms. These two atoms are trapped in each other's presence; they can get no further apart than r_2, where $E = E_p$. For Point d, $E > 0$, and the atoms are not trapped in each other's presence. Point d corresponds to two hydrogen atoms not combined in a molecule; Point c corresponds to a hydrogen molecule with vibrational energy. The motions of this vibration involve changes of interatomic distance from r_2 to r_3 to r_2 to r_3, and so on, the total energy remaining constant, as shown by the horizontal line, but the potential and kinetic energies changing continuously. The distance between Points b and c represents the kinetic energy of the vibrating atoms at c (separation $= r_1$).

In Figure 10–2, the dotted horizontal line, $E = 0$, corresponds to the potential energy of infinitely separated atoms. Not all energies in the shaded region below this line *are possible.* According to quantum mechanics, only certain definite vibrational energies are allowed. In the region of the minimum of the potential-energy curve (where the curve is approximately parabolic, $E_p = E_0 + \frac{1}{2}k(r - r_0)^2$, with E_0 the potential energy at r_0, the separation at the minimum), the allowed vibrational energies are given by the equation

$$E_v = h\nu_0(v + \tfrac{1}{2})$$

in which h is Planck's constant, ν_0 is the fundamental vibrational frequency of the diatomic molecule, and v is the vibrational quantum number that can have only the values:

$$v = 0, 1, 2, \cdots$$

The fundamental vibrational frequency ν_0 is related to a force constant k, which measures the springiness of the bond, and the masses of bonded atoms; the relationship is

$$\nu_0 = \frac{1}{2\pi}\sqrt{\frac{k}{\mu}}$$

in which μ (called the reduced mass of the system) is given by

$$\mu = \frac{m_1 m_2}{m_1 + m_2}$$

in which m_1 and m_2 are the masses of the two bonded atoms. Of particular interest at this point is the fact that even with $v = 0$, the lowest possible value of v, the molecule is vibrating. The smallest allowed amount of vibrational energy,

$$E_{v=0} = \frac{1}{2}h\left(\frac{1}{2\pi}\sqrt{\frac{k}{\mu}}\right)$$

is called the *zero-point energy*. Figure 10–3 shows the potential-energy curve

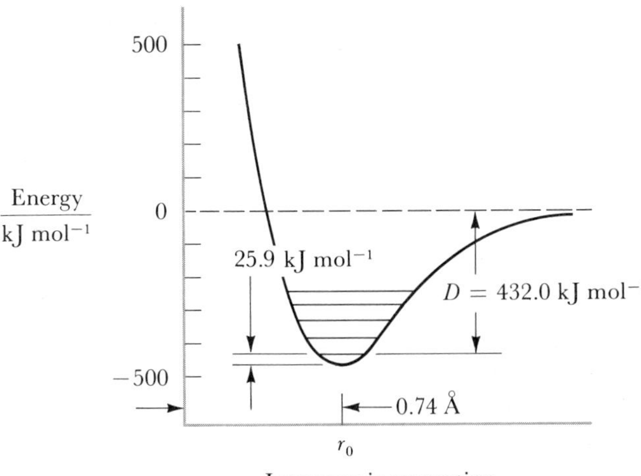

FIGURE 10–3
The potential energy for two hydrogen atoms bonded to give a hydrogen molecule. Horizontal lines give energies of allowed vibrational states with $v = 0, 1, 2, 3$, and 4.

for hydrogen, with the lowest few allowed vibrational levels drawn in. As indicated in the equation for E_v, the vibrational states are equally spaced in the vicinity of the minimum. At higher vibrational energies, where the potential-energy curve is not parabolic, the spacing of the levels is closer.

This graph has other features that deserve attention. The interatomic separation corresponding to the minimum of the curve commonly is designated r_0, and it is this distance that is taken as the interatomic separation in the molecule. Values of r_0 for certain diatomic molecules are presented in Table 10–1. However, even in the lowest possible vibrational energy state, the atoms are moving with respect to one another; therefore they are not separated by a fixed distance r_0. For each of the isotopic hydrogen molecules, $^1H_2$ or $^2H_2$ (deuterium), the value of r_0 is 0.74 Å, but

TABLE 10–1
Bond-Dissociation Energies and Interatomic Distances in Some Diatomic Molecules

Molecule	$\frac{r_0}{\text{Å}}$	$\frac{D}{\text{kJ mol}^{-1}}$
$^1H_2$	0.74	432.0
$^2H_2$	0.74	439.6
F_2	1.44	154
Cl_2	1.99	239
Br_2	2.28	190
I_2	2.67	149
O_2	1.21	494
N_2	1.10	942
HF	0.92	565
HCl	1.27	427
HBr	1.42	363
HI	1.61	295

the separation of atoms in the lowest vibrational state ($v = 0$) has a different range in the two molecules, ~0.64 to ~0.84 Å in $^1H_2$ and ~0.66 to ~0.83 Å in $^2H_2$. We see in this table also that for a series of related molecules, for example, the halogens F_2, Cl_2, Br_2, and I_2, or the hydrogen halides HF, HCl, HBr, and HI, the interatomic separation increases as the principal quantum number of the valence shell of one or both of the bonded atoms increases.

The difference between the energy for the lowest vibration state of the molecule ($v = 0$) and that corresponding to the atoms separated by an infinite distance (but with no excess kinetic energy), represented by D in Figure 10–3, is the dissociation energy of the diatomic molecule. It is the minimum energy that must be added to a diatomic molecule in the lowest vibration state to dissociate the molecule into its constituent atoms. Values for D for a selection of diatomic molecules are given in Table 10–1. The table includes entries for the two different isotopic hydrogen molecules $^1H_2$ and $^2H_2$; the difference between 439.6 kJ mol^{-1} for $^2H_2$ and 432.0 kJ mol^{-1} for $^1H_2$ (7.6 kJ mol^{-1}) is a significant amount of energy. The zero-point energy for H_2 (shown in Figure 10–3) is 25.9 kJ mol^{-1}. Because the reduced mass of the deuterium molecule is twice as large as that of the normal hydrogen molecule, the zero-point energy of deuterium is lower by a factor of $\sqrt{2}$ (see the equation relating ν_0 to reduced mass μ):

$$E_{v=0}(^2H_2) = E_{v=0}(^1H_2) \times \frac{1}{\sqrt{2}}$$

$$= 25.9 \text{ kJ mol}^{-1} \times \frac{1}{\sqrt{2}} = 18.3 \text{ kJ mol}^{-1}$$

The potential-energy curve for the deuterium molecule ($^2H_2$) is identical to that for the normal hydrogen molecule ($^1H_2$), but there is a difference in the spacing of the vibrational energy levels and a difference of the zero-point energies. For this reason the dissociation energies differ. As a consequence there is a difference in the chemical and physical properties of deuterium and hydrogen, both in the elemental state (the diatomic molecules being discussed) and in compounds. The percentage differences between the masses of isotopes of other elements are not as great as that between the isotopes of hydrogen. We usually assume that the chemical properties of isotopically different molecules of particular elements are identical. Because the effects are small, except for molecules containing the lightest elements, this assumption generally is valid.

10–3 Molecules of the Nonmetals

The discussion until now and Table 10–1 show that hydrogen, the halogens, nitrogen, and oxygen all exist in the elemental state as diatomic molecules. Atoms of each halogen have a valence shell with seven electrons, just as in the valence shell of the fluorine atom, and the Lewis structure for each diatomic halogen molecule is the same as that for fluorine:

$$:\ddot{\underset{..}{X}}:\ddot{\underset{..}{X}}: \quad \text{or} \quad :\ddot{\underset{..}{X}}—\ddot{\underset{..}{X}}:$$

in which a short line represents a shared pair of electrons with opposite spin.

The diatomic molecules O_2 and N_2 have dramatically higher dissociation energies and appreciably shorter bonds than F_2, as is shown in Figure 10–4. There are fewer valence electrons in these molecules, twelve ($2 \times 6 = 12$) in O_2 and ten ($2 \times 5 = 10$) in N_2, and it is not possible for these molecules to have the valence-electron distribution just shown for the fourteen electron molecule X_2 (X = a halogen). For nitrogen, with an electronic configuration

			1s	2s	2p
N	$1s^22p^22p^3$	N	(⇅)	(⇅)	(↑)(↑)(↑)

the principles already described suggest that two nitrogen atoms can share three pairs of electrons. The resulting diatomic molecule with three shared

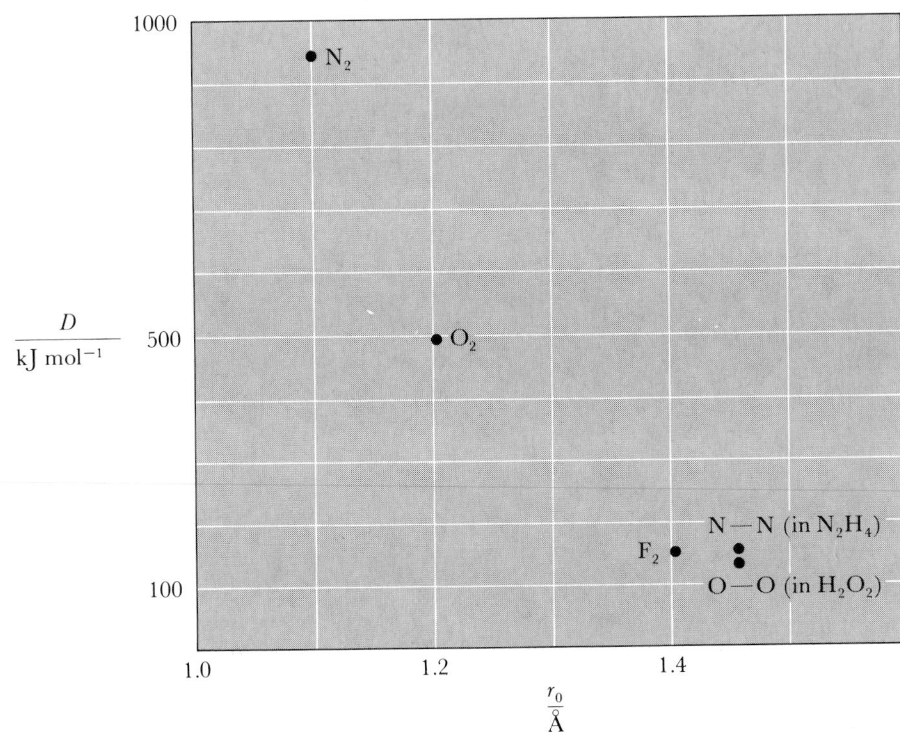

FIGURE 10–4
The bond-dissociation energies and bond lengths in N_2, O_2, and F_2. Also shown are these properties for single-bonded N—N and O—O units in H_2NNH_2 (hydrazine) and HOOH (hydrogen peroxide).

pairs of electrons is represented as

$$:\mathrm{N}:::\mathrm{N}: \quad \text{or} \quad :\mathrm{N}{\equiv}\mathrm{N}:$$

The bond involving three shared pairs of electrons is called a *triple bond.* The enormous bond energy in N_2 (942 kJ mol^{-1}) and the short bond distance (1.10 Å) are attributable to the triple bond.

The simple pictures of diatomic fluorine, F_2, involving a single bond and diatomic nitrogen, N_2, involving a triple bond suggest that bonding in diatomic oxygen, O_2, would involve a *double bond* (two shared pairs of electrons). The structure

$$\ddot{\underset{\cdot\cdot}{\mathrm{O}}}::\ddot{\underset{\cdot\cdot}{\mathrm{O}}} \quad \text{or} \quad \ddot{\underset{\cdot\cdot}{\mathrm{O}}}{=}\ddot{\underset{\cdot\cdot}{\mathrm{O}}}$$

employs the twelve electrons available from the valence shells of two oxygen atoms, which have the electronic structure

	1*s*	2*s*	2*p*
O	(⇅)	(⇅)	(⇅)(↑)(↑)

A double-bonded structure for O_2,

$$\ddot{\underset{\cdot\cdot}{\mathrm{O}}}::\ddot{\underset{\cdot\cdot}{\mathrm{O}}}$$

implies, however, that all of the electrons are paired, with no resultant spin angular momentum for the molecule. But oxygen is attracted into a magnetic field, as is shown in Figure 10–5, and this property, known as *paramagnetism*, is associated with unpaired electrons. An atomic or molecular species with unpaired electrons has a magnetic moment. This means that it acts like a little magnet and its potential energy is lowered in a magnetic field. Thus there is a net force attracting a paramagnetic substance, like

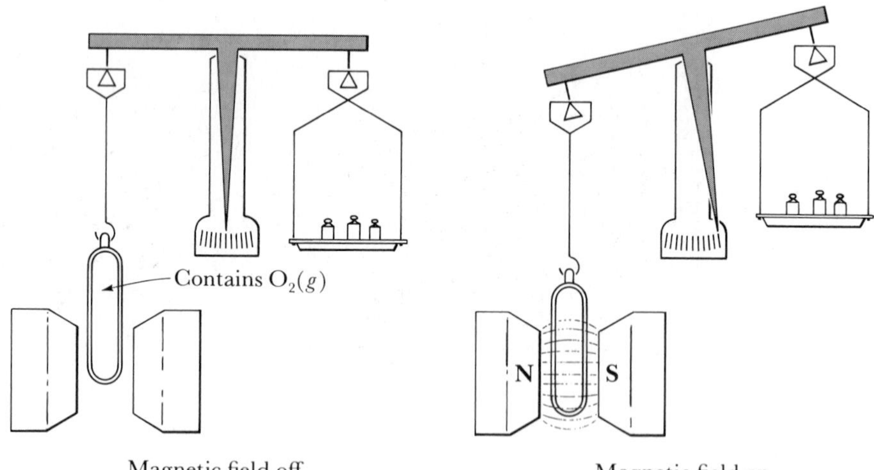

FIGURE 10–5
The paramagnetism of oxygen. The sample of oxygen is attracted into the region of high magnetic field.

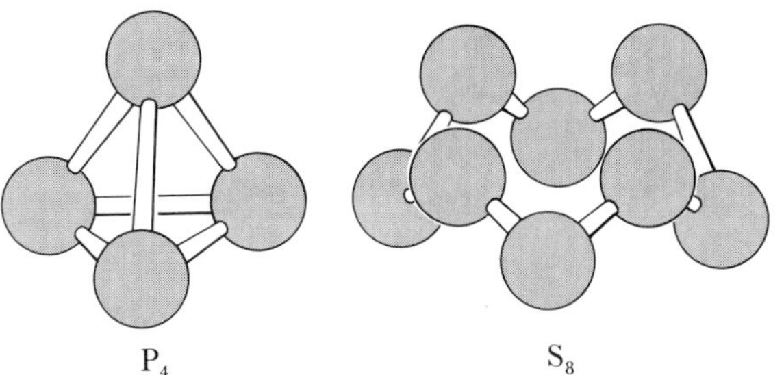

FIGURE 10–6
The structures of molecules of phosphorus, P_4, and sulfur, S_8, shown with ball-and-stick models. Bond angles: $\angle$ PPP = 60°; $\angle$ SSS = 107.9°.

molecular oxygen, into a magnetic field.[2] (Molecules with all electrons paired are repelled slightly by a magnetic field; such molecules are said to be *diamagnetic*.) That oxygen has approximately the bond strength and bond length expected for a molecule with a double bond, but has unpaired electrons, will be explained later in this chapter in a discussion of the molecular orbital theory of the bonding of diatomic molecules.

Because the stable form of nitrogen, the diatomic molecule N_2, involves a nitrogen–nitrogen triple bond, an analogous situation might be expected for elemental phosphorus, the element below nitrogen in the periodic table. Phosphorus, like nitrogen, has three half-filled p orbitals in its valence shell

	$3s$	$3p$
P(Ne core)	(↑↓)	(↑)(↑)(↑)

but phosphorus does not form a stable P_2 molecule under ordinary conditions. Elemental phosphorus exists in several different forms, one of which, white phosphorus, is made up of discrete molecules containing four phosphorus atoms. The geometry of the P_4 molecule is tetrahedral; in the structure, shown in Figure 10–6, the twenty valence electrons are distributed as follows:

$6 \times 2 = 12$ electrons as P–P single bonds represented by the six edges of the tetrahedron

$4 \times 2 = 8$ electrons as unshared pairs, one on each phosphorus atom

This distribution gives each phosphorus atom an octet of electrons in its valence shell. The structure involves single bonds joining phosphorus atoms, and is more stable than that of diatomic P_2, involving a triple bond. Hence the tendency for multiple bonding between phosphorus atoms is much smaller than that between nitrogen atoms. In fact, the comparison is a more general one: other examples will show that the tendency for multiple bonding is lower for atoms with a valence shell having $n \geq 3$ than for atoms with a valence shell having $n = 2$.

The stable forms of elemental sulfur also show the lessened tendency for third-row elements to form multiple bonds. Under ordinary conditions the molecular form of this element is a cyclic S_8 molecule involving single

[2]The paramagnetism of oxygen is utilized in an instrument (Taylor–Servomex Paramagnetic Oxygen Analyzer, cost ~1500 dollars) for analysis of oxygen.

bonds. The puckered ring, which resembles a crown (with $\angle SSS = 107.9°$) is shown in Figure 10–6. In this structure the forty-eight valence electrons are distributed as follows:

$8 \times 2 = 16$ electrons as S–S single bonds

$8 \times 4 = 32$ electrons as unshared pairs, two on each sulfur atom

This distribution gives each sulfur atom an octet of electrons in its valence shell. Each of the two common crystalline modifications, rhombic sulfur and monoclinic sulfur (see Figure 6–12 for a phase diagram), contains the S_8 molecule; the packing of the molecules is different in the two kinds of crystals.

In Lewis structures for the molecules discussed so far, each atom has, by the sharing of electrons, acquired an octet of electrons in its valence shell (or for hydrogen, a pair of electrons). This octet of electrons corresponds to the electronic structure of a noble gas, and an octet of electrons in a valence shell is given almost magic significance in simple discussions of covalent bonding. The presence of an octet in the valence shells of the bonded atoms of the most stable molecules is easy enough to explain. Consider the two molecules F_2, with a single bond, and N_2, with a triple bond.

If the electrons belonging to each of the two fluorine atoms are distinguished for the purposes of this discussion, the two single fluorine atoms:

		1s	2s	2p
Atom 1	F	$\binom{x}{x}$	$\binom{x}{x}$	$\binom{x}{x}\binom{x}{x}(^{x})$
Atom 2	F	$(:)$	$(:)$	$(:)(:)(\cdot)$

acquire in the diatomic molecule the structures

		1s	2s	2p
Atom 1	F	$\binom{x}{x}$	$\binom{x}{x}$	$\binom{x}{x}\binom{x}{x}\binom{x}{\cdot}$
Atom 2	F	$(:)$	$(:)$	$(:)(:)\binom{x}{\cdot}$

With the sharing of one pair of electrons, all orbitals of the $n = 2$ shell for each atom are occupied. Thus additional sharing of electrons in low-energy orbitals (orbitals with $n = 1$ or 2) is blocked.

For nitrogen we will first consider the structures of the free atoms and the atoms in a hypothetical diatomic molecule with a single bond. The nitrogen atoms

		1s	2s	2p
Atom 1	N	$\binom{x}{x}$	$\binom{x}{x}$	$(^{x})(^{x})(^{x})$
Atom 2	N	$(:)$	$(:)$	$(_{\cdot})(_{\cdot})(_{\cdot})$

by sharing one pair of electrons acquire the structures:

		1s	2s	2p
Atom 1	N	$\binom{x}{x}$	$\binom{x}{x}$	$\binom{x}{\cdot}(^{x})(^{x})$
Atom 2	N	$(:)$	$(:)$	$\binom{x}{\cdot}(_{\cdot})(_{\cdot})$

The single-bonded molecule depicted in this way,

$$:\dot{\underset{\bullet}{\mathrm{N}}}:\dot{\underset{\bullet}{\mathrm{N}}}:$$

should be stable with respect to the separated atoms even though each nitrogen atom does not have an octet in this molecule. But still greater stability (a more negative potential energy) is possible if three pairs of electrons are shared in N_2:

		1*s*	2*s*	2*p*
Atom 1	N	$(\overset{x}{x})$	$(\overset{x}{x})$	$(\overset{x}{\bullet})(\overset{x}{\bullet})(\overset{x}{\bullet})$
Atom 2	N	(:)	(:)	$(\overset{x}{\bullet})(\overset{x}{\bullet})(\overset{x}{\bullet})$

With the sharing of three pairs of electrons, all orbitals with $n = 1$ and 2 are occupied, and no more sharing of electrons in low-energy orbitals is possible. As in the case of F_2 with a single shared pair of electrons, this maximum sharing of electrons, and therefore maximum stability, occurs when each atom has the *s* and *p* orbitals of its valence shell completely occupied (i.e., has an octet).

Although the discussion of electronic structure of molecules in terms of Lewis electron-dot formulas serves us very well, there are other ways that are even more appropriate in some situations. The molecular-orbital description of bonding is one such alternative.

10–4 Molecular-Orbital Descriptions of Diatomic Molecules

The description of the bond in diatomic hydrogen as an electron-pair bond leaves unanswered many questions regarding the nature of such a bond. Just as quantum mechanics provides a clearer picture of the electronic structure of the atom, so also it provides a clearer picture of the electronic structure of molecules, However, the mathematical problems of applying quantum mechanics to a system of two or more nuclei and one or more electrons are much more complicated than those encountered in solving the hydrogen-atom problem. The modern high-speed electronic computer has eased the calculation problem, and progress is being made on more and more complicated systems. Here we shall not make detailed calculations, but shall present a simplified picture of the molecular orbitals in diatomic molecules. In Chapter 8 we described an orbital by a mathematical equation which defines an allowed electron distribution; Table 8–3 gives these mathematical equations for ψ, which, upon being squared, give the electron distributions for the 1*s*, 2*s*, $2p_x$, $2p_y$, and $2p_z$ orbitals in a one-electron atom. For an atom with one electron, the energy of the system can be calculated accurately using quantum mechanics. For atoms with more than one electron, approximate procedures are used. The same facets of electronic structure are to be explored for diatomic molecules. What are the allowed electron distributions? What are the allowed energies? And in addition there is the question that did not arise in connection with atomic structure: What is the geometrical arrangement of the nuclei in the most stable electronic state of the molecule? For a diatomic molecule, the question of geometrical arrangement simplifies to the question of the value of r_0, the separation of the atoms.

MOLECULES INVOLVING HYDROGEN AND HELIUM

The simplest diatomic molecule is the hydrogen molecule-ion H_2^+, which has the interatomic distance $r_0 = 1.06 \times 10^{-8}$ cm, and the dissociation energy $D(H\text{–}H^+) = 255.5$ kJ mol^{-1}. The equation for coulombic potential energy for a system of one electron and two nuclei has three terms, two for the attraction between the electron and each of the two nuclei and the third term for the repulsion between the two nuclei. This is shown in Figure 10–7. This equation for potential energy would be used in the Schrödinger equation applied to the hydrogen molecule-ion. The quantum-mechanical problem associated with the hydrogen molecule-ion has been solved satisfactorily with calculated values of r_0 and D that agree with the experimental values. The electron distribution for the ground state of the hydrogen molecule-ion is shown as a probability contour map in Figure 10–8a, and also in Figure 10–8b, in which the contour map is sliced along the interatomic axis. In this figure the electron distribution is compared with the radial distribution of an electron in a 1*s* orbital of each hydrogen atom. Compared to a 1*s* electron in each hydrogen atom, the electron in the hydrogen molecule-ion has a higher probability of being between the two nuclei and a lower probability of being on the periphery of the molecule. The attraction of the electron for each proton results in binding the two nuclei into a molecular unit, H_2^+, if the electron is located between the two nuclei.

There also is another solution to the Schrödinger equation applied to the hydrogen molecule-ion. This solution corresponds to an electron distribution that is less stable than the ground state. Figure 10–9 gives the potential-energy curve and electron distribution for this state, in which the nuclei repel one another at all separations. There is a lower probability of the electron being between the nuclei and a higher probability of its being in the peripheral regions. The attraction of the electron for each of the two

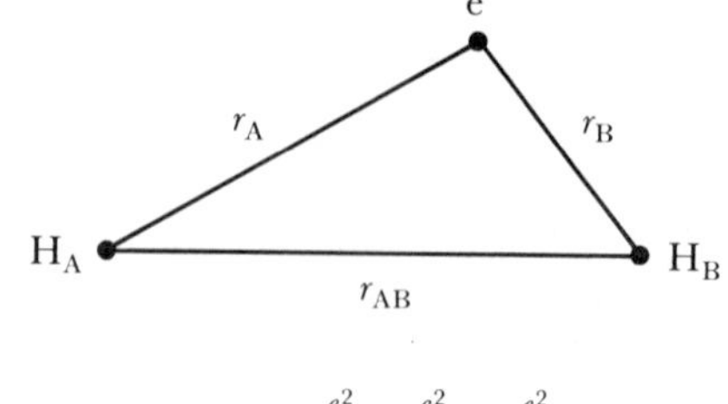

$$U = -\frac{e^2}{r_A} - \frac{e^2}{r_B} + \frac{e^2}{r_{AB}}$$

FIGURE 10–7
The coulombic potential energy of two protons and one electron (*e* is the magnitude of the charge on the electron and the proton).

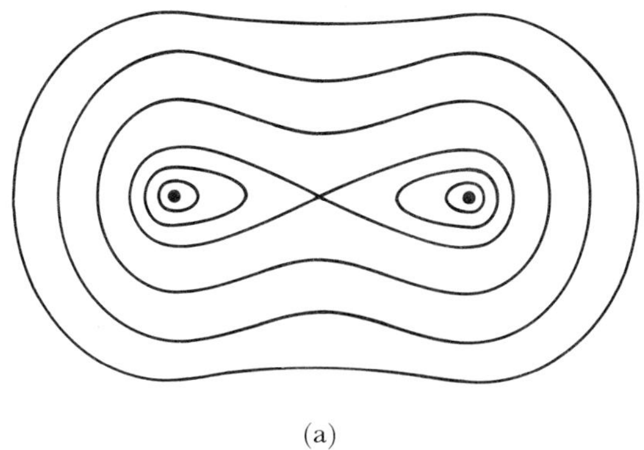

(a)

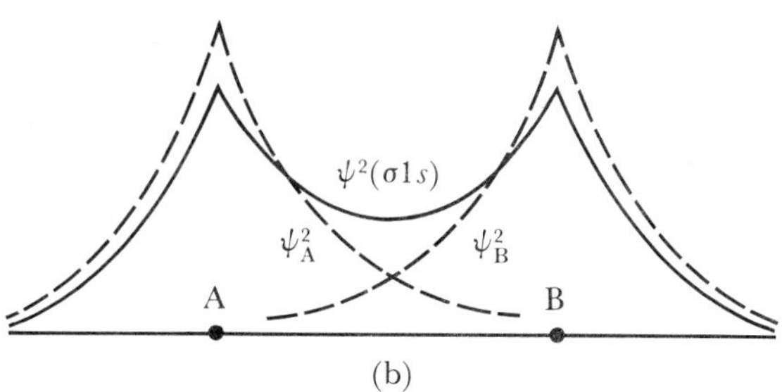

(b)

FIGURE 10–8
The electron distribution for the ground state of the hydrogen molecule-ion; the electron is in the $\sigma 1s$ orbital, a bonding molecular orbital. (a) Contours of constant electron density (i.e., constant values of ψ^2). The nuclei are separated by 1.06×10^{-8} cm. (b) The solid line shows values of ψ^2 along the axis of the two nuclei, labeled A and B. The dashed lines show ψ^2 for individual atoms, account taken of there being only one electron.

nuclei does not result in binding if the electron spends most of its time outside of the region between the two nuclei.

The wave functions $\psi(1s)$, $\psi(2s)$, $\psi(2p_x)$, and so on, which arose as solutions to the Schrödinger equation applied to a one-electron atom, correspond to electron distributions that are called *atomic orbitals*. Similarly, the electron distributions shown in Figures 10–8 and 10–9 are called *molecular orbitals*. That given in Figure 10–8 is called a *bonding molecular orbital* and that in Figure 10–9 an *antibonding molecular orbital*. (An asterisk is used to indicate that a molecular orbital is antibonding.)

Simple equations for ψ as a function of coordinates such as those presented in Table 8–3 for the one-electron atom cannot be presented for the molecular orbitals pictured in Figure 10–8 and 10–9. However, they can be approximated by appropriate combinations of atomic orbitals belonging to each of the two atoms. The cylindrically symmetric electron distribution of Figure 10–8, the bonding molecular orbital, is similar to that obtained by adding the wave functions for the $1s$ orbital of each hydrogen atom, and that for the antibonding molecular orbital, also cylindrically symmetric (shown in Figure 10–9), by subtracting the wave functions for the $1s$ orbital of each hydrogen atom. That is,

$$\sigma 1s = \psi A + \psi B \text{ (bonding MO)}$$
$$\sigma^* 1s = \psi A - \psi B \text{ (antibonding MO)}$$

in which ψA and ψB are the wave functions for the $1s$ orbitals of hydrogen

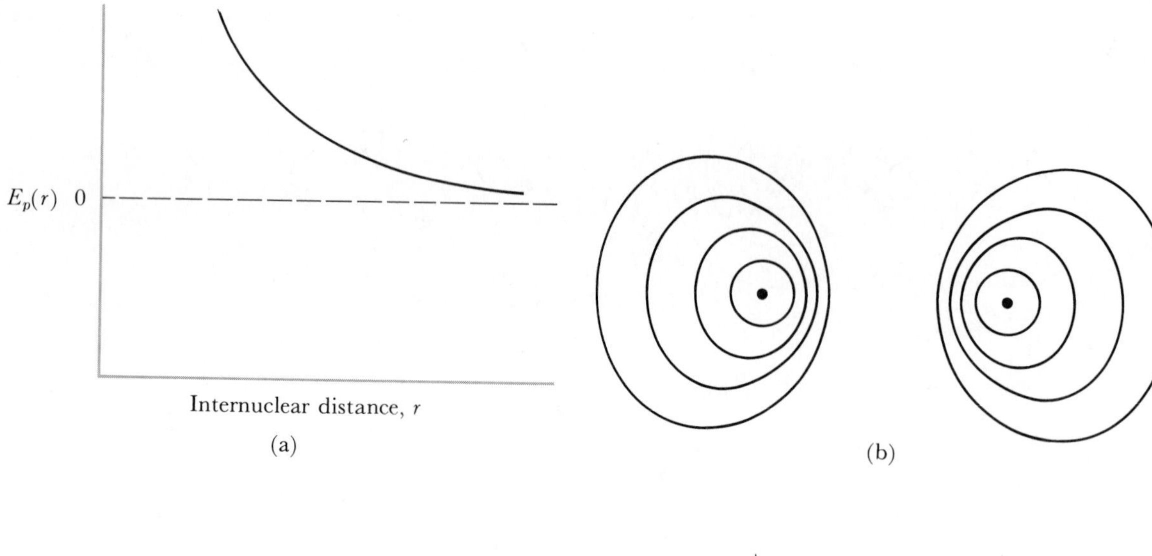

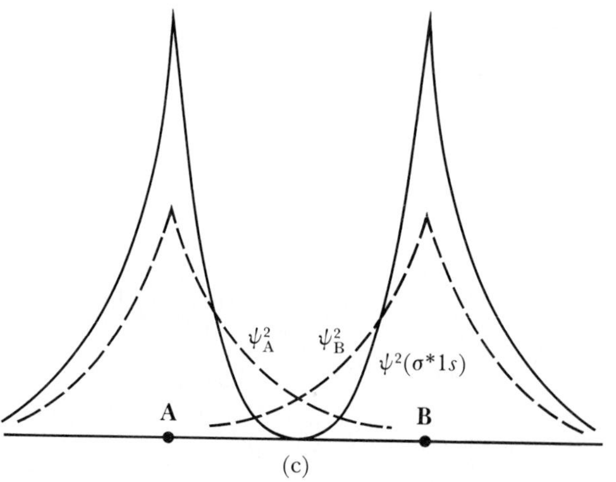

FIGURE 10–9
The excited state for the hydrogen molecule-ion; the electron is in the σ^*1s orbital, an antibonding orbital. (a) E_p versus internuclear distance; nuclei repel one another at all distances. (b) Contours of constant electron density, with nuclei separated by the internuclear distance observed for the ground state. (c) The solid line shows values of ψ^2 along the axis of the two nuclei, labeled A and B. The dashed lines show ψ^2 for individual atoms, account taken of there being only one electron.

atoms identified as A and B, respectively. The designation of these molecular orbitals as sigma (σ) orbitals denotes the cylindrical symmetry of the electron distribution. A bond involving a sigma molecular orbital is called a *sigma bond*.

The relative energies of these two sigma molecular orbitals and the atomic orbitals of the separated atoms are depicted in Figure 10–10. The

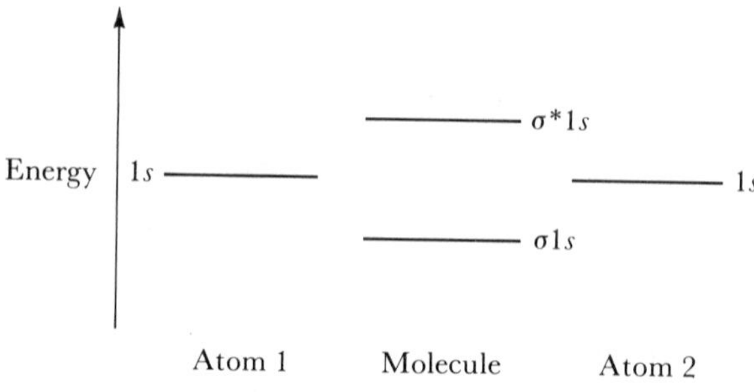

FIGURE 10–10
The energies of molecular orbitals for homonuclear diatomic molecules involving atoms with a valence shell having $n = 1$. The two $1s$ orbitals, one on each of the two atoms, become at this internuclear separation two molecular orbitals, one more stable and one less stable than the atomic orbitals.

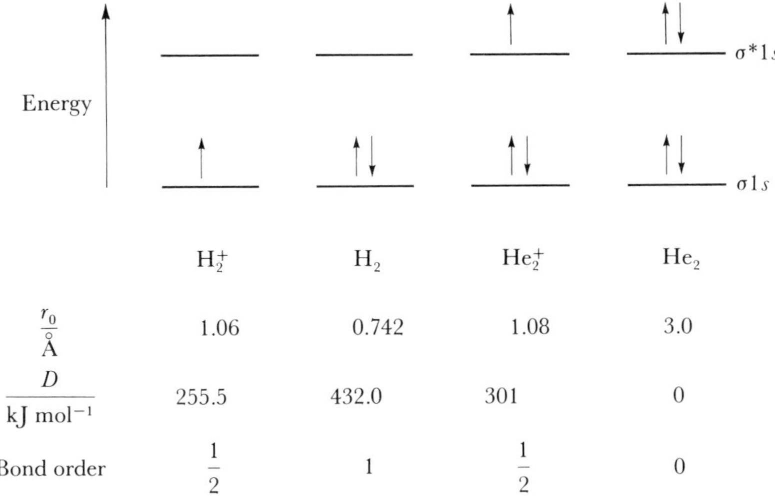

	H_2^+	H_2	He_2^+	He_2
$\frac{r_0}{\text{Å}}$	1.06	0.742	1.08	3.0
$\frac{D}{\text{kJ mol}^{-1}}$	255.5	432.0	301	0
Bond order	$\frac{1}{2}$	1	$\frac{1}{2}$	0

FIGURE 10–11
Occupancy of molecular orbitals in the homonuclear diatomic molecules of hydrogen and helium.

bonding molecular orbital, $\sigma 1s$, is more stable and the antibonding molecular orbital, $\sigma^* 1s$, is less stable than the $1s$ orbitals of the separated atoms. The electronic structures of known and hypothetical homonuclear diatomic molecules involving the $\sigma 1s$ and $\sigma^* 1s$ molecular orbitals are derived by placing the electrons into these orbitals. Pauli's principle applies here also: two electrons can occupy the same molecular orbital only if they have m_s values of the opposite sign, that is, have opposite spins. Figure 10–11 shows the occupancy of these molecular orbitals in H_2^+, H_2, He_2^+, and He_2.

In this simple picture, there is bonding of the two atoms into a molecule if more electrons are in bonding molecular orbitals than in antibonding orbitals. In molecular orbital theory, the bond order, which is one factor determining the bond energy, is defined in terms of the occupancy of bonding and antibonding molecular orbitals:

Bond order $= \frac{1}{2} \times$ (the number of electrons in bonding molecular orbitals minus the number of electrons in antibonding molecular orbitals)

The bond orders for H_2^+, H_2, He_2^+, and He_2, calculated in this way, are 1/2, 1, 1/2, and 0. We see that there is a correlation for these closely related molecules between the calculated bond orders and the experimentally determined values of the dissociation energies and bond lengths. The larger the bond order the greater is the dissociation energy and the smaller is the bond distance. For He_2, the bond order is zero, and this species is not a stable diatomic molecule. (The value of r_0 given for He_2 is derived from the Lennard-Jones parameters for helium, analogous to those given in Table 4–7 for other noble gases.)

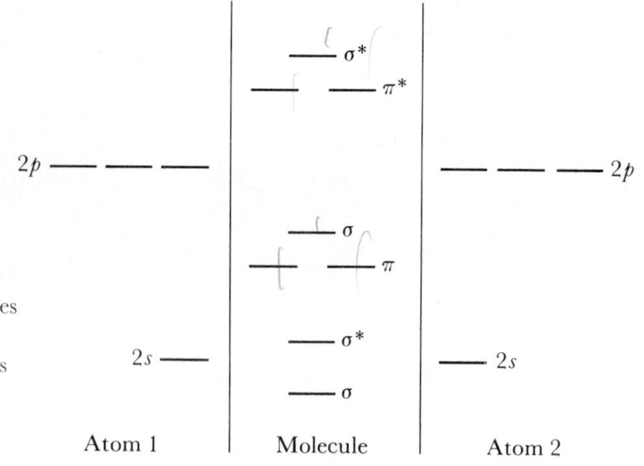

FIGURE 10–12
Molecular orbitals formed by combinations of 2s and 2p orbitals on two identical atoms. This order of the energies of the molecular orbitals is correct for the homonuclear second-row diatomic molecules Li_2 through N_2, and it is adequate for considering O_2 and F_2.

DIATOMIC MOLECULES INVOLVING SECOND-ROW ELEMENTS

The electronic structures of diatomic molecules involving elements with atomic numbers 3 to 10 can be described in terms of the occupancy of molecular orbitals, just as we have described H_2^+, H_2, and He_2^+. These molecular orbitals will be considered combinations of 2s and 2p atomic orbitals. The relative energies of the eight molecular orbitals formed by combinations of 2s, $2p_x$, $2p_y$, and $2p_z$ orbitals on the two atoms depend upon the number of electrons occupying these orbitals and the nuclear charges on the bonded atoms. We shall consider primarily the relative energies that explain the properties that we will discuss; this order is given in Figure 10–12.

There are four bonding molecular orbitals and four antibonding molecular orbitals. Of the four bonding molecular orbitals, two are σ orbitals, which can be viewed as arising from combinations of the 2s and $2p_z$ orbitals on each of the two atoms. (The internuclear axis is assumed to lie in the z direction.) The remaining two bonding molecular orbitals are pi (π) orbitals. A π molecular orbital is one with a high probability on either side of a plane containing the bonded atoms, but zero probability in that plane, which is called a *nodal plane*. The $2p_x$ atomic orbitals on each atom have zero electron density in the yz plane, and molecular orbitals formed by combination of these atomic orbitals are π orbitals with zero electron density in the yz plane. Figure 10–13 shows the regions of high electron density in the π and π^* orbitals formed by combining p_x atomic orbitals. The regions of high electron density in the π and π^* molecular orbitals formed from p_y atomic orbitals are identical to these, but they have zero probability in the xz plane. The energies of the two π orbitals are the same, and the energies of the two π^* orbitals are the same. Molecular orbitals that have identical energies are said to be *degenerate*. Thus the two π orbitals are degenerate, and the two π^* orbitals are degenerate. We see in the π^* orbital shown in Figure 10–13 a nodal plane bisecting the inter-

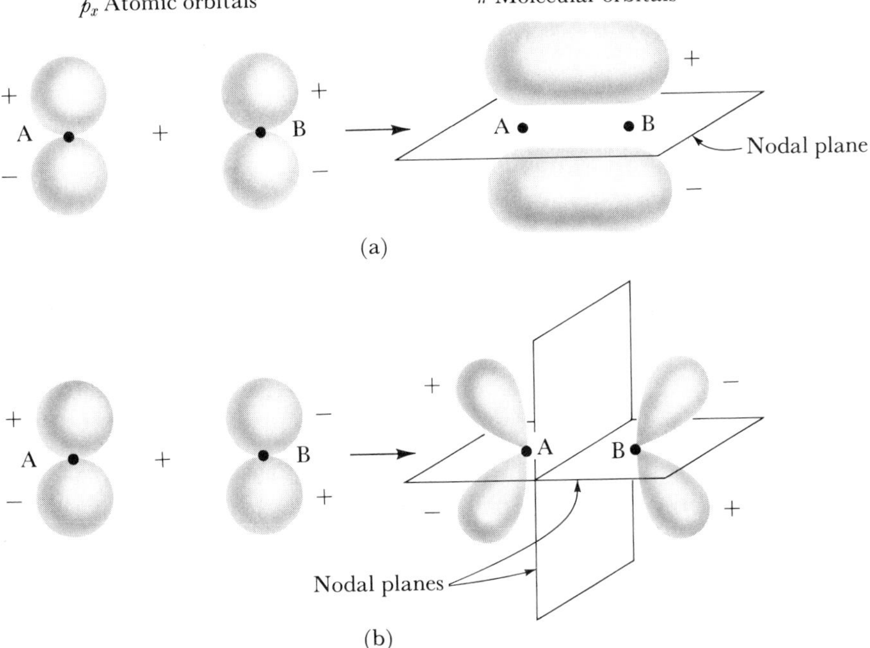

FIGURE 10–13
Electron density in bonding and antibonding π molecular orbitals. (a) Bonding molecular orbital formed by combining $2p_x$ orbitals on each atom. The yz plane containing the atoms is a nodal plane. (b) Antibonding molecular orbital formed by combining $2p_x$ orbitals on each atom. The yz plane is a nodal plane, as is the plane that bisects the internuclear axis.

nuclear axis. A molecular orbital with a nodal plane separating two atoms does not contribute to the bonding of those atoms.

An experimental result showing that the order of energies of orbitals given in Figure 10–12 is correct for B_2 is the paramagnetism of this molecule, which has been studied in the vapor phase. The situation is illustrated in Figure 10–14a, which shows the two least tightly held electrons in degenerate π molecular orbitals. The system is more stable with these two electrons in different orbitals and having the same spin. Hund's rule for atoms

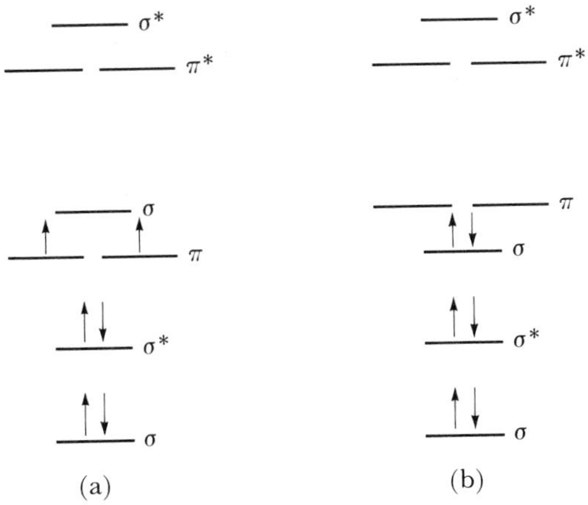

FIGURE 10–14
Possible orders of energies of molecular orbitals for B_2, showing the six valence electrons. (a) The order shown in Figure 10–12. On the basis of this occupancy (2 unpaired electrons), B_2 is expected to be paramagnetic, which is consistent with the observed magnetic behavior of this molecule. (b) An alternative order, from which B_2 is expected to be diamagnetic.

is valid also for molecular systems. An alternate order of the energies of the σ and π orbitals, shown in Figure 10–14b, would have the two least tightly held electrons paired in a sigma orbital; this electronic structure would make B_2 diamagnetic, contrary to what is observed.

Degeneracy of the π^* orbitals explains the paramagnetism of molecular oxygen, as is shown in the upper right part of Figure 10–15. There are twelve electrons in the $n = 2$ shells of two oxygen atoms. With ten electrons in the five most stable molecular orbitals, there are two electrons to go into the two degenerate π^* orbitals; these unpaired electrons make oxygen gas paramagnetic. The bond order associated with this occupancy of orbitals is

$$\tfrac{1}{2}(8 - 4) = 2$$

which is consistent with the strength of the bond in O_2.

Table 10–2 gives the orbital occupancy for various diatomic species involving second-row elements; it also gives the values of bond-dissociation energy (D) and interatomic separation (r_0), if these data are available. The table includes diatomic species in which the two atoms are different. For such species the atomic orbitals for the two atoms do not have the same

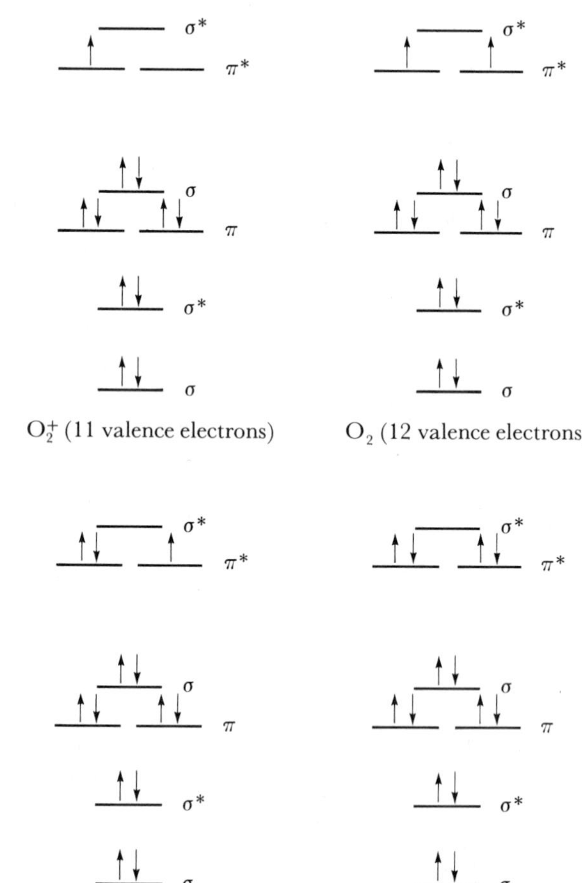

FIGURE 10–15
Orbital occupancy in the ground states of diatomic molecular and ionic species of oxygen. The order of the highest-energy orbitals, the σ^* and π^* orbitals, is given correctly. The order of the next orbitals, the σ and π orbitals, is that appropriate for molecules with atoms having lower Z (i.e., N_2), but the order of these orbitals, all filled, is not pertinent to our discussion of these oxygen species.

TABLE 10–2
Properties of Diatomic Species Involving Second-Row Elements

	Orbital occupancy (number of electrons)				
Species	Bonding MO's	Antibonding MO's	Bond order	r_0 / Å	D / kJ mol^{-1}
Li_2^+	1	0	0.5	—	149
Li_2	2	0	1	2.67	108
B_2	4	2	1	1.59	289
C_2	6	2	2	1.31	628
BN	6	2	2	1.28	385
CO^+	7	2	2.5	1.12	—
CN	7	2	2.5	1.17	787
N_2	8	2	3	1.10	942
CO	8	2	3	1.13	1070
NO	8	3	2.5	1.15	678
O_2^+	8	3	2.5	1.12 (gas)	—
				1.17 ± 0.17 (salts)	—
O_2	8	4	2	1.21	494
O_2^-	8	5	1.5	1.32–1.35	—
O_2^{2-}	8	6	1	1.49	139
F_2	8	6	1	1.42	154

energy, but our general discussion is not marred by considering Figure 10–12 to be a diagram of orbital energies associated with all species given in Table 10–2. The table reveals some interesting trends. If we compare only the neutral homonuclear diatomic species Li_2, B_2, C_2, N_2, O_2, and F_2, we see a correlation of the bond-dissociation energy and bond length with the calculated bond order. This is shown in Figure 10–16. However, there is not a quantitative correlation. Species with bond order 1, Li_2, B_2, and F_2, have a range of bond-dissociation energies (108–289 kJ mol^{-1}); C_2 and O_2, with bond order 2, have different bond-dissociation energies (628 kJ mol^{-1} and 494 kJ mol^{-1}).

Clearly, factors other than bond order affect dissociation energies. One factor is the nuclear charge. Its effect can be seen by comparing isoelectronic species. Peroxide ion, O_2^{2-}, is isoelectronic with molecular fluorine, F_2, and the increased nuclear charges in the fluorine molecule shrink the electron cloud and make the bond stronger and shorter.

The diatomic oxygen species provide an interesting sequence, in which there is a qualitative correlation of bond length and bond order. The oxygen species O_2^+ [occurring in $O_2(PtF_6)$], O_2, O_2^- (occurring in KO_2), and O_2^{2-} (occurring in Na_2O_2) have bond order $2\frac{1}{2}$, 2, $1\frac{1}{2}$, and 1. The bond distances in these species are: 1.12 Å, 1.21 Å, 1.34 Å, and 1.49 Å.

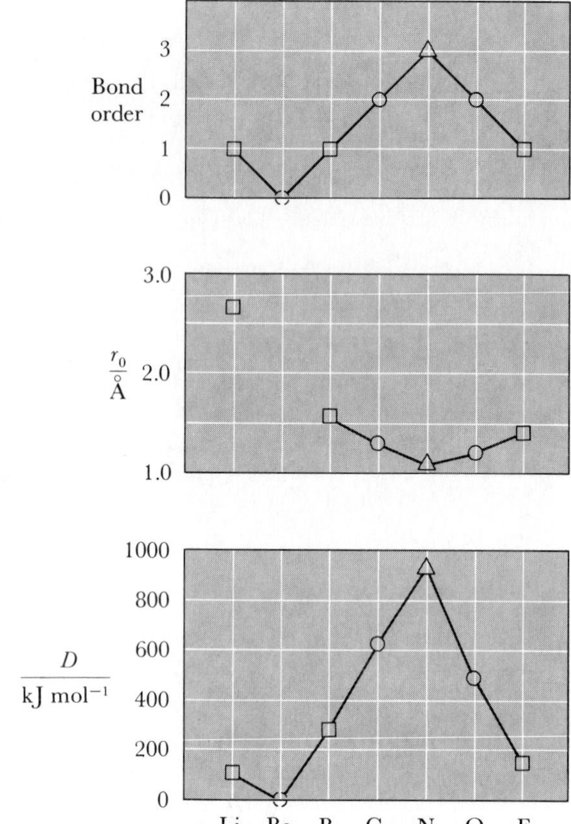

FIGURE 10–16
Properties of diatomic homonuclear species: bond order, bond distance, r_0, and bond-dissociation energy, D. The type of points indicate the bond order:
◌ 0 □ 1 ○ 2 △ 3

One comparison in this table is not even qualitatively consistent with the simple correlation of bond energy with bond order. The bond in Li_2^+ (bond order 1/2) is stronger than the bond in Li_2 (bond order 1). This complete failure of the simple theory confirms that factors not included in the simple concept of bond order can play dominant roles in determining bond energy.

10–5 Electronic Structures and Geometry of Polyatomic Molecules

For polyatomic molecules the geometry is an important characteristic. It is redundant to say that the diatomic nitrogen molecule, N_2, is linear, but it is not redundant to say that the triatomic water molecule, H_2O, is nonlinear. Now we will consider simple aspects of the electronic structures of polyatomic species (molecules and ions) that are conveyed by electron-dot formulas and a way in which these simple formulations allow accurate predictions of molecular geometry.

LEWIS ELECTRON-DOT REPRESENTATIONS OF SOME SIMPLE MOLECULES

Although molecular-orbital descriptions of bonding in covalent diatomic molecules, just described, are more informative than electron-dot formulas, the simplicity of Lewis structures makes them useful. This is especially true for polyatomic molecules, the molecular-orbital descriptions of which are more complex.

First, let us consider the covalently bonded molecules of hydrogen with carbon, nitrogen, oxygen, and fluorine. Atoms of these elements have four, five, six, and seven valence-shell electrons, respectively. To complete an octet of electrons in the $n = 2$ shell by sharing electrons with hydrogen (each atom of which has one electron to share), an atom of each of these elements must bond to four, three, two, and one hydrogen atoms, respectively:

$$\begin{array}{cccc} \mathrm{H} & \mathrm{H} & \mathrm{H} & \\ \mathrm{H:\overset{..}{\underset{..}{C}}:H} & \mathrm{:\overset{..}{\underset{..}{N}}:H} & \mathrm{:\overset{..}{\underset{..}{O}}:H} & \mathrm{:\overset{..}{\underset{..}{F}}:H} \\ \mathrm{H} & \mathrm{H} & & \end{array}$$

These substances, methane, CH_4, ammonia, NH_3, water, H_2O, and hydrogen fluoride, HF, are stable chemical compounds.

The number of half-filled orbitals in the ground electronic states of nitrogen, oxygen, and fluorine atoms (shown in Table 8–5) is the number of bonds formed to hydrogen in each of these compounds. For carbon, the situation is different. The ground electronic state of the carbon atom,

	1*s*	2*s*	2*p*
C	(↑↓)	(↑↓)	(↑)(↑)( )

with two half-filled orbitals, may suggest that carbon forms CH_2 rather than CH_4. That CH_4 is the stable hydrocarbon with one carbon atom is explained by considering an excited state of the carbon atom, C*,

	1*s*	2*s*	2*p*
C*	(↑↓)	(↑)	(↑)(↑)(↑)

in which an electron is promoted to the empty 2*p* orbital from the doubly occupied 2*s* orbital. This atom can form bonds to four hydrogen atoms.

There are many stable covalent compounds in which an atom forms as many bonds as the maximum number of unpaired electrons in its valence shell, not just the number of unpaired electrons in the valence shell in the ground electronic state of the atom. Examples are the fluorides of iodine, IF, IF_3, IF_5, and IF_7.[3] In these molecules, each fluorine atom contributes one electron to a shared pair, as it did in F_2, which we discussed previously. In contrast, the iodine atom contributes one, three, five, and seven electrons

[3]Although each of these is known, only IF_5 and IF_7 are stable enough to be studied in detail.

to shared electron pairs in covalently bonded representations of IF, IF_3, IF_5 and IF_7, respectively. To explain this we must consider not only the ground state of the iodine atom,

	5*s*	5*p*	5*d*
I	(↑↓)	(↑↓)(↑↓)(↑)	( )( )( )( )( )

(in which electrons in orbitals with $n \leq 4$ are not shown), but also the excited states,

	5*s*	5*p*	5*d*
I	(↑↓)	(↑↓)(↑)(↑)	(↑)( )( )( )( )
I	(↑↓)	(↑)(↑)(↑)	(↑)(↑)( )( )( )
I	(↑)	(↑)(↑)(↑)	(↑)(↑)(↑)( )( )

in which there are three, five, and seven half-filled orbitals. If IF_3, IF_5, and IF_7 are considered to involve three, five, and seven electron-pair bonds the central iodine atom has ten, twelve, and fourteen electrons in its valence shell. We see that the electrons in excess of eight are accommodated in valence-shell *d* orbitals.

The concept of isoelectronic species, presented already in connection with monatomic species (F^-, Ne, and Na^+ are isoelectronic) and diatomic species (O_2^{2-} and F_2 are isoelectronic), is also useful in considering polyatomic species. Series of isoelectronic species with three to five atoms are:

1. CO_2, NO_2^+, N_2O, and N_3^- (each with 16 valence electrons, and a total of 22 electrons)
2. NH_3, H_3O^+ (each with 8 valence electrons, and a total of 10 electrons)
3. CO_3^{2-}, NO_3^-, BF_3 (each with 24 valence electrons, and a total of 32 electrons)
4. BH_4^-, CH_4, and NH_4^+ (each with 8 valence electrons, and a total of 10 electrons).

The polyatomic species in each of these groups have the same number of electrons; they are *isoelectronic*. But as the term is used here, to be isoelectronic also requires that the species have the same number of atoms.[4] The concept of isoelectronic polyatomic species is useful because such species have the same geometrical structure, and also because the concept may help us predict the existence of unusual chemical compounds.

The relationship of geometrical structure to electronic structure is an important topic, which we mentioned in discussing the water molecule in Chapter 1, but have not yet developed. The electron-dot formula for

[4]Other definitions of isoelectronic species are useful in discussing some problems. In one alternate definition, hydrogen atoms are not counted in reckoning the number of atoms, and in another definition the total number of electrons need not be the same, although the number of valence electrons must be the same. The former relaxation of the rigid definition allows classification of HN_3 and N_3^- as isoelectronic or H_3BCO and CO_2 as isoelectronic, and the latter relaxation of the definition allows classification of SO_2 and O_3 as isoelectronic or CH_4 and SiH_4 as isoelectronic.

ammonia or isoelectronic hydronium ion,

```
    H            H +
    ••           ••
  : N : H   or : O : H
    ••           ••
    H            H
```

implies nothing about whether these species are planar or pyramidal. The octet of electrons in the outer shell of the nitrogen or oxygen atom does, in fact, impose pyramidal geometry upon these species. Now we will discuss the basis for this fact and for the geometry of other covalent polyatomic molecules.

THE VALENCE-SHELL ELECTRON-PAIR REPULSION MODEL

We will consider the geometry of polyatomic molecules of the type XY_n in which a central nonmetal X atom is covalently bonded to *n* Y atoms. You can predict the geometry of such molecules simply by considering the mutual repulsion of the electron pairs around the X atom. The basis for this repulsion is *both* the coulombic repulsion of like charges *and* the repulsion of electrons with the same spin, repulsion due to the Pauli exclusion principle. The total number of valence-shell electrons around the X atom in XY_n is the sum of valence-shell electrons of X plus the electrons donated to the X—Y bonds by the Y atoms. The pairs of electrons around the central atom are proposed to have an angular distribution that keeps them as far away from each other as possible. Because the total number of valence-shell electron pairs around X may exceed the number of Y atoms bonded to X, the geometry of a molecule does not follow directly from its formula: the total number of valence-shell electron pairs must be considered.

This theory, now commonly called the "valence-shell electron-pair repulsion" model (VSEPR), was developed by SIDGWICK and NYHOLM and has been extended by GILLESPIE. The most stable angular distributions of two through six electron pairs around a central atom are summarized in Table 10–3.

We will use this table to explain the structures of two XY_3 molecules, BF_3 and NF_3, in each of which a central nonmetal atom is bonded to three fluorine atoms. The geometries of these two molecules are different: boron trifluoride is planar but nitrogen trifluoride is pyramidal, as shown in Figure 10–17. The important difference between these two molecules is in the number of valence-shell electron pairs around the central atom. Three electrons are donated to the central atom by the three fluorine atoms in each case. The nitrogen atom has five valence-shell electrons, and in nitrogen trifluoride, the central nitrogen atom has four electron pairs $[\frac{1}{2}(5 + 3) = 4]$. The boron atom has three valence-shell electrons, and in a single-bonded representation of boron trifluoride the central boron atom has only three electron pairs $[\frac{1}{2}(3 + 3) = 3]$. To be as far away from each other as possible, the four pairs of electrons around the nitrogen atom are

TABLE 10–3
The Arrangements of Valence-Shell Electron Pairs Around an Atom

Number of electron pairs	Angle between nucleus–electron pair axes	Figure defined by electron pairs
2	180°	line
3	120°	triangle
4	109.5°	tetrahedron[a]
5	120° and 90°	trigonal bipyramid
6	90°	octahedron

[a] If you study the figure given below, which shows a point in the center of a cube and four points that define a tetrahedron inscribed in the cube, you will see that the tetrahedral angle θ is two times the angle for which the cosine is $(1/\sqrt{3})$:

$$\cos(\tfrac{1}{2}\theta) = 1/\sqrt{3} = 0.577 \qquad \tfrac{1}{2}\theta = 54.76^\circ$$
$$\theta = 109.5^\circ$$

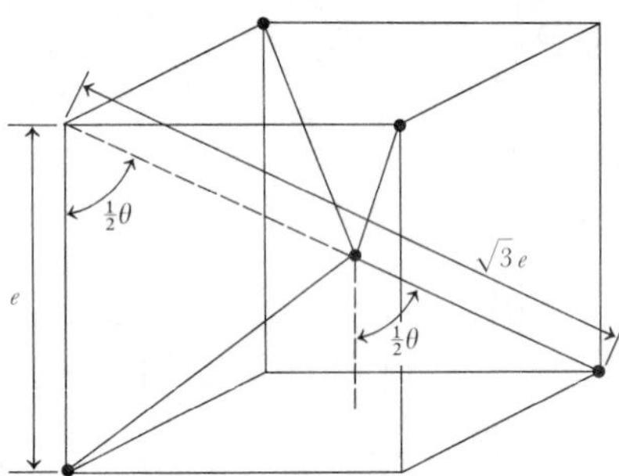

expected to have high probability along axes radiating tetrahedrally from the nitrogen atom. Three of these electron pairs are involved in bonding with fluorine atoms and the fourth pair is a lone pair. If the angular distribution of the electron pairs around the nitrogen atom were exactly tetrahedral, the angle ∠FNF would be 109.5°. The angle, found experimentally, is 102.5°. This difference is explained in the framework of the present theory by assuming that there is greater repulsion between a lone pair of electrons and a bonding pair than between two bonding pairs of electrons. The bonding pairs of electrons repel one another less because a bonding pair is under the influence of two nuclei. In contrast, a lone pair is under the influence of only one nucleus and it occupies more space. To be as far

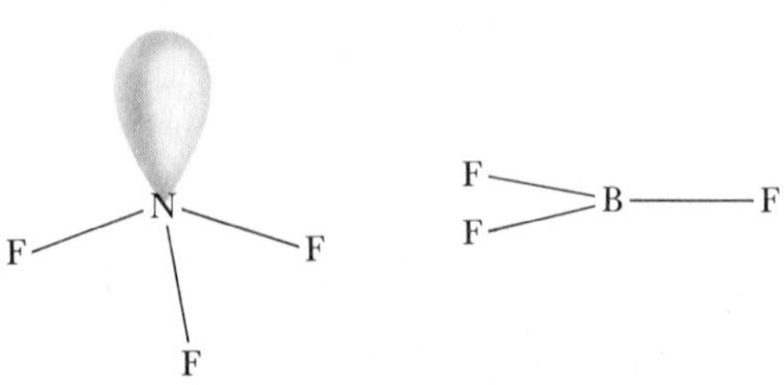

FIGURE 10–17
Two XY_3 molecules: pyramidal NF_3 with central atom having four pairs of valence electrons, and planar BF_3 with central atom having three pairs of valence electrons. The lines represent bonding pairs of electrons; the shaded region represents a lone pair of electrons (as in Figure 1–7).

away from each other as possible, the three pairs of electrons around the boron atom in boron trifluoride are expected to have high probability along axes radiating from the boron atom in a planar trigonal orientation with $\angle$ FBF = 120°. The boron trifluoride molecule has the expected regular planar geometry. (We will return to the electronic structure of boron trifluoride when we discuss resonance.)

The compounds of hydrogen with carbon, nitrogen, oxygen, and fluorine, already mentioned, have geometries expected on the basis of the simple principle just outlined. Each central nonmetal atom in these compounds has an octet of electrons in its valence shell. According to the electron-pair repulsion model, these eight electrons (four with one spin and four with the opposite spin) are grouped in four pairs (each pair with one electron having each of the two possible spin orientations) with maximum probability approximately along the tetrahedral directions. The geometry of each of these molecules, as characterized by the angle $\angle$HXH which deviates mildly from the 109.5° for all except CH_4, is consistent with the order of electron-pair repulsions:

lone pair–lone pair repulsion > lone pair–bonding pair repulsion
> bonding pair–bonding pair repulsion

Molecules of these substances are pictured in Figure 10–18. In addition to the trend in angles, the trend in bond distances,

C—H, 1.09 Å > N—H, 1.01 Å > O—H, 0.96 Å > F—H, 0.92 Å

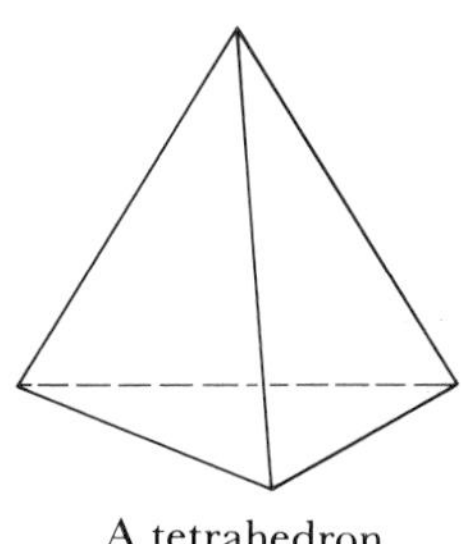
A tetrahedron

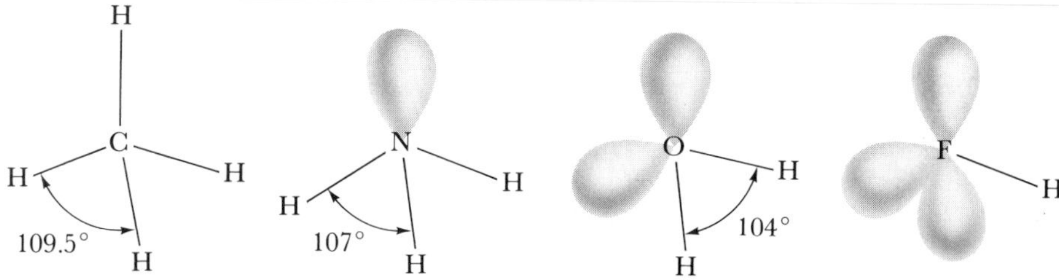

FIGURE 10–18
The molecular geometry of nonmetal hydride with a central atom having four pairs of valence electrons. Valence-shell electron bookkeeping:

CH_4 $4 + 4 \times 1 = 8$ OH_2 $6 + 2 \times 1 = 8$
NH_3 $5 + 3 \times 1 = 8$ FH $7 + 1 \times 1 = 8$

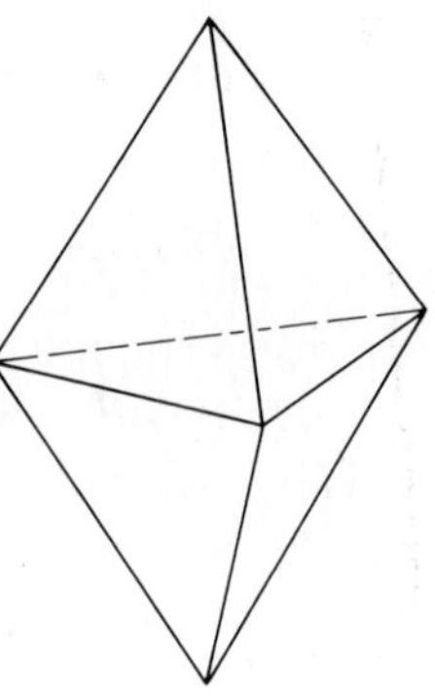

A trigonal bipyramid

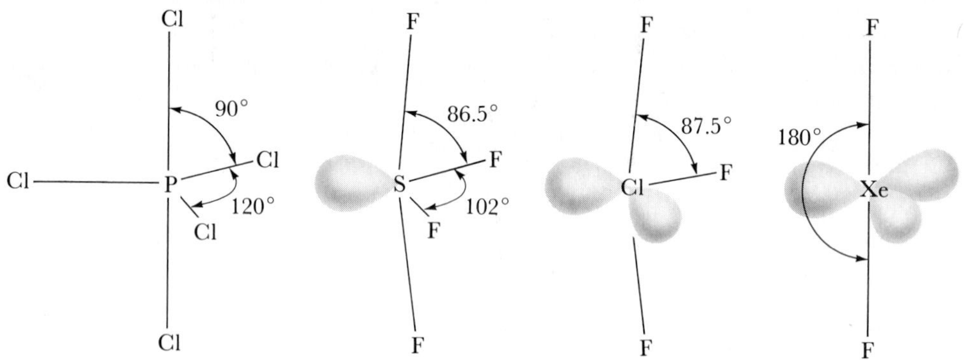

FIGURE 10–19
Trigonal bipyramidal coordination of five chlorine atoms around phosphorus in $PCl_5(g)$. Also shown are the geometries of other molecules in which the central atom has five pairs of valence electrons. Valence-shell electron bookkeeping:

PCl_5	$5 + 5 \times 1 = 10$	ClF_3	$7 + 3 \times 1 = 10$
SF_4	$6 + 4 \times 1 = 10$	XeF_2	$8 + 2 \times 1 = 10$

is that expected; the increase in nuclear charge in going from left to right (carbon to fluorine) causes the electron pairs and the bonded hydrogen atoms to be pulled closer to the central nonmetal atom.

The geometrical figure defined by the preferred arrangement of five electron pairs around a central atom is the trigonal bipyramid (shown in Figure 10–19). Gaseous phosphorus pentachloride, PCl_5, in which the central phosphorus atom has five electron pairs $[\frac{1}{2}(5 + 5) = 5]$, has this geometry. Unlike the triangular and tetrahedral orientations of three and four electron pairs in which the axes of maximum electron density are equivalent, the five axes along which there is high electron density in the trigonal bipyramid orientation are not equivalent. This being so, it is especially interesting to examine the structures of sulfur tetrafluoride (SF_4) and chlorine trifluoride (ClF_3), each a molecule having a central atom with five electron pairs but fewer than five coordinated atoms.

Figure 10–19 also shows the experimentally observed structures of these molecules. These observed structures have fewer unfavorable electron-pair repulsions at ~90° than there are in alternate configurations, as summarized in Figure 10–20. Notice that the structures of SF_4 and ClF_3, shown in Figure 10–19, are mild distortions of the idealized geometry, distortions that help minimize the potential energy of the different types of electron-pair repulsions.

The geometrical figure defined by the preferred arrangement of six pairs of electrons around a central atom is the octahedron (shown in Figure 10–21). An XY_6 molecule with six pairs of electrons in the valence shell of X is sulfur hexafluoride, SF_6. As expected, this molecule has a regular octahedral geometry. Examples of molecules with fewer than six atoms bonded to the central atom, but with six electron pairs in the valence shell of the central atom, are iodine pentafluoride (IF_5) and xenon tetrafluoride (XeF_4), which have the structures shown in Figure 10–21. The geometry of each molecule is related to the octahedral geometry of sulfur hexafluoride,

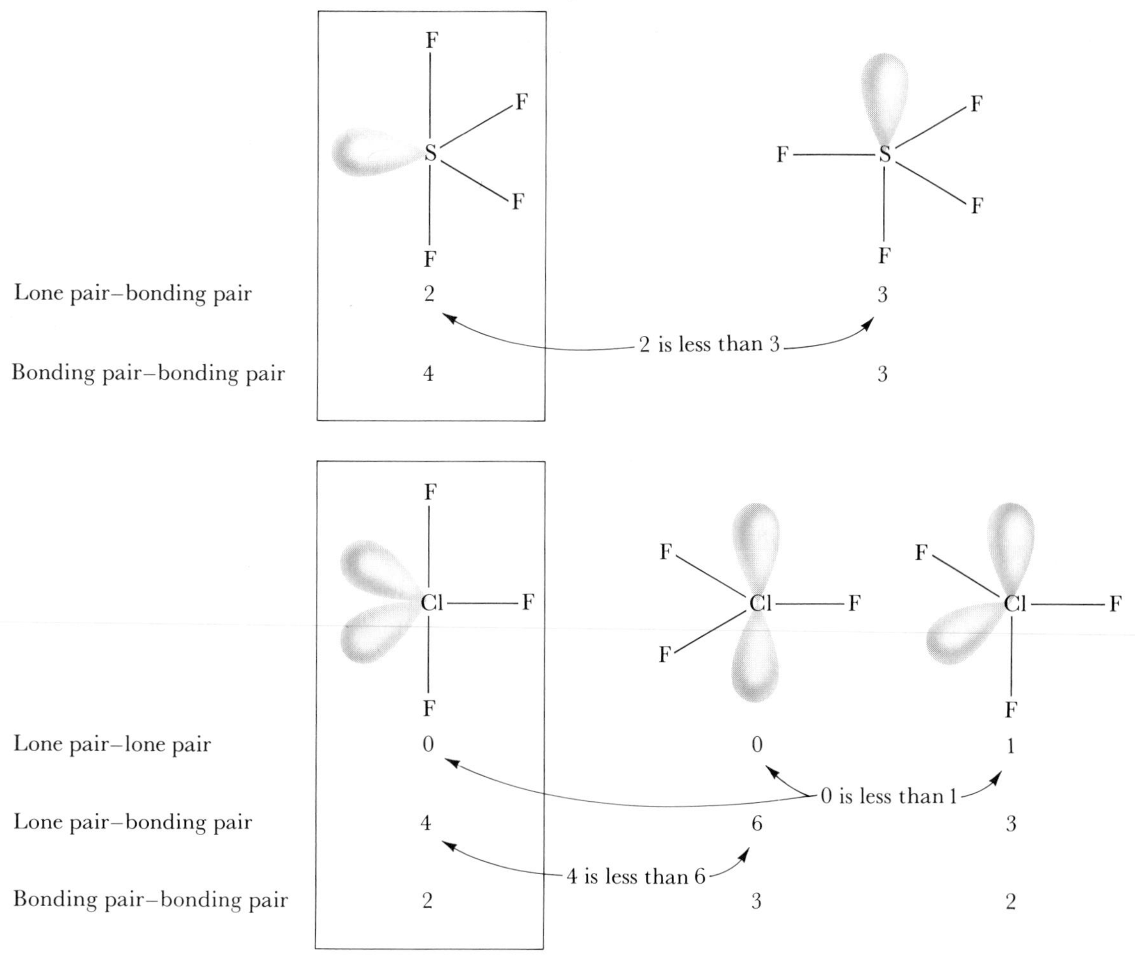

FIGURE 10–20
The numbers of electron-pair interactions at ~90° in alternate structures for SF_4 and ClF_3. The observed structures are enclosed in boxes.

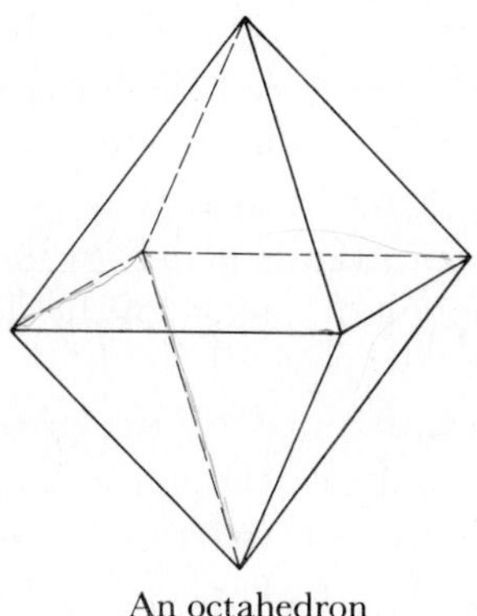

An octahedron

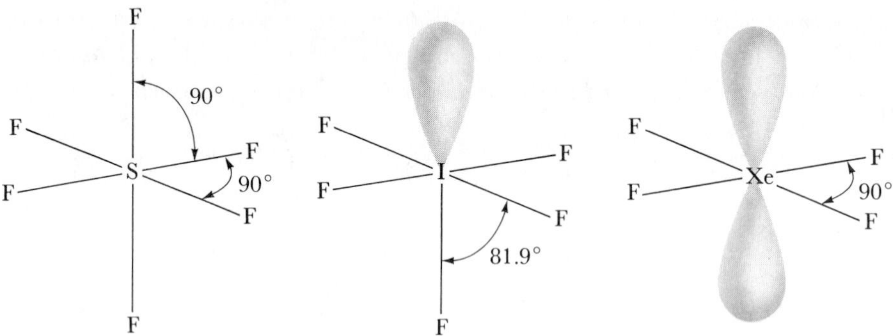

FIGURE 10–21

Octahedral coordination of six fluorine atoms around sulfur in SF_6. Also shown are the geometries of other molecules in which the central atom has six pairs of valence electrons. Valence-shell electron bookkeeping:

SF_6 $6 + 6 \times 1 = 12$ IF_5 $7 + 5 \times 1 = 12$ XeF_4 $8 + 4 \times 1 = 12$

with lone pairs of electrons occupying one and two of the octahedron's corners.

Among the substances discussed in presenting the valence-shell electron-pair repulsion model was phosphorus pentachloride (PCl_5); in the gas phase molecules of this substance have the expected structure. In the solid state the structure is very different. The solid with the composition PCl_5 is an ionic substance $(PCl_4^+)(PCl_6^-)$. In each of these ions, the phosphorus–chlorine bonds are essentially covalent. In applying the valence-shell electron-pair repulsion model to these ionic species (and to all other ionic species), we will take into account the surplus of electrons for negative ions or the deficiency of electrons for positive ions:

	PCl_4^+	PCl_6^-
Electrons from P	5	5
Electrons from Cl	4×1	6×1
Extra electrons	-1	$+1$
Total	8	12
Number of electron pairs around P	4	6

Tetrachlorophosphonium cation (PCl_4^+) has a regular tetrahedral geometry, and hexachlorophosphate anion (PCl_6^-) has a regular octahedral geometry. Each of these ions, in addition to the gaseous molecule (PCl_5), has a geometry consistent with the simple principles outlined in this section; however, simple theory does not allow prediction that this substance is monomeric PCl_5 in the gas phase but is ionic (PCl_4^+)(PCl_6^-) in the solid phase. The coulombic lattice energy stabilizes this ionic form in the solid relative to a solid containing the neutral PCl_5 molecules, which would have no such stabilizing energy.

The molecules and ions that we have considered so far have only single bonds. In applying the valence-shell electron-pair repulsion model to species with double bonds or triple bonds, we consider the two electron pairs of a double bond or the three electron pairs of a triple bond to be a single group of electrons. This point can be illustrated with the molecules H_2CO (formaldehyde), CO_2 (carbon dioxide), and HCN (hydrogen cyanide). Lewis formulas for these compounds are

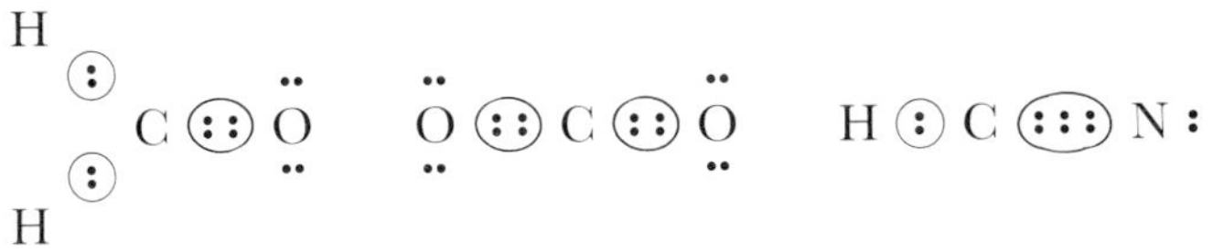

H (:) C (::) Ö: with second H (:) — O (::) C (::) O — H (:) C (:::) N:

In these formulas, each *group* of electrons is encircled. Although the carbon atom in each of these molecules has an octet of electrons in its valence shell, we will deduce their geometries by considering not the number of electron pairs in the valence shell around carbon, but rather the number of groups of electrons. These are:

H_2CO 2 groups of 1 electron pair each plus 1 group of 2 electron pairs = 3 groups; therefore the molecule is planar.

CO_2 2 groups of 2 electron pairs each = 2 groups; therefore the molecule is linear.

HCN 1 group of 1 electron pair plus 1 group of 3 electron pairs = 2 groups; therefore the molecule is linear.

The observed geometries of these molecules are consistent with these predictions. The structure of formaldehyde reveals an additional aspect of the subject. The bond angles in this molecule are

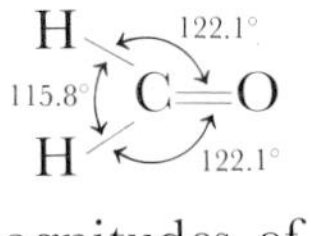

H–C=O angles: ∠HCO 122.1°, ∠HCH 115.8°, ∠HCO 122.1°

From the observed relative magnitudes of these angles, that is, ∠HCO greater than ∠HCH, we conclude that the two electron pairs of a double bond take more space than the one pair of a single bond. This conclusion is confirmed by data for other molecules.

RESONANCE

Carbonate ion (CO_3^{2-}) and nitrate ion (NO_3^-) are isoelectronic species with twenty-four valence electrons. Structural studies show that each of these ions is planar with $\angle OCO$ or $\angle ONO = 120°$; in CO_3^{2-} there are three equal C—O bond lengths, 1.31 Å, and in NO_3^- there are three equal N—O bond lengths, 1.24 Å. An electronic structure

$$\left[\begin{array}{ccc} :\!\ddot{O}\!: & & \\ & X::\ddot{\underset{..}{O}} & \\ :\!\ddot{O}\!: & & \end{array}\right]^{n-}$$

with $n = 1$ if X is N and $n = 2$ if X is C, is consistent with the observed planar structures. (There are three groups of electrons around X in this structure.) But this structure implies that lengths of the X—O bonds would not be the same. This unsatisfactory feature is eliminated by considering the electronic structure to be an average of three equivalent structures:

$$\left[\begin{array}{cc} :\!\ddot{O}\!: & \\ & X::\ddot{\underset{..}{O}} \\ :\!\ddot{O}\!: & \end{array}\right]^{n-} \longleftrightarrow \left[\begin{array}{cc} :\!\ddot{O}\!: & \\ & X:\ddot{\underset{..}{O}}: \\ :\!\ddot{O}\!: & \end{array}\right]^{n-} \longleftrightarrow \left[\begin{array}{cc} :\!\ddot{O}\!: & \\ & X:\ddot{\underset{..}{O}}: \\ :\!\ddot{O}\!: & \end{array}\right]^{n-}$$

Each X–O bonds has one third double-bond character and two thirds single-bond character; no single Lewis structure explains the properties of carbonate ion or nitrate ion, but an "average" of these three structures does. The structure is said to be a *resonance hybrid* of the three equivalent Lewis structures. The concept of resonance, made popular by L. PAULING, is very useful because there are many molecular and ionic species with properties that cannot be explained by a single Lewis structure, but which can be explained in terms of a resonance hybrid of two or more Lewis structures.

There is more to the theory of resonance than is brought out in this example. A molecule (or ion) with an electronic structure that can be described as a resonance hybrid of two or more Lewis structures is more stable than would be predicted on the basis of any one of the contributing structures. (This point will be documented with numerical data in Chapter 11.)

A molecule that is isoelectronic with carbonate ion and nitrate ion is boron trifluoride, BF_3. We already have discussed this molecule in terms of the single-bonded structure

$$\begin{array}{cc} :\!\ddot{F}\!: & \\ & B:\ddot{\underset{..}{F}}: \\ :\!\underset{..}{F}\!: & \end{array}$$

which does not give the boron atom an octet of electrons. Structures such as those proposed for carbonate ion and nitrate ion,

$$\begin{array}{cc} :\!\ddot{F}\!: & \\ & B::\ddot{\underset{..}{F}} \\ :\!\ddot{F}\!: & \end{array} \longleftrightarrow \begin{array}{cc} :\!\ddot{F}\!: & \\ & B:\ddot{\underset{..}{F}}: \\ :\!\ddot{F}\!: & \end{array} \longleftrightarrow \begin{array}{cc} :\!\ddot{F}\!: & \\ & B:\ddot{\underset{..}{F}}: \\ :\!\ddot{F}\!: & \end{array}$$

do give the boron atom an octet of electrons, and these structures are equally consistent with the planarity of boron trifluoride. Other data (bond lengths and chemical reactivity) also suggest that these equivalent structures contribute to the structure of this molecule. The structure of boron trifluoride is a resonance hybrid of all four of the structures.

HYBRID ATOMIC ORBITALS

In our discussion of the geometry of XY_n molecules, we have not used information regarding the angular distribution of electron density in atoms (i.e., the dependence of ψ on θ and ϕ), which was presented in Chapter 8. You should not interpret this omission to mean that the bonding and geometry of an XY_n molecule cannot be related to the angular distribution of the s and p orbitals (and d orbitals, if necessary) of the X atom. It means simply that we can predict geometry no more reliably by starting from the dependence of ψ upon θ and ϕ than by using the valence-shell electron-pair repulsion approach. The angular distribution predictions based upon the dependence of ψ upon θ and ϕ may, in fact, be less reliable unless they take appropriate account of *orbital hybridization.*

Figure 8–10 shows the angular distribution of electron density in the $2s$ and the three $2p$ orbitals. This picture plus that inferred earlier for the orbital occupancy in methane (CH_4),

	1s	2s	2p
C	(:)	(ẋ)	(ẋ) (ẋ) (ẋ)

with x's for electrons donated by hydrogen atoms, does not predict methane's tetrahedral geometry. To explain this discrepancy, we must recognize that the equations for $\psi(2s)$, $\psi(2p_x)$, $\psi(2p_y)$, and $\psi(2p_z)$ given in Table 8–3 are solutions to a differential equation, the Schrödinger equation. These solutions are appropriate for the free atom, but they are not the appropriate solutions if the bonding of other atoms to the central atom is being considered. There are combinations of these equations [the equations for $\psi(2s)$, $\psi(2p_x)$, $\psi(2p_y)$, and $\psi(2p_z)$] that are solutions of the original differential equation, and Pauling showed in 1931 that four particular independent combinations of the $2s$ and three $2p$ orbitals give maximum probabilities along the tetrahedral axes. In addition, these four new orbitals, which are called *hybrid orbitals*, allow for maximum overlap with the orbitals of four atoms bonded to the central atom. Because the energy of a covalent bond is assumed to increase with the extent of this overlap, these hybrid orbitals (called sp^3 hybrid orbitals) are more appropriate for discussion of the bonding than are the individual s and p atomic orbitals. A pictorial representation of the mathematics behind the transformation of one $2s$ orbital plus three $2p$ orbitals into four sp^3 hybrid orbitals is given in Figure 10–22. Thus the geometry and bonding in methane (CH_4) can be described in terms of four tetrahedrally oriented bonds that result from strong overlap of the $1s$ orbital of each of the hydrogen atoms with the four sp^3 hybrid orbitals of the carbon atom.

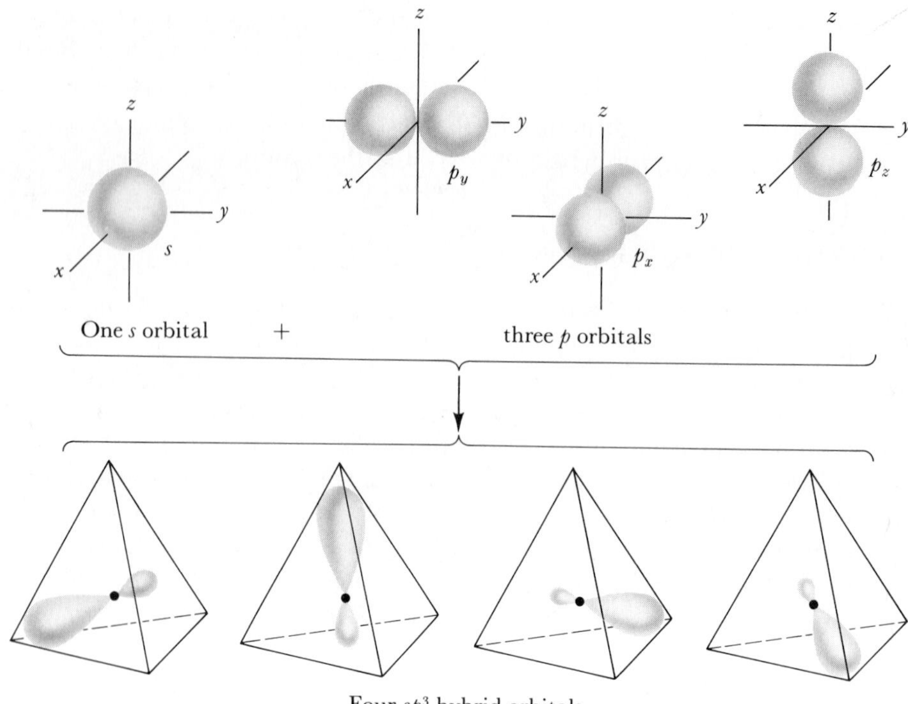

FIGURE 10–22
The sp^3 hybrid orbitals formed by combining the one s and the three p orbitals. The shaded regions represent the angular probability distribution functions for an electron, as in Figure 8–10.

In discussing the molecular-orbital approach to the bonding in diatomic molecules, we considered molecular orbitals to be linear combinations of atomic orbitals. A molecular-orbital description of the bonding in methane could involve linear combinations of the 1s orbitals of the hydrogen atoms and the four sp^3 hybrid orbitals of the carbon atom.

If the four atoms bonded to carbon are not the same and if the angles $\angle$ XCY are not exactly the tetrahedral angle, the four hybrid orbitals do not have the same relative amounts of s-orbital and p-orbital contribution. The same is true for the XY_n molecules with $n \leq 3$; in such molecules the hybrid orbitals used in bonding are not equivalent to those occupied by unshared pairs of electrons.

Hybrid orbitals also can be formulated for a central atom with any other number of pairs of electrons. If more than four pairs of electrons are present in the valence shell of the central atom, d orbitals (as well as s and p orbitals) must be involved in the formulation. Discussion of the covalent bonding in molecules of the nonmetals, XY_n, of whatever observed geometry can be accommodated in terms of hybrid orbitals just as it can be in terms of the valence-shell electron-pair repulsion model. The latter approach, because of its greater simplicity, is described in some detail in this chapter. If you extend your study of chemistry to a full course in organic chemistry, it is very likely that you will learn much more about hybrid orbitals.

10–6 Polarity of Covalent Molecules; Electronegativity

Chapter 9 introduced you to the concept of polarity in its discussion of gaseous diatomic ionic compounds. That a gaseous molecule containing a positive ion and a negative ion is electrically unsymmetrical is not a surprise. It also should not be a surprise to learn that homonuclear diatomic molecules, for example, H_2 or F_2, are not polar. It is less certain what to expect for heteronuclear diatomic molecules such as HF. We already learned in this chapter that HF is polar, with a dipole moment of 1.82 D, which suggests an appreciable asymmetry in the sharing of the bonding pair of electrons in the molecule:

$$\mathrm{H:\ddot{\underset{\cdot\cdot}{F}}:}$$

The dipole moments of other diatomic molecules are given in Table 10–4. These data and other types of experimental data, as well as theoretical calculations, are consistent with the idea that atoms in a covalently bonded molecule have a characteristic attraction for electrons. This characteristic attraction by atoms for electrons in a molecule is called the *electronegativity* of the element.

We have been introduced to measures of the attraction of an atom for electrons that are amenable to strict, quantitative description. The ionization energy (IE) of an element X is the value of ΔU for the process

$$\mathrm{X}(g) = \mathrm{X}^+(g) + \mathrm{e}^- \qquad IE = \Delta U$$

and the electron affinity (EA) of an element X is the negative of ΔU for the process

$$\mathrm{X}(g) + \mathrm{e}^- = \mathrm{X}^-(g) \qquad EA = -\Delta U$$

No such strict, quantitative description of electronegativity is possible, but scientists have proposed a number of different measures of this quantity, denoted by χ. One proposed by R. S. MULLIKEN involves ionization energy and electron affinity:

$$\chi = \frac{IE + EA}{2}$$

TABLE 10–4
Dipole Moments of Covalently Bonded Molecules

Diatomic molecules				Polyatomic molecules			
	p/D		p/D		p/D		p/D
HF	1.82	ClF	0.88	CH_4	0	CO_2	0
HCl	1.08	NO	0.153	CH_3Cl	1.87	SO_2	1.63
HBr	0.82	CO	0.112	CH_2Cl_2	1.60	SO_3	0
HI	0.44			$CHCl_3$	1.01	NH_3	1.47
				CCl_4	0	NF_3	0.24
				H_2O	1.85	SF_6	0

TABLE 10–5
Electronegativities of Certain Nonmetals[a]

Element	χ	Element	χ
H	2.2	Si	1.8
B	2.0	P	2.1
C	2.6	S	2.5
N	3.0	Cl	3.0
O	3.5	Br	2.8
F	4.0	I	2.4

[a]For our purposes, the electronegativity values will be considered to be dimensionless. The electronegativities of metals range from 1.0 ± 0.1 (alkali metals) to 1.8 ± 0.1 (Cu, Cd, Sn, Ag, Au).

This method of evaluating the electronegativity of an element involves measurements only on that uncombined element. Other methods involve studying compounds; these methods generally give differences of the electronegativities of the bonded atoms. Because there is no operational definition of electronegativity, there is no quantitative agreement among the various proposed measures. Further, it is found that application of any one method of estimating the electronegativity of a particular element in compounds with different structures, for example, phosphorus in $PCl_3(g)$ and $PCl_5(g)$, give different values of χ. Because of this uncertainty, the values of electronegativity in Table 10–5 are given only to two significant figures. An approximately linear correlation exists between the values of $\delta+$ (and $\delta-$), derived from values of the dipole moments of the hydrogen halides and the differences of electronegativities of the bonded atoms. Because the electronegativity of hydrogen is smaller than the electronegativities of each halogen, the polarity of each hydrogen halide molecule is assumed to be the same as already suggested for hydrogen fluoride. The correlation of $\delta-$ with $(\chi_X - \chi_H)$ is shown in Figure 10–23.

Each hydrogen halide is formed exothermically from the constituent elements (with both reactants present in the gaseous state). This means

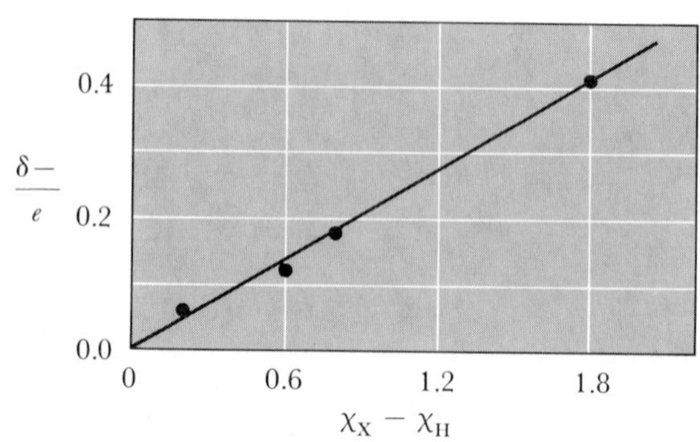

FIGURE 10–23
Correlation of $\delta-$, the partial charge on the halogen atom in the hydrogen halides with $(\chi_X - \chi_H)$. ($\delta-$ values are in units of the charge on the electron, e, and they were calculated by the method used in Chapter 9.)

that the bond-dissociation energy of each hydrogen halide is greater than the average of the bond-dissociation energies of the diatomic reactant molecules, H_2 and X_2. A particularly simple view of covalent bond formation proposed by Pauling suggested that the bond energy of an X–Y bond exceeds the average of the bond energies of X–X and Y–Y bonds to an extent determined by the ionic character of the bond.

Figure 10–24 shows for the hydrogen halides a plot of $\Delta = D(\text{H—X}) - \frac{1}{2}[D(\text{X—X}) + D(\text{H—H})]$ versus $(\chi_X - \chi_H)$. The plot's curvature suggests a dependence of Δ upon $(\chi_X - \chi_H)$ raised to a power greater than one. Pauling proposed a second power dependence, $\Delta = 96.2 \text{ kJ mol}^{-1} \times (\chi_X - \chi_H)^2$, which is given as the line in Figure 10–24. Although we may use this relationship in quantitative calculations, the principal message we will take from this correlation is:

> The formation of covalent bonds between unlike atoms occurs exothermically because in such a chemical change the *more electronegative* element *gains* in its share of electrons.

This is an important principle that will help you in understanding the chemistry of the elements.

The observed dipole moment of a polyatomic molecule results from unequal sharing of the bonding electrons by atoms with different electronegativities plus a contribution to electrical asymmetry from any unshared electron pairs in the molecule. The role of unshared electrons is illustrated by a comparison of ammonia and nitrogen trifluoride, which have dipole moments 1.47 D and 0.24 D, respectively. The pyramidal geometries of the two molecules are similar, the angles being $\angle$ HNH = 106.8° and $\angle$ FNF = 102.5°. The electronegativity differences of the bonded atoms are similar for the two molecules ($\chi_N - \chi_H = 0.8$, and $\chi_F - \chi_N = 1.0$), but there is an enormous difference between the net dipole moments of the two.

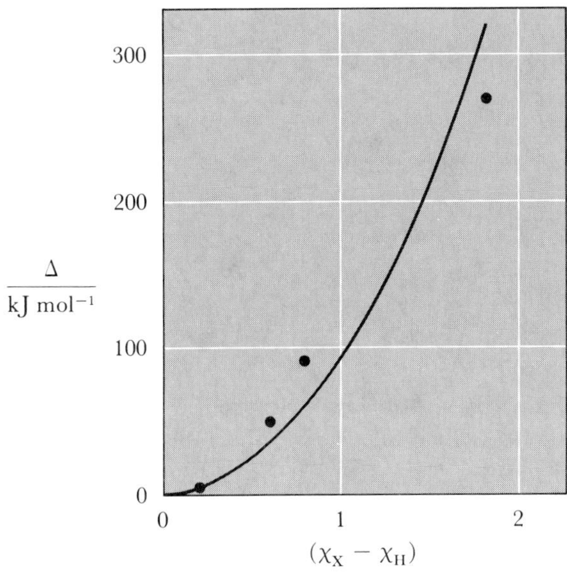

FIGURE 10–24
Correlation of bond-energy enhancement (Δ) with the difference of electronegativity of the bonded atoms, $\Delta = D(\text{H—X}) - \frac{1}{2}[D(\text{X—X}) + D(\text{H—H})]$. The solid line is $\Delta = 96.2 \text{ kJ mol}^{-1} \times (\chi_X - \chi_H)^2$.

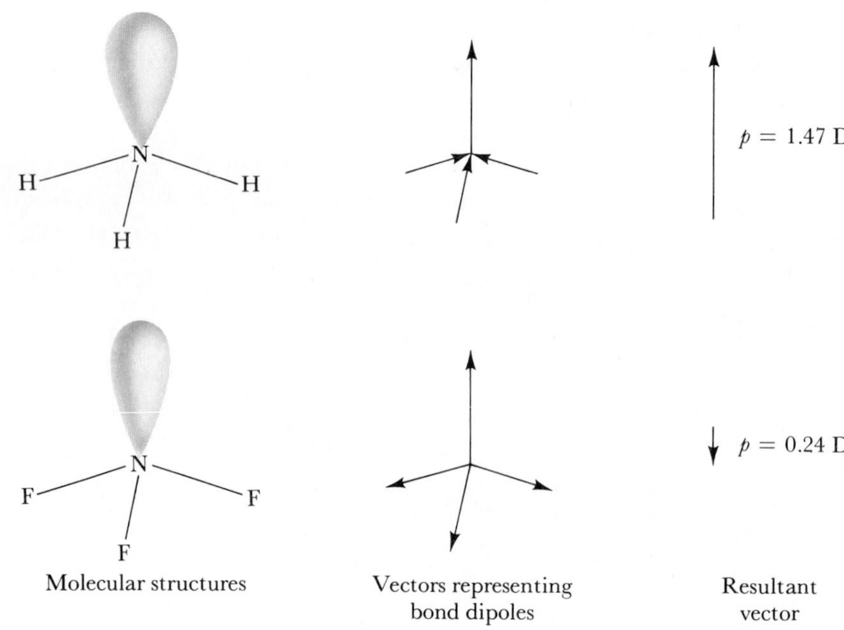

FIGURE 10–25
The contribution of unshared electron pairs to the dipole moment of molecules; a comparison of ammonia and nitrogen trifluoride. Dipole vectors point toward the negative end of the dipole.

A reasonable explanation of this is shown in Figure 10–25. In ammonia the N–H bond dipole moments and the electrical asymmetry due to the unshared electron pair add to give the net dipole moment; in nitrogen trifluoride these contributions to the net dipole moment are opposed. In the diagrams in Figure 10–25 each contribution to the dipole moment is represented by a vector; the net dipole is represented by the resultant vector.

A polyatomic molecule is nonpolar despite the polarity of the bonds if it has the appropriate symmetry. Methane and carbon tetrachloride are nonpolar because bond dipole moments cancel each other exactly in these regular tetrahedral molecules. Cancellation of bond dipole moments occurs also for sulfur hexafluoride, which has a regular octahedral shape (shown in Figure 10–21). Since the carbon–hydrogen and carbon–chlorine bond dipole moments are unequal, the molecules $CH_{4-n}Cl_n$ ($n = 1$, 2, and 3), given in Table 10–4, have net dipole moments.

10–7 More About Bond Energies and Bond Lengths

In Section 10–2 a discussion of the potential-energy curve for the interaction of two hydrogen atoms as a function of interatomic separation introduced the concepts of bond-dissociation energy (D in Figure 10–3) and r_0, the internuclear separation at the minimum of the potential-energy curve.

Similar potential-energy curves can be drawn for other diatomic molecules. The quantity D can be identified with ΔU for the reaction

$$XY(g, 0\ K) = X(g, 0\ K) + Y(g, 0\ K)$$

(Specifying 0 K as the temperature is a way of specifying that all XY molecules are in the lowest vibrational state.) For a polyatomic molecule more than one interatomic distance is needed to define positions of the atoms relative to one another, and we cannot draw a simple pictorial representation of the potential energy of the system as a function of all of the atomic positions. The concepts introduced in Figure 10–3 are useful, nonetheless. The energy required to convert the molecule in the lowest vibrational state of the ground electronic state to separated atoms (in their ground electronic states) is the sum of the dissociation energies of all of the bonds in the molecule.

Let us consider methane, CH_4. The carbon–hydrogen bonds in this molecule are very strong. For the atomization reaction,

$$CH_4(g) = C(g) + 4H(g)$$

the value of ΔU is $+1652$ kJ mol^{-1}. This is the energy necessary to break the four equivalent carbon–hydrogen single bonds, and one fourth of this (413 kJ mol^{-1}) is taken as the *bond energy* of the carbon–hydrogen single bond. Although the four carbon–hydrogen single bonds in methane are equivalent, the values of ΔU for the stepwise dissociation of the four bonds are not equal:

	$\frac{\Delta U}{\text{kJ mol}^{-1}}$	
$CH_4(g) = CH_3(g) + H(g)$	435	$D(H_3C—H)$
$CH_3(g) = CH_2(g) + H(g)$	453	$D(H_2C—H)$
$CH_2(g) = CH(g) + H(g)$	425	$D(HC—H)$
$CH(g) = C(g) + H(g)$	339	$D(C—H)$
Sum $CH_4(g) = C(g) + 4H(g)$	1652	
or $\frac{1}{4}CH_4(g) = \frac{1}{4}C(g) + H(g)$	413	$E(C—H)$

These values of ΔU for the individual steps, called *bond-dissociation energies*, are represented by D. These quantities are not identical to *bond energies*, represented by E, which are averaged quantities for all bonds of a particular type in a molecule. (Each quantity is useful, but you must be careful not to mistake one of them for the other.)

A value for the carbon–carbon single bond energy can be derived from the energy required to break all of the bonds in ethane, C_2H_6,

```
H     H
 \   /
H—C—C—H
 /   \
H     H
```

$$C_2H_6(g) = 2C(g) + 6H(g) \qquad \Delta U = 2825\ \text{kJ mol}^{-1}$$

This energy of atomization is the sum of all of the bond energies:

$$\Delta U = E(\text{C—C}) + 6E(\text{C—H})$$

If it is assumed that the average carbon–hydrogen bond energy in ethane is the same as that in methane, the carbon–carbon single bond energy can be calculated:

$$\begin{aligned} E(\text{C—C}) &= \Delta U - 6E(\text{C—H}) \\ &= 2825 \text{ kJ mol}^{-1} - 6 \times 413 \text{ kJ mol}^{-1} \\ &= 347 \text{ kJ mol}^{-1} \end{aligned}$$

By procedures analogous to this, the average bond energies for bonds between atoms of most of the common elements can be obtained. Table 10–6 summarizes some of these quantities.

We can use average bond energies to estimate the thermochemistry of gas-phase chemical reactions involving reactants and products for which the values of ΔH_f^0 are not known. Consider the unknown molecule, ammonia oxide, H_3NO,

```
     H
     ..  ..
  H : H : O :
     ..  ..
     H
```

which has the same composition as the known molecule, hydroxylamine, H_2NOH,

```
     H
     ..  ..
  H : N : O : H
     ..  ..
```

TABLE 10–6
Average Bond Energies[a]

Single bonds			
H—H	432	N—C	305
C—H	413	O—C	358
N—H	391	F—C	485
O—H	467	Cl—C	339
S—H	347	N—N	160
F—H	565	O—O	146
Cl—H	427	F—F	154
Br—H	363	Cl—Cl	239
I—H	295	Br—Br	190
C—C	347	I—I	149
Multiple bonds			
C=O	745	C≡C	837
N=N	418	N≡N	942
C=C	614		

[a]Bond energies, E, are in kJ mol^{-1}.

The value of ΔH^0 for the reaction

$$\mathrm{H_3N{-}\ddot{\underset{..}{O}}\!:} = \mathrm{H_2N{-}\ddot{\underset{..}{O}}{-}H}$$

can be estimated by using the average bond energies tabulated in Table 10–6:

$$\begin{aligned}\Delta H^0 &= -[2E(\mathrm{N{-}H}) + E(\mathrm{N{-}O}) + E(\mathrm{O{-}H})\\ &\quad - 3E(\mathrm{N{-}H}) - E(\mathrm{N{-}O})]\\ &= E(\mathrm{N{-}H}) - E(\mathrm{O{-}H})\\ &= 391\ \mathrm{kJ\ mol^{-1}} - 467\ \mathrm{kJ\ mol^{-1}}\\ &= -76\ \mathrm{kJ\ mol^{-1}}\end{aligned}$$

Thus we can say that ammonia oxide is unstable with respect to hydroxylamine primarily because the O–H bond is stronger than the N–H bond. (The entropy change in this reaction, in which one gaseous molecule gives one gaseous molecule, is expected to be small.)

Features of molecular structure which enhance or diminish stability are revealed by comparing the experimentally measured heat of formation of the compound with that calculated using average bond energies. Cyclopropane, C_3H_6, is a molecule with strained bond angles:

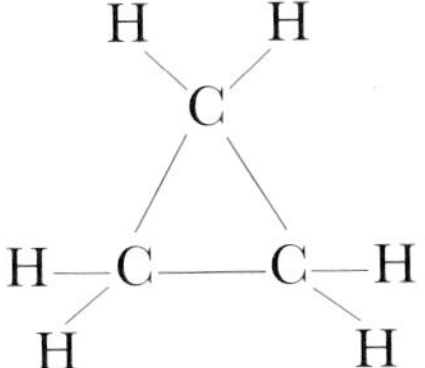

Instead of the tetrahedral angle (109.5°), the angles in the ring of this molecule are $\angle \mathrm{CCC} = 60°$. Experimental data allow calculation of the energy change for the process:

$$\mathrm{C_3H_6}(g) = 3\mathrm{C}(g) + 6\mathrm{H}(g) \qquad \Delta U = +3404\ \mathrm{kJ\ mol^{-1}}$$

Using the average C—C and C—H bond energies derived from data on "strain-free" molecules, one calculates for this reaction:

$$\begin{aligned}\Delta U\ (\text{calcd}) &= 3E(\mathrm{C{-}C}) + 6E(\mathrm{C{-}H})\\ &= 3 \times 347\ \mathrm{kJ\ mol^{-1}} + 6 \times 413\ \mathrm{kJ\ mol^{-1}}\\ &= +3519\ \mathrm{kJ\ mol^{-1}}\end{aligned}$$

Less energy is required to dissociate cyclopropane into its constituent atoms than is predicted in this calculation; the difference ($115\ \mathrm{kJ\ mol^{-1}}$) is attributed to the *strain* in cyclopropane.

It is convenient in some contexts to think of covalently bonded molecules as being made up of atoms with particular bonding radii, even though bond lengths vary during a vibration. The distance of separation of atoms at the potential-energy minimum, r_0, in Figure 10–3, can be taken as the sum of the covalent radii for the bonded atoms. The values of r_0 for diatomic halogen molecules, from Table 10–1, allow us to calculate the covalent radii for fluorine, chlorine, bromine, and iodine. These values are 0.72 Å, 0.99 Å, 1.14 Å, and 1.33 Å, respectively. Let us use these covalent radii to predict the interatomic distances in some diatomic interhalogen molecules, X—Y. The calculated and experimentally determined values are:

	$(r_X + r_Y)$ Å	r_0(X–Y) Å
ClF	1.71	1.63
BrF	1.86	1.76
BrCl	2.13	2.14
ICl	2.32	2.32

The agreement is good for bromine chloride and iodine chloride, but it is not adequate for the molecules containing fluorine. The bond between fluorine atoms in the diatomic molecule F_2 is weaker than expected, as already mentioned in Chapter 8, and there is a correspondingly long bond in F_2. For better agreement of calculated with observed interatomic distances in other covalent fluorides, a radius of 0.64 Å should be used for fluorine.

The carbon–carbon bond distances in diamond and in a number of hydrocarbons (in which single bonds join carbon atoms) range between 1.53 Å and 1.54 Å. These data lead to a value of 0.77 Å for the single bond radius for carbon. With this covalent radius and those for the halogens, we can calculate the expected C—X interatomic distances in methyl halides, H_3CX:

	$(r_C + r_X)$ Å	r_0(C–X) Å
H_3CF	1.41	1.39
H_3CCl	1.76	1.77
H_3CBr	1.91	1.94
H_3CI	2.10	2.14

The agreement is reasonable, but not perfect.

Ethylene, C_2H_4, and acetylene, C_2H_2, have structures involving carbon–carbon double and triple bonds:

H₂C=CH₂ (C=C distance 1.344 Å) and

H—C≡C—H (C≡C distance 1.204 Å)

TABLE 10–7
Covalent Radii[a] for Some Common Elements

Single-bond radii				Double-bond radii			
H	0.28[b]	C	0.77	C	0.67	O	0.62
F	0.64[b]	O	0.66	N	0.62		
Cl	0.99	S	1.04	Triple-bond radii			
Br	1.14	N	0.70				
I	1.33	P	1.10	C	0.60	N	0.55

[a]Radii are in angstroms.
[b]The single-bond radii for H and F given here are not one half of the bond distances in H_2 and F_2. Rather, they are values that best fit the observed bond distances in various H—Y and Z—F molecules.

Studies of these and other closely related hydrocarbons lead to values 0.67 Å for the double-bond covalent radius for carbon and 0.60 Å for the triple-bond covalent radius for carbon. A compilation of covalent radii is given in Table 10–7. The single-bond covalent radii, plotted as a function of atomic number in Figure 10–26, decrease as the nuclear change increases for elements in the same row of the periodic table. Electrons are attracted more closely to the nucleus as the nuclear charge increases. Values of the covalent radii for elements in a particular column of the periodic table increase as the principal quantum number of the valence shell increases.

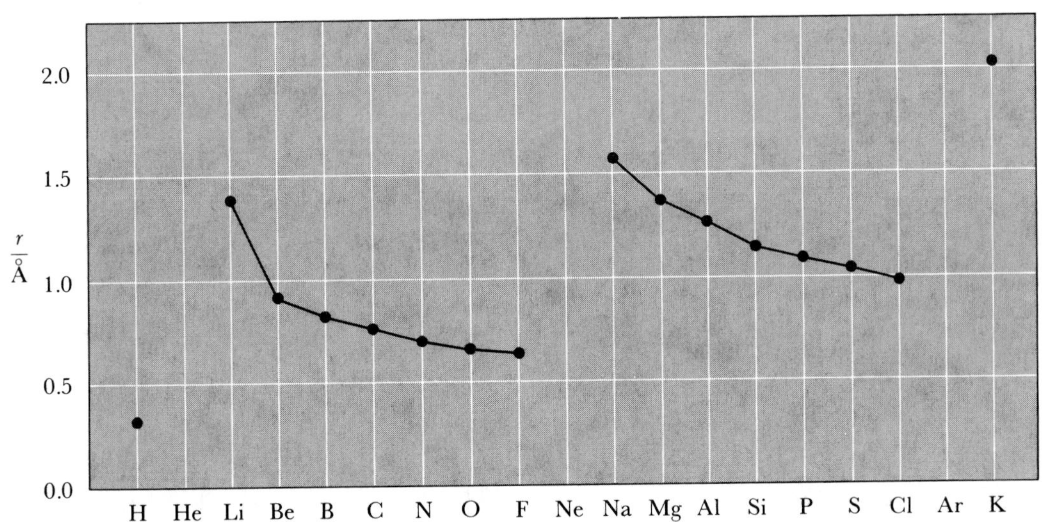

FIGURE 10–26
Single-bond covalent radii for elements H ($Z = 1$) through K ($Z = 19$).

Biographical Notes

RONALD J. GILLESPIE (1924–) was born and educated in England (B.Sc., Ph.D., and D.Sc. degrees from University College, London). He has been on the faculty of McMaster University, Hamilton, Ontario since 1960. In addition to working on molecular geometry, he has done important work on the behavior of solutions in very strongly acidic solvents.

ROBERT S. MULLIKEN (1895–), an American chemical physicist, was educated at the Massachusetts Institute of Technology (B.S.) and the University of Chicago (Ph.D.). For many years he was Professor of Physics at the latter institution. He is particularly known for his work on the molecular-orbital description of bonding in covalent molecules, for which he received the Nobel Prize in Chemistry in 1966.

RONALD S. NYHOLM (1917–1971) was born in Australia and received his bachelor's degree there. He received his Ph.D. and D.Sc. degrees from University College, London, an institution at which he was later Professor of Chemistry. He made outstanding contributions to the chemistry of transition metals.

LINUS C. PAULING (1901–), American chemist, was educated at Oregon State College and the California Institute of Technology. with which he was associated for over 40 years. He now is director of the Linus Pauling Institute for Science and Medicine in Menlo Park, California. He made many important contributions to the theory of chemical bonding and molecular structure, including the structure of proteins. For this work, he received the Nobel Prize in Chemistry in 1954. His efforts to stop the testing of nuclear weapons were recognized by the Nobel Peace Prize in 1962.

NEVIL V. SIDGWICK (1873–1952), an English chemist, was educated at Oxford and Tubingen, Germany. He was Professor of Chemistry at Oxford from 1933 to 1945. He did much early work in valency theory.

Problems and Questions

10–1 Which of the following molecules or ions is paramagnetic: NO_2, N_2O, Cl_2O, ClO_2, SF_6^-, S_2F_{10}?

10–2 Draw electron-dot formulas for OH^- (hydroxide ion), NH_2^- (amide ion), and CO_4^{4-} (orthocarbonate ion, an unstable species). What neutral molecule is isoelectronic with each of these species?

10–3 Draw electron-dot formulas for HONO (nitrous acid), H_2CO (formaldehyde), HOClO (chlorous acid), and HOCl (hypochlorous acid).

10–4 Give the formulas of a cation and an anion that are isoelectronic with CH_4.

10–5 Draw an electron-dot formula for formate ion, HCO_2^-, in which the two C–O bonds are equally long. Is this species planar, or pyramidal?

10–6 The two S–O bonds in sulfur dioxide have the same length. Draw an electron-dot formula for SO_2. Is this molecule linear?

10–7 Consider regular polygons, planar n-sided figures with sides of equal length. Derive an equation for the angle between adjacent sides of such polygons. For what value of n is this angle closest to the regular tetrahedral angle (109.5°)? For values of n larger than this, is the angle between adjacent sides of the figure made smaller, or larger, by allowing the figure to become nonplanar? Use this calculation to explain why S_8 is not planar.

10–8 White phosphorus (the form of phosphorus composed of discrete P_4 molecules) is very reactive. Is there a feature of its structure that explains this property?

10–9 Consider the vibration of a diatomic molecule in its lowest vibrational state. What interatomic separation maximizes the kinetic energy of the atoms? What separation minimizes it?

10–10 The potential-energy curves for isotopic molecules are the same. Thus the potential-energy curves for $^2H_2$ and $^3H_2$ are the same as that for $^1H_2$, shown in Figure 10–3. The force constant k is the same for each of these molecules. What are the energy differences between the $v = 1$ and $v = 0$ states in $^1H_2$, $^2H_2$, and $^3H_2$? What are the dissociation energies for $^2H_2$ and $^3H_2$?

10–11 The force constant k for diatomic chlorine is $k = 3.2 \times 10^{-3}$ N m^{-1}. (Recall 1 N = 1 kg m s^{-2}.) What is the separation of vibrational energy levels in $^{35}Cl_2$? What are the zero-point energies for $^{35}Cl_2$ and $^{37}Cl_2$? If the dissociation energy of $^{35}Cl_2$ is 239 kJ mol^{-1}, what is the dissociation energy of $^{37}Cl_2$?

10–12 The bond energy in H_2, with bond order = 1, is not two times the bond energy in H_2^+, with bond order = 1/2. Suggest reasons why we cannot expect a quantitative correlation, $D \propto$ bond order, for related species such as H_2^+ and H_2.

10–13 Explain the trends of interatomic distances in the pairs of isoelectronic species given in Table 10–3: CO^+ and CN, NO and O_2^+, and O_2^{2-} and F_2.

10–14 The diatomic ions NO^+ and CN^- are isoelectronic with N_2. Predict the relative bond lengths in these three diatomic species.

10–15 There are two diamagnetic excited electronic states of O_2 that have more energy than the ground-state molecule. Use an orbital diagram like those in Figure 10–13 to suggest what the electronic structures of these molecules are.

10–16 Which do you expect to have the larger bond energy, F_2 or F_2^+? What is the basis for your prediction? Use this prediction to help you decide whether F or F_2 has the larger ionization energy.

10–17 Predict geometries of the following ions: SF_5^-, SF_3^+, PF_6^-, SiF_6^{2-}, I_3^-, I_3^+, ICl_4^+, and IF_4^+.

10–18 In our discussion of XY_n molecules with four pairs of electrons in the valence shell of X (Figure 10–18), we used as examples molecules in which Y was hydrogen. Predict the geometries of CF_4, NF_3, and OF_2, in each of which the central atom has an octet of electrons.

10–19 The electron-dot formula for thiocyanate ion, SCN^-, is $:\ddot{\underset{..}{S}}:C:::N:^-$. Is this ion linear? Both HSCN and HNCS are known. Which angle, ∠HSC or ∠HNC, is 180°?

10–20 If gaseous hydrogen chloride were 100% ionic, what would you expect the dipole moment to be? Using a procedure from Chapter 9, calculate the percent ionic character for hydrogen chloride.

10–21 Using average bond energies, estimate the values of ΔH^0 for the following reactions:

$$\begin{array}{c} \begin{array}{ccccc} & H & H & & \\ & | & | & & \\ H- & C- & C- & O- & H \\ & | & | & & \\ & H & H & & \end{array} = \begin{array}{ccccc} & H & & H & \\ & | & & | & \\ H- & C- & O- & C- & H \\ & | & & | & \\ & H & & H & \end{array} \\ \text{Ethyl alcohol} \qquad\qquad \text{Dimethyl ether} \end{array}$$

$$\begin{array}{c} \begin{array}{cccc} & H & H & \\ & \diagdown & \diagup & \\ & & C & \\ & \diagup & \diagdown & \\ H- & C- & C- & H \\ & | & | & \\ & H & H & \end{array} = \begin{array}{cccc} & H & & H \\ & | & & \diagup \\ H- & C= & C- & C-H \\ & & | & \diagdown \\ & & H & H \end{array} \\ \text{Cyclopropane} \qquad\qquad \text{Propylene} \end{array}$$

Which of these estimations do you expect to be incorrect? Why?

10–22 Using average bond energies, estimate the values of ΔU for the reactions:

$$\begin{array}{ccc} H & & H \\ & \diagdown \; \diagup & \\ & C{=}C & \\ & \diagup \; \diagdown & \\ H & & H \end{array} + \begin{array}{c} H \\ \diagdown \\ O-H \end{array} = \begin{array}{ccccc} & H & H & & \\ & | & | & & \\ H- & C- & C- & O- & H \\ & | & | & & \\ & H & H & & \end{array}$$

$$\begin{array}{ccc} H & & H \\ & \diagdown \; \diagup & \\ & C{=}C & \\ & \diagup \; \diagdown & \\ H & & H \end{array} + H-Cl = \begin{array}{cccc} & H & H & \\ & | & | & \\ H- & C- & C- & Cl \\ & | & | & \\ & H & H & \end{array}$$

$$\begin{array}{ccc} H & & H \\ & \diagdown \; \diagup & \\ & C{=}C & \\ & \diagup \; \diagdown & \\ H & & H \end{array} + Cl-Cl = \begin{array}{cccc} & H & H & \\ & | & | & \\ Cl- & C- & C- & Cl \\ & | & | & \\ & H & H & \end{array}$$

$$\begin{array}{cc} H & \\ \diagdown & \\ & C{=}O \\ \diagup & \\ H & \end{array} + H-H = \begin{array}{cccc} & H & & \\ & | & & \\ H- & C- & O- & H \\ & | & & \\ & H & & \end{array}$$

Structures of Simple Carbon Compounds

11–1 Introduction

The living world is based upon the compounds of carbon, and the large area of carbon chemistry is called organic chemistry. At one time it was thought that compounds of carbon originating from living sources (plant or animal) had special characteristics because of their origin. This idea has been dead for one hundred and fifty years now: carbon compounds obtained from living sources also can be obtained by synthesis from inorganic reactants. Organic chemists synthesize many thousands of new carbon compounds each year. However, some compounds of carbon are not within the domain of organic chemistry: for example, the oxides of carbon and carbonates.

In this chapter we will consider the structures of simple compounds of carbon. These structures will expand and further illustrate the principles outlined in Chapter 10. Central to any discussion of organic compounds is the ability of a carbon atom to form four covalent bonds, an ability that is consistent with an excited electronic state of carbon with four unpaired electrons:

	1*s*	2*s*	2*p*
C (excited state)	(↑↓)	(↑)	(↑)(↑)(↑)

The tetrahedral orientation of four covalent bonds to carbon is consistent with either of the points of view presented in the last chapter, the valence-shell electron-pair repulsion model or the hydridization of the one $2s$ and three $2p$ orbitals to give four hybrid sp^3 bonds.

Much of our discussion in this chapter will deal with hydrocarbons, compounds of carbon and hydrogen, but we will learn also about organic compounds with particular groups of atoms, called *functional groups*, which bestow on the molecules particular chemical properties. The present chapter will provide part of the basis for studying in Chapter 20 more complex structures, including many with important roles in biology.

11–2 Carbon and Its Simple Inorganic Species

Although this chapter deals primarily with organic compounds, we will start by considering the elemental forms of carbon and simple inorganic species of carbon. In the first discussion we will see structural features that will recur in our later consideration of organic molecules. In the latter discussion we will learn a simple way to describe the distribution of charge in molecules and ions.

DIAMOND AND GRAPHITE

Two allotropic crystalline forms of carbon are diamond and graphite. Some of the properties of these substances differ dramatically:

Diamond	Graphite
Colorless	Black
Very hard	Soft
An electrical insulator	An electrical conductor

These differences are explained by the structures, which are shown in Figure 11–1.

In diamond, each carbon atom is equidistant from four nearest neighbor carbon atoms. These four carbon atoms define a regular tetrahedron, which has a carbon atom in its center. The array continues in all directions. Examine the diamond structure in Figure 11–1, and you will see that it is a face-centered cubic array of atoms with additional atoms in one half of the tetrahedral sites. (The location of tetrahedral sites in a face-centered cubic lattice was shown in Figure 9–7.) Each line between adjacent carbon atoms in the diamond structure is intended to represent an electron-pair bond. One electron of each bonding pair is donated by each of the two bonded carbon atoms. The hardness of diamond can be attributed to the three-dimensional network of covalent carbon–carbon bonds. That diamond is not a conductor of electricity is consistent with this electronic structure: the valence electrons are all involved in strong tetrahedrally directed covalent bonds. (Recall the point already made, in Section 5–4, that in

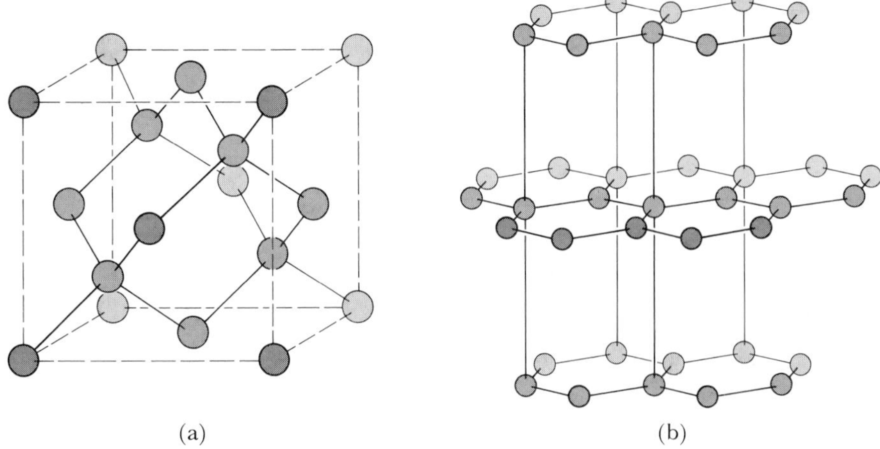

FIGURE 11–1
The structures of diamond and graphite. (a) Diamond, $d_{C-C} = 1.54$ Å. (b) Graphite, $d_{C-C} = 1.42$ Å in plane; $d_{C-C} = 3.35$ Å between adjacent planes.

free elements coordination numbers less than those found in metals, eight and twelve, imply strong directional bonding.)

In graphite, the carbon atoms are arranged in a hexagonal array in planar sheets, each atom surrounded in the plane by three other atoms that define an equilateral triangle. The distances between nearest-neighbor carbon atoms in a plane are all equal (1.42 Å); this distance is shorter than the carbon–carbon single bond distance in diamond (1.54 Å), but it is longer than the carbon–carbon double bond distance in ethylene (1.344 Å). Accordingly, the bonding between carbon atoms is judged to be intermediate between a single bond and a double bond. The distance between the planar sheets of carbon atoms is relatively large (3.35 Å), and no covalency is assumed in the attraction between carbon atoms in adjacent planes.

To explain the equivalence of all carbon–carbon bond lengths in the sheets of atoms in graphite, as well as the value of this bond length, we will use the concept of resonance to which you were introduced in Section 10–5. Figure 11–2 shows a number of arrangements of four bonds to each carbon atom in a segment of a sheet of atoms in graphite. The electronic structure of graphite, from the point of view of resonance theory, is a hybrid of these and other equivalent structures. Of the six structures shown in Figure 11–2, two show double bonding between the adjacent carbon atoms designated with black dots and four show single bonding between these carbon atoms. If all possible resonance structures were considered, there would be this same relative contribution of double and single bonding for every pair of adjacent carbon atoms. That is, each carbon–carbon bond would be one third double bond and two thirds single bond; this corresponds to a bond-order of one and one third:

$$(\tfrac{1}{3} \times 2) + (\tfrac{2}{3} \times 1) = 1\tfrac{1}{3}$$

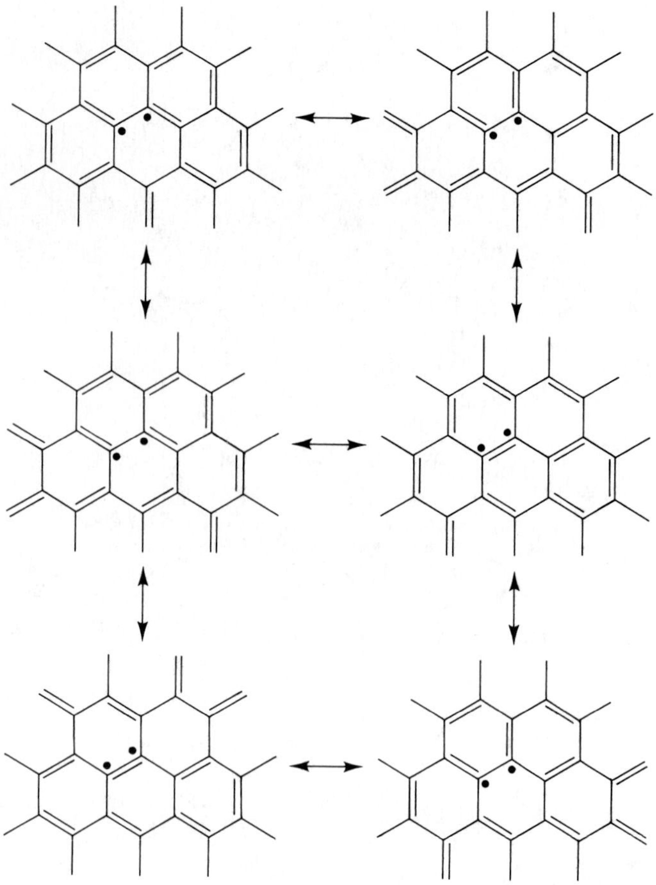

FIGURE 11–2
Resonance structures for the bonding of carbon atoms in graphite. The double-headed arrows denote resonance. The two dots simply label a particular pair of atoms; they do not denote electrons.

The bond order 1 1/3 also follows from reasoning that simply divides the total number of bonds formed by each carbon atom (4) by the total number of equivalent atoms to which each carbon atom is bonded (3):

$$\text{Bond order} = \tfrac{4}{3} = 1\tfrac{1}{3}$$

This bond-order is consistent with the observed carbon–carbon bond distances in substances with carbon–carbon bond orders 1, 2, and 3, as shown in Figure 11–3.

The electronic structure of graphite also can be described in terms of molecular orbitals. If the sheet of atoms defines the *xy* plane, the bonding involves a network of sigma bonding molecular orbitals made up of *s*, p_x, and p_y orbitals of the carbon atoms plus pi bonding molecular orbitals made up of the p_z orbitals of the carbon atoms. Electrons in the pi molecular orbitals belonging to the entire array of atoms have a mobility that is consistent with the electrical conductance of graphite. The attraction between the sheets of carbon atoms is weak, and graphite is easily cleaved along these planes. This easy cleavage makes graphite useful as a lubricant.

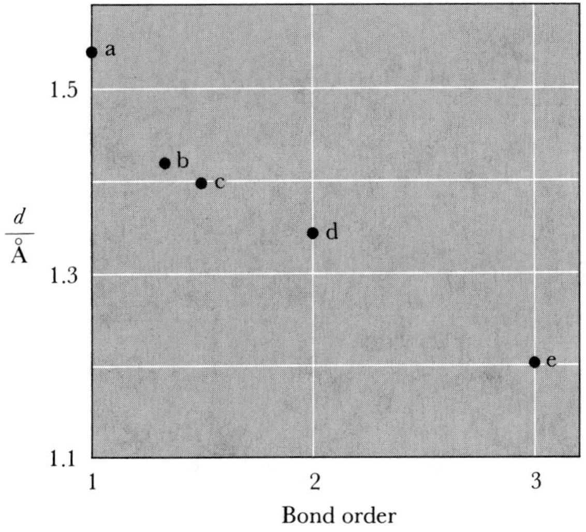

FIGURE 11–3
The length of carbon–carbon bonds as a function of the bond order.
(a) C—C in diamond or ethane (bond order 1)
(b) C---C in graphite (bond order 1.33)
(c) C---C in benzene (bond order 1.5)
(d) C═C in ethylene (bond order 2)
(e) C≡C in acetylene (bond order 3)

At ordinary temperatures and pressures, diamond is unstable with respect to graphite:

$$\text{C(diamond, 1 atm)} = \text{C(graphite, 1 atm)}$$
$$\Delta G^0(298.2\ \text{K}) = -2.87\ \text{kJ mol}^{-1}$$

However, the rate of this reaction is immeasurably low under these conditions, and the spontaneous conversion of diamond to graphite does not occur at 1 atm and 298.2 K. The density of diamond ($3.51\ \text{g cm}^{-3}$) is greater than that of graphite ($2.22\ \text{g cm}^{-3}$), and equilibrium in the conversion reaction can be shifted to diamond at sufficiently high pressures. The phase diagram for carbon is shown in Figure 11–4. The successful conversion of graphite to diamond, a goal sought for almost one hundred years, was achieved in 1953 and 1954.[1] Synthetic diamonds are now produced commercially: 40% of industrial diamonds are man-made.

Because graphite is more stable at 1 atm and 298.2 K, this form of carbon is conventionally selected as the standard state for the element. You saw in Table 3–2 that

$$\Delta H_f^0\ \text{(C, graphite)} = 0$$
$$\Delta H_f^0\ \text{(C, diamond)} = 1.88\ \text{kJ mol}^{-1}$$

The entry in this table,

$$\Delta H_f^0[CO_2(g)] = -393.5\ \text{kJ mol}^{-1}$$

[1]The synthesis of diamond was first achieved in the United States by H. Tracy Hall (1919–) and associates in December 1954 at the General Electric Research Laboratory. In an article describing his first synthesis of diamond [*J. Chem. Educ.* **38**, 484 (1961)], Dr. Hall cites claims by Swedish scientists (H. Liander and E. Lundblad) to have synthesized diamonds in February 1953.

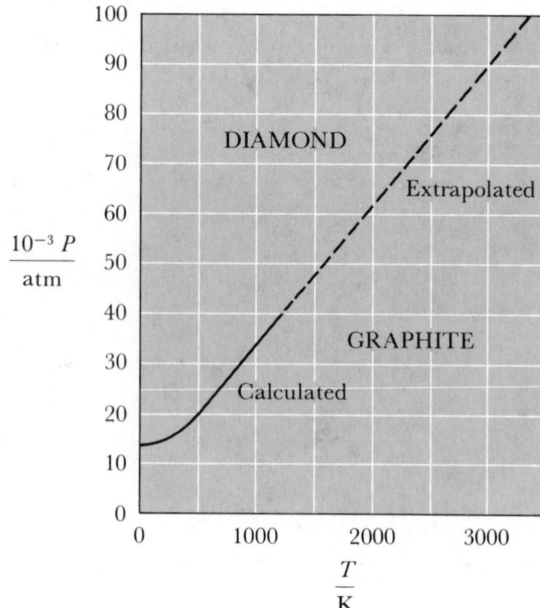

FIGURE 11–4
The phase diagram for carbon.

pertains to the reaction

$$C(s, \text{graphite}) + O_2(g) = CO_2(g)$$

Analysis of coal shows it to contain, in addition to impurities, almost as much hydrogen as carbon (on an atomic basis); the carbon–hydrogen composition of a typical bituminous coal is $CH_{0.8 \pm 0.3}$. The framework of carbon atoms in coal includes segments having the structural features of graphite with peripheral hydrogen atoms.

OXIDES OF CARBON

You already have learned something about the principal oxides of carbon—carbon monoxide, CO, and carbon dioxide, CO_2. Each is formed exothermically from its elements:

$$2C(s, \text{graphite}) + O_2(g) = 2CO(g) \qquad \Delta H^0 = -221.0 \text{ kJ mol}^{-1}$$
$$C(s, \text{graphite}) + O_2(g) = CO_2(g) \qquad \Delta H^0 = -393.5 \text{ kJ mol}^{-1}$$

as we explained in our discussion of chemistry and energy in Chapter 3. Carbon monoxide, with atoms having four and six valence electrons, is isoelectronic with diatomic nitrogen, N_2. A Lewis electronic structure analogous to that for N_2, with a triple bond,

:C:::O:

explains the extraordinary strength of the carbon–oxygen bond [$D(\text{C–O}) = 1070.3 \text{ kJ mol}^{-1}$], but this structure implies an unreasonable distribution of charge in the molecule, as we shall show. The distribution of charge for a Lewis formula of a covalently bonded molecule may be obtained by calculating the *formal charges* on each atom. These are obtained by an accounting of valence-shell electrons that gives each bonded atom its lone-pair electrons plus one half of the electrons it shares with other atoms. The negative charge of this share of electrons is compared to the positive charge on the atom's *kernel*, which is the neutral atom *minus* its valence electrons. Thus in triple-bonded carbon monoxide, the formal charge on each atom is:

	Kernel charge	+	Electron charge	=	Formal charge
C	+4		−5	=	−1
O	+6		−5	=	+1

Because oxygen is more electronegative than carbon, a charge distribution in which the carbon atom is negative and the oxygen atom positive is unlikely. Another difficulty with this structure is its inconsistency with the low measured dipole moment, $p(\mathrm{CO}) = 0.10$ D. The correct structure is believed to be a resonance hybrid of the structures:

$$\overset{-}{:\mathrm{C}}:::\overset{+}{\mathrm{O}}: \longleftrightarrow :\mathrm{C}::\ddot{\underset{\cdot\cdot}{\mathrm{O}}} \longleftrightarrow :\overset{+}{\mathrm{C}}:\ddot{\underset{\cdot\cdot}{\mathrm{O}}}:^{-}$$

in which the formal charges in the first and third structure are shown. (We see in this example that the sum of the formal charges on all of the atoms in a neutral molecule is zero.) The extraordinary strength of the carbon–oxygen bond is attributed in part to this resonance, and the small dipole moment is attributed to opposition of the dipole moments of the two polar structures.

The linear structure of carbon dioxide is consistent with the electron-dot formula,

$$\ddot{\underset{\cdot\cdot}{\mathrm{O}}}::\mathrm{C}::\ddot{\underset{\cdot\cdot}{\mathrm{O}}}$$

if you consider the carbon atom to be surrounded by two groups of electrons with two pairs of electrons in each group. In this structure the formal charge on each atom is zero.

SIMPLE SPECIES OF CARBON WITH HYDROGEN, OXYGEN, AND NITROGEN

The description of carbon oxides will help us understand some other simple carbon species with which they are isoelectronic. We can formulate such species easily if we recognize that N^- and C^{2-} are isoelectronic with O. Species that are isoelectronic with carbon monoxide (2 atoms and 10 valence electrons) are:

$$CN^- \quad (HCN) \qquad C_2^{2-} \quad (HCCH)$$

Both cyanide ion, CN^-, and acetylide ion, C_2^{2-}, and the molecules formed by protonating these ions are known species.

Isoelectronic with carbon dioxide (3 atoms and 16 valence electrons) are *isomeric* species with two different compositions:

$$NCO^- \quad CNO^- \quad CON^- \quad CNN^{2-} \quad NCN^{2-}$$

The species with the same composition and molecular weight but with different structure are called *isomers*. Here we can use the principle that structures with large formal charges (|formal charge| > 1) are unstable to decide which isomer is the stable structure for each composition.[2] The double-bonded structure for a triatomic sixteen-electron species has the indicated charges due to the valence electrons:

$$\underset{\cdot\cdot}{\overset{\cdot\cdot}{X}} :: Y :: \underset{\cdot\cdot}{\overset{\cdot\cdot}{Z}}$$
$$-6 \quad -4 \quad -6$$

and a structure involving one triple bond and one single bond has the indicated charges due to the valence electrons:

$$: X ::: Y : \underset{\cdot\cdot}{\overset{\cdot\cdot}{Z}} :$$
$$-5 \quad -4 -7$$

If we match these charges with kernel charges, the structures with minimum development of formal charges are the stable species:

NCO^-, cyanate ion

$$\underset{\cdot\cdot}{\overset{\cdot\cdot}{N}} :: C :: \underset{\cdot\cdot}{\overset{\cdot\cdot}{O}}{}^- \longleftrightarrow : N ::: C : \underset{\cdot\cdot}{\overset{\cdot\cdot}{O}} :^-$$
$$-1 \quad 0 \quad 0 \qquad\qquad 0 \quad 0 - 1$$

NCN^{2-}, cyanamide ion

$$\underset{\cdot\cdot}{\overset{\cdot\cdot}{N}} :: C :: \underset{\cdot\cdot}{\overset{\cdot\cdot}{N}}{}^{2-}$$
$$-1 \quad 0 \; -1$$

(Notice that the sum of the formal charges is the net charge on the ion.) The ions CON^- and CNN^{2-} are not known; each of these structures matched to the electron distributions given in the XYZ formulas have formal charges of −2 on one or more atoms. The ion CNO^- is known; it is fulminate ion. The formal charges of the structures

$$\underset{\cdot\cdot}{\overset{\cdot\cdot}{C}} :: N :: \underset{\cdot\cdot}{\overset{\cdot\cdot}{O}}{}^- \longleftrightarrow : C ::: N : \underset{\cdot\cdot}{\overset{\cdot\cdot}{O}} :^-$$
$$-2 \; +1 \quad 0 \qquad\qquad -1 \quad +1 -1$$

suggest that the second structure is a better representation of the electron

[2]This is essentially an approach proposed by Pauling in 1926.

distribution in fulminate ion, but that even this structure with each atom having a formal charge of ± 1 may be unstable. In fact, fulminates are explosively unstable.

11–3 Hydrocarbons

In the discussion of covalent bonding in Chapter 10, you were introduced to methane, CH_4, the simplest hydrocarbon, and also to hydrocarbons with two carbon atoms joined by a single bond, ethane (C_2H_6), a double bond, ethylene (C_2H_4), and a triple bond, acetylene (C_2H_2). The ability of carbon atoms to form strong covalent bonds with each other, as in diamond or ethane, and with hydrogen atoms, as in methane or ethane, as well as their ability to form double bonds and triple bonds with other carbon atoms, is the basis for the enormous number of hydrocarbon molecules, C_nH_m. The ratio of hydrogen to carbon in these molecules (m/n) varies from a maximum of 4 in CH_4 to 1 or less than 1: for example, 1.0 in C_6H_6 (benzene) or C_2H_2 (acetylene), 0.8 in $C_{10}H_8$ (naphthalene), or 0.71 in $C_{14}H_{10}$ (phenanthrene). We now will consider the structures of these diverse molecules.

Although we will not deal at length with molecules in which carbon atoms are bonded to halogen atoms, you will remember from the preceding chapter that fluorine atoms (and atoms of other halogens) can, like hydrogen, form one single covalent bond, as in hydrogen fluoride, HF. The discussion of polarity of polyatomic molecules included the series of chloromethanes, CH_3Cl, CH_2Cl_2, $CHCl_3$, and CCl_4, in which the hydrogen atoms of methane were replaced by chlorine atoms. So as you study hydrocarbons, remember that there are many molecules in which one or more hydrogen atoms of a hydrocarbon molecule are replaced by halogen atoms.

SATURATED HYDROCARBONS

Hydrocarbons that contain no multiple bonds are called saturated hydrocarbons. Methane is the simplest, and the next in order of molecular weight are ethane, C_2H_6, and propane, C_3H_8. The structures of these molecules are shown in Figure 11–5. Each of the bond angles ∠HCH, ∠HCC, and ∠CCC in these molecules is very close to the tetrahedral angle. The formulas and structures of methane, ethane, and propane can be viewed as a number of connected CH_2 units (called methylene groups) with hydrogen atoms at each of the two ends:

$$CH_4 \text{ is } H(CH_2)_1H$$
$$C_2H_6 \text{ is } H(CH_2)_2H$$
$$C_3H_8 \text{ is } H(CH_2)_3H$$

The general formula is $(CH_2)_nH_2$ or C_nH_{2n+2}, and hydrocarbons with

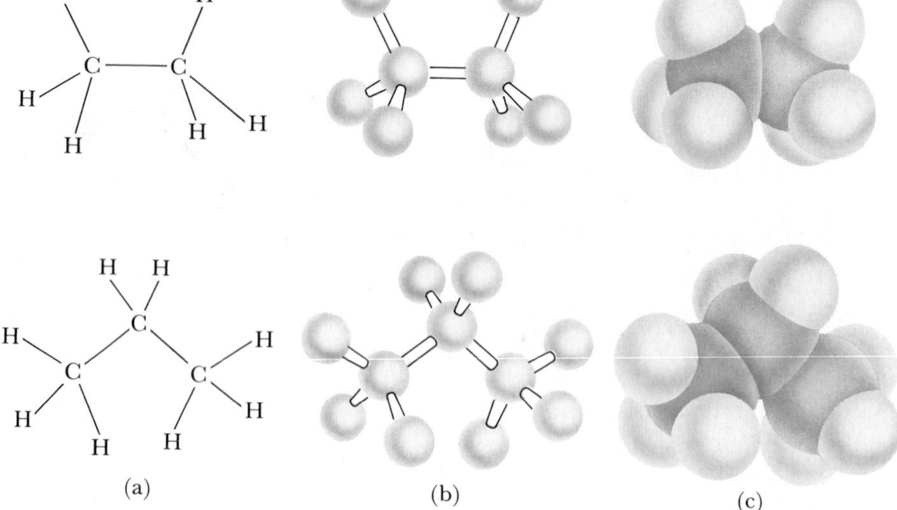

FIGURE 11–5
Structures of ethane (C_2H_6) and propane (C_3H_8). Representation with: (a) chemical symbols, (b) ball-and-stick models, and (c) space-filling models. Ethane is shown in eclipsed conformation; propane is shown in staggered conformation.

this composition make up the class of compounds called alkanes or saturated hydrocarbons.

The saturated hydrocarbons with four and five carbon atoms are butane, C_4H_{10}, and pentane, C_5H_{12}. There are isomers for each of these compositions, two for C_4H_{10} and three for C_5H_{12}. The arrangement of atoms in the isomers of butane and pentane are shown in Table 11–1. We see that each isomer has characteristic melting and boiling points.[3] Chemical composition and molecular weight are the only respects in which these isomers are the same. Isomers such as these are different chemical substances, and they differ in their chemical and physical properties. The continuous-chain hydrocarbons are named with the prefix *normal*, and the prefix *n*- in a name designates this structure. Two names are given for each of the branched hydrocarbons. The names 2-methylbutane and 2,2-dimethylpropane identify the structures unambiguously, as the names isopentane and neopentane do not. These unambiguous names are based upon a system of nomenclature recommended by the International Union of Pure and Applied Chemistry (IUPAC). In this system of nomenclature,

[3]The relatively high melting point of neopentane is particularly striking. Neopentane is a highly symmetric molecule (it could be called tetramethylmethane), and highly symmetric nonpolar molecules have high melting points. This is explained in terms of the smaller entropy increase accompanying the melting of highly symmetric molecules. For highly symmetric molecules, many different indistinguishable orientations of the molecule are possible in the crystalline solid. Because the randomness corresponding to these different indistinguishable orientations already is present in the solid, the entropy "driving force" that causes melting to occur is smaller, and a higher temperature is needed for $T\Delta S$ to equal ΔH, the enthalpy change upon melting.

TABLE 11–1
The Isomeric Butanes and Pentanes[a]

Butanes (C_4H_{10})		mp °C	bp °C
n-Butane	H H H H | | | | H—C—C—C—C—H | | | | H H H H	−138.3	−0.5
Isobutane (2-methylpropane)	H H H | | | H—C—C—C—H / | \\ H H H—C—H | H	−159.6	−11.7
Pentanes (C_5H_{12})			
n-Pentane	H H H H H | | | | | H—C—C—C—C—C—H | | | | | H H H H H	−129.7	36.1
Isopentane (2-methylbutane)	H H H H | | | | H—C—C—C—C—H | | | | H H C H / | \\ H H H	−159.9	27.9
Neopentane (2,2-dimethylpropane)	H | H—C—H H | H H—C—C—C—H H | H H—C—H | H	−16.6	9.5

[a]In these structures, the zigzag nature of the chain of carbon atoms is not shown.

a substance is named as a derivative of the longest continuous-chain hydrocarbon. The locations of substituent hydrocarbon groups on this longest continuous chain (methyl groups in the examples under consideration) are designated by the numbers of the carbon atoms on which substituents occur.

Each isomeric pentane has the same number of each particular kind of chemical bond. There are four carbon–carbon single bonds and twelve carbon–hydrogen bonds. If the energy of a particular type of bond were independent of the molecule in which it occurred, the enthalpy changes for combustion of all isomeric saturated hydrocarbons of a particular composition would be the same. For the isomeric pentanes, values of ΔH^0 associated with the reactions $C_5H_{12}(g) + 8O_2(g) = 5CO_2(g) + 6H_2O(l)$ at 298.2 K are:

n-Pentane	$\Delta H^0 = -3536.1\ \text{kJ mol}^{-1}$
Isopentane	$\Delta H^0 = -3528.2\ \text{kJ mol}^{-1}$
Neopentane	$\Delta H^0 = -3516.5\ \text{kJ mol}^{-1}$

The chemical equations for the oxidation reactions for the isomeric pentanes can be subtracted from one another to obtain equations for the isomerization reactions:

$$n\text{-}C_5H_{12}(g) = iso\text{-}C_5H_{12}(g) \qquad \frac{\Delta H^0}{\text{kJ mol}^{-1}} = -3536.1 - (-3528.2)$$

$$\Delta H^0 = -7.9\ \text{kJ mol}^{-1}$$

$$n\text{-}C_5H_{12}(g) = neo\text{-}C_5H_{12}(g) \qquad \frac{\Delta H^0}{\text{kJ mol}^{-1}} = -3536.1 - (-3516.5)$$

$$\Delta H^0 = -19.6\ \text{kJ mol}^{-1}$$

The relative thermodynamic stabilities of these isomeric substances are determined by these values of ΔH^0 and the values of ΔS^0 for these reactions. From such data, it can be calculated that at 298.2 K *at equilibrium* the relative amounts of the isomeric pentanes are 3% *n*-C_5H_{12}, 44% *iso*-C_5H_{12}, and 53% *neo*-C_5H_{12}. These data for pentane illustrate a general point: a branched hydrocarbon is thermodynamically more stable than the continuous-chain isomer. Hydrocarbons do not reach isomerization equilibrium rapidly under ordinary conditions, but catalysts, reagents that increase reaction velocity, are known for the isomerization of hydrocarbons. Such catalyzed reactions are important in the petroleum industry because branched hydrocarbons are better fuels than the normal hydrocarbons.

TABLE 11–2
Names and Numbers of Possible Isomeric Alkanes, C_nH_{2n+2}

n	Formula	Name	Number of isomers
1	CH_4	Methane	1
2	C_2H_6	Ethane	1
3	C_3H_8	Propane	1
4	C_4H_{10}	Butane	2
5	C_5H_{12}	Pentane	3
6	C_6H_{14}	Hexane	5
7	C_7H_{16}	Heptane	9
8	C_8H_{18}	Octane	18
9	C_9H_{20}	Nonane	35
10	$C_{10}H_{22}$	Decane	75

The names of the alkanes (saturated hydrocarbons) with up to ten carbon atoms are given in Table 11–2. As you would expect, the number of isomers increases rapidly as the number of carbon atoms in the molecule increases.

THE CONFORMATIONS OF SATURATED HYDROCARBONS

An aspect of the structures of saturated hydrocarbon molecules is not covered by either the representations in Figure 11–5 or in Table 11–1. This is the internal rotation of parts of the molecule relative to one another. We already have learned that an increase in a molecule's energy of rotation as temperature increases contributes to the heat capacity of a substance. This is true both for rotation of an entire molecule and for internal rotation of parts of a molecule relative to one another. Heat-capacity data for ethane show that the two methyl groups rotate freely relative to one another at ordinary temperatures.

Possible relative orientations of the hydrogen atoms of the two methyl groups of ethane are shown in Figure 11–6, which shows also the potential energy of the system as a function of the angle of rotation. The potential energy is a minimum when the hydrogen atoms on one methyl group are staggered 60° relative to those on the other methyl group. The potential energy is a maximum if the hydrogen atoms on the two methyl groups are lined up with one another. The structures corresponding to various values of the angle ϕ in Figure 11–6 are called *conformations*. The conformation corresponding to $\phi = 60°$ is called the *staggered* conformation, and that

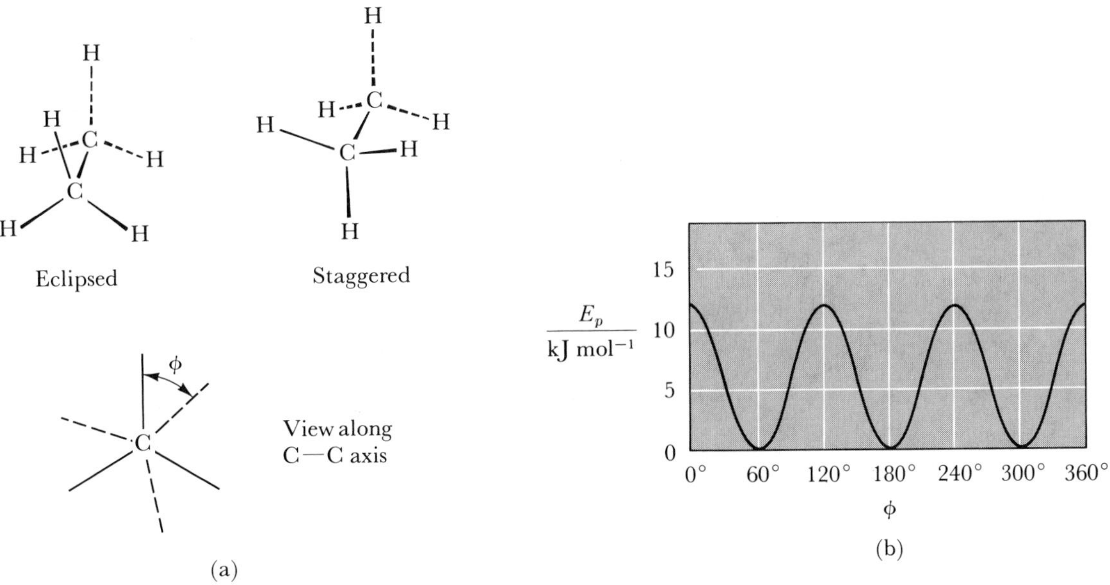

FIGURE 11–6
Internal rotation of methyl groups in ethane. (a) Different conformations of ethane. (b) Potential energy as a function of angle ϕ. $E_p = 0$ for staggered conformation ($\phi = 60°$, 180°, and 300°); $E_p \cong 12$ kJ mol^{-1} for eclipsed conformation ($\phi = 0°$, 120°, and 240°).

with $\phi = 0^\circ$ is called the *eclipsed* conformation. The difference between the potential energies of the staggered and eclipsed conformations is called the barrier to internal rotation. In ethane, this barrier is ~ 12 kJ mol^{-1}. This energy is larger than RT (~ 2.5 kJ mol^{-1} at 298 K), but not enormously larger. Consequently most ethane molecules at 298 K have the staggered conformation, but the methyl groups rotate rapidly relative to one another (i.e., ϕ changing from 60° to 180° to 300° to 60°). The barrier to internal rotation is greater than ~ 12 kJ mol^{-1} if atoms or groups larger than hydrogen are bonded to the carbon atoms in ethane. Thus if two hydrogens on one of the carbon atoms of ethane are replaced by methyl groups (giving 2-methylpropane), the barrier to internal rotation is raised to 16.3 kJ mol^{-1}.

Just as there are six hydrogen atoms bonded to the two carbon atoms in ethane, there are six carbon atoms bonded to a two-carbon-atom segment in diamond (see Figure 11–1). In the diamond structure the three carbon atoms bonded to one carbon atom have a *staggered conformation* relative to the three carbon atoms bonded to the other carbon atom of the two-carbon segment. Various conformations of hydrocarbons can be visualized by focusing attention on the diamond lattice. Figure 11–7 shows some distinguishable conformations of the carbon atoms of normal hexane inscribed in a diamond lattice. Because of the relatively free rotation around single bonds at ordinary temperatures, the various conformations are rapidly transformed into one another in the gas phase and in the liquid phase, and molecules with many conformations are present in a sample. In the solid state, however, the molecules can be packed more closely together if they all have the same conformation.

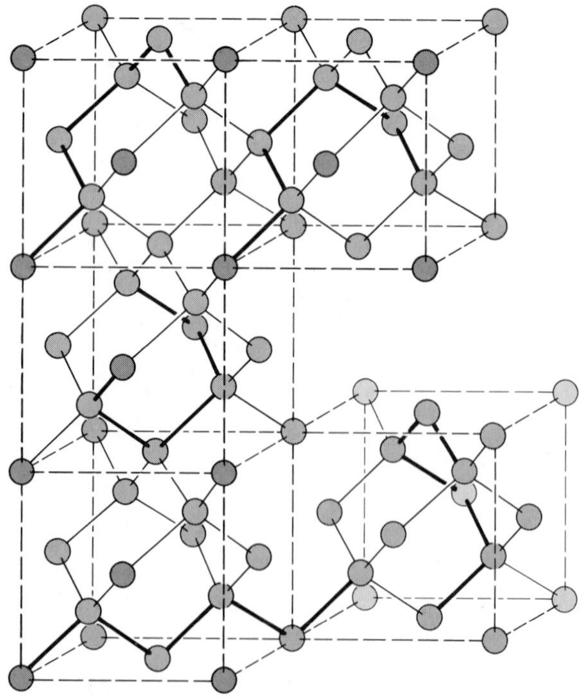

FIGURE 11–7
Various conformations of the continuous chain of carbon atoms in normal hexane inscribed in a diamond lattice. The atoms bonded to each carbon atom are either carbon atoms, which are part of the chain (shown), or hydrogen atoms, which are not part of the chain. In each of these conformations there is a staggered conformation of the atoms bonded to adjacent carbon atoms. The conformation in the lower left is the fully extended conformation.

For a saturated hydrocarbon with the general chemical formula C_nH_{2n+2}, the ratio of the numbers of hydrogen atoms to carbon atoms is $[2 + (2/n)]$ to 1. As n becomes very large, this ratio approaches 2 to 1. Hydrocarbons in which this ratio is 2 to 1 or smaller have structural features not present in alkanes. You already have been introduced to molecules (C_nH_m) with $m/n \leq 2$: C_2H_4, ethylene (the simplest alkene, also called ethene), a molecule with a double bond, C_2H_2, acetylene (the simplest alkyne, also called ethyne), a molecule with a triple bond, and C_3H_6, cyclopropane (the simplest cycloalkane), a molecule containing a ring of carbon atoms.

If a hydrocarbon has the molecular formula $C_nH_{2n+2-2a}$, the significance of a is

$$a = d + 2t + r$$

in which d = number of double bonds, t = number of triple bonds, and r = number of rings. That is, compared to the chemical formula for an alkane, C_nH_{2n+2}, each deficiency of two hydrogen atoms means one double bond or one ring, and a deficiency of four hydrogen atoms means a triple bond or two of the other features just mentioned. Hydrocarbons with double or triple bonds are called *unsaturated*.

Unsaturated Hydrocarbons—Alkenes and Alkynes. The presence of one or more multiple bonds in a hydrocarbon molecule gives the molecule some unusual characteristics. Unlike the carbon–carbon single bond, around which the six bonded atoms or groups of atoms can rotate relatively freely, the carbon–carbon double bond is a structural unit which does not allow free rotation. This is shown by the contrasting numbers of isomers of dichloroethane (two) and dichloroethene (three), the structures of which are shown in Figure 11–8. Each isomeric dichloroethene is planar (as we would predict from principles discussed in Chapter 10). The rigidity of this planar structure is consistent with a model of two carbon atoms joined by a double bond being depicted as two tetrahedra sharing an edge:

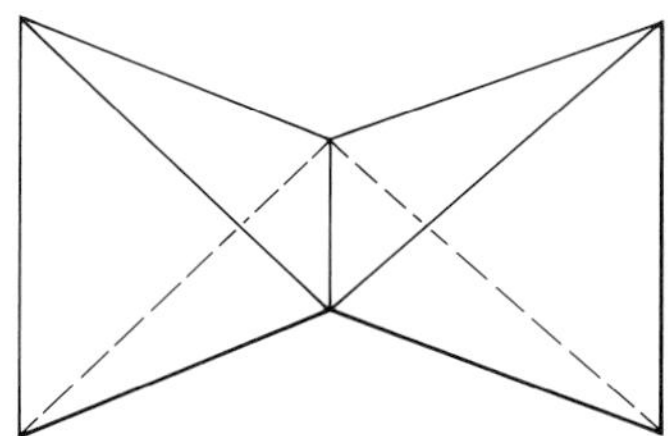

The two carbon atoms are in the centers of the tetrahedra. In this same type of picture, two carbon atoms joined by a triple bond is depicted as two tetrahedra sharing a face:

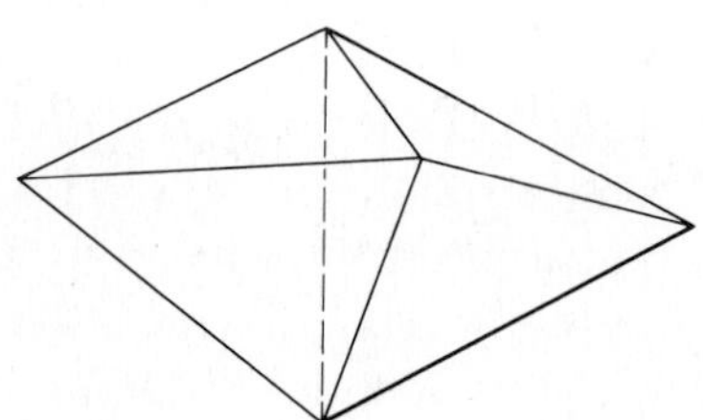

Like the valence-shell electron-pair repulsion model, this model predicts a linear geometry for acetylene (ethyne).

The next simplest alkene is propylene or propene, $H_3C{-}CH{=}CH_2$, and the next simplest alkyne is methyl acetylene or propyne, $H_3C{-}C{\equiv}CH$. Carbon–carbon double bonds and triple bonds are found in many other molecules, and the geometrical arrangements of atoms bonded to the carbon atoms of these substances are those just outlined. The properties of long-chain hydrocarbons with periodic double bonds depend upon whether there is a cis or trans configuration at each double bond. This point will be discussed in Chapter 20, where the molecular structure of rubber will be discussed.

We saw in the preceding chapter (Table 10–6) that multiple bonds are stronger than single bonds, but carbon–carbon multiple bonds are not strong enough to make these molecules unreactive. The reaction of

1,1-Dichloroethane
mp −97.4°C
bp 57.3°C

1,2-Dichloroethane
mp −35.3°C
bp 83.5°C

(a) $C_2H_4Cl_2$

1,1-Dichloroethene
mp −122.1°C
bp 37.0°C

cis-1,2-Dichloroethene
mp −80.5°C
bp 60.3°C

trans-1,2-Dichloroethene
mp −50.0°C
bp 47.7°C

(b) $C_2H_2Cl_2$

FIGURE 11–8
The isomers of $C_2H_4Cl_2$ and $C_2H_2Cl_2$. (a) There is relatively free rotation around the C–C bond in each dichloroethane isomer. (b) The isomeric dichloroethenes are planar molecules; free rotation does not occur.

ethene with hydrogen,

$$C_2H_4(g) + H_2(g) = C_2H_6(g)$$

is exothermic ($\Delta H^0 = -137$ kJ mol^{-1} at 298.2 K), and a summary of the bonds present in the reactants and product, along with the average bond energies from Table 10–6, shows why this is the case:

Reactants	Product
1 C=C bond 614 kJ mol^{-1}	1 C—C bond 347 kJ mol^{-1}
4 C—H bonds $4 \times 413 = 1652$ kJ mol^{-1}	6 C—H bonds $6 \times 413 = 2478$ kJ mol^{-1}
1 H—H bond 432 kJ mol^{-1}	

The energy of the two C–H bonds formed offsets the energy of the H–H bond broken as well as the energy difference between the carbon–carbon double and single bonds. Using these bond energies we can calculate for this reaction $\Delta U = -127$ kJ mol^{-1}; correction to a value of ΔH^0 at 298.2 K gives $\Delta H^0 = -129.5$ kJ mol^{-1}, in fair agreement with the experimental value. The exothermicity of this reaction is an indication of reactivity that is general for unsaturated molecules, a reactivity which renders them useful reagents in chemical synthesis.

Cycloalkanes. The geometrical arrangements imposed upon atoms in cyclic hydrocarbons may deviate from the stable arrangements of atoms in the alkanes, in which there are approximately tetrahedral bond angles and staggered (rather than eclipsed) conformations. We already have seen that the enthalpy of atomization of cyclopropane is less positive than predicted for a hydrocarbon with three C–C bonds and six C–H bonds. The instability imposed on cyclopropane by the bond angles $\angle CCC = 60°$ and the eclipsed hydrogen atoms shows itself also in the enthalpy change in the isomerization reaction producing propylene (also called propene):

$$cyclo\text{-}C_3H_6(g) = CH_3CH{=}CH_2(g) \qquad \Delta H^0 = -32.9 \text{ kJ mol}^{-1}$$

A cyclic hydrocarbon that has received much attention is cyclohexane, C_6H_{12}. The conformation of the six-atom saturated ring in this compound is a model for the conformation of six-atom saturated rings in other compounds, for example, sugars and hormones, to be discussed in Chapter 20. If the six carbon atoms in cyclohexane defined a regular planar hexagon, the bond angles $\angle CCC$ would be 120° and all of the hydrogen atoms would be eclipsed, making the conformation an unstable one. A nonplanar conformation is more stable. Two possibilities with $\angle CCC = 109.5°$ (the tetrahedral angle) are shown in Figure 11–9. These two conformations, called *chair* and *boat* conformations, are not equally stable: >99% of cyclohexane molecules have the chair conformation at room temperature.

In the boat conformation of cyclohexane, the eight hydrogen atoms on the sides of the boat are eclipsed, and the two hydrogen atoms that

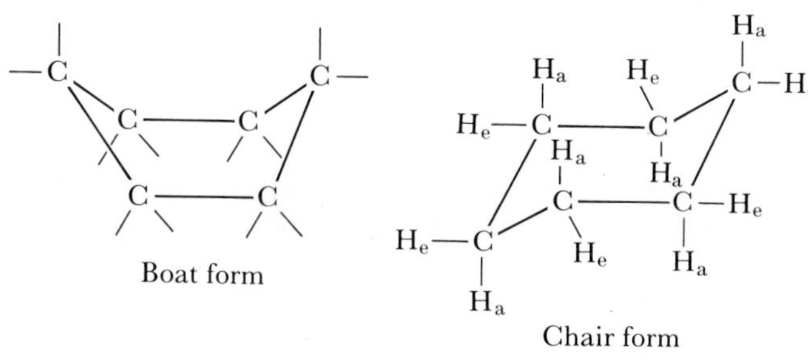

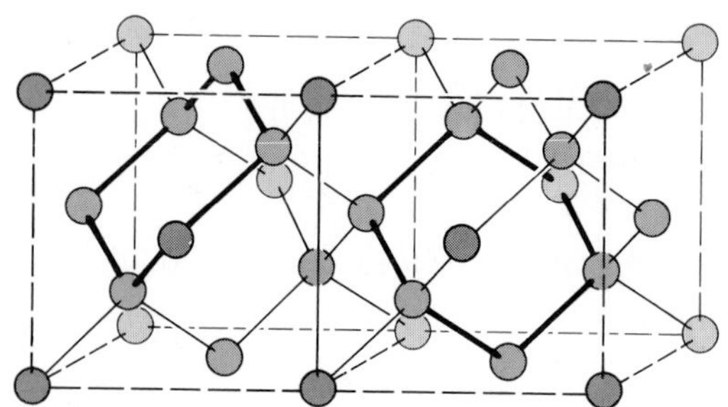

FIGURE 11–9
The boat and chair conformations of cyclohexane. Hydrogen atoms in the chair conformation, which are labeled equatorial (e) or axial (a), are staggered relative to one another. Also shown is the chair conformation of the six-atom ring inscribed in a diamond lattice (two views).

project upward at the front and back of the boat are forced close to one another. Each of these features makes this conformation unstable. In contrast, hydrogen atoms on adjacent carbon atoms in the chair form are staggered relative to one another. There are two different types of hydrogen atoms in the chair form of cyclohexane: six are slightly above or below the average plane of the six carbon atoms and the other six project above and below this plane. Hydrogen atoms of the former group are called *equatorial* hydrogen atoms and those of the latter group are called *axial* hydrogen atoms. There are smaller repulsive interactions between a substituent on a cyclohexane ring and hydrogen atoms bonded to other carbon atoms of the ring if the substituent is in an equatorial position. The relative lack of strain in cyclohexane is shown in the enthalpy change for the isomerization producing 1-hexene (with the double bond joining the first and second carbon atoms),

$$\textit{cyclo}\text{-}C_6H_{12}(g) = C_4H_9\text{—}CH\text{=}CH_2(g) \qquad \Delta H^0 = +81 \text{ kJ mol}^{-1}$$

This endothermic isomerization of a relatively strain-free cyclic hydrocarbon can be contrasted with the exothermic isomerization of cyclopropane. Simply using values of the bond energies given in Table 10–6, we may predict that the value of ΔH for a reaction in which two carbon–carbon single bonds become one carbon–carbon double bond would be $+80$ kJ mol^{-1}.

Benzene. Benzene is a naturally occurring hydrocarbon with the molecular formula C_6H_6. It has eight fewer hydrogen atoms than the saturated hydrocarbon with six carbon atoms (hexane, C_6H_{14}). According to the formula presented before, this deficiency of four pairs of hydrogen atoms means that benzene contains either:

4 rings
3 rings + 1 double bond
2 rings + 2 double bonds
2 rings + 1 triple bond
1 ring + 3 double bonds
1 ring + 1 double bond + 1 triple bond
0 rings + four double bonds
0 rings + two double bonds + 1 triple bond
0 rings + 2 triple bonds

In 1865, KEKULÉ proposed a cyclic structure with three double bonds and three single bonds:

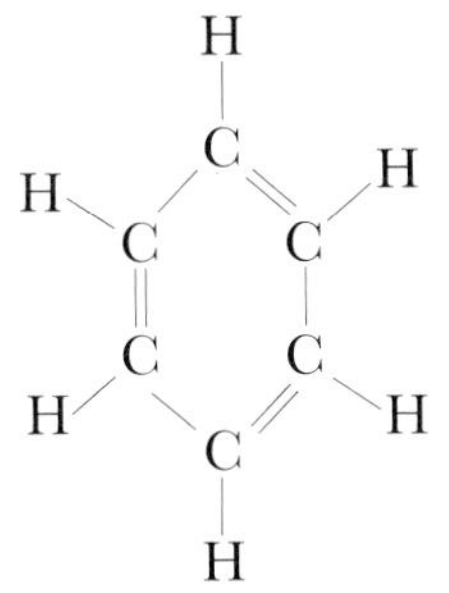

However, this simple valence-bond picture is inadequate because all carbon–carbon bonds in the planar benzene molecule are equal in length (1.397 Å). The single Kekulé structure suggests alternating longer carbon–carbon single bonds (~1.54 Å, as found in diamond or ethane) and shorter carbon–carbon double bonds (~1.344 Å, as found in ethylene). The actual electronic structure of the benzene molecule can be described in terms of a resonance hybrid of two equivalent valence-bond structures:

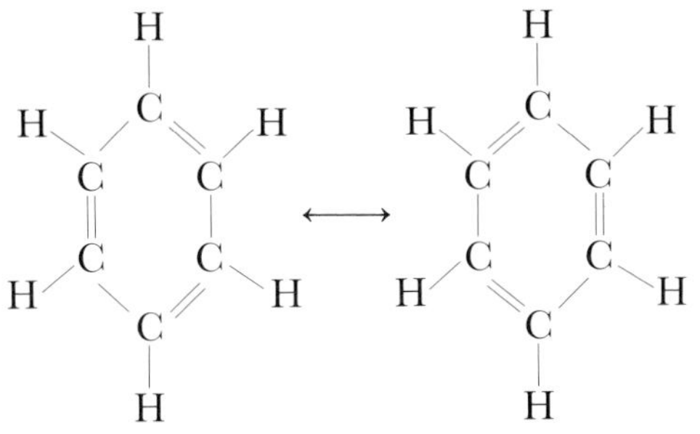

Each carbon–carbon bond is intermediate between a single bond and a double bond, and the observed bond length is consistent with this, as shown in Figure 11–3. We can say that three of the pairs of electrons are *delocalized*. As was mentioned in Chapter 10, a molecule with an electronic structure that is a resonance hybrid of two or more structures is more stable than predicted on the basis of any one of the valence-bond structures contributing to the actual structure. Now we can check this point with a quantitative calculation.

The heat of hydrogenation of benzene to give cyclohexane has been measured:

$$C_6H_6(g) + 3H_2(g) = C_6H_{12}(g) \qquad \Delta H^0 = -208.4 \text{ kJ mol}^{-1}$$

Benzene Cyclohexane

This can be compared with the heat of hydrogenation of three moles of cyclohexene (a molecule with one double bond, in which there would be no resonance):

$$C_6H_{10}(g) + H_2(g) = C_6H_{12}(g) \qquad \Delta H^0 = -119.7 \text{ kJ mol}^{-1}$$

Cyclohexene Cyclohexane

or -359.1 kJ for three moles. This comparison is appropriate because the final product, cyclohexane, is the same in each case. The comparison shows that benzene is more stable than expected on the basis of a structure with three independent double bonds; less heat is evolved in the reaction in which benzene is the reactant. The difference of these changes of enthalpy, 150.7 kJ mol^{-1}, is interpreted as a quantitative measure of the resonance stabilization of benzene.

The extra stability of benzene due to resonance is reflected in its chemical reactivity. Although a halogen (e.g., chlorine) *adds* to the double bond in an alkene (e.g., ethylene),

$$CH_2{=}CH_2 + Cl_2 = CH_2ClCH_2Cl$$

converting the molecule to a saturated substance, the reaction of benzene and chlorine,

$$C_6H_6 + Cl_2 = C_6H_5Cl + HCl$$

results in *substitution* of chlorine for hydrogen. The basic benzene structure representable by a resonance hybrid of the two Kekulé structures is *not* altered in this substitution reaction, but addition of one mole of chlorine to a mole of benzene would give a product $C_6H_6Cl_2$ with only two double bonds and no resonance.

The replacement of hydrogen in benzene by chlorine can be carried to complete substitution of chlorine in C_6Cl_6, hexachlorobenzene. We can study the problem of isomerism in substituted benzenes through the examples of the various chlorobenzenes, $C_6H_{6-n}Cl_n$, which are intermediate in composition between C_6H_6 and C_6Cl_6. Because the structure of benzene is a planar, regular hexagon, all carbon atoms are equivalent, and there is but a single monochlorobenzene:

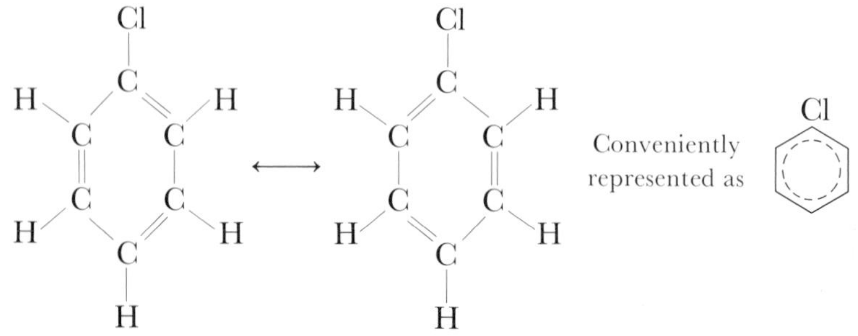

Cl Cl
H C H H C H
C C ⟷ C C Conveniently represented as Cl
C C C C
H C H H C H
H H

The dashed circle represents the delocalized electrons of the benzene ring. The introduction of a second chlorine leads to the possibility of isomerism, and there are three isomeric dichlorobenzenes:

ortho-Dichlorobenzene
mp −17.0°C
bp 180.5°C

meta-Dichlorobenzene
mp −24.7°C
bp 173°C

para-Dichlorobenzene
mp 53.1°C
bp 174°C

and three isomeric trichlorobenzenes:

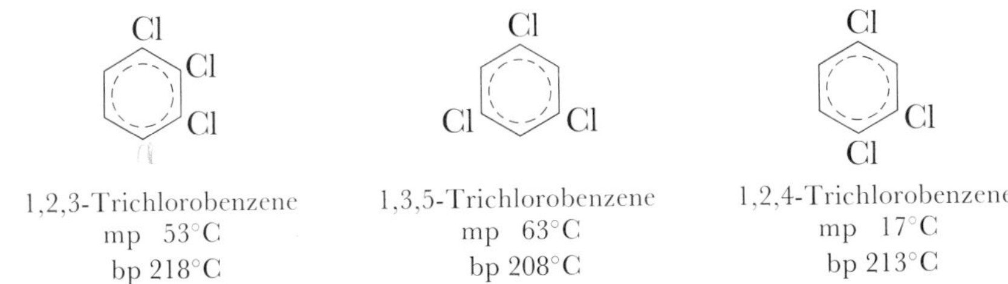

1,2,3-Trichlorobenzene
mp 53°C
bp 218°C

1,3,5-Trichlorobenzene
mp 63°C
bp 208°C

1,2,4-Trichlorobenzene
mp 17°C
bp 213°C

It is interesting to notice that the melting points of both the isomeric dichlorobenzenes and trichlorobenzenes conform to the generalization that highly symmetric molecules melt at relatively high temperatures. Only the most symmetric dichlorobenzene is a solid at room temperature; this substance, *para*-dichlorobenzene, is used as moth crystals.

An interesting method for establishing the structures of polysubstituted benzenes is that devised by WILHEIM KÖRNER, although it has long been superseded by other techniques. In Körner's method, attention is paid to the number of isomeric trisubstituted benzenes that are formed from a particular disubstituted benzene. Studying the structures of the dichlorobenzenes and the trichlorobenzenes, we find the following possible interconversions:

$$C_6H_4Cl_2 + Cl_2 = C_6H_3Cl_3 + HCl$$

- *o*-$C_6H_4Cl_2$ → 1,2,3-$C_6H_3Cl_3$
- *o*-$C_6H_4Cl_2$ → 1,2,4-$C_6H_3Cl_3$
- *m*-$C_6H_4Cl_2$ → 1,2,3-$C_6H_3Cl_3$
- *m*-$C_6H_4Cl_2$ → 1,3,5-$C_6H_3Cl_3$
- *m*-$C_6H_4Cl_2$ → 1,2,4-$C_6H_3Cl_3$
- *p*-$C_6H_4Cl_2$ → 1,2,4-$C_6H_3Cl_3$

Each particular isomeric disubstituted benzene gives a different number of isomeric trisubstituted benzenes. This fact can be used to identify isomers. If, for instance, a pure isomeric dichlorobenzene of unknown structure were found upon reaction with chlorine to give two different isomeric trichlorobenzenes, the reactant would be identified as *ortho*-dichlorobenzene.

One or more hydrogen atoms of benzene also can be replaced by a hydrocarbon radical (to be discussed more fully in the next section). Here, we simply will introduce one example, toluene, $C_6H_5CH_3$,

CH_3

an important chemical, which is much less toxic than benzene, but still is a hazardous chemical. The effect of molecular symmetry upon melting point, already mentioned, is illustrated in a comparison of benzene (mp 5.5°C) and toluene (mp −95°C). Solutions of benzene and toluene were considered at length in Chapter 7.

11–4 Derivatives of Hydrocarbons

In our discussion of hydrocarbons, we have mentioned related compounds in which one or more hydrogen atoms are replaced by halogen atoms. If one hydrogen atom of methane is replaced by a chlorine atom, the resulting molecule, CH_3Cl, is called methyl chloride or chloromethane. The simplest saturated hydrocarbon radicals (a hydrocarbon minus a hydrogen atom) are called *alkyl groups*, and are designated as R in many formulas. The names of a few of these groups are given in Table 11–3. The name methyl chloride may remind you of the names of metal chloride salts (e.g., sodium chloride), but this resemblance is misleading: methyl chloride is not ionic. It is a covalent compound, with the structure given in Figure 11–10; we see that the bond angles in methyl chloride deviate slightly from the regular tetrahedral angle. This generally is observed for molecules in which the atoms or groups of atoms bonded to a carbon atom are different. Also given in Figure 11–10 are the structures of other chlorine-substituted hydrocarbons, which introduce some new points in our discussion. In Figure

TABLE 11–3
Alkyl Groups ($—C_nH_{2n+1}$) Derived From Alkanes (C_nH_{2n+2})

Hydrocarbon		Alkyl group	
Methane	CH_4	Methyl	$—CH_3$
Ethane	H_3CCH_3	Ethyl	$—CH_2CH_3$
Propane	$H_3CCH_2CH_3$	*n*-Propyl	$—CH_2CH_2CH_3$
		i-Propyl	$—CH(CH_3)_2$
Butane[a]	C_4H_{10}	Butyl[a,c]	$—C_4H_9$
Pentane[b]	C_5H_{12}	Pentyl[b,c]	$—C_5H_{11}$

[a]There are two isomeric butanes and four isomeric butyl groups.
[b]There are three isomeric pentanes and eight isomeric pentyl groups.
[c]Naming derivatives of some of the isomeric butanes and pentanes according to IUPAC rules avoids the necessity of using a special name for each isomeric butyl or pentyl group.

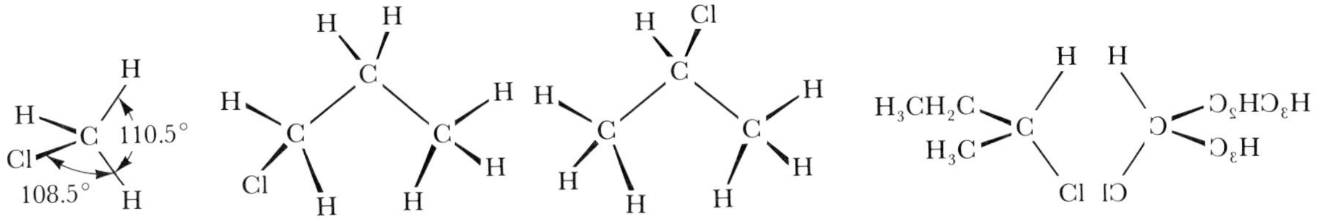

(a) (b) *n*-Propyl chloride (1-chloropropane) Isopropyl chloride (2-chloropropane) (c)

FIGURE 11–10
The structures of some alkyl chlorides. (a) Methyl chloride, CH_3Cl; notice the deviations of the bond angles from 109.5°. (b) Isomeric propyl chlorides. (c) Optical isomers of 2-chlorobutane; these chiral molecules are nonsuperimposable mirror images.

11–10b we see two isomers of propyl chloride. Although propane, C_3H_8, does not exist in isomeric forms, propyl chloride, C_3H_7Cl, exists in two structural isomeric forms. These differ in the point of substitution of the chlorine atom. In normal propyl chloride (also called 1-chloropropane), the chlorine atom is bonded to an end carbon atom; in isopropyl chloride (also called 2-chloropropane), the chlorine atom is bonded to the middle carbon atom. (In examining the alternate names given here, you are learning more about the IUPAC system of naming organic compounds.)

In 2-chlorobutane there is a new type of isomerism. In this molecule a carbon atom is bonded to four different atoms or groups. These atoms and groups are hydrogen, chlorine, the methyl group, and the ethyl group:

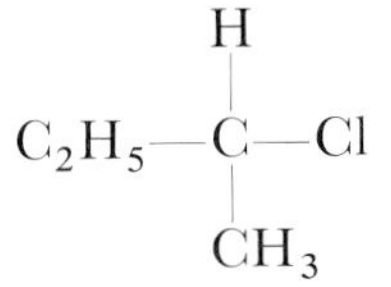

$$\begin{array}{c} H \\ | \\ C_2H_5-C-Cl \\ | \\ CH_3 \end{array}$$

A molecule such as this, in which four different groups are attached to a particular carbon atom, cannot be superimposed upon its mirror image. This is shown in Figure 11–10c. The situation is analogous to that of a right-handed glove and a left-handed glove: two such gloves are not superimposable. Molecules with this characteristic, this "handedness," are called *chiral*, from the Greek word for hand. Isomers having this property are called *optical isomers*. The plane of polarized light is rotated by passing through a vessel containing one or the other of these isomers; each isomer is said to exhibit *optical activity*. The direction in which the polarized light is rotated differs for the two isomers.

If 2-chlorobutane is prepared by a reaction of nonchiral reagents, for example, by adding hydrogen chloride to the double bond in 2-butene,

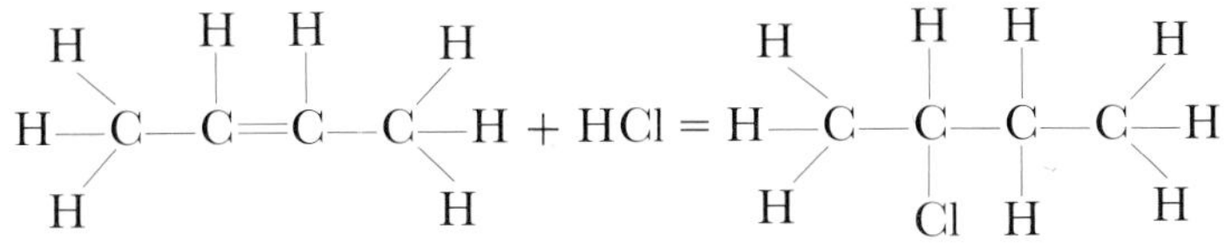

$$\begin{array}{ccccccccccc} & H & H & H & H & & & H & H & H & H \\ H & -C- & C= & C- & C-H & + HCl = & H- & C- & C- & C- & C-H \\ & H & & & H & & & H & Cl & H & H \end{array}$$

the two optical isomers are formed in equal amounts. This mixture containing equal amounts of the two isomers does not cause any net rotation of the plane of polarized light. The rotation in one direction caused by the one isomer is offset by an equal and opposite rotation caused by the other isomer.

Optical isomers have identical properties, such as melting point and boiling point. Each of the optical isomers has the same chemical reactivity toward a nonchiral reagent but has different chemical reactivity toward a chiral reagent, just as right-hand and left-hand gloves react identically with a spherical ball (a nonchiral object), but nonidentically with a right hand (a chiral object). Many natural organic substances (from plant or animal sources) are optically active.

OXYGEN-CONTAINING DERIVATIVES

A large variety of oxygen-containing compounds can be viewed as derivatives of hydrocarbons in which:

1. a hydrogen atom is replaced by a hydroxyl group (a water molecule minus a hydrogen atom):

$$\cdot\mathrm{H} \text{ is replaced by } \cdot\ddot{\underset{\cdot\cdot}{\mathrm{O}}}\!:\!\mathrm{H}$$

2. a pair of hydrogen atoms on the same carbon atom is replaced by an oxygen atom:

$$2\ \cdot\mathrm{H} \text{ are replaced by } :\ddot{\underset{\cdot\cdot}{\mathrm{O}}}$$

Let us consider first the stepwise replacement of the four hydrogen atoms of methane with hydroxyl groups. The molecules generated in this way are given in the first column of Table 11–4.[4]

Some molecules that have two hydroxyl groups bonded to one carbon atom are unstable. In the reaction

$$\underset{\text{Dihydroxymethane}}{\mathrm{H_2C(OH)_2}(g)} = \underset{\text{Formaldehyde}}{\mathrm{H_2C{=}O}(g)} + \mathrm{H_2O}(g)$$

the number of bonds of each kind in reactant and products is:

	Dihydroxymethane	Formaldehyde + water
C—H	2	2
C—O	2	0
O—H	2	2
C=O	0	1

[4] Do not think of the transformation

$$\mathrm{H_3C\!:\!H} + \cdot\ddot{\underset{\cdot\cdot}{\mathrm{O}}}\!:\!\mathrm{H} = \mathrm{H_3C\!:\!}\ddot{\underset{\cdot\cdot}{\mathrm{O}}}\!:\!\mathrm{H} + \cdot\mathrm{H}$$

as a realizable chemical reaction. This is simply a way of developing a sequence of related molecules containing one carbon atom.

TABLE 11–4
Molecules Generated From Methane by Replacement of a Hydrogen Atom With a Hydroxyl Group[a]

Parent molecule	Oxidation state of carbon	Parent molecule *minus* one water molecule	Parent molecule *minus* two water molecules
CH_4 (structural formula: C bonded to four H) Methane	−4		
CH_3OH (C bonded to three H and OH) Methyl alcohol	−2		
[$CH_2(OH)_2$] (C bonded to H, H, OH, HO) Dihydroxymethane	0	H–CH=O Formaldehyde	
[$CH(OH)_3$] (C bonded to H, HO, OH, HO) Trihydroxymethane	+2	H–C(=O)–OH Formic acid	
[$C(OH)_4$] (C bonded to OH, HO, OH, HO) Tetrahydroxymethane	+4	OH–C(=O)–OH Carbonic acid	O=C=O Carbon dioxide

[a]Brackets enclose molecules that are not stable.

Therefore this reaction involves the transformation of two carbon–oxygen single bonds into one carbon–oxygen double bond:

$$\rangle C\langle^{O}_{O} \longrightarrow \rangle C{=}O$$

We can estimate the value of ΔU for this reaction in the gas phase by using the bond energies given in Table 10–6:

$$\Delta U = 2D(\text{C—O}) - D(\text{C═O})$$

$$\frac{\Delta U}{\text{kJ mol}^{-1}} = 2 \times 358 - 745 = -29$$

$$\Delta U = -29 \text{ kJ mol}^{-1}$$

Loss of water with the concomitant conversion of two carbon–oxygen single bonds into a carbon–oxygen double bond is an exothermic change; the strength of the carbon–oxygen double bond is responsible for the low stability of molecules with two hydroxyl groups bonded to the same carbon atom.[5] It is this same type of transformation that converts hypothetical tetrahydroxymethane, $C(OH)_4$, into carbonic acid, $OC(OH)_2$, and carbonic acid into carbon dioxide. This latter transformation is an important reaction in aqueous solution,

$$OC(OH)_2(aq) = H_2O(l) + CO_2(aq)$$

and it will be discussed further in Chapter 13. This reaction also establishes the relationship between an acid (here, carbonic acid) and its anhydride (here, carbon dioxide). Later we will see this same type of reaction converting other acids into their anhydrides.

The molecules in Table 11–4 have been generated by replacing hydrogen atoms bonded to carbon with hydroxyl groups. These molecules could be produced from the parent methane by oxidation. The series of chemical reactions involving oxygen as the oxidizing reactant is:

$$CH_4 + \tfrac{1}{2}O_2 = CH_3OH$$
$$CH_3OH + \tfrac{1}{2}O_2 = H_2CO + H_2O$$
$$H_2CO + \tfrac{1}{2}O_2 = HCOOH$$
$$HCOOH + \tfrac{1}{2}O_2 = CO_2 + H_2O$$

Each of these molecules can be characterized by the *oxidation state* of the carbon, values of which are given in the table. The oxidation state of carbon in a compound is calculated by application of simple rules. In its compounds, hydrogen is assigned an oxidation state of +1 (except in ionic hydrides, introduced in Chapter 9), and oxygen is assigned an oxidation state of −2 (except in peroxides or superoxides, mentioned in Chapter 10). The oxidation state of carbon in each of the compounds given in Table 11–4 is assigned the value that makes the sum of the oxidation states of all elements

[5]With appropriate electron-withdrawing groups in the compound, substances with this structure are known, and are called *gem*-diols. Trichloroacetaldehyde, CCl_3CHO, forms a *gem*-diol chloral hydrate, $CCl_3CH(OH)_2$. In aqueous solution, an equilibrium is established between the aldehyde form and the *gem*-diol, for example, for formaldehyde $H_2O + H_2C{=}O \rightleftarrows H_2C(OH)_2$. In addition to the bond energies, the solvation energies of reactants and products play a role in this reaction in solution.

in the compound equal to the net charge on the molecule or ion; this is zero for each of these molecules. Let us designate the oxidation state of carbon in each of these compounds as x and calculate its values:

CH_4	$x + 4(+1) = 0$	$x = -4$
CH_3OH	$x + 4(+1) + (-2) = 0$	$x = -2$
CH_2O	$x + 2(+1) + (-2) = 0$	$x = 0$
CHOOH	$x + 2(+1) + 2(-2) = 0$	$x = +2$
CO_2	$x + 2(-2) = 0$	$x = +4$

In each step of the oxidation of methane to carbon dioxide, the oxidation state of carbon increases by two units. (We will find the concept of oxidation state very useful in our study of electrochemistry and the chemistry of other elements. A more detailed discussion of this concept will be presented in Chapter 16.)

We have been considering the molecular structure of methyl alcohol to be methane with a hydrogen atom replaced by a hydroxyl radical. We also can consider methyl alcohol to be water with a hydrogen atom replaced by a methyl radical:

```
H                   H                        H
|                  /                        /
O—H minus •H plus •C—H gives H—O—C—H
                   \                        \
                    H                        H
```

From this point of view, we also can substitute methyl groups for both hydrogen atoms of water to give dimethyl ether. Thus we have the series of molecules:

	H_2O	CH_3OH	$(CH_3)_2O$
mp:	0.0°C	−97.7°C	−138.5°C
bp:	100.0°C	64.5°C	−23.0°C

in which the role of hydrogen bonding in determining the attractive interactions in the solid and liquid states is seen clearly in the melting and boiling points. There is not appreciable hydrogem bonding between molecules of dimethyl ether, hence the relatively low boiling and melting points. (A hydrogen atom bonded to carbon generally does not form hydrogen bonds, although, compared to the hydrogen of a methyl group, the hydrogen atom of chloroform is made more positive by the three electronegative chlorine atoms, and we saw in Chapter 7 that chloroform forms hydrogen bonds with acetone.)

The functional groups and the hydrocarbon derivatives that we have been discussing are summarized in Table 11–5.

The common names of the simplest aldehydes and carboxylic acids, which are given in Table 11–6, are related to one another, but they do not follow in a simple way from the names of the corresponding alcohols. (You

TABLE 11–5
Classes of Organic Compounds Containing Hydroxyl Groups (—OH), Alkoxy Groups (—OR), and Carbonyl Groups (>C=O)

Number of each group			General formula	Name of class
—O—H	—O—R	>C=O		
1	0	0	R—O—H	Alcohol
0	1	0	R—O—R	Ether[a]
0	0	1	R—C(=O)—H	Aldehyde[b]
0	0	1	R—C(=O)—R′	Ketone[a]
1	0	1	R—C(=O)—O—H	Carboxylic acid
0	1	1	R—C(=O)—O—R′	Ester[a]

[a]The two alkyl groups (R) in an ether, a ketone, or an ester can be the same or different.
[b]The simplest aldehyde is formaldehyde, in which R = H.

TABLE 11–6
Names of Aldehydes and Carboxylic Acids

Aldehydes, R—C(H)=O		Carboxylic acids, R—C(=O)—OH	
Formula	Name	Formula	Name
H—C(H)=O	Formaldehyde	H—C(=O)—OH	Formic acid
H—C(H)(H)—C(H)=O	Acetaldehyde	H—C(H)(H)—C(=O)—OH	Acetic acid
H—C(H)(H)—C(H)(H)—C(H)=O	Propionaldehyde	H—C(H)(H)—C(H)(H)—C(=O)OH	Propionic acid
H—C(H)(H)—C(H)(H)—C(H)(H)—C(H)=O	Butyraldehyde	H—C(H)(H)—C(H)(H)—C(H)(H)—C(=O)OH	Butyric acid

should notice that the aldehyde and acid with a particular number of carbon atoms contains one fewer than this number of carbon atoms in its alkyl group, a point implied in Table 11–5).

As you probably know from previous study (and as we will consider at length in the next chapter), an acid is a substance capable of donating a proton to another molecule or ion, called a base. Acetic acid in water solution reacts to a small extent to give ions in the dissociation reaction

$$\underset{\text{Acetic acid}}{CH_3CO_2H} + H_2O = \underset{\text{Acetate ion}}{CH_3CO_2^-} + H_3O^+$$

in which a proton is transferred from the hydroxyl group of acetic acid to a water molecule. This reaction occurs to an extent of $\sim 1.3\%$ in a solution with a stoichiometric concentration of acetic acid of 0.10 mol L^{-1}. An alcohol also contains a hydroxyl group, but alcohols are much less acidic than carboxylic acids. The reaction

$$CH_3CH_2OH + H_2O = CH_3CH_2O^- + H_3O^+$$

occurs to an extent less than $10^{-8}\%$ in dilute aqueous solution. An explanation for the different extents of these two reactions, in each of which

$$\geqslant C\text{—}OH + H_2O = \geqslant C\text{—}O^- + H_3O^+$$

occurs, is to be found in the structures of the anions resulting from proton dissociation. In ethoxide ion, $CH_3CH_2O^-$, formed by proton dissociation from ethyl alcohol, the resulting negative charge is localized on the oxygen. (Draw an electron-dot formula for ethoxide ion, and calculate the formal charge on each atom.) In acetate ion, $CH_3CO_2^-$, formed by proton dissociation from acetic acid, the two oxygen atoms are equivalent, and the one unit of negative charge is distributed between the two atoms. There is structural evidence for this equivalence of the two oxygen atoms: The two carbon–oxygen bond lengths are equal. This is consistent with a structure that is a resonance hybrid:

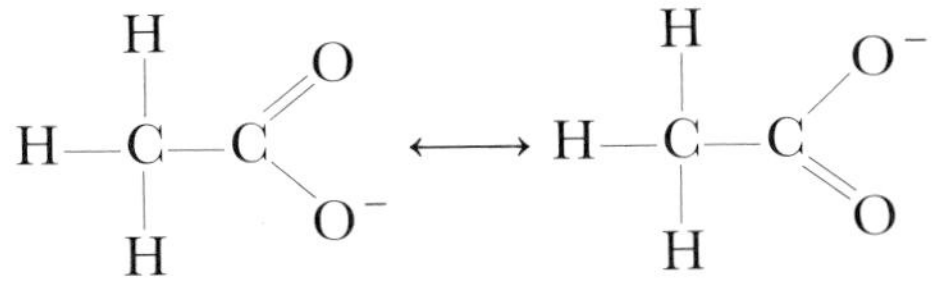

```
     H       O                 H       O⁻
     |     //                  |      /
 H—C—C          ⟷    H—C—C
     |     \                   |      \\
     H       O⁻                H       O
```

The delocalization of negative charge implied by this resonance, which also can be represented by the dashed line in

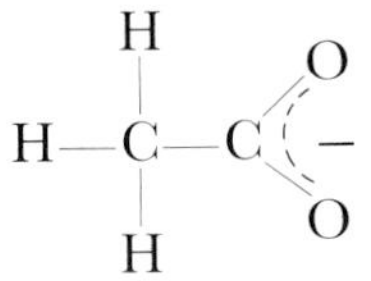

```
     H       O
     |     //
 H—C—C ( −
     |     \\
     H       O
```

is a factor contributing to the relative stability of acetate ion.

Replacement of a hydrogen atom in the methyl group of acetic acid by an atom of the more electronegative element chlorine ($\chi_H = 2.2$, $\chi_{Cl} =$

3.0) dramatically changes the extent of acid dissociation in aqueous solution:

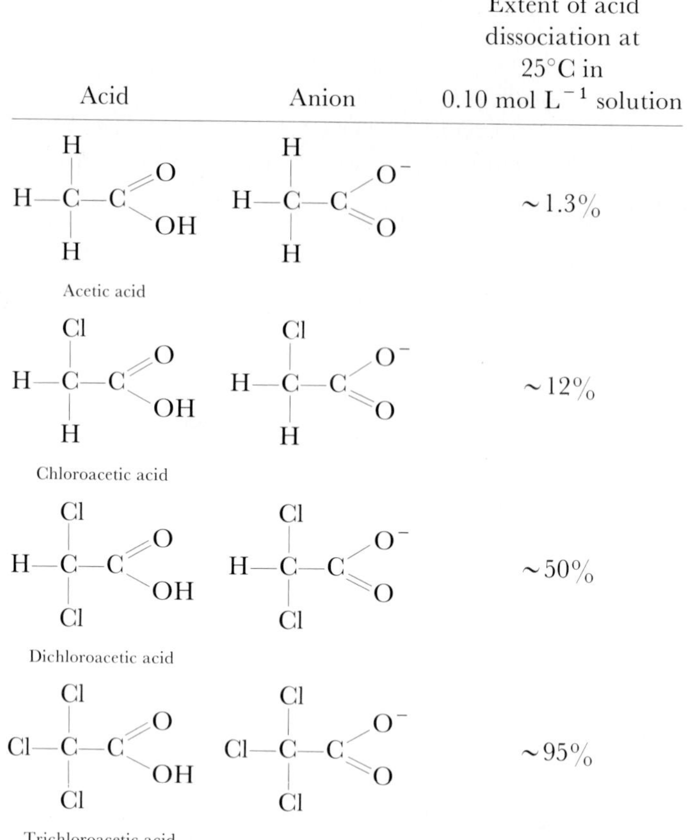

Acid
Anion
Extent of acid dissociation at 25°C in 0.10 mol L⁻¹ solution
Acetic acid
~1.3%
Chloroacetic acid
~12%
Dichloroacetic acid
~50%
Trichloroacetic acid
~95%

The more electronegative chlorine atom attracts electrons from the carboxyl group in the molecule, making

1. dissociation of the proton from the carboxyl group easier, or
2. the anionic species more stable.

As you would expect, the influence of the chlorine-for-hydrogen substitution is greater the closer the substitution is to the site of proton dissociation. This is illustrated by the extent of acid dissociation of butyric acid and its monochloro derivatives in aqueous solutions with a concentration of 0.10 mol L^{-1}:

Acid	Extent of acid dissociation at 25°C in 0.10 mol L^{-1} solution
H—C(H)(H)—C(H)(H)—C(H)(H)—C(=O)OH Butyric acid	1.2%

(*Continued*)

$$\mathrm{H{-}\underset{H}{\overset{Cl}{C}}{-}\underset{H}{\overset{H}{C}}{-}\underset{H}{\overset{H}{C}}{-}C\begin{matrix}{\scriptstyle =O}\\{\scriptstyle -OH}\end{matrix}}\qquad 1.7\%$$

γ-Chlorobutyric acid[6]

$$\mathrm{H{-}\underset{H}{\overset{H}{C}}{-}\underset{H}{\overset{Cl}{C}}{-}\underset{H}{\overset{H}{C}}{-}C\begin{matrix}{\scriptstyle =O}\\{\scriptstyle -OH}\end{matrix}}\qquad 2.9\%$$

β-Chlorobutyric acid[6]

$$\mathrm{H{-}\underset{H}{\overset{H}{C}}{-}\underset{H}{\overset{H}{C}}{-}\underset{H}{\overset{Cl}{C}}{-}C\begin{matrix}{\scriptstyle =O}\\{\scriptstyle -OH}\end{matrix}}\qquad 11\%$$

α-Chlorobutyric acid[6]

These examples illustrate the influence that a substituent can have on the electron distribution in a molecule. Chemists have developed many aspects of the theory of organic chemistry by observing the effects of substituents on the extent of dissociation of acids.

In the sequence of oxidation reactions by which methane was converted to carbon dioxide, the oxidation of methyl alcohol gave formaldehyde. However, not all alcohols give aldehydes upon oxidation. The product of oxidation depends upon the number of alkyl groups and hydrogen atoms bonded to the carbon atom to which the hydroxyl group is bonded. The relationship is summarized in Table 11–7. The simplest secondary alcohol is isopropyl alcohol; upon oxidation it forms acetone, the simplest ketone:

$$\mathrm{H_3C{-}\underset{CH_3}{\overset{H}{C}}{-}OH} + \tfrac{1}{2}\,\mathrm{O_2} = \mathrm{\begin{matrix}H_3C\\H_3C\end{matrix}\!\!>\!C{=}O} + \mathrm{H_2O}$$

Isopropyl alcohol — Acetone (dimethyl ketone)

Ketones are named simply by identifying the two alkyl groups attached to the carbon of the carbonyl group, $\mathrm{{-}\overset{\overset{\displaystyle O}{\|}}{C}{-}}$. The two alkyl groups need not be

[6]In this system of nomenclature the position of a substituent (chlorine in this example) relative to the principal functional group (the carboxyl group in this example) is designated α (alpha) if the substituent is on the first carbon, β (beta) if the substituent is on the second carbon, and so on.

TABLE 11–7

Types of Alcohols and Their Oxidation Products[a]

Alcohol		Oxidation product
Primary alcohols		
$\mathrm{H{-}\overset{\displaystyle H}{\underset{\displaystyle H}{C}}{-}OH} \xrightarrow{[O]}$	$\mathrm{\overset{\displaystyle H}{\underset{H\diagup}{C}}{=}O}$	Formaldehyde
$\mathrm{H{-}\overset{\displaystyle H}{\underset{\displaystyle R}{C}}{-}OH} \xrightarrow{[O]}$	$\mathrm{\overset{\displaystyle H}{\underset{R\diagup}{C}}{=}O}$	Aldehyde
Secondary alcohols		
$\mathrm{R'{-}\overset{\displaystyle H}{\underset{\displaystyle R}{C}}{-}OH} \xrightarrow{[O]}$	$\mathrm{\overset{\displaystyle R}{\underset{R'\diagup}{C}}{=}O}$	Ketone
Tertiary alcohol		
$\mathrm{R'{-}\overset{\displaystyle R''}{\underset{\displaystyle R}{C}}{-}OH} \xrightarrow{[O]}$	breaking of C–C bonds	

[a]The symbol $\xrightarrow{[O]}$ denotes reaction with some oxidizing agent, not necessarily oxygen.

the same. Isopropyl methyl ketone has the structure

$$\mathrm{H{-}\overset{\displaystyle H}{\underset{\displaystyle H}{C}}{-}\overset{\displaystyle H}{\underset{H-\overset{|}{C}-H \atop \overset{|}{H}}{C}}{-}\overset{\displaystyle O}{\overset{\|}{C}}{-}\overset{\displaystyle H}{\underset{\displaystyle H}{C}}{-}H}$$

Tertiary alcohols are not easily oxidized. The simplest of them is tertiary butyl alcohol:

$$\mathrm{H{-}\overset{\displaystyle H}{\underset{\displaystyle H}{C}}{-}\overset{H-\overset{\overset{\displaystyle H}{|}}{C}-H}{\underset{H-\underset{\underset{\displaystyle H}{|}}{C}-H}{C}}{-}O{-}H}$$

With very strong oxidizing agents, such alcohols react to produce oxidized molecules with smaller numbers of carbon atoms.

A simple class of organic nitrogen compounds is the *amines*. These compounds can be viewed as derivatives of ammonia, with one or more hydrogen atoms replaced by alkyl groups:

NH_3	ammonia	mp	−78°C
		bp	−33°C
$NH_2(CH_3)$	methylamine	mp	−93.5°C
		bp	−6.3°C
$NH(CH_3)_2$	dimethylamine	mp	−92.2°C
		bp	6.9°C
$N(CH_3)_3$	trimethylamine	mp	−117.3°C
		bp	2.9°C

Each of these molecules is pyramidal because the nitrogen atom is surrounded by four pairs of electrons, one lone pair plus three bonding pairs. (It is the lone pair of electrons on the nitrogen atom that gives ammonia and substituted amines the ability to accept a proton from an acid, for example,

$$CH_3CO_2H + NH_3 = CH_3CO_2^- + NH_4^+$$

This ability to accept protons from an acid defines a base, as will be explained in the next chapter.)

The NH_2 group (ammonia, NH_3, minus a hydrogen atom) is analogous to the hydroxyl group, OH (water, H_2O, minus a hydrogen atom), and there are many organic nitrogen compounds containing NH_2 groups and substituted NH_2 groups analogous to the hydroxyl and alkoxyl compounds already discussed:

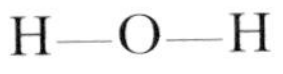

H—O—H

Water

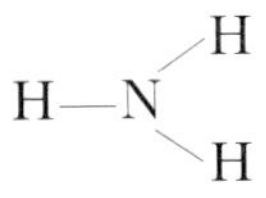

H—N(—H)—H

Ammonia

R—O—H

Alcohol

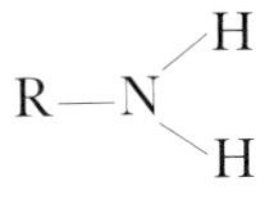

R—N(—H)—H

Primary amine

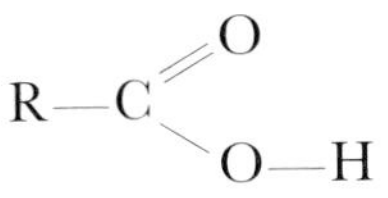

R—C(=O)—O—H

Acid

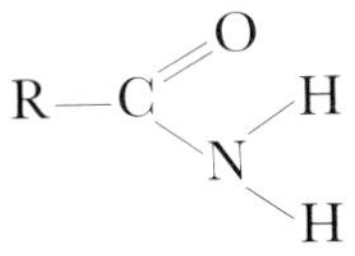

R—C(=O)—N(—H)—H

Amide

O=C(—O—H)—O—H

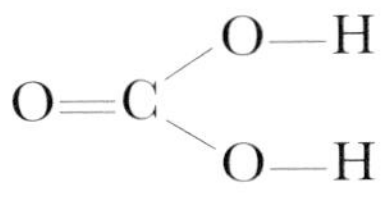

Carbonic acid (unstable)

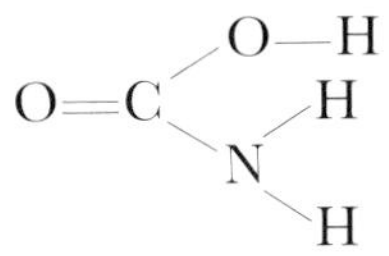

O=C(—O—H)—N(—H)—H

Carbamic acid (unstable)

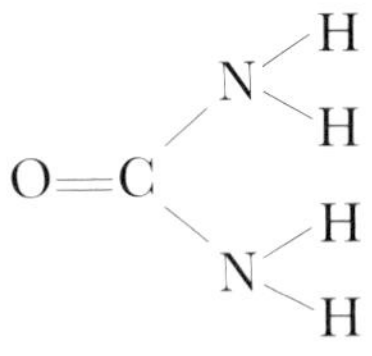

O=C(—N(—H)—H)—N(—H)—H

Urea

Other simple organic nitrogen compounds can be generated by replacement of successive hydrogen atoms in ammonia with alkyl groups:

$R{-}N{<}^{H}_{H}$	$R{-}N{<}^{R'}_{H}$	$R{-}N{<}^{R'}_{R''}$
Primary amine	Secondary amine	Tertiary amine

Each of these amines has, like ammonia, an unshared pair of electrons, and each is pyramidal. It is possible for nitrogen to be bonded to four atoms as in:

NH_4^+	$N(CH_3)_4^+$
Ammonium ion [isoelectronic with CH_4]	Tetramethylammonium ion [isoelectronic with $C(CH_3)_4$]

An NH_2 group may replace a hydrogen atom in the hydrocarbon parts of molecules that also contain other functional groups. A particularly important class of organic compounds with this feature is the α-amino acid,[7]

$$R{-}CH(NH_2){-}CO_2H$$

a carboxylic acid with an NH_2 group substituted on the α-carbon atom. These are important compounds because proteins, the building blocks of biological matter, are built of α-amino acids (proteins will be discussed in Chapter 20). The two simplest α-amino acids are glycine, α-aminoacetic

[7] In Chapter 13 it will be shown that the predominant electrically neutral form of an amino acid is that with the separation of positive and negative charges:

$$R{-}CH(NH_3^+){-}CO_2^-$$

This point will be ignored at this time in writing the formulas of amino acids.

acid, and alanine, α-aminopropionic acid:

```
     H                 H   H
     |                 |   |
H—C—CO2H        H—C—C—CO2H
     |                 |   |
    NH2                H  NH2
  Glycine            Alanine
```

There is an important difference between the pattern of substituents on the α carbon atom of glycine and on that of alanine. Two of the groups bonded to the α carbon atom of glycine are the same, the hydrogen atoms. However, the α carbon atom of alanine has four different atoms or groups bonded to it: a hydrogen atom, a methyl group, an amino group, and a carboxylic acid group. Therefore alanine and other more complicated α-amino acids, but not glycine, are chiral, and they exist as optical isomers. Virtually all of the optically active α-amino acids occurring as constituents of proteins have the same configuration at the α carbon atom:

```
        CO2H                  CO2H
       /  |  \                  ▲
H2N ———+——— H          H2N ► C ◄ H
       \  |  /                  ▼
         R                      R
```

The twenty α-amino acids from which most naturally occurring proteins are made are listed in Table 20–1.

Biographical Notes

F. A. KEKULÉ (1829–1896) was Professor of Chemistry at Ghent and Bonn. He worked on the structure of benzene, and also did important pioneering work in synthetic dyes.

WILHEIM KÖRNER (1839–1925) was born in Germany and was Professor at the University of Milan. His work on polysubstituted benzenes led him to propose the use of the terms *ortho*, *meta*, and *para* to designate the isomeric C_6H_4XY molecules.

Problems and Questions

11–1 An unstable oxide of carbon, not discussed in this chapter, is carbon suboxide, C_3O_2. Using the rules for octets of electrons, formal charges, and stability, suggest the structure of this molecule.

11–2 Calculate the formal charges on the atoms in carbonate ion, CO_3^{2-}.

11–3 Draw carbon skeleton formulas for the five isomeric hexanes and the nine isomeric heptanes, and name each.

11–4 What is the simplest chiral saturated hydrocarbon? (Here, "simplest" means having the smallest number of carbon atoms.)

11–5 Draw carbon skeleton formulas for the octanes that are chiral, and name them.

11–6 Draw structural formulas for each of the following: 2,3-dimethylpentane, 2,3,4-trimethylhexane, and 3,3-diethylpentane.

11–7 Give the distance between the centers of the terminal carbon atoms in *n*-pentane, in the most extended conformation that staggers all C–H bonds for adjacent carbon atoms. Compare this distance with the sum of the four C–C bond lengths in this chain of atoms.

11–8 Draw the structural formula 2,2,3,3-tetramethylbutane. This is an isomer of what continuous-chain hydrocarbon? Look up the melting points of these two compounds, and explain the contrasting values.

11–9 What is the simplest chiral alcohol?

11–10 Draw carbon skeleton formulas for the isomeric pentyl chlorides. Which of these molecules are chiral?

11–11 Tartaric acid has the chemical formula

$$\mathrm{HO{-}\overset{\overset{\displaystyle O}{\|}}{C}{-}\underset{\underset{\displaystyle OH}{|}}{\overset{\overset{\displaystyle H}{|}}{C}}{-}\underset{\underset{\displaystyle OH}{|}}{\overset{\overset{\displaystyle H}{|}}{C}}{-}\overset{\overset{\displaystyle O}{\|}}{C}{-}OH}$$

Is this molecule chiral? Comment on any special features of this compound that play roles in its chirality.

11–12 What ether has the same molecular composition as *n*-propyl alcohol? Look up the boiling points of these substances and explain the contrast.

11–13 Suppose that rotation about the carbon–carbon single bond were prevented by a very high potential-energy barrier. In substituted ethanes the various different staggered conformations would become "isomers." How many such isomers would there be for each of the following?

1,1-difluoroethane	CHF_2CH_3
1,2-difluoroethane	CH_2FCH_2F
1,1-chlorofluoro-2-chloroethane	$CHFClCH_2Cl$

11–14 Which of the isomeric dichloroethenes has zero dipole moment?

11–15 Neither cyclobutane, C_4H_8, nor cyclopentane, C_5H_{10}, is exactly planar. To what factor(s) can this be attributed?

11–16 Draw electron-dot formulas for 2,2-dimethylpropane and an isoelectronic molecule $(H_3C)_3BNH_3$. Calculate the formal charges on the atoms in each of these molecules. In what important respects do you expect these substances to differ?

11–17 Boron and nitrogen form compounds with the composition BN that have the structures of diamond and graphite. What are the formal charges for the boron and nitrogen atoms in these compounds?

11–18 The reactions of acetylene and nitrogen with hydrogen,

$$C_2H_2(g) + 2H_2(g) = C_2H_6(g)$$
$$N_2(g) + 2H_2(g) = N_2H_4(g)$$

are analogous, in the sense that a triple bond between nonmetal atoms (—C≡C— and N≡N) is replaced by four nonmetal–hydrogen single bonds plus a single bond between nonmetal atoms. Yet the first reaction is exothermic ($\Delta H^0 = -312$ kJ mol^{-1}), and the second reaction is endothermic ($\Delta H^0 = +95$ kJ mol^{-1}). Check the relevant bond energies and discuss the factors responsible for this difference.

11–19 Draw one isomer of benzene with each of the possible sets of values of d, t, and r (double bonds, triple bonds, and rings) itemized in this chapter. Draw an isomer of benzene in which the carbon atoms are in a straight line (i.e., all angles ∠CCC = 180°).

11–20 For a rainy day: Draw all of the isomers of benzene (there are over 100). In drawing these isomers, you should not concern yourself with possible strain in the molecule. Each isomer should conform only to the valence rules for carbon (4 bonds) and hydrogen (1 bond).

11–21 Draw the isomeric dichlorofluorobenzenes ($C_6H_3Cl_2F$).

11–22 Do either dichloroethane isomers have zero dipole moment?

11–23 Explain the positive value of ΔS^0 for the conversion of cyclopropane to propene, $\Delta S^0 = +30$ J K^{-1} mol^{-1}.

11–24 Given the following standard heats of formation (at 298.2 K):

C(g)	715.0 kJ mol^{-1}	C_2H_4(g)	+52.5 kJ mol^{-1}
CH_4(g)	−74.9 kJ mol^{-1}	C_2H_2(g)	+226.7 kJ mol^{-1}
H(g)	+216.0 kJ mol^{-1}		

Assume that the carbon–hydrogen bond energy in ethylene and in acetylene is the same as in methane. On this basis, estimate the carbon–carbon double-bond energy and the carbon–carbon triple-bond energy.

11–25 Naphthalene, $C_{10}H_8$ (a planar molecule), has the structure:

```
           H     H
           |     |
           C     C
        //   \ /   \\
   H—C        C       C—H
      |       ||      |
   H—C        C       C—H
        \\   / \   //
           C     C
           |     |
           H     H
```

Draw each of the valence-bond structures of which the true structure is a resonance hybrid. Predict whether all of the C–C bonds have the same length.

11–26 Draw each isomeric pentyl alcohol. Classify each as a primary, secondary, or tertiary alcohol.

11–27 The oxidation of what alcohol gives ethyl methyl ketone?

11–28 You were told that chloroacetic acid ($ClCH_2CO_2H$) dissociates 12% in an aqueous solution with $C(ClCH_2CO_2H) = 0.10$ mol L^{-1}. Do you expect the dissociation of fluoroacetic acid (FCH_2CO_2H) to be larger, or smaller, than this?

11–29 What hydrocarbon is isoelectronic with diethylammonium ion?

11–30 This chapter has shown a relationship between the structures of carbonic acid, carbamic acid, and urea. Can you suggest the chemical formula of an additional carbon-nitrogen-hydrogen molecule that can be considered a member of this series?

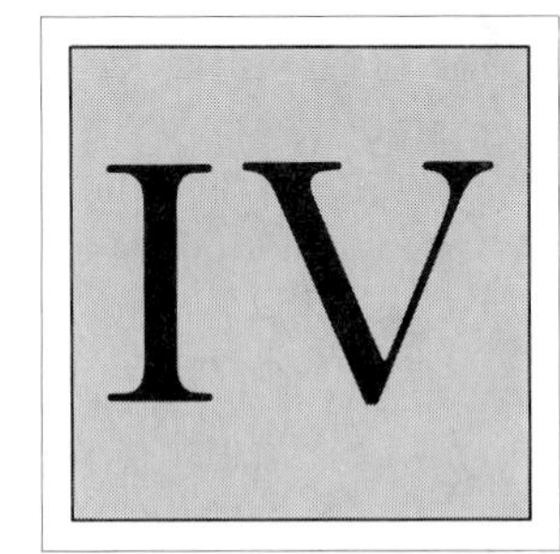

PRINCIPLES OF CHEMICAL REACTIONS

The next five chapters deal with aspects of chemical reactions, a subject introduced in Chapter 2. The concept of chemical equilibrium will be developed thoroughly in Chapters 12, 13, and 14. The reactions of compounds of hydrogen will be the focus of much of the discussions, which will consider acid–base reactions in both the gas phase and in liquid solution. The laboratory work that accompanies your classroom study of chemistry may involve acid–base reactions in aqueous solution; there is a close relationship between the material in Chapter 13 and that with which you will be working in the laboratory. Other types of equilibria that are related to your laboratory work, solubility equilibria and complex-ion equilibria, are considered in Chapter 14. In discussing a chemical equilibrium, we do not ask how rapidly the change from reactants to products occurs. Chemical thermodynamics, upon which quantitative treatment of chemical equilibrium is based, cannot answer questions about rates of chemical reactions.

But the rate at which equilibrium is established is an important characteristic of a chemical reaction, and this subject is presented in Chapter 15. There you will learn also that most chemical reactions occur through a sequence of steps. The sum of these steps is the overall chemical change. Such a sequence of steps is called a *reaction mechanism*. You will use the understanding of reaction rates and reaction mechanisms you gain in this chapter in later discussions of the chemistry of the elements.

Chemical reactions involving gain and loss of electrons by the reactants are classified as oxidation–reduction reactions. Such reactions are the basis for *galvanic cells* (batteries), in which a spontaneous chemical change can be harnessed to do useful work. An external source of electrical potential difference can be used in *electrolytic cells* to bring about a nonspontaneous chemical change. In addition to considering these types of electrical cells, Chapter 16 provides the background for discussion of the oxidation–reduction properties of the various elements. Reactions involving oxidation and reduction pervade all of chemistry and biology.

12

Hydrogen, Equilibrium, and Acid–Base Chemistry

12–1 Introduction

In this chapter we will develop two very important topics, *chemical equilibrium* and *acid-base reactions*, in a discussion of some compounds of hydrogen, the simplest element. Despite its simplicity, hydrogen forms a number of different types of compounds. These types include:

1. Ionic hydrides with metals. Saltlike compounds involving H^-, such as sodium hydride, NaH (considered in Chapter 9), are formed with the most electropositive elements. With some transition metals, hydrogen forms nonstoichiometric compounds, e.g., $ZrH_{1.9}$, in which the bonding probably is not strictly ionic. There also are metal hydrides in which the bonding has much covalent character (e.g., in $LiAlH_4$ the AlH_4^- species is isoelectronic with covalently bonded SiH_4 and PH_4^+). Although chemists often mention hydrogen ion, H^+, a cation, this species does not persist as such in aqueous solution, nor, as we will learn, does it persist as such in the gas phase in the presence of any neutral or anionic species. When a chemist speaks of hydrogen ion, he usually is referring to the cationic species in which hydrogen ion is bonded to a solvent molecule, for example, H_3O^+, formed in the union of H^+ and H_2O.
2. Covalent hydrides with nonmetals. The hydrocarbons and the hydrogen halides, water, and ammonia are covalent; we considered the first groups in Chapter 11 and the second in Chapter 10. Covalent hydrides

in which hydrogen is bonded to an electronegative element are acids, and these compounds will be the subject of much of the discussion in this chapter.

3. Compounds in which a hydrogen atom acts as a bridge between two atoms. These fall into two classes: (1) the compounds involving conventional hydrogen bonding, such as KHF_2 containing the hydrogen-bonded anion, FHF^-, and (2) the electron-deficient boron hydrides, of which the simplest example is diborane, B_2H_6; two of the hydrogen atoms in diborane are bridging atoms. Both of these classes of compounds will be discussed in this chapter.

This simple classification does not cover all types of hydrogen compounds, but it does provide an introduction to a broad area spanning from classical compounds (e.g., water, H_2O) to electron-deficient compounds (e.g., pentaborane(9), B_5H_9).

12–2 Chemical Equilibrium in the Gas Phase

We started our study of chemical thermodynamics in Chapter 3, which described the Gibbs free energy criterion for whether a process occurs:

$\Delta G > 0$	process does not occur
$\Delta G = 0$	process is at equilibrium
$\Delta G < 0$	process occurs

Chapter 3 also explored the way in which the temperature and the molar volumes of gases involved in a process influence the value of ΔG. The molar volume of a gas involved in a process is relevant to the question of whether the process occurs because the molar entropy of a gas depends upon its volume. Now we will carry the discussion further to show that the dependence of molar entropy upon molar volume is the root of the *equilibrium-constant* equation, an exceedingly useful principle. We shall show this dependence by considering reactions of hydrogen with nonmetals:

	ΔH^0 (kJ mol^{-1})	ΔS^0 (J K^{-1} mol^{-1})	$\Delta G^0_{298.2\,K}$ (kJ mol^{-1})
$3H_2(g) + N_2(g) = 2NH_3(g)$	−91.8	−198.1	−32.7
$H_2(g) + Br_2(g) = 2HBr(g)$	−103.8	+21.2	−110.1

The bond energies in these two nonmetal hydrides are similar (see Table 10–6):

$$E(\text{N—H}) = 391 \text{ kJ mol}^{-1}$$
$$E(\text{Br—H}) = 363 \text{ kJ mol}^{-1}$$

but the values of ΔH^0 in the reactions that form one mole of hydrogen–

nonmetal bonds differ by a larger amount:

$$\mathrm{NH_3} \quad \Delta H_f^0 = -15.3\ \mathrm{kJ}\ \ (\mathrm{mol\ N{-}H\ bonds})^{-1}$$
$$\mathrm{HBr} \quad \Delta H_f^0 = -51.9\ \mathrm{kJ}\ \ (\mathrm{mol\ Br{-}H\ bonds})^{-1}$$

The reason for this difference is the enormously high bond energy in diatomic nitrogen; this is shown in Figure 12–1.

But the contrasting values of ΔS^0 are even more important than the values of ΔH^0 in the comparison of these two reactions. The large difference in the values of ΔS^0 is due to the difference in the values of Δn:

In forming 2 mol NH_3, $\Delta n = 2 - 4 = -2$,

4 mol reactant gases → 2 mol product gases

In forming 2 mol HBr, $\Delta n = 2 - 2 = 0$,

2 mol reactant gases → 2 mol product gases

For the reaction in which there is a net disappearance of gas, the value of ΔS^0 is very negative ($\Delta S^0 = -198.1\ \mathrm{J\ K^{-1}\ mol^{-1}}$), but for the reaction in which $\Delta n = 0$, the value of ΔS^0 is small ($\Delta S^0 = +21.2\ \mathrm{J\ K^{-1}\ mol^{-1}}$).

These values of ΔS^0 were derived from values of the molar entropies, which can be obtained from heat-capacity measurements or theoretical calculations related to the equation $S = R \ln \Omega$, discussed in Chapter 5. For the substances involved in the reactions being considered, values of the molar entropies are:

Substance	S^0 / $\mathrm{J\ K^{-1}\ mol^{-1}}$
$H_2(g)$	130.6
$N_2(g)$	191.5
$Br_2(g)$	245.4
$NH_3(g)$	192.6
$HBr(g)$	198.6

and the values of ΔS^0 calculated using these values are:

$$\Delta S^0 = 2S^0[\mathrm{NH_3}(g)] - 3S^0[\mathrm{H_2}(g)] - S^0[\mathrm{N_2}(g)]$$

$$\frac{\Delta S^0}{\mathrm{J\ K^{-1}\ mol^{-1}}} = 2 \times 192.6 - 3 \times 130.6 - 191.5 = -198.1$$

$$\Delta S^0 = 2S^0[\mathrm{HBr}(g)] - S^0[\mathrm{H_2}(g)] - S^0[\mathrm{Br_2}(g)]$$

$$\frac{\Delta S^0}{\mathrm{J\ K^{-1}\ mol^{-1}}} = 2 \times 198.6 - 130.6 - 245.4 = +21.2$$

The values of ΔS^0 pertain to the reaction under standard conditions; now we will consider the change of the value of ΔS as the reaction is carried out under nonstandard conditions. The principles will be developed in connection with the synthesis of ammonia.

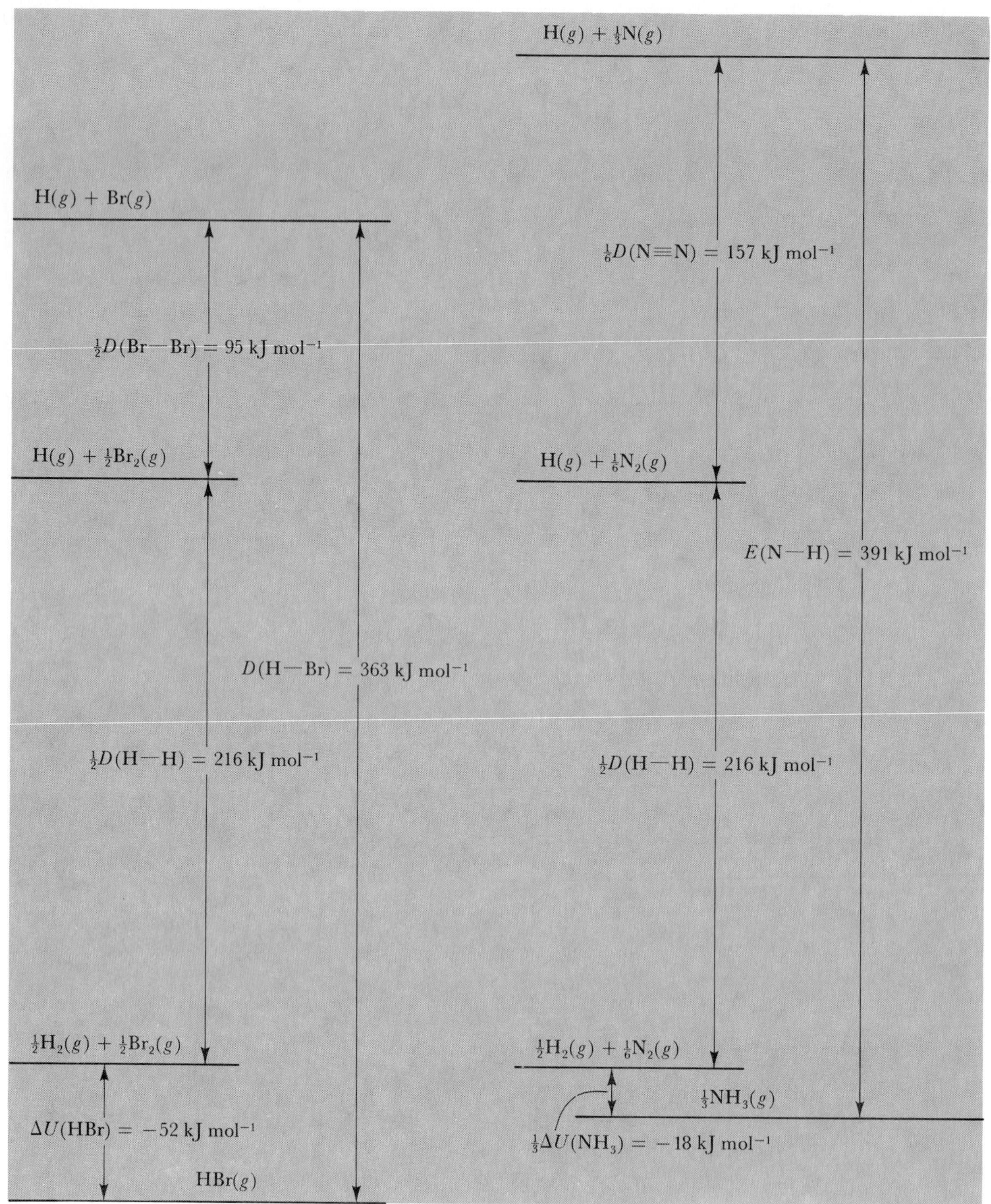

FIGURE 12–1
The formation of one mole of hydrogen–nitrogen bonds and one mole of hydrogen–bromine bonds. Notice the relative contributions of $\frac{1}{2}D$(Br—Br) and $\frac{1}{6}D$(N≡N) in determining the value of ΔU. [The value of ΔU(HBr) is equal to the value of ΔH given in the text because $\Delta n = 0$; the value of $\frac{1}{3}\Delta U(NH_3)$ is less negative than the value of ΔH given in the text because $\Delta n = -\frac{1}{3}$.]

In the discussion of the ammonia-synthesis reaction in Chapter 3, each of the gaseous reactants and products was in its standard state,

$$P^0_{N_2} = P^0_{H_2} = P^0_{NH_3} = \text{exactly 1 atm (ideal gases)}$$

and the values of ΔH^0 and ΔS^0 just presented for the reaction pertain to these conditions. Under these conditions

$$\Delta G^0 = -91{,}800 \text{ J mol}^{-1} + 198.1 \text{ J K}^{-1} \text{ mol}^{-1} \times T$$

and we can calculate that

$$\Delta G^0 = 0 \text{ at } 463 \text{ K (see footnote 1)}$$

$$\text{with } \Delta G^0 > 0 \text{ at } T > 463 \text{ K} \qquad \text{and } \Delta G^0 < 0 \text{ at } T < 463 \text{ K}$$

Thus we calculate that standard conditions are equilibrium conditions at 463 K. But we are interested in the equilibrium conditions at other temperatures: to what extent does the reaction go to establish equilibrium when the reactants are brought together at some nonstandard set of pressure conditions at a particular temperature. To find this extent, we will take the equation $\Delta G = \Delta H - T\Delta S$ for a particular temperature and two different sets of pressure conditions, one of them standard conditions, and subtract:

$$\begin{array}{r} \Delta G = \Delta H - T\Delta S \\ -(\Delta G^0 = \Delta H^0 - T\Delta S^0) \\ \hline (\Delta G - \Delta G^0) = (\Delta H - \Delta H^0) - T(\Delta S - \Delta S^0) \end{array}$$

We will assume the gases are ideal because this will allow us to use what we have already learned in Chapter 4 about ideal gases. The internal energy and enthalpy of an *ideal gas* do not depend upon the pressure at a particular temperature. Therefore, $H - H^0 \equiv 0$ for each gas, and $\Delta H - \Delta H^0 \equiv 0$ for the reaction.

In Chapters 3 and 4 we learned that the entropy of an ideal gas depends upon its molar volume. For an isothermal change the molar volume is inversely proportional to the pressure ($V/n = RT/P$), and we have for each gas

$$S - S^0 = R \ln\left(\frac{V}{V^0}\right) = -R \ln\left(\frac{P}{P^0}\right)$$

which gives for $P^0 = 1$ atm,

$$S - S^0 = -R \ln\left(\frac{P}{1 \text{ atm}}\right)$$

[1]This calculation is based upon the assumption that the values of ΔH^0 and ΔS^0 do not depend upon temperature. In truth, the temperature at which $\Delta G^0 = 0$ is 456 K. (The small error in the calculation is due to incorrectness in the assumption.)

For the reaction being considered,

$$\begin{aligned}\Delta S - \Delta S^0 &= 2(S_{NH_3} - S^0_{NH_3}) - (S_{N_2} - S^0_{N_2}) - 3(S_{H_2} - S^0_{H_2})\\ &= 2\left[-R\ln\left(\frac{P_{NH_3}}{1\text{ atm}}\right)\right] - \left[-R\ln\left(\frac{P_{N_2}}{1\text{ atm}}\right)\right] - 3\left[-R\ln\left(\frac{P_{H_2}}{1\text{ atm}}\right)\right]\\ &= -R\ln\left[\left(\frac{P^2_{NH_3}}{P_{N_2}\times P^3_{H_2}}\right)\times \text{atm}^2\right]\end{aligned}$$

with $\Delta H - \Delta H^0 \equiv 0$, the equation for $(\Delta G - \Delta G^0)$ for the reaction becomes:

$$\begin{aligned}\Delta G - \Delta G^0 &= -T(\Delta S - \Delta S^0)\\ &= RT\ln\left[\left(\frac{P^2_{NH_3}}{P_{N_2}\times P^3_{H_2}}\right)\times \text{atm}^2\right]\end{aligned}$$

Figure 12–2 gives a plot of $\Delta G/RT$ versus $\ln\{[P^2_{NH_3}/(P_{N_2}\times P^3_{H_2})]\times \text{atm}^2\}$ for $T = 298.2$ K. The plot is linear with slope 1. With each gas in its standard state ($P = 1$ atm with gas ideality), the logarithmic term becomes zero, $\ln 1 = 0$, and the intercept on the ordinate is $\Delta G^0/RT$. With the value of $\Delta G = 0$, the value of the quotient of pressures has its equilibrium

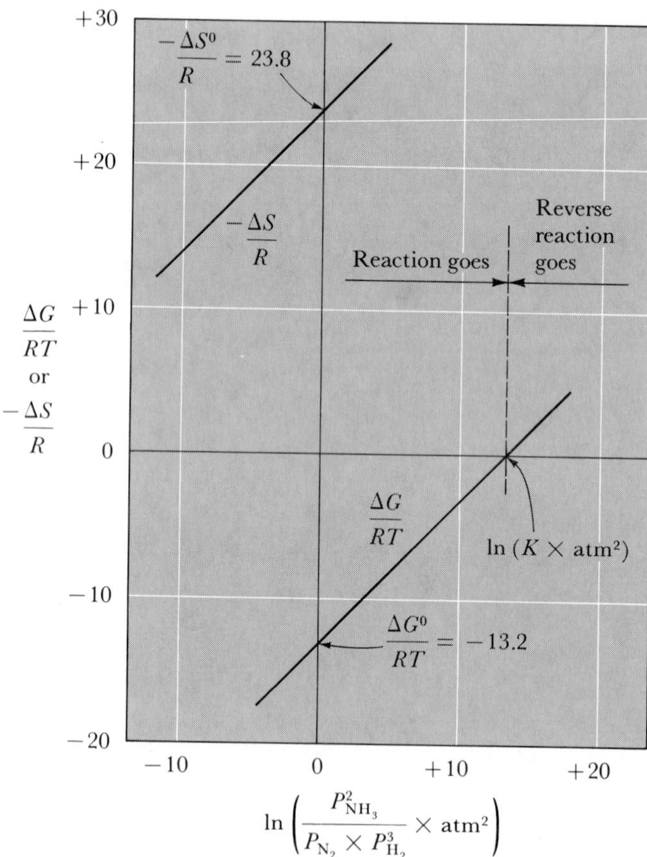

FIGURE 12–2
A plot of

$$\frac{\Delta G}{RT} \text{ versus } \ln\left[\left(\frac{P^2_{NH_3}}{P_{N_2}\times P^3_{H_2}}\right)\times \text{atm}^2\right]$$

for the reaction

$$N_2(g) + 3H_2(g) = 2NH_3(g)$$

For values of the quotient of pressures less than K ($K = 5.3\times 10^5$ atm^{-2} at 298.2 K) the reaction occurs as written; for values greater than K, the reverse reaction occurs. (The upper line gives $-\Delta S/R$ as a function of the same variable.)

value, represented by K, commonly called the *equilibrium constant* for the reaction,

$$K = \left(\frac{P^2_{NH_3}}{P_{N_2} \times P^3_{H_2}}\right)_{eq}$$

The intercept on the abscissa is $\ln (K \times atm^2)$. Also shown in this figure is a plot of $-\Delta S/R$ versus the same logarithmic function of the pressures of reactants and products. We see the two straight lines are parallel; it is the linear dependence of $-\Delta S/R$ upon $\ln\{[P^2_{NH_3}/(P_{N_2} \times P^3_{H_2})] \times atm^2\}$ that is responsible for the linear dependence of $\Delta G/RT$ upon this same variable.

The value of K tells us about the extent to which a reaction goes to establish equilibrium; for equilibrium, $\Delta G = 0$ and K is related to the value of ΔG^0,

$$-\Delta G^0 = RT \ln (K \times atm^2)$$

or, expressed in exponential form,

$$K \times atm^2 = e^{-\Delta G^0/RT}$$

Using the value of ΔG^0, already given for 298.2 K, the value of K can be calculated:

$$K \times atm^2 = e^{-\Delta G^0/RT}$$

$$= e^{-\left(\frac{-32{,}700 \text{ J mol}^{-1}}{8.315 \text{ J K}^{-1} \text{ mol}^{-1} \times 298.2 \text{ K}}\right)} = 5.3 \times 10^5$$

$$K = 5.3 \times 10^5 \text{ atm}^{-2}$$

At 298.2 K a mixture of nitrogen, hydrogen, and ammonia is at chemical equilibrium if

$$\frac{P^2_{NH_3}}{P_{N_2} \times P^3_{H_2}} = 5.3 \times 10^5 \text{ atm}^{-2}$$

If the value of this quotient for a mixture of these gases is less than 5.3×10^5 atm^{-2}, the reaction that occurs in approaching equilibrium consumes nitrogen and hydrogen and produces more ammonia. If the value of this quotient for existing conditions is greater than 5.3×10^5 atm^{-2}, ammonia decomposes to produce additional nitrogen and hydrogen. At 298.2 K the rate of establishing chemical equilibrium in this system is immeasurably low. However, this does not prevent us from making calculations pertaining to equilibrium at this temperature.

EXAMPLE: A sample of pure ammonia gas (initial pressure = 0.0100 atm) in a container of fixed volume is allowed to come to equilibrium at 298.2 K. What are the partial pressures of hydrogen, nitrogen, and ammonia at equilibrium?

Study of the balanced chemical equation leads to the following assignment of partial pressures at equilibrium:

$$P_{N_2} = \tfrac{1}{3}P_{H_2} \qquad P_{NH_3} = 0.0100 \text{ atm} - \tfrac{2}{3}P_{H_2}$$

Substitution of these quantities into the equilibrium-constant equation gives

$$\frac{(0.0100\ \text{atm} - \frac{2}{3}P_{H_2})^2}{\frac{1}{3}P_{H_2} \times P_{H_2}^3} = 5.3 \times 10^5\ \text{atm}^{-2}$$

Multiplying each side of the equation by 1/3 and taking the square root gives

$$\frac{(0.0100\ \text{atm} - \frac{2}{3}P_{H_2})}{P_{H_2}^2} = 4.2 \times 10^2\ \text{atm}^{-1}$$

which upon rearrangement gives the quadratic equation:

$$420\ \text{atm}^{-1} \times P_{H_2}^2 + \tfrac{2}{3}P_{H_2} - 0.0100\ \text{atm} = 0$$

Using

$$x = \frac{(-b \pm \sqrt{b^2 - 4ac})}{2a}$$

as the solution to the equation $ax^2 + bx + c = 0$, we get

$$P_{H_2} = \frac{-2/3 + \sqrt{4/9 + 4 \times 420\ \text{atm}^{-1} \times 0.01000\ \text{atm}}}{2 \times 420\ \text{atm}^{-1}}$$

$$P_{H_2} = 4.15 \times 10^{-3}\ \text{atm}$$

Although three significant figures are not justified by the data (a value of ΔG^0 given to 100 J mol^{-1} leads to a value of K given to two significant figures), we will keep the three significant figures for each pressure until we have completed the calculation. (The quadratic equation has two solutions; the other solution is negative, $P_{H_2} = -5.7 \times 10^{-3}$ atm, and is a physically meaningless solution to the equation.) Therefore the other equilibrium pressures are:

$$P_{N_2} = \tfrac{1}{3}P_{H_2} = 1.38 \times 10^{-3}\ \text{atm}$$

$$P_{NH_3} = 0.01000\ \text{atm} - \tfrac{2}{3}P_{H_2}$$

$$P_{NH_3} = 0.01000\ \text{atm} - 2.77 \times 10^{-3}\ \text{atm} = 0.00723\ \text{atm}$$

We can check our answers by substituting them into the equilibrium-constant equation:

$$K = \frac{P_{NH_3}^2}{P_{N_2} \times P_{H_2}^3} = \frac{(0.00723\ \text{atm})^2}{1.38 \times 10^{-3}\ \text{atm} \times (4.15 \times 10^{-3}\ \text{atm})^3}$$

$$= 5.30 \times 10^5\ \text{atm}^{-2}$$

which is the correct value of K. (It always is desirable to check an equilibrium calculation by substituting the derived pressures into the equilibrium-constant equation.) At this point we can round the calculated pressures to an appropriate number of significant figures:

$$P_{H_2} = 0.0042\ \text{atm} \qquad P_{N_2} = 0.0014\ \text{atm} \qquad P_{NH_3} = 0.0072\ \text{atm}$$

EXAMPLE: The gases in a mixture at 298.2 K have the partial pressures $P_{NH_3} = 0.0146$ atm, $P_{N_2} = 0.00340$ atm, and $P_{H_2} = 0.00560$ atm. Is this mixture at chemical equilibrium with respect to the reaction $N_2(g) + 3H_2(g) = 2NH_3(g)$? If not, in what direction does reaction occur to establish equilibrium? What will be the partial pressure of each gas after equilibrium is established?

We can calculate for these initial conditions the value of the quotient:

$$\frac{P_{NH_3}^2}{P_{N_2} \times P_{H_2}^3}$$

It is

$$\frac{(0.0146 \text{ atm})^2}{(3.40 \times 10^{-3} \text{ atm})(5.60 \times 10^{-3} \text{ atm})^3} = 3.57 \times 10^5 \text{ atm}^{-2}$$

Because this is smaller than 5.3×10^5 atm^{-2}, the reaction must occur to the right (i.e., more ammonia must be formed) to establish equilibrium. If we represent by x the increase in the pressure of ammonia necessary to establish equilibrium, we have as the partial pressures at equilibrium:

$$P_{NH_3} = 0.0146 \text{ atm} + x$$
$$P_{N_2} = 0.00340 \text{ atm} - \tfrac{1}{2}x$$
$$P_{H_2} = 0.00560 \text{ atm} - \tfrac{3}{2}x$$

Substituting these quantities into the equilibrium-constant equation, we get

$$\frac{(0.0146 \text{ atm} + x)^2}{(0.00340 \text{ atm} - \frac{1}{2}x)(0.00560 \text{ atm} - \frac{3}{2}x)^3} = 5.3 \times 10^5 \text{ atm}^{-2}$$

Here we have a quartic equation that we cannot simplify by taking the square root of each side as we did in the preceding example. We can determine the value of x by a procedure called *the method of successive approximations*. The procedure, which we will use in other problems as well, consists for this problem of the steps:

1. assuming a value of x,
2. calculating the value of the quotient that is the left side of the preceding equilibrium-constant equation,
3. comparing the calculated value with the correct equilibrium value, 5.3×10^5 atm^{-2}, and
4. repeating Steps 1, 2, and 3 with other assumed values of x until the calculated value of the quotient is the equilibrium value (within the justified accuracy).

In assuming a value for x we will recognize the limits placed on its value by our preceding calculation. With $x = 0$ atm, the quotient is slightly too small, and an upper limit is $x = 0.0037$ atm, for which the quotient would approach infinity because $(0.0056 \text{ atm} - \frac{3}{2}x) \cong 0$. Using this reasoning let us take as a first approximation $x = 5.0 \times 10^{-4}$ atm, which gives

$$P_{NH_3} = 0.0146 \text{ atm} + 0.00050 \text{ atm} = 0.0151 \text{ atm}$$
$$P_{N_2} = 0.00340 \text{ atm} - \tfrac{1}{2} \times 0.00050 \text{ atm} = 0.00315 \text{ atm}$$
$$P_{H_2} = 0.00560 \text{ atm} - \tfrac{3}{2} \times 0.00050 \text{ atm} = 0.00485 \text{ atm}$$

Substitution of these partial pressures into the quotient that is the left side of the equation gives

$$\frac{P^2_{NH_3}}{P_{N_2} \times P^3_{H_2}} = \frac{(0.0151 \text{ atm})^2}{(0.00315 \text{ atm})(0.00485 \text{ atm})^3} = 6.34 \times 10^5 \text{ atm}^{-2}$$

This value is larger than the value of K, suggesting a second approximation in the range:

$$0 < x < 5.0 \times 10^{-4} \text{ atm}$$

Since the first assumed value for x ($x = 5.0 \times 10^{-4}$ atm) is almost correct, a reasonable second approximation is $x = 3.0 \times 10^{-4}$ atm, which gives

$$P_{NH_3} = 0.0146 \text{ atm} + 0.00030 \text{ atm} = 0.0149 \text{ atm}$$
$$P_{N_2} = 0.00340 \text{ atm} - \tfrac{1}{2} \times 0.00030 \text{ atm} = 0.00325 \text{ atm}$$
$$P_{H_2} = 0.00560 \text{ atm} - \tfrac{3}{2} \times 0.00030 \text{ atm} = 0.00515 \text{ atm}$$

Substitution of these partial pressures into the quotient gives

$$\frac{P^2_{NH_3}}{P_{N_2} \times P^3_{H_2}} = \frac{(0.0149 \text{ atm})^2}{(0.00325 \text{ atm})(0.00515 \text{ atm})^3} = 5.00 \times 10^5 \text{ atm}$$

which is 6% smaller than the value of K. Further refinement of the answer probably is not justified, but if you wish to perform one more cycle of calculations, you now know that

$$3.0 \times 10^{-4} \text{ atm} < x < 5.0 \times 10^{-4} \text{ atm}$$

Because the second approximation is closer than the first, a value of $x = 3.5 \times 10^{-4}$ atm would be a reasonable third approximation.[2]

These examples illustrate the usefulness of the equilibrium-constant equation, which is called the *law of mass action.* With an appropriate value of the constant (i.e., the value that is correct for the temperature and other relevant conditions), we can calculate the composition of the system at equilibrium for all possible combinations of initial values of the partial pressures. In these examples the partial pressures were so low that each gas could be assumed to be an ideal gas. As we will learn, nonideality of reactant and product gases alters the value of the equilibrium constant.

---

[2]With this value of x, the calculated value of the quotient is $5.31 \times 10^5 \text{ atm}^{-2}$.

Consideration of the dependence of ΔS upon the pressures of the gaseous reactants and products in the reaction

$$N_2(g) + 3H_2(g) = 2NH_3(g)$$

led us to the relationship

$$\Delta G^0 = -RT\ln\left[\left(\frac{P^2_{NH_3}}{P_{N_2} \times P^3_{H_2}}\right)_{eq} \times \text{atm}^2\right]$$

$$= -RT\ln(\text{K} \times \text{atm}^2)$$

Now we will review and generalize the relationship between the change of free energy for the reaction under standard conditions and the equilibrium constant for the reaction.

For a reaction

$$a\text{A}(g) + b\text{B}(g) = c\text{C}(g) + d\text{D}(g)$$

a derivation for $(\Delta S - \Delta S^0)$ just like that used in the ammonia-synthesis reaction gives

$$\Delta S - \Delta S^0 = -R\ln\left[\left(\frac{P^c_C \times P^d_D}{P^a_A \times P^b_B}\right) \times \text{atm}^{-\Delta n}\right]$$

in which $\Delta n = c + d - a - b$. Combining this with $(\Delta H - \Delta H^0) \equiv 0$ (a consequence of the assumption of gas ideality) and $\Delta G = 0$ for equilibrium conditions gives

$$\Delta G^0 = -RT\ln\left[\left(\frac{P^c_C \times P^d_D}{P^a_A \times P^b_B}\right)_{eq} \times \text{atm}^{-\Delta n}\right]$$

$$= -RT\ln(K \times \text{atm}^{-\Delta n})$$

in which

$$K = \left(\frac{P^c_C \times P^d_D}{P^a_A \times P^b_B}\right)_{eq}$$

The equilibrium constant for a reaction involving ideal gases is the product of the partial pressures of the gaseous products divided by the product of the partial pressures of the gaseous reactants, each partial pressure being raised to a power equal to the coefficient of that species in the chemical equation.

The magnitudes of ΔG^0 and K are changed if a chemical equation is multiplied by a numerical factor. Thus, at 298.2 K, for

$$N_2(g) + 3H_2(g) = 2NH_3(g) \qquad \Delta G^0 = -32{,}700 \text{ J mol}^{-1}$$
$$K = 5.3 \times 10^5 \text{ atm}^{-2}$$

and for

$$\tfrac{1}{2}N_2(g) + \tfrac{3}{2}H_2(g) = NH_3(g) \qquad \Delta G^0 = \tfrac{1}{2}(-32{,}700\ \text{J mol}^{-1}) = -16{,}350\ \text{J mol}^{-1}$$
$$K = (5.3 \times 10^5\ \text{atm}^{-2})^{1/2} = 7.3 \times 10^2\ \text{atm}^{-1}$$

The equilibrium constant

$$K = \frac{P_{NH_3}^2}{P_{N_2} \times P_{H_2}^3}$$

has the dimension $(\text{atm})^{-2}$ if the partial pressures are expressed in atmospheres. Therefore the product $(K \times \text{atm}^2)$ is dimensionless, and its logarithm can be taken. The factor $(\text{atm})^2$ enters the mathematical relationship because for each molecule appearing in the chemical equation, there is a term

$$S - S^0 = -R \ln\left(\frac{P}{1\ \text{atm}}\right)$$

in the equation for $\Delta S - \Delta S^0$, and for the ammonia-synthesis reaction there are four molecules on the reactant side and two molecules on the product side ($\Delta n = -2$). For the general case

$$\Delta G^0 = -RT \ln(K \times \text{atm}^{-\Delta n}) \qquad \text{or} \qquad K \times \text{atm}^{-\Delta n} = e^{-\Delta G^0/RT}$$

For the ammonia-synthesis reaction and for any other gas-phase reaction with $\Delta n = 0$, the numerical values of K and ΔG^0 depend upon the dimensions in which the pressures are expressed. This is equivalent to a dependence upon the choice of the standard state for each gaseous reactant and product. Let us convert the value of the equilibrium constant for the ammonia synthesis at 298.2 K, $K = 5.3 \times 10^5\ \text{atm}^{-2}$, to the values for the pressures expressed in torr and pascal; we will use the conversion factors

$$1\ \text{atm} \equiv 760\ \text{torr} \qquad 1\ \text{atm} \equiv 101{,}325\ \text{Pa}$$

We make this conversion simply by multiplying the value of K by the appropriate unit conversion factor:

$$K = 5.3 \times 10^5\ \text{atm}^{-2} \times \left(\frac{1\ \text{atm}}{760\ \text{torr}}\right)^2$$

$$K = 0.92\ \text{torr}^{-2}$$

$$K = 5.3 \times 10^5\ \text{atm}^{-2} \times \left(\frac{1\ \text{atm}}{101\ 325\ \text{Pa}}\right)^2$$

$$K = 5.2 \times 10^{-5}\ \text{Pa}^{-2}$$

We see that inclusion of dimensions with the value of the equilibrium constant tells us unambiguously which pressure unit is used. If $\Delta n = 0$, the equilibrium constant is dimensionless.

THE DEPENDENCE OF EQUILIBRIUM CONSTANTS ON TEMPERATURE

Now we will consider the equilibrium constant's dependence upon temperature. Substituting $\Delta G^0 = \Delta H^0 - T\,\Delta S^0$ into the equation

$$\Delta G^0 = -RT \ln (K \times \text{atm}^{-\Delta n})$$

and than dividing by RT, we get

$$\ln (K \times \text{atm}^{-\Delta n}) = -\frac{\Delta H^0}{RT} + \frac{\Delta S^0}{R}$$

Taking the antilogarithm of each side of the equation, we obtain the exponential form:

$$K \times \text{atm}^{-\Delta n} = e^{-\Delta H^0/RT} \times e^{\Delta S^0/R}$$

which shows that the numerical value of an equilibrium constant is made up of two factors. One factor, $e^{-\Delta H^0/RT}$, involves the enthalpy change in the reaction, and this factor varies with temperature. The other factor, involving the standard entropy change, $e^{\Delta S^0/R}$, does not vary with temperature. The equation in the logarithmic form shows that a plot of $\ln (K \times \text{atm}^{-\Delta n})$ versus $1/T$ will produce a linear graph if ΔH^0 and ΔS^0 are independent of temperature. Figure 12–3 shows this plot for the ammonia-synthesis reaction. The data do not fall precisely on a straight line; the

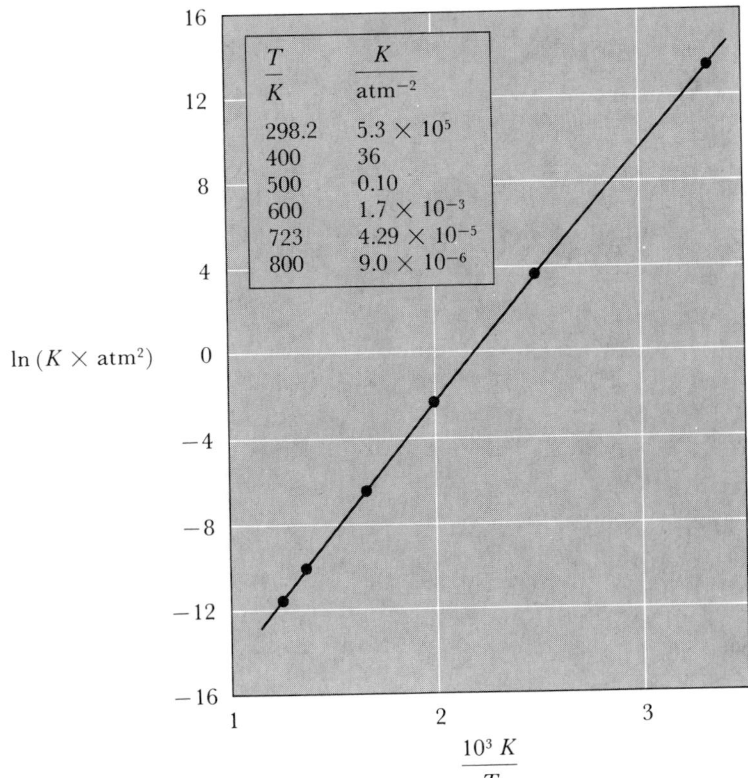

$\frac{T}{K}$	$\frac{K}{\text{atm}^{-2}}$
298.2	5.3×10^5
400	36
500	0.10
600	1.7×10^{-3}
723	4.29×10^{-5}
800	9.0×10^{-6}

FIGURE 12–3
The dependence of K upon temperature for the reaction $N_2(g) + 3H_2(g) = 2NH_3(g)$. The plot is $\ln (K \times \text{atm}^2)$ versus $10^3 K/T$.

value of ΔH^0 depends upon temperature because the value of ΔC_p is not exactly zero.

Chapter 6 outlined the types of calculations that use the equation

$$\ln\left(\frac{P}{\text{atm}}\right) = -\frac{\Delta H^0_{\text{vap}}}{RT} + \frac{\Delta S^0_{\text{vap}}}{R}$$

These are:

1. calculation of ΔH^0_{vap}, given the vapor pressure, P, at two temperatures,
2. calculation of P at some temperature, given the value of ΔH^0_{vap} and the value of P at some other temperature, and
3. calculation of the temperature at which the value of P has some particular value, given the same information as in 2.

We can make analogous calculations using the relationship

$$\ln(K \times \text{atm}^{-\Delta n}) = -\frac{\Delta H^0}{RT} + \frac{\Delta S^0}{R}$$

EXAMPLE: Values of the equilibrium constant for the reaction $N_2O_4(g) = 2NO_2(g)$ as a function of temperature are 25.0°C, $K =$ 0.1426 atm, and 45.0°C, $K = 0.6706$ atm. What are the values of ΔH^0 and ΔS^0? Estimate the value of K at 65.0°C.

Writing the equation relating K, ΔH^0, and ΔS^0 for the two temperatures, T_1 and T_2, and then subtracting these equations, we have

$$\ln\frac{K_1}{K_2} = -\frac{\Delta H^0}{R}\left(\frac{1}{T_1} - \frac{1}{T_2}\right)$$

or

$$\ln\left(\frac{0.1426}{0.6706}\right) = -\frac{\Delta H^0}{8.31\ \text{J K}^{-1}\ \text{mol}^{-1}} \times \left(\frac{1}{298.2\ \text{K}} - \frac{1}{318.2\ \text{K}}\right)$$

Upon rearrangement of this equation, we obtain

$$\Delta H^0 = -8.31\ \text{J K}^{-1}\ \text{mol}^{-1} \times \frac{1}{\left(\frac{1}{298.2\ \text{K}} - \frac{1}{318.2\ \text{K}}\right)} \times \ln\left(\frac{0.1426}{0.6706}\right)$$

$$= +61.0\ \text{kJ mol}^{-1}$$

With the values of ΔH^0 ($+61.0$ kJ mol^{-1}) and the equilibrium constant at some temperature ($K = 0.1426$ atm at 298.2 K), the value of ΔS^0 can be calculated:

$$\Delta S^0 = +\frac{\Delta H^0}{T} + R\ln(K \times \text{atm}^{-\Delta n})$$

$$= +\frac{61{,}000\ \text{J mol}^{-1}}{298.2\ \text{K}} + 8.31\ \text{J K}^{-1}\ \text{mol}^{-1} \times \ln(0.1426\ \text{atm} \times \text{atm}^{-1})$$

$$= 188\ \text{J K}^{-1}\ \text{mol}^{-1}$$

In this reaction one gas molecule gives two gas molecules, $\Delta n = +1$. We see that under standard conditions the reaction is accompanied by a large increase in entropy, as we would expect.

We can estimate the value of K at some other temperature if we are willing to assume that ΔH^0 and ΔS^0 do not depend upon temperature. This almost certainly is not a correct assumption, but for the problem at hand we will make it in order to obtain an approximate answer. (With values of K at only two temperatures, we have no data to show the quality of this assumption.) We can calculate the value of K at 65.0°C directly from the already derived values of ΔH^0 and ΔS^0:

$$\ln(K \times \text{atm}^{-1}) = -\frac{\Delta H^0}{RT} + \frac{\Delta S^0}{R}$$

$$= -\frac{61{,}000 \text{ J mol}^{-1}}{8.31 \text{ J K}^{-1} \text{ mol}^{-1} \times 338.2 \text{ K}} + \frac{188 \text{ J K}^{-1} \text{ mol}^{-1}}{8.31 \text{ J K}^{-1} \text{ mol}^{-1}} = 0.919$$

$$K = 2.51 \text{ atm}$$

There are other ways to answer this question. We can plot $\ln(K \times \text{atm}^{-1})$ versus T^{-1}, draw a straight line through the two experimental points, and read the value of $\ln(K \times \text{atm}^{-1})$ at $T^{-1} = (338.2 \text{ K})^{-1} = 2.957 \times 10^{-3} \text{ K}^{-1}$.

NONIDEALITY IN THE AMMONIA-SYNTHESIS REACTION

The choice of conditions for practical production of ammonia from nitrogen and hydrogen is influenced by the extremely low rate of the reaction at ordinary temperatures where the value of the equilibrium constant is large. To increase the reaction rate, higher temperatures and catalysts are employed.[3] But at higher temperatures, the value of K is smaller. At 723 K (a temperature in the range employed in the Haber–Bosch process), the value of K for low total pressures is 4.29×10^{-5} atm^{-2}. Although this value of K is small, the yield of ammonia increases with an increase of the total pressure. Using this value of K, we can calculate the equilibrium yield of ammonia as a function of the total pressure in a gaseous mixture containing nitrogen and hydrogen in the stoichiometric relative amounts 1 mol N_2 per 3 mol H_2. The results, given as the solid line in Figure 12–4, were obtained in a calculation based upon the assumption that each gas was an ideal gas. This assumption is approximately correct if the total pressure is less than ~30 atm, but certainly is incorrect if the total pressure is >300 atm. Nonideality is taken into account in an equilibrium constant by introducing correction factors called *activity coefficients*, which are represented by γ. For gases, a pressure is multiplied by an activity coefficient to give a quantity called *activity*, represented by a, which is dimensionless. Thus if $a = P \times \gamma$, and P is expressed in atm, the dimensions of γ must be atm^{-1}.

[3]You will learn about reaction rates and catalysis in Chapter 15.

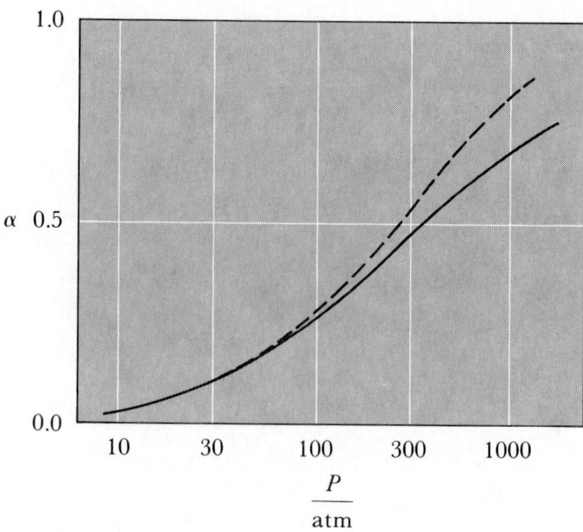

FIGURE 12–4
The fractional yield of ammonia, α ($\alpha = 3P_{NH_3}/2P^0_{H_2}$), at 723 K as a function of the total pressure, with logarithmic abscissa. ($P^0_{H_2}$ is the initial pressure of hydrogen; $P^0_{N_2} = \frac{1}{3}P^0_{H_2}$.) The solid line is based upon $K = 4.29 \times 10^{-5}$ atm^{-2}. The dashed line is based upon

$$K = 4.29 \times 10^{-5} \times \frac{\gamma_{N_2}\gamma^3_{H_2}}{\gamma^2_{NH_3}}$$

For this reaction the equilibrium-constant equation including activity coefficients, represented by K^0, is

$$K^0 = \frac{a^2_{NH_3}}{a_{N_2} \times a^3_{H_2}} = \frac{P^2_{NH_3}}{P_{N_2} \times P^3_{H_2}} \times \frac{\gamma^2_{NH_3}}{\gamma_{N_2} \times \gamma^3_{H_2}}$$

At low pressures the gases are almost ideal, and each activity coefficient has the value ~ 1.00 atm^{-1}, $\gamma_{NH_3} = \gamma_{N_2} = \gamma_{H_2} \cong 1.00$ atm^{-1}. Therefore the value of K^0 is numerically equal to the value of K that was used in plotting the solid line in Figure 12–4:

$$K^0 = K \times \frac{\gamma^2_{NH_3}}{\gamma_{N_2} \times \gamma^3_{H_2}}$$

$$= 4.29 \times 10^{-5} \text{ atm}^{-2} \times \frac{(1.00 \text{ atm}^{-1})^2}{1.00 \text{ atm}^{-1} \times (1.00 \text{ atm}^{-1})^3} = 4.29 \times 10^{-5}$$

For conditions under which the quotient of activity coefficients is not approximately equal to 1.0 atm^2, we can rearrange this equation to obtain the value of K that is appropriate for the conditions at hand:

$$K = \frac{P^2_{NH_3}}{P_{N_2} \times P^3_{H_2}} = 4.29 \times 10^{-5} \times \frac{\gamma_{N_2}\gamma^3_{H_2}}{\gamma^2_{NH_3}}$$

With increasing pressure this quotient of activity coefficients increases, therefore the value of K increases. This is shown in Table 12–1, and the consequences of increasing the total pressure are shown as the dashed line in Figure 12–4.

For the ammonia-formation reaction, the equilibrium yield in any particular mixture of nitrogen and hydrogen increases with increasing total pressure because:

1. At low pressures, this increase in the extent of reaction (increase in α in Figure 12–4) is necessary to maintain the quotient $P^2_{NH_3}/(P_{N_2} \times P^3_{H_2})$ at the correct value ($K = 4.29 \times 10^{-5}$ atm^{-2} at 723 K).

TABLE 12–1
The Effect of Gas Nonideality Upon the Equilibrium Constant[a]

$$K = \frac{P^2_{NH_3}}{P_{N_2}P^3_{H_2}} = K^0 \times \frac{\gamma^2_{N_2}\gamma^3_{H_2}}{\gamma^2_{NH_3}} \text{ at } T = 723 \text{ K}$$

P(total) atm	$\gamma_{N_2}\gamma^3_{H_2}/\gamma^2_{NH_3}$ atm^{-2}	$10^5\ K$ atm^{-2}
0	1.000	4.29
10	1.017	4.36
30	1.051	4.51
50	1.089	4.67
100	1.193	5.12
300	1.825	7.83
600	3.866	16.59
1000	12.14	52.08

[a]From H. F. Gibbard and M. R. Emptage, *J. Chem. Educ.* **53**, 218 (1976).
[b]The product of a pressure and an activity coefficient is dimensionless: $P_{NH_3} \times \gamma_{NH_3} \equiv 1$. Thus γ has the dimensions of atm^{-1} and the dimensions of $(\gamma_{N_2}\gamma^3_{H_2}/\gamma^2_{NH_3})$ are atm^{-2}.

2. At higher pressures, this increase in the extent of reaction is necessary to maintain the quotient $P^2_{NH_3}/(P_{N_2} \times P^3_{H_2})$ at a value that increases appreciably (by a factor of >10 to 5.2×10^{-4} atm $^{-2}$ at 1000 atm).

In most of our calculations we will assume that gases are ideal, but you should keep in mind that gas nonideality may play a role in establishing the extent to which a gaseous reaction occurs at equilibrium if the conditions involve high pressures. This obviously must be considered in any practical situation; in the design of a chemical plant for production of ammonia, account would be taken of the value of K that is appropriate for the high-pressure conditions of the process.

THE HYDROGEN HALIDES

Now we will continue our discussion of covalent hydrides of nonmetals and chemical equilibrium by considering hydrogen bromide (HBr) and the other hydrogen halides, HF, HCl, and HI, certain properties of which are summarized in Table 12–2. The hydrogen–fluorine bond has the largest bond energy of any single bond (except for the silicon–fluorine bond, for which the average bond energy is 582 kJ mol^{-1}). The hydrogen–halogen single bonds all are stronger than the single bonds in the diatomic halogen molecules, but only the hydrogen–fluorine bond is stronger than the single bond in the diatomic hydrogen molecule (432 kJ mol^{-1}). In interpreting values for the standard enthalpy of formation, you should remember that in most compilations the standard state for bromine is the liquid, $Br_2(l)$,

TABLE 12–2
Properties of the Hydrogen Halides

	mp/K	bp/K	T_c[a]/K	ΔH_f^0(298.2 K)[b]/kJ mol^{-1}	D(H—X)/kJ mol^{-1}	S^0[e]/J K^{-1} mol^{-1}
HF	189.8	292.7	461	−271	565	173.7
HCl	158.9	188.1	324.6	−92.3	427	186.8
HBr	186.2	206.4	363.2	−36.4[c]	363	198.6
HI	222.3	237.8	423.2	+26.5[d]	295	206.5

[a] T_c = critical temperature.
[b] ΔH_f^0 for formation of the gaseous hydrogen halide.
[c] The standard state for Br_2 is $Br_2(l)$.
[d] The standard state for I_2 is $I_2(s)$.
[e] S^0 values for gas (at 1 atm).

and that for iodine is the solid, $I_2(s)$. For correlation with the values for hydrogen fluoride and hydrogen chloride, it is more appropriate to consider the formation of the following hydrogen halides from the gaseous halogen:

Reaction	ΔH^0/kJ mol^{-1}
$\frac{1}{2}H_2(g) + \frac{1}{2}Br_2(g) = HBr(g)$	−51.9
$\frac{1}{2}H_2(g) + \frac{1}{2}I_2(g) = HI(g)$	−4.6

We see, in particular, that formation of hydrogen iodide from the gaseous constituent elements is slightly exothermic.

Now we will consider the formation of hydrogen bromide in more detail. Unlike the reaction for the formation of ammonia, the formation of hydrogen bromide from gaseous reactants has values of ΔH^0 and ΔS^0 with opposite sign, and there is no temperature at which the value of ΔG^0 is zero. Therefore solving for T in the equation (with values of ΔH^0 and ΔS^0 given previously):

$$\Delta G^0 = \Delta H^0 - T\Delta S^0$$
$$= -103{,}800 \text{ J mol}^{-1} - 21.2 \text{ J K}^{-1} \text{ mol}^{-1} \times T$$

would give a negative value. This is shown in Figure 12–5; Line **2** does not cross the $\ln K = 0$ axis. Line **2** can be contrasted with Line **4**, the $\ln (K \times \text{atm}^2)$ versus $1/T$ plot for the ammonia synthesis, which crosses the $\ln K = 0$ line at $T^{-1} = (463 \text{ K})^{-1} = 2.16 \times 10^{-3} \text{ K}^{-1}$.

The reaction forming hydrogen bromide serves to illustrate other points. At temperatures below 332.6 K (the normal boiling point of bromine), the reaction involving liquid bromine as a reactant is the appropriate representation. At temperatures above 1770 K, where atomic bromine has a pressure of 1 atm in equilibrium with diatomic bromine at 1 atm, the reaction involving atomic bromine as a reactant is the appropriate representation. For these reactions, the values of ΔH^0 and ΔS^0,

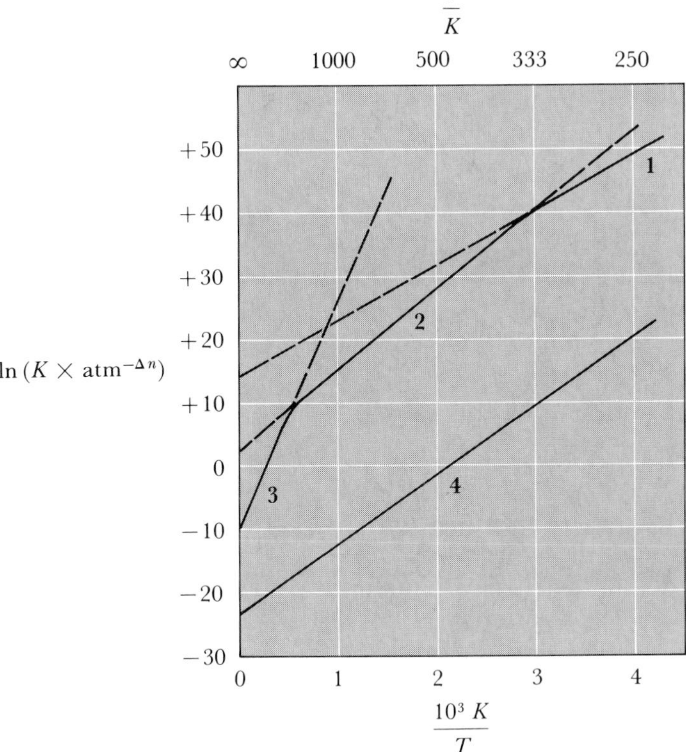

FIGURE 12–5
Plots of $\ln(K \times \mathrm{atm}^{-\Delta n})$ versus $1/T$ for the reactions:

1	$H_2(g) + Br_2(l) = 2HBr(g)$	$T < 357$ K, bp of $Br_2(l)$
2	$H_2(g) + Br_2(g) = 2HBr(g)$	$357\ \mathrm{K} < T < 1770$ K
3	$H_2(g) + 2Br(g) = 2HBr(g)$	$T > 1770$ K, at which $K = P^2_{Br}/P_{Br_2} = 1$ atm
4	$3H_2(g) + N_2(g) = 2NH_3(g)$	

Each straight line is drawn with a slope of $\Delta H^0/R$ and an intercept of $\Delta S^0/R$, using the values of ΔH^0 and ΔS^0 for 298.2 K. (Standard state for all gases, 1 atm.) The continuous solid line represents the reaction involving the predominant form of bromine.

assumed to be temperature-independent, are

Reaction	$\frac{\Delta H^0}{\mathrm{kJ\ mol^{-1}}}$	$\frac{\Delta S^0}{\mathrm{J\ K^{-1}\ mol^{-1}}}$
$H_2(g) + Br_2(l) = 2HBr(g)$	−72.9	+114.4
$H_2(g) + 2Br(g) = 2HBr(g)$	−296.6	−83.2

Lines **1** and **3** in Figure 12–5 show the dependence of $\ln(K \times \mathrm{atm}^{-\Delta n})$ for these reactions upon temperature. The solid line connecting segments of Lines **1**, **2**, and **3** gives the value of $(K \times \mathrm{atm}^{-\Delta n})$ for the reaction that involves bromine in the most stable form at the temperature in question. The values of Δn for Reactions **1**, **2**, and **3** are +1, 0, and −1, and this is reflected in the values of ΔS^0, $+114.4\ \text{J K}^{-1}\ \text{mol}^{-1}$, $+21.2\ \text{J K}^{-1}\ \text{mol}^{-1}$, and $-83.2\ \text{J K}^{-1}\ \text{mol}^{-1}$.

Under conditions where Reaction **3** is the appropriate representation, the values of ΔH^0 and ΔS^0 have opposite signs, the value of ΔG^0 is zero at a positive temperature, and Line **3** crosses the ln (K × atm) = 0 line in Figure 12–5. The crossing occurs at $T = 3560$ K ($T^{-1} = 2.81 \times 10^{-4}\ \text{K}^{-1}$).

This detailed consideration of the formation of hydrogen bromide shows the necessity of keeping track of the predominant form in which each reactant and product exists. The stability of hydrogen bromide relative to hydrogen and bromine is diminished at temperatures where the predominant form of bromine is atomic bromine. [It would be further diminished at still higher temperatures (>3750 K) where atomic hydrogen is the predominant form of hydrogen.]

THE STANDARD GIBBS FREE ENERGY OF FORMATION

We saw in Chapter 3 that a tabulation of values of ΔH_f^0, the change in enthalpy for the formation of a substance from constituent elements in their standard states, is very useful for calculating the values of ΔH^0 for other reactions of these substances. A tabulation of values of the standard free energy of formation of substances is even more useful because from this tabulation we can calculate equilibrium constants. From the values of ΔH^0 and ΔS^0 for the ammonia-synthesis reaction already considered [$3H_2(g) + N_2(g) = 2NH_3(g)$], we can calculate the value of ΔG_f^0 for a particular temperature. For 298.2 K,

$$\begin{aligned}\Delta G^0 &= -91{,}800\ \text{J mol}^{-1} - 298.2\ \text{K} \times (-198.1\ \text{J K}^{-1}\ \text{mol}^{-1}) \\ &= -32{,}700\ \text{J mol}^{-1} \\ \Delta G_f^0 &= -16.4\ \text{kJ mol}^{-1}\ \text{(for 1 mol } NH_3)\end{aligned}$$

This value can be combined with values of ΔG_f^0 for other substances to obtain values of ΔG^0 for reactions by procedures analogous to those already studied in Chapter 4. From the following tabulation for 298.2 K,

Substance	ΔG_f^0 / kJ mol^{-1}
$NH_3(g)$	−16.4
$NO(g)$	86.6
$H_2O(g)$	−228.6
$O_2(g)$	0

we can calculate the value of ΔG^0 for

$$4NH_3(g) + 5O_2(g) = 4NO(g) + 6H_2O(g)$$

which is a reaction in the process by which ammonia is converted to nitric acid. Substitution into the equation

$$\Delta G^0 = 4\,\Delta G_f^0[\mathrm{NO}(g)] + 6\,\Delta G_f^0[\mathrm{H_2O}(g)] - 4\,\Delta G_f^0[\mathrm{NH_3}(g)] - 5\,\Delta G_f^0[\mathrm{O_2}(g)]$$

gives ($\Delta G_f^0[\mathrm{O_2}(g)] = 0$ because $O_2(g)$ is the standard state for oxygen):

$$\frac{\Delta G^0}{\mathrm{kJ\ mol}} = 4 \times 86.6 + 6 \times (-228.6) - 4 \times (-16.4) = -959.6$$

$$\Delta G^0 = -959.6\ \mathrm{kJ\ mol^{-1}}$$

Therefore for this reaction, with $\Delta n = +1$, the equilibrium constant for 298.2 K can be calculated:

$$\begin{aligned} K \times \mathrm{atm}^{-1} &= e^{-\Delta G^0/RT} \\ &= e^{+959{,}600\ \mathrm{J\ mol^{-1}}/(8.315\ \mathrm{J\,K^{-1}\ mol^{-1}} \times 298.2\ \mathrm{K})} \\ &= 1.2 \times 10^{168} \end{aligned}$$

(Appendix 3 tabulates values of ΔG_f^0 for a number of substances.)

12–3 The Proton, A Very Special Chemical Species

The proton, H^+, is unlike all other atomic or molecular species encountered in ordinary chemistry. It is a species with zero electrons, a bare nucleus.[4] Its size is $\sim 10^{-5}$ times the size of most atoms and ions ($\sim 10^{-13}$ cm compared to $\sim 10^{-8}$ cm, the radius of a typical atom or monatomic ion). Free protons exist as such only in the absence of other molecules. In the gas phase the proton reacts exothermically with any atom, molecule, or anion to form a protonated species. Techniques have been developed recently to study gas-phase protonation reactions, and there now is a rich supply of thermodynamic data for these processes. Knowledge of the energetics of the gas-phase reactions is proving a benefit also to those studying protonation reactions in liquid solution where solvation effects introduce diverse complications. Comparison of the gas-phase and liquid-solution data help us understand these complications.

GAS-PHASE PROTON-ASSOCIATION REACTIONS

Let us start with some examples of gas-phase proton association and the energy changes accompanying these reactions. Proton association occurs with anionic species to form familiar neutral molecules, for example,

$$\mathrm{H^+}(g) + \mathrm{Cl^-}(g) = \mathrm{HCl}(g) \qquad \Delta U = -1394\ \mathrm{kJ\ mol^{-1}}$$

---

[4]Other species with zero electrons are He^{2+}, Li^{3+}, and so on, but these species are not encountered in ordinary chemistry. The alpha particle is He^{2+}, and you will encounter this species in our discussion of the nucleus and radioactivity in Chapter 22.

Proton association also occurs with molecular species to form familiar cations, such as

$$H^+(g) + NH_3(g) = NH_4^+(g) \qquad \Delta U = -866 \text{ kJ mol}^{-1}$$

and also with noble-gas atoms, for example,

$$H^+(g) + Ar(g) = HAr^+(g) \qquad \Delta U = -343 \text{ kJ mol}^{-1}$$

There also is proton association with molecular species that at first sight would seem to have no ability to associate with an additional atom, examples are

$$H^+(g) + H_2(g) = H_3^+(g) \qquad \Delta U = -440 \text{ kJ mol}^{-1}$$
$$H^+(g) + CH_4(g) = CH_5^+(g) \qquad \Delta U = -527 \text{ kJ mol}^{-1}$$

Here, proton association forms unfamilar cationic species. In both diatomic hydrogen and methane, all of the valence-shell electrons are involved in bonding, but each of these gaseous molecules reacts exothermically with a proton. Clearly the bonding in the product cations, H_3^+ and CH_5^+, involves principles that were not outlined in Chapters 10 and 11.

The negative energy change ($\Delta U < 0$) in a gas-phase proton-association reaction generally is given as a positive *proton affinity* (PA); $PA = -\Delta U$. Thus $PA(Cl^-) = 1394$ kJ mol^{-1}, $PA(NH_3) = 866$ kJ mol^{-1}, $PA(CH_4) = 527$ kJ mol^{-1}, $PA(H_2) = 440$ kJ mol^{-1}, and $PA(Ar) = 343$ kJ mol^{-1}. These values can be used to estimate the equilibrium extents of gas-phase protonation reactions.

Consider the species in this group with the smallest proton affinity, Ar. Since our goal is an estimate, not a refined calculation, we will assume for the process

$$H^+(g) + Ar(g) = HAr^+(g)$$

$\Delta H^0 \cong \Delta U = -343$ kJ mol^{-1}, and $\Delta S^0 \cong -93$ J K^{-1} mol^{-1} (the known value for the analogous reaction of chloride ion). Thus, at 298.2 K,

$$\begin{aligned}\Delta G^0 &= -343{,}000 \text{ J mol}^{-1} - 298.2 \text{ K} \times (-93 \text{ J K}^{-1} \text{ mol}^{-1})\\ &= -315 \text{ kJ mol}^{-1}\end{aligned}$$

and

$$K \times \text{atm} = e^{-\left(\frac{-315{,}000 \text{ J mol}^{-1}}{8.31 \text{ J K}^{-1} \text{ mol}^{-1} \times 298.2 \text{ K}}\right)} = 1.6 \times 10^{55}$$
$$K \cong 2 \times 10^{55} \text{ atm}^{-1}$$

Because each proton affinity generally is known to an accuracy no better than ~ 5 kJ mol^{-1}, an estimated equilibrium constant can be known to an accuracy no better than a factor of ~ 10; a factor of 10 in an equilibrium constant corresponds at 298.2 K to 8.31 J K^{-1} mol^{-1} $\times$ 298.2 K $\times$ ln 10 = 5.7 kJ mol^{-1}.

The qualitative meaning of this value of K is clear: only an essentially perfect vacuum allows protons to exist as such at equilibrium. However, it is possible to establish measurable equilibria between different protonated species in the gas phase if these species have similar proton affinities. For instance, the proton affinities of acetone, $(CH_3)_2CO$, and methyl cyanide, CH_3CN, are:

$$PA[(CH_3)_2CO] = 795 \text{ kJ mol}^{-1}$$
$$PA[CH_3CN] = 778 \text{ kJ mol}^{-1}$$

In a gaseous mixture of acetone and methyl cyanide in which some protons are produced (for instance, by the bombardment of hydrogen-containing molecules with high-energy electrons), equilibrium is established in the reaction

$$CH_3CNH^+(g) + (CH_3)_2CO(g) \rightleftarrows CH_3CN(g) + (CH_3)_2COH^+(g)$$

Study of this equilibrium as a function of temperature gives $\Delta H^0 = -17 \text{ kJ mol}^{-1}$. This value is the difference between the proton affinities of these species:

$$\Delta H^0 = \Delta U = -PA[(CH_3)_2CO] + PA[CH_3CN]$$
$$= -795 \text{ kJ mol}^{-1} + 778 \text{ kJ mol}^{-1} = -17 \text{ kJ mol}^{-1}$$

(Recall from Section 12–2 that the temperature dependence of the equilibrium constant depends upon the value of ΔH.) By combining information from proton-transfer reactions with absolute determinations of the proton affinities of some reference molecules or anions, chemists now have found a large number of proton affinity values. Some of these values are given in Table 12–3.

TABLE 12–3
Proton Affinities (*PA*) of Various Atoms, Molecules, and Anions[a]

Species	*PA* kJ mol^{-1}	Species	*PA* kJ mol^{-1}
H^-	1675	H_2O	686
He	176	NH_3	866
F^-	1548	CH_4	527
Ne	201	C_2H_4	669
Cl^-	1394	C_2H_2	636
Ar	343	CH_3OH	753
OH^-	1630	$(CH_3)_2O$	778
NH_2^-	1690	NH_2CH_3	904
CH_3^-	>1690	$NH(CH_3)_2$	931
H_3O^+	<0	$N(CH_3)_3$	949
H_2	440	Cl_3CCH_2OH	769
HF	548	CH_3CH_2OH	801
HCl	586		

[a]The uncertainty in each of these values is $\pm 5 \text{ kJ mol}^{-1}$

Comparison of isoelectronic pairs of species, for example, F^- and Ne, OH^- and HF, and NH_2^- and H_2O, shows the enormous effect of ionic charge. The proton affinity of each anion in the table is much larger than the proton affinity of any of the neutral molecules. We can assume that the association of a proton by cationic species to form species of charge +2 (e.g., by H_3O^+ to give H_4O^{2+}, which is isoelectronic with methane) is endothermic in the gas phase.

Other more subtle effects of changing the electron distribution within a molecule are revealed in proton affinity values. The effect of substituting a methyl group for hydrogen is seen in two series of related molecules given in Table 12–3: $PA[(CH_3)_2O] > PA(CH_3OH) > PA(H_2O)$, and $PA[(CH_3)_3N] > PA[(CH_3)_2NH] > PA(CH_3NH_2) > PA(NH_3)$. Using these data we can conclude that a methyl group attracts electrons within a molecule less strongly than a hydrogen atom does. In view of the trends in dissociation of chlorine-substituted acetic acid given in Chapter 11, we would expect substitution of chlorine for hydrogen to have the opposite effect. This is confirmed by the comparison $PA(CH_3CH_2OH) > PA(Cl_3CCH_2OH)$. The study of protonation processes in the gas phase promises to be extremely fruitful.

We have not used the terms *acid* and *base* in this discussion, but you certainly remember them from your previous study of chemistry, and recall the role of the proton in these classifications. Acids and bases will be our next subject for study.

BRØNSTED ACIDS AND BASES

We have just considered recent studies of proton association with various species, neutral and anionic, in the gas phase. This field of study has developed in the last two decades. But the type of reaction that is the focus of attention in these recent studies, $XH^+ + Z = X + ZH^+$, is a reaction that has been a central concern of chemists throughout the history of chemistry. This is an *acid-base reaction.* Relatively recently (1923), BRØNSTED has given us a particularly useful definition of acids and bases: *An acid is a proton donor; a base is a proton acceptor.* Thus in the reaction just given, XH^+ and ZH^+ are Brønsted acids and X and Z are Brønsted bases. (Most of the acids and bases we will study are acids and bases in the sense defined by Brønsted, and the prefix Brønsted will be omitted in most of our discussion.) In the framework of these definitions, each acid is related to a base, and each base is related to an acid:

Acid	Base
XH^+	X
ZH^+	Z

It is customary to call X the conjugate base of the acid XH^+ and XH^+ the conjugate acid of the base X; analogous statements can be made about ZH^+ and Z.

The experimental data presented in our discussion of exothermic gas-phase proton-association reactions, suggest that:

> *Any neutral or anionic gaseous atom or molecule is a base.* No neutral or anionic atom or molecule has so little basicity that it will not react exothermically with a proton in the gas phase.

If we consider potentially basic species in liquid solution, no such statement can be made. In solution, the basicity of a species is demonstrable only if the species has a basicity approximately equal to or greater than that of the solvent. Solvation of a base and its conjugate acid influences the extent of protonation of a base in solution. Thus chloride ion in aqueous solution generally is not classified as a base, despite its high proton affinity in the gas phase, because it does not compete effectively with the solvent for protons. The reaction

$$Cl^-(aq) + H_3O^+(aq) = HCl(aq) + H_2O(l)$$

does not occur to an appreciable extent. Ammonia, which has a smaller proton affinity than chloride ion in the gas phase (see Table 12–3), does have demonstrable basicity in aqueous solution. The reaction

$$NH_3(aq) + H_2O(l) \rightleftarrows NH_4^+(aq) + OH^-(aq)$$

does occur to an appreciable extent ($\sim 1.3\%$ in $\sim 0.10 M\ NH_3$), and the reaction

$$NH_3(aq) + H_3O^+(aq) \rightleftarrows NH_4^+(aq) + H_2O(l)$$

occurs almost to completion. The hydration of ions is more exothermic than that of neutral molecules; this is the main reason why chloride ion and ammonia have different relative basicities in aqueous solution and in the gas phase. The greater hydration energy of chloride ion compared to that of hydrogen chloride makes chloride ion a weaker base in aqueous solution; the greater hydration energy of ammonium ion compared to that of ammonia makes ammonia a stronger base in aqueous solution.

Before we can understand the reactions of acids and bases in aqueous solution, we must consider the acid–base relationships of water, which is both an acid and a base:

Acid	Base
H_3O^+	H_2O
H_2O	OH^-

The ionic species related to water by acid–base relationships are hydroxide ion, OH^-, which is isoelectronic with hydrogen fluoride, HF, and hydronium ion, H_3O^+, which is isoelectronic with ammonia, NH_3. The structures of these ions and the isoelectronic neutral molecules are shown in Figure 12–6.

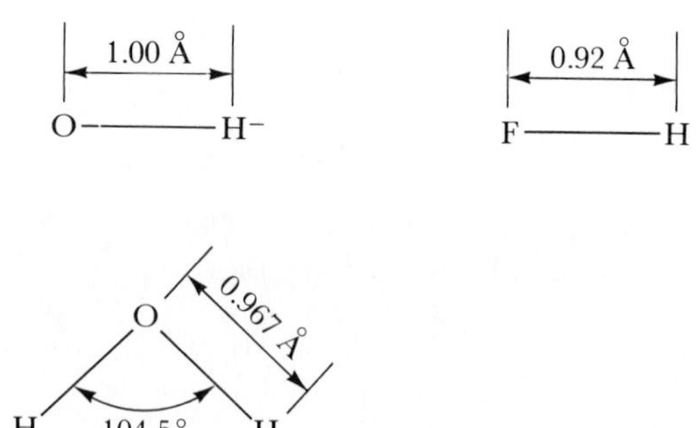

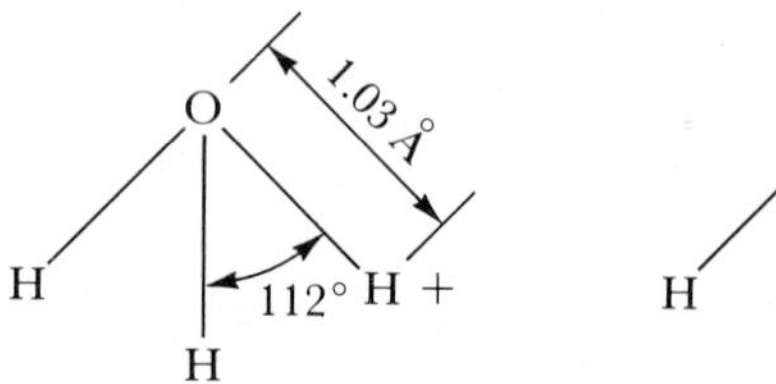

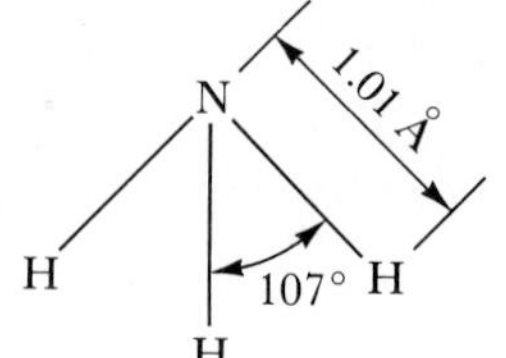

FIGURE 12–6
The structures of water, hydronium ion, and hydroxide ion (and neutral molecules isoelectronic with the ions). Hydronium ion (and ammonia) are pyramidal.

There are many stable crystalline ionic hydroxides, for example, $Na^+(OH^-)$, $K^+(OH^-)$, and $Ba^{2+}(OH^-)_2$.[5] There are some crystalline ionic compounds containing hydronium ion: for example, the crystalline hydrates of hydrogen chloride and hydrogen perchlorate, $HCl \cdot H_2O$ and $HOClO_3 \cdot H_2O$, are actually ionic compounds $(H_3O^+)(Cl^-)$ and $(H_3O^+)(ClO_4^-)$. Hydronium ion and the analogous sulfonium ion, H_3S^+, are found in salts with hexafluoroantimonate ion, $(H_3O^+)(SbF_6^-)$ and $(H_3S^+)(SbF_6^-)$. When ionic substances containing hydroxide ion or hydronium ion dissolve in water, the basic or acidic species of the solvent is formed directly:

$$Na^+(OH^-)(s) = Na^+(aq) + OH^-(aq)$$
$$(H_3O^+)(ClO_4^-)(s) = H_3O^+(aq) + ClO_4^-(aq)$$

Proton hydration in the gas phase involves a series of hydrates, $H(H_2O)_n^+$ with $n = 1, 2, 3, \cdots$, that is, H_3O^+, $H_5O_2^+$, $H_7O_3^+$, $H_9O_4^+$, and so on. Figure 12–7 shows a thermochemical cycle for the hydration of the proton. We see that the largest energy is associated with the first step:

$$H^+(g) + H_2O(g) = H_3O^+(g) \qquad \Delta U = -PA(H_2O) = -686 \text{ kJ mol}^{-1}$$

The hydration energy of hydronium ion in liquid water,

$$H_3O^+(g) = H_3O^+(aq) \qquad \Delta U = -442 \text{ kJ mol}^{-1}$$

is similar to that of other small cations of charge +1 (Li^+, -510 kJ mol^{-1}; Ag^+, -469 kJ mol^{-1}; Na^+, -402 kJ mol^{-1}). The figure shows that this

[5]Here the charges on the constituent ions are shown; however, this is not the convention generally used for the formulas of ionic compounds.

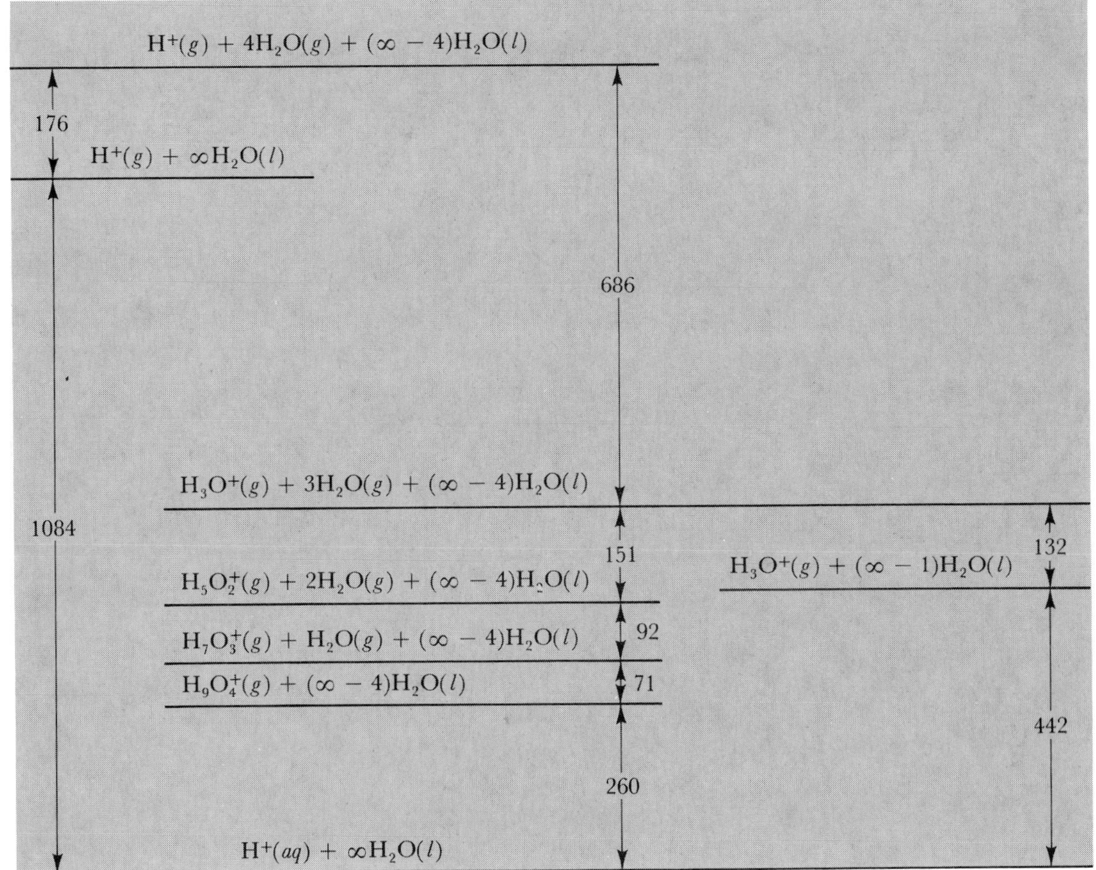

FIGURE 12–7
The hydration of gaseous hydrogen ion. Gaseous hydrates of hydrogen ion with more than four water molecules also are known. (Energies are in kJ mol^{-1}.)

hydration energy of hydronium ion can be resolved into the change in energy for the addition of each water molecule to the growing cluster. However, for most purposes it will not be necessary formally to recognize the hydration of hydronium ion, and the chemical formula used for hydrated hydrogen ion in aqueous solution will be that of hydronium ion, H_3O^+.

The hydration both of hydrogen ion and the basic species affect the dissociation of acids in aqueous solution. This is shown in Figure 12–8, in which the reaction

$$HCl(g) = H^+(aq) + Cl^-(aq)$$

is dissected. In the absence of strongly exothermic solvation steps, hydrogen chloride is undissociated, as it is in the gas phase, in the pure liquid, and in nonpolar solvents (e.g., cyclohexane, see Table 7–3).

Analogous studies of the interaction of hydroxide ion with water in the gas phase reveal species $OH(H_2O)_n^-$, with $n = 1$–5, inclusive. The enthalpy changes are much smaller than they are for hydration of hydrogen

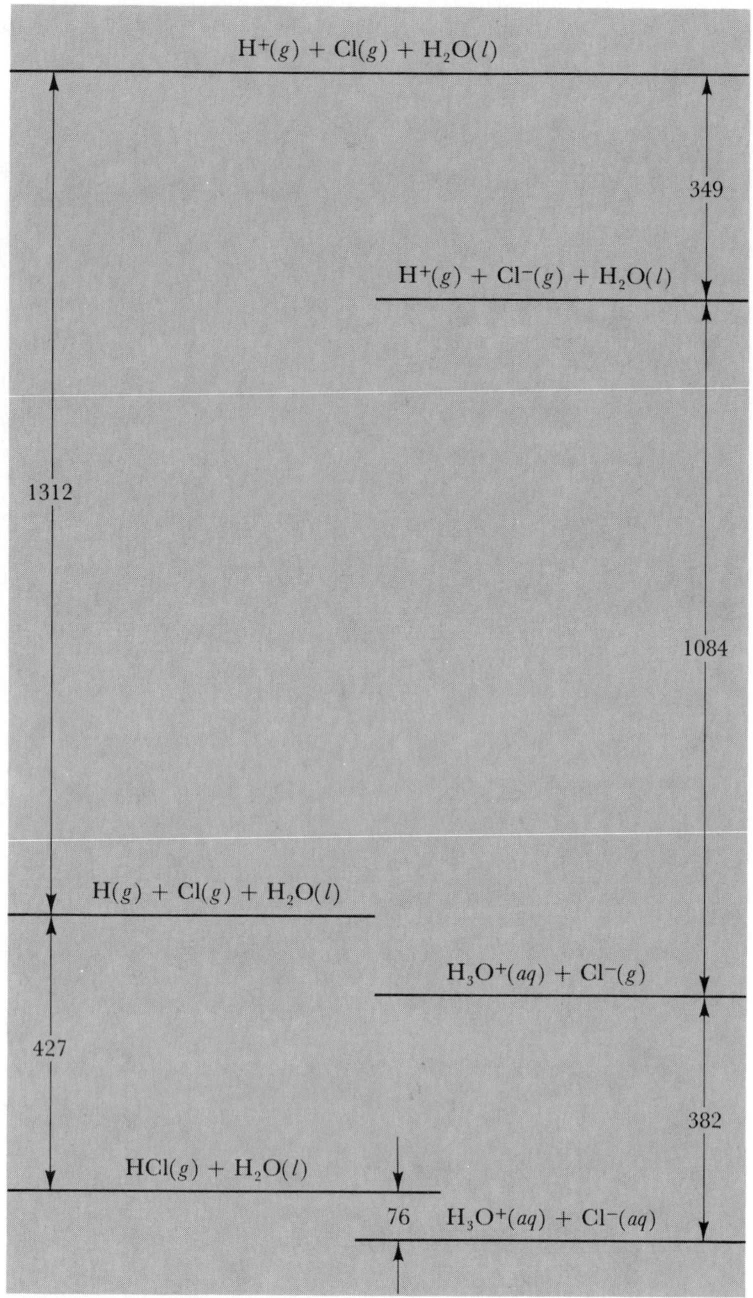

FIGURE 12–8
The thermochemical cycle for the reaction $HCl(g) = H^+(aq) + Cl^-(aq)$. (Energies are in kJ mol^{-1}.)

ion, ranging from $\Delta H = -94$ kJ mol^{-1} for the first step, $OH^-(g) + H_2O(g) \rightarrow OH(H_2O)^-(g)$, to $\Delta H = -59$ kJ mol^{-1} for the fourth and fifth steps. (The values of ΔH for the stepwise hydration reactions of hydroxide ion in the gas phase are approximately the same as those for fluoride ion.)

Table 12–4 presents the chemical formulas of a number of acids and their conjugate bases. The acids in this table are arranged in order of decreasing proton-donating ability. (This ability will be discussed in

TABLE 12–4
Some Acids and Their Conjugate Bases

Acid	Base
$HOClO_3$	ClO_4^-
$(HO)_2SO_2$	$HOSO_3^-$
HCl	Cl^-
H_3O^+	H_2O
$HONO_2$	NO_3^-
$HOSO_3^-$	SO_4^{2-}
HF	F^-
$H_3^+NCH_2CO_2H$ (glycine cation)	$H_3^+NCH_2CO_2^-$ (glycine zwitterion)
$Al(OH_2)_6^{3+}$	$Al(OH_2)_5OH^{2+}$
CH_3CO_2H	$CH_3CO_2^-$
H_2S	HS^-
HCN	CN^-
$H_3^+NCH_2CO_2^-$	$H_2NCH_2CO_2^-$ (glycine anion)
NH_4^+	NH_3
HS^-	S^{2-}
H_2O	OH^-
NH_3	NH_2^-
OH^-	O^{2-}

quantitative terms in the next chapter.) For the acids above hydronium ion (e.g., HCl), the acid–base reaction

$$\underset{\text{Acid}}{HCl(aq)} + \underset{\text{Base}}{H_2O(l)} = \underset{\text{Acid}}{H_3O^+(aq)} + \underset{\text{Base}}{Cl^-(aq)}$$

goes essentially to completion. For bases below hydroxide ion (e.g., amide ion, NH_2^-), the reaction

$$\underset{\text{Base}}{NH_2^-(aq)} + \underset{\text{Acid}}{H_2O(l)} = \underset{\text{Base}}{OH^-(aq)} + \underset{\text{Acid}}{NH_3(aq)}$$

goes essentially to completion. For acids below hydronium ion and above water and for bases above hydroxide ion and below water in the table, the acid–base reactions with water, the solvent, go to measurable but incomplete extents.

There are acid–base reactions in solution that are called by different special names, even though each is an example of a reaction which can be represented:

$$\underset{\text{Acid}}{XH^+} + \underset{\text{Base}}{Z} = \underset{\text{Base}}{X} + \underset{\text{Acid}}{ZH^+}$$

These catagories of acid–base reactions are: solvent self-dissociation (or solvent autoprotolysis), acid dissociation, base dissociation, neutralization, and hydrolysis. Examples and further description of each of these categories follow.

Solvent Self-Dissociation. This type of acid–base reaction occurs in all solvents with appreciable polarity containing molecules in which hydrogen is bonded to an electronegative element. For water, ammonia, and acetic acid the self-dissociation reactions, which occur to measurable extents, are:

Acid		Base		Base		Acid
H_2O	+	H_2O	=	OH^-	+	H_3O^+
NH_3	+	NH_3	=	NH_2^-	+	NH_4^+
CH_3CO_2H	+	CH_3CO_2H	=	$CH_3CO_2^-$	+	$CH_3CO_2H_2^+$

However, in nonpolar cyclohexane the reaction

$$C_6H_{12} + C_6H_{12} = C_6H_{11}^- + C_6H_{13}^+$$

does not occur to a measurable extent. In the autoprotolysis reactions of water, ammonia, and acetic acid, the product ionic species are produced in equal amounts. Therefore these species are present at equal concentrations, both in the pure solvent and in solutions of solutes that do not react preferentially with one of these acidic or basic species derived from the solvent. Solutions with this equality of concentration (e.g., $[H_3O^+] = [OH^-]$) are said to be *neutral.* Solutions in a particular solvent are said to be *acidic* if the concentration of the acidic species derived from the autoprotolysis reaction is greater than the concentration of the basic species so derived. The solutions are said to be *basic* if the opposite inequality holds. The meaning of these terms, applied to each of three particular solvents, water, ammonia, and acetic acid, is summarized:

Acidic	Neutral	Basic
$[H_3O^+] > [OH^-]$	$[H_3O^+] = [OH^-]$	$[H_3O^+] < [OH^-]$
$[NH_4^+] > [NH_2^-]$	$[NH_4^+] = [NH_2^-]$	$[NH_4^+] < [NH_2^-]$
$[CH_3CO_2H_2^+] > [CH_3CO_2^-]$	$[CH_3CO_2H_2^+] = [CH_3CO_2^-]$	$[CH_3CO_2H_2^+] < [CH_3CO_2^-]$

The terms "acidic," "neutral," and "basic" as defined here refer to solutions in a *particular solvent.*

Acid Dissociation. In this type of reaction, an acidic solute donates a proton to the solvent. For instance, the acid-dissociation reactions of hydrochloric acid and perchloric acid:

Acid		Base		Base		Acid
$HCl(aq)$	+	$H_2O(l)$	=	$Cl^-(aq)$	+	$H_3O^+(aq)$
$HOClO_3(aq)$	+	$H_2O(l)$	=	$ClO_4^-(aq)$	+	$H_3O^+(aq)$

go essentially to completion. If the dissociation reaction of an acid HY goes essentially to completion at moderate concentrations [$C(HY) \geq 0.1$ mol L^{-1}], the acid is called a *strong acid.* Some acids that are strong in aqueous solution are given in Table 12–5. In this list we see familiar hydrochloric acid (HCl), nitric acid (HNO_3), and sulfuric acid (H_2SO_4). But

TABLE 12–5
Strong Acids in Aqueous Solution[a]

HCl
HBr
HI
$HClO_4$ $(HOClO_3)$[b]
H_2SO_4 $[(HO)_2SO_2]$[b] dissociation of first proton only
HNO_3 $(HONO_2)$[b]
HSO_3CF_3 $[F_3CSO_2(OH)]$[b]
HSO_3F $(HOSO_2F)$[b]

[a]For these acids the reaction $HX(aq) + H_2O(l) = H_3O^+(aq) + X^-(aq)$ goes to an extent > 97% at 0.10–1.00 mol L^{-1}.
[b]These line formulas correctly represent the fact that hydrogen is bonded to oxygen in these acids.

we also see the other hydrohalic acids, hydrobromic acid and hydroiodic acid (but not hydrofluoric acid, which is not a strong acid), perchloric acid ($HClO_4$), and two acids that are derivatives of sulfuric acid with one hydroxyl group replaced by a trifluoromethyl group ($—CF_3$) or by a fluorine atom. The names of these acids are trifluoromethylsulfuric acid and fluorosulfuric acid. The dissociation of acetic acid, CH_3CO_2H (sometimes represented as HOAc),

$$\underset{\text{Acid}}{CH_3CO_2H(aq)} + \underset{\text{Base}}{H_2O(l)} = \underset{\text{Base}}{CH_3CO_2^-(aq)} + \underset{\text{Acid}}{H_3O^+(aq)}$$

occurs only to a slight extent (in a 0.1 mol L^{-1} solution, ~1.3% of the acetic acid molecules are dissociated). If an acid-dissociation reaction does not go essentially to completion at moderate concentrations ($C \cong 0.1$ mol L^{-1}), the acid is called a *weak acid.*

Base Dissociation. This is a convenient term because of its similarity to the term acid dissociation, but the term is slightly misleading. The "base dissociation" of ammonia,

$$\underset{\text{Acid}}{H_2O} + \underset{\text{Base}}{NH_3} = \underset{\text{Base}}{OH^-} + \underset{\text{Acid}}{NH_4^+}$$

does not involve dissociation of the base, but is an acid–base reaction of the type being discussed. This reaction occurs only to a slight extent, and ammonia is called a *weak base.* The common *strong bases* are alkali-metal hydroxides and alkaline-earth hydroxides. The dissolving of these substances in water is a dissociation, but direct interaction with water is not shown in the conventional representation. For example, we would write the dissociation of sodium hydroxide in water

$$Na^+(OH^-)(s) = Na^+(aq) + OH^-(aq)$$

Table 12–6 gives the solubilities of these strong bases in water.

TABLE 12–6
Strong Bases in Aqueous Solution[a]

	Concentration of saturated solution at 25°C in mol L^{-1}
LiOH	5.1
NaOH	20.7
KOH	14.8
$Ca(OH)_2$	0.020
$Ba(OH)_2$	0.273

[a] Dissociation is essentially complete at $C = 0.10$ mol L^{-1}; therefore $[OH^-] \cong C_{MOH}$ or $2C_{M(OH)_2}$.

Neutralization. This term describes an acid–base reaction in which a solute acid and a solute base react essentially to completion. The neutralization of an aqueous solution of hydrochloric acid (a strong acid) with an aqueous solution of sodium hydroxide (a strong base) is represented by the chemical equation

$$\underset{\text{Acid}}{H_3O^+} + \underset{\text{Base}}{OH^-} = \underset{\text{Base}}{H_2O} + \underset{\text{Acid}}{H_2O}$$

The neutralization of an aqueous solution of acetic acid (a weak acid) with an aqueous solution of sodium hydroxide is represented by the chemical equation

$$\underset{\text{Acid}}{CH_3CO_2H} + \underset{\text{Base}}{OH^-} = \underset{\text{Base}}{CH_3CO_2^-} + \underset{\text{Acid}}{H_2O}$$

The neutralization of an aqueous solution of hydrochloric acid with an aqueous solution of ammonia (a weak base) is represented by the chemical equation

$$\underset{\text{Acid}}{H_3O^+} + \underset{\text{Base}}{NH_3} = \underset{\text{Base}}{H_2O} + \underset{\text{Acid}}{NH_4^+}$$

In each of these chemical equations representing a neutralization reaction, the reactant solute acid and base are represented by chemical formulas for the acidic and basic species that are predominant in the solutions prior to reaction. (The conventions for writing chemical equations for the net reaction were described in Chapter 2; these conventions should be reviewed.)

The neutralization of the weak acid acetic acid and the weak base ammonia is represented by the chemical equation

$$\underset{\text{Acid}}{CH_3CO_2H} + \underset{\text{Base}}{NH_3} = \underset{\text{Base}}{CH_3CO_2^-} + \underset{\text{Acid}}{NH_4^+}$$

Despite the fact that both acid and base are weak, this reaction occurs to more than 99% in aqueous solution, and "neutralization" is an appropriate term for it. If, however, the solute acid and the solute base are both

TABLE 12–7
Types of Acid–Base Reactions[a]

Type	Acid	+	Base	=	Base	+	Acid
Acid dissociation	CH_3CO_2H	+	H_2O	=	$CH_3CO_2^-$	+	H_3O^+
	NH_4^+	+	H_2O	=	NH_3	+	H_3O^+
Autoprotolysis	H_2O	+	H_2O	=	OH^-	+	H_3O^+
	CH_3OH	+	CH_3OH	=	CH_3O^-	+	$CH_3OH_2^+$
Base dissociation	H_2O	+	NH_3	=	OH^-	+	NH_4^+
	H_2O	+	$CH_3CO_2^-$	=	OH^-	+	CH_3CO_2H
Neutralization	H_3O^+	+	OH^-	=	H_2O	+	H_2O
	CH_3CO_2H	+	OH^-	=	$CH_3CO_2^-$	+	H_2O
	H_3O^+	+	NH_3	=	H_2O	+	NH_4^+
Hydrolysis	H_2O	+	$CH_3CO_2^-$	=	OH^-	+	CH_3CO_2H
	NH_4^+	+	H_2O	=	NH_3	+	H_3O^+

[a] Brønsted acid + Brønsted base = Brønsted base + Brønsted acid.

very weak, the extent of the reaction may be so small that we cannot say neutralization has occurred. A solution of ammonium chloride (containing the acid NH_4^+) reacts with one of sodium acetate (containing the base $CH_3CO_2^-$) to less than 1%:

$$\underset{\text{Acid}}{NH_4^+} + \underset{\text{Base}}{CH_3CO_2^-} = \underset{\text{Base}}{NH_3} + \underset{\text{Acid}}{CH_3CO_2H}$$

It can hardly be said that this acid and base have neutralized one another. (Notice that this reaction is the reverse of the one given previously.)

Hydrolysis. The term "hydrolysis" has been used to describe the reaction of certain salts with water to produce an acidic or basic solution. Sodium carbonate is said to hydrolyze in solution; solutions of this salt contain hydroxide ion due to the reaction

$$\underset{\text{Acid}}{H_2O} + \underset{\text{Base}}{CO_3^{2-}} = \underset{\text{Base}}{OH^-} + \underset{\text{Acid}}{HOCO_2^-}$$

(This reaction makes sodium carbonate, or washing soda, useful as a detergent.) The reaction producing hydroxide ion can equally well be viewed as the acid–base reaction of water (an acid) with carbonate ion (a base). Solutions of ammonium salts are slightly acidic: ammonium chloride, NH_4Cl, is an example. This salt, the salt of a strong acid (hydrochloric acid) and a weak base (ammonia), is said to hydrolyze to give an acidic solution. The reaction

$$\underset{\text{Acid}}{NH_4^+} + \underset{\text{Base}}{H_2O} = \underset{\text{Base}}{NH_3} + \underset{\text{Acid}}{H_3O^+}$$

is, in fact, simply an acid–base reaction, and it could equally well be called the acid–dissociation reaction of the weak acid, ammonium ion.

Aqueous solutions of salts of many metal ions with a charge $\geq +2$ are acidic. Although it can be said that salts of aluminum(III) ion hydrolyze, it is preferable to recognize that aluminum(III) ion in aqueous solution is present as hexaaquaaluminum(III) ion (see Chapters 14 and 18), and this species is an acid. The reaction responsible for the acidity of solutions of salts of aluminum(III) ion is[6]

$$\underset{\text{Acid}}{Al(OH_2)_6^{3+}} + \underset{\text{Base}}{H_2O} = \underset{\text{Base}}{Al(OH_2)_5OH^{2+}} + \underset{\text{Acid}}{H_3O^+}$$

The several types of acid–base reactions that we have just discussed are summarized in Table 12–7.

12–4 The Hydrogen Bond

We first discussed the hydrogen bond in Chapter 1, in our description of the structures of liquid and solid water. In our later discussion of liquids, we also saw manifestations of hydrogen bonding; recall that liquid water and liquid ethyl alcohol have abnormally high values for the change of entropy on vaporization (Table 6–4). Now we shall consider additional properties of matter that show the effects of hydrogen bonding.

Hydrogen bonding, X----H----Y, may occur with atoms of X and Y that have a range of values of electronegativity, but the bonding is strongest if X and Y are relatively electronegative atoms in the second row of the periodic table (i.e., N, O, and F). As a gas, hydrogen fluoride deviates greatly from ideal gas behavior, the value of PV/nRT being much less than unity except at high temperatures and low pressures. The dependence of vapor density upon pressure has suggested the equilibria

$$n\mathrm{HF}(g) \rightleftarrows (\mathrm{HF})_n(g)$$

in which $n = 2$ through 6. The hexamer $(HF)_6$ is the predominant species, and a structural study has shown this molecule to have a puckered ring structure (Figure 12–9).

A very strong hydrogen bond is formed between fluoride ion and hydrogen fluoride. The enthalpy change in the reaction

$$\mathrm{F^-}(g) + \mathrm{HF}(g) = \mathrm{FHF^-}(g)$$

has been estimated at $-\Delta H \cong 150$–160 kJ mol^{-1}, thus making this the strongest known hydrogen bond. Other hydrogen dihalide ions are known, generally in salts containing large cations of charge +1, for example, $N(C_4H_9)_4^+$ (tetra-*n*-butylammonium ion). Data for these ions are summarized in Table 12–8. We see that the distance between the two halogen atoms in hydrogen dihalide ions is smaller than the closest distance between two halide ions in ionic crystals (see Table 9–1). This bonding effect is

[6]Depending upon the concentration of the aluminum(III) salt, other reactions are possible.

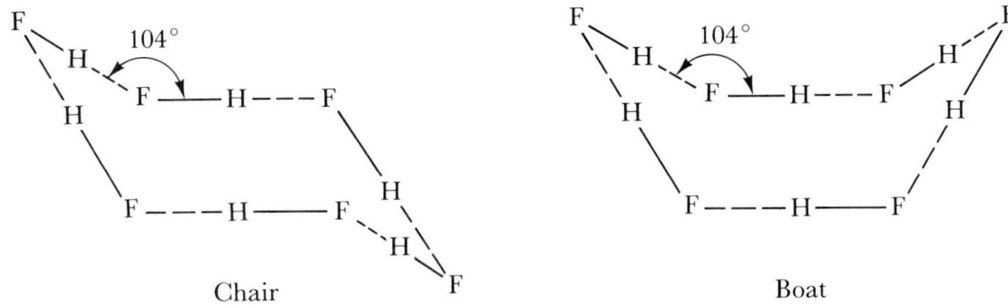

FIGURE 12–9
The structure of the hexamer of hydrogen fluoride, $(HF)_6$. This molecule is not rigid, and the chair ⇄ boat transformation may occur rapidly.

understandable if one considers the hydrogen dihalide ion as two halide ions in contact with a proton inserted at the midpoint of the axis connecting the two halide ions, as depicted in Figure 12–10 for HCl_2^-. The proton pulls the two chloride ions closer together.

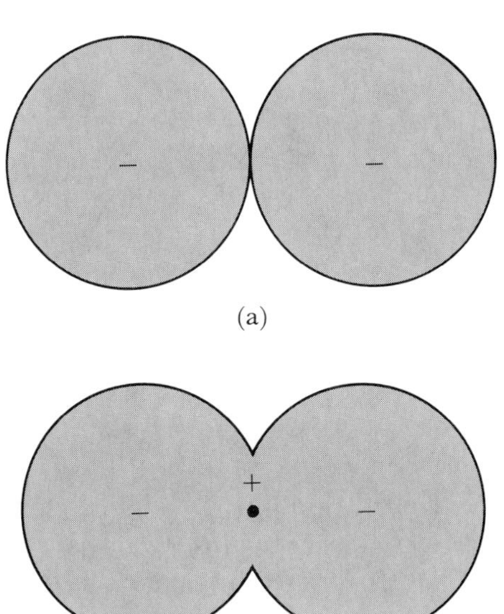

FIGURE 12–10
The hydrogen dichloride ion, $ClHCl^-$. (a) Two chloride ions in contact [$r_{Cl^-} = 1.70$ Å, $d(Cl^-—Cl^-) = 3.40$ Å]. (b) A proton is inserted between the two chloride ions [$d(Cl^-—Cl^-) = 3.14$ Å].

TABLE 12–8
Hydrogen Dihalide Ions, X—H—X$^-$

Ion	X—X distance / Å	$2 \times r_{X^-}$[a] / Å	$-\Delta H^b$ / kJ mol^{-1}
FHF^-	2.26	2.38	150–160
$ClHCl^-$	3.14	3.40	60
$BrHBr^-$	3.35	3.74	50
IHI^-	3.80	4.24	50

[a] r_{X^-} is the ionic radius for halide ion, from Table 9–1.
[b] $-\Delta H$ estimated for process $HX(g) + X^-(g) = XHX^-(g)$.

For most hydrogen bonds, X---H---Y, the hydrogen nucleus is not located midway between the X and Y atoms. There are exceptions, hydrogen difluoride ion being one, and the hydrogen bond in sodium hydrogen diacetate, $NaH(CH_3CO_2)_2$, being another. In this salt, acetate ions are hydrogen-bonded together.

The hydrogen bonding of acetate units is known also for acetic acid in the gas phase, and in solutions in nonpolar solvents (e.g., hydrocarbons) in which there is an equilibrium between monomeric acid and dimeric acid:

$$2CH_3CO_2H \rightleftarrows (CH_3CO_2H)_2$$

The dimer has a cyclic structure with two hydrogen bonds:

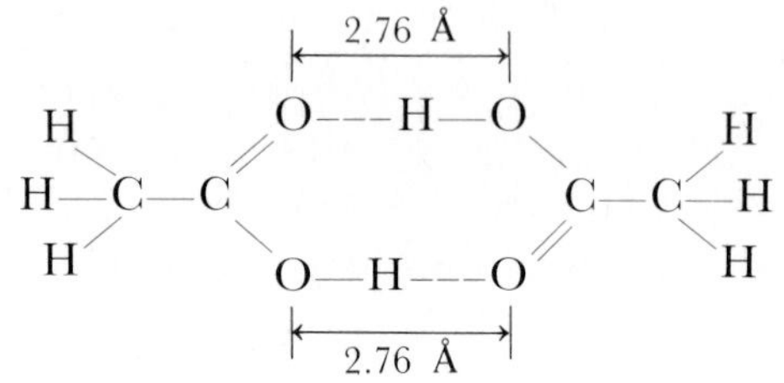

```
            |← 2.76 Å →|
   H        O---H—O          H
    \      //      \        /
 H—C—C              C—C—H
    /      \       //       \
   H        O—H---O          H
            |← 2.76 Å →|
```

The oxygen–oxygen distance in this dimer (2.76 ± 0.05 Å) is the same as that in ice (2.75 Å). This dimerization does not occur to an appreciable extent in aqueous solution because the monomeric solute molecules are hydrogen-bonded preferentially to the solvent molecules, which are in large excess.

In the valence-shell electron-pair repulsion model for the structure of water (see Figure 10–18), the two lone pairs of electrons are oriented relative to each other and to the two O–H bonds at angles that are approximately tetrahedral. This orientation is consistent with the structure of ice (see Figures 1–7 or 5–1), in which each hydrogen atom is covalently bonded to one oxygen atom and hydrogen-bonded to another. (In ice, the hydrogen atoms are not midway between the oxygen atoms.) The site of this hydrogen bonding is an unshared electron pair of the oxygen atom. The structure of the water molecule, with two hydrogen atoms and two unshared pairs of electrons per molecule, is particularly suitable for a three-dimensional network of hydrogen bonds such as is found in ice. A particularly simple way to estimate the energy of the hydrogen bonds in ice is to attribute the entire change of internal energy upon sublimation of ice (48.8 kJ mol^{-1}) to the energy of the hydrogen bonds. When one mole of ice is vaporized to give separated molecules of H_2O, two moles of hydrogen bonds are broken; thus this estimate of the energy of the bond

$$\mathrm{O{-}{-}{-}H{-}O}$$

is ~24 kJ mol^{-1}. This hydrogen bond between neutral molecules is appreciably weaker than that in hydrogen dihalide ions (see Table 12–8) or in $H(OH_2)_2^+$ (see Figure 12–7):

$$H_2O(g) + H_3O^+(g) = (H_2O)H(OH_2)^+(g) \qquad \Delta U = -151 \text{ kJ mol}^{-1}$$

The hydrogen-bonded structure of ice has much open space. When ice melts, this open structure partially collapses, thereby making liquid water more dense than ice. With a further increase of temperature above the melting point, the density first increases and then decreases, the temperature of maximum density being 4°C; this is shown in Figure 12–11. The conflicting factors that give rise to this maximum density at 4°C are the increasing thermal motion of the molecules, which would make the density decrease with increasing temperature, and the breaking of some hydrogen bonds with increasing temperature, which causes the open structure to collapse, thereby making the density greater.

Hydrogen bonding does not necessarily involve the interaction of a hydrogen atom in a single X—H molecule with an unshared pair of electrons in Y. In solid ammonia, NH_3, each nitrogen atom is surrounded by six hydrogen atoms. Three are close, at a distance nearly the same as in an isolated ammonia molecule (1.01 Å), and three are farther away (2.37 Å). These latter three hydrogen atoms belong to three different ammonia molecules. The need for efficient packing of molecules in the solid may be responsible for this arrangement.

Hydrogen bonding, which plays an important role in determining the properties of liquid water, also plays an important role in determining solubilities in water as a solvent. Substances that can form hydrogen bonds with water are more soluble than otherwise might be expected. With dipole moments of 2.04 D and 1.68 D, *n*-propyl iodide and *n*-propyl alcohol provide an interesting contrast. Despite its higher polarity, *n*-propyl iodide is less soluble in water than is *n*-propyl alcohol. The solubility of *n*-propyl

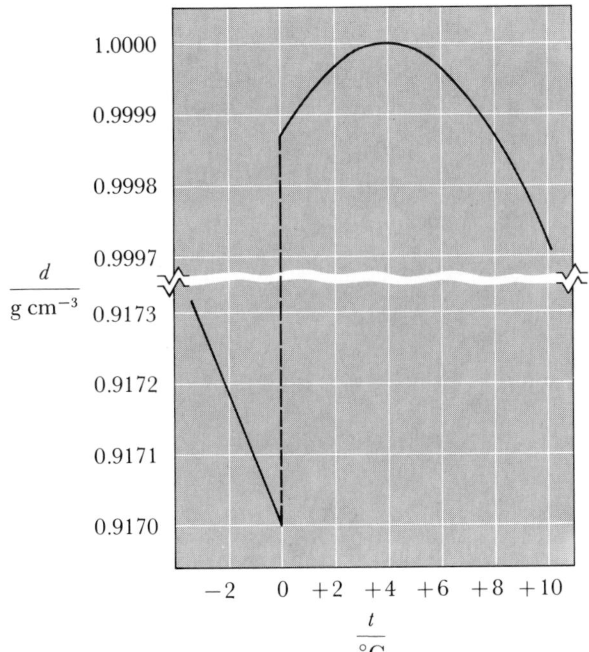

FIGURE 12–11
The density of water as a function of temperature.

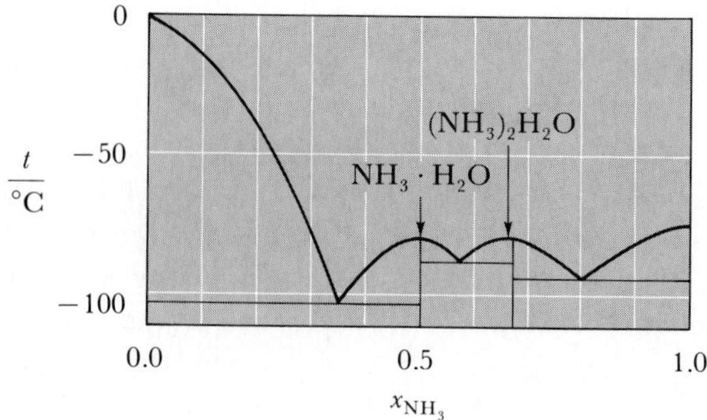

FIGURE 12–12
A freezing-temperature diagram for the system NH_3–H_2O.

iodide, with little tendency for hydrogen bonding, at 20°C is only 0.0054 mol L^{-1}, but *n*-propyl alcohol, with a hydroxyl group for hydrogen bonding, is miscible with water in all proportions.

The high solubility of ammonia in water ($x_{NH_3} = 0.48$ at $P_{NH_3} =$ 1.00 atm, 0.0°C) results from the hydrogen bonding between molecules of water and ammonia. The freezing-point diagram for this system (given in Figure 12–12) shows the existence of solid hydrates of ammonia $NH_3 \cdot H_2O$ and $(NH_3)_2H_2O$. However, the low melting points of these compounds (~ -80°C) and the vapor-pressure data for the solutions at ordinary temperatures suggest that discrete ammonia hydrate molecules do not exist as such in solution at room temperature. We will represent ammonia in aqueous solution as $NH_3(aq)$, not as ammonium hydroxide, NH_4OH.

A quantitative comparison of the solubilities of different liquids in water does not give the relative hydrogen bonding of water with the different substances directly. Different extents of hydrogen bonding in the pure liquid solutes also play a role in determining the solubilities. This can be illustrated by comparing the solubilities of isoamyl alcohol and ethyl propionate given in Table 12–9. At first sight, it will seem surprising that the alcohol is only slightly more soluble than the ester because molecules of the alcohol have a much greater ability to form hydrogen bonds with water molecules. But, compared to molecules of the ester, molecules of the alcohol also have a greater tendency to form hydrogen bonds with each

TABLE 12–9
Solubilities in Water at 20°C[a]

Solute	Solution saturated with pure liquid solute	Solution saturated with solute at pressure = 1 torr
Isoamyl alcohol[b]	0.32 mol L^{-1}	0.14 mol L^{-1}
Ethyl propionate[c]	0.23 mol L^{-1}	0.0083 mol L^{-1}

[a]From J. H. Hildebrand and R. L. Scott, "The Solubility of Non-Electrolytes", 3rd ed., Dover Publishing Co., New York, 1964, p. 264.
[b]The vapor pressure of pure isoamyl alcohol at 20°C is 2.3 torr.
[c]The vapor pressure of pure ethyl propionate at 20°C is 27.8 torr.

other in the pure liquid state. This is reflected in the much lower vapor pressure of the alcohol. A more rational comparison of the solubilities of isoamyl alcohol and ethyl propionate in water is a comparison in which the partial pressure of each solute is the same. (The solubility at any particular pressure can be calculated from the solubility of the pure liquid solute of known vapor pressure, under the assumption that Henry's law is obeyed for solutions of this solute.) We see in Table 12–9 that comparison of the solubilities when the partial pressure of solute vapor is 1 torr shows the solubility of the alcohol is 16.7-fold greater. This comparison shows the effect of hydrogen bonding between the solute alcohol and the solvent water, which comparison of solubilities of the liquids did not show.

12–5 Hydrogen-Bridged Electron-Deficient Molecules

Our discussion of the proton affinity of molecules presented the fact that molecular hydrogen, H_2, reacts exothermically with a proton:

$$H_2(g) + H^+(g) = H_3^+(g) \qquad \Delta U = -440 \text{ kJ mol}^{-1}$$

The product species of this process, H_3^+, is an interesting molecular entity; it has a total of two electrons, and it has three nuclei. Theoretical studies indicate that the structure of this species is an equilateral triangle. This structure is the prototype of a chemical bond called a *three-center bond*, three atoms bonded together by one pair of electrons. The species H_3^+ can be called *electron-deficient* because there are fewer valence-shell electrons (2) than there are stable valence-shell orbitals (3) in the atoms involved.

A large class of electron-deficient molecules are the boron hydrides. Before discussing these compounds, we should recognize that boron atoms, which have three valence electrons,

$$\text{B} \qquad 1s^2 2s^2 2s^1$$

form some compounds that have conventional electron-pair bonds. The planar structure of boron trifluoride is consistent with an electronic structure using only conventional two-center bonds. The correct structure probably is a resonance hybrid between a structure in which the boron atom shares three pairs of electrons (shown in Figure 10–17) and three equivalent structures in which the boron atom shares four pairs of electrons (the carbonate ion structure, see Section 10–5). Boron trifluoride reacts with alkali-metal fluoride to give ionic compounds containing tetrafluoroborate anion, BF_4^-,

$$Na^+F^-(s) + BF_3(g) = Na^+BF_4^-(s)$$

which is a tetrahedral ion, isoelectronic with carbon tetrafluoride. An analogous borohydride ion, BH_4^-, is isoelectronic with methane and ammonium ion. Neither of the anions BF_4^- or BH_4^- is electron-deficient.

Boron trifluoride and sodium borohydride react to give a boron hydride, B_2H_6, called diborane:

$$3NaBH_4 + 4BF_3 = 3NaBF_4 + 2B_2H_6$$

Diborane is the simplest stable boron hydride; a monoborane, BH_3, has only transitory existence. The correspondence between the formulas of diborane (B_2H_6) and ethane (C_2H_6) first led chemists to suggest that diborane had the ethane structure. In ethane there are fourteen valence electrons, $(2 \times 4) + (6 \times 1)$, and each line in the formula

```
    H  H
    |  |
 H—C—C—H
    |  |
    H  H
```

represents a two-center electron-pair bond. In electron-deficient diborane there are only twelve valence electrons, $(2 \times 3) + (6 \times 1)$. If diborane had the ethane molecular geometry, clearly it could not have the ethane electronic structure. But speculation regarding the ethanelike structure for diborane stopped in the 1940s, when investigations showed that diborane has the structure given in Figure 12–13. This structure involves two three-center bonds. The six pairs of valence electrons in diborane are used in bonding as follows:

4 pairs in 4 B—H sigma bonds

2 pairs in 2 B–H–B three-center bonds

Diborane and other boron hydrides, itemized in Table 12–10, all electron-deficient, were synthesized by ALFRED STOCK and his associates in the period 1912–1936. The original preparation of the boron hydrides involved the reaction of magnesium boride, MgB_2, with acid; for example,

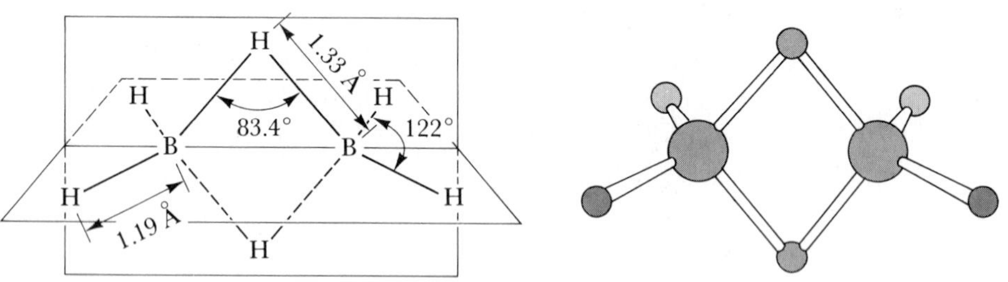

FIGURE 12–13
Two representations of the structure of diborane. Valence-electron bookkeeping:

Electrons available: $(2 \times 3) + (6 \times 1) = 12$

4 B—H bonds: $4 \times 2 = 8$
2 B—H—B bonds: $2 \times 2 = 4$
12

TABLE 12–10
The Hydrides of Boron Synthesized by Stock

Formula	Name	mp °C	bp °C
B_2H_6	Diborane(6)	−164.9	−92.6
B_4H_{10}	Tetraborane(10)	−120	18
B_5H_9	Pentaborane(9)	−46.6	48
B_5H_{11}	Pentaborane(11)	−123	63
B_6H_{10}	Hexaborane(10)	−62.3	108
$B_{10}H_{14}$	Decaborane(14)	+99.7	213

for hexaborane(10) the balanced equation is

$$7MgB_2 + 14H_3O^+ = 2B_6H_{10}(g) + 7Mg^{2+} + 2B(OH)_3 + 8H_2O$$

This reaction is not the only one occurring in the reaction mixture; other boron hydrides also are produced under these conditions. The boron atoms in magnesium boride are arranged in negatively charged graphitelike sheets. (The boron atoms in magnesium boride MgB_2 have an average charge of -1; B^- is isoelectronic with C, and the sheets of boron atoms in a hexagonal array have the same structure as the sheets of carbon atoms in graphite.) These sheets of anionic boron atoms are attacked by hydronium ion through a mechanism that has not been explained, but which results in a variety of products.

Figure 12–14 shows the structures of pentaborane(9) and decaborane-(14). If you count the lines (the atomic connections) in these structures, you will find that the number exceeds the number of electron pairs available for bonding. For pentaborane(9), there are twenty-four valence electrons, $(5 \times 3) + (9 \times 1)$. These twelve electron pairs hold the molecule together as follows:

5 pairs in 5 B—H sigma bonds (one for each boron atom)

4 pairs in 4 B(H)B three-center bonds (one for each edge of the base of the pyramid of boron atoms)

3 pairs in 1 B(B)B three-center bond plus 2 B—B sigma bonds, which bond the top boron atom to the four basal atoms. (Four equivalent structures can be drawn and the true structure is a resonance hybrid of these.)

LIPSCOMB studied this compound and other boron hydrides through x-ray crystallography, and did much to describe their electronic structures.

In decaborane(14) the framework of boron atoms is much the same as in the dodecahedron of twelve boron atoms in the tetragonal form of elemental boron, shown in Figure 5–18. Two adjacent boron atoms of the dodecahedron are replaced by four hydrogen atoms that are involved in

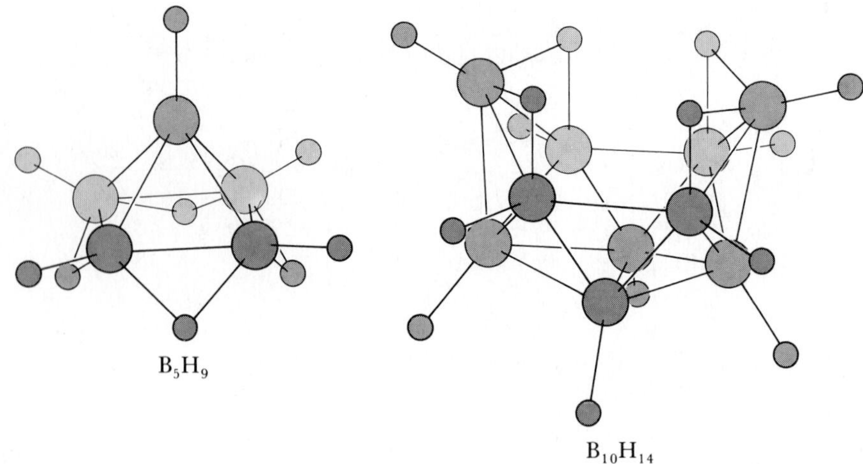

FIGURE 12–14
The structures of pentaborane(9) and decaborane(14). Valence-electron bookkeeping:

	B_5H_9	$B_{10}H_{14}$
Electrons available:	$(5 \times 3) + (9 \times 1) = 24$	$(10 \times 3) + (14 \times 1) = 44$
B—H bonds:	$5 \times 2 = 10$	$10 \times 2 = 20$
B—H—B bonds:	$4 \times 2 = 8$	$4 \times 2 = 8$
B atom framework:	6	16
	24	44

three-center bonding. In tetragonal boron, each atom in a dodecahedron also is bonded to a boron atom that is not part of that dodecahedron. In decaborane(14), the ten peripheral hydrogen atoms satisfy this bonding affinity of boron.

There are other electron-deficient compounds that involve bridging hydrogen atoms. In beryllium borohydride, $Be(BH_4)_2$, and aluminum borohydride, $Al(BH_4)_3$, there are three-center bonds, M⟋H⟍B (M = Be or Al).

The diversity of hydrogen chemistry has been illustrated in this chapter, and also you have been introduced to quantitative calculations regarding chemical equilibrium. You will continue such calculations in the next chapter.

Biographical Notes

JOHANNES N. BRØNSTED (1879–1947), Professor of Physical Chemistry in the University of Copenhagen, made important contributions to many areas of chemistry, particularly to the study of acids and bases in catalysis of reactions in solution.

WILLIAM N. LIPSCOMB (1919–), an American physical chemist, was educated at the University of Kentucky (B.S., 1941) and the California Institute of Technology (Ph.D., 1946). He has been on the faculties of the University of Minnesota and Harvard University. He received the Nobel Prize in Chemistry in 1976 for his work on boron hydrides, and he has done outstanding work on the structure of proteins (see Figure 5–8).

ALFRED E. STOCK (1876–1946), a German chemist, was born in Danzig. He studied in Berlin under Emil Fischer, the outstanding organic chemist of his day. Stock's pioneering use of high-vacuum techniques made possible the study of air-sensitive compounds, such as the boron hydrides.

Problems and Questions

12–1 Using data available in Chapters 8 and 10, devise a cycle that allows you to evaluate the energy change for the process $H_2(g) = H^+(g) + H^-(g)$.

12–2 A nonstoichiometric hydride of lanthanum has the composition $LaH_{2.87}$. If this compound contains both La^{2+} and La^{3+}, what fraction of the lanthanum ions are La^{2+}?

12–3 The enthalpy change in the reaction $2H_2(g) + O_2(g) = 2H_2O(g)$ at 298.2 K is $\Delta H^0 = -483.7$ kJ mol^{-1}. What is the value of ΔU^0?

12–4 In which of the following gas-phase equilibria is the yield of products increased by increasing the total pressure on the reaction mixture?

$$CO(g) + H_2O(g) \rightleftarrows CO_2(g) + H_2(g)$$

$$2NO(g) + Cl_2(g) \rightleftarrows 2NOCl(g)$$

$$2SO_3(g) \rightleftarrows 2SO_2(g) + O_2(g)$$

12–5 For the reaction $PCl_5(g) = PCl_3(g) + Cl_2(g)$ the values of relevant thermodynamic quantities for 298.2 K for reactants and products are:

	$\frac{\Delta H_f^0}{\text{kJ mol}^{-1}}$	$\frac{S^0}{\text{J K}^{-1}\text{ mol}^{-1}}$
$PCl_5(g)$	−398.94	352.7
$PCl_3(g)$	−306.35	311.7
$Cl_2(g)$	0	222.9

(The standard state of each gas is the gas at a pressure of 1 atm.) (a) What is the value of ΔG^0 at 298.2 K? (b) What is the value of K at 298.2 K? Give the value of K using each of the pressure units atm, torr, and pascals. (c) If the values of ΔH^0 and ΔS^0 for this reaction do not vary with temperature, at what temperature will $K = 1$ atm?

12–6 A 0.100-mol sample of $PCl_5(g)$ is heated to 480 K in a vessel with a volume of 12.0 L. What are the partial pressures of each gas at equilibrium? [As in Problem 12–5(c) assume that ΔH^0 and ΔS^0 do not depend upon temperature.]

12–7 Instead of placing 0.100 mol PCl_5 in a 12.0-L vessel (Problem 12–6), the sample is placed in a 3.0-L vessel. What is the partial pressure of each gas at equilibrium? What is the total pressure in each of these vessels, and what fraction of the PCl_5 is dissociated in each?

12–8 For the reaction $2NOCl(g) \rightleftarrows 2NO(g) + Cl_2(g)$ the value of the equilibrium constant at 298.2 K is 4.40×10^{-8} atm, and the value of ΔH^0 is 76.0 kJ mol^{-1}. What is the value of ΔS^0 for the reaction? If the values of molar entropies for chlorine and nitric oxide at 298.2 K are 223 J K^{-1} mol^{-1} and 211 J K^{-1} mol^{-1}, respectively, what is the molar entropy of nitrosyl chloride (NOCl) at this temperature?

12–9 Construct a graph of $\Delta G/RT$ versus $\ln[(P_{Cl_2} \times P_{NO}^2/P_{NOCl}^2) \times atm^{-1}]$ for the reaction $2NOCl(g) = 2NO(g) + Cl_2(g)$ analogous to Figure 12–2. The necessary data are given in Problem 12–8. Locate on this graph $\Delta G^0/RT$ and $\ln(K \times atm^{-1})$. What is the value of ΔG for this reaction if each reactant and product is present at a pressure of 1.00 torr?

12–10 At 298.2 K the value of K for the ammonia-synthesis reaction is 5.3×10^5 atm^{-2}. A sample of pure ammonia is placed in a constant-volume vessel. What should the initial pressure of ammonia be if it is 50% decomposed at equilibrium?

12–11 At 298.2 K a gaseous mixture has the initial partial pressures $P_{NH_3} =$ 0.0200 atm, $P_{N_2} = 0.0100$ atm, and $P_{H_2} = 0.00500$ atm. The mixture is allowed to reach equilibrium in a constant-volume vessel. What are the partial pressures at equilibrium?

12–12 At high temperatures the reaction $P_4(g) = 2P_2(g)$ occurs to a measurable extent; values of the equilibrium constant as a function of temperature are 1173 K, $K = 0.00645$ atm; 1273 K, $K = 0.0375$ atm; 1373 K, $K =$ 0.1671 atm; and 1473 K, $K = 0.6118$ atm. From these data determine the values of ΔH^0 and ΔS^0_{atm} for this range of temperature. At 1373 K with a total pressure of 1.00 atm, what fraction of P_4 is dissociated?

12–13 A sample of gaseous nitrosyl bromide (NOBr) with a pressure of 0.0515 atm at 298.2 K has a density of 0.1861 g L^{-1}. This density is lower than expected for pure NOBr; the explanation lies in the dissociation reaction,

$$2NOBr(g) \rightleftarrows 2NO(g) + Br_2(g)$$

From the gas-density data, calculate the equilibrium constant for this reaction.

12–14 To take into account the state of iodine, the equation for the formation of hydrogen iodide from the constituent elements should be written as one of the following, depending upon the temperature:

(a) $H_2(g) + I_2(s) = 2HI(g)$ (c) $H_2(g) + I_2(g) = 2HI(g)$
(b) $H_2(g) + I_2(l) = 2HI(g)$ (d) $H_2(g) + 2I(g) = 2HI(g)$

(a) at $T < 387$ K, where I_2 is a solid, (b) at 387 K $< T <$ 456 K, where I_2 is a liquid, (c) at 456 K $< T <$ 1500 K, where I_2 is a gas, and (d) at $T >$ 1500 K, where I_2 has dissociated into atoms. Predict the relative magnitude of ΔH^0 and ΔS^0 for each of these reactions.

12–15 The gas-phase species HeH^+, NeH^+, ArH^+, KrH^+, and XeH^+ are isoelectronic with what neutral species?

12–16 What is the enthalpy change in each of the gas-phase proton-transfer reactions?

$$(CH_3)_2OH^+(g) + CH_3CH_2OH(g) = (CH_3)_2O(g) + CH_3CH_2OH_2^+(g)$$
$$F^-(g) + H_2(g) = HF(g) + H^-(g)$$
$$CH_3^-(g) + NH_3(g) = CH_4(g) + NH_2^-(g)$$

12–17 The energy required to dissociate a hydrogen-containing compound in the gas phase into a hydrogen atom and a neutral radical, $HY(g) = H(g) + Y(g)$, is less than that required to produce ions, $HY(g) = H^+(g) + Y^-(g)$. Why is this? Despite this, an acid in aqueous solution dissociates to give ions, not neutral species. Why?

12–18 Write balanced chemical equations for each of the following reactions: (a) the solvent self-dissociation for methyl alcohol (CH_3OH), (b) the hydrolysis of cyanide ion, (c) the base dissociation of methylamine (H_3CNH_2) in water, and (d) the neutralization of methylamine with hydrochloric acid.

12–19 What are the conjugate bases for each of the following Brønsted acids: NH_4^+, $Fe(OH_2)_6^{3+}$, $HOSO_3^-$, H_3O^+, and HOCl?

12–20 What are the conjugate acids for each of the following Brønsted bases: $CH_3CO_2^-$, NH_2^-, H_2O, $HOSO_3^-$, and S^{2-}?

12–21 Consider Figure 12–12, the phase diagram for the water–ammonia system. The phases present in nine regions of this diagram are not labeled. Label them. (Study Figure 7–23; this may help you.) How many eutectic compositions are there in this system?

12–22 Commercially available antacid preparations for reducing the acidity of gastric liquid are $NaHCO_3$, $CaCO_3$, and Al_2O_3. Write balanced chemical equations for each of these solid compounds reacting with excess hydrogen ion (H^+ or H_3O^+). What weight of each reacts with one mole of hydrogen ion?

12–23 The species H_4O^{2+} is isoelectronic with NH_4^+. This species is not known in dilute aqueous solutions of strong acids. Can you suggest a reason? If experiments to detect this species (H_4O^{2+}) were to be conducted, what type of solutions should be examined?

12–24 Balance equations for the following reactions (all reactants and products are given):

$$B_2H_6 = B_5H_{11} + H_2$$

$$B_2H_6 + NaH = NaBH_4$$

$$B_2H_6 + H_2O = B(OH)_3 + H_2$$

$$MgB_2 + H^+ + H_2O = B_5H_9 + B(OH)_3 + Mg^{2+}$$

12–25 The structure of $B_{10}H_{14}$ involves ten B—H sigma bonds and four B–H–B (H bridging two B) three-center bonds (see Figure 12–14). How many electron pairs are there to form B—B sigma bonds and B–B–B (B bridging two B) three-center bonds?

12–26 In the mass-spectrometric analysis of a vapor containing $H_2O(g)$ and $CH_3OH(g)$, in which protons have reacted with water and methanol, peaks are observed at $m/e = 69, 83, 97, 101, 115$, and 129. What species cause these peaks?

13

Acid–Base Reactions in Solution

13–1 Introduction; Equilibrium in Solution

In this chapter we will consider quantitatively the acid-dissociation reactions mentioned in the previous two chapters. These quantitative considerations will involve calculations using the equilibrium constants for acid–base reactions. For a gas-phase reaction, for example, the ammonia-synthesis reaction,

$$N_2(g) + 3H_2(g) = 2NH_3(g)$$

the equilibrium-constant equation developed in the preceding chapter is

$$K = \frac{P^2_{NH_3}}{P_{N_2} \times P^3_{H_2}}$$

This equation involves the pressure of each gaseous reactant and product. The concentration, n/V, of an ideal gas is proportional to the pressure of the gas, $n/V = P/RT$, and the equilibrium constant for the ammonia-synthesis reaction also could be written in terms of concentrations,

$$K' = \frac{[NH_3]^2}{[N_2] \times [H_2]^3}$$

in which a chemical formula in square brackets is the concentration of that species, for example, $[NH_3] = n_{NH_3}/V$. The value of K' for this reaction is related to the value of the previously defined K, which involved the pressures of the gases:

$$K' = \frac{[NH_3]^2}{[N_2][H_2]^3} = \frac{P^2_{NH_3}}{P_{N_2} \times P^3_{H_2}} \times \frac{\left(\frac{[NH_3]}{P_{NH_3}}\right)^2}{\left(\frac{[N_2]}{P_{N_2}}\right) \times \left(\frac{[H_2]}{P_{H_2}}\right)^3}$$

But each quotient of concentration and pressure is equal to $(RT)^{-1}$; therefore, for this reaction,

$$K' = K \times (RT)^2$$

For the ammonia-synthesis reaction, $K = 5.3 \times 10^5$ atm^{-2} at 298.2 K. The value of K' at this temperature is

$$\begin{aligned} K' &= 5.3 \times 10^5 \text{ atm}^{-2} \times (0.0821 \text{ L atm mol}^{-1} \text{ K}^{-1} \times 298.2 \text{ K})^2 \\ &= 3.2 \times 10^8 \text{ L}^2 \text{ mol}^{-2} \end{aligned}$$

In the general case the power to which RT is raised in converting K to K' is $-\Delta n$:

$$K' = K \times (RT)^{-\Delta n}$$

(For the equation for the ammonia-synthesis reaction, $\Delta n = -2$.)

For reactions in solution the equilibrium-constant equation involves concentrations, as in this alternate expression of the equilibrium constant for the ammonia-synthesis reaction. Generally we will employ molar concentrations for solute species and mole fraction units for the solvent component. Because the mole fraction of the solvent component of a dilute solution is close to one, this convention seems to omit the concentration of solvent from the equilibrium-constant equation. For the dissociation of acetic acid (represented as HOAc) in water,

$$\text{HOAc}(aq) + H_2O(l) \rightleftarrows H_3O^+(aq) + \text{OAc}^-(aq)$$

the equilibrium-constant equation is

$$K = \frac{[H_3O^+][\text{OAc}^-]}{[\text{HOAc}]x_{H_2O}}$$

which is equal to

$$K = \frac{[H_3O^+][\text{OAc}^-]}{[\text{HOAc}]}$$

for $x_{H_2O} \cong 1$. (But if the solution being considered is not dilute and the mole fraction of the solvent is appreciably less than 1, the factor $x^{-1}_{H_2O}$ should be included in the equation. There are other complications in concentrated solution, and these also would have to be taken into account.)

In our discussion we will learn of the complexities of ionic equilibria in solution and also of the simplifications that make many calculations easy. These calculations have a range of application, for instance, from learning the effect of substituting a chlorine for a hydrogen in acetic acid (mentioned in Chapter 11) to correlating the acid–base properties of components of biological fluids with their ability to control the acidity of these fluids.

13–2 Strong Acids and Bases

We saw in the preceding chapter that certain acids and bases dissociate completely or almost completely in aqueous solution. Solutions of these acids and bases find much use both in industry and in the laboratory. The study of solutes that dissociate completely also provides the basis for judging the extent of dissociation of weak acids and bases. Before we consider aqueous solutions of strong acids and bases, we must consider the dissociation of water itself.

THE SELF-DISSOCIATION OF WATER

Water from a municipal water supply or water taken from a mountain stream conducts an electrical current. This electrical conductance is due largely to ionic solutes in the water. Painstaking chemical treatment and multiple distillation can remove these ionic solutes, but the electrical conductance of the water does not then become zero. Pure water has an electrical conductance caused by hydronium ion and hydroxide ion resulting from the dissociation:

$$2H_2O(l) \rightleftarrows H_3O^+ + OH^- \quad \text{or} \quad H_2O(l) \rightleftarrows H^+ + OH^- \quad \text{(see footnote 1)}$$

Careful work by KOHLRAUSCH and others has provided the conductance data that can be interpreted to give the extent of this self-dissociation reaction. At 298.2 K the concentrations of ions in pure water are:

$$[H^+] = 1.00 \times 10^{-7} \text{ mol L}^{-1}$$
$$[OH^-] = 1.00 \times 10^{-7} \text{ mol L}^{-1}$$

(The balanced chemical equation shows that the numbers of hydrogen ions and hydroxide ions from the dissociation reaction are equal.) These concentrations correspond to an extremely small extent of dissociation. The concentration of water molecules in liquid water is $\sim 55.3 \text{ mol L}^{-1}$. [At 298.2 K the density of water is 0.997 g cm^{-3}; thus the concentration of water is $0.997 \text{ g cm}^{-3} \times 1000 \text{ cm}^3 \text{ L}^{-1} \times (1 \text{ mol}/18.02 \text{ g}) = 55.3 \text{ mol L}^{-1}$.] The fraction of water molecules which are dissociated is

$$f_{\text{dissoc}} = \frac{[H^+]}{[H_2O]} = \frac{1.00 \times 10^{-7} \text{ mol L}^{-1}}{55.3 \text{ mol L}^{-1}} = 1.81 \times 10^{-9}$$

that is, ~ 2 molecules in a billion.

---

[1]As explained in Chapter 12, hydrogen ion (H^+, a proton) does not exist as such in solution, but we will use this symbol to stand for the solvated hydrogen ion (H_3O^+ and its hydrates) wherever we can do so without introducing error.

Under the conventions presented previously in this chapter, the equilibrium constant for water self-dissociation is

$$K_w = [H^+][OH^-]$$

For 298.2 K, the value of K_w can be calculated from the equilibrium values of the concentrations of hydrogen ion and hydroxide ion:

$$K_w = (1.00 \times 10^{-7}\ \text{mol L}^{-1})^2 = 1.00 \times 10^{-14}\ \text{mol}^2\ \text{L}^{-2}$$

If we had developed this equilibrium-constant equation as we did those in Chapter 12, we would have derived the equation for the change of free energy under standard conditions as

$$\Delta G^0 = -RT \ln [K_w \times (\text{mol L}^{-1})^{-2}]$$

For 298.2 K,

$$\begin{aligned} -\Delta G^0 &= 8.31\ \text{J K}^{-1}\ \text{mol}^{-1} \times 298.2\ \text{K} \\ &\quad \times \ln [1.00 \times 10^{-14}\ \text{mol}^2\ \text{L}^{-2} \times (\text{mol L}^{-1})^{-2}] \\ \Delta G^0 &= +79{,}900\ \text{J mol}^{-1} \end{aligned}$$

We cannot directly measure the value of ΔH^0 for the self-dissociation of water, but we can measure the value of ΔH^0 for the reverse reaction,

$$H^+ + OH^- = H_2O$$

This is simply the enthalpy change in the neutralization of a strong acid with a strong base. This value, measured calorimetrically, is

$$\Delta H^0 = -55.9\ \text{kJ mol}^{-1}$$

For the self-dissociation, therefore,

$$\Delta H^0 = +55.9\ \text{kJ mol}^{-1}$$

With the values of ΔG^0 and ΔH^0 known, the value of ΔS^0 can be calculated:

$$\begin{aligned} \Delta G^0 &= \Delta H^0 - T\Delta S^0 \\ \Delta S^0 &= \frac{\Delta H^0 - \Delta G^0}{T} \\ &= \frac{(55{,}900 - 79{,}900)\ \text{J mol}^{-1}}{298.2\ \text{K}} \\ &= -80.5\ \text{J K}^{-1}\ \text{mol}^{-1} \end{aligned}$$

Thus for the self-dissociation reaction at 298.2 K under standard conditions:

$$\begin{aligned} &H_2O(l) \rightleftarrows H^+(1\ \text{mol L}^{-1}) + OH^-(1\ \text{mol L}^{-1}) \\ &\quad \Delta G^0 = +79.9\ \text{kJ mol}^{-1} \\ &\quad \Delta H^0 = +55.9\ \text{kJ mol}^{-1} \\ &\quad \Delta S^0 = -80.5\ \text{J K}^{-1}\ \text{mol}^{-1} \end{aligned}$$

At first sight you may be surprised that the standard change of entropy in a reaction in which one molecular species gives two could be negative. If a gaseous reaction were being considered, for example,

$$PCl_5(g) = PCl_3(g) + Cl_2(g)$$

your feeling for the relationship of the value of ΔS^0 and the value of Δn would be correct: the value of ΔS^0 for this gas-phase reaction is $+181.9\ \text{J K}^{-1}\ \text{mol}^{-1}$. The water self-dissociation reaction has a negative value of ΔS^0 because the strong attractive interaction of the ions H^+ and OH^- with the solvent causes water molecules to lose randomness when these ions are formed. (We will see other examples of ion-producing dissociation reactions in aqueous solution for which $\Delta S^0 < 0$.)

For the self-dissociation reaction under equilibrium conditions at 298.2 K,

$$H_2O(l) \rightleftarrows H^+(1.00 \times 10^{-7}\ \text{mol L}^{-1}) + OH^-(1.00 \times 10^{-7}\ \text{mol L}^{-1})$$

values of the thermodynamic quantities are:

$$\Delta G = 0$$

$$\Delta H = \Delta H^0 = +55{,}900\ \text{J mol}^{-1}$$

$$\Delta S = \frac{\Delta H}{T} = \frac{+55{,}900\ \text{J mol}^{-1}}{298.2\ \text{K}} = +187.5\ \text{J K}^{-1}\ \text{mol}^{-1}$$

As explained before (Section 7–3), the larger the volume available to a solute species the more positive is its molar entropy; under these equilibrium conditions the volume available to the one mole of each solute species, the reaction products H^+ and OH^-, is 10^7-fold greater than in the reaction under standard conditions.

Because the self-dissociation of water is endothermic, the value of K_w increases with an increase of temperature. Values of K_w as a function of temperature are given in Table 13–1. We see that this equilibrium constant increases by a factor of 443 over the temperature range 0–100°C.

TABLE 13–1
Values of the Equilibrium Constant for Self-Dissociation of Water[a]

$\frac{t}{°C}$	$\frac{10^{14} \times K_w}{(\text{mol L}^{-1})^2}$	$\frac{t}{°C}$	$\frac{10^{14} \times K_w}{(\text{mol L}^{-1})^2}$
0	0.114	50	5.37
10	0.292	60	9.33
20	0.678	70	15.3
25	1.00	80	23.9
30	1.46	90	35.5
40	2.87	100	50.5

[a] $H_2O(l) \rightleftarrows H^+ + OH^-$
$K_w = [H^+][OH^-]$

DILUTE SOLUTIONS OF STRONG ACIDS AND BASES; THE pH SCALE

The equilibrium-constant equation for the self-dissociation of water,

$$K_w = [H^+][OH^-] = 1.00 \times 10^{-14} \text{ mol}^2 \text{ L}^{-2}$$

is applicable also to dilute solutions of acids, bases, and other solutes in water at 298.2 K. Consider solutions prepared by successive 100-fold dilutions of solutions of hydrochloric acid and sodium hydroxide. Before dilution, the two solutions have the concentrations[2]

$$C_{HCl} = 0.0100 \text{ mol L}^{-1}$$

$$C_{NaOH} = 0.0100 \text{ mol L}^{-1}$$

This acid and this base are each completely dissociated strong electrolytes in aqueous solution, which means that the reactions

$$HCl(aq) = H^+ + Cl^- \quad \text{and} \quad NaOH(aq) = Na^+ + OH^-$$

go to completion, and the reaction

$$H_2O(l) \rightleftarrows H^+ + OH^-$$

goes to an extent dictated by $[H^+][OH^-] = 1.00 \times 10^{-14}$ mol^2 L^{-2}. The concentration of hydrogen ion from the strong acid in the hydrochloric acid solution is enormously greater than that from the water dissociation, making

$$[H^+] = C_{HCl} = 0.0100 \text{ mol L}^{-1}$$

and

$$[OH^-] = \frac{K_w}{[H^+]} = \frac{1.00 \times 10^{-14} \text{ mol}^2 \text{ L}^{-2}}{1.00 \times 10^{-2} \text{ mol L}^{-1}} = 1.00 \times 10^{-12} \text{ mol L}^{-1}$$

Because the dissociation of water produces equal amounts of hydrogen ion and hydroxide ion, the concentration of hydrogen ion from the water dissociation is the same as this concentration of hydroxide ion, 1.00×10^{-12} mol L^{-1}, which is minute compared to the concentration of hydrogen ion derived from the solute. For the original, undiluted solution of sodium hydroxide, we have

$$[OH^-] = C_{NaOH} = 1.00 \times 10^{-2} \text{ mol L}^{-1}$$

and

$$[H^+] = \frac{K_w}{[OH^-]} = \frac{1.00 \times 10^{-14} \text{ mol}^2 \text{ L}^{-2}}{1.00 \times 10^{-2} \text{ mol L}^{-1}} = 1.00 \times 10^{-12} \text{ mol L}^{-1}$$

---

[2]Here and elsewhere we will use the convention introduced in Chapter 7 in which C stands for a stoichiometric concentration, and the substance generally is designated, as in C_{HCl} and C_{NaOH}, to avoid possible ambiguity.

TABLE 13–2
Solutions of Strong Acid (HCl) and Solutions of Strong Base (NaOH)

	Concentrations[a] from solution composition		Concentrations[a] calculated from $[H^+][OH^-] = 1.00 \times 10^{-14}$ mol^2 L^{-2}	pH[b]
C_{HCl}	$[H^+]$	$[Cl^-]$	$[OH^-]$	
0.0100	0.0100	0.0100	1.00×10^{-12}	2.00
1.00×10^{-4}	1.00×10^{-4}	1.00×10^{-4}	1.00×10^{-10}	4.00
1.00×10^{-6}	1.00×10^{-6}	1.00×10^{-6}	1.00×10^{-8}	6.00
C_{NaOH}	$[OH^-]$	$[Na^+]$	$[H^+]$	pH[b]
0.0100	0.0100	0.0100	1.00×10^{-12}	12.00
1.00×10^{-4}	1.00×10^{-4}	1.00×10^{-4}	1.00×10^{-10}	10.00
1.00×10^{-6}	1.00×10^{-6}	1.00×10^{-6}	1.00×10^{-8}	8.00

[a]All concentrations in mol L^{-1}.
[b]Calculated using $\text{pH} = -\log([H^+] \times \text{L mol}^{-1})$.

The concentrations of species in these solutions are summarized in Table 13–2. In this table we consider only solutions in the range of concentrations

$$C_{HCl} = 1.00 \times 10^{-2} \text{ to } 1.00 \times 10^{-6} \text{ mol L}^{-1}$$
$$C_{NaOH} = 1.00 \times 10^{-2} \text{ to } 1.00 \times 10^{-6} \text{ mol L}^{-1}$$

We impose this limitation for two reasons:

1. At C_{HCl} or $C_{NaOH} > 1.00 \times 10^{-2}$ mol L^{-1}, solution nonideality has an influence on the value of K_w.
2. At C_{HCl} or $C_{NaOH} < 1.00 \times 10^{-6}$ mol L^{-1}, the concentrations of both the hydrogen ion and the hydroxide ion from the water dissociation are appreciable compared to those provided by the strong electrolyte.

We will discuss each of these complications, but first let us consider the column of Table 13–2 labeled pH. The pH of a solution is a measure of its "acidity." The equation employed to calculate the pH of each solution in Table 13–2 is

$$\text{pH} = -\log\left(\frac{[H^+]}{\text{mol L}^{-1}}\right)$$

We will use this equation in all calculations of this type.[3] The pH is a quantity that also can be measured experimentally, usually by electrochemical methods, although other methods can be used. In each of these methods, solution nonideality influences the experimental measurement.

[3]Analogous equations can be used for expressing other concentrations on a logarithmic scale, for example, $\text{pOH} = -\log([OH^-]/\text{mol L}^{-1})$, $\text{pCl} = -\log([Cl^-]/\text{mol L}^{-1})$, and so on. We will not use this method for expressing the concentrations of species other than hydrogen ion.

The electrochemical method, with appropriate calibration, measures a quantity that includes a factor for nonideality:

$$\mathrm{pH} = -\log\left([\mathrm{H}^+]\gamma_{\mathrm{H}^+}\right)$$

in which γ_{H^+} is the activity coefficient of hydrogen ion in the solution. The subject of activities is, in fact, rather complicated, and we will not enter into all of the complications. Rather, in all our calculations we will use the equation

$$\mathrm{pH} = -\log\left(\frac{[\mathrm{H}^+]}{\mathrm{mol\ L^{-1}}}\right)$$

but we will recognize that a pH meter (see Chapter 16) does not measure the concentration of hydrogen ion exactly:

$$[\mathrm{H}^+] \neq 10^{-\mathrm{pH\ (measd)}}$$

if $\gamma_{\mathrm{H}^+} \neq 1\ \mathrm{L\ mol^{-1}}$.

Now let us consider a solution in which the amount of hydrogen ion furnished by the strong electrolyte hydrochloric acid is not enormously greater than that provided by dissociation of the solvent. What are the concentrations of hydrogen ion and hydroxide ion and what is the pH for a solution of hydrochloric acid with $C_{\mathrm{HCl}} = 3.00 \times 10^{-7}\ \mathrm{mol\ L^{-1}}$? The mathematical relationships that allow us to calculate these quantities are:

1. $[\mathrm{Cl}^-] = C_{\mathrm{HCl}} = 3.00 \times 10^{-7}\ \mathrm{mol\ L^{-1}}$
2. $[\mathrm{H}^+][\mathrm{OH}^-] = 1.00 \times 10^{-14}\ \mathrm{mol^2\ L^{-2}}$
3. $[\mathrm{H}^+] = [\mathrm{Cl}^-] + [\mathrm{OH}^-]$

Equation 1 is simply a statement that all of the chloride ion comes from complete dissociation of the solute, Equation 2 is the equilibrium-constant equation, and Equation 3 is a statement of the electroneutrality condition: The sum of the concentrations of cations must equal the sum of concentrations of anions, appropriate account having been taken of the charges on these ions.[4] The electroneutrality condition for the solution being considered,

$$[\mathrm{H}^+] = [\mathrm{Cl}^-] + [\mathrm{OH}^-]$$

also can be viewed as a statement that the hydrogen ion comes from two sources, hydrochloric acid $[\mathrm{H}^+]_{\mathrm{HCl}}$ and water $[\mathrm{H}^+]_{\mathrm{H_2O}}$, with $[\mathrm{H}^+]_{\mathrm{HCl}} = [\mathrm{Cl}^-]$ and $[\mathrm{H}^+]_{\mathrm{H_2O}} = [\mathrm{OH}^-]$:

$$[\mathrm{H}^+] = [\mathrm{H}^+]_{\mathrm{HCl}} + [\mathrm{H}^+]_{\mathrm{H_2O}} = [\mathrm{Cl}^-] + [\mathrm{OH}^-]$$

Equation 3 can be converted to an equation in only one unknown, $[\mathrm{H}^+]$, by use of Equations 1 and 2:

$$[\mathrm{H}^+] = C_{\mathrm{HCl}} + \frac{K_{\mathrm{w}}}{[\mathrm{H}^+]}$$

[4]The electroneutrality condition for an aqueous solution of sulfuric acid $(\mathrm{HO})_2\mathrm{SO}_2$ is $[\mathrm{H}^+] = 2[\mathrm{SO_4^{2-}}] + [\mathrm{HOSO_3^-}] + [\mathrm{OH}^-]$.

This equation is a quadratic equation,

$$[H^+]^2 - C_{HCl}[H^+] - K_w = 0$$

for which the solution is[5]

$$[H^+] = \frac{C_{HCl} + (C_{HCl}^2 + 4K_w)^{1/2}}{2}$$

For our problem, the concentration of hydrogen ion is

$$[H^+] = \frac{3.00 \times 10^{-7} \text{ mol L}^{-1} + [(3.00 \times 10^{-7} \text{ mol L}^{-1})^2 + 4 \times 1.00 \times 10^{-14} \text{ mol}^2 \text{ L}^{-2}]^{1/2}}{2}$$

$$[H^+] = 3.30 \times 10^{-7} \text{ mol L}^{-1}$$

$$[OH^-] = \frac{1.00 \times 10^{-14} \text{ mol}^2 \text{ L}^{-2}}{3.30 \times 10^{-7} \text{ mol L}^{-1}} = 3.03 \times 10^{-8} \text{ mol L}^{-1}$$

$$\text{pH} = -\log\left(\frac{3.30 \times 10^{-7} \text{ mol L}^{-1}}{\text{mol L}^{-1}}\right) = 6.48$$

We see in this example that we would have made an error of ~9% if we had calculated the concentration of hydrogen ion as we did previously for concentrations in the range

$$1.00 \times 10^{-2} \text{ mol L}^{-1} > C_{HCl} > 1.00 \times 10^{-6} \text{ mol L}^{-1}$$

that is,

$$[H^+] = C_{HCl} \qquad (13\text{–}1)^6$$

Instead we obtained the correct answer by using the equation

$$[H^+] = C_{HCl} + \frac{K_w}{[H^+]} \qquad (13\text{–}2)$$

At 298.2 K, for solutions in which the concentration of ionic solutes are greater than ~0.0100 mol L^{-1}, there is a problem connected with using the expression

$$[H^+][OH^-] = 1.00 \times 10^{-14} \text{ mol}^2 \text{ L}^{-2}$$

Equilibrium in the reaction

$$H_2O(l) \rightleftarrows H^+ + OH^-$$

is shifted to the right by ionic solutes at low to moderate concentrations (~0.01 mol L^{-1} to 1.0 mol L^{-1}). The long-range ion-atmosphere effects mentioned in Chapter 9 are responsible. Studies have been made of the

---

[5]Recall that for a quadratic equation $ax^2 + bx + c = 0$ the solutions are

$$x = \frac{-b \pm (b^2 - 4ac)^{1/2}}{2a}$$

In general, only one root of the equation gives a physically meaningful answer to a problem such as that being considered.

[6]We will number certain equations in this chapter because this will make it easier to convey relationships between them.

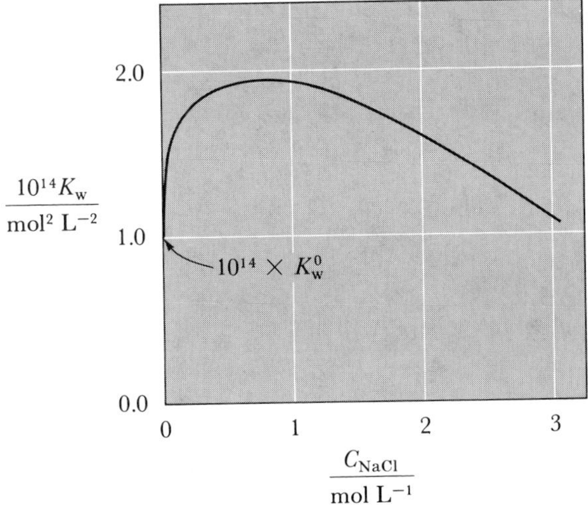

FIGURE 13–1
$K_w = [H^+][OH^-]$ at 298.2 K as a function of the concentration of sodium chloride. $K_w = K_w^0 \times (\gamma_{H^+}\gamma_{OH^-})^{-1}$; variation in K_w with concentration of sodium chloride is due to variation of activity coefficients.

dependence of the equilibrium constant

$$K_w = [H^+][OH^-]$$

upon the concentration of ionic solutes. Figure 13–1 shows the dependence for sodium chloride. We see that K_w increases and then decreases as the sodium chloride concentration increases to ~ 3 mol L^{-1}. The maximum value of K_w is almost twice its value in electrolyte-free solution. The customary approach to this problem (and corresponding problems for other ionic equilibria) is analogous to that used in the equilibrium constant for the ammonia-synthesis reaction at high pressures: the incorporation of an activity coefficient for each concentration term. Thus the equilibrium constant for water dissociation becomes

$$K^0 = \frac{[H^+][OH^-]\gamma_{H^+}\gamma_{OH^-}}{x_{H_2O}\gamma_{H_2O}}$$

For very dilute solutions at 298.2 K, in which $x_{H_2O}\gamma_{H_2O} = 1.00$ and the ionic species behave ideally, $\gamma_{H^+} \cong \gamma_{OH^-} \cong 1.00$ L mol^{-1},

$$K_w^0 = 1.00 \times 10^{-14}\ \text{mol}^2\ \text{L}^{-2} \times 1.00\ \text{L}^2\ \text{mol}^{-2} = 1.00 \times 10^{-14}$$

The equation for K^0 can be rearranged to give

$$K_w = [H^+][OH^-] = K_w^0 \times (\gamma_{H^+}\gamma_{OH^-})^{-1} = 1.00 \times 10^{-14} \times (\gamma_{H^+}\gamma_{OH^-})^{-1}$$

(Here the factor $x_{H_2O}\gamma_{H_2O}$ is assumed to be 1.00.)

THE NEUTRALIZATION OF A STRONG ACID BY A STRONG BASE

As we already have stated, the neutralization of a strong acid by a strong base,

$$H_3O^+ + OH^- \rightleftarrows H_2O + H_2O$$

is an example of the general acid–base reaction

$$\text{Acid} + \text{Base} = \text{Base} + \text{Acid}$$

To be considered a neutralization, the reactant solute acid and reactant solute base must be so strong that the neutralization reaction goes essentially to completion when equivalent amounts of acid and base are mixed. This prerequisite is met if solutions of the strong acid, hydrochloric acid, and the strong base, sodium hydroxide, are mixed. The chemical equation for this reaction, which also can be written

$$H^+ + OH^- = H_2O$$

does not involve chloride ion or sodium ion because in aqueous solution each of the substances—hydrochloric acid, sodium hydroxide, and sodium chloride—is a completely dissociated strong electrolyte. (This was explained in Chapter 2.) When exactly equivalent amounts of hydrochloric acid and sodium hydroxide are mixed, the resulting solution is identical in all respects with one prepared by adding an appropriate amount of pure sodium chloride to water. Principles outlined in Chapter 7 allow us to calculate the volume of base of a known concentration required to neutralize a certain amount of acid.

EXAMPLE: What volume of $0.1147M$ NaOH is needed to neutralize $20.00\ \text{cm}^3$ of $0.2407M$ HCl?

The balanced equation shows that 1 mol of hydronium ion (provided by 1 mol of hydrochloric acid) reacts with 1 mol of hydroxide ion (provided by 1 mol of sodium hydroxide). The amount of acid,

$$n_{\text{H}^+} = \frac{20.00\ \text{cm}^3}{1000\ \text{cm}^3\ \text{L}^{-1}} \times 0.2407\ \text{mol L}^{-1}$$

must, therefore, be equal to the amount of base that is in a volume V of $0.1147M$ NaOH:

$$n_{\text{OH}^-} = \frac{V}{1000\ \text{cm}^3\ \text{L}^{-1}} \times 0.1147\ \text{mol L}^{-1}$$

Equating these amounts of acid and base, $n_{\text{H}^+} = n_{\text{OH}^-}$, we have

$$V = \frac{20.00\ \text{cm}^3 \times 0.2407\ \text{mol L}^{-1}}{0.1147\ \text{mol L}^{-1}} = 41.97\ \text{cm}^3$$

EXAMPLE: A more practical question is the following. A $20.00\ \text{cm}^3$ portion of hydrochloric acid solution of unknown concentration is neutralized by $36.43\ \text{cm}^3$ of $0.1103M$ sodium hydroxide, in an experiment in which the hydrochloric acid solution is measured with a pipet and the sodium hydroxide solution is measured with a buret (see Figure 7–1). What is the concentration of the hydrochloric acid solution?

TABLE 13–3
The Concentrations of Hydrogen Ion at Various Stages in the Titration of 20.00 cm^3 of 0.2009*M* HCl with 0.1103*M* NaOH

Volume of NaOH added cm^3	Total volume[a] cm^3	Percent neutralization	$[H^+]$ $mol\ L^{-1}$	$[OH^-]$ $mol\ L^{-1}$	pH[c]
0.00	20.00	0	0.2009		0.697
10.00	30.00	27.45	0.0973		1.012
20.00	40.00	54.90	0.0453		1.344
30.00	50.00	82.35	0.01418		1.848
36.00	56.00	98.82	8.47×10^{-4}		3.072
36.20	56.20	99.37	4.51×10^{-4}		3.346
36.40	56.40	99.92	5.87×10^{-5}		4.23
36.43	56.43	100.00	1.00×10^{-7}	1.00×10^{-7}	7.00
36.46	56.46	100.08	1.71×10^{-10}[b]	5.86×10^{-5}	9.77
36.70	56.70	100.74	1.90×10^{-11}[b]	5.25×10^{-4}	10.72
36.90	56.90	101.29	1.10×10^{-11}[b]	9.11×10^{-4}	10.96
37.50	57.50	102.94	4.87×10^{-12}[b]	2.05×10^{-3}	11.31
39.00	59.00	107.05	2.08×10^{-12}[b]	4.80×10^{-3}	11.68
45.00	65.00	123.52	6.88×10^{-13}[b]	1.45×10^{-2}	12.16

[a]Volumes are assumed to be additive, a good assumption for these dilute solutions.
[b]Calculated from $[H_3O^+] = 1.00 \times 10^{-14}\ mol^2\ L^{-2}/[OH^-]$, the value of $[OH^-]$ being calculated $(n_{OH^-} - n_{H^+})/V$.
[c]Calculated using $pH = -\log([H^+] \times L\ mol^{-1})$.

Using the same reasoning as in the preceding example,

$$n_{H^+} = n_{OH^-}$$

$$\frac{20.00\ cm^3}{1000\ cm^3\ L^{-1}} \times C_{HCl} = \frac{36.43\ cm^3}{1000\ cm^3\ L^{-1}} \times 0.1103\ mol\ L^{-1}$$

which becomes

$$C_{HCl} = \frac{36.43\ cm^3}{20.00\ cm^3} \times 0.1103\ mol\ L^{-1} = 0.2009\ mol\ L^{-1}$$

At exact neutralization, the concentration of hydrogen ion in the reaction mixture is the same as in sodium chloride solution, which is essentially the same as in pure water.[7] Table 13–3 summarizes the concentration of hydrogen ion during the titration in the example; it shows that the last drop (0.03 cm^3) of sodium hydroxide solution added before the equivalence point changes the concentration of hydrogen ion by a factor of ~590. This factor can be contrasted with that of ~2 for the addition of the first 10 cm^3 of sodium hydroxide.

[7]Recall, however, the discussion of the effect of electrolyte upon the equilibrium $H_2O \leftrightharpoons H^+ + OH^-$.

In the course of the titration before enough base is added for exact neutralization, the reaction

$$H^+ + OH^- \rightleftarrows H_2O$$

occurs essentially to completion. The limited amount of hydroxide ion is consumed by the excess acid. For instance, after addition of 35.00 cm^3 of base, the total volume of the solution is 55.00 cm^3, and the total amounts of acid and base present before reaction are:

$$n_{HCl} = \frac{20.00 \text{ cm}^3}{1000 \text{ cm}^3 \text{ L}^{-1}} \times 0.2009 \text{ mol L}^{-1} = 4.018 \times 10^{-3} \text{ mol}$$

$$n_{NaOH} = \frac{35.00 \text{ cm}^3}{1000 \text{ cm}^3 \text{ L}^{-1}} \times 0.1103 \text{ mol L}^{-1} = 3.861 \times 10^{-3} \text{ mol}$$

At this point the extent of neutralization is

$$100 \times \frac{3.861 \times 10^{-3} \text{ mol}}{4.018 \times 10^{-3} \text{ mol}} = 96.1\%$$

and the amount of excess acid is

$$\begin{aligned} n_{H^+} &= n_{HCl} - n_{NaOH} \\ &= 4.018 \times 10^{-3} \text{ mol} - 3.861 \times 10^{-3} \text{ mol} = 0.157 \times 10^{-3} \text{ mol} \end{aligned}$$

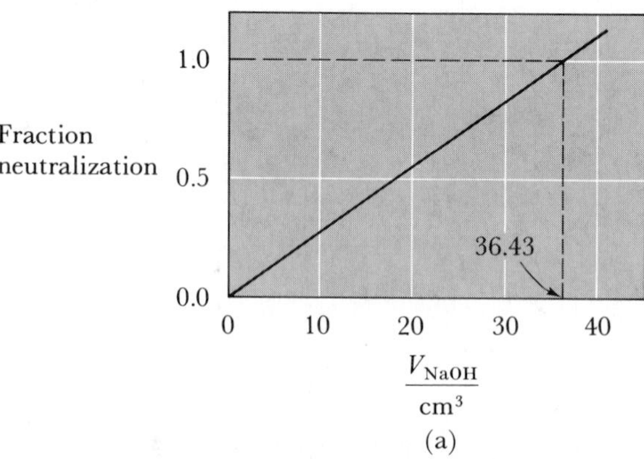

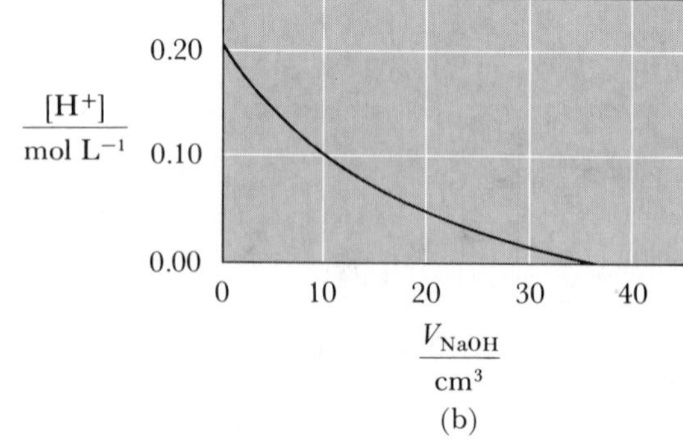

FIGURE 13–2
The neutralization of hydrochloric acid (20.00 cm^3 of 0.2009*M* HCl) with sodium hydroxide ($C_{NaOH} = 0.1103$ mol L^{-1}). (a) Fraction neutralized versus volume of sodium hydroxide solution. (b) [H$^+$] versus volume of sodium hydroxide solution.

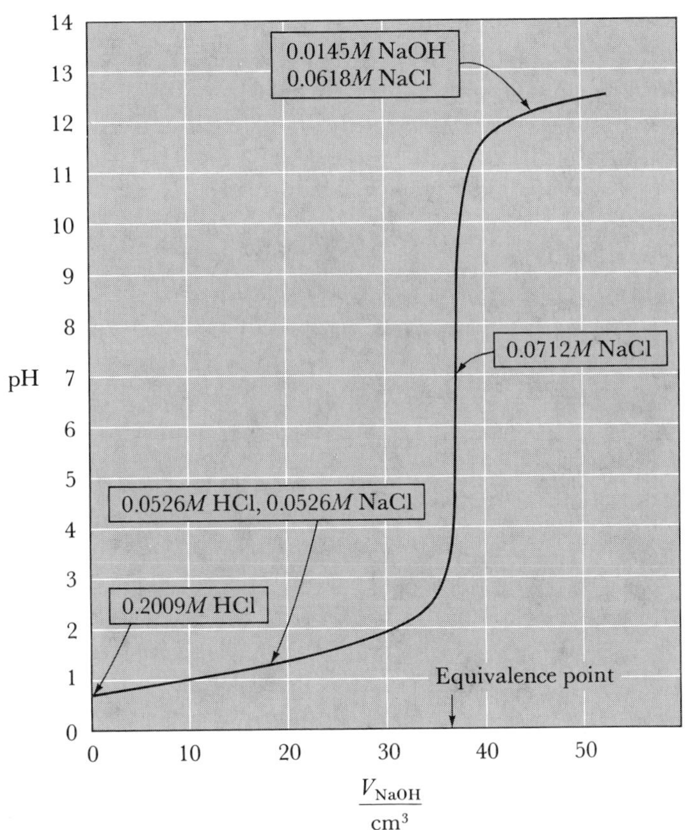

FIGURE 13–3
The neutralization of 20.00 cm^3 of 0.2009*M* HCl with 0.1103*M* NaOH. The equivalence point is at 36.43 cm^3 of added solution of NaOH. Boxes show the stoichiometric concentrations at addition of 0, 50, 100, and 123.5% of the base needed for exact neutralization.

The concentration of hydrogen ion at this point is

$$[H^+] = \frac{n_{H^+}}{V} = \frac{0.157 \times 10^{-3}\ \text{mol}}{0.05500\ \text{L}} = 2.85 \times 10^{-3}\ \text{mol L}^{-1}$$

and the calculated value of pH is

$$\begin{aligned} \text{pH} &= -\log([H^+] \times \text{L mol}^{-1}) \\ &= -\log(2.85 \times 10^{-3}) = 2.545 \end{aligned}$$

The linear relationship between the fraction of the acid neutralized and the volume of base added in the titration is shown in Figure 13–2a. The relationship between the concentration of hydrogen ion and the amount of added base, shown in Figure 13–2b, is not quite linear because the solution changes in volume as titration proceeds. This latter figure does not show clearly the dramatic variation in the relative concentration of hydrogen ion near the equivalence point; Figure 13–3, in which the pH is plotted versus the volume of base, does show this.

13–3 Weak Acids and Bases

There are many naturally occurring compounds, both organic and inorganic, that are weak acids or their conjugate bases. They occur in vegetables (e.g., oxalic acid in spinach), in fruits (e.g., citric acid in oranges or malic acid in apples), or in biological fluids (e.g., phosphates in blood). The incomplete dissociation of each of these acids is governed by an equilibrium-constant equation, and now we will learn how to include such equations among those that must be satisfied by the equilibrium concentrations.

EQUILIBRIUM IN AQUEOUS SOLUTIONS OF WEAK ACIDS

The extent of dissociation of a weak acid in aqueous solution,

$$HX(aq) + H_2O(l) = H_3O^+(aq) + X^-(aq)$$

which usually we will represent

$$HX = H^+ + X^-$$

is characterized by an equilibrium constant

$$K_a = \frac{[H^+][X^-]}{[HX]}$$

As with the equilibrium constant for water dissociation, the complete equation involves an activity-coefficient factor,

$$K_a^0 = \frac{[H^+][X^-]}{[HX]} \times \frac{\gamma_{H^+}\gamma_{X^-}}{\gamma_{HX}}$$

but generally we will not concern ourselves with complexities due to nonideality.

Now we will learn to calculate equilibrium concentrations for aqueous solutions of weak acids containing no other solutes. In such solutions there is the equilibrium

$$H_2O \rightleftarrows H^+ + OH^-$$

in addition to the acid-dissociation equilibrium. In an aqueous solution containing the acid HX, there are four solute species—HX, H^+, X^-, and OH^-—and we will calculate the concentration of each of these species in a solution with a particular stoichiometric concentration of acid, C. Therefore we must find four equations relating the four concentrations [HX], $[H^+]$, $[X^-]$, and $[OH^-]$. The four equations are:

$$[HX] + [X^-] = C$$

$$[H^+] = [X^-] + [OH^-]$$

$$\frac{[H^+][X^-]}{[HX]} = K_a$$

$$[H^+][OH^-] = K_w$$

The first equation is a simple conservation equation: The sum of the concentrations of all of the species containing X must be equal to the total stoichiometric concentration of acid, HX. (The pure HX added to make up the solution with stoichiometric concentration C must be present in the solution either as X^- or as HX.) The second equation is a statement of the electroneutrality condition. The ionic species in the solution are

Cation: H^+ Anions: OH^- and X^-

The third and fourth equations are the equilibrium-constant equations appropriate to this problem.

Now we will convert these four equations in four unknowns to one equation in one unknown, $[H^+]$. Elimination of [HX] between the first and third equations gives

$$\frac{[H^+][X^-]}{C - [X^-]} = K_a$$

In this equation, $[X^-]$ can be expressed in terms of $[H^+]$ by use of the second and fourth equations:

$$[X^-] = [H^+] - [OH^-]$$
$$= [H^+] - \frac{K_w}{[H^+]}$$

This gives us the complete equation, a cubic equation in $[H^+]$:

$$\frac{[H^+]\left([H^+] - \dfrac{K_w}{[H^+]}\right)}{C - \left([H^+] - \dfrac{K_w}{[H^+]}\right)} = K_a \qquad (13\text{–}3)^8$$

When we must use the cubic equation, we may solve it conveniently by a method of successive approximations. However, most practical problems do not involve a cubic equation because it can be simplified to different equations, each a quadratic, by recognizing the consequences of certain inequalities:

1. The inequality

$$[H^+] \gg \frac{K_w}{[H^+]}$$

is equivalent to $[H^+] \gg [OH^-]$, which is true for solutions with $[H^+] > 1 \times 10^{-6}$ mol L^{-1}. The consequence of this inequality is

$$\left([H^+] - \frac{K_w}{[H^+]}\right) \cong [H^+]$$

[8]That this is a cubic equation in $[H^+]$ can be shown by cross-multiplication followed by multiplication through by $[H^+]$; the result is

$$[H^+]^3 + K_a[H^+]^2 - (K_w + K_aC)[H^+] - K_wK_a = 0$$

2. The inequality

$$C \gg \left([\mathrm{H}^+] - \frac{K_w}{[\mathrm{H}^+]}\right)$$

is a statement that the extent of dissociation is slight. The quantity

$$\left([\mathrm{H}^+] - \frac{K_w}{[\mathrm{H}^+]}\right) \quad \text{or} \quad ([\mathrm{H}^+] - [\mathrm{OH}^-])$$

is the concentration of the conjugate base $[X^-]$; if this quantity is very small compared to the stoichiometric concentration, C, the extent of dissociation is slight. The consequence of this inequality is

$$C - \left([\mathrm{H}^+] - \frac{K_w}{[\mathrm{H}^+]}\right) \cong C$$

We can impose each of these inequalities individually upon the complete equation (Equation 13–3) or we can impose both together. We then get three different simplified equations, Equations 13–4, 13–4′, and 13–5, which are given in Figure 13–4. The solutions of these quadratic equations for the concentration of hydrogen ion are designated 13–4s, 13–4′s, and 13–5s.

We also can obtain each of these simpler equations by beginning with the approximations. If we consider only the dissociation of the weak acid,

$$\mathrm{HX} = \mathrm{H}^+ + \mathrm{X}^-$$

the equation for K_a,

$$K_a = \frac{[\mathrm{H}^+][\mathrm{X}^-]}{[\mathrm{HX}]}$$

becomes for solutions with $[\mathrm{X}^-] = [\mathrm{H}^+]$,

$$K_a = \frac{[\mathrm{H}^+]^2}{C - [\mathrm{H}^+]}$$

which is Equation 13–4. Rearrangement of this equation gives the quadratic equation,

$$[\mathrm{H}^+]^2 + K_a[\mathrm{H}^+] - K_aC = 0$$

for which the positive root is

$$[\mathrm{H}^+] = \frac{-K_a + (K_a^2 + 4K_aC)^{1/2}}{2} \tag{13–4s}$$

If dissociation of the acid is slight and water dissociation contributes appreciably to the concentration of hydrogen ion, the electroneutrality condition,

$$[\mathrm{H}^+] = [\mathrm{OH}^-] + [\mathrm{X}^-]$$

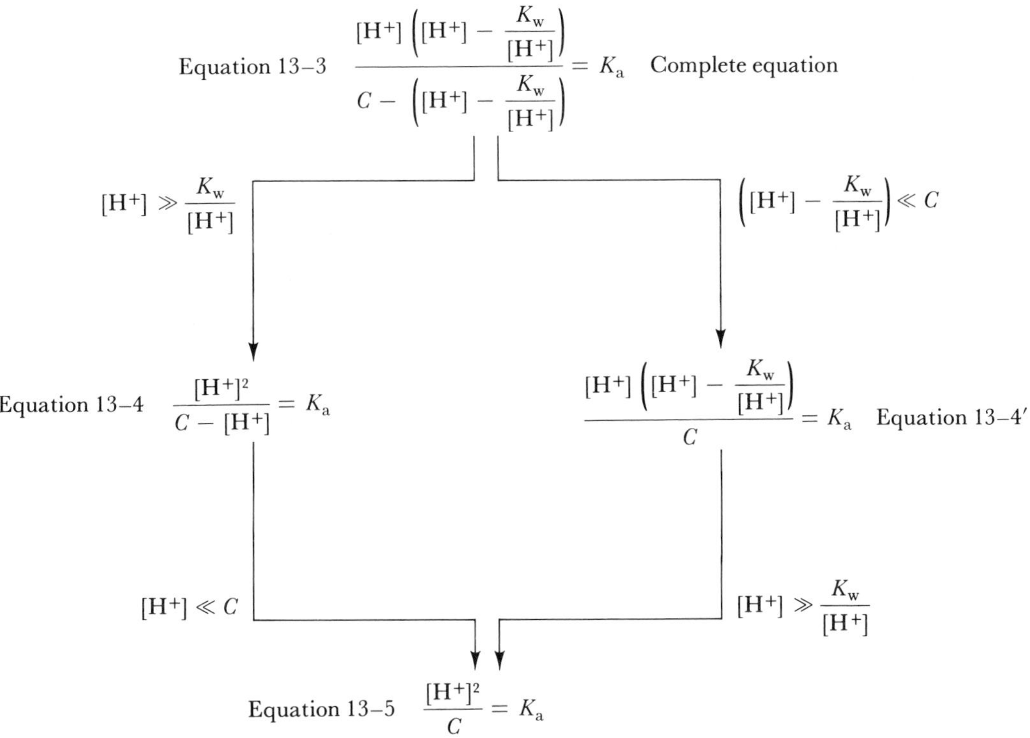

Equation 13–3 $\dfrac{[H^+]\left([H^+] - \dfrac{K_w}{[H^+]}\right)}{C - \left([H^+] - \dfrac{K_w}{[H^+]}\right)} = K_a$ Complete equation
$[H^+] \gg \dfrac{K_w}{[H^+]}$
$\left([H^+] - \dfrac{K_w}{[H^+]}\right) \ll C$
Equation 13–4 $\dfrac{[H^+]^2}{C - [H^+]} = K_a$
$\dfrac{[H^+]\left([H^+] - \dfrac{K_w}{[H^+]}\right)}{C} = K_a$ Equation 13–4′
$[H^+] \ll C$
$[H^+] \gg \dfrac{K_w}{[H^+]}$
Equation 13–5 $\dfrac{[H^+]^2}{C} = K_a$

FIGURE 13–4
Ways of simplifying the complete equation for the concentration of hydrogen ion in solutions of a weak acid. The equations solved for $[H^+]$ are:

$$[H^+] = \frac{-K_a + (K_a^2 + 4K_aC)^{1/2}}{2} \qquad (13\text{–}4s)$$

$$[H^+] = (K_w + K_aC)^{1/2} \qquad (13\text{–}4's)$$

$$[H^+] = (K_aC)^{1/2} \qquad (13\text{–}5s)$$

gives us, with $C \cong [HX]$,

$$[H^+] = \frac{K_w}{[H^+]} + K_a \frac{C}{[H^+]}$$

which can be rearranged to

$$\frac{[H^+]\left([H^+] - \dfrac{K_w}{[H^+]}\right)}{C} = K_a \qquad (13\text{–}4')$$

This is a simple quadratic equation,

$$[H^+]^2 = K_w + K_aC$$

which gives Equation 13–4′s:

$$[H^+] = (K_w + K_aC)^{1/2} \qquad (13\text{–}4's)$$

We may derive the simplest equation, Equation 13–5, if we consider only the weak acid dissociation,

$$HX \rightleftarrows H^+ + X^-$$

and assume that the extent of dissociation is slight, that is, $[HX] \cong C$; we then have

$$K_a = \frac{[H^+][X^-]}{[HX]} \cong \frac{[H^+]^2}{C} \qquad (13\text{–}5)$$

Solving this equation for $[H^+]$ we have an equation that is very simple to use:

$$[H^+] = (KC)^{1/2} \qquad (13\text{–}5s)$$

We will study the use of these equations by considering solutions of two weak acids with very different values of K_a, acetic acid ($K_a = 1.75 \times 10^{-5}$ mol L^{-1}) and ammonium chloride (for ammonium ion, $K_a = 5.69 \times 10^{-10}$ mol L^{-1}). (These values of K_a and values for other weak acids are given in Table 13–4.) In each calculation, we will start by using the simplest equation relating the concentration of hydrogen ion to the stoichiometric concentration of acid (Equation 13–5s).

Sample Calculations for Acetic Acid. Two solutions, 0.1000*M* HOAc and $1.000 \times 10^{-4}M$ HOAc will be considered; this will show the effect of dilution.

For 0.1000*M* HOAc

By Equation 13–5s, the calculated concentrations for this solution are:

$$[H^+] = (K_aC)^{1/2} \qquad (13\text{–}5s)$$

$$[H^+] = (1.75 \times 10^{-5} \text{ mol L}^{-1} \times 0.1000 \text{ mol L}^{-1})^{1/2}$$
$$= 1.32 \times 10^{-3} \text{ mol L}^{-1}$$

$$[OAc^-] = [H^+] = 1.32 \times 10^{-3} \text{ mol L}^{-1}$$

$$[HOAc] = C - [OAc^-]$$
$$= 0.1000 \text{ mol L}^{-1} - 1.32 \times 10^{-3} \text{ mol L}^{-1}$$
$$= 0.0987 \text{ mol L}^{-1}$$

We see that the assumption that $C \gg [H^+]$ is accurate to within 1.3% for this solution. We now will calculate the concentration of hydroxide ion:

$$[OH^-] = \frac{K_w}{[H^+]} = \frac{1.00 \times 10^{-14} \text{ mol}^2 \text{ L}^{-2}}{1.32 \times 10^{-3} \text{ mol L}^{-1}}$$
$$= 7.6 \times 10^{-12} \text{ mol L}^{-1}$$

which is very small compared to the concentration of hydrogen ion. Within the justified accuracy, these calculated concentrations are correct. All of these concentrations are consistent with the inequalities which give the simple equation: $[H^+] \gg [OH^-]$, and $C \gg [H^+]$. But now consider a more dilute solution of acetic acid.

TABLE 13–4
Acid-Dissociation Constants for $HX \rightleftarrows H^+ + X$ in Water at 25°C[a]

$$K_a = \frac{[H^+][X]}{[HX]}$$

Acid	K_a / mol L^{-1}	Acid	K_a / mol L^{-1}
CH_3CO_2H	1.754×10^{-5}	$(HO)_3PO$	7.52×10^{-3}
$ClCH_2CO_2H$	1.39×10^{-3}	$(HO)_2PO_2^-$	6.23×10^{-8}
Cl_2CHCO_2H	5.0×10^{-2}	$(HO)PO_3^{2-}$	2.2×10^{-13}
Cl_3CCO_2H	1.3×10^{-1}		
CO_2	4.45×10^{-7}	H_2S	1.02×10^{-7}
$HOCO_2^-$	5.69×10^{-11}	HS^-	1.29×10^{-13}
HCN	6.2×10^{-10}	$HOSO_3^-$	1.03×10^{-2}
$HONO_2$	25	SO_2	1.3×10^{-2}
HONO	4.5×10^{-4}	$HOSO_2^-$	5.6×10^{-8}
HN_3	1.8×10^{-5}	HOClO	1.1×10^{-2}
NH_4^+	5.69×10^{-10}	HOCl	2.95×10^{-8}
HF	6.8×10^{-4}	$Fe(OH_2)_6^{3+}$	6.3×10^{-3}

[a] The charges are omitted from HX and X; the charge on X is one unit less positive than that on HX.

For $1.000 \times 10^{-4} M$ HOAc

By Equation 13–5s, the calculated concentrations for this solution are:

$$[H^+] = (K_a C)^{1/2} \qquad (13\text{–}5s)$$

$$[H^+] = (1.75 \times 10^{-5} \text{ mol L}^{-1} \times 1.000 \times 10^{-4} \text{ mol L}^{-1})^{1/2}$$
$$= 4.18 \times 10^{-5} \text{ mol L}^{-1}$$

$$[OAc^-] = [H^+] = 4.18 \times 10^{-5} \text{ mol L}^{-1}$$

But this calculated concentration of hydrogen ion cannot be correct because the inequality $C \gg [H^+]$ is not satisfied. If we study Figure 13–4, we see that failure of this inequality takes us from Equation 13–5 to Equation 13–4 for solution of this problem:

$$[H^+] = \frac{-K_a + (K_a^2 + 4K_a C)^{1/2}}{2}$$

Using this equation for $1.000 \times 10^{-4} M$ HOAc, we have:

$$\frac{[H^+]}{\text{mol L}^{-1}} = \frac{-1.75 \times 10^{-5} + [(1.75 \times 10^{-5})^2 + 4 \times 1.75 \times 10^{-5} \times 1.000 \times 10^{-4}]^{1/2}}{2}$$

(Continued)

$$[H^+] = 3.40 \times 10^{-5}\ \text{mol L}^{-1}$$
$$[OAc^-] = [H^+] = 3.40 \times 10^{-5}\ \text{mol L}^{-1}$$
$$[HOAc] = C - [OAc^-] = 1.000 \times 10^{-4}\ \text{mol L}^{-1} - 3.40 \times 10^{-5}\ \text{mol L}^{-1}$$
$$= 6.60 \times 10^{-5}\ \text{mol L}^{-1}$$

$$[OH^-] = \frac{K_w}{[H^+]} = \frac{1.00 \times 10^{-14}\ \text{mol}^2\ \text{L}^{-2}}{3.40 \times 10^{-5}\ \text{mol L}^{-1}} = 2.94 \times 10^{-10}\ \text{mol L}^{-1}$$

We see that the simplest equation (Equation 13–5s in Figure 13–4) is satisfactory for calculations for 0.1000*M* HOAc, but is not satisfactory for calculations for $1.000 \times 10^{-4}M$ HOAc. The error that results from using Equation 13–5 when Equation 13–4 should be used depends upon the value of the dimensionless ratio C/K_a, and Figure 13–5 shows the error as a function of this variable.[9] This error is 1% if $C/K_a \cong 2500$, is 3% if $C/K_a \cong 300$, and is 10% if $C/K_a \cong 28$. You can use this graph for help in choosing which equation to use to calculate the concentration of hydrogen ion in solutions containing a single weak acid. For example, you can see that for acetic acid, with $K_a = 1.75 \times 10^{-5}$ mol L^{-1}, the equation $[H^+] = (K_aC)^{1/2}$ is accurate to closer than 1% if $C > 2500 \times 1.75 \times 10^{-5}$ mol L^{-1} or $C > 0.044$ mol L^{-1}. (In attempting to decide which equation to use to calculate the hydrogen ion concentration, consider the accuracy with which the values of K_a and C are known.)

The calculated concentrations of species in acetic acid solutions are tabulated in Table 13–5. These calculations illustrate the effect of dilution upon an equilibrium. The extent of dissociation of acetic acid increases as the concentration is lowered. For the equilibrium

$$HOAc \rightleftarrows H^+ + OAc^-$$

with two solute particles on the product side of the equation and one solute particle on the reactant side, dilution will cause reaction to occur from left to right. The basis for this prediction is not altered by writing the equilibrium as

$$HOAc + H_2O \rightleftarrows H_3O^+ + OAc^-$$

because in the relatively dilute solutions being considered the mole fraction of the solvent is essentially constant ($x \cong 1$) as the concentration of acetic acid changes from 0.1000 mol L^{-1} to 1.000×10^{-4} mol L^{-1}.

---

[9] The error in using Equation 13–5s can be defined:

$$\text{Error} = \frac{(K_aC)^{1/2} - [H^+]}{[H^+]}$$

Use of Equation 13–4s for $[H^+]$ gives:

$$\text{Error} = \frac{\left(1 + 4\dfrac{C}{K_a}\right)^{1/2} - 1 - \left(4\dfrac{C}{K_a}\right)^{1/2}}{1 - \left(1 + 4\dfrac{C}{K_a}\right)^{1/2}}$$

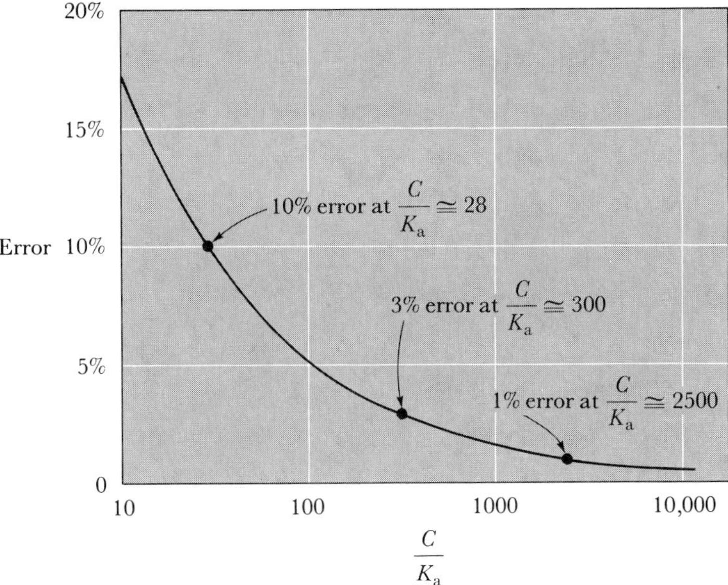

FIGURE 13–5
The error in the concentration of hydrogen ion calculated using Equation 13–5s compared to the correct value calculated using Equation 13–4s (notice logarithmic abscissa).

$$[H^+] = (K_aC)^{1/2} \qquad (13\text{–}5s)$$

$$[H^+] = \frac{-K_a + (K_a^2 + 4K_aC)^{1/2}}{2} \qquad (13\text{–}4s)$$

That the relative number of solute species on each side of the chemical equation determines the effect of dilution can be shown by a simple example. Suppose that a solution of 0.1000*M* HOAc is diluted by a factor of ten, and that no shift in equilibrium occurs upon dilution. Each concentration will be reduced to one tenth of its value before dilution. The situa-

TABLE 13–5
The Concentrations of Species in Aqueous Acetic Acid

$\frac{C}{\text{mol L}^{-1}}$	$\frac{(K_aC)^{1/2}}{\text{mol L}^{-1}}$	$\frac{[H^+] = [OAc^-]^b}{\text{mol L}^{-1}}$	$\frac{[HOAc]}{\text{mol L}^{-1}}$	Percent dissociation
0.1000	1.32×10^{-3}	1.314×10^{-3}	0.0987	1.3
0.01000	4.18×10^{-4}	4.10×10^{-4}	9.59×10^{-3}	4.1
0.001000	1.32×10^{-4}	1.24×10^{-4}	8.76×10^{-4}	12.4
1.000×10^{-4}	4.18×10^{-5}	3.40×10^{-5}	6.60×10^{-5}	34.0
1.000×10^{-5}	$(1.32 \times 10^{-5})^c$	7.11×10^{-6}	2.89×10^{-6}	71.1
1.000×10^{-6}	$(4.18 \times 10^{-6})^c$	9.49×10^{-7}	5.14×10^{-8}	94.9

[a] At $t = 25°C$; $K_a = 1.75 \times 10^{-5}$ mol L^{-1}.
[b] Calculated using Equation 13–4s,

$$[H^+] = \tfrac{1}{2}[-K_a + (K_a^2 + 4K_aC)^{1/2}]$$

[c] The concentration of hydrogen ion cannot exceed the stoichiometric concentration of acid unless $C \le 10^{-7}$ mol L^{-1}. Thus these first approximations are certainly incorrect.

tion would be:

Before dilution

$$\frac{[H^+][OAc^-]}{[HOAc]} = \frac{(1.314 \times 10^{-3}\ \text{mol L}^{-1})^2}{(0.0987\ \text{mol L}^{-1})} = 1.75 \times 10^{-5}\ \text{mol L}^{-1}$$

After dilution, with no shift in equilibrium

$$\frac{[H^+][OAc^-]}{[HOAc]} = \frac{(1.314 \times 10^{-4}\ \text{mol L}^{-1})^2}{(0.00987\ \text{mol L}^{-1})} = 1.75 \times 10^{-6}\ \text{mol L}^{-1}$$

But this value of the quotient of concentrations is not the value that corresponds to equilibrium. It is too small. The equilibrium value of the quotient of concentrations will be reached if the value of the numerator increases and the value of the denominator decreases. An increase in the relative value of the numerator results from an increase in the degree of dissociation with dilution. An additional 2.78% (4.10% − 1.32%) of the acetic acid molecules dissociate when equilibrium is established after dilution. The true situation is:

After dilution, with a shift in equilibrium

$$\frac{[H^+][OAc^-]}{[HOAc]} = \frac{(4.10 \times 10^{-4}\ \text{mol L}^{-1})^2}{(0.00959\ \text{mol L}^{-1})} = 1.75 \times 10^{-5}\ \text{mol L}^{-1}$$

Equation 13–4 can be rearranged to obtain an equation for the extent of dissociation, which is $[H^+]/C$:

$$\frac{[H^+]}{C} = \frac{1}{C} \times \left[\frac{-K_a + (K_a^2 + 4K_aC)^{1/2}}{2}\right]$$

$$= \frac{1}{2}\left\{-\left(\frac{K_a}{C}\right) + \left[\left(\frac{K_a}{C}\right)^2 + 4\left(\frac{K_a}{C}\right)\right]^{1/2}\right\}$$

This equation shows that the extent of dissociation for a weak acid depends only upon the dimensionless ratio K_a/C if the solution is one to which Equation 13–4 is applicable. Figure 13–6 gives a plot of $[H^+]/C$

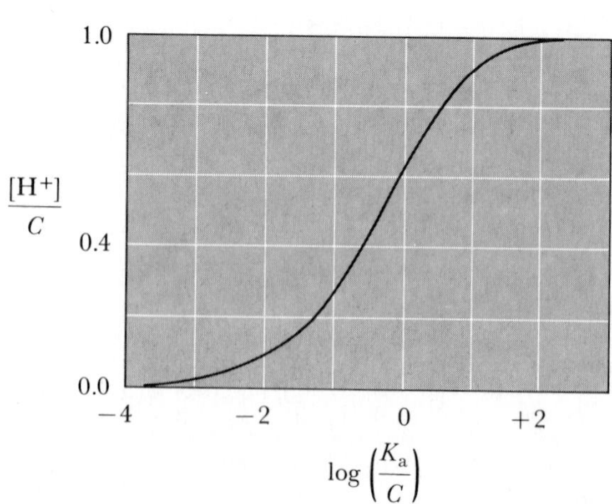

FIGURE 13–6
The extent of dissociation of a weak acid as a function of K_a/C:

$$\frac{[H^+]}{C} = \frac{1}{2}\left\{-\left(\frac{K_a}{C}\right) + \left[\left(\frac{K_a}{C}\right)^2 + 4\left(\frac{K_a}{C}\right)\right]^{1/2}\right\}$$

If $[H^+] < 10^{-6}$ mol L^{-1}, this relationship, based upon Equation 13–4, does not hold.

K_a/C	$[H^+]/C$	K_a/C	$[H^+]/C$
1.00×10^{-3}	0.031	1.00	0.618
1.00×10^{-2}	0.095	10.0	0.916
0.100	0.270	100	0.990

versus log (K_a/C). We see that the larger the value of K_a/C the greater the extent of dissociation: $[H^+]/C = 0.95$ at $K_a/C = 18$, and $[H^+]/C = 0.05$ at $K_a/C = 0.0026$.

Sample Calculations for Ammonium Chloride. Now we will continue our sample calculations by considering solutions of ammonium chloride. Some new features will be revealed by study of a very weak acid, an acid with a value of $K_a < 1 \times 10^{-8}$ mol L^{-1}. For ammonium ion at 298.2 K,

$$K_a = \frac{[H^+][NH_3]}{[NH_4^+]} = 5.69 \times 10^{-10} \text{ mol L}^{-1}$$

For 0.1000M NH_4^+

By Equation 13–5s, the calculated concentrations are:

$$[H^+] = (5.69 \times 10^{-10} \text{ mol L}^{-1} \times 0.1000 \text{ mol L}^{-1})^{1/2}$$
$$= 7.54 \times 10^{-6} \text{ mol L}^{-1}$$
$$[NH_3] = [H^+] = 7.54 \times 10^{-6} \text{ mol L}^{-1}$$
$$[NH_4^+] = C - [NH_3]$$
$$= 0.1000 \text{ mol L}^{-1} - 7.54 \times 10^{-6} \text{ mol L}^{-1}$$
$$= 0.1000 \text{ mol L}^{-1}$$
$$[OH^-] = \frac{K_w}{[H^+]} = \frac{1.00 \times 10^{-14} \text{ mol}^2 \text{ L}^{-2}}{7.54 \times 10^{-6} \text{ mol L}^{-1}} = 1.33 \times 10^{-9} \text{ mol L}^{-1}$$

These concentrations, calculated as in the 0.1000M HOAc example, are consistent with the inequalities $[H^+] \gg [OH^-]$ and $C \gg [H^+]$, which are necessary for use of Equation 13–5s (Figure 13–4).

For $1.000 \times 10^{-4}M$ NH_4^+

By Equation 13–5s, the calculated concentrations are:

$$[H^+] = (5.69 \times 10^{-10} \text{ mol L}^{-1} \times 1.000 \times 10^{-4} \text{ mol L}^{-1})^{1/2}$$
$$= 2.39 \times 10^{-7} \text{ mol L}^{-1}$$

This calculated value of $[H^+]$ conforms to the approximation $C \gg [H^+]$, but it does not conform to $[H^+] \gg [OH^-]$. According to Figure 13–4, failure of this inequality leads us back to Equation 13–4′ or 13–4′s,

$$[H^+] = (K_w + K_a C)^{1/2}$$

For the present problem, this gives

$$[H^+] = (1.00 \times 10^{-14} \text{ mol}^2 \text{ L}^{-2} + 5.69 \times 10^{-10} \times 1.000 \times 10^{-4} \text{ mol}^2 \text{ L}^{-2})^{1/2}$$
$$= 2.59 \times 10^{-7} \text{ mol L}^{-1}$$

which is 7.7% larger than the value calculated incorrectly by using Equation 13–5s. Because an appreciable amount of this hydrogen ion comes from self-dissociation of the solvent, the concentration of ammonia is not equal to the concentration of hydrogen ion. We can calculate the value of $[NH_3]$ quite accurately:

$$[NH_3] = \frac{K_a[NH_4^+]}{[H^+]} \cong \frac{K_aC}{[H^+]}$$

We may then substitute C for $[NH_4^+]$, assuming that the extent of dissociation is slight, an assumption which will prove to be correct:

$$[NH_3] = \frac{5.69 \times 10^{-10} \text{ mol L}^{-1} \times 1.000 \times 10^{-4} \text{ mol L}^{-1}}{2.59 \times 10^{-7} \text{ mol L}^{-1}}$$

$$= 2.20 \times 10^{-7} \text{ mol L}^{-1}$$

$$[NH_4^+] = C - [NH_3]$$

$$= 1.000 \times 10^{-4} \text{ mol L}^{-1} - 2.20 \times 10^{-7} \text{ mol L}^{-1}$$

$$= 9.98 \times 10^{-5} \text{ mol L}^{-1}$$

$$[OH^-] = \frac{K_w}{[H^+]} = \frac{1.00 \times 10^{-14} \text{ mol}^2 \text{ L}^{-2}}{2.59 \times 10^{-7} \text{ mol L}^{-1}}$$

$$= 3.86 \times 10^{-8} \text{ mol L}^{-1}$$

After performing so complex a calculation, you should check the results by substituting the derived concentrations into the equations you have not used individually. For the present example, these are:

$$\frac{[H^+][NH_3]}{[NH_4^+]} = \frac{(2.59 \times 10^{-7} \text{ mol L}^{-1})(2.20 \times 10^{-7} \text{ mol L}^{-1})}{(9.98 \times 10^{-5} \text{ mol L}^{-1})}$$

$$= 5.71 \times 10^{-10} \text{ mol L}^{-1}$$

This result "agrees" with $K_a = 5.69 \times 10^{-10}$ mol L^{-1}. The check on electroneutrality is a second check of our derived concentrations. The net concentration of positive charge is

$$[H^+] + [NH_4^+] - [OH^-] - [Cl^-]$$

$$= 2.59 \times 10^{-7} \text{ mol L}^{-1} + 9.98 \times 10^{-5} \text{ mol L}^{-1}$$

$$- 3.86 \times 10^{-8} \text{ mol L}^{-1} - 1.000 \times 10^{-4} \text{ mol L}^{-1}$$

$$= +0.20 \times 10^{-7} \text{ mol L}^{-1}$$

This net charge is small compared to the concentration of ammonium ion, the species that has primary control over the acid–base equilibria in this solution. (If additional significant figures had been used for the various concentrations, the final calculated net charge would have been smaller.) We see that the simplest equation (Equation 13–5s in Figure 13–4) is satisfactory for calculations for $0.1000M$ NH_4^+ but is not satisfactory for calculations for $1.000 \times 10^{-4}M$ NH_4^+. The error in using Equation 13–5s when Equation 13–4′s should be used depends upon the dimensionless quotient K_aC/K_w, and Figure 13–7 shows the error as a function of this

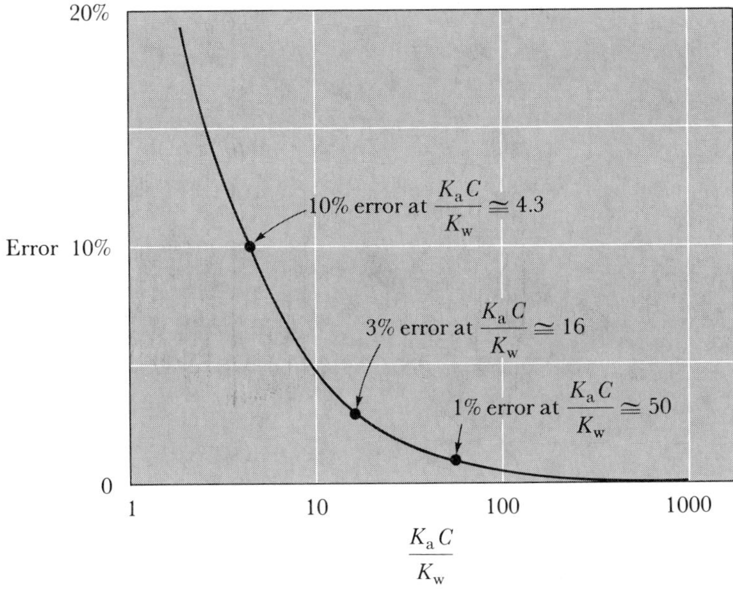

FIGURE 13–7
The error in the concentration of hydrogen ion calculated using Equation 13–5s compared to the correct value calculated using Equation 13–4′s (notice logarithmic abscissa).

$$[H^+] = (K_a C)^{1/2} \qquad (13\text{–}5s)$$
$$[H^+] = (K_w + K_a C)^{1/2} \qquad (13\text{–}4's)$$

variable.[10] This error is 1% if $K_a C/K_w \cong 50$, is 3% if $K_a C/K_w \cong 16$, and is 10% if $K_a C/K_w \cong 4.3$. Thus for ammonium ion with $K_a = 5.69 \times 10^{-10}$ mol L^{-1}, the equation $[H^+] = (K_a C)^{1/2}$ is accurate to within 1% if $C > 8.8 \times 10^{-4}$ mol L^{-1}.

The concentrations of species in ammonium chloride solutions are summarized in Table 13–6. A point worth noting in this table is the fact that the extent of acid dissociation does not continue to increase with dilution in solutions in which the hydrogen ion is produced primarily from dissociation of the solvent. At high dilution of ammonium ion ($C < 1 \times 10^{-5}$ mol L^{-1}), the concentration of hydrogen ion closely approaches the value for pure water. For the limiting value of $[H^+]$, 1.00×10^{-7} mol L^{-1},

---

[10]The error in using Equation 13–5s can be defined:

$$\text{Error} = \frac{[H^+] - (K_a C)^{1/2}}{[H^+]}$$

Using this definition in Equation 13–4′ for $[H^+]$, we get

$$\text{Error} = 1 - \frac{(K_a C/K_w)^{1/2}}{(1 + K_a C/K_w)^{1/2}}$$

TABLE 13–6
The Concentrations of Species in Ammonium Chloride Solutions[a]

$\frac{C}{\text{mol L}^{-1}}$	$\frac{(K_aC)^{1/2}}{\text{mol L}^{-1}}$	$\frac{[H^+]^b}{\text{mol L}^{-1}}$	$\frac{[NH_3]^c}{\text{mol L}^{-1}}$	$\frac{[NH_4^+]}{\text{mol L}^{-1}}$	Percent dissociation
0.1000	7.54×10^{-6}	7.54×10^{-6}	7.54×10^{-6}	0.1000	7.5×10^{-3}
0.01000	2.39×10^{-6}	2.39×10^{-6}	2.39×10^{-6}	0.01000	2.4×10^{-2}
0.001000	7.54×10^{-7}	7.61×10^{-7}	7.48×10^{-7}	9.99×10^{-4}	7.5×10^{-2}
1.000×10^{-4}	2.39×10^{-7}	2.59×10^{-7}	2.20×10^{-7}	9.98×10^{-5}	0.22
1.000×10^{-5}	$(7.54 \times 10^{-8})^d$	1.25×10^{-7}	4.55×10^{-8}	9.95×10^{-6}	0.46
1.000×10^{-6}	$(2.39 \times 10^{-8})^d$	1.03×10^{-7}	5.52×10^{-9}	9.94×10^{-7}	0.55

[a]At $t = 25°C$; $K_a = 5.69 \times 10^{-10}$ mol L^{-1}.
[b]Calculated using the equation

$$[H^+] = (K_w + K_aC)^{1/2}$$

[c]Calculated using the equation

$$[NH_3] = [H^+] - \frac{K_w}{[H^+]} \quad (\text{for } C \geq 10^{-4} \text{ mol L}^{-1}) \quad \text{or}$$

$$[NH_3] = \frac{K_a[NH_4^+]}{[H^+]} = \frac{K_aC}{[H^+]} \quad (\text{for } C \leq 10^{-5} \text{ mol L}^{-1})$$

[d] An incorrectly calculated concentration of hydrogen ion. A solution of NH_4^+, an acid, cannot have $[H^+] \ll 1.00 \times 10^{-7}$ mol L^{-1}.

the extent of dissociation becomes

$$\frac{[NH_3]}{C} \cong \frac{[NH_3]}{[NH_4^+]} = \frac{K_a}{[H^+]}$$

$$= \frac{5.69 \times 10^{-10} \text{ mol L}^{-1}}{1.00 \times 10^{-7} \text{ mol L}^{-1}} = 0.00569 \text{ or } 0.569\%$$

Six different equations can be used to calculate the concentration of hydrogen ion in an aqueous solution of acid, and we have used five of these in our calculations. Which should be used depends upon the values of C and K_a, as is summarized in Figure 13–8. The borderlines are drawn where use of a simpler equation introduces a 2% error in the calculated value of $[H^+]$. We have not used the complete equation (Equation 13–3); the figure shows that it is needed only in a range of values of K_a and C that is unlikely to arise in practical problems.

The figure also shows that dilute solutions of an acid with a value of K_a that is not too small ($K_a > 10^{-4}$ mol L^{-1}) can be treated as if the acid were completely dissociated. That is, Equation 13–1, $[H^+] = C$, can be used to calculate the concentration of hydrogen ion.

This can be illustrated with a calculation for a solution of dichloroacetic acid, $1.00 \times 10^{-4}M$ Cl_2CHCO_2H ($K_a = 5.0 \times 10^{-2}$ mol L^{-1}):

$$[H^+] \cong C = 1.00 \times 10^{-4} \text{ mol L}^{-1}$$

$$[Cl_2CHCO_2^-] = [H^+] = 1.00 \times 10^{-4} \text{ mol L}^{-1}$$

The concentration of undissociated dichloroacetic acid can be calculated

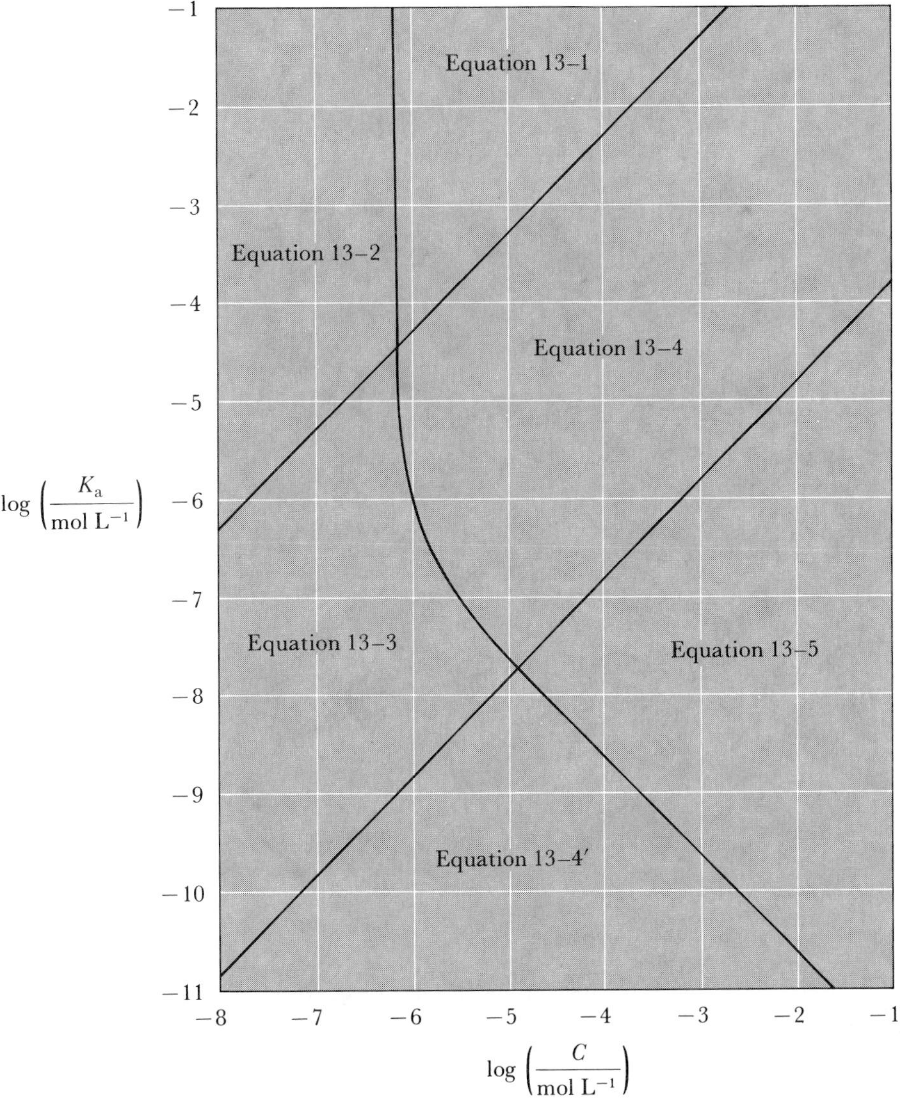

FIGURE 13–8
The ranges of values of K_a and C in which the various equations can be used to calculate $[H^+]$. The borderlines are drawn where use of a simpler equation introduces a 2% error in the calculated value of $[H^+]$. [Based upon figure in J. D. Burke, *J. Chem. Educ.* **53**, 79 (1976)].

using the equilibrium-constant equation, appropriately rearranged:

$$[Cl_2CHCO_2H] = \frac{[H^+][Cl_2CHCO_2^-]}{K_a}$$

$$= \frac{(1.00 \times 10^{-4}\ mol\ L^{-1})^2}{(5.0 \times 10^{-2}\ mol\ L^{-1})} = 2.0 \times 10^{-7}\ mol\ L^{-1}$$

This is 0.2% of the stoichiometric concentration. The extent of dissociation is 99.8%, and our use of Equation 13–1 to obtain $[H^+]$ was justified.

We can use Figure 13–8 also to draw some rough borderlines in nomenclature based upon the intersection of the four regions at $K_a \cong 10^{-8}$ mol L^{-1} and $C \cong 10^{-5}$ mol L^{-1}. These are:

Weak acid:[11]	$18C > K_a > 10^{-8}$ mol L^{-1}
Very weak acid:	$K_a < 10^{-8}$ mol L^{-1}
Dilute solution:	$C > 10^{-5}$ mol L^{-1}
Very dilute solution:	$C < 10^{-5}$ mol L^{-1}

THE NEUTRALIZATION OF A WEAK ACID BY A STRONG BASE

We have just considered solutions of weak acids containing no other acidic or basic solutes. Now we will consider solutions containing both a weak acid and a strong base; in such solutions, this neutralization reaction

$$HX + OH^- = X^- + H_2O$$

occurs. We will consider the course of this reaction in the titration of a definite amount of acid with a solution of base. At each point in the titration we can calculate concentrations of the species by using equilibrium principles. Each calculation depends upon making an appropriate assignment of the stoichiometric concentration at that point.

Stoichiometric Concentrations. Let us consider the titration of 100.0 cm^3 of $0.1000M$ HOAc with a solution of $0.2500M$ NaOH. The amount of acetic acid in the initial solution is

$$n_{HOAc} = 100.0\ cm^3 \times \frac{1\ L}{1000\ cm^3} \times 0.1000\ mol\ L^{-1} = 1.000 \times 10^{-2}\ mol$$

The amount of base needed to neutralize this acid is 1.000×10^{-2} mol, which will be contained in

$$V = \frac{n_{OH^-}}{C_{NaOH}}$$

$$= \frac{1.000 \times 10^{-2}\ mol}{0.2500\ mol\ L^{-1}} = 0.04000\ L = 40.00\ cm^3$$

Before addition of any base the solution has the composition

$$C_{HOAc} = 0.1000\ mol\ L^{-1}$$

At the equivalence point, the reaction

$$HOAc + OH^- = OAc^- + H_2O$$

---

[11] If $K_a/C = 18$, $[H^+]/C = 0.95$, and the acid is almost completely dissociated (see Figure 13–6).

has occurred essentially to completion; we will assume it has occurred to completion to assign the stoichiometric composition. The solution at the equivalence point has a volume of 140.0 cm^3, and it contains 1.000×10^{-2} mol of sodium acetate.[12] At the equivalence point the solution prepared by mixing solutions of acetic acid and sodium hydroxide is identical to one prepared with water and sodium acetate. The stoichiometric concentration of sodium acetate in this solution is

$$C_{\text{NaOAc}} = \frac{n}{V}$$

$$= \frac{0.01000 \text{ mol}}{0.1400 \text{ L}} = 0.0714 \text{ mol L}^{-1}$$

In a volume of 120.0 cm^3, the solution at the midpoint of the titration contains, before reaction,

$$0.01000 \text{ mol HOAc} \quad \text{and} \quad 0.00500 \text{ mol NaOH}$$

Again for the purpose of calculating the stoichiometric concentrations, we will consider the reaction to have gone to completion, thereby making the amounts of solutes present equal to

$$0.00500 \text{ mol HOAc} \quad \text{and} \quad 0.00500 \text{ mol NaOAc}$$

In a volume of 120.0 cm^3 (0.1200 L), the stoichiometric concentrations are:

$$C_{\text{HOAc}} = \frac{0.00500 \text{ mol}}{0.1200 \text{ L}} = 0.0417 \text{ mol L}^{-1}$$

$$C_{\text{NaOAc}} = \frac{0.00500 \text{ mol}}{0.1200 \text{ L}} = 0.0417 \text{ mol L}^{-1}$$

Thus to calculate the concentration of hydrogen ion in the solution at various stages of the titration, we start with the stoichiometric concentrations:

At 0% titration, $C_{\text{HOAc}} = 0.1000 \text{ mol L}^{-1}$
At 100% titration (the equivalence point), $C_{\text{NaOAc}} = 0.0714 \text{ mol L}^{-1}$
At 50.0% titration, $C_{\text{HOAc}} = 0.0417 \text{ mol L}^{-1}$ and $C_{\text{NaOAc}} = 0.0417 \text{ mol L}^{-1}$

We already have calculated the concentration of hydrogen ion in 0.1000M HOAc.

The Equivalence Point; Base Dissociation. For a solution of 0.0714M NaOAc, we will start by recognizing that this is simply an aqueous solution of a Brønsted base, OAc^-, in which the principal equilibrium is

$$OAc^- + H_2O \rightleftarrows HOAc + OH^-$$

[12] It is reasonable to assume additivity of volumes for mixing of these dilute solutions.

We can decide that this is the principal equilibrium if we recall that a weak Brønsted base will interact in this way with water, a weak Brønsted acid, to give an alkaline solution (see the previous discussion in Section 12–3). We also should recall that the equilibrium

$$H_2O \rightleftarrows H^+ + OH^-$$

contributes a lesser amount of hydroxide ion than does the principal equilibrium. The equilibrium constant for the reaction of the Brønsted base, OAc^-, with water to produce hydroxide ion,

$$K_b = \frac{[HOAc][OH^-]}{[OAc^-]}$$

is related to the equilibrium constant for acid dissociation of the conjugate acid. This can be shown by multiplying the numerator and denominator of this quotient by the concentration of hydrogen ion:

$$K_b = \frac{[HOAc][OH^-]}{[OAc^-]} \times \frac{[H^+]}{[H^+]}$$

and then by grouping the concentration terms:

$$K_b = \frac{[HOAc]}{[H^+][OAc^-]} \times [H^+][OH^-]$$

$$= \frac{1}{K_a} \times K_w = \frac{K_w}{K_a}$$

Thus the equilibrium constant for this hydroxide ion-forming reaction of acetate ion at 298.2 K is

$$K_b = \frac{1.00 \times 10^{-14}\ \text{mol}^2\ \text{L}^{-2}}{1.754 \times 10^{-5}\ \text{mol L}^{-1}} = 5.70 \times 10^{-10}\ \text{mol L}^{-1}$$

Table 13–7 presents values of K_b for the conjugate bases of all acids for which values of K_a were given in Table 13–4. However, you should recognize that Table 13–7 is superfluous. Since a base-dissociation constant is obtainable directly from the acid-dissociation constant for the conjugate acid ($K_b = K_w/K_a$), a table of acid-dissociation constants such as Table 13–4 also provides values of the equilibrium constants for the conjugate bases reacting with the solvent at the same temperature. The numerical relationship between the acid-dissociation constant and base-dissociation constant for a particular conjugate acid–base pair also depends on the autoprotolysis constant of the solvent.

It is interesting to observe the mathematical relationship of the combinations of chemical equations and the combinations of equilibrium constants. The chemical equations for some reactions may be obtained by *subtracting* chemical equations for two reactions. For example,

$$OAc^- + H_2O = HOAc + OH^-$$

is

TABLE 13–7
Equilibrium Constants for $H_2O + B \rightleftarrows HB + OH^-$ in Water at 25°C[a]

$$K_b = \frac{[HB][OH^-]}{[B]}$$

Base[b]	K_b / mol L^{-1}	Base[b]	K_b / mol L^{-1}
$CH_3CO_2^-$	5.70×10^{-10}	$(HO)_2PO_2^-$	1.33×10^{-12}
$ClCH_2CO_2^-$	7.19×10^{-12}	$(HO)PO_3^{2-}$	1.61×10^{-7}
$Cl_2CHCO_2^-$	2.0×10^{-13}	PO_4^{3-}	4.6×10^{-2}
$Cl_3CCO_2^-$	7.7×10^{-14}	HS^-	9.8×10^{-8}
$HOCO_2^-$	2.25×10^{-8}	S^{2-}	7.8×10^{-2}
CO_3^{2-}	1.76×10^{-4}	SO_4^{2-}	9.7×10^{-13}
CN^-	1.6×10^{-5}	$HOSO_2^-$	7.7×10^{-13}
NO_3^-	4.0×10^{-16}	SO_3^{2-}	1.79×10^{-7}
NO_2^-	2.2×10^{-11}	ClO_2^-	9.0×10^{-13}
N_3^-	5.6×10^{-10}	ClO^-	3.39×10^{-7}
NH_3	1.76×10^{-5}	$Fe(OH_2)_5OH^{2+}$	1.59×10^{-12}
F^-	1.47×10^{-11}		

[a]The charges have been omitted from HB and B; the charge on B is one unit more negative than that on HB.
[b]The bases are the conjugate bases for the acids given in Table 13–4. These values of K_b are obtained from the values of K_a:

$$K_b = \frac{K_w}{K_a} = \frac{1.00 \times 10^{-14}\ \text{mol}^2\ \text{L}^{-2}}{K_a}$$

$$H_2O = H^+ + OH^-$$

minus

$$HOAc = H^+ + OAc^-$$

The equilibrium constant for the first reaction is the *quotient* of equilibrium constants for the last two reactions. (If the chemical equation for a reaction is obtained by *adding* chemical equations for two reactions, the equilibrium constant for this reaction is the *product* of the equilibrium constants for the two reactions. Other illustrations of this procedure will appear from time to time.)

Now to continue with the numerical calculation of base dissociation, we recognize that we can solve this problem by using an equation analogous to Equation 13–5s for the calculation of the concentration of hydrogen ion in ammonium ion solutions with $C > 10^{-3}$ mol L^{-1}. (The value of K_b for acetate ion is very close to the value of K_a for ammonium ion.) With

$$[OH^-] = [HOAc] \quad \text{and} \quad [OAc^-] = C - [HOAc^-] \cong C$$

we have

$$K_b = \frac{[OH^-]^2}{C}$$

(*Continued*)

and

$$[OH^-] = (K_bC)^{1/2}$$
$$[OH^-] = (5.70 \times 10^{-10}\ \text{mol L}^{-1} \times 0.0714\ \text{mol L}^{-1})^{1/2}$$
$$= 6.38 \times 10^{-6}\ \text{mol L}^{-1}$$
$$[HOAc] = 6.38 \times 10^{-6}\ \text{mol L}^{-1}$$
$$[OAc^-] = 0.0714\ \text{mol L}^{-1} - 6.38 \times 10^{-6}\ \text{mol L}^{-1} = 0.0714\ \text{mol L}^{-1}$$
$$[H^+] = \frac{K_w}{[OH^-]} = \frac{1.00 \times 10^{-14}\ \text{mol}^2\ \text{L}^{-2}}{6.38 \times 10^{-6}\ \text{mol L}^{-1}} = 1.57 \times 10^{-9}\ \text{mol L}^{-1}$$
$$\text{pH} = -\log\left(\frac{[H^+]}{\text{mol L}^{-1}}\right) = -\log(1.57 \times 10^{-9}) = 8.80$$

(This calculation will help you with analogous calculations for solutions of other weak bases.) Depending upon the stoichiometric concentration of the base and the value of K_b, calculations for base dissociation involve equations analogous to Equations 13–1 through 13–5 presented in our detailed discussion of acids. These equations with $[H^+]$ replaced by $[OH^-]$ and with K_a replaced by K_b (i.e., K_w/K_a) plus a set of figures analogous to Figures 13–4 through 13–8 provide you with the basis for doing all possible calculations pertaining to base dissociation. If you have learned how to do equilibrium calculations for acids, you also can do them for bases. For many solutions of weak bases it will be appropriate to use the equation analogous to Equation 13–5s, $[OH^-] = (K_bC)^{1/2}$.

The Titration Midpoint. Now we will consider the midpoint of the titration, the solution with the stoichiometric composition

$$C_{HOAc} = 0.0417\ \text{mol L}^{-1} \quad \text{and} \quad C_{NaOAc} = 0.0417\ \text{mol L}^{-1}$$

For a partially neutralized weak acid the principal equilibrium is the acid-dissociation reaction; for the present problem it is

$$HOAc \rightleftarrows H^+ + OAc^-$$

This is a reaction by which the species HOAc and OAc^- determine the concentration of hydrogen ion. The concentrations of these species at equilibrium are the stoichiometric concentrations altered slightly by the reaction that produces hydrogen ion:

$$[OAc^-] = C_{NaOAc} + [H^+]$$
$$= 0.0417\ \text{mol L}^{-1} + [H^+]$$
$$[HOAc] = C_{HOAc} - [H^+]$$
$$= 0.0417\ \text{mol L}^{-1} - [H^+]$$

That is, in the half-neutralized solution, the concentrations of the species HOAc and OAc^- are equal, except for a slight inequality introduced by dissociation of the acid: every hydrogen ion formed converts an acetic acid molecule into an acetate ion. Substitution of these relationships into the equilibrium-constant equation gives

$$\frac{[H^+](0.0417 \text{ mol L}^{-1} + [H^+])}{(0.0417 \text{ mol L}^{-1} - [H^+])} = 1.75 \times 10^{-5} \text{ mol L}^{-1}$$

Examination of this equation suggests that $[H^+] \ll 0.0417 \text{ mol L}^{-1}$, making

$$0.0417 \text{ mol L}^{-1} + [H^+] \cong 0.0417 \text{ mol L}^{-1}$$

and

$$0.0417 \text{ mol L}^{-1} - [H^+] \cong 0.0417 \text{ mol L}^{-1}$$

The simplified equation leads to

$$[H^+] = 1.75 \times 10^{-5} \text{ mol L}^{-1} \times \frac{0.0417 \text{ mol L}^{-1}}{0.0417 \text{ mol L}^{-1}} = 1.75 \times 10^{-5} \text{ mol L}^{-1}$$

(Because 1.75×10^{-5} is small compared to 0.0417, the approximations upon which this calculation is based are valid.) This interesting result is generally valid: *The concentration of hydrogen ion in a half-neutralized solution of a weak monobasic acid is equal in value to the acid-dissociation constant.*

The Complete Titration Curve. Within the framework of the assumptions we used in calculating the concentration of hydrogen ion in a half-neutralized solution of acetic acid, we can derive a simple equation for the concentration of hydrogen ion as a function of the fraction of neutralization, f, of the weak acid HZ:

$$[HZ] = C_{HZ}(1 - f)$$
$$[Z^-] = C_{HZ} f$$

which gives

$$[H^+] = K_{HZ} \times \frac{[HZ]}{[Z^-]} = K_{HZ} \times \frac{C_{HZ}(1 - f)}{C_{HZ} f} = K_{HZ} \times \frac{1 - f}{f}$$

Using this equation to calculate the concentration of hydrogen ion in solutions of acetic acid that are 30%, 40%, 50%, 60%, and 70% neutralized, we find:

f	$\frac{1-f}{f}$	$\frac{[H^+]}{\text{mol L}^{-1}}$
0.30	2.33	4.08×10^{-5}
0.40	1.50	2.63×10^{-5}
0.50	1.00	1.75×10^{-5}
0.60	0.67	1.17×10^{-5}
0.70	0.43	7.5×10^{-6}

With these values and those already calculated, we can calculate the complete titration curve for acetic acid. Such a graph is given in Figure

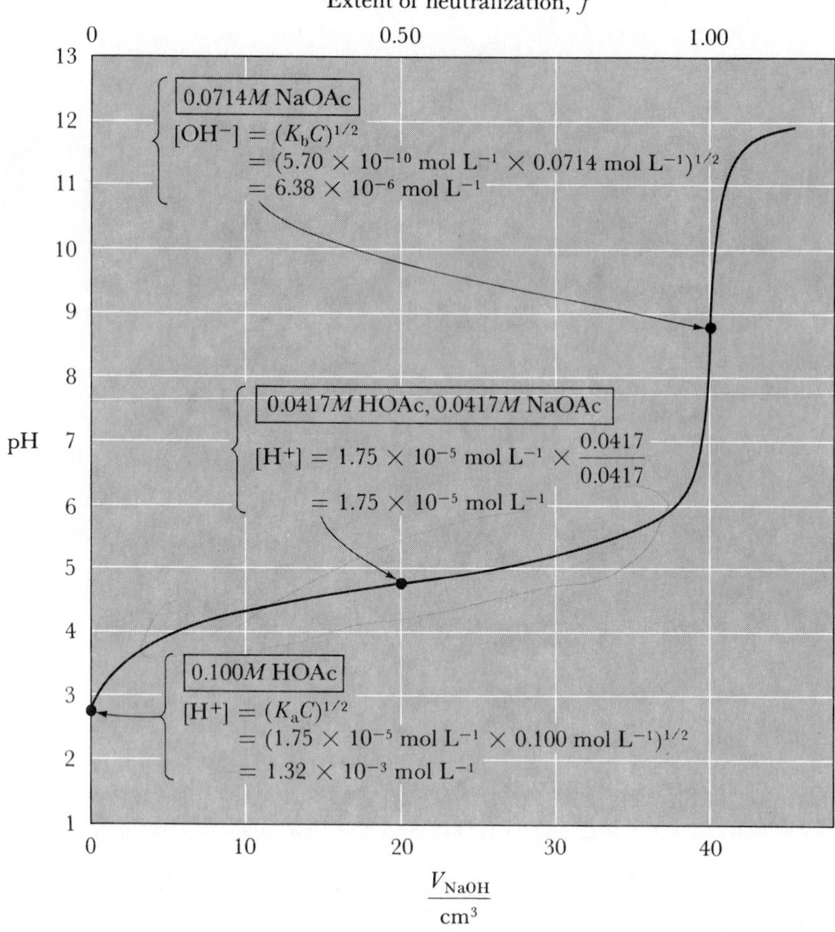

FIGURE 13–9
The neutralization of acetic acid with sodium hydroxide. A 100-cm^3 sample of 0.1000*M* HOAc is titrated with 0.2500*M* NaOH; the equivalence point is at 40.00 cm^3 of added NaOH. Boxes show stoichiometric concentrations at 0, 50%, and 100% neutralization.

13–9. For points appreciably past the equivalence point ($f = 1.00$), the concentration of hydroxide ion is that provided by the excess base. For instance, at 0.80 cm^3 of base added after the equivalence point ($f = 1.020$, $V = 140.80$ cm^3), the stoichiometric composition of the solution is

$$C_{\mathrm{NaOH}} = 0.2500 \text{ mol L}^{-1} \times \frac{0.80 \text{ cm}^3}{140.80 \text{ cm}^3} = 1.42 \times 10^{-3} \text{ mol L}^{-1}$$

$$C_{\mathrm{NaOAc}} = 0.1000 \text{ mol L}^{-1} \times \frac{100.00 \text{ cm}^3}{140.80 \text{ cm}^3} = 0.0710 \text{ mol L}^{-1}$$

The concentration of hydroxide ion is equal to the stoichiometric concentration of sodium hydroxide:

$$[\mathrm{OH}^-] = C_{\mathrm{NaOH}} = 1.42 \times 10^{-3} \text{ mol L}^{-1}$$

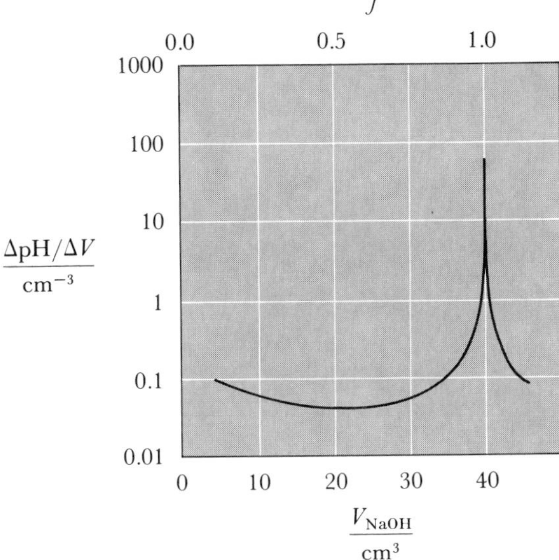

FIGURE 13–10
The slope of the titration curve $\Delta pH/\Delta V_{NaOH}$ versus V_{NaOH} for titration of 100 cm^3 of $0.1000M$ HOAc with $0.2500M$ NaOH. (The titration curve is shown in Figure 13–9; notice logarithmic scale for ordinate.)

The reaction of acetate ion with water,

$$OAc^- + H_2O = HOAc + OH^-$$

would produce additional hydroxide ion, but it occurs to a negligible extent. This can be shown by calculating the concentration of acetic acid:

$$[HOAc] = K_b \frac{[OAc^-]}{[OH^-]}$$

$$[HOAc] = 5.70 \times 10^{-10} \text{ mol L}^{-1} \times \frac{0.0710 \text{ mol L}^{-1}}{1.42 \times 10^{-3} \text{ mol L}^{-1}}$$

$$= 2.85 \times 10^{-8} \text{ mol L}^{-1}$$

which is very small compared to 1.42×10^{-3} mol L^{-1}.

Particularly worth noting in Figure 13–9 is the dependence of the slope of the line $\Delta pH/\Delta V_{NaOH}$ upon the extent of neutralization. Figure 13–10, a plot of $\Delta pH/\Delta V_{NaOH}$ versus V_{NaOH}, shows the large slope in the vicinity of the equivalence point ($f = 1.00 \pm 0.03$); this helps us find the equivalence point in an acid–base titration by use of an indicator or by electrometric means. The small slope at half-neutralization ($f = 0.5 \pm 0.2$) is relevant in the use of partially neutralized weak acids or weak bases as buffer solutions.

BUFFER SOLUTIONS

A solution is said to be buffered if its pH remains relatively constant even though small amounts of acid or base are added to the solution. Such solutions are of considerable importance. The normal pH of blood is 7.35–7.45, and it is maintained near this value because carbonate species, phosphate species, plasma proteins, and hemoglobin (all acid–base systems) are components of blood. Generally, solutions containing partially neutralized weak acids have buffering capacity. A solution containing a partially neutralized weak acid is one containing both a weak acid and its conjugate base.

The way in which the mixture of a weak acid and its conjugate base maintains relatively constant pH will be illustrated by comparing two solutions, each initially with $[H^+] = 1.75 \times 10^{-5}$ mol L^{-1} (pH = 4.76). One is a solution of nitric acid, a strong acid, and the other is a solution of acetic acid and its salt, sodium acetate (containing the conjugate base, acetate ion). The stoichiometric concentrations of these two solutions are

1. Nitric acid solution:

$$C_{HNO_3} = 1.75 \times 10^{-5} \text{ mol L}^{-1}$$

2. Acetic acid–sodium acetate solution:

$$C_{HOAc} = 1.00 \times 10^{-2} \text{ mol L}^{-1}$$
$$C_{NaOAc} = 1.00 \times 10^{-2} \text{ mol L}^{-1}$$

Consider the change in pH of the first solution when 2.000×10^{-4} mol of hydrogen chloride, a strong acid, is added to 1.00 L of the solution. The hydrogen chloride dissociates completely upon being dissolved in water, giving 2.00×10^{-4} mol of hydrogen ion. This hydrogen ion, added to that from the nitric acid present initially, gives 2.175×10^{-4} mol of hydrogen ion:

H^+ from HNO_3	1.75×10^{-5} mol
H^+ from HCl	20.00×10^{-5} mol
Total H^+	21.75×10^{-5} mol

The hydrogen ion concentration is 2.175×10^{-4} mol L^{-1}, or 12.43 times the concentration before addition of the hydrogen chloride. Clearly, in this unbuffered solution with an initial pH = 4.76, the addition of 2.0×10^{-4} mol of hydrogen chloride has a substantial effect on the concentration of hydrogen ion and on the pH, which is decreased to 3.66.

The situation is quite different when 2.00×10^{-4} mol of hydrogen chloride is added to 1.00 L of solution with the composition 0.0100*M* HOAc, 0.0100*M* NaOAc. Before addition of the hydrochloric acid, the hydrogen ion concentration of this solution is 1.75×10^{-5} mol L^{-1}, as calculated in the preceding section for another solution in which $C_{HOAc} = C_{NaOAc}$. The addition of 2.00×10^{-4} mol of hydrogen chloride to 1.00 L of the solution converts 2.00×10^{-4} mol of acetate ion to 2.00×10^{-4} mol of acetic acid, making the stoichiometric concentrations 0.0102*M* HOAc, 0.0098*M* NaOAc, and 0.00020*M* NaCl. As in the calculations already given for a half-neutralized acid, we can assume provisionally that

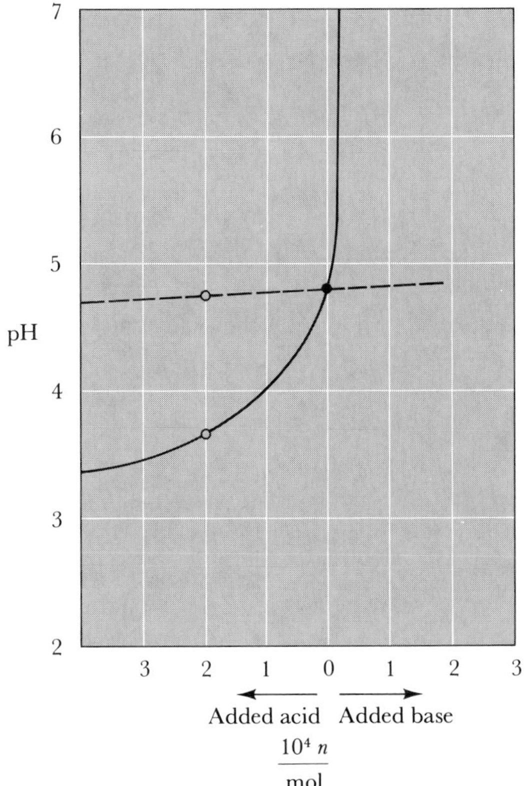

FIGURE 13–11
Titration curves for nitric acid and 0.0100*M* HOAc in the region where $[H^+] = 1.75 \times 10^{-5}$ mol L^{-1}.

Nitric acid ————

Acetic acid – – – – –

● Solutions with pH = 4.76

○ Solutions formed by addition of 2.0×10^{-4} mol of strong acid to 1.00 L of solution.

the concentrations of HOAc and OAc^- are stoichiometric:

$$[HOAc] = C_{HOAc} - [H^+] \cong C_{HOAc} = 0.01020 \text{ mol L}^{-1}$$
$$[OAc^-] = C_{NaOAc} + [H^+] \cong C_{NaOAc} = 0.0098 \text{ mol L}^{-1}$$

The equilibrium-constant equation then becomes

$$\frac{[H^+] \times 0.0098 \text{ mol L}^{-1}}{0.0102 \text{ mol L}^{-1}} = 1.75 \times 10^{-5} \text{ mol L}^{-1}$$

or

$$[H^+] = 1.75 \times 10^{-5} \text{ mol L}^{-1} \times \frac{0.0102}{0.0098} = 1.82 \times 10^{-5} \text{ mol L}^{-1}$$

The calculated concentration of hydrogen ion is small compared to 0.010 mol L^{-1}; therefore our provisional assumption is valid. This concentration of hydrogen ion is 1.04 times the concentration before addition of the hydrochloric acid; the pH has decreased only slightly, from 4.76 to 4.74.

The contrast between these two situations is striking. Addition of 2.000×10^{-4} mol of hydrogen chloride to the liter of unbuffered solution of pH = 4.76 lowers the pH to 3.66, but the same amount of acid added to the buffered solution lowers the pH only to 4.74.

The contrasting properties of the buffered and unbuffered solutions can be understood by comparing the titration curves for acetic acid and nitric acid (or any other strong acid) in the region of pH = 4.76. The relevant sections of the titration curves, taken from Figures 13–3 and 13–9, are given in Figure 13–11.

If a buffer solution is so concentrated that the approximations

$$[HX] = C_{HX} - [H^+] = C_{HX} \quad \text{and} \quad [X^-] = C_{NaX} + [H^+] = C_{NaX}$$

are valid, dilution of the buffer solution does not change the pH appreciably; the concentration of hydrogen ion depends only upon the ratio of concentrations,

$$[H^+] = K_a \frac{[HX]}{[X^-]} \cong K_a \frac{C_{HX}}{C_{NaX}}$$

which is not altered by dilution.

INDICATORS AND THEIR USE

Many organic substances are weak bases, and some of these weak bases also are intensely colored. If protonation of the basic site (an unshared pair of electrons) alters the relative energies of the ground and excited electronic states of the molecule appreciably, the color of the conjugate acid (i.e., the protonated species) may be different from that of the base. Such a substance is useful as an indicator of the concentration of hydrogen ion, for it is the concentration of hydrogen ion that determines the relative amounts of indicator in each of the two differently colored forms.

The indicator, a conjugate acid–base pair, is involved in an equilibrium

$$BH^+ \rightleftarrows B + H^+$$

for which there is an equilibrium constant:

$$K_a = \frac{[B][H^+]}{[BH^+]}$$

Because indicators are intensely colored, they are used at very low concentrations; therefore it is the other acid–base systems present at higher concentrations that are largely responsible for establishing the concentration of hydrogen ion. The indicator system comes to equilibrium with this concentration of hydrogen ion, and the relative concentrations of the two differently colored forms are

$$\frac{[BH^+]}{[B]} = \frac{[H^+]}{K_a}$$

This relationship of the concentrations of the protonated and unprotonated forms of a conjugate acid–base pair of species is applicable to all such pairs, not just to indicators.

An equation for the fraction of the base present in each of the two forms is easy to derive. The fraction, f_B, of the acid–base conjugate pair B and BH^+ which is present as base B is

$$\text{Fraction present as B} = f_B = \frac{[B]}{[B] + [BH^+]}$$

This equation is simply a definition of the fraction of the substance present in the basic form. Dividing numerator and denominator by [B] gives

$$f_B = \frac{\frac{[B]}{[B]}}{\frac{[B]}{[B]} + \frac{[BH^+]}{[B]}}$$

But $[BH^+]/[B]$ can be expressed in terms of K_a and $[H^+]$,

$$\frac{[BH^+]}{[B]} = \frac{[H^+]}{K_a}$$

which gives

$$f_B = \frac{1}{1 + \frac{[H^+]}{K_a}} = \frac{K_a}{K_a + [H^+]}$$

Correspondingly, the fraction present as BH^+ can be derived:

$$\text{Fraction present as } BH^+ = f_{BH^+} = \frac{[H^+]}{K_a + [H^+]}$$

The sum of these two fractions is one.

These relationships for f_B and f_{BH^+} show that conversion of an indicator base (or any other base) to its conjugate acid occurs over an approximately 100-fold range of hydrogen ion concentration (a range of approximately 2 pH units). At $[H^+] = 0.1K_a$, $[BH^+]/[B] = 0.1$, and at a 100-fold higher concentration, that is, $[H^+] = 10K_a$, $[BH^+]/[B] = 10$. At the lower of these concentrations of hydrogen ion, the base is ~91% present as B, and at the higher concentration the base is ~9% present as B. For these and other relative concentrations of hydrogen ion, the situation is shown in Figure 13-12 and is summarized here:

$\frac{[H^+]}{K_a}$	$\frac{[BH^+]}{[B]}$	$\frac{[BH^+]}{[BH^+] + [B]}$
0.0100	0.0100	0.0099
0.100	0.100	0.091
0.300	0.300	0.231
1.00	1.00	0.500
3.00	3.00	0.750
10.0	10.0	0.909
100	100	0.990

The actual range of concentration of hydrogen ion at which a particular indicator changes from being predominantly in the basic form $[([BH^+]/[B]) < 0.1]$ to being predominantly in the acidic form $[([BH^+]/[B]) > 10]$ depends upon the value of K_a. Depending upon their chemical structures, different indicator bases have different values of K_a.

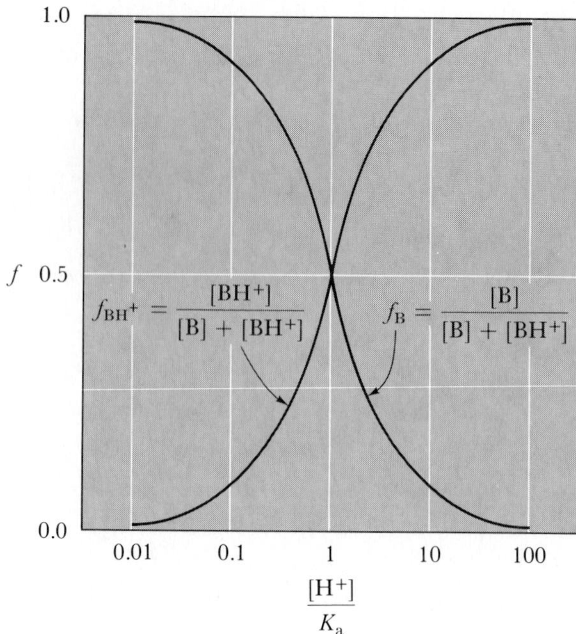

FIGURE 13–12
The relative concentrations of a weak acid (BH^+) and its conjugate base (B) as a function of the quotient $[H^+]/K_a$ (notice logarithmic scale for abscissa).

Table 13-8 gives the values of K_a for some commonly used indicators. You should be careful in choosing an indicator to determine the equivalence point in an acid–base titration. The indicator should have a value of K_a that is approximately equal to the concentration of hydrogen ion at the equivalence point, a quantity you can calculate as already explained. Thus for the titration of acetic acid, we have calculated that if $C_{NaOAc} = 0.0714$

TABLE 13–8
Common Indicators

Name	$\frac{K_a{}^a}{\text{mol L}^{-1}}$	Useful pH range and colors	
Methyl orange	5.6×10^{-4}	pH = 3	red
		pH = 4	orange
		pH = 5	yellow
Methyl red	5.0×10^{-6}	pH = 5	red
		pH = 6	orange
		pH = 7	yellow
Bromthymol blue	1.0×10^{-7}	pH = 6	yellow
		pH = 7	green
		pH = 8	blue
Phenolphthalein	1.0×10^{-9}	pH = 8	colorless
		pH = 9	pink
		pH = 10	red

[a]At 25°C.

mol L^{-1}, the equivalence point occurs at

$$[OH^-] = (K_b C_{NaOAc})^{1/2} = 6.38 \times 10^{-6} \text{ mol L}^{-1}$$

The concentration of hydrogen ion in this solution,

$$[H^+] = 1.57 \times 10^{-9} \text{ mol L}^{-1}$$

is close to K_a for phenolphthalein ($K_a = 1.0 \times 10^{-9}$ mol L^{-1}).

The matching of the value of K_a for the indicator and the concentration of hydrogen ion at the equivalence point need not be perfect because of the typically large values of $\Delta pH/\Delta V$ at the equivalence point. But clearly the use of methyl red ($K_a = 5.0 \times 10^{-6}$ mol L^{-1}) in the titration of acetic acid (or any other weak acid) with sodium hydroxide would be unsound. For the titration depicted in Figure 13-9, methyl red would change color in the range of fraction neutralization $f = 0.26$ to $f = 0.97$.

However, methyl red is a useful indicator in the titration of weak bases (e.g., ammonia) with strong acids (e.g., hydrochloric acid). At the equivalence point in such a titration the predominant species is the conjugate acid of the base being titrated (e.g., ammonium ion). In Table 13-6, you see that the concentration of hydrogen ion in a solution with the composition $C_{NH_4Cl} = 0.100$ mol L^{-1} is $[H^+] = 7.54 \times 10^{-6}$ mol L^{-1}, a value close to K_a for this indicator.

CONCENTRATION OF HYDROGEN ION; BASE-NEUTRALIZING CAPACITY

At this stage of our discussion we can summarize certain points about the concentration of hydrogen ion and the base-neutralizing capacity of solutions of a strong acid, hydrochloric acid, and a weak acid, acetic acid. Appropriately calibrated, a pH-measuring device or an indicator measures the concentration of hydrogen ion with some accuracy.[13] Titration with a strong base, in contrast, measures the stoichiometric concentration of the acid. The situation is summarized in Table 13-9. The two solutions,

TABLE 13–9
A Comparison of $[H^+]$ and C_{HX} for Three Solutions[a]

	100 cm^3 0.0100*M* HCl	100 cm^3 0.0100*M* HOAc	100 cm^3 4.10×10^{-4}*M* HCl
$\frac{[H^+]}{\text{mol L}^{-1}}$	0.0100	4.10×10^{-4}	4.10×10^{-4}
Volume of 0.0200*M* NaOH[a]	50.0	50.0	2.05

[a] $[H^+]$ is measured by an indicator. C_{HX} is measured by titration with 0.0200*M* NaOH.

[13]Recall, however, our mention of nonideality in Section 13–3.

0.0100*M* HCl and 0.0100*M* HOAc, have different concentrations of hydrogen ion, but have equal capacities for neutralizing bases. The two solutions, 0.0100*M* HOAc and $4.10 \times 10^{-4}M$ HCl, have different capacities for neutralizing bases, but have equal concentrations of hydrogen ion.

DIBASIC ACIDS; GENERAL DERIVATIONS

To this point, our discussion of equilibria in solutions of weak acids has dealt with acids having one acidic proton per molecule (monobasic acids). Now we will consider weak acids with two acidic protons per molecule (dibasic acids). We shall develop certain additional principles, useful for problems involving weak acids having three or more acidic protons, which we shall mention only briefly.

Equilibrium in an aqueous solution of a dibasic acid H_2Y and/or its salt(s), for instance NaHY and Na_2Y, is governed by two equilibrium-constant expressions, those corresponding to the stepwise acid-dissociation reactions,

$$H_2Y + H_2O \rightleftarrows H_3O^+ + HY^- \qquad K_1 = \frac{[H^+][HY^-]}{[H_2Y]}$$

$$HY^- + H_2O \rightleftarrows H_3O^+ + Y^{2-} \qquad K_2 = \frac{[H^+][Y^{2-}]}{[HY^-]}$$

in addition to the equilibrium constant for the self-dissociation of water, $K_w = [H^+][OH^-]$. We could begin to calculate the concentrations of species present in solutions of a dibasic acid by writing five equations involving the five unknown concentrations, $[H^+]$, $[OH^-]$, $[H_2Y]$, $[HY^-]$ and $[Y^{2-}]$, as we did for monobasic acids. The equation we finally derived would be a quartic equation in $[H^+]$. However, few practical problems need the complete equation for their solution, and our approach here will be to simplify the chemical problem before attacking the mathematical problem.

If the acid-dissociation constants K_1 and K_2 differ by several powers of 10, which is true for most common dibasic acids, the neutralization of the dibasic acid with a strong base can be described in terms of two separate reactions,

$$H_2Y + OH^- = HY^- + H_2O \quad \text{and} \quad HY^- + OH^- = Y^{2-} + H_2O$$

the first of which occurs almost to completion before the second occurs to an appreciable extent. At any particular concentration of hydrogen ion, each of the equations

$$\frac{[HY^-]}{[H_2Y]} = \frac{K_1}{[H^+]} \quad \text{and} \quad \frac{[HY^-]}{[Y^{2-}]} = \frac{[H^+]}{K_2}$$

is valid. If the concentration of hydrogen ion is intermediate between K_1

and K_2, we have the inequalities

$$K_1 \gg [H^+] \gg K_2 \quad \text{or} \quad \frac{K_1}{[H^+]} \gg 1 \quad \text{and} \quad \frac{[H^+]}{K_2} \gg 1$$

These inequalities correspond to

$$\frac{[HY^-]}{[H_2Y]} \gg 1 \quad \text{and} \quad \frac{[HY^-]}{[Y^{2-}]} \gg 1$$

That is, the monoprotonated form of the dibasic weak acid is the predominant form in solutions in the range of hydrogen ion concentration specified by these inequalities. This is the range of concentrations near the equivalence point in the reaction of *one* mole of H_2Y with *one* mole of hydroxide ion. Clearly, a particular concentration of hydrogen ion can be very small compared to K_1 and very large compared to K_2 only if K_1 is very large compared to K_2.

In performing equilibrium calculations for solutions of a weak dibasic acid and/or its salts, it is useful to think in terms of the principal equilibrium existing in the solution. A solution will fall into one of three regions, and there is a different principal equilibrium in each region. The regions are:

Region 1. Solutions of H_2Y, or H_2Y and NaHY, or H_2Y and HX (HX is a strong acid).
Region 2. Solutions of NaHY, or NaHY with a relatively small amount of H_2Y or Na_2Y.
Region 3. Solutions of Na_2Y, or Na_2Y and NaHY, or Na_2Y and NaOH.

These three regions correspond roughly to the extents of neutralization of both acidic protons of the acid H_2Y:

Region 1. <0 to 45%
Region 2. 45–55%
Region 3. 55 to $>100\%$

The point already has been made that for an acid with $K_1 \gg K_2$ the neutralization can be considered to occur in two essentially independent stages. Calculation of the concentration of hydrogen ion in solutions in Regions 1 and 3 follows the same pattern as that already outlined for solutions containing a monobasic acid at various stages of neutralization.

The principal equilibrium in Region 1 is the first stage of dissociation of the weak acid:

$$H_2Y + H_2O \rightleftarrows H_3O^+ + HY^-$$

In the range of composition that makes $[H_2Y]/[HY^-] > 0.1$, the concentration of hydrogen ion is very large compared to K_2, and the species Y^{2-} is not a dominant species; the principal equilibrium does not involve Y^{2-}.

The principal equilibrium in Region 3 is the base-dissociation reaction of Y^{2-}:

$$Y^{2-} + H_2O \rightleftarrows OH^- + HY^-$$

In the range of composition that makes $[Y^{2-}]/[HY^-] > 0.1$, the concentration of hydrogen ion is very small compared to K_1, and the species H_2Y is not a dominant species. (For many weak dibasic acids, $K_2 < 10^{-7}$ mol L^{-1}, and solutions in the composition range in which $[Y^{2-}]/[HY^-] > 0.1$ are basic; therefore the equation for the principal equilibrium is written to involve hydroxide ion, not hydrogen ion.)

The principal equilibrium in Region 2 for most dibasic acids is

$$2HY^- \rightleftarrows H_2Y + Y^{2-}$$

a reaction that involves neither hydrogen ion nor hydroxide ion. As we will learn, for many dibasic acids, solutions in this region of composition are close enough to neutral (i.e., pH $= 7 \pm 3$) that calculations assuming this to be the principal equilibrium are correct.

Here are some examples of calculations for solutions in Region 1.

EXAMPLE: Calculate the concentration of each species in a $0.2000M$ solution of a weak acid H_2Y ($K_1 = 4.00 \times 10^{-7}$ mol L^{-1}, and $K_2 = 6.00 \times 10^{-11}$ mol L^{-1}).

Using the reasoning described earlier (see Figure 13–5), we can conclude from the value of C/K_1 for this solution ($C/K_1 = 5 \times 10^5$) that Equation 13–5s,

$$[H^+] = (K_1 C)^{1/2}$$

will be accurate for calculating the concentration of hydrogen ion. This equation gives

$$[H^+] = (0.2000 \text{ mol L}^{-1} \times 4.00 \times 10^{-7} \text{ mol L}^{-1})^{1/2}$$
$$= 2.83 \times 10^{-4} \text{ mol L}^{-1}$$

The concentrations of other species involved in the principal equilibrium are

$$[HY^-] = [H^+] = 2.83 \times 10^{-4} \text{ mol L}^{-1}$$
$$[H_2Y] = C - [HY^-] = 0.2000 \text{ mol L}^{-1} - 2.83 \times 10^{-4} \text{ mol L}^{-1}$$
$$= 0.1997 \text{ mol L}^{-1}$$

The assumption that $[H^+] \ll C$, upon which our use of the simple equation was based, is a good one. We now can calculate the concentrations of species not involved in the principal equilibrium, $[Y^{2-}]$ and $[OH^-]$, using the concentrations established by the principal equilibrium and the appropriate equilibrium-constant equations. To calculate $[Y^{2-}]$, we use

$$[Y^{2-}] = K_2 \times \frac{[HY^-]}{[H^+]}$$

$$= 6.00 \times 10^{-11} \text{ mol L}^{-1} \times \frac{2.83 \times 10^{-4} \text{ mol L}^{-1}}{2.83 \times 10^{-4} \text{ mol L}^{-1}}$$

$$= 6.00 \times 10^{-11} \text{ mol L}^{-1}$$

This calculated concentration of Y^{2-} supports the assumption made in choosing the principal equilibrium. The concentration of hydrogen ion provided by the first acid-dissociation step ($[H^+] = 2.83 \times 10^{-4}$ mol L^{-1}) is very large compared to K_2 (6.00×10^{-11} mol L^{-1}), and the second acid-dissociation step is suppressed. The amount of hydrogen ion produced in the second stage of dissociation is equal to the amount of Y^{2-}, which is only $\sim 2 \times 10^{-7}$ times that produced in the first stage. To calculate the concentration of hydroxide ion, we use

$$[OH^-] = \frac{K_w}{[H^+]}$$

$$= \frac{1.00 \times 10^{-14}\ \text{mol}^2\ \text{L}^{-2}}{2.83 \times 10^{-4}\ \text{mol L}^{-1}} = 3.53 \times 10^{-11}\ \text{mol L}^{-1}$$

Each of these concentrations is much smaller than the concentrations of species involved in the principal equilibrium. Notice an interesting feature of this calculation: in a solution of H_2Y containing no other acids or bases, and in which $[H^+] \cong [HY^-]$, the concentration of Y^{2-} is equal to the value of K_2.

Example: Calculate the concentrations of species present in a buffer solution with the stoichiometric composition $0.1500M$ H_2Y, $0.0500M$ NaHY. (This solution can be prepared from the solution described in the preceding example by addition of 0.250 mol of NaOH per mole of H_2Y.)

This solution is one in which $[H_2Y]/[HY^-] > 0.1$, therefore we are still concerned with the first stage of acid dissociation as the principal equilibrium:

$$H_2Y \leftrightarrows H^+ + HY^-$$

We can use the approach we used previously in discussing partially neutralized acetic acid to calculate the concentration of hydrogen ion:

$$[H^+] = K_1 \frac{[H_2Y]}{[HY^-]} = K_1 \frac{C_{H_2Y} - [H^+]}{C_{NaHY} + [H^+]} \cong K_1 \frac{C_{H_2Y}}{C_{NaHY}}$$

$$= 4.00 \times 10^{-7}\ \text{mol L}^{-1} \times \frac{0.1500\ \text{mol L}^{-1}}{0.0500\ \text{mol L}^{-1}}$$

$$= 1.20 \times 10^{-6}\ \text{mol L}^{-1}$$

$$\text{pH} = -\log\left(\frac{[H^+]}{\text{mol L}^{-1}}\right) = -\log(1.20 \times 10^{-6}) = 5.92$$

The assumption that $[H^+]$ is small compared to C_{H_2Y} and C_{NaHY} is seen to be valid, and the concentrations of H_2Y and HY^- are essentially the stoichiometric concentrations:

$$[H_2Y] = 0.1500\ \text{mol L}^{-1} - 1.20 \times 10^{-6}\ \text{mol L}^{-1} = 0.1500\ \text{mol L}^{-1}$$
$$[HY^-] = 0.0500\ \text{mol L}^{-1} + 1.20 \times 10^{-6}\ \text{mol L}^{-1} = 0.0500\ \text{mol L}^{-1}$$

The concentrations of species not involved in the principal equilibrium can be calculated as in the preceding example:

$$[Y^{2-}] = K_2 \frac{[HY^-]}{[H^+]}$$

$$= 6.00 \times 10^{-11} \text{ mol L}^{-1} \times \frac{0.0500 \text{ mol L}^{-1}}{1.20 \times 10^{-6} \text{ mol L}^{-1}}$$

$$= 2.50 \times 10^{-6} \text{ mol L}^{-1}$$

$$[OH^-] = \frac{K_w}{[H^+]}$$

$$= \frac{1.00 \times 10^{-14} \text{ mol}^2 \text{ L}^{-2}}{1.20 \times 10^{-6} \text{ mol L}^{-1}}$$

$$= 8.33 \times 10^{-9} \text{ mol L}^{-1}$$

Although the calculated concentration of Y^{2-} is not small compared to the calculated concentration of hydrogen ion, it is small compared to the concentrations of H_2Y and HY^-, which are the concentrations that establish the concentrations of hydrogen ion.

Here is an example of a calculation for a solution in Region 3.

EXAMPLE: What are the concentrations of species present in a solution prepared by dissolving 0.1000 mol of Na_2Y in 350 cm^3 of water? (This solution can be considered to result from quantitative neutralization of a solution of H_2Y.)

Before starting the equilibrium calculations, we must calculate the stoichiometric concentration of Na_2Y:

$$C_{Na_2Y} = \frac{0.1000 \text{ mol}}{0.350 \text{ L}} = 0.286 \text{ mol L}^{-1}$$

The first part of this calculation is the same as that already discussed for a solution of sodium acetate; the principal equilibrium is that of the Brønsted base Y^{2-} reacting with water:

$$Y^{2-} + H_2O \leftrightharpoons HY^- + OH^-$$

The equilibrium constant for this reaction is related to equilibrium constants K_w and K_2:

$$K = \frac{[HY^-][OH^-]}{[Y^{2-}]} = \frac{[HY^-]}{[H^+][Y^{2-}]} \times [H^+][OH^-] = \frac{1}{K_2} \times K_w$$

$$= \frac{1.00 \times 10^{-14} \text{ mol}^2 \text{ L}^{-2}}{6.00 \times 10^{-11} \text{ mol L}^{-1}} = 1.67 \times 10^{-4} \text{ mol L}^{-1}$$

In this equilibrium the concentrations of HY^- and OH^- are equal, and they can be calculated using the equilibrium constant just calculated:

$$K = \frac{[HY^-][OH^-]}{[Y^{2-}]} = \frac{[OH^-]^2}{0.286\text{ mol L}^{-1} - [OH^-]}$$
$$= 1.67 \times 10^{-4}\text{ mol L}^{-1}$$

If we assume that $[OH^-] \ll 0.286\text{ mol L}^{-1}$, we have

$$\frac{[OH^-]^2}{0.286\text{ mol L}^{-1}} = 1.67 \times 10^{-4}\text{ mol L}^{-1}$$

$$[OH^-] = (1.67 \times 10^{-4}\text{ mol L}^{-1} \times 0.286\text{ mol L}^{-1})^{1/2}$$
$$= 6.91 \times 10^{-3}\text{ mol L}^{-1}$$

This concentration is 2.4% of the stoichiometric concentration of Y^{2-}.

Refinement of the calculation would be justified if the stoichiometric concentration of Na_2Y and the value of K_2 were known fairly accurately. This refinement can be made by solving the quadratic equation in which the approximation $[OH^-] \ll 0.286\text{ mol L}^{-1}$ is not used:

$$[OH^-]^2 + 1.67 \times 10^{-4}\text{ mol L}^{-1} \times [OH^-] - 4.78 \times 10^{-5}\text{ mol}^2\text{ L}^{-2} = 0$$

$$\frac{[OH^-]}{\text{mol L}^{-1}} = \frac{-1.67 \times 10^{-4} + (2.79 \times 10^{-8} + 1.912 \times 10^{-4})^{1/2}}{2}$$

$$[OH^-] = 6.83 \times 10^{-3}\text{ mol L}^{-1}$$

Or a second approximation to the hydroxide ion concentration can be calculated using

$$[Y^{2-}] = C_{Na_2Y} - [OH^-]_{\text{1st approx.}}$$
$$= 0.286\text{ mol L}^{-1} - 6.91 \times 10^{-3}\text{ mol L}^{-1}$$
$$= 0.279\text{ mol L}^{-1}$$

This gives

$$[OH^-]_{\text{2nd approx.}} = [1.67 \times 10^{-4}\text{ mol L}^{-1} \times 0.279\text{ mol L}^{-1}]^{1/2}$$
$$= 6.83 \times 10^{-3}\text{ mol L}^{-1}$$

The concentrations of other species involved in the principal equilibrium are:

$$[HY^-] = [OH^-] = 6.83 \times 10^{-3}\text{ mol L}^{-1}$$
$$[Y^{2-}] = 0.286\text{ mol L}^{-1} - 0.00683\text{ mol L}^{-1} = 0.279\text{ mol L}^{-1}$$

The concentrations of species not involved in the principal equilibrium and the pH are:

$$[H^+] = \frac{K_w}{[OH^-]} = \frac{1.00 \times 10^{-14}\text{ mol}^2\text{ L}^{-2}}{6.83 \times 10^{-3}\text{ mol L}^{-1}}$$

$$= 1.46 \times 10^{-12}\text{ mol L}^{-1}$$

(Continued)

$$\mathrm{pH} = -\log\left(\frac{[\mathrm{H^+}]}{\mathrm{mol\ L^{-1}}}\right) = -\log(1.46 \times 10^{-12}) = 11.84$$

$$[\mathrm{H_2Y}] = \frac{[\mathrm{H^+}][\mathrm{HY^-}]}{K_1}$$

$$= \frac{1.46 \times 10^{-12}\ \mathrm{mol\ L^{-1}} \times 6.83 \times 10^{-3}\ \mathrm{mol\ L^{-1}}}{4.00 \times 10^{-7}\ \mathrm{mol\ L^{-1}}}$$

$$= 2.49 \times 10^{-8}\ \mathrm{mol\ L^{-1}}$$

The predominant species in solutions corresponding to Region 2 is HY^-. This species is a weak acid, the equilibrium associated with this property being

$$\mathrm{HY^- + H_2O \rightleftarrows Y^{2-} + H_3O^+}$$

with $K = K_2$. This species also is a weak base, the equilibrium associated with this property being

$$\mathrm{HY^- + H_2O \rightleftarrows H_2Y + OH^-}$$

with $K = K_w/K_1$. If the equilibrium constants for these two reactions were exactly the same (which is quite unlikely), a solution of NaHY would be exactly neutral (pH = 7.00). The acidity of the species HY^- would exactly balance its basicity, and the principal equilibrium in solutions of HY^- would be

$$\mathrm{2HY^- \rightleftarrows H_2Y + Y^{2-}}$$

a reaction that neither produces nor consumes hydrogen ion. Fortunately for those concerned with calculations for such solutions, *this reaction is the principal equilibrium* for solutions of NaHY even if K_2 and K_w/K_1 differ by as much as several powers of ten. Under such conditions the solution has a pH within a few units of 7. The equilibrium constant for this principal equilibrium is

$$K = \frac{[\mathrm{H_2Y}][\mathrm{Y^{2-}}]}{[\mathrm{HY^-}]^2}$$

which is related to K_1 and K_2; multiplying numerator and denominator by $[H^+]$ gives

$$K = \frac{[\mathrm{H_2Y}][\mathrm{Y^{2-}}][\mathrm{H^+}]}{[\mathrm{HY^-}][\mathrm{HY^-}][\mathrm{H^+}]}$$

Grouping the concentration factors gives

$$K = \frac{[\mathrm{H_2Y}]}{[\mathrm{HY^-}][\mathrm{H^+}]} \times \frac{[\mathrm{H^+}][\mathrm{Y^{2-}}]}{[\mathrm{HY^-}]} = \frac{1}{K_1} \times K_2$$

Considering the balanced equation for the principal equilibrium, you will see that the concentrations of H_2Y and Y^{2-} are equal:

$$[\mathrm{H_2Y}] = [\mathrm{Y^{2-}}]$$

Therefore,

$$\frac{K_2}{K_1} = \frac{[H_2Y][Y^{2-}]}{[HY^-]^2} = \left(\frac{[H_2Y]}{[HY^-]}\right)^2$$

which gives

$$\frac{[H_2Y]}{[HY^-]} = \left(\frac{K_2}{K_1}\right)^{1/2}$$

To calculate the concentration of hydrogen ion in this solution, we will use the equilibrium-constant expression involving K_1:

$$[H^+] = K_1 \frac{[H_2Y]}{[HY^-]}$$

Substituting the already derived value for $[H_2Y]/[HY^-]$, we get

$$[H^+] = K_1 \times \left(\frac{K_2}{K_1}\right)^{1/2} = \sqrt{K_1 K_2}$$

This result, $[H^+] = \sqrt{K_1 K_2}$, is an interesting one: The concentration of hydrogen ion at the equivalence point in the first stage of neutralization of a weak dibasic acid is the geometric mean of the two acid-dissociation constants. Notice also that this equation for the concentration of hydrogen ion in a solution of NaHY *does not* involve the stoichiometric concentration of NaHY. In the absence of effects due to solution nonideality, the concentration of hydrogen ion is independent of the concentration of NaHY. This result is consistent with simple reasoning about the effect of dilution upon the principal equilibrium. For the reaction

$$2HY^- \rightleftarrows H_2Y + Y^{2-}$$

with the same number of solute particles on each side of the equation, dilution does not shift the equilibrium. A ten-fold dilution of an equilibrium mixture lowers the concentration of each of the species by a factor of ten, and the value of the quotient of concentrations,

$$\frac{[H_2Y][Y^{2-}]}{[HY^-]^2}$$

remains unchanged. Similarly, values of each of the quotients of concentrations upon which the concentration of hydrogen ion depends, namely,

$$\frac{[HY^-]}{[Y^{2-}]} \quad \text{and} \quad \frac{[H_2Y]}{[HY^-]}$$

do not change with dilution. However, there is a limit to the usefulness of this approach. With continued dilution of a solution of NaHY, the concentration of hydrogen ion will change in the limit of infinite dilution to the value for pure water,

$$[H^+] = 1.00 \times 10^{-7} \text{ mol L}^{-1}$$

In the derivation that gave

$$[H^+] = \sqrt{K_1 K_2}$$

it was tacitly assumed that the inequality of the concentrations of hydrogen ion and hydroxide ion did not appreciably upset the electroneutrality condition. We will learn about the validity of this approximation through an example. Following is an example of a calculation of the concentrations of species in a solution in Region 2.

EXAMPLE: What are the concentrations of species present in a solution in which the stoichiometric concentration of NaHY is 0.1500 mol L^{-1}?

The concentration of hydrogen ion in this solution is approximately equal to $(K_1 K_2)^{1/2}$:

$$\begin{aligned}[H^+] &= (4.00 \times 10^{-7}\ \text{mol L}^{-1} \times 6.00 \times 10^{-11}\ \text{mol L}^{-1})^{1/2} \\ &= 4.90 \times 10^{-9}\ \text{mol L}^{-1}\end{aligned}$$

The concentration of hydroxide ion is

$$\begin{aligned}[OH^-] &= \frac{K_w}{[H^+]} \\ &= \frac{1.00 \times 10^{-14}\ \text{mol}^2\ \text{L}^{-2}}{4.90 \times 10^{-9}\ \text{mol L}^{-1}} = 2.04 \times 10^{-6}\ \text{mol L}^{-1}\end{aligned}$$

The concentrations of H_2Y, HY^-, and Y^{2-} can be calculated from the concentration of hydrogen ion and the stoichiometric concentration of the salt, C_{NaHY}. The concentration of each of these species is the product of the stoichiometric concentration and f, the fraction of the total Y present in that particular form:

$$\begin{aligned}[H_2Y] &= C_{NaHY} \times f_{H_2Y} \\ [HY^-] &= C_{NaHY} \times f_{HY^-} \\ [Y^{2-}] &= C_{NaHY} \times f_{Y^{2-}}\end{aligned}$$

The equation for each fraction, f, a function of $[H^+]$, K_1, and K_2, follows simply from the definition of that fraction. For f_{H_2Y} we have

$$f_{H_2Y} \equiv \frac{[H_2Y]}{[H_2Y] + [HY^-] + [Y^{2-}]}$$

Dividing numerator and denominator by $[H_2Y]$, we have

$$f_{H_2Y} = \frac{1}{1 + \dfrac{[HY^-]}{[H_2Y]} + \dfrac{[Y^{2-}]}{[H_2Y]}}$$

In this equation each quotient of concentrations can be expressed in terms of $[H^+]$, K_1, and K_2:

$$\frac{[HY^-]}{[H_2Y]} = \frac{K_1}{[H^+]}$$

$$\frac{[Y^{2-}]}{[H_2Y]} = \frac{[Y^{2-}]}{[HY^-]} \times \frac{[HY^-]}{[H_2Y]} = \frac{K_2}{[H^+]} \times \frac{K_1}{[H^+]}$$

Thus for f_{H_2Y} we have

$$f_{H_2Y} = \frac{1}{1 + \dfrac{K_1}{[H^+]} + \dfrac{K_1K_2}{[H^+]^2}}$$

In the same manner, the equations for f_{HY^-} and $f_{Y^{2-}}$ can be derived:

$$f_{HY^-} = \frac{\dfrac{K_1}{[H^+]}}{1 + \dfrac{K_1}{[H^+]} + \dfrac{K_1K_2}{[H^+]^2}}$$

$$f_{Y^{2-}} = \frac{\dfrac{K_1K_2}{[H^+]^2}}{1 + \dfrac{K_1}{[H^+]} + \dfrac{K_1K_2}{[H^+]^2}}$$

For this problem:

$$[H^+] = 4.90 \times 10^{-9} \text{ mol L}^{-1}$$

$$\frac{K_1}{[H^+]} = \frac{4.00 \times 10^{-6} \text{ mol L}^{-1}}{4.90 \times 10^{-9} \text{ mol L}^{-1}} = 816$$

$$\frac{K_1K_2}{[H^+]^2} = 1 \quad \text{because} \quad [H^+] = (K_1K_2)^{1/2}$$

and

$$[H_2Y] = 0.1500 \text{ mol L}^{-1} \times \frac{1}{1 + 816 + 1} = 1.83 \times 10^{-4} \text{ mol L}^{-1}$$

$$[HY^-] = 0.1500 \text{ mol L}^{-1} \times \frac{816}{1 + 816 + 1} = 0.1496 \text{ mol L}^{-1}$$

$$[Y^{2-}] = [H_2Y] = 1.83 \times 10^{-4} \text{ mol L}^{-1}$$

To check the validity of these calculated concentrations, we will check the electroneutrality condition. The complete electroneutrality condition is

$$[H^+] + [Na^+] = [HY^-] + 2[Y^{2-}] + [OH^-]$$

This can be transformed into an equation for the net concentration of positive charge in the solution by using the relationship

$$C_{NaHY} = [Na^+] = [H_2Y] + [HY^-] + [Y^{2-}]$$

The net concentration of positive charge is

$$[H^+] - [OH^-] + [H_2Y] - [Y^{2-}] = [H^+] - [OH^-]$$

For the solution under consideration, the net concentration of positive charge is

$$4.90 \times 10^{-9} \text{ mol L}^{-1} - 2.04 \times 10^{-6} \text{ mol L}^{-1} = -2.04 \times 10^{-6} \text{ mol L}^{-1}$$

If the calculated net concentration of positive charge is small compared to the calculated concentrations of the species that control the concentration of hydrogen ion, that is, $[HY^-]$ and $[H_2Y]$ or $[HY^-]$ and $[Y^{2-}]$, the calculated concentrations are approximately correct. This is true for this example:

$$2.04 \times 10^{-6} \text{ mol L}^{-1} \ll 0.1496 \text{ mol L}^{-1} \quad \text{and}$$
$$2.04 \times 10^{-6} \text{ mol L}^{-1} \ll 1.83 \times 10^{-4} \text{ mol L}^{-1}$$

EXAMPLE: Calculate the concentrations of species in $1.500 \times 10^{-4}M$ NaHY by the methods of the preceding example, and check the electroneutrality condition. (This solution is a 1000-fold dilution of the solution just considered.)

For this diluted solution, the methods of the previous example give calculated concentrations of hydrogen ion and hydroxide ion that are the same as in $0.1500M$ NaHY:

$$[H^+] = 4.90 \times 10^{-9} \text{ mol L}^{-1}$$
$$[OH^-] = 2.04 \times 10^{-6} \text{ mol L}^{-1}$$

The calculated concentration of each Y-containing species would be 1/1000 of the value in $0.1500M$ NaHY:

$$[H_2Y] = 1.83 \times 10^{-7} \text{ mol L}^{-1}$$
$$[HY^-] = 1.496 \times 10^{-4} \text{ mol L}^{-1}$$
$$[Y^{2-}] = 1.83 \times 10^{-7} \text{ mol L}^{-1}$$

The calculated net concentration of charge using these conditions is the same as in the preceding example:

$$\begin{aligned}\text{Net concentration of charge} &= [H^+] - [OH^-] \\ &= 4.90 \times 10^{-9} \text{ mol L}^{-1} \\ &\quad - 2.04 \times 10^{-6} \text{ mol L}^{-1} \\ &= -2.04 \times 10^{-6} \text{ mol L}^{-1}\end{aligned}$$

However, in this more dilute solution, the magnitude of this net charge is not small compared to both of the concentrations, $[HY^-]$ (1.496×10^{-4} mol L^{-1}) and $[H_2Y]$ (1.83×10^{-7} mol L^{-1}). Therefore we can conclude that this value, $[H^+] = 4.90 \times 10^{-9}$ mol L^{-1}, is wrong.[14]

A rigorous approach to problems involving very dilute solutions of NaHY can be developed along the lines used previously in this chapter, but we shall not pursue this matter here.

[14] The correct value of $[H^+]$ for this solution is 7.14×10^{-9} mol L^{-1}

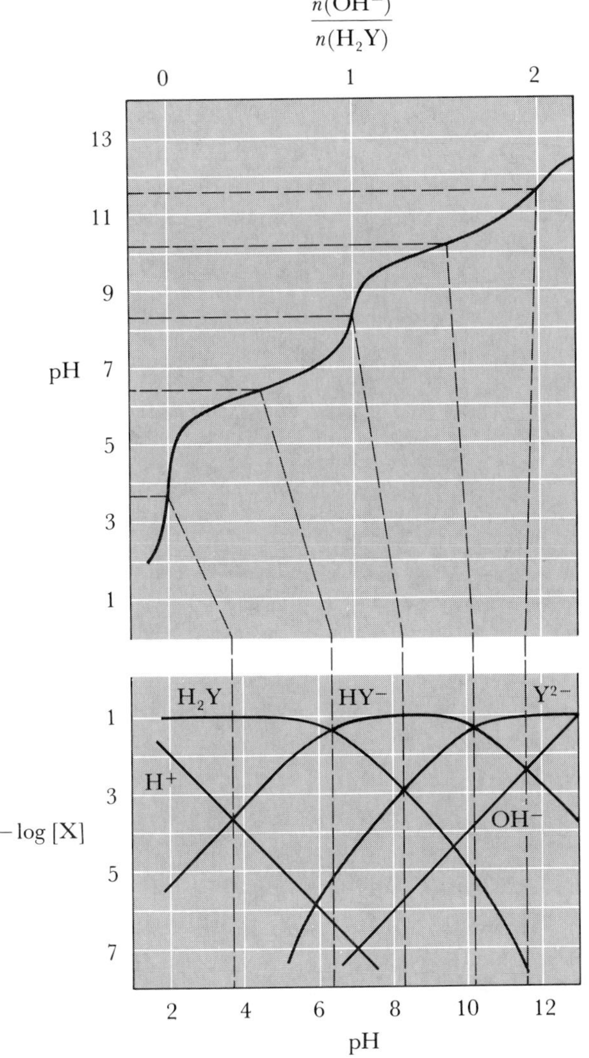

FIGURE 13–13
A titration curve for a dibasic weak acid, H_2Y ($K_1 = 4.0 \times 10^{-7}$ mol L^{-1}, $K_2 = 6.0 \times 10^{-11}$ mol L^{-1}); initial concentration is 0.1000 mol L^{-1}. (a) Upper figure, titration curve; pH versus added base. (b) Lower figure, concentrations of the various species, X, versus pH. Tie lines connect corresponding points in the two figures. (Tie lines drawn at 0, 0.5, 1.0, 1.5, and 2.0 mol OH^-/mol H_2Y.)

Figure 13–13 shows the titration curve for a dibasic acid with $K_1 = 4.0 \times 10^{-7}$ mol L^{-1} and $K_2 = 6.0 \times 10^{-11}$ mol L^{-1} (values of equilibrium constants close to those for carbon dioxide) and also a graph showing the distribution of species as a function of the concentration of hydrogen ion; lines connect corresponding parts of each graph. As in a monobasic-acid system, the distribution of species in a dibasic-acid system is a function of the concentration of hydrogen ion.

The reaction that is the sum of the stepwise dissociation reactions for H_2Y,

$$H_2Y \rightleftarrows 2H^+ + Y^{2-}$$

has an equilibrium constant that is the product K_1K_2:

$$K = \frac{[H^+]^2[Y^{2-}]}{[H_2Y]} = \frac{[H^+][HY^-]}{[H_2Y]} \times \frac{[H^+][Y^{2-}]}{[HY^-]} = K_1K_2$$

The equilibrium concentrations of H^+, H_2Y, and Y^{2-} in solution satisfy this equilibrium-constant expression, but the corresponding chemical equation is *not* the principal equilibrium for any solution, be it in Region 1, 2, or 3. (If we thought this equation alone were the principal equilibrium for a solution of H_2Y, we would conclude that $[H_3O^+] = 2[Y^{2-}]$, which is wrong.) However, this equilibrium-constant equation is useful for solving problems dealing with the concentration of Y^{2-} in a solution containing H_2Y and strong acid, for example hydrochloric acid. In such a solution $[H^+] \cong C_{HCl}$, and $[H_2Y] \cong C_{H_2Y}$, and the concentration of Y^{2-} can be calculated:

$$[Y^{2-}] = K_1K_2\frac{[H_2Y]}{[H^+]^2} \cong K_1K_2\frac{C_{H_2Y}}{C_{HCl}^2}$$

EXAMPLE: What is the concentration of Y^{2-} in a solution with the stoichiometric composition $0.100M$ H_2Y, $0.250M$ HCl?

$$[Y^{2-}] = K_1K_2 \times \frac{[H_2Y]}{[H^+]^2} \cong K_1K_2 \times \frac{C_{H_2Y}}{C_{HCl}^2}$$

$$= 2.40 \times 10^{-17}\ \text{mol}^2\ \text{L}^{-2} \times \frac{0.100\ \text{mol L}^{-1}}{(0.250\ \text{mol L}^{-1})^2}$$

$$= 3.84 \times 10^{-17}\ \text{mol L}^{-1}$$

DIBASIC ACIDS; SPECIFIC EXAMPLES

There are a number of common dibasic acids that are especially interesting. We will consider hydrogen sulfide, H_2S, carbon dioxide or carbonic acid, $CO_2(aq)$, and glycine cation, $H_3^+NCH_2CO_2H$. Values of the acid-dissociation constants at 25°C for these acids are

	$\frac{K_1}{\text{mol L}^{-1}}$	$\frac{K_2}{\text{mol L}^{-1}}$	$\frac{K_1}{K_2}$
H_2S	1.02×10^{-7}	1.29×10^{-13}	7.91×10^5
$CO_2(aq)$	4.45×10^{-7}	5.69×10^{-11}	7.82×10^3
$H_3\overset{+}{N}CH_2CO_2H$	4.46×10^{-3}	1.66×10^{-10}	2.69×10^7

For each of these acids $K_1 \ggg K_2$, and we can apply the general procedures that we have just developed with the acid H_2Y ($K_1 = 4.00 \times 10^{-7}$ mol L^{-1}, $K_2 = 6.00 \times 10^{-11}$ mol L^{-1}, $K_1/K_2 = 6.67 \times 10^3$).

Hydrogen Sulfide. Although hydrogen sulfide is an analogue of water, its physical properties are rather different from those of water. The abnormally high melting point and boiling point of water are due to the strong hydrogen bonding involving the very electronegative element oxygen. Sulfur is less electronegative, and hydrogen bonding between hydrogen sulfide molecules is much weaker. Pure hydrogen sulfide is a gas at ordinary temperatures and pressures; its melting point is −82.9°C and its boiling point is −59.6°C. At 25°C and 1.00 atm pressure, the solubility of hydrogen sulfide in water is 0.10 mol L^{-1}. This system obeys Henry's law reasonably well, so we can calculate the concentration of H_2S in solution from the partial pressure over the solution by using Henry's law:

$$[H_2S] = \frac{P_{H_2S}}{k} = \frac{P_{H_2S}}{10 \text{ atm L mol}^{-1}}$$
$$= 0.10 \text{ mol L}^{-1} \text{ atm}^{-1} \times P_{H_2S}$$

Equilibria involving hydrogen sulfide play roles in determining the solubility of sulfide minerals. The large range of solubilities of various metal sulfides is exploited in analysis, and you may use sulfide precipitation procedures in your laboratory work. For this reason we will make calculations for several different sulfide-containing solutions.

The principal equilibrium in a solution prepared by dissolving hydrogen sulfide in water containing no other acids or bases is

$$H_2S + H_2O \rightleftarrows HS^- + H_3O^+$$

Using the methods already described, we can calculate the concentrations of species in 0.100*M* H_2S. The values of K_a (1.02×10^{-7} mol L^{-1}) and C (0.100 mol L^{-1}) indicate that Equation 13–5s is appropriate:

$$[H^+] = (K_aC)^{1/2}$$
$$= (1.02 \times 10^{-7} \text{ mol L}^{-1} \times 0.100 \text{ mol L}^{-1})^{1/2}$$
$$= 1.01 \times 10^{-4} \text{ mol L}^{-1}$$
$$[HS^-] = [H^+] = 1.01 \times 10^{-4} \text{ mol L}^{-1}$$

The very low concentration of sulfide ion is established by the equilibrium

$$HS^- + H_2O \rightleftarrows S^{2-} + H_3O^+$$
$$[S^{2-}] = K_2 \times \frac{[HS^-]}{[H_3O^+]}$$
$$[S^{2-}] = 1.29 \times 10^{-13} \text{ mol L}^{-1} \times \frac{[HS^-]}{[H_3O^+]}$$

However, because $[H_3O^+] = [HS^-]$ in this solution,

$$[S^{2-}] = 1.29 \times 10^{-13} \text{ mol L}^{-1}$$

(This result we already have seen: $[Y^{2-}] = K_2$ in a solution containing H_2Y with no added acidic or basic solutes.)

The concentration of sulfide ion is even lower than this in solutions of hydrogen sulfide and a strong acid such as hydrochloric acid. Consider a solution that is 0.100*M* HCl, and 0.100*M* H_2S. As already indicated, this situation is one in which the equilibrium constant for the overall dissociation reaction,

$$H_2S + 2H_2O \rightleftarrows 2H_3O^+ + S^{2-} \qquad K = K_1K_2 = 1.32 \times 10^{-20} \text{ mol}^2 \text{ L}^{-2}$$

can be used to calculate the concentration of sulfide ion:

$$[S^{2-}] = K_1K_2 \times \frac{[H_2S]}{[H^+]^2} \cong K_1K_2 \frac{C_{H_2S}}{C^2_{HCl}}$$

$$= 1.32 \times 10^{-20}\ \text{mol}^2\ \text{L}^{-2} \frac{0.100\ \text{mol L}^{-1}}{(0.100\ \text{mol L}^{-1})^2}$$

$$= 1.32 \times 10^{-19}\ \text{mol L}^{-1}$$

Solutions of sodium sulfide and ammonium sulfide with the same stoichiometric concentration have very different concentrations of sulfide ion. The acidity of ammonium ion causes greater conversion of sulfide ion to hydrogen sulfide ion in ammonium sulfide solution.

The principal equilibrium in sodium sulfide solution is

$$S^{2-} + H_2O \rightleftarrows HS^- + OH^-$$

with an equilibrium constant:

$$K = \frac{[HS^-][OH^-]}{[S^{2-}]} = \frac{[HS^-]}{[H^+][S^{2-}]} \times [H^+][OH^-] = \frac{K_w}{K_2}$$

$$= \frac{1.00 \times 10^{-14}\ \text{mol}^2\ \text{L}^{-2}}{1.29 \times 10^{-13}\ \text{mol L}^{-1}} = 7.75 \times 10^{-2}\ \text{mol L}^{-1}$$

In ammonium sulfide solution, the principal equilibrium is

$$S^{2-} + NH_4^+ \rightleftarrows HS^- + NH_3$$

with an equilibrium constant:

$$K = \frac{[HS^-][NH_3]}{[S^{2-}][NH_4^+]} = \frac{[HS^-]}{[H^+][S^{2-}]} \times \frac{[NH_3][H^+]}{[NH_4^+]} = \frac{K_a(NH_4^+)}{K_2(H_2S)}$$

$$= \frac{5.69 \times 10^{-10}\ \text{mol L}^{-1}}{1.29 \times 10^{-13}\ \text{mol L}^{-1}} = 4.41 \times 10^3$$

With an equilibrium constant of 4.41×10^3, the acid–base reaction between ammonium ion and sulfide ion occurs to a large extent, and it really is more appropriate to describe a solution of ammonium sulfide as a solution of ammonia and ammonium hydrogen sulfide.

We can use equilibrium constants for these principal equilibria in numerical calculations. Comparison of the results for $0.100M$ Na_2S and $0.100M$ $(NH_4)_2S$ (better described as $0.100M$ NH_4HS, $0.100M$ NH_3) shows the point already made, that the sulfide ion concentration is much larger in sodium sulfide solution than in ammonium sulfide solution. A summary of results of these calculations is:

Stoichiometric composition	$\frac{[S^{2-}]}{\text{mol L}^{-1}}$	$\frac{[OH^-]}{\text{mol L}^{-1}}$	$\frac{[HS^-]}{\text{mol L}^{-1}}$	$\frac{[H_2S]}{\text{mol L}^{-1}}$
$0.100M$ Na_2S	0.0426	0.0574	0.0574	9.8×10^{-8}
$\{0.100M$ NH_4HS, $0.100M$ $NH_3\}$	2.27×10^{-5}	1.76×10^{-5}	0.100	5.57×10^{-4}

The much greater concentration of sulfide ion in sodium sulfide solution makes it a preferable reagent for dissolving insoluble sulfides of metals that form soluble thiometal anions (e.g., AsS_2^- or AsS_3^{3-}). Even the extremely insoluble mercury(II) sulfide dissolves appreciably in sodium sulfide solution to form HgS_2^{2-}, but it does not dissolve in ammonium sulfide solution. We will learn more about the interaction of acid–base equilibria with solubility equilibria and complex-ion equilibria in the next chapter, which also will present a graph (Figure 14–14) showing the results of these calculations for sulfide solutions.

Carbon Dioxide or Carbonic Acid. Pure carbon dioxide is a gas at ordinary temperatures and pressures. From the phase diagram (Figure 6–12b), we learn that at 1.00 atm pressure, the solid is in equilibrium with the gas at −78.5°C, and the liquid and solid are in equilibrium at 5.1 atm and −56.7°C. The linear carbon dioxide molecule is nonpolar, and the attractive forces between molecules are not strong. The solubility of carbon dioxide in water is 0.034 mol L^{-1} at 25°C if the pressure of carbon dioxide is 1 atm. The solubility at other pressures can be calculated using Henry's law; values of the Henry's law constant for other temperatures were given in Table 7–4. The interaction of carbon dioxide with water is important in many areas, including the chemistry of natural waters (rivers, lakes, oceans) and the chemistry of respiration. Carbonate species have roles as buffering agents in each of these areas.

Aqueous solutions of carbon dioxide are acidic; the acid-dissociation equilibria are:

$$CO_2(aq) + 2H_2O \rightleftarrows H_3O^+ + HOCO_2^-$$
$$HOCO_2^- + H_2O \rightleftarrows H_3O^+ + CO_3^{2-}$$

For these equilibria, the conventionally defined equilibrium constants at 25°C are

$$K_1 = \frac{[H^+][HOCO_2^-]}{[CO_2(aq)]} = 4.45 \times 10^{-7} \text{ mol L}^{-1}$$

$$K_2 = \frac{[H^+][CO_3^{2-}]}{[HOCO_2^-]} = 5.69 \times 10^{-11} \text{ mol L}^{-1}$$

There are special points of interest regarding carbonic acid. Both the acid, $(HO)_2CO$, and its anhydride, CO_2, exist in aqueous solution; the reaction by which they are interconverted,

$$\underset{\text{Acid anhydride}}{CO_2} + H_2O = \underset{\text{Acid}}{(HO)_2CO}$$

is slow (half time for the reaction ~23 seconds at 25°C), and equilibrium favors the anhydride,

$$K = \frac{[(HO)_2CO]}{[CO_2]} = 0.0016 \qquad \text{at } 25°\text{C}$$

(In these respects carbon dioxide differs from some acid anhydrides. For instance, N_2O_5, the anhydride of nitric acid, is rapidly and quantitatively converted to nitric acid, $HONO_2$. We will discuss nonmetal oxides as acid anhydrides further in Chapter 17.) In the chemical equation given for the first acid-dissociation equilibrium and in the corresponding equilibrium-constant equation, $CO_2(aq)$ stands for the sum of the two neutral species CO_2 and $(HO)_2CO$. But the *predominant* uncharged carbonate species in aqueous solution is carbon dioxide:

$$f_{CO_2} = \frac{[CO_2]}{[CO_2] + [(HO)_2CO]} = \frac{1}{1 + \dfrac{[(HO)_2CO]}{[CO_2]}}$$

$$= \frac{1}{1 + 0.0016} = 0.9984$$

Because this fraction is so close to one, in almost all equilibrium calculations we can take

$$[CO_2] = [CO_2] + [(HO)_2CO]$$

(You may have to solve problems about reaction mechanisms in which you need to know the concentration of carbonic acid, $(HO)_2CO$. In such problems you must remember that there are two neutral forms of carbon dioxide in aqueous solution.)

Glycine Cation. Protonated glycine, $H_3^+NCH_2CO_2H$, is a dibasic acid, the complete neutralization of which yields the anion $H_2NCH_2CO_2^-$; the chemical equation for complete neutralization by a strong base is

$$H_3\overset{+}{N}CH_2CO_2H + 2OH^- = H_2NCH_2CO_2^- + 2H_2O$$

The stepwise neutralization is describable in terms of two equations analogous to those already given for the model weak dibasic acid H_2Y.

Glycine is especially important because it is the simplest α-amino acid (see Section 11–4), and it is a constituent of proteins. Of special interest are the two isomeric species with zero net charge, which have the structures

```
  H        :O:              H        :O:
   \       //                \       //
H—C—C                     H—C—C
   /       \                 /       \
  N:       :O—H            N+        :O:⁻
 / |                      / | \
H  H                     H  H  H
```

The first structure is the one suggested by the name, α-aminoacetic acid, which denotes acetic acid with an amino group in place of a hydrogen atom on the alpha carbon atom. But this structure has an acidic site, the carboxylic acid group, —COOH, and a basic site, the amino group, —$\ddot{N}H_2$. Isomeric with the first structure is the second structure, one in which an extra proton is associated with the amino group and no proton is associated with the carboxylate group. The charges shown in this structure are formal charges on the nitrogen atom and on an oxygen atom of the carboxylate group. (Recall, however, that the oxygen atoms in a carboxylate anion are equivalent, and that the structure given here is one of the two equivalent resonance forms.) A structure of this type is called a *zwitterion*, which means a species carrying both positive and negative charges.

By simple reasoning we may conclude that the zwitterion is the predominant form, a conclusion which is supported by independent data. The glycine cation,

```
      H           O
        \        //
      H—C—C
        /        \
     N+           O—H
   /  |  \
  H   |   H
      H
```

contains two nonequivalent acidic groups, one a substituted ammonium ion, RNH_3^+, and the other a carboxylic acid, $R'CO_2H$. We can estimate the relative extents of proton dissociation from the two acid groups of the glycine cation by considering the acidic properties of monobasic acids with approximately the same structures. Appropriate models for the two weakly acidic groups of the glycine cation are the acids

$HOCH_2CH_2NH_3^+$	CH_3CO_2H
Ethanolammonium ion	Acetic acid

with acid-dissociation constants at 25°C:

$$K = \frac{[H^+][HOC_2H_4NH_2]}{[HOC_2H_4NH_3^+]} = 3.2 \times 10^{-10} \text{ mol L}^{-1}$$

$$K = \frac{[H^+][CH_3CO_2^-]}{[CH_3CO_2H]} = 1.75 \times 10^{-5} \text{ mol L}^{-1}$$

Although equilibrium constants for the stepwise acid dissociation of the glycine cation cannot be assumed the same as those for these model monobasic acids, this difference of five powers of ten is so great that it lets us assume one thing with certainty: the carboxylic acid group ionizes preferentially in the glycine cation,

$$H_3\overset{+}{N}CH_2CO_2H + H_2O \rightleftarrows H_3\overset{+}{N}CH_2CO_2^- + H_3O^+$$

thereby giving the zwitterion, $H_3\overset{+}{N}CH_2CO_2^-$. This species with spatially separated formal charges has a very large dipole moment, $p = 12.2$ D; its moment can be contrasted with $p = 1.74$ D for acetic acid, which does not have formal charges.

Molecules of a dibasic weak acid for which the completely protonated form is a cation and the completely deprotonated form is an anion will at some concentration of hydrogen ion have an average net charge of zero. This concentration of hydrogen ion is called the *isoelectric point*, and it is a quantity of interest in the study of amino acids, such as glycine. In relatively acidic solutions, in which glycine (or another similar amino acid) is a cation, this species migrates in solution toward the negatively charged electrode of an electrolytic cell; in relatively basic solution glycine is an anion, and this species migrates in solution toward the positive electrode. In solutions of some particular intermediate concentration of hydrogen ion, glycine migrates in neither direction in an electric field. The average charge on glycine is zero if concentrations of the cationic and anionic forms are equal, $[H_3^+NCH_2CO_2H] = [H_2NCH_2CO_2^-]$. We have already explained that concentrations of the fully protonated and deprotonated forms of a dibasic weak acid are equal in a solution in which the dibasic acid is one half neutralized. In this solution of HY^- ($H_3^+NCH_2CO_2^-$ in this example), the concentration of hydrogen ion was shown to be

$$[H^+] = (K_1K_2)^{1/2}$$

which, therefore, defines the isoelectric point. For glycine, the isoelectric point for 25°C is

$$[H^+] = [(4.46 \times 10^{-3} \text{ mol L}^{-1}) \times (1.66 \times 10^{-10} \text{ mol L}^{-1})]^{1/2}$$
$$= 8.60 \times 10^{-7} \text{ mol L}^{-1}$$
$$\text{pH} = 6.07$$

POLYBASIC ACIDS

There are several polybasic weak acids which you may encounter in your study or in everyday life; with their acid-dissociation constants, they are:

Phosphoric acid, $(HO)_3PO$
$K_1 = 7.52 \times 10^{-3}$ mol L^{-1}
$K_2 = 6.23 \times 10^{-8}$ mol L^{-1}
$K_3 = 2.2 \times 10^{-13}$ mol L^{-1}

Pyrophosphoric acid, $(HO)_2OPOPO(OH)_2$
$K_1 = 0.14$ mol L^{-1}
$K_2 = 1.1 \times 10^{-2}$ mol L^{-1}
$K_3 = 2.69 \times 10^{-7}$ mol L^{-1}
$K_4 = 2.40 \times 10^{-10}$ mol L^{-1}

Citric acid, $HOC(CH_2CO_2H)_2CO_2H$
$K_1 = 7.45 \times 10^{-4}$ mol L^{-1}
$K_2 = 1.73 \times 10^{-5}$ mol L^{-1}
$K_3 = 4.02 \times 10^{-7}$ mol L^{-1}

Ethylenediaminetetraacetic acid (EDTA),
$(HO_2CCH_2)_2NCH_2CH_2N(CH_2CO_2H)_2$

$$K_1 = 1.02 \times 10^{-2} \text{ mol L}^{-1}$$
$$K_2 = 2.14 \times 10^{-3} \text{ mol L}^{-1}$$
$$K_3 = 6.92 \times 10^{-7} \text{ mol L}^{-1}$$
$$K_4 = 5.89 \times 10^{-11} \text{ mol L}^{-1}$$

A point worth noting about phosphoric acid is the magnitude of the third dissociation constant, $K_3 = 2.2 \times 10^{-13}$ mol L^{-1}. Because hydrogen phosphate ion, $HOPO_3^{2-}$, is such a weak acid, its conjugate base, phosphate ion, PO_4^{3-}, is a moderately strong base. For the reaction

$$PO_4^{3-} + H_2O \rightleftarrows HOPO_3^{2-} + OH^-$$

the equilibrium constant is

$$K = \frac{K_w}{K_3} = \frac{1.00 \times 10^{-14} \text{ mol}^2 \text{ L}^{-2}}{2.2 \times 10^{-13} \text{ mol L}^{-1}} = 4.5 \times 10^{-2} \text{ mol L}^{-1}$$

This is 1000-fold larger than the equilibrium constant for the reaction in which ammonia acts as a base. The base-dissociation constant for phosphate ion is so large that at low stoichiometric concentrations ($C_{Na_3PO_4} < 1 \times 10^{-4}$ mol L^{-1}) the reaction just given occurs essentially to completion. The detergent action of solutions of trisodium phosphate (TSP) is caused by this hydroxide ion-producing equilibrium.

Pyrophosphoric acid (also called diphosphoric acid), a partially dehydrated phosphoric acid,

$$2(HO)_3PO = (HO)_2OPOPO(OH)_2 + H_2O$$

is the simplest polyphosphoric acid, of which there are many. These acids, their anions, and their organic derivatives play important roles in areas of inorganic and organic chemistry, and in the chemistry of biological systems. Chapter 20 will introduce the chemistry of adenosine triphosphate (ATP) and adenosine diphosphate (ADP).

The ratios of successive acid-dissociation constants K_n/K_{n+1} are very large for the dibasic acids considered in the previous section ($K_1/K_2 = 10^4$ to 10^7). We see that for pyrophosphoric acid, for citric acid, and for ethylenediaminetetraacetic acid, some of these ratios are much smaller, 10^3 or less. The smaller ratios follow from the greater separation of the acidic groups in these acids. The negative charge developed by loss of a proton is farther from the site of dissociation of the next proton. The structure of pyrophosphoric acid is

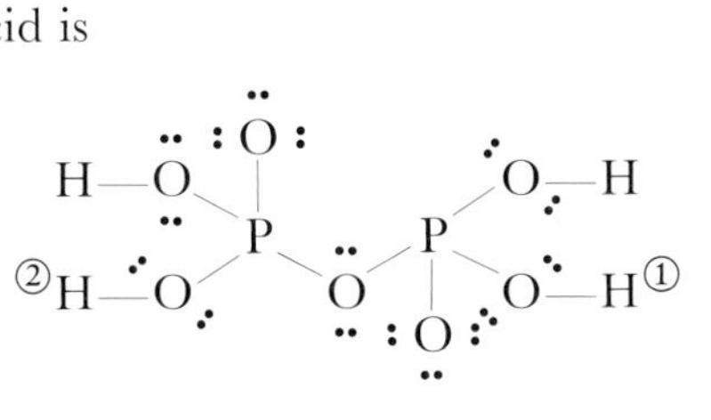

H—O, O—H, P, P, ②H—O, O, O—H①, :O:, :O:

and the ratios of the successive acid-dissociation constants, $K_1/K_2 = 13$, $K_2/K_3 = 4.1 \times 10^4$, and $K_3/K_4 = 1.1 \times 10^3$, are reasonable if the order of dissociation of the protons is that indicated by the numbers given with the structure. The second proton to dissociate comes from the end of the molecule which has not developed a negative charge, but the third proton to dissociate must come from an end of the molecule that already has one unit of negative charge. The fourth proton to dissociate comes from an end of the molecule with one unit of negative charge, and the nearby other end of the molecule has two units of negative charge.

Some equilibrium calculations for solutions of dibasic or polybasic acids with $K_n/K_{n+1} < 10^3$ are more complex than the ones which we have studied. The neutralization of polybasic acids with smaller values of K_n/K_{n+1} does not proceed in well defined separate stages. For instance, in a solution of partially neutralized citric acid at pH = 4.0, ~13% of the molecules have dissociated two acidic protons and ~10% have not dissociated any protons.

13–4 Acids in Very Concentrated Aqueous Solution

Our discussion to this point has dealt with relatively dilute solutions of acids and bases. In such solutions, values of the activity coefficients of the solute species and the solvent are close to one. However, many common strong acids are very soluble in water, and in the concentrated solutions the activity coefficients of the solute species and the solvent differ greatly from one. Because of this, by any of a number of measurements, concentrated solutions of strong acids are much more acidic than one would expect from the molarity of hydrogen ion. For instance, as measured by the ability of the solution to protonate certain neutral indicator bases, a solution of 10.0*M* HCl is 38,000 times more acidic than a solution of 0.100*M* HCl. For this change of stoichiometric concentration, the concentration of hydrogen ion increases by a factor of 10^2, but this measure of the acidity increases by a factor of over 10^4.

A partial explanation for this very high acidity is the strong interaction of hydrogen ion with water molecules. For simplicity, we will assume that the predominant hydrated proton species is the tetrahydrate (see Figure 12–7), and that the hydration of the protonated indicator and its uncharged conjugate base involve approximately the same number of molecules of water. With these assumptions, the chemical equation for protonation of an indicator base is

$$H(H_2O)_4^+ + B \rightleftarrows BH^+ + 4H_2O$$

(The specific hydration of B and BH^+ is omitted because we are assuming that these species are hydrated by approximately equal numbers of water molecules.) If we were employing the conventions used for dilute solutions, the equation for the ratio $[BH^+]/[B]$ would be

$$\frac{[BH^+]}{[B]} = K[H^+]$$

But if we recognize the nonideality of these solutions and also the participation of water, the equilibrium-constant equation is

$$K^0 = \frac{[BH^+]\gamma_{BH^+}(x_{H_2O}\gamma_{H_2O})^4}{[B]\gamma_B[H_9O_4^+]\gamma_{H_9O_4^+}}$$

This can be rearranged:

$$\frac{[BH^+]}{[B]} = K^0 \frac{[H_9O_4^+]\dfrac{\gamma_{H_9O_4^+}\gamma_B}{\gamma_{BH^+}}}{(x_{H_2O}\gamma_{H_2O})^4}$$

Like the simple equation $([BH^+]/[B]) = K[H^+]$, this complete equation can be written

$$\frac{[BH^+]}{[B]} = K^0 h_0$$

in which

$$h_0 = \frac{[H_9O_4^+] \times \dfrac{\gamma_{H_9O_4^+}\gamma_B}{\gamma_{BH^+}}}{(x_{H_2O}\gamma_{H_2O})^4}$$

The 38,000-fold increase in h_0 between $0.100M$ HCl and $10.0M$ HCl can be dissected into the changes of the three factors in h_0:

	$\dfrac{[H_9O_4^+]}{\text{mol L}^{-1}}$	$\dfrac{\gamma_{H_9O_4^+}\gamma_B/\gamma_{BH^+}}{\text{L mol}^{-1}}$	$(x_{H_2O}\gamma_{H_2O})^{-4}$
$0.100M$ HCl	0.100	~1.0	~1.0
$10.0M$ HCl	10.0	~2.7	~140
Ratio	100	~2.7	~140

We see that the hydration of the proton that introduced the factor $(x_{H_2O}\gamma_{H_2O})^{-4}$ into the equation for $[BH^+]/[B]$ plays the most important role in making $10.0M$ HCl ~38,000-fold more acidic than $0.100M$ HCl.

The quantity h_0, defined by the ratio $[BH^+]/[B]$,

$$h_0 = \frac{[BH^+]}{[B]} \times \frac{1}{K^0}$$

generally is expressed in a logarithmic form analogous to the expression of $[H^+]$ in terms of pH:

$$\text{pH} = -\log([H^+] \times \text{L mol}^{-1})$$
$$H_0 = -\log h_0$$

H_0 is called the *Hammett acidity function*, named for LOUIS P. HAMMETT. It is appropriate to consider H_0 values as being on a pH scale extended to very concentrated solutions. One unit of pH and one unit of H_0 each correspond to a ten-fold change of acidity. A more negative value of H_0 denotes a greater acidity. Table 13–10 gives the H_0 values of some dilute and some concentrated solutions of strong acids. We see in this table that concentrated solutions of sulfuric and perchloric acid are even more acidic than hydrochloric acid solutions of the same molarity. A ten-fold increase in the molarity of perchloric acid (from 1.00*M* to 10.00*M*) decreases the value of H_0 by 5.57 units, or increases the acidity by a factor of $10^{5.57}$ (or 370,000).

A little bit more needs to be said about the phrase "increases the acidity by a factor of 370,000" in the preceding sentence. Because neutral indicator bases were used in the evaluation of H_0, the phrase means that the extent of protonation of a neutral indicator base as measured by the value of $[BH^+]/[B]$ is 370,000-fold greater in the more acidic solution. When Hammett first developed his acidity function scale, in 1932, many scientists hoped that the scale would be generally useful for correlating the ability of concentrated solutions to protonate neutral bases of all sorts. This hope has not been realized completely, but the H_0 scale is widely used nonetheless.

The value of H_0 for 100% sulfuric acid is −11.93. Although this liquid is enormously acidic, the acidity can be raised still further. The addition of sulfur trioxide to pure sulfuric acid causes the formation of $H_2S_2O_7$, called pyrosulfuric acid or disulfuric acid:

$$(HO)_2SO_2 + SO_3 = (HO)O_2SOSO_2(OH)$$

TABLE 13–10
The Acidity of Various Solutions of Strong Acids at 25°C

Solution	$\frac{[H^+]}{\text{mol L}^{-1}}$	pH[a]	H_0[b]
$1.00 \times 10^{-2}M$ HCl	1.00×10^{-2}	2.00	2.00
$1.00 \times 10^{-2}M$ $HClO_4$	1.00×10^{-2}	2.00	2.00
1.00*M* HCl	1.00	—	−0.20
1.00*M* $HClO_4$	1.00	—	−0.22
5.00*M* HCl	5.00	—	−1.76
5.00*M* $HClO_4$	5.00	—	−2.23
5.00*M* H_2SO_4[c]	—	—	−2.28
10.00*M* HCl	10.00	—	−3.60
10.00*M* $HClO_4$	10.00	—	−5.79
10.00*M* H_2SO_4[c]	—	—	−4.89
100% H_2SO_4	—	—	−11.93

[a] Calculated using $pH = -\log([H^+] \times \text{L mol}^{-1})$.
[b] Measured values.
[c] The second stage of dissociation of H_2SO_4 is incomplete; $2C > [H^+] > C$.

This liquid is even more acidic than pure sulfuric acid. A solution containing equal amounts of sulfuric acid and sulfur trioxide (which corresponds in composition to pure pyrosulfuric acid) has an H_0 value of -14.44. Another very strong acid is fluorosulfuric acid, $HOSO_2F$, in which one hydroxyl group of sulfuric acid is replaced by a fluorine atom. The value of H_0 for pure fluorosulfuric acid has been estimated as -15.1. Mixtures of SbF_5 and $HOSO_2F$, called *magic acid*, have still higher proton-donating abilities. For a solution with 11.4 mol % SbF_5, the value of H_0 has been estimated as -18.

Biographical Notes

LOUIS P. HAMMETT (1894–), an American chemist, was a pioneer in introducing physical chemical approaches into the study of organic chemistry. He was educated at Harvard (A.B., 1916) and Columbia (Ph.D., 1923), and he was on the faculty of Columbia from 1920 until his retirement in 1961.

FRIEDRICH W. G. KOHLRAUSCH (1840–1910), a German physicist, did much work in electricity and magnetism, particularly in the development of measuring devices. His studies of electrical conductance led to the definition of the ohm and the establishment of conductances of individual ions in solution.

Problems and Questions

13–1 What are the values of ΔG^0 for water dissociation

$$H_2O(l) = H^+(1.00 \text{ mol L}^{-1}) + OH^-(1.00 \text{ mol L}^{-1})$$

at 10°C and 70°C? Calculate these values from the values of K_w given in Table 13–1.

13–2 What is the concentration of hydroxide ion in each of the following solutions: 0.0543M HCl, 0.147M NaOH, and 0.0042M $Ba(OH)_2$?

13–3 What is the pH of each of the following solutions: 0.0150M HCl, 0.00470M HCl, $7.20 \times 10^{-4}M$ HBr, 0.0310M NaOH, and $3.40 \times 10^{-4}M$ NaOH?

13–4 A solution with what concentration of hydrochloric acid or sodium hydroxide has each of the following values of pH: 2.09, 3.76, 10.83, and 12.41?

13–5 The concentration of a solution of hydrochloric acid has an uncertainty of 0.5%; the concentration is 0.01630 ± 0.00008 mol L^{-1}. What is the pH of this solution? Include the uncertainty.

13–6 Equal volumes of solutions of $1.00 \times 10^{-3}M$ HCl and $1.00 \times 10^{-4}M$ NaOH are mixed. What is the concentration of hydrogen ion and the pH of the resulting solution?

13–7 A solution is prepared by mixing 50.0 cm^3 of 0.450M HCl and 1.00 g CsOH(s) and diluting the resulting solution to a final volume of 100.0 cm^3. What reagent was in excess, the acid or the base? This solution can be labeled as containing HCl and CsCl or CsCl and CsOH. Which of these is appropriate, and what concentration of each of the two solutes should be given on the label?

13–8 What volume of $0.0514M$ NaOH is required to neutralize exactly $25.00\ cm^3$ of $0.0942M$ HCl?

13–9 What volume of $1.50 \times 10^{-2}M$ NaOH must be added to $500\ cm^3$ of $0.200M$ HCl to give a solution of pH = 2.15?

13–10 Calculate the concentration of hydrogen ion and the pH at points in a titration of $25.00\ cm^3$ of $0.0834M$ HCl with $0.1042M$ NaOH. (Take into account the actual volume at each point.) Make calculations for 30, 60, 90, 95, 98, 99, 99.5, and 100.5% neutralization, and plot your results as pH versus volume of base.

13–11 What is the pH of $7.00 \times 10^{-7}\ M$ HCl?

13–12 Chapter 11 gave the extent of dissociation of various carboxylic acids at a concentration of $0.100\ mol\ L^{-1}$. From these data calculate the value of K_a for each acid.

13–13 A solution is prepared by dissolving 1.00 g of pure formic acid (HCO_2H) in water and diluting the solution to a final volume of $500\ cm^3$. By the use of indicators, the concentration of hydrogen ion is found to be $2.7 \times 10^{-3}\ mol\ L^{-1}$. What is the concentration of undissociated formic acid in this solution? What is the numerical value of the acid-dissociation constant? What is the base-dissociation constant for formate ion?

13–14 What are the concentrations of all of the species present in $0.0340M$ HOAc at 25°C? ($K_a = 1.75 \times 10^{-5}\ mol\ L^{-1}$.) What is the pH of the solution?

13–15 What are the concentrations of all species in $0.0200M$ $ClCH_2CO_2H$? ($K_a = 1.39 \times 10^{-3}\ mol\ L^{-1}$.)

13–16 What are the concentrations of all of the species present in $0.0450M$ NH_3? [$K_a(NH_4^+) = 5.69 \times 10^{-10}\ mol\ L^{-1}$.]

13–17 To 1.00 L of a solution of $0.150M$ CH_3CO_2H, enough sodium acetate ($NaO_2C_2H_3$) is added to make the pH = 4.10. What weight of sodium acetate is required?

13–18 To $100\ cm^3$ of a buffer solution with the stoichiometric composition $0.080M$ HOAc, $0.100M$ NaOAc, is added 1.5×10^{-4} mol NaOH. What is the pH of the solution before and after the addition of base? What weight of sodium hydroxide must be added to the original solution to make the pH = 6.08?

13–19 A solution of ammonia ($C = 0.400\ mol\ L^{-1}$) is to be titrated with a solution of hydrochloric acid. At the equivalence point the total volume will be ~1.5-fold larger than the original volume. At what pH will the equivalence point occur? What use can be made of the results of this calculation in the choice of an indicator?

13–20 Borax, $Na_2B_4O_7 \cdot 10H_2O$, dissolves in water to give equal amounts of boric acid, $B(OH)_3$, and borate ion, $B(OH)_4^-$, if the solution is dilute; the reaction is

$$Na_2B_4O_7 \cdot 10H_2O = 2Na^+ + 3H_2O + 2B(OH)_3 + 2B(OH)_4^-$$

The acid-dissociation constant for boric acid is $1.0 \times 10^{-9}\ mol\ L^{-1}$. What is the pH of $0.015M$ $Na_2B_4O_7$? Although the pH of a sodium carbonate solution changes with dilution, the pH of a solution of sodium tetraborate, $Na_2B_4O_7$, does not change with moderate dilution. Explain.

13–21 A solution of acetic acid, in water containing no other solutes, has a pH = 2.68 at 25°C. What volume of $0.0975M$ NaOH is required to neutralize $25.00\ cm^3$ of the solution?

13–22 If the solution with pH = 2.68 (see Problem 13–21) had been a hydrochloric acid solution, what volume of base (0.0975*M* NaOH) would have been required in the neutralization?

13–23 Consider solutions of the weak acids CH_3CO_2H ($K_a = 1.75 \times 10^{-5}$ mol L^{-1}), HONO ($K_a = 4.5 \times 10^{-4}$ mol L^{-1}), and HCN ($K_a = 6.2 \times 10^{-10}$ mol L^{-1}). At what stoichiometric concentration of each of these acids is there a 5% error in the concentration of hydrogen ion calculated using Equation 13–5s? (You can read approximate answers from figures, or can calculate more accurate values using the appropriate equation given in Footnotes 9 or 10.)

13–24 The acid-dissociation constants of ammonium ion and acetic acid are given in Table 13–4. From these values calculate the value of the equilibrium constant for the reaction $NH_4^+ + CH_3CO_2^- \rightleftarrows NH_3 + CH_3CO_2H$.

13–25 Ammonia gas is very soluble in water. (At 0°C an aqueous solution of ammonia with $x_{NH_3} = 0.481$ is in equilibrium with ammonia gas at $P = 1$ atm.) A sample of ammonia gas (10.00 L at a pressure of 1.00 atm at 20.0°C) is brought in contact with 2.00 L of hydrochloric acid ($C_{HCl} = 0.100$ mol L^{-1}). What are the concentrations of species in the final solution?

13–26 A solution contains both hydrochloric acid and acetic acid: 0.0500*M* HCl, 0.100*M* HOAc. What is the concentration of acetate ion in this solution?

13–27 Calculate the value of ΔG^0 for the dissociation of each of these weak acids:

$ClCH_2CO_2H$	$K = 1.39 \times 10^{-3}$ mol L^{-1}
NH_4^+	$K = 5.69 \times 10^{-10}$ mol L^{-1}
H_2O	$K_w = 1.00 \times 10^{-14}$ mol^2 L^{-2}

13–28 Smelling salts, once used to revive a person after a fainting spell, depend for their action on the sharp odor of ammonia. This solid contains ammonium ion and carbonate ion. For simplicity, assume it is pure ammonium carbonate. What is the balanced chemical equation for the production of ammonia?

13–29 An aqueous solution at 25°C contains two weak acids, formic acid ($K_a = 1.77 \times 10^{-4}$ mol L^{-1}) and propionic acid ($K_a = 1.34 \times 10^{-5}$ mol L^{-1}); the concentrations are 0.0500*M* HCOOH and 0.150*M* CH_3CH_2COOH. What are the concentrations of all species in this solution?

13–30 The indicator bromphenol blue is a weak acid, with $K_a = 1.6 \times 10^{-4}$ mol L^{-1}. Over what range of pH does this indicator change from 95% BH^+ to 95% B?

13–31 List the concentrations that are equal at each pH at which there is a tie line connecting features of Figure 13–13a and 13–13b.

13–32 What is the concentration of hydroxide ion in a solution of 0.500*M* Na_2CO_3?

13–33 To 2.50 L of 0.105*M* Na_2CO_3 at 25°C is added 0.135 mol of $CO_2(g)$. What is the pH of the resulting solution?

13–34 What are the concentrations of all species in a solution prepared by adding 0.150 mol of HCl(*g*) to 1.00 L of 0.400*M* Na_2CO_3? [No $CO_2(g)$ is allowed to escape.]

13–35 A buffer with pH = 10.15 is to be prepared by adding solid sodium hydrogen carbonate ($NaHCO_3$) to 1.00 L of 0.300*M* Na_2CO_3. What weight is required?

13–36 Consider solutions of glycine ($K_1 = 4.46 \times 10^{-3}$ mol L^{-1}; $K_2 = 1.66 \times 10^{-10}$ mol L^{-1} at 298.2 K). What are the relative concentrations of all of the species of glycine in a buffered solution with pH = 4.27?

13–37 The stepwise dissociation constants for the cation of serine at 25°C are $K_1 = 6.52 \times 10^{-3}$ mol L^{-1} and $K_2 = 6.19 \times 10^{-10}$ mol L^{-1}. What is the isoelectric point of this amino acid? (Express it both as a concentration of hydrogen ion and as a pH.)

13–38 Explain the pattern of relative values of K_1, K_2, K_3, and K_4 for ethylenediaminetetraacetic acid.

13–39 Both nitrous acid (HONO) and nitric acid ($HONO_2$) have basic properties; in what type of solution would you expect the species NO^+ and NO_2^+ to be present at the highest possible concentrations? With what neutral molecules are these two cations isoelectronic?

14

Solubility and Complex-Ion Equilibria

14–1 Introduction

In Chapter 13 we considered equilibria for acid–base reactions in aqueous solution. Now we will extend this discussion to include both qualitative and quantitative aspects of other equilibria:

1. solubility equilibria involving ionic solids and saturated solutions containing the constituent ions of the solids,
2. equilibria involving a hydrated metal ion and metal ion species called *complex ions*; in these species ligands are bonded to the metal ion,[1]
3. solubility equilibria involving a metal salt or hydroxide and a saturated solution in which complex ions of the metal with ligands other than water are the predominant species, and
4. heterogeneous equilibria in which a solute is distributed between two different solvents that are relatively insoluble in one another.

Scientists in chemistry and related fields must know how to do equilibrium calculations for systems of these types. Many acid–base, solubility, and complex-ion equilibria are involved in the chemistry of natural waters, of many metallurgical processes, and of biological fluids. A very important type of question in each of these areas asks what reaction or reactions occur

[1] "Ligand" comes from the Latin, *ligo*, to bind.

when certain reagents are brought together, and how far the reaction goes in the establishment of equilibrium. In this chapter you will learn how to answer this question by quantitative application of equilibrium principles.

14–2 Solubility Equilibria

We devoted most of our introduction to solubility equilibria in Chapter 7 to the solubility of solid iodine, a nonelectrolyte, in perfluoroheptane. Now we will consider the solubility of solid compounds that dissolve in water to give ionic solute species. For solutes MX and MY_2, the equilibria between the solid and the saturated solutions are represented by the chemical equations

$$MX(s) \rightleftarrows M^+(aq) + X^-(aq) \quad \text{and} \quad MY_2(s) \rightleftarrows M^{2+}(aq) + 2Y^-(aq)$$

For these reactions the equilibrium constants, called *solubility-product* constants or simply solubility products, are

$$K_s = [M^+][X^-] \quad \text{and} \quad K_s = [M^{2+}][Y^-]^2$$

This formulation of equilibrium-constant equations follows the principles already developed. The concentration of a component in the solid phase is represented in an equilibrium constant by its mole fraction. Because we are considering the solubilities of pure solid phases, $MX(s)$ and $MY_2(s)$, the mole fraction of each of these components is one,

$$x_{MX} = 1 \quad \text{and} \quad x_{MY_2} = 1$$

and these factors are not apparent in the solubility product, K_s. As in other ionic equilibria in solution, appreciable concentrations of ionic solutes cause ion-atmosphere effects, which make the numerical value of the activity-coefficient factor in the equilibrium constant deviate from one. For the solubility equilibria being considered, complete equilibrium-constant equations are

$$K_s^0 = [M^+][X^-]\gamma_{M^+}\gamma_{X^-} \quad \text{and} \quad K_s^0 = [M^{2+}][Y^-]^2\gamma_{M^{2+}}\gamma_{Y^-}^2$$

Therefore we have

$$K_s = [M^+][X^-] = K_s^0 \times \frac{1}{\gamma_{M^+}\gamma_{X^-}}$$

$$K_s = [M^{2+}][Y^-]^2 = K_s^0 \times \frac{1}{\gamma_{M^{2+}}\gamma_{Y^-}^2}$$

Most tabulations of solubility-product constants give the values of K_s^0. In our calculations we will use these as the values of K_s for the concentration conditions in question; that is, we will assume the numerical value of the activity-coefficient factor is one. This is a good assumption in calculations for solutions in which the total concentration of ions is very low (<0.001 mol L^{-1}), but it is a poor assumption for solutions in which the concentration of electrolyte is moderate or high (>0.10 mol L^{-1}). To illustrate this point, we will consider the solubility of calcium iodate hexahydrate, $Ca(IO_3)_2 \cdot 6H_2O$, in aqueous solutions of sodium chloride. Like the ion-producing dissociation of liquid water, discussed in Chapter 13,

$$H_2O(l) \rightleftarrows H^+(aq) + OH^-(aq)$$

the ion-producing dissolution of solid calcium iodate hexahydrate,

$$Ca(IO_3)_2 \cdot 6H_2O(s) \rightleftarrows Ca^{2+}(aq) + 2IO_3^-(aq) + 6H_2O(l)$$

occurs to a greater extent in solutions containing moderate concentrations of sodium chloride (a strong electrolyte) than it does in pure water; this is shown in Figure 14–1. Study of the solubility of calcium iodate has not been carried to a high enough concentration of sodium chloride to reveal whether the value of K_s goes through a maximum, as does the value of K_w, shown in Figure 13–1. Figures 13–1 and 14–1 show the errors associated with the assumption that the numerical value of the activity-coefficient factor for an ionic reaction is one in solutions containing appreciable concentrations of electrolyte.

More important than the influence of added electrolyte upon the solubility product of an ionic solute due to ion-atmosphere effects are the effects upon the solubility due to chemical participation of an ion of the added electrolyte in the solubility equilibrium. An added ion can participate as a *common ion*, or as a reagent which interacts with one of the constituent ions of the solid. The common-ion effect upon the solubility of our model salts $MX(s)$ and $MY_2(s)$ results if a soluble salt of M^+ or X^- is added to a solution saturated with $MX(s)$, or if a soluble salt of M^{2+} or Y^- is added to a solution saturated with $MY_2(s)$. Qualitatively, the common-ion effect can be considered a demonstration of Le Chatelier's principle. The addition of NaX to a solution saturated with $MX(s)$ causes the reaction to occur that consumes X^-, which is the reverse of the reaction

$$MX(s) \rightleftarrows M^+(aq) + X^-(s)$$

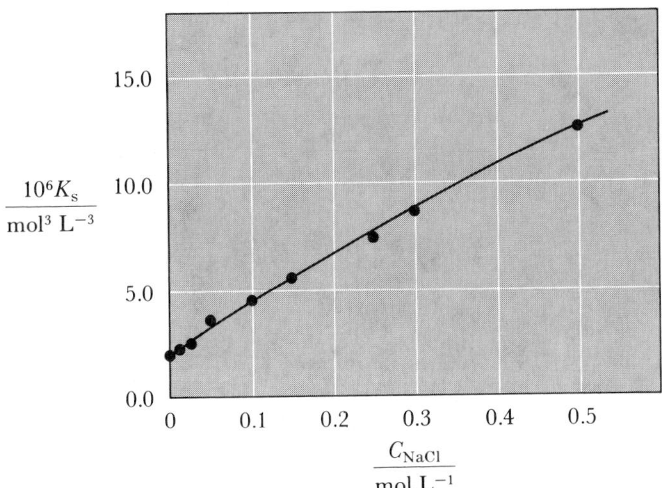

FIGURE 14–1
The value of the solubility-product constant for calcium iodate hexahydrate, $K_s = [Ca^{2+}][IO_3^-]^2$, in aqueous solutions of sodium chloride at 25.0°C. (Because only dilute solutions are considered, $x_{H_2O} \cong 1.0$ and $x_{H_2O}^6$ is omitted from the equation for K_s.)

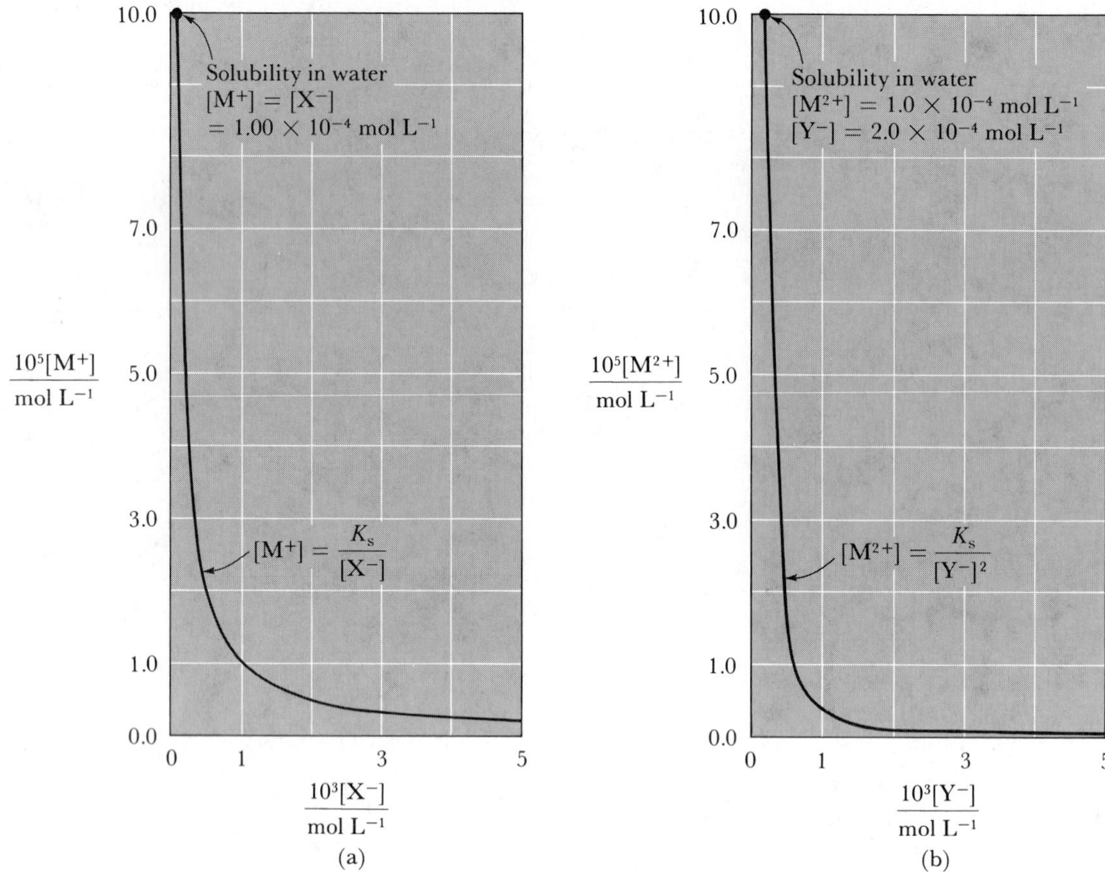

FIGURE 14–2
The effect of a common ion on the solubility of MX and MY_2.

(a) MX: $K_s = [M^+][X^-] = 1.00 \times 10^{-8}$ mol^2 L^{-2}
(b) MY_2: $K_s = [M^{2+}][Y^-]^2 = 4.00 \times 10^{-12}$ mol^3 L^{-3}

Consequently the concentration of M^+ is decreased by the addition of an X^--containing solute. Quantitative aspects of the common-ion effect are presented in Figure 14–2. These data also can be presented in linear plots, as is shown in Figure 14–3:

$$\log([M^+]/\text{mol L}^{-1}) \text{ versus } \log([X^-]/\text{mol L}^{-1})$$
$$\log([M^{2+}]/\text{mol L}^{-1}) \text{ versus } \log([Y^-]/\text{mol L}^{-1})$$

The basis for the linearity of the log–log plots is simple enough; if

$$[M^+] = K_s[X^-]^{-1} \quad \text{and} \quad [M^{2+}] = K_s[X^-]^{-2}$$

taking the logarithm of each side of these equations gives

$$\log[M^+] = \log K_s - 1 \times \log[X^-] \quad \text{and}$$
$$\log[M^{2+}] = \log K_s - 2 \times \log[Y^-]$$

The slopes of these plots, $\log[M^+]$ versus $\log[X^-]$ and $\log[M^{2+}]$ versus

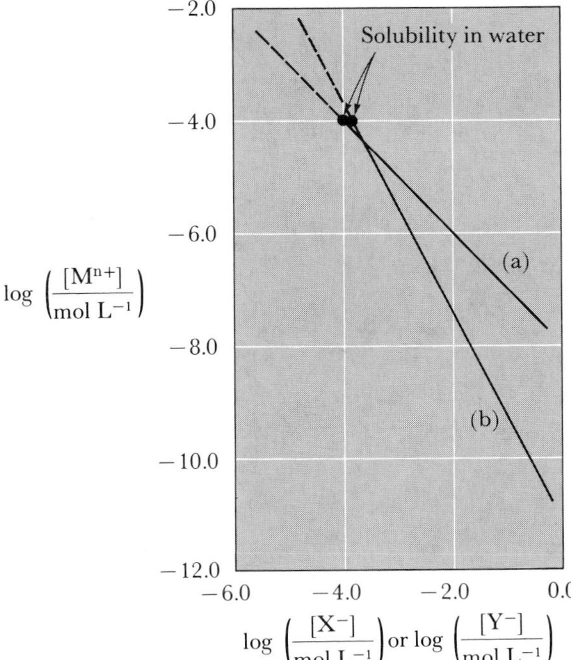

FIGURE 14–3
The effect of a common ion on the solubility of MX and MY_2.

(a) MX: $K_s = [M^+][X^-] = 1.00 \times 10^{-8}\ mol^2\ L^{-2}$

(b) MY_2: $K_s = [M^{2+}][Y^-]^2 = 4.00 \times 10^{-12}\ mol^3\ L^{-3}$

Slopes of these linear plots, −1 and −2, give the dependences of $[M^{n+}]$ upon $[X^-]$ and $[Y^-]$: $[M^+] \propto [X^-]^{-1}$, and $[M^{2+}] \propto [Y^-]^{-2}$. Plots for $[M^+] > 1.00 \times 10^{-4}$ mol L^{-1} are realized if an excess of metal ion is present.

log $[Y^-]$, are −1 and −2, respectively. Each slope is the negative of the exponent to which the anion concentration is raised in the solubility-product equation. In this example and in others we will encounter, the use of log–log plots is informative.

The solubility of a salt is lowered by increasing the concentration of a common ion, but the solubility is increased by addition of a reagent that reacts with an ion of the solid. If HX were a weak acid, the addition of nitric acid to a solution saturated with MX would increase the solubility; the reaction

$$MX(s) + H^+(aq) \rightleftarrows M^+(aq) + HX(aq)$$

increases the concentration of M^+ with increasing concentration of hydrogen ion.

THERMODYNAMICS OF DISSOLUTION REACTIONS

The solubility-product constant, like other equilibrium constants, is related to the thermodynamic quantities ΔG^0, ΔH^0, and ΔS^0 by relationships with which you are familiar:

$$-RT \ln [K_s \times (\text{mol L}^{-1})^{-\Delta n}] = \Delta G^0 = \Delta H^0 - T\Delta S^0$$

which leads to

$$\ln [K_s \times (\text{mol L}^{-1})^{-\Delta n}] = -\frac{\Delta H^0}{RT} + \frac{\Delta S^0}{R}$$

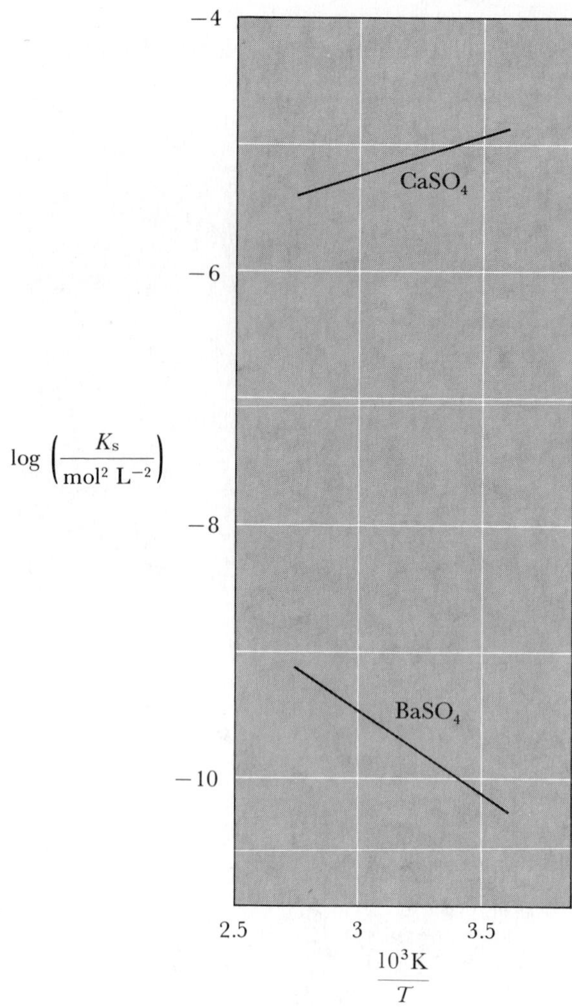

FIGURE 14–4
The dependence of the solubility-product constants for $CaSO_4$ and $BaSO_4$ upon temperature.

As in other examples that we have discussed, the dependence upon temperature of this equilibrium constant is determined by the standard enthalpy change accompanying the dissolution reaction: the slope of a plot of $\ln [K_s \times (\text{mol L}^{-1})^{-\Delta n}]$ versus T^{-1} is $-\Delta H^0/R$. Figure 14–4 shows such plots for two substances, calcium sulfate ($CaSO_4$) and barium sulfate ($BaSO_4$); the changes of enthalpy and entropy for these reactions under standard conditions are:

Reaction	$\frac{\Delta H^0}{\text{kJ mol}^{-1}}$	$\frac{\Delta S^0}{\text{J K}^{-1}\text{ mol}^{-1}}$
$CaSO_4(s) \rightleftarrows Ca^{2+}(aq) + SO_4^{2-}(aq)$	−11.8	−136
$BaSO_4(s) \rightleftarrows Ba^{2+}(aq) + SO_4^{2-}(aq)$	+25.8	−104

The dissolution reaction for calcium sulfate is exothermic, and that for barium sulfate is endothermic. The dominant factor in making the dissolution of calcium sulfate exothermic is the greater solvation energy of the smaller calcium ion (see Table 9–11). With an increase of temperature, the value of K_s for calcium sulfate decreases and that for barium sulfate increases.

For each of these reactions the change of entropy under standard conditions is negative. As we pointed out previously, many reactions producing ions in aqueous solution are accompanied by a decrease of entropy because hydration of the ions reduces the water molecules' randomness of location. The influence of ionic charge upon the value of ΔS^0 of a dissolution reaction, shown previously, can be seen also in comparison of isoelectronic substances, potassium perchlorate and calcium sulfate, and potassium chloride and calcium sulfide:

Reaction	$\frac{\Delta S^0}{\text{J K}^{-1}\text{ mol}^{-1}}$
$KClO_4(s) = K^+(aq) + ClO_4^-(aq)$	+133
$CaSO_4(s) = Ca^{2+}(aq) + SO_4^{2-}(aq)$	−136
$KCl(s) = K^+(aq) + Cl^-(aq)$	+75
$CaS(s) = Ca^{2+}(aq) + S^{2-}(aq)$	−138

The greater the charge on the ions, the more negative is the value of ΔS^0 in the dissolution reaction.

Plots of the solubilities of some compounds as a function of temperature show discontinuities, but for other compounds there are no such discontinuities. Figure 14–5 gives data for an example of each type. Over the range 0–60°C, the solubility of sodium chloride (on the molal scale) increases very slightly (from 6.097 mol kg^{-1} to 6.413 mol kg^{-1}) with no discontinuity. Over this same range of temperature, the solubility of sodium sulfate increases with increasing temperature until 32.4°C, and above this temperature the solubility decreases with increasing temperature. The

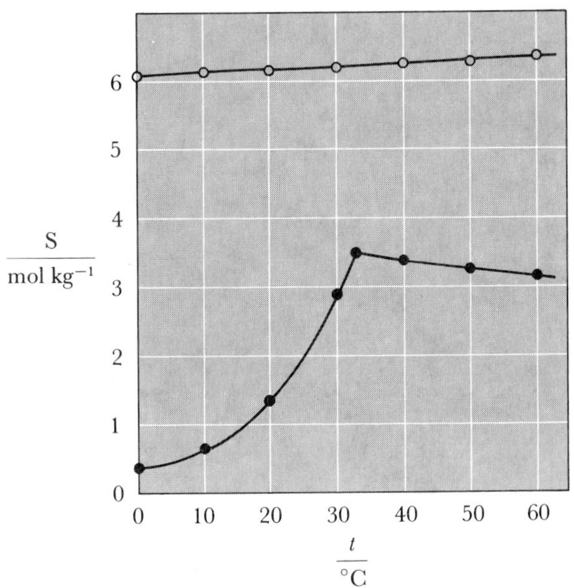

FIGURE 14–5
The solubility of sodium chloride and sodium sulfate in water as a function of temperature.

○ Sodium chloride, solid phase NaCl

● Sodium sulfate, solid phase:

$Na_2SO_4 \cdot 10H_2O$ $\quad t < 32.4°C$

Na_2SO_4 $\quad t > 32.4°C$

explanation for a discontinuity of this type is a *change of composition of the solid phase*. The solid phase in equilibrium with a saturated aqueous solution is $Na_2SO_4 \cdot 10H_2O$ below 32.4°C and is the anhydrous salt Na_2SO_4 above this temperature. The one process is endothermic:

$$Na_2SO_4 \cdot 10H_2O(s) = Na_2SO_4 \text{ (saturated solution)} \qquad \Delta H > 0$$

The other process is exothermic:

$$Na_2SO_4(s) = Na_2SO_4 \text{ (saturated solution)} \qquad \Delta H < 0$$

The solubility of sodium sulfate decahydrate increases with increasing temperature because the dissolution process absorbs heat, and the solubility of anhydrous sodium sulfate decreases with increasing temperature because the dissolution process evolves heat. This variation in the enthalpy change is reasonable. A factor that helps make the dissolution of an ionic salt exothermic is the hydration energy of the ions in solution. This factor is certainly important in making the anhydrous salt dissolve with the evolution of heat: the sodium ion and sulfate ion are hydrated in the saturated solution. The different behavior of sodium sulfate decahydrate can be explained by the fact that these ions already are hydrated in the solid, $Na_2SO_4 \cdot 10H_2O$.

The endothermic solid-phase transition,

$$\begin{aligned} Na_2SO_4 \cdot 10H_2O(s) &= Na_2SO_4(s) \\ &\quad + 10H_2O \text{ } (l\text{, saturated with sodium sulfate}) \\ \Delta H &= +67 \text{ kJ mol}^{-1} \end{aligned}$$

has been proposed as a means of solar-energy storage. Exposed to a temperature above the transition temperature, the system (a vat of water and excess sodium sulfate) absorbs energy from the surroundings (the solar collector) and the reaction goes to the right. When the temperature of the surroundings is less than the transition temperature, the reaction goes to the left, and heat flows from the system to the surroundings (e.g., the circulating water of a heating system).

SOLUBILITY-PRODUCT CALCULATIONS

Many kinds of numerical calculations are based upon the solubility-product equation. Now we will do calculations pertaining to two slightly soluble compounds, lead(II) sulfate ($PbSO_4$), with a solubility of 0.0382 g L^{-1} in water at 25°C, and magnesium hydroxide ($Mg(OH)_2$), with a solubility of 0.00855 g L^{-1} in water at 25°C. The molecular weights of lead sulfate and magnesium hydroxide are 303.26 and 58.33, respectively. Therefore values of the molar solubility, S, are

$$PbSO_4 \qquad S = \frac{0.0382 \text{ g L}^{-1}}{303.26 \text{ g mol}^{-1}} = 1.260 \times 10^{-4} \text{ mol L}^{-1}$$

$$Mg(OH)_2 \qquad S = \frac{0.00855 \text{ g L}^{-1}}{58.33 \text{ g mol}^{-1}} = 1.466 \times 10^{-4} \text{ mol L}^{-1}$$

The only species present at appreciable concentrations in a saturated aqueous solution of lead(II) sulfate are lead(II) ion (Pb^{2+}) and sulfate ion (SO_4^{2-}). The solubility equilibrium is

$$PbSO_4(s) \rightleftarrows Pb^{2+} + SO_4^{2-}$$

and the solubility product has the form

$$K_s = [Pb^{2+}][SO_4^{2-}]$$

Because dissolving one mole of solid lead sulfate produces one mole of lead(II) ion plus one mole of sulfate ion in solution, the concentration of each of these ions in the saturated solution is equal to the solubility, S:

$$[Pb^{2+}] = [SO_4^{2-}] = S = 1.260 \times 10^{-4} \text{ mol L}^{-1}$$

and

$$\begin{aligned} K_s &= [Pb^{2+}][SO_4^{2-}] \\ &= (1.260 \times 10^{-4} \text{ mol L}^{-1})(1.260 \times 10^{-4} \text{ mol L}^{-1}) \\ &= 1.59 \times 10^{-8} \text{ mol}^2 \text{ L}^{-2} \end{aligned}$$

The only species present at appreciable concentrations in a saturated solution of magnesium hydroxide are magnesium ion (Mg^{2+}) and hydroxide ion (OH^-). The solubility equilibrium is

$$Mg(OH)_2(s) \rightleftarrows Mg^{2+} + 2OH^-$$

and the solubility product has the form

$$K_s = [Mg^{2+}][OH^-]^2$$

Dissolving one mole of solid magnesium hydroxide gives one mole of magnesium ion:

$$[Mg^{2+}] = S = 1.466 \times 10^{-4} \text{ mol L}^{-1}$$

However, it gives *two* moles of hydroxide ion:

$$[OH^-] = 2 \times S = 2.932 \times 10^{-4} \text{ mol L}^{-1}$$

The value of the solubility product is found by substituting these concentrations into the equation

$$\begin{aligned} K_s &= [Mg^{2+}][OH^-]^2 \\ &= (1.466 \times 10^{-4} \text{ mol L}^{-1})(2.932 \times 10^{-4} \text{ mol L}^{-1})^2 \\ &= 1.260 \times 10^{-11} \text{ mol}^3 \text{ L}^{-3} \end{aligned}$$

These calculations are simple, and they are correct because the constituent ions of the solid phase are, in fact, the predominant species in the saturated solution.[2]

[2]The lead sulfate and magnesium hydroxide solubility examples are used in Appendix 2 in a discussion of significant figures.

A measurement of the solubility of a substance in pure water does not, however, disclose whether the constituent ions are the predominant species in the saturated solution. If the constituent ions are not the predominant species in solution, it is meaningless to calculate solubility-product constant as we just have done. The solubility of mercury(II) chloride in water at 25°C is $S = 68.9 \text{ g L}^{-1} = 0.254 \text{ mol L}^{-1}$. The predominant species in the solution is $HgCl_2$, which dissociates only very, very slightly to give ions. Therefore, the concentration of mercury(II) ion *is not* equal to the solubility, the concentration of chloride ion *is not* equal to two times the solubility, and the value of the solubility-product constant *is not* obtained as in the calculation for magnesium hydroxide:

$$K_s(HgCl_2) \neq S(2S)^2$$

We must know more than their solubilities in pure water to show that lead(II) sulfate and magnesium(II) hydroxide give predominantly the constituent ions in the saturated solution, and that mercury(II) chloride gives predominantly the neutral molecule in the saturated solution. A measurement of electrical conductance would shed some light on the question. Saturated solutions of lead(II) sulfate and magnesium(II) hydroxide have conductance values expected for solutions of strong electrolytes with the measured concentrations; a saturated solution of mercury(II) chloride has an electrical conductance only slightly greater than that of pure water.

Before we attempt quantitative calculations involving the solubility-product constant, we will review certain points. The solubility-product constant for lead(II) sulfate has the value $1.59 \times 10^{-8} \text{ mol}^2 \text{ L}^{-2}$ at 25°C:

$$K = [Pb^{2+}][SO_4^{2-}] = 1.59 \times 10^{-8} \text{ mol}^2 \text{ L}^{-2}$$

This means that an aqueous solution *saturated* with lead(II) sulfate at 25°C has a concentration of lead(II) ion and a concentration of sulfate ion such that their product is equal to $1.59 \times 10^{-8} \text{ mol}^2 \text{ L}^{-2}$. A solution containing lead(II) ions and sulfate ions at such concentrations that their product is less than $1.59 \times 10^{-8} \text{ mol}^2 \text{ L}^{-2}$ is a solution in which additional lead(II) sulfate can dissolve: it is a solution that is *not saturated* with lead(II) sulfate. A solution containing lead(II) ions and sulfate ions at such concentrations that their product is greater than $1.59 \times 10^{-8} \text{ mol}^2 \text{ L}^{-2}$ is a solution which is *supersaturated*. In time, the supersaturation will be relieved: lead(II) sulfate will precipitate from the solution until the product of concentrations, $[Pb^{2+}][SO_4^{2-}]$, decreases to the value corresponding to equilibrium, that is,

$$[Pb^{2+}][SO_4^{2-}] = 1.59 \times 10^{-8} \text{ mol}^2 \text{ L}^{-2}$$

Here are several solubility-product problems of varied difficulty.

EXAMPLE: The solubility product for barium sulfate at 25°C is $K_s = [Ba^{2+}][SO_4^{2-}] = 1.10 \times 10^{-10} \text{ mol}^2 \text{ L}^{-2}$. What is the solubility of barium sulfate in water at 25°C in moles per liter and in grams per liter? Assume that the predominant species in the saturated aqueous solution are barium ion Ba^{2+} and sulfate ion SO_4^{2-}.

The equation $BaSO_4(s) \rightleftarrows Ba^{2+} + SO_4^{2-}$ shows that one mole of barium sulfate dissolves to give one mole of barium ion and one mole of sulfate ion. The molarity of each of these ions in the saturated solutions is equal to the molar solubility, S:

$$[Ba^{2+}] = [SO_4^{2-}] = S \qquad K = [Ba^{2+}][SO_4^{2-}] = S^2$$

$$S^2 = 1.10 \times 10^{-10}\ \text{mol}^2\ \text{L}^{-2}$$

$$S = (1.10 \times 10^{-10}\ \text{mol}^2\ \text{L}^{-2})^{1/2} = 1.05 \times 10^{-5}\ \text{mol L}^{-1}$$

$$S = 1.05 \times 10^{-5}\ \text{mol L}^{-1} \times 233.4\ \text{g mol}^{-1} = 2.45 \times 10^{-3}\ \text{g L}^{-1}$$

EXAMPLE: The solubility-product constant for mercury(I) chloride (Hg_2Cl_2) at 25°C is $K_s = [Hg_2^{2+}][Cl^-]^2 = 1.32 \times 10^{-18}$ mol³ L⁻³. What is the solubility of mercury(I) chloride in water at 25°C? Assume that the predominant species in the saturated solution are Hg_2^{2+} and Cl^-. (Notice that mercury(I) ion is the dimeric species Hg_2^{2+}. This species will be explained in terms of electronic structure in Chapter 18. Notice also that unlike mercury(II) chloride, which is relatively soluble but does not dissociate to give ions in solution, mercury(I) chloride is relatively insoluble, and the dissolved compound dissociates to give ions.)

The equation $Hg_2Cl_2(s) \rightleftarrows Hg_2^{2+} + 2Cl^-$ shows that one mole of mercury(I) chloride (Hg_2Cl_2) dissolves to give one mole of mercury(I) ion Hg_2^{2+} and two moles of chloride ion. The relationships of the concentrations to the solubility, S, are

$$[Hg_2^{2+}] = S \qquad [Cl^-] = 2S$$

Therefore the solubility-product equation involving the solubility S is

$$K_s = [Hg_2^{2+}][Cl^-]^2 = S \times (2S)^2 = 4S^3$$

We solve for S as follows:

$$4S^3 = 1.32 \times 10^{-18}\ \text{mol}^3\ \text{L}^{-3}$$

$$S^3 = 3.30 \times 10^{-19}\ \text{mol}^3\ \text{L}^{-3}$$

$$S = 6.91 \times 10^{-7}\ \text{mol L}^{-1}$$

EXAMPLE: What is the solubility of magnesium hydroxide at 25°C in a solution containing 0.150*M* NaOH? The solubility product for magnesium hydroxide is 1.260×10^{-11} mol³ L⁻³.

The solubility equilibrium, $Mg(OH)_2(s) \rightleftarrows Mg^{2+} + 2OH^-$, allows us to assign the equilibrium concentrations

$$[Mg^{2+}] = S \quad \text{and} \quad [OH^-] = C_{NaOH} + 2S = 0.150\ \text{mol L}^{-1} + 2S$$

Because the additional hydroxide ion from the sodium hydroxide lowers the solubility of magnesium hydroxide ion below its value in water (1.466×10^{-4} mol L⁻¹), we can make the approximation that $C_{NaOH} \gg 2S$, which gives

$$[OH^-] = 0.150\ \text{mol L}^{-1}$$

Substituting into the solubility-product equation and solving for S, we have:

$$S \times (0.150 \text{ mol L}^{-1})^2 = 1.260 \times 10^{-11} \text{ mol}^3 \text{ L}^{-3}$$

$$S = \frac{1.260 \times 10^{-11} \text{ mol}^3 \text{ L}^{-3}}{(0.150 \text{ mol L}^{-1})^2} = 5.60 \times 10^{-10} \text{ mol L}^{-1}$$

Although each mole of magnesium ion going into solution is accompanied from the solid by two moles of hydroxide ion, the hydroxide ion added in this way is very small compared to that from the sodium hydroxide:

$$[OH^-] \text{ from } Mg(OH)_2 = 2 \times [Mg^{2+}] = 1.12 \times 10^{-9} \text{ mol L}^{-1}$$

and the solution of the problem based upon the approximation $[OH^-] = 0.150 \text{ mol L}^{-1}$ is valid.

EXAMPLE: What is the solubility at 25°C of magnesium hydroxide in $1.00 \times 10^{-4} M$ NaOH?

Given the solubility in pure water (1.466×10^{-4} mol L^{-1}), it seems likely that the solubility in this solution is large enough to make the equilibrium concentration of hydroxide ion appreciably greater than that of the sodium hydroxide. With the solubility represented by S, the concentrations are

$$[Mg^{2+}] = S \quad \text{and} \quad [OH^-] = 2S + 1.00 \times 10^{-4} \text{ mol L}^{-1}$$

Substitution into the solubility-product equation gives

$$S(2S + 1.00 \times 10^{-4} \text{ mol L}^{-1})^2 = 1.260 \times 10^{-11} \text{ mol}^3 \text{ L}^{-3}$$

which is a cubic equation in S. This cubic equation can be solved simply by the method of successive approximations. In the present problem this procedure involves the following steps:

1. Rearrange the above equation to give

$$S = \frac{1.260 \times 10^{-11} \text{ mol}^3 \text{ L}^{-3}}{(2S + 1.00 \times 10^{-4} \text{ mol L}^{-1})^2}$$

   (In using the method of successive approximations for equilibrium calculations, you often will be able to rearrange the equation to the form $x = f(x)$, in which $f(x)$ involves a polynomial function of x.)

2. Assume a value of S, and substitute this assumed value into the right side of the equation. This allows you to calculate a value of S.
3. Try successive assumed values of S until the calculated value is the same as the assumed value. This is easier than it might seem. The correct value of S is between the assumed and calculated values in each step.

The procedure will be illustrated by assuming as a first approximation $S = 1.0 \times 10^{-4}$ mol L^{-1}, which is somewhat smaller than the value in pure water (1.466×10^{-4} mol L^{-1}):

Approx. number	$\frac{S(\text{assumed})}{\text{mol L}^{-1}}$	$\frac{\left[\frac{1.260 \times 10^{-11}\ \text{mol}^3\ \text{L}^{-3}}{(2S + 1.00 \times 10^{-4}\ \text{mol L}^{-1})^2}\right]}{\text{mol L}^{-1}}$
1	1.00×10^{-4}	1.40×10^{-4}
2	1.18×10^{-4}	1.12×10^{-4}
3	1.15×10^{-4}	1.16×10^{-4}

In each approximation after the first, the value of S was assumed to be the geometric mean of the assumed and calculated values from the previous approximations. The third approximation yields the correct value of the product $[Mg^{2+}][OH^-]^2$, within the accuracy justified by the data.

In the dependence of the solubility of magnesium hydroxide on the concentration of sodium hydroxide, we see an illustration of the common-ion effect. Addition of a common ion, hydroxide ion, shifts equilibrium in the solubility reaction to lower the concentration of magnesium ion:

$\frac{C_{NaOH}}{\text{mol L}^{-1}}$	$\frac{S}{\text{mol L}^{-1}}$
0	1.466×10^{-4}
1.00×10^{-4}	1.16×10^{-4}
0.150	5.60×10^{-10}

EXAMPLE: What is the solubility of copper(II) oxide in water? Assume the solubility equilibrium for this substance to be

$$CuO(s) + H_2O \rightleftarrows Cu^{2+} + 2OH^-$$

The equilibrium constant for the dissolution equilibrium for CuO(s) is

$$K_s = 2.19 \times 10^{-20}\ \text{mol}^3\ \text{L}^{-3}$$

If we were to use the approach that we used in the magnesium hydroxide example,

$$[Cu^{2+}] = S \quad \text{and} \quad [OH^-] = 2S$$

we would have

$$4S^3 = 2.19 \times 10^{-20}\ \text{mol}^3\ \text{L}^{-3}$$
$$S = 1.76 \times 10^{-7}\ \text{mol L}^{-1}$$

This approach would give

$$[OH^-] = 2S = 3.52 \times 10^{-7}\ \text{mol L}^{-1}$$

which is only slightly larger than the concentration of hydroxide ion from the dissociation of water. Thus in an accurate calculation we must consider the equilibrium

$$H_2O(l) \rightleftarrows H^+ + OH^-$$

together with the solubility equilibrium. The electroneutrality condition applied to the saturated solution is

$$[H^+] + 2[Cu^{2+}] = [OH^-]$$

With the substitutions

$$[Cu^{2+}] = S$$

$$[OH^-] = \left(\frac{K_s}{[Cu^{2+}]}\right)^{1/2} = \left(\frac{K_s}{S}\right)^{1/2}$$

$$[H^+] = \frac{K_w}{[OH^-]} = \frac{K_w}{\left(\dfrac{K_s}{S}\right)^{1/2}}$$

the electroneutrality equation gives

$$\frac{K_w}{\left(\dfrac{K_s}{S}\right)^{1/2}} + 2S = \left(\frac{K_s}{S}\right)^{1/2}$$

or

$$S = \frac{1}{2}\left[\left(\frac{K_s}{S}\right)^{1/2} - \frac{K_w}{\left(\dfrac{K_s}{S}\right)^{1/2}}\right]$$

This is an equation that can be solved by successive approximations. For the first approximation, we will use the value of S calculated previously under the assumption that $2S \gg 1 \times 10^{-7}$ mol L^{-1}.

$\dfrac{S(\text{assumed})}{\text{mol L}^{-1}}$	$\dfrac{\frac{1}{2}\left[\left(\dfrac{2.19 \times 10^{-20}\ \text{mol}^3\ \text{L}^{-3}}{S}\right)^{1/2} - \dfrac{1.00 \times 10^{-14}\ \text{mol}^2\ \text{L}^{-2}}{(2.19 \times 10^{-20}\ \text{mol}^3\ \text{L}^{-3}/S)^{1/2}}\right]}{\text{mol L}^{-1}}$
1.76×10^{-7}	1.62×10^{-7}
1.69×10^{-7}	1.66×10^{-7}
1.67×10^{-7}	1.67×10^{-7}

Thus, the concentrations of species in the saturated solution are:

$$[Cu^{2+}] = 1.67 \times 10^{-7}\ \text{mol L}^{-1}$$

$$[OH^-] = \left(\frac{2.19 \times 10^{-20}\ \text{mol}^3\ \text{L}^{-3}}{1.67 \times 10^{-7}\ \text{mol L}^{-1}}\right)^{1/2} = 3.62 \times 10^{-7}\ \text{mol L}^{-1}$$

$$[H^+] = \frac{1.00 \times 10^{-14}\ \text{mol}^2\ \text{L}^{-2}}{3.62 \times 10^{-7}\ \text{mol L}^{-1}} = 2.76 \times 10^{-8}\ \text{mol L}^{-1}$$

The concentration of copper(II) ion derived in the refined calculation is 5.4% lower than the value derived in the initial calculation. Depending upon your objective and the accuracy of your data (the stoichiometric concentrations and equilibrium constants), you may sometimes choose to refine calculations and sometimes choose to let approximate calculation do. In the example we have just considered, we assumed that species other than Cu^{2+}, H^+, and OH^- were not present in solution at appreciable concentrations. In fact, in slightly alkaline solution, copper(II) ion forms the hydroxo species $CuOH^+$ to an appreciable extent. A highly accurate calculation should take this into account. We will explore the effect of hydroxo-metal ion species upon the solubility of metal hydroxides in a discussion of zinc hydroxide later in this chapter.

Table 14–1 gives the solubility-product constants for several relatively insoluble substances. In comparing values of these equilibrium constants, remember that a direct comparison can be made only if the same number of solute ions are produced in solution (e.g., two ions: $AgCl(s) \rightleftarrows Ag^+ + Cl^-$, $K_s = 1.78 \times 10^{-10}$ mol^2 L^{-2}; and $PbSO_4(s) \rightleftarrows Pb^{2+} + SO_4^{2-}$, $K_s = 1.59 \times 10^{-8}$ mol^2 L^{-2}; or three ions: $Mg(OH)_2(s) \rightleftarrows Mg^{2+} + 2OH^-$, $K_s = 1.26 \times 10^{-11}$ mol^3 L^{-3}; and $Ag_2SO_4(s) \rightleftarrows 2Ag^+ + SO_4^{2-}$, $K_s = 1.59 \times 10^{-5}$ mol^3 L^{-3}). For substances giving the same number of ions in solution, the solubility-product constants have the same dimensions. Among substances that can be legitimately compared, a substance with a smaller solubility-product constant is less soluble in water. No such simple

TABLE 14–1
Values of the Solubility-Product Constants for Relatively Insoluble Substances[a] at 25°C

$MX(s) \rightleftarrows M^{n+} + X^{n-}$		$MY_2 \rightleftarrows M^{2+} + 2Y^-$ or $M_2Z \rightleftarrows 2M^+ + Z^{2-}$	
Substance	K_s / mol^2 L^{-2}	Substance	K_s / mol^3 L^{-3}
$CaSO_4$	9.2×10^{-6}	Ag_2SO_4	1.59×10^{-5}
$PbSO_4$	1.59×10^{-8}	CaF_2	4.0×10^{-11}
$CaCO_3$	7.22×10^{-9}	$Mg(OH)_2$	1.26×10^{-11}
$BaCO_3$	5.13×10^{-9}	$Fe(OH)_2$	8.0×10^{-16}
AgCl	1.78×10^{-10}	Hg_2Cl_2[b]	1.32×10^{-18}
$BaSO_4$	1.10×10^{-10}	CuO[c]	2.19×10^{-20}
AgBr	5.25×10^{-13}		
AgI	8.32×10^{-17}	$MW_3 \rightleftarrows M^{3+} + 3W^-$ or $M_3V \rightleftarrows 3M^+ + V^{3-}$	
$AlPO_4$	5.75×10^{-19}	Substance	K_s / mol^4 L^{-4}
		$Al(OH)_3$	1.1×10^{-33}
		$Fe(OH)_3$	6.0×10^{-38}
		Ag_3PO_4	1.3×10^{-20}

[a]Solubility-product constants for sulfides are given in Table 14–3.
[b]Gives Hg_2^{2+} and $2Cl^-$ in aqueous solution.
[c]Gives $Cu^{2+} + 2OH^-$ in aqueous solution (see example in the text).

comparison of solubility-product constants is possible for substances giving different numbers of ions in solution. This point already has been illustrated: lead sulfate is less soluble in water than magnesium hydroxide (1.26×10^{-4} mol L^{-1} compared to 1.466×10^{-4} mol L^{-1}). However, the numerical value of the solubility-product constant for lead sulfate is larger than that for magnesium hydroxide.

The values of K_s given in the table are appropriate for an electrolyte-free aqueous solution at a particular temperature, 25°C; for another temperature, there are different values of K_s. (Only if the enthalpy change for the solid dissolving in the saturated solution is fortuitously close to zero will the value of K_s be approximately independent of temperature.) The values cannot be used in highly accurate calculations for solutions with appreciable concentrations of electrolyte. As we already have learned, values of the activity-coefficient factor should be included in accurate equilibrium calculations.

Table 14–1 presents only data for relatively insoluble substances. The solubilities of substances with moderate to high solubilities generally are not discussed in terms of the solubility-product concept. Rather, compilations of solubilities generally give data in weight or moles of solute per unit volume or per unit weight of pure solvent. Relatively soluble substances have a large range of solubilities (e.g., the solubilities of alkali-metal hydroxides given in Table 12–6). In saturated solutions of such substances, the activity coefficients of ions are very different from 1 L mol^{-1}, and they change appreciably with addition of other solutes. Thus calculations of the type we have been making are more complex for relatively soluble substances, and generally such calculations must be based upon uncertain assumptions about the variation of activity coefficients with concentration. We will not venture into this area.

Solubility-product constants will tell us how to separate from a mixture of two ions the ion that forms the less-soluble compound. As an example, let us consider the separation of iron(II) ion and magnesium(II) ion by precipitation of iron(II) hydroxide. Figure 14–6 shows the concentration of each of these metal ions in equilibrium with the solid metal hydroxides as a function of the concentration of hydroxide ion. Suppose a solution with the initial stoichiometric composition

$$C_{FeCl_2} = 0.0300 \text{ mol L}^{-1} \quad \text{and} \quad C_{MgCl_2} = 0.0200 \text{ mol L}^{-1}$$

is treated with concentrated sodium hydroxide solution. (We will assume that the concentration of this sodium hydroxide solution is so high that the volume of the solution barely changes as the sodium hydroxide is added.) The hydroxide ion concentration at which the precipitation of iron(II) hydroxide starts can be calculated:

$$[OH^-] = \left(\frac{K_s}{[Fe^{2+}]}\right)^{1/2}$$

$$[OH^-] = \left(\frac{8.0 \times 10^{-16} \text{ mol}^3 \text{ L}^{-3}}{0.030 \text{ mol L}^{-1}}\right)^{1/2} = 1.6 \times 10^{-7} \text{ mol L}^{-1}$$

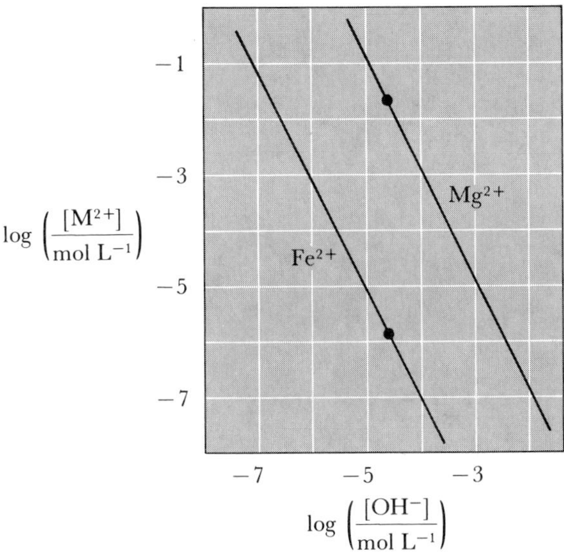

FIGURE 14–6
The concentrations of iron(II) ion and magnesium(II) ion in equilibrium with the solid hydroxides:

$$[Fe^{2+}][OH^-]^2 = 8.0 \times 10^{-16} \text{ mol}^3 \text{ L}^{-3}$$
$$[Mg^{2+}][OH^-]^2 = 1.26 \times 10^{-11} \text{ mol}^3 \text{ L}^{-3}$$

(The points show the concentration of each metal ion at $[OH^-] = 2.5 \times 10^{-5}$ mol L^{-1}.)

We will get the optimum separation of these two metal ions if we add sodium hydroxide only to the point at which the concentration of hydroxide ion is infinitesimally smaller than the concentration that would cause the precipitation of magnesium hydroxide to start. This concentration can be calculated:

$$[OH^-] = \left(\frac{K_s}{[Mg^{2+}]}\right)^{1/2}$$
$$= \left(\frac{1.26 \times 10^{-11} \text{ mol}^3 \text{ L}^{-3}}{0.020 \text{ mol L}^{-1}}\right)^{1/2} = 2.51 \times 10^{-5} \text{ mol L}^{-1}$$

At this concentration of hydroxide ion, the concentration of iron(II) ion is very much lower than it was at the start because most of it has been converted to the solid hydroxide:

$$[Fe^{2+}] = \frac{K_s}{[OH^-]^2}$$
$$= \frac{8.0 \times 10^{-16} \text{ mol}^3 \text{ L}^{-3}}{(2.51 \times 10^{-5} \text{ mol L}^{-1})^2} = 1.27 \times 10^{-6} \text{ mol L}^{-1}$$

Thus the fraction of iron(II) remaining in solution at this point is

$$\frac{[Fe^{2+}]}{C_{FeCl_2}} = \frac{1.27 \times 10^{-6} \text{ mol L}^{-1}}{0.0300 \text{ mol L}^{-1}} = 4.2 \times 10^{-5}$$

That is, 99.996% of the iron(II) has precipitated from the solution, and none of the magnesium(II) has precipitated.

Now let us calculate the concentrations of all species in the final solution. The concentrations of magnesium ion and chloride ion are the same as presented initially:

$$[Mg^{2+}] = C_{MgCl_2} = 0.0200 \text{ mol L}^{-1}$$
$$[Cl^-] = 2C_{MgCl_2} + 2C_{FeCl_2} = 0.1000 \text{ mol L}^{-1}$$

The concentrations of iron(II) ion and hydroxide ion have the values that we have calculated:

$$[Fe^{2+}] = 1.27 \times 10^{-6} \text{ mol L}^{-1} \quad \text{and} \quad [OH^-] = 2.51 \times 10^{-5} \text{ mol L}^{-1}$$

The concentration of sodium ion can be calculated using the electroneutrality condition,

$$[Na^+] + [H^+] + 2[Fe^{2+}] + 2[Mg^{2+}] = [Cl^-] + [OH^-]$$

or

$$[Na^+] = [Cl^-] + [OH^-] - [H^+] - 2[Fe^{2+}] - 2[Mg^{2+}]$$

Because the concentrations of the other ions are very small,

$$\begin{aligned}[Na^+] &\cong [Cl^-] - 2[Mg^{2+}] \\ &\cong 0.100 \text{ mol L}^{-1} - 0.040 \text{ mol L}^{-1} = 0.060 \text{ mol L}^{-1}\end{aligned}$$

This separation plan is based upon two assumptions:

1. A solid phase containing both iron(II) ion and magnesium(II) ion does not form. (If mixed crystals formed in this system, each metal ion would be less soluble than we have calculated.)
2. We can find the point at which we should stop adding sodium hydroxide. We can do this by monitoring the addition with a pH meter (see Section 16–4), and stopping at pH $\cong$ 9.40, or we can get the right concentration of hydroxide ion by using an appropriate ammonia–ammonium ion buffer. For ammonium ion $K_a = 5.69 \times 10^{-10}$ mol L^{-1}, and a solution with $[NH_4^+]/[NH_3] \cong 0.70$ has a pH $\cong$ 9.40.

14–3 Metal–Ligand Complexes

Metal ions in solution are coordinated to ligands, which can be molecules of the solvent or other neutral or anionic species. You learned in Chapter 9 that coordination numbers in ionic solids depend upon the ratio of radii of cation and anion, and this radius-ratio factor also plays a role in determining the coordination number of a metal ion in solution. In some well-behaved systems the coordination number remains constant through a series of species [e.g., $Be(OH_2)_{4-n}F_n^{2-n}$ with $n = 0$–4, inclusive] and in other systems the coordination number changes as one ligand replaces another [e.g., $Cd(OH_2)_6^{2+}$ going to CdI_4^{2-}]. If the concentrations of ligands other than the solvent are low, the metal ion will be coordinated exclusively to solvent molecules, and because water is the solvent in which we are most interested we will start this discussion with hydrated metal ions.

HYDRATED METAL IONS

As you already have learned, the hydration of metal ions in solution is partly responsible for the ability of a crystalline ionic salt to dissolve in water. The hydration of a cation and the enthalpy change in the process,

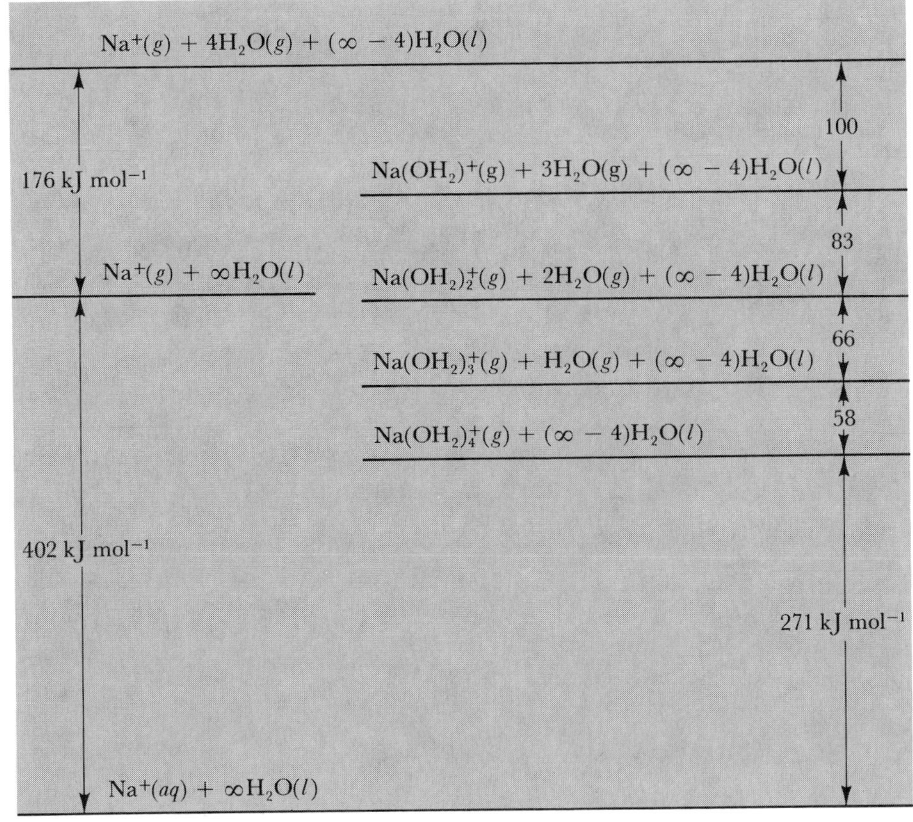

FIGURE 14–7
The stepwise hydration of gaseous sodium ion. (There are data suggesting that hydrated sodium ion in aqueous solution is $Na(OH_2)_4^+$.)

for example, for sodium ion,

$$Na^+(g) = Na^+(aq) \qquad \Delta H^0 = -402 \text{ kJ mol}^{-1}$$

can be dissected into a series of steps based upon a combination of gas-phase studies, which characterize the successive hydration steps,

$$Na(OH_2)_{n-1}^+(g) + H_2O(g) = Na(OH_2)_n^+(g)$$

and solution studies, which reveal the hydrated sodium ion to be

$$Na(OH_2)_4^+$$

This dissection is presented in Figure 14–7, which can be compared with Figure 12–7, an analogous plot for the hydration of the proton. In the line-formula representation of the hydrated species, $Na(OH_2)_4^+$, the fact that bonding is between the metal atom and the oxygen atom is indicated. This bonding for most metal ions is largely the electrostatic attraction between the positive charge of the cation and the negative charge of the oxygen end of the water dipole. Other metal ions interact with water to form hydrated ions with various compositions, geometries, and bond energies, which depend upon the charge on the ion, its electronic structure, size, and

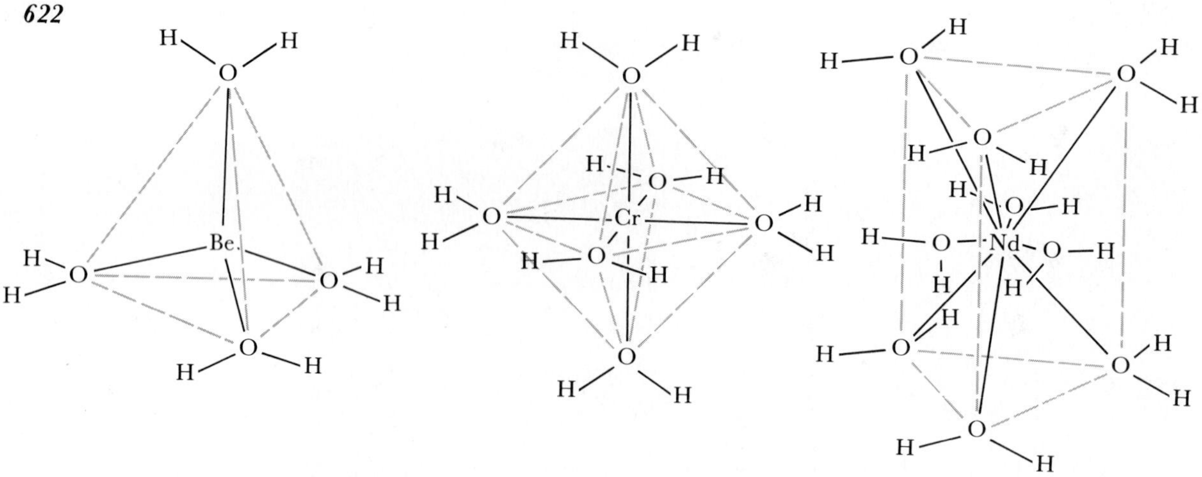

FIGURE 14–8
The configurations of various hydrated metal ions:

$Be(OH_2)_4^{2+}$	tetrahedral
$Cr(OH_2)_6^{3+}$	octahedral
$Nd(OH_2)_9^{3+}$	triangular prism with water molecule at each corner and on each rectangular face

polarizability. Figure 14–8 shows the geometrical structures of some hydrated ions: $Be(OH_2)_4^{2+}$, $Cr(OH_2)_6^{3+}$, and $Nd(OH_2)_9^{3+}$.

The solvation of chromium(III) ion has been characterized also in other polar solvents, and in these it is solvated by six molecules. For instance, solvated chromium(III) ion in methyl alcohol is $Cr(OHCH_3)_6^{3+}$.

In this chapter we will discuss the equilibria in aqueous solutions of metal ions and those ligands that can displace a water molecule from the hydrated metal ion. In a displacement reaction, for example,

$$Cr(OH_2)_6^{3+}(aq) + Cl^- \rightleftarrows Cr(OH_2)_5Cl^{2+}(aq) + H_2O(l)$$

a metal–ligand bond is broken (Cr–O in this example) and a metal–ligand bond is formed (Cr–Cl in this example). The change of enthalpy in these reactions may be small despite the energy of the bonds because the overall process involves both bond-making and bond-breaking. For this chromium-(III) reaction, $\Delta H^0 = +25$ kJ mol^{-1}.

Some ligands displace one water molecule from the hydration shell around a metal ion; these are called *monodentate ligands*. Other ligands with two or more basic coordination sites can displace two or more water molecules; these are called *polydentate ligands*. A complexed metal ion involving a polydentate ligand is called a chelated metal ion or simply *chelate*.[3] The interactions of metal ions with biological molecules generally involve chelation, and synthetic chelating agents are very useful in chemical analysis and water treatment. Table 14–2 gives some common ligands.

[3]From *chele*, a Greek word, meaning claw.

TABLE 14–2
Common Ligands

Monatomic, monodentate ligands	Polyatomic, monodentate ligands	
F^-	NH_3	N_3^-
Cl^-	H_2O	SCN^-
Br^-	CH_3OH	O_2
I^-	$(CH_3)_2SO$	CO

Bidentate chelating ligands

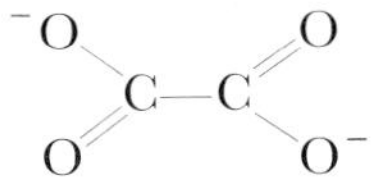

Oxalate ion, $C_2O_4^{2-}$ (one resonance form)

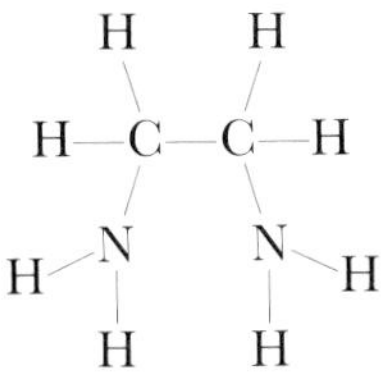

Ethylenediamine, $C_2H_4(NH_2)_2$

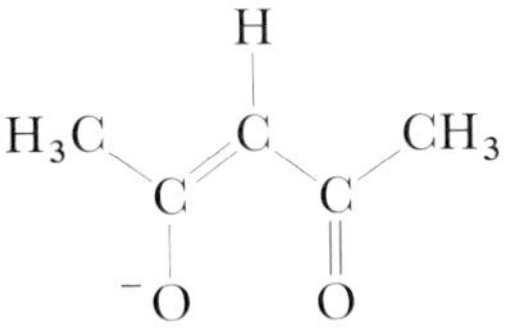

Acetylacetonate ion (acac), $C_5H_7O_2^-$ (one resonance form)

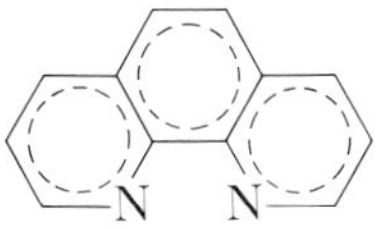

1, 10-Phenanthroline (phen), $C_{12}H_8N_2$ (isoelectronic with phenanthrene, $C_{14}H_{10}$)

Polydentate chelating ligands

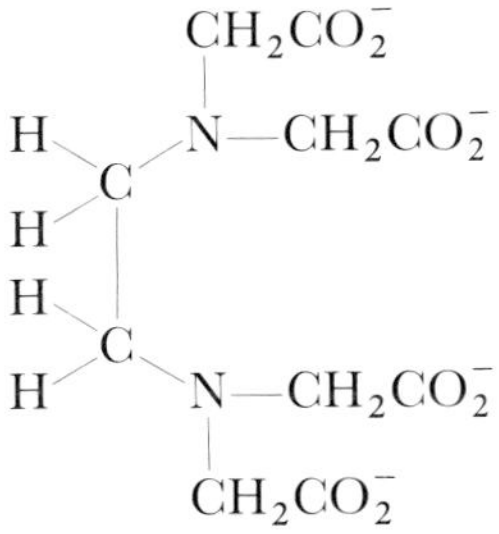

Ethylenediaminetetraacetate (EDTA), $C_2H_4[N(CH_2CO_2)_2]_2^{4-}$

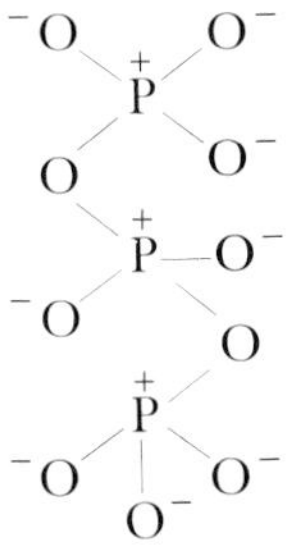

Tripolyphosphate ion, $P_3O_{10}^{5-}$ (related to ATP)

COMPLEX-ION EQUILIBRIA

In the reaction of a monodentate ligand with a hydrated metal ion, the fully complexed species generally forms in a series of stepwise ligand replacement reactions. For example, for the fluoride complexes of beryllium-(II) ion, the stepwise reactions are:

1. $Be(OH_2)_4^{2+} + F^- \rightleftarrows Be(OH_2)_3F^+ + H_2O$
2. $Be(OH_2)_3F^+ + F^- \rightleftarrows Be(OH_2)_2F_2 + H_2O$
3. $Be(OH_2)_2F_2 + F^- \rightleftarrows Be(OH_2)F_3^- + H_2O$
4. $Be(OH_2)F_3^- + F^- \rightleftarrows BeF_4^{2-} + H_2O$

Showing the waters of hydration in these chemical formulas makes it clear that each coordinated fluoride ion in the fluoride complexes has replaced a water molecule of the hydrated beryllium(II) ion, $Be(OH_2)_4^{2+}$. For these four stepwise reactions at 25.0°C in aqueous solution, the equilibrium constants are:[4]

$$K_1 = \frac{[BeF^+]}{[Be^{2+}][F^-]} = 7.9 \times 10^4 \text{ L mol}^{-1}$$

$$K_2 = \frac{[BeF_2]}{[BeF^+][F^-]} = 5.8 \times 10^3 \text{ L mol}^{-1}$$

$$K_3 = \frac{[BeF_3^-]}{[BeF_2][F^-]} = 6.1 \times 10^2 \text{ L mol}^{-1}$$

$$K_4 = \frac{[BeF_4^{2-}]}{[BeF_3^-][F^-]} = 2.7 \times 10 \text{ L mol}^{-1}$$

(In representing concentrations in equilibrium-constant equations, the hydration need not be included with the chemical formulas.) Just as the relative concentrations at equilibrium of species containing different numbers of protons depend upon the concentration of hydrogen ion in the solution, the relative concentrations at equilibrium of species containing different numbers of fluoride ions depend upon the concentration of fluoride ion in solution. Rearrangement of the equations for K_1 and K_2 gives

$$\frac{[BeF^+]}{[Be^{2+}]} = K_1[F^-] = 7.9 \times 10^4 \text{ L mol}^{-1} [F^-]$$

$$\frac{[BeF_2]}{[BeF^+]} = K_2[F^-] = 5.8 \times 10^3 \text{ L mol}^{-1} [F^-]$$

A combination of these equations gives another ratio of relative concentrations:

$$\frac{[BeF_2]}{[Be^{2+}]} = \frac{[BeF_2]}{[BeF^+]} \times \frac{[BeF^+]}{[Be^{2+}]} = K_1K_2[F^-]^2 = 4.58 \times 10^8 \text{ L}^2 \text{ mol}^{-2} [F^-]^2$$

Notice that the relative concentrations of species that differ in composition by one fluoride ion depend upon the first power of the concentration of

[4]Notice that these equilibrium constants are for association reactions. For the stepwise association of protons with a base, the convention is to consider the dissociation reactions, as we did in Chapter 13.

fluoride ion, and, if the compositions differ by two fluoride ions, the ratio of concentrations depends upon the square of the concentration of fluoride ion.

An equation for the fraction of beryllium(II) in the form of each particular species is easily derived from the equations for the relative concentrations of pairs of species. For instance, the fraction of beryllium(II) present as trifluoroberyllium(II) ion, $f_{BeF_3^-}$, is

$$f_{BeF_3^-} = \frac{[BeF_3^-]}{[Be^{2+}] + [BeF^+] + [BeF_2] + [BeF_3^-] + [BeF_4^{2-}]}$$

Dividing the numerator and denominator by the concentration of uncomplexed beryllium(II) ion, $[Be^{2+}]$, gives

$$f_{BeF_3^-} = \frac{\dfrac{[BeF_3^-]}{[Be^{2+}]}}{1 + \left(\dfrac{[BeF^+]}{[Be^{2+}]}\right) + \left(\dfrac{[BeF_2]}{[Be^{2+}]}\right) + \left(\dfrac{[BeF_3^-]}{[Be^{2+}]}\right) + \left(\dfrac{[BeF_4^{2-}]}{[Be^{2+}]}\right)}$$

The equilibrium-constant equations allow the fraction of beryllium(II) present as trifluoroberyllium(II) ion to be given as a function of the concentration of free fluoride ion:

$$f_{BeF_3^-} = \frac{K_1K_2K_3[F^-]^3}{1 + K_1[F^-] + K_1K_2[F^-]^2 + K_1K_2K_3[F^-]^3 + K_1K_2K_3K_4[F^-]^4}$$

Analogous equations can be derived from the fraction of beryllium(II) present in each of the forms. Substitution of a particular concentration of fluoride ion into these equations gives a numerical value for each fraction at that concentration of fluoride ion. At $[F^-] = 0.00150$ mol L^{-1}, the magnitude of each term in this quotient is:

$$\frac{[BeF^+]}{[Be^{2+}]} = K_1[F^-]$$
$$= 7.9 \times 10^4 \text{ L mol}^{-1} \times 0.00150 \text{ mol L}^{-1} = 1.19 \times 10^2$$

$$\frac{[BeF_2]}{[Be^{2+}]} = K_1K_2[F^-]^2$$
$$= 4.58 \times 10^8 \text{ L}^2 \text{ mol}^{-2} \times (0.00150 \text{ mol L}^{-1})^2 = 1.03 \times 10^3$$

$$\frac{[BeF_3^-]}{[Be^{2+}]} = K_1K_2K_3[F^-]^3$$
$$= 2.80 \times 10^{11} \text{ L}^3 \text{ mol}^{-3} \times (0.00150 \text{ mol L}^{-1})^3 = 9.5 \times 10^2$$

$$\frac{[BeF_4^{2-}]}{[Be^{2+}]} = K_1K_2K_3K_4[F^-]^4$$
$$= 7.5 \times 10^{12} \text{ L}^4 \text{ mol}^{-4} \times (0.00150 \text{ mol L}^{-1})^4 = 38$$

and the value of $f_{BeF_3^-}$ is

$$f_{BeF_3^-} = \frac{9.5 \times 10^2}{1 + 1.19 \times 10^2 + 1.03 \times 10^3 + 9.5 \times 10^2 + 38} = 0.44$$

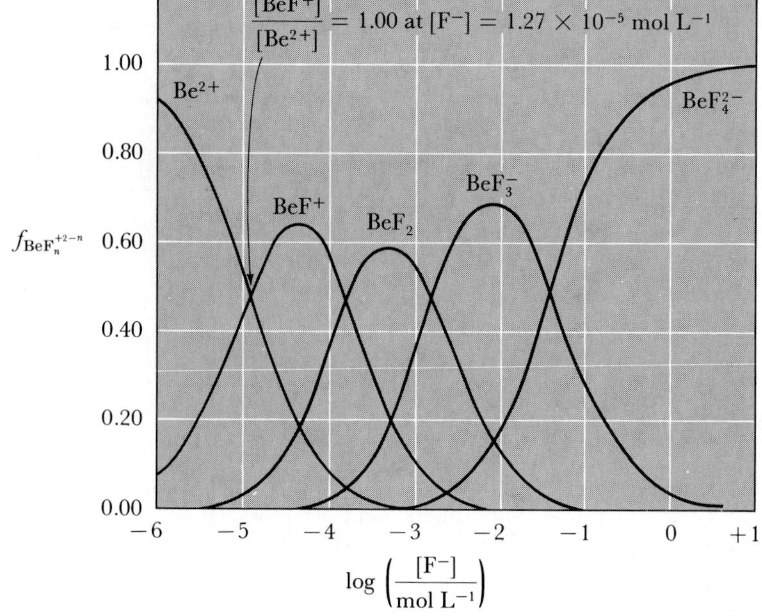

FIGURE 14–9
The relative concentrations of beryllium(II) fluoride complexes in aqueous solution at 25°C as a function of the concentration of fluoride ion. [From R. E. Mesmer and C. F. Baes, Jr., *Inorg. Chem.* **8**, 618 (1969).]

A graph of the fraction of beryllium(II) present in each form as a function of the concentration of fluoride ion, calculated in this way, is given in Figure 14–9. Notice how the relative concentration of each species with an intermediate number of fluoride ions (i.e., 1, 2, or 3) first increases with increasing concentration of fluoride ion and then decreases. Also notice the concentration of fluoride ion at which there are equal concentrations of species with compositions differing by one fluoride ion. Because

$$\frac{[BeF^+]}{[Be^{2+}]} = K_1[F^-]$$

the ratio $[BeF^+]/[Be^{2+}]$ is 1.00 at $[F^-] = K_1^{-1} = 1.27 \times 10^{-5}$ mol L^{-1}:

$$\frac{[BeF^+]}{[Be^{2+}]} = K_1[F^-]$$
$$= 7.9 \times 10^4 \text{ L mol}^{-1} \times 1.27 \times 10^{-5} \text{ mol L}^{-1} = 1.00$$

Although our discussion has involved primarily the stepwise formation of each of the several complexes, BeF_n^{2-n}, in which $n = 1, 2, 3,$ and 4, there are circumstances in which one can use the equilibrium constant for the formation of the completely complexed metal ion from the free metal ion:

$$Be^{2+} + 4F^- \rightleftarrows BeF_4^{2-}$$

This reaction is the sum of the four stepwise reactions, and the equilibrium constant,

$$K = \frac{[BeF_4^{2-}]}{[Be^{2+}][F^-]^4}$$

is the product of the equilibrium constants for the four stepwise reactions:

$$\begin{aligned} K &= K_1K_2K_3K_4 \\ &= (7.9 \times 10^4 \text{ L mol}^{-1})(5.8 \times 10^3 \text{ L mol}^{-1}) \\ &\quad \times (6.1 \times 10^2 \text{ L mol}^{-1})(27 \text{ L mol}^{-1}) \\ &= 7.5 \times 10^{12} \text{ L}^4 \text{ mol}^{-4} \end{aligned}$$

A reaction such as this is not the principal equilibrium in any solution, but this equilibrium constant is useful in calculating the concentration of uncomplexed beryllium(II) ion in a solution in which beryllium(II) is predominantly tetrafluoroberyllium(II) ion. Thus at a concentration of free fluoride ion of 1.00 mol L^{-1}, the only beryllium(II) species present at appreciable relative concentrations are BeF_3^- and BeF_4^{2-}, with over 96% of beryllium present as BeF_4^{2-}. Thus in a solution with $[F^-] = 1.00$ mol L^{-1}, the relative concentration of uncomplexed beryllium(II) ion can be calculated:

$$\begin{aligned} \frac{[Be^{2+}]}{C_{Be(II)}} &\cong \frac{[Be^{2+}]}{[BeF_4^{2-}]} = \frac{1}{K[F^-]^4} \\ &= \frac{1}{7.5 \times 10^{12} \text{ L}^4 \text{ mol}^{-4} \times (1.00 \text{ mol L}^{-1})^4} = 1.3 \times 10^{-13} \end{aligned}$$

The considerations that went into the construction of Figure 14–9 and the features it displays are pertinent to many aspects of the solution chemistry of metals. Analytical procedures for many metal ions depend upon measurement of light absorption by some particular complex ion having strong light absorption. A graph such as Figure 14–9 shows the concentration of free ligand necessary to cause formation of the species with the desired light-absorption qualities. An analysis for copper(II) ion can be based upon the light absorption by the deeply blue colored species $Cu(NH_3)_4^{2+}$. Figure 14–10 shows that a concentration of free ammonia greater than ~0.32 mol L^{-1} is necessary if $f_{Cu(NH_3)_4^{2+}}$ is to be >0.98.

Table 14–2 lists the diatomic neutral molecules carbon monoxide and oxygen as ligands. Hemoglobin is an important system in which these molecules act as ligands. Hemoglobin, the oxygen-carrying component of blood, is a complex protein containing four iron atoms, each of which can bind an oxygen molecule (or a carbon monoxide molecule). We will consider the structure of this molecule in Chapter 20; here we can represent the four stepwise equilibria (Hb_4 represents the hemoglobin molecule):

1. $Hb_4(aq) + O_2(g) \rightleftarrows Hb_4O_2(aq)$
2. $Hb_4O_2(aq) + O_2(g) \rightleftarrows Hb_4(O_2)_2(aq)$
3. $Hb_4(O_2)_2(aq) + O_2(g) \rightleftarrows Hb_4(O_2)_3(aq)$
4. $Hb_4(O_2)_3(aq) + O_2(g) \rightleftarrows Hb_4(O_2)_4(aq)$

Values of the equilibrium constants for these four reactions for sheep

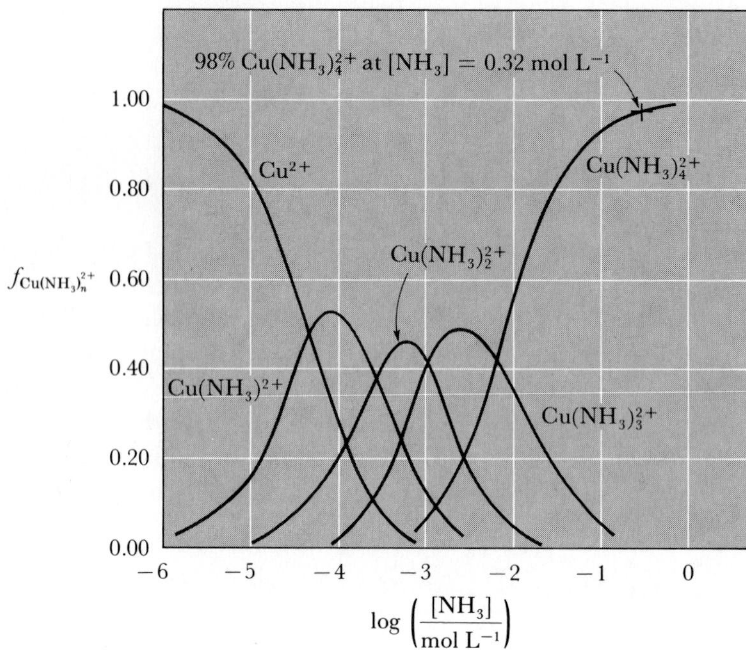

FIGURE 14–10
The relative concentrations of copper(II)–ammonia complexes in aqueous solution at 25°C as a function of the concentration of ammonia. The equilibrium constants upon which this graph is based are $K_1 = 1.86 \times 10^4$ L mol^{-1}, $K_2 = 3.88 \times 10^3$ L mol^{-1}, $K_3 = 1.00 \times 10^3$ L mol^{-1}, and $K_4 = 1.55 \times 10^2$ L mol^{-1}.

hemoglobin[5] at 19°C in an aqueous solution at pH = 9.1 (maintained with a boric acid–sodium borate buffer) are:

$$K_1 = \frac{[Hb_4O_2]}{[Hb_4]P_{O_2}} = 0.112 \text{ torr}^{-1}$$

$$K_2 = \frac{[Hb_4(O_2)_2]}{[Hb_4O_2]P_{O_2}} = 0.12 \text{ torr}^{-1}$$

$$K_3 = \frac{[Hb_4(O_2)_3]}{[Hb_4(O_2)_2]P_{O_2}} = 0.15 \text{ torr}^{-1}$$

$$K_4 = \frac{[Hb_4(O_2)_4]}{[Hb_4(O_2)_3]P_{O_2}} = 2.00 \text{ torr}^{-1}$$

These equilibrium constants are written in terms of the partial pressure of oxygen over the solution; they could be rewritten in terms of the molarity of oxygen in the solution by using the Henry's law constant for 19°C:

$$P_{O_2} = 5.39 \times 10^5 \text{ torr L mol}^{-1} \times [O_2]$$

which gives

$$[O_2] = 1.86 \times 10^{-6} \text{ mol L}^{-1} \text{ torr}^{-1} \times P_{O_2}$$

[5]Sheep hemoglobin was used because the hemoglobin has to be free of carbon monoxide, which contaminates the hemoglobin of smokers.

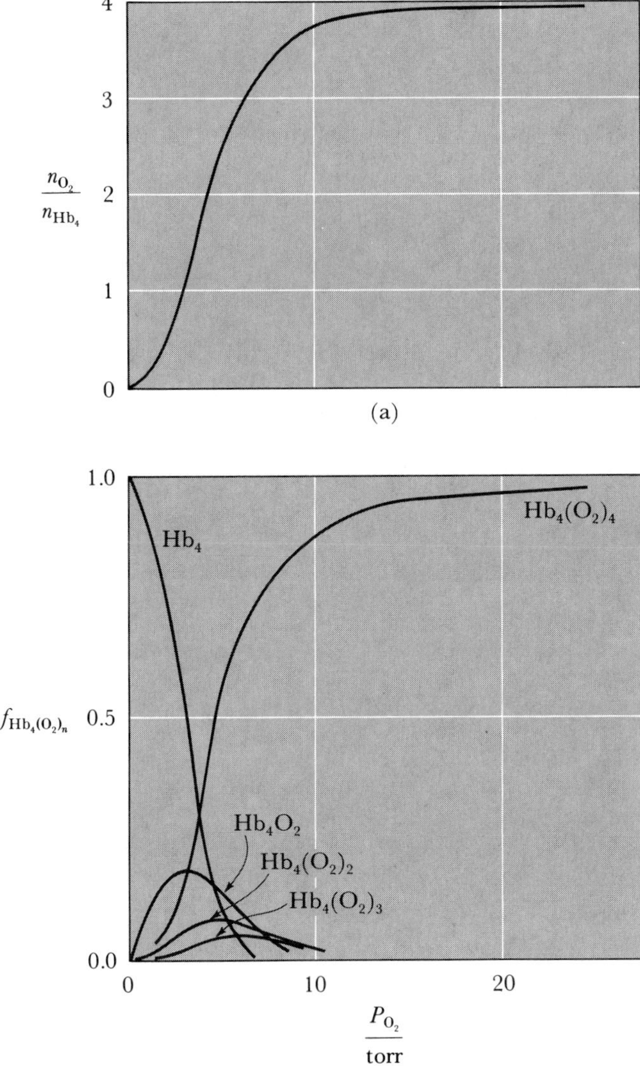

FIGURE 14–11
The oxygenation of sheep hemoglobin (symbol Hb_4). pH = 9.1 at 19°C. (a) Oxygen-binding, n_{O_2}/n_{Hb_4}. (b) Relative concentrations of Hb_4, Hb_4O_2, $Hb_4(O_2)_2$, $Hb_4(O_2)_3$, and $Hb_4(O_2)_4$.

Unlike the beryllium(II)–fluoride and copper(II)–ammonia systems, in which the intermediate species reach moderately high relative concentrations, the hemoglogin–oxygen system is one in which the intermediate species are not predominant at any pressure of oxygen. The binding of oxygen by hemoglobin is said to be *cooperative*; the binding of oxygen at one iron-coordination site in hemoglobin increases the affinity of the other sites for oxygen. This is shown in Figure 14–11. The steepness of the plot in Figure 14–11a shows the efficiency with which oxygen can be transported by hemoglobin. Without this cooperativity, a three-fold change of oxygen pressure would change the fractional binding by ~0.27 or less; with cooperativity, a three-fold change of oxygen pressure can change the fractional binding by as much as ~0.80. Oxygen is taken up by hemoglobin in the lungs where the partial pressure and concentration are high, and is delivered to organs where the concentration is low. These concentrations differ by a factor of 3 to 5. An important characteristic of an oxygen carrier is the amount of oxygen it can give up between these concentrations of free oxygen.

Carbon monoxide binds to the iron atoms in hemoglobin much more strongly than does oxygen, and an iron atom cannot bind both of these molecules at the same time. Carbon monoxide is so lethal because it converts hemoglobin to a species incapable of carrying oxygen.

14–4 The Influence of Other Equilibria on Solubility Equilibria

A solubility equilibrium is influenced by species that interact with one (or more) of the solute species produced by dissolution of the solute compound. This principle will be illustrated by specific examples, which include both nonelectrolytes and electrolytes as solutes. Many of the principles illustrated in the examples provide the basis for separation of metal ions by combinations of precipitation reactions and complex-ion reactions.

THE INFLUENCE OF IODIDE ION ON THE SOLUBILITY OF IODINE

The solubility of solid iodine in water at 25°C is 1.331×10^{-3} mol L^{-1}. Because iodine and iodide ion react to form triiodide ion,

$$I_2(aq) + I^-(aq) \rightleftarrows I_3^-(aq) \qquad K = \frac{[I_3^-]}{[I_2][I^-]} = 710 \text{ L mol}^{-1} \text{ (at 25°C)}$$

the total solubility of iodine in an aqueous solution of potassium iodide is greater than its solubility in water. We can derive an equation for the total solubility of iodine, S, determined by a titration with thiosulfate ion, that involves the value of the equilibrium constant, just given, and the stoichiometric concentration of iodide, C_I. The derivation is

$$S = [I_2] + [I_3^-] = [I_2] + K[I_2][I^-]$$

We now must find a relationship between free iodide ion concentration and the stoichiometric concentration of potassium iodide; this relationship follows from

$$C_I = [I^-] + [I_3^-] = [I^-] + K[I_2][I^-]$$

which gives

$$[I^-] = \frac{C_I}{1 + K[I_2]}$$

Substitution of this expression into the equation for S gives

$$S = [I_2] + \frac{K[I_2]}{1 + K[I_2]} \times C_I$$

Using $K = 710 \text{ L mol}^{-1}$ and $[I_2] = 1.331 \times 10^{-3} \text{ mol L}^{-1}$, we find

$$S = 1.331 \times 10^{-3} \text{ mol L}^{-1} + \frac{0.945}{1.945} C_I = 1.331 \times 10^{-3} \text{ mol L}^{-1} + 0.486 C_I$$

This relationship is shown graphically in Figure 14–12.

Although we have pursued this example by starting with knowledge of the equilibrium constant, it is clear that an experimental value of the dependence of solubility upon the stoichiometric concentration of potassium iodide ($\Delta S/\Delta C_I = 0.486$) allows evaluation of K: rearrangement of

$$\frac{\Delta S}{\Delta C_I} = \frac{1.331 \times 10^{-3} \text{ mol L}^{-1} \times K}{1 + 1.331 \times 10^{-3} \text{ mol L}^{-1} \times K}$$

gives

$$K = \frac{\Delta S/\Delta C_I}{1.331 \times 10^{-3} \text{ mol L}^{-1} (1 - \Delta S/\Delta C_I)}$$

$$= \frac{0.486}{1.331 \times 10^{-3} \text{ mol L}^{-1} (1 - 0.486)} = 710 \text{ L mol}^{-1}$$

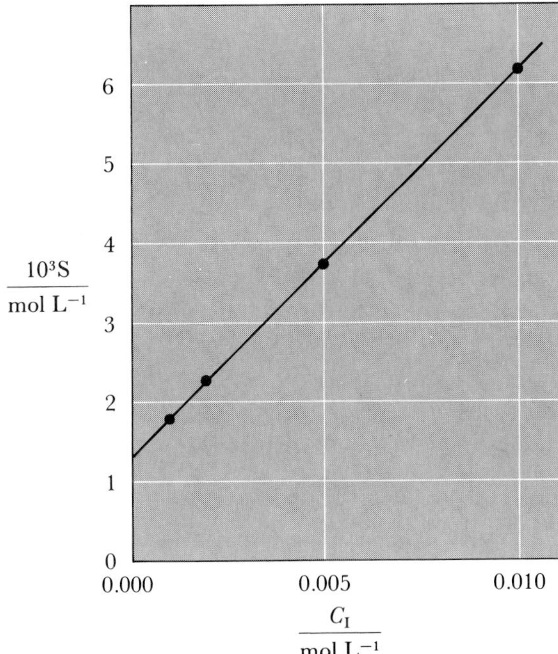

FIGURE 14–12
The solubility of iodine in aqueous solution as a function of the concentration of potassium iodide at 25°C. The experimental values of solubility are shown as points. The slope of the line is $\Delta S/\Delta C_I = 0.486$.

THE INFLUENCE OF AMMONIUM ION ON THE SOLUBILITY OF MAGNESIUM HYDROXIDE

If magnesium hydroxide is precipitated from solution with aqueous ammonia solution, the equation for the net reaction is

$$Mg^{2+} + 2NH_3 + 2H_2O \rightleftarrows Mg(OH)_2(s) + 2NH_4^+$$

This also is the principal equilibrium existing in the system consisting of excess solid magnesium hydroxide and a saturated solution containing ammonia and ammonium ion. The equilibrium-constant equation for this reaction is

$$K = \frac{[NH_4^+]^2}{[Mg^{2+}][NH_3]^2}$$

We can obtain the value of this equilibrium constant by combining values of the equilibrium constants for other reactions. The reaction in question is

$$2 \times (NH_3 + H_2O \rightleftarrows NH_4^+ + OH^-) \quad \textit{minus} \quad Mg(OH)_2(s) \rightleftarrows Mg^{2+} + 2OH^-$$

Therefore the equilibrium constant is

$$K = \frac{K_b^2}{K_s} = \left(\frac{[NH_4^+][OH^-]}{[NH_3]}\right)^2 \times \frac{1}{[Mg^{2+}][OH^-]^2}$$

$$= \frac{[NH_4^+]^2}{[Mg^{2+}][NH_3]^2}$$

From Tables 13–7 and 14–1, we have values for K_b and K_s:

$$K_b = 1.76 \times 10^{-5} \text{ mol L}^{-1}$$

$$K_s = 1.26 \times 10^{-11} \text{ mol}^3 \text{ L}^{-3}$$

Thus

$$K = \frac{[NH_4^+]^2}{[Mg^{2+}][NH_3]^2} = \frac{(1.76 \times 10^{-5} \text{ mol L}^{-1})^2}{1.26 \times 10^{-11} \text{ mol}^3 \text{ L}^{-3}} = 24.6 \text{ L mol}^{-1}$$

It is clear from both the chemical equation and the equilibrium-constant expression that increasing the concentration of ammonium ion, a weak acid, increases the concentration of magnesium ion in equilibrium with excess solid magnesium hydroxide. It is easy to calculate the solubility of magnesium hydroxide in a solution containing both ammonia and ammonium chloride at high concentrations because the final equilibrium concentrations of ammonia and ammonium ion are essentially the same as the initial stoichiometric concentrations. This is the case for the solubility in a solution with the composition 2.00*M* NH_3, 0.50*M* NH_4Cl:

$$[Mg^{2+}] = \frac{1}{K} \times \frac{[NH_4^+]^2}{[NH_3]^2} \cong \frac{1}{K} \times \frac{C_{NH_4^+}^2}{C_{NH_3}^2}$$

$$= \frac{1}{24.6 \text{ L mol}^{-1}} \times \left(\frac{0.500 \text{ mol L}^{-1}}{2.000 \text{ mol L}^{-1}}\right)^2$$

$$= 2.54 \times 10^{-3} \text{ mol L}^{-1}$$

This solubility also can be calculated by first finding the concentration of hydroxide ion in this relatively concentrated buffer solution; this can be done using the value of K_b:

$$[OH^-] = K_b \times \frac{[NH_3]}{[NH_4^+]} \cong K_b \times \frac{C_{NH_3}}{C_{NH_4^+}}$$

$$= 1.76 \times 10^{-5}\ \text{mol L}^{-1} \times \frac{2.00\ \text{mol L}^{-1}}{0.50\ \text{mol L}^{-1}} = 7.04 \times 10^{-5}\ \text{mol L}^{-1}$$

and

$$[Mg^{2+}] = \frac{K_s}{[OH^-]^2} = \frac{1.26 \times 10^{-11}\ \text{mol}^3\ \text{L}^{-3}}{(7.04 \times 10^{-5}\ \text{mol L}^{-1})^2} = 2.54 \times 10^{-3}\ \text{mol L}^{-1}$$

If the buffer solution is appreciably more dilute, we must recognize that every mole of magnesium hydroxide that dissolves will convert two moles of ammonium ion into two moles of ammonia. We allow for this fact automatically by considering the balanced chemical equation

$$Mg^{2+} + 2NH_3 + 2H_2O \rightleftarrows Mg(OH)_2(s) + 2NH_4^+$$

If no reaction:	C_{NH_3}	$C_{NH_4^+}$
At equilibrium:	$(C_{NH_3} + 2[Mg^{2+}])$	$(C_{NH_4^+} - 2[Mg^{2+}])$

The equilibrium-constant equation gives

$$[Mg^{2+}] = \frac{1}{K} \times \frac{[NH_4^+]^2}{[NH_3]^2} = \frac{1}{K} \times \frac{(C_{NH_4^+} - 2[Mg^{2+}])^2}{(C_{NH_3} + 2[Mg^{2+}])^2}$$

We can solve this cubic equation for the concentration of magnesium ion by successive approximations as in previous examples.

THE INFLUENCE OF COMPLEXING LIGANDS ON THE SOLUBILITY OF SILVER CHLORIDE

Silver(I) forms complex ions with many different ligands. In most of these systems there are complex species containing one or two ligands per metal ion. The presence of a reagent that forms such complex ions increases the solubility of a silver(I) salt. The complex ions of interest in this example are the ammonia complexes $AgNH_3^+$ and $Ag(NH_3)_2^+$, the thiosulfate complexes $AgS_2O_3^-$ and $Ag(S_2O_3)_2^{3-}$, and the chloride complexes $AgCl(aq)$ and $AgCl_2^-$. Chemical equations and experimentally determined equilibrium constants for the stepwise formation of these species in aqueous solution at 25°C are:

$Ag^+ + NH_3 \rightleftarrows AgNH_3^+$	$K_1 = 2.1 \times 10^3\ \text{L mol}^{-1}$
$AgNH_3^+ + NH_3 \rightleftarrows Ag(NH_3)_2^+$	$K_2 = 8.2 \times 10^3\ \text{L mol}^{-1}$
$Ag^+ + S_2O_3^{2-} \rightleftarrows AgS_2O_3^-$	$K_1 = 7.4 \times 10^8\ \text{L mol}^{-1}$
$AgS_2O_3^- + S_2O_3^{2-} \rightleftarrows Ag(S_2O_3)_2^{3-}$	$K_2 = 3.9 \times 10^4\ \text{L mol}^{-1}$
$Ag^+ + Cl^- \rightleftarrows AgCl(aq)$	$K_1 = 2.0 \times 10^3\ \text{L mol}^{-1}$
$AgCl(aq) + Cl^- \rightleftarrows AgCl_2^-$	$K_2 = 8.9 \times 10\ \text{L mol}^{-1}$

In these three systems there is wide range in the values of K_2/K_1, which determine the relative stabilities of the one-to-one complexes, that is, $AgNH_3^+$, $AgS_2O_3^-$, and $AgCl(aq)$. For the reaction

$$2AgX(aq) \rightleftarrows Ag^+ + AgX_2^-$$

the equilibrium constant is the quotient K_2/K_1:

$$K = \frac{[Ag^+][AgX_2^-]}{[AgX]^2} = \left(\frac{[Ag^+][X^-]}{[AgX]}\right)\left(\frac{[AgX_2^-]}{[AgX][X^-]}\right) = \frac{K_2}{K_1}$$

Here the ligand is represented as having a charge of -1, but X^- stands for each ligand we are considering, NH_3, $S_2O_3^{2-}$, and Cl^-. Values of the equilibrium constants for these three systems are:

$$Ag^+\text{–}NH_3 \qquad K = \frac{8.2 \times 10^3 \text{ L mol}^{-1}}{2.1 \times 10^3 \text{ L mol}^{-1}} = 3.9$$

$$Ag^+\text{–}S_2O_3^{2-} \qquad K = \frac{3.9 \times 10^4 \text{ L mol}^{-1}}{7.4 \times 10^8 \text{ L mol}^{-1}} = 5.3 \times 10^{-5}$$

$$Ag^+\text{–}Cl^- \qquad K = \frac{89 \text{ L mol}^{-1}}{2.0 \times 10^3 \text{ L mol}^{-1}} = 0.045$$

These reactions are analogous to that of the principal equilibrium in a solution of HY^-, the intermediate species of the dibasic acid H_2Y:

$$2HY^- \rightleftarrows H_2Y + Y^{2-} \qquad K = \frac{K_{a2}}{K_{a1}}$$

The calculations for the complex-ion system are analogous to some we already have made.

The relative concentration of the one-to-one complex, $AgX(aq)$, is a maximum in a solution in which the average number of X^- ligands bound by silver(I) ion is one; in this solution, $C_{Ag} = [Ag^+] + [AgX] + [AgX_2^-]$

$$[AgX_2^-] = [Ag^+] = \tfrac{1}{2}(C_{Ag} - [AgX])$$

Substitution in the equilibrium-constant equation gives

$$K = \frac{[Ag^+][AgX_2^-]}{[AgX]^2} = \frac{[\frac{1}{2}(C_{Ag} - [AgX])]^2}{[AgX]^2}$$

which can be converted to an equation for the relative concentration of the one-to-one complex:

$$\frac{[AgX]}{C_{Ag}} = \frac{1}{1 + 2K^{1/2}}$$

For the systems being considered, the values for this quotient are

$$\frac{[AgNH_3^+]}{C_{Ag}} = 0.202 \qquad \frac{[AgS_2O_3^-]}{C_{Ag}} = 0.986 \qquad \frac{[AgCl(aq)]}{C_{Ag}} = 0.702$$

We see that the one-to-one complex $AgNH_3^+$ is not a dominant species, but that $AgS_2O_3^-$ is. A factor that causes this striking difference is the different charges on the species. Addition of thiosulfate ion to $AgS_2O_3^-$ builds up the negative charge on the metal complex to -3; addition of an ammonia molecule to $AgNH_3^+$ does not change the charge. There is no electrostatic repulsion working against formation of the complex containing two ammonia molecules.

Let us consider the effect of either ammonia or thiosulfate ion on the solubility of silver(I) chloride. If the concentration of the complexing agent is large compared to $1/K_2$, the predominant soluble silver(I) species will have two molecules of complexing agent per silver(I) ion, $Ag(NH_3)_2^+$ or $Ag(S_2O_3)_2^{3-}$. For these conditions the principal equilibrium in an aqueous solution of ammonia saturated with AgCl is:

$$AgCl(s) + 2NH_3 \rightleftarrows Ag(NH_3)_2^+ + Cl^-$$

$$\begin{aligned} K = \frac{[Ag(NH_3)_2^+][Cl^-]}{[NH_3]^2} &= \frac{[Ag(NH_3)_2^+]}{[Ag^+][NH_3]^2} \times ([Ag^+][Cl^-]) \\ &= K_1K_2 \times K_s \\ &= (2.1 \times 10^3 \text{ L mol}^{-1})(8.2 \times 10^3 \text{ L mol}^{-1}) \\ &\quad \times (1.78 \times 10^{-10} \text{ mol}^2 \text{ L}^{-2}) \\ &= 3.1 \times 10^{-3} \end{aligned}$$

For a thiosulfate solution, the corresponding equilibrium is

$$AgCl(s) + 2S_2O_3^{2-} \rightleftarrows Ag(S_2O_3)_2^{3-} + Cl^-$$

$$\begin{aligned} K = \frac{[Ag(S_2O_3)_2^{3-}][Cl^-]}{[S_2O_3^{2-}]^2} &= \frac{[Ag(S_2O_3)_2^{3-}]}{[Ag^+][S_2O_3^{2-}]^2} \times ([Ag^+][Cl^-]) \\ &= K_1K_2 \times K_s \\ &= (7.4 \times 10^8 \text{ L mol}^{-1})(3.9 \times 10^4 \text{ L mol}^{-1}) \\ &\quad \times (1.78 \times 10^{-10} \text{ mol}^2 \text{ L}^{-2}) \\ &= 5.1 \times 10^3 \end{aligned}$$

Even a superficial examination of the values of these equilibrium constants discloses useful information. The solubility of silver(I) chloride is increased by the presence of ammonia, but a moderately high concentration is needed to make silver(I) chloride dissolve appreciably. In contrast, even a small excess of thiosulfate ion dissolves silver(I) chloride.

These qualitative conclusions can be supported by calculating the concentration of complexing agent X^- needed to produce a total dissolved silver concentration of 0.100 mol L^{-1} in equilibrium with silver(I) chloride. If the reaction

$$AgCl + 2X^- \rightleftarrows AgX_2^- + Cl^-$$

gives $[AgX_2^-] = 0.100$ mol L^{-1}, the concentration of chloride ion also will be 0.100 mol L^{-1}. Therefore

$$K = \frac{(0.100 \text{ mol L}^{-1})(0.100 \text{ mol L}^{-1})}{[X^-]^2}$$

or

$$[X^-] = \frac{0.100 \text{ mol L}^{-1}}{K^{1/2}}$$

For the ammonia and thiosulfate systems being considered:

$$[NH_3] = \frac{0.100 \text{ mol L}^{-1}}{(3.1 \times 10^{-3})^{1/2}} = 1.80 \text{ mol L}^{-1}$$

and

$$[S_2O_3^{2-}] = \frac{0.100 \text{ mol L}^{-1}}{(5.1 \times 10^3)^{1/2}} = 1.40 \times 10^{-3} \text{ mol L}^{-1}$$

Each of these calculated concentrations is the concentration of free complexing agent needed to make the solubility of silver(I) chloride 0.100 mol L^{-1}. To calculate the stoichiometric concentration of complexing agent, we must take into account the amount bound in the complex. This is two times the concentration of dissolved silver ion; therefore the stoichiometric concentrations of ammonia and thiosulfate ion needed are

$$C_{NH_3} = 1.80 \text{ mol L}^{-1} + 2 \times 0.100 \text{ mol L}^{-1} = 2.00 \text{ mol L}^{-1}$$
$$C_{S_2O_3^{2-}} = 1.40 \times 10^{-3} \text{ mol L}^{-1} + 2 \times 0.100 \text{ mol L}^{-1} = 0.201 \text{ mol L}^{-1}$$

Silver(I) iodide is much less soluble than silver(I) chloride ($K_s = [Ag^+][I^-] = 8.32 \times 10^{-17}$ mol^2 L^{-2}), so we expect that higher concentrations of ammonia and thiosulfate ion are needed to dissolve this substance. The principal solubility equilibria in solutions of ammonia and sodium thiosulfate saturated with silver iodide are

$$AgI + 2NH_3 \rightleftharpoons Ag(NH_3)_2^+ + I^-$$
$$AgI + 2S_2O_3^{2-} \rightleftharpoons Ag(S_2O_3)_2^{3-} + I^-$$

The equilibrium constants for these reactions, calculated as in the silver chloride examples, are

$$K = K_1K_2 \times K_s$$

Hence, for $AgI + 2NH_3$:

$$K = (2.1 \times 10^3 \text{ L mol}^{-1})(8.2 \times 10^3 \text{ L mol}^{-1})(8.32 \times 10^{-17} \text{ mol}^2 \text{ L}^{-2})$$
$$= 1.4 \times 10^{-9}$$

and for $AgI + 2S_2O_3^{2-}$:

$$K = (7.4 \times 10^8 \text{ L mol}^{-1})(3.9 \times 10^4 \text{ L mol}^{-1})(8.32 \times 10^{-17} \text{ mol}^2 \text{ L}^{-2})$$
$$= 2.4 \times 10^{-3}$$

Using the procedure we have just presented, we can calculate the concentrations of ammonia and thisulfate ion needed to raise the solubility of silver iodide to 0.10 mol L^{-1}:

$$[NH_3] = \frac{0.10 \text{ mol L}^{-1}}{(1.4 \times 10^{-9})^{1/2}} = 2.7 \times 10^3 \text{ mol L}^{-1}$$

$$[SO_3^{2-}] = \frac{0.10 \text{ mol L}^{-1}}{(2.4 \times 10^{-3})^{1/2}} = 2.0 \text{ mol L}^{-1}$$

This calculated concentration of ammonia is completely unrealistic: the concentration of concentrated aqueous ammonia is ~15 mol L^{-1}. Thus silver(I) iodide is insoluble in aqueous ammonia. The calculated concentration of sodium thiosulfate is less than the solubility of this salt ($S \cong 3.9$ mol L^{-1}); thus the solubility of silver(I) iodide is raised to moderate values in available solutions of sodium thiosulfate.

The dissolving of silver(I) chloride in aqueous ammonia is a step in a common analytical procedure for separating silver(I) from mercury(I), which does not form a soluble ammonia complex. The dissolving of silver(I) chloride (and silver(I) bromide) in thiosulfate solution is the basis for fixing exposed photographic film. In this step of the developing process, the "hypo" (sodium thiosulfate) removes from the film the silver(I) halide without reacting with the metallic silver.

The example of solutions containing excess chloride ion that are saturated with silver(I) chloride exposes new facets of this subject because chloride ion is both the anionic constituent of the solid and the ligand associated with silver(I) ion in the solution. The total solubility of silver(I) chloride, represented by S, is the sum of the concentrations of the species Ag^+, $AgCl(aq)$, and $AgCl_2^-$:

$$S = [Ag^+] + [AgCl] + [AgCl_2^-] = [Ag^+]\left(1 + \frac{[AgCl]}{[Ag^+]} + \frac{[AgCl_2^-]}{[Ag^+]}\right)$$

This equation can be converted into one involving the concentration of free chloride ion by making the substitutions

$$[Ag^+] = \frac{K_s}{[Cl^-]} \qquad \frac{[AgCl]}{[Ag^+]} = K_1[Cl^-] \qquad \frac{[AgCl_2^-]}{[Ag^+]} = K_1K_2[Cl^-]^2$$

This gives

$$S = \frac{K_s}{[Cl^-]}(1 + K_1[Cl^-] + K_1K_2[Cl^-]^2)$$
$$= K_s[Cl^-]^{-1} + K_sK_1 + K_sK_1K_2[Cl^-]$$

With values for equilibrium constants already given, this equation becomes

$$S = 1.78 \times 10^{-10} \text{ mol}^2 \text{ L}^{-2} [Cl^-]^{-1} + 3.6 \times 10^{-7} \text{ mol L}^{-1} + 3.2 \times 10^{-5} [Cl^-]$$

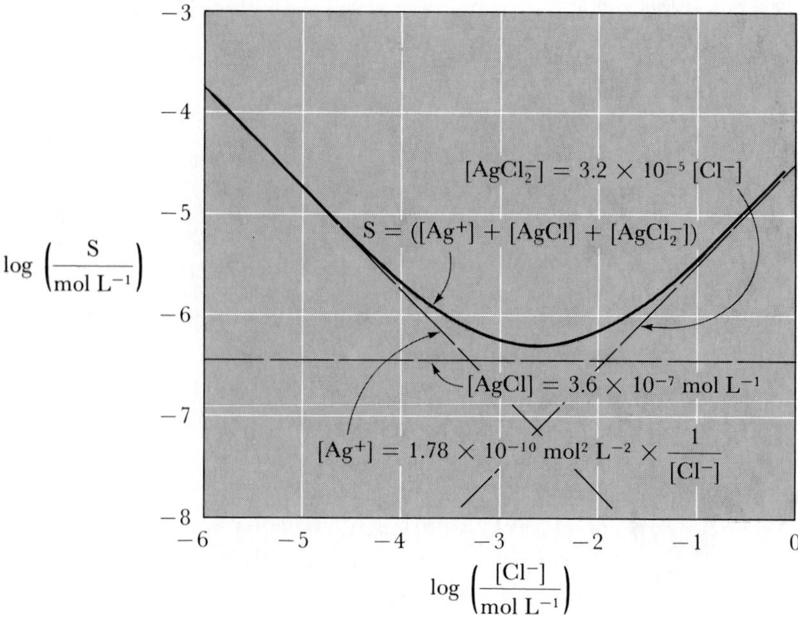

FIGURE 14–13
The solubility of silver(I) chloride in chloride-containing solutions at 25°C. The solid line gives total solubility; the dashed lines give concentrations of individual species.

Figure 14–13 is a plot of the calculated solubility as a function of the concentration of chloride ion; experimental values agree closely with the calculations. Also shown are the calculated concentrations of the individual species. The different dependence of the concentration of each species on the concentration of chloride ion is shown clearly by the slopes of the straight lines in this log–log plot: $[Ag^+] \propto [Cl^-]^{-1}$, $[AgCl] \propto [Cl^-]^0$, and $[AgCl_2^-] \propto [Cl^-]$. The exponent in each case is numerically equal to the number of chloride ions that must be added to the solid to give the species in question:

$$\begin{aligned} Ag^+ &= AgCl - 1\,Cl^- \\ AgCl &= AgCl + 0\,Cl^- \\ AgCl_2^- &= AgCl + 1\,Cl^- \end{aligned}$$

The line with slope -1 gives the concentration of Ag^+ calculated using the solubility-product equation $K_s = [Ag^+][Cl^-] = 1.78 \times 10^{-10}\ \text{mol}^2\ \text{L}^{-2}$. The graph shows that Ag^+ is the dominant soluble species at low concentrations of chloride ion: $[Ag^+]/S \geq 0.98$ if $[Cl^-] \leq 1.02 \times 10^{-5}\ \text{mol L}^{-1}$.

The stability of silver(I) complexes involving basic ligands depends upon the acidity of the solution. Thus addition of nitric acid to a solution containing chlorosilver(I) species *does not* shift the complex-ion equilibria appreciably because chloride ion does not associate with hydrogen ion in aqueous solution. In contrast, the addition of an excess of nitric acid to a solution of diamminesilver(I) ion causes the decomposition reaction,

$$Ag(NH_3)_2^+ + 2H^+ = Ag^+ + 2NH_4^+$$

In a solution with the concentrations $[NH_4^+] = 1.0 \text{ mol L}^{-1}$ and $[H^+] = 1.0 \text{ mol L}^{-1}$, the concentration of free ammonia is

$$[NH_3] = K_a \frac{[NH_4^+]}{[H^+]}$$

$$= 5.69 \times 10^{-10} \text{ mol L}^{-1} \times \frac{1.0 \text{ mol L}^{-1}}{1.0 \text{ mol L}^{-1}}$$

$$= 5.69 \times 10^{-10} \text{ mol L}^{-1}$$

At this concentration of free ammonia, hydrated silver ion is not converted appreciably into ammonia complexes. If a solution of diamminesilver(I) ion also contains chloride ion, the acidification with nitric acid causes precipitation of silver chloride:

$$Ag(NH_3)_2^+ + Cl^- + 2H^+ = AgCl(s) + 2NH_4^+$$

THE INFLUENCE OF ACIDITY AND SULFIDE ION ON THE SOLUBILITY OF METAL SULFIDES

Most metal sulfides are relatively insoluble. However, the range of numerical values of the solubility-product constants given in Table 14–3 is large, and adjustment of the concentration of sulfide ion allows a correspondingly large range of measured solubilities. We can adjust the concentration of sulfide ion by choosing the right concentration of hydrogen ion.

We already have discussed the stepwise acid dissociation of hydrogen sulfide (Section 13–3). Figure 14–14 shows the concentration of sulfide ion as a function of the concentration of hydrogen ion over the range of

TABLE 14–3
The Solubility-Product Constants for Various Metal Sulfides at 25°C

Metal sulfide	K_s / $\text{mol}^3 \text{ L}^{-3}$	Metal sulfide	K_s / $\text{mol}^2 \text{ L}^{-2}$
Cu_2S	1×10^{-48}	CdS	7.1×10^{-29}
Ag_2S	7.1×10^{-50}	PbS	8×10^{-30}
		CuS	7.9×10^{-36}
	K_s / $\text{mol}^2 \text{ L}^{-2}$	HgS	3.0×10^{-52}
		PtS	8×10^{-73}
MnS	1×10^{-11}		
ZnS^a	7.1×10^{-26}		K_s / $\text{mol}^5 \text{ L}^{-5}$
SnS	1×10^{-26}		
CoS^a	1.5×10^{-27}	La_2S_3	2×10^{-13}
NiS^a	2×10^{-28}	Bi_2S_3	1×10^{-96}

[a]These sulfides exist in more than one form. The listed values of K_s are for the less-soluble form.

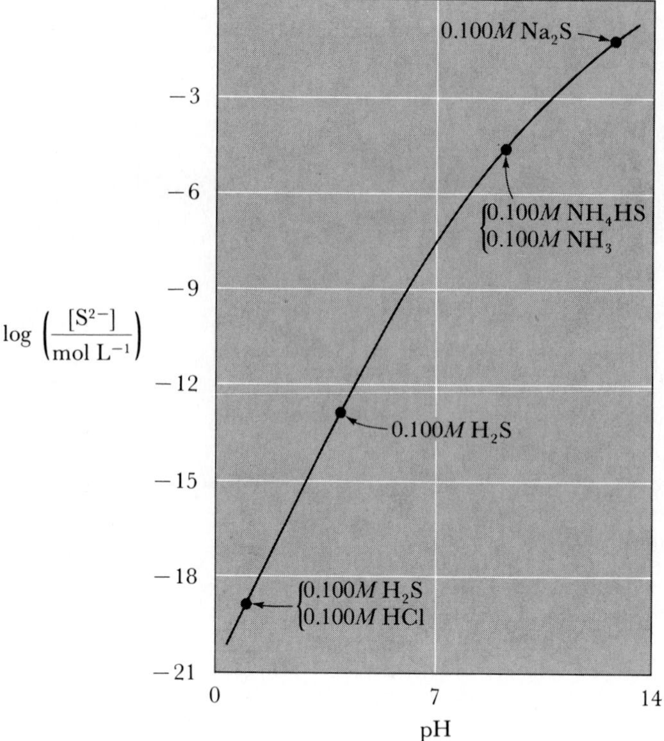

FIGURE 14–14
The concentration of sulfide ion as a function of the pH (25°C); the total concentration of sulfide is 0.10 mol L^{-1}. The four indicated points correspond to the four solutions discussed in Chapter 13: 0.100*M* HCl–0.100*M* H_2S; 0.100*M* H_2S; 0.100*M* NH_4HS–0.100*M* NH_3; and 0.100*M* Na_2S. The calculations are based upon $K_1 = 1.02 \times 10^{-7}$ mol L^{-1}, $K_2 = 1.29 \times 10^{-13}$ mol L^{-1}.

pH = 0–14 for solutions in which the total concentration of sulfide is 0.10 mol L^{-1}. Notice that the concentration of sulfide ion is inversely proportional to the square of the concentration of hydrogen ion in solutions in which the predominant species is H_2S (pH < 6), but inversely proportional to the first power of the concentration of hydrogen ion in solutions in which the predominant species is HS^- (8 < pH < 12).

For a divalent metal sulfide, MS, the principal equilibria determining its solubility to give the hydrated metal ion in solutions with pH < 12 are:

At $[H^+] > K_1$: $MS(s) + 2H^+ \rightleftarrows M^{2+} + H_2S$

At $K_1 > [H^+] > K_2$: $MS(s) + H^+ \rightleftarrows M^{2+} + HS^-$

For the first of these reactions the equilibrium constant is

$$K = \frac{[M^{2+}][H_2S]}{[H^+]^2} = [M^{2+}][S^{2-}] \times \frac{[H_2S]}{[H^+]^2[S^{2-}]}$$

$$= \frac{K_s}{K_1 K_2} = \frac{K_s}{1.32 \times 10^{-20}\ \text{mol}^2\ \text{L}^{-2}}$$

This leads to

$$[M^{2+}] = 7.6 \times 10^{19}\ L^2\ mol^{-2} \times K_s \times \frac{[H^+]^2}{[H_2S]}$$

For the region of hydrogen ion concentration in which the second reaction is the principal equilibrium, the equilibrium constant is

$$K = \frac{[M^{2+}][HS^-]}{[H^+]} = [M^{2+}][S^{2-}] \times \frac{[HS^-]}{[H^+][S^{2-}]}$$

$$= \frac{K_s}{K_2} = \frac{K_s}{1.29 \times 10^{-13}\ mol\ L^{-1}}$$

This leads to

$$[M^{2+}] = 7.8 \times 10^{12}\ L\ mol^{-1} \times K_s \times \frac{[H^+]}{[HS^-]}$$

Figure 14–15 shows the solubility of several divalent metal sulfides calculated as a function of pH for the range of pH in which the hydrated metal ion is the predominant species.

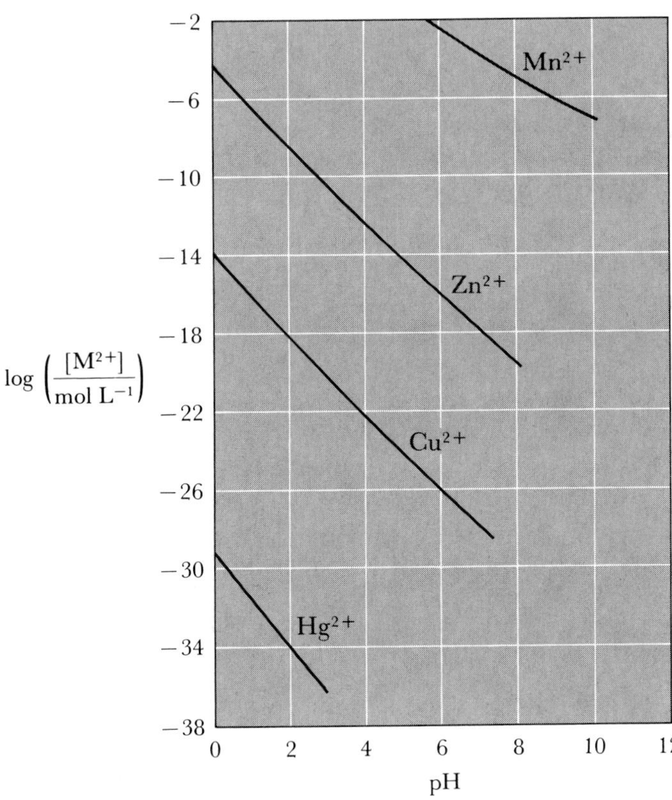

FIGURE 14–15
The solubility of various metal sulfides as a function of pH at 25°C. The total concentration of sulfide in all forms is 0.10 mol L^{-1}. The lines extend over the pH range in which hydroxo-metal species do not form to an appreciable extent.

The dependences of solubility on hydrogen and sulfide ions are altered if the predominant form of the metal in solution is not the simple hydrated ion. We can use arsenic(III) to illustrate two different points. In acidic solutions with very low concentrations of sulfide ion, the predominant species of arsenic(III) is arsenious acid, $As(OH)_3$ ($K_{a1} = 6 \times 10^{-10}$ mol L^{-1}). However, in solutions containing a high concentration of sulfide ion the predominant arsenic(III) species are AsS_2^- and AsS_3^{3-}. At conditions intermediate between these extremes, solutions probably will have species containing both bound sulfide ion and hydroxide ion [e.g., $AsS(OH)_2^-$] or both bound sulfide ion and hydrogen sulfide ion [e.g., AsS(SH)]. The arsenic system has not been characterized completely, but the principal equilibria between the solid phase As_2S_3 and the predominant species at the extremes of the different conditions illustrate points not brought out in the previous discussions. Where the principal solubility equilibrium is

$$As_2S_3(s) + 3H_2O \rightleftarrows 3H_2S + 2As(OH)_3$$

the solubility of arsenic(III) sulfide does not depend upon the concentration of hydrogen ion. Where the principal solubility equilibrium is

$$As_2S_3(s) + HS^- + OH^- \rightleftarrows 2AsS_2^- + H_2O$$

or

$$As_2S_3(s) + 3HS^- + 3OH^- \rightleftarrows 2AsS_3^{3-} + 3H_2O$$

the solubility increases with increasing concentration of hydrogen sulfide ion and hydroxide ion.

THE INFLUENCE OF HYDROXIDE ION ON THE SOLUBILITY OF AN AMPHOTERIC METAL HYDROXIDE

The solubility of zinc hydroxide is low ($K_s = 3.2 \times 10^{-17}$ mol^3 L^{-3}), and it is increased by either excess acid or excess base. A substance that behaves this way is both a base and an acid; such substances are said to be *amphoteric*.[6] Equilibria that play a role in determining the solubility of zinc hydroxide as a function of the pH of the solution, and the associated equilibrium constants determined from solubility measurements at 25°C, are:

$$Zn(OH)_2(s) + 2H^+ \rightleftarrows Zn^{2+} + 2H_2O \qquad K = 3.2 \times 10^{11} \text{ L mol}^{-1}$$

$$Zn(OH)_2(s) + H^+ \rightleftarrows ZnOH^+ + H_2O \qquad K = 3 \times 10^{2}$$

$$Zn(OH)_2(s) + OH^- \rightleftarrows Zn(OH)_3^- \qquad K = 3 \times 10^{-3}$$

$$Zn(OH)_2(s) + 2OH^- \rightleftarrows Zn(OH)_4^{2-} \qquad K = 4 \times 10^{-2} \text{ L mol}^{-1}$$

(The concentration of soluble $Zn(OH)_2$ is so low that we cannot evaluate its stability accurately.) Using these equilibrium constants, we can calculate

---

[6] From the Greek word *amphoteros*, meaning both.

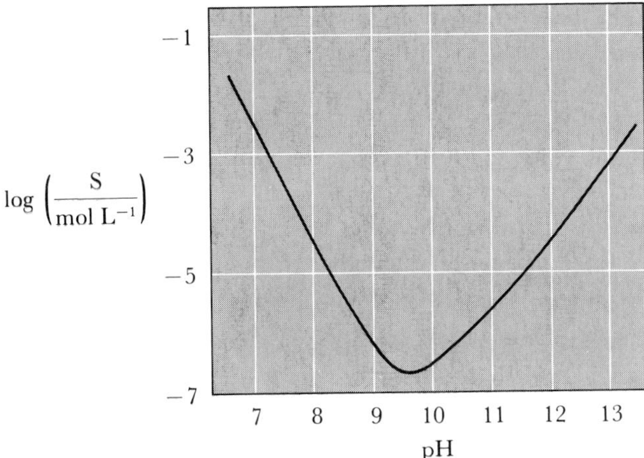

FIGURE 14–16
The solubility of zinc hydroxide as a function of the pH of the solution. $S = [Zn^{2+}] + [ZnOH^{+}] + [Zn(OH)_3^-] + [Zn(OH)_4^{2-}]$. $Zn(OH)_2(aq)$ is not an important species. The negative of slope of the line, $-\Delta \log (S/\text{mol L}^{-1})/\Delta pH$, gives the average charge on the zinc(II) species.

the solubility of zinc hydroxide as a function of pH; the results of such calculations are graphed in Figure 14–16.

Zinc hydroxide is but one example of a relatively insoluble amphoteric metal hydroxide. Even metal hydroxides commonly considered to show only basic properties may form anionic hydroxo species at very high concentrations of hydroxide ion. For instance, copper(II) hydroxide dissolves appreciably in 6*M* NaOH to form $Cu(OH)_3^-$ and $Cu(OH)_4^{2-}$. One possible complication in preparing solutions of anionic hydroxo-metal complex ions in concentrated solutions of sodium hydroxide (or another alkali-metal hydroxide) is precipitation of the sodium salt of the hydroxo-metal complex [e.g., $Na_2Cu(OH)_4$]. Although increasing the concentration of sodium hydroxide increases the solubility of copper(II) hydroxide,

$$Cu(OH)_2(s) + OH^- \rightleftarrows Cu(OH)_3^-$$

this concentration change decreases the solubility of sodium tetrahydroxocuprate(II),

$$Na_2Cu(OH)_4(s) \rightleftarrows 2Na^+ + Cu(OH)_3^- + OH^-$$

The solid phase $Cu(OH)_2$ is stable at low concentrations of sodium hydroxide, and the solid phase $Na_2Cu(OH)_4$ is stable at high concentrations. There is a particular intermediate concentration of sodium hydroxide at which the two solid phases are in equilibrium:

$$Cu(OH)_2(s) + 2Na^+(aq) + 2OH^-(aq) = Na_2Cu(OH)_4(s)$$

$$K = \frac{1}{[Na^+]^2[OH^-]^2}$$

At this intermediate concentration of sodium hydroxide, $C_{NaOH} = K^{-1/4}$, the concentration of dissolved copper(II) in the saturated solutions is at a maximum.

14–5 Distribution Equilibria

In our discussion of liquid solutions (Chapter 7), we learned that a volatile solute is distributed between the liquid phase and the vapor phase. The concentrations of the solute, X, in the two phases are proportional to one another; Henry's law gives this proportionality:

$$P_X = k[X]_l$$

in which $[X]_l$ is the concentration of X in the liquid phase. If we assume that X is an ideal gas, its concentration in the gas phase $[X]_g$ is related to its pressure:

$$[X]_g = \frac{n_X}{V} = \frac{P_X}{RT}$$

Hence we can write a concentration distribution coefficient, D,

$$\frac{[X]_g}{[X]_l} = \frac{\dfrac{P_X}{RT}}{\dfrac{P_X}{k}} = \frac{k}{RT} = D$$

This coefficient governs the distribution of the substance X between the two phases.

Another example of the distribution of a substance between different phases is the distribution of a solute between relatively immiscible liquid phases. A simple example is the distribution of iodine between water and carbon tetrachloride. Iodine is slightly soluble in water ($S = 1.331 \times 10^{-3}$ mol L^{-1} at 25°C) and considerably more soluble in carbon tetrachloride ($S = 0.119$ mol L^{-1} at 25°C). Each of these saturated solutions is in equilibrium with pure solid iodine. If these two saturated solutions are prepared separately and then, with the excess solid iodine removed, the two liquid phases are mixed thoroughly (water and carbon tetrachloride do not dissolve appreciably in one another), the concentration of iodine in each of the liquid phases will remain unchanged.[7] The two liquid phases, each of which is in equilibrium with solid iodine, are in equilibrium with one another. This is an important principle. The equilibrium distribution of the solute between the two liquid phases can be treated mathematically using a distribution coefficient, D, which is the equilibrium constant for

$$I_2(\text{in } H_2O) \rightleftarrows I_2(\text{in } CCl_4)$$

with

$$D = \frac{[I_2]_{CCl_4}}{[I_2]_{H_2O}} = \frac{\text{Solubility of } I_2 \text{ in } CCl_4}{\text{Solubility of } I_2 \text{ in } H_2O}$$

$$= \frac{0.119 \text{ mol L}^{-1}}{1.331 \times 10^{-3} \text{ mol L}^{-1}} = 89.4 \text{ (at 25°C)}$$

[7] This would not be true if the two solvents were appreciably soluble in one another. In such a case, the two solvent phases would not have the same composition after mixing as before.

This value of D is approximately correct for all concentrations of iodine, from low values to the saturation limit, since in each of these solvents iodine obeys Henry's law, approximately.

Two calculations will illustrate the removal of iodine from an aqueous phase by extraction with carbon tetrachloride. Consider two 100-cm^3 portions of aqueous iodine solution with a particular initial concentration. One of these portions is extracted with an equal volume (100 cm^3) of carbon tetrachloride. The other is extracted with five successive 20-cm^3 portions of carbon tetrachloride. What fraction of the iodine remains in each aqueous phase after these procedures? In calculating distribution equilibria with different (or the same) volumes of the two phases, we should keep track of the amount of the solute, rather than its concentration. Using the notation

n = total amount of iodine in an extraction

x = amount of iodine in aqueous phase at equilibrium

we have for a single extraction

$$D = 89.4 = \frac{\dfrac{(n-x)}{0.100\ \text{L}}}{\dfrac{x}{0.100\ \text{L}}}$$

which gives

$$1 - \left(\frac{x}{n}\right) = 89.4\left(\frac{x}{n}\right) \quad \text{or} \quad \frac{x}{n} = \frac{1}{90.4} = 0.011$$

Thus this single extraction removes 98.9% of the iodine from the aqueous phase. For each of the successive extractions with 20 cm^3 of carbon tetrachloride, we have

$$D = 89.4 = \frac{\dfrac{(n-x)}{0.020\ \text{L}}}{\dfrac{x}{0.100\ \text{L}}}$$

which gives

$$1 - \left(\frac{x}{n}\right) = \frac{89.4}{5}\left(\frac{x}{n}\right) \quad \text{or} \quad \left(\frac{x}{n}\right) = \frac{1}{18.9} = 0.053$$

This is the fraction of iodine remaining in the aqueous phase after each extraction. After five successive extractions, the fraction of the iodine in the aqueous phase is $(0.053)^5 = 4.2 \times 10^{-7}$. Thus use of the 100 cm^3 of carbon tetrachloride in five 20-cm^3 portions removes 99.99996% of the iodine from the aqueous solution. This calculation illustrates the greater efficiency of an extraction procedure if a particular volume of extracting reagent is used in several smaller portions.

Problems and Questions

14–1 Consider the equations for the dissolution of the following isoelectronic compounds:

$$KNO_3(s) = K^+(aq) + NO_3^-(aq)$$
$$CaCO_3(s) = Ca^{2+}(aq) + CO_3^{2-}(aq)$$

For which reaction is ΔS^0 more positive? Explain the reason for your answer.

14–2 What is the value of ΔG for the process

$$Na_2SO_4 \cdot 10H_2O(s) = Na_2SO_4(s) + 10H_2O(l, \text{saturated with sodium sulfate})$$

at 32.4°C? What is the sign of ΔH for the above process?

14–3 A solution of sodium sulfate contains 3.25 mol per kg of water at 40°C. A precipitate forms both when this solution is cooled and when it is heated. Explain.

14–4 What are the solubilities of silver bromide and silver iodide in water at 298.2 K? (See Table 14–1 for K_s values.)

14–5 The value of K_s for calcium iodate, $Ca(IO_3)_2 \cdot 6H_2O$, at 298.2 K is 2.03×10^{-6} mol³ L⁻³. What is the solubility of this salt in (a) water, (b) 0.100M $NaIO_3$, and (c) 0.00500M $NaIO_3$?

14–6 The value of K_s for calcium fluoride, CaF_2, at 298.2 K is 4.00×10^{-11} mol³ L⁻³. What is the solubility of this salt in (a) water and (b) 3.00M NaF?

14–7 What is the concentration of iodide ion in a solution of $1.50 \times 10^{-4}M$ $AgNO_3$ that is saturated with silver iodide at 298.2 K? [$K_s(AgI) = 8.32 \times 10^{-17}$ mol² L⁻².]

14–8 Thorium(IV) iodate, $Th(IO_3)_4$, has a very low solubility, but the dissolved salt is completely dissociated: at 25°C, $K_s = [Th^{4+}][IO_3^-]^4 = 2.40 \times 10^{-15}$ mol⁵ L⁻⁵. What is the solubility of this salt in water at 25°C? What is the solubility in 0.0500M $NaIO_3$? Compare these two solubilities by calculating the quotient $\Delta \log [Th^{4+}]/\Delta \log [IO_3^-]$. What is the significance of this value?

14–9 Equal volumes of barium chloride solution ($C_{BaCl_2} = 0.0150$ mol L⁻¹) and sodium sulfate solution ($C_{Na_2SO_4} = 0.0100$ mol L⁻¹) are mixed. What are the concentrations of species present in the final solution after barium sulfate has precipitated?

14–10 Calculate the equilibrium constant for the reaction $AgBr(s) + Cl^-(aq) \rightleftarrows AgCl(s) + Br^-(aq)$.

14–11 Calculate the equilibrium constant for each of the following reactions:

$$CaCO_3(s) + 2H^+(aq) \rightleftarrows Ca^{2+}(aq) + CO_2(aq) + H_2O$$
$$CaCO_3(s) + CO_2(aq) + H_2O \rightleftarrows Ca^{2+}(aq) + 2HCO_3^-(aq)$$

For what situations would each of these equations be the principal equilibrium?

14–12 The solubility product for calcium phosphate is $K_s = [Ca^{2+}]^3[PO_4^{3-}]^2 = 1 \times 10^{-26}$ mol⁵ L⁻⁵ at 25°C. Calculate the equilibrium constant for the reaction $Ca_3(PO_4)_2(s) + 2H_2O(l) \rightleftarrows 3Ca^{2+} + 2HPO_4^{2-} + 2OH^-$.

14–13 Write balanced equations for the principal equilibrium in the dissolution of silver(I) phosphate Ag_3PO_4 in buffered solutions in the pH ranges: (a) pH = 1–2, (b) pH = 4–6, (c) pH = 8–12, and (d) pH = 14. What is the equilibrium constant for each of these reactions?

14–14 Both silver(I) oxide (Ag_2O) and silver(I) chloride (AgCl) are relatively insoluble. The former is converted to the latter by hydrochloric acid. Write a balanced equation for this reaction. What increase in weight of the precipitate results if an excess of silver(I) oxide is treated with 45.4 cm^3 of 0.274*M* HCl? What weight of pure silver(I) oxide must be added to 500.0 cm^3 of 0.1500*M* HCl to make the pH of the final solution equal to 1.85?

14–15 If calcium sulfate and barium sulfate do not form mixed crystals, what is the maximum fraction of the barium ion that can be precipitated as the sulfate from a solution with the initial composition $C_{BaCl_2} = 0.080$ mol L^{-1}, $C_{CaCl_2} = 0.100$ mol L^{-1}, before precipitation of calcium sulfate starts?

14–16 What is the concentration of ammonia in a solution that is 0.500*M* NH_4Cl and in which the solubility of $Mg(OH)_2$ is 1.10×10^{-3} mol L^{-1}?

14–17 What is the principal equilibrium in a system containing solid calcium carbonate and an aqueous solution of ammonium chloride? What is the equilibrium constant for this reaction? What is the solubility of calcium carbonate in 0.650*M* NH_4Cl?

14–18 What fraction of beryllium(II) is present as BeF_4^{2-} in a solution in which the concentration of free fluoride ion is (a) 0.0015 mol L^{-1} and (b) 0.015 mol L^{-1}?

14–19 By examining Figure 14–9 you will see that the relative concentration of BeF^+ is at a maximum at approximately the same concentration of fluoride ion at which the concentrations of Be^{2+} and BeF_2 are equal. What is this concentration of fluoride ion? At what concentrations of fluoride ion are the concentrations of BeF_2 and BeF_3^- each at a maximum?

14–20 Derive an equation for the number of moles of oxygen bound per mole of hemoglobin, Hb_4, as a function of the pressure of oxygen, P_{O_2}.

14–21 From the equilibrium constants for the stepwise reactions in the copper(II)–ammonia system (given in the legend to Figure 14–10), calculate the equilibrium constant for $Cu^{2+} + 4NH_3 \rightleftarrows Cu(NH_3)_4^{2+}$. Calculate the equilibrium constant for the reaction $CuO(s) + H_2O + 4NH_3 \rightleftarrows Cu(NH_3)_4^{2+} + 2OH^-$. Above what concentration of free ammonia is this the principal equilibrium in an ammonia solution saturated with copper(II) oxide? (Answer by calculating the concentration of ammonia at which $[Cu(NH_3)_4^{2+}]/[Cu(NH_3)_3^{2+}] \cong 20$.)

14–22 What is the concentration of uncomplexed copper(II) ion, Cu^{2+}, in a solution in which the total copper(II) is 0.150 mol L^{-1} and the stoichiometric concentration of ammonia is 3.00 mol L^{-1}?

14–23 Silver chloride dissolves in 2*M* NH_3 but does not dissolve in 2*M* NH_4NO_3. Explain.

14–24 What is the solubility of silver(I) bromide in a solution with the composition 0.200*M* NH_3 at 298.2 K?

14–25 What total concentration of thiosulfate ion is needed to make the total concentration of dissolved silver(I) in equilibrium with solid silver(I) chloride equal to 0.0080 mol L^{-1}? What is the principal soluble silver(I) species present in this solution?

14–26 A solution with a concentration of free bromide ion of 1.50×10^{-5} mol L^{-1} is in equilibrium with a mixture of silver(I) bromide and silver(I) iodide. What is the concentration of iodide ion in this solution?

14–27 *For students who have had calculus*: Derive an equation for the concentration of chloride ion at which the solubility of silver(I) chloride is at a minimum.

14–28 The solubility of gold(III) in sodium hydroxide solution has a very sharp maximum at $0.42M$ NaOH. The solubility is lower at concentrations of sodium hydroxide above and below this value. Explain.

14–29 The distribution coefficient for bromine between carbon tetrachloride and water at 0°C is 27.0, that is, $D = [Br_2]_{CCl_4}/[Br_2]_{H_2O} = 27$. What volume of carbon tetrachloride is needed to extract 95% of the bromine in 100 cm^3 of water in a single extraction?

14–30 What fraction of the iodine is extracted from 200 cm^3 of aqueous solution by four successive 30-cm^3 portions of carbon tetrachloride?

14–31 What fraction of the dissolved iodine is present as triodide ion in a solution with a concentration of free iodide ion of 0.0250 mol L^{-1}?

14–32 Equal volumes of carbon tetrachloride and an aqueous solution of potassium iodide ($C_{KI} = 0.0500$ mol L^{-1}) are in contact with one another. A trace of iodine (I_2) is added, and the heterogeneous system is equilibrated by shaking. What fraction of the iodine is present in the aqueous phase at equilibrium? [Neither potassium iodide (KI) nor potassium triodide (KI_3) is soluble in carbon tetrachloride.]

14–33 The value of K_s for CuS is 7.9×10^{-36} mol^2 L^{-2}. Calculate the equilibrium constant for the reaction $CuS + H^+ \rightleftarrows Cu^{2+} + HS^-$. In what range of pH is this equilibrium the principal equilibrium between the solid and the saturated solution?

14–34 What is the solubility of bismuth(III) sulfide in a solution which contains $0.100M$ HCl and is saturated with hydrogen sulfide gas?

14–35 What is the solubility of zinc hydroxide in water? What is the solubility in $0.100M$ NaOH? What are the concentrations of the various soluble species in this latter solution?

14–36 In a simple adiabatic calorimetric experiment, 100 cm^3 of $0.100M$ $AgNO_3$ is mixed with 100 cm^3 of $0.0800M$ NaCl. The energy effect in this exothermic reaction measured by the temperature increase of the system is 525.5 J. From this information and the value of K_s at 25°C, calculate the value of K_s at 50°C.

15

Rates and Mechanisms of Chemical Reactions

15–1 Introduction

Our previous discussions of chemical reactions have focused on the extent to which a reaction occurs in establishing equilibrium. We have learned that the extent of a reaction is determined by the concentration conditions and by the magnitude of the equilibrium constant, which in turn is determined by the changes of enthalpy and entropy for the reaction occurring under standard conditions:

$$K = e^{-\Delta G^0/RT} = e^{-\Delta H^0/RT} \times e^{\Delta S^0/R}$$

The enthalpy change, ΔH^0, and entropy change, ΔS^0, for the reaction under standard conditions are determined by properties of the *reactants* and *products* of the reaction. These properties are bond energies (Chapter 10) and the spacing of rotational and vibrational energy levels (Chapter 5 and 21). If the reaction is one for which $\Delta n \neq 0$, the value of ΔS^0 depends also upon the standard states of the gases and solute species (i.e., the pressure of each gas and the concentration of each solute species in its standard state). For reactions in solution, interactions of reactant and product species with solvent influence the values of both ΔH^0 and ΔS^0.

However, these factors do not establish the velocity with which the reaction occurs. Similar reactions with similar equilibrium constants can have reaction velocities that are very different. For instance, the complex-ion formation reactions

$$Cr(OH_2)_6^{3+} + SCN^- \rightleftarrows Cr(OH_2)_5NCS^{2+} + H_2O$$
$$Fe(OH_2)_6^{3+} + SCN^- \rightleftarrows Fe(OH_2)_5NCS^{2+} + H_2O$$

have equilibrium constants that are almost the same: at 25°C,

$$K = \frac{[CrNCS^{2+}]}{[Cr^{3+}][SCN^-]} = 1.2 \times 10^3 \text{ L mol}^{-1}$$

$$K = \frac{[FeNCS^{2+}]}{[Fe^{3+}][SCN^-]} = 9.3 \times 10^2 \text{ L mol}^{-1}$$

But the velocities of these reactions differ dramatically. At 25°C, the chromium(III) reaction occurs over a period of months and the iron(III) reaction occurs in a fraction of a second. We can cite other examples to show that reaction velocity and equilibrium extent of reaction are not necessarily correlated. Hydrogen gas can escape into an air-filled room and the reaction

$$2H_2(g) + O_2(g) = 2H_2O(l)$$

for which (at 25°C)

$$K = \frac{1}{P_{O_2} \times P_{H_2}^2} = 1.3 \times 10^{83} \text{ atm}^{-3}$$

will not occur at an appreciable rate. However, if a flame is present reaction may occur with explosive violence. It should be clear from these examples that we learn about the velocity of chemical reactions by direct study of reaction velocity and not simply by a study of the equilibrium properties of the system.

A number of aspects of the study of reaction velocity will concern us. Some are practical:

1. What mathematical relationships describe the way in which the concentrations of reactants and products change with time?
2. What time is required for the reaction to go to a particular extent (e.g., 98% to completion)?
3. How is the rate of reaction influenced by temperature?
4. Is the reaction subject to catalysis? (In a catalyzed reaction the rate is increased by the presence of a substance that does not appear in the balanced equation for the overall reaction.)

Some are theoretical:

1. Does the overall reaction occur in a single step (an elementary reaction), or does it occur in a sequence of steps, the net result of which is the overall reaction? What are the chemical formulas of the reactants and products of each elementary reaction in the sequence?
2. How is the rate of an elementary reaction dependent upon the structures of the reactants and products of the reaction, and what unstable arrangements of atoms form along the pathway between reactants and products in each elementary reaction?

3. What is the relationship between the velocity of the reaction and the bond energies and spacing of energy levels in the reactants, products, and these intermediate unstable arrangements of atoms?
4. In what elementary reaction of a catalyzed pathway does the catalyst participate? Why is the catalyst effective?

The formation and breaking of chemical bonds are key features of most chemical reactions. In the decomposition of cyclobutane to give ethylene in the gas phase,

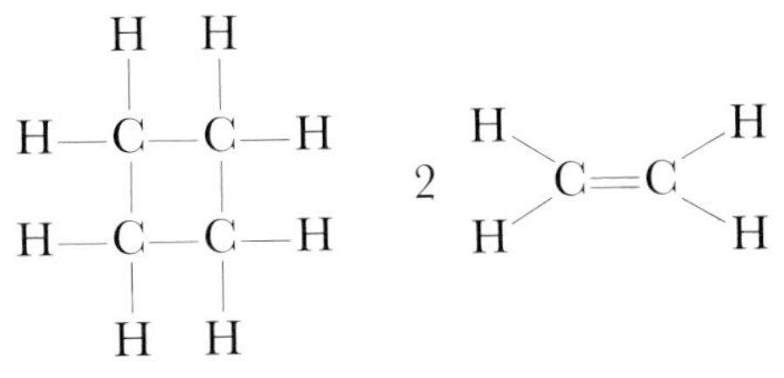

```
    H   H
    |   |
H — C — C — H        H         H
    |   |        2     \     /
H — C — C — H           C = C
    |   |              /     \
    H   H            H         H
```

two carbon–carbon single bonds are broken completely, and two carbon–carbon single bonds are transformed into two carbon–carbon double bonds. The process of breaking chemical bonds, as in this reaction, requires energy, and an important factor in determining the rate of a chemical reaction is its energy requirement and the time scale for the energy transfer. In this gas-phase decomposition reaction, the energy needed is provided by collisions of molecules. Therefore the rate of such a reaction clearly is dependent upon the frequency of the fruitful collisions.

Collisions also play a role in reactions in which atoms are exchanged between molecules in converting reactants to products, as in the reaction of nitric oxide with ozone,

$$NO(g) + O_3(g) = NO_2(g) + O_2(g)$$

a reaction that contributes to urban air pollution. Collisions between two or more molecules or atoms are needed to convert nitric oxide to nitrogen dioxide, whether this reaction goes in a single step or in a sequence of steps.

In liquid solutions, where many important chemical reactions occur, the collision process is more complex than in the gas phase. Solute species (neutral molecules or ions) collide at high frequency with neighboring solvent molecules, but the surrounding solvent lowers the rate of collision of the initially separated reactant solute species. In time, however, reactant solute species diffuse into contact with one another; this is called an encounter within a solvent cage. The envelope of solvent molecules then helps keep the reactant species together for a while.

15–2 Rates of Chemical Reactions

As most chemists use the term, *reaction rate* means the rate of change of concentration with time. If a gas-phase reaction is carried out in a vessel of constant volume or if the volume of the liquid solution in which a reaction occurs does not change appreciably with the reaction, the changes in con-

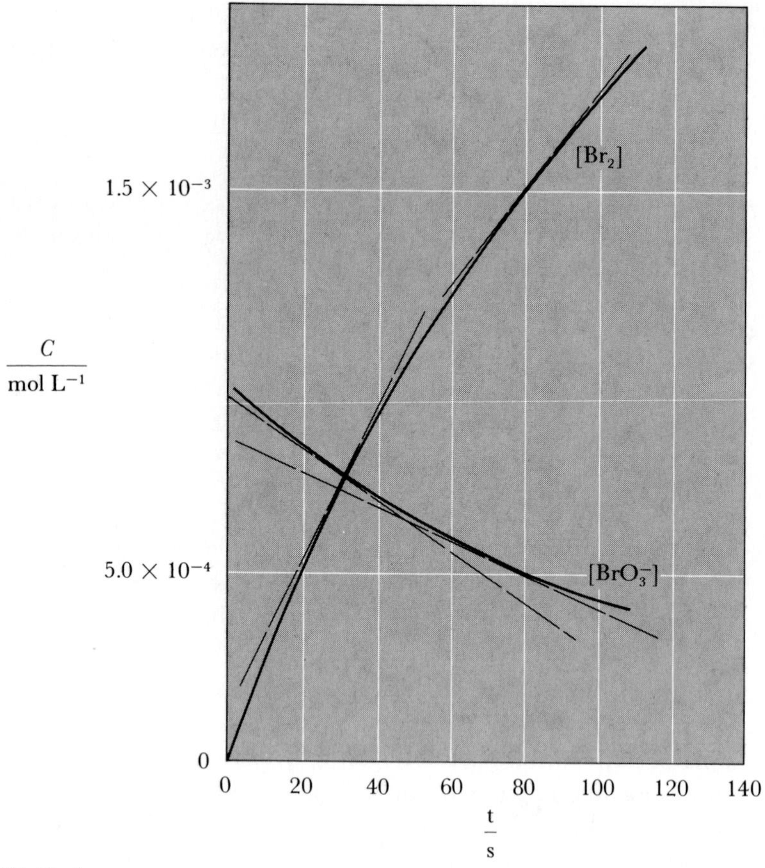

FIGURE 15–1
The concentrations of bromine and bromate ion as a function of time in the reaction

$$BrO_3^- + 5Br^- + 6H^+ = 3Br_2 + 3H_2O$$

carried out with the initial concentration conditions $[BrO_3^-]_0 = 1.00 \times 10^{-3}$ mol L^{-1}, $[Br^-]_0 = 0.1000$ mol L^{-1}, and $[H^+]_0 = 0.1010$ mol L^{-1}. The straight lines show the slopes $d[Br_2]/dt$ and $d[BrO_3^-]/dt$ at 30 s and 80 s.

centrations of reactants and products with time are due to the changes of amounts of reactants and products. In a plot of concentration versus time, the reaction rate is related to the slope of the line. Figure 15–1 gives an example of such a plot, a plot for the reaction of bromate ion and bromide ion in acidic solution,

$$BrO_3^- + 5Br^- + 6H^+ = 3Br_2 + 3H_2O$$

Equilibrium in this reaction lies far to the right in an experiment with the initial concentration conditions:

$$[Br^-]_0 = 0.1000 \text{ mol L}^{-1}$$
$$[H^+]_0 = 0.1010 \text{ mol L}^{-1}$$
$$[BrO_3^-]_0 = 1.00 \times 10^{-3} \text{ mol L}^{-1}$$
$$[Br_2]_0 = 0$$

With increasing time, the concentration of bromate ion decreases from

0.00100 mol L^{-1} almost to zero, and the concentration of bromine increases almost to 0.00300 mol L^{-1}. The rate of this reaction is defined as the rate of decrease of bromate ion concentration:[1]

$$\text{Reaction rate} \equiv -\frac{d[BrO_3^-]}{dt}$$

This quantity, the negative of the slope of a tangent to the curve, $[BrO_3^-]$ versus time, decreases as the reaction proceeds. Tangents are drawn at two points in Figure 15–1:

At t = 30 s

$$[BrO_3^-] = 7.6 \times 10^{-4} \text{ mol L}^{-1} \qquad -\frac{d[BrO_3^-]}{dt} = 6.7 \times 10^{-6} \text{ mol L}^{-1}\text{ s}^{-1}$$

At t = 80 s

$$[BrO_3^-] = 5.0 \times 10^{-4} \text{ mol L}^{-1} \qquad -\frac{d[BrO_3^-]}{dt} = 4.2 \times 10^{-6} \text{ mol L}^{-1}\text{ s}^{-1}$$

(Notice that we have defined the reaction rate to be a positive quantity; because the concentration of bromate ion decreases with time, $-d[BrO_3^-]/dt$ is positive.)

As we will learn in later discussions, this reaction occurs in a sequence of steps, but the concentrations of reaction intermediates (species of bromine in intermediate oxidation states) do not build up appreciably during the course of the reaction. Under such circumstances, at all times during the reaction the rate of decrease of the concentration of bromate ion $(-d[BrO_3^-]/dt)$ and the rate of increase of the concentration of bromine $(+d[Br_2]/dt)$ are related by coefficients in the balanced chemical equation:

$$1BrO_3^- \quad \text{gives} \quad 3Br_2$$

$$-\frac{d[BrO_3^-]}{dt} = +\frac{1}{3}\frac{d[Br_2]}{dt}$$

From the slope of these tangents, we see that the reaction rate decreases as the concentration of bromate ion decreases. Because the initial concentrations of the other reactants, bromide ion and hydrogen ion, were very large compared to the initial concentration of bromate ion, the concentrations of these species have changed only by ~1–3%:

Time:	t = 0	t = 30 s	t = 80 s
$\frac{[BrO_3^-]}{\text{mol L}^{-1}}$:	1.00×10^{-3}	7.6×10^{-4}	5.0×10^{-4}
$\frac{[Br^-]}{\text{mol L}^{-1}}$:	0.1000	0.0988	0.0975

(Table continued)

[1] A review of the calculus employed in this chapter is given in Appendix 2.

Time:	t = 0	t = 30 s	t = 80 s
$\frac{[H^+]}{\text{mol L}^{-1}}$:	0.1010	0.0996	0.0980
$\frac{[Br_2]}{\text{mol L}^{-1}}$:	0	7.2×10^{-4}	1.50×10^{-3}

The bromine concentration has changed greatly during this time, but we will assume provisionally that the decrease in rate with increasing time is due to the decrease in concentration of bromate ion, and not to the increase in the concentration of bromine. In fact, we can show that the reaction rate is directly proportional to the concentration of bromate ion, using the values of $-d[BrO_3^-]/dt$ already given. We show this by calculating the ratio of the rate to the concentration of bromate ion:

At $[BrO_3^-] = 7.6 \times 10^{-4}$ mol L^{-1}

$$\frac{-\left(\dfrac{d[BrO_3^-]}{dt}\right)}{[BrO_3^-]} = \frac{6.7 \times 10^{-6}\ \text{mol L}^{-1}\ \text{s}^{-1}}{7.6 \times 10^{-4}\ \text{mol L}^{-1}} = 8.8 \times 10^{-3}\ \text{s}^{-1}$$

At $[BrO_3^-] = 5.0 \times 10^{-4}$ mol L^{-1}

$$\frac{-\left(\dfrac{d[BrO_3^-]}{dt}\right)}{[BrO_3^-]} = \frac{4.2 \times 10^{-6}\ \text{mol L}^{-1}\ \text{s}^{-1}}{5.0 \times 10^{-4}\ \text{mol L}^{-1}} = 8.4 \times 10^{-3}\ \text{s}^{-1}$$

Finding this correlation of reaction rate and the concentration of bromate ion ($-d[BrO_3^-]/dt \div [BrO_3^-] \cong$ a constant) is the first step in establishing an equation relating the reaction rate to all of the concentration factors that influence it. Such an equation is called a *rate law*. A thorough experimental study of the rate of the reaction being considered reveals the rate law to be

$$-\frac{d[BrO_3^-]}{dt} = k[BrO_3^-][Br^-][H^+]^2$$

This equation summarizes the proportionality between the rate of the reaction and the product of the first powers of the concentrations of bromate ion and bromide ion and the second power of the concentration of hydrogen ion. The exponent of a concentration in a rate law of this type is called the *order* of the reaction for the species in question; this reaction is first order in bromate ion, first order in bromide ion, and second order in hydrogen ion. The sum of these exponents is four, so this rate law is called a fourth-order rate law. *The form of this rate law could not have been predicted from the equation for the overall reaction*; rather, it was established by an appropriate experimental study of the reaction rate.

As we will learn, the form of a rate law, that is, the reaction order for each substance that influences the reaction rate, gives us information about the reaction mechanism. Although many different types of data shed light on the mechanism of a reaction, the experimentally determined rate law is an important source of information.

The proportionality constant k in this rate law is called the *rate constant.* For the reaction we have been studying the rate constant at 25°C has the value

$$k = 9.0 \text{ L}^3 \text{ mol}^{-3} \text{ s}^{-1}$$

Notice that this rate constant has dimensions such that multiplying it by the product of concentrations in the rate law, we get the rate, with the dimensions

$$-\frac{d[\text{BrO}_3^-]}{dt} \equiv \text{mol L}^{-1} \text{ s}^{-1}$$

At zero time in the reaction described in Figure 15–1, the rate of reaction is

$$\begin{aligned} -\frac{d[\text{BrO}_3^-]}{dt} &= 9.0 \text{ L}^3 \text{ mol}^{-3} \text{ s}^{-1} \times (1.00 \times 10^{-3} \text{ mol L}^{-1}) \\ &\quad \times (0.100 \text{ mol L}^{-1}) \times (0.101 \text{ mol L}^{-1})^2 \\ &= 9.2 \times 10^{-6} \text{ mol L}^{-1} \text{ s}^{-1} \end{aligned}$$

If each concentration in a rate law of this form is exactly 1 mol L^{-1}, the rate is numerically equal to k; for this reason, another name used for the rate constant is the *specific reaction rate.*

The value of a rate constant for a particular rate law is constant only at a particular temperature, and, for reactions in solution, only in a particular solvent. An increase of temperature generally increases the value of a rate constant, and a change of solvent generally changes the value of a rate constant. (A drastic change in the nature of the solvent, such as a change from water to a hydrocarbon, may change the form of a rate law, or even the identity of the products coming from certain reactants.)

We already have learned that gas nonideality at high pressures and solute nonideality at high concentrations change the values of equilibrium constants. Gas and solute nonideality also change the values of rate constants, but we will not explore this aspect of reaction kinetics.

THE INTEGRATED RATE LAWS FOR FIRST-ORDER AND SECOND-ORDER REACTIONS

We have just seen that correlation of the rate of reaction, $-d[\text{BrO}_3^-]/dt$, with the concentration of bromate ion, $[\text{BrO}_3^-]$, in an experiment in which the other concentrations did not change appreciably during reaction showed the reaction to be first order in bromate ion. This same conclusion can be drawn more simply by fitting experimental data ($[\text{BrO}_3^-]$ as a function of time) to the equation that results from integration of a first-order law.

Before applying this procedure to the specific reactions, let us consider the reaction

$$A = B$$

governed by a first-order rate law

$$-\frac{d[A]}{dt} = k_1[A]$$

and the reaction

$$E = F$$

governed by a second-order rate law

$$-\frac{d[E]}{dt} = k_2[E]^2$$

The mathematical operations that convert these equations involving derivatives into equations that involve no derivatives are:

1. Multiply each side of the rate law by dt and divide each side by the concentration factor in the simple rate law.
2. Integrate both sides of the resulting equation.
3. Evaluate the constant of integration by substituting the initial concentration conditions.

Performing these operations on our two examples, we have:

	First order	Second order
	$-\frac{d[A]}{dt} = k_1[A]$	$-\frac{d[E]}{dt} = k_2[E]^2$
1.	$-\frac{d[A]}{[A]} = k_1 dt$	$-\frac{d[E]}{[E]^2} = k_2 dt$
2.	$-\ln [A] = k_1 t + C_1$	$+\frac{1}{[E]} = k_2 t + C_2$
3.	$-\ln [A]_0 = k_1 \times 0 + C_1$	$+\frac{1}{[E]_0} = k_2 \times 0 + C_2$
	$C_1 = -\ln [A]_0$	$C_2 = \frac{1}{[E]_0}$
	$\ln [A] = \ln [A]_0 - k_1 t$	$\frac{1}{[E]} = \frac{1}{[E]_0} + k_2 t$
	$\ln \frac{[A]}{[A]_0} = -k_1 t$	$\frac{1}{[E]} - \frac{1}{[E]_0} = k_2 t$

Each of these integrated rate laws suggests a particular graphical treatment of experimental data. The linearity of a particular plot discloses

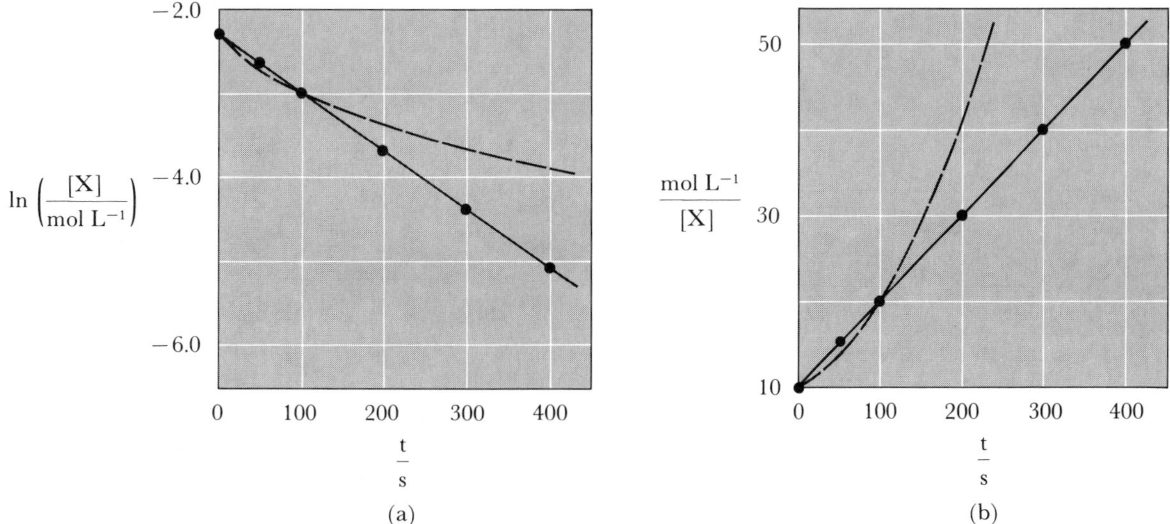

FIGURE 15–2
Plots of data that conform to first-order and second-order rate laws.
The data:

$\frac{t}{s}$	$\frac{[A]}{mol\ L^{-1}}$	$\ln\left(\frac{[A]}{mol\ L^{-1}}\right)$	$\frac{[E]}{mol\ L^{-1}}$	$\frac{mol\ L^{-1}}{[E]}$
0	0.1000	−2.303	0.1000	10.00
50	0.0707	−2.650	0.0667	15.00
100	0.0500	−2.996	0.0500	20.00
200	0.0250	−3.689	0.0333	30.00
300	0.0125	−4.382	0.0250	40.00
400	0.00625	−5.075	0.0200	50.00
$k_1 = 6.93 \times 10^{-3}\ s^{-1}$			$k_2 = 0.100\ L\ mol^{-1}\ s^{-1}$	

The plots:

In each ordinate, [X] may be [A] or [E].
(a) The solid line is a linear plot of ln ([A]/mol L^{-1}) versus time; the dashed line is a nonlinear plot of ln ([E]/mol L^{-1}) versus time.
(b) The solid line is a linear plot of 1/[E] versus time; the dashed line is a nonlinear plot of 1/[A] versus time.

both the order of the reaction and the value of the rate constant. In Figure 15–2 such graphical presentations are given. If the data conform to a first-order rate law, the plot of ln ([A]/mol L^{-1}) versus time is linear. If the data conform to a second-order rate law, the plot of 1/[E] versus time is linear. The rate constants k_1 and k_2 can be evaluated from the slopes of these lines. If we consider the tabulated concentrations at t = 0 and t = 400 s, the calculated values of k_1 and k_2 are:

$$k_1 = \frac{-\ln\left(\dfrac{[A]}{[A]_0}\right)}{t}$$

$$= \frac{-\ln\left(\dfrac{0.00625 \text{ mol L}^{-1}}{0.1000 \text{ mol L}^{-1}}\right)}{400 \text{ s}} = 6.93 \times 10^{-3} \text{ s}^{-1}$$

$$k_2 = \frac{\dfrac{1}{[E]} - \dfrac{1}{[E]_0}}{t}$$

$$= \frac{50.0 \text{ L mol}^{-1} - 10.0 \text{ L mol}^{-1}}{400 \text{ s}} = 0.100 \text{ L mol}^{-1} \text{ s}^{-1}$$

Each graph in Figure 15–2 also presents the nonlinear plot of data not conforming to the requirements for linearity. Thus if data conforming to a second-order rate law are plotted as $\ln([E]/\text{mol L}^{-1})$ versus time, the plot is not linear, and if data conforming to a first-order rate law are plotted as $[A]^{-1}$ versus time, the plot is not linear. (These plots show the reaction going to an extent $>75\%$; if we collect data for only a small extent of reaction, experimental uncertainty can make it hard for us to determine the reaction order by judging which type of plot is more nearly linear.)

Linearity in a plot of $\ln([A]/\text{mol L}^{-1})$ versus time corresponds to a constant value of $([A]_{t_2}/[A]_{t_1})$ with each constant unit of time $(t_2 - t_1)$. In the example shown in Figure 15–2, the value of $([A]_{t_2}/[A]_{t_1})$ for $(t_2 - t_1) = 50.0$ s is 0.707, and the value of $[A]_{400\text{ s}}/[A]_0$ is $(0.707)^4$:

$$\frac{[A]_{400\text{ s}}}{[A]_0} = \frac{[A]_{50\text{ s}}}{[A]_0} \times \frac{[A]_{100\text{ s}}}{[A]_{50\text{ s}}} \times \frac{[A]_{150\text{ s}}}{[A]_{100\text{ s}}} \times \frac{[A]_{200\text{ s}}}{[A]_{150\text{ s}}}$$

$$= \left(\frac{[A]_{50\text{ s}}}{[A]_0}\right)^4 = (0.707)^4 = 0.250$$

This is the observed value; the reaction goes three fourths to completion in 200 s. The rate of some changes governed by first-order rate laws (e.g., the disintegration of radioactive nuclides, to be discussed in Chapter 22) is usually characterized by giving the half-time (or half-life), $t_{1/2}$, rather than the rate constant. The half-time is related to the rate constant

$$\ln\left(\frac{1}{2}\right) = -kt_{1/2}$$

which gives

$$t_{1/2} = \frac{\ln 2}{k} = \frac{0.693}{k}$$

For the example shown in Figure 15–2, the value of $t_{1/2}$, calculated in this way, is

$$t_{1/2} = \frac{0.693}{6.93 \times 10^{-3}\ s^{-1}} = 100\ s$$

This value readily is discerned in the tabulation of data given with the figure.

We also see in the plots in Figure 15–2 that a second-order reaction slows down more with the occurrence of reaction than does a first-order one. This point can be confirmed by calculation. Each of the two sets of data plotted in Figure 15–2 corresponds to the reaction going 50% to completion in 100 s. Now let us calculate the time required for each reaction to go 98% to completion. For the first-order reaction with $k_1 = 6.93 \times 10^{-3}$ s^{-1}, we have

$$\ln\frac{[A]}{[A]_0} = -k_1 t \quad \text{or} \quad t = -\frac{1}{k_1} \times \ln\frac{[A]}{[A]_0}$$

With $[A] = 0.020[A]_0$,

$$t = -\frac{1}{6.93 \times 10^{-3}\ s^{-1}} \times \ln(0.020) = 564\ s$$

For the second-order reaction with $k_2 = 0.100\ L\ mol^{-1}\ s^{-1}$, we have

$$\frac{1}{[E]} - \frac{1}{[E]_0} = k_2 t \quad \text{or} \quad t = \frac{1}{k_2}\left(\frac{1}{[E]} - \frac{1}{[E]_0}\right)$$

With $[E]_0 = 0.100\ mol\ L^{-1}$ and $[E]_t = 0.0020\ mol\ L^{-1}$, this equation becomes

$$t = \frac{1}{0.100\ L\ mol\ s^{-1}} \times \left(\frac{1}{0.0020\ mol\ L^{-1}} - \frac{1}{0.100\ mol\ L^{-1}}\right) = 4900\ s$$

Many higher-order reactions can be studied under conditions that show the logarithmic decrease in concentration of a reactant characteristic of a first-order reaction. Consider the fourth-order bromate–bromide reaction with the rate law

$$-\frac{d[BrO_3^-]}{dt} = k[BrO_3^-][Br^-][H^+]^2$$

under the initial concentration conditions

$$[BrO_3^-]_0 \ll [Br^-]_0,\ [BrO_3^-]_0 \ll [H^+]_0$$

Under such conditions the concentrations of bromide ion and hydrogen ion do not change appreciably as the bromate ion concentration decreases from its initial value to zero. The essentially constant values of these two concentrations can be incorporated into an apparent rate constant for this particular experiment:

$$-\frac{d[BrO_3^-]}{dt} \cong (k[Br^-]_0[H^+]_0^2)\,[BrO_3^-]$$

or

$$-\frac{d[BrO_3^-]}{dt} = k'[BrO_3^-]$$

in which $k' = k[Br^-]_0[H^+]_0^2$. In the experiment depicted in Figure 15–1 with the initial concentration conditions $[Br^-]_0 = 0.1000$ mol L^{-1}, $[H^+]_0 = 0.1010$ mol L^{-1}, and $[BrO_3^-]_0 = 1.00 \times 10^{-3}$ mol L^{-1}, the concentrations of bromide ion and hydrogen ion decrease only by 5% and 6%, respectively, as the concentration of bromate ion goes from 1.00×10^{-3} mol L^{-1} to zero.

Figure 15–3 shows an approximately linear plot of the logarithm of the concentration of bromate ion versus time for this experiment. In a situation such as this, in which the concentration of a reactant follows approximately a first-order rate law even though the complete rate law is higher order, the reaction is said to be pseudo first order. The slope of the logarithmic plot is related to the pseudo first-order rate constant, k'; if

$$-\frac{d[BrO_3^-]}{dt} = k'[BrO_3^-]$$

then

$$k' = -\frac{d\ln\left(\frac{[BrO_3^-]}{\text{mol L}^{-1}}\right)}{dt} = -\frac{\Delta\ln\left(\frac{[BrO_3^-]}{\text{mol L}^{-1}}\right)}{\Delta t}$$

If different sets of relatively high initial concentrations of bromide ion and/or hydrogen ion were employed in other experiments, each such experiment

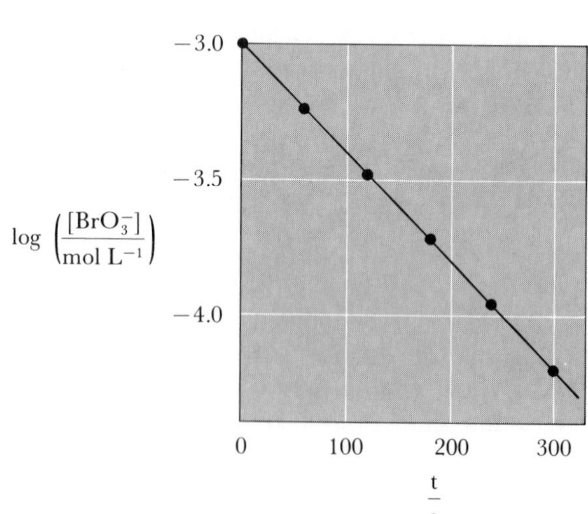

FIGURE 15–3
The concentration of bromate ion as a function of time, log $([BrO_3^-]/\text{mol L}^{-1})$ versus time. The course of the reaction conforms approximately to first-order kinetics, even though the complete rate law is a fourth-order rate law:

$$-\frac{d[BrO_3^-]}{dt} = k[BrO_3^-][Br^-][H^+]^2$$

This happens because the concentrations of bromide ion and hydrogen ion do not change appreciably during the run. Initial concentrations: $[BrO_3^-]_0 = 1.00 \times 10^{-3}$ mol L^{-1}, $[Br^-]_0 = 0.1000$ mol L^{-1}, and $[H^+]_0 = 0.1010$ mol L^{-1}.

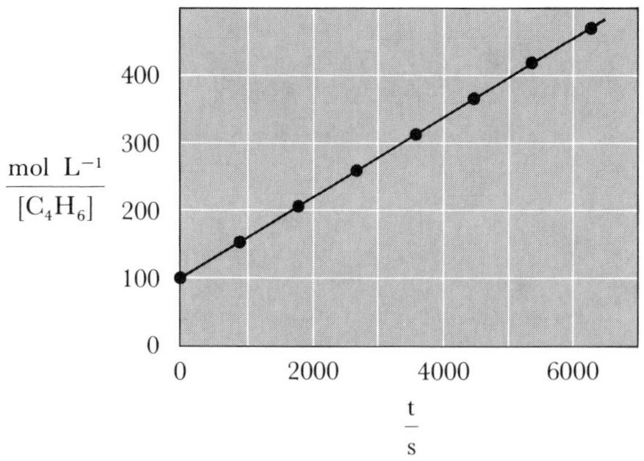

FIGURE 15–4
The reciprocal of the concentration of butadiene as a function of time during the reaction $2C_4H_6 = C_8H_{12}$, which is governed by the second-order rate law,

$$-\frac{d[C_4H_6]}{dt} = k_2[C_4H_6]^2$$

For the experiment shown:

$$T = 620\ \text{K} \qquad [C_4H_6]_0 = 0.010\ \text{mol L}^{-1}$$
$$k_2 = 5.9 \times 10^{-2}\ \text{L mol}^{-1}\ \text{s}^{-1}$$

would have a rate determined by the particular value of k' for the average concentrations of bromide ion and hydrogen ion existing during that run:

$$k' = k[Br^-]_{av}[H^+]^2_{av}$$

(If we studied the variation of a pseudo first-order rate constant with the concentration of each reagent, then we could determine the order of the reaction with respect to that reagent.)

Figure 15–4 shows a plot of $[C_4H_6]^{-1}$ versus time for a dimerization reaction, the DIELS–ALDER condensation of butadiene in the gas phase to give 4-ethenylcyclohexene:

$$2\ H_2C{=}CH{-}CH{=}CH_2 \longrightarrow \text{(4-ethenylcyclohexene)}$$

From a previous derivation we have

$$k_2 = \frac{\dfrac{1}{[C_4H_6]_{t_2}} - \dfrac{1}{[C_4H_6]_{t_1}}}{t_2 - t_1}$$

Using the data points at 0 and 6300 s, we can calculate k_2:

$$k_2 = \frac{470\ \text{L mol}^{-1} - 100\ \text{L mol}^{-1}}{6300\ \text{s}} = 5.9 \times 10^{-2}\ \text{L mol}^{-1}\ \text{s}^{-1}$$

EXAMPLE: In carbon tetrachloride solution at 45°C, the decomposition of dinitrogen pentaoxide,

$$2N_2O_5 = 4NO_2 + O_2(g)$$

can be followed by measuring of the volume of oxygen evolved. (Oxygen has a very low solubility in carbon tetrachloride.) Because the molar

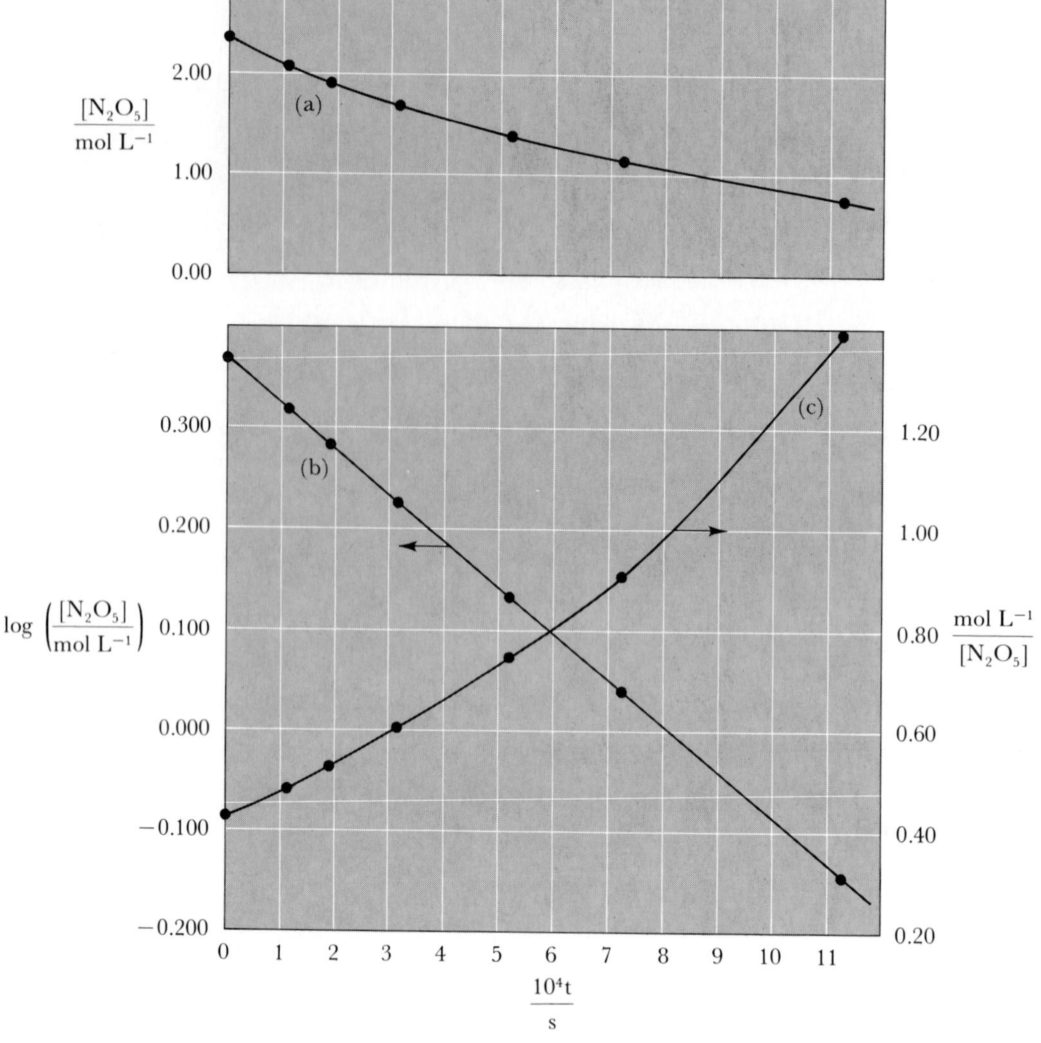

FIGURE 15–5

Plots of various functions of the concentration of dinitrogen pentaoxide versus time: (a) $[N_2O_5]$ versus time, (b) log $\left(\frac{[N_2O_5]}{\text{mol L}^{-1}}\right)$ versus time, and (c) $[N_2O_5]^{-1}$ versus time.

volume of oxygen at 45.0°C and 1.00 atm is 26,110 cm^3 mol^{-1}, the concentration of dinitrogen pentaoxide at a time when the volume of evolved oxygen is v is given by

$$[N_2O_5] = [N_2O_5]_0 - 2 \times \frac{v}{26{,}110 \text{ cm}^3 \text{ mol}^{-1}} \times \frac{1}{V}$$

in which V is the volume of the solution. Values of concentration of dinitrogen pentaoxide as a function of time, calculated from the volume of evolved oxygen, are:

$\frac{t}{s}$	$\frac{[N_2O_5]}{mol\ L^{-1}}$	$\frac{t}{s}$	$\frac{[N_2O_5]}{mol\ L^{-1}}$
0	2.33	5.20×10^4	1.35
1.10×10^4	2.08	7.19×10^4	1.11
1.91×10^4	1.91	11.3×10^4	0.72
3.16×10^4	1.67		

What is the reaction order, and what is the value of the rate constant?

Figure 15–5 shows these data plotted in three ways: $[N_2O_5]$ versus time, log ($[N_2O_5]/\text{mol L}^{-1}$) versus time, and $[N_2O_5]^{-1}$ versus time. This reaction is judged to be first order because the logarithmic plot is linear, and the other two plots are not. The value of k can be obtained from the value of the half-time read from the graph (6.66×10^4 s):

$$k = \frac{0.693}{t_{1/2}} = \frac{0.693}{6.66 \times 10^4 \text{ s}} = 1.04 \times 10^{-5} \text{ s}^{-1}$$

THE TEMPERATURE COEFFICIENT OF REACTION VELOCITY (EMPIRICAL)

The point already has been made that most rate constants depend upon temperature. The usual dependence is an increase with an increase of temperature. For the gas-phase decomposition of cyclobutane to give ethylene,

$$C_4H_8(g) = 2C_2H_4(g)$$

the first-order rate constant increases by a factor of ~20 over the temperature range 693–741 K. This is shown in Figure 15–6, which also shows the temperature dependence of the second-order rate constant for the Diels–Alder condensation of butadiene. The logarithm of the rate constant for each of these reactions is a linear function of the reciprocal of the absolute temperature:

$$\ln\left[\frac{k}{(\text{L mol}^{-1})^{n-1}\ \text{s}^{-1}}\right] = \alpha + \beta \times \frac{1}{T}$$

in which n is the reaction order. (We need this cubersome notation if we are to take the logarithm of a dimensionless quantity.) The lines in Figure 15–6 resemble other linear plots of the logarithm of a temperature-dependent quantity versus the reciprocal of the temperature. You have seen such plots involving the vapor pressure of water (Figure 6–5) and the solubility-product constants for calcium sulfate and barium sulfate (Figure 14–4). The slope of each of these plots was identified as an energy quantity divided

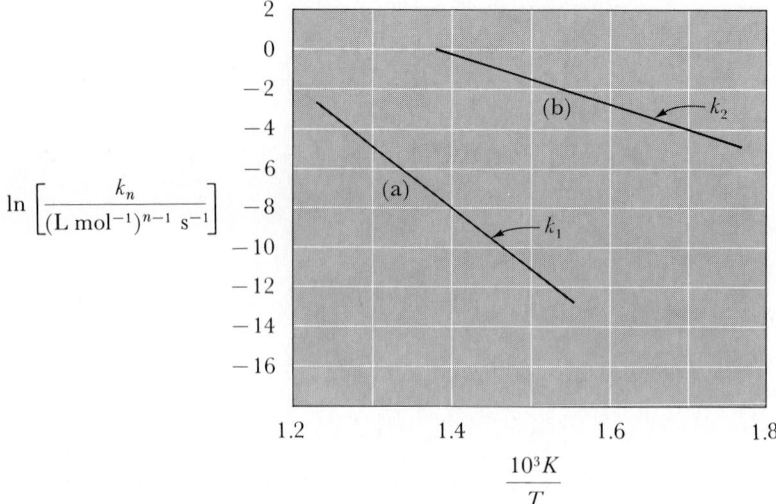

FIGURE 15–6
The temperature dependence of the rate constants for (a) the decomposition of cyclobutane:

$$-\frac{d[C_4H_8]}{dt} = k_1[C_4H_8] \qquad E_a = 261 \text{ kJ mol}^{-1}$$

(b) the condensation of butadiene:

$$-\frac{d[C_4H_6]}{dt} = k_2[C_4H_6]^2 \qquad E_a = 99.2 \text{ kJ mol}^{-1}$$

by the gas constant,

$$\frac{d\ln\left(\frac{P_{\text{vap}}}{\text{torr}}\right)}{d\left(\frac{1}{T}\right)} = -\frac{\Delta H_{\text{vap}}}{R}$$

$$\frac{d\ln\left(\frac{K_s}{\text{mol}^2\ \text{L}^{-2}}\right)}{d\left(\frac{1}{T}\right)} = -\frac{\Delta H_s}{R}$$

in which ΔH_{vap} is the change of enthalpy for the process $H_2O(l) \rightleftarrows H_2O(g)$, and ΔH_s is the change of enthalpy for the reaction $MSO_4(s) \rightleftarrows M^{2+}(aq) + SO_4^{2-}(aq)$, M = Ca or Ba. The slope of the plot of $\ln(k/(\text{L mol})^{n-1}\ \text{s}^{-1})$ versus T^{-1} also is identified as an energy quantity, the activation energy of the reaction, E_a, divided by the gas constant,

$$\text{Slope} = \beta = \frac{d\ln\left(\frac{k}{(\text{L mol}^{-1})^{n-1}\ \text{s}^{-1}}\right)}{d\left(\frac{1}{T}\right)} = -\frac{E_a}{R}$$

The temperature dependence of a rate constant usually is expressed in an equation involving an exponential rather than in one involving a logarithm.

The equation

$$\ln\left(\frac{k}{(\mathrm{L\ mol^{-1}})^{n-1}\ \mathrm{s^{-1}}}\right) = \alpha + \beta \times \frac{1}{T}$$

is transformed into

$$\frac{k}{(\mathrm{L\ mol^{-1}})^{n-1}\ \mathrm{s^{-1}}} = e^{\alpha} \times e^{\beta/T}$$

by taking the inverse logarithm of each side of the equation. With $\beta = -E_a/R$ and with $e^{\alpha} = A/(\mathrm{L\ mol^{-1}})^{n-1}\ \mathrm{s^{-1}}$, this equation becomes

$$k = Ae^{-E_a/RT}$$

in which form it is known as the Arrhenius equation. Graphs such as Figure 15–6 are known as Arrhenius plots. The Arrhenius equation for the cyclobutane decomposition is

$$k_1 = 4.0 \times 10^{15}\ \mathrm{s^{-1}} \times e^{-261\ \mathrm{kJ\ mol^{-1}}/RT}$$

That for the condensation of butadiene is

$$k_2 = 1.3 \times 10^{7}\ \mathrm{L\ mol^{-1}\ s^{-1}} \times e^{-99.2\ \mathrm{kJ\ mol^{-1}}/RT}$$

Later in this chapter we will say more about the significance of the parameters of the Arrhenius equation. For the present, we will notice only the factor by which a rate constant increases for a 10 K increase of temperature. To compare the value of a rate constant at 303.2 K with the value at 293.2 K, we take

$$\frac{k(303.2\ \mathrm{K})}{k(293.2\ \mathrm{K})} = \frac{Ae^{-E_a/(303.2\ \mathrm{K} \times 8.31\ \mathrm{J\ K^{-1}\ mol^{-1}})}}{Ae^{-E_a/(293.2\ \mathrm{K} \times 8.31\ \mathrm{J\ K^{-1}\ mol^{-1}})}}$$

$$= e^{\frac{E_a}{8.31\ \mathrm{J\ K\ mol^{-1}}}\left(\frac{1}{293.2\ \mathrm{K}} - \frac{1}{303.2\ \mathrm{K}}\right)}$$

$$= e^{+E_a/73.9\ \mathrm{kJ\ mol^{-1}}}$$

TABLE 15–1
The Temperature Dependence of Reaction Velocity

$\frac{E_a}{\mathrm{kJ\ mol^{-1}}}$	$\frac{k(303.2\ \mathrm{K})}{k(293.2\ \mathrm{K})}$	$\frac{k(603.2\ \mathrm{K})}{k(593.2\ \mathrm{K})}$
20	1.31	1.07
40	1.72	1.14
60	2.25	1.22
80	2.95	1.31
100	3.87	1.40
120	5.07	1.50
140	6.64	1.60
160	8.70	1.71
200	14.94	1.96
240	25.66	2.24
280	44.07	2.56

Values for $[k(303.2\text{ K})/k(293.2\text{ K})]$ and $[k/(603.2\text{ K})/k(593.2\text{ K})]$ for various values of E_a between 20 kJ mol^{-1} and 280 kJ mol^{-1} are given in Table 15–1. We see that the larger the activation energy, the greater the dependence of the rate constant upon temperature. This also was shown in Figure 15–6. We notice also that a temperature increase of 10 K influences the rate constant less at higher temperatures.

SOME COMPLEX EMPIRICAL RATE LAWS

In our discussion to this point, we have considered several empirical rate laws:

$$-\frac{d[BrO_3^-]}{dt} = k[BrO_3^-][Br^-][H^+]^2 \qquad -\frac{d[N_2O_5]}{dt} = k[N_2O_5]$$

$$-\frac{d[C_4H_6]}{dt} = k[C_4H_6]^2 \qquad -\frac{d[C_4H_8]}{dt} = k[C_4H_8]$$

Each of these rate laws consists of one term with one concentration factor or with the product of two or more concentration factors, each raised to a positive power. There are empirical rate laws that display additional features. By mentioning some of these rate laws, we will set the stage for later discussion of the mechanistic significance of rate laws.

There are one-term rate laws that involve one or more concentration factors raised to negative powers. An example is the rate law for the oxidation of iodide ion by hypochlorite ion in strongly alkaline solution ($[OH^-] > 0.25$ mol L^{-1}),

$$OCl^- + I^- = OI^- + Cl^-$$

The rate law under these conditions,

$$-\frac{d[OCl^-]}{dt} = k\frac{[OCl^-][I^-]}{[OH^-]}$$

involves an inverse dependence of reaction rate upon the concentration of hydroxide ion, which does not appear in the balanced chemical equation.

In addition to one-term rate laws, there are empirical rate laws that involve more than a single term. For reactions governed by such rate laws, the order of reaction with respect to one or more species varies with the concentration conditions. The empirical rate laws which correlate such data have either

1. a sum of terms, or
2. a quotient with a denominator having a sum of terms.

Rate laws of the first type include that for the reaction of hydrogen peroxide and iodide ion in acidic solutions,

$$H_2O_2 + 3I^- + 2H^+ = I_3^- + 2H_2O$$

which is governed by the rate law

$$-\frac{d[H_2O_2]}{dt} = k_1[H_2O_2][I^-] + k_2[H_2O_2][I^-][H^+]$$

The order of this reaction with respect to hydrogen ion *increases* from zero at low concentrations of hydrogen ion ($[H^+] \ll k_1/k_2$) to one at high concentrations of hydrogen ion ($[H^+] \gg k_1/k_2$). You will recognize the relevance of these inequalities if we write the empirical rate law as

$$-\frac{d[H_2O_2]}{dt} = k_2[H_2O_2][I^-]\left(\frac{k_1}{k_2} + [H^+]\right)$$

At 25°C, values of the rate constants are

$$k_1 = 1.15 \times 10^{-2}\ \text{L mol}^{-1}\ \text{s}^{-1} \qquad k_2 = 1.75 \times 10^{-1}\ \text{L}^2\ \text{mol}^{-2}\ \text{s}^{-1}$$

and their ratio has the value

$$\frac{k_1}{k_2} = 0.0657\ \text{mol L}^{-1}$$

At $[H^+] = k_1/k_2 = 0.0657$ mol L^{-1}, the two rate law terms are equal.

EXAMPLE: At what concentration of hydrogen ion does the reaction pathway corresponding to the term $k_1[H_2O_2][I^-]$ contribute 80.0% of the total rate of production of triiodide ion?

If the term $k_1[H_2O_2][I^-]$ contributes 80.0% of the total rate ($k_1[H_2O_2][I^-] + k_2[H_2O_2][H^+]$), this term must be 80.0/(100.0 − 80.0) times as large as the other term at the acidity in question:

$$1.15 \times 10^{-2}\ \text{L mol}^{-1}\ \text{s}^{-1}\ [H_2O_2][I^-] = 4.00 \times (1.75 \times 10^{-1}\ \text{L}^2\ \text{mol}^{-2}\ \text{s}^{-1}\ [H_2O_2][I^-][H^+])$$

Therefore

$$[H^+] = \frac{1.15 \times 10^{-2}\ \text{L mol}^{-1}\ \text{s}^{-1}}{4.00 \times (1.75 \times 10^{-1}\ \text{L}^2\ \text{mol}^{-2}\ \text{s}^{-1})} = 0.0164\ \text{mol L}^{-1}$$

The changing order with respect to hydrogen ion is summarized in Figure 15–7, a plot of the logarithm of the apparent second-order rate constant

$$k' = \frac{1}{[H_2O_2][I^-]} \times \left(-\frac{d[H_2O_2]}{dt}\right)$$

versus the logarithm of the concentration of hydrogen ion. We see that the slope of this line is ~0 at $[H^+] < 6.6 \times 10^{-3}$ mol L^{-1} and is ~1 at $[H^+] > 0.66$ mol L^{-1}. The slope in a $\log k'$ versus $\log [H^+]$ plot such as this tells us the reaction order with respect to hydrogen ion. The reaction is zero order with respect to hydrogen ion if the rate-law term $k_1[H_2O_2][I^-]$ is dominant, and is first order with respect to hydrogen ion if the rate-law term $k_2[H_2O_2][I^-][H^+]$ is dominant.

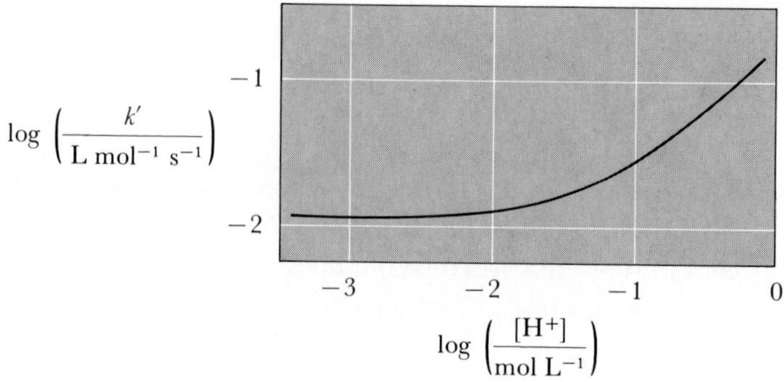

FIGURE 15–7
The dependence of the apparent second-order rate constant,

$$k' = \frac{1}{[H_2O_2][I^-]} \times \left(-\frac{d[H_2O_2]}{dt}\right)$$

upon the concentration of hydrogen ion. At $[H^+] = 0.0657$ mol L^{-1}, slope = 0.50; at $[H^+] < 0.0066$ mol L^{-1}, $0.00 < \text{slope} < 0.09$; at $[H^+] > 0.66$ mol L^{-1}, $0.91 < \text{slope} < 1.00$.

A rate law of this same type governs the hydration of carbon dioxide to give carbonate species.[2] In the acidity range pH ≅ 7 to pH ≅ 11, the reaction is governed by the two-term rate law

$$-\frac{d[CO_2]}{dt} = k_1[CO_2] + k_2[CO_2][OH^-]$$

with $k_1 = 3.9 \times 10^{-2}\ s^{-1}$ and $k_2 = 8.5 \times 10^3$ L $mol^{-1}\ s^{-1}$ at 25°C. The concentration of hydroxide ion at which the two rate-law terms contribute equally to the rate is:

$$k_1 = k_2[OH^-]$$

$$[OH^-] = \frac{k_1}{k_2}$$

$$= \frac{3.9 \times 10^{-2}\ s^{-1}}{8.5 \times 10^3\ \text{L mol}^{-1}\ s^{-1}} = 4.6 \times 10^{-6}\ \text{mol L}^{-1}$$

Under these conditions, where the predominant carbonate species is hydrogen carbonate ion, the net chemical change is

$$CO_2 + OH^- = HOCO_2^-$$

and the half-time for the reaction can be obtained from the pseudo first-order rate constant, k':

$$t_{1/2} = \frac{0.693}{k'} = \frac{0.693}{k_1 + k_2[OH^-]}$$

$$= \frac{0.693}{0.039\ s^{-1} + 8.5 \times 10^3\ \text{L mol}^{-1}\ s^{-1} \times 4.6 \times 10^{-6}\ \text{mol L}^{-1}} = 8.9\ s$$

[2]This subject is reviewed by D. M. Kern, *J. Chem. Educ.* **37**, 14 (1960).

At the pH of physiological fluids (pH = 7.0 ± 0.5), the half-time is even longer; at $[OH^-] = 1.0 \times 10^{-7}$ mol L^{-1}:

$$t_{1/2} = \frac{0.693}{0.039\ s^{-1} + 8.5 \times 10^3\ L\ mol^{-1}\ s^{-1} \times 1.0 \times 10^{-7}\ mol\ L^{-1}} = 17.4\ s$$

Metabolic processes that depend upon carbon dioxide hydration would not be possible if the reaction were as slow as this calculation implies. The reaction is subject to catalysis by an enzyme, carbonic anhydrase, which is present in physiological fluids, and the rate of the enzyme-catalyzed reaction is over 10^6-fold greater than the rate of the uncatalyzed reaction. (We will learn more about enzyme catalysis later in this chapter.)

The gas-phase reaction of hydrogen with bromine,

$$H_2(g) + Br_2(g) = 2HBr(g)$$

is governed by a rate law with a sum of terms in the denominator:

$$-\frac{d[Br_2]}{dt} = \frac{k[H_2][Br_2]^{3/2}}{[Br_2] + K[HBr]}$$

This law can be simplified to two different limiting forms, depending upon the relative magnitudes of $[Br_2]$ and $K[HBr]$. If $[Br_2] \gg K[HBr]$,

$$-\frac{d[Br_2]}{dt} \cong k[H_2][Br_2]^{1/2}$$

If $[Br_2] \ll K[HBr]$,

$$-\frac{d[Br_2]}{dt} \cong \left(\frac{k}{K}\right)\frac{[H_2][Br_2]^{3/2}}{[HBr]}$$

The orders with respect to bromine and hydrogen bromide change as the concentrations of these reagents change. At a constant concentration of hydrogen bromide, the order of the reaction with respect to bromine *decreases* from 3/2 to 1/2 as the bromine concentration *increases*. At a constant concentration of bromine, the order of the reaction with respect to hydrogen bromide *decreases* from 0 to −1 as the hydrogen bromide concentration *increases*.

We see in these examples an important difference between the ways in which reaction orders may change with increasing concentration. If the sum of terms is in the numerator of the rate law, a changing reaction order is an *increase* with an *increase* of the concentration of that species. If the sum of terms is in the denominator, a changing order is a *decrease* with an *increase* of the concentration of that species.

In presenting these empirical rate laws, we have said nothing about their meaning in terms of *reaction mechanisms*. Although an empirical rate law rarely provides proof for a reaction mechanism, a proposed reaction mechanism must be consistent with an empirical rate law. Now we will consider the rate laws that are associated with particular mechanisms. Seeing these relationships, you will be more able to answer the difficult question of what mechanisms are consistent with particular empirical rate laws.

15–3 Mechanisms of Chemical Reactions

In the chemical equation for an overall reaction, it is the net chemical change that is shown. Examples of such equations are:

1. The gas-phase conversion of cyclobutane to ethylene,

$$C_4H_8(g) = 2C_2H_4(g)$$

2. The hydrolysis of sucrose ($C_{12}H_{22}O_{11}$) to glucose ($C_6H_{12}O_6$) and fructose (also $C_6H_{12}O_6$),

$$C_{12}H_{22}O_{11} + H_2O = C_6H_{12}O_6 + C_6H_{12}O_6$$

3. The oxidation of iron(II) by chromium(VI) in acidic aqueous solution,

$$HOCrO_3^- + 3Fe^{2+} + 7H^+ = Cr^{3+} + 3Fe^{3+} + 4H_2O$$

4. The combustion of methane,

$$CH_4(g) + 2O_2(g) = CO_2(g) + 2H_2O(g)$$

5. The reaction of nitric oxide with ozone,

$$NO(g) + O_3(g) = NO_2(g) + O_2(g)$$

In most reactions, even ones with relatively simple stoichiometry, the net chemical change is the result of a sequence of elementary reactions involving reactive intermediate species. The sequence of *elementary reactions* is called the *reaction mechanism* or the reaction pathway. Because the rates of the elementary chemical reactions determine the rate of the overall net chemical change, reaction rates, empirical rate laws, the mechanisms of chemical change, and the identity and reactivity of unstable intermediate species are closely related topics.

Each elementary reaction involves some atomic rearrangements in the species that come together in that step. Relatively stable reactant species go through relatively unstable configurations enroute to relatively stable product species. The least stable of the atomic configurations on the pathway for an elementary reaction is called the *activated complex* for that reaction step.

Elementary reactions can be classified according to the number of molecular species that come together in the reaction. The elementary reactions

$$A \rightarrow B + C$$
$$D + E \rightarrow F + G$$
$$H + I + J \rightarrow K + L$$

are classified as *unimolecular*, *bimolecular*, and *termolecular*, respectively. This *molecularity* is a *theoretical concept* describing postulated elementary reactions.[3] Molecularity is to be contrasted with empirical reaction *order*, which per-

[3]Some experimental approaches allow chemists to study specific elementary reactions directly. We shall discuss one such approach, the use of molecular beams.

tains to an overall reaction and is a quantity that is measured experimentally. If a reaction occurs in a single step, the total reaction order is the same as the molecularity. The rate laws for the elementary reactions just given are

$$-\frac{d[\mathrm{A}]}{dt} = k_1[\mathrm{A}] \qquad -\frac{d[\mathrm{E}]}{dt} = k_2[\mathrm{D}][\mathrm{E}] \qquad -\frac{d[\mathrm{J}]}{dt} = k_3[\mathrm{H}][\mathrm{I}][\mathrm{J}]$$

That is, we can write the rate law for an elementary reaction simply by inspection. (These simple relationships are based upon more elaborate derivations of the concentration dependences of collision rates in the gas phase or encounter rates in the liquid phase.) Notice that the reaction orders of the rate laws for these elementary reactions give the compositions of the activated complexes for the reactions. Thus the activated complexes for these three reactions have the compositions

$$\mathrm{A}^{\ddagger} \qquad \mathrm{DE}^{\ddagger} \qquad \mathrm{HIJ}^{\ddagger}$$

in which ‡ denotes the activated complex. This same relationship between reaction orders and activated-complex composition holds for empirical rate laws. We will proceed to explore this relationship further.

THE MEANING OF AN EMPIRICAL RATE LAW

The simplest interpretation of the reaction orders in an empirical rate law identifies these concentration dependences with the composition of the activated complex for the slow step of the reaction. Thus the first-order rate law for decomposition of cyclobutane,

$$-\frac{d[\mathrm{C_4H_8}]}{dt} = k[\mathrm{C_4H_8}]$$

discloses that the activated complex has the composition

$$\mathrm{C_4H_8}^{\ddagger}$$

That is, the reaction is first order in C_4H_8; therefore the activated complex for the slow step of the reaction contains *one* unit of composition C_4H_8. A reasonable reaction sequence consistent with this rate law is:

Slow step

```
 H  H                 H           H   ‡          H   H   H   H
 |  |                 |           |              |   |   |   |
H—C—C—H   ——→    H—C-----------C—H   ——→   ·C—C—C—C·
  |  |                 \         /               |   |   |   |
H—C—C—H              H—C——C—H              H   H   H   H
 |  |                   |   |
 H  H                   H   H               Biradical intermediate
```

Fast step

```
 H  H  H  H
 |  |  |  |              H       H
·C—C—C—C·  ——→ 2     \     /
 |  |  |  |               C=C
 H  H  H  H              /     \
                         H       H
Biradical intermediate
```

The activated complex has one C–C single bond stretched to the breaking point.

For the rate law

$$-\frac{d[BrO_3^-]}{dt} = k[BrO_3^-][Br^-][H^+]^2$$

an analogous interpretation is possible. The reaction is first order in bromate ion, first order in bromide ion, and second order in hydrogen ion; therefore the activated complex for the reaction contains the atoms in one bromate ion, one bromide ion, and two hydrogen ions. This interpretation leads to the representation

$$BrO_3^- + Br^- + 2H^+ \pm nH_2O \rightarrow \{Br_2O_3H_2 \pm nH_2O\}^\ddagger \rightarrow$$
Species that undergo further reaction(s) to give the final products

The activated complex, $\{Br_2O_3H_2 \pm (H_2O)_n\}^\ddagger$, exists along the pathway for this reaction. We do not know the number of water molecules in the activated complex because we cannot determine the order of the reaction with respect to the solvent. (We find the reaction order by determining the variation of the rate with variation of concentration; in dilute solutions, the concentration of the solvent is essentially constant.) Because the reaction orders in this reaction are not the same as the coefficients in the balanced equation for the overall reaction, the reaction must occur in a sequence of steps. The products of decomposition of the activated complex, $\{Br_2O_3H_2 \pm (H_2O)_n\}^\ddagger$, must react with four additional bromide ions and four additional hydrogen ions before the final products bromine, Br_2, and water, H_2O, are produced. A rate law discloses nothing regarding the nature of the reactions following the slow step, nor does it disclose how many fast steps precede the slow step. For a system as complex as this one, much speculation about detailed mechanisms is possible. Sensible speculation involves only mechanisms with a step having an activated complex of the composition $\{Br_2O_3H_2 \pm nH_2O\}^\ddagger$, with this step being of such a nature that it reasonably may be slow.

In some cases supplementary data may help in deciding between alternate mechanisms consistent with a particular empirical rate law. The rate law for the oxidation of iodide ion by hypochlorite ion in strongly alkaline solution,

$$OCl^- + I^- = OI^- + Cl^-$$

$$-\frac{d[OCl^-]}{dt} = k\frac{[OCl^-][I^-]}{[OH^-]}$$

indicates as the simplest possibility that the activated complex has the composition $\{OClI^{2-} + H_2O - OH^-\}^\ddagger$ or $HOClI^{-\ddagger}$. An activated complex with this composition is a feature in two mechanisms:

Mechanism A

1. $OCl^- + H_2O \rightleftarrows HOCl + OH^-$ A rapid equilibrium
2. $HOCl + I^- \rightarrow HOI + Cl^-$ The slow step
3. $HOI + OH^- \rightleftarrows OI^- + H_2O$ A rapid equilibrium

Mechanism B

1. $I^- + H_2O \rightleftarrows HI + OH^-$ A rapid equilibrium
2. $HI + OCl^- \rightarrow HOI + Cl^-$ The slow step
3. $HOI + OH^- \rightleftarrows OI^- + H_2O$ A rapid equilibrium

Each of these postulated mechanisms has a rapid acid–base equilibrium as the first step. But equilibrium in the postulated first step of Mechanism B is extremely unfavorable. Hydroiodic acid (HI) is a very strong acid ($K_a(\text{HI}) \gg 1$ mol L^{-1}); therefore iodide ion is very weak base. However, hypochlorous acid is a weak acid ($K_a(\text{HOCl}) = 2.95 \times 10^{-8}$ mol L^{-1} at 25°C), and the equilibrium

$$OCl^- + H_2O \rightleftarrows HOCl + OH^-$$

would produce an appreciable concentration of hypochlorous acid, the reactant in the slow step, the *rate-determining step*, of Mechanism A.

We can derive a rate law for Mechanism A to verify that it has the form of the empirical rate law. Since the second step is the slow step, the overall rate is given by the rate of this step,

$$-\frac{d[OCl^-]}{dt} = +\frac{d[Cl^-]}{dt} = k_2[HOCl^-][I^-]$$

but the concentration of hypochlorous acid is established by equilibrium in the first step,

$$[HOCl] = K_b\frac{[OCl^-]}{[OH^-]}$$

Thus,

$$-\frac{d[OCl^-]}{dt} = +\frac{d[Cl^-]}{dt} = k_2K_b\frac{[OCl^-][I^-]}{[OH^-]}$$

which has the form of the observed rate law. This derivation identifies the empirical rate constant k with the product of k_2, the second-order rate constant for the rate-determining step, and K_b, the base-dissociation constant for hypochlorite ion. Keep in mind a point developed in this derivation: *a simple inverse concentration dependence* indicates that an equilibrium preceding the rate-determining step produces the species in question as well as one or more species that react in the rate-determining step. In our example the species in question is hydroxide ion, and the species reacting in the rate-determining step is hypochlorous acid.

A rate law that is a sum of terms indicates that the reaction has as many pathways as there are terms in the rate law. The two-term rate law for the reaction of hydrogen peroxide with iodide ion indicates that the activated complexes for the slow steps of the two pathways have the compositions $H_2O_2I^{-\ddagger}$ and $H_3O_2I^{\ddagger}$. We can speculate regarding the mechanism for each pathway. Our speculation is guided by our knowledge of the chemistry of iodine and hydrogen peroxide and by the fact that the activated complexes must have these compositions. Relevant chemical information includes the following facts:

1. One molecule of hydrogen peroxide containing two oxygen atoms loses two units of oxidation state in becoming two molecules of water.
2. There is a known +1 oxidation state of iodine (in hypoiodous acid, HOI, and hypoiodite ion, OI^-); this state is two units of oxidation state more positive than iodide ion (−1 oxidation state).
3. Hypoiodous acid rapidly oxidizes iodide ion to iodine, which rapidly forms triiodide ion in the presence of excess iodide ion.

Reasonable mechanisms, consistent with the two-term rate law given previously and these facts, are

The $k_1[H_2O_2][I^-]$ term

1.	$H_2O_2 + I^- \rightarrow H_2O + OI^-$	Slow
2.	$OI^- + H^+ \rightleftarrows HOI$	Rapid
3.	$HOI + I^- + H^+ \rightleftarrows I_2 + H_2O$	Rapid
4.	$I_2 + I^- \rightleftarrows I_3^-$	Rapid
Sum:	$H_2O_2 + 3I^- + 2H^+ = I_3^- + 2H_2O$	

The $k_2[H_2O_2][I^-][H^+]$ term

1.	$H_2O_2 + I^- + H^+ \rightarrow H_2O + HOI$	Slow
2.	$HOI + I^- + H^+ \rightleftarrows I_2 + H_2O$	Rapid
3.	$I_2 + I^- \rightleftarrows I_3^-$	Rapid
Sum:	$H_2O_2 + 3I^- + 2H^+ = I_3^- + 2H_2O$	

THE STEADY-STATE DERIVATION OF RATE LAWS

For the multistep mechanisms considered thus far, one particular step has been rate-determining. However, there are multistep reaction mechanisms in which different steps may be rate-determining, depending upon the concentration conditions. This situation may be described by a rate law with a sum of terms in the denominator. We saw such a rate law for the hydrogen–bromine reaction:

$$-\frac{d[Br_2]}{dt} = \frac{k[H_2][Br_2]^{3/2}}{[Br_2] + K[HBr]}$$

This rate law indicates activated complexes with the compositions

$$\{H_2Br_3 - Br_2\}^{\ddagger} = H_2Br^{\ddagger} \quad \text{and} \quad \{H_2Br_3 - HBr\}^{\ddagger} = HBr_2^{\ddagger}$$

A mechanism incorporating these features is

1. $Br_2 \overset{K_1}{\rightleftarrows} 2Br$ A rapid equilibrium with $[Br] = K_1^{1/2}[Br_2]^{1/2}$
2. $Br + H_2 \underset{k_{-2}}{\overset{k_2}{\rightleftarrows}} HBr + H$ A slow reaction in each direction
3. $Br_2 + H \overset{k_3}{\rightarrow} HBr + Br$ A slow step; for this reaction $K \cong 3 \times 10^{30}$, and the reverse reaction is unimportant

To derive a rate law for this mechanism, we will focus attention on Step 3, which consumes Br_2; the rate law for this step is

$$-\frac{d[Br_2]}{dt} = k_3[Br_2][H]$$

However, to make this rate law useful we must express the concentration of atomic hydrogen, an unstable intermediate, in terms of the concentrations of stable reactant and product species, Br_2, H_2, and HBr. We do this by postulating that the atomic hydrogen is maintained at an approximately steady concentration by a balance of its rate of formation (in the k_2 step),

$$k_2[Br][H_2] = k_2K_1^{1/2}[Br_2]^{1/2}[H_2]$$

and its rate of consumption (in the k_{-2} and k_3 steps),

$$k_{-2}[HBr][H] + k_3[Br_2][H]$$

Equating these rates,

$$k_2K_1^{1/2}[Br_2]^{1/2}[H_2] = k_{-2}[HBr][H] + k_3[Br_2][H]$$

and solving the resulting equation for [H], we have

$$[H] = \frac{k_2K_1^{1/2}[Br_2]^{1/2}[H_2]}{k_{-2}[HBr] + k_3[Br_2]}$$

In this quotient, which is the equation for the steady-state concentration of hydrogen atoms, the numerator is the rate of the process that forms hydrogen atoms, and the denominator is the sum of the pseudo first-order rate constants for processes that consume hydrogen atoms. Upon substituting this concentration of hydrogen atoms into the equation for $-d[Br_2]/dt$, we obtain

$$-\frac{d[Br_2]}{dt} = \frac{k_3k_2K_1^{1/2}[Br_2]^{3/2}[H_2]}{k_{-2}[HBr] + k_3[Br_2]}$$

which has the form of the empirical rate law.

EXAMPLE: Compare the empirical rate law having two parameters, k and K, with the derived rate law. Identify these empirical parameters with constants of the proposed mechanism.

If the numerator and denominator of the derived rate law are divided by k_3, we obtain

$$-\frac{d[Br_2]}{dt} = \frac{k_2K_1^{1/2}[Br_2]^{3/2}[H_2]}{[Br_2] + \left(\frac{k_{-2}}{k_3}\right)[HBr]}$$

Thus

$$k = k_2K_1^{1/2} \quad \text{and} \quad K = \frac{k_{-2}}{k_3}$$

RATE LAWS AND MECHANISMS FOR REVERSIBLE REACTIONS

The discussion in Chapters 3 and 12 showed that a chemical reaction is at equilibrium if the concentration conditions correspond to $\Delta G = 0$. Nothing was said about reaction rates at equilibrium. Now you will recognize that a chemical reaction is at equilibrium if the net reaction rate is zero at the given concentration conditions. By *net reaction rate* we mean the difference between the rate of the forward reaction and the rate of the reverse reaction.

First let us consider an elementary reaction

$$A + B \underset{k_{-1}}{\overset{k_1}{\rightleftarrows}} 2C$$

The rate of the forward reaction is

$$-\frac{d[A]}{dt} = k_1[A][B]$$

and the rate of the reverse reaction is

$$+\frac{d[A]}{dt} = k_{-1}[C]^2$$

The net rate of reaction is given by a rate law that takes into account both the forward and reverse reactions:

$$-\frac{d[A]}{dt} = k_1[A][B] - k_{-1}[C]^2$$

At equilibrium the net reaction rate is zero:

$$-\frac{d[A]}{dt} = 0$$

This condition exists when the forward and reverse rates are equal:

$$k_1[A][B] = k_{-1}[C]^2$$

Rearrangement of this equation gives

$$\frac{k_1}{k_{-1}} = \frac{[C]^2}{[A][B]}$$

which relates the concentrations of reactants and products at equilibrium. This relationship of concentrations has the same form as that already derived for the equilibrium constant:

$$K = \frac{[\mathrm{C}]^2}{[\mathrm{A}][\mathrm{B}]}$$

so we can make the identification

$$K = \frac{k_1}{k_{-1}}$$

Notice the dynamic nature of chemical equilibrium: Although the net reaction rate is zero at equilibrium, neither $k_1[\mathrm{A}][\mathrm{B}]$ nor $k_{-1}[\mathrm{C}]^2$ is zero. At chemical equilibrium, reactant molecules are being converted to product molecules and product molecules are being converted to reactant molecules.

In this hypothetical example of an elementary reaction, the form of the rate law follows directly from the chemical equation. The exponents of concentrations in the rate law are the same as the coefficients in the balanced chemical equation. For real chemical reactions that may occur via multistep mechanisms, the situation is more complex. We already have stated that the form of the rate law cannot be predicted from the chemical equation for the overall reaction. However, this does not mean that there is no relationship between the form of the complete rate law (including both positive and negative terms) and the concentration conditions appropriate for equilibrium. An example will illustrate the problem.

The reaction of triiodide ion with arsenic(III),

$$\mathrm{As(OH)_3 + I_3^- + H_2O = OAs(OH)_3 + 3I^- + 2H^+}$$

occurs almost to completion at $[\mathrm{H^+}] < 10^{-5}$ mol L^{-1}, and this reaction is used in quantitative analysis for arsenic(III) or triiodide ion. The rate law for this reaction when it is far from equilibrium is

$$-\frac{d[\mathrm{I_3^-}]}{dt} = k_f \frac{[\mathrm{As(OH)_3}][\mathrm{I_3^-}]}{[\mathrm{H^+}][\mathrm{I^-}]^2}$$

In solutions of high acidity, arsenic(V) oxidizes iodide ion (the reverse reaction); the rate law for this reaction is

$$\frac{d[\mathrm{I_3^-}]}{dt} = k_r[\mathrm{OAs(OH)_3}][\mathrm{I^-}][\mathrm{H^+}]$$

If we assume that there is a single predominant pathway for this reaction over the entire range of pH, 0–5, we can find the net rate of decrease of the concentration of triiodide ion for solutions of intermediate acidity by combining these rate laws:

$$-\frac{d[\mathrm{I_3^-}]}{dt} = k_f \frac{[\mathrm{As(OH)_3}][\mathrm{I_3^-}]}{[\mathrm{H^+}][\mathrm{I^-}]^2} - k_r[\mathrm{OAs(OH)_3}][\mathrm{I^-}][\mathrm{H^+}]$$

If the equilibrium condition $-d[\mathrm{I_3^-}]/dt = 0$ is applied to this rate law, the resulting relationship of concentrations is

$$k_f \frac{[As(OH)_3][I_3^-]}{[H^+][I^-]^2} = k_r[OAs(OH)_3][I^-][H^+]$$

This equation can be rearranged to give

$$\frac{k_f}{k_r} = \frac{[OAs(OH)_3][H^+]^2[I^-]^3}{[As(OH)_3][I_3^-]}$$

which has the form of the equilibrium-constant equation

$$K = \frac{[OAs(OH)_3][H^+]^2[I^-]^3}{[As(OH)_3][I_3^-]}$$

As in the simple example, the ratio of rate constants k_f/k_r is the equilibrium constant:

$$\frac{k_f}{k_r} = K$$

The experimentally determined numerical values of k_f and k_r and the independently determined value of K are consistent with this; for 0°C,

$$k_f = 1.57 \times 10^{-5}\ \text{mol}^2\ \text{L}^{-2}\ \text{s}^{-1}$$
$$k_r = 1.05 \times 10^{-4}\ \text{L}^2\ \text{mol}^{-2}\ \text{s}^{-1}$$

$$\frac{k_f}{k_r} = \frac{1.57 \times 10^{-5}\ \text{mol}^2\ \text{L}^{-2}\ \text{s}^{-1}}{1.05 \times 10^{-4}\ \text{L}^2\ \text{mol}^{-2}\ \text{s}^{-1}} = 0.15\ \text{mol}^4\ \text{L}^{-4}$$

The independently determined value of K is

$$K = 0.16\ \text{mol}^4\ \text{L}^{-4}$$

Although the forms of rate laws for the forward and reverse reactions cannot be predicted from the chemical equation for the overall reaction, they are related to one another so as to yield the equilibrium-constant expression when the net reaction rate is zero.

The positive and negative terms in the complete rate law for a reversible reaction have forms that give the same composition of activated complex. For this example, the composition is

$$\underline{\text{The } k_f \frac{[As(OH)_3][I_3^-]}{[H^+][I^-]^2} \text{ term}}$$

$$\{As(OH)_3I_3^- - H^+ - 2I^- \pm nH_2O\}^\ddagger = \{AsIO_2 + n'H_2O\}^\ddagger \quad (n' = n + 1)$$

$$\underline{\text{The } k_r[OAs(OH)_3][I^-][H^+] \text{ term}}$$

$$\{OAs(OH)_3IH \pm mH_2O\}^\ddagger = \{AsIO_2 + m'H_2O\}^\ddagger \quad (m' = m + 2)$$

A reasonable mechanism consistent with this composition of activated complex with $n' = m' = 2$ [i.e., activated complex = $As(OH)_4I^{\ddagger}$] is

1. $I_3^- + H_2O \rightleftarrows HOI + 2I^- + H^+$ A rapid equilibrium
2. $HOI + As(OH)_3 \rightleftarrows OAs(OH)_3 + I^- + H^+$ The slow reaction

This is the mechanism both for the forward reaction and for the reverse reaction. It is an important principle that each pathway for a chemical reaction going in one direction is also a pathway for the reaction going in the opposite direction. At equilibrium the rates of forward and reverse reactions along each particular pathway are the same. This is the principle of *microscopic reversibility*.

This principle can be used to predict the form of the complete rate law, given the form of the rate-law term for the forward reaction. We will discuss this prediction for the decomposition of nitrous acid,

$$3HONO = H^+ + NO_3^- + 2NO + H_2O$$

for which the experimentally determined rate law for the forward reaction is

$$\frac{d[NO_3^-]}{dt} = k_f \frac{[HONO]^4}{P_{NO}^2}$$

We can write the complete rate law as

$$\frac{d[NO_3^-]}{dt} = k_f \frac{[HONO]^4}{P_{NO}^2} - k_r f(C)$$

in which $f(C)$ is a function of the concentrations of the species in the reaction mixture. We wish to find the form of $f(C)$. If this rate is equated to zero and the equation is solved for k_f/k_r, we obtain

$$\frac{k_f}{k_r} = \frac{P_{NO}^2 f(C)}{[HONO]^4}$$

Let us compare this equation, which holds at equilibrium, with the equilibrium-constant equation,

$$K = \frac{[H^+][NO_3^-]P_{NO}^2}{[HONO]^3}$$

Equating these two quotients of concentrations and solving for $f(C)$ we find

$$\frac{P_{NO}^2 f(C)}{[NONO]^4} = \frac{[H^+][NO_3^-]P_{NO}^2}{[HONO]^3}$$

$$f(C) = [H^+][NO_3^-][HONO]$$

Thus we expect the complete rate law to be

$$\frac{d[NO_3^-]}{dt} = k_f \frac{[HONO]^4}{P_{NO}^2} - k_r[H^+][NO_3^-][HONO]$$

This reaction has been studied under conditions in which each term is important, and the experimentally observed rate law for the reverse reaction is the one we have just predicted.

If a reaction can go by two or more independent pathways, the principle of microscopic reversibility holds for each pathway. Thus a complete rate law has a negative term for each positive term. Each such pair of terms can be combined to give an equilibrium-constant equation.

We can see the relationship between the activation energies of the forward reaction, E_{af}, and the reverse reaction, E_{ar}, and the enthalpy change for the overall reaction, ΔH^0, by expressing k_f, k_r, and K in the following equations:

$$k_f = A_f e^{-E_{af}/RT} \qquad k_r = A_r e^{-E_{ar}/RT} \qquad K = e^{\Delta S^0/R} e^{-\Delta H^0/RT}$$

We combine these to give

$$\frac{k_f}{k_r} = K \qquad \frac{A_f e^{-E_{af}/RT}}{A_r e^{-E_{ar}/RT}} = e^{\Delta S^0/R} \times e^{-\Delta H^0/RT}$$

For this relationship to hold over a range of temperatures, the temperature-independent factors must be equal on each side of the equation:

$$\frac{A_f}{A_r} = e^{\Delta S^0/R}$$

as well as the temperature-dependent factors:

$$e^{-(E_{af} - E_{ar})/RT} = e^{-\Delta H^0/RT}$$

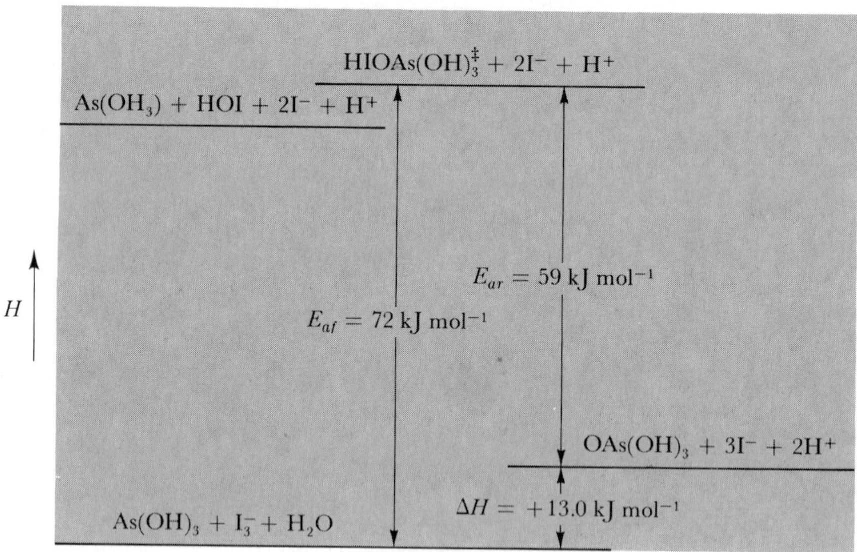

FIGURE 15–8
The enthalpy content (H) of reactants, products, intermediate (HOI), and transition state ($HIOAs(OH)_3^‡$) in the reaction

$$As(OH)_3 + I_3^- + H_2O = OAs(OH)_3 + 3I^- + 2H^+$$

The latter equation shows the relationship of E_{af}, E_{ar}, and ΔH^0 to be

$$E_{af} - E_{ar} = \Delta H^0$$

a relationship that also is shown in Figure 15–8, where these quantities are given for the triiodide ion–arsenious acid reaction. We see in this figure that the activation energy for the forward reaction is the difference of the energies of the reactants and the activated complex, and that the activation energy for the reverse reaction is the difference of the energies of the products and the activated complex (the same activated complex).

CATALYSIS; MECHANISMS OF CATALYZED REACTIONS

Many chemical reactions occur by efficient pathways that involve species not included in the chemical equation for the overall reaction. The presence of such a species in a system makes a new pathway accessible, and, as a consequence, increases the reaction rate. This phenomenon is called *catalysis*, and species responsible for these additional reaction pathways are called *catalysts*. Catalysis is extremely important in all areas of chemistry and biology. Many reactions of enormous practical importance occur at useful rates only in the presence of a catalyst. Examples of such reactions are:

1. the synthesis of ammonia from nitrogen and hydrogen, discussed in Chapter 12,
2. the production of methyl alcohol from products of the water-gas reaction (Sections 3–3 and 3–5), hydrogen and carbon monoxide,

$$CO + 2H_2 = H_3COH$$

3. the conversion of sulfur dioxide to sulfur trioxide (which is the anhydride of sulfuric acid),

$$2SO_2(g) + O_2(g) = 2SO_3(g)$$

In some systems, there are many different reactions for which $\Delta G < 0$, all of which are extremely slow. A particular catalyst may make one of the many possible reactions occur at a reasonable rate without providing new pathways for the other reactions. Nowhere is the specificity of catalysis more striking than in living organisms, where a certain enzyme may catalyze only a single reaction among many for a particular reactant.

There is no single mechanism for catalysis, but rather many different types of mechanisms varying with the nature of the reaction; only a few types will be discussed here.

An early study of catalysis dealt with the acid-catalyzed hydrolysis of sucrose to give glucose and fructose (see Chapter 20),

$$\underset{\text{Sucrose}}{C_{12}H_{22}O_{11}} + H_2O = \underset{\text{Glucose}}{C_6H_{12}O_6} + \underset{\text{Fructose}}{C_6H_{12}O_6}$$

In this reaction a naturally occurring disaccharide, ordinary cane sugar, gives its two constituent monosaccharide molecules.

This reaction involves chiral reactant and products, and the sign of the angle of rotation of polarized light changes during the reaction; the process is called *inversion.* The product mixture is called invert sugar. (Honey is largely invert sugar.) The inversion reaction is extremely slow in neutral solution, but it is increased by the presence of acid. (This reaction, which is the first stage in the metabolism of sucrose, also is catalyzed by an enzyme, invertase.) For the acid-catalyzed pathway in dilute solutions of strong acids, the rate law is

$$-\frac{d[C_{12}H_{22}O_{11}]}{dt} = k[H^+][C_{12}H_{22}O_{11}]$$

Hydrogen ion is not consumed in this reaction; therefore its concentration remains constant as the sucrose concentration decreases. The rate law is a second-order rate law, but the kinetic course of reaction in each experiment is first order,

$$-\frac{d[C_{12}H_{22}O_{11}]}{dt} = k'[C_{12}H_{22}O_{11}]$$

in which $k' = k[H^+]$. Because the ether linkage (C—O—C) connecting the two monosaccharides is being broken in this step, it is likely that protonation of the weakly basic oxygen atom of this linkage is responsible for the catalysis. A mechanism consistent with this rate law and the nature of the reactant and product molecules is (only the ether linkage is shown):

1. $>\!C{-}O{-}C\!< + H^+ \rightleftharpoons\ >\!C{-}\overset{H}{\underset{+}{O}}{-}C\!<$ Rapid equilibrium

2. $>\!C{-}\overset{H}{\underset{+}{O}}{-}C\!< \longrightarrow\ >\!C{-}OH + {}^{+}C\!<$ Slow

3. $H_2O + {}^{+}C\!< \longrightarrow\ >\!C{-}OH + H^+$ Fast

Net change: $>\!C{-}O{-}C\!< + H_2O = 2\ >\!C{-}OH$

Hydrogen ion is regenerated in the last reaction step; hence it is a catalyst, not a reactant. This regeneration is a general characteristic of a catalytic pathway. The catalyst is used in one reaction step and regenerated in a later step.

In concentrated aqueous solutions of acid, the rate of this reaction is greater than we might expect: its speed is not proportional to the hydrogen ion concentration as given by the rate law already presented. A 10-fold increase in the concentration of hydrogen ion increases the rate by a factor of over 100. The pseudo first-order rate constant k', defined by the rate law

$$-\frac{d[C_{12}H_{22}O_{11}]}{dt} = k'[C_{12}H_{22}O_{11}]$$

depends upon the Hammett acidity function, H_0, introduced in Chapter 13; $-\Delta \log k'/\Delta H_0 \cong 1.00$, as is shown in Figure 15–9. If $k' = kh_0$

$$\log k' = \log k + \log h_0 = \log k - H_0$$

which is the dependence displayed in the figure. This correlation is reasonable. The function H_0 measures the ability of the solution to protonate a neutral indicator base. Sucrose is a weak organic base, and its reaction rate in acidic aqueous solution is determined by its extent of protonation, which in turn is determined by the value of H_0 of the solution.

Many, many organic reactions are subject to catalysis by acids. The presence of basic groups (R—O—R, R—OH, R—NH_2, etc.) in the reactant molecules makes it easy to visualize the role of protonation.

Among the most powerful catalysts are enzymes, high-molecular-weight proteins that catalyze reactions in living matter. Some reactions that are catalyzed by enzymes also are catalyzed by acid or base. As already mentioned, the hydrolysis of sucrose also is catalyzed by invertase, an enzyme. Since proteins contain amino acids with acidic and basic groups, this suggests the possibility that some enzyme catalysis may be nothing more than acid or base catalysis. This oversimplified view cannot be the complete answer because catalysis by most enzymes is enormously more effective than is catalysis by hydrogen ion. For instance, the hydrolysis of adenosine triphosphate (ATP; see Table 20–3), a substance that plays a role in many metabolic reactions, can be catalyzed by hydrogen ion or by an enzyme, myosin, a macromolecule of molecular weight 470,000. The hydrolysis of this triphosphate, $AOP_3O_9H^{3-}$, to adenosine diphosphate (ADP), represented as $AOP_2O_6H^{2-}$, can be written

$$AOP_3O_9H^{3-} + H_2O = AOP_2O_6H^{2-} + (HO)_2PO_2^-$$

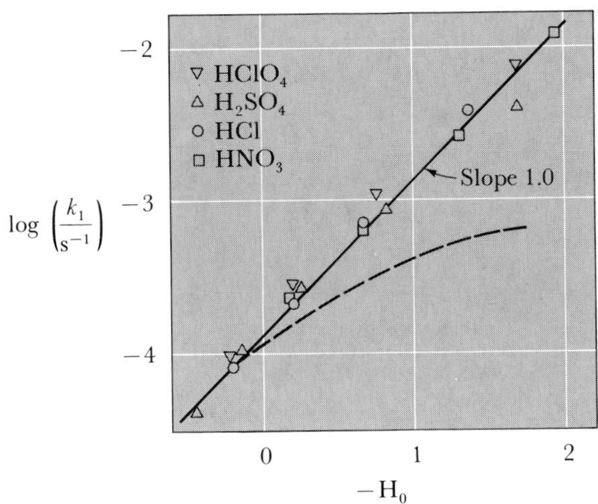

FIGURE 15–9
The rate constant for the hydrolysis of sucrose, k', as a function of the Hammett acidity function (H_0):

$$-\frac{d[C_{12}H_{22}O_{11}]}{dt} = k'[C_{12}H_{22}O_{11}]$$

$$\log k' = \log k - H_0 \quad \text{or} \quad k' = kh_0$$

(If the reaction rate were proportional to the concentration of hydrogen ion, $k' = k[H^+]$, the data for perchloric acid would fall on the dashed line.)

This hydrolysis is analogous to the hydrolysis of simple diphosphate ion,

$$P_2O_7H_2^{2-} + H_2O = 2(HO)_2PO_2^-$$

a reaction that goes faster in highly acidic solutions. The second-order rate constant for the hydrolysis of ATP catalyzed by myosin, represented as E,

$$-\frac{d[\text{ATP}]}{dt} = k[\text{E}][\text{ATP}]$$

is over 10^{12}-fold larger than the corresponding second-order rate constant for the hydrolysis catalyzed by hydrogen ion,

$$-\frac{d[\text{ATP}]}{dt} = k[\text{H}^+][\text{ATP}]$$

It is believed that the incredible efficiency of some enzymes is due to the conformation of the macromolecule, which allows the reactant molecule(s) to fit tightly on the active site of the enzyme, that part of the protein molecule that brings about the catalysis. Figure 15–10 depicts the active site of an enzyme, with its ability to hold the reactant molecule(s), called the substrate(s), in an orientation that facilitates reaction. This picture implies that the active site has a specific size and shape, which gives it a lock-and-key relationship with the substrate. This model, proposed in 1890 by EMIL FISCHER, has proved to be very useful, and it is believed to be essentially correct for many enzyme systems. However, recent structural studies have shown that the enzyme carboxypeptidase-A (see Figure 5–8)

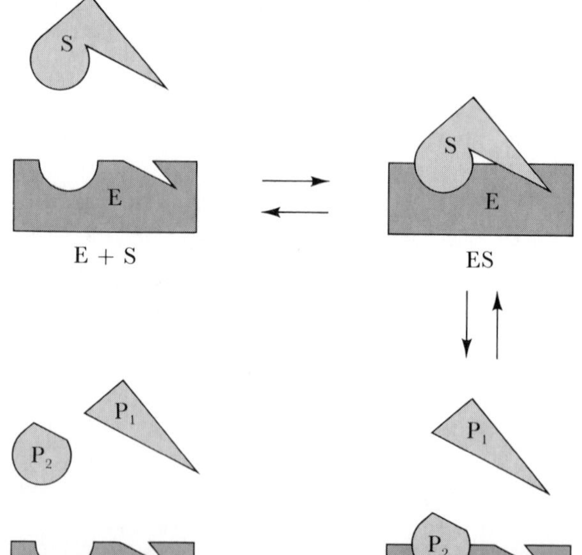

FIGURE 15–10
A pictorial representation of the role of the active site of an enzyme in catalyzing a reaction.

Net reaction: $S = P_1 + P_2$

Mechanism: $E + S \rightleftarrows ES$

$ES \rightleftarrows EP_2 + P_1$

$EP_2 \rightleftarrows E + P_2$

modifies its conformation when it binds the substrate glycyltyrosine. To fit the substrate, the enzyme's amino acid groups that are associated with the substrate in the enzyme–substrate complex have moved from their positions in the free enzyme. This result suggests that the rigid structure of the active site implied in Figure 15–10 is an oversimplification. The active site in some enzymes may change conformations to accommodate the substrate(s).

The second-order rate law for an enzyme-catalyzed reaction is a limiting form of a more complex rate law. For the simplest type of reaction, S = P (standing for substrate = product), the common type of rate law for the forward reaction is

$$-\frac{d[\mathrm{S}]}{dt} = \frac{kC_{\mathrm{E}}[\mathrm{S}]}{K_m + [\mathrm{S}]}$$

in which C_{E} is the total stoichiometric concentration of enzyme. The generally accepted mechanism corresponding to rate laws of this form was proposed by Michaelis and Menten in 1913; and the constant K_m is called the MICHAELIS constant. This mechanism incorporates formation of an enzyme–substrate complex ES,

$$\mathrm{E} + \mathrm{S} \rightleftarrows \mathrm{ES}$$

followed by its decomposition, which regenerates the enzyme,

$$\mathrm{ES} \rightarrow \mathrm{E} + \mathrm{P}$$

This mechanism and the associated rate law indicate that the reaction order for the substrate goes from first at low concentrations of substrate to zero at high concentrations; the rate law

$$-\frac{d[\mathrm{S}]}{dt} = \frac{kC_{\mathrm{E}}[\mathrm{S}]}{K_m + [\mathrm{S}]}$$

becomes

$$-\frac{d[\mathrm{S}]}{dt} = \frac{k}{K_m} C_{\mathrm{E}}[\mathrm{S}]$$

for $[\mathrm{S}] \ll K_m$, and

$$-\frac{d[\mathrm{S}]}{dt} = kC_{\mathrm{E}}$$

for $[\mathrm{S}] \gg K_m$.

This is shown in Figure 15–11. The meaning of the limiting rate at high concentrations of substrate is simple enough. At high concentrations of substrate, all of the active sites in the enzyme are associated with substrate, and further increase of the concentration of substrate cannot, therefore, increase the reaction rate. The figure also shows that the value of the Michaelis constant is equal to the concentration of substrate at which the

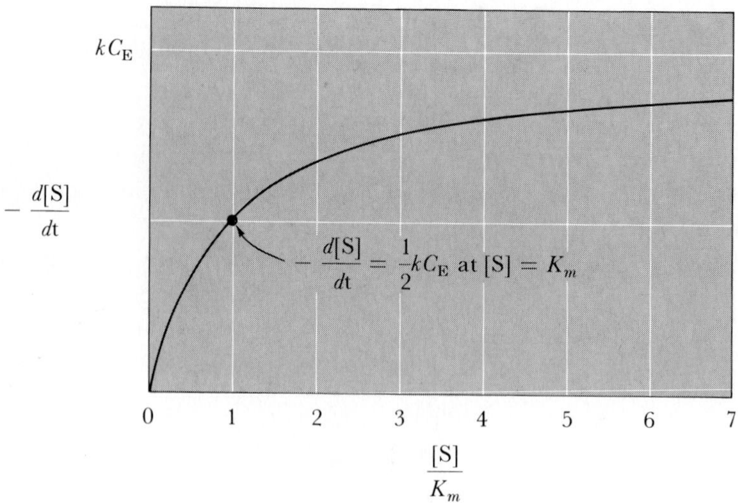

FIGURE 15–11
The dependence of the rate of an enzyme-catalyzed reaction upon the concentration of substrate:

$$-\frac{d[S]}{dt} = \frac{kC_E[S]}{K_m + [S]}$$

rate is equal to one half of the limiting value; the rate law

$$-\frac{d[S]}{dt} = \frac{kC_E[S]}{K_m + [S]}$$

becomes

$$-\frac{d[S]}{dt} = \frac{kC_E K_m}{K_m + K_m} = \tfrac{1}{2}kC_E$$

at $[S] = K_m$.

An important area of catalysis is that in which a solid material catalyzes a reaction of species in the gas or liquid phases. The mechanism for such *heterogeneous catalysis* involves the adsorption of the one or more reactants onto the solid surface through a specific chemical interaction. The nature of the solid surface influences the catalytic power and, indeed, can determine what reaction occurs. On an alumina (Al_2O_3) surface, ethyl alcohol decomposes to give ethylene and water:

$$C_2H_5OH = C_2H_4 + H_2O$$

However, on a copper surface the decomposition gives acetaldehyde and hydrogen:

$$C_2H_5OH = CH_3CHO + H_2$$

For surface reactions involving a single reactant, such as the reactions of ethyl alcohol just considered, the dependence of the reaction rate on the pressure of the reactant, given in Figure 15–12, resembles the substrate concentration dependence of an enzyme-catalyzed reaction given in

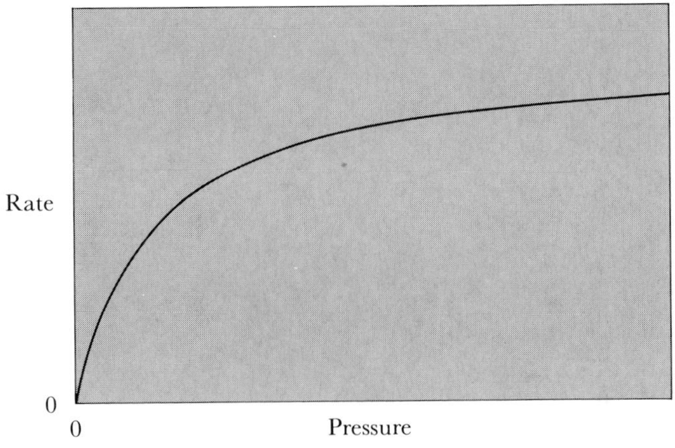

FIGURE 15–12
The dependence of the rate for a surface-catalyzed reaction on the pressure of the reactant.

Figure 15–11. Just as the rate of an enzyme-catalyzed reaction will not increase further as the concentration of substrate becomes so high that the active sites of the enzyme are saturated, so also the rate of a surface-catalyzed reaction will not increase further as the pressure of the gas becomes so high that the surface of the solid catalyst is covered completely.

In many cases of heterogeneous catalysis a reactant molecule is dissociated on the surface. The interaction of hydrogen with metallic platinum and palladium is strong enough to dissociate the hydrogen molecule despite its strong bond [$D(\text{H—H}) = 432 \text{ kJ mol}^{-1}$]. There are palladium hydrides, and they undoubtedly are involved as intermediates in the catalysis by palladium of reactions of hydrogen, for example, in the addition of hydrogen to a carbon–carbon double bond:

$$H_2 + H_2C{=}CH_2 \rightarrow H_3C{-}CH_3$$

One last point must be made about catalysis. The presence of a catalyst in a reaction mixture does not shift the equilibrium in a reaction. *A catalyst for the forward reaction also is a catalyst for the reverse reaction.* This point also can be made in the framework of the principle of microscopic reversibility, already discussed. If the rate law for the forward reaction includes a term with the concentration of a catalyst (which does not appear in the equilibrium-constant equation), the rate law for the reverse reaction also must contain a term with the same catalyst concentration factor.

CHAIN REACTIONS

The two steps of the postulated mechanism for the reaction of triiodide ion with arsenious acid (the unstable intermediate species enclosed) are:

$$I_3^- + H_2O \rightleftarrows \boxed{HOI} + 2I^- + H^+$$

$$\boxed{HOI} + As(OH)_3 \rightleftarrows OAs(OH)_3 + I^- + H^+$$

Let us compare these steps with the two product-producing steps of the postulated mechanism for the hydrogen-bromine reaction:

$$\boxed{Br} + H_2 \rightleftarrows HBr + \boxed{H}$$

$$\boxed{H} + Br_2 \rightleftarrows HBr + \boxed{Br}$$

The comparison brings out an important similarity and an important difference. In each case the sum of the two reactions is the overall reaction: this is the similarity. In the triiodide ion–arsenious acid reaction, a single unstable intermediate is produced in the first step and consumed in the second step. But in the hydrogen–bromine reaction, two unstable intermediates are involved. One is consumed and one is formed in each step, hence the difference between the two reactions. Each intermediate in the hydrogen–bromine reaction is analogous to a catalyst: it is consumed in one step and regenerated in another. But here, the "catalyst" is generated from the reaction mixture itself. Reactions with this feature are called *chain reactions*.

Some chain reactions occur with explosive violence. The mechanisms for these exothermic reactions generally involve one or more steps in which one reactive intermediate produces two (or more) reactive intermediates. This process is called *branching*, and it causes the rate to increase as reaction occurs; if this rate increase is coupled with the rapid evolution of heat, the reaction may get out of hand. The complex mechanism for the hydrogen–oxygen reaction involves unstable intermediate species, H and OH in the chain-propagating step,

$$H_2 + OH \rightarrow H_2O + H$$

and O in the branching steps,

$$H + O_2 \rightarrow OH + O$$
$$O + H_2 \rightarrow OH + H$$

In each of these branching steps one unstable intermediate gives two unstable intermediates. The net result of these two branching steps is conversion of two reactant molecules into two chain-propagating intermediates,

$$H_2 + O_2 \rightarrow 2OH$$

15–4 Theoretical Aspects of Reaction Kinetics

Even the simplest elementary reaction is a composite of reactant molecules with a range of energies reacting to give product molecules with a range of energies. The reaction rate in a bulk sample of matter is a weighted average of the rates of reaction of molecules with a Boltzmann distribution of energies. Although new techniques allow one to study the rate of a reaction involving reactants with specific energies, we are concerned primarily with reaction rates in bulk gaseous or liquid samples.

THE TEMPERATURE COEFFICIENT OF REACTION VELOCITY (THEORETICAL)

The empirical Arrhenius equation relating a rate constant and temperature,

$$k = Ae^{-E_a/RT}$$

was illustrated previously with data for two reactions:

$$2C_4H_6(g) = C_8H_{12}(g)$$

a second-order reaction with $A = 1.3 \times 10^7$ L mol^{-1} s^{-1} and $E_a = 99.2$ kJ mol^{-1}, and

$$C_4H_8(g) = 2C_2H_4(g)$$

a first-order reaction with $A = 4.0 \times 10^{15}$ s^{-1} and $E_a = 261$ kJ mol^{-1}. Now we will consider the theoretical significance of the Arrhenius parameters for these reactions in more detail.

For an elementary bimolecular gas-phase reaction,

$$A + B \rightarrow C$$

with a second-order rate law

$$-\frac{d[A]}{dt} = k[A][B]$$

the proportionality constant k is determined by several factors. These are:

1. the frequency factor for collisions of A and B,
2. a steric factor, the fraction of the collisions of A and B that have the spatial orientation that allows A and B to be transformed into the product, and
3. the fraction of the collisions with the correct orientation that have the energy needed to break certain chemical bonds and thus accomplish the reaction.

For a bimolecular reaction the rate constant can be expressed in a form that resembles the Arrhenius equation:

$$k = PZe^{-E^*/RT}$$

in which P is the steric factor for the reaction and Z is the frequency factor. The energy quantity E^* in this equation is the minimum relative translational energy of the colliding molecules needed for reaction to occur. The exponential factor gives the fraction of the collisions that have this energy:

$$\frac{\text{number of collisions with } E > E^*}{\text{all collisions}} \propto e^{-E^*/RT}$$

(If the energy is expressed per molecule, the Boltzmann constant k appears in the equation; if the energy is expressed per mole, the gas constant R

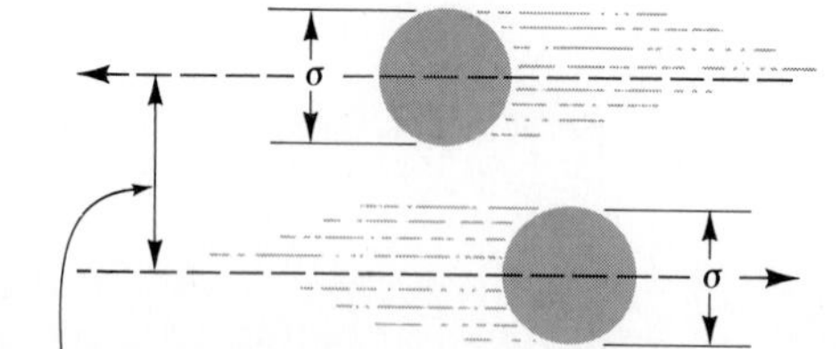

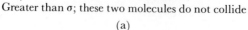

Greater than σ; these two molecules do not collide

(a)

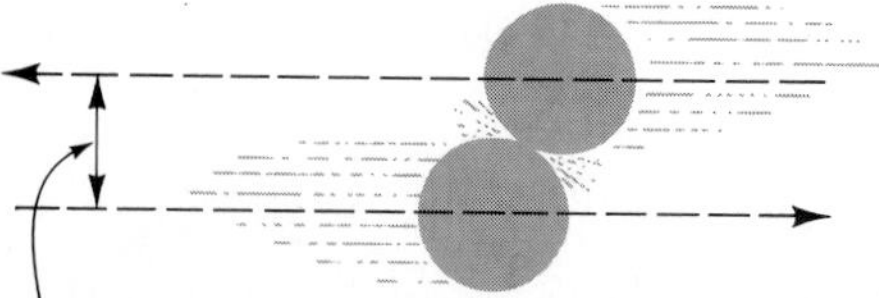

Less than σ; these two molecules do collide

(b)

FIGURE 15–13
The approach of two molecules to one another. (a) Molecules do not approach within σ of one another; there is no collision. (b) Molecules do approach within σ of one another; the molecules collide.

appears in the equation.) The frequency factor depends on the temperature and on the size of the colliding molecules:

$$Z = N_A \sigma_{A,B}^2 \left(\frac{8\pi RT}{M}\right)^{1/2}$$

In this equation, $\sigma_{A,B}$ is the mean collision diameter of molecules A and B; it is the distance within which the molecules must approach to collide, as is shown in Figure 15–13. The quantity M in this equation is a function of the masses of A and B, $M = M_A M_B/(M_A + M_B)$. The temperature dependence of this factor, $T^{1/2}$, is simply the kinetic-molecular-theory dependence relating the average translational velocity of gaseous molecules and temperature ($\frac{1}{2}m\overline{u^2} \propto T$). Each of the two factors Z and $e^{-E^*/RT}$ helps to increase the rate of a gas-phase reaction with increasing temperature, but the latter factor is very much more important. If only the dependence of average collision frequency upon temperature were responsible for the temperature dependence of a rate constant, the dependence would be much smaller than it usually is. An increase of temperature from 20°C to 30°C (293.2 K to 303.2 K) would increase the average collision frequency by a factor of

$$\left(\frac{303.2}{293.2}\right)^{1/2} = 1.017$$

and an increase from 320.0°C to 330.0°C (593.2 K to 603.2 K) would increase the average collision frequency by a factor of

$$\left(\frac{603.2}{593.2}\right)^{1/2} = 1.008$$

But many reactions are speeded up by a factor of two or more for such an increase of temperature. Table 15–1 presented the factor by which the rate constant increases for a temperature increase of 10 K. For the Diels–Alder condensation of butadiene, an increase of temperature from 593.2 K to 603.2 K increases the rate by a factor of 1.40.

The steric factors for bimolecular reactions in the gas phase vary over wide limits. For simple reactions, values approaching unity have been observed. For the reaction $Br + CH_4 \rightarrow CH_3 + HBr$, a step in the mechanism for the reaction of bromine with methane, $P \cong 0.2$. The reaction $NO + O_3 \rightarrow NO_2 + O_2$, which plays a role in air pollution, has a steric factor $P \cong 0.008$. For the bimolecular reaction of more complicated molecules, the value of P can be much smaller. For instance, in the Diels–Alder condensation of butadiene, $P \cong 3 \times 10^{-5}$.

In a reaction in which two molecules combine to give one molecule, for example, $2C_4H_6(g) = C_8H_{12}(g)$, the role of molecular collisions is clear enough. Two molecules of butadiene must come together in some step of the mechanism in order to give the product. In a reaction in which one molecule decomposes to give two molecules, for example, $C_4H_8(g) = 2C_2H_4(g)$, the role of molecular collisions may be less obvious. But collisions are necessary in this case also. The activation energy for this first-order reaction, 261 kJ mol^{-1}, is the difference between the average energy of the molecules that do react (the activated molecules) and the average energy of all of the reactant molecules. The molecules with the requisite energy to react get that energy from molecular collisions. For the cyclobutane decomposition, in which two carbon–carbon single bonds are broken, it is reasonable to consider the activation energy to be the energy required to stretch one of the single bonds to the breaking point, as shown previously. The activation energy, 261 kJ mol^{-1}, can be compared with the average carbon–carbon single-bond energy, 347 kJ mol^{-1}. Since there is strain in the reactant molecule, it takes less energy to produce the activated complex than we predict from the average bond energy.

Although the mechanism for decomposition of cyclobutane is called a unimolecular mechanism, bimolecular collisions provide the requisite activation energy. We will consider this puzzle with reference to a decomposition reaction like that of cyclobutane:

$$A = 2P$$

The simplest unimolecular mechanism for this reaction is one proposed independently in 1921 and 1922 by CHRISTIANSEN and LINDEMANN:

$$A + M \underset{k_{-1}}{\overset{k_1}{\rightleftarrows}} A^* + M$$

$$A^* \overset{k_2}{\rightarrow} 2P$$

Here M is a molecule of any gaseous substance, A, P, or a foreign gas, and A* is an activated molecule with sufficient energy to react. Applying the steady-state method to this mechanism, A* being the unstable intermediate with a "steady" concentration, we obtain

$$k_1[A][M] = k_{-1}[A^*][M] + k_2[A^*]$$

This equation gives

$$[A^*] = \frac{k_1[A][M]}{k_{-1}[M] + k_2}$$

for this steady-state concentration. By using the equation for the steady-state concentration of A*, we can convert the rate of the reaction,

$$-\frac{d[A]}{dt} = \frac{1}{2}\frac{d[P]}{dt} = k_2[A^*]$$

into a rate law involving the total concentration of A. This rate law is

$$-\frac{d[A]}{dt} = \frac{k_1 k_2[A][M]}{k_{-1}[M] + k_2}$$

Like our earlier steady-state rate law with a two-term denominator, this rate law can be simplified to two different limiting forms.

If $k_{-1}[M] \gg k_2$, the probable fate of A* is deactivation to give the reactant,

$$A^* + M \xrightarrow{k_{-1}} A + M$$

and this complete rate law becomes a first-order law for this limiting case:

$$-\frac{d[A]}{dt} = \left(\frac{k_1}{k_{-1}}\right) k_2\,[A]$$

The empirical first-order rate constant is related to the rate constants involved in this mechanism:

$$k = \left(\frac{k_1}{k_{-1}}\right) k_2 = K_1 k_2$$

in which K_1 is the equilibrium constant for the activating reaction. In this limit, the activated molecules are maintained essentially at the concentration corresponding to thermal equilibrium.

An interesting feature of this mechanism develops if $k_2 \gg k_{-1}[M]$: under such limiting conditions the probable fate of A* is not deactivation, but is decomposition to give products, and the rate law becomes

$$-\frac{d[A]}{dt} = k_1[M][A]$$

That is, the rate law for the unimolecular decomposition of a molecule A is the second-order rate law that gives the rate of the activating collisions. All gas-phase unimolecular reactions show limiting second-order behavior if the pressure is low enough.

Understanding this mechanism, you will be able to see why a first-order rate law governs the decomposition of cyclobutane in the presence of a foreign gas,

$$-\frac{d[C_4H_8]}{dt} = k[C_4H_8]$$

whereas second-order rate laws govern the decomposition of nitrogen dioxide and the dissociation of iodine at high temperatures in the presence of a foreign gas M,

$$2NO_2 = 2NO + O_2 \qquad I_2 = 2I$$

The rate laws are

$$-\frac{d[NO_2]}{dt} = k[NO_2][M] \quad \text{and} \quad -\frac{d[I_2]}{dt} = k[I_2][M]$$

The key to this difference between the two decompositions is the rate of the process by which the activated molecule A* becomes the activated complex for the production of product(s),

$$A^* \rightarrow A^\ddagger \rightarrow 2P$$

The more complex the molecule, the slower is this process of localizing the extra energy in the bond being broken. This process for cyclobutane,

$$C_4H_8{}^* \xrightarrow{k_2} 2C_2H_4$$

is so slow that $k_{-1}[M] \gg k_2$, and the first-order rate law results. For a molecule as simple as nitrogen dioxide with three atoms, the reaction in which the requisite energy is redistributed to break a nitrogen–oxygen bond,

$$NO_2{}^* \xrightarrow{k_2} NO + O$$

occurs very rapidly, making $k_2 \gg k_{-1}[M]$, and resulting in the second-order rate law.

The limit of simplicity in molecules is a diatomic molecule, for which energizing and decomposition occur essentially in one bimolecular step. If a collision between M and I_2 has sufficient energy to dissociate the iodine molecule, the molecule dissociates at once,

$$M + I_2 \rightarrow M + I + I$$

With only one bond, the bond which is being broken, an appropriately energized iodine molecule has no way to store the energy acquired in the collision, so the bond breaks in the time of one vibration.

Since a simple molecule, such as I_2, requires a bimolecular reaction for dissociation,

$$M + I_2 \rightarrow M + I + I$$

the recombination of atoms requires a termolecular reaction,

$$M + I + I \rightarrow M + I_2$$

In this collision the third body, as M is called, takes away the energy lost when the bond between iodine atoms is formed. (Recall Figure 3–2.) The situation is pictured in another way in Figure 15–14, which shows the potential-energy diagram for iodine. An activating collision with M gives an iodine molecule, initially in the lowest vibrational state, enough energy to result in its dissociation (Figure 15–14a). The kinetic energy of the iodine

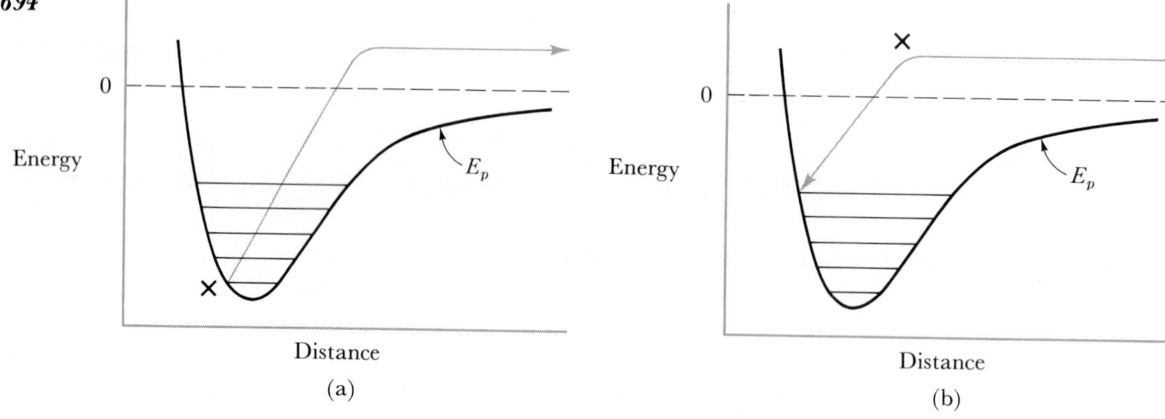

FIGURE 15–14
A potential-energy diagram for I_2. (a) A collision of an I_2 molecule with a molecule M at X. (Initially, the I_2 molecule is in the ground vibrational state.) (b) A collision of two I atoms with a molecule M at X. (The I_2 molecule is formed in an excited vibrational state.)

atoms relative to one another exceeds the dissociation energy, and they escape from one another. If two iodine atoms approach one another head-on without the third body M, the two iodine atoms simply bounce off of one another in a collision resembling that of two argon atoms as depicted in Figure 4–17. With a third body participating in the collision of the iodine atoms, some energy is lost to the third body, allowing an *iodine molecule* to form in an excited vibrational state, and giving this third body M more translational energy (and rotational and vibrational energy also, if it contains two or more atoms) than it brought into the collision.

THE ELEMENTARY REACTIONS OF MOLECULES WITH SPECIFIC ENERGIES; MOLECULAR BEAMS

In a typical bulk experiment involving the reaction of gases, the system in thermal equilibrium consists of molecules with a Boltzmann distribution of translational energies. The molecules also have rotational and vibrational energies, the number of rotational and vibrational states occupied depending upon the temperature and the spacing of the energy states (as shown in Figure 5–21). A range of fruitful collisions is possible, and the measured reaction rate in a bulk experiment is the sum of all of these. A bulk experiment discloses little about the unfruitful collisions. Since the middle 1950s a whole new field of chemical kinetics has developed; it has become possible to observe the consequences of individual collisions in experiments that involve beams of molecules.

In a molecular-beam apparatus, the reactant molecules or atoms collide in a region of highly evacuated space, and the scattering of the molecules in the bimolecular collisions can be studied. Three types of scattering can occur. *Elastic scattering* results from a billiard-ball collision in which translational energy is not converted to rotational or vibrational energy. *Inelastic scattering* results from a collision in which chemical reaction does not occur, but translational energy is converted to rotational and/or vibrational energy. *Reactive scattering* is the name for a collision that produces chemical reaction. Because these collisions involve pairs of molecules, the processes being studied are elementary bimolecular processes. The probability of each type of scattering depends upon the energy and orientation of the colliding molecules, and with appropriate experimental apparatus it is possible to study the collisions of molecules with specific energies and orientations. Figure 15–15 is a diagram of a molecular-beam apparatus in which potassium atoms of a particular speed collide with methyl iodide molecules of a particular orientation. The reaction that occurs in the reactive collisions produces potassium iodide and methyl radicals:

$$K + CH_3I \rightarrow KI + CH_3$$

(The methyl radicals produced do not persist indefinitely, but their fate is not a concern in this experiment.) The probability of each kind of scattering can be characterized by a cross section; in the simplest picture, the cross section is $\pi\sigma^2$, where σ is the collision diameter defined in Figure 15–13. The cross section for scattering can be determined by studying the intensity of the scattered potassium atoms or potassium iodide molecules as a function

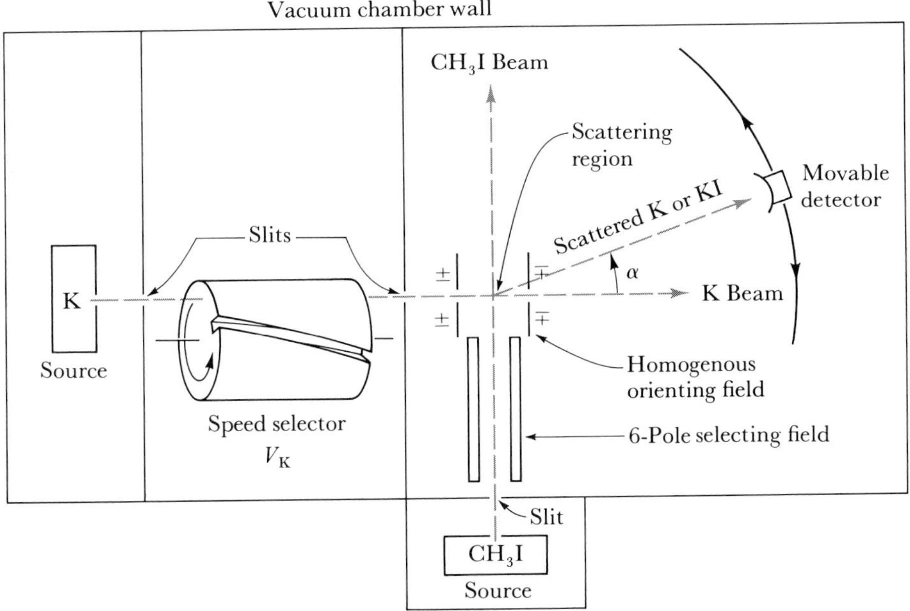

FIGURE 15–15
A diagram of a molecular-beam apparatus for study of the reaction $K + CH_3I = KI + CH_3$. Although the potassium source emits atoms with a Boltzmann range of energies, the rotating grooved cylinder (the speed selector) allows passage of atoms with only a particular narrow range of energies. [From E. F. Greene and J. Ross, *Science* **159**, 587 (1968).]

of the scattering angle (see Figure 15–15). For the reaction in question, the cross section for reactive scattering is 30 Å^2 (which corresponds to $\sigma \cong 3$ Å, a reasonable value).

For some reactions, the reaction cross section is much greater than is consistent with the sizes of the atoms or molecules involved. For instance, the reaction

$$K + Br_2 \rightarrow KBr + Br$$

has a cross section of ~150 Å^2. To explain this large cross section, a mechanism called the "harpoon" mechanism has been proposed. In the harpoon mechanism, a potassium atom transfers an electron to a bromine molecule from a great distance,

$$K + Br_2 \rightarrow K^+ + Br_2^-$$

(This gives the process a large cross section.) The electrostatic attraction of the oppositely charged ions draws the reactants together, and the reaction is completed with

$$K^+ + Br_2^- \rightarrow KBr + Br$$

The transfer of the electron is the throwing of the harpoon, and the electrostatic attraction is the drawing in of the catch.

With the molecular-beam technique, the intimate details of many gas-phase reactions are being revealed.

15–5 Fast Reactions in Solution

For a unimolecular reaction $A \rightarrow P$ in solution, the necessary energizing collisions are made with surrounding solvent molecules. The rate of such a reaction will be governed by the first-order rate law,

$$-\frac{d[A]}{dt} = k[A]$$

over the entire range of concentration of A. The equilibrium concentration of activated reactant molecules, A*, is maintained by collisions with the solvent molecules, which are present at very high concentrations.

A reaction in solution that occurs via a bimolecular mechanism, $A + B \rightarrow P$, can occur no faster than the molecules of A and B diffuse into contact with one another. Having diffused together, the molecules A and B will, on the average, bounce together many times with varying energies as they rattle around in the cage of surrounding solvent molecules. Eventually, they will bounce together with appropriate energy and orientation for reaction, or they will escape from each other's presence before they react. This act of rattling around together in the same solvent cage is called an *encounter*. If the reaction of A and B has an activation energy much larger than RT, most encounters end fruitlessly, and the rate of reaction is smaller than the rate of encounter of A and B. If the reaction itself has an activation energy close to zero, the temperature dependence of the second-order rate constant is determined by the activation energy for diffusion, which is ~4–13 kJ mol^{-1}. Reactions that have a very low activation energy and that occur upon each encounter are called *diffusion-controlled* reactions. These reactions occur very rapidly.

Among diffusion-controlled reactions is the neutralization reaction

$$H_3O^+ + OH^- \rightarrow 2H_2O$$

for which the second-order rate constant at 25°C is

$$k = (1.3 \pm 0.2) \times 10^{11} \text{ L mol}^{-1} \text{ s}^{-1}$$

This is one of the largest known second-order rate constants for a reaction in aqueous solution. If it were possible to mix equal volumes of 0.10*M* HCl and 0.10*M* NaOH instantaneously, the neutralization reaction would be 99% over in ~10^{-6} s.

The reactions of hydronium ion with most other Brønsted bases occur almost as rapidly as this; examples are:

$$H_3O^+ + F^- \rightarrow HF + H_2O \qquad k = 1.0 \times 10^{11} \text{ L mol}^{-1} \text{ s}^{-1}$$

$$H_3O^+ + CH_3CO_2^- \rightarrow CH_3CO_2H + H_2O \qquad k = 5.1 \times 10^{10} \text{ L mol}^{-1} \text{ s}^{-1}$$

Because the rate constants for reaction of hydronium ion and various Brønsted bases are very similar, the varying strengths of the acids must result from the variation of the rates of dissociation of the acids (i.e., the rates of reaction of the acids with water). For a series of reactions

$$HB + H_2O \underset{k_2}{\overset{k_1}{\rightleftharpoons}} H_3O^+ + B^-$$

with values of $K_a = k_1/k_2$, the values of k_1 must differ for different compounds if for these compounds the values of K_a vary but the values of k_2 are all approximately the same. The weaker the acid HB, the smaller is k_1, the rate constant for its dissociation. Table 15–2 presents data for four acids

TABLE 15–2
Rate Constants for Acid-Dissociation Reactions[a] at 25°C

X^-	$\frac{10^{-10} \times k_2}{\text{L mol}^{-1} \text{ s}^{-1}}$	$\frac{k_1}{\text{s}^{-1}}$	$\frac{K_a}{\text{mol L}^{-1}}$
F^-	10	6.8×10^7	6.8×10^{-4}
$CH_3CO_2^-$	5.1	8.9×10^5	1.75×10^{-5}
HS^-	6.3	6.4×10^3	1.02×10^{-7}
NH_3	4.0	2.3×10	5.69×10^{-10}

[a]For the equilibrium

$$HX + H_2O \underset{k_2}{\overset{k_1}{\rightleftharpoons}} H_3O^+ + X^-$$

with dissociation constants varying by a factor of 10^6. In this series there is a 3×10^6-fold change in the value of k_1, but the reaction of each conjugate base with hydrogen ion occurs with a rate close to that for a diffusion-controlled reaction.

Another second-order reaction that occurs at approximately the diffusion-controlled rate is

$$I_2(aq) + I^-(aq) \rightleftarrows I_3^-(aq)$$

for which $k_f = (4.1 \pm 0.4) \times 10^{10}$ L mol^{-1} s^{-1}.

Conventionally, a reaction in solution is started by mixing the two or more solutions containing different reactants. Depending upon the mixing technique used, this takes 0.5 s to 5 s. If the reaction occurs essentially to completion in the time of mixing, nothing is learned about the kinetics of the reaction other than a lower limit for the rate. Until recently it was customary to refer to such reactions as being *immeasurably fast*. Now there are many techniques for studying kinetics of reactions that are faster than the limit imposed by conventional mixing procedures.

Before discussing these techniques for studying very fast reactions, it is important to remark that some reactions with large rate constants can still be studied by conventional methods. A second-order reaction occurs more slowly the lower the concentration. If a reaction governed by the second-order rate law

$$-\frac{d[A]}{dt} = k[A][B]$$

goes one half to completion in 2 s when the initial concentrations of A and B are each 0.10 mol L^{-1}, it will go one half to completion in 2000 s if the initial concentrations are each 1.0×10^{-4} mol L^{-1}. A sensitive analytical method makes it possible to study fast second-order reactions at low concentrations using conventional batch techniques, which may obviate the need for use of fast reaction methods.

Two general methods for studying fast reactions are (1) rapid-mixing flow methods, and (2) relaxation methods. In the rapid-mixing flow methods, the mixing time can be reduced to $\sim 10^{-3}$ s. Mixing is avoided entirely in the studies that involve relaxation methods. In these, a system at chemical equilibrium is disturbed by a very rapid perturbation, and the rate at which the system approaches a new position of equilibrium is observed.

In the simplest flow procedure, the reagents are driven into a mixing chamber from syringes, and the mixed solution then flows down an observation tube, as shown in Figure 15–16. The time between mixing and observation depends upon the flow velocity of the mixed solution and the distance between the mixing chamber and the observation point. Each of these may be varied. In an experiment with continuous flow at a constant rate and an observation point a particular distance from the mixing chamber, the concentrations of reactant and product species at the observation point do

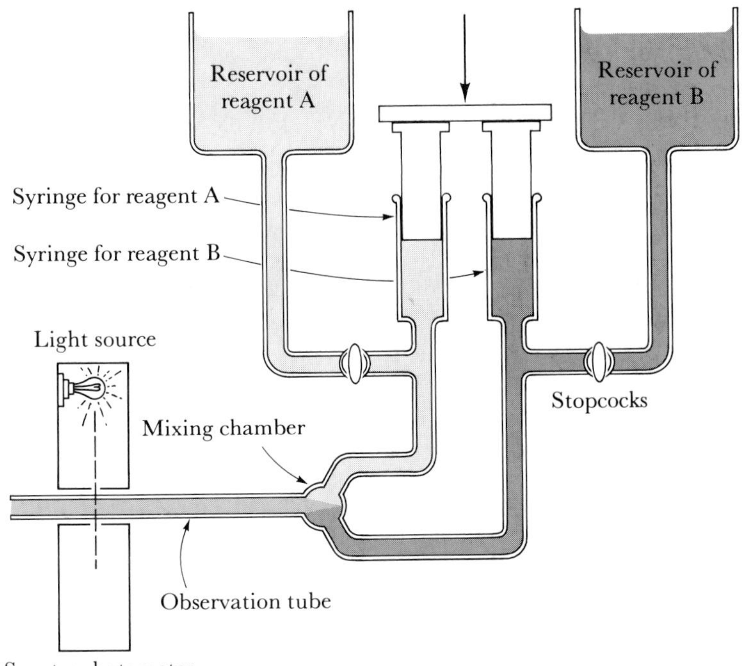

FIGURE 15–16
A diagram of equipment for study of a fast reaction between A and B. The light absorption of the mixed solution is observed a few milliseconds after mixing.

not change with time. Therefore the monitoring spectrophotometer need not respond quickly to changes in light absorption. However, continuous-flow methods are wasteful of reagents, often highly purified, and these methods have been superseded by stopped-flow methods. In a stopped-flow experiment, the mechanism driving the syringes stops when the mixed solution has come into view at the observation point. The spectrophotometer follows the change of light absorption with time at this point, and its readings are displayed on an oscilloscope screen.

Because the reaction for dissociation of water is endothermic,

$$2H_2O = H_3O^+ + OH^- \qquad \Delta H^0 = +55.9 \text{ kJ mol}^{-1}$$

the extent of this reaction increases with temperature. A relaxation method called the T-jump (temperature jump) method has been used to study the rate of this reaction. If a high-voltage condenser is discharged through a cell containing water, the temperature is raised very rapidly. In experiments developed by EIGEN and his associates, a temperature increase of 2–10 K occurs in $\sim 10^{-6}$ s. The conductance of the water in the cell, which increases to its new equilibrium value at the higher temperature, can be determined as a function of time by appropriate equipment. An observation that the half-time in which the conductance approaches its new equilibrium value is $\sim 25 \times 10^{-6}$ s allows calculation that the second-order rate constant for the bimolecular reaction of hydronium ion and hydroxide ion is 1.3×10^{11} L mol^{-1} s^{-1}.

Biographical Notes

KURT ALDER (see OTTO DIELS).

J. A. CHRISTIANSEN (1888–), a Danish physical chemist, was Professor at the University of Copenhagen from 1931–1959. He did much work in the field of chemical kinetics.

OTTO P. H. DIELS (1876–1954) and KURT ALDER (1902–1958), German chemists, received the Nobel Prize in Chemistry in 1950 for their discovery of the reaction of a diene with an unsaturated compound to give a cyclic compound. This general reaction, in which noncyclic molecules produce cyclic molecules, has been very useful in synthesis.

MANFRED EIGEN (1927–), a German physical chemist, was educated at the University of Göttingen. Since 1953 he has been associated with the Max Planck Institute for Physical Chemistry in Göttingen. At this institute he developed methods for studying very fast reactions, work for which he received the Nobel Prize in Chemistry in 1967.

EMIL FISCHER (1852–1919), German chemist, was the outstanding organic chemist of his day. He received the Nobel Prize in Chemistry in 1902 for his work in synthetic organic chemistry. He was particularly interested in sugars, polypeptides, and purines.

F. A. LINDEMANN (1886–1957), a British physicist, received his Ph.D. from the University of Berlin. He was a professor at Oxford, and was science adviser to Winston Churchill during World War II.

L. MICHAELIS (1875–1949), a physician and biochemist, was born and educated in Germany. Before coming to the United States in 1926, he taught in Germany and Japan. He was associated with the Rockefeller Institute from 1929 until his retirement in 1940.

Problems and Questions

15–1 For the reaction of bromate ion with bromide ion in acidic solution, the value of $-d[BrO_3^-]/dt$ in a reaction mixture at a particular instant is 1.47×10^{-3} mol L^{-1}s^{-1}. What are the values of $-d[Br^-]/dt$, $-d[H^+]/dt$, and $d[Br_2]/dt$ at this same instant?

15–2 A first-order reaction has a half-time of 17.3 min. What time is required for the reaction to go 98.0% to completion? What is the extent of reaction in 42.0 min?

15–3 Two isomeric substances, A and B, each decompose in first-order reactions with $k_A = 4.50 \times 10^{-4}$ s^{-1} and $k_B = 3.72 \times 10^{-3}$ s^{-1}. If at t = 0, a mixture contains equal amounts of A and B, $[A]_0 = [B]_0$, at what time is [A] = 4.00 [B]?

15–4 The rate of a reaction A + B = C + D is determined by measuring the concentration of A as a function of time in an experiment with $[A]_0 = [B]_0$. The data are:

$\frac{\text{time}}{\text{min}}$	0	10	20	30
$\frac{[A]}{\text{mol L}^{-1}}$	0.100	0.063	0.0397	0.025

What is the rate law for the forward reaction? What is the rate constant for the forward reaction? What is the form of the rate law for the reverse reaction? (There may be more than one possible rate law.)

15–5 A reaction $2A = C$ is governed by a second-order rate law, $d[C]/dt = k[A]^2$. In a reaction mixture with $[A]_0 = 0.184 \text{ mol L}^{-1}$, $[C]_0 = 0$, the concentration of C at $t = 550 \text{ s}$ is 0.047 mol L^{-1}. (Equilibrium in the reaction lies far to the right. The reverse reaction does not occur to a measurable extent.) What is the value of k and what is the concentration of A at 1000 s?

15–6 The second-order rate constant for the dimerization of butadiene at 620 K is $5.9 \times 10^{-2} \text{ L mol}^{-1} \text{ s}^{-1}$. In a sample with an initial concentration of butadiene of 0.180 mol L^{-1}, what is the reaction rate after 105 min?

15–7 Detailed kinetic studies have shown the reaction $2C_4H_6(g) = C_8H_{12}(g)$ to be governed by a second-order rate law, $d[C_8H_{12}]/dt = k[C_4H_6]^2$. At a particular temperature, concentration-time data are:

t	$[C_4H_6]$
0	$0.0116 \text{ mol L}^{-1}$
4000 s	$0.0070 \text{ mol L}^{-1}$

What is the value of k, and what time is required for the reaction to go 90.0% to completion (i.e., to $[C_4H_6] = 1.16 \times 10^{-3} \text{ mol L}^{-1}$)?

15–8 The fourth-order rate constant for the bromate ion–bromide ion reaction at 25°C is $9.0 \text{ L}^3 \text{ mol}^{-3} \text{ s}^{-1}$. An experiment is carried out with the initial concentration conditions $[BrO_3^-]_0 = 1.00 \times 10^{-3} \text{ mol L}^{-1}$, $[Br^-]_0 = 0.150 \text{ mol L}^{-1}$, $[H^+] = 0.151 \text{ mol L}^{-1}$. Under these conditions the course of the reaction is essentially first order. Why? What is the value of k', the pseudo first-order rate constant defined by the equation

$$-\frac{d[BrO_3^-]}{dt} = k'[BrO_3^-]$$

under these conditions? What is the half-time for the reaction under these conditions?

15–9 The fast reaction of iodide ion with hypochlorite ion,

$$I^- + OCl^- = OI^- + Cl^-$$

has been studied at 25°C in $1M$ NaOH. The initial concentrations of the reactants are the same ($0.00200M$), and they decrease with time:

$\frac{\text{time}}{\text{s}}$	0	3	6	14
$\frac{[OCl^-]}{\text{mol L}^{-1}}$	0.00250	0.00158	0.00123	0.00073

What is the order of this reaction, and what is the value of the rate constant for this experiment?

15–10 The first-order rate constant for a reaction $A \rightarrow P$ shows the following dependence upon temperature:

$\frac{t}{°C}$	50.0	70.1	89.4	101.0
$\frac{k}{s^{-1}}$	0.0108	0.0734	0.454	1.38

What is the activation energy for this reaction?

15–11 Many textbooks suggest that a 10 K increase of temperature (from 20°C to 30°C) doubles the velocity of a reaction. This statement is valid only for reactions with a particular value of the activation energy. What is this value of E_a?

15–12 The activation energy for the decomposition of cyclobutane is 262 kJ mol^{-1}. By what factor does the rate of this reaction change in going from 700 K to 710 K?

15–13 A reactant can undergo either of two first-order reactions,

$$A \xrightarrow{k_1} P \qquad A \xrightarrow{k_2} Q$$

The value of k_1 and k_2 are equal at 70°C, and the activation energies associated with the two rate constants are 64.4 kJ mol^{-1} (k_1) and 85.8 kJ mol^{-1} (k_2). At what temperature is the value of $k_2/k_1 = 2.00$?

15–14 Using numerical values given in the text, calculate the concentration of hydrogen ion at which the rate law term $k_1[H_2O_2][I^-]$ is two times as large as the term $k_2[H_2O_2][I^-][H^+]$.

15–15 Explain the large difference in activation energy for the decomposition of nitrogen dioxide,

$$2NO_2 = 2NO + O_2$$

via the activated complex $(NO_2)_2^{\ddagger}$ (113.8 kJ mol^{-1} in the range 600–650 K) and the activated complex $NO_2 \cdot Ar^{\ddagger}$ (247 kJ mol^{-1} in the range 1400–2300 K).

15–16 The rate of formation of phosgene,

$$CO(g) + Cl_2(g) = COCl_2(g)$$

is governed by the rate law

$$\frac{d[COCl_2]}{dt} = k_f[CO][Cl_2]^{3/2}$$

What is the composition of the activated complex for this reaction? Predict the form of the rate law for the reverse reaction.

15–17 A mechanism proposed for the gas-phase decomposition of ozone, $2O_3(g) = 3O_2(g)$, is

$$O_3 + M \rightleftarrows O_2 + O + M$$
$$O + O_3 \rightleftarrows 2O_2$$

in which M is a third body. Treat atomic oxygen as an unstable intermediate to which the steady-state approximation can be applied, and derive the rate law.

15–18 The decomposition reaction for gaseous dinitrogen pentaoxide is first order, but the mechanism is not a simple unimolecular reaction. The mechanism proposed for this reaction, $2N_2O_5 = 4NO_2 + O_2$, is

1. $M + N_2O_5 \underset{k_2}{\overset{k_1}{\rightleftarrows}} NO_3 + NO_2 + M$

2. $NO_3 + NO_2 \xrightarrow{k_3} NO + O_2 + NO_2$

3. $NO_3 + NO \xrightarrow{k_4} 2NO_2$

Treat NO_3 and NO as unstable intermediates to which the steady-state approximation can be applied, and derive a rate law consistent with this mechanism.

15–19 The rate law for the hydration of carbon dioxide is

$$-\frac{d[CO_2]}{dt} = k_1[CO_2] + k_2[CO_2][OH^-]$$

with (at 25.0°C) $k_1 = 3.9 \times 10^{-2}\ s^{-1}$ and $k_2 = 8.5 \times 10^3\ L\ mol^{-1}\ s^{-1}$. At what concentration of hydroxide ion is the fraction of the reaction going by the k_1 pathway 0.90, 0.50, and 0.10?

15–20 The rate of decomposition of hyponitrite ion ($N_2O_2^{2-}$) in alkaline solution ($[OH^-] > 10^{-2}$ mol L^{-1}),

$$N_2O_2^{2-} + H_2O = N_2O(g) + 2OH^-$$

is governed by the rate law

$$-\frac{d[N_2O_2^{2-}]}{dt} = k\frac{[N_2O_2^{2-}]}{[OH^-]^2}$$

Interpret this rate law in terms of a reaction mechanism. (For hyponitrous acid, $K_{a1} = 9 \times 10^{-8}$ mol L^{-1} and $K_{a2} = 1.0 \times 10^{-11}$ mol L^{-1}).

15–21 The rate of oxidation of iodide ion by iron(III) ion,

$$2Fe^{3+} + 2I^- = I_2 + 2Fe^{2+}$$

is governed at high concentrations of iron(II) ion by the rate law

$$\frac{d[I_2]}{dt} = \frac{k[Fe^{3+}]^2[I^-]^2}{[Fe^{2+}]}$$

What is the composition of the activated complex for this reaction under the concentration conditions that give this limiting rate law? Suggest a mechanism consistent with this rate law.

15–22 A reaction A = B is subject to catalysis by hydrogen ion; the rate law is $-d[A]/dt = k[H^+][A]$. At $[H^+] = 0.043$ mol L^{-1}, the reaction goes 80% to completion in 93.4 min. What is the value of k? What is the half-time for reaction at $[H^+] = 0.071$ mol L^{-1}?

15–23 Explain the fallacy in the proposal that a catalyst may increase the rate of a reaction going in one direction without increasing the rate of the reaction going in the other direction.

16

Oxidation–Reduction Reactions and Electrochemistry

16–1 Introduction

Electrons are a constituent of matter, and chemical reactions involve changes of electronic structure within atoms of the substances involved. The changes are not dramatic in some reactions; for instance, in the neutralization reaction

$$H_3O^+(aq) + OH^-(aq) = 2H_2O(l)$$

the number of valence-shell electrons involved in bonding a hydrogen atom to an oxygen atom is the same in each reactant and in the product. In other reactions the changes of electronic structure are dramatic; for instance, in the reaction of sodium metal with chlorine gas to produce sodium chloride (discussed in Chapter 9),

$$2Na(s) + Cl_2(g) = 2NaCl(s)$$

the valence-shell electron of each sodium atom is transferred to a chlorine atom, and the compound produced consists of ions (Na^+ and Cl^-). This latter reaction involving electron transfer is called an *oxidation–reduction reaction*. This chapter deals with oxidation–reduction reactions, a subject with many facets.

The electron transfer is explicit in oxidation–reduction reactions that occur in electrolysis. An external electrical potential causes the flow of an electrical current through an electrolytic cell, and electron transfer is forced to occur at each electrode. The large-scale commercial production of some materials (e.g., aluminum) is accomplished by electrolysis.

A chemical system in which a spontaneous chemical change can occur is a system that may provide useful electrical energy in a galvanic cell (a battery). The oxidation–reduction reaction of lead and lead dioxide in sulfuric acid is a spontaneous reaction that is harnessed to produce electrical energy in the common lead storage battery:

$$Pb + PbO_2 + 2H^+ + 2HOSO_3^- = 2PbSO_4 + 2H_2O$$

Quantitative study of the electromotive force (emf) of galvanic cells also gives us information about the equilibrium constants for oxidation–reduction reactions.

16–2 Oxidation–Reduction Reactions

The term "oxidation" suggests reaction with oxygen, and reactions of oxygen do involve oxidation. In its reaction with oxygen, magnesium metal is oxidized:

$$2Mg(s) + O_2(g) = 2MgO(s)$$

The resulting compound, magnesium oxide, is an ionic compound containing magnesium ion, Mg^{2+}, and oxide ion, O^{2-}, in the sodium chloride lattice (see Figure 9–3). The same transformation of magnesium occurs in its reaction with chlorine gas,

$$Mg(s) + Cl_2(g) = MgCl_2(s)$$

(magnesium chloride is an ionic compound containing magnesium ion, Mg^{2+}, and chloride ion, Cl^-). Again, the same transformation occurs when magnesium reacts with a solution of acid (e.g., hydrochloric acid),

$$Mg(s) + 2H^+(aq) = Mg^{2+}(aq) + H_2(g)$$

In each of these reactions, magnesium metal is changed to magnesium ion by the loss of two electrons per atom: this is shown in the half-reaction

$$Mg(s) = Mg^{2+} + 2e^-$$

The term *oxidation* describes the loss of electrons that magnesium has undergone whether this loss of electrons has occurred through the reaction with oxygen, with chlorine, or with hydrogen ion. Each reaction in which one or more electrons have been lost by an atom, an ion, or a molecular species is a reaction in which electrons have been gained by another atom, ion, or molecular species. The term *reduction* is used to describe the gain of electrons. In the three reactions just considered,

1. oxygen gas gains electrons and is reduced to oxide ion,

$$O_2(g) + 4e^- = 2O^{2-}$$

2. chlorine gas gains electrons and is reduced to chloride ion,

$$Cl_2(g) + 2e^- = 2Cl^-$$

3. hydrogen ion gains electrons and is reduced to hydrogen gas,

$$2H^+ + 2e^- = H_2(g)$$

Just as each magnesium atom loses electrons in its oxidation to magnesium ion, so each molecule in the series of compounds with one carbon atom discussed in Chapter 11 loses electrons in its oxidation:

$$CH_4 \rightarrow CH_3OH \rightarrow CH_2O \rightarrow HCOOH \rightarrow CO_2$$

Each of these transformations can be described in a half-reaction:

$$CH_4 + H_2O = CH_3OH + 2H^+ + 2e^-$$
$$CH_3OH = CH_2O + 2H^+ + 2e^-$$
$$CH_2O + H_2O = HCOOH + 2H^+ + 2e^-$$
$$HCOOH = CO_2 + 2H^+ + 2e^-$$

Therefore each of these changes corresponds to our description of oxidation. In our previous discussion of these changes (Section 11–4), we introduced the concept of oxidation state, and we showed that values of the oxidation state of carbon in the five compounds of this series are -4 (CH_4), -2 (CH_3OH), 0 (CH_2O), $+2$ ($HCOOH$), and $+4$ (CO_2). Thus definitions of oxidation and reduction may involve the oxidation states of the elements. Either kind of definition is useful; our definition may be based either on loss or gain of electrons or on the increase or decrease of oxidation state:

1. A substance or species is said to be *oxidized* if it loses electrons, or if the oxidation state of one of its constituent elements increases.
2. A substance or species is said to be *reduced* if it gains electrons, or if the oxidation state of one of its constituent elements decreases.

A substance or species that is oxidized acts as a *reducing agent*. A substance or species that is reduced acts as an *oxidizing agent*.

OXIDATION STATES

We introduced the concept of oxidation state in Chapter 11, but did not develop it fully at that point. Now full development is necessary. In this chapter and in later chapters dealing with the chemistry of the elements, we will make much use of the concept.

Applying a set of rules, we can find the oxidation state of an element in the elemental form, in a compound (whether ionic or covalent), or in a monatomic or polyatomic ion:

1. *In the elemental form, the oxidation state of an element is zero.* The oxidation state is zero for an element in each allotropic form; for example, oxygen

has the oxidation state zero both in the diatomic molecule O_2 and in the triatomic molecule ozone, O_3.

2. *In simple binary ionic compounds, the oxidation state of each element is the same as the ionic charge on atoms of that element.* The oxidation states of iron and chlorine in ferrous chloride ($FeCl_2$, containing Fe^{2+} and Cl^-) are $+2$ and -1, respectively, and in ferric chloride ($FeCl_3$, containing Fe^{3+} and Cl^-) are $+3$ and -1, respectively. [The preferred nomenclature for these compounds specifies the oxidation state of the metal, iron(II) chloride and iron(III) chloride.] By the rule just given, the oxidation state of oxygen in ionic oxides (e.g., MgO and K_2O) is -2, and the oxidation state of oxygen in ionic peroxides is, by this rule, -1 (e.g., Na_2O_2 containing Na^+ and O_2^{2-}.) Notice that the oxidation state is calculated per atom; the oxidation state of oxygen in peroxide ion is -1.
3. *The oxidation state of oxygen in covalent compounds is assigned the same value as in its ionic compounds*, that is -2; however, in peroxides the oxidation state of oxygen is assigned the value -1. (If we do not know the structure of the compound, we customarily assign an oxidation state of -2 to oxygen, but, as we will see, this can lead us to give highly artificial oxidation state values to other elements.)
4. *The sum of the oxidation states of all elements in a molecule or ion is the net charge on the molecule or ion.* This rule, coupled with the convention that the oxidation state of oxygen in covalent compounds is assigned in the same way as in ionic compounds, allows us to assign oxidation states to elements in many covalently bonded species. Some examples are:

   In water, H_2O, the oxidation state of hydrogen (x) can be calculated:

   $$2x + (-2) = 0 \qquad 2x = +2$$
   $$x = +1$$

   In carbon dioxide, CO_2, the oxidation state of carbon (x) can be calculated:

   $$x + 2(-2) = 0 \qquad x = +4$$

   In perchlorate ion, ClO_4^-, the oxidation state of chlorine (x) can be calculated:

   $$x + 4(-2) = -1 \qquad x = +7$$

   Chromium pentaoxide, CrO_5, is a peroxide, $Cr(O_2)_2O$, with two peroxide units and one oxide unit. The oxidation state of chromium (x) is

   $$x + 4(-1) + (-2) = 0 \qquad x = +6$$

   We made this calculation knowing the structure of chromium pentaoxide derived from an x-ray diffraction study. If we knew nothing about this structure, and we assumed the oxygen to be entirely oxide

(O^{2-}), we would calculate the oxidation state of chromium to be

$$x + 5(-2) = 0 \qquad x = +10$$

(As we shall see, although this calculated oxidation state is based upon the incorrect assumption that all oxygen atoms are equivalent, it allows us to find the true oxidation capacity of this species.)

5. *In its covalent compounds with nonmetals, hydrogen is assigned oxidation state* $+1$, *and in its covalent compounds with metals, hydrogen is assigned oxidation state* -1. (In ionic hydrides, such as sodium hydride NaH, containing Na^+ and H^-, Rule 2 assigns -1 as the oxidation state of hydrogen.) In Chapter 11, this rule coupled with Rule 4 for the oxidation state of oxygen allowed us to assign an oxidation state to carbon in compounds ranging from methane, CH_4, to carbon dioxide, CO_2. These rules also allow us to assign the oxidation state of nitrogen, x, in its compounds with hydrogen and oxygen, some of which are:

Ammonia, NH_3	$x + 3(+1) = 0$	$x = -3$
Ammonium ion, NH_4^+	$x + 4(+1) = +1$	$x = -3$
Hydrazine, N_2H_4	$2x + 4(+1) = 0$	$x = -2$
Hydroxylamine, H_2NOH	$x + 3(+1) + 1(-2) = 0$	$x = -1$

A broad range of recently synthesized compounds contains metal hydrogen bonds, for example, $RhHCl_5^{3-}$. The oxidation state assigned to rhodium (x) in this species is based upon oxidation state -1 assigned both to chlorine and to hydrogen:

$$x + (-1) + 5(-1) = -3 \qquad x = +3$$

The oxidation state for an element calculated by application of these rules is an average oxidation state, which need not be an integer. For instance in Fe_3O_4, the average oxidation state of iron (x) is calculated as follows:

$$3x + 4(-2) = 0 \qquad x = +2\tfrac{2}{3}$$

This average, $+2\frac{2}{3}$, is the consequence of two thirds of the iron atoms' being iron(III) and one third being iron(II):

$$\tfrac{2}{3}(+3) + \tfrac{1}{3}(+2) = +2\tfrac{2}{3}$$

The structures of iron oxides, including Fe_3O_4, are given in Table 9–3. We can see how to calculate average oxidation states in covalently bonded species containing nonequivalent atoms of a particular element by considering thiosulfate ion ($S_2O_3^{2-}$) and tetrathionate ion ($S_4O_6^{2-}$). In $S_2O_3^{2-}$, the average oxidation state of sulfur (x) can be calculated:

$$2x + 3(-2) = -2 \qquad x = +2$$

In $S_4O_6^{2-}$, the average oxidation state of sulfur (x) can be calculated:

$$4x + 6(-2) = -2 \qquad x = +2\tfrac{1}{2}$$

These two species are related; in a common analytical procedure in which thiosulfate ion is oxidized by iodine (or triiodide ion), tetrathionate ion is the product:

$$2S_2O_3^{2-} + I_2 = S_4O_6^{2-} + 2I^-$$

The structures of thiosulfate ion and tetrathionate ion (given as electron-dot formulas) show the nonequivalence of sulfur atoms in each of these species:

```
                    2−                                 2−
       ..                         ..        ..
      :O:                        :O:       :O:
 ..    ..  ..                ..   ..   ..   ..   ..
:O  :  S : S:               :O : S  :  S :  S :  O:
 ..    ..  ..                ..   ..   ..   ..   ..
      :O:                        :O:       :O:
       ..                         ..        ..
```

Thiosulfate ion — Tetrathionate ion

There are situations in which we may not be able to assign oxidation states to all of the elements in a substance or species by using these rules. What we should do in such cases depends upon the problem at hand. If oxidation states are to be used in balancing a chemical equation for an oxidation–reduction reaction, a consistent but arbitrary assignment is satisfactory. If the oxidation state is to be used to describe the electron distribution in the substance, data in addition to chemical composition may be needed. For instance, we use the magnetic properties of the species $FeNO^{2+}(aq)$ to decide that this species, formed by reaction of nitric oxide, NO, and iron(II) ion, is a species of iron(I) and NO^+, not a species of iron(II) and NO. (Notice that NO^+ is isoelectronic with CO, a ligand that also forms complex species with many metals in low oxidation states.)

The oxidation states calculated using these rules for elements in most covalent molecules or ions assign values which are the charges each atom would have if the shared electrons were held wholly by the more electronegative element. For instance, nitrogen is more electronegative than hydrogen, $\chi_N = 3.0$, $\chi_H = 2.2$ (see Table 10–5), and this allocation of electrons in ammonia gives the charges

```
             3−
           ( .. )
    1+H   (: N :)  H1+
           ( .. )
            H1+
```

which are equal to the oxidation states already assigned (N, −3, and H, +1).

HALF-REACTIONS

In showing the reduction of oxygen, chlorine, and hydrogen ion in their reactions with magnesium, we used equations in which electrons appeared on the reactant side of the chemical equation, for example,

$$O_2(g) + 4e^- = 2O^{2-} \text{ (in MgO)}$$

A chemical reaction of this type is called a *half-reaction*. Half-reactions are used for many purposes:

1. to show the chemical changes occurring at the electrodes in both electrolytic and galvanic cells,
2. to dissect an oxidation–reduction into the two changes (oxidation and reduction), and
3. to balance chemical equations for oxidation-reduction reactions.

A single half-reaction does not occur in isolation.[1] The reaction of magnesium metal with oxygen to produce magnesium oxide can be dissected into the two half-reactions:

$$Mg(s) = Mg^{2+} \text{ (in MgO)} + 2e^-$$
$$O_2(g) + 4e^- = 2O^{2-} \text{ (in MgO)}$$

The complete reaction,

$$2Mg(s) + O_2(g) = 2Mg^{2+} \text{ (in MgO)} + 2O^{2-} \text{ (in MgO)}$$

or simply

$$2Mg(s) + O_2(g) = 2MgO(s)$$

is two times the first half-reaction plus the second half-reaction. The electrons are canceled. This complete reaction was obtained by *adding* two half-reactions, one written as an oxidation [$Mg(s) \rightarrow Mg^{2+}$] and the other written as a reduction [$O_2(g) \rightarrow 2O^{2-}$]. However, all half-reactions can be written in one particular way, as oxidations or as reductions. For most purposes, we will employ the convention of writing half-reactions as reductions. Under this convention, the half-reaction for magnesium, in our example would be written

$$Mg^{2+} \text{ (in MgO)} + 2e^- = Mg(s)$$

and the complete reaction would be obtained by *subtracting* the two half-reactions (with appropriate multiplication of the coefficients in the equations, if necessary).

BALANCING OXIDATION–REDUCTION EQUATIONS

A balanced chemical equation for an oxidation–reduction reaction is subject to the same rules as other chemical equations: atoms of each element must be conserved, and the sum of the charges on species on each side of the equation must be the same. Chemical equations for many oxidation–reduction reactions are relatively difficult to balance by inspection, and special methods have been developed for balancing such equations. Before considering these, we shall illustrate an approach that uses only the conservation rules already mentioned.

---

[1] We are not considering here the processes studied in gas-phase electron-affinity measurements, for example, $Cl(g) + e^- = Cl^-(g)$; this is not a half-reaction.

The Algebraic Method. In acidic solution manganese(VII) oxidizes iron(II) to iron(III) and is reduced to manganese(II):

$$MnO_4^- + Fe^{2+} \quad \text{gives} \quad Mn^{2+} + Fe^{3+}$$

To balance oxygen, water is required on the product side, and to balance hydrogen, hydrogen ion is required on the reactant side. With undetermined coefficients, the chemical equation is

$$MnO_4^- + aFe^{2+} + bH^+ = cMn^{2+} + dFe^{3+} + eH_2O$$

the coefficient of MnO_4^- having been chosen arbitrarily as one. (This will prove to be an appropriate choice, but it is not a necessary one. After balancing a chemical equation, fractional coefficients can be eliminated by multiplying all coefficients by an appropriate factor.) Because only relative values of the coefficients are physically meaningful in an equation intended only to convey stoichiometric relationships, one coefficient can be chosen arbitrarily. There are five unknown coefficients that we must determine. The five equations relating these five unknowns are:

$$a = d \text{ (conservation of iron atoms)}$$

$$b = 2e \text{ (conservation of hydrogen atoms)}$$

$$c = 1 \text{ (conservation of manganese atoms)}$$

$$e = 4 \text{ (conservation of oxygen atoms)}$$

$$-1 + 2a + b = 2c + 3d \text{ (conservation of charge)}$$

The solution comes from appropriate substitutions:

$$b = 2e = 8$$

$$-1 + 2a + 8 = 2 + 3a$$

which gives

$$a = 5$$

Therefore the balanced equation is

$$MnO_4^- + 5Fe^{2+} + 8H^+ = Mn^{2+} + 5Fe^{3+} + 4H_2O$$

Although this may not be the quickest method of balancing equations for oxidation–reduction reactions, it is presented first to emphasize that special methods *are not needed* to balance equations for most oxidation–reduction reactions.

The Half-Reaction Method. You should learn the half-reaction method because of the role of half-reactions in galvanic and electrolytic cells. (Equations for half-reactions also tell us the meaning of oxidation-state changes.) To illustrate the half-reaction method, we will use this same reaction:

$$Fe^{2+} + MnO_4^- \quad \text{gives} \quad Fe^{3+} + Mn^{2+}$$

One half-reaction involves the reactant and product species of manganese, and the other involves the reactant and product species of iron. For a half-reaction occurring in acidic aqueous solution, oxygen is balanced by addition of water, hydrogen is balanced by addition of hydrogen ion, and charge is balanced by addition of electrons, *in that order*. A start at the half-reaction for reduction of permanganate ion is

$$MnO_4^- = Mn^{2+}$$

First, oxygen is balanced by adding water to the right side; four molecules of water are needed:

$$MnO_4^- = Mn^{2+} + 4H_2O$$

Next, hydrogen is balanced by adding hydrogen ion to the left side; eight hydrogen ions are needed:

$$MnO_4^- + 8H^+ = Mn^{2+} + 4H_2O$$

Finally, charge is balanced by adding electrons to the left side, which needs five units of negative charge:

$$MnO_4^- + 8H^+ + 5e^- = Mn^{2+} + 4H_2O$$

For the half-reaction for oxidation of iron(II), a start is

$$Fe^{2+} = Fe^{3+}$$

Because neither oxygen nor hydrogen is contained in either of these species, only charge needs to be balanced. This is done by adding one electron to the right side:

$$Fe^{2+} = Fe^{3+} + e^-$$

In the balanced equation for the overall reaction *there are no electrons* because electrons are neither produced nor consumed, only transferred. Such an equation results if five times the iron(II)–iron(III) half-reaction is added to the manganese(VII)–manganese(II) half-reaction:

$$\begin{array}{r} MnO_4^- + 8H^+ + 5e^- = Mn^{2+} + 4H_2O \\ 5Fe^{2+} = 5Fe^{3+} + 5e^- \\ \hline MnO_4^- + 5Fe^{2+} + 8H^+ = Mn^{2+} + 5Fe^{3+} + 4H_2O \end{array}$$

This is the result we already have obtained.

Another example is the oxidation of sulfur dioxide by triodide ion to give sulfate ion and iodide ion in acidic solution:

$$SO_2(aq) + I_3^- \quad \text{gives} \quad SO_4^{2-} + I^-$$

Incomplete equations for the half-reactions are

$$I_3^- = I^- \quad \text{and} \quad SO_2 = SO_4^{2-}$$

In the first of these half-reactions iodine is balanced by setting the coefficient of iodide ion to be three:

$$I_3^- = 3I^-$$

Then charge is balanced by adding two electrons to the left side:

$$2e^- + I_3^- = 3I^-$$

To balance oxygen in the second half-reaction, water is added to the left side, then to balance hydrogen, hydrogen ion is added to the right side:

$$2H_2O + SO_2 = SO_4^{2-} + 4H^+$$

This equation needs two electrons on the right side to balance charge:

$$2H_2O + SO_2 = SO_4^{2-} + 4H^+ + 2e^-$$

The two balanced half-reactions, each of which involves two electrons, can be added directly to give the balanced equation for the overall reaction:

$$I_3^- + SO_2 + 2H_2O = 3I^- + SO_4^{2-} + 4H^+$$

Elements other than hydrogen, oxygen, and the element undergoing oxidation or reduction may participate in a half-reaction. To illustrate the balancing of a half-reaction for such a system, we will consider the oxidation of mercury(II) sulfide in the presence of chloride ion to give sulfur and tetrachloromercury(II) ion. The steps in balancing this half-reaction start with the chemical formulas of the mercury and sulfur species:

$$HgS = HgCl_4^{2-} + S$$

Chlorine is balanced with chloride ion:

$$4Cl^- + HgS = HgCl_4^{2-} + S$$

Then charge is balanced with electrons:

$$4Cl^- + HgS = HgCl_4^{2-} + S + 2e^-$$

We use the same rules for balancing half-reactions involving organic species as we have used for those involving inorganic species. Let us consider the oxidation of ethyl alcohol to acetic acid in acidic solution. We start with

$$C_2H_5OH = CH_3CO_2H$$

then balance oxygen with H_2O,

$$H_2O + C_2H_5OH = CH_3CO_2H$$

next balance hydrogen with hydrogen ion,

$$H_2O + C_2H_5OH = CH_3CO_2H + 4H^+$$

and finally balance charge with electrons,

$$H_2O + C_2H_5OH = CH_3CO_2H + 4H^+ + 4e^-$$

This is the balanced half-reaction.

For a reaction in alkaline solution, we balance hydrogen and oxygen by adding hydroxide ion and water to the appropriate sides of the equation. Because each of these species (H_2O and OH^-) involves both hydrogen and oxygen, balancing a half-reaction in alkaline solution is not as simple as balancing one in acidic solution. We can make it relatively simple, however, by using the rules we have used for reactions in acidic solution. Then we modify the equation by recognizing that

$$H^+ \quad \text{equals} \quad H_2O \quad \text{minus} \quad OH^-$$

We shall illustrate the process by considering the oxidation of iron(III) hydroxide to ferrate(VI) ion by hypochlorite ion, which is reduced to chloride ion. The incomplete half-reactions involving the *chemical formulas appropriate for strongly alkaline solution* are

$$Fe(OH)_3 = FeO_4^{2-} \quad \text{and} \quad OCl^- = Cl^-$$

The first step in each case is to balance oxygen by adding water to the side of the equation needing oxygen:

$$H_2O + Fe(OH)_3 = FeO_4^{2-} \qquad OCl^- = Cl^- + H_2O$$

The next step is to balance hydrogen by adding hydrogen ion:

$$H_2O + Fe(OH)_3 = FeO_4^{2-} + 5H^+ \qquad 2H^+ + OCl^- = Cl^- + H_2O$$

The next step is to balance charge with electrons on the side of the equation needing negative charge:

$$H_2O + Fe(OH)_3 = FeO_4^{2-} + 5H^+ + 3e^-$$
$$2e^- + 2H^+ + OCl^- = Cl^- + H_2O$$

The final modification involves replacing hydrogen ions with a like number of hydroxide ions on the opposite side of the equation and adjusting the number of water molecules:

$$5OH^- + Fe(OH)_3 = FeO_4^{2-} + 4H_2O + 3e^-$$
$$2e^- + OCl^- + H_2O = Cl^- + 2OH^-$$

The balanced equation for the overall reaction is obtained by adding two times the iron(III) hydroxide–ferrate(VI) ion half-reaction (which involves three electrons) to three times the hypochlorite ion–chloride ion half-reaction (which involves two electrons):

$$2[5OH^- + Fe(OH)_3 = FeO_4^{2-} + 4H_2O + 3e^-]$$
$$3[2e^- + OCl^- + H_2O = Cl^- + 2OH^-]$$
$$\overline{2Fe(OH)_3 + 3OCl^- + 4OH^- = 2FeO_4^{2-} + 3Cl^- + 5H_2O}$$

We can summarize the procedure for balancing a half-reaction equation as follows:

1. Write the correct formulas for the reactant and product species of the element being oxidized or reduced, with coefficients that balance the atoms of that element.
2. Balance elements other than oxygen and hydrogen that are incorporated in the species being oxidized or reduced but which themselves are not so changed.
3. Balance oxygen with H_2O.

4. Balance hydrogen with H^+.
5. Balance charge with electrons.
6. If the half-reaction occurs in alkaline solution, replace hydrogen ions with a like number of hydroxide ions on the other side of the equation and balance water appropriately.
7. Check the equation to be sure it is balanced.

You should notice that in balancing these half-reaction equations, *we needed no reckoning of the oxidation states*. But the number of electrons appearing in the balanced half-reaction equation is the change in oxidation state of the element involved. For the reactions considered, the changes are:

$MnO_4^- + 5e^- \rightarrow Mn^{2+}$; the oxidation state of manganese decreases from +7 to +2, a change of 5

$Fe^{2+} \rightarrow Fe^{3+} + e^-$; the oxidation state of iron increases from +2 to +3, a change of 1

$2e^- + I_3^- \rightarrow 3I^-$; the average oxidation state of iodine decreases from −1/3 to −1, a change of 2/3 per atom or 2 per three atoms

$SO_2 \rightarrow SO_4^{2-} + 2e^-$; the oxidation state of sulfur increases from +4 to +6, a change of 2

$2e^- + OCl^- \rightarrow Cl^-$; the oxidation state of chlorine decreases from +1 to −1, a change of 2

$Fe(OH)_3 \rightarrow FeO_4^{2-} + 3e^-$; the oxidation state of iron increases from +3 to +6, a change of 3

The Oxidation-State-Change Method. A simple method of balancing complete equations for oxidation–reduction reactions involves balancing the changes of oxidation-state of the elements being oxidized and reduced. The steps in this procedure are:

1. Write the correct chemical formulas for the reactant and product species, and calculate the oxidation states of the elements in these species.
2. Calculate the changes in oxidation state for the elements in each species, and select coefficients that make the total increase in oxidation state of the element(s) being oxidized equal to the total decrease in oxidation state of the element(s) being reduced.
3. Balance any elements other than oxygen or hydrogen that are not oxidized or reduced.
4. Balance charge by (a) adding H^+ to the side of the equation needing positive charge *if the reaction occurs in acidic solution*, (b) adding OH^- to the side of the equation needing negative charge *if the reaction occurs in alkaline solution*, or (c) adding H^+ or OH^- to the product side of the equation *if the reaction occurs in an initially neutral solution.*
5. Balance hydrogen and oxygen by adding water where needed.
6. Check the equation to be sure that it is balanced.

This procedure will be illustrated with several examples.

Zinc is oxidized by nitric acid to zinc ion, with the nitrate ion being reduced to ammonium ion:

$$Zn + NO_3^- = Zn^{2+} + NH_4^+$$

The oxidation states of the elements involved in the oxidation and reduction are Zn(0), $Zn^{2+}(+2)$, NO_3^-(N, +5), NH_4^+(N, −3). The changes of oxidation state, an increase of 2 for zinc and a decrease of 8 for nitrogen, are balanced by the coefficients 4 for Zn and 4 for Zn^{2+} and 1 for NO_3^- and 1 for NH_4^+:

$$4Zn + NO_3^- = 4Zn^{2+} + NH_4^+$$

An inspection of this equation shows that positive charge is needed on the left side. This positive charge is provided by hydrogen ions:

$$hH^+ + 4Zn + NO_3^- = 4Zn^{2+} + NH_4^+$$

An equation for the balance of charge is

$$h - 1 = 4(+2) + (+1) \qquad h = 8 + 1 + 1 = 10$$

This gives

$$10H^+ + 4Zn + NO_3^- = 4Zn^{2+} + NH_4^+$$

To balance hydrogen, water is added to the right side, a procedure that also balances the oxygen:

$$10H^+ + 4Zn + NO_3^- = 4Zn^{2+} + NH_4^+ + 3H_2O$$

Chromium dioxide (CrO_2) is not stable in acidic solution; upon dissolution in acid, this oxide, in which the oxidation state of chromium is +4, reacts to give chromium in a higher state [chromium(VI) in $HOCrO_3^-$] and a lower state [chromium(III) in Cr^{3+}]:

$$CrO_2 = Cr^{3+} + HOCrO_3^-$$

(A reaction of this type, in which a particular species is both the oxidizing agent and the reducing agent, is called *disproportionation.*) For each hydrogen chromate ion formed, the oxidation state of chromium increases by two units (+4 to +6), and for each chromium(III) ion formed, the oxidation state of chromium decreases by one unit (+4 to +3). Balance is achieved because two chromium(III) ions and one hydrogen chromate ion are produced from three molecules of chromium dioxide:

$$3CrO_2 = 2Cr^{3+} + HOCrO_3^-$$

To balance charge, we will add hydrogen ion to the left side of the equation:

$$5H^+ + 3CrO_2 = 2Cr^{3+} + HOCrO_3^-$$

To balance hydrogen, we add water to the right side, a procedure that also balances oxygen:

$$5H^+ + 3CrO_2 = 2Cr^{3+} + HOCrO_3^- + 2H_2O$$

In alkaline solution an alkene is oxidized to a carboxylate anion by permanganate ion, which is reduced to manganese dioxide; using ethylene as the alkene, we start with

$$C_2H_4 + MnO_4^- = CH_3CO_2^- + MnO_2$$

The oxidation states of the elements involved in the oxidation and reduction are C_2H_4(C, −2), $CH_3CO_2^-$ (C, average oxidation state = 0), MnO_4^- (Mn, +7), and MnO_2(Mn, +4). The oxidation state of carbon increases by 2 units per atom or 4 units per C_2H_4 molecule; the oxidation state of manganese decreases by 3 units. These oxidation state changes are balanced with the coefficients:

$$3C_2H_4 + 4MnO_4^- = 3CH_3CO_2^- + 4MnO_2$$

Charge is balanced with hydroxide ion:

$$3C_2H_4 + 4MnO_4^- = 3CH_3CO_2^- + 4MnO_2 + OH^-$$

Then hydrogen is balanced with water, a procedure that also balances oxygen:

$$3C_2H_4 + 4MnO_4^- = 3CH_3CO_2^- + 4MnO_2 + OH^- + H_2O$$

There are situations in which oxidation states assigned in the usual way are quite artificial, but if we assign these states consistently, we can use them in balancing equations for oxidation–reduction reactions. Consider the blue peroxochromium(VI) species CrO_5, already discussed; CrO_5 contains chromium(VI), two peroxide units O_2^{2-}, and an oxide unit, O^{2-}. The reduction of this species to chromium(III) and water,

$$CrO_5 = Cr^{3+} + H_2O$$

involves the following changes of oxidation states:

	Change
Cr(VI) → Cr(III) (+6 → +3)	−3
$2O_2^{2-} \rightarrow H_2O$ [4 × (−1 → −2)]	−4
Total change:	−7

This is the same change we would calculate if we did not know the structure of this compound and we assigned +10 as the oxidation state of chromium; in this case we would calculate the change

	Change
Cr(X) → Cr(III) (+10 → +3)	−7

This also is the number of electrons appearing in the balanced equation for the half-reaction:

$$7e^- + CrO_5 + 10H^+ = Cr^{3+} + 5H_2O$$

Let us consider another example, the oxidation of thiocyanate ion to sulfate ion, carbonate ion, and nitrate ion by hypochlorite ion in alkaline solution:

$$OCl^- + SCN^- = SO_4^{2-} + CO_3^{2-} + NO_3^- + Cl^-$$

There is no rational way to assign oxidation states to the constituent elements of thiocyanate ion, but we can balance this equation by assigning in an arbitrary way to carbon and to nitrogen the same oxidation states as in the products (C, +4, and N, +5) and making the oxidation state for sulfur whatever value gives the correct net charge on the species:

$$x + 4 + 5 = -1 \qquad x = -10$$

The increase in oxidation state of sulfur from −10 to +6, 16 units, is balanced by the reduction of 8 hypochlorite ions:

$$8OCl^- + SCN^- = SO_4^{2-} + CO_3^{2-} + NO_3^- + 8Cl^-$$

Charge is balanced with 4 hydroxide ions:

$$4OH^- + 8OCl^- + SCN^- = SO_4^{2-} + CO_3^{2-} + NO_3^- + 8Cl^-$$

Then hydrogen is balanced with water, a procedure that also balances oxygen:

$$4OH^- + 8OCl^- + SCN^- = SO_4^{2-} + CO_3^{2-} + NO_3^- + 8Cl + 2H_2O$$

Thus, the equation is balanced by the use of a consistent, but arbitrary, assignment of oxidation states.

16–3 Electrolysis

Electrolysis is the process in which an electrical current makes an oxidation–reduction reaction occur. The vessel in which the reaction occurs is called an *electrolytic cell*; a diagram of the essential parts of an electrolytic cell for the electrolysis of water is given in Figure 16–1. An external source of electrical potential (a battery or a generator) provides electrons to one electrode and removes them from the other. Each electrode is an interface between the external circuit, where the electrical current is a flow of electrons, and the chemical substance or mixture of substances present in the cell. For current to flow in this circuit, chemical changes (half-reactions) that convert electrons into ions or vice versa must occur at each electrode; the electrical current carriers in the cell are ions. We will consider several electrolytic processes to illustrate various aspects of electrolysis.

ELECTROLYSIS OF MOLTEN SODIUM CHLORIDE

In terms of the chemistry involved, a particularly simple type of electrolysis involves a molten salt. At temperatures above 800°C, sodium chloride is a liquid made up of positively charged sodium ions and negatively charged

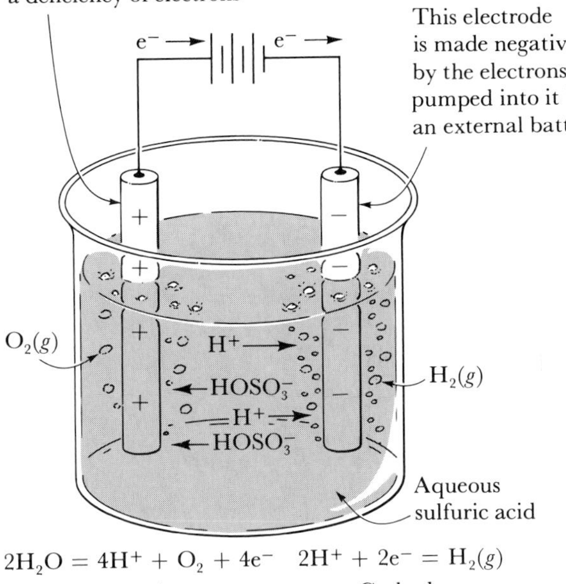

FIGURE 16–1
An electrolytic cell containing aqueous sulfuric acid in which the decomposition of water,

$$2H_2O(l) = 2H_2(g) + O_2(g)$$

is the net chemical change. (If this cell is operated at high current density, another reaction occurs; see Chapter 17.)

chloride ions.[2] If an electric field is imposed upon this liquid, as when two conducting electrodes attached by wires to the terminals of a battery or a generator are placed in the liquid, the ions tend to move under the influence of the electric field. Motion of ions, positive ions moving in one direction and negative ions moving in the other direction, is a current of electricity, but the electrical circuit is not complete without half-reactions occurring at the two electrodes. The half-reactions accomplish the chemical change that consumes and produces the current carriers, the electrons in the external circuit and the ions in the molten salt of the electrolytic cell. At one electrode metallic sodium is produced (it is a liquid at this temperature: the melting point of sodium is 97.5°C, and its boiling point is 892°C), and at the other electrode chlorine gas is produced. Figure 16–2 shows a cell for the electrolysis of molten sodium chloride.

To analyze the electrolysis of molten sodium chloride, let us first consider the electrode into which the battery or generator pumps electrons. With the accumulation of a relatively small excess of electrons in the electrode, the resulting *negative charge*

1. attracts positively charged sodium ions,
2. repels negatively charged chloride ions, and
3. causes a chemical reaction that consumes electrons and sodium ions: $e^- + Na^+ = Na$.

[2]The melting point of sodium chloride is 800.4°C.

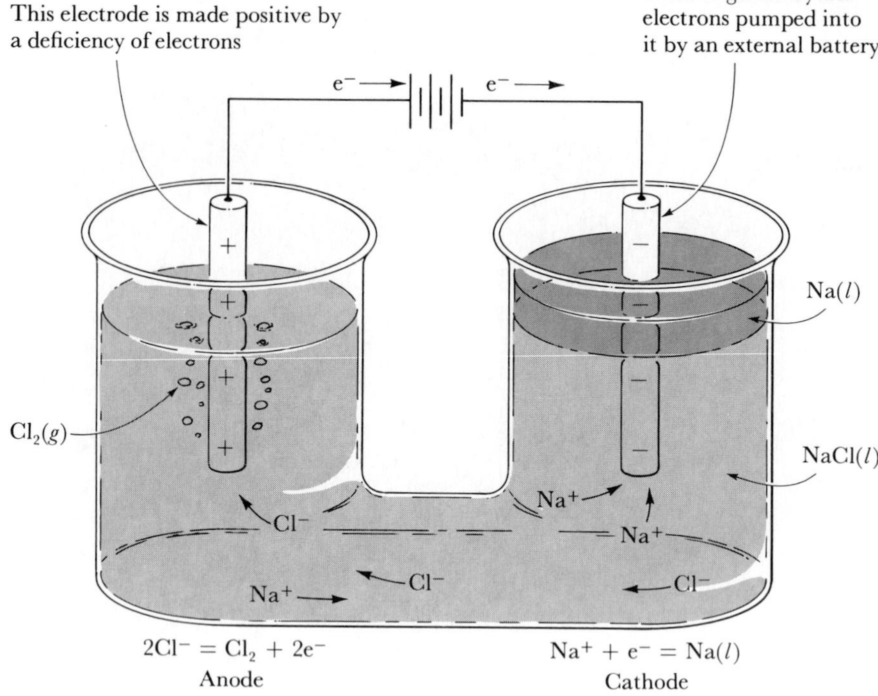

FIGURE 16–2
A cell for the electrolysis of molten sodium chloride: $Na(l, d = 1.0\ g\ cm^{-3})$ floats on $NaCl(l, d = 2.2\ g\ cm^{-3})$. Solid electrodes can be made of any unreactive electrical conductor (e.g., graphite or platinum can be used).

This electrode, which is made negative by the electrons pumped into it from the external source, is called the *cathode*, the electrode that attracts the *cations*. It is the *cathode at which reduction occurs*. At the other electrode, the electron pump is removing electrons, giving the electrode a *positive charge* which

1. attracts negatively charged chloride ions,
2. repels positively charged sodium ions, and
3. causes a chemical reaction that consumes chloride ion and produces electrons: $Cl^- = \frac{1}{2}Cl_2 + e^-$.

This electrode is called the *anode*, the electrode that attracts *anions*. It is the *anode at which oxidation occurs*.

The equation for the net reaction that occurs as electrons are pumped through this electrolytic cell is obtained by adding the equations for the two half-reactions; in the electrolysis of molten sodium chloride, the net reaction is

$$2NaCl(l) = 2Na(l) + Cl_2(g)$$

This reaction does not occur spontaneously [ΔG (823°C) $\cong$ 670 kJ mol^{-1}]; sodium chloride does not spontaneously decompose to give its constituent elements at this temperature. The reverse reaction, the combination of sodium metal and chlorine gas, does go spontaneously. For this reason the products of the two electrode reactions must be prevented from coming together. One preventive method is illustrated in Figure 16–2. In trying to prevent the products of the two electrode reactions from mixing, we *must not* close the two electrode compartments off from one another. If we isolated these two compartments from one another, the electrical circuit would not be complete, and current would not flow.

STOICHIOMETRIC RELATIONSHIPS IN ELECTROCHEMISTRY

The stoichiometric relationships implied by the coefficients for chemical species and electrons in balanced equations for half-reactions are analogous to those we have used in stoichiometric calculations for complete reactions. The use of these relationships, called *Faraday's laws*, can be illustrated by the half-reactions occurring in the electrolysis of acidified water (see Figure 16–1),

$$2H^+ + 2e^- = H_2(g)$$
$$2H_2O = O_2(g) + 4H^+ + 4e^-$$

The balanced equations indicate that the flow of one mole of electrons causes production of one half mole of hydrogen gas and one quarter mole of oxygen gas. One mole of electrons is that number of electrons equal to the number of atoms in exactly 12 g of carbon-12. It is 6.022×10^{23} electrons.[3] This number of electrons corresponds to a specific amount of electrical charge, which we can express in coulombs, the SI unit of electrical charge:

$$1 \text{ coulomb} = 1 \text{ ampere second}$$
$$1\ \text{C} = 1\ \text{A s}$$

The charge on the electron is

$$1\ e = 1.602 \times 10^{-19}\ \text{C}$$

Therefore, the factor converting moles of electrons to coulombs, called *Faraday's constant* (symbol $\mathscr{F}$), is

$$\mathscr{F} = 6.022 \times 10^{-23}\ \text{mol}^{-1} \times 1.602 \times 10^{-19}\ \text{C}$$
$$= 9.65 \times 10^{4}\ \text{C mol}^{-1} \text{ (see footnote 4)}$$

or, because $1\ \text{J} = 1\ \text{V C}$,

$$\mathscr{F} = 9.65 \times 10^{4}\ \text{J V}^{-1}\ \text{mol}^{-1}$$

(In calories, the conversion factor is $\mathscr{F} = 2.306 \times 10^{4}$ cal V^{-1} mol^{-1}.)

---

[3] The most accurate value of Avogadro's constant is $6.0220943 \times 10^{23}\ \text{mol}^{-1}$.

[4] If the best values of e and N_A are used, the calculated value of $\mathscr{F}$ is $96{,}486.7 \pm 0.5\ \text{C mol}^{-1}$.

EXAMPLE: What weight of metallic copper is produced at the cathode when a current of 6.43 A flows for 5.00 h if the half-reaction is $Cu^{2+} + 2e^- = Cu$?

The amount of electrical charge that flows during this period is

$$6.43 \text{ A} \times 5.00 \text{ h} \times \frac{3600 \text{ s}}{1 \text{ h}} \times \frac{1 \text{ C}}{1 \text{ A s}} = 1.157 \times 10^5 \text{ C}$$

This is equal to

$$1.157 \times 10^5 \text{ C} \times \frac{1}{9.65 \times 10^4 \text{ C mol}^{-1}} = 1.199 \text{ mol e}^-$$

which will produce

$$1.199 \text{ mol e}^- \times \frac{0.5 \text{ mol Cu}}{1 \text{ mol e}^-} \times \frac{63.55 \text{ g Cu}}{1 \text{ mol Cu}} = 38.11 \text{ g Cu}$$

EXAMPLE: A current of 10.40 A flows in an electrolytic cell in which aluminum metal is produced:

$$Al^{3+} + 3e^- = Al$$

What time is required to produce 1.000 pound of metal?

The amount of aluminum in 1.000 pound is

$$1.000 \text{ lb} \times \frac{453.6 \text{ g}}{1 \text{ lb}} \times \frac{1 \text{ mol}}{26.98 \text{ g}} = 16.81 \text{ mol}$$

The number of moles of electrons needed is three times this:

$$3 \times 16.81 \text{ mol} = 50.43 \text{ mol}$$

The number of coulombs of electrical charge is

$$9.65 \times 10^4 \text{ C mol}^{-1} \times 50.43 \text{ mol} = 4.87 \times 10^6 \text{ C} = 4.87 \times 10^6 \text{ A s}$$

With a current of 10.40 A, the time required for the flow of this amount of electrical charge is

$$t = \frac{4.87 \times 10^6 \text{ A s}}{10.40 \text{ A}} \times \frac{1 \text{ h}}{3600 \text{ s}} = 130.1 \text{ h}$$

ELECTROLYSIS OF AQUEOUS SODIUM CHLORIDE

In the example that we have discussed (the electrolysis of molten sodium chloride), only *one* possible reduction reaction can occur at the cathode: this is the reduction of sodium ion. In addition, only *one* possible oxidation reaction can occur at the anode: this is the oxidation of chloride ion. (We assumed that the electrode material was inert; later we shall consider examples in which the electrode takes part in the reaction.) The situation becomes more complicated when more than one reducible substance and/or more than one oxidizable substance are present in the electrolytic cell.

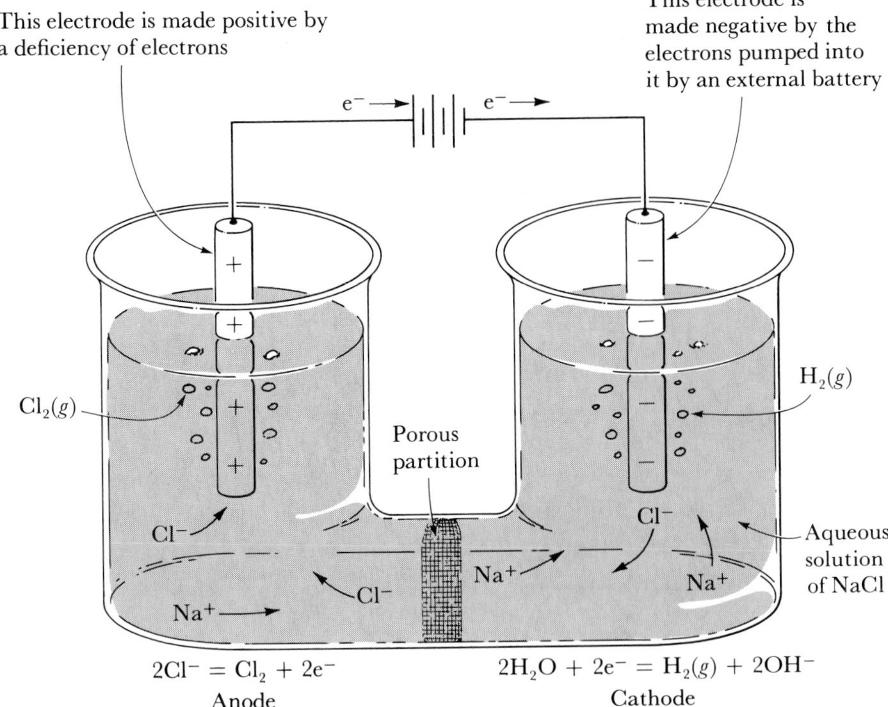

FIGURE 16–3
A cell for the electrolysis of an aqueous solution of sodium chloride.

At anode: $2Cl^- = Cl_2(g) + 2e^-$ occurs instead of
$2H_2O = O_2(g) + 4H^+ + 4e^-$

At cathode: $2H_2O + 2e^- = H_2 + 2OH^-$ occurs instead of
$Na^+ + e^- = Na$

Under such conditions, one asks which reducible substances will be reduced at the cathode and which oxidizable substances will be oxidized at the anode.

To illustrate such a situation, consider the electrolysis of an aqueous solution of sodium chloride. Figure 16–3 shows a cell for this process. The solution contains

At relatively large concentrations: H_2O, Na^+, Cl^-

At extremely low concentrations: H^+, OH^-

Although hydrogen ion and hydroxide ion may play roles in the mechanisms of the oxidation and reduction reactions of the solvent, water, the equations for the net reactions occurring at the electrodes should be written with water, a major species, as the reactant. The possible reduction half-reactions are

$$e^- + Na^+ = Na \qquad 2e^- + 2H_2O = H_2 + 2OH^-$$

In practice, the second reaction occurs at the cathode in the electrolysis of an aqueous solution of sodium chloride. We might expect this, knowing that sodium is more active chemically than hydrogen; witness:

1. the relative values of the electrode potentials for the two half-reactions (a topic we will discuss later in this chapter), and
2. the known rapid reaction of sodium and water (see Chapter 18).

If the possible half-reactions have similar tendencies to occur as measured by values of the electrode potentials, their kinetic properties may determine which occurs. This is illustrated by the situation at the anode. The possible oxidation half-reactions are

$$2Cl^- = Cl_2 + 2e^- \qquad 2H_2O = 4H^+ + O_2 + 4e^-$$

On the basis of the electrode potential values we would expect oxygen to be evolved at the anode. This does not happen; in practice, the rate at which oxygen is evolved at the anode is very low unless a high voltage is used. The excess voltage over that required on the basis of the electrode potentials is called the *overvoltage*. The overvoltage for the evolution of oxygen is much larger than the overvoltage for the evolution of chlorine, and in practice, with concentrated salt solution the product of oxidation at the anode is largely chlorine. The overall reaction occurring in the cell, the sum of the two half-reactions, is

$$2Cl^- + 2H_2O = Cl_2 + H_2 + 2OH^-$$

This reaction *does not occur spontaneously*: the reaction is brought about by the passage of the electric current provided by an external source. As in the electrolysis of molten sodium chloride, the products of the two electrode reactions must be kept apart because they tend to react. However, the electrolysis of an aqueous solution of sodium chloride presents an additional problem. Not only is the anode product, chlorine, able to react with one of the cathode products, hydrogen, but also chlorine reacts with hydroxide ion, the other cathode product, to give chloride ion and hypochlorite ion:

$$Cl_2 + 2OH^- = Cl^- + OCl^- + H_2O$$

Thus if solutions from the anode and cathode regions of the cell are mixed, the final oxidation product is hypochlorite ion rather than chlorine gas. The result of electrolysis under such conditions is a solution containing sodium hypochlorite and sodium chloride, which is widely used as an inexpensive disinfectant and bleaching agent.

ELECTROLYTIC PRODUCTION OF ALUMINUM

An electrolytic method is used to produce aluminum metal commercially. Like sodium ion in the previous example, aluminum ion is less easily reduced than water, and electrolysis of an aqueous aluminum salt solution does not give the metal. In 1886, HALL in the United States and HEROULT in France discovered that aluminum metal could be produced by electrolysis of a solution of aluminum oxide dissolved in molten cryolite [Na_3AlF_6, a salt containing sodium ion and the complex ion hexafluoroaluminate(III) ion AlF_6^{3-}, a species isoelectronic with sulfur hexafluoride]. At ~1300 K, where cryolite is a liquid, the reaction at the cathode (the carbon lining of an iron vessel) is

$$Al^{3+} + 3e^- = Al(l)$$

The reaction at the anode (carbon rods immersed in the molten salt) is

$$2O^{2-} = O_2(g) + 4e^-$$

but at the temperature of the operation, oxygen oxidizes the carbon electrode, which therefore must be replenished. Because neither sodium ion nor fluoride ion [which is coordinated to aluminum(III) ion] is consumed in the electrode reactions, the molten cryolite solvent need not be replenished. Aluminum(III) oxide is simply added to the liquid as needed. However, this aluminum oxide must be very pure. To purify aluminum oxide from the impurities with which it is found (oxides of iron, titanium, and silicon), advantage is taken of its amphoteric nature. Bauxite (an aluminum oxide mineral) is treated with concentrated sodium hydroxide producing aluminate ion:[5]

$$Al_2O_3(s) + 2OH^- = 2AlO_2^- + H_2O$$

This reaction occurs under conditions that do not allow iron(III) oxide, titanium(IV) oxide, or silicon(IV) oxide to dissolve appreciably. The solution containing sodium ion, hydroxide ion, and aluminate ion is separated from the sludge containing these insoluble oxides. Lowering the concentration of hydroxide ion by adding carbon dioxide,

$$CO_2(g) + OH^- = HOCO_2^-$$

reprecipitates hydrated aluminum(III) oxide,

$$2CO_2(g) + 2AlO_2^- + (n+1)H_2O = 2HOCO_2^- + Al_2O_3 \cdot nH_2O$$

This oxide is dehydrated by heating prior to being added to the molten electrolytic bath.

Large amounts of electrical energy are used in producing aluminum, so recycling of aluminum is now being fostered to conserve energy.

16–4 Galvanic Cells

Many combinations of oxidizing agents and reducing agents react when they are brought together. The reaction may be useful in analysis, or the product(s) formed in the reaction may have a use. If either is the case, the energy change accompanying the reaction is not a primary concern. However, a spontaneous oxidation–reduction reaction can be the basis

[5]Aluminate ion is $Al(OH_2)_2(OH)_4^-$; AlO_2^- is the anhydrous representation.

for a galvanic cell (a battery) that produces electrical energy. These electrochemical cells can be contrasted with electrolytic cells, in which nonspontaneous reactions occur by means of electrical current produced by an external source.

CONSTRUCTION OF GALVANIC CELLS; ION MIGRATION

To introduce you to galvanic cells, we will consider the reaction of iron(III) ion with copper in acidic solution to produce copper(II) ion and iron(II) ion:

$$2Fe^{3+} + Cu(s) = 2Fe^{2+} + Cu^{2+}$$

The equilibrium constant for this reaction is very large; at 25°C,

$$K = \frac{[Cu^{2+}][Fe^{2+}]^2}{[Fe^{3+}]^2} = 4.7 \times 10^{14} \text{ mol L}^{-1}$$

and the reaction goes essentially to completion when the reactants are brought together. We can show this by calculating the concentrations at equilibrium in a solution prepared by adding excess copper to an acidic solution of iron(III) perchlorate, with $C_{Fe(ClO_4)_3} = 0.100 \text{ mol L}^{-1}$.[6] Before reaction the solution has the concentrations

$$[Fe^{3+}]_0 = 0.100 \text{ mol L}^{-1} \qquad [Fe^{2+}]_0 = 0 \qquad [Cu^{2+}]_0 = 0$$

At equilibrium the concentrations of all species can be expressed in terms of the equilibrium concentration of iron(III):

$$[Fe^{2+}] = 0.100 \text{ mol L}^{-1} - [Fe^{3+}]$$

$$[Cu^{2+}] = \tfrac{1}{2}[Fe^{2+}] = 0.050 \text{ mol L}^{-1} - 0.5[Fe^{3+}]$$

Substituting these concentrations into the equilibrium-constant equation and rearranging it, we get

$$[Fe^{3+}]^2 = \frac{1}{K} \times [Fe^{2+}]^2[Cu^{2+}]$$

$$= \frac{(0.100 \text{ mol L}^{-1} - [Fe^{3+}])^2(0.050 \text{ mol L}^{-1} - 0.5[Fe^{3+}])}{4.7 \times 10^{14} \text{ mol L}^{-1}}$$

If $[Fe^{3+}] \ll 0.050 \text{ mol L}^{-1}$, this equation can be simplified to

$$[Fe^{3+}]^2 = \frac{(0.100 \text{ mol L}^{-1})^2(0.050 \text{ mol L}^{-1})}{4.7 \times 10^{14} \text{ mol L}^{-1}}$$

$$[Fe^{3+}] = 1.03 \times 10^{-9} \text{ mol L}^{-1}$$

Therefore our assumption that $[Fe^{3+}] \ll 0.050 \text{ mol L}^{-1}$ is valid, and we see that only $10^{-6}\%$ of the iron(III) remains in solution at equilibrium.

---

[6]The presence of a low concentration of an acid (e.g., 0.010*M* $HClO_4$) suppresses the acid dissociation of hydrated iron(III) ion.

To make this spontaneous reaction occur without the iron(III) coming in contact with the metallic copper, we will construct a galvanic cell made up of two half-cells. The first step is to divide the reaction into the two half-reactions. We will write each half-reaction as a reduction:

1. $Fe^{3+} + e^- = Fe^{2+}$
2. $Cu^{2+} + 2e^- = Cu$

The complete reaction is obtained by subtracting Half-reaction 2 from twice Half-reaction 1. Each half-cell should contain all of the species present as reactants and products in the equation for the half-reaction in question. In addition, each half-cell must have a conducting electrode to allow flow of electrons to or from the site of the half-reaction. In the copper–copper(II) half-cell, the copper metal acts as the conductor; in the iron(II)–iron(III) half-cell, an inert metal (e.g., platinum) dipping into the solution fills this role. If each half-cell is constructed in a separate beaker and the two electrodes are connected to each other with an ammeter in the circuit, as shown in Figure 16–4a, *current does not flow.*

For current to flow, the circuit must be complete. It can be completed with a salt bridge between the two half-cells (Figure 16–4b). To understand the need for a complete circuit, you should consider one half-cell in isolation, for instance that involving the half-reaction

$$Cu^{2+} + 2e^- = Cu$$

If this half-reaction occurred to the right, the solution surrounding the copper electrode would acquire a negative charge from the surplus anions present, and the electrode would become positively charged from the deficiency of electrons. If the half-reaction occurred to the left, the solution surrounding the electrode would acquire a positive charge from the surplus copper(II) ions present, and the electrode would become negatively charged from the surplus electrons. Each of these hypothetical changes would develop a state of higher potential energy due to the separation of unlike charges. This half-reaction (or any other half-reaction) cannot occur to a measurable extent *in isolation*. However, with the external circuit closed and the salt bridge in place, current can flow. Electrons will flow through the external circuit *from* the electrode at which there is the greater tendency for oxidation to occur *to* the electrode at which there is the greater tendency for reduction to occur.

In the present example, electrons flow through the external circuit from the copper–copper(II) half-cell to the iron(II)–iron(III) half-cell. At the copper–copper(II) electrode, the half-reaction that gives electrons to the external circuit occurs:

$$Cu = Cu^{2+} + 2e^-$$

The copper(II) ions go into solution, where they

1. migrate from the electrode region, and
2. attract anions (ClO_4^-) into the electrode region.

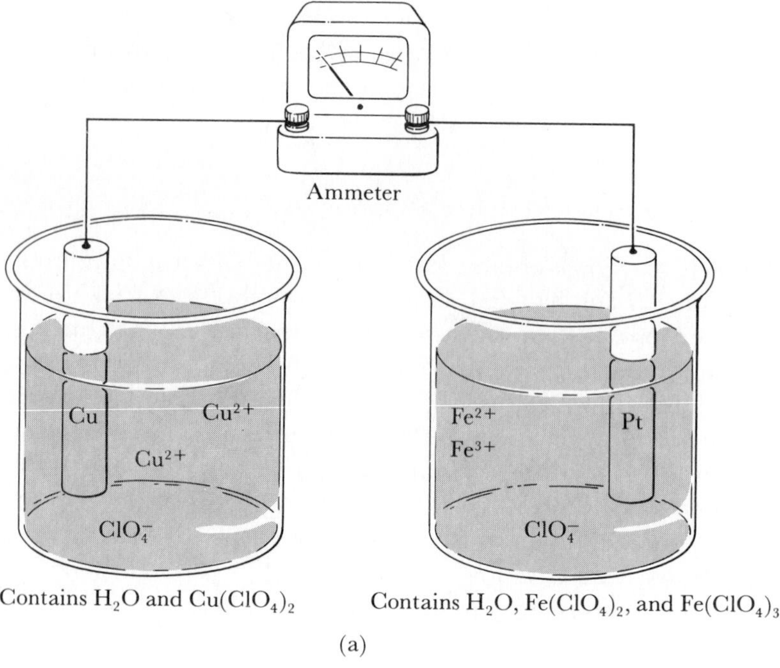

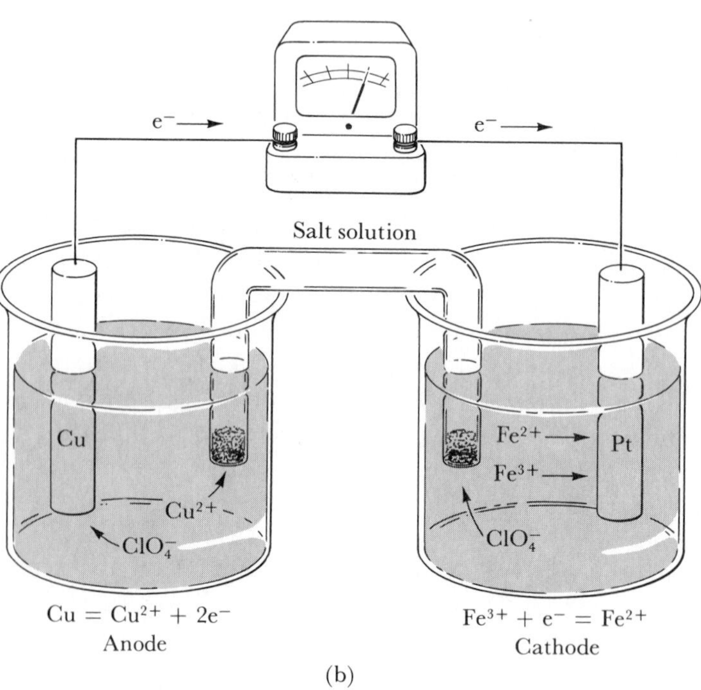

FIGURE 16–4

A galvanic cell in which the reaction $2Fe^{3+} + Cu = 2Fe^{2+} + Cu^{2+}$ occurs. (a) The separate half-cells; circuit not complete. (b) The complete circuit, including a salt bridge. The directions of electron and net ion migration are shown. An electrical current results because the two half-reactions have unequal tendencies to occur.

At the inert metal electrode in the solution containing iron(II) and iron(III), the following half-reaction occurs:

$$Fe^{3+} + e^- = Fe^{2+}$$

taking electrons from the external circuit. Because the electrode reaction converts an ion of charge +3 into an ion of charge +2, cations migrate into the electrode region and anions (ClO_4^-) migrate from the electrode region. We have come to the same conclusions about the direction of ion migration by considering the half-reaction at each electrode: positive ions migrate toward the inert electrode in the iron(III)–iron(II) half-cell, and negative ions migrate toward the copper electrode. Figure 16–4 shows this:

Cations move to the right; that is, they migrate toward the electrode at which *reduction* occurs; this electrode is called the *cathode*.
Anions move to the left; that is, they migrate toward the electrode at which *oxidation* occurs; this electrode is called the *anode*.

All ions present in both half-cells *and* in the salt bridge migrate as indicated. (The transfer of electrolyte due to electrode reactions is, of course, accompanied by normal diffusion. We will not consider this diffusion, but it should be obvious that over a long enough time period, iron(III) will diffuse through the salt bridge to the copper half-cell and there react directly with the copper.)

Notice that the terms "cathode" and "anode" (introduced in our discussion of electrolysis) also can be used for the electrodes in galvanic cells. For both types of cells, these terms imply the same direction of ion migration and the same nature of the electrode reaction:

At the anode, *oxidation* occurs; anions are attracted to this electrode, and cations are repelled from this electrode.
At the cathode, *reduction* occurs; cations are attracted to this electrode, and anions are repelled from this electrode.

If a galvanic cell is to be used as a practical battery, the electrodes are labeled to show the direction of electron flow. A negative label is placed on the electrode from which electrons flow into the external circuit as the battery delivers electrical current; this is the copper electrode in the galvanic cell being discussed (Figure 16–4). A positive label is placed on the electrode into which the electrons flow from the external circuit, the platinum electrode in this cell. (Notice that the labels on the electrodes *do not* predict the direction of ion migration in a galvanic cell: these labels simply tell the user of the battery the direction of electron flow in the external circuit.)

EMF OF GALVANIC CELLS; STANDARD EMF VALUES FOR HALF-REACTIONS

The simple galvanic cell pictured in Figure 16–4 can be made to do useful work: for instance, the flow of electrons in the external circuit can drive a small electrical motor. The amount of electrical work a reaction in a battery can do depends upon the voltage of the battery and the amount of current that flows. The flow of 1.00 mol of electrons through a potential difference of 0.500 V corresponds to an energy of

$$\begin{aligned}\text{Energy} &= 0.500\ \text{V} \times 1.00\ \text{mol} \times 9.65 \times 10^4\ \text{C mol}^{-1}\\ &= 4.82 \times 10^4\ \text{V C} = 4.82 \times 10^4\ \text{J}\end{aligned}$$

The emf of the battery depends upon the chemical reaction that occurs upon discharge, the concentration conditions in the battery, and upon the current that flows. Each of these factors is important. The current determines the potential drop due to the internal resistance of the battery. Such losses are unavoidable if a galvanic cell is to be used as a practical battery. However, if we wish to use a galvanic cell

1. as an emf standard,
2. as a means of characterizing the thermodynamic properties of an oxidation–reduction reaction, or
3. as an instrument in certain types of chemical analysis,

we must reduce the current to zero. We can do this by opposing the emf of the battery with a variable emf that can be adjusted to allow infinitesimal flow of current in either direction. If you recall from Chapter 4 the definition of *reversibility* in the expansion of a gas, you will realize that the emf of a galvanic cell measured by balancing it with an opposing emf to give zero current is the voltage corresponding to reversible occurrence of the reaction. Under these idealized circumstances with no flow of electrical current to cause a voltage drop within the cell, the voltage of the cell is a maximum. This voltage, $\mathscr{E}$, is related to the value of ΔG for the spontaneous reaction that would occur if current were allowed to flow; at constant temperature and pressure,

$$-\Delta G = n\mathscr{F}\mathscr{E}$$

in which n is the number of moles of electrons flowing per mole of chemical change, as defined by the chemical equation associated with ΔG, and $\mathscr{F}$ is the faraday. The values of ΔG and $\mathscr{E}$ depend upon the concentration conditions in the cell. We will focus attention first on the *standard conditions*,

$$-\Delta G^0 = n\mathscr{F}\mathscr{E}^0$$

This equation allows $\mathscr{E}^0$ to be related to the equilibrium constant for the reaction that occurs in the cell. Because

$$-\Delta G^0 = RT \ln K^0$$

the relationship between K^0 and $\mathscr{E}^0$ is

$$\mathscr{E}^0 = \frac{RT}{n\mathscr{F}} \ln K^0$$

For 298.2 K, this equation becomes

$$\mathscr{E}^0 = \frac{8.315 \text{ J K}^{-1} \text{ mol}^{-1} \times 298.2 \text{ K}}{n \times 9.65 \times 10^4 \text{ J V}^{-1} \text{ mol}^{-1}} \times \ln K^0$$

$$= \frac{0.0257 \text{ V}}{n} \ln K^0$$

This equation generally is written with decadic logarithms; for 298.2 K

$$\mathscr{E}^0 = \frac{0.0592 \text{ V}}{n} \log K^0$$

or the equation can be rearranged to

$$K^0 = 10^{n\mathscr{E}^0/0.0592 \text{ V}}$$

which shows that at 298.2 K one power of ten in an equilibrium constant corresponds to a value of 0.0592 V/n in $\mathscr{E}^0$:

$\mathscr{E}^0$	K^0	$\mathscr{E}^0$	K^0
$\frac{0.0592 \text{ V}}{n}$	$10^{1.00}$	$\frac{0.1776 \text{ V}}{n}$	$10^{3.00}$
$\frac{0.1184 \text{ V}}{n}$	$10^{2.00}$	$\frac{0.5920 \text{ V}}{n}$	10^{10}

In this equation, K^0 is the dimensionless equilibrium constant that has incorporated into it the activity-coefficient factor. (In most of our calculations we have ignored solution nonideality, and have focused attention on equilibrium constants which have dimensions.)

The standard states conventionally used to define $\mathscr{E}^0$ values are:

1. each solute species is present at a concentration of one molal (which for the aqueous solutions being considered here will be taken to be the same as one molar), and
2. the effects of nonideality are appropriately taken into account.

For the reaction being considered at 298.2 K,

$$2Fe^{3+} + Cu = 2Fe^{2+} + Cu^{2+}$$

with $n = 2$ and $K = 4.7 \times 10^{14}$ mol L^{-1} ($K^0 = 4.7 \times 10^{14}$), the values of $\mathscr{E}^0$ and ΔG^0 are:

$$\mathscr{E}^0 = \frac{0.0592 \text{ V}}{2} \log (4.7 \times 10^{14}) = 0.434 \text{ V}$$

$$\Delta G^0 = -8.315 \text{ J K}^{-1} \text{ mol}^{-1} \times 298.2 \text{ K} \times \ln (4.7 \times 10^{14})$$
$$= -83.8 \text{ kJ mol}^{-1}$$

A positive value of $\mathscr{E}^0$ and a negative value of ΔG^0 indicate a reaction that occurs spontaneously under standard conditions.

This value of $\mathscr{E}^0$ for the complete reaction can be measured experimentally, but the value of $\mathscr{E}^0$ for each of the constituent half-reactions cannot be measured. However, this does not prevent formulation of a useful scale of $\mathscr{E}^0$ values for individual half-reactions. As we will see, arbitrary assignment of a value of $\mathscr{E}^0$ to one particular half-reaction fixes values of $\mathscr{E}^0$ for all half-reactions.

In our development of this scale of $\mathscr{E}^0$ values, we will continue to consider the reaction

1. $2Fe^{3+} + Cu = 2Fe^{2+} + Cu^{2+}$

We can think of this reaction as a combination of two other reactions, each involving hydrogen gas being oxidized to hydrogen ion; each of these reactions can be made the basis for a galvanic cell, and the values of $\mathscr{E}^0$ can be determined. The reactions and experimentally determined values of $\mathscr{E}^0$ are:

2. $2Fe^{3+} + H_2(g) = 2Fe^{2+} + 2H^+$ $\qquad \mathscr{E}^0_2 = 0.771$ V
3. $Cu^{2+} + H_2(g) = Cu(s) + 2H^+$ $\qquad \mathscr{E}^0_3 = 0.337$ V

Reaction 2 minus Reaction 3 is Reaction 1. We can consider these three reactions also as combinations of three half-reactions:

a. $Fe^{3+} + e^- = Fe^{2+}$
b. $Cu^{2+} + 2e^- = Cu(s)$
c. $2H^+ + 2e^- = H_2(g)$

The combinations are:

Reaction 1 = 2 × Half-reaction *a* minus Half-reaction *b*
Reaction 2 = 2 × Half-reaction *a* minus Half-reaction *c*
Reaction 3 = Half-reaction *b* minus Half-reaction *c*

If, arbitrarily, we assign $\mathscr{E}^0$ = exactly 0 V to the hydrogen ion–hydrogen gas half-reaction, the values of $\mathscr{E}^0$ for the other two half-reactions follow directly from the experimentally determined values of $\mathscr{E}^0$ for Reactions 2 and 3. As we have discussed in Chapter 12, the values of ΔG^0 can be combined in Hess-law-type calculations. Because

$$\text{Reaction 2} = 2 \times \text{Half-reaction } a \text{ minus Half-reaction } c$$
$$\Delta G^0_2 = 2 \times \Delta G^0_a - \Delta G^0_c$$
$$-n_2\mathscr{F}\mathscr{E}^0_2 = -2 \times (n_a\mathscr{F}\mathscr{E}^0_a) + n_c\mathscr{F}\mathscr{E}^0_c$$

The values of n for Reaction 2, Half-reaction a, and Half-reaction c are 2, 1, and 2, respectively. This equation becomes

$$-2\mathscr{F}\mathscr{E}^0_2 = -2 \times (1\mathscr{F}\mathscr{E}^0_a) + 2\mathscr{F}\mathscr{E}^0_c;$$

with $\mathscr{E}^0_c \equiv 0$, we have

$$\mathscr{E}^0_a = \frac{2\mathscr{F}\mathscr{E}^0_2}{2\mathscr{F}} = \mathscr{E}^0_2 = +0.771 \text{ V}$$

Similarly,

Reaction 3 = Half-reaction *b* minus Half-reaction *c*

gives the result $\mathscr{E}_b^0 = \mathscr{E}_3^0 = +0.337$ V.

A summary of values of $\mathscr{E}^0$ and ΔG^0 for these reactions and half-reactions is given in Table 16–1. We can derive all the entires in this table from the voltages of the galvanic cells in which the chemical changes upon discharge are Reactions 2 and 3. The two entries for the iron(III)–iron(II) half-reaction show that the value of ΔG^0 changes if a reaction or half-reaction is multiplied by a factor, but the value of $\mathscr{E}^0$ does not. This further illustrates the difference between an extensive property and an intensive property, defined in Chapter 1. The standard change of Gibbs free energy, ΔG^0, is an extensive property of the reaction. But the standard change of Gibbs free energy per coulomb is an intensive property of the reaction; this is $-\mathscr{E}^0$, $-\mathscr{E}^0 = \Delta G^0/n\mathscr{F}$.

The hydrogen electrode assembly shown in Figure 16–5 plays a central role in defining values of $\mathscr{E}^0$ for half-reactions. This assembly allows hydrogen gas and a solution of acid saturated with hydrogen gas to be in contact with an inert electrode, generally platinum. With the convention that the $\mathscr{E}^0$ value for the hydrogen electrode is exactly zero, the half-reactions

$$Fe^{3+} + e^- = Fe^{2+} \qquad \mathscr{E}^0 = +0.771 \text{ V}$$

$$Cu^{2+} + 2e^- = Cu \qquad \mathscr{E}^0 = +0.337 \text{ V}$$

are *equivalent* to

$$2Fe^{3+} + H_2(g) = 2Fe^{2+} + 2H^+ \qquad \mathscr{E}^0 = +0.771 \text{ V}$$

and

$$Cu^{2+} + H_2(g) = Cu + 2H^+ \qquad \mathscr{E}^0 = +0.337 \text{ V}$$

TABLE 16–1
The Thermodynamic Quantities, ΔG^0 and $\mathscr{E}^0$, for Reactions and Half-Reactions at 25.0°C

Reaction	n	ΔG^0 / kJ mol^{-1}	$\mathscr{E}^0$ / V
1. $2Fe^{3+} + Cu(s) = 2Fe^{2+} + Cu^{2+}$	2	−83.8	+0.434
2. $2Fe^{3+} + H_2(g) = 2Fe^{2+} + 2H^+$	2	−148.8	+0.771
3. $Cu^{2+} + H_2(g) = Cu + 2H^+$	2	−65.0	+0.337
Half-Reaction			
a. $Fe^{3+} + e^- = Fe^{2+}$	1	−74.4	+0.771
a. $2Fe^{3+} + 2e^- = 2Fe^{2+}$	2	−148.8	+0.771
b. $Cu^{2+} + 2e^- = Cu(s)$	2	−65.0	+0.337
c. $2H^+ + 2e^- = H_2(g)$	2	exactly 0	exactly 0

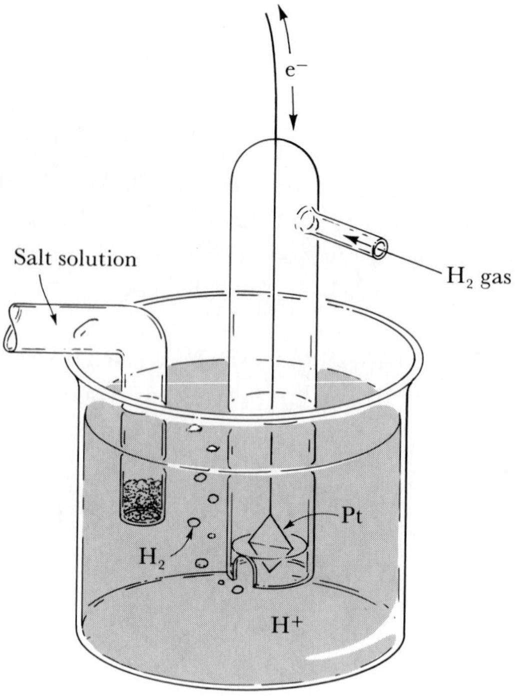

FIGURE 16–5
The hydrogen electrode:

$$\mathscr{E} = 0.000\text{ V} - \frac{0.0592\text{ V}}{2}\log\left(\frac{P_{H_2}}{[H^+]^2} \times \text{mol}^2\text{ L}^{-2}\text{ atm}^{-1}\right)$$

Therefore in this sense the electron shown in conventional half-reactions associated with values of $\mathscr{E}^0$ stands for

$$e^- \equiv \tfrac{1}{2}H_2\,(g,\ 1\text{ atm})\ \textit{minus}\ H^+\ (1\text{ molar, ideal solution})$$

and it is not to be confused with the solvated electron that can exist in certain solvents (to be discussed in Chapter 18).

Positive values of $\mathscr{E}^0$ for the reaction of iron(III) with hydrogen gas and the reaction of copper(II) with hydrogen gas indicate that both iron(III) and copper(II) are better oxidizing agents than hydrogen ion. Because the value of $\mathscr{E}^0$ for the iron(III)–iron(II) half-reaction is more positive than that for the copper(II)–copper(0) half-reaction, iron(III) ion is a better oxidizing agent than is copper(II) ion. For half-reactions written in this way, with the oxidized form on the left side of the equation, the greater the oxidizing power of the oxidized form of the couple the more positive is the value of $\mathscr{E}^0$.[7]

Table 16–2 presents $\mathscr{E}^0$ values for a number of half-reactions; some of these values are presented graphically in Figure 16–6. In Figure 16–6

[7]When using a table of $\mathscr{E}^0$ values, be certain that you understand the conventions adopted for the values presented. In some places, half-reactions are written in a way opposite to the way given here. Also, in books on biochemistry, the values given may pertain to solutions with pH = 7.00, an acidity closer to those encountered in biological and physiological studies. These conventions will be discussed later in this chapter.

TABLE 16–2
Standard Potentials for Half-Reactions in Acidic Aqueous Solution at 25.0°C

Half-Reaction	$\mathscr{E}^0$/V
$F_2(g) + 2e^- = 2F^-$	+2.87
$H_2O_2 + 2H^+ + 2e^- = 2H_2O$	+1.77
$Ce^{4+} + e^- = Ce^{3+}$	+1.70
$Cl_2(aq) + 2e^- = 2Cl^-$	+1.358
$O_2(g) + 4H^+ + 4e^- = 2H_2O$	+1.229
$ClO_4^- + 2H^+ + 2e^- = ClO_3^- + H_2O$	+1.19
$2IO_3^- + 12H^+ + 10e^- = I_2 + 6H_2O$	+1.19
$Br_2 + 2e^- = 2Br^-$	+1.08
$VO_2^+ + 2H^+ + e^- = VO^{2+} + H_2O$	+1.00
$Pu^{4+} + e^- = Pu^{3+}$	+0.982
$Ag^+ + e^- = Ag$	+0.7994
$Fe^{3+} + e^- = Fe^{2+}$	+0.771
$O_2 + 2H^+ + 2e^- = H_2O_2$	+0.69
$I_2(aq) + 2e^- = 2I^-$	+0.621
$Cu^{2+} + 2e^- = Cu$	+0.337
$CH_3CHO + 2H^+ + 2e^- = CH_3CH_2OH$	+0.19
$SO_4^{2-} + 4H^+ + 2e^- = SO_2 + 2H_2O$	+0.17
$S_4O_6^{2-} + 2e^- = 2S_2O_3^{2-}$	+0.169
$2H^+ + 2e^- = H_2(g)$	0.000
$CH_3CO_2H + 2H^+ + 2e^- = CH_3CHO + H_2O$	−0.12
$CO_2(g) + 2H^+ + 2e^- = CO(g) + H_2O$	−0.12
$CO_2(g) + 2H^+ + 2e^- = HCO_2H$	−0.20
$Fe^{2+} + 2e^- = Fe$	−0.44
$2CO_2(g) + 2H^+ + 2e^- = H_2C_2O_4$	−0.49
$Cr^{3+} + e^- = Cr^{2+}$	−0.50
$Zn^{2+} + 2e^- = Zn$	−0.763
$Mn^{2+} + 2e^- = Mn$	−1.18
$V^{2+} + 2e^- = V$	−1.2
$Al^{3+} + 3e^- = Al$	−1.66
$Mg^{2+} + 2e^- = Mg$	−2.37
$Na^+ + e^- = Na$	−2.713

the vertical distance between the levels corresponding to two half-reactions gives the value of $\mathscr{E}^0$ for a galvanic cell involving those two half-reactions.

Now we will prove that the value of $\mathscr{E}^0$ for a complete reaction is obtained by subtraction of $\mathscr{E}^0$ values for the component half-reactions. The value of ΔG^0 for an overall reaction is the difference of the values of ΔG^0 for the half-reactions that are subtracted to give the overall reaction.

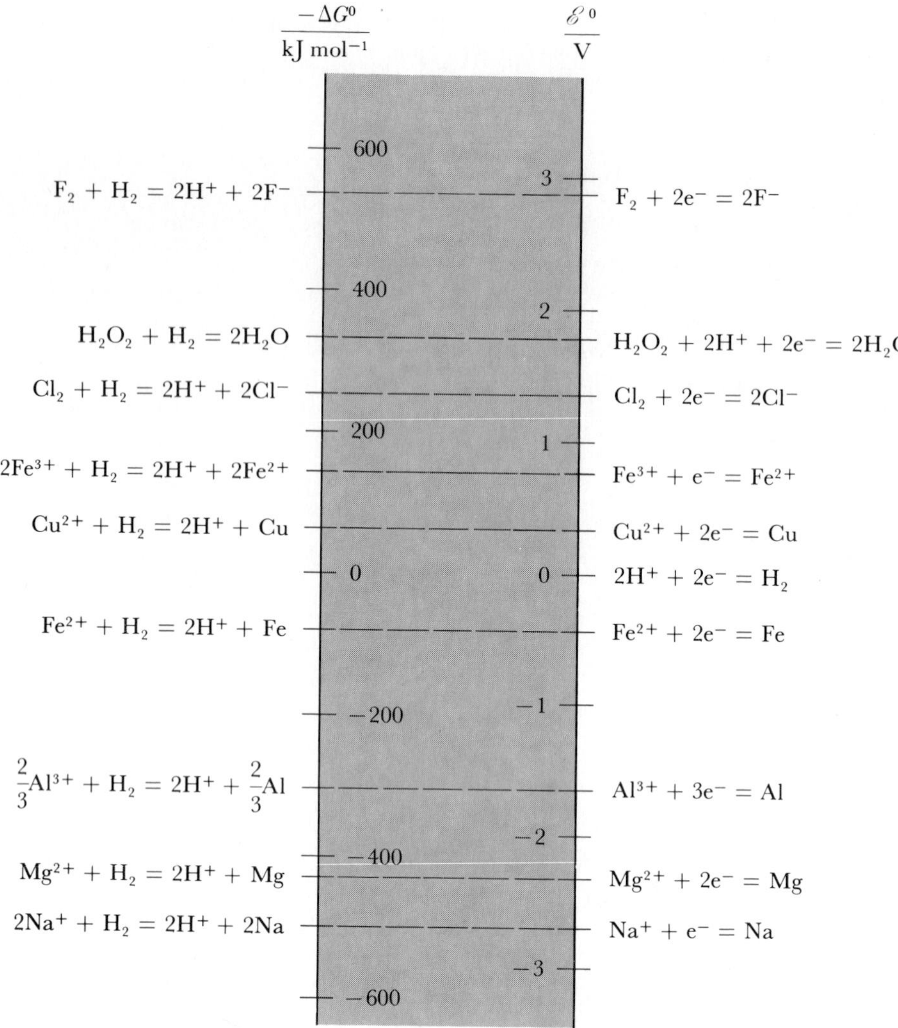

FIGURE 16–6
Free energy changes in oxidation–reduction reactions in acidic solution at 298.2 K:

$$\Delta G^0 \text{ for reaction } (\tfrac{2}{n})\text{Ox} + H_2 = (\tfrac{2}{n})\text{Red} + 2H^+$$
$$\mathscr{E}^0 \text{ for half-reaction Ox} + ne^- = \text{Red}$$

For each complete reaction as written, $n = 2$; $-\Delta G^0 = 2 \times \mathscr{F}\mathscr{E}^0 = 193 \text{ kJ mol}^{-1} \times \mathscr{E}^0$

For the value of ΔG^0, we can substitute the value of $(-n\mathscr{F}\mathscr{E}^0)$. Multiplication of the coefficients in a half-reaction by a factor does not change the value of $\mathscr{E}^0$, but it does change the value of $n\mathscr{F}\mathscr{E}^0$:

	$\mathscr{E}^0$	$-n\mathscr{F}\mathscr{E}^0$
$Fe^{3+} + e^- = Fe^{2+}$	0.771 V	$-1\mathscr{F} \times 0.771$ V
$2Fe^{3+} + 2e^- = 2Fe^{2+}$	0.771 V	$-2\mathscr{F} \times 0.771$ V

We subtract the two half-reactions and associated energy quantities (values of $-n\mathscr{F}\mathscr{E}^0$) to give a complete reaction as follows:

	$\mathscr{E}^0$	$-n\mathscr{F}\mathscr{E}^0$
$2Fe^{3+} + 2e^- = 2Fe^{2+}$	$+0.771$ V	$-2\mathscr{F} \times 0.771$ V
$Cu^{2+} + 2e^- = Cu$	$+0.337$ V	$-2\mathscr{F} \times 0.337$ V
$2Fe^{3+} + Cu = 2Fe^{2+} + Cu^{2+}$		$-2\mathscr{F} \times (0.771\text{ V} - 0.337\text{ V})$

For the overall reaction:

$$\Delta G^0 = -n\mathscr{F}\mathscr{E}^0 = -2\mathscr{F}\mathscr{E}^0 = -2\mathscr{F} \times (0.771\text{ V} - 0.337\text{ V})$$

therefore the $\mathscr{E}^0$ for the overall reaction (the $\mathscr{E}^0$ measured in the galvanic cell) is:

$$\mathscr{E}^0 = \frac{2\mathscr{F} \times (0.771\text{ V} - 0.337\text{ V})}{2\mathscr{F}}$$

$$\mathscr{E}^0 = 0.771\text{ V} - 0.337\text{ V} = 0.434\text{ V}$$

This is an important result. The $\mathscr{E}^0$ value for an overall reaction is the difference of the $\mathscr{E}^0$ values for the component half-reactions. Although the number of electrons, n, comes into the equation for the energy change in each half-reaction, it is canceled in the calculation of $\mathscr{E}^0$ for the overall reaction. The value of n for the overall reaction necessarily is the same as the n value in each of the half-reactions as they are subtracted.

To determine the value of $\mathscr{E}^0$ for a half-reaction from the values of $\mathscr{E}^0$ for other half-reactions, the value of n must be taken into account explicitly. (The values of n do not cancel out as in the example just presented.) Consider the relationship between the values of $\mathscr{E}^0$ for the half-reactions

a. $Fe^{3+} + e^- = Fe^{2+}$ $\quad \mathscr{E}^0_a = +0.771$ V
d. $Fe^{2+} + 2e^- = Fe$ $\quad \mathscr{E}^0_d = -0.44$ V

and the half-reaction that is the sum of these two:

e. $Fe^{3+} + 3e^- = Fe$ $\quad \mathscr{E}^0_e = ?$

The values of ΔG^0 (i.e., $-n\mathscr{F}\mathscr{E}^0$) are additive:

$$\Delta G^0_e = \Delta G^0_a + \Delta G^0_d$$
$$3\mathscr{F}\mathscr{E}^0_e = 1\mathscr{F}\mathscr{E}^0_a + 2\mathscr{F}\mathscr{E}^0_d$$

therefore

$$\mathscr{E}^0_e = \tfrac{1}{3}(\mathscr{E}^0_a + 2\mathscr{E}^0_d)$$
$$= \tfrac{1}{3}(+0.771\text{ V} - 0.88\text{ V}) = -0.036\text{ V}$$

We see that the value of $\mathscr{E}^0$ for the half-reaction

$$Fe^{3+} + 3e^- = Fe$$

is the weighted average of the values of $\mathscr{E}^0$ for the half-reactions that are added to give this half-reaction; the weighting factors are the values of n.

In combining half-reactions to give another half-reaction, the values of n are not canceled in calculation of the value of $\mathscr{E}^0$.

THE THERMODYNAMICS OF GALVANIC CELLS

We already have stated that the reaction

$$2Fe^{3+}(aq) + Cu = 2Fe^{2+}(aq) + Cu^{2+} \qquad \mathscr{E}^0 = 0.434 \text{ V}$$

can be carried out irreversibly by adding metallic copper to a solution containing an iron(III) salt. Or the reaction can be carried out reversibly in a galvanic cell. In the irreversible reaction, heat is evolved ($q < 0$); in a experiment at constant pressure,

$$q = \Delta H^0 = -15.9 \text{ kJ mol}^{-1}$$

In the reversible reaction, useful electrical work can be done by harnessing the battery. The magnitude of this useful work to be done on the surroundings ($-w$) is the negative of the value of ΔG; for carrying out the reaction reversibly under standard conditions (at 1.00 atm and 298.2 K),

$$\begin{aligned} -w = -\Delta G^0 &= n\mathscr{F}\mathscr{E}^0 \\ &= 2 \times 9.65 \times 10^4 \text{ J V}^{-1} \text{ mol}^{-1} \times 0.434 \text{ V} \\ &= 83.8 \text{ kJ mol}^{-1} \end{aligned}$$

The change of internal energy of the system, ΔU, is essentially the same in the irreversible and reversible situations, and is approximately equal to ΔH^0 because only condensed phases are involved [$\Delta(PV) \cong 0$]. The first law of thermodynamics,

$$\Delta U = q + w$$

applied to the reversible reaction in the galvanic cell operating under standard conditions gives

$$q_{\text{rev}} = \Delta U - w = \Delta H^0 - \Delta G^0$$

But ($\Delta H^0 - \Delta G^0$) is equal to the product of the temperature and the standard entropy change

$$\Delta H^0 - \Delta G^0 = T\Delta S^0$$

or

$$q_{\text{rev}} = T\Delta S^0$$

This relationship is one we already have introduced in discussing the reversible expansion of gases (Chapter 4). For the Fe(III)–Cu reaction,

$$\begin{aligned} q_{\text{rev}} = \Delta H^0 - \Delta G^0 &= -15.9 \text{ kJ mol}^{-1} - (-83.8 \text{ kJ mol}^{-1}) \\ &= +67.9 \text{ kJ mol}^{-1} \end{aligned}$$

$$\Delta S^0 = \frac{67{,}900 \text{ J mol}^{-1}}{298.2 \text{ K}} = +228 \text{ J K}^{-1} \text{ mol}^{-1}$$

When this reaction occurs irreversibly, the energy that the chemical system loses goes to heat the surroundings:

$$q_{\text{irrev}} = \Delta H^0 = -15{,}900 \text{ J mol}^{-1}$$

When the reaction occurs reversibly, heat is absorbed from the surroundings:

$$q_{\text{rev}} = +67{,}900 \text{ J mol}^{-1}$$

Here useful work is done at the expense of this heat and the decrease of energy of the chemical system. The useful work obtained from the reaction's occurring reversibly is the maximum possible; but remember that with reversibility we also would have infinite slowness.

THE DEPENDENCE OF $\mathscr{E}$ ON CONCENTRATION; THE NERNST EQUATION

In discussion of the ammonia-synthesis reaction in Chapter 12, we derived

$$\Delta G = \Delta G^0 + RT \ln\left(\frac{P_{\text{NH}_3}^2}{P_{\text{N}_2} \times P_{\text{H}_2}^3} \times \text{atm}^2\right)$$

An analogous equation holds for an oxidation–reduction reaction in solution, with the quotient of pressures replaced by a quotient of concentrations. For the oxidation of metallic copper by iron(III) ion:

$$2\text{Fe}^{3+} + \text{Cu}(s) = 2\text{Fe}^{2+} + \text{Cu}^{2+}$$

$$\Delta G = \Delta G^0 + RT \ln\left(\frac{[\text{Fe}^{2+}]^2[\text{Cu}^{2+}]}{[\text{Fe}^{3+}]^2} \times \text{L mol}^{-1}\right)$$

(In this equation, the activity-coefficient factors have been omitted. For the purpose of the present discussion, solution ideality is assumed.) This equation can be written in terms of values of $\mathscr{E}$ rather than values of ΔG; with

$$-n\mathscr{F}\mathscr{E} = \Delta G \quad \text{and} \quad -n\mathscr{F}\mathscr{E}^0 = \Delta G^0$$

the equation becomes

$$-n\mathscr{F}\mathscr{E} = -n\mathscr{F}\mathscr{E}^0 + RT \ln\left(\frac{[\text{Fe}^{2+}]^2[\text{Cu}^{2+}]}{[\text{Fe}^{3+}]^2} \times \text{L mol}^{-1}\right)$$

or, because $n = 2$,

$$\mathscr{E} = \mathscr{E}^0 - \frac{RT}{2\mathscr{F}} \ln\left(\frac{[\text{Fe}^{2+}]^2[\text{Cu}^{2+}]}{[\text{Fe}^{3+}]^2} \times \text{L mol}^{-1}\right)$$

At 298.2 K and with decadic logarithms, the equation becomes

$$\mathscr{E} = \mathscr{E}^0 - \frac{0.0592 \text{ V}}{2} \log\left(\frac{[\text{Fe}^{2+}]^2[\text{Cu}^{2+}]}{[\text{Fe}^{3+}]^2} \times \text{L mol}^{-1}\right)$$

A decrease in the concentration of a product of the reaction, copper(II) or iron(II), *or* an increase in the concentration of the reactant, iron(III), makes the value of $\mathscr{E}$ more positive. Substitution of 1.00 mol L^{-1} for each of the concentrations (i.e. standard states) gives $\mathscr{E} = \mathscr{E}^0$, which, of course, it must.

For a battery made up of these two half-cells with the concentration conditions

$$[Fe^{3+}] = 0.0100 \text{ mol L}^{-1} \qquad [Fe^{2+}] = 0.100 \text{ mol L}^{-1}$$
$$[Cu^{2+}] = 0.0500 \text{ mol L}^{-1}$$

the value of $\mathscr{E}$ is

$$\mathscr{E} = 0.434 \text{ V} - \frac{0.0592 \text{ V}}{2} \log\left[\frac{(0.100 \text{ mol L}^{-1})^2(0.0500 \text{ mol L}^{-1})}{(0.0100 \text{ mol L}^{-1})^2} \times \text{L mol}^{-1}\right]$$

$$= 0.434 \text{ V} - \frac{0.0592 \text{ V}}{2} \log 5.00 = 0.434 \text{ V} - 0.0207 \text{ V} = 0.413 \text{ V}$$

For the general case of a reaction $aA + bB = cC + dD$ with each of the species A, B, C, and D present in solution, the equation for the concentration dependence of $\mathscr{E}$ at 298.2 K is

$$\mathscr{E} = \mathscr{E}^0 - \frac{0.0592 \text{ V}}{n} \log\left(\frac{[C]^c[D]^d}{[A]^a[B]^b} \times (\text{mol L}^{-1})^{-\Delta n}\right)$$

in which $\Delta n = c + d - a - b$. This equation is known as the *Nernst equation.*

THE SILVER(I)–SILVER COUPLE

For a particular pair of oxidation states, we may write more than a single half-reaction if one or both of these states exists in different forms, depending on what other reagents are present. We shall illustrate the relationship of the $\mathscr{E}^0$ values for such related half-reactions with examples involving the silver(I)–silver couple. First, let us consider the dependence of $\mathscr{E}$ for the silver(I) ion–silver half-reaction at 25.0°C upon the concentration of silver(I) ion; with $\mathscr{E}^0$ given to 0.001 V, we have

$$Ag^+ + e^- = Ag(s) \qquad \mathscr{E}^0 = 0.799 \text{ V}$$

$$\mathscr{E} = +0.799 \text{ V} - 0.0592 \text{ V} \log\left(\frac{\text{mol L}^{-1}}{[Ag^+]}\right)$$

$$= +0.799 \text{ V} + 0.0592 \text{ V} \log([Ag^+] \times \text{L mol}^{-1})$$

We can use this equation to calculate the values of $\mathscr{E}$ as a function of the concentration of silver(I); for instance, at $[Ag^+] = 1.00 \times 10^{-2}$ mol L^{-1},

$$\mathscr{E} = 0.799 \text{ V} + 0.0592 \text{ V} \times (-2.00) = 0.799 \text{ V} - 0.118 \text{ V} = 0.681 \text{ V}$$

Calculations of this type yield the values of $\mathscr{E}$:

$\frac{[Ag^+]}{mol\ L^{-1}}$	$\frac{\mathscr{E}}{V}$
1.00	+0.799
0.100	+0.740
0.0100	+0.681
1.00×10^{-5}	+0.503
1.00×10^{-9}	+0.266
1.00×10^{-10}	+0.207

The linear plot of $\mathscr{E}$ versus the logarithm of the concentration of silver(I) ion is given in Figure 16–7.

If the solution in contact with the silver electrode is 1.00*M* HCl and and is saturated with silver(I) chloride ($K_s = 1.78 \times 10^{-10}$ mol^2 L^{-2}), the concentration of silver(I) ion is

$$[Ag^+] = \frac{K_s}{[Cl^-]} = \frac{1.78 \times 10^{-10}\ mol^2\ L^{-2}}{1.00\ mol\ L^{-1}} = 1.78 \times 10^{-10}\ mol\ L^{-1}$$

The electromotive force of the silver(I)–silver electrode under these conditions is

$$\begin{aligned}\mathscr{E} &= 0.799\ V + 0.0592\ V \log(1.78 \times 10^{-10}) \\ &= 0.799\ V - 0.577\ V = 0.222\ V\end{aligned}$$

But this situation, a silver electrode in contact with a solution containing chloride ion with $[Cl^-] = 1.00$ mol L^{-1} and saturated with silver(I) chloride, is that which defines $\mathscr{E}^0$ for the half-reaction $AgCl + e^- = Ag + Cl^-$. Therefore $\mathscr{E}^0$ for this reaction is 0.222 V. The conventional silver(I) chloride–silver electrode involves a coating of silver(I) chloride on metallic silver; the $\mathscr{E}$ value for a half-cell in which this half-reaction occurs at 25°C is

$$\mathscr{E} = +0.222\ V - 0.0592\ V \log([Cl^-] \times L\ mol^{-1})$$

That is, the value of $\mathscr{E}$ depends only upon the concentration of chloride ion: it is this concentration that determines the concentration of silver(I) ion in equilibrium with solid silver(I) chloride.

We can consider each half-reaction involving silver metal and a compound of silver(I) as a combination of the half-reaction

$$Ag^+ + e^- = Ag$$

and the reaction in which Ag^+ is formed from the species of silver(I), for example,

$$AgCl(s) \rightleftarrows Ag^+ + Cl^-$$

Thus the half-reaction

$$AgCl(s) + e^- = Ag + Cl^-$$

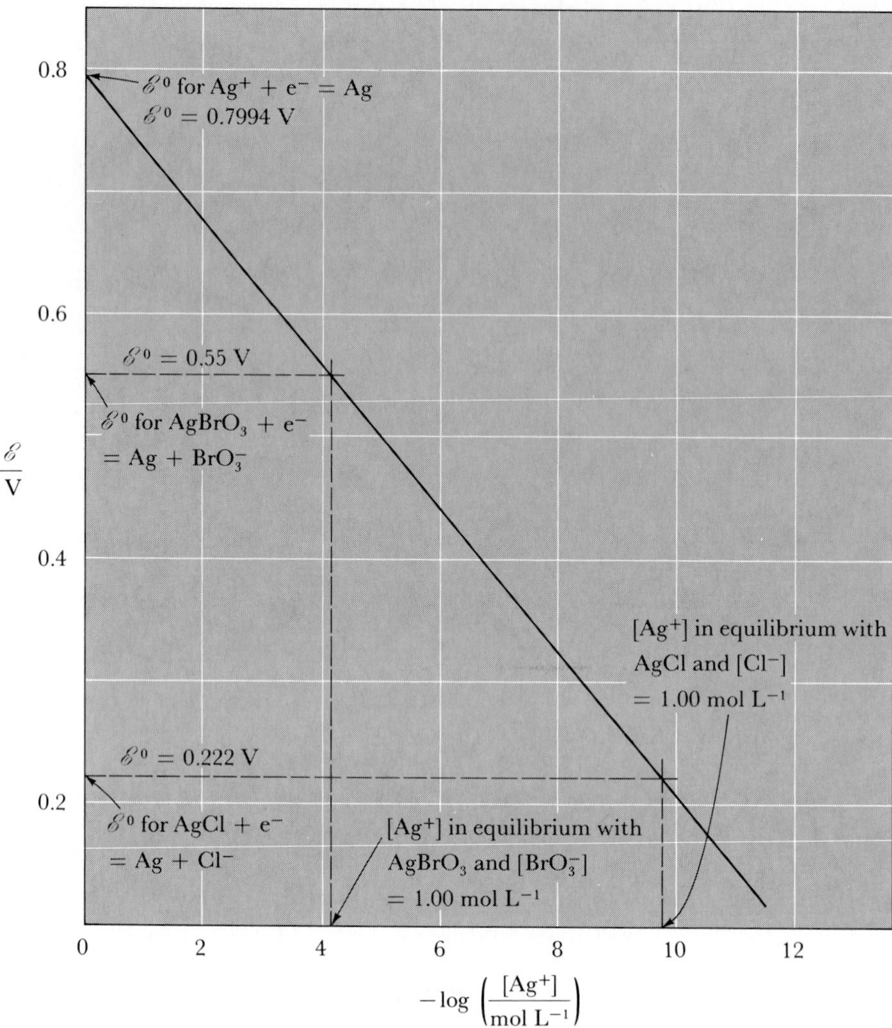

FIGURE 16–7
The voltage of the half-reaction $Ag^+ + e^- = Ag$ as a function of the concentration of silver(I) ion at 298.2 K:

$$\mathscr{E} = 0.7994\ V + 0.0592\ V \log\left(\frac{[Ag^+]}{mol\ L^{-1}}\right)$$

is the sum of

$$AgCl(s) \rightleftarrows Ag^+ + Cl^- \qquad \Delta G^0 = -RT \ln K_s^0 = -RT \ln (1.78 \times 10^{-10})$$

and

$$Ag^+ + e^- = Ag \qquad \Delta G^0 = -1 \times \mathscr{F}\mathscr{E}^0 = -1\mathscr{F} \times 0.799\ V$$

The value of $\mathscr{E}^0$ for the half-reaction

$$AgCl(s) + e^- = Ag + Cl^-$$

therefore is

$$\mathscr{E}^0 = -\frac{\Delta G^0}{\mathscr{F}} = -\frac{1}{\mathscr{F}}[-RT\ln(1.78 \times 10^{-10}) - 0.799\ \text{V} \times \mathscr{F}]$$

$$= -0.577\ \text{V} + 0.799\ \text{V} = +0.222\ \text{V}$$

which is the value already calculated.

Table 16–3 gives the values of $\mathscr{E}^0$ for several half-reactions of the type

$$\text{AgX} + \text{e}^- = \text{Ag} + \text{X}^-$$

and corresponding values of the solubility-product constant for AgX. Clearly, if we measure the value of $\mathscr{E}^0$ for a galvanic cell that involves the AgX–Ag half-cell, we can calculate the solubility-product constant of the compound AgX. This indirect method of determining the solubility product is useful, especially for compounds of very low solubility.

It is worth noting that a battery can be constructed using half-cells based upon the two half-reactions

$$\text{Ag}^+ + \text{e}^- = \text{Ag} \qquad \mathscr{E}^0 = +0.799\ \text{V}$$
$$\text{AgCl}(s) + \text{e}^- = \text{Ag} + \text{Cl}^- \qquad \mathscr{E}^0 = +0.222\ \text{V}$$

In such a battery, illustrated in Figure 16–8, the spontaneous chemical change that occurs, obtained by subtracting the second half-reaction from the first, is

$$\text{Ag}^+ + \text{Cl}^- = \text{AgCl}(s)$$

This overall reaction *does not* involve oxidation and reduction, but its occurrence in this battery produces electrical energy.

TABLE 16–3
Values of $\mathscr{E}^0$ for Half-Reactions

$$\text{AgX}(s) + \text{e}^- = \text{Ag} + \text{X}^-(aq)$$

and Values of K^0 for

$$\text{AgX}(s) \rightleftarrows \text{Ag}^+(aq) + \text{X}^-(aq)$$

at 25°C

AgX(s)	$\frac{\mathscr{E}^0}{\text{V}}$	K^0
AgOAc	+0.643	2.3×10^{-3}
$AgBrO_3$	+0.55	5.4×10^{-5}
$AgIO_3$	+0.35	3.1×10^{-8}
AgCl	+0.222	1.78×10^{-10}
AgBr	+0.071	5.0×10^{-13}
AgCN	−0.144	1.2×10^{-16}
AgI	−0.152	8.5×10^{-17}

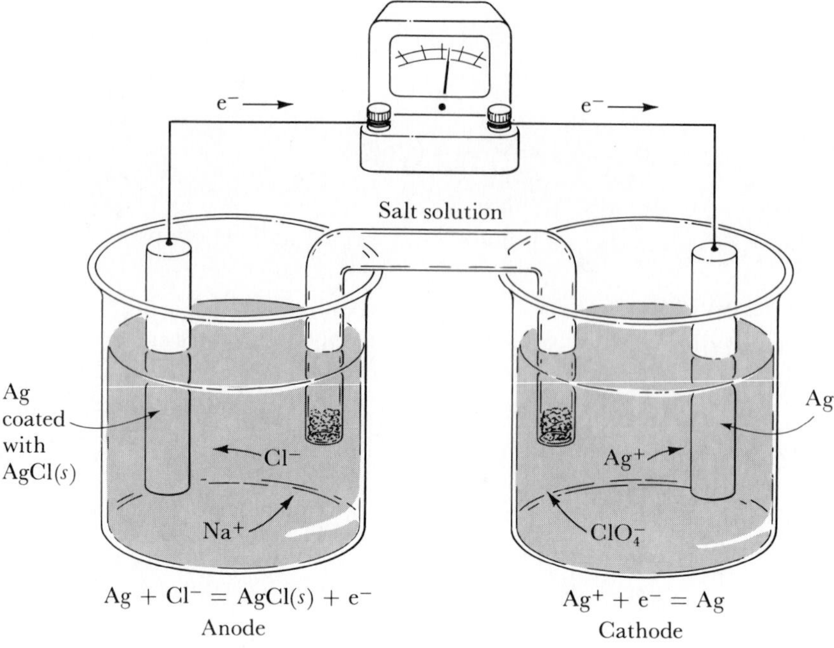

FIGURE 16–8

A battery in which the net chemical change does not involve oxidation and reduction:

$$\begin{array}{rl} & Ag^{+} + e^{-} = Ag \\ & -[AgCl(s) + e^{-} = Ag + Cl^{-}] \\ \hline \text{Net reaction:} & Ag^{+} + Cl^{-} = AgCl(s) \\ & \mathscr{E}^{0} = +0.799\text{ V} - (0.222\text{ V}) = 0.577\text{ V (298.2 K)} \end{array}$$

In discussing the value of $\mathscr{E}$ for the half-reaction $Ag^{+}(aq) + e^{-} = Ag(s)$, we have focused attention on changes brought about by changing the concentration of silver(I) in solution. But the value of $\mathscr{E}$ also can be changed by using as an electrode a solid solution of silver(0), for example, a solid solution of silver and gold. This system was discussed in Chapter 7, where we stated that solid solutions in this system were close to ideal. If this is strictly correct, the value of $\mathscr{E}$ for this half-reaction is given by the Nernst equation,

$$\mathscr{E} = \mathscr{E}^{0} - \frac{0.0592\text{ V}}{1} \log\left(\frac{x_{Ag}}{[Ag^{+}] \times \text{L mol}^{-1}}\right)$$

in which the mole fraction of silver in the solid solution, x_{Ag}, is taken into account.

THE DEPENDENCE OF EMF ON ACIDITY; THE pH METER

The electromotive force of a galvanic cell may depend on the concentration of hydrogen ion in the constituent half-cells. This dependence arises because either

1. hydrogen ion is a reactant or product in the overall reaction, or
2. with a change of hydrogen ion concentration, the form of one or more of the reactants and/or products changes.

Many of the half-reactions tabulated in Table 16–2 involve hydrogen ion. One is that involving oxygen gas (oxidation state of oxygen = 0) and water (oxidation state of oxygen = −2):

$$O_2(g) + 4H^+ + 4e^- = 2H_2O \qquad \mathscr{E}^0 = 1.229 \text{ V}$$

The value of $\mathscr{E}$ for this half-reaction with $P_{O_2} = 1.00$ atm but with a non-standard concentration of hydrogen ion, represented as $\mathscr{E}^{0\prime}$, is given by (for 25.0°C)[8]:

$$\mathscr{E}^{0\prime} = \mathscr{E}^0 - \frac{0.0592 \text{ V}}{4} \log\left(\frac{\text{mol}^4 \text{ L}^{-4}}{[H^+]^4}\right)$$

or

$$\mathscr{E}^{0\prime} = 1.229 \text{ V} + 0.0592 \text{ V} \log([H^+] \times \text{L mol}^{-1})$$

At $[H^+] = 1.00 \times 10^{-7}$ mol L^{-1},

$$\mathscr{E}^{0\prime} = 1.229 \text{ V} + 0.0592 \text{ V} \times (-7.000)$$
$$= 0.815 \text{ V}$$

At $[H^+] = 1.00 \times 10^{-14}$ mol L^{-1} ($[OH^-] = 1.00$ mol L^{-1}),

$$\mathscr{E}^{0\prime} = 1.229 \text{ V} + 0.0592 \text{ V} \times (-14.000)$$
$$= 0.400 \text{ V}$$

This value of $\mathscr{E}^{0\prime}$ at $[OH^-] = 1.00$ mol L^{-1} is the value of $\mathscr{E}^0$ for the oxygen(0)–oxygen(−2) half-reaction that involves hydroxide ion:

$$O_2(g) + 2H_2O + 4e^- = 4OH^- \qquad \mathscr{E}^0 = 0.400 \text{ V}$$

The dependence of $\mathscr{E}^{0\prime}$ for the oxygen (0)–oxygen (−2) couple is given in Figure 16–9. A single continuous straight line gives the dependence; the line can be represented by the equation

$$\mathscr{E}^{0\prime} = 0.400 \text{ V} - 0.0592 \text{ V} \log([OH^-] \times \text{L mol}^{-1})$$

which goes with the half-reaction written for basic solutions, or by the equation

$$\mathscr{E}^{0\prime} = 1.229 \text{ V} + 0.0592 \text{ V} \log([H^+] \times \text{L mol}^{-1})$$

which goes with the half-reaction written for acidic solution.

Another half-reaction that involves hydrogen ion is

$$2H^+ + 2e^- = H_2(g)$$

For this reaction, the dependence of $\mathscr{E}^{0\prime}$ ($P_{H_2} = 1.00$ atm) on the concentra-

[8]The symbol $\mathscr{E}^{0\prime}$ generally is used to represent the electromotive force for reactions and half-reactions in which the concentrations of all species except hydrogen ion have the standard value.

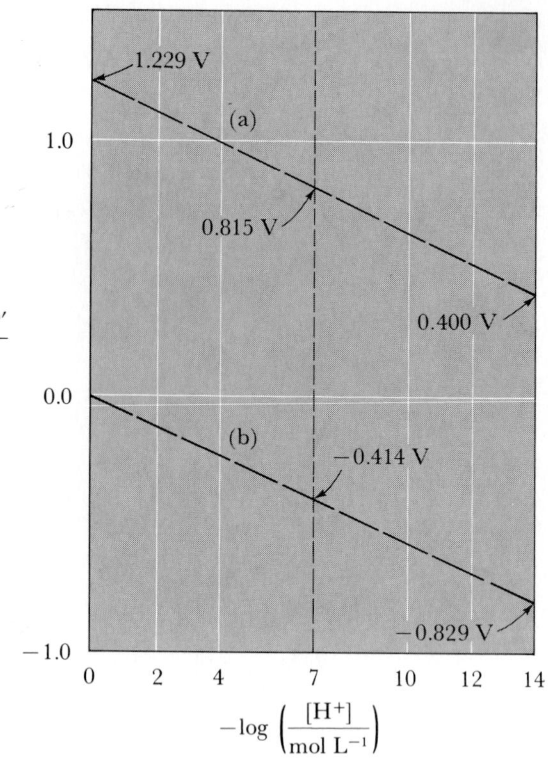

FIGURE 16–9
The dependences of $\mathscr{E}^{0\prime}$ for the half-reactions upon the concentration of hydrogen ion at 298.2 K.

(a) $O_2(g) + 4H^+ + 4e^- = 2H_2O$ $\quad \mathscr{E}^0 = 1.229$ V

or

$O_2(g) + 2H_2O + 4e^- = 4OH^-$ $\quad \mathscr{E}^0 = 0.400$ V

(b) $2H^+ + 2e^- = H_2(g)$ $\quad \mathscr{E}^0 = 0.000$ V

or

$2H_2O + 2e^- = 2OH^- + H_2(g)$ $\quad \mathscr{E}^0 = -0.829$ V

tion of hydrogen ion at 25.0°C is

$$\mathscr{E}^{0\prime} = \mathscr{E}^0 - \frac{0.0592 \text{ V}}{2} \log\left(\frac{\text{mol}^2 \text{ L}^{-2}}{[H^+]^2}\right)$$

or

$$\mathscr{E}^{0\prime} = 0.000 \text{ V} + 0.0592 \text{ V} \log([H^+] \times \text{L mol}^{-1})$$

At $[H^+] = 1.00 \times 10^{-7}$ mol L^{-1},

$$\mathscr{E}^{0\prime} = 0.000 \text{ V} + 0.0592 \text{ V} \times (-7.00)$$
$$= -0.414 \text{ V}$$

[At $[H^+] = 1.00 \times 10^{-14}$ mol L^{-1} ($[OH^-] = 1.00$ mol L^{-1}),

$$\mathscr{E}^{0\prime} = 0.000 \text{ V} + 0.0592 \text{ V} \times (-14.00)$$
$$= -0.829 \text{ V}$$

This value of $\mathscr{E}^{0\prime}$ for the half-reaction at $[H^+] = 1.00 \times 10^{-14}$ mol L^{-1} is the value of $\mathscr{E}^0$ for the hydrogen (+1)–hydrogen(0) half-reaction that involves hydroxide ion:

$$2H_2O + 2e^- = H_2(g) + 2OH^- \qquad \mathscr{E}^0 = -0.829 \text{ V}$$

The dependence of $\mathscr{E}^{0\prime}$ upon the concentration of hydrogen ion for this couple also is plotted in Figure 16–9. We see that the linear dependences of $\mathscr{E}^{0\prime}$ on $\log([H^+] \times \text{L mol}^{-1})$ for these two half-reactions are parallel. In the balanced equation for each of these half-reactions in acidic solution, the coefficient of H^+ is equal to the coefficient of e^-. It is the quotient of these coefficients that determines the slope, $\Delta\mathscr{E}/\Delta \log([H^+] \times L\ mol^{-1})$.

For a half-reaction

$$A + hH^+ + ne^- = B$$

the dependence of $\mathscr{E}^{0\prime}$ ($[A] = 1.00$ mol L^{-1} and $[B] = 1.00$ mol L^{-1}) at 25.0°C on the concentration of hydrogen ion is

$$\mathscr{E}^{0\prime} = \mathscr{E}^0 - \frac{0.0592\ V}{n} \log\left[\frac{(mol\ L^{-1})^h}{[H^+]^h}\right]$$

$$= \mathscr{E}^0 + 0.0592\ V \times \frac{h}{n} \times \log([H^+] \times L\ mol^{-1})$$

The dependence is determined by the ratio of the coefficients of hydrogen ion and electron, h/n, in the balanced half-reaction. For this same half-reaction involving hydroxide ion,

$$hH_2O + A + ne^- = B + hOH^-$$

the dependence of $\mathscr{E}^{0\prime}$ on the concentration of hydrogen ion is the same as the one already derived. In the equation relating $\mathscr{E}^{0\prime}$ and $\mathscr{E}^0$, h is the coefficient of hydrogen ion on the electron side of the half-reaction, or the coefficient of hydroxide ion on the opposite side.

This dependence is shown by the slopes of the lines for the several half-reactions given in Figure 16–10. The intersections of the lines in this figure with the $[H^+] = 1.00$ mol L^{-1} (pH = 0) axis are the values of $\mathscr{E}^0$ for the half-reactions involving hydrogen ion. The intersection of the lines with the $[H^+] = 1.00 \times 10^{-14}$ mol L^{-1} ($[OH^-] = 1.00$ mol L^{-1}) axis are the values of $\mathscr{E}^0$ for half-reactions involving hydroxide ion. Table 16–4 presents these values of $\mathscr{E}^0$ and others for alkaline solution.

The intersection of the lines in Figure 16–10 with the $[H^+] = 1.00 \times 10^{-7}$ mol L^{-1} line are the values of $\mathscr{E}^{0\prime}$ for neutral solutions. These values of $\mathscr{E}^{0\prime}$ for pH = 7.00 are of particular interest in biology, physiology, and biochemistry because many physiological processes occur in approximately neutral aqueous solutions.

The lines in Figure 16–10 do not extend into a region of hydrogen ion concentration in which one or both of the oxidation states involved in the couple are unstable. For instance, the line for the half-reaction

$$Cl_2 + 2e^- = 2Cl^-$$

does not extend below $[H^+] \cong 10^{-4}$ mol L^{-1}; at this acidity, equilibrium in the disproportionation reaction

$$Cl_2 + H_2O = H^+ + Cl^- + HOCl$$

has shifted to the right (for $[Cl^-] = 1.00$ mol L^{-1}), and it is not useful to discuss the chlorine(0)–chlorine(−1) couple at $[H^+] < 10^{-4}$ mol L^{-1}.

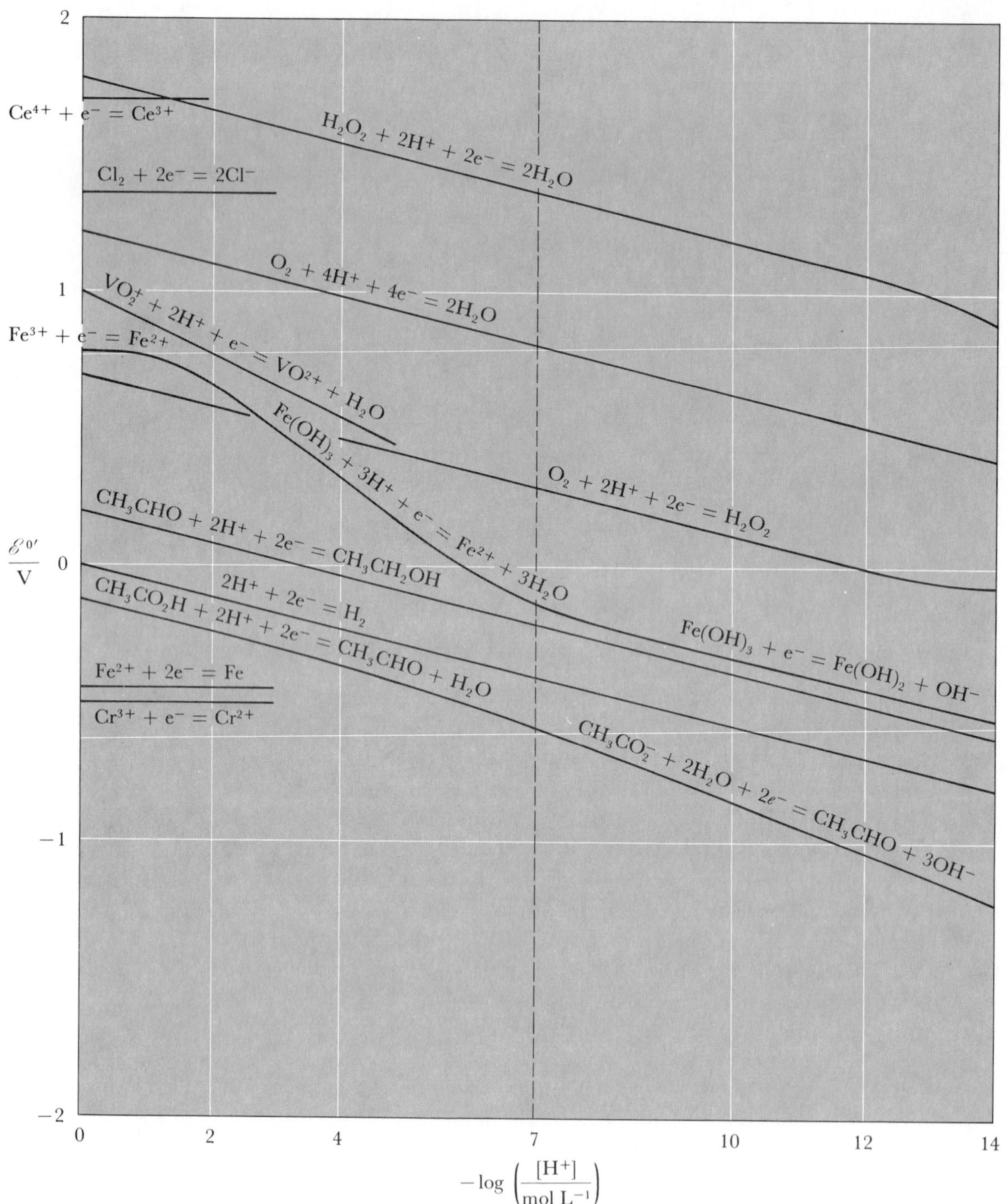

FIGURE 16–10

The dependence of $\mathscr{E}^{0\prime}$ upon the concentration of $[H^+]$. The intersections of lines with the left ordinate ($[H^+] = 1.00$ mol L^{-1}) are $\mathscr{E}^0$ values for acid half-reactions; intersections with dashed line at $[H^+] = 1.00 \times 10^{-7}$ mol L^{-1} are $\mathscr{E}^{0\prime}$ values used by biochemists and biologists; intersections with right ordinate are $\mathscr{E}^0$ values for base half-reactions (written involving OH^-). The curvature at pH = 12 to 14 in the lines for half-reactions involving H_2O_2 is due to the formation of HO_2^- in this range.

TABLE 16–4
Standard Potentials for Half-Reactions in Alkaline Aqueous Solution at 25.0°C

Half-Reaction	$\frac{\mathscr{E}^0}{V}$
$ClO^- + H_2O + 2e^- = Cl^- + 2OH^-$	+0.89
$HO_2^- + H_2O + 2e^- = 3OH^-$	+0.89
$MnO_4^- + 2H_2O + 3e^- = MnO_2 + 4OH^-$	+0.585
$O_2 + 2H_2O + 4e^- = 4OH^-$	+0.400
$Ag_2O + H_2O + 2e^- = 2Ag + 2OH^-$	+0.344
$S_4O_6^{2-} + 2e^- = 2S_2O_3^{2-}$	+0.169
$O_2 + H_2O + 2e^- = HO_2^- + OH^-$	−0.076
$Fe(OH)_3 + e^- = Fe(OH)_2 + OH^-$	−0.56
$CH_3CHO + 2H_2O + 2e^- = CH_3CH_2OH + 2OH^-$	−0.63
$2H_2O + 2e^- = H_2 + 2OH^-$	−0.819
$SO_4^{2-} + H_2O + 2e^- = SO_3^{2-} + 2OH^-$	−0.93
$CH_3CO_2^- + 2H_2O + 2e^- = CH_3CHO + 3OH^-$	−1.22
$AlO_2^- + 2H_2O + 3e^- = Al + 4OH^-$	−2.35

For some couples, one or both of the oxidation states are involved in acid–base equilibria. In this case the predominant species of the oxidized and/or reduced forms of the couple change in the ranges of hydrogen ion concentration where the protonation equilibria shift. An example is the acetate–acetaldehyde couple. For acidic solution, the half-reaction and associated value of $\mathscr{E}^0$ are

$$CH_3CO_2H + 2H^+ + 2e^- = CH_3CHO + H_2O \qquad \mathscr{E}^0 = -0.12\ V$$

For alkaline solution, the half-reaction should be written with hydroxide ion, not hydrogen ion, and acetate ion, not acetic acid. [Because $K_a(HOAc) = 1.75 \times 10^{-5}$ mol L^{-1}, the predominant form of acetate at $[H^+] < 1.00 \times 10^{-7}$ mol L^{-1} is acetate ion.] The half-reaction for basic solution,

$$CH_3CO_2^- + 2H_2O + 2e^- = CH_3CHO + 3OH^-$$

is the algebraic combination of the acetate–acetaldehyde half-reaction for acidic solution and the acid-dissociation reactions for water and acetic acid. The value of $\mathscr{E}^0$ for this half-reaction can be obtained from the value of ΔG^0, which is the algebraic combination of values of ΔG^0 for the three reactions being combined; each of these values of ΔG^0 is either $\Delta G^0 = -n\mathscr{F}\mathscr{E}^0$ or $\Delta G^0 = -RT \ln K^0$:

Reaction	ΔG^0
$CH_3CO_2H + 2H^+ + 2e^- = CH_3CHO + H_2O$	$-2\mathscr{F} \times (-0.12\ V)$
$+3 \times (H_2O = H^+ + OH^-)$	$-3RT \ln(1.00 \times 10^{-14})$
$-(CH_3CO_2H = CH_3CO_2^- + H^+)$	$+RT \ln(1.75 \times 10^{-5})$

Thus, the value of $\mathscr{E}^0$ for this half-reaction is

$$\mathscr{E}^0 = \frac{-\Delta G^0}{2\mathscr{F}}$$

$$= -\frac{1}{2\mathscr{F}}[-2\mathscr{F} \times (-0.12 \text{ V}) - 3RT \ln(1.00 \times 10^{-14}) + RT \ln(1.75 \times 10^{-5})]$$

Expressed in decadic logarithms, this equation becomes

$$\mathscr{E}^0 = -0.12 \text{ V} + \frac{3 \times 0.0592 \text{ V}}{2} \times \log(1.00 \times 10^{-14}) - \frac{0.0592 \text{ V}}{2} \times \log(1.75 \times 10^{-5})$$

$$= -0.12 \text{ V} - 1.24 \text{ V} + 0.14 \text{ V} = -1.22 \text{ V}$$

This is the value of $\mathscr{E}^{0\prime}$ in Figure 16–10 at the intersection of the line for this couple with the vertical line for $[H^+] = 1.00 \times 10^{-14}$ mol L^{-1}. Notice that the slope of this line changes over the range of concentrations of hydrogen ion in which acetic acid becomes predominantly acetate ion, $10K_a > [H^+] > 0.1K_a$, that is, 1.75×10^{-4} mol $L^{-1} > [H^+] > 1.75 \times 10^{-6}$ mol L^{-1}. For $[H^+] > 1.75 \times 10^{-4}$ mol L^{-1}, the predominant form is CH_3CO_2H and $h/n = 1$; for $[H^+] < 1.75 \times 10^{-6}$ mol L^{-1}, the predominant form is $CH_3CO_2^-$ and $h/n = 1.5$.

The iron(III)–iron(II) couple provides another example; we see in Figure 16–10 that the line for this couple has three linear segments. At $[H^+] > \sim 0.1$ mol L^{-1}, the simple hydrated ions are the predominant species, and the appropriate half-reaction is

$$Fe^{3+} + e^- = Fe^{2+} \qquad \frac{h}{n} = 0$$

At 1×10^{-3} mol $L^{-1} > [H^+] > 1 \times 10^{-6}$ mol L^{-1}, iron(III) hydroxide is the predominant form of iron(III) but simple hydrated iron(II) ion is the predominant form of the +2 oxidation state, and the appropriate half-reaction is

$$Fe(OH)_3 + e^- + 3H^+ = Fe^{2+} + 3H_2O \qquad \frac{h}{n} = 3$$

At $[H^+] < 1 \times 10^{-8}$ mol L^{-1}, both iron(II) and iron(III) are present as hydroxides, and the appropriate half-reaction is

$$Fe(OH)_3 + e^- = Fe(OH)_2 + OH^- \qquad \frac{h}{n} = 1$$

The values of h/n for these three half-reactions, 0, 3, and 1, are the relative values of the slopes of the three linear segments of the line for the iron(III)–iron(II) couple.

Whether the value of $\mathscr{E}^{0\prime}$ for a complete reaction varies with the concentration of hydrogen ion depends on the relative values of h/n for the two half-reactions that are combined to give the overall reaction. This point can be illustrated by three reactions involving the following four half-reactions:

Half-reaction	$\frac{h}{n}$	$\frac{\mathscr{E}^0}{\text{V}}$
$VO_2^+ + 2H^+ + e^- = VO^{2+} + H_2O$	2	+1.00
$O_2 + 4H^+ + 4e^- = 2H_2O$	1	+1.23
$ClO_4^- + 2H^+ + 2e^- = ClO_3^- + H_2O$	1	+1.19
$Pu^{4+} + e^- = Pu^{3+}$	0	+0.982

Balanced chemical equations for oxidation of the reduced form by oxygen in acidic solution are:

Reaction[9]	$\frac{h}{n}$
$O_2 + 4VO^{2+} + 2H_2O = 4VO_2^+ + 4H^+$	−1
$O_2 + 2ClO_3^- = 2ClO_4^-$	0
$O_2 + 4Pu^{3+} + 4H^+ = 4Pu^{4+} + 2H_2O$	+1

Although an examination of the oxygen half-reaction alone suggests that oxygen would be a better oxidizing agent (in the equilibrium sense) the greater the concentration of hydrogen ion, this is not a useful conclusion. You should examine the balanced equations for the overall reactions. The balanced equations for these three reactions, each with oxygen a reactant, involve hydrogen ion as a product, no hydrogen ion, and hydrogen ion as a reactant. Only in the last of the reactions does the equilibrium extent of the reaction increase with an increase of the concentration of hydrogen ion.

The dependence of the electromotive force of the hydrogen electrode on the concentration of hydrogen ion, shown in Figure 16–9 and 16–10, is the basis for the pH scale and for electrochemical devices which measure pH. If nonideality of solution behavior is taken into account, the equation given earlier for $\mathscr{E}^{0\prime}$ for the half-reaction

$$2H^+ + 2e^- = H_2(g)$$

must be modified to include the activity coefficient of hydrogen ion, γ_{H^+}; the resulting equation for 25.0°C is

$$\mathscr{E}^{0\prime} = 0.000 \text{ V} + 0.0592 \text{ V} \times \log([H^+]\gamma_{H^+})$$

The idealized definition of pH is

$$\text{pH} = -\log([H^+]\gamma_{H^+})$$

[9] The value of n is 4 in each of these reactions; two atoms of oxygen each change oxidation state by two units (from zero to minus two) in each reaction.

TABLE 16–5
pH Values of Reference Solutions at 25.0°C

Solution	$[H^+] \times L\ mol^{-1}$	$-\log([H^+] \times L\ mol^{-1})$	pH
0.1000*M* HCl	0.1000	1.000	1.088
{0.01000*M* HCl, 0.09000*M* KCl}	0.01000	2.000	2.098
{0.1000*M* HOAc, 0.1000*M* NaOAc}	—	—	4.652
{0.0250*M* KH_2PO_4, 0.0250*M* Na_2HPO_4}	—	—	6.865

Therefore the value of pH is related to the value of $\mathscr{E}^{0\prime}$ by

$$pH = -\frac{\mathscr{E}^{0\prime}}{0.0592\ V} \qquad \text{(see footnote 10)}$$

It would seem that measurement of the electromotive force of a galvanic cell involving this half-reaction would give the pH of the solution in which the hydrogen electrode is immersed. However, there are problems that mar this simple approach:

1. The galvanic cell involves two half-cells, and there is a junction between the solutions in the two half-cells. There is a small unknown potential at this interface. This is called the *liquid junction potential.*
2. The activity coefficient of a single ion (e.g., γ_{H^+} in the present example) cannot be measured. Therefore it is not possible to know the exact value of $[H^+]\gamma_{H^+}$ for a solution.

These two problems have been studied intensively, and in the framework of reasonable assumptions a useful set of standard solutions with assigned values of pH has been proposed by the National Bureau of Standards. Some of these standards are given in Table 16–5. The first two entries show the significant difference between $-\log[H^+]$ and pH ($pH = -\log[H^+]\gamma_{H^+}$). It simply is *not correct* to assume that an electrometric method of finding solution acidity measures the concentration of hydrogen ion exactly. Solution nonideality influences the voltage of a galvanic cell. Nonetheless, we will continue to calculate the value of pH from the concentration of hydrogen ion using the equation

$$pH_{calcd} = -\log([H^+] \times L\ mol^{-1})$$

[10]Recall that the factor 0.0592 V is the value of $RT/\mathscr{F}$ for 298.2 K. For other temperatures, the value is different. The temperature-compensating adjustment on a direct reading pH meter, generally a knob to be turned, corrects this proportionality between pH and $\mathscr{E}^{0\prime}$ for the temperature of the study.

In practice, it is not convenient to use the hydrogen electrode to evaluate the pH of a solution. The physical characteristics of this electrode (Figure 16–5) and its need for hydrogen gas make it impossible to construct a rough-and-ready pH meter involving the hydrogen electrode. Other obstacles may come from chemical complications: because hydrogen gas is a relatively good reducing agent, use of a hydrogen electrode in a solution containing an oxidizing agent may result in a chemical reaction. For instance, if the solution contains copper(II) ion, the following reaction will occur:

$$H_2 + Cu^{2+}(aq) = 2H^+(aq) + Cu$$

It was discovered early in this century that an electrical potential is generated across a thin glass membrane separating solutions of different acidity. This potential is determined by the difference between values of the pH of the two solutions. In the 1920s and 1930s, rugged pH meters using the glass electrode were developed. Figure 16–11 shows a glass electrode. On the enclosed side of the glass membrane is a solution of hydrochloric acid; in this solution is a silver(I) chloride–silver electrode. The contribution of this half-reaction to the measured emf is constant because the composition of the solution inside the glass electrode does not change. On the other side of the membrane is a solution, the acidity of which is being measured. The galvanic cell is completed by a salt bridge

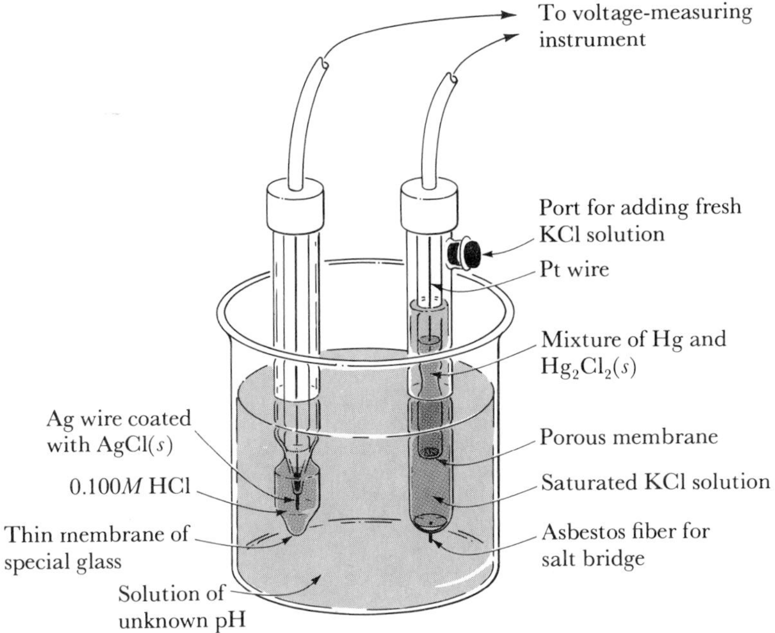

FIGURE 16–11
The glass electrode–calomel electrode pH meter. Solution nonideality influences the value of $\mathscr{E}$ for this cell, and the measured pH involves an activity-coefficient factor:

$$pH = -\log([H^+] \times \gamma_{H^+})$$

connecting this half-cell with another half-cell, conventionally the saturated calomel electrode, for which the half-reaction is

$$Hg_2Cl_2(s) + 2e^- = Hg(l) + 2Cl^- \text{ (in satd KCl soln)} \qquad \mathscr{E} = 0.241 \text{ V}$$

With the glass electrode immersed in different solutions, the complete galvanic cell has different voltages, but the only thing that is changed in going from one solution to another is the contribution to the voltage generated at the acid–glass interface. Thus, by appropriate calibration, the pH meter with a glass electrode can be made to give directly the pH of the solution being studied.

The range of pH over which a glass electrode responds as if it were a hydrogen electrode depends upon the composition of the glass. For a glass commonly used in glass electrodes (72.2 mol % SiO_2, 6.4 mol % CaO, and 21.4 mol % Na_2O), the range is pH = 1 to 9.5. For solutions more alkaline than pH = 9.5, this glass electrode responds to the sodium ion in the solution.

There has been much recent development of rugged, serviceable specific ion electrodes, analogous to the glass electrode, which respond over a large concentration range specifically to some particular ion (e.g., Na^+, K^+, NH_4^+, F^-, Cl^-). The silver chloride–silver electrode is a monitor of the concentration of chloride ion:

$$AgCl(s) + e^- = Ag + Cl^- \qquad \mathscr{E}^0 = +0.222 \text{ V}$$

$$\mathscr{E} = \mathscr{E}^0 + 0.0592 \text{ V} \log([Cl^-]\gamma_{Cl^-}) \text{ at } 298.2 \text{ K}$$

This electrode is based directly upon the principles that have been developed in this chapter. A fluoride-ion-specific electrode has been developed that resembles the glass electrode. The membrane separating the internal reference solution and electrode from the unknown fluoride-containing solution is a small single crystal of lanthanum fluoride (LaF_3). This electrode is used to determine concentrations of fluoride ion in city water supplies.

16–5 Some Practical Batteries; Fuel Cells

Some galvanic cells used as practical batteries cannot be recharged. The nature of the electrode reactions and the physical construction of the battery make this impractical. These cells are called *primary cells.* The common dry cell (or the Leclanché cell, named for GEORGE LECLANCHÉ) is a member of this class. For other galvanic cells, the spontaneous chemical reaction that occurs as current is produced can be reversed by an applied potential. Batteries based upon such cells, which are called *secondary cells,* can be recharged. Common storage batteries are members of this class. If the spontaneous oxidation–reduction reaction occurring in the battery involves reactants that can be fed continuously into the cell, the device for obtaining electrical energy directly from a chemical reaction is called a *fuel cell.* Now we will consider examples of each of these kinds of cells.

THE DRY CELL

The common dry cell uses zinc as the reducing agent and manganese dioxide as the oxidizing agent. The zinc can, which is the container for the battery, is the anode, and the cathode is a carbon rod that dips into a wet mixture of manganese dioxide, carbon, and ammonium chloride. The net chemical change that occurs when this cell discharges is

$$Zn(s) + 2MnO_2(s) + 2NH_4^+ = Zn(NH_3)_2^{2+} + 2MnO(OH)(s)$$

[It is an oversimplification to represent the zinc(II) product as the diammine complex; some association with chloride ion probably occurs in the moistened ammonium chloride, and $Zn(NH_3)_2Cl_2$ probably is precipitated as zinc is converted to zinc(II).] A fresh dry cell based upon this reaction has an electromotive force of ~1.55 V.

The same oxidizing and reducing agents used in the Leclanché cell are used in the common alkaline dry cells. The electrolyte present in this cell is aqueous sodium hydroxide, and the net reaction occurring upon discharge is

$$Zn(s) + 2MnO_2(s) + 2H_2O + 2OH^- = Zn(OH)_4^{2-} + 2MnO(OH)(s)$$

The stronger construction needed to prevent the concentrated sodium hydroxide solution from leaking is responsible for the greater cost of these batteries.

THE LEAD STORAGE BATTERY

Most automobiles have lead storage batteries to start their engines and to provide electricity when the car is not running. The net reaction occurring upon discharge,

$$PbO_2(s) + Pb + 4H^+ + 2SO_4^{2-} = 2PbSO_4(s) + 2H_2O$$

is accompanied by a large decrease in Gibbs free energy,

$$\Delta G^0 = -392 \text{ kJ mol}^{-1}$$

The cell voltage corresponding to this value of ΔG° is

$$\mathscr{E}^0 = \frac{-\Delta G^0}{n\mathscr{F}}$$

$$\mathscr{E}^0 = \frac{392{,}000 \text{ J mol}^{-1}}{2 \times 9.65 \times 10^4 \text{ J V}^{-1} \text{ mol}^{-1}} = 2.03 \text{ V}$$

In a twelve-volt storage battery, there are six cells in series; in a six-volt battery, there are three. The actual emf of a storage battery depends upon the concentration of sulfuric acid. When the battery is fully charged, the concentration of sulfuric acid is ~5 mol L^{-1} or higher, and the emf is ~2.1 V. (The emf depends also upon the temperature.)

The chemical equation for the reaction in the lead storage battery shows that only the sulfuric acid and water are present in the solution in the cell. The lead, lead dioxide, and lead sulfate are pure solid phases. The insolubility of lead sulfate ($K_s = [Pb^{2+}][SO_4^{2-}] = 1.59 \times 10^{-8}\ mol^2\ L^{-2}$) makes the value of ΔG^0 for this reaction more negative than for the corresponding reaction involving simple lead(II) ion:

$$PbO_2(s) + Pb + 4H^+ = 2Pb^{2+} + 2H_2O \qquad \Delta G^0 = -305\ kJ\ mol^{-1}$$
$$\mathscr{E}^0 = 1.581\ V$$

The half-reactions that comprise the two oxidation–reduction reactions being considered are:

Lead(IV)–Lead(II) Couple

$$PbO_2 + 4H^+ + SO_4^{2-} + 2e^- = PbSO_4 + 2H_2O \qquad \mathscr{E}^0 = +1.68\ V$$
$$PbO_2 + 4H^+ + 2e^- = Pb^{2+} + 2H_2O \qquad \mathscr{E}^0 = +1.455\ V$$

Lead(II)–Lead Couple

$$PbSO_4 + 2e^- = Pb + SO_4^{2-} \qquad \mathscr{E}^0 = -0.351\ V$$
$$Pb^{2+} + 2e^- = Pb \qquad \mathscr{E}^0 = -0.126\ V$$

Comparison of the two half-reactions for reduction of lead dioxide shows that it is a better oxidizing agent in the presence of sulfuric acid, in which it is reduced to lead(II) sulfate, than in an acidic medium (e.g., perchloric acid) in which lead(II) ion is soluble. Similarly, comparison of the half-reactions for reduction of lead(II) sulfate and lead(II) ion shows that metallic lead is a better reducing agent (i.e., is more easily oxidized) in the presence of sulfuric acid, in which it is oxidized to lead(II) sulfate, than in an acid in which it is oxidized to lead(II) ion.

When a lead storage battery is charged the reverse reaction occurs: electrolysis oxidizes lead(II) sulfate to lead dioxide and reduces lead(II) sulfate to metallic lead. Cycles of discharge and charge cannot, however, be repeated indefinitely. In time, the physical structure of the electrodes deteriorates, and internal short circuits develop. During the charging of a lead storage battery, electrode reactions other than the desired ones take place. The electrolysis of water occurs; this involves the half-reactions

$$2H_2O = O_2 + 4H^+ + 4e^- \quad \text{and} \quad 2H^+ + 2e^- = H_2$$

which occur at lower values of emf than do the transformations of lead sulfate into lead dioxide and metallic lead. However, these gas-evolution reactions are sluggish and do not prevent concurrent occurrence of the desired charging reactions.

In the new, sealed lead storage batteries, the hydrogen and oxygen produced in the charging process combine to reform water in a catalyzed reaction. Thus there is no net loss of water and none need be replaced. The need to add water periodically to a regular lead storage battery is due in part to the loss of water by electrolysis.

FUEL CELLS

When a fuel is burned and the resulting heat is used to run an electrical generator, the conversion of chemical energy to electrical energy is inherently inefficient because of the thermodynamic limitations of a heat power engine. The inefficiency can be eliminated if the combustion reaction occurs directly in a galvanic cell. Such galvanic cells are called *fuel cells*. The most practical fuel cell, developed for use in the Apollo spacecraft, employs the reaction

$$2H_2(g) + O_2(g) = 2H_2O(l)$$

A single cell operating at 25°C with the reactant gases at 1.0 atm has an electromotive force of 1.23 V, and multiples of this voltage are realized by having cells in series. A 2 kW system involving 31 cells was developed for the Apollo spacecraft. The aqueous electrolyte used in these fuel cells is concentrated potassium hydroxide; therefore the half-reactions are those appropriate for alkaline solution:

Anode (where H_2 is oxidized)

$$H_2(g) + 2OH^-(aq) = 2H_2O(l) + 2e^-$$

Cathode (where O_2 is reduced)

$$O_2(g) + 2H_2O(l) + 4e^- = 4OH^-(aq)$$

It was not easy to develop this simple oxidation–reduction chemistry into a reliable fuel cell. The practical fuel cell using this reaction must be run at elevated temperatures, and the water produced is distilled from the cell for use by the astronauts.

One possible future use of this fuel cell is energy storage. Water could be electrolyzed at off-peak power-usage periods, and the hydrogen and oxygen produced employed in the fuel cells during peak power usage.

Scientists are studying other oxidation–reduction reactions for use in fuel cells. Among those they are considering are the oxidation of hydrocarbons (e.g., methane) and methanol:

$$CH_4(g) + 2O_2(g) = CO_2(g) + 2H_2O(l)$$
$$2CH_3OH(l) + 3O_2(g) = 2CO_2(g) + 4H_2O(l)$$

Thermochemical data for these substances allow us to calculate the values of $\mathscr{E}^0$ for these reactions: the values are +1.09 V and +1.22 V, respectively.

16–6 Oxidation–Reduction Reactions and $\mathscr{E}^0$

The values of $\mathscr{E}^0$ for half-reactions help us think about oxidation–reduction reactions. As an example let us consider the reaction

$$Br_2 + SO_2 + 2H_2O = SO_4^{2-} + 2Br^- + 4H^+$$

The two half-reactions that give this reaction and their $\mathscr{E}^0$ values are

$$Br_2 + 2e^- = 2Br^- \qquad \mathscr{E}^0 = +1.08\ V$$
$$4H^+ + SO_4^{2-} + 2e^- = SO_2 + 2H_2O \qquad \mathscr{E}^0 = +0.17\ V$$

We can conclude from these values of $\mathscr{E}^0$ that in a solution with $[H^+] = 1.00\ mol\ L^{-1}$, bromine, Br_2, is a better oxidizing agent than sulfate ion, SO_4^{2-}, or, stated differently, sulfur dioxide, SO_2, is a better reducing agent than bromide ion, Br^-. Therefore, the reaction goes as written because its reactants are bromine, the better oxidizing agent, and sulfur dioxide, the better reducing agent. The reverse reaction involves the poorer oxidizing agent (sulfate ion) and the poorer reducing agent (bromide ion); this reverse reaction does not go spontaneously.

When considering the solution chemistry of an element in an intermediate oxidation state (an oxidation state less positive than some and more positive than others), we must ask how stable the intermediate oxidation state is with respect to disproportionation. We can dissect the disproportionation reactions for iron(II) and for hydrogen peroxide into the half-reactions involving these species:

$$H_2O_2 + 2H^+ + 2e^- = 2H_2O \qquad \mathscr{E}^0 = +1.77\ V$$
$$Fe^{3+} + e^- = Fe^{2+} \qquad \mathscr{E}^0 = +0.771\ V$$
$$O_2 + 2H^+ + 2e^- = H_2O_2 \qquad \mathscr{E}^0 = +0.69\ V$$
$$Fe^{2+} + 2e^- = Fe \qquad \mathscr{E}^0 = -0.44\ V$$

Examining these values of $\mathscr{E}^0$, we can conclude that the disproportionation of iron(II),

$$3Fe^{2+} = Fe + 2Fe^{3+} \qquad \mathscr{E}^0 = -0.44\ V - (+0.771\ V) = -1.21\ V$$

does not go. The reactant in this disproportionation reaction is the poorer of the oxidizing agents [iron(II) is a poorer oxidizing agent than iron(III)] and the poorer of the reducing agents [iron(II) is a poorer reducing agent than metallic iron]. From the values of $\mathscr{E}^0$ for the half-reactions involving hydrogen peroxide, we can conclude that its disproportionation,

$$2H_2O_2 = 2H_2O + O_2 \qquad \mathscr{E}^0 = +1.77\ V - (+0.69\ V) = +1.08\ V$$

does go. In this reaction, hydrogen peroxide, a better oxidizing agent than oxygen, reacts with hydrogen peroxide, a better reducing agent than water. Although hydrogen peroxide is unstable, decomposing into water and oxygen, this decomposition is *very* slow in the absence of a catalyst.

In each of these examples we decided which was the better oxidizing agent by examining of the half-reactions and associated $\mathscr{E}^0$ values with the guiding principle: the more positive the $\mathscr{E}^0$ value, the greater the oxidizing power of the oxidized form of the couple. (The decision as to which is the better reducing agent is redundant once you have decided which is the better oxidizing agent. The better reducing agent is the reduced form of the poorer oxidizing agent.)

The values of $\mathscr{E}^0$ for the half-reactions of an element with several oxidation states can be summarized concisely in a potential diagram. Consider the half-reactions for acidic solutions involving the several oxidation states of vanadium:

$$VO_2^+ + 2H^+ + e^- = VO^{2+} + H_2O \qquad \mathscr{E}^0 = 1.00\text{ V}$$
$$VO^{2+} + 2H^+ + e^- = V^{3+} + H_2O \qquad \mathscr{E}^0 = 0.36\text{ V}$$
$$V^{3+} + e^- = V^{2+} \qquad \mathscr{E}^0 = -0.25\text{ V}$$
$$V^{2+} + 2e^- = V(s) \qquad \mathscr{E}^0 = -1.2\text{ V}$$

All the information in these half-reactions is summarized in the diagram

$$VO_2^+ \xrightarrow{1.00\text{ V}} VO^{2+} \xrightarrow{0.36\text{ V}} V^{3+} \xrightarrow{-0.25\text{ V}} V^{2+} \xrightarrow{-1.2\text{ V}} V$$

The diagram does not show explicitly the protons (hydronium ions) involved in each half-reaction, but we can deduce whether hydrogen ion is involved in a half-reaction by inspecting the formulas of the species. Hydrogen ion and water are involved in the VO_2^+–VO^{2+} and VO^{2+}–V^{3+} half-reactions because different numbers of oxide ions are coordinated per vanadium in the oxidized and reduced forms. Similarly, hydrogen ion and water are not involved in the V^{3+}–V^{2+} and V^{2+}–V half-reactions because oxide ion is not coordinated in these vanadium species. Applying principles outlined previously in this chapter, you can balance any half-reaction if you know the formulas of the oxidized and reduced forms.

An examination of the potential diagram shows that each oxidation state of vanadium is a better oxidizing agent than is its reduced form. Therefore, no intermediate oxidation state of vanadium is unstable with respect to disproportionation in a solution with $[H^+] = 1.00\text{ mol L}^{-1}$. For instance, for the disproportionation reaction of vanadium(IV),

$$2VO^{2+} = VO_2^+ + V^{3+}$$

the value of $\mathscr{E}^0$ is

$$\mathscr{E}^0 = 0.36\text{ V} - (1.00\text{ V}) = -0.64\text{ V}$$

therefore

$$K^0 = 10^{-0.64\text{ V}/0.0592\text{ V}} = 1.5 \times 10^{-11}$$

By calculation, we can confirm that an equilibrium constant as small as this corresponds to very little disproportionation.

Example: What is the concentration of vanadium(III) in a solution with $[H^+] = 1.00\text{ mol L}^{-1}$ and $[VO^{2+}]_0 = 0.100\text{ mol L}^{-1}$?

From considering the balanced equation for disproportionation, we can express the equilibrium concentrations:

$$[VO^{2+}] = 0.100\text{ mol L}^{-1} - 2[V^{3+}]$$
$$[VO_2^+] = [V^{3+}]$$

Substituting these concentrations into

$$K = \frac{[VO_2^+][V^{3+}]}{[VO^{2+}]^2} = 1.5 \times 10^{-11}$$

we get

$$\frac{[V^{3+}]^2}{(0.100 \text{ mol L}^{-1} - 2[V^{3+}])^2} = 1.5 \times 10^{-11}$$

Taking the square root of each side of the equation, we get

$$\frac{[V^{3+}]}{0.100 \text{ mol L}^{-1} - 2[V^{3+}]} = 3.9 \times 10^{-6}$$

Because $2[V^{3+}] \ll 0.100 \text{ mol L}^{-1}$,

$$[V^{3+}] = 0.100 \text{ mol L}^{-1} \times 3.9 \times 10^{-6} = 3.9 \times 10^{-7} \text{ mol L}^{-1}$$

Thus less than $10^{-3}\%$ of the vanadium(IV) disproportionates in this acidic solution.

The potential diagram for manganese in acidic solution,

$$MnO_4^- \xrightarrow{0.564 \text{ V}} MnO_4^{2-} \xrightarrow{2.26 \text{ V}} MnO_2 \xrightarrow{0.95 \text{ V}} Mn^{3+} \xrightarrow{1.51 \text{ V}} Mn^{2+} \xrightarrow{-1.18 \text{ V}} Mn$$

(MnO_4^- to MnO_2: 1.69 V; MnO_2 to Mn^{2+}: 1.23 V)

shows two oxidation states (+6, MnO_4^{2-}; and +3, Mn^{3+}) that are unstable with respect to disproportionation in a solution with $[H^+] = 1.0 \text{ mol L}^{-1}$. You may wonder how the $\mathscr{E}^0$ values are obtained for half-reactions involving an unstable oxidation state. For some substances (e.g., hydrogen peroxide), the disproportionation reaction does not reach equilibrium rapidly, and the unstable oxidation state may be characterized without difficulty. The two oxidation states of manganese in question can be stabilized by appropriate changes in the concentration of hydrogen ion. The disproportionation reaction for manganese(VI),

$$3MnO_4^{2-} + 4H^+ = 2MnO_4^- + MnO_2 + 2H_2O$$
$$\mathscr{E}^0 = 2.26 \text{ V} - (0.564 \text{ V}) = +1.70 \text{ V}$$

is sensitive to the concentration of hydrogen ion. Rearrangement of the equilibrium constant equation gives

$$[MnO_4^{2-}] = \left(\frac{[MnO_4^-]^2}{K[H^+]^4}\right)^{1/3}$$

In strongly alkaline solution manganate ion is stable. There are pure crystalline manganate salts (e.g., K_2MnO_4), and if we dissolve such a salt in a concentrated solution of base, we obtain a solution of green manganate ion that can be studied with little interference from disproportionation. The disproportionation of manganese(III),

$$2Mn^{3+} + 2H_2O = Mn^{2+} + MnO_2 + 4H^+$$
$$\mathscr{E}^0 = 1.51\ V - (0.95\ V) = +0.56\ V$$

also has a strong dependence upon the concentration of hydrogen ion:

$$[Mn^{3+}] \propto [H^+]^2$$

(at constant $[Mn^{2+}]$), so it is possible to prepare and study highly acidic solutions containing an appreciable concentration of manganese(III).

The manganese potential diagram shows $\mathscr{E}^0$ values for couples which bypass an unstable intermediate oxidation state. In many situations, such $\mathscr{E}^0$ values are more useful in calculations than those for the stepwise half-reactions. Using the manganese(VII)–manganese(IV) couple as an example, we shall derive the relationship of the $\mathscr{E}^0$ values. The half-reaction

$$MnO_4^- + 4H^+ + 3e^- = MnO_2 + 2H_2O$$

is the sum of two other half-reactions:

$$MnO_4^- + e^- = MnO_4^{2-} \qquad \mathscr{E}^0 = +0.564\ V$$
$$MnO_4^{2-} + 4H^+ + 2e^- = MnO_2 + 2H_2O \qquad \mathscr{E}^0 = +2.26\ V$$

As we have shown, the value of $\mathscr{E}^0$ for the half-reaction involving three electrons is the weighted average of the values of $\mathscr{E}^0$ for the two half-reactions that combine to give it:

$$\mathscr{E}^0 = \tfrac{1}{3}(0.564\ V + 2 \times 2.26\ V) = 1.69\ V$$

We already have learned that the value of $\mathscr{E}^0$ for a particular couple in alkaline solution may be different from the value in acidic solution. For the correlation of oxidation–reduction reactions in alkaline solution, you should use half-reactions and values of $\mathscr{E}^0$ for such solutions. Potential diagrams can be constructed for alkaline solution just as for acidic solution; that for manganese at 25°C is:

$$MnO_4^- \xrightarrow{0.564\ V} MnO_4^{2-} \xrightarrow{0.596\ V} MnO_2 \xrightarrow{-0.23\ V} Mn(OH)_3 \xrightarrow{+0.13\ V} Mn(OH)_2 \xrightarrow{-1.56\ V} Mn$$

(MnO_4^- to MnO_2: 0.585 V; MnO_2 to $Mn(OH)_2$: −0.05 V)

Consider the lines in Figures 16–9 and 16–10 for the half-reactions

$$O_2 + 4H^+ + 4e^- = 2H_2O \qquad \mathscr{E}^0 = +1.23\ V$$
$$2H^+ + 2e^- = H_2 \qquad \mathscr{E}^0 = 0.00\ V$$

These half-reactions are relevant to the stabilities of aqueous solutions of oxidizing and reducing agents. The oxidizing agent (Ox) in a couple,

$$Ox + ne^- = Red$$

with $\mathscr{E}^0 > \sim 1.23$ V is capable (at $[H^+] = 1.00$ mol L^{-1}) of oxidizing water:

$$\tfrac{4}{n}\text{Ox} + 2H_2O = O_2 + \tfrac{4}{n}\text{Red}$$

The reducing agent (Red) in a couple,

$$\text{Ox} + ne^- = \text{Red}$$

with $\mathscr{E}^0 < \sim 0.00$ V is capable (at $[H^+] = 1.00$ mol L^{-1}) of reducing hydrogen ion:

$$\tfrac{2}{n}\text{Red} + 2H^+ = H_2 + \tfrac{2}{n}\text{Ox}$$

The situation at other acidities is described in Figure 16–10. Half-reactions inside the region defined by lines for the hydrogen and oxygen half-reactions involve an oxidized form that is not capable of oxidizing water or hydroxide ion and a reduced form that is not capable of reducing water or hydrogen ion. The iron(III)–iron(II) couple is in this region; neither of the reactions occurs:

$$4Fe^{3+} + 2H_2O \neq 4Fe^{2+} + O_2 + 4H^+$$
$$2Fe^{2+} + 2H^+ \neq 2Fe^{3+} + H_2$$

Half-reactions outside of this region of the graph involve an oxidized form capable of oxidizing water or hydroxide ion or a reduced form capable of reducing water or hydrogen ion. Aqueous solutions of such reagents, for instance acidic solutions of cerium(IV) or chromium(II), are unstable in the thermodynamic sense. The values of $\mathscr{E}^0$ for the half-reactions

$$Ce^{4+} + e^- = Ce^{3+} \qquad \mathscr{E}^0 = +1.70 \text{ V}$$
$$Cr^{3+} + e^- = Cr^{2+} \qquad \mathscr{E}^0 = -0.50 \text{ V}$$

show that equilibrium in each of the reactions

$$4Ce^{4+} + 2H_2O = 4Ce^{3+} + 4H^+ + O_2$$
$$\mathscr{E}^0 = +1.70 \text{ V} - 1.23 \text{ V} = +0.47 \text{ V}$$

$$2Cr^{2+} + 2H^+ = 2Cr^{3+} + H_2$$
$$\mathscr{E}^0 = 0.00 \text{ V} - (-0.50 \text{ V}) = +0.50 \text{ V}$$

is far to the right. But neither of these reactions occurs rapidly. Solutions of cerium(IV) perchlorate in perchloric acid, a useful reagent, are sold commercially. The solutions can be kept for years with the concentration of cerium(IV) decreasing only very slowly. Acidic solutions of chromium(II) are less easily kept, but their instability is due to the oxidation of chromium(II) by oxygen of the air, not by hydrogen ion.

16–7 Oxidation–Reduction in Analysis

Oxidation–reduction reactions are employed in the analysis of many substances. Permanganate ion, a versatile analytical reagent, is a powerful oxidizing agent in acidic solution, and its color (deep purple) allows us to see the end point in a titration without using an indicator. Before we consider some analytical reactions of permanganate ion, let us consider the stability of its aqueous stock solutions in which this reaction may occur:

$$4MnO_4^- + 2H_2O = 3O_2(g) + 4MnO_2(s) + 4OH^-$$

The equilibrium constant for this reaction can be calculated from values of $\mathscr{E}^0$ for the relevant half-reactions, which are

$$MnO_4^- + 2H_2O + 3e^- = MnO_2 + 4OH^- \qquad \mathscr{E}^0 = +0.585\ V$$
$$O_2 + 2H_2O + 4e^- = 4OH^- \qquad \mathscr{E}^0 = +0.400\ V$$

Subtraction of these values of $\mathscr{E}^0$ gives for the overall reaction (for which $n = 12$)

$$\mathscr{E}^0 = +0.585\ V - 0.400\ V = +0.185\ V$$

and

$$K^0 = 10^{0.185\ V/(0.0592\ V \div 12)} = 3.2 \times 10^{37}$$
$$K = 3.2 \times 10^{37}\ atm^3$$

This corresponds essentially to complete reduction of permanganate ion in an initially neutral solution. But despite this, solutions of potassium permanganate can be kept for a long time if no catalyst for the reaction is present. (Manganese dioxide, a product of the decomposition reaction, increases the rate of decomposition. For this reason, solutions of potassium permanganate should be prepared with highly purified water, and after standardization they can be kept only until the first traces of manganese dioxide appear.)

Established analytical procedures for iron(II), hydrogen peroxide, and oxalic acid ($H_2C_2O_4$) involve oxidation of each of these substances by permanganate ion. The relevant half-reactions and values of $\mathscr{E}^0$ are:

$$Fe^{3+} + e^- = Fe^{2+} \qquad \mathscr{E}^0 = +0.771\ V$$
$$O_2 + 2H^+ + 2e^- = H_2O_2 \qquad \mathscr{E}^0 = +0.69\ V$$
$$2CO_2 + 2H^+ + 2e^- = H_2C_2O_4 \qquad \mathscr{E}^0 = -0.49\ V$$

If you consider these values of $\mathscr{E}^0$ and those for the MnO_4^-–MnO_2 couple ($\mathscr{E}^0 = +1.69$ V) and the MnO_2–Mn^{2+} couple ($\mathscr{E}^0 = +1.23$ V), you will realize that the identity of the manganese product depends upon the stoichiometric proportions used. If the reducing agent is in excess, manganese(VII) is reduced to manganese(II). This is the case when permanganate ion is added from a buret to a solution of the reducing agent. With this order of addition, the reducing agent is in excess up to the end point, and the net reactions for oxidation of these reducing agents are:

$$MnO_4^- + 5Fe^{2+} + 8H^+ = Mn^{2+} + 5Fe^{3+} + 4H_2O$$
$$2MnO_4^- + 5H_2O_2 + 6H^+ = 2Mn^{2+} + 5O_2 + 8H_2O$$
$$2MnO_4^- + 5H_2C_2O_4 + 6H^+ = 2Mn^{2+} + 10CO_2 + 8H_2O$$

The end point in each of these titrations comes when the purple color of a slight excess of permanganate ion appears, but this color fades in time. The reaction responsible for this fading of color is

$$2MnO_4^- + 3Mn^{2+} + 2H_2O = 5MnO_2 + 4H^+$$

a reaction between the slight excess of permanganate ion and manganese(II) ion produced in the original reaction. If we add reagents in the opposite order, adding each of these reducing agents to a solution of permanganate ion, manganese dioxide will form initially, and will not be reduced to manganous ion until all of the permanganate ion is consumed. Even an excess of reducing agent will not dissolve the manganese dioxide rapidly. This order of addition, with the reducing agent in the buret, does not lend itself to the development of useful analytical procedures involving permanganate ion.

With reagents less intensely colored than permanganate ion, the end point of an oxidation–reduction titration must be detected in another way. Measurements of emf may be employed, and there also are indicators that are useful. To monitor the course of an oxidation–reduction reaction by emf measurements, the oxidation–reduction reaction is carried out in a titration vessel containing an inert electrode (e.g., metallic platinum). This vessel is one half-cell; it is connected by a salt bridge to the other half-cell containing a reference electrode (e.g., the saturated calomel electrode).

Let us consider a titration half-cell in which vanadium(IV) is being oxidized by cerium(IV) in the presence of acid ($[H^+] = 1.00 \text{ mol L}^{-1}$). The net reaction is

$$Ce^{4+} + VO^{2+} + H_2O = Ce^{3+} + VO_2^+ + 2H^+$$

The value of $\mathscr{E}^0$ for this reaction ($\mathscr{E}^0 = +0.70$ V) is obtained from the values of $\mathscr{E}^0$ for the component half-reactions; for 25°C,

$$Ce^{4+} + e^- = Ce^{3+} \qquad \mathscr{E}^0 = +1.70 \text{ V}$$
$$VO_2^+ + 2H^+ + e^- = VO^{2+} + H_2O \qquad \mathscr{E}^0 = +1.00 \text{ V}$$

The equilibrium constant for the reaction (with $n = 1$) is very large, and the reaction occurs rapidly:

$$K^0 = 10^{0.70 \text{ V}/0.0592 \text{ V}} = 10^{11.8} = 6.7 \times 10^{11}$$
$$K = 6.7 \times 10^{11} \text{ mol}^2 \text{ L}^{-2}$$

Each added increment of cerium(IV) is converted essentially completely to cerium(III) by the reaction with excess VO^{2+}. The concentration ratio

$[VO^{2+}]/[VO_2^+]$ decreases during the course of the titration, and the electromotive force of this half-cell increases; the value of $\mathscr{E}$ is given by

$$\mathscr{E} = +1.00\ \text{V} - \frac{0.0592\ \text{V}}{1}\log\left(\frac{[VO^{2+}]\ \text{mol}^2\ \text{L}^{-2}}{[VO_2^+][H^+]^2}\right)$$

which for $[H^+] = 1.00\ \text{mol L}^{-1}$, becomes

$$\mathscr{E} = 1.00\ \text{V} - \frac{0.0592\ \text{V}}{1}\log\left(\frac{[VO^{2+}]}{[VO_2^+]}\right)$$

At 50% reaction, $[VO^{2+}] = [VO_2{}^+]$, $\mathscr{E} = \mathscr{E}^0 = 1.00$ V. This is shown in Figure 16–12, which gives the value of $\mathscr{E}$ throughout the course of the titration. The value of $\mathscr{E}$ is related to the concentration ratio $[VO^{2+}]/[VO_2^+]$ as already shown, but it also is related to the concentration ratio $[Ce^{3+}]/[Ce^{4+}]$:

$$\mathscr{E} = +1.70\ \text{V} - \frac{0.0592\ \text{V}}{1}\log\left(\frac{[Ce^{3+}]}{[Ce^{4+}]}\right)$$

At each point in the titration, equilibrium is established in the overall reaction, and the values of $\mathscr{E}$ for each of the two half-reactions are the same. We make use of both of these equations in calculating the value of $\mathscr{E}$ at the equivalence point, where

$$[Ce^{4+}] = [VO^{2+}] \quad \text{and} \quad [Ce^{3+}] = [VO_2^+]$$

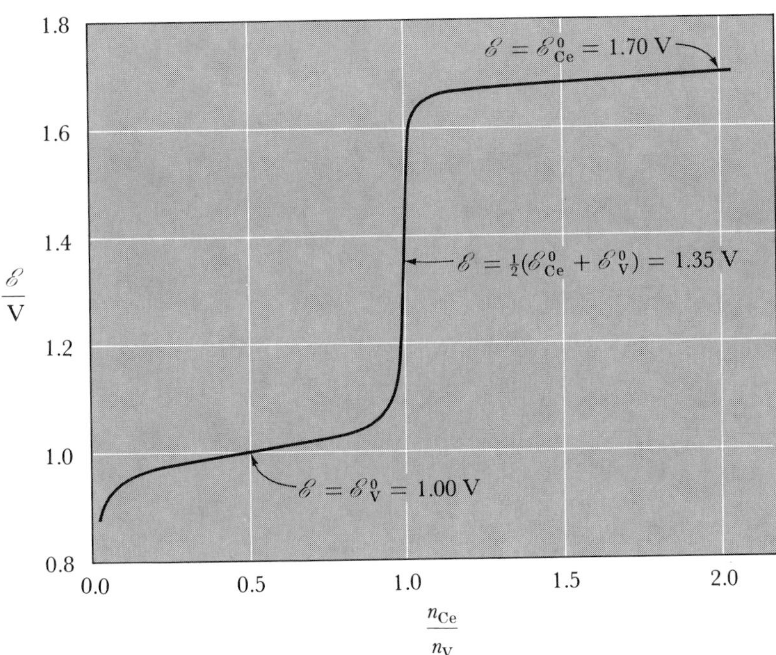

FIGURE 16–12
The voltage of a galvanic cell. In one half-cell the titration of VO^{2+} by Ce^{4+} is being carried out; the other half-cell has $\mathscr{E} = 0$ V.

(Study the chemical equation for the reaction to convince yourself that this is the relationship of the concentrations when exactly one mole Ce(IV) has been added per mole of VO^{2+}.) The emf of the titration half-cell at this point can be calculated by adding the equations for $\mathscr{E}$ for each of the two half-reactions

$$\mathscr{E} = 1.70\ \text{V} - 0.0592\ \text{V} \log\left(\frac{[Ce^{3+}]}{[Ce^{4+}]}\right)$$

$$\mathscr{E} = 1.00\ \text{V} - 0.0592\ \text{V} \log\left(\frac{[VO^{2+}]}{[VO_2^+]}\right)$$

$$2\mathscr{E} = 1.00\ \text{V} + 1.70\ \text{V} - 0.0592\ \text{V} \log\left(\frac{[Ce^{3+}][VO^{2+}]}{[Ce^{4+}][VO_2^+]}\right)$$

Cancellation of the concentrations that are equal gives

$$2\mathscr{E} = (1.00\ \text{V} + 1.70\ \text{V}) - 0.0592\ \text{V} \log 1.00$$

$$\mathscr{E} = \tfrac{1}{2}(1.00\ \text{V} + 1.70\ \text{V}) = 1.35\ \text{V}$$

This is the average of the $\mathscr{E}^0$ values for the two half-reactions.

We see in Figure 16–12 that the value of $\mathscr{E}$ changes very dramatically at the equivalence point. At 99.80% reaction,

$$\frac{[VO^{2+}]}{[VO_2^+]} = \frac{0.20}{99.80} = 2.0 \times 10^{-3}$$

and

$$\mathscr{E} = 1.00\ \text{V} - 0.0592\ \text{V} \log (2.0 \times 10^{-3}) = 1.16\ \text{V}$$

At 0.20% past the equivalence point,

$$\frac{[Ce^{3+}]}{[Ce^{4+}]} = \frac{100}{0.20} = 500$$

and

$$\mathscr{E} = +1.70\ \text{V} - 0.0592\ \text{V} \log 500 = +1.54\ \text{V}$$

Thus the value of $\mathscr{E}$ changes by 0.38 V (1.54V – 1.16V) at the equivalence point ±0.2%. Figure 16–12 shows the value of $\mathscr{E}$ well past the equivalence point. If a two-fold excess of cerium (IV) is added, the half-cell contains equal concentrations of cerium(IV) and cerium(III), and

$$\mathscr{E} = 1.70\ \text{V} - 0.0592\ \text{V} \log 1.00 = 1.70\ \text{V}$$

The equivalence point in the titration of vanadium(IV) by cerium(IV) also can be determined by using an appropriate oxidation–reduction indicator, for example, *tris*-phenanthrolineiron(II)–(III) (see Table 14–2) for the structure of phenanthroline, which will be abbreviated phen):

$$\underset{\text{Pale blue}}{Fe(phen)_3^{3+}} + e^- = \underset{\text{Blood red}}{Fe(phen)_3^{2+}} \qquad \mathscr{E}^0 = 1.12\ \text{V}$$

When this system is used as an indicator at very low concentrations, the ratio $[Fe(phen)_3^{2+}]/[Fe(phen)_3^{3+}]$ is determined by the value of $\mathscr{E}$ that is established by the couples involved in the oxidation–reduction titration. The equation

$$\mathscr{E} = 1.12\text{ V} - 0.0592\text{ V}\log\left\{\frac{[Fe(phen)_3^{2+}]}{[Fe(phen)_3^{3+}]}\right\}$$

can be converted to

$$\frac{[Fe(phen)_3^{2+}]}{[Fe(phen)_3^{3+}]} = 10^{(1.12\text{ V}-\mathscr{E})/0.0592\text{V}}$$

Thus in the titration depicted in Figure 16–12, the value of this concentration ratio changes dramatically in the vicinity of the equivalence point. At 99.8% reaction, $\mathscr{E} = 1.16$ V so

$$\frac{[Fe(phen)_3^{2+}]}{[Fe(phen)_3^{3+}]} = 10^{-0.04\text{V}/0.0592\text{V}} = 0.2$$

and at the equivalence point, $\mathscr{E} = 1.35$ V and

$$\frac{[Fe(phen)_3^{2+}]}{[Fe(phen)_3^{3+}]} = 10^{-0.23\text{V}/0.0592\text{V}} = 1.3 \times 10^{-4}$$

Thus at 99.8% reaction, the relative concentration of the deeply colored iron(II) species is high enough to make the solution red. With the addition of another drop of cerium(IV), the relative concentration of the iron(II) species becomes very low, and the color of the solution becomes very pale blue, almost colorless.

Biographical Notes

CHARLES M. HALL (1863–1914) was a student at Oberlin College (Oberlin, Ohio) when he became interested in finding a method for producing aluminum. He invented the electrolytic process described in the text in the year after his graduation.

PAUL L. T. HEROULT (1863–1914) was a French chemist, who not only discovered independently the successful electrolytic method for producing aluminum, but also invented an electric furnace for producing steel.

GEORGE LECLANCHE′ (1839–1882), a French chemist, developed the dry cell battery as well as other practical electrochemical devices.

Problems and Questions

16–1 What is the oxidation state of chromium in each of the substances or species CrO, $Cr(OH)_2$, $Cr(OH)_3$, $NaCrO_2$, CrO_2, Na_2CrO_4, $K_2Cr_2O_7$? What is the oxidation state of sulfur in Na_2SO_4, $S_2O_3^{2-}$, $S_2O_6^{2-}$, $S_2O_4^{2-}$, $HOSO_2^-$?

16–2 Balance the following equations for half-reactions:

For acidic solution	For alkaline solution
$ClO_4^- = ClO_3^-$	$Ni(OH)_3 = Ni(OH)_2$
$MnO_4^- = MnO_2$	$H_3PO_2 = P_4$
$SO_4^{2-} = SO_2$	$ZnO_2^{2-} = Zn$
$NO_3^- = NH_4^+$	$Sn(OH)_6^{2-} = SnO_2^{2-}$
$PuO_2^{2+} = Pu^{4+}$	$AgO = Ag_2O$
$VO_2^+ = V^{3+}$	$PbO_2 = PbO$
$N_2H_5^+ = NH_4^+$	$MnO_4^- = MnO_2$
$ICl_2^- = I_2$	$N_3^- = NH_3$

16–3 For each of the balanced half-reaction equations in Problem 16–2 show that the change in oxidation state and the number of electrons involved in the half-reaction are numerically equal.

16–4 Balance the following equations for reactions (use each of the methods described in Section 16–2 for two examples in each category):

For acidic solution	For alkaline solution
$HCrO_4^- + H_2O_2 = Cr^{3+} + O_2$	$P_4 = H_2PO_3^- + PH_3$
$BrO_3^- + Br^- = Br_2$	$OCl^- + MnO_2 = MnO_4^{2-} + Cl^-$
$ICl_2^- = IO_3^- + I_2$	$Br_2 = Br^- + BrO_3^-$
$MnO_4^- + H_2C_2O_4 = CO_2 + Mn^{2+}$	$Hg_2Cl_2 + NH_3 = HgNH_2Cl + Hg$
$HgS + NO_3^- = HgCl_4^{2-} + SO_4^{2-} + NO$	$Al = AlO_2^- + H_2$
$SCN^- + S_2O_8^{2-} = SO_4^{2-} + CO_2 + NO_3^-$	$AlH_4^- = AlO_2^- + H_2$
$Fe^{2+} + NO_3^- = FeNO^{2+} + Fe^{3+}$	$CH_3CO_2^- + MnO_4^- = CO_3^{2-} + MnO_2$

16–5 If the electrolytic cell shown in Figure 16–1 were divided into two compartments by a porous membrane that prevented mixing but did not prevent ion migration, at which electrode would the solution become more concentrated in sulfuric acid?

16–6 Slow electrolysis of dilute sulfuric acid results in the decomposition of water, $2H_2O = 2H_2(g) + O_2(g)$ (illustrated in Figure 16–1). However, in electrolysis at higher current density, less oxygen gas is evolved than corresponds to the stoichiometry of this equation. Can you suggest a reason?

16–7 An electrolytic plant produces 50,000 tons of aluminum per year. If the plant is in operation at all times (even Christmas day), what current must flow to give this production?

16–8 Two electrolytic cells are in series. In one there is a solution of copper(II) sulfate ($CuSO_4$) and in the other there is a solution of copper(I) chloride in hydrochloric acid (the complex ion $CuCl_2^-$ is the predominant species). How much copper metal will be deposited at the cathode in each cell if a current of 6.00 amperes flows for 2.00 hours?

16–9 Diagram a galvanic cell based upon each of the following oxidation–reduction reactions:

$$H_2(g) + Br_2(aq) = 2H^+ + 2Br^-$$
$$2Ag(s) + I_3^-(aq) = 2AgI(s) + I^-$$
$$VO_2^+ + V^{3+} = 2VO^{2+}$$

For each cell write equations for the two half-reactions, give the value of $\mathscr{E}^0$ for 298.2 K, and indicate the direction of migration of all ions in the solutions.

16–10 What would be the electromotive force of the hydrogen–bromine battery in Problem 16–9 under the following conditions? $T = 298.2$ K, $P_{H_2} = 3.00$ atm, $[Br_2] = 2.00 \times 10^{-3}$ mol L^{-1}, $[H^+] = 0.500$ mol L^{-1}, $[Br^-] = 0.0400$ mol L^{-1}.

16–11 Diagram a galvanic cell based upon the oxidation–reduction reaction in alkaline solution

$$H_2O + ClO^- + 2Fe(OH)_2 = 2Fe(OH)_3 + Cl^-$$

What are the half-reactions, the value of $\mathscr{E}^0$, the direction of ion migration during discharge, and the value of $\mathscr{E}$ at $[OCl^-] = 5.00 \times 10^{-3}$ mol L^{-1}, $[Cl^-] = 1.00$ mol L^{-1}, $[OH^-] = 0.100$ mol L^{-1}? Which electrode is the anode during discharge? What side reaction may occur in the cell?

16–12 The value of $\mathscr{E}^0$ for the half-reaction $AgN_3 + e^- = Ag + N_3^-$ is +0.29 V. What is the value of K_s for silver azide?

16–13 The reaction $AgCl(s) + I^-(aq) = AgI(s) + Cl^-(aq)$ does not involve oxidation–reduction. However, this reaction can be the basis for a galvanic cell. Diagram this cell; what is the value of $\mathscr{E}^0$?

16–14 From values of $\mathscr{E}^0$ given in Table 16–2 calculate values of $\mathscr{E}^0$ for the half-reactions

$$CH_3CO_2H + 4H^+ + 4e^- = CH_3CH_2OH + H_2O$$
$$ClO_4^- + H_2O + 2e^- = ClO_3^- + 2OH^-$$

16–15 From the two values of $\mathscr{E}^0$ for the peroxide–water half-reactions,

$$H_2O_2 + 2H^+ + 2e^- = 2H_2O \qquad \mathscr{E}^0 = +1.77 \text{ V}$$
$$HO_2^- + H_2O + 2e^- = 3OH^- \qquad \mathscr{E}^0 = +0.89 \text{ V}$$

calculate the first acid-dissociation constant for hydrogen peroxide.

16–16 From the $\mathscr{E}^0$ for the half-reaction $Fe^{2+} + 2e^- = Fe$ ($\mathscr{E}^0 = -0.44$ V) and the value of K_s for $Fe(OH)_2$ ($K_s = 8.0 \times 10^{-16}$ mol^2 L^{-2}), calculate the value of $\mathscr{E}^0$ for the iron(II)–iron(0) half-reaction in alkaline solution, $Fe(OH)_2 + 2e^- = Fe + 2OH^-$.

16–17 For each of the half-reactions given in Figure 16–10, the value of $\mathscr{E}^{0\prime}$ either is independent of pH or decreases with increasing pH. What characteristics in an oxidation–reduction couple would cause the value of $\mathscr{E}^{0\prime}$ to show the opposite trend (an increase in $\mathscr{E}^{0\prime}$ with an increase of pH)?

16–18 What reaction occurs when an acidic solution of vanadium(II) is treated with excess vanadium(V)? What reaction occurs when an acidic solution of vanadium(V) is treated with excess vanadium(II)?

16–19 Use the potential diagram (Section 16–6) for manganese in alkaline solution to determine equilibrium constants for the reactions:

$$3MnO_4^{2-} + 2H_2O = 2MnO_4^- + MnO_2 + 4OH^-$$
$$2MnO_4^- + 3Mn(OH)_2 = 5MnO_2 + 2OH^- + 2H_2O$$
$$4MnO_4^- + 2H_2O = 4MnO_2 + 3O_2 + 4OH^-$$

16–20 Is acetaldehyde stable with respect to disproportionation in acidic solution? (Use $\mathscr{E}^0$ values in Table 16–2 to answer.) Solutions of this reagent are available. What does your answer coupled with this information tell you about reaction rates?

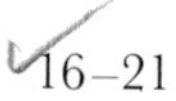

16–21 Using $\mathscr{E}^0$ values given in Table 16–2 and in the text, calculate the equilibrium constant at 298.2 K for each of these reactions:

$$I_2(aq) + 2S_2O_3^{2-} = S_4O_6^{2-} + 2I^-$$
$$IO_3^- + 5I^- + 6H^+ = 3I_2(aq) + 3H_2O$$
$$2Cl_2(g) + 2H_2O = 4H^+ + 4Cl^- + O_2(g)$$

16–22 An acidic solution of vanadium(II) ion is treated with excess silver(I) ion. What reaction occurs, and what is the equilibrium constant for this reaction?

16–23 Write the balanced equation for the reaction that occurs when a lead storage battery is charged. What other reaction occurs during the charging process? Compare the values of $\mathscr{E}^0$ for each of these reactions, and explain why the desired charging reaction is the one that occurs predominantly.

16–24 Aqueous solutions of iron(II) sulfate that are exposed to air maintain their reducing capacity better at high acidity. Is this due to equilibrium or to kinetic factors?

16–25 What is the equilibrium partial pressure of $O_2(g)$ over a solution of 0.100*M* NaOH containing a mixture of $Ag_2O(s)$ and $Ag(s)$?

16–26 A sample of pure Fe_3O_4 weighing 0.4532 g is dissolved in dilute sulfuric acid and then titrated with 0.02430*M* $KMnO_4$. What volume of permanganate solution is required?

16–27 For the reaction $Ce^{4+} + Fe^{2+} = Ce^{3+} + Fe^{3+}$, the value of $\mathscr{E}^0$ is +0.93 V. What is the value of the equilibrium constant for this reaction? What is the value of $\mathscr{E}$ for the iron(III)–iron(II) half-reaction in a reaction mixture in which a solution of iron(II) is 90% titrated to iron(III) by the addition of cerium(IV) ion? What is the ratio of the concentrations of cerium(IV) to cerium(III) in this solution? What is the value of $\mathscr{E}$ in this half-cell at the equivalence point?

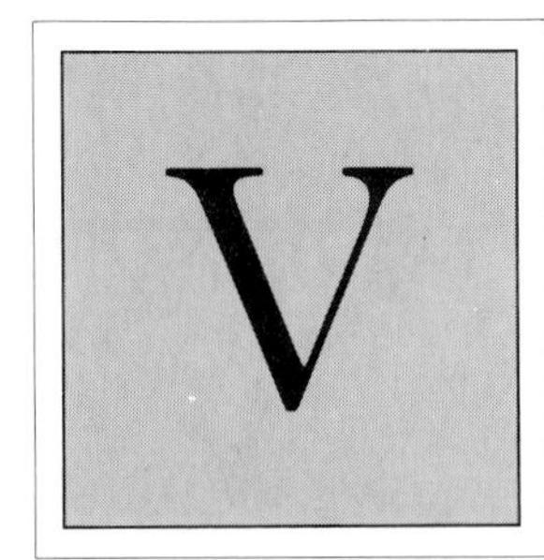

CHEMISTRY OF THE ELEMENTS

The next three chapters deal with the properties and reactions of selected elements. With the background provided by previous chapters, we can consider:

1. *The electronic structures of compounds.* Both covalent and ionic compounds will be encountered. In many of the ionic compounds, one or both of the ions are polyatomic covalently bonded units. For most polyatomic molecules and ions, we will find it possible to draw satisfactory electron-dot formulas, with the concept of resonance involved in some of the structures. When we discuss transition-metal compounds we will treat aspects of chemical bonding that were not introduced in previous chapters.
2. *The geometrical arrangement of atoms in molecules.* For polyatomic molecules and ions of nonmetals, the valence-shell electron-pair repulsion model will be the basis for most of our discussion of structure. For ionic compounds the role of relative ionic radii of the cation and anion in determining coordination numbers has been covered thoroughly in Chapter 9. But the principles outlined in that chapter treated all ions as spheres; when we introduce the possibility of some directional covalency in the bonding of compounds of the metals the discussion of structure goes beyond what we learned previously.
3. *Acid–base, oxidation–reduction, solubility, and complex-ion formation reactions.* We will consider the position of equilibrium in these reactions (using

values of equilibrium constants and of the thermodynamic quantities ΔH_f^0, ΔG_f^0, and S^0 for the reactant and product substances), and also their rate laws and mechanisms. We will see additional examples of the way in which the empirical (the rate law) and the theoretical (the mechanism) are correlated.

The survey will not deal with each element of the periodic table, but will consider some elements from each class—nonmetals and metals, both nontransition and transition metals. A brief discussion of the oxidation–reduction chemistry of transuranium elements will appear in the chapter on radioactivity and the nucleus (Chapter 22).

17

The Nonmetals

17–1 Introduction

The elements in the segment of the periodic table given in Figure 17–1 generally are classified as nonmetals. These elements show a range of chemical reactivity: fluorine is one of the most reactive elements, whereas the noble gases helium, neon, argon, krypton, xenon (and radon) are very unreactive. (Until 1962, the noble gases were thought by many to be completely unreactive.) No one set of properties is exhibited by all nonmetals and by no metals. Nonetheless, some properties help distinguish nonmetals from metals:

1. *Melting points and boiling points in the elemental state.* Nonmetals generally have lower melting points and lower boiling points than metals. Ten nonmetals shown in Figure 17–1 and hydrogen are gases at standard temperature and pressure, and one is a liquid (bromine). No metals are gases at standard temperature and pressure; one metal is a liquid (mercury).
2. *The nature of elemental molecules.* Many nonmetals in the elemental state exist as simple discrete molecules containing a small number of atoms. The noble gases exist as monatomic units; the halogens, oxygen, and nitrogen occur as diatomic molecules; phosphorus forms a molecule containing four atoms, and sulfur forms a molecule containing eight atoms. Because nonmetals form simple discrete molecules, their melting and boiling points are relatively so low. A nonmetal that exists in each

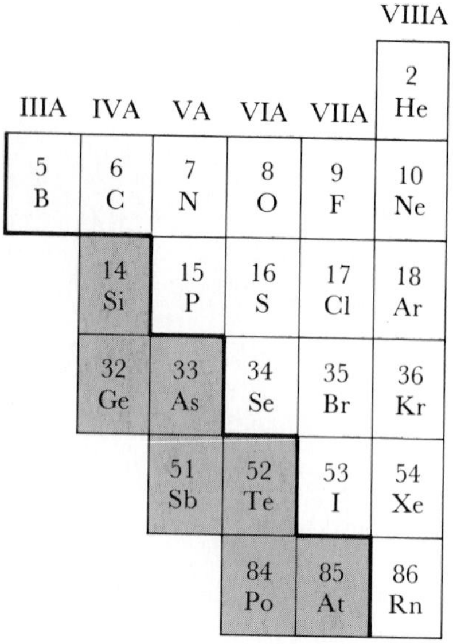

IIIA	IVA	VA	VIA	VIIA	VIIIA
					2 He
5 B	6 C	7 N	8 O	9 F	10 Ne
	14 Si	15 P	16 S	17 Cl	18 Ar
	32 Ge	33 As	34 Se	35 Br	36 Kr
		51 Sb	52 Te	53 I	54 Xe
			84 Po	85 At	86 Rn

FIGURE 17–1
The nonmetals. The shaded elements are metalloids, elements on the borderline between metals and nonmetals.

state (gas, liquid, and solid) as the same discrete molecule melts and vaporizes without the breaking of covalent bonds.

3. *Electrical conductivity*. Nonmetals (except carbon as graphite) are very poor conductors of electricity.
4. *The acidity and basicity of the oxide*. The oxides of most nonmetals are acidic: nonmetal oxides react with water to give Brønsted acids. In contrast, most metal oxides are basic; however, some metal oxides also are acidic so this property is not adequate to characterize a nonmetal.
5. *The attraction of the atom for electrons*. Atoms of nonmetals generally have larger ionization energies, electron affinities, and electronegativities than do atoms of metals. (See Figure 17–2.)
6. *The nature of the compounds with nonmetals*. Most compounds of nonmetals with metals are ionic solids; most compounds of nonmetals with nonmetals are discrete covalently bonded molecules.

The simplest principle of covalent chemical bonding is that an atom tends to share as many electrons as possible with other atoms to fill its valence-shell s and p orbitals. This principle is successful in explaining the composition and structure of most compounds of nonmetals. For nonmetal atoms having a valence shell with $n \geq 3$, d orbitals also may play a role in chemical bonding, and more than an octet of electrons may be present in the valence shell of such a nonmetal atom. The discussion of covalent bonding in Chapter 10 included examples of molecules in which nonmetal atoms had more than an octet of electrons [e.g., $PCl_5(g)$ and SF_6].

Now we will generate the formulas of some molecules containing nitrogen, oxygen, and fluorine, in much the same way that we generated a series of carbon compounds in Table 11–4. To acquire an octet of electrons by combining with hydrogen, an atom with five, six, or seven valence electrons (e.g., nitrogen, oxygen, and fluorine) adds three, two, or one

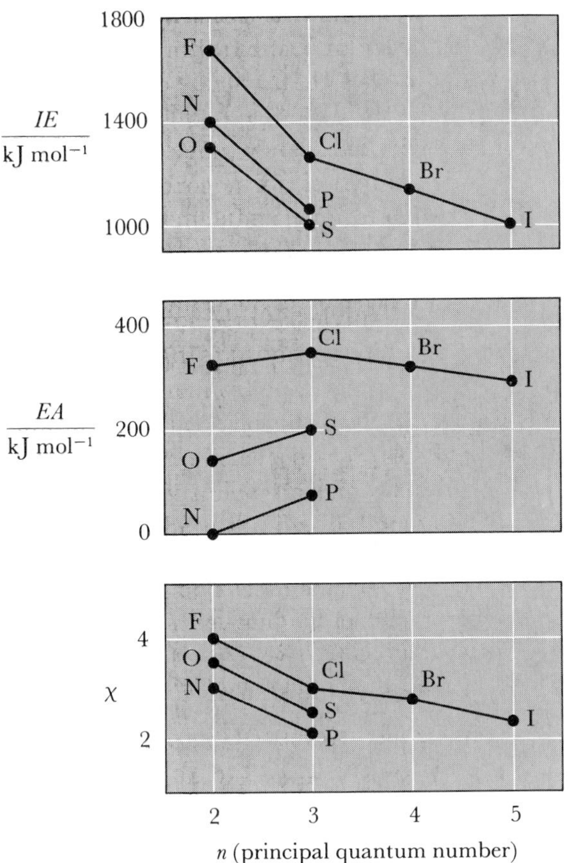

FIGURE 17–2
Some properties of nonmetal atoms: ionization energy, electron affinity, electronegativity. Comparison with metals: Only one metal (Hg) has a value of $IE > 1000$ kJ mol^{-1}; only two metals (Pt and Au) have values of $EA > 200$ kJ mol^{-1}; most metals have values of $\chi < 1.5$.

hydrogen atoms, respectively. This is the situation in ammonia, water, and hydrogen fluoride, each of which is a known stable substance:

$$\cdot\ddot{\text{N}}\cdot + 3\text{H}\cdot \longrightarrow \text{NH}_3 \;(\ddot{\text{N}} \text{ bonded to H, H, H})$$

$$:\ddot{\text{O}}\cdot + 2\text{H}\cdot \longrightarrow :\ddot{\text{O}}\text{—H} \;(\text{O bonded to H, H})$$

$$:\ddot{\underset{\cdot\cdot}{\text{F}}}\cdot + \text{H}\cdot \longrightarrow :\ddot{\underset{\cdot\cdot}{\text{F}}}\text{—H}$$

The formulas of other known compounds of nitrogen, oxygen, and fluorine can be generated by replacing the hydrogen atoms of ammonia, water, and hydrogen fluoride by the hydroxyl radical. This process is summarized in Table 17–1. As we showed in our development of CH_aO_b molecules (Table 11–4), a molecule with two hydroxyl groups bonded to the same nonmetal atom is unstable in that it may lose a water molecule. In the series now being considered, the molecules $O(OH)_2$, $HN(OH)_2$, and $N(OH_3)$ have more than one hydroxyl groups bonded to the same nonmetal atom and none of these compounds is stable. Their decomposition

TABLE 17–1

Molecules Derived from Hydrogen Fluoride, Water, and Ammonia by Replacing Hydrogen Atoms with Hydroxyl Radicals

Parent molecule	Parent molecule minus ·H plus ·ÖH		Anhydride
:F̤̈—H	:F̤̈—Ö̤H Hypofluorous acid	⟶	:F̤̈—Ö̤—F̤̈: Oxygen difluoride
H—Ö̤—H	H—Ö̤—Ö̤H Hydrogen peroxide		
	HÖ̤—Ö̤—Ö̤H Dihydrogen trioxide[a]	⟶	O_2
H—N̈—H | H	H—N̈—Ö̤H | H Hydroxylamine		
	HÖ̤—N̈—Ö̤H | H Unknown	⟶	HÖ̤—N̈=N̈—Ö̤H Hyponitrous acid ↓ :N̈=N=Ö: Nitrous oxide
	HÖ̤—N̈—Ö̤H | :Ö: H Unknown	⟶	HÖ̤—N̈=Ö Nitrous acid ↓ :Ö=N̈—Ö̤—N̈=Ö: Dinitrogen trioxide

[a]Unstable; the deuterium (^{2}H) compound, which is more stable than the normal hydrogen (^{1}H) compound, decomposes in a first-order reaction with a half-time of 139 s at 0°C in 0.027 *M* $DClO_4$ in heavy water, D_2O.

forms anhydrides, which involve double bonds, a structural feature that is particularly stable for atoms of nonmetals with a valence shell having $n = 2$. If the valence shell of the nonmetal has $n \geq 3$, double-bonded structures are relatively less stable, and molecules with two or more hydroxyl groups bonded to a particular nonmetal atom are stable. For the $n = 3$

nonmetals, such known molecules include orthosilicic acid [$Si(OH)_4$], orthophosphoric acid [$OP(OH)_3$], and sulfuric acid [$O_2S(OH)_2$].

This approach to compounds of nitrogen does not generate some important known molecules:

NO	NO_2	$HONO_2$
Nitric oxide	Nitrogen dioxide	Nitric acid

The first two of these, the oxides NO and NO_2, have an odd number of electrons, and they cannot be made to fit into a series of molecules in which the parent molecule contains an even number of electrons. The last of these molecules, nitric acid, $HONO_2$, involves nitrogen in the +5 oxidation state. To generate the more positive oxidation states of the nonmetals, we must start with some substance other than the hydrides. We will do so by considering the molecules in which the nonmetal atom is bonded to a maximum number of oxygen atoms by single bonds plus a maximum number of hydroxyl radicals by single bonds. For a nitrogen atom with five electrons in the $n = 2$ shell, this bonding situation gives orthonitric acid:

```
                                        H
                                        ··
                                       :O:
      ·        ··        ··        ··   ··   ··
    : N · + 1 O : + 3 · O : H ——→ : O : N : O : H
      ·        ··        ··        ··   ··   ··
                                       :O:
                                        ··
                                        H
```

Orthonitric acid

Orthonitric acid is not stable; a molecule of orthonitric acid loses one molecule of water to become the stable form of nitric acid ($HONO_2$, generally written HNO_3):

```
         H
         ··
        :O:                    :O:
   ··    ··   ··               ··
  :O  :  N  : O : H  ——→   : O : N        + H2O
   ··    ··   ··           ··     ·O·
        :O:                       ··
         ··                        H
         H
```

Nitric acid

Additional loss of water results in dinitrogen pentaoxide (N_2O_5), the anhydride of nitric acid:

```
      :O:                        ·O·              ·O·
       ··                          ··   ··        ··
  2 : O : N          ——→            N : O : N           + H2O
      ··    ·O·                    ··   ··        ··
             ··                  ·O·              ·O·
              H
```

Dinitrogen pentaoxide

If the structures for nitric acid and dinitrogen pentaoxide are to be described

TABLE 17–2

Maximum and Minimum Oxidation States Exhibited by Nonmetals in Compounds with Hydrogen and Oxygen, and in Simple Ions

C	N	O	F
(−4) C^{4-}, CH_4	(−3) N^{3-}, NH_3	(−2) O^{2-}, H_2O	(−1) F^-, HF
(+4) $(HO)_2CO$, CO_3^{2-}, CO_2	(+5) $HONO_2$, NO_3^-, N_2O_5		
	P	**S**	**Cl**
	(−3) P^{3-}, PH_3	(−2) S^{2-}, H_2S	(−1) Cl^-, HCl
	(+5) $(HO)_3PO$, PO_4^{3-}, P_4O_{10}	(+6) $(HO)_2SO_2$, SO_4^{2-}, SO_3	(+7) $HOClO_3$, ClO_4^-, Cl_2O_7
			Br
			(−1) Br^-, HBr
			(+7) $HOBrO_3$, BrO_4^-
			I
			(−1) I^-, HI
			(+7) $HOIO_3$, IO_4^-, I_2O_7 $(HO)_5IO$, $(HO)_4IO_2^-$

with electron-dot formulas, as they are here, resonance is needed. Only a single contributing structure is shown here for each of these molecules.

Table 17–2 gives the maximum and minimum oxidation states for the nonmetals that we will be studying.

17–2 The Free Elements

Of the nonmetals, only nitrogen, oxygen, sulfur, and the noble gases occur in nature as the free elements. Air, the composition of which is given in Table 17–3, is the principal source of nitrogen, oxygen, and the noble gases except helium. Because of differences in boiling points, it is possible to separate these elements from one another by fractional distillation of liquid air. Figure 17–3 shows the boiling point–composition diagram for liquid solutions of nitrogen and oxygen. We see from this figure that distillation of liquid air gives an enrichment of nitrogen in the early distillate and of oxygen in the residual liquid. Fractional distillation is used commercially to prepare pure liquid nitrogen and pure liquid oxygen; for 1978 the production in the United States was 360 billion cubic feet of $N_2(l)$ and 420 billion cubic feet of $O_2(l)$.

TABLE 17–3
The Composition of Dry Air

Substance		Volume percent	Boiling point, K
Nitrogen	N_2	78.084 ± 0.004	77.4
Oxygen	O_2	20.946 ± 0.002	90.2
Carbon dioxide	CO_2	0.033 ± 0.001	—
Argon	Ar	0.934 ± 0.001	87.5
Neon	Ne	$(18.18 \pm 0.04) \times 10^{-4}$	27.2
Helium	He	$(5.24 \pm 0.004) \times 10^{-4}$	4.6
Krypton	Kr	$(1.14 \pm 0.01) \times 10^{-4}$	120.9
Xenon	Xe	$(0.087 \pm 0.001) \times 10^{-4}$	166
Hydrogen	H_2	$\sim 5 \times 10^{-5}$	20
Methane	CH_4	$\sim 2 \times 10^{-4}$	111.7
Nitrous oxide	N_2O	$\sim 5 \times 10^{-5}$	185

Argon, the first noble gas discovered, was found in the atmosphere after STRUTT and RAMSAY had observed that the density of the unreactive gas in air, presumed initially to be pure nitrogen, was greater than that of nitrogen gas obtained by decomposing pure nitrogen compounds. This heavier "nitrogen" contained argon, which makes up $\sim 1\%$ of air. The difference between nitrogen from pure compounds and nitrogen from air was ap-

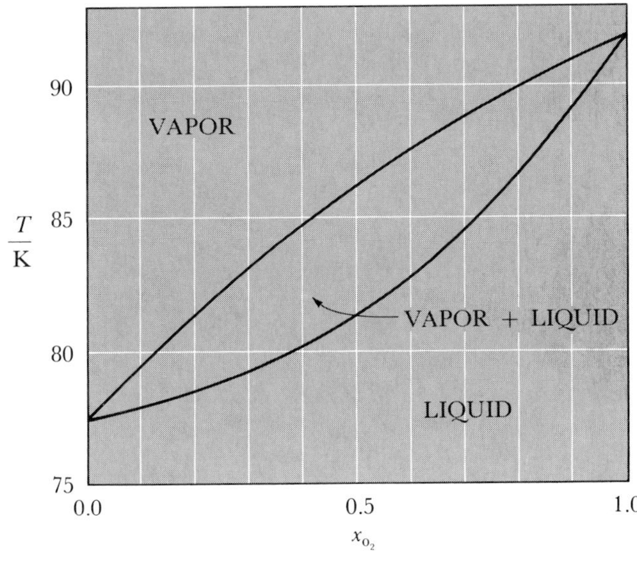

FIGURE 17–3
Boiling point–composition diagram for liquid solutions of nitrogen and oxygen (P = 1 atm).

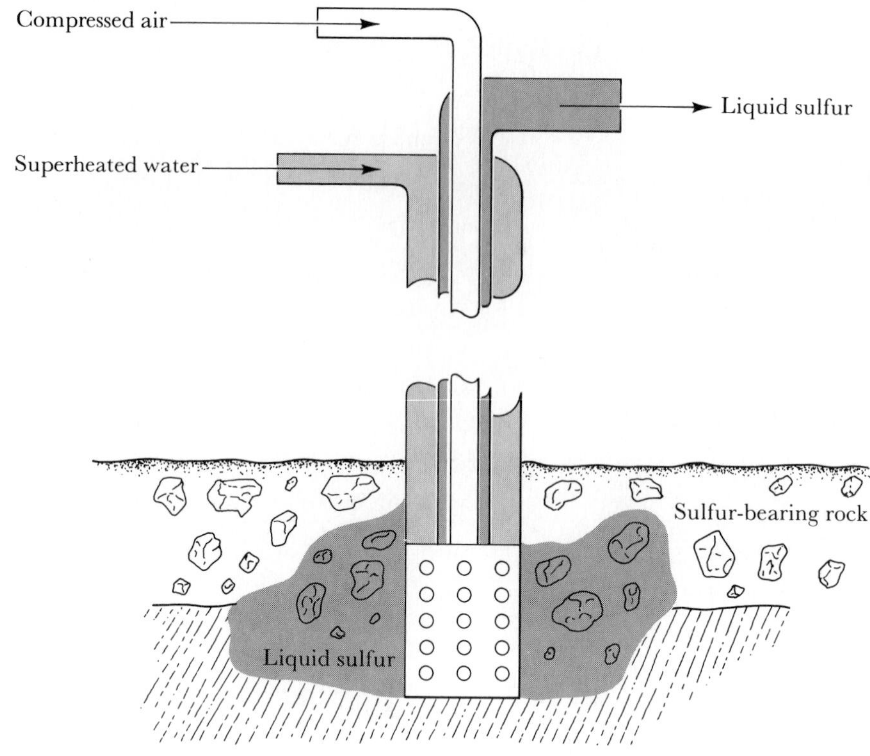

FIGURE 17–4
The Frasch process for mining sulfur.

preciable. At 1 atm and 0°C:

Pure N_2

$$d = \frac{2 \times 14.007 \text{ g mol}^{-1}}{22.415 \text{ L mol}^{-1}} = 1.2498 \text{ g L}^{-1}$$

78.084 vol% N_2 + 0.934 vol% Ar

$$d = \frac{78.084 \times 2 \times 14.007 \text{ g mol}^{-1} + 0.934 \times 39.948 \text{ g mol}^{-1}}{(78.084 + 0.934) \times 22.415 \text{ L mol}^{-1}}$$

$$= 1.2561 \text{ g L}^{-1}$$

This is a difference of 0.50%.

Elemental sulfur (with a melting point of ~110°C) is found in the United States primarily in large underground deposits in Louisiana.[1] It is mined by injection of high-pressure hot water (~170°C), which melts it, allowing compressed air to force it to the surface of the Earth. Figure 17–4 is a diagram showing how sulfur is mined in this way; this process is called the FRASCH process.

[1]Sulfur exists in different crystalline modifications, and the molten liquid contains molecular species that are not in the crystalline solid. These species lower the equilibrium melting point.

TABLE 17–4
The Enthalpy Changes[a] for Dissociation of Molecules of Nonmetals at 25°C

Reaction	ΔH^0	ΔH^0 / Atom formed	ΔH^0 / Attachment broken
$N_2(g) = 2N(g)$	945	473	945
$O_2(g) = 2O(g)$	498	249	498
$P_4(g) = 4P(g)$	1207	302	201
$S_8(g) = 8S(g)$	2127	266	266
$F_2(g) = 2F(g)$	157	79	157
$Cl_2(g) = 2Cl(g)$	242	121	242
$Br_2(g) = 2Br(g)$	193	97	193
$I_2(g) = 2I(g)$	152	76	152

[a] ΔH^0 in kJ mol^{-1}.

Oxygen and sulfur exist in nature also in the form of compounds. Oxygen is the most abundant element in the Earth's crust, making up 85.8% by weight of ocean water and a large fraction of the aluminosilicate minerals of the solid crust.[2] (The weight percentage of oxygen in silica, SiO_2, is 53.3% and in alumina, Al_2O_3, is 47.1%.) The solid crust of the Earth is 46.6% oxygen. The compounds of sulfur that are relatively abundant are pyrites, FeS_2, and gypsum, $CaSO_4 \cdot 2H_2O$.

The chemical reactivity of the remaining nonmetals prevents them from occurring in nature in the elemental state. Phosphorus occurs primarily as orthophosphate minerals, $Ca_3(PO_4)_2$ and apatites with the composition $Ca_5Cl(PO_4)_3$ and $Ca_5F(PO_4)_3$, the latter of which also is an important source of fluorine. Fluorine occurs as fluorite, CaF_2, and cryolite, Na_3AlF_6, in addition to fluoroapatite. Chlorine, bromine, and iodine occur in sea water and in underground brines as halide ions.

The diminished tendency for double bonding in the second-row elements (phosphorus and sulfur) compared to the first-row elements (nitrogen and oxygen) is responsible for the stability of the molecules P_4 and S_8, which involve single covalent bonds, instead of diatomic molecules such as nitrogen N_2 with bond order 3 and oxygen O_2 with bond order 2.[3] Table 17–4 gives values of ΔH^0 for reactions in which the free atoms are formed from the gaseous molecules. Before these values are compared one to another to learn about bond strengths, account should be taken of the number of atomic attachments that are broken in each atomization reaction—one in nitrogen and one in oxygen, six in the tetrahedral P_4 molecule, and eight in the ring molecule S_8. This is done in the last column of the table.

[2] The figure for ocean water differs from the percentage in pure water (88.81% by weight) because of the dissolved salts in the ocean.

[3] There are other discrete cyclic sulfur molecules, S_n, with $n = 5, 6, 7, 9, 10, 11, 12, 18$, and 20. These molecules are less stable than S_8.

The enthalpy required to produce phosphorus atoms from the P_4 molecule is not a good measure of the strength of a normal phosphorus–phosphorus single bond because of the strain in the tetrahedral P_4 molecule with its PPP angle of 60°. This strain makes white phosphorus (containing discrete P_4 molecules) unstable with respect to other allotropic modifications called red and black phosphorus, in which there exist three-dimensional networks of covalently bonded atoms.

Other examples of allotropy are found in the nonmetals being considered. Ozone, an unstable allotropic form of oxygen,

$$3O_2(g) = 2O_3(g) \qquad \Delta H^0 = +285.3 \text{ kJ mol}^{-1}$$

was discussed in Sections 2–4 and 3–3; it is one of the most powerful oxidizing agents available:

$$O_3 + 2H^+ + 2e^- = O_2 + H_2O \qquad \mathscr{E}^0 = +2.07 \text{ V}$$

It is possible to use ozone instead of chlorine to purify municipal water supplies. An advantage is the fact that the reduction products of ozone are molecular oxygen and water, both innocuous.

We will consider the nonmetal elements in this order: halogens and noble gases, oxygen, nitrogen, and sulfur and phosphorus. Our discussion of phosphate chemistry will include some material that is pertinent to biological processes, which we shall discuss in Chapter 20.

17–3 The Halogens

The halogens are familiar to all. A few decades ago the familiarity might have been limited to sodium chloride (table salt), silver bromide in photographic film, and tincture of iodine. But now everyone has learned something also about fluorine (in toothpaste and drinking water as fluoride ion and bonded to carbon in Teflon and Freons) and also about the presence of chlorine and bromine atoms in molecules that are pesticides, defoliants, flame retardants, and so on. All these substances are useful, but have very bad effects if accidentally ingested by man or animal.

THE ELEMENTS

The physical properties of the free halogens are summarized in Table 17–5.[4] These substances exist as diatomic molecules in each of the states of matter. Bromine and iodine exist in a condensed state under ordinary conditions because of the strong van der Waals attractions among their molecules. (The van der Waals constant a for iodine is 15×10^{-6} m^6 atm mol^{-2}, over two times the largest value given in Table 4–6, that for chlorine.)

[4] The fifth halogen, astatine ($Z = 85$), does not occur in nature. All isotopes of astatine are unstable. The longest-lived isotope is ^{210}At, with a half-life of 8.4 hours.

TABLE 17–5
Physical Properties of Halogens

	Mp / K	Bp / K	$T_c{}^a$ / K	$\Delta H_{vap}{}^b$ / kJ mol^{-1}	Color and state^c
F_2	54	85	—	3.3	Pale yellow gas
Cl_2	172	239	417	11.9	Pale green gas
Br_2	266	331	584	15.0	Dark red liquid
I_2	387	455	785	20.8	Dark purple solid

$^a T_c$ = critical temperature.
^{b}For vaporization of liquid at its normal boiling point.
^{c}State at 25°C and P = 1 atm.

The free elements can be produced from naturally occurring halide salts electrochemically (we described this process for chlorine in Chapter 16) or by appropriate chemical reactions. Chlorine is capable of oxidizing bromide ion, and it is used to produce bromine from sea water, where the concentration of bromide ion is 65 ppm (8.3×10^{-4} mol L^{-1}).[5] These reactions produce elemental bromine from the very dilute solution of bromide ion:

$$Cl_2(g) + 2Br^-(aq) = 2Cl^-(aq) + Br_2(aq)$$

$$Br_2(aq) \xrightarrow[\text{air}]{\text{stream of}} Br_2(g)$$

$$3Br_2(g) + 6CO_3^{2-} + 3H_2O = 5Br^- + BrO_3^- + 6HOCO_2^-$$
$$5Br^- + BrO_3^- + 6H^+ = 3Br_2(g) + 3H_2O$$
$$Br_2(g) = Br_2(l)$$

The Henry's-law constant for bromine in aqueous solution (at 20°C, $k = P_{Br_2}/[Br_2] = 6.4 \times 10^2$ torr L mol^{-1}) is large enough so that bromine can be swept from the dilute solution by a stream of air.[6] This gaseous mixture of air and bromine vapor then is passed through a solution of sodium carbonate, a useful base in which the bromine is trapped by the disproportionation reaction. Treatment of this solution with sulfuric acid reverses the disproportionation reaction, and bromine is liberated.

This $Br_2 \rightarrow BrO_3^- + Br^-$ reaction is an additional illustration of how a change of hydrogen ion concentration can shift an equilibrium.[7] For the reaction at 25°C,

$$BrO_3^- + 5Br^- + 6H^+ = 3Br_2(aq) + 3H_2O$$

[5]We converted 65 ppm to 8.3×10^{-4} mol L^{-1} in Section 7–2.
[6]The constant is calculated from data at 20°C: solubility of liquid bromine in water is 0.21 mol L^{-1}; vapor pressure of liquid bromine is 133.5 torr.
[7]Recall the discussion in Chapter 16 of the effect of the concentration of hydrogen ion upon the stability of manganese in the +6 and +3 oxidation states.

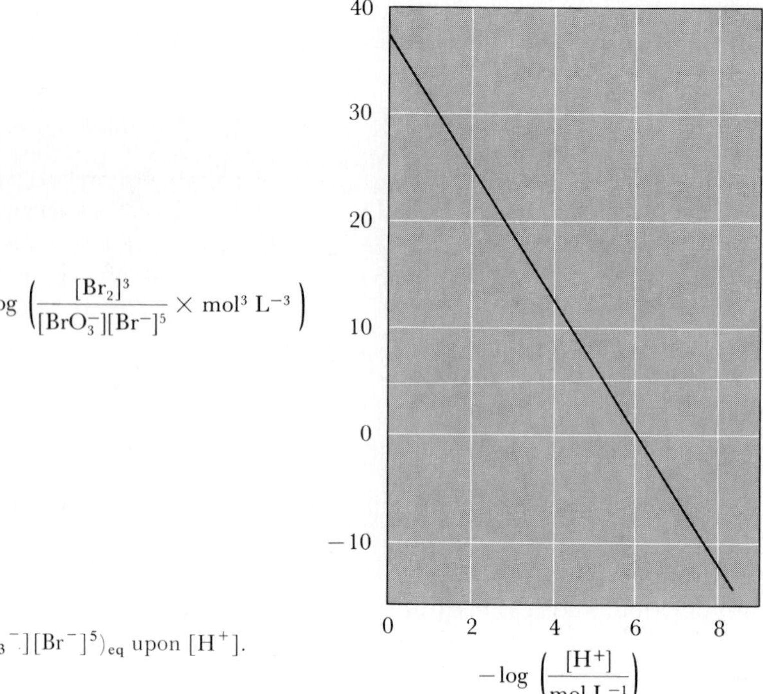

FIGURE 17–5
Dependence of the quotient $([Br_2]^3/[BrO_3^-][Br^-]^5)_{eq}$ upon $[H^+]$.

the equilibrium constant has the value

$$K = \frac{[Br_2]^3}{[BrO_3^-][Br^-]^5[H^+]^6} = 1.45 \times 10^{37}\ L^9\ mol^{-9}$$

Rearrangement of this equation gives

$$\left(\frac{[Br_2]^3}{[BrO_3^-][Br^-]^5}\right)_{eq} = 1.45 \times 10^{37}\ L^9\ mol^{-9} \times [H^+]^6$$

Figure 17–5 shows the dramatic dependence of this quotient upon the concentration of hydrogen ion. Because of the sixth power dependence upon $[H^+]$, the quotient of concentrations, which is a measure of the position of equilibrium, has an enormous value in a solution of moderate acidity and a very small value in alkaline solution. Thus at $[H^+] = 6 \times 10^{-11}$ mol L^{-1} (the value for a carbonate ion–hydrogencarbonate buffer solution in which $[CO_3^{2-}] \cong [HOCO_2^-]$), this quotient of concentrations is

$$\left(\frac{[Br_2]^3}{[BrO_3^-][Br^-]^5}\right)_{eq} \cong 7 \times 10^{-25}\ L^3\ mol^{-3}$$

The disproportionation of bromine in this alkaline solution is essentially complete. Reacidification then causes stoichiometric formation of bromine from bromide ion and bromate ion.

OXIDATION STATES AND OXIDATION–REDUCTION CHEMISTRY

Each of the common halogens except fluorine has oxidation states between −1 and +7. Fluorine forms compounds in which nominally it would be assigned a +1 oxidation state, for example, F_2O and HOF. However, this assignment is misleading since fluorine ($\chi_F = 4.0$) is more electronegative than oxygen ($\chi_O = 3.5$). Table 17–6 summarizes the positive oxidation states of the halogens; both the oxides and the oxoanions are shown. The oxides (except ClO_2) are acid anhydrides, for example, hypochlorous acid forms when dichlorine monoxide, Cl_2O, dissolves in water,

$$Cl_2O + H_2O = 2HOCl$$

(The trend in acid strengths of the chlorine oxoacids and other oxoacids will be discussed later in this chapter.)

Whether iodine(VII) has coordination number 4 or 6 in aqueous solution depends upon the concentration of hydrogen ion of the solution. This situation arises because metaperiodic acid, $HOIO_3$, is a strong acid, but orthoperiodic acid, $(HO)_5IO$, is a weak acid; for 25°C,

$$HOIO_3(aq) \rightleftarrows H^+ + IO_4^- \qquad K \gg 1 \text{ mol L}^{-1}$$
$$(HO)_5IO(aq) \rightleftarrows H^+ + (HO)_4IO_2^- \qquad K = 1.0 \times 10^{-3} \text{ mol L}^{-1}$$

At acidities where the predominant species of iodine(VII) has charge −1 ($[H^+] < 10^{-4}$ mol L^{-1}), metaperiodate is more stable:

$$(HO)_4IO_2^- \rightleftarrows IO_4^- + 2H_2O \qquad K = 29$$

TABLE 17–6
Oxides and Oxoanions of the Halogens[a]

Oxidation state	Oxide	Anion	Acid[b]	Comment
+1	X_2O	XO^-	HOX	
+3	X_2O_3	XO_2^-	HOXO	Not known for iodine.
+4	XO_2	—	—	An odd-electron molecule; unstable, but ClO_2 is a useful reagent.
+5	X_2O_5	XO_3^-	$HOXO_2$	
+7	X_2O_7	XO_4^-	$HOXO_3$	Iodine(VI) oxoanion, $OI(OH)_5^-$, in which iodine(VII) has coordination number 6, also is known.

[a]Represented by X.
[b]The names of the chlorine oxoacids are HOCl, hypochlorous acid; HOClO, chlorous acid; $HOClO_2$, chloric acid; and $HOClO_3$, perchloric acid.

At $[H^+] = 1.00\ mol\ L^{-1}$, the predominant species is the uncharged orthoperiodic acid. At this acidity the relative concentrations are calculated as follows:

$$\frac{[(HO)_4IO_2^-]}{[(HO)_5IO]} = \frac{1.0 \times 10^{-3}\ mol\ L^{-1}}{[H^+]} = 1.0 \times 10^{-3}$$

$$\frac{[IO_4^-]}{[(HO)_5IO]} = \frac{[IO_4^-]}{[(HO)_4IO_2^-]} \times \frac{[(HO)_4IO_2^-]}{[(HO)_5IO]}$$

$$= 29 \times (1.0 \times 10^{-3}) = 0.029$$

Therefore the relative concentrations at this acidity are

$$[(HO)_4IO_2^-]:[IO_4^-]:[(HO)_5IO] = 1.0 \times 10^{-3}:0.029:1.00$$

Approximately 97% of iodine(VII) is present as uncharged orthoperiodic acid in a solution with $[H^+] = 1.00\ mol\ L^{-1}$.

The oxidation–reduction properties of the halogens in aqueous solution are summarized in the potential diagrams giving $\mathscr{E}^0$ values for acidic solution:

$$F_2(g) \xrightarrow{2.87\ V} F^- \qquad F_2(g) \xrightarrow{3.06\ V} HF$$

$$ClO_4^- \xrightarrow{1.19\ V} ClO_3^- \xrightarrow{1.21\ V} HOClO \xrightarrow{1.64\ V} HOCl \xrightarrow{1.63\ V} Cl_2(g) \xrightarrow{1.358\ V} Cl^-$$

$$BrO_3^- \xrightarrow{1.50\ V} HOBr$$

$$BrO_4^- \xrightarrow{1.76\ V} BrO_3^- \xrightarrow{1.49\ V} HOBrO \xrightarrow{1.51\ V} HOBr \xrightarrow{1.60\ V} Br_2(aq) \xrightarrow{1.08\ V} Br^-$$

$$BrO_3^- \xrightarrow{1.52\ V} Br_2(aq)$$

$$(HO)_5IO \xrightarrow{1.6\ V} IO_3^- \xrightarrow{1.13\ V} HOI \xrightarrow{1.44\ V} I_2(aq) \xrightarrow{0.621\ V} I^-$$

$$IO_3^- \xrightarrow{1.19\ V} I_2(aq) \qquad I_3^- \xrightarrow{0.537\ V} I^-$$

The range of values of $\mathscr{E}^0$ for the halogen–halide ion couples is enormous, +2.87 V for fluorine to +0.621 V for iodine. Before using the tabulated values, however, you should notice certain points:

1. Since hydrofluoric acid is a weak acid ($K_a = 6.8 \times 10^{-4}\ mol\ L^{-1}$), the half-reaction involving $HF(aq)$ and not $F^-(aq)$ is the one that should be used in practical calculations in acidic solution ($pH < 3$).
2. The half-reactions for bromine and iodine given in the potential diagrams involve $X_2(aq)$, not $Br_2(l)$ and $I_2(s)$, the usual standard states for these elements.

The tabulated $\mathscr{E}^0$ values are related to values of ΔG^0 for the reaction

$$H_2(g) + X_2(g) = 2H^+(aq) + 2X^-(aq)$$

The values are:

	$\frac{\Delta G^0}{\text{kJ mol}^{-1}}$		$\frac{\Delta G^0}{\text{kJ mol}^{-1}}$
Fluorine	−554	Bromine	−208
Chlorine	−262	Iodine	−123

It is the value for fluorine that stands out. We can learn the factor responsible for this by dissecting the reaction into the steps:

1. $H_2(g) = 2H(g)$
2. $H(g) = H(g)^+ + e^-$
3. $H^+(g) = H^+(aq)$

4. $X_2(g) = 2X(g)$
5. $X(g) + e^- = X^-(g)$
6. $X^-(g) = X^-(aq)$

Steps 1–3 pertain to hydrogen and, therefore, are the same for the reaction of each halogen. The enthalpy changes for Steps **4–6** for each halogen are shown in Figure 17–6. We see in this figure that the major factor responsible for the oxidizing power of fluorine in aqueous solution is the hydration energy of fluoride ion, $\Delta H_{\text{hyd}}(F^-) = -506$ kJ mol^{-1}. This is 125 kJ mol^{-1} more negative than the hydration energy of chloride ion. (See the values tabulated in Table 9–11.)

The values of $\mathscr{E}^0$ are determined by the values of ΔG^0, not ΔH^0, and standard changes of entropy, ΔS^0, also play a role. The molar entropies of the halogens and halide ions in aqueous solution are[8]:

Halogen	$\frac{S^0[X_2(g)]}{\text{J K}^{-1}\text{ mol}^{-1}}$	$\frac{S^0[X^-(aq)]}{\text{J K}^{-1}\text{ mol}^{-1}}$	$\frac{2S^0[X^-(aq)] - S^0[X_2(g)]}{\text{J K}^{-1}\text{ mol}^{-1}}$
Fluorine	202.9	−13.8	−230.5
Chlorine	223.0	56.5	−110.0
Bromine	245.4	82.4	−80.6
Iodine	260.6	105.9	−48.8

The trend in the values of the entropy differences shows the effect of ionic size on the entropy of an ionic solute species in aqueous solution: the value $S^0[X^-(aq)]$ is less positive the smaller the ion. The net effect of this trend in the values of $2S^0[X^-(aq)] - S^0[X_2(g)]$ is to make the range of oxidizing power of the halogens smaller than it would be if it depended only upon the enthalpy changes shown in Figure 17–6.

The potential diagrams can be scanned to reveal which intermediate oxidation states are unstable with respect to disproportionation at $[H^+] = 1.00$ mol L^{-1}. The clearest cases for disproportionation are:

[8] Entropies of ions in solution are based upon the assignment $S^0[H^+(aq)] = 0$. This is discussed in Appendix 3.

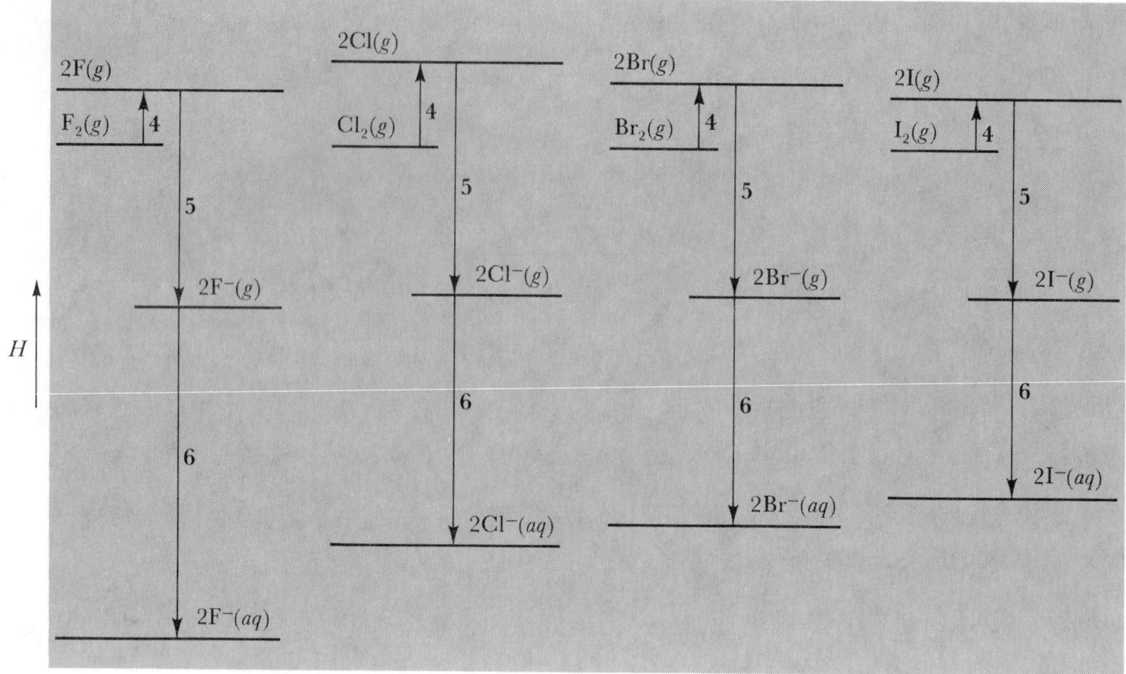

FIGURE 17–6
Dissection of the enthalpy change for the half-reaction

$$X_2(g) + 2e^- = 2X^-(aq)$$

into the enthalpy changes for the component steps:

4. $X_2(g) = 2X(g)$
5. $2X(g) + 2e^- = 2X^-(g)$
6. $2X^-(g) = 2X^-(aq)$

chlorous acid, HOClO,

$$2HOClO = HOCl + ClO_3^- + H^+$$
$$\mathscr{E}^0 = +1.64\ V - 1.21\ V = +0.43\ V$$

hypobromous acid, HOBr,

$$5HOBr = 2Br_2 + BrO_3^- + H^+ + 2H_2O$$
$$\mathscr{E}^0 = +1.60\ V - 1.50\ V = 0.10\ V$$

and hypoiodous acid, HOI,

$$5HOI = 2I_2 + IO_3^- + H^+ + 2H_2O$$
$$\mathscr{E}^0 = 1.44\ V - 1.13\ V = +0.31\ V$$

As we have already seen in discussing the production of bromine, these disproportionation equilibria can be influenced by the concentration of hydrogen ion.

The potential diagram suggests that chlorine is stable with respect to disproportionation:

$$Cl_2(g) + H_2O \rightleftarrows H^+ + Cl^- + HOCl$$

$$\mathscr{E}^0 = +1.358\ V - 1.63\ V = -0.27\ V$$

$$K^0 = 10^{+n\mathscr{E}^0/0.0592\ V} = 10^{-1 \times 0.27\ V/0.0592\ V} = 2.7 \times 10^{-5}$$

$$\text{or} \quad K = 2.7 \times 10^{-5}\ mol^3\ L^{-3}\ atm^{-1}$$

Although disproportionation is unfavorable in a solution with $[H^+] = 1.00\ mol\ L^{-1}$, it occurs to a greater extent in an initially neutral solution, and to a still greater extent in a solution buffered at pH = 7.00.

It is possible to prepare a solution containing hypochlorous acid but free of hydrochloric acid if the equilibrium

$$Cl_2 + H_2O \rightleftarrows H^+ + Cl^- + HOCl$$

is shifted to the right by a reagent that consumes both hydrogen ion and chloride ion. A reagent containing oxide ion or hydroxide ion combined with a metal ion that forms an insoluble chloride meets these specifications. Silver(I) oxide is such a reagent. The net reaction occurring when an aqueous solution of chlorine is treated with silver(I) oxide is

$$Ag_2O(s) + 2Cl_2 + H_2O = 2AgCl(s) + 2HOCl$$

It was thought until recently that perbromate ion could not exist, but work in 1969 led to its preparation by the oxidation of bromate in alkaline solution (5*M* NaOH) with fluorine:

$$F_2 + BrO_3^- + 2OH^- = BrO_4^- + 2F^- + H_2O$$

Although perbromate ion is an extremely powerful oxidizing agent in acidic aqueous solution, it reacts slowly with most reducing agents. In this respect it resembles perchlorate ion. The reaction

$$2Cr^{2+} + ClO_4^- + 2H^+ = 2Cr^{3+} + ClO_3^- + H_2O$$

with an equilibrium constant $K = 1.2 \times 10^{57}\ L^2\ mol^{-2}$ at 25°C does not occur at an appreciable rate at ordinary temperatures. But in concentrated solution perchloric acid or perchlorate ion reacts with organic matter explosively. Even in the absence of reducing agents, some metal perchlorates can detonate. The reaction

$$AgClO_4(s) = AgCl(s) + 2O_2(g)$$

is exothermic, and has been known to occur explosively.

Many of the values of $\mathscr{E}^0$ given in potential diagrams for the halogens are more positive than +1.23 V, the value of $\mathscr{E}^0$ for the half-reaction

$$O_2 + 4H^+ + 4e^- = 2H_2O$$

As we learned in Chapter 16, the oxidized form of a couple for which $\mathscr{E}^0 > 1.23$ V is capable of oxidizing water under standard conditions; therefore an aqueous solution of such a reagent will evolve oxygen gas and lose its oxidizing ability. For fluorine, the reaction

$$2F_2 + 2H_2O = O_2 + 4HF \quad \mathscr{E}^0 = 3.06\ V - 1.23\ V = +1.83\ V$$

is overwhelmingly favorable and also occurs rapidly. In addition, acidic solutions of bromic acid, hypochlorous acid, or chlorine evolve oxygen; the reactions are:

$$4H^+ + 4BrO_3^- = 5O_2 + 2Br_2 + 2H_2O$$
$$\mathscr{E}^0 = +1.52\ V - 1.23\ V = +0.29\ V$$
$$4HOCl = O_2 + 2Cl_2 + 2H_2O$$
$$\mathscr{E}^0 = +1.63\ V - 1.23\ V = +0.40\ V$$
$$2Cl_2 + 2H_2O = O_2 + 4H^+ + 4Cl^-$$
$$\mathscr{E}^0 = +1.358\ V - 1.23\ V = 0.13\ V$$

However, these reactions occur slowly.

INTERHALOGEN MOLECULES AND IONS

In our previous discussion of covalent bonding (Section 10–5) we considered iodine fluorides IF_n ($n = 1$, 3, 5, and 7) to illustrate the possible involvement of valence-shell d orbitals of iodine in the molecules with $n \geq 3$. Now we shall consider interhalogen molecules and ions of this type to illustrate certain additional points regarding chemical bonding and molecular structure.

The gaseous diatomic interhalogen molecules are formed exothermically from the gaseous diatomic elements, for example,

$$F_2(g) + Cl_2(g) = 2ClF(g) \qquad \Delta H^0 = -99.7\ \text{kJ mol}^{-1}\ \text{(at 298.2 K)}$$

The two atoms in a diatomic interhalogen molecule have different electronegativities, hence bonding in the molecule has some ionic character. Pauling postulated that this ionic character makes the X–Y bond stronger than the average of the X–Y and Y–Y bonds. We learned previously (see Figure 10–24) that for the hydrogen halides, there is approximate proportionality between the quantity Δ,

$$\Delta = D(\text{H—X}) - \tfrac{1}{2}[D(\text{X—X}) + D(\text{H—H})]$$

and $(\chi_X - \chi_H)^2$; Figure 17–7 shows an analogous correlation for the four known diatomic interhalogen molecules. Two of the possible diatomic interhalogen molecules are not stable because they can decompose to products other than the free halogens. For example, the reactions

$$3BrF = BrF_3 + Br_2 \quad \text{and} \quad 5IF = IF_5 + 2I_2$$

are responsible for the instability of BrF and IF.

There are many polyhalogen ions, both cations and anions. You are familiar with triiodide ion, I_3^-; this species was discussed in Chapter 14, where the increased solubility of iodine in potassium iodide solution was

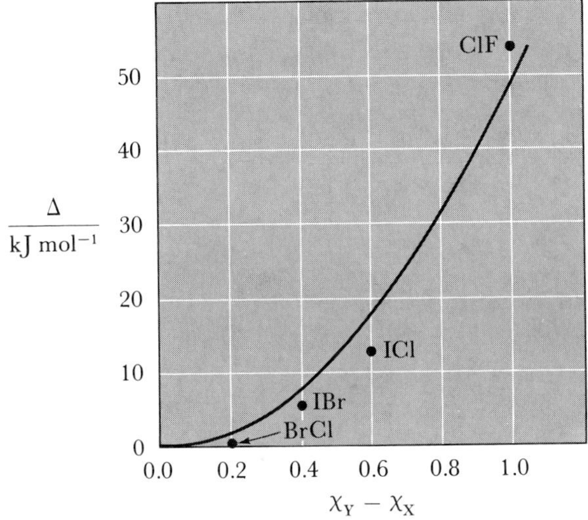

FIGURE 17–7
Correlation of bond-energy enhancement (Δ) with the difference of electronegativities of the bonded atoms.

$$\Delta = D(\text{X—Y}) - \tfrac{1}{2}[D(\text{X—X}) + D(\text{Y—Y})]$$

$$\Delta = -\Delta H \text{ for the reaction}$$

$$\tfrac{1}{2}\text{X}_2(g) + \tfrac{1}{2}\text{Y}_2(g) = \text{XY}(g)$$

The solid line is $\Delta = 47 \text{ kJ mol}^{-1} \times (\chi_Y - \chi_X)^2$

correlated with formation of triiodide ion. You probably also have used solutions of this reagent in analysis. (This use will be discussed later in this chapter.) But for our present discussion, we will focus attention on chlorine fluoride and iodine fluoride ions. A cationic species XF_{n-1}^+ can be formed by removing fluoride ion from a neutral molecule XF_n, and an anionic species XF_{n+1}^- can be formed by adding fluoride ion to a neutral molecule XF_n. These relationships, involving known ions, are shown in Table 17–7. The geometry of each ion in this compilation is predicted correctly by the valence-shell electron-pair repulsion model (see Chapter 10). For ClF_4^-, the model predicts a planar geometry:

$$ClF_4^-, \; 7 + (4 \times 1) + 1 = 12 \text{ valence-shell electrons}$$
$$= 6 \text{ pairs of electrons}$$

TABLE 17–7
Formulas and Structures of Iodine Fluoride and Chlorine Fluoride Molecules and Ions

Number of valence-shell electrons on X	XF_{n-1}^+ ← $-F^-$	XF_n	$+F^-$ → XF_{n+1}^-
8	ClF_2^+	ClF	
10	IF_4^+	ClF_3	ClF_2^-
12	IF_6^+	IF_5	ClF_4^-
14		IF_7	IF_6^-
16			(IF_8^-)[a]

[a]Not known.

with octahedral orientation of electron pairs about the central Cl atom. The two unshared pairs are as far as possible from one another, which places the fluorine atoms in the same plane. This ion has the same number of valence-shell electrons as does XeF_4, the structure of which was shown in Figure 10–21.

A metal fluoride, M^+F^-, can be used to form the anionic species XF_{n+1}^- from neutral XF_n:

$$M^+F^-(s) + XF_n(l \text{ or } g) = M^+XF_{n+1}^-(s)$$

In this reaction an X–F bond is formed, but the lattice energy of M^+F^- is lost and the lattice energy of $M^+XF_{n+1}^-$ is gained. The lattice-energy change is a factor that makes the reaction tend not to occur because $U(M^+XF_{n+1}^-) < U(M^+F^-)$. But this inequality of lattice energies is smaller for compounds containing large cations. Thus reactions of the type we are considering occur with cesium fluoride but not with lithium fluoride. For instance,

$$Cs^+F^-(s) + ClF_3(l) = Cs^+ClF_4^-(s)$$

occurs, but

$$Li^+F^-(s) + ClF_3(l) = Li^+ClF_4^-(s)$$

does not occur.

To form the cationic species, the neutral XF_n is allowed to react with a substance having a high affinity for fluoride ion, for example, antimony pentafluoride:

$$SbF_5 + ClF_3 = ClF_2^+SbF_6^-$$

The formation of tetrafluorochlorate(III) ion, ClF_4^-, from chlorine trifluoride, ClF_3, and fluoride ion, F^-, is a Lewis acid–Lewis base reaction. A Lewis acid is an *electron-pair acceptor*; chlorine trifluoride is an example. A Lewis base is an *electron-pair donor*; fluoride ion is an example. In the reaction $ClF_3 + F^- = ClF_4^-$, fluoride ion "donates" an electron pair for bonding to chlorine in chlorine trifluoride.

OTHER NONMETAL FLUORIDES, INCLUDING XENON FLUORIDES

Many nonmetal molecules and ions, like the halogen fluoride molecules and ions just discussed, involve a central nonmetal atom bonded to fluorine. Several of these (e.g., BF_3, NF_3, SF_4, and SF_6) were used in Chapter 10 to illustrate the VSEPR model for predicting molecular geometry.

Now let us consider the ionic species NF_4^+ (isoelectronic with BF_4^- and CF_4), which is particularly interesting because it was long thought to be incapable of existence. We cannot prepare tetrafluoronitrogen(V) ion, NF_4^+, by taking F^- away from NF_5 because NF_5 does not exist, but we can imagine preparing the compound by reacting NF_3 with F_2:

$$NF_3 + F_2 = NF_4^+F^-$$

However, this reaction *does not occur* under conditions that have been tried. To form NF_4^+ from NF_3 requires the addition of F^+ to NF_3. The generation of F^+ from F_2 is aided by a reagent with a large affinity for fluoride ion. We have just learned that antimony(V) fluoride is such a reagent and is capable of producing ClF_2^+ from ClF_3; this reagent works here also and this reaction does occur:

$$NF_3 + F_2 + SbF_5 = NF_4^+SbF_6^-$$

As expected, NF_4^+ is a regular tetrahedral ion.

The existence and structure of NF_4^+ are reasonable when it is considered a member of the isoelectronic series BF_4^-, CF_4, and NF_4^+. Other nonmetal fluorides also fit nicely into isoelectronic series: for instance, AlF_6^{3-}, SiF_6^{2-}, PF_6^-, SF_6, and ClF_6^+ (which does not exist). (The nonexistence of ClF_6^+ may be due to the steric repulsion, that is, crowding, of six fluorines around a chlorine; IF_6^+ is known, but an iodine atom is much larger than a chlorine atom. It is possible also that no one has tried the appropriate reaction to prepare ClF_6^+.)

Isoelectronic with the iodine fluoride anions IF_2^-, IF_4^-, and IF_6^- are neutral xenon fluorides XeF_2, XeF_4, and XeF_6. But until 1962, no stable compound containing covalently bonded noble-gas atoms had been prepared.[9] In that year, BARTLETT ended the prevailing notion that the noble-gas elements, with completely filled *s* and *p* orbitals of the valence shell, did not form compounds. He prepared the first compound from xenon and platinum hexafluoride, a reaction he conceived as analogous to the reaction of oxygen with platinum hexafluoride,

$$O_2 + PtF_6 = O_2^+PtF_6^-$$

which he had been studying. His basis for this analogy was the similarity of the ionization energies of xenon and oxygen:

$$Xe(g) = Xe^+(g) + e^- \qquad IE = \Delta U = 1170.3 \text{ kJ mol}^{-1}$$
$$O_2(g) = O_2^+(g) + e^- \qquad IE = \Delta U = 1165.2 \text{ kJ mol}^{-1}$$

A crystalline yellow compound forms when xenon and platinum hexafluoride react, but its composition (probably a mixture of $XeF^+PtF_6^-$ and $XeF^+Pt_2F_{11}^-$) is not strictly analogous to that of the oxygen compound.

Discovery of the compound of xenon with platinum hexafluoride stimulated much additional study, and the binary fluorides XeF_2, XeF_4, and XeF_6 were prepared soon after. The standard enthalpies of formation of the xenon fluorides and their molecular geometries, which can be predicted by use of the VSEPR model, are given in Table 17–8. The average bond energies in these compounds can be calculated from the standard heats of formation. For instance, in xenon(VI) fluoride, the average bond energy,

$$E(Xe—F) = 126 \text{ kJ mol}^{-1}$$

[9]Recall from Chapter 12, however, the gas-phase protonation of noble-gas atoms to give cationic species, for example, HNe^+, isoelectronic with HF.

TABLE 17–8
Properties of Xenon Fluorides

Fluorides	Number of valence-shell electrons around xenon	Molecular geometry	ΔH_f^0 / kJ mol^{-1}
XeF_2	10	Linear	-108.4[a]
XeF_4	12	Square planar	-215.5[a]
XeF_6	14	Distorted octahedron	-294.6[a]

[a]These are standard heats of formation of the gaseous compounds. These fluorides are solids at room temperature; the melting points are XeF_2, 129°C; XeF_4, 117°C; and XeF_6, 49.6°C.

is calculated from the standard heat of formation:

$$\mathrm{Xe}(g) + 3\mathrm{F}_2(g) = \mathrm{XeF}_6(g)$$
$$-\Delta H_f \cong 6E(\mathrm{Xe{-}F}) - 3D(\mathrm{F{-}F})$$
$$-6E(\mathrm{Xe{-}F}) \cong \Delta H_f - 3D(\mathrm{F{-}F})$$
$$E(\mathrm{Xe{-}F}) \cong \tfrac{1}{6}(294.6 \text{ kJ mol}^{-1} + 3 \times 154 \text{ kJ mol}^{-1})$$
$$= 126 \text{ kJ mol}^{-1}$$

The xenon–fluorine bond is not strong compared to most covalent bonds, but it is almost as strong as the fluorine–fluorine bond in F_2.

We already have noted that many nonmetal fluorides form anionic species with fluoride ion. Xenon(VI) fluoride does this also; the reaction

$$2\mathrm{CsF} + \mathrm{XeF}_6 = \mathrm{Cs}_2\mathrm{XeF}_8$$

produces a compound which is stable to ~400°C. The octafluoroxenate(VI) ion has 18 electrons in the valence shell of the central xenon atom. The geometry of this ion has been found [in $(NO)_2XeF_8$] to be a slightly distorted Archimedian antiprism (the figure generated if one face of a cube is rotated in its plane by 45° relative to the opposite face).

17–4 Oxygen and Oxygen-Containing Compounds of Nonmetals

A good place to begin our discussion of oxygen is to mention the natural processes that consume and produce it. Our lives depend upon oxygen, and people have worried that man's activities may be diminishing the concentration of oxygen in the atmosphere. This seems not to be the case. There is an enormous reservoir of oxygen in the atmosphere: 60,000 moles of oxygen gas cover each square meter of the Earth's surface. The natural consumption and production of oxygen, which is coupled with the production and consumption of carbon dioxide, can be represented by the oxidation of a carbohydrate,

$$C_6H_{12}O_6(s) + 6O_2(g) = 6CO_2(g) + 6H_2O(l)$$

and its reverse, the photosynthesis reaction,

$$6CO_2(g) + 6H_2O(l) = C_6H_{12}O_6(s) + 6O_2(g)$$

These reactions annually turn over approximately eight moles of oxygen per square meter of the Earth's surface.[10] This is a very mild perturbation of the vast amount of oxygen in the atmosphere. In addition, there appear to be other regulatory mechanisms that keep the oxygen level approximately constant.

The photosynthesis that preceded the formation of fossil fuels occurred over a long period eons ago, and the burning of these fuels is not balanced by a process that consumes carbon dioxide. This is a potentially serious problem. Our combustion of fossil fuels is increasing the amount of carbon monoxide and carbon dioxide in the atmosphere. Carbon monoxide is a poison and carbon dioxide plays a role in the heat balance of radiant energy coming to and leaving the Earth.

The oxygen atom O $1s^22s^22p^6$ in the ground state has the following distribution of valence-shell electrons:

	2s	2p
O	(⇅)	(⇅)(↑)(↑)

It forms compounds with most other elements: only helium, neon, and possibly argon do not form oxides. Several different types of compounds are formed by oxygen with other elements:

1. There are compounds containing oxide ion, O^{2-}, in which the oxygen atom has an octet of electrons in the $n = 2$ shell (the neon structure); generally the two extra electrons are gained from an electropositive element (i.e., a metal).
2. There are compounds containing covalently bonded oxygen with single bonds between an oxygen atom and two atoms of other elements (e.g., water, H_2O, with single bonds to hydrogen atoms, or dimethyl ether, $(H_3C)_2O$, with single bonds to carbon atoms, or hypochlorous acid, HOCl, with single bonds to a hydrogen atom and a chlorine atom).
3. There are compounds containing covalently bonded oxygen with one double bond between an atom of oxygen and an atom of another element (e.g., formaldehyde, H_2CO, with a double bond to a carbon atom).
4. There are compounds in which an oxygen atom accepts two electrons from one other atom to form a polar single bond (e.g., an amine oxide, $R_3N\text{—}\ddot{\underset{..}{O}}\text{:}$, in which the bonding electrons are provided by the nitrogen

[10] For the combustion of glucose, $C_6H_{12}O_6$, the values of ΔH^0 and ΔG^0 are -2816 kJ mol^{-1} and -2879 kJ mol^{-1}, respectively. The production of oxygen and glucose from carbon dioxide and water is a process for which $\Delta G > 0$; the process occurs only because the needed energy is provided by the light absorbed by chlorophyll.

atom, giving nitrogen and oxygen formal charges of +1 and −1, respectively).

5. There are compounds containing diatomic or polyatomic oxygen ions or radicals, for example, hydrogen peroxide, HOOH, and sodium peroxide, Na_2O_2 (containing O_2^{2-}), cesium superoxide, CsO_2 (containing O_2^-), and cesium ozonide, CsO_3 (containing O_3^-).

OXIDES AND OXOACIDS

Nonmetals form many compounds with oxygen. The tendency to form several oxides of different composition is shown strikingly by nitrogen and chlorine, for which six and five different oxides have been characterized:

N_2O	Nitrous oxide or dinitrogen oxide	Cl_2O	Dichlorine oxide
NO	Nitric oxide	ClO_2	Chlorine dioxide
N_2O_3	Dinitrogen trioxide	Cl_2O_4	Dichlorine tetraoxide
NO_2	Nitrogen dioxide	Cl_2O_6	Dichlorine hexaoxide
N_2O_4	Dinitrogen tetraoxide	Cl_2O_7	Dichlorine heptaoxide
N_2O_5	Dinitrogen pentaoxide		

Some of these compounds (NO, NO_2, and ClO_2) are odd-electron molecules (i.e., the molecules have odd numbers of electrons), and most are thermodynamically unstable relative to the constituent elements. Of these oxides only dinitrogen pentaoxide forms from its elements exothermically. The slowness of the decomposition reactions for some of these oxides allows us to study them, but some (e.g., ClO_2) decompose explosively. Ways in which a thermodynamically unstable compound can be prepared will be illustrated later with the formation of xenon trioxide and nitric oxide.

Most nonmetal oxides are acid anhydrides; we mentioned this previously in our discussion of carbon dioxide and carbonic acid (Section 13–3). Equations for conversion of the acid anhydrides dinitrogen trioxide and dinitrogen pentaoxide to the corresponding acids are

$$N_2O_3 + H_2O = \underset{\text{Nitrous acid}}{2HONO} \qquad N_2O_5 + H_2O = \underset{\text{Nitric acid}}{2HONO_2}$$

Nitrous acid in aqueous solution is a weak acid; the acid-dissociation reaction,

$$HONO + H_2O \rightleftarrows H_3O^+ + NO_2^-$$

is governed by the equilibrium constant $K_a = 4.5 \times 10^{-4}$ mol L^{-1}. In contrast, nitric acid is a strong acid: it dissociates completely in dilute solution. For the reaction

$$HONO_2 + H_2O \rightleftarrows H_3O^+ + NO_3^-$$

the equilibrium constant is $K_a \cong 25$ mol L^{-1} at 25°C. Only in concentrated solution (>6 mol L^{-1}) is molecular nitric acid present at appreciable concentration.

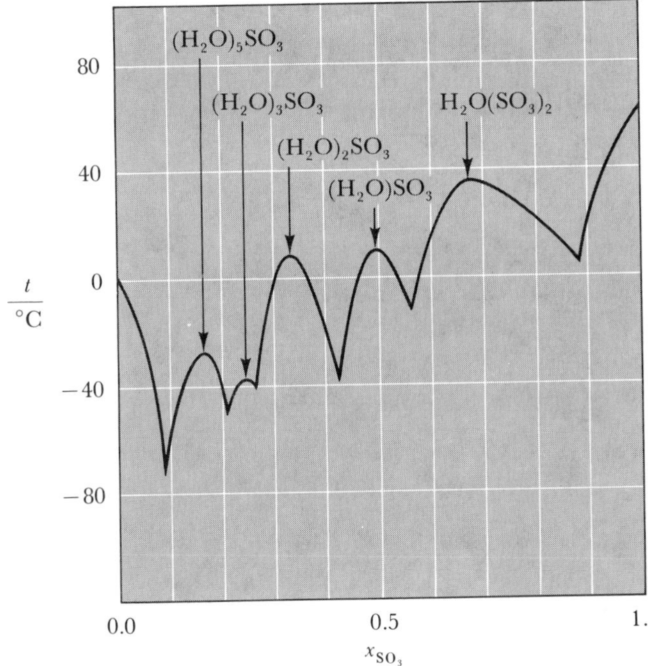

FIGURE 17–8
A freezing temperature–composition diagram for the system SO_3–H_2O. The maxima occur at the compositions and temperatures:

$(H_2O)(SO_3)_2$ or $H_2S_2O_7$	36°C
$(H_2O)(SO_3)$ or H_2SO_4	10.3°C
$(H_2O)_2(SO_3)$ or $(H_3O^+)(HSO_4^-)$	8.5°C
$(H_2O)_3(SO_3)$ or $H_2SO_4{\cdot}2H_2O$	−39.5°C
$(H_2O)_5(SO_3)$ or $H_2SO_4{\cdot}4H_2O$	−28.3°C

For many nonmetal oxide–water systems, definite hydrates form in addition to the simple acids. This is illustrated by the sulfur trioxide–water system; the existence of several different hydrates is revealed in the freezing temperature–composition diagram given in Figure 17–8. As we learned in Chapter 7, a maximum in a freezing temperature–composition diagram at a particular composition discloses a compound with that composition. Maxima are observed at mole ratios of water to sulfur trioxide of 0.50, 1.0, 2.0, 3.0, and 5.0; these compositions correspond to the following chemical formulas:

$\frac{n_{H_2O}}{n_{SO_3}}$	Formula and name
0.50	$H_2S_2O_7$, $(HOSO_2)_2O$, Disulfuric acid or pyrosulfuric acid
1.0	H_2SO_4, $(HO)_2SO_2$, Sulfuric acid
2.0	$H_2SO_4 \cdot H_2O$, $(H_3O^+)(HOSO_3^-)$, Sulfuric acid monohydrate or hydronium hydrogensulfate
3.0	$H_2SO_4 \cdot 2H_2O$, Sulfuric acid dihydrate
5.0	$H_2SO_4 \cdot 4H_2O$, Sulfuric acid tetrahydrate

(The dihydrate and tetrahydrates undoubtedly are hydrates of hydronium hydrogensulfate, but the structures of these low-melting solids are not known.) Sulfuric acid is, as we already have learned, a relatively strong acid:

$$(HO)_2SO_2 \rightleftarrows H^+ + HOSO_3^- \qquad K_1 \gg 1 \text{ mol L}^{-1}$$
$$HOSO_3^- \rightleftarrows H^+ + SO_4^{2-} \qquad K_2 = 1.03 \times 10^{-2} \text{ mol L}^{-1} \text{ (at 25°C)}$$

Some nonmetal oxides do not react completely with water. One example is carbon dioxide, discussed in Chapter 13, and another is sulfur dioxide. As was true for carbon dioxide, the predominant neutral form of sulfur dioxide in water is SO_2, not $(HO)_2SO$. The acid-dissociation equilibria are (K_a values for 25°C):

$$SO_2(aq) + H_2O \rightleftarrows H^+ + HOSO_2^- \qquad K_1 = 1.3 \times 10^{-2} \text{ mol L}^{-1}$$
$$HOSO_2^- \rightleftarrows H^+ + SO_3^{2-} \qquad K_2 = 5.6 \times 10^{-8} \text{ mol L}^{-1}$$

Thus we see that the oxoacid of sulfur(IV) is much weaker than the oxoacid of sulfur(VI).

This trend also is observed for the oxoacids of nitrogen and most other nonmetals. Consider the oxoacids of chlorine (K_a values for 25°C):

HOCl	$K_a = 2.95 \times 10^{-8}$ mol L^{-1}	$HOClO_2$	$K_a > 1$ mol L^{-1}
HOClO	$K_a = 1.1 \times 10^{-2}$ mol L^{-1}	$HOClO_3$	$K_a \gg 1$ mol L^{-1}

This trend in the values of K_a easily is correlated either (1) with the oxidation state of the central atom, or (2) with the formal charge on the central atom, because these two quantities each increase in the series of chlorine oxoacids. Electron-dot formulas for single-bonded representations of the chlorine oxoacids allow us to calculate for each acid the formal charge on the chlorine atom:

Chlorine oxoacids	Oxidation state of Cl	Formal charge on Cl
H : Ö : C̈l : (each O and Cl with full octet) Hypochlorous acid	+1	0
H : Ö : C̈l : Ö : Chlorous acid	+3	+1
H : Ö : C̈l : Ö : with : Ö : below Cl Chloric acid	+5	+2
H : Ö : C̈l : Ö : with : Ö : above and below Cl Perchloric acid	+7	+3

In the reactions,

$$H_2O + HOClO_{n-1} \rightleftarrows H_3O^+ + ClO_n^- \qquad (n = 1\text{–}4)$$

the proton is being dissociated from an oxygen atom bonded to the central chlorine atom. It is reasonable that increasing the positive formal charge on chlorine increases the ease of removal of the proton. The electron cloud on the oxygen atom to which the hydrogen atom is bonded is attracted by positive charge on the chlorine atom.

In the series of phosphorus oxoacids, $(HO)OPH_2$, $(HO)_2OPH$, and $(HO)_3PO$, the formal charge on the phosphorus atom is the same despite variation in the oxidation state. The electron-dot formulas for single-bonded representations of these acids, the formal charges on the central phosphorus atom, its oxidation state, and the values of K_a for 25°C are:

Phosphorus oxoacid	Oxidation state of P	Formal charge on P	$\frac{K_a}{\text{mol L}^{-1}}$
·· :O: ·· ·· H : P : O : H ·· ·· H Hypophosphorous acid	+1	+1	1.0×10^{-2}
·· :O: ·· H : P : O : H ·· ·· :O: ·· H Phosphorous acid	+3	+1	1.6×10^{-2}
H ·· :O: ·· ·· ·· H : O : P : O : H ·· ·· ·· :O: ·· Phosphoric acid	+5	+1	7.52×10^{-3}

Because of the structures of these acids (information on which is hidden if the line formulas are written H_3PO_2, H_3PO_3, and H_3PO_4), the formal charge on phosphorus is the same (+1) in each acid even though the oxidation state is different.

The relationship between formal charge on the central atom and acid-dissociation constant seems to hold even if different central atoms are involved. The chlorine oxoacid in which the formal charge on chlorine is +1 (chlorous acid, HOClO) has $K_a = 1.1 \times 10^{-2}$ mol L^{-1}, a value very similar to that for the three phosphorus oxoacids, in each of which the formal charge on the phosphorus is +1. This view is oversimplified, but it is useful in discussing the approximate magnitude of the acid-dissociation constants for inorganic oxoacids. Other examples to which we can apply this reasoning are the iodine(VII) oxoacids, mentioned previously. In metaperiodic acid, $HOIO_3$, the formal charge on iodine is +3, the same as the formal charge on chlorine in perchloric acid, and metaperiodic acid is a strong acid ($K_a \gg 1$ mol L^{-1}). In orthoperiodic acid, $(HO)_5IO$, the formal charge on iodine is +1 and the acid-dissociation constant of this acid is $K_a = 1.0 \times 10^{-3}$ mol L^{-1}, which is close (within a factor of 10) to values for the phosphorus oxoacids, in which the formal charge on the phosphorus atom is +1.

The success of these correlations, which involve single-bonded representations of the bonding between the central nonmetal atom and oxygen, should not be interpreted as proof that these representations adequately explain other properties of these species. The oxygen–nonmetal bonds in oxoanions generally are shorter than expected for single bonds, and structures involving double bonds probably also contribute to the structures of these species. For simplicity we will continue to use single-bonded representations that give the central atom an octet of electrons. (Recall that a single-bonded representation does not satisfy this requirement for nitric acid.)

Xenon forms oxides that are unstable with respect to the constituent elements, as are the oxides of chlorine and nitrogen (except dinitrogen pentaoxide). The xenon–oxygen bond energy in xenon trioxide is only 96 kJ mol^{-1}, a value ~75% of the Xe–F bond energy in xenon hexafluoride. This weak bond, coupled with the strong bond in diatomic oxygen, makes xenon trioxide unstable with respect to the constituent elements:

$$2XeO_3(s) = 2Xe(g) + 3O_2(g) \qquad \Delta H^0 = -800 \text{ kJ mol}^{-1}$$

(The value of ΔG^0 is even more negative than this because the value of ΔS^0 for this gas-producing reaction is very positive.) Clearly xenon trioxide *cannot* be prepared by mixing xenon and oxygen. It is prepared by coupling the formation of unstable xenon trioxide with the formation of stable hydrogen fluoride. The reaction

$$XeF_6 + 3H_2O = XeO_3 + 6HF$$

goes because the hydrogen–fluorine bond is much stronger than the hydrogen–oxygen bond:

$$D(\text{H—F}) = 565 \text{ kJ mol}^{-1} \qquad E(\text{H—O}) = 467 \text{ kJ mol}^{-1}$$

This system provides an example of coupled reactions. Consider the sequence of three favorable reactions

$$3H_2(g) + \tfrac{3}{2}O_2(g) = 3H_2O(l)$$
$$Xe(g) + 3F_2(g) = XeF_6(s)$$
$$XeF_6(s) + 3H_2O(l) = XeO_3(s) + 6HF(aq)$$

for each of which $\Delta G^0 < 0$. The sum of these reactions is

$$3H_2(g) + \tfrac{3}{2}O_2(g) + Xe(g) + 3F_2(g) = XeO_3(s) + 6HF(aq)$$

But this reaction also is the sum of two reactions:

$$3H_2(g) + 3F_2(g) = 6HF(aq)$$

and

$$Xe(g) + \tfrac{3}{2}O_2(g) = XeO_3(s)$$

the first of which is a reaction with $\Delta G^0 < 0$ and the second of which is a reaction with $\Delta G^0 > 0$. The sequence of three favorable reactions couples the formation of an unstable compound (XeO_3) to the formation of a stable compound (HF).

Xenon also forms the oxide XeO_4. One means of preparing xenon(VIII) oxide is the reaction of an oxoanion of xenon(VI) with ozone in dilute alkaline solution,

$$3OH^- + O_3 + \underset{\text{Hydrogen-xenate ion}}{HOXeO_3^-} = \underset{\text{Perxenate ion}}{XeO_6^{4-}} + O_2 + 2H_2O$$

Reaction of barium perxenate Ba_2XeO_6 with concentrated sulfuric acid gives the oxide,

$$Ba_2XeO_6(s) + 2H_2SO_4 = 2BaSO_4 + 2H_2O + XeO_4(g)$$

Both XeO_3 and XeO_4 [or its hydrate, $(HO)_4XeO_2$] are powerful oxidizing agents in acidic aqueous solution:

$$(HO)_4XeO_2 \xrightarrow{2.36\ V} XeO_3 \xrightarrow{2.12\ V} Xe$$

Oxygen is evolved from acidic solutions of either and each decomposes explosively in the absence of water.

Simple electron-dot formulas coupled with valence-shell electron-pair repulsion will allow you to deduce the structures of most nonmetal oxides and oxoanions. Simple structures involving only single bonds given already for the chlorine oxoacids show four pairs of electrons around the central chlorine atom in each species. As predicted from this feature of the electronic structure, ClO_2^- is nonlinear, ClO_3^- is nonplanar, and ClO_4^- (and the isoelectronic species SO_4^{2-} and PO_4^{3-}) is tetrahedral. The xenon oxides XeO_3 and XeO_4 are isoelectronic with the iodine oxoanions IO_3^- and IO_4^-, and, like the analogous chlorine oxoanions, these iodine oxoanions and xenon oxides are nonplanar and tetrahedral, respectively.

OXIDATION–REDUCTION CHEMISTRY OF OXYGEN; SUPEROXIDES AND PEROXIDES

The quantitative aspects of equilibria in reactions of oxygen acting as an oxidizing agent in aqueous solution are described by the half-reaction

$$O_2 + 4H^+ + 4e^- = 2H_2O(l) \qquad \mathscr{E}^0 = 1.23\ V$$

To consider only this change, $O_2 \rightarrow 2H_2O$, which involves four electrons, is to omit all aspects of the mechanism of oxidation. The reduction of oxygen at a cathode or in reaction with a chemical reagent probably involves two or more reaction steps and not the direct production of water (or oxide ion) by addition of four electrons to an oxygen molecule. Known species are produced if fewer than four electrons are added to a molecule of oxygen:

Reaction	$\frac{\Delta H}{\text{kJ mol}^{-1}}$
$O_2(g) + e^- = O_2^-(g)$	-63
$O_2^-(g) + e^- = O_2^{2-}(g)$	$+670$
$O_2^{2-}(g) + e^- = O^-(g) + O^{2-}(g)$	$+624$

The molecular-orbital descriptions of the electronic structures of the diatomic oxygen species O_2, O_2^-, and O_2^{2-} were given in Chapter 10 (see Figure 10–15). Stable crystalline compounds are known that contain O_2^-, superoxide ion, and O_2^{2-}, peroxide ion. These diatomic anionic species (or their protonated forms) and the hydroxyl radical (OH) play roles in the reduction of oxygen in solution.[11] The potential diagram for oxygen in acidic aqueous solution, including the protonated species HO_2, hydrogen superoxide, H_2O_2, hydrogen peroxide, and OH, hydroxyl radical, is

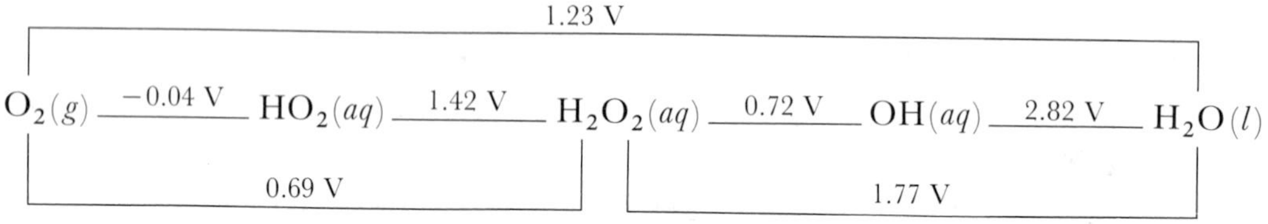

$$O_2(g) \xrightarrow{-0.04\text{ V}} HO_2(aq) \xrightarrow{1.42\text{ V}} H_2O_2(aq) \xrightarrow{0.72\text{ V}} OH(aq) \xrightarrow{2.82\text{ V}} H_2O(l)$$

1.23 V ($O_2(g)$ to $H_2O(l)$); 0.69 V ($O_2(g)$ to $H_2O_2(aq)$); 1.77 V ($H_2O_2(aq)$ to $H_2O(l)$)

Because all of the protonated species are weak acids [$K_a(HO_2) = 1.3 \times 10^{-5}$ mol L^{-1}, $K_{a1}(H_2O_2) = 2.2 \times 10^{-12}$ mol L^{-1}, and $K_a(OH) = 1.3 \times 10^{-12}$ mol L^{-1}], these weak acids are present predominantly in the protonated form in solutions with $[H^+] = 1.00$ mol L^{-1}.

We see in this potential diagram that both hydrogen superoxide and hydrogen peroxide are unstable with respect to disproportionation:

$$2HO_2 = H_2O_2 + O_2(g) \qquad \mathscr{E}^0 = +1.42\text{ V} - (-0.04\text{ V}) = 1.46\text{ V}$$
$$K^0 = 10^{1.46\text{ V}/0.0592\text{ V}} = 4.6 \times 10^{24} \text{ at } 25.0°\text{C}$$
$$K = 4.6 \times 10^{24} \text{ atm L mol}^{-1}$$
$$2H_2O_2 = 2H_2O + O_2(g) \qquad \mathscr{E}^0 = 1.77\text{ V} - 0.69\text{ V} = 1.08\text{ V}$$
$$K^0 = 10^{2\times 1.08\text{ V}/0.0592\text{ V}} = 3.1 \times 10^{36} \text{ at } 25.0°\text{C}$$
$$K = 3.1 \times 10^{36} \text{ atm L}^2 \text{ mol}^{-2}$$

The disproportionation of hydrogen superoxide is rapid; the rate depends upon the pH, with the second-order rate constant being a maximum at pH $\cong$ 3.9. Study of the rate as a function of pH shows the predominant

[11]The cationic species O_2^+ has not been studied in aqueous solution. Equilibrium in the reaction $4O_2^+ + 2H_2O = 5O_2(g) + 4H^+$ lies far to the right.

reaction pathway at this pH to be

$$HO_2 + O_2^- \rightarrow O_2 + HO_2^-$$

The rate of disproportionation of hydrogen peroxide is low in the absence of catalysts. However, the enormous value of the equilibrium constant for this reaction conveys some messages:

1. An attempt to prepare hydrogen peroxide by bubbling oxygen gas through water will fail.
2. The disporportionation will develop an enormous pressure of oxygen gas. (Bottles of 30% hydrogen peroxide reagent are vented to allow escape of oxygen. Bottles of 3% hydrogen peroxide, obtainable at drugstores, contain an inhibitor that retards the decomposition. These bottles are not vented.)

Hydroxyl radicals do not persist in aqueous solution, but they do play roles as reaction intermediates. (When we discuss radiation chemistry in Chapter 22, you again will encounter hydroxyl radicals.)

Many reactions of oxygen, although favored thermodynamically, are slow, a slowness that is caused by:

1. an unfavorable polymolecular mechanism if the oxidation occurs in one step, or
2. an unfavorable equilibrium in the first step of the reaction if this first step involves one unit of oxidation-state change.

We can explore these causes in some detail by considering the reaction of oxygen with iron(II) ion. Equilibrium in the overall reaction

$$O_2(g) + 4H^+ + 4Fe^{2+} = 2H_2O + 4Fe^{3+}$$

is very favorable:

$$\mathscr{E}^0 = 1.23\ \text{V} - 0.77\ \text{V} = 0.46\ \text{V}$$
$$K^0 = 10^{4 \times 0.46\ \text{V}/0.0592\ \text{V}} = 1.2 \times 10^{31}$$
$$K = 1.2 \times 10^{31}\ \text{L}^4\ \text{mol}^{-4}\ \text{atm}^{-1}$$

but a one-step mechanism for this reaction of nine species having a total charge of +12 is not reasonable. If the first step of the reaction were a one-electron process,

$$O_2 + H^+ + Fe^{2+} \rightarrow HO_2 + Fe^{3+}$$

this first step would be a deterrent to the reaction because equilibrium in this step is very unfavorable:

$$\mathscr{E}^0 = -0.04\ \text{V} - 0.77\ \text{V} = -0.81\ \text{V}$$
$$K^0 = 10^{-0.81\ \text{V}/0.0592\ \text{V}} = 2.1 \times 10^{-14}$$
$$K = 2.1 \times 10^{-14}\ \text{atm}^{-1}$$

If the first step were a two-electron process,

$$O_2 + 2H^+ + 2Fe^{2+} \rightarrow H_2O_2 + 2Fe^{3+}$$

the situation would be more favorable:

$$\mathscr{E}^0 = 0.69 \text{ V} - 0.77 \text{ V} = -0.08 \text{ V}$$
$$K^0 = 10^{-2 \times 0.08 \text{ V}/0.0592 \text{ V}} = 2.0 \times 10^{-3}$$
$$K = 2.0 \times 10^{-3} \text{ L mol}^{-1} \text{ atm}^{-1}$$

The experimentally determined rate law for the reaction of oxygen with iron(II) in acidic solution is

$$-\frac{d[O_2]}{dt} = k[Fe^{2+}]^2[O_2]$$

This demonstrates that the activated complex for the slow reaction step has the composition $\{Fe_2O_2^{4+}\}^{\ddagger}$, and it is reasonable to propose the sequence of two bimolecular steps,

$$1.\ Fe^{2+} + O_2 \rightleftarrows FeO_2^{2+} \qquad \text{(fast)}$$
$$2.\ FeO_2^{2+} + Fe^{2+} \rightarrow FeO_2Fe^{4+} \qquad \text{(slow)}$$

which produces a dimeric complex of iron(III) with peroxide ion. Such a species is known. The overall reaction is completed by the rapid reaction of this peroxide species with additional iron(II) ion and hydrogen ion; the equation

$$FeO_2Fe^{4+} + 2Fe^{2+} + 4H^+ = 4Fe^{2+} + 2H_2O$$

expresses the sum of the several rapid reactions that complete the net change.

Reactions of oxygen with some reagents also may involve excited electronic states of diatomic oxygen. These excited electronic states, unlike the paramagnetic ground state, are diamagnetic. The molecular-orbital energy-level diagrams for these species, called singlet oxygen, are presented in Figure 17–9, which includes the relative energies of these species. These excited electronic states can be produced by the action of an appropriately powerful oxidizing agent on hydrogen peroxide. Hypochlorite ion in alkaline solution is such a reagent; the rapid reaction

$$OCl^- + H_2O_2 = H_2O + Cl^- + O_2^*$$

produces the more stable of the singlet oxygen molecules, represented as O_2^*. If an appropriate reactant is present, singlet oxygen may react rapidly in a reaction ordinary oxygen will not perform. One such reaction is the addition of oxygen to butadiene to form a cyclic peroxide:

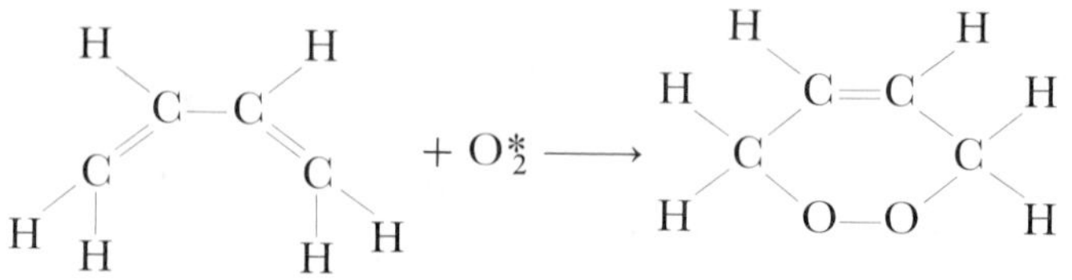

```
H         H                      H           H
 \       /                H       \         /      H
  C — C                     \      C = C         /
 //     \\                   C  /        \  C
C         C     + O2*  --->     |          |
/ |        | \               H   \ O — O /   H
H  H       H   H                              
```

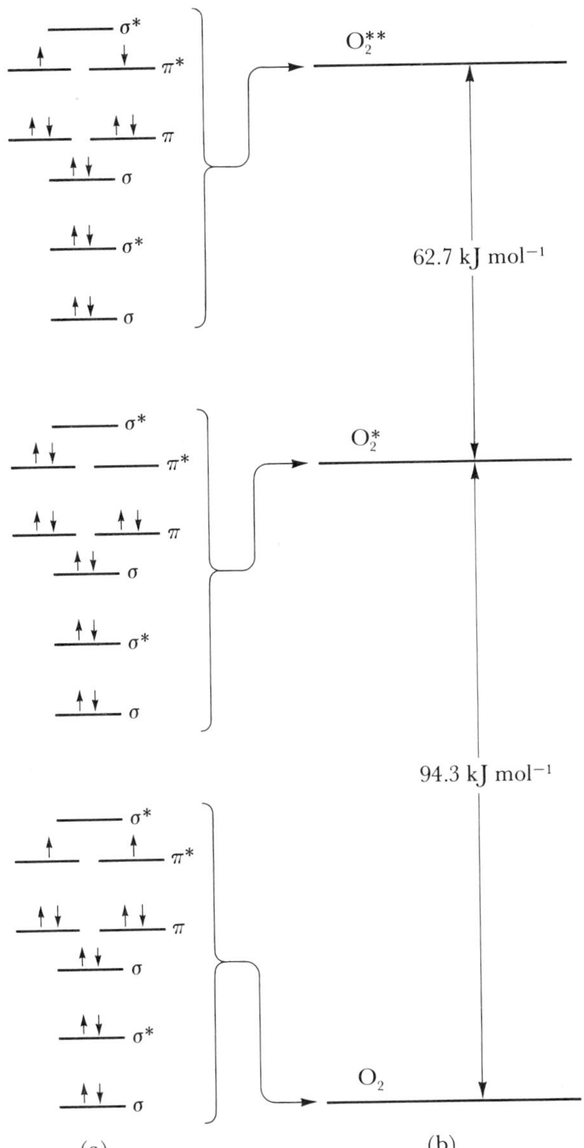

FIGURE 17–9
The ground triplet state and excited singlet states of molecular oxygen. (a) Occupancy of molecular orbitals (the order of orbitals is given correctly; in Chapter 10 the order appropriate for lighter second-row elements was used). (b) The energies of these states.

If an appropriate reactant is not present, the singlet oxygen molecules lose their energy in a process in which two molecules produce one photon of red light[12]:

$$2O_2^* \rightarrow 2O_2 + h\nu \qquad \Delta E = -2 \times 94.3 \text{ kJ mol}^{-1}$$

We can calculate the wavelength of light expected in this process using the

[12]This can be shown in a lecture demonstration; see B. Z. Shakashiri and L. G. Williams, *J. Chem. Educ.* **53**, 358 (1976).

Planck relationship (see Section 8–3):

$$\lambda = \frac{hc}{|\Delta E|}$$

$$= \frac{6.626 \times 10^{-34}\ \text{J s} \times 3.00 \times 10^{8}\ \text{m s}^{-1}}{\dfrac{2 \times 9.43 \times 10^{4}\ \text{J mol}^{-1}}{6.02 \times 10^{23}\ \text{mol}^{-1}}}$$

$$= 6.34 \times 10^{-7}\ \text{m} = 634\ \text{nm}$$

(The observed spectrum is consistent with this calculation, but in a complete solution of the problem, the vibrational energy of each of the two states of oxygen (O_2^* and O_2) would have to be taken into account.)

The decomposition of hydrogen peroxide to water and oxygen may be catalyzed by many different substances, for example, iron(II) salts, bromine or bromide ion, or solid manganese dioxide. The mechanism of the catalytic action depends upon the specific substance responsible for catalysis. The action of bromine or bromide ion illustrates one mode of catalysis. With the value of $\mathscr{E}^0 = 1.08$ V, the half-reaction

$$Br_2(aq) + 2e^- = 2Br^-$$

involves a reduced form, Br^-, capable of reducing hydrogen peroxide to water, and an oxidized form, Br_2, capable of oxidizing hydrogen peroxide to oxygen. The reactions are

$$H_2O_2 + 2Br^- + 2H^+ = Br_2 + 2H_2O$$
$$\mathscr{E}^0 = 1.77\ \text{V} - 1.08\ \text{V} = 0.69\ \text{V} \qquad K = 2.1 \times 10^{23}\ \text{L}^4\ \text{mol}^{-4}$$
$$H_2O_2 + Br_2 = 2H^+ + 2Br^- + O_2$$
$$\mathscr{E}^0 = 1.08\ \text{V} - 0.69\ \text{V} = 0.39\ \text{V} \qquad K = 1.5 \times 10^{13}\ \text{atm mol}^2\ \text{L}^{-2}$$

With concentration conditions under which these two reactions occur at the same rate, the net chemical change is the sum of the two chemical reactions:

$$2H_2O_2 = 2H_2O + O_2$$

Under these conditions the bromine–bromide ion couple acts as a catalyst for this chemical change.

Ionic superoxides (compounds containing O_2^-) and ionic peroxides (compounds containing O_2^{2-}) are produced in the reaction of excess oxygen with very active metals. With potassium, rubidium, and cesium, the superoxide is formed under ordinary conditions; for potassium the reaction is

$$K(s) + O_2(g) = KO_2(s)$$

With sodium, the peroxide is formed,

$$2Na(s) + O_2(g) = Na_2O_2(s)$$

With lithium, the oxide is formed

$$4Li(s) + O_2(g) = 2Li_2O(s)$$

The relative stabilities of superoxides, peroxides, and oxides are related to their lattice energies, values of which are given in Table 17–9. There are two important points to notice in this table:

1. The expected approximate dependence of lattice energy upon the square of the charge of the ions. Compare the alkali-metal superoxides ($M^+O_2^-$) with the isoelectronic alkaline-earth peroxides ($M^{2+}O_2^{2-}$) (e.g., compare $K^+O_2^-$ with $Ca^{2+}O_2^{2-}$).
2. The smaller range of lattice energy values for the peroxides, ionic compounds containing a large anion, compared to the corresponding values for oxides, ionic compounds containing a small anion:

$$U_{Li_2O} - U_{Rb_2O} = 2910 - 2150 \text{ kJ mol}^{-1} = 760 \text{ kJ mol}^{-1}$$
$$U_{Li_2O_2} - U_{Rb_2O_2} = 2420 - 1940 \text{ kJ mol}^{-1} = 480 \text{ kJ mol}^{-1}$$
$$U_{CaO} - U_{BaO} = 3510 - 3100 \text{ kJ mol}^{-1} = 410 \text{ kJ mol}^{-1}$$
$$U_{CaO_2} - U_{BaO_2} = 2960 - 2660 \text{ kJ mol}^{-1} = 300 \text{ kJ mol}^{-1}$$

The stabilities of peroxides are limited by their decomposition to oxygen and the metal oxide. For instance,

$$2Li_2O_2(s) = 2Li_2O(s) + O_2(g) \quad \text{and} \quad 2CaO_2(s) = 2CaO(s) + O_2(g)$$

are the decomposition reactions for lithium peroxide and calcium peroxide. Peroxides of smaller cations are less stable than those of larger cations because the lattice energy of the oxide, a product in the decomposition reaction, is relatively larger for the small cations. Lithium peroxide and calcium peroxide decompose below 400°C, but cesium peroxide and barium peroxide do not decompose until the temperature is raised above 650°C.

TABLE 17–9
Lattice Energies[a] for Superoxides, Peroxides, and Oxides

Metal ion	Superoxides	Peroxides		Oxides	
	MO_2	M_2O_2	MO_2	M_2O	MO
Li^+		2420		2910	
Na^+	800	2230		2520	
K^+	720	2020		2230	
Rb^+	690	1940		2150	
Cs^+	670	1860			
Ca^{2+}			2960		3510
Sr^{2+}			2780		3300
Ba^{2+}			2660		3100

[a] Values are $\dfrac{U}{\text{kJ mol}^{-1}}$.

One use of potassium superoxide follows from its ability to yield oxygen in reaction with carbon dioxide and water vapor,

$$4KO_2 + CO_2 + H_2O = K_2CO_3 + 2KOH + 3O_2$$

This reaction, in which the superoxide ion disproportionates,

$$4O_2^- + 2H_2O = 4OH^- + 3O_2$$

is the basis for the kind of respiration apparatus used in high-mountain expeditions, in fire fighting and mine rescue operations, and by astronauts. Exhaled water vapor and carbon dioxide react with potassium superoxide to give oxygen gas (plus solid products).

There are a number of important covalently bonded peroxides. Some of these compounds can be viewed as acids in which a hydroxyl group is replaced by a hydrogenperoxo group. Examples are:

$CH_3CO(OOH)$	Peroxoacetic acid
O_2NOOH	Peroxonitric acid
$(HO)O_2SOOH$	Peroxomonosulfuric acid
$(HO)O_2SOOSO_2(OH)$	Peroxodisulfuric acid

The last two of these compounds are involved in the large-scale production of hydrogen peroxide. In the electrolysis of aqueous sulfuric acid or ammonium hydrogensulfate at high current densities, the predominant anode reaction is

$$2SO_4^{2-} = S_2O_8^{2-} + 2e^-$$

not the production of oxygen gas,

$$2H_2O = 4H^+ + O_2(g) + 4e^-$$

which occurs at low current density. The anode product, peroxodisulfate ion, with the structure

$$\left[\begin{array}{ccccccc} & \ddot{O} & & & \ddot{O} & \\ :\ddot{O}: & S & :\ddot{O}:\ddot{O}: & S & :\ddot{O}: \\ & \ddot{O} & & & \ddot{O} & \end{array}\right]^{2-}$$

reacts slowly with water in two stages to give hydrogen peroxide. The first stage,

$$S_2O_8^{2-} + H_2O = HOSO_3^- + HOOSO_3^-$$

gives hydrogensulfate ion and hydrogenperoxomonosulfate ion, which in turn reacts with water,

$$HOOSO_3^- + H_2O = H_2O_2 + HOSO_3^-$$

giving hydrogen peroxide. This is removed from the electrolyte solution by distillation at reduced pressure. Hydrogen peroxide also is produced industrially by the reduction of oxygen with isopropyl alcohol:

$$(CH_3)_2CHOH + O_2 = (CH_3)_2CO + H_2O_2$$

The rate at which peroxodisulfate ion hydrolyzes in water is low enough so that a solution of this species can be prepared and then used as a powerful oxidizing agent:

$$S_2O_8^{2-} + 2e^- = 2SO_4^{2-} \qquad \mathscr{E}^0 = 2.0 \text{ V}$$

Many reactions of this oxidizing agent are catalyzed by metal ions. For instance, the oxidation of either vanadium(IV) or chromium(III),

$$2VO^{2+} + S_2O_8^{2-} + 2H_2O = 2VO_2^+ + 4H^+ + 2SO_4^{2-}$$
$$2Cr^{3+} + 3S_2O_8^{2-} + 8H_2O = 2HCrO_4^- + 14H^+ + 6SO_4^{2-}$$

is catalyzed by silver(I). The rate law for each of the reactions is

$$-\frac{d[S_2O_8^{2-}]}{dt} = k[S_2O_8^{2-}][Ag^+]$$

and the value of k is the same in each case. The rate does not depend upon the concentration of the reducing agent, vanadium(IV) or chromium(III). This rate law indicates that the rate-determining step for each of the reactions is the same, the reaction of peroxodisulfate ion and silver(I) ion:

$$S_2O_8^{2-} + Ag^+ \rightarrow Ag^{3+} + 2SO_4^{2-}$$

The assumed product, silver(III), then oxidizes the reducing agent (VO^{2+} or Cr^{3+}) in a rapid reaction sequence in which silver(III) is transformed back to silver(I):

1. $Ag^{3+} + VO^{2+} + H_2O \rightarrow Ag^{2+} + VO_2^+ + 2H^+$
2. $Ag^{2+} + VO^{2+} + H_2O \rightarrow Ag^+ + VO_2^+ + 2H^+$

Silver(I) thereby conforms to the classical definition of a catalyst, a reagent that increases a reaction velocity without being permanently transformed.

17–5 Nitrogen and Its Compounds

Nitrogen constitutes 78% of the atmosphere, its compounds are very important in technology and agriculture, and it is present in the molecules of life: amino acids, which are the building blocks of proteins (see Chapter 20), and purine and pyrimidine bases, which are components of nucleic acids, the building blocks of deoxyribonucleic acid (DNA) (see Chapter 20). The chemistry of this element is wonderfully diverse and complex, and only some of its simplest features will be presented here.

The nitrogen atom, N $1s^22s^22p^3$, in its ground state has the following distribution of valence-shell electrons

	$2s$	$2p$
N	(↑↓)	(↑)(↑)(↑)

As we already have learned, the elemental form of nitrogen is the diatomic molecule, N_2, with one of the strongest known covalent chemical bonds, $D(N{\equiv}N) = 942\ kJ\ mol^{-1}$. The electron-dot formulation with a triple bond,

$$:N:::N:$$

or the molecular-orbital description with six more electrons in bonding molecular orbitals than in antibonding molecular orbitals, giving a bond order of three,

$$N_2(\sigma 2s)^2(\sigma^* 2s)^2(\pi 2p_x)^2(\pi 2p_y)^2(\sigma 2p_z)^2$$

explains the extraordinarily strong bond. Because of this particularly stable structure, molecular nitrogen is relatively unreactive, and many compounds of nitrogen are thermodynamically unstable with respect to the constituent elements. However, kinetic inertness is characteristic of many nitrogen compounds, and many thermodynamically unstable compounds can be characterized and studied despite their inherent instability.

Simple binary ionic compounds of nitrogen are the nitrides, containing N^{3-} (which is isoelectronic with neon) and azides, containing N_3^- (which is isoelectronic with carbon dioxide). The electron affinity of nitrogen is nearly zero, but high lattice energies make nitrides of the most active metals stable.

COMPOUNDS OF NITROGEN WITH HYDROGEN AND OXYGEN

Covalently bonded molecules containing hydrogen, oxygen, and one atom of nitrogen, generated from ammonia by substituting hydroxyl for hydrogen, were discussed in Section 17–1. Because of the way in which oxidation states are assigned, the substitution of hydroxyl for hydrogen increases the oxidation state by two units. Even though there is little physical significance to oxidation states assigned to elements in covalent substances, they are useful in surveying the range of covalent compounds of an element such as nitrogen, as well as in balancing equations for oxidation–reduction reactions (see Chapter 16). Each of the oxidation states -3 to $+5$ is shown by nitrogen in its compounds with hydrogen and oxygen; these are summarized in Table 17–10.

The four pairs of electrons in the valence shell of the nitrogen atom in ammonia and substituted ammonia make the shape of this molecule pyramidal. The pyramidal shape for a substituted ammonia with three different groups attached to nitrogen, for example, methylethylamine, $(CH_3)(C_2H_5)NH$, gives the molecule chirality (see Section 11–4). However, it is not possible to isolate the optical isomers because of rapidly occurring inversion:

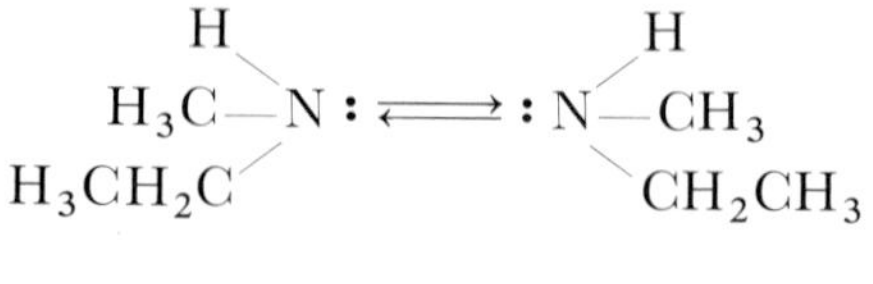

$$\begin{matrix} & H & & & H & \\ H_3C & — N: & \rightleftharpoons & :N & — CH_3 \\ H_3CH_2C & & & & CH_2CH_3 \end{matrix}$$

TABLE 17–10
Compounds of Nitrogen with Hydrogen and Oxygen

Chemical formula	Oxidation state of nitrogen	Name[a]
NH_3	−3	Ammonia
N_2H_4	−2	Hydrazine
NH_2OH	−1	Hydroxylamine
N_2	0	Nitrogen
N_2O	+1	Dinitrogen oxide or nitrous oxide
$H_2N_2O_2$ (HONNOH)	+1	Hyponitrous acid
NO	+2	Nitric oxide
N_2O_3 (ONONO)	+3	Dinitrogen trioxide
HNO_2 (HONO)	+3	Nitrous acid
NO_2	+4	Nitrogen dioxide
N_2O_5 (O_2NONO_2)	+5	Dinitrogen pentaoxide[b]
HNO_3 ($HONO_2$)	+5	Nitric acid

[a]The oxides of nitrogen also can be named by indicating the oxidation state of nitrogen: N_2O, nitrogen(I) oxide; NO, nitrogen(II) oxide; N_2O_3, nitrogen(III) oxide; NO_2, nitrogen(IV) oxide; and N_2O_5, nitrogen(V) oxide.
[b]Often called nitrogen pentaoxide.

The inversion, a motion like that of an umbrella flipping inside out, converts the molecule to its mirror image. For ammonia the frequency of this inversion is greater than $10^{10}\ s^{-1}$.

The unshared pair of electrons on the nitrogen atom in many of its compounds makes these compounds basic; ammonia, hydrazine, and hydroxylamine all are basic, but hydroxylamine and hydrazine are less basic than ammonia (see Table 17–11). Hydroxylamine is ammonia with one hydrogen atom replaced by a hydroxyl group. The oxygen atom (more electronegative than hydrogen) attracts electrons to it more strongly, making the unshared electron pair less available for donation to a proton than does the unshared electron pair in ammonia.

The replacement of a hydrogen atom of ammonia by an amido group ($—NH_2$) to give hydrazine reduces the basicity compared to ammonia for the same reason. This molecule has two basic nitrogen atoms, but an unfavorable electrostatic factor prevents the formation of appreciable amounts of the doubly protonated hydrazinium ion, $N_2H_6^{2+}$, in aqueous solution; however, crystalline salts of this cation (e.g., $N_2H_6SO_4$) are known. The unshared pairs of electrons on the nitrogen atoms of the hydrazine molecule are believed to be responsible for the relative weakness of the nitrogen–nitrogen single bond in this substance. The carbon–carbon bond energy in ethane is 347 kJ mol^{-1}, but the nitrogen–nitrogen bond in hydrazine has a bond energy of only 160 kJ mol^{-1}. The repulsion of the unshared pairs of

TABLE 17–11
The Basicities of Nitrogen Bases in Water at 25°C

Acid	Base	K_a / mol L^{-1}	K_b / mol L^{-1}
NH_4^+	NH_3	5.69×10^{-10}	1.78×10^{-5}
$N_2H_6^{2+}$	$N_2H_5^+$	~ 10	$\sim 1 \times 10^{-15}$
$N_2H_5^+$	N_2H_4	1.0×10^{-7}	1.0×10^{-7}
$HONH_3^+$	$HONH_2$	6.6×10^{-8}	1.53×10^{-7}
$H_3CNH_3^+$	H_3CNH_2	2.4×10^{-11}	4.2×10^{-4}
$(H_3C)_2NH_2^+$	$(H_3C)_2NH$	1.7×10^{-11}	5.9×10^{-4}
$(H_3C)_3NH^+$	$(H_3C)_3N$	1.6×10^{-10}	6.3×10^{-5}
$(HOCH_2)_3CNH_3^+$	$(HOCH_2)_3CNH_2$[a]	8.39×10^{-9}	1.19×10^{-6}
$C_5H_5NH^+$[b]	C_5H_5N	5.6×10^{-6}	1.8×10^{-9}
$C_6H_5NH_3^+$[c]	$C_6H_5NH_2$	2.6×10^{-5}	3.8×10^{-10}
$C_{12}H_{14}O_2N_2H^+$	$C_{12}H_{14}O_2N_2$[d]	$<10^{-14}$	>1

[a]Tris(hydroxymethyl)aminomethane, commonly called "tris" or "tham," is widely used as a buffer in physiological studies.
[b]Pyridinium ion; isoelectronic with benzene (see Chapter 11).
[c]Anilinium ion; isoelectronic with toluene (see Chapter 11).
[d]Structure given in text.

electrons on the two nitrogen atoms is undoubtedly a factor in this dramatic difference of ~ 200 kJ mol^{-1} in the bond energies of the central bonds in these related molecules.

We saw previously (Table 12–3) that gas-phase proton affinities for the methylamines CH_3NH_2, $(CH_3)_2NH$, and $(CH_3)_3N$ are greater than that for ammonia. A comparison of the range of these proton-affinity values with the range of values of ΔH^0 for the base-dissociation reactions in aqueous solution,

$$(CH_3)_aNH_{3-a} + H_2O \rightleftarrows (CH_3)_aNH_{4-a}^+ + OH^- \qquad (a = 0\text{–}3)$$

will help us see the role of solvation in determining the relative basicities of these nitrogen bases in solution. The comparison is:

Base	PA / kJ mol^{-1}	$-\Delta H^0$ / kJ mol^{-1}
NH_3	866	52.8
CH_3NH_2	904	60.6
$(CH_3)_2NH$	931	61.5
$(CH_3)_3N$	949	55.9

We see a much smaller range of values of ΔH^0 for the reactions in solution than for the gas-phase protonation reactions. In solution, the reactant nitrogen base is hydrogen-bonded to water, as is the protonated product.

The net effect that replacing hydrogen atoms in ammonia with methyl groups has on the solvation energies of reactant and product species is to make the range of basicity exhibited by these molecules smaller than would be expected on the basis of the range of proton-affinity values.

If a nitrogen atom is bonded to a benzene ring, as in aniline, or is part of a resonance-stabilized ring, as in pyridine, it is appreciably less basic than in ammonia or in simple alkyl-substituted amines. The basicities of these molecules are given in Table 17–11, as is that for a remarkable neutral nitrogen base, $C_{12}H_{14}O_2N_2$, 1,8-diamino-2,7-dimethoxynapthalene,

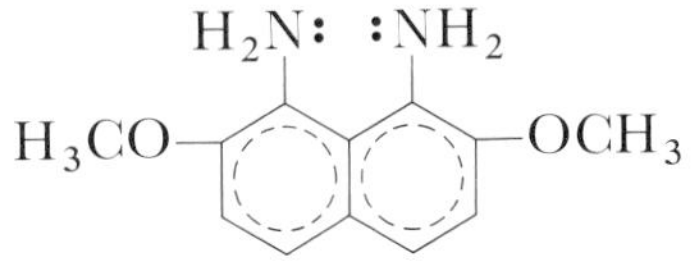

H_2N: :NH_2

H_3CO— —OCH_3

which is a strong base. The extremely high basicity of this molecule is attributed to the proximity of the two basic nitrogen atoms, which allows a single proton to become coordinated to both. Molecules of this type with very high basicity are called "proton sponges."

REACTIONS OF NITROGEN; THE FIXATION OF ATMOSPHERIC NITROGEN

Because elemental nitrogen is available in the atmosphere, and because nitrogen compounds are so important, reactions by which elemental nitrogen is converted to compounds deserve our attention. Table 17–12 indicates the difficulties associated with the problem of nitrogen fixation. Only ammonia, hydroxylamine, and dinitrogen pentaoxide are formed exo-

TABLE 17–12
The Standard Changes of Enthalpy and Free Energy for the Formation of Certain Nitrogen Compounds[a]

	ΔH_f^0 kJ mol^{-1}	ΔG_f^0 kJ mol^{-1}
$NH_3(g)$	−46.2	−16.7
$N_2H_4(l)$	+50.4	+149.0
$NH_2OH(s)$	−106.7	—
$N_2O(g)$	+81.6	+103.6
$NO(g)$	+90.3	+86.7
$N_2O_3(g)$	+83.7	+140.1
$NO_2(g)$	+33.8	+51.8
$N_2O_4(g)$	+9.7	+98.3
$N_2O_5(s)$	−41.8	+133.9

[a]The values for the standard changes of enthalpy and free energy are at $T = 298.2$ K.

thermically from the elements. Only ammonia is formed from its elements in a reaction with an equilibrium constant greater than 1 at 298.2 K. Therefore the production of most nitrogen compounds from the constituent elements is not a thermodynamically favored process at 298.2 K. A change of temperature can alter the situation, as will be shown.

The synthesis of ammonia from the constituent elements is an important commercial process for nitrogen fixation. The reaction

$$N_2(g) + 3H_2(g) \rightleftarrows 2NH_3(g)$$

which has been discussed at length in Chapter 12, provides an example of the balancing of factors that influence both the equilibrium yield of the products in an exothermic reaction and the rate of formation of the products. A high temperature increases the rate of reaction but decreases the extent of reaction at equilibrium. A high pressure increases the rate of reaction and, for this reaction with $\Delta n < 0$, also increases the extent of reaction at equilibrium, but the strength of reaction vessels limits the pressures that can be used in practice.

There is an enormous market for ammonia: in 1978, 1.7×10^7 tons were produced. Ammonia is used as a fertilizer, and it also can be converted to nitric oxide by oxidation,

$$4NH_3(g) + 5O_2(g) = 4NO(g) + 6H_2O(g)$$

a reaction for which $K \cong 10^{168}$ atm. The sequence of reactions

$$N_2(g) + 3H_2(g) = 2NH_3(g)$$
$$4NH_3(g) + 5O_2(g) = 4NO(g) + 6H_2O(g)$$

results in the formation of an unstable compound, nitric oxide (NO); its formation is coupled to the formation of a stable compound, water (H_2O). The sum of the coupled reactions is

$$2N_2(g) + 6H_2(g) + 5O_2(g) = 4NO(g) + 6H_2O(g)$$

This example resembles a previously discussed example, the coupled reactions that produce unstable xenon trioxide.

The oxidation of ammonia by oxygen also can yield nitrogen,

$$4NH_3(g) + 3O_2(g) = 2N_2(g) + 6H_2O(g)$$

a reaction with $K \cong 10^{228}$ atm. The desired reaction, that producing nitric oxide, is catalyzed by metallic platinum, and with this catalyst present at 1000 K, nitric oxide can be produced at a practical rate. The undesired reaction producing nitrogen is not catalyzed by platinum, and it does not occur at an appreciable rate under these conditions.

Equilibrium in the direct reaction of nitrogen with oxygen to form nitric oxide,

$$N_2(g) + O_2(g) = 2NO(g)$$

TABLE 17–13
The Equilibrium Constant[a] for
$N_2(g) + O_2(g) = 2NO(g)$

$\frac{T}{K}$	K	$\frac{T}{K}$	K
298.1	4.5×10^{-31}	2100	6.73×10^{-4}
1000	7.5×10^{-9}	2200	1.08×10^{-3}
1500	1.06×10^{-5}	2300	1.65×10^{-3}
1900	2.24×10^{-4}	2400	2.45×10^{-3}
2000	4.00×10^{-4}	2500	3.51×10^{-3}

[a] $K = \frac{P_{NO}^2}{P_{N_2}P_{O_2}}$; $\Delta H^0 = 180.6$ kJ mol^{-1} at 298.2 K.

does not produce an appreciable yield of nitric oxide at ordinary temperatures:

$$K = \frac{P_{NO}^2}{P_{N_2}P_{O_2}} = 4.5 \times 10^{-31} \text{ at } 298.2 \text{ K}$$

Nevertheless, the reaction is endothermic ($\Delta H^0 = 180.6$ kJ mol^{-1} at 298.2 K), and the equilibrium becomes more favorable as the temperature increases. Table 17–13 and Figure 17–10 give values of the equilibrium constant as a function of temperature. This reaction, with $\Delta n = 0$ and consequently a small value of ΔS^0 ($\Delta S^0 = +24.8$ J K^{-1} mol^{-1} at 298.2 K, $\Delta S^0 = +22.5$ J K^{-1} mol^{-1} at 6000 K), does not have a large value of K even at extreme temperatures ($K = 0.51$ at 6000 K). At temperatures of $\sim$2200 K, the yield of nitric oxide is appreciable, but a process for nitrogen fixation which depends on this reaction at high temperatures must allow for rapid cooling of the high temperature equilibrium mixture. With slow cooling, equilibrium in the mixture will be reestablished at temperatures lower than the maximum temperature reached. Therefore it is necessary to cool the mixture of gases (nitrogen, oxygen, and nitric oxide) very quickly to realize a yield like that given by the equilibrium at the highest temperature reached. The reaction under consideration has the same number of gaseous molecules on the reactant and product sides of the chemical equation ($\Delta n = 0$); therefore the equilibrium yield of nitric oxide is not increased by an increase of the pressure. A commercial process based on this reaction, the Birkeland–Eyde process, passes air through an electric arc. Although conditions in the arc are ill-defined, there is a region in the arc in which the temperature is high, and the process yields 1% to 2% nitric oxide. However, the high cost of the electric energy needed in this process has made it uneconomic, and it is no longer used.

The reaction of nitrogen and oxygen to form nitric oxide in internal combustion engines is the first step in a sequence that helps form photochemical smog. Although the yield of nitric oxide is not high at the temperatures existing in the cylinders of automobile engines, the total volume

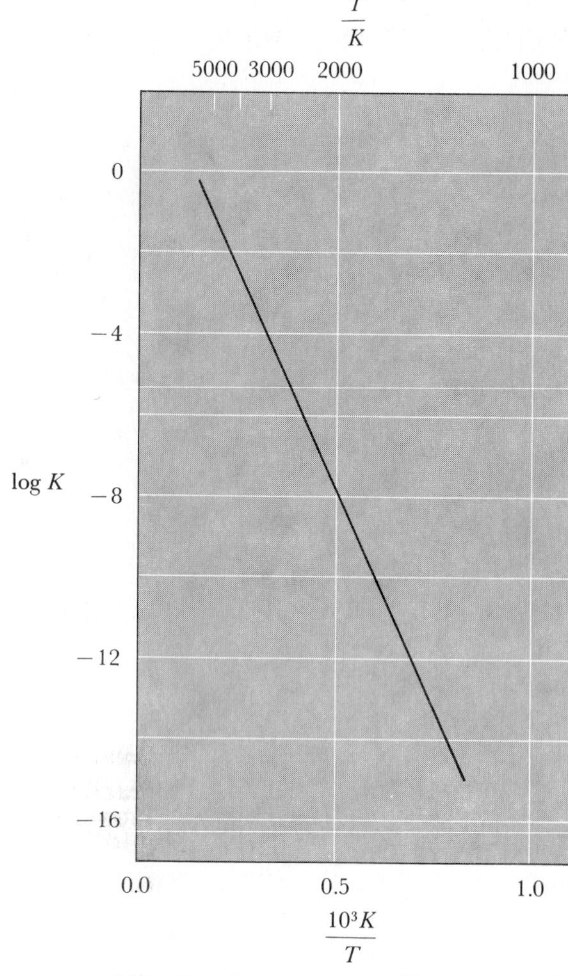

FIGURE 17–10

The temperature dependence of $K = \dfrac{P_{NO}^2}{P_{N_2}P_{O_2}}$ at high temperature. The range of temperature shown in this graph is 1200 K to 6000 K.

of air "processed" by all the automobiles in a large metropolitan center is enough to create a problem.

It is possible to fix atmospheric nitrogen with certain reactive reagents. Whether such reactions are practical for commercial production of nitrogen compounds depends upon the expense of producing the reactive reagent itself. Most active metals (e.g., Li or Ca) form ionic nitrides (containing N^{3-}, isoelectronic with neon):

$$6Li(s) + N_2(g) = 2Li_3N(s) \qquad \Delta H^0 = -390 \text{ kJ mol}^{-1}$$
$$3Ca(s) + N_2(g) = Ca_3N_2(s) \qquad \Delta H^0 = -430 \text{ kJ mol}^{-1}$$

Upon hydrolysis, these nitrides form ammonia, for example,

$$Ca_3N_2 + 6H_2O = 3Ca(OH)_2 + 2NH_3$$

However, these reactions are not economically practical for fixing atmospheric nitrogen.

The cyanamide process is a practical method for fixing nitrogen with an active reducing agent less expensive than calcium or lithium. This process involves calcium carbide, CaC_2, an ionic compound containing calcium ions Ca^{2+} and carbide ions, C_2^{2-}. (Carbide ion is isoelectronic with cyanide ion, CN^-, and carbon monoxide, CO.) Calcium carbide is produced from inexpensive reagents, limestone and coke:

$$CaCO_3(s) = CaO(s) + CO_2(g) \qquad \Delta H^0 = +178 \text{ kJ mol}^{-1}$$
$$CaO(s) + 3C(s) = CaC_2(s) + CO(g) \qquad \Delta H^0 = +460 \text{ kJ mol}^{-1}$$

Each of the two reactions is strongly endothermic, but each reaction produces a gas and can be made to occur essentially to completion by removing the product gas. The equilibria are made more favorable and the reaction rate is increased by working at elevated temperatures. The production of calcium carbide from lime (CaO) and coke (C) is carried out at 3000 K. Calcium cyanamide, Ca(NCN), an ionic compound containing calcium ion, Ca^{2+}, and cyanamide ion, NCN^{2-}, which is isoelectronic with carbon dioxide, OCO, cyanate ion, OCN^-, and azide ion, N_3^-, is produced in the reaction:

$$CaC_2(s) + N_2(g) = Ca(NCN)(s) + C(s) \qquad \Delta H^0 = -289 \text{ kJ mol}^{-1}$$

Hydrolysis of calcium cyanamide produces calcium carbonate and ammonia:

$$Ca(NCN)(s) + 3H_2O(l) = 2NH_3(g) + CaCO_3(s)$$
$$\Delta H^0 = -48.5 \text{ kJ mol}^{-1}$$

If the balanced equations for each reaction in the cyanamide process are added, the resulting equation is

$$2C(s) + N_2(g) + 3H_2O(l) = CO(g) + CO_2(g) + 2NH_3(g)$$

This equation reveals that it is elemental carbon that is oxidized to allow nitrogen to be reduced.

OXIDATION–REDUCTION CHEMISTRY

With oxidation states from −3 to +5, nitrogen has a complex oxidation–reduction chemistry. Following is a summary of the standard reduction potentials at 25°C for half-reactions involving the various oxidation states in acidic solution:

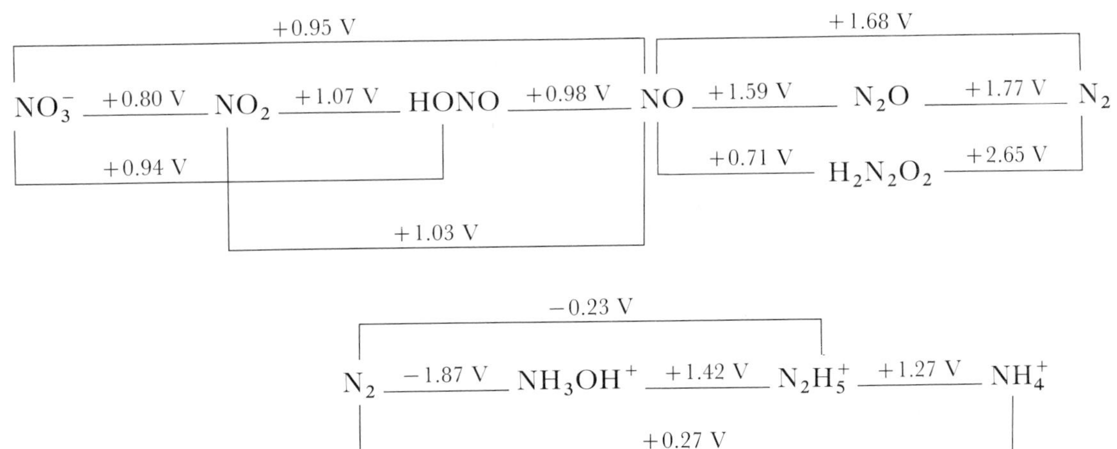

This diagram shows that many intermediate oxidation states of nitrogen are unstable with respect to disproportionation. Nonetheless, these states persist for varying lengths of time in acid solution because most of the disproportionation reactions are slow.

A disproportionation reaction is a step in the industrial production of nitric acid. We already saw in this chapter that fixation of atmospheric nitrogen could be carried to nitric oxide:

$$N_2 \xrightarrow{H_2} NH_3 \xrightarrow{O_2} NO$$

Now we will consider the reactions that convert nitric oxide to nitric acid. These reactions are oxidation,

$$2NO(g) + O_2(g) = 2NO_2(g)$$

and disproportionation,

$$3NO_2(g) + H_2O = 2H^+ + 2NO_3^- + NO(g)$$

The nitric oxide formed in this step is recycled. The net oxidation of nitrogen(II) in nitric oxide, NO, to nitrogen(V) in nitrate ion, NO_3^-, is brought about in this sequence by oxygen.

Nitric acid is a relatively powerful oxidizing agent regardless of the oxidation state to which the nitrogen is reduced. For some reactions the product formed depends upon the concentration of the acid. For instance, the oxidation of copper to copper(II) ion by nitrate ion in acidic solution gives nitric oxide in dilute acid solution,

$$3Cu + 8H^+ + 2NO_3^- = 3Cu^{2+} + 2NO(g) + 4H_2O$$

and nitrogen dioxide in concentrated acid solution,

$$Cu + 4H^+ + 2NO_3^- = Cu^{2+} + 2NO_2(g) + 2H_2O$$

Under a range of experimental conditions, both oxides are produced in the reduction of nitric acid by copper. The reaction of nitric acid with zinc, a more powerful reducing agent than copper,

$$Zn^{2+} + 2e^- = Zn \qquad \mathscr{E}^0 = -0.763 \text{ V}$$

yields ammonium ion:

$$10H^+ + NO_3^- + 4Zn = 4Zn^{2+} + NH_4^+ + 3H_2O$$

If we consider the half-reactions

$$Fe^{3+} + e^- = Fe^{2+} \qquad \mathscr{E}^0 = +0.77 \text{ V}$$
$$MnO_2 + 4H^+ + 2e^- = Mn^{2+} + 2H_2O \qquad \mathscr{E}^0 = +1.23 \text{ V}$$

along with those summarized in the diagram for nitrogen, we see why manganese(II) is stable in aqueous nitric acid, but iron(II) is not stable in aqueous nitric acid. Nitrate ion is not capable of oxidizing manganese(II) in acidic solution, but it is capable of oxidizing iron(II).

The decomposition of nitrous acid in acidic solution by disproportionation,

$$3HONO = 2NO(g) + H^+ + NO_3^- + H_2O$$

is reversible under easily realized concentration conditions, and we have discussed in Chapter 15 the rate law for this reaction. For the reaction as written, the equilibrium constant is

$$K = \frac{[H^+][NO_3^-]P_{NO}^2}{[HONO]^3} = 28.7 \text{ atm}^2 \text{ L mol}^{-1}$$

and we would expect the extent of disproportionation to increase as the concentration of hydrogen ion decreases:

$$\frac{[NO_3^-]P_{NO}^2}{[HONO]^3} \propto \frac{1}{[H^+]}$$

But this prediction is based on a chemical equation that is appropriate only for solutions in which the predominant species of nitrogen(III) is undissociated nitrous acid. Because the acid-dissociation constant for nitrous acid is $K_a = 4.5 \times 10^{-4}$ mol L^{-1}, the chemical equation already given is appropriate for $[H^+] > 4.5 \times 10^{-3}$ mol L^{-1}. For acidic solutions in which nitrite ion is the predominant species of nitrogen(III) (1.0×10^{-7} mol L^{-1} $< [H^+] < 4.5 \times 10^{-5}$ mol L^{-1}), the appropriate chemical equation is

$$3NO_2^- + 2H^+ = 2NO(g) + NO_3^- + H_2O$$

For alkaline solutions the appropriate chemical equation is

$$3NO_2^- + H_2O = 2NO(g) + NO_3^- + 2OH^-$$

These two chemical equations show that nitrogen(III) becomes more stable with respect to disproportionation with decreasing concentration of hydrogen ion at low acidity. This relationship is illustrated in Figure 17–11, which shows the dependence of the quotient

$$\left\{\frac{[NO_3^-]P_{NO}^2}{([HONO] + [NO_2^-])^3}\right\}_{eq}$$

upon the concentration of hydrogen ion. This quotient, with the concentrations (or partial pressures) of the nitrogen-containing products of disproportionation in the numerator, is a measure of the "extent" of disproportionation of nitrogen(III) at equilibrium. From the graph we see that nitrogen(III) is stable with respect to disproportionation in alkaline solution; this quotient is less than 1.00 atm^2 L^2 mol^{-2} for $[H^+] < 1.8 \times 10^{-6}$ mol L^{-1}.

Although both hydroxylamine and hydrazine are unstable with respect to disproportionation in acidic solution,

$$3NH_3OH^+ = N_2 + NH_4^+ + 2H^+ + 3H_2O$$

$$3N_2H_5^+ + H^+ = N_2 + 4NH_4^+$$

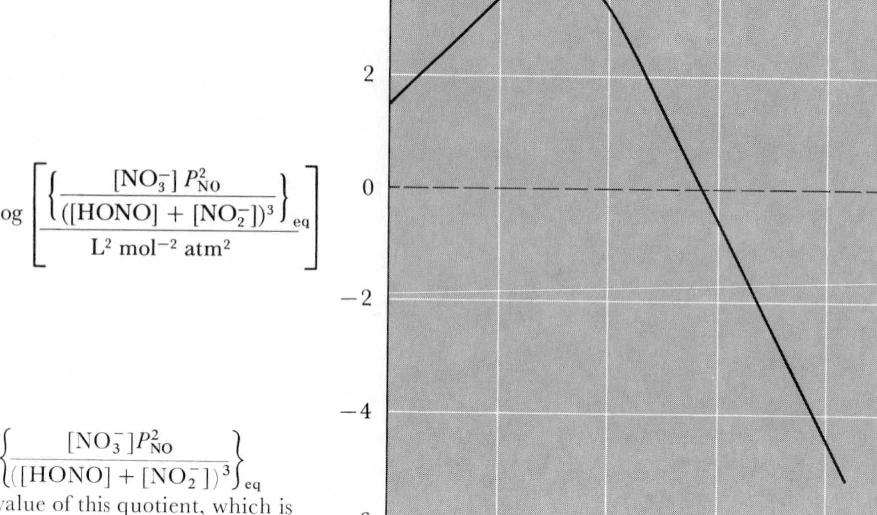

FIGURE 17–11

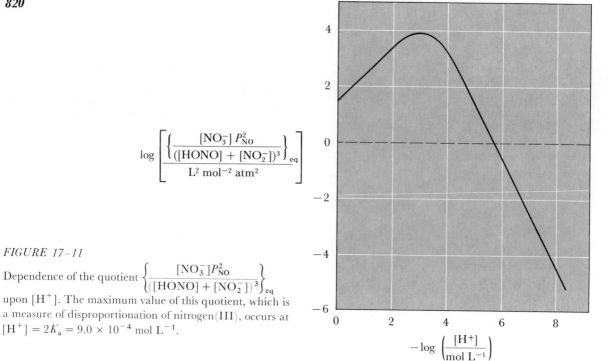

Dependence of the quotient $\left\{\frac{[NO_3^-]P_{NO}^2}{([HONO]+[NO_2^-])^3}\right\}_{eq}$ upon $[H^+]$. The maximum value of this quotient, which is a measure of disproportionation of nitrogen(III), occurs at $[H^+] = 2K_a = 9.0 \times 10^{-4}$ mol L^{-1}.

these species of nitrogen in negative oxidation states do not decompose rapidly is solution. Each can act both as an oxidizing agent and as a reducing agent. For instance, hydrazine in acidic solution oxidizes titanium(III) ion,

$$N_2H_5^+ + 2Ti^{3+} + 2H_2O = 2NH_4^+ + 2TiO^{2+} + H^+$$

and reduces bromine,

$$N_2H_5^+ + 2Br_2 = N_2 + 4Br^- + 5H^+$$

17–6 Sulfur and Phosphorus

Sulfur and phosphorus have the electronic structures

	1s	2s	2p	3s	3p
S	(⇅)	(⇅)	(⇅)(⇅)(⇅)	(⇅)	(⇅)(↑)(↑)
P	(⇅)	(⇅)	(⇅)(⇅)(⇅)	(⇅)	(↑)(↑)(↑)

Their valence-shell electron distributions are the same as those for oxygen and nitrogen, respectively. The properties of these elements with an $n = 3$ valence shell are quite different from those of the corresponding elements with an $n = 2$ valence shell. This is the result of differences of ionization energies and electron affinities (see Figure 17–2), different capacities of these valence shells,

$n = 2$ Capacity 8 electrons; $2s$ and $2p$ orbitals

$n = 3$ Capacity 18 electrons; $3s$, $3p$, and $3d$ orbitals

and different tendencies to form double bonds.

The dominant factor that makes sulfur and phosphorus less nonmetallic than oxygen and nitrogen is their different ionization energies; this difference offsets an unexpected trend in the electron affinities:

$$IE(\text{S}) - IE(\text{O}) = -314 \text{ kJ mol}^{-1}$$
$$IE(\text{P}) - IE(\text{N}) = -339 \text{ kJ mol}^{-1}$$
$$EA(\text{S}) - EA(\text{O}) = 59.3 \text{ kJ mol}^{-1}$$
$$EA(\text{P}) - EA(\text{N}) = 72 \text{ kJ mol}^{-1}$$

By itself, the difference of the electron affinities would make sulfur and phosphorus more nonmetallic. Recall that one proposed measure of electronegativity is the average of the ionization energy and electron affinity. The electronegativity differences for these elements,

$$\chi(\text{S}) - \chi(\text{O}) = -1.0 \qquad \chi(\text{P}) - \chi(\text{N}) = -0.9$$

are dominated by the differences of ionization energies.

Table 17–14 summarizes the thermodynamic properties of some compounds of sulfur and phosphorus with hydrogen and oxygen. These are in dramatic contrast with the corresponding quantities for oxygen and nitrogen compounds. Hydrogen sulfide is less stable than water, and the hydride of phosphorus, phosphine (PH_3), is less stable than the hydride of nitrogen, ammonia. But the oxides of sulfur and phosphorus are much more stable than the oxides of nitrogen.

TABLE 17–14
The Standard Changes of Enthalpy and Free Energy for the Formation of Certain Compounds of Sulfur and Phosphorus[a]

	ΔH_f^0 kJ mol^{-1}	ΔG_f^0 kJ mol^{-1}
$H_2S(g)$	−20.6	−33.6
$SO_2(g)$	−296.8	−300.2
$SO_3(g)$	−395.7	−371.1
$PH_3(g)$	+5.4	+13.4
$P_4O_6(s)$	−1640	—
$P_4O_{10}(s)$	−2984	−2698

[a]The values for the standard changes of enthalpy and free energy are at $T = 298.2$ K.

Sulfur and phosphorus exhibit oxidation states over the entire range expected for elements with their electronic configurations. The extreme oxidation states -2 and $+6$ for sulfur and -3 and $+5$ for phosphorus are shown in Table 17–2. The potential diagram for sulfur in acidic solution, showing only the more important species, is

$$SO_4^{2-} \xrightarrow{+0.17\ V} SO_2 \xrightarrow{+0.48\ V} S_4O_6^{2-} \xrightarrow{+0.17\ V} S_2O_3^{2-} \xrightarrow{+0.50\ V} S \xrightarrow{+0.14\ V} H_2S$$

$$SO_2 \xrightarrow{+0.40\ V} S_2O_3^{2-}$$

Using this potential diagram, we can evaluate many features of the diverse oxidation–reduction chemistry of this element.

For instance, we can calculate the equilibrium constant for an important reaction of analytical chemistry, the reaction of thiosulfate ion with triiodide ion:

$$I_3^- + 2S_2O_3^{2-} = 3I^- + S_4O_6^{2-} \qquad \mathcal{E}^0 = +0.537\ V - 0.17\ V = 0.37\ V$$

$$K^0 = 10^{2\times 0.37\ V/0.0592\ V} = 3.2 \times 10^{12}$$

$$K = 3.2 \times 10^{12}\ mol\ L^{-1}$$

The end point in this reaction is detected by the deep blue color of a complex of iodine(0) with starch. This blue color is developed by the presence of triodide ion at a concentration of $\sim 1 \times 10^{-5}$ mol L^{-1}. Let us see what concentration of unreacted thiosulfate ion exists in equilibrium with the other species in a solution with a composition typical for an end point in such a titration:

$$[I_3^-] = 1.0 \times 10^{-5}\ mol\ L^{-1}$$

$$[I^-] = 0.150\ mol\ L^{-1}$$

$$[S_4O_6^{2-}] = 0.030\ mol\ L^{-1}$$

$$[S_2O_3^{2-}] = \left\{\frac{[I^-]^3[S_4O_6^{2-}]}{K \times [I_3^-]}\right\}^{1/2}$$

$$= \left\{\frac{(0.150\ mol\ L^{-1})^3(0.030\ mol\ L^{-1})}{(3.2 \times 10^{12}\ mol\ L^{-1})(1.0 \times 10^{-5}\ mol\ L^{-1})}\right\}^{1/2}$$

$$= 1.8 \times 10^{-6}\ mol\ L^{-1}$$

Thus at the appearance of the blue starch-iodine color:

1. the concentration of unreacted thiosulfate ion is very low ($\sim 2 \times 10^{-6}$ mol L^{-1}), and
2. the concentration of unreacted triiodide ion ($\sim 1 \times 10^{-5}$ mol L^{-1}) is larger than the equivalent concentration of unreacted thiosulfate ion,

$$[I_3^-] - \tfrac{1}{2}[S_2O_3^{2-}] = 9 \times 10^{-6}\ mol\ L^{-1}$$

but this difference is very small compared to typical initial concentrations. Thus the end point occurs essentially at the equivalence point.

The potential diagram for sulfur shows the disproportionation of thiosulfate ion to be a favorable reaction at $[H^+] = 1.00$ mol L^{-1}:

$$S_2O_3^{2-} + 2H^+ = SO_2 + S(s) + H_2O$$

$$\mathscr{E}^0 = +0.50 \text{ V} - 0.40 \text{ V} = 0.10 \text{ V}$$

$$K^0 = 10^{2 \times 0.10 \text{ V}/0.0592 \text{ V}} = 2.4 \times 10^3$$

$$K = 2.4 \times 10^3 \text{ L}^2 \text{ mol}^{-2}$$

This disproportionation equilibrium becomes less favorable as the acidity of the solution is lowered. At $[H^+] = 1.00 \times 10^{-7}$ mol L^{-1} in the presence of solid sulfur,

$$\frac{[SO_2]}{[S_2O_3^{2-}]} = 2.4 \times 10^3 \text{ L}^2 \text{ mol}^{-2} \times [H^+]^2 = 2.4 \times 10^{-11}$$

Thus thiosulfate ion is stable with respect to disproportionation at $[H^+] = 1.0 \times 10^{-7}$ mol L^{-1}, but is appreciably unstable ($[SO_2]/[S_2O_3^{2-}] > 0.01$) at $[H^+] > 2.0 \times 10^{-3}$ mol L^{-1}.[13]

The potential diagram gives values of $\mathscr{E}^0$ which are appropriate for *standard conditions*. Therefore the value of $\mathscr{E}^0$ for the sulfate ion–sulfur dioxide couple does not reveal the oxidizing ability of concentrated sulfuric acid. Let us consider the oxidation–reduction reaction of sulfate ion with bromide ion in acidic solution:

$$SO_4^{2-} + 2Br^- + 4H^+ = SO_2(aq) + Br_2(aq) + 2H_2O$$

$$\mathscr{E}^0 = +0.17 \text{ V} - 1.08 \text{ V} = -0.91 \text{ V}$$

$$K^0 = 10^{2 \times (-0.91 \text{ V})/0.0592 \text{ V}} = 1.8 \times 10^{-31}$$

$$K = 1.8 \times 10^{-31} \text{ L}^5 \text{ mol}^{-5}$$

Bromide ion is not oxidized by dilute sulfuric acid, which is consistent with this small value of K. But concentrated sulfuric acid does oxidize bromide ion. Rearrangement of the equilibrium-constant equation, with *the activity-coefficient factors* included, gives

$$[Br_2] = K^0 \times \frac{[H^+]^4[SO_4^{2-}][Br^-]^2\gamma_{H^+}^4\gamma_{SO_4^{2-}}\gamma_{Br^-}^2}{[SO_2]x_{H_2O}^2\gamma_{SO_2}\gamma_{H_2O}^2\gamma_{Br_2}}$$

When the concentration of sulfuric acid is increased to large values (e.g., from 1 mol L^{-1} to ~ 18.0 mol L^{-1}) the value of $([H^+]^4[SO_4^{2-}]\gamma_{H^+}^4\gamma_{SO_4^{2-}}\gamma_{Br^-}^2)$ increases greatly, and the factor $(x_{H_2O}^2\gamma_{H_2O}^2)$ decreases greatly. The quotient of these factors increases enormously, which makes it possible for sulfuric acid to oxidize bromide ion to an appreciable extent. This particular point is relevant to the problem of preparing gaseous hydrogen halides from the corresponding salts. It is convenient to do this by treating a halide salt with concentrated sulfuric acid because hydrogen halides are

---

[13] Add dilute hydrochloric acid to a small portion of thiosulfate solution and you will see a milky precipitate of sulfur form.

not very soluble in this liquid. This reaction involving sodium chloride is

$$H^+ + HOSO_3^- + NaCl(s) = HCl(g) + NaHOSO_3(s)$$

Sulfuric acid cannot oxidize chloride ion, so a side reaction producing chlorine does not occur. In contrast, the reaction for sodium bromide,

$$H^+ + HOSO_3^- + NaBr(s) = HBr(g) + NaHOSO_3(s)$$

is accompanied by the side reaction producing bromine:

$$H^+ + 3HOSO_3^- + 2NaBr(s) = Br_2(g) + SO_2(g) + 2NaHOSO_3(s) + H_2O$$

This example shows that activity-coefficient factors, assumed to have a magnitude of ~ 1.00 in many of our calculations, can have a magnitude very different from 1.00 in concentrated solutions of sulfuric acid.

Sulfur dioxide is thermodynamically unstable with respect to disproportionation; for example,

$$4SO_2(aq) + 3H_2O = 2SO_4^{2-} + S_2O_3^{2-} + 6H^+$$
$$\mathscr{E}^0 = +0.40\ V - 0.17\ V = +\ 0.23\ V$$

But neither this reaction nor other possible disproportionation reactions of this species occur at an appreciable rate under ordinary conditions. The gas-phase disproportionation reaction,

$$3SO_2(g) = 2SO_3(g) + S(s)$$

is unfavorable ($K = 2 \times 10^{-28}\ atm^{-1}$ at 298.2 K).

The potential diagram for phosphorus in acidic solution is[14]

$$H_3PO_4 \underline{\ -0.28\ V\ } H_3PO_3 \underline{\ -0.50\ V\ } H_3PO_2 \underline{\ -0.51\ V\ } P_4 \underline{\ +0.06\ V\ } PH_3$$

None of the positive oxidation states is a good oxidizing agent. Rather, even mild oxidizing agents can oxidize compounds of phosphorus in lower oxidation states ($\leq +3$) all the way to the $+5$ oxidation state; for example, the reaction of hypophosphorous acid with triiodide ion goes through both stages:

$$I_3^- + H_3PO_2 + H_2O = H_3PO_3 + 3I^- + 2H^+$$
$$I_3^- + H_3PO_3 + H_2O = H_3PO_4 + 3I^- + 2H^+$$

The rate of the first stage of this reaction at moderate concentrations of triiodide ion is governed by the rate law

$$-\frac{d[H_3PO_2]}{dt} = k[H_3PO_2][H^+]$$

This rate law does not involve the concentration of triiodide ion. It has been interpreted to mean that the slow step in the reaction under these

[14]Because the line formulas for phosphorous acid and hypophosphorous acid that disclose their structures are cumbersome, we use conventional formulation in this diagram and in some chemical equations of this section.

conditions is conversion of phosphorus(I) from a tetrahedral species to a trigonal bipyramidal species:

$$(HO)OPH_2 + H^+ \leftrightarrows (HO)_2PH_2^+ \qquad K \ll 1 \text{ L mole}^{-1}$$

$$\mathrm{(HO)(OH^+)PH_2} + H_2O \longrightarrow \mathrm{HO{-}PH_2(OH)_2} + H^+ \qquad \text{Slow}$$

The phosphorus atom in this pentacoordinated intermediate, $(HO)_3PH_2$, has bonded to it the three oxygen atoms and one hydrogen atom which are bonded in the final product $(HO)_2OPH$. This product is produced in the rapid oxidation step, which removes one of the hydrogen atoms bonded to phosphorus in the intermediate species, and produces phosphorous acid:

$$I_3^- + (HO)_3PH_2 \rightarrow 3I^- + 2H^+ + (HO)_2OPH$$

At high temperatures, hypophosphite ion decomposes in alkaline solution to produce phosphite ion and hydrogen:

$$H_2PO_2^- + OH^- = HPO_3^{2-} + H_2$$

With a strong reducing agent, such as zinc, hypophosphorous acid will act as an oxidizing agent; reduction takes the phosphorus to phosphine:

$$H_3PO_2 + 2Zn + 4H^+ = PH_3 + 2Zn^{2+} + 2H_2O$$

(Phosphine is not basic enough to be protonated in dilute acidic solution.)

The slight oxidizing ability of phosphorus(V) makes it possible to use concentrated phosphoric acid as a source of protons to produce hydrogen bromide and hydrogen iodide from salts. That is, the reaction

$$NaBr(s) + H_3PO_4 \text{ (conc.)} = HBr(g) + NaH_2PO_4(s)$$

occurs without the occurrence of a side reaction involving oxidation and reduction. This is to be contrasted with the reaction of sulfuric acid already discussed.

However, phosphorus(V) does act as an oxidizing agent in the reaction by which elemental phosphorus is produced from common ores of phosphorus, for example, calcium orthophosphate, $Ca_3(PO_4)_2$. The reducing agent is carbon. The reaction

$$2Ca_3(PO_4)_2(s) + 10C(s) = 6CaO(s) + 10CO(g) + P_4(g)$$

does not occur, but this same oxidation–reduction process can be made to occur if calcium oxide, a basic oxide, is converted to a salt. Silicon dioxide, an acidic oxide, does this. With silicon dioxide present, the net reaction at elevated temperatures is

$$2Ca_3(PO_4)_2(s) + 10C(s) + 6SiO_2(s) = 6CaSiO_3(s) + 10CO(g) + P_4(g)$$

This reaction is carried out on a large scale using phosphate rock [$Ca_3(PO_4)_2$], coke (C), and sand (SiO_2) in an electric furnace, which gives temperatures of $>800°C$. The reaction is made essentially irreversible by removing the carbon monoxide and phosphorus, which distill from the furnace. The phosphorus is collected under water to keep it from being oxidized by air.

Much of the elemental phosphorus produced in this way then is reoxidized to produce phosphoric acid. Although simple treatment of calcium orthophosphate with concentrated sulfuric acid produces phosphoric acid, the reaction

$$3H_2SO_4(\text{conc.}) + Ca_3(PO_4)_2(s) = 2H_3PO_4 + 3CaSO_4(s)$$

does not remove some of the impurities present in the phosphate ore. Thus the more circuitous route to pure phosphoric acid is necessary. However, the treatment with a limited amount of sulfuric acid is used to convert insoluble calcium orthophosphate to a mixture of calcium sulfate and soluble calcium dihydrogenphosphate:

$$Ca_3PO_4(s) + 2H_2SO_4 = Ca(H_2PO_4)_2 + CaSO_4(s)$$

This mixture is called superphosphate of lime and is used as a fertilizer.

OXIDES AND OXOACIDS

Much about the chemistry of the oxides and acids of sulfur and phorphorus already has appeared in our general discussion of nonmetal oxides and oxoacids and in the discussion of the oxidation–reduction chemistry of these elements. Now we will consider some points that we have not covered.

Each reaction in the sequence

$$S(s) + O_2(g) = SO_2(g)$$

and

$$2SO_2(g) + O_2(g) = 2SO_3(g)$$

is thermodynamically favored, $\Delta G^0 \ll 0$, but the second reaction is very slow. The large-scale production of sulfuric acid involves this reaction, and a number of catalysts for it are known (V_2O_5, Pt, NO). The sulfur trioxide formed is readily absorbed in concentrated sulfuric acid to form disulfuric acid ($H_2S_2O_7$),

$$SO_3(g) + (HO)_2SO_2 = H_2S_2O_7$$

which then is treated with water to give sulfuric acid,

$$H_2S_2O_7 + H_2O = 2(HO)_2SO_2$$

These two reactions involve moving along the abscissa of Figure 17–8 from $x_{SO_3} = 0.50$ to 0.667 and then back to 0.50.

One or both hydroxyl groups of sulfuric acid can be replaced by groups derived from elements adjacent to oxygen in the periodic table. The analogous groups are

$$\cdot\ddot{\underset{\cdot\cdot}{N}}:H \quad \cdot\ddot{\underset{\cdot\cdot}{O}}:H \quad \cdot\ddot{\underset{\cdot\cdot}{F}}:$$
$$\text{H}$$

The molecules that can be generated by these substitutions are

$H_2NSO_2(OH)$ $K_a = 0.100 \text{ mol L}^{-1}$.
Amidosulfuric acid

$FSO_3(OH)$ $K_a > 1 \text{ mol L}^{-1}$
Fluorosulfuric acid

The oxides of phosphorus(III) and phosphorus(V) have the molecular formulas P_4O_6 and P_4O_{10}; these formulas contrast with the molecular formulas N_2O_3 and N_2O_5 for the corresponding oxides of nitrogen. The structure of P_4O_6 can be described as a tetrahedron with a phosphorus atom at each of its four corners and an oxygen atom slightly off the middle of each of its six edges, as shown in Figure 17–12. The structure of phosphorus(V) oxide, P_4O_{10}, is related to that of phosphorus(III) oxide; a peripheral oxygen atom is bonded to each phorphorus atom without altering the P_4O_6 tetrahedron (Figure 17–12). These oxides of phosphorus are the anhydrides of phosphorous acid [H_3PO_3, preferably written $(HO)_2OPH$] and phosphoric acid [H_3PO_4, preferably written $(HO)_3PO$]. The affinity of P_4O_{10} for water is enormous: it is used as a dessicant. The vapor pressure of water over mildly hydrated P_4O_{10} is less than 3×10^{-5} torr.

Like sulfuric acid, which forms disulfuric acid by partial dehydration, phosphoric acid forms diphosphoric acid, $H_4P_2O_7$ (also called pyrophosphoric acid), and triphosphoric acid, $H_5P_3O_{10}$, in this way. The suc-

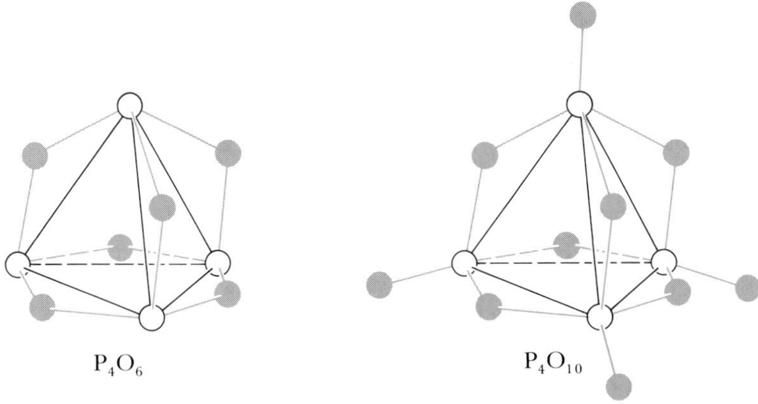

FIGURE 17–12
The oxides of phosphorus. Each bridging oxygen atom is bonded to two phorphorus atoms with single bonds. In phosphorus(III) oxide (P_4O_6), the resulting formal charge on each atom is zero. In phosphorus(V) oxide (P_4O_{10}), the bonds between phosphorus atoms and peripheral oxygen atoms have some double-bond character.

cessive acid-dissociation constants for diphosphoric acid were discussed in Section 13–3, where the structure of the acid was given. Linear polyphosphate ions with the composition $P_nO_{3n+1}^{-(n+2)}$, with n having values as high as 10, are chains of atoms alternating oxygen with phosphorus; each phosphorus atom also has two additional oxygen atoms bonded to it. Thus $P_5O_{16}^{7-}$ has the structure

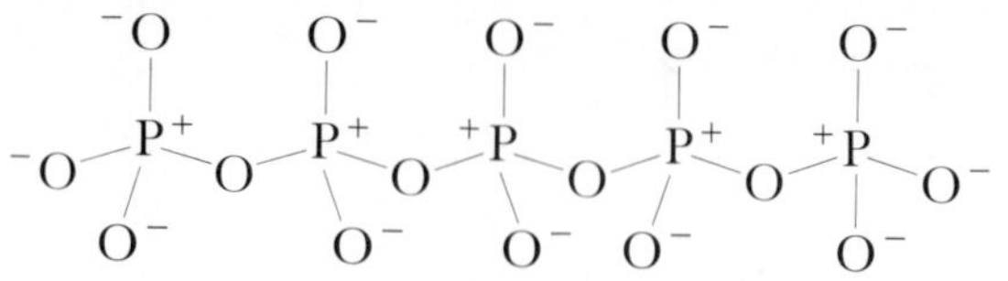

$$\begin{array}{c} {}^{-}O \quad O^{-} \quad O^{-} \quad O^{-} \quad O^{-} \\ {}^{-}O{-}P^{+}{-}O{-}P^{+}{-}O{-}{}^{+}P{-}O{-}P^{+}{-}O{-}{}^{+}P{-}O^{-} \\ O^{-} \quad O^{-} \quad O^{-} \quad O^{-} \quad O^{-} \end{array}$$

with the formal charges shown that would be associated with an electronic structure having only single bonds. There is a net charge of -2 associated with each terminal —PO_3 group $(+1 - 3 = -2)$ and a net charge of -1 associated with each nonterminal $>PO_2$ group $(+1 - 2 = -1)$. If a strong acid is added to a solution of a linear polyphosphate ion, each of the two terminal —PO_3^{2-} groups will add one proton at high pH (pH > 6); the remaining n units of net negative charge are neutralized by associating protons at lower pH (pH < 3). Thus the predominant form of the fully protonated acid, $H_{n+2}P_nO_{3n+1}$, has two protons on oxygen atoms of each terminal —PO_3^{2-} group and one proton on an oxygen atom of each nonterminal $>PO_2^-$ group.

Linear polyphosphate species can lose water to form cyclic polyphosphate species; for example,

$$\underset{\text{Dihydrogen-triphosphate ion}}{H_2P_3O_{10}^{3-}} = \underset{\text{Trimetaphosphate ion}}{P_3O_9^{3-}} + H_2O$$

The structure of trimetaphosphate ion is cyclic:

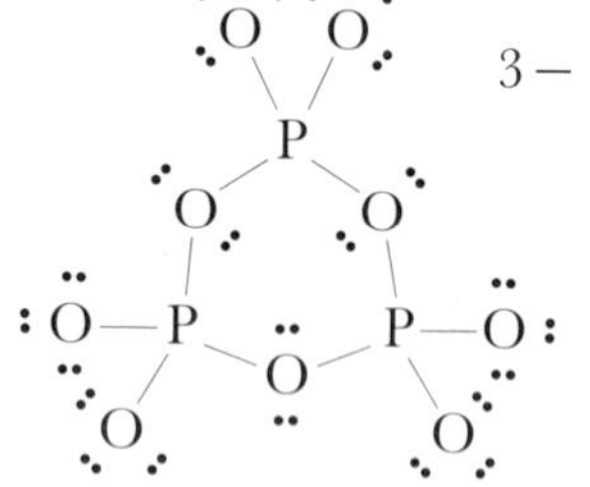

$$\left[\begin{array}{c} :\ddot{O}\quad\ddot{O}: \\ \diagdown\;\diagup \\ P \\ \diagup\quad\diagdown \\ :\ddot{O}\qquad\ddot{O}: \\ | \qquad\quad | \\ :\ddot{O}{-}P \quad :\ddot{O}: \quad P{-}\ddot{O}: \\ | \qquad\qquad\quad | \\ :\ddot{O}: \qquad\quad :\ddot{O}: \end{array}\right]^{3-}$$

Trimetaphosphate ion

The conversion of metaphosphates to orthophosphate ion in aqueous solution is a slow process, occurring over a period of days to months (depending upon the temperature and concentration of the hydrogen ion). The polymetaphosphates are the phosphate constituent of detergents, in which they have many functions, including provision of mild alkalinity (in the anionic form, metaphosphate ions are Brønsted bases) and the complexing of calcium ion to soften the water.

Calcium(II) ion forms a number of compounds of low solubility with phosphate ions. In addition, ion-pair complex ions involving calcium ion and phosphate ions, for example, $Ca^{2+}O_3POH^{2-}$, are known. It is important to study this rather complex field because calcium phosphates are constituents of teeth and bones, and the natural regulation of the phosphate concentration in lakes and rivers is related to solubilities and rates of precipitation of calcium phosphates. Slightly soluble calcium phosphates (with their mineralogical names) are:

$Ca_5(PO_4)_3OH$, Hydroxyapatite
$Ca(HOPO_3) \cdot 2H_2O$, Brushite
$Ca(HOPO_3)$, Monetite
$Ca_8(HOPO_3)_2(PO_4)_4 \cdot 5H_2O$, Octacalcium phosphate
$Ca_3(PO_4)_2$, Whitlockite

The solubility of each of these compounds varies in a different way with variation of the acidity of the solution. For instance, in solutions in which the principal phosphate species is dihydrogenphosphate ion, $(HO)_2PO_2^-$ (i.e., a solution with $K_1 > [H^+] > K_2$), the solubility equilibria for $Ca_5(PO_4)_3OH$, $Ca_3(PO_4)_2$, and $Ca(HOPO_3)$ are:

$$Ca_5(PO_4)_3OH(s) + 7H^+ \rightleftarrows 5Ca^{2+} + 3(HO)_2PO_2^- + H_2O \qquad S \propto [H^+]^{7/8}$$
$$Ca_3(PO_4)_2(s) + 4H^+ \rightleftarrows 3Ca^{2+} + 2(HO)_2PO_2^- \qquad S \propto [H^+]^{4/5}$$
$$Ca(HOPO_3)(s) + H^+ \rightleftarrows Ca^{2+} + (HO)_2PO_2^- \qquad S \propto [H^+]^{1/2}$$

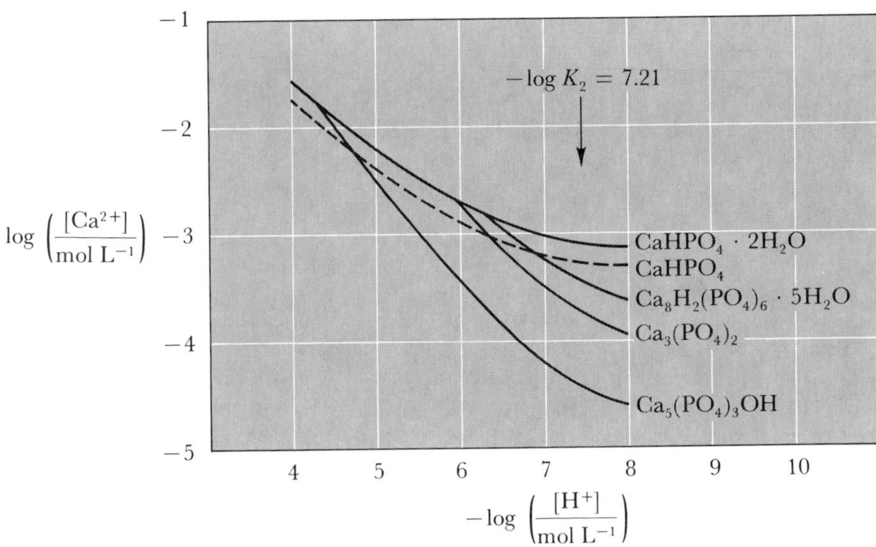

FIGURE 17–13
The concentration of calcium ion in equilibrium with various calcium phosphates as a function of pH. At $[H^+] > 6 \times 10^{-7}$ mol L^{-1}, $H_2PO_4^-$ is the predominant form. At $[H^+] < 6 \times 10^{-9}$ mol L^{-1}, HPO_4^{2-} is the predominant form. Notice the change of slopes of lines in this range of $[H^+]$.

(The exponent of $[H^+]$ in each case is the coefficient of hydrogen ion on the reactant side of the balanced chemical equation divided by the sum of the coefficients of calcium ion and dihydrogenphosphate ion on the product side of the balanced chemical equation.) Because of the different dependences of solubility upon concentration of hydrogen ion, the relative solubilities of the compounds depend upon the solution acidity, as is shown in Figure 17–13, which indicates the values of the pH at which the solubility curves cross. If all equilibria between the solution phase and the various solid phases were established rapidly, the solid phase with the lowest solubility at a particular pH would be the only solid phase capable of existing in equilibrium with a solution of that pH. Because the rates of precipitation of the various compounds differ, a solution with concentrations exceeding those allowed by the solubility-product constant may persist without precipitation occurring.

PHOSPHATE ESTERS

An ester of a carboxylic acid, discussed in Chapter 11, is simply a carboxylic acid with an alkyl group instead of an acidic hydrogen:

```
     H    O              H    O
     |   //              |   //
  H—C—C               H—C—C
     |   \               |   \
     H    O—H            H    O—CH3
```

Acetic acid   Methyl acetate

Inorganic oxoacids also form esters, and their structures are related in this same way. Here we are concerned with esters of phosphoric acid and polyphosphoric acid. The structures of phosphoric acid and methyl phosphate are:

```
         H                    H
        /                    /
       O                    O
       |     H              |     CH3
    O—P—O/              O—P—O/
  H/   |              H/   |
       O                    O
```

Phosphoric acid   Methyl phosphate

These substances interest us because various phosphate esters play important roles in biology. We shall mention some of these roles in Chapter 20; here we shall be concerned primarily with the dependence of the stability of the ester linkage (P—O—C) and the polyphosphate linkage (P—O—P) on the nature of the alkyl group.

Phosphate esters have acidic protons, and the extent to which these are dissociated in solution is determined by the concentration of hydrogen ion

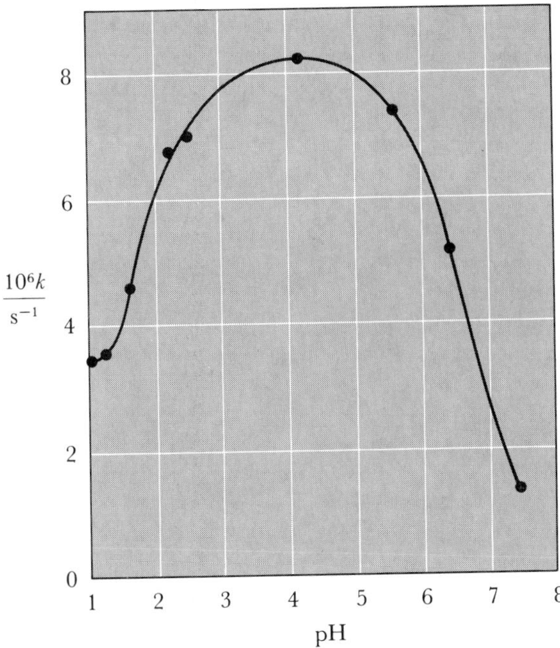

FIGURE 17–14
The rate constant for hydrolysis of methyl phosphate at 100.1°C as a function of pH (in the range $1.0 < pH < 7.5$). [From C. A. Bunton, D. R. Llewellyn, K. G. Oldham, and C. A. Vernon, *J. Chem. Soc.*, 3574 (1958).]

in solution. The two acid-dissociation constants for methyl phosphate are (for 22°C):

$$K_1 = \frac{[H^+][H_3COPO_3H^-]}{[H_3COPO_3H_2]} = 0.030 \text{ mol L}^{-1}$$

$$K_2 = \frac{[H^+][H_3COPO_3^{2-}]}{[H_3COPO_3H^-]} = 2.6 \times 10^{-7} \text{ mol L}^{-1}$$

These values are each ~4-fold larger than the values of K_1 and K_2 for orthophosphoric acid at 25°C (given in Section 13–3).

The two types of reactions in which we are interested are (with charges appropriate for pH ≅ 5):

Ester hydrolysis

$$ROPO_3H^- + H_2O = ROH + (HO)_2PO_2^-$$

Polyphosphate hydrolysis (e.g., a triphosphate)

$$RO(PO_3)_2PO_3H^{3-} + H_2O = ROPO_3PO_3H^{2-} + (HO)_2PO_2^-$$

First let us consider the ester hydrolysis reaction. How does the reaction rate depend on pH and how does the extent of reaction depend on the identity of the R— group? An answer to the first of these questions is provided by data for methyl phosphate presented in Figure 17–14, which shows that the rate is highest in the range of pH in which the ion of charge -1, $H_3COPO_3H^-$, is predominant. Therefore the kinetic data show that this species is more reactive than the neutral species, $H_3COPO_3H_2$, or the ion of charge -2, $H_3COPO_3^{2-}$.

Equilibrium constants for hydrolysis of two phosphate esters are (for pH = 7 and 25°C):

Ester	$\frac{K}{\text{mol L}^{-1}}$
$CH_3C(=O)\text{—}OPO_3H^-$ Acetyl phosphate	2.5×10^7
$HOCH_2CH_2OHCH_2OPO_3H^-$ Glycerol 1-phosphate[15]	41

Among these two substances, acetyl phosphate has the greater tendency to transfer its phosphate group to water. As a consequence, the reaction in which acetyl phosphate transfers a phosphate group to glycerol,

Acetyl phosphate + Glycerol = Glycerol 1-phosphate + Acetate ion

has an equilibrium constant $K > 1$:

$$K = \frac{2.5 \times 10^7 \text{ mol L}^{-1}}{41 \text{ mol L}^{-1}} = 6.1 \times 10^5$$

Like the ester hydrolysis reactions just considered, the hydrolysis of a polyphosphate produces phosphate ion. But in this case, bond-breaking occurs in the P—O—P bonding system. For the hydrolysis of adenosine triphosphate (ATP) to give adenosine diphosphate (ADP) (see Table 20–3 for the structures), the equilibrium constant (at pH = 7 and 25°C) is $K = 2.2 \times 10^5 \text{ mol L}^{-1}$, a value intermediate between those given for hydrolysis of the two monophosphates. Thus by phosphate-transfer reactions acetyl phosphate can transform ADP to ATP, and ATP can transform glycerol to glycerol 1-phosphate.

Many metabolic processes involve phosphate-transfer reactions analogous to those just considered. There are mechanisms, generally catalyzed by one or more enzymes, that allow a favorable phosphate-transfer reaction (i.e., one for which $K > 1$) to be coupled to some other reaction that is unfavorable, thereby causing the unfavorable reaction (with $K < 1$) to occur. This introduction will help you to understand some of the discussion in Chapter 20.

[15] We shall describe the structure of glycerol more completely in our discussion of fats in Chapter 20. Glycerol is propane with a hydroxyl group substituted for one hydrogen atom on each carbon atom.

Biographical Notes

NEIL BARTLETT (1932–), now Professor of Chemistry at the University of California (Berkeley), was educated in England and Canada. He was a faculty member at the University of British Columbia when he synthesized the first compound of xenon.

HERMAN FRASCH (1851–1914) was born in Germany, but he emigrated to the United States as a youth. He attended the Philadelphia College of Pharmacy. Between 1894 and 1901, he developed the process for mining sulfur that bears his name.

WILLIAM RAMSEY (1852–1916), English chemist, was the discoverer by himself or with others of all of the noble gases. He received the Nobel Prize for Chemistry in 1904.

JOHN W. STRUTT (Lord Rayleigh) (1842–1919), an English mathematician and physicist, did much basic research on light. He developed a theory of the scattering of light by gas molecules, now known as Rayleigh scattering. He received the Nobel Prize for Physics in 1904.

Problems and Questions

17–1 From Figure 17–3, what is the boiling point of liquid air at $P = 1$ atm and what is the composition of the vapor in equilibrium with liquid air?

17–2 What is the equilibrium constant for the reaction $O_3 + H_2O + 2F^- = O_2 + F_2 + 2OH^-$ at 25°C? Can fluorine be produced by bubbling a 50%–50% mixture of ozone and oxygen through an aqueous solution of sodium fluoride?

17–3 Do you expect cyclic S_6 to be planar, or nonplanar? Explain.

17–4 Iodine(VII) forms species with coordination numbers 4 (IO_4^-) and 6 [$(HO)_5IO$ and $(HO)_4IO_2^-$]. At what concentration of hydrogen ion is $[IO_4^-]/([(HO)_5IO] + [(HO)_4IO_2^-]) = 1.00$?

17–5 What is the equilibrium partial pressure of oxygen over a solution with the composition: $[H^+] = 2.50$ mol L^{-1}, $[IO_3^-] = 0.100$ mol L^{-1}, $[I_2] = 1.00 \times 10^{-4}$ mol L^{-1}, at 25.0°C?

17–6 Aqueous solutions of hydrobromic acid and hydroiodic acid become colored in contact with air. Explain.

17–7 Write balanced equations for the reactions expected to occur when the nonmetal fluorides PF_5, SF_6, and IF_7 react with excess sodium hydroxide.

17–8 Calculate $\mathscr{E}^0$ values for the following half-reactions from those given in this chapter [$K(HOCl) = 3.2 \times 10^{-8}$ mol L^{-1}]:

$$HOCl + 2e^- = Cl^- + OH^-$$

$$OCl^- + H_2O + 2e^- = Cl^- + 2OH^-$$

17–9 Calculate the value of $\mathscr{E}^0$ for the half-reactions

$$BrO_4^- + H_2O + 2e^- = BrO_3^- + 2OH^-$$

$$BrO_4^- + 7H^+ + 6e^- = HOBr + 3H_2O$$

17–10 If before 1969 you were asked to produce hitherto unknown perbromate ion (BrO_4^-) from bromate ion (BrO_3^-) in aqueous solution, what oxidizing agents would you have tried? With the BrO_4^-—BrO_3^- potential now known, which of your proposed reactions certainly would not have occurred?

17–11 What geometry do you expect for each of the following nonmetal fluorine species: PF_4^+, SF_4, SF_3^+, IF_2^-, XeF_2?

17–12 What geometry do you expect for each of the following nonmetal oxygen species: XeO_3, TeO_3^{2-}, XeO_6^{4-}, IO_2^+, SiO_4^{4-}, F_2O?

17–13 What geometry do you expect for each of the following nonmetal oxohalide species: SOF_2, SO_2F_2, SO_2F^-, BrO_3F, NF_2O^+, NO_2F, $XeOF_4$?

17–14 Predict the sign (positive, negative, or close to zero) of ΔS^0 for each of the following reactions:

$$2ClF(g) = Cl_2(g) + F_2(g)$$
$$3BrF(g) = BrF_3(s) + Br_2(g)$$
$$ClF_5(g) = ClF_3(g) + F_2(g)$$

17–15 Nitric acid forms two hydrates: $HNO_3 \cdot H_2O$ (mp $-37.68°C$), $HNO_3 \cdot 3H_2O$ (mp $-18.47°C$). The melting point of pure nitric acid is $-41.64°C$. From this information sketch a phase diagram for the system H_2O–HNO_3 under the assumption that no solid solutions form in this system.

17–16 Give chemical equations for the formation of each of the following acids from the corresponding acid anhydride: H_2SO_4, H_3PO_3, HNO_3, H_4SiO_4, H_5IO_6, and $HClO_4$. Also rewrite the line formulas of each of these acids in a way that shows their structures more clearly.

17–17 Using the correlation between first-dissociation constants for acids $(HO)_aXO_b$ and the formal charge on X shown by the chlorine oxoacids, predict the value of K_1 for $(HO)_4XeO_2$.

17–18 For the half-reaction $XeO_3 + 6H^+ + 6e^- = Xe + 6H_2O$, $\mathscr{E}^0 = +2.12$ V. What is the equilibrium constant for the oxidation of bromate ion (BrO_3^-) to perbromate ion (BrO_4^-) by xenon trioxide in acidic aqueous solution?

17–19 Xenon(VI) oxide cannot be prepared from the constituent elements. Ozone and atomic oxygen are more powerful oxidizing agents than oxygen; is it likely that xenon(VI) oxide could be formed from xenon and ozone or xenon and atomic oxygen? ($\Delta H_f^0(O_3) = +143$ kJ mol^{-1} and $\Delta H_f^0(O) = +249$ kJ mol^{-1})

17–20 Peroxide ion, O_2^{2-}, is formed endothermically from O_2 in the gas phase, yet ionic peroxides (e.g., BaO_2) are stable. Explain.

17–21 Potassium superoxide, KO_2, has a density of 2.14 g cm^{-3} at 25°C. Its reaction with water and carbon dioxide gives 3/4 mole O_2 per mole KO_2. At what pressure would oxygen gas have to be stored to have the same density of usable oxygen? (Initially, assume ideal gas behavior in your calculation. From your answer, do you believe that this assumption is valid?)

17–22 If singlet oxygen O_2^* lost its energy by emission of light, with one molecule giving one photon, what would be the wavelength of this light?

17–23 Draw the electron-dot formulas for peroxonitric acid, O_2NOOH, and peroxodisulfuric acid, $(HO)O_2SOOSO_2(OH)$.

17–24 What half-reactions can be combined to give the overall reaction $2O_3(g) = 3O_2(g)$? From the values of $\mathscr{E}^0$ for these half-reactions, calculate K for the above reaction. (The value of n is not obvious when you examine the overall reaction, but it will be when you consider the half-reactions that combine to give this reaction.)

17–25 Contrast the thermodynamics of the two nitrogen-fixing reactions, $N_2 + 3H_2 = 2NH_3$ (see Figure 12–3), and $N_2 + O_2 = 2NO$ (see Figure 17–10). Show that the value of ΔH^0 determines the dependence of K upon temperature, but that the value of ΔS^0 determines the limiting value of K^0 approached at very high temperatures.

17–26 A reaction that produces relatively pure nitric oxide is the reduction of nitrous acid by iodide ion:

$$2HONO + 3I^- + 2H^+ = 2NO(g) + I_3^- + 2H_2O$$

Using $\mathscr{E}^0$ values given in this chapter, calculate the equilibrium constant for this reaction.

17–27 Although the iodine–thiosulfate ion reaction, which produces tetrathionate ion, is an analytically useful reaction, side reactions may interfere with it. Write a balanced chemical equation for each of the reactions to produce $S_4O_6^{2-}$ (tetrathionate ion), HSO_3^- (hydrogensulfite ion), and SO_4^{2-} (sulfate ion) in an initially neutral solution. What is the equilibrium constant for each of these reactions?

17–28 Sulfur dissolves to an appreciable extent in aqueous sodium sulfide solution to form polysulfide ions S_n^{2-}. The simplest such ion, S_2^{2-}, is isoelectronic with what neutral molecule? If a solution containing polysulfide ion is acidified with hydrochloric acid, sulfur is precipitated. Write a balanced equation for the reaction (using S_2^{2-} as the polysulfide ion). What does the occurrence of this reaction tell you about the relative acid strengths of HS^- and HS_2^-?

17–29 There is a trimer of sulfur trioxide, $(SO_3)_3$; this is isoelectronic with a phosphorus species discussed in this chapter. What species is this? Using this relationship, suggest a structure for $(SO_3)_3$.

17–30 These reactions are discussed in this chapter:

$$H_2PO_2^- + OH^- = HPO_3^{2-} + H_2$$
$$H_3PO_2 + 2Zn + 4H^+ = PH_3 + 2Zn^{2+} + 2H_2O$$

Dissect these reactions into their component half-reactions.

17–31 Draw structures for the linear polyphosphate anions $P_3O_{10}^{5-}$ and $P_4O_{13}^{6-}$.

17–32 Diphosphorus pentaoxide (actually P_4O_{10}, which is $2P_2O_5$) absorbs water to form phosphoric acid [H_3PO_4, actually $(HO)_3PO$]. If 10.0 L of air saturated with water vapor at 25.0°C were dried with phosphorus pentaoxide, by how much would the weight of the drying agent be increased?

18

The Metals

18–1 Introduction

Most elements in the periodic table are metals (Figure 18–1). Although metals exhibit a wide range of properties, some properties are common to most metals. The contrasting properties of metals and nonmetals were itemized at the start of Chapter 17. Only a few additional points need be added here:

1. *Melting points and boiling points in the elemental state.* Although one metal is a liquid at ordinary room temperature (Hg, mp −38.87°C) and two other metals melt slightly above room temperature (Cs, mp 28.40°C; Ga, mp 29.78°C), many metals have extremely high melting points and boiling points. Among the nontransition metals only beryllium has a melting temperature above 1000°C (Be, mp 1280°C, bp 2480°C), but many transition metals have much higher melting temperatures: for example, chromium, molybdenum, and tungsten (each in family VIB) have melting temperatures of 1900°C, 2610°C, and 3410°C, respectively.
2. *The nature of the elemental molecules.* Most metals do not form discrete small molecules. The diatomic alkali-metal molecules (e.g., Li_2; see Table 10–2) are the exception. For a metal under ordinary conditions, the molecule is the crystal with its closest-packed or body-centered arrangement of atoms. There is strong attraction of the atoms for one another in

IA	IIA	IIIB	IVB	VB	VIB	VIIB	VIIIB			IB	IIB	IIIA	IVA	VA	VIA	VIIA
3 Li	4 Be															
11 Na	12 Mg											13 Al	14 Si			
19 K	20 Ca	21 Sc	22 Ti	23 V	24 Cr	25 Mn	26 Fe	27 Co	28 Ni	29 Cu	30 Zn	31 Ga	32 Ge	33 As		
37 Rb	38 Sr	39 Y	40 Zr	41 Nb	42 Mo	43 Tc	44 Ru	45 Rh	46 Pd	47 Ag	48 Cd	49 In	50 Sn	51 Sb	52 Te	
55 Cs	56 Ba	57- 71	72 Hf	73 Ta	74 W	75 Re	76 Os	77 Ir	78 Pt	79 Au	80 Hg	81 Tl	82 Pb	83 Bi	84 Po	85 At
87 Fr	88 Ra	89- 103	104 Rf	105 Ha												

57 La	58 Ce	59 Pr	60 Nd	61 Pm	62 Sm	63 Eu	64 Gd	65 Tb	66 Dy	67 Ho	68 Er	69 Tm	70 Yb	71 Lu
89 Ac	90 Th	91 Pa	92 U	93 Np	94 Pu	95 Am	96 Cm	97 Bk	98 Cf	99 Es	100 Fm	101 Md	102 No	103 Lr

FIGURE 18–1
The metals. (The shaded elements are the metalloids, elements on the border line between metals and nonmetals.)

these arrangements, as demonstrated by the high melting points and boiling points just discussed. But the easy malleability and ductility of metals indicates that the strong attraction of the atoms for one another is not appreciably lessened as they slide over one another. (Diamond with strong directional bonds is neither malleable nor ductile.)

3. *Electrical conductivity*. Some metals are very good conductors of electricity. In this property, copper, silver, and gold excel, but less expensive aluminum has a conductivity that is ~60% of that of copper.
4. *Acidity and basicity of the oxides*. Most metal oxides are basic but not acidic, a few are acidic but not basic, and some are both acidic and basic. The ions of a particular metal may be in different categories, depending on the oxidation state. The conditions of acidity or basicity to which a metal oxide is subjected also influence the category into which it falls.
5. *The attraction of atoms for electrons*. In Chapter 8 we learned of trends of the ionization energies within the periodic table, and in Chapter 9 we used the ionization energies of metals in Born–Haber cycles for a number of ionic compounds. Gaseous metal atoms may add electrons exothermically; although uncommon, compounds containing metal anions are known.

6. *The nature of compounds of metals.* The saltlike ionic compounds of metals with nonmetals have been the focus of much of our discussion, but metals also form compounds with one another (e.g., the compounds of copper, CuZn, Cu_3Al, Cu_5Sn, Cu_5Zn_8, Cu_9Al_4, and $Cu_{31}Sn_8$), and with organic molecules or radicals [e.g., $Cr(C_6H_6)_2$].

We will consider many aspects of the chemistry of metals in this chapter and the next. This chapter deals primarily with the nontransition metals, but the electronic structures of atoms of both nontransition and transition metals are discussed. The next chapter deals with transition metals, including an extensive discussion of coordination compounds.

18–2 Electronic Structures of Metal Atoms

The neutral atoms of most metals have one to three electrons in the *s* and *p* orbitals of their valence shells. Tin and lead, with neutral atoms having four electrons in their valence shells are metallic, and antimony and bismuth, with atoms having five electrons in their valence shells, exhibit some properties of metals. Electrons in *d* and *f* orbitals of shells beneath the valence shell are involved in the chemistry of the transition metals, the lanthanide metals, and the actinide metals.

Table 8–5 presented the electronic structures of elements with atomic numbers 1–18. You will recall that at argon ($Z = 18$), all of the orbitals with $n = 1$ and 2 are filled, and the 3*s* and 3*p* orbitals also are filled. As we explained in Chapter 8, the ground states of the next two elements, potassium ($Z = 19$) and calcium ($Z = 20$) involve occupancy of the 4*s* orbital:

$$\text{K (argon core)} \quad 4s^1 \qquad \text{Ca (argon core)} \quad 4s^2$$

Occupancy of the 3*d* orbitals first occurs at scandium ($Z = 21$):

$$\text{Sc (argon core)} \quad 3d^1 4s^2$$

Table 18–1, giving the electronic configuration of elements with atomic number 19–36, has a number of important features. From scandium ($Z = 21$) through zinc ($Z = 30$), the five 3*d* orbitals are being filled.[1] As we learned previously, it is energetically favorable for electrons with the same values of n and l to occupy different orbitals and to have parallel spins. The observed electronic configurations shown in Table 18–1 confirm this. For example, the ground state of the titanium atom has two electrons with parallel spins in two different 3*d* orbitals, rather than having them with opposite spins in the same 3*d* orbital.

[1] Recall that *d* orbitals correspond to the quantum number l having the value 2. The quantum number m_l can have values from $+l$ to $-l$. With $l = 2$, the possible values of m_l are 2, 1, 0, -1, and -2. The orbitals with $n = 3$, $l = 2$, and these five different values of m_l are the five 3*d* orbitals.

The ground state electronic configurations of chromium and copper are somewhat surprising. The configuration

$$\text{Cr (argon core)} \quad 3d^5 4s^1$$

is more stable (i.e., has a lower energy) than the configuration

$$\text{Cr (argon core)} \quad 3d^4 4s^2$$

because the six electrons outside of the argon core repel one another less if they occupy six orbitals (five *d* and one *s*) of closely similar energy than if they occupy five orbitals (four *d* and one *s*). The configuration

$$\text{Cu (argon core)} \quad 3d^{10} 4s^1$$

is more stable than the configuration

$$\text{Cu (argon core)} \quad 3d^9 4s^2$$

also for reasons related to electron–electron repulsion and the energies of 4*s* and 3*d* orbitals, but the greater stability of the observed configuration is harder to explain because the same number of orbitals is occupied in each of these states.

TABLE 18–1
Electronic Configuration and Orbital Occupancy for the Ground States of Elements K Through Kr

Element		Configuration	Orbital occupancy									IE_1
Z	Symbol	(outside Ar core)	3*d*	3*d*	3*d*	3*d*	3*d*	4*s*	4*p*	4*p*	4*p*	kJ mol^{-1}
19	K	$4s^1$						(↑)				418.8
20	Ca	$4s^2$						(⇅)				589.5
21	Sc	$3d^1 4s^2$	(↑)					(⇅)				633
22	Ti	$3d^2 4s^2$	(↑)	(↑)				(⇅)				661
23	V	$3d^3 4s^2$	(↑)	(↑)	(↑)			(⇅)				649
24	Cr	$3d^5 4s^1$	(↑)	(↑)	(↑)	(↑)	(↑)	(↑)				653
25	Mn	$3d^5 4s^2$	(↑)	(↑)	(↑)	(↑)	(↑)	(⇅)				717
26	Fe	$3d^6 4s^2$	(⇅)	(↑)	(↑)	(↑)	(↑)	(⇅)				761
27	Co	$3d^7 4s^2$	(⇅)	(⇅)	(↑)	(↑)	(↑)	(⇅)				757
28	Ni	$3d^8 4s^2$	(⇅)	(⇅)	(⇅)	(↑)	(↑)	(⇅)				736
29	Cu	$3d^{10} 4s^1$	(⇅)	(⇅)	(⇅)	(⇅)	(⇅)	(↑)				745
30	Zn	$3d^{10} 4s^2$	(⇅)	(⇅)	(⇅)	(⇅)	(⇅)	(⇅)				906
31	Ga	$3d^{10} 4s^2 4p^1$	(⇅)	(⇅)	(⇅)	(⇅)	(⇅)	(⇅)	(↑)			577
32	Ge	$3d^{10} 4s^2 4p^2$	(⇅)	(⇅)	(⇅)	(⇅)	(⇅)	(⇅)	(↑)	(↑)		761
33	As	$3d^{10} 4s^2 4p^3$	(⇅)	(⇅)	(⇅)	(⇅)	(⇅)	(⇅)	(↑)	(↑)	(↑)	946
34	Se	$3d^{10} 4s^2 4p^4$	(⇅)	(⇅)	(⇅)	(⇅)	(⇅)	(⇅)	(⇅)	(↑)	(↑)	941
35	Br	$3d^{10} 4s^2 4p^5$	(⇅)	(⇅)	(⇅)	(⇅)	(⇅)	(⇅)	(⇅)	(⇅)	(↑)	1142
36	Kr	$3d^{10} 4s^2 4p^6$	(⇅)	(⇅)	(⇅)	(⇅)	(⇅)	(⇅)	(⇅)	(⇅)	(⇅)	1351

Following copper comes zinc; in a zinc atom, all of the $3d$ orbitals and the $4s$ orbital are filled. With the next element, gallium, occupancy of the three $4p$ orbitals starts, and at krypton, the three $4p$ orbitals are fully occupied.

The relative energies of an electron in $4s$ and $3d$ orbitals depend upon the occupancy of other orbitals as well as on the nuclear charge of the atom. In a one-electron atom, each orbital with $n = 3$ (including the $3d$ orbitals) is more stable than each orbital with $n = 4$. In an atom with more than one electron, the energy of an electron depends upon its interaction both with the nucleus and with the other electrons. This complex situation is simplified by considering the positive nuclear charge as being partially screened from the electron in question by the other electrons. In this simplified picture, we speak of the penetration of an electron through the screening effect of other electrons. Because of the greater penetration of s orbitals compared to d orbitals, the $4s$ orbital becomes relatively more stable than the $3d$ orbitals as electrons fill orbitals with $n = 1$, 2, and 3. For an atom with $Z = 19$ and with nineteen electrons (a neutral potassium atom), the $4s$ orbital is more stable than the $3d$ orbitals, with the result that potassium has an electron in the $4s$ orbital but none in $3d$ orbitals. With occupancy of the $4s$ orbital, the $3d$ orbitals become relatively more stable because electrons in the $4s$ orbital do not screen the $3d$ orbitals effectively. For an atom with $Z = 21$ and twenty-one electrons (a neutral scandium atom), a $3d$ orbital is more stable than a $4p$ orbital, with the result that scandium has an electron in a $3d$ orbital but none in $4p$ orbitals.

The observed electronic structures of atoms in their ground states provide a simple picture of the relative energies of electrons in the various orbitals. The order of energy derived from studying the electronic configuration of ground states of successive neutral atoms is:

$$1s < 2s < 2p < 3s < 3p < 4s < 3d < 4p < 5s < 4d < 5p < 6s < 5d \cong 4f < 6p < 7s < 6d \cong 5f < 7p$$

The relationship given above, $3p < 4s < 3d$, is derived from the ground states of neutral atoms of argon, potassium, calcium, and scandium:

Ar (Ne core)	$3s^2 3p^6$
K (Ne core)	$3s^2 3p^6 4s^1$
Ca (Ne core)	$3s^2 3p^6 4s^2$
Sc (Ne core)	$3s^2 3p^6 3d^1 4s^2$

Presenting an order of orbital energies may imply that the orbitals have particular relative energies independent of nuclear charge and the occupancy of other orbitals, but this implication is not correct. We can learn more by considering the relative energies of the ground state and several excited states of

1. the isoelectronic atomic species K, Ca^+, Sc^{2+}, and Ti^{3+}, each of which has one electron in addition to the argon core, and
2. the neutral atoms K, Ca, and Sc, which have one, two, and three electrons, respectively, in addition to the argon core.

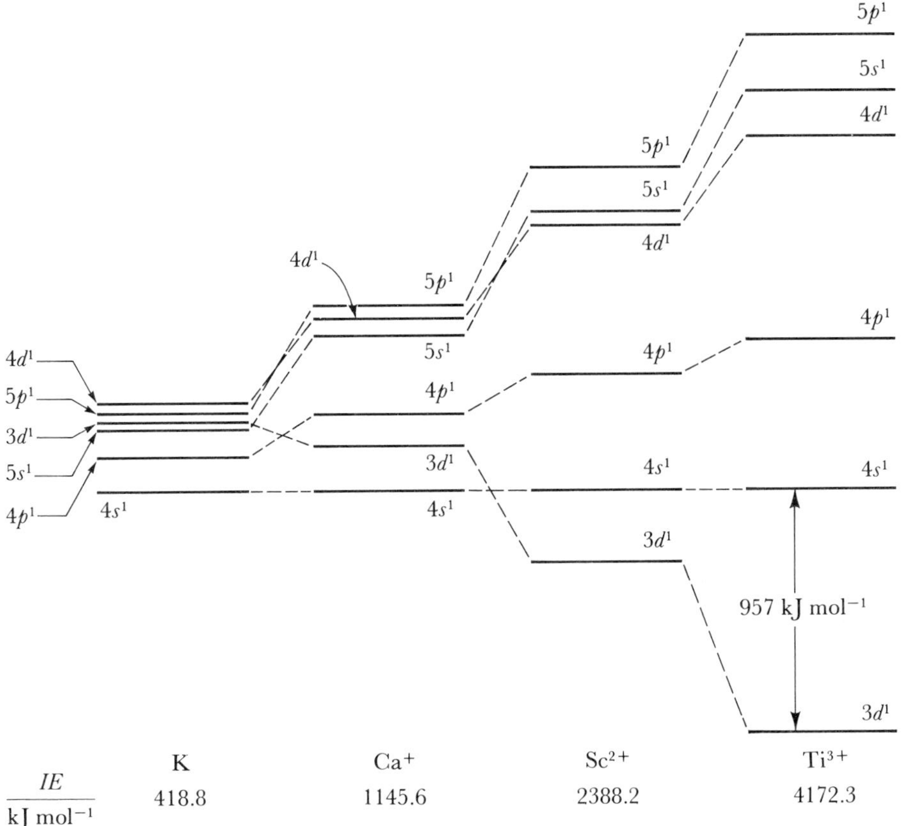

FIGURE 18–2
The relative energies of various states of isoelectronic atomic species with 19 electrons: K, Ca^+, Sc^{2+}, and Ti^{3+}. In this figure, the energy of the state with the $4s^1$ configuration for each species is drawn at the same level. The energies of the most stable states relative to the ionization limit to give K^+, Ca^{2+}, Sc^{3+}, and Ti^{4+} are given. The 18 most tightly held electrons occupy the $1s$, $2s$, $2p$, $3s$, and $3p$ orbitals, the argon structure.

Figure 18–2 shows the energies of the six most stable states of isoelectronic species with nineteen electrons but varying nuclear charge (from $Z = 19$ to 22). We see in this figure that a change of nuclear charge influences the energies of different states differently. The smaller the value of n for the one electron outside of the argon core, the greater is the effect of increasing the nuclear charge. Notice how the energy change (ΔU) for the following $4s^1$ to $3d^1$ configuration change varies:

Process	$\dfrac{\Delta U}{\text{kJ mol}^{-1}}$
$K(4s^1) = K(3d^1)$	+258
$Ca^+(4s^1) = Ca^+(3d^1)$	+164
$Sc^{2+}(4s^1) = Sc^{2+}(3d^1)$	−305
$Ti^{3+}(4s^1) = Ti^{3+}(3d^1)$	−957

With increasing nuclear charge, the energy of the orbital with $n = 3$ decreases more than does the energy of the orbital with $n = 4$.

The situation is more complex for the series of neutral atoms, which includes atomic species with two and three electrons outside of the argon core. The relative energies for states of these atomic species, shown in Figure 18–3, depend upon both the relative spins of the electrons and the orbitals that are occupied. We encountered this dependence in Chapter 8 when we considered the ground state and excited states of helium. As in the helium example, the energy of a state in which electrons have parallel spins is lower than that of a state with the same orbitals occupied, but in which the electron spins are paired.

In Table 18–1 and Figure 18–3, the five $3d$ orbitals have no labels. As long as we simply are considering the energies of gaseous atoms in the absence of magnetic or electric fields, the five $3d$ orbitals have the same energy, and distinguishing them from one another is unimportant. But we will have occasion in the next chapter to consider aspects of transition-metal chemistry in which the angular distribution of the five $3d$ orbitals is very important. Like the three p orbitals discussed in Chapter 8, the five d orbitals have different dependences of ψ upon θ and ϕ. The angular dis-

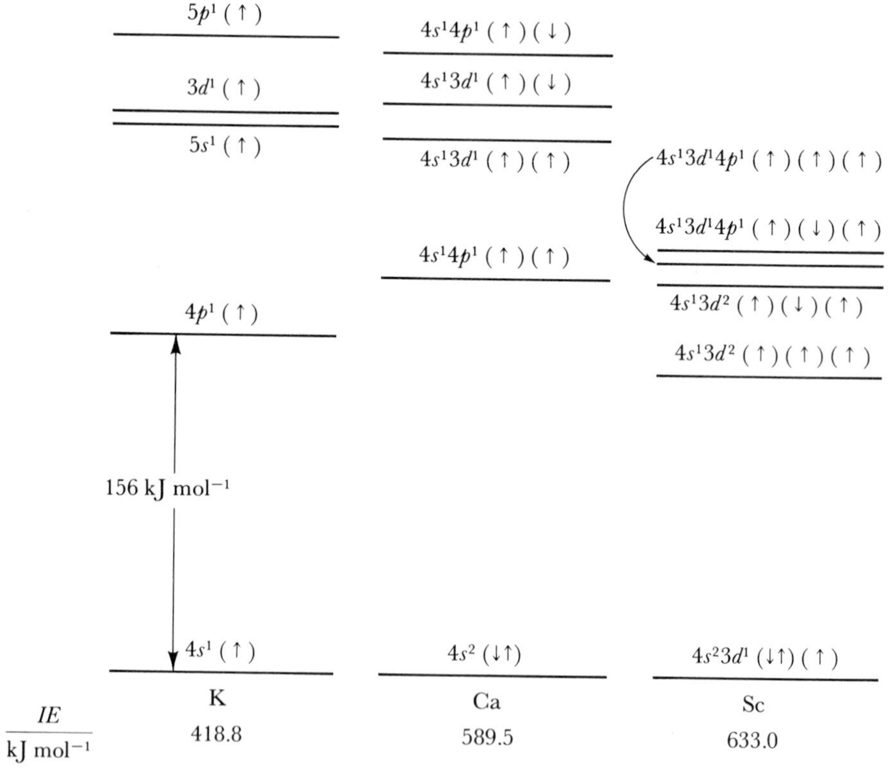

FIGURE 18–3

The energies of states of neutral atoms of K, Ca, and Sc relative to the ground state. The eighteen-electron argon core, $1s^22s^22p^63s^23p^6$, underlies each configuration. The spin orientations are shown for each configuration. In this figure, the energy of the ground state for each element is drawn at the same level.

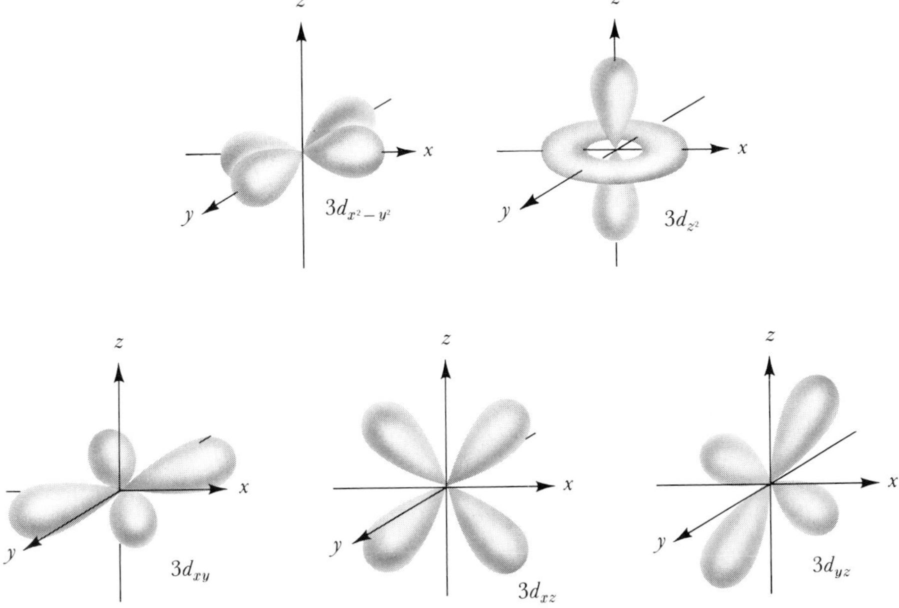

FIGURE 18–4
The angular distribution of the five $3d$ orbitals. Nodal surfaces for these orbitals are:

$d_{x^2-y^2}$	2 planes defined by $x = y$ and $x = -y$
d_{z^2}	2 cones with angle 54.73° to z axis
d_{xy}	2 planes defined by $x = 0$ (yz plane) and $y = 0$ (xz plane)
d_{xz}	2 planes defined by $x = 0$ (yz plane) and $z = 0$ (xy plane)
d_{yz}	2 planes defined by $y = 0$ (xz plane) and $z = 0$ (xy plane)

tributions for the five d orbitals are pictured in Figure 18–4, a figure analogous to Figure 8–10 for the three different p orbitals.

The electronic configurations of ground states of the eighteen elements following krypton ($Z = 36$) are analogous to those given in Table 18–1, but the situation is more complex for the elements following xenon ($Z = 54$) and radon ($Z = 86$). Following each of these noble gases (as after argon for $n = 4$), there is the sequence for $n = 5$, 6, and 7:

Noble gas	$(n-1)s^2(n-1)p^6$
Alkali metal	$(n-1)s^2(n-1)p^6ns^1$
Alkaline earth	$(n-1)s^2(n-1)p^6ns^2$
First transition element	$(n-1)s^2(n-1)p^6(n-1)d^1ns^2$

For these values of n, the noble gas, the following alkali metal and alkaline earth, and the first transition element are:

$n = 5$	Kr	Rb	Sr	Y
$n = 6$	Xe	Cs	Ba	La
$n = 7$	Rn	Fr	Ra	Ac

The ten elements starting with yttrium and ending with cadmium are, like the sequence scandium–zinc, a series of metals in which *d* orbitals of the shell beneath the valence shell are being filled. However, the situation is not analogous for the elements following lanthanum and actinium. In lanthanum, with the electronic configuration

$$\text{La (Xe core)} \quad 5d^1 6s^2$$

there are unoccupied orbitals with $n = 4$, which is two units less than that of the valence shell ($n = 6$): the $4f$ orbitals are unfilled. Because of efficient screening by electrons with $n = 1$, 2, and 3, the energies of electrons in the nonpenetrating $4f$ orbitals are relatively uninfluenced by increasing nuclear charge until electrons occupy shells with *n* greater than four. At lanthanum, eleven electrons are in the $n = 5$ and $n = 6$ shells, and the energy of an electron in a $4f$ orbital is similar to that of an electron in a $5d$ orbital. One more unit of nuclear charge makes the $4f$ orbital more stable, and the next element, cerium, has the electronic configuration

$$\text{Ce (Xe core)} \quad 4f^2 6s^2$$

(For *f* orbitals, $l = 3$, and m_l can have seven values ranging from -3 to $+3$.) The seven $4f$ orbitals, with $m_l = -3, -2, -1, 0, +1, +2$, and $+3$, are filled in the succeeding elements. These elements, cerium through lutecium, are called the *rare earths* or *lanthanides*.

The electronic configurations of the lanthanide elements are given in Table 18–2. The energy of an electron in a $5d$ orbital remains close to that

TABLE 18–2

Electronic Configuration and Orbital Occupancy for the Ground States of the Lanthanide Elements

Element		Configuration	Orbital Occupancy								
Z	Symbol	(outside Xe core)	$4f$	$4f$	$4f$	$4f$	$4f$	$4f$	$4f$	$5d$	$6s$
57	La	$5d^1 6s^2$								(↑)	(⇅)
58	Ce	$4f^2 6s^2$	(↑)	(↑)							(⇅)
59	Pr	$4f^3 6s^2$	(↑)	(↑)	(↑)						(⇅)
60	Nd	$4f^4 6s^2$	(↑)	(↑)	(↑)	(↑)					(⇅)
61	Pm	$4f^5 6s^2$	(↑)	(↑)	(↑)	(↑)	(↑)				(⇅)
62	Sm	$4f^6 6s^2$	(↑)	(↑)	(↑)	(↑)	(↑)	(↑)			(⇅)
63	Eu	$4f^7 6s^2$	(↑)	(↑)	(↑)	(↑)	(↑)	(↑)	(↑)		(⇅)
64	Gd	$4f^7 5d^1 6s^2$	(↑)	(↑)	(↑)	(↑)	(↑)	(↑)	(↑)	(↑)	(⇅)
65	Tb	$4f^9 6s^2$	(⇅)	(⇅)	(↑)	(↑)	(↑)	(↑)	(↑)		(⇅)
66	Dy	$4f^{10} 6s^2$	(⇅)	(⇅)	(⇅)	(↑)	(↑)	(↑)	(↑)		(⇅)
67	Ho	$4f^{11} 6s^2$	(⇅)	(⇅)	(⇅)	(⇅)	(↑)	(↑)	(↑)		(⇅)
68	Er	$4f^{12} 6s^2$	(⇅)	(⇅)	(⇅)	(⇅)	(⇅)	(↑)	(↑)		(⇅)
69	Tm	$4f^{13} 6s^2$	(⇅)	(⇅)	(⇅)	(⇅)	(⇅)	(⇅)	(↑)		(⇅)
70	Yb	$4f^{14} 6s^2$	(⇅)	(⇅)	(⇅)	(⇅)	(⇅)	(⇅)	(⇅)		(⇅)
71	Lu	$4f^{14} 5d^1 6s^2$	(⇅)	(⇅)	(⇅)	(⇅)	(⇅)	(⇅)	(⇅)	(↑)	(⇅)

of an electron in a $4f$ orbital in this series of elements. This is shown by the electronic configuration of gadolinium, Gd ($Z = 64$). In europium, Eu ($Z = 63$), there is one electron in each of the seven $4f$ orbitals and zero electrons in $5d$ orbitals. If one more electron were added to the $4f$ orbitals, it would go into an orbital that already contains an electron. Instead, the electron added at gadolinum goes into a $5d$ orbital.

There is an analogous situation after actinium: the energies of the $5f$ orbitals are low enough to let these orbitals fill after one or two electrons are added to the $6d$ orbitals. At lawrencium Lr ($Z = 103$), the seven $5f$ orbitals are filled, and there is one electron in a $6d$ orbital. This sequence of elements, called the *actinides*, includes eleven elements that do not occur in nature but are man-made (neptunium through lawrencium).

With the increase in atomic number in the lanthanide series of elements, the radii of ions of these elements decrease because of the increased attraction the increased nuclear charge exerts on the outermost electrons. Figure 18–5 gives the radii of +3 ions of the lanthanide elements as a function of atomic number. The contraction of atomic and ionic size in this series, called the *lanthanide contraction*, strongly affects the properties of elements in the third series of transition elements, as compared with those of the second. Following are the ionic radii for some isoelectronic ions:

Kr structure	Sr^{2+}	1.3 Å	Y^{3+}	0.9 Å	Zr^{4+}	0.8 Å
Xe structure	Ba^{2+}	1.4 Å	La^{3+}	1.1 Å	Hf^{4+}	0.8 Å

Hafnium(IV) has the xenon structure plus 14 electrons in the seven $4f$ orbitals. Instead of having a radius ~20% larger than that of the corresponding ion of the element above it in the periodic table (as is true in a comparison of La^{3+} and Y^{3+}), Hf^{4+} has a radius equal to that of Zr^{4+}. Because of this, many of the chemical properties of hafnium resemble very closely those of zirconium. There is no such close resemblance between lanthanum and yttrium.

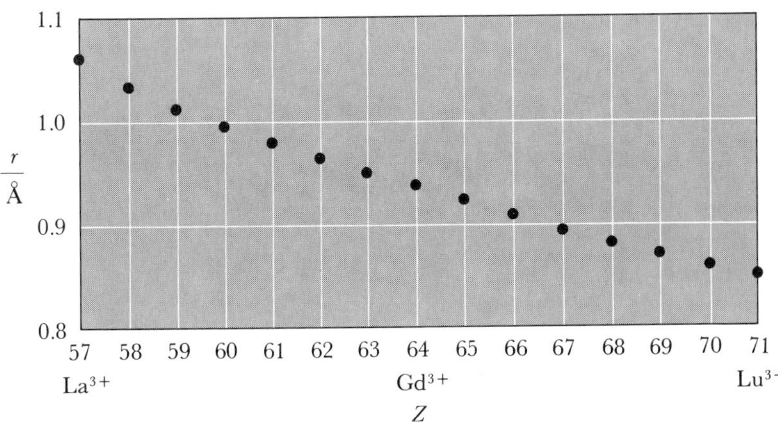

FIGURE 18–5
The ionic radii of +3 lanthanide ions. Each ion has a xenon core plus zero (for La^{3+}) to fourteen (for Lu^{3+}) electrons in $4f$ orbitals.

18–3 The Metallic State

We learned in Chapter 5 that most metals crystallize in one of the arrangements for the closest packing of spheres of identical size, hexagonal closest packing or cubic closest packing. However, nothing was said then about the nature of the bonding in metals. Now we shall consider that subject. In a metal with a closest-packed arrangement of atoms, each atom has twelve nearest neighbor atoms; this coordination number is larger than either the number of stable valence-shell orbitals or the number of valence electrons in the atom. This situation contrasts with that of diamond, in which each carbon atom has four nearest-neighbor atoms, four stable valence-shell orbitals, and four valence electrons. And diamond, because of its structure with strong directional covalent bonds, does not have the properties of a metal.

As a contrast with diamond, consider magnesium, a metal that crystallizes with a hexagonal closest-packed arrangement of atoms. Each magnesium atom has twelve nearest-neighbor atoms, four relatively stable valence-shell orbitals (the $3s$ and $3p$ orbitals; the $3d$ orbitals have appreciably higher energy at this point), and two valence electrons. Whatever the nature of the bonding, it holds the atoms together strongly; the boiling point of magnesium is 1400 K. The concept of resonance can be applied to the description of the electronic structure of metals. In this approach, a magnesium atom in magnesium metal is described as being bonded to each of its twelve neighbors by one sixth of an electron-pair bond. That is, the structure is described as a resonance hybrid of structures in which the magnesium atoms are bonded together by electron-pair bonds, but with only one sixth of the contributing structures having an electron-pair bond between a particular pair of magnesium atoms. The delocalization of electrons in such a structure would account for the electrical conductivity of magnesium.

A generally more useful approach to the theory of bonding in metals is an extension of the free-electron model developed by LORENTZ in the early part of this century. In this model a metal is viewed as a crystalline array of cations held together by a sea of valence electrons, essentially a free-electron gas enveloping the lattice of cations. However, this simple picture predicts an incorrect value of the heat capacity for metals. If a mole of magnesium consisted of a mole of magnesium(II) ions and two moles of electron gas, we would expect the molar heat capacity at constant volume to be

$$C_v = 3R + 2\left(\frac{3R}{2}\right) = 6R$$

(The term $3R$ arises from the vibration of the atoms, and the term $2(3R/2)$ arises from the translation of two moles of electron gas.)[2] However, the

[2]See Chapters 3 and 4 for a discussion of the molar heat capacity of a gas.

heat capacity of magnesium at 25°C is 24.9 J K^{-1} mol^{-1}, close to $3R$ (25.0 J K^{-1} mol^{-1}), the value expected if translational motion of the electrons made no contribution to the heat capacity.

This failure of the free-electron model is rectified by quantum mechanics. Just as two hydrogen atoms close together have a bonding molecular orbital that can be viewed as a combination of the 1*s* orbitals on the isolated hydrogen atoms, so also a lattice of metal atoms has bonding molecular orbitals that can be viewed as combinations of the valence-shell atomic orbitals on the separated atoms. A crystalline sample of one mole of magnesium metal has 3×10^{23} bonding orbitals resulting from combining the 3*s* orbital on each of the 6×10^{23} atoms, plus 9×10^{23} bonding orbitals resulting from combining the three 3*p* orbitals on each of the 6×10^{23} atoms. (There also are an equal number of antibonding orbitals.) The energies of the bonding molecular orbitals resulting from combinations of these atomic orbitals are very close to one another; such groups of closely spaced energy levels are commonly called bands of energy levels. For magnesium, the band of levels resulting from combinations of 3*s* atomic orbitals has a range of energies that overlaps the energies of the band of levels resulting from combination of 3*p* atomic orbitals. Each of these molecular orbitals, which extends throughout the entire crystal, can hold two electrons. Only electrons in highest occupied energy levels of the band can be excited to higher unoccupied (or half-occupied) levels by an increase of temperature. This is a very, very small proportion of the total number of electrons, and the contributions of the electrons to the molar heat capacity is very small.

The trend in cohesive energy in metals is consistent with the band theory just outlined. For example, in the sequence sodium, magnesium, and aluminum, all of the valence electrons are in bonding orbitals of the metal, and with an increasing number of electrons so situated, the cohesive energy increases. This is shown in Table 18–3.

TABLE 18–3
Correlation of Cohesive Energy of Metals with the Number of Valence Electrons

	Number of valence electron	ΔH^a_{vap} / kJ mol^{-1}
Na	1	108
Mg	2	149
Al	3	324

[a] ΔH for process M(*s*) = M(*g*) at 298.2 K.

18–4 The Occurrence of Metals in Nature

The chemical forms in which metals occur in nature depend upon a number of factors. These include

1. the tendency for the metal to be oxidized,
2. the solubility in water of common geologically important compounds of the metal, for example, the oxide, the sulfide, and the carbonate, and
3. the tendency for atoms of the rarer metals to be substituted for atoms of common metals in their minerals.

The relevance of these factors can be illustrated with examples. For gold, the first ionization energy is relatively high (890 kJ mol^{-1}), and the oxide (Au_2O_3) would be formed endothermically:

$$4Au(s) + 3O_2(g) = 2Au_2O_3(s) \qquad \Delta H^0 = 81 \text{ kJ mol}^{-1}$$
$$\Delta S^0 = -540 \text{ J K}^{-1} \text{ mol}^{-1}$$

Both the enthalpy and entropy factors are against stability for gold(III) oxide. As a result of this small tendency to be oxidized, gold occurs in nature in the free state. (However, gold does occur as the telluride, $AuTe_2$.) Like gold, the other elements that generally occur in nature in the free state (ruthenium, rhodium, palladium, osmium, iridium, and platinum) have high ionization energies and oxides that are relatively unstable.

The oxidation of gold by oxygen does not occur. In contrast, the corresponding reaction of iron does occur:

$$4Fe(s) + 3O_2(g) = 2Fe_2O_3(s) \qquad \Delta H^0 = -1651 \text{ kJ mol}^{-1}$$
$$\Delta S^0 = -543 \text{ J K}^{-1} \text{ mol}^{-1}$$

The first ionization energy for iron is 761 kJ mol^{-1}, 130 kJ mol^{-1} less than that for gold. For the oxidation of iron the favorable change of enthalpy makes equilibrium in this reaction very favorable. Iron(III) oxide is found in nature; it is responsible for the red color of many rocks.

The alkali metals are oxidized easily, and most of their compounds are relatively soluble. These elements are found in the oceans and in inland lakes. Soluble compounds of the alkali metals also are found in surface deposits in arid regions; the deposits of sodium nitrate in Chili are an example.

Some of the alkaline-earth elements (e.g., magnesium) also are found in sea water. However, compounds of these metals generally are less soluble than those of the alkali metals, and the carbonates and sulfates, compounds of low solubility, are important minerals (e.g., magnesite, $MgCO_3$; limestone, $CaCO_3$; gypsum, $CaSO_4 \cdot 2H_2O$; and barite, $BaSO_4$).

Although the chemistry of thallium differs in many respects from that of potassium (or the other alkali metals), the similarity of the ionic radii of thallium(I) ion (1.54 Å) and potassium(I) ion (1.44 Å) allows thallium(I) to coprecipitate in potassium silicate minerals. Thallium(I) ion, Tl^+, occupies a very small fraction of the sites normally occupied by potassium(I) ion, K^+, in these minerals.

The high stability of many metal sulfides and the relative abundance of sulfur in the Earth's crust are responsible for the occurrence of many metals as sulfide minerals.[3] We have learned already (Chapter 14) that most metal sulfides have a very low solubility.

18–5 The Nontransition Metals

The nontransition metals are metals whose atoms both in the free state and in compounds have either empty or completely filled sets of *d* orbitals or *f* orbitals. This defines the elements shown in Figure 18–6 as nontransition metals. Notice that this classification places zinc, cadmium, and mercury in the nontransition metal class because they have a completely filled set of *d* orbitals in both the 0 and +2 oxidation states. This classification also places scandium, yttrium, lanthanum, and actinium in the transition-metal class because their neutral atoms have one *d* electron in the $(n-1)$ shell. But the +3 oxidation state of each of these elements has a noble-gas electronic configuration, with no partially filled *d* orbitals; hence these elements in the +3 oxidation state are more like nontransition metals. No completely rational borderline between nontransition and transition metals can be drawn.

We can introduce the nontransition metals by considering their oxidation states. In compounds, each alkali metal and each alkaline-earth metal has the oxidation state that corresponds to removal of all electrons from the valence shell:

$$+1 \quad Li^+ \quad Na^+ \quad K^+ \quad Rb^+ \quad Cs^+ \quad Fr^+$$
$$+2 \quad Be^{2+} \quad Mg^{2+} \quad Ca^{2+} \quad Sr^{2+} \quad Ba^{2+} \quad Ra^{2+}$$

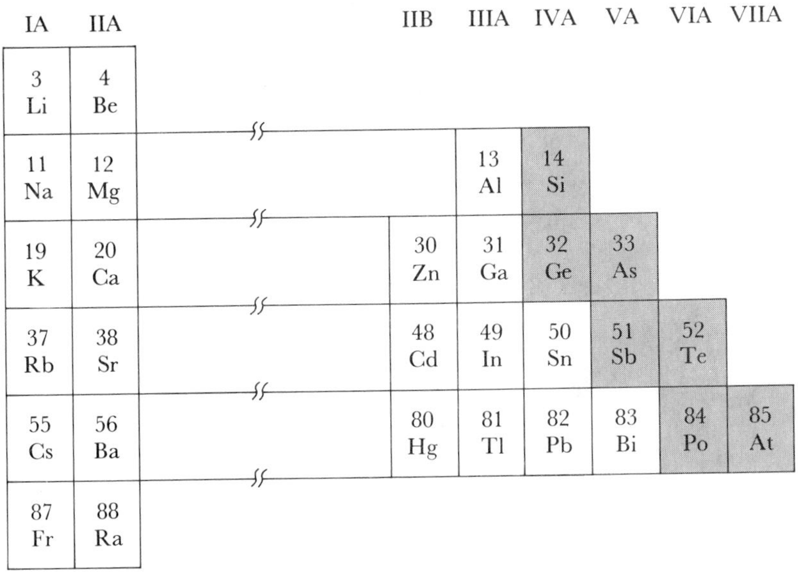

IA	IIA	IIB	IIIA	IVA	VA	VIA	VIIA
3 Li	4 Be						
11 Na	12 Mg		13 Al	14 Si			
19 K	20 Ca	30 Zn	31 Ga	32 Ge	33 As		
37 Rb	38 Sr	48 Cd	49 In	50 Sn	51 Sb	52 Te	
55 Cs	56 Ba	80 Hg	81 Tl	82 Pb	83 Bi	84 Po	85 At
87 Fr	88 Ra						

FIGURE 18–6
The nontransition metals. (The shaded elements are metalloids, elements on the border line between metals and nonmetals.)

[3]Of the nonmetals, only oxygen, fluorine, and phosphorus are more abundant than sulfur in the Earth's crust.

These simple ions and the simple ion of aluminum(III), Al^{3+}, have noble-gas electronic configurations.

In forming the +2 oxidation state, atoms of zinc, cadmium, and mercury acquire an electronic configuration with filled *s*, *p*, and *d* orbitals of the shell beneath the valence shell of the neutral atom:

Zn^{2+} (Ne core)	$3s^2 3p^6 3d^{10}$
Cd^{2+} (Ar core)	$4s^2 4p^6 4d^{10}$
Hg^{2+} (Kr core + $4f^{14}$)	$5s^2 5p^6 5d^{10}$

In the zero oxidation state, atoms of these elements have two electrons in the 4*s*, the 5*s*, and the 6*s* orbital, respectively.

The maximum oxidation states of the remaining nontransitional metals have configurations with eighteen electrons in the shell beneath the valence shell of the neutral atom. Simple ions of these elements, which are isoelectronic with Zn^{2+}, Cd^{2+}, and Hg^{2+}, are:

Zn^{2+} structure	Ga^{3+}	
Cd^{2+} structure	In^{3+}	Sn^{4+}
Hg^{2+} structure	Tl^{3+}	Pb^{4+}

Although gallium, indium, and thallium are below aluminum in the periodic table, the chemistry of their +3 ions does not closely resemble that of aluminum(III) ion; these ionic species have eighteen-electron kernels, but aluminum(III) ion has a noble-gas kernel.

Some of these elements also exhibit oxidation states with electronic configurations that are isoelectronic with neutral atoms of zinc, cadmium, and mercury:

Zn structure	Ga^+		
Cd structure	In^+	Sn^{2+}	
Hg structure	Tl^+	Pb^{2+}	Bi^{3+}

Each of these species has two electrons in the *s* orbital of the valence shell. Because these oxidation states are intermediate between known higher and lower oxidation states, we may wonder about their tendency to disproportionate. The potential diagrams for indium and thallium in acidic solution,

$$In^{3+} \xrightarrow{-0.43\ V} In^+ \xrightarrow{-0.16\ V} In$$

$$Tl^{3+} \xrightarrow{+1.27\ V} Tl^+ \xrightarrow{-0.336\ V} Tl$$

show that in solution, indium(I) is unstable with respect to disproportionation, whereas thallium(I) is stable. A particular oxidation state may be unstable in aqueous solution but stable in another solvent or in the solid state. Indium(I) is an example; the compound $In(AlCl_4)$ containing In^+ and $AlCl_4^-$ is known.

For the nontransition metals we have discussed so far, each oxidation state (other than zero) corresponds to an even number of electrons in the atom. There are oxidation states corresponding to atoms that have an odd number of electrons, but these oxidation states do not persist as monatomic species. Two examples are

Hg(I) $\cdot Hg^{+}$ dimerizes to $Hg:Hg^{2+}$, the predominant form of mercury(I)

Tl(II) $\cdot Tl^{2+}$ can be formed photochemically, but it rapidly disproportionates to Tl^{+} and Tl^{3+}

It is interesting to consider the possibility that there are negative oxidation states for some metals, a possibility which is suggested by the high values of the electron affinity for some metal atoms (see Table 8–6). However, the exothermic formation of a metal anion in the gas phase does not mean that formation of compounds with metal anions will be exothermic.

THE ALKALI AND ALKALINE-EARTH METALS

The alkali metals (Li, Na, K, Rb, and Cs) and alkaline-earth metals (Be, Mg, Ca, Sr, and Ba) are found in the Groups IA and IIA of the periodic table; certain of their properties are given in Table 18–4.[4] The electron or electrons in the *s* orbital of the valence shell of atoms of these elements are more easily removed than are valence-shell electrons from atoms of

TABLE 18–4
Properties of Alkali Metals and Alkaline-Earth Metals

	Mp	Bp	Density[a]	Ionization energy[b] MJ mol^{-1}			$r^{c}_{M^{n+}}$	D(M—M)
	°C	°C	g cm^{-3}	1st	2nd	3rd	Å	kJ mol^{-1}
Li	180	1330	0.53	0.520	7.30	11.82	0.86	108
Na	98	892	0.97	0.496	4.57	6.92	1.12	71
K	64	760	0.86	0.419	3.08	4.4	1.44	50
Rb	39	688	1.53	0.409	2.66	3.9	1.58	46
Cs	29	690	1.87	0.382	2.43	3.4	1.84	44
Be	1280	2480	1.85	0.899	1.763	14.86	0.70	
Mg	650	1110	1.74	0.738	1.451	7.74	0.87	
Ca	838	1490	1.55	0.590	1.146	4.94	1.18	
Sr	770	1380	2.60	0.556	1.07	4.1	1.32	
Ba	714	1640	3.50	0.509	0.969	3.4	1.49	

[a]Density of solid at 25.0°C.
[b]1 MJ = 10^6 J.
[c]Ionic radius.

[4]A heavier alkali metal, francium, Fr ($Z = 87$), and a heavier alkaline-earth metal, radium, Ra ($Z = 88$), are highly radioactive. We will not consider the chemistry of these two elements.

any other element. Because of this, the alkali metals and alkaline-earth metals form many ionic compounds; some of these compounds served as examples in our early discussion of the Born–Haber cycle (see Chapter 9).

As the preceding section of this chapter explained, the number of valence electrons helps determine the interatomic attraction in the metallic state. Because atoms of the alkali metals have only one valence electron, these metals are soft and they have low melting points. On the MOHS scale of hardness where diamond has a hardness of 10 and talc, $Mg_3Si_4O_{10}(OH)_2$, a hardness of 1, sodium metal has a hardness of 0.4. The hardness of magnesium on this scale is 2.0. Magnesium and beryllium are used as structural metals, but the other alkaline-earth elements and the alkali metals are not.

The alkali metals' relatively low melting points make it possible for solutions of some of these metals in one another to have eutectic temperatures below room temperature. The freezing temperature–composition diagram for solutions of sodium and potassium is given in Figure 18–7. Because the eutectic mixture of these metals has a large temperature range over which it is liquid, and also a high specific heat, it possibly may be used as a coolant in nuclear reactors. The figure also reveals the existence of an intermetallic compound, KNa_2. (The melting point of this compound is so much lower than that of sodium that there is no maximum in the freezing-point curve at this composition. The details of this phase diagram came from interpretation of cooling curves, as described in Chapter 7.)

At their boiling points, the alkali metals in the vapor phase tend to form diatomic molecules, such as Na_2. Because an alkali-metal atom has a single valence electron, the bonding in the diatomic molecule is easily visualized in terms of a pair of electrons in a bonding sigma orbital or an ordinary electron-pair bond:

$$Na_2(\sigma 3s)^2 \text{ or } Na{:}Na$$

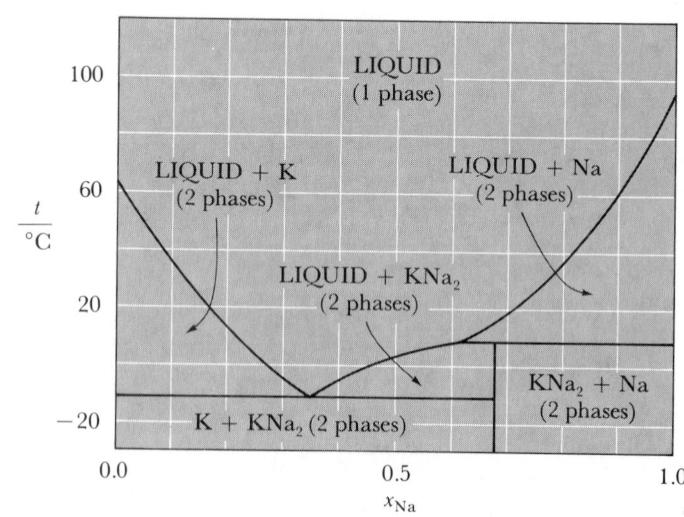

FIGURE 18–7
The freezing temperature–composition diagram for solutions of sodium and potassium. The eutectic has the composition 34 mole % Na, 66 mole % K, and it melts at −12.3°C. This system has an intermetallic compound, KNa_2. The vertical line at $x_{Na} = 0.67$ is at the composition of this compound.

However, the bond is weak compared with most covalent bonds; the Na–Na bond energy is 71 kJ mol^{-1}. The extent to which sodium is dimerized in the vapor depends upon the temperature and the pressure. At 1165 K, the normal boiling point, $K = [Na_2]/[Na]^2 \cong 0.29$ atm^{-1}; at a total pressure of 1 atm, the partial pressure of Na_2 is ~0.19 atm.

The Alkali Metals as Reducing Agents; Solvated Electrons. We already have stated (Chapter 8) that the trend in values of the change of enthalpy for the reaction

$$M(s) + H^+(aq) = M^+(aq) + \tfrac{1}{2}H_2(g)$$

is the consequence of a number of factors. If we dissect this reaction for lithium and cesium and compare the enthalpy changes, we find

Reaction	$\frac{\Delta H}{\text{kJ mol}^{-1}}$ Li	Cs
$M(s) = M(g)$	161	80
$M(g) = M^+(g) + e^-$	520	376
$M^+(g) = M^+(aq)$	−510	−255

We see that the enormous hydration energy of lithium ion is the factor that makes lithium the best reducing agent of the alkali metals, as measured by the values of $\mathscr{E}^0$, which are (for 25.0°C):

Half-reaction	$\frac{\mathscr{E}^0}{V}$
$Li^+(aq) + e^- = Li(s)$	−3.03
$Na^+(aq) + e^- = Na(s)$	−2.713
$K^+(aq) + e^- = K(s)$	−2.925
$Rb^+(aq) + e^- = Rb(s)$	−2.93
$Cs^+(aq) + e^- = Cs(s)$	−2.92

The difference of entropy, $S^0_{M(s)} - S^0_{M^+(aq)}$, also contributes to these values of $\mathscr{E}^0$; these values are 14, −8, −39, −47, and −50 J K^{-1} mol^{-1}, respectively. Thus the trend in entropy changes, which reflects variation in the hydration of the ions in solution, would make lithium the least good reducing agent.

These very negative values of $\mathscr{E}^0$ show that these metals can reduce hydrogen ion or water to hydrogen, and analogous reactions occur with ammonia and alcohols (e.g., methyl alcohol). The reactions for sodium are:

$$2Na(s) + 2H_2O(l) = 2Na^+(aq) + 2OH^-(aq) + H_2(g)$$
$$2Na(s) + 2NH_3(l) = 2NaNH_2(s) + H_2(g)$$
$$2Na(s) + 2CH_3OH(l) = 2Na^+ + 2OCH_3^- + H_2(g)$$

The values of $\mathscr{E}^0$ neither tell us about the rates of these reactions nor disclose information about possible intermediate stages of the reactions. The intermediate stage has been studied most completely for the reaction of alkali metals with liquid ammonia.

When sodium metal is dissolved in liquid ammonia, a blue solution is formed, and, in the absence of a catalyst, hydrogen gas is not evolved immediately. The reaction occurring initially is

$$\mathrm{Na}(s) = \mathrm{Na}^{+}(\mathrm{in\ NH_3}) + \mathrm{e}^{-}\ (\mathrm{in\ NH_3})$$

Because alkali-metal ions are colorless, and the same blue color appears in dilute solutions of different alkali metals in liquid ammonia, the color has been attributed to electrons solvated by ammonia. The blue color of the solutions (hence of the electrons) persists for long periods of time, and such solutions are useful as reducing agents in organic synthesis. The solvated electron in liquid ammonia reduces an alkyne to a *trans*-alkene, for example, with butyne-2 as the reactant, *trans*-butene-2 is formed:

$$2\mathrm{e}^{-}(\mathrm{in\ NH_3}) + 2\mathrm{NH_3} + \mathrm{CH_3C{\equiv}CCH_3}$$
$$= 2\mathrm{NH_2^-} + trans\text{-}\mathrm{CH_3CH{=}CHCH_3}$$

In the absence of a solute that can be reduced, the solvated electrons slowly reduce the solvent to give hydrogen gas,

$$2\mathrm{e}^{-}\ (\mathrm{in\ NH_3}) + 2\mathrm{NH_3}(l) = 2\mathrm{NH_2^-} + \mathrm{H_2}(g)$$

and the blue color disappears.

When an alkali metal dissolves in liquid ammonia, the volume of the solution is greater than the volume of the components. The density of liquid ammonia at $-33°\mathrm{C}$ is 0.683 g $\mathrm{cm^{-3}}$, but the density of a saturated solution (15.66 molal) of lithium metal in liquid ammonia at this temperature is only 0.477 g $\mathrm{cm^{-3}}$, the lowest known density for a liquid at ordinary temperatures. From these data, one concludes that the solvated electron occupies a relatively large volume in solution.

The reaction of hydrated electrons with water is very fast, and when an alkali metal is dissolved in water an intermediate stage of blue coloration does not appear. However, high-energy radiation does produce hydrated electrons in water, and study of such solutions is an active field of research. In pure water the hydrated electron has a half-life of $\sim 2.3 \times 10^{-4}$ s, with each of the following paths for destruction being important at $\mathrm{pH} = 7$:

$$\mathrm{e}^{-}(aq) + \mathrm{H_3O^+} \rightarrow \mathrm{H_2O} + \mathrm{H} \quad \text{and} \quad \mathrm{e}^{-}(aq) + \mathrm{H_2O} \rightarrow \mathrm{OH^-} + \mathrm{H}$$

The first of these reactions is dominant at $\mathrm{pH} < 5.4$; its rate is governed by the second-order rate constant $k = 2.1 \times 10^{10}\ \mathrm{L\ mol^{-1}\ s^{-1}}$, which is similar to the very large rate constants given in Chapter 15 for some diffusion-controlled reactions. The hydrogen atoms produced in these initial steps rapidly yield hydrogen gas.

The reducing ability of the hydrated electron can be characterized by an $\mathscr{E}^0$ value for the reaction

$$2\mathrm{e}^{-}(aq) + 2\mathrm{H}^{+}(aq) = \mathrm{H_2}(g) \qquad \mathscr{E}^0 = +2.67\ \mathrm{V}$$

which is equivalent to the conventionally written half-reaction:

$$e^- = e^-(aq) \qquad \mathscr{E}^0 = -2.67\ \text{V}$$

Thus the hydrated electron is an extremely powerful reducing agent.

The properties of concentrated solutions of alkali metals in liquid ammonia (solutions with a copper color and a metallic luster) show that these solutions are more complex than the dilute solutions. The predominant solute species in concentrated solutions may be the solvated alkali-metal cations and alkali-metal anions:

$$Na(NH_3)_n^+ \qquad \text{and} \qquad Na(NH_3)_m^-$$

We shall say more about alkali-metal anions later in this section.

Production of the Metals. The production of metallic sodium by electrolysis of molten sodium chloride was discussed in Chapter 16. Analogous electrolytic procedures would seem to be possible for the other alkali metals. They are for lithium, but *not* for the heavier alkali metals. The melting point of potassium chloride is 1043 K, 10 K higher than the *boiling* point for metallic potassium. Clearly, grave technical difficulties prevent us from producing metallic potassium by electrolysis of molten potassium chloride. Potassium is produced by reaction of the molten salt with sodium:

$$Na(g) + KCl(l) = K(g) + NaCl(l)$$

Of the alkaline-earth metals, magnesium is produced on the largest scale. Electrolysis of a molten mixture of halides ($MgCl_2 + CaCl_2 + NaCl$) produces magnesium at the cathode, for magnesium is the most easily reduced of these metals. (Use of a molten salt mixture allows the electrolysis to take place at a temperature lower than the melting point of pure magnesium chloride.) Magnesium metal also can be produced at high temperatures by reduction of the oxide with carbon. Equilibrium in this reaction is very unfavorable at temperatures where magnesium metal is a solid. But if magnesium is gaseous, the large increase of entropy in this reaction ($\Delta S^0 = +314\ \text{J K}^{-1}\ \text{mol}^{-1}$),

$$MgO(s) + C(s) = Mg(g) + CO(g)$$

makes the equilibrium favorable at high temperatures. (Notice that if magnesium is gaseous, the reaction produces two moles of gas from two moles of solid.)

Alkali-Metal Ions and Cage Ligands; Sodium Anions. Sodium ion and potassium ion play important roles in biology. In human blood plasma their concentrations are

$$[Na^+] \cong 0.15\ \text{mol L}^{-1} \qquad \text{and} \qquad [K^+] \cong 0.005\ \text{mol L}^{-1}$$

but in the fluid inside cells the relative concentrations are reversed:

$$[Na^+] \cong 0.005\ \text{mol L}^{-1} \qquad \text{and} \qquad [K^+] \cong 0.16\ \text{mol L}^{-1}$$

Salts of these cations control the osmotic pressure in biological fluids, and each has specific functions in metabolic processes. The functions are different, however, and the differences can be attributed to the difference of radii of the two ions (see Table 9–1):

$$r_{Na^+} = 1.12\ \text{Å} \quad \text{and} \quad r_{K^+} = 1.44\ \text{Å}$$

The size of an ion helps determine its tendency to lose its solvation shell of water molecules and form complex ions; it has been found that potassium ion forms more stable complexes with some important biological ligands than sodium ion does. One such ligand is the antibiotic valinomycin, a cyclic molecule containing both hydroxy acids and amino acids joined by ester and amide linkages:

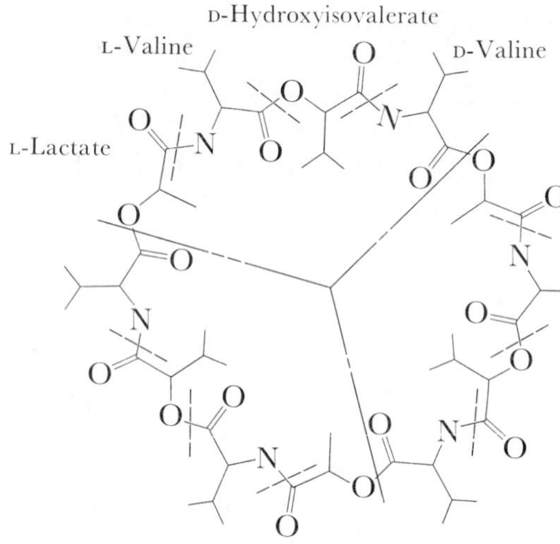

D-Hydroxyisovalerate
L-Valine
D-Valine
L-Lactate

With this ligand, represented as **C**, the association reactions of sodium ion and potassium ion are

$$Na(OH_2)_4^+ + \mathbf{C} = Na\mathbf{C}^+ + 4H_2O \qquad K(OH_2)_4^+ + \mathbf{C} = K\mathbf{C}^+ + 4H_2O$$

For the alkali-metal ion to be coordinated to the cyclic ligand (a cage), it must fit the internal cavity, and to do so the ion must lose its hydration shell. Quantitative study of reactions in methyl alcohol at 25°C give equilibrium constants for the association reactions:

$$K = \frac{[M\mathbf{C}^+]}{[M^+][\mathbf{C}]}$$
$$= 3.7\ \text{L mol}^{-1}\ (\text{for Na}^+)$$
$$= 6.3 \times 10^4\ \text{L mol}^{-1}\ (\text{for K}^+)$$

A complex ion with a ligand such as valinomycin has an organic periphery, and such a species can penetrate hydrocarbonlike cell membranes that are not permeable to simple hydrated ions. Thus complexes such as these are believed to help transport potassium ion into the interior of cells.

Chemists are very much interested in synthetic cyclic ethers, such as 2,5,8,15,18,21-hexaoxatricyclohexacosane; the structure of this ligand, with each —CH or —CH_2 group represented as a bend in a line, is;

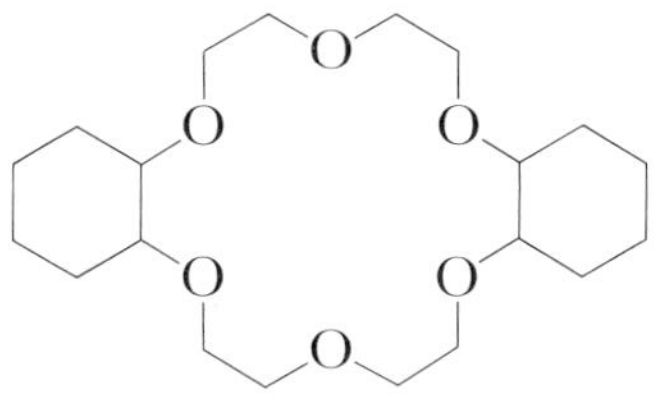

This structure resembles some of the biological ligands. The synthetic ether just given (represented here by **C**) forms a stable complex with potassium ion in aqueous solution:

$$K(OH_2)_4^+ + \mathbf{C} \rightleftarrows K\mathbf{C}^+ + 4H_2O \qquad K = 1.0 \times 10^2 \text{ L mol}^{-1}$$

In addition to serving as models for the biological ligands, these synthetic cyclic ethers (called crown ethers) that form cage complexes have practical uses. Potassium permanganate is a useful oxidizing agent, but its solubility in organic solvents of low polarity, such as ethers, is too low for such solutions to be of practical use. In the presence of a crown ether, potassium permanganate is much more soluble:

$$KMnO_4(s) + \mathbf{C} \rightleftarrows K\mathbf{C}^+MnO_4^-$$

Such solutions are useful also in preparative organic chemistry.

Molecules analogous to crown ethers were involved in the synthesis of the first compound containing an alkali-metal anion. We already have learned that in the gas phase, alkali-metal atoms react exothermically with electrons to form anions. However, the electron affinity for each alkali metal is smaller than the first ionization energy, and a process of electron transfer, such as that for sodium,

$$2Na(g) = Na^+(g) + Na^-(g)$$

is endothermic. In the solid, the coulombic lattice energy helps stabilize an ionic structure but this stabilization is not sufficient to make solid sodium an ionic solid, Na^+Na^-. We would expect stabilization of the cation by complexation with a cage ligand to help stabilize an ionic structure. It has proved to do so. Sodium has a very low solubility ($<10^{-6}$ mol L^{-1}) in ethylamine, but the solubility in the presence of an excess of a cage ligand, that is, a bicyclic polyoxadiamine,

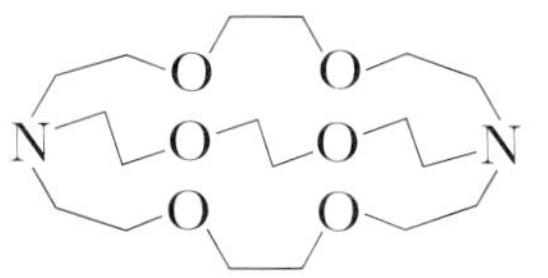

is enormously larger, as high as 0.4 mol L^{-1}. From such solutions of sodium and ligand (also represented as **C**), a crystalline solid of the composition $Na_2\mathbf{C}$ can be obtained. Studying the crystal structure of this compound, and comparing it with the crystal structure of the ionic compound NaC^+I^-, we can conclude that the compound Na_2C is NaC^+Na^-. From this study, we also can conclude that the ionic radius of Na^- is 2.17 Å, similar to that of I^- (2.12 Å).

Alkaline-Earth Compounds. The alkaline-earth metals form compounds in which the metal is present in the +2 oxidation state. Removing two electrons from an alkaline-earth atom requires a large amount of energy: for magnesium,

$$Mg(g) = Mg^{2+}(g) + 2e^- \qquad \Delta U = 2189 \text{ kJ mol}^{-1}$$

This energy can be contrasted with that needed to form the isoelectronic ion of charge +1:

$$Na(g) = Na^+(g) + e^- \qquad \Delta U = 495.8 \text{ kJ mol}^{-1}$$

Nevertheless, ionic compounds of the alkaline-earth ions of charge +2 are stable because of their large lattice energies. Recall the Born–Haber cycles for sodium fluoride and magnesium oxide in Chapter 9 (see Figure 9–12). Because it requires only slightly more energy to remove one electron from an alkaline-earth atom (e.g., $IE_1(\text{Mg}) = 737.6 \text{ kJ mol}^{-1}$) than from an alkali-metal atom, it may seem reasonable to you that ionic compounds of halogens with alkaline-earth metals in the +1 oxidation state should be stable relative to the constituent elements. Indeed, for magnesium(I) fluoride we can estimate the lattice energy to be $\sim 960 \text{ kJ mol}^{-1}$, and using a Born–Haber cycle we can calculate that the following reaction is exothermic:

$$2Mg(s) + F_2(g) = 2MgF(s) \qquad \Delta H^0 \cong -650 \text{ kJ mol}^{-1}$$

But magnesium(I) fluoride is not a known compound. The explanation lies in its *instability with respect to disproportionation*,

$$2MgF(s) = Mg(s) + MgF_2(s)$$

for which

$$\begin{aligned}\Delta H^0 &= \Delta H_f^0[MgF_2(s)] - 2\Delta H_f^0[MgF(s)] \\ &= -1124 \text{ kJ mol}^{-1} - 2(-325 \text{ kJ mol}^{-1}) = -474 \text{ kJ mol}^{-1}\end{aligned}$$

The same arguments hold for the other alkaline-earth metals, and no stable crystalline compound of these metals in the +1 oxidation is known.

The strong coulombic attraction of alkaline-earth cations and oxide ions makes their compounds highly refractory. The melting points of the alkaline-earth oxides are:

BeO	2803 K	SrO	2703 K
MgO	3073 K	BaO	2193 K
CaO	2843 K		

(Bricks for the lining of furnaces can be made from magnesium oxide.)

These oxides, except for beryllium oxide, react with water to form the corresponding hydroxides:

$$MO(s) + H_2O(g) = M(OH)_2(s)$$

Values of the equilibrium constant for this reaction, $K = P_{H_2O}^{-1}$ at 298.2 K, are Mg, 2.0×10^6 atm^{-1}; Ca, 3.6×10^{11} atm^{-1}; Sr, 1.5×10^{14} atm^{-1}; and Ba, 2.1×10^{17} atm^{-1}. These equilibrium constants decrease with increasing temperature: for instance, the value of K for calcium becomes 1.0 atm^{-1} at ~820 K.

The solubilities of the alkaline-earth hydroxides at 25°C increase from $Mg(OH)_2$ (discussed at length in Chapter 14) to $Ba(OH)_2$:

$Mg(OH)_2$	$S = 1.47 \times 10^{-4}$ mol L^{-1}
$Ca(OH)_2$	$S = 0.020$ mol L^{-1}
$Sr(OH)_2$	$S = 0.065$ mol L^{-1}
$Ba(OH)_2$	$S = 0.273$ mol L^{-1}

These hydroxides are largely dissociated in aqueous solution, and solutions of relatively soluble barium hydroxide are used as a convenient source of hydroxide ion. Because calcium hydroxide is only slightly soluble, a slurry of this hydroxide in a saturated solution, known as milk of lime, can be used as a base.

For some purposes, barium hydroxide is preferable as a base compared to alkali-metal hydroxides. Solutions of alkali-metal hydroxides that are exposed to air absorb carbon dioxide, and convert it to carbonate ion:

$$CO_2(g) + 2OH^- = H_2O + CO_3^{2-}$$

The same reaction occurs in barium hydroxide solution, but the low solubility of barium carbonate [$K_s(BaCO_3) = 5.1 \times 10^{-9}$ mol^2 L^{-2}] causes the carbonate ion, so formed, to precipitate:

$$CO_2(g) + 2OH^- + Ba^{2+} = BaCO_3(s) + H_2O$$

Thus solutions of barium hydroxide are relatively free of carbonate ion.

Unlike the oxides and hydroxides of magnesium, calcium, strontium, and barium, which are basic and show no acidic properties, the oxide of beryllium is amphoteric. Although its solubility in water is very low, beryllium oxide dissolves both in acid and in base:

$$BeO(s) + 2H_3O^+ + H_2O = Be(OH_2)_4^{2+}$$
$$BeO(s) + 2OH^- + H_2O = Be(OH)_4^{2-}$$

The tetrahydroxoberyllium(II) ion, $Be(OH)_4^{2-}$, is isoelectronic with tetrafluoroberyllium(II) ion, BeF_4^{2-}; Chapter 14 described the stepwise formation of the latter from tetraaquaberyllium(II) ion, $Be(OH_2)_4^{2+}$.

Both magnesium and calcium ions play biological roles, calcium ion being particularly important. Skeletal material is composed largely of compounds of calcium: for example, calcium phosphates (see Figure 17-13) in the bones and teeth of vertebrates, and calcium carbonate as shells of shellfish. Calcium ion is involved in enzyme-catalyzed biological reactions and also in certain physiological processes, such as muscle contraction. The interaction of calcium ion with high-molecular-weight biopolymers (see Chapter 20) involves the same types of ligand groups as can be studied in experiments with simple model ligands, for example,

Carboxylate groups, a model being acetate ion:

$$Ca^{2+} + CH_3CO_2^- \rightleftarrows CaO_2CCH_3^+ \qquad K = 6 \text{ L mol}^{-1} \text{ (25.0°C)}$$

Polyphosphate groups, a model being diphosphate ion:

$$Ca^{2+} + P_2O_7^{4-} \rightleftarrows CaP_2O_7^{2-} \qquad K = 1 \times 10^5 \text{ L mol}^{-1} \text{ (25.0°C)}$$

Amino acid groups, a model being glycine:

$$Ca^{2+} + H_2NCH_2CO_2^- \rightleftarrows CaO_2CCH_2NH_2^+ \qquad K = 25 \text{ L mol}^{-1} \text{ (25.0°C)}$$

Natural water containing calcium ion, called hard water, has several undesirable characteristics. If the hardness is due to the presence of calcium hydrogencarbonate, the hardness is said to be temporary, and calcium carbonate precipitates if the water is boiled. Boiling expels carbon dioxide, which causes the reaction

$$Ca^{2+} + 2HOCO_2^- = CO_2 + CaCO_3(s) + H_2O$$

to go to completion, thereby removing the calcium ion from solution. The precipitated calcium carbonate can deposit in pipes and in boilers, thereby causing loss of heat-transfer efficiency. If other anions are present in the water and there is less hydrogencarbonate ion than needed in the reaction just considered, the water is said to be permanently hard. Boiling does not cause complete precipitation of calcium ion from such water.

Another undesirable characteristic of hard water is the precipitation by calcium ion of the anion of soap, for example, stearate ion, $C_{17}H_{35}CO_2^-$. The precipitation reaction is

$$Ca^{2+} + 2C_{17}H_{35}CO_2^- = Ca(C_{17}H_{35}CO_2)_2$$

Although the effect of calcium ion in water can be suppressed with a reagent that forms a soluble complex with calcium ion, for example, a polyphosphate, a more effective water softening procedure is *ion exchange*, in which the calcium ion is replaced by sodium ion (see Figure 18–8). Ion exchange applied to the softening of water involves a cation exchanger, an insoluble polymeric ionic material with its negative charges on atoms of the polymer network. The system's electroneutrality is maintained with simple cations, which are free to be displaced by other cations. The polymer network may be a zeolite, which is a natural aluminosilicate (to be discussed in the next section), or a synthetic hydrocarbon resin with covalently bonded anionic groups (e.g., sulfonate groups,—SO_3^-). When hard water flows through a column containing a cation-exchange material in the sodium

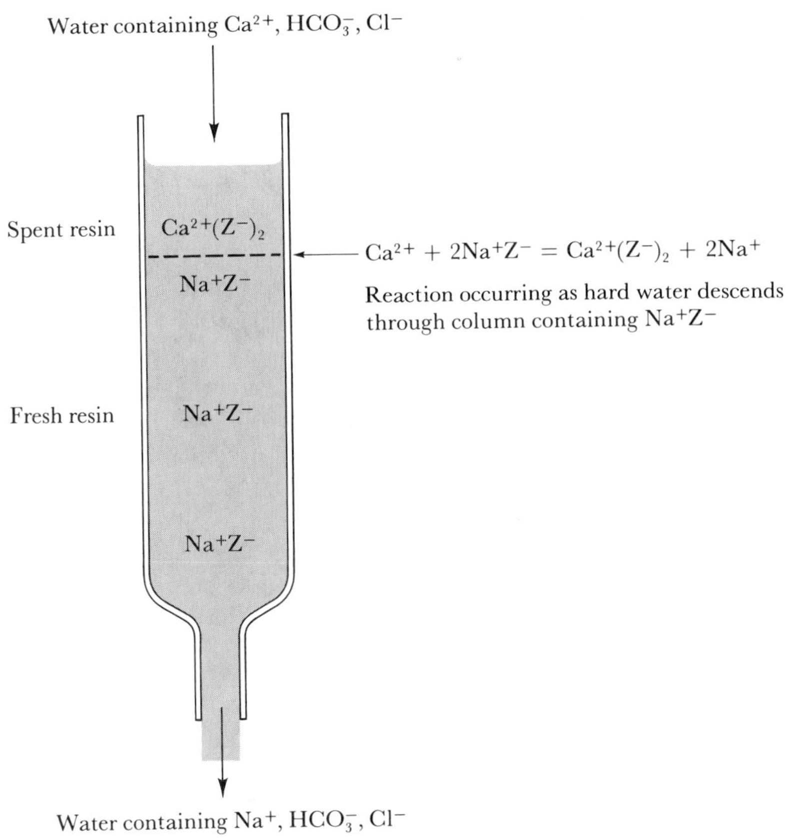

Water containing Ca^{2+}, HCO_3^-, Cl^-

Spent resin $Ca^{2+}(Z^-)_2$

$Ca^{2+} + 2Na^+Z^- = Ca^{2+}(Z^-)_2 + 2Na^+$

Reaction occurring as hard water descends through column containing Na^+Z^-

Na^+Z^-

Fresh resin Na^+Z^-

Na^+Z^-

Water containing Na^+, HCO_3^-, Cl^-

FIGURE 18–8
A water softener that involves cation exchange. The column contains resin beads; Z^- is stationary. There must be cations to balance the charge of Z^-; the cation is Ca^{2+} in the spent resin and Na^+ in the fresh resin.

ion form, represented as Na^+Z^-, this reaction occurs:

$$Ca^{2+}(aq) + 2Na^+Z^- = Ca^{2+}(Z^-)_2 + 2Na^+(aq)$$

The water coming from the column contains electrolytes, but these electrolytes are sodium salts, not salts of calcium ion or magnesium ion. After enough hard water passes through a water softener of this type, all of the sodium ion is displaced, and the capacity of the softener is exhausted. The softener may be regenerated by passing concentrated sodium chloride through the column; a reaction occurs that is the reverse of the reaction just given:

$$2Na^+(aq) + Ca^{2+}(Z^-)_2 = Ca^{2+}(aq) + 2Na^+Z^-$$

The effluent from this regeneration is discarded, and the residual salt solution is rinsed away, leaving the cation exchanger in the sodium ion form, just as it was originally.

Although ion exchange of the type just outlined exchanges one cation for another, it does not remove electrolyte from the water. However, complete deionization of water can be accomplished by a combination of a cation exchanger in the hydrogen ion form (H^+Z^-) and an anion exchanger in the hydroxide ion form (R^+OH^-). Passage of a solution of an electrolyte, for instance sodium chloride, through a column of these exchangers has the net result of removing the ions completely:

$$Na^+(aq) + Cl^-(aq) + H^+Z^- + R^+OH^- = H_2O + Na^+Z^- + R^+Cl^-$$

Ion exchange is very useful in analysis and in chemical separations.

Our introduction to complex ions showed the structure of the polydentate ligand ethylenediaminetetraacetate, EDTA (Table 14–2), and Chapter 13 gave the acid-dissociation constants for the protonated forms. An analysis for calcium ion (or magnesium ion) is based upon the great stability of complexes of EDTA (represented as Y^{4-}):

$$Ca^{2+} + Y^{4-} \rightleftarrows CaY^{2+} \qquad K = 1.0 \times 10^{11}\ \mathrm{L\ mol^{-1}}\ (25°C)$$

This great stability is attributed to the ligand's ability to coordinate to the metal ion with six basic coordination groups, four carboxylate groups and two amino groups:

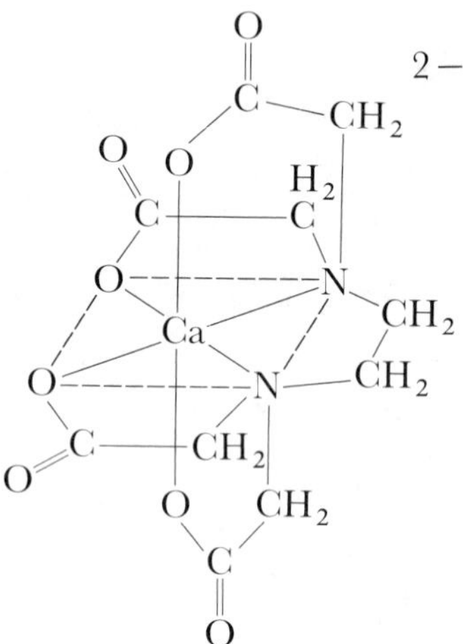

The analysis involves titration of the unknown (e.g., well water being analyzed for hardness) with a standard solution of EDTA. The indicator, represented as HIn^{2-}, forms a colored complex with calcium ion, $CaIn^-$. At the endpoint, the reaction

$$CaIn^- + HY^{3-} = CaY^{2-} + HIn^{2-}$$

converts the colored calcium complex to the colorless complex with EDTA (CaY^{2-}) plus differently colored free indicator.

ALUMINUM AND SILICON; SILICATES AND ALUMINOSILICATES

Next to oxygen, silicon and aluminum are the most abundant elements in the Earth's crust. The weight percentage abundances are oxygen, 46.60%; silicon, 27.72%; and aluminum, 8.13%. Aluminum and silicon are adjacent in the periodic table; the electronic structures of the neutral atoms are:

$$\text{Al(Ne core)} \quad 3s^2 3p^1$$
$$\text{Si(Ne core)} \quad 3s^2 3p^2$$

The dividing line separating metals and metalloids is between these elements. The properties of silicon definitely are borderline, but aluminum is predominately metallic.

The crystal structure of one form of silicon dioxide, β-cristobalite, is shown in Figure 9–3 and is repeated in Figure 18–9. Each silicon atom is surrounded tetrahedrally by four oxygen atoms, and each oxygen atom is surrounded by two silicon atoms. The bonding in silicon dioxide is partially ionic and partially covalent; an estimate of 50%–50% is based upon the difference between the electronegativities of silicon and oxygen, $(\chi_O - \chi_{Si}) = 1.7$. Even though the bonding is not strictly ionic, the partial replacement of silicon in silicon dioxide by aluminum can be viewed as replacement of Si^{4+} by Al^{3+}. Such a replacement would give an electrically charged lattice. However, another cation is incorporated concomitantly into an open space of the silicon dioxide framework. Thus in albite, a feldspar mineral $NaAlSi_3O_8$, one fourth of the silicon of silicon dioxide is replaced by aluminum, and charge neutrality is maintained by the presence of one sodium ion per aluminum ion:

$$1\ Si^{4+} \text{ replaced by } 1\ Al^{3+} \textit{ plus } 1\ Na^+$$

Not all silicate minerals maintain the three-dimensional framework of crystalline silicon dioxide. Let us consider some other possibilities, starting with the products of reaction of a basic metal oxide with acidic silicon dioxide. Reaction of excess sodium oxide with silicon dioxide gives

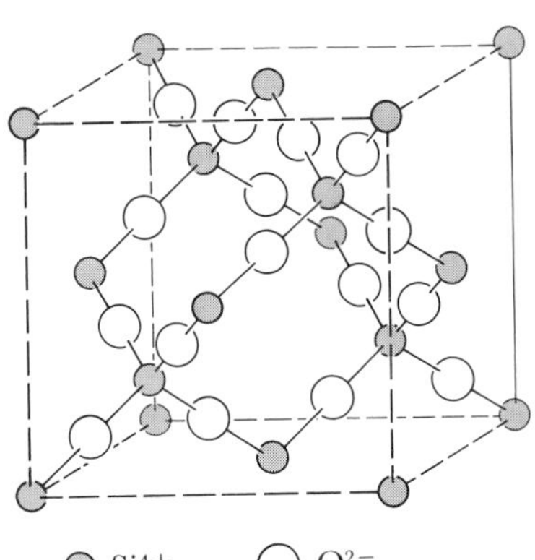

FIGURE 18–9
The atomic arrangement in the β-cristobalite form of SiO_2. Coordination numbers: for Si^{4+}, 4; for O^{2-}, 2.

sodium orthosilicate,

$$2Na_2O + SiO_2 = Na_4SiO_4$$

which contains tetrahedral orthosilicate ion SiO_4^{4-}; this ion is isoelectronic with phosphate ion, PO_4^{3-}, sulfate ion, SO_4^{2-}, and perchlorate ion, ClO_4^-. The coordination about each silicon atom in orthosilicate ion is the same tetrahedral arrangement as found in crystalline silicon dioxide and in these isoelectronic ions. Orthosilicate ion is not common in minerals, although it is present in some, for example in zircon, $ZrSiO_4$. If less metal oxide is present per mole of silicon dioxide, for example, 1.5 mol Na_2O per mole of SiO_2,

$$3Na_2O + 2SiO_2 = Na_6Si_2O_7$$

the product is a salt containing disilicate ion, $Si_2O_7^{6-}$ (isoelectronic with diphosphate ion $P_2O_7^{4-}$ and disulfate ion $S_2O_7^{2-}$). A mineral containing disilicate ion is thortveitite, $Sc_2Si_2O_7$. The next most complex arrangements of silicate anions are found in infinite chain anions of composition $(SiO_3)_n^{2n-}$, double-chain anions of composition $(Si_4O_{11})_n^{6n-}$, and sheets of composition $(Si_2O_5)_n^{2n-}$. These structures are shown in Figure 18–10.

Aluminum(III) ion also can replace silicon(IV) in these chain or sheet structures; each such replacement causes the net charge of the aluminosilicate structure to be one unit more negative per aluminum(III) ion. Many minerals contain these infinite chains or sheets, and some properties of a mineral are a consequence of this feature of its structure. Thus asbestos, $(MgOH)_6Si_4O_{11}$, with a fibrous character, contains the double-chain silicate anions, and mica, $Na(AlOH)_2Si_3AlO_{10}$, easily cleaved into thin transparent sheets, contains aluminosilicate sheet anions. Like mica, clay minerals have sheet structures containing the charged sheets $(Si_4O_{10})_n^{4n-}$ or $(AlSi_3O_{10})_n^{5n-}$. These sheets and the counter ions located between the sheets, for example, Na^+, K^+, $AlOH^{2+}$, and $FeOH^{2+}$, have a high affinity for water, thereby giving clay one of the properties needed in a soil if it is to support plant growth.

Although aluminosilicate minerals are very abundant, they have not yet proved to be usable sources of aluminum because it is not easy to separate aluminum from silicon in these minerals. Base dissolves both aluminum oxide and silicon dioxide:

$$Al_2O_3 + 3H_2O + 2OH^- = 2Al(OH)_4^-$$
$$SiO_2 + 4OH^- = SiO_4^{4-} + 2H_2O$$

and dilute acid does not leach aluminum(III) ion from aluminosilicates, even though aluminum(III) oxide is soluble in solutions of intermediate acidity, in which the solubility of silicon dioxide is not appreciable.

Aluminum metal is widely used despite its tendency to be oxidized:

$$4Al + 3O_2(g) = 2Al_2O_3$$
$$\Delta H^0 = -3352 \text{ kJ mol}^{-1} \qquad \Delta S^0 = -625 \text{ J K}^{-1} \text{ mol}^{-1}$$
$$\Delta G^0_{298.2\,K} = -3166 \text{ kJ mol}^{-1}$$

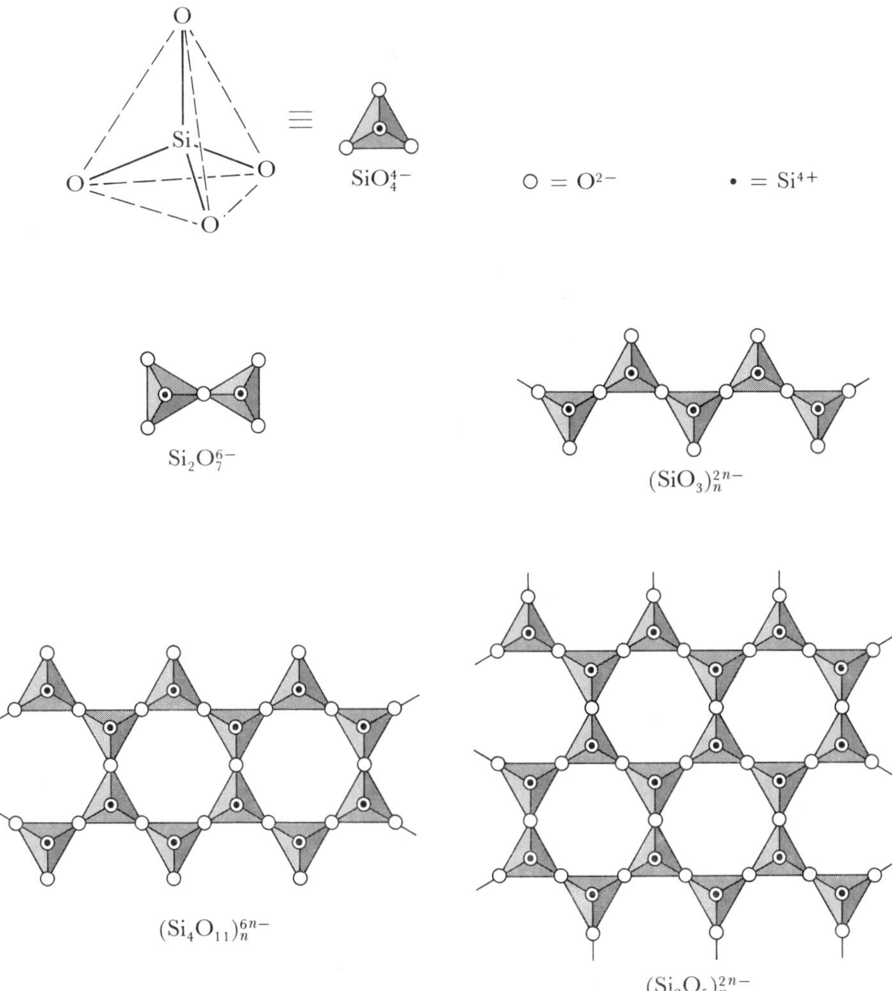

FIGURE 18–10
Silicate anions:

SiO_4^{4-}, orthosilicate ion (in zircon, $ZrSiO_4$)
$Si_2O_7^{6-}$, disilicate ion (or pyrosilicate ion; in thortveitite, $Sc_2Si_2O_7$)
$(SiO_3)_n^{2n-}$, an infinite chain [in enstatite, $MgSiO_3$, and spodumene, $LiAl(SiO_3)_2$]
$(Si_4O_{11})_n^{6n-}$, a double-stranded infinite chain [in asbestos, $(MgOH)_6(Si_4O_{11})$]
$(Si_2O_5)_n^{2n-}$, an infinite sheet [in talc, $Mg_3(OH)_2(Si_2O_5)_2$]

Metallic aluminum in contact with air rapidly acquires a coating of oxide. This film prevents oxidation of the bulk metal, and aluminum objects in contact with air may not deteriorate over periods of years.

Aluminum(III) ion in aqueous solution is hydrated by six molecules of water, and this species is a weak acid:

$$Al(OH_2)_6^{3+} \rightleftarrows H^+ + Al(OH_2)_5OH^{2+} \qquad K = 1.1 \times 10^{-5} \text{ mol L}^{-1} \ (25°C)$$

There are successive reactions by which cationic hexaaquaaluminum(III) ion is converted to anionic species [e.g., $Al(OH_2)_2(OH)_4^-$] by reaction with hydroxide ion. However, in solutions of intermediate pH the maximum concentration of aluminum(III) ion is low because of the insolubility of the hydroxide, $K_s = [Al^{3+}][OH^-]^3 = 1.1 \times 10^{-33}$ mol^4 L^{-4}.

The six water molecules in hexaaquaaluminum(III) ion can be replaced by fluoride ions, in stepwise reactions analogous to those already discussed (Chapter 14) for tetraaquaberyllium(II) ion. At concentrations of fluoride ion above 0.1 mol L^{-1}, $Al(OH_2)F_5^{2-}$ and AlF_6^{3-} are the predominant species in solution. The fluoride complexes of silicon(IV) have been less thoroughly studied, but the species SiF_6^{2-} is well known, and the dissolving of glass (here represented as SiO_2) in hydrofluoric acid is due to its formation:

$$SiO_2 + 6HF = SiF_6^{2-} + 2H_2O + 2H^+$$

TIN AND LEAD

The column of the periodic table headed by carbon displays the gradation from predominantly nonmetallic properties to predominantly metallic properties with increasing atomic size. Tin and lead have many of the properties of metals, although some of their properties show their closeness to the dividing line separating metals from nonmetals. Elemental tin has two crystalline modifications, one is a distorted close-packed lattice, and the other, gray tin, stable below 18°C, has the covalently bonded diamond structure. Tin(IV) chloride is not saltlike; it is a liquid at ordinary temperatures, with a melting point of −33°C and a normal boiling point of 114.1°C. In aqueous solution, tin(IV) exists as complex ions (e.g., $SnCl_n^{-n+4}$) or hydrolyzed species [e.g., $Sn(OH)_n^{-n+4}$], but not as a simple hydrated ion of charge 4+.

There is a marked contrast between the oxidation–reduction properties associated with the M(IV)–M(II) couple of these two elements. The potential diagrams for acidic solution are:

$$Sn(IV) \xrightarrow{+0.15\ V} Sn^{2+} \xrightarrow{-0.14\ V} Sn$$

$$PbO_2 \xrightarrow{1.455\ V} Pb^{2+} \xrightarrow{-0.126\ V} Pb$$

Tin(IV) is a very poor oxidizing agent, but lead(IV) oxide is an extremely good oxidizing agent. The role of the oxidation states of lead in the lead storage battery has already been discussed in Chapter 16. The oxidation of tin(II) in solution by atmospheric oxygen goes readily. Either solutions of tin(II) must be kept away from oxygen, or metallic tin must be added to keep the tin in the +2 oxidation state. Equilibrium in the reaction

$$Sn(IV) + Sn = 2Sn^{2+}$$

lies to the right. (In this equation, as well as in the potential diagram just given, the designation of tin(IV) as Sn(IV), not Sn^{4+}, indicates uncertainty about the species of tin(IV) present in acidic solution.) Tin(II) is useful as a reducing agent in analysis, for example, in the reduction of iron(III):

$$2Fe^{3+} + Sn^{2+} = 2Fe^{2+} + Sn(IV)$$

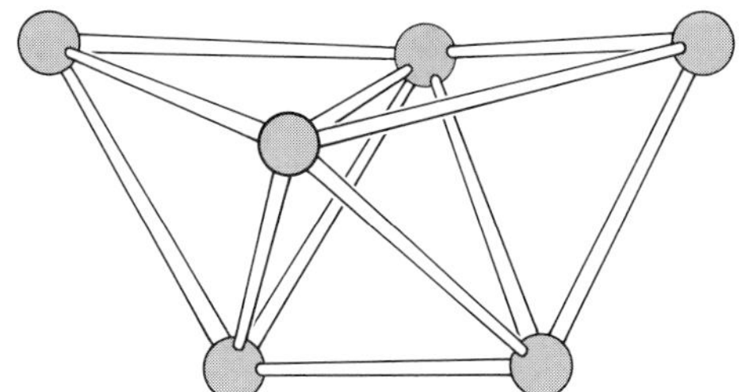

FIGURE 18–11
The structure of the hexameric ion $Pb_6O(OH)_6^{4+}$, found in $[Pb_6O(OH)_6](ClO_4)_4 \cdot H_2O$. Only lead atoms are shown. These atoms define three tetrahedra, which share faces. The oxide ion is near the center of the middle tetrahedron; the six hydroxide ions are near the exposed faces of the end tetrahedra.

It also is useful in preparative organic chemistry, such as in the reduction of nitrobenzene to aniline:

$$6H^+ + \underset{\text{Nitrobenzene}}{C_6H_5NO_2} + 3Sn^{2+} = \underset{\text{Aniline}}{C_6H_5NH_2} + 3Sn(IV) + 2H_2O$$

The hydroxides of both tin(II) and lead(II) are amphoteric. The chemical equations for lead,

$$Pb(OH)_2 + 2H^+ \rightleftarrows Pb^{2+} + 2H_2O \quad \text{and} \quad Pb(OH)_2 + OH^- \rightleftarrows Pb(OH)_3^-$$

oversimplify a more complex situation. If the lead(II) concentration is moderate, polymeric cationic species form. Among the polymeric species present in solution is the hexamer, $Pb_6O(OH)_6^{4+}$, known in a crystalline basic perchlorate to have the unusual structure given in Figure 18–11.

The great toxicity of lead has caused it to be removed from household paints, and chemists are working hard to find substitutes for tetraethyl lead, $Pb(C_2H_5)_4$, as an antiknock agent in the internal combustion engine.

ZINC, CADMIUM, AND MERCURY

Since atoms of these elements have completely filled *d* orbitals both in the free elements and in their compounds, they are classified as nontransition elements. Nonetheless, they exhibit a rich and varied coordination chemistry and oxidation–reduction chemistry. The neutral atoms have two electrons in the *s* orbital of the valence shell and a filled set of five *d* orbitals in the next inner shell. In the +2 oxidation state, common to each of these elements, the atom has lost the two electrons from its valence shell. The energy required to produce the +2 ions in the gaseous state is appreciably greater for these elements than for the alkaline earth elements (e.g., Ca):

Process	$\frac{IE_1 + IE_2}{\text{kJ mol}^{-1}}$
$Ca(g) = Ca^{2+}(g) + 2e^-$	1736
$Zn(g) = Zn^{2+}(g) + 2e^-$	2639
$Cd(g) = Cd^{2+}(g) + 2e^-$	2499
$Hg(g) = Hg^{2+}(g) + 2e^-$	2816

In addition, this family of elements does not display the usual trend whereby the ionization energy decreases with the increasing value of *n*, the valence-shell quantum number.

These elements are oxidized by oxygen, but the reactions are much less exothermic than those for the alkaline-earth metals. At 298.2 K the values of $\Delta H_f/\text{kJ mol}^{-1}$ are:

CaO	−635	ZnO	−348
SrO	−603	CdO	−258
BeO	−599	HgO	−90

Particularly striking is the marginal stability of mercury(II) oxide relative to the constituent elements. The decomposition of mercury(II) oxide upon being heated:

$$2HgO(s) = 2Hg(l) + O_2(g) \qquad \Delta H^0 = +180 \text{ kJ mol}^{-1}$$
$$\Delta S^0 = +217 \text{ J K}^{-1} \text{ mol}^{-1}$$

is a convenient laboratory source of pure oxygen. This reaction was one of those used by LAVOISIER, PRIESTLEY, and SCHEELE in their studies of the nature of combustion and the composition of air.

Like aluminum, in contact with air zinc acquires an oxide coating that retards further oxidation. A coating of zinc protects iron from corrosion. The protection of iron coated with zinc, called *galvanized iron*, persists even if the coating is not complete. This is a general phenomenon in which a more active metal protects a less active metal from oxidation simply by being in contact with it. If oxygen is reduced at a metal surface covered by a film of moisture containing dissolved carbon dioxide, the net half-reaction is

$$O_2(g) + 4CO_2 + 2H_2O + 4e^- = 4HOCO_2^-$$

The electrons required for reaction come from the metal; for a metal M that forms an ion M^{2+}, the half-reaction providing the electrons is

$$M(s) = M^{2+} + 2e^-$$

If two metals, both exposed to air, are in contact with one another, and if their surface is covered by a film of moisture containing carbon dioxide, it is the *more active* of the two metals that is oxidized, *regardless* of the site of oxygen reduction. The electrons flow throughout the metal conductor in response to the tiny potential gradients developed by occurrence of these half-reactions. The film of water plays a role in this electrochemical process. The production of anionic species at one site ($HOCO_2^-$ in the oxygen half-reaction) and cations at another site (M^{2+} in the metal half-reaction) would produce spatial electrical imbalance were it not for the ion migration that occurs in the solution film. This ion migration is analogous to the ion migration in galvanic cells already discussed.

Consideration of the half-reactions

$$Sn^{2+} + 2e^- = Sn \qquad \mathscr{E}^0 = -0.14 \text{ V}$$
$$Fe^{2+} + 2e^- = Fe \qquad \mathscr{E}^0 = -0.44 \text{ V}$$
$$Zn^{2+} + 2e^- = Zn \qquad \mathscr{E}^0 = -0.763 \text{ V}$$

shows that zinc is more easily oxidized than iron. Therefore, when the zinc coating of galvanized iron is broken, protection of the iron from corrosion continues: zinc is oxidized when oxygen is reduced. In contrast, when iron is protected by a coating of tin, a metal *less active* than iron, the tin protects the iron only so long as it covers it completely. Once the tin coating is broken, the exposed iron corrodes rapidly because the exposed iron then protects the tin, just as zinc protects iron. (You probably have noticed the rust along scratches on a "tin can," an iron can coated with tin.) These situations are contrasted in Figure 18–12. [This is not a complete discussion of iron corrosion; further oxidation of iron(II) ion produces iron(III) oxide, Fe_2O_3.]

The lower stability of mercury(II) oxide compared to zinc(II) oxide makes for a difference between the methods of claiming the metals from sulfide ores. When mercury(II) sulfide is heated in air, the reaction

$$HgS(s) + O_2(g) = Hg(g) + SO_2(g)$$

occurs; when zinc(II) sulfide is heated in air, the metal oxide is formed:

$$2ZnS(s) + 3O_2(g) = 2ZnO(s) + 2SO_2(g)$$

To obtain metallic zinc, it is necessary to heat the oxide with carbon (coal) at a temperature above the boiling point of the metal (907°C):

$$ZnO(s) + C(s) = Zn(g) + CO(g)$$

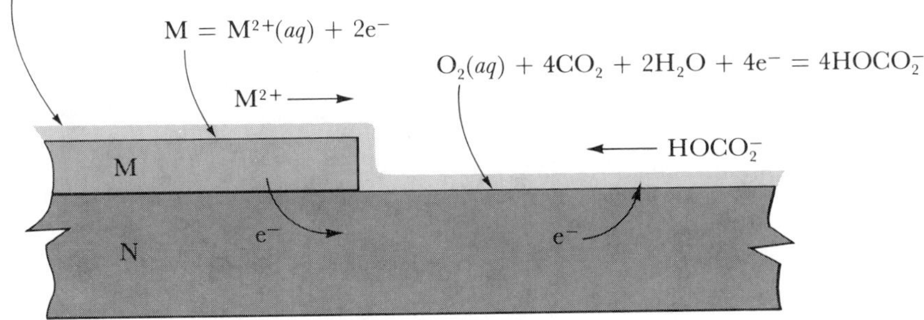

FIGURE 18–12
The protection of a less-active metal, N, by a more active metal, M.

Cathode half-reaction:

$$O_2 + 4CO_2 + 2H_2O + 4e^- = 4HOCO_2^-$$

occurs over entire surface, including surface of N.

Anode half-reaction:

$$M = M^{2+} + 2e^-$$

occurs with more active metal. For zinc–iron, M = Zn, N = Fe; for tin–iron, M = Fe, N = Sn.

The smaller tendency of mercury to be oxidized also is disclosed by the potential diagrams:

$$Zn^{2+} \xrightarrow{-0.763\text{ V}} Zn$$

$$Cd^{2+} \xrightarrow{-0.403\text{ V}} Cd$$

$$Hg^{2+} \xrightarrow{+0.920} Hg_2^{2+} \xrightarrow{+0.789\text{ V}} Hg \quad (Hg^{2+} \to Hg: 0.855\text{ V})$$

The +1 oxidation state is stable in aqueous solution only for mercury. [Solid compounds of cadmium(I) are known: for example, $(Cd_2^{2+})(AlCl_4^-)_2$. Mercury(I), in the solid state or in solution, is a dimeric ion, Hg_2^{2+}. The value of $\mathscr{E}^0$ for the disproportionation reaction,

$$Hg_2^{2+} \rightleftarrows Hg^{2+} + Hg(l) \qquad \mathscr{E}^0 = +0.789\text{ V} - 0.920\text{ V} = -0.131\text{ V}$$

can be used to calculate the equilibrium constant:

$$K = 10^{-0.131\text{ V}/0.0592\text{ V}} = 6.1 \times 10^{-3}$$

Thus mercury(I) ion in aqueous solution is stable with respect to disproportionation. An aqueous solution of mercury(I) perchlorate (or similarly for compounds with any other noncomplexing anion) can be prepared by adding excess metallic mercury to an acidified solution of mercury(II) perchlorate. The reaction

$$Hg^{2+} + Hg(l) = Hg_2^{2+}$$

occurs to the extent dictated by the equilibrium constant already given:

$$\frac{[Hg_2^{2+}]}{[Hg^{2+}]} = \frac{1}{6.1 \times 10^{-3}} = 163$$

Thus in such a solution the fraction of dissolved mercury which is present as mercury(I) is ~0.997.

The disproportionation equilibrium can be shifted by reagents that have a high affinity for one or the other of the oxidation states. Mercury(II) cyanide is very stable, and in the presence of this ion mercury(I) disproportionates completely:

$$Hg_2^{2+} + 2CN^- = Hg(l) + Hg(CN)_2(aq)$$

Disproportionation also occurs if hydrogensulfide ion or hydroxide ion is added to a solution of mercury(I):

$$Hg_2^{2+} + HS^- + OH^- = HgS(s) + Hg(l) + H_2O$$
$$Hg_2^{2+} + 2OH^- = HgO(s) + Hg(l) + H_2O$$

In a standard scheme of qualitative analysis, mercury(I) and silver(I) are precipitated together as relatively insoluble chlorides. Addition of aqueous ammonia to this precipitate causes silver(I) chloride to dissolve [to form $Ag(NH_3)_2^+$], and the presence of mercury(I) is disclosed when the white precipitate [the original mixture of $AgCl(s)$ and $Hg_2Cl_2(s)$] becomes black. The reaction that occurs is disproportionation,

$$Hg_2Cl_2(s) + 2NH_3 = Hg(l) + HgNH_2Cl(s) + NH_4^+ + Cl^-$$

and the black color is due to the metallic mercury dispersed throughout the amidomercury(II) chloride.

Shifting the disproportionation equilibrium further toward mercury(I) is not easy because it does not form many stable complexes. With pyrophosphate ion in approximately neutral solution mercury(I) forms $Hg_2(P_2O_7)OH^{3-}$, and pyrophosphate ion shifts the disproportionation equilibrium to stabilize mercury(I).

Mercury(II) forms very stable halide complexes, but the relative stabilities of its successive complexes are very different from those of most systems. Equilibrium constants for the following reactions at 25°C allow us to calculate the distribution curves in Figure 18–13:

Reaction	$\frac{K}{\text{L mol}^{-1}}$
$Hg^{2+} + Cl^- \rightleftarrows HgCl^+$	5.5×10^6
$HgCl^+ + Cl^- \rightleftarrows HgCl_2$	3.0×10^6
$HgCl_2 + Cl^- \rightleftarrows HgCl_3^-$	8.9
$HgCl_3^- + Cl^- \rightleftarrows HgCl_4^{2-}$	11.2

Mercury(II) chloride is a linear molecule, and in aqueous solution hydrated mercury(II) ion is believed to be a linear diaquamercury(II) ion, $(H_2O)Hg(OH_2)^{2+}$. In formation of the monochloro- and dichloromercury(II) species in solution, these two water molecules of hydration are displaced. The neutral molecule $HgCl_2$ is very stable, and in solutions of this compound there is very little dissociation to give ions (Hg^{2+}, $HgCl^+$, and Cl^-). However, at moderate concentrations of free chloride ion the coordination sphere around mercury(II) is disrupted and additional chloride ion becomes coordinated; tetrachloromercurate(II) ion, $HgCl_4^{2-}$, is a tetrahedral species. It is for systems such as mercury(II)–chloride, in which a change of coordination number occurs, that an irregular pattern of relative stabilities (Figure 18–13) results. This pattern is to be contrasted

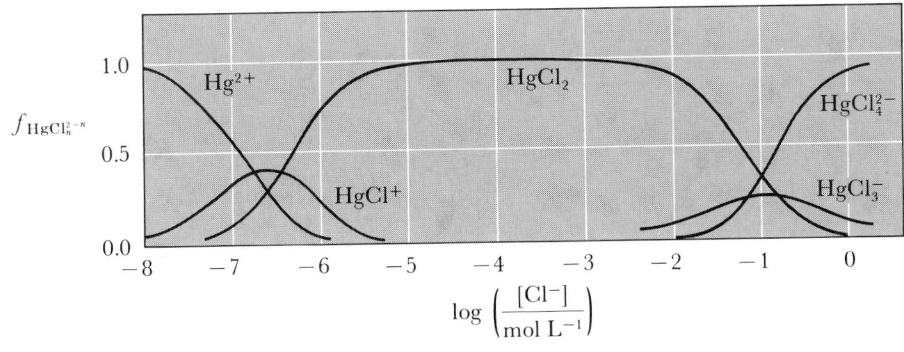

FIGURE 18–13
The distribution of mercury(II) among various chloromercury(II) species, $HgCl_n^{2-n}$, as a function of the concentration of chloride ion. $f_{HgCl_n^{2-n}}$ = fraction of mercury(II) in the form of $HgCl_n^{2-n}$.

with the more regular pattern of stepwise one-to-one replacement of water in $M(OH_2)_n^{m+}$ by a ligand **L** to give $M(OH_2)_{n-p}\mathbf{L}_p^{m-p}$ (if **L** has -1 charge) with $p = 1, 2, 3, \cdots, n$. Such a pattern was shown by systems discussed earlier: beryllium(II)–fluoride (Figure 14–9) and copper(II)–ammonia (Figure 14–10).

Hydrated mercury(II) ion, $Hg(OH_2)_2^{2+}$ is a much stronger acid than expected for a +2 hydrated metal ion,

$$Hg(OH_2)_2^{2+} \rightleftarrows Hg(OH_2)(OH)^+ + H^+ \qquad K_1 = 1.6 \times 10^{-3} \text{ mol L}^{-1}$$

This value of K is 10^5-fold larger than the value for any hydrated +2 transition-metal ion given in Table 19–2, and it is $\sim 10^6$-fold larger than the values for the hydrated ions of zinc(II) and cadmium(II). Mercury(II) ion and gold(I) (isoelectronic species) are appreciably more polarizable than other metal ions we have discussed, and bonding in their complex ions has appreciable covalent character.

This trait (covalency) is responsible for the contrasting stabilities of mercury(II)–halide complexes and zinc(II)–halide complexes. Let us consider the formation of the monohalide complexes:

$$K_1 = \frac{[MX^+]}{[M^{2+}][X^-]}$$

X	$M^{2+} = Zn^{2+}$	$M^{2+} = Hg^{2+}$
F^-	18 L mol^{-1}	3.6×10^1 L mol^{-1}
Cl^-	2.7 L mol^{-1}	5.5×10^6 L mol^{-1}
Br^-	~ 1 L mol^{-1}	8.7×10^8 L mol^{-1}
I^-	<1 L mol^{-1}	7.4×10^{12} L mol^{-1}

We see two significant differences:

1. the chloride, bromide, and iodide complexes of mercury(II) are enormously more stable than those of zinc(II), and
2. the orders of stabilities differ:

$$\text{For } Zn^{2+} \qquad F^- > Cl^- > Br^- > I^-$$
$$\text{For } Hg^{2+} \qquad I^- > Br^- > Cl^- > F^-$$

The order of stabilities of halide complexes has been used to classify metals. The order displayed by zinc(II) is typical of highly charged ions with low polarizability, for example, Be(II), Al(III), Cr(III), or Fe(III). The order displayed by mercury(II) is typical of ions having small charge with high polarizability, for example Ag^+ or Cu^+. The former group of metal ions is called the "Type *a*" or "hard" metal ions and the latter group is called the "Type *b*" or "soft" metal ions. The classifications allow prediction and correlation of more than simply the stability of metal halide complexes. For instance, Type *a* metal ions form more stable complexes with alkyl amines (R_3N) than with alkyl phosphines (R_3P), and the reverse is true for Type *b* metal ions.

The discussion of metals continues in the next chapter, where we focus attention on transition metals. The same general topics, oxidation–

reduction chemistry and complex-ion chemistry, will be pursued, and many facets of these topics are especially rich for transition-metal chemistry. Many transition metals have several oxidation states in addition to zero, and structural aspects of the complex-ion chemistry of these elements have been explored rather thoroughly.

Biographical Notes

ANTOINE L. LAVOISIER (1743–1794), a French chemist, generally is considered to be the founder of modern chemistry. He introduced quantitative methods and helped lay the foundations of thermochemistry. He was guillotined in Paris.

HENDRICK A. LORENTZ (1853–1928), a Dutch physicist, made many contributions to theoretical physics, particularly in areas of electromagnetic theory. He shared the Nobel Prize in Physics for 1902 with P. Zeeman, another Dutch physicist. Lorentz was Professor of Theoretical Physics at the University of Leiden from 1878 to 1923.

FRIEDRICH MOHS (1773–1839), a German mineralogist, in 1822 devised the scale of mineral hardness that bears his name.

JOSEPH PRIESTLEY (1733–1804), an English chemist, was an ordained minister. He was a prolific writer on many subjects, including politics, religion, and science. He moved to the United States in 1794, where he settled in Northumberland, Pennsylvania.

CARL W. SCHEELE (1742–1786), a self-taught Swedish chemist, discovered oxygen independently of Lavoisier and Priestley, but his work was not published until later. He also discovered chlorine and many inorganic and organic acids.

Problems and Questions

18–1 Why do 3*d* orbitals in neutral atoms become more stable relative to 4*s* orbitals as the atomic number increases beyond 20?

18–2 Of the lanthanide elements, only europium (Eu) and ytterbium (Yb) exhibit a relatively stable +2 oxidation state. Explain this phenomenon in terms of the electronic structures of atoms of these elements.

18–3 Explain the order of stability of the several states of the neutral scandium atom shown in Figure 18–3.

18–4 Predict which element has the higher normal boiling point, potassium or calcium.

18–5 Using Figure 18–7, find the phases that are present at 20°C in mixtures of sodium and potassium containing 10.0 mol % Na, 47.0 mol % Na, and 90.0 mol % Na.

18–6 The enthalpy change for vaporization of liquid sodium to give Na(*g*) is +108 kJ mol^{-1} and the bond-dissociation energy of $Na_2(g)$ is 71 kJ mol^{-1}. Does the ratio (P_{Na_2}/P_{Na}) for the vapor in equilibrium with liquid sodium increase, or decrease, with an increase of temperature?

18–7 Explain why an alkali metal (e.g., sodium) may dissolve in a dilute solution of liquid ammonia to give

$$Na(NH_3)_n^+ + e^- \text{ (solvated by } NH_3)$$

but in concentrated solution gives

$$Na(NH_3)_n^+ + Na(NH_3)_m^-$$

18–8 Explain the trend in values of the equilibrium constants for the reaction $MO(s) + H_2O(g) = M(OH)_2(s)$ for the alkaline-earth metals.

18–9 From analogies you may draw with the thermal stabilities of alkaline-earth hydroxides, predict the relative thermal stabilities of $CaCO_3$ and $BaCO_3$ in the reaction $MCO_3(s) = MO(s) + CO_2(g)$.

18–10 Show how temporary hardness is imparted to an underground water containing carbon dioxide when it comes in contact with limestone ($CaCO_3$).

18–11 The equilibrium constant for association of calcium ion and acetate ion at 25°C is small:

$$K = \frac{[CaOAc^+]}{[Ca^{2+}][OAc^-]} = 6 \text{ L mol}^{-1}$$

What is the concentration of $CaOAc^+$ in $0.100M$ $Ca(OAc)_2$? Does your calculation support the idea that calcium acetate is a strong electrolyte?

18–12 To analyze hard water for calcium ion by an EDTA titration we must use a buffered solution with pH $\cong$ 10. Why?

18–13 Examine the structures for the silicate chain and sheet anions, $(SiO_3)_n^{2n-}$, $(Si_4O_{11})_n^{6n-}$, and $(Si_2O_5)_n^{2n-}$, given in Figure 18–10. In each structure pick a segment having the indicated composition.

18–14 An ionic species isoelectronic with silane, SiH_4, is AlH_4^-. From what you know about the stability of aluminum hydroxide, predict what reaction occurs when $NaAlH_4$ is dissolved in water.

18–15 Write the balanced equation for the reaction that occurs when a solution of sodium nitrate flows through a column of cation-exchange resin in the hydrogen ion form. How can this reaction be used to simplify analysis of solutions of sodium nitrate?

18–16 What is the concentration of hydrogen ion in a solution of aluminum(III) chloride [$C(AlCl_3) = 0.050$ mol L^{-1}] if formation of $(H_2O)_5AlOH^{2+}$ is the only hydrolytic reaction that can occur? Is the solubility product for aluminum hydroxide exceeded in this solution?

18–17 Tin(IV) bromide is a known compound, but lead(IV) bromide is not known. Explain this.

18–18 A solution of thallium(I) perchlorate ($[Tl^+] = 0.050$ mol L^{-1}) is in contact with metallic thallium. What is the equilibrium concentration of thallium(III)?

18–19 To what temperature must mercury(II) oxide be heated in order for decomposition to occur in contact with air ($P_{O_2} = 0.20$ atm)?

18–20 Mercury(II) oxide is more soluble in $0.050M$ HCl than it is in $0.050M$ HNO_3. Explain.

18–21 From data in this chapter and the solubility product for mercury(I) chloride ($K_s = [Hg_2^{2+}][Cl^-]^2 = 1.32 \times 10^{-18}$ mol^3 L^{-3}), calculate the equilibrium constant for the reaction

$$Hg_2Cl_2(s) + 2Cl^- \rightleftarrows HgCl_4^{2-} + Hg(l)$$

18–22 Write a balanced equations for the reactions that occur when

(a) $Al(OH)_3(s)$ is treated with a solution of $AlCl_3$.

(b) A solution of $HgCl_2$ is treated with enough sodium chloride to make $[Cl^-] = 1.00$ mol L^{-1}.

(c) A solution of tetrahydroxoaluminate ion, $Al(OH)_4^-$, is treated with ammonium chloride.

(d) Potassium metal is added to methyl alcohol.

(e) Beryllium metal is treated with sodium hydroxide solution to produce a gas.

18–23 You wish to attempt preparation of zinc(I) chloride. What experimental conditions would favor its formation?

18–24 With the help of Figure 18–13, justify the classification of mercury(II) chloride as a weak electrolyte.

19

Transition Metals and Coordination Chemistry

19–1 Introduction

Atoms of the transition metals have partially filled sets of *d* or *f* orbitals in the free state and in compounds. In most discussions of transition metals, the classification is subdivided into:

1. the 3*d* transition metals (elements with partially filled 3*d* orbitals),
2. the heavier transition metals (elements with partially filled 4*d* or 5*d* orbitals),
3. the lanthanide metals, also called the rare earths (lanthanum and the elements with 1 to 14 electrons in 4*f* orbitals), and
4. the actinide metals (actinium and the elements with 1 to 14 electrons in 5*f* orbitals). The lanthanides and actinides also are called inner transition elements.

Figure 19–1 shows the portions of the periodic table that contain these fifty-six metals. Most of our discussion will deal with the elements of the 3*d* transition series. All of the actinide metals are radioactive; in Chapter 22 we shall discuss the chemical properties of some of these elements, as well as their nuclear properties.

The transition metals have high melting and high boiling points; they are good conductors of heat and electricity. As pure metals many are hard and strong, and these characteristics are improved in the many

IIIB	IVB	VB	VIB	VIIB	VIIIB			IB
21 Sc	22 Ti	23 V	24 Cr	25 Mn	26 Fe	27 Co	28 Ni	29 Cu
39 Y	40 Zr	41 Nb	42 Mo	43 Tc	44 Ru	45 Rh	46 Pd	47 Ag
57-71	72 Hf	73 Ta	74 W	75 Re	76 Os	77 Ir	78 Pt	79 Au
89-103	104 Rf	105 Ha						

57 La	58 Ce	59 Pr	60 Nd	61 Pm	62 Sm	63 Eu	64 Gd	65 Tb	66 Dy	67 Ho	68 Er	69 Tm	70 Yb	71 Lu
89 Ac	90 Th	91 Pa	92 U	93 Np	94 Pu	95 Am	96 Cm	97 Bk	98 Cf	99 Es	100 Fm	101 Md	102 No	103 Lr

FIGURE 19–1
The transition metals.

alloys formed by the transition metals. Most of these elements form compounds in more than one oxidation state, and many of the compounds are colored.

Many transition metals have been known since antiquity, for example, iron, copper, silver, and gold. Of the transition metals, iron is the most abundant in the Earth's crust. On a weight percentage basis, its abundance is exceeded only by the abundances of oxygen, silicon, and aluminum. Although the lanthanide metals are called the rare earths, some of them are relatively abundant in the Earth's crust. The abundance of cerium (~ 60 g ton^{-1}) exceeds that of many elements considered to be common, such as arsenic or tin (~ 2 g ton^{-1} for each).

Chapter 18 introduced the electronic structures of the transition metals. In the first sections of this chapter we will classify compounds of the $3d$ transition series according to the number of electrons in their $3d$ orbitals, without reference to the role these electrons play in chemical bonding. In the last section of this chapter, we will consider in greater detail the electronic structures of atoms of these elements in various compounds.

19–2 Compounds of the Transition Metals

We learned in Chapter 18 that when one or more electrons are removed from gaseous atoms of scandium and titanium the resulting ionic species have fewer 4*s* electrons than the neutral atom; the ground states of sucessively ionized atoms of scandium and titanium are:

$$\begin{array}{ll}
\text{Sc(Ar core)} \quad 3d^1 4s^2 & \text{Ti(Ar core)} \quad 3d^2 4s^2 \\
\quad\downarrow -1\,e^- & \quad\downarrow -1\,e^- \\
\text{Sc}^+\text{(Ar core)} \quad 3d^1 4s^1 & \text{Ti}^+\text{(Ar core)} \quad 3d^2 4s^1 \\
\quad\downarrow -1\,e^- & \quad\downarrow -1\,e^- \\
\text{Sc}^{2+}\text{(Ar core)} \quad 3d^1 & \text{Ti}^{2+}\text{(Ar core)} \quad 3d^2 \\
\quad\downarrow -1\,e^- & \quad\downarrow -1\,e^- \\
\text{Sc}^{3+}\text{(Ar core)} & \text{Ti}^{3+}\text{(Ar core)} \quad 3d^1 \\
 & \quad\downarrow -1\,e^- \\
 & \text{Ti}^{4+}\text{(Ar core)}
\end{array}$$

The energy required to remove each successive electron is larger, but there is no marked discontinuity after removal of the two 4*s* electrons. The discontinuity comes when an electron is removed from an underlying noble-gas core. This is true whether a transition metal or a nontransition element is being considered. To illustrate this, the successive ionization energies for titanium and germanium are compared in Figure 19–2. For atoms of each of these elements, removal of a fifth electron requires appreciably more energy; this electron comes from an orbital of the shell beneath the valence shell of the neutral atom. Each 3*d* transition metal is like scandium and titanium; the 4*s* electrons are removed most easily. Each of these elements except scandium exhibits the +2 oxidation state in aqueous

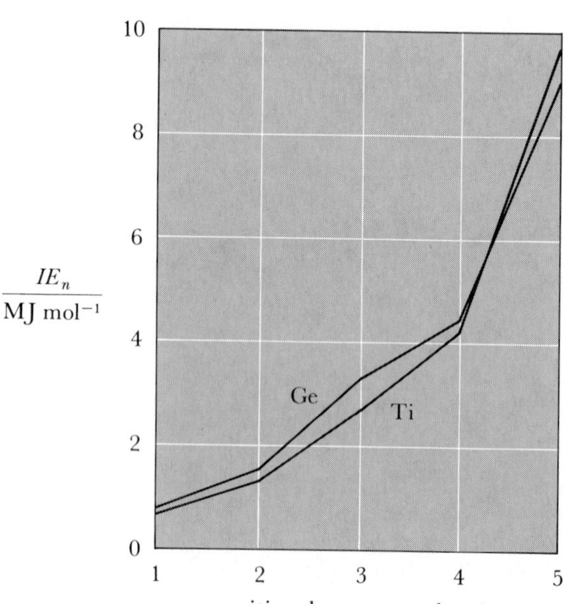

FIGURE 19–2

The ionization energies for removal of successive electrons from atoms of titanium (a 3*d* transition metal) and germanium (a nontransition element). For titanium: IE_1–IE_4 are for removal of the 4*s* and 3*d* electrons; IE_5 is for removal of a 3*p* electron (i.e., an electron in the Ar core). For germanium: IE_1–IE_4 are for removal of the 4*p* and 4*s* electrons; IE_5 is for removal of a 3*d* electron (i.e., an electron in the Zn^{2+} core).

solution and in simple compounds (e.g., halides and oxides); in this oxidation state, both 4*s* electrons are removed.

The maximum oxidation state for each element of the first half of the 3*d* transition metals (scandium through manganese) corresponds to complete removal of both the 4*s* and 3*d* electrons, but this statement is misleading if it suggests that this leads to simple ions of high charge. Halides and oxides of 3*d* transition metals in +2 and +3 oxidation states are predominantly ionic, but simple ions of charge ≥ +4 are not known for these elements. Let us consider the chlorides of titanium:

$TiCl_2$ is a saltlike substance that decomposes at 475°C. Titanium(II) ions occupy one half of the octahedral sites in a hexagonal closest-packed lattice of chloride ions.

$TiCl_3$ is a saltlike substance that decomposes at 440°C. Titanium(III) ions occupy one third of the octahedral sites in a hexagonal closest-packed lattice of chloride ions.

$TiCl_4$ is not saltlike. It melts at −23°C and boils at 136°C. The titanium atom is surrounded tetrahedrally by four chlorine atoms in each state; no strong cation–anion interactions are broken in changes of state. The small, highly charged titanium(IV) ion [$r(Ti^{4+}) = 0.68$ Å] has an enormous polarizing effect on the electron clouds of the four surrounding chloride ions (Figure 9–18 depicts the polarization of one ion by another), and bonding in $TiCl_4$ has much covalent character.

In many common compounds of the 3*d* transition metals in high oxidation states, the transition metal is present as an oxometal anion: for example,

V(V)	Na_3VO_4	Sodium orthovanadate
Cr(VI)	Na_2CrO_4	Sodium chromate
Mn(VII)	$NaMnO_4$	Sodium permanganate

The anions in these compounds, VO_4^{3-}, CrO_4^{2-}, and MnO_4^-, are isoelectronic and have regular tetrahedral geometry. The next member of this isoelectronic series would be the oxide of iron(VIII), FeO_4, which is not known. (Ruthenium and osmium, elements under iron in the periodic table, form tetraoxides, OsO_4 and RuO_4.)

The oxidation states exhibited by a transition metal depend on the other element(s) with which the transition metal is combined. For some cases, the reason for this dependence is simple enough. For instance, a high oxidation state will not be exhibited by a transition metal in combination with an easily oxidized anion. Thus nickel(IV) exists in combination with fluoride ion in K_2NiF_6, but analogous compounds of nickel(IV) and the more easily oxidized halides (chloride, bromide, and iodide) are not known. In combination with carbon monoxide oxidation state zero is common (e.g., nickel(0) in $Ni(CO)_4$ and chromium(0) in $Cr(CO)_6$), and some transition metals even exhibit negative oxidation states (e.g., vanadium(−I) in $V(CO)_6^-$).

OXIDATION STATES OF 3*d* TRANSITION METALS IN AQUEOUS SOLUTION

A summary of the oxidation states exhibited by the 3*d* transition metals in acidic solution is given in Table 19–1. In this table the species are classified according to the number of electrons that would be in *d* orbitals of the metal atom if it existed as the simple ion. Thus the +2, +3, and +4 oxidation states of titanium are classified as d^2, d^1, and d^0 species, respectively. Oxidation states with a d^0 configuration are the highest known for each of the elements through manganese, but the highest oxidation state known for iron is iron(VI), a d^2 species. Table 19–1 includes formulas of some species (in parenthesis) that are unstable or have marginal stability in dilute acidic solutions. In some cases a species is stable in basic solution even if it is not stable in acidic solution; this is true for manganate ion, MnO_4^{2-} (discussed in Chapter 16). The absence of an entry for a particular oxidation state means that this oxidation state is not stable enough in dilute acidic solution to allow characterization of it. Chromium(IV) is an example. Compounds of chromium(IV) are known; the oxide CrO_2, has the TiO_2 structure (see Figure 9–3). This compound with a d^2 metal ion is used in the manufacture of high-fidelity magnetic tapes. Treatment of this oxide with an acidic aqueous solution results in rapid disproportionation, giving only chromium(III) and chromium(VI):

$$3CrO_2 + 5H^+ = 2Cr^{3+} + HCrO_4^- + 2H_2O$$

TABLE 19–1
Oxidation States of the 3*d* Transition Metals in Acidic Aqueous Solution[a]

Number of 3*d* electrons	Oxidation state +1	+2	+3	+4	+5	+6	+7
d^0		Ca^{2+}[b]	Sc^{3+}	TiO^{2+}	VO_2^+	$HOCrO_3^-$	MnO_4^-
d^1			Ti^{3+}	VO^{2+}		(MnO_4^{2-})[d]	
d^2		Ti^{2+}	V^{3+}			(FeO_4^{2-})[e]	
d^3		V^{2+}	Cr^{3+}	$MnO_2(s)$[c]			
d^4		Cr^{2+}	Mn^{3+}				
d^5		Mn^{2+}	Fe^{3+}				
d^6		Fe^{2+}	Co^{3+}				
d^7		Co^{2+}					
d^8		Ni^{2+}					
d^9		Cu^{2+}					
d^{10}	Cu^+	Zn^{2+}[b]					

[a]The hydration of species is not shown. The formula given is that which is predominant at $[H^+] = 0.1 - 1 \text{ mol L}^{-1}$.
[b]Neither calcium nor zinc is a transition element in our classification.
[c]Manganese(IV) oxide has very low solubility in both acidic and basic solutions.
[d]Unstable with respect to disproportionation in acidic solution.
[e]Unstable with respect to oxidation of water.

Species of chromium(IV) exist in solution as reaction intermediates, but the coordination number of the metal and the net charge of such species have not been characterized with certainty.

For conciseness in Table 19–1, the formulas of ions with the net charge equal to the oxidation state are given without the hydration shell of coordinated water molecules. As we already have learned (Section 14–3), metal ions in aqueous solution are strongly solvated by water molecules. For most transition-metal ions, the solvation results in a well-defined species, for example, hexaaquachromium(III) ion, $Cr(OH_2)_6^{3+}$. Although simple hydrated ions of charge +4 are known [in acidic solution, thorium(IV) with a large ionic radius, 1.02 Å, exists as $Th(OH_2)_n^{4+}$, with n equal to 8 or 9], no simple hydrated ions of charge +4 are known for the 3d transition metals. The metal ions of charge ≥ +4 are so small [e.g., $r(Ti^{4+}) = 0.68$ Å and $r(V^{5+}) = 0.59$ Å] that a proton dissociates from a coordinated water molecule even in highly acidic solutions. For both titanium(IV) and vanadium(IV), the hypothetical hydrated species of charge +4 revert in acidic solution to oxometal ions of charge +2 by proton dissociation:

$$Ti(OH_2)_6^{4+} + 2H_2O \rightarrow (H_2O)_5TiO^{2+} + 2H_3O^+$$
$$V(OH_2)_6^{4+} + 2H_2O \rightarrow (H_2O)_5VO^{2+} + 2H_3O^+$$

If the oxidation state is +6 or +7, the tendency to coordinate oxide ion is so great that anionic species predominate even in acidic solution:

Chromium(VI)	$HOCrO_3^-$ and $Cr_2O_7^{2-}$
Manganese(VII)	MnO_4^-

The chemical formulas for the several oxidation states of a particular element in Table 19–1 convey information about acidity and basicity of the oxides. Thus the formulas Cr^{2+}, Cr^{3+}, and $HOCrO_3^-$, the predominant species of chromium(II), chromium(III), and chromium(VI) in solutions with $[H^+] \cong 0.10$ mol L^{-1}, indicate that:

CrO and $Cr(OH)_2$ are basic $\quad Cr(OH)_2 + 2H^+ = Cr^{2+} + 2H_2O$

Cr_2O_3 and $Cr(OH)_3$ are basic $\quad Cr(OH)_3 + 3H^+ = Cr^{3+} + 3H_2O$

CrO_3 is acidic $\quad CrO_3 + H_2O = H^+ + HOCrO_3^-$

Whether the hydroxides of chromium(II) and chromium(III) also have acidic properties is not disclosed by formulas of the species present in acidic solution. That chromium(II) hydroxide is not appreciably acidic and chromium(III) hydroxide is acidic is disclosed by the formulas $Cr(OH)_2(s)$ and $Cr(OH)_4^-$ for the predominant species in an alkaline solution ($[OH^-] \cong$ 0.10 mol L^{-1}).

THE ACIDITY AND BASICITY OF OXIDES AND HYDROXIDES OF 3d TRANSITION METALS

The acidity and basicity of metal oxides can be revealed in a number of ways. A metal oxide (e.g., CrO_3) shows acidic character by reacting with a basic metal oxide,

$$BaO(s) + CrO_3(s) = BaCrO_4(s)$$

or by forming an anionic species in alkaline solution,

$$2OH^- + CrO_3(s) = H_2O + CrO_4^{2-}$$

or by reacting with liquid water to give an acidic solution,

$$H_2O(l) + CrO_3(s) = H^+ + HOCrO_3^-$$

A metal oxide (e.g., MnO) shows basic character by reacting with carbon dioxide gas,

$$CO_2(g) + MnO(s) = MnCO_3(s)$$

or by reacting with water vapor,

$$H_2O(g) + MnO(s) = Mn(OH)_2(s)$$

or by reacting with an aqueous solution of acid,

$$2H^+(aq) + MnO(s) = Mn^{2+}(aq) + H_2O$$

We also can discuss the acid–base properties of a hydrated transition-metal ion in terms of acid-dissociation equilibria. The first stage in the acid dissociation of a hydrated metal ion (assumed to be a hexaaquametal ion),

$$H_2O(l) + M(OH_2)_6^{m+}(aq) = M(OH_2)_5OH^{m-1}(aq) + H_3O^+(aq)$$

has been studied for many of the 3*d* transition-metal ions. Values of the acid-dissociation constants are given in Table 19–2. Accurate measurement of these acid-dissociation constants is difficult, and some of these values are uncertain. It is clear, however, that hydrated transition-metal ions of charge

TABLE 19–2
Acid-Dissociation Constants for Hydrated Transition-Metal Ions

M	$K_a[M(OH_2)_6^{2+}]$ / mol L^{-1}	$K_a[M(OH_2)_6^{3+}]$ / mol L^{-1}
Sc	—	5×10^{-5}
Ti	—	6×10^{-3}
V	—	5×10^{-3}
Cr	—	1×10^{-4}
Mn	3×10^{-11}	—
Fe	3×10^{-10}	6×10^{-3}
Co	2×10^{-10}	—
Ni	1×10^{-10}	—
Cu	1×10^{-8}	—

[a]For the equilibrium at 25°C

$$H_2O(l) + M(OH_2)_6^{m+}(aq) \rightleftarrows M(OH_2)_5OH^{m-1}(aq) + H_3O^+(aq)$$

+2 are only slightly acidic, but those of charge +3 are much more acidic. All of the $M(OH_2)_6^{3+}$ ions are stronger acids than acetic acid, and the strongest of these acids, $Ti(OH_2)_6^{3+}$ and $Co(OH_2)_6^{3+}$, have acid-dissociation constants as large as that for hydrogensulfate ion, $HOSO_3^-$.

Dissociation of successive protons by a hexaaquametal(III) ion would result in the formation of a sequence of species:

$$M(OH_2)_6^{3+} \rightleftarrows M(OH_2)_5OH^{2+} \rightleftarrows M(OH_2)_4(OH)_2^+ \rightleftarrows$$
$$M(OH_2)_3(OH)_3 \rightleftarrows M(OH_2)_2(OH)_4^- \rightleftarrows M(OH_2)(OH)_5^{2-} \rightleftarrows M(OH)_6^{3-}$$

For no +3 metal ion can this complete sequence of reactions be characterized. Many of the aquahydroxometal(III) ions tend to form a range of polymeric species, the simplest of which are dimeric species, such as $Fe_2(OH)_2^{4+}$:

$$2FeOH^{2+} \rightleftarrows Fe_2(OH)_2^{4+} \qquad K = 27\ \text{L mol}^{-1}\ (25^\circ C)$$

This dimeric species involves two hydroxide ions bridging between the two metal ions:

$$\begin{array}{ccccccc} & H_2O & H & OH_2 & & & 4+ \\ H_2O & & O & & OH_2 & & \\ & Fe & & Fe & & & \\ H_2O & & O & & OH_2 & & \\ & H_2O & H & OH_2 & & & \end{array}$$

Whether the various hydroxometal species can attain an appreciable concentration in aqueous solution of variable hydrogen ion concentration depends on the solubility of the metal hydroxide and the solubilities of salts of the hydroxometal anion, as was shown in the discussion of zinc hydroxide and copper hydroxide in Chapter 14. Of the 3*d* transition-metal(III) ions, chromium(III) shows the greatest tendency to form anionic hydroxometal species:

$$2H_2O + Cr(OH)_3(s) + OH^- \rightleftarrows Cr(OH_2)_2(OH)_4^- \qquad K \cong 0.4$$

The +4 oxidation state is exhibited by several of the 3*d* transition-metal ions, and the oxides of two of these elements, titanium and vanadium, are basic. The other common 3*d* transition-metal dioxide, MnO_2, is neither acidic nor basic, and it is not soluble to an appreciable extent either in dilute acid or in dilute base. However, manganese dioxide does react with hydrochloric acid solution in an oxidation–reduction reaction:

$$MnO_2(s) + 4H^+ + 2Cl^- = Cl_2(g) + Mn^{2+} + 2H_2O$$

The +5 oxidation state of vanadium is represented by V_2O_5, which exhibits both acidic and basic properties. The simple monomeric ions VO_2^+ and VO_4^{3-} are known, and these form when a small amount of the oxide is dissolved in strongly acidic and strongly basic solution:

$$V_2O_5(s) + 2H^+ \rightleftarrows 2VO_2^+ + H_2O$$
$$V_2O_5(s) + 6OH^- \rightleftarrows 2VO_4^{3-} + 3H_2O$$

However, this system is enormously complicated, and depending on the concentrations of hydroxide ion and vanadium(V), the species $V_2O_7^{4-}$, $V_4O_{12}^{4-}$, and $V_{10}O_{28}^{6-}$ form, together with partially protonated forms of these species. The structure of the polymeric anion $V_{10}O_{28}^{6-}$, shown in Figure 19–3, is a network of ten octahedra, each with a vanadium(V) atom surrounded by six oxide ions. Protonation of this species starts in dilute acidic solution; between pH = 5 and pH = 6 the species is converted to $V_{10}O_{28}H^{5-}$. Complete protonation to neutral $V_{10}O_{28}H_6$ is not observed because in solutions of high acidity the simple monomeric oxocation VO_2^+ is formed.

There also are concentration-dependent equilibria for oxometal anions of chromium(VI):

$$(HO)_2CrO_2 \rightleftarrows H^+ + HOCrO_3^- \qquad K > 1 \text{ mol L}^{-1}$$
$$HOCrO_3^- \rightleftarrows H^+ + CrO_4^{2-} \qquad K = 7.4 \times 10^{-7} \text{ mol L}^{-1}$$
$$2HOCrO_3^- \rightleftarrows Cr_2O_7^{2-} + H_2O \qquad K = 35 \text{ L mol}^{-1}$$

Thus at $[H^+] < 10^{-7}$ mol L^{-1}, the predominant form of chromium(VI) is CrO_4^{2-}, but at $[H^+] > 10^{-5}$ mol L^{-1}, the predominant form is $HOCrO_3^-$ or $Cr_2O_7^{2-}$, depending on the concentration of chromium(VI). We can demonstrate this dependence by calculating the fraction of chromium(VI) present as monomer $HOCrO_3^-$ and dimer $Cr_2O_7^{2-}$ in solutions with different total concentrations of chromium(VI).

EXAMPLE: What fraction of chromium(VI) in aqueous solution with pH = 3 is present as $Cr_2O_7^{2-}$ if the total concentration of chromium(VI) is 0.00100 mol L^{-1}; also if the concentration is 0.0200 mol L^{-1}?

Let $[HOCrO_3^-] = C - 2[Cr_2O_7^{2-}]$, which upon substitution into the equilibrium-constant equation gives

$$[Cr_2O_7^{2-}] = 35 \text{ L mol}^{-1} \times (C - 2[Cr_2O_7^{2-}])^2$$

or

$$4[Cr_2O_7^{2-}]^2 - \left(4C + \frac{1}{35 \text{ L mol}^{-1}}\right)[Cr_2O_7^{2-}] + C^2 = 0$$

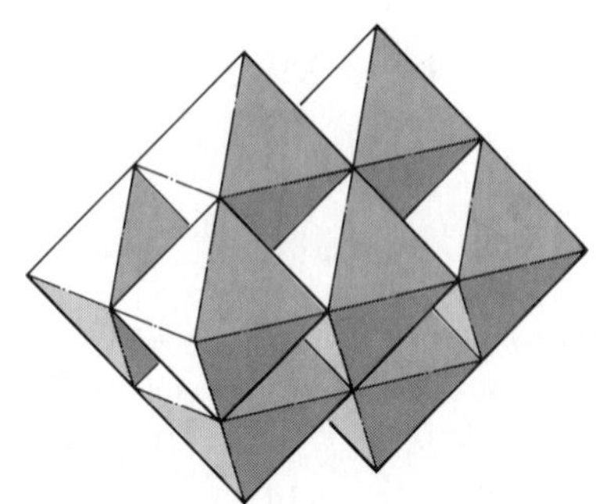

FIGURE 19–3
Structure of decavanadate ion, $V_{10}O_{28}^{6-}$. The structure consists of 10 octahedra sharing edges. The 38 atoms lie in 5 planes, which contain:

1. top of structure $2O^{2-}$
2. next plane $6O^{2-}$ and $2V^{5+}$
3. next plane $12O^{2-}$ and $6V^{5+}$
4. next plane $6O^{2-}$ and $2V^{5+}$
5. bottom of structure $2O^{2-}$

This structure is found in solid compounds and in solution, where it can be protonated to give $V_{10}O_{28}H^{5-}$ and $V_{10}O_{28}H_2^{4-}$

Solving this quadratic equation gives

$$[Cr_2O_7^{2-}] = \frac{4C + \dfrac{1}{35\ \text{L mol}^{-1}} - \left[\dfrac{8C}{35\ \text{L mol}^{-1}} + \dfrac{1}{(35\ \text{L mol}^{-1})^2}\right]^{1/2}}{8}$$

which gives:

For $C = 0.00100$ mol L^{-1}, $[Cr_2O_7^{2-}] = 3.1 \times 10^{-5}$ mol L^{-1}
For $C = 0.0200$ mol L^{-1}, $[Cr_2O_7^{2-}] = 4.4 \times 10^{-3}$ mol L^{-1}

The twenty-fold increase in stoichiometric concentration of chromium(VI) increases the fraction of chromium(VI) present as dichromate ion from 6.2% to 44%.

The oxoanion of manganese(VII) is not basic enough to be protonated at $[H^+] \cong 1$ mol L^{-1}:

$$HOMnO_3 \rightleftarrows H^+ + MnO_4^- \qquad K \gg 1\ \text{mol L}^{-1}$$

but concentrated sulfuric acid causes formation of the unstable anhydride, Mn_2O_7, which decomposes explosively to give MnO_2:

$$2H^+ + 2MnO_4^- = Mn_2O_7(s) + H_2O$$
$$2Mn_2O_7(s) = 4MnO_2(s) + 3O_2(g) \qquad \Delta H^0 \cong -630\ \text{kJ mol}^{-1}$$
$$\Delta S^0 \cong +630\ \text{J K}^{-1}\ \text{mol}^{-1}$$

Potassium permanganate must not be mixed with concentrated sulfuric acid!

OXIDATION–REDUCTION CHEMISTRY OF SOME TRANSITION METALS

We saw in Table 19–1 that most of the 3*d* transition elements exist in several oxidation states. The oxidation–reduction chemistry of these elements is a rich field. We considered the potential diagrams for vanadium and manganese in Chapter 16. Here we first will consider the diagrams for chromium and iron:

$$HOCrO_3^- \xrightarrow{+1.20\ V} Cr^{3+} \xrightarrow{-0.50\ V} Cr^{2+} \xrightarrow{-0.86\ V} Cr$$
$$FeO_4^{2-} \xrightarrow{+2.2\ V} Fe^{3+} \xrightarrow{+0.771\ V} Fe^{2+} \xrightarrow{-0.44\ V} Fe$$

Like Table 19–1, these diagrams do not include the unstable +4 and +5 oxidation states of chromium and iron. The oxidation states given in these two diagrams include:

1. a good oxidizing agent, $HOCrO_3^-$ (or its anhydride, $Cr_2O_7^{2-}$), which is very useful in preparative chemistry and in analysis,
2. a very good reducing agent, Cr^{2+}; we already have learned (Section 16–6) that chromium(II) persists for long periods of time in oxygen-free

aqueous acid, despite an equilibrium in the reaction that favors Cr^{3+}

$$2Cr^{2+} + 2H^+ = 2Cr^{3+} + H_2$$

3. an extremely powerful oxidizing agent, FeO_4^{2-}, which is stable in strongly alkaline solution but does not persist in acidic aqueous solution for an appreciable time. The reaction

$$4FeO_4^{2-} + 20\,H^+ = 4Fe^{3+} + 3O_2 + 10H_2O$$

occurs rapidly.
4. a couple, $Fe^{3+} - Fe^{2+}$, which plays important roles in some biological oxidation–reduction reactions.

In analytical chemistry an important reaction that involves chromium(VI) is the oxidation of iron(II) ion; if the chromium(VI) concentration is low ($< 10^{-3}$ mol L^{-1}) the appropriate equation is

$$HOCrO_3^- + 3Fe^{2+} + 7H^+ = 3Fe^{3+} + Cr^{3+} + 4H_2O$$

and if the chromium concentration is higher ($>2 \times 10^{-2}$ mol L^{-1}), the appropriate equation is

$$Cr_2O_7^{2-} + 6Fe^{2+} + 14H^+ = 6Fe^{3+} + 2Cr^{3+} + 7H_2O$$

In this reaction a one-electron reducing agent, iron(II), reacts with a three-electron oxidizing agent, chromium(VI). That this reaction must occur by a multistep reaction mechanism is suggested by the complexity of each of these chemical equations. Studies of reaction kinetics suggest that this reaction proceeds via a sequence of three one-electron steps. With only the oxidation state (and not the complete formula) of each of the chromium species given, the sequence of steps comprising the proposed mechanism is

1. $Cr(VI) + Fe^{2+} \underset{k_2}{\overset{k_1}{\rightleftarrows}} Cr(V) + Fe^{3+}$

2. $Cr(V) + Fe^{2+} \xrightarrow{k_3} Cr(IV) + Fe^{3+}$

3. $Cr(IV) + Fe^{2+} \xrightarrow{k_5} Cr(III) + Fe^{3+}$

The second reaction in this three-step sequence is the slow step, and only for Step 1 is it necessary to include the reverse reaction.

The experimental basis for this conclusion is the form of the empirical rate law,

$$-\frac{d[Cr(VI)]}{dt} = \frac{k'[HCrO_4^-][Fe^{2+}]^2}{[Fe^{3+}]}$$

(This rate law is appropriate for dilute solutions of chromium(VI) with constant acidity. The rate of this reaction does depend on the concentration of hydrogen ion, that is, k' depends upon $[H^+]$.) The inverse dependence of reaction rate on the concentration of iron(III) indicates that iron(III) and a reactive intermediate [in this case, chromium(V)] are produced in a rapid equilibrium prior to the rate-determining step. Because the second reaction, that with rate constant k_3, is rate-determining, the rate is given by the rate law

$$-\frac{d[Fe^{2+}]}{dt} = 3k_3[Fe^{2+}][Cr(V)]$$

The factor 3 is included because a total of 3 mol of iron(II) ions are consumed per 1 mol of iron(II) reacting in Step 2. The chromium(V) that is the reactant in the rate-determining step is produced in Step 1, and its concentration is determined by equilibrium in this rapid step:

$$[Cr(V)] = \frac{k_1}{k_2}\frac{[Cr(VI)][Fe^{2+}]}{[Fe^{3+}]}$$

Substitution of this into the rate law for the slow reaction gives

$$-\frac{d[Fe^{2+}]}{dt} = \frac{3k_1k_3}{k_2}\frac{[HCrO_4^-][Fe^{2+}]^2}{[Fe^{3+}]}$$

a rate law that has the form of the empirical rate law. The empirical rate constant k' is identified in this mechanism as $(3k_1k_3/k_2)$. Reactions of chromium(VI) with some other one-electron reducing agents [e.g., vanadium(IV)] are governed by rate laws with analogous forms, and these reactions are assumed to occur by an analogous sequence of three one-electron steps. To explain the relative slowness of the chromium(V) → chromium(IV) step in each of these reactions, it has been postulated that this is the step in which the metal's coordination number (CN) changes:

$$\underset{[\text{e.g., }(HO)_3CrO]}{Cr(V)\ (CN\ 4)} \rightarrow \underset{[\text{e.g., }Cr(OH_2)_6^{4+}]}{Cr(IV)\ (CN\ 6)}$$

In this step, chromium–oxygen bonds are formed.

Reducing agents that do not undergo one-electron reactions easily are constrained to react with chromium(VI) by other mechanisms. This is the situation in the chromium(VI) oxidation of alcohols. Chromium(VI) is a useful reagent for the laboratory-scale oxidation of isopropyl alcohol to acetone:

$$2HOCrO_3^- + 3(CH_3)_2CHOH + 8H^+ = 2Cr^{3+} + 3(CH_3)_2CO + 8H_2O$$

This reaction has been studied carefully, and it has been shown that the initial step of the reaction forms an ester of chromic acid, isopropyl hydrogenchromate:

$$H^+ + (CH_3)_2CHOH + HOCrO_3^- \rightleftarrows (CH_3)_2CHOCrO_3H + H_2O$$

This reaction involves no changes of oxidation state: the chromate ester is a species of chromium(VI). This chromate ester reacts with water in a two-electron step to give acetone and a species of chromium(IV). Additional steps, which have not been thoroughly characterized, result in the net stoichiometry: two hydrogenchromate ions oxidize three molecules of isopropyl alcohol.

Recently it has been found that in the presence of both isopropyl alcohol and another reducing agent (glycolic acid, $HOCH_2CO_2H$), there is no intermediate formation of chromium(IV). In one step, chromium(VI) is converted to chromium(III), the three electrons coming from two different molecules: two from isopropyl alcohol, and one from glycolic acid.

Chromium(II) ion is a very good reducing agent. Acidic aqueous solutions of this species are prepared easily by reducing chromium(III) ion with zinc,

$$Zn + 2Cr^{3+} = Zn^{2+} + 2Cr^{2+}$$

or by dissolving pure metallic chromium in dilute perchloric acid,

$$2H^+ + Cr = Cr^{2+} + H_2$$

However, these solutions must be kept free of air because oxygen rapidly oxidizes chromium(II) to chromium(III). The process is, in fact, a complicated one, and simple hexaaquachromium(III) ion is not produced. Rather, a dimeric hydroxo species, $Cr_2(OH)_2^{4+}$, analogous to $Fe_2(OH)_2^{4+}$ is produced; the net change is

$$O_2 + 4Cr^{2+} + 2H_2O = 2Cr_2(OH)_2^{4+}$$

but the mechanism consists of several steps.

That iron(VI) can be prepared in alkaline solution, where it is stable, despite its instability in acidic solution, is a consequence of the strong dependence of $\mathscr{E}$ on acidity for each of the half-reactions:

$$FeO_4^{2-} + 8H^+ + 3e^- = Fe^{3+} + 4H_2O \qquad \left(\frac{h}{n} = 2\tfrac{2}{3}\right)$$

which is appropriate at $[H^+] > \sim 10^{-2}$ mol L^{-1}, and

$$FeO_4^{2-} + 4H_2O + 3e^- = Fe(OH)_3 + 5OH^- \qquad \left(\frac{h}{n} = 1\tfrac{2}{3}\right)$$

which is appropriate at $[H^+] < \sim 10^{-7}$ mol L^{-1}. By using the principles outlined in Chapter 16 and the solubility-product constant,

$$K_s = [Fe^{3+}][OH^-]^3 = 6 \times 10^{-38} \text{ mol}^4 \text{ L}^{-4}$$

we can show that the value of $\mathscr{E}^0$ for the half-reaction in acid, $\mathscr{E}^0 = +2.2$ V, leads to the value of $\mathscr{E}^0$ for the half-reaction in base, $\mathscr{E}^0 = +0.7$ V. Combining the half-reaction for base with that for the hypochlorite–chloride couple in base,

$$OCl^- + H_2O + 2e^- = Cl^- + 2OH^- \qquad \mathscr{E}^0 = +0.89 \text{ V}$$

we get the equation

$$3OCl^- + 2Fe(OH)_3(s) + 4OH^- = 3Cl^- + 2FeO_4^{2-} + 5H_2O$$

with

$$\mathscr{E}^0 = +0.89\text{ V} - 0.7\text{ V} \cong 0.2\text{ V}$$

which gives ($n = 6$):

$$K^0 = 10^{6 \times 0.2\text{ V}/0.059\text{ V}}$$
$$K = 10^{20}\text{ L}^2\text{ mol}^{-2}$$

Thus hypochlorite ion in base is capable of oxidizing iron(III) hydroxide to ferrate(VI)ion.

The iron(III)–iron(II) couple,

$$Fe^{3+} + e^- = Fe^{2+} \qquad \mathscr{E}^0 = +0.77\text{ V}$$

was the basis for one half-cell in the galvanic cell discussed in Chapter 16. Both iron(III) and iron(II) form many complex ions, and half-reactions can be written for couples involving these complex iron(III) and iron(II) species. With cyanide ion as the ligand, iron(III) forms the more stable complex, and hexacyanoferrate(III) ion is a less good oxidizing agent than hexaaquairon(III) ion:

$$Fe(CN)_6^{3-} + e^- = Fe(CN)_6^{4-} \qquad \mathscr{E}^0 = +0.42\text{ V}$$

With phenanthroline (see Table 14–2) as ligand, iron(II) forms the more stable complex, and tris(phenanthroline)iron(III) ion is a better oxidizing agent than hexaaquairon(III) ion:

$$Fe(phen)_3^{3+} + e^- = Fe(phen)_3^{2+} \qquad \mathscr{E}^0 = +1.12\text{ V}$$

As discussed in Chapter 16, iron(II) phenanthroline is a useful indicator for oxidation–reduction titrations.

In biological systems there are many iron-containing proteins in which the iron exists in the +3 or +2 oxidation states. The role of these systems in biological oxidation–reduction reactions is dependent upon the values of $\mathscr{E}^0$ for the half-reactions. The dependence of $\mathscr{E}^0$ on the identity of the protein–ligand system is illustrated by a few couples for pH = 7.0:

Half-Reaction	$E^{0\prime}$/V
Methemoglobin + e^- = Hemoglobin	0.17
Peroxidase–Fe(III) + e^- = Peroxidase–Fe(II)	−0.17
Spinach ferredoxin–Fe(III) + e^- = Spinach ferredoxin–Fe(II)	−0.43

There are enormous effects. In spinach ferredoxin, a sulfur-containing protein, the $\mathscr{E}^{0\prime}$ value differs by more than 1.0 V from the value for the simple hydrated ions. Each of these $\mathscr{E}^0$ values indicates a ligand stabilization of the iron(III) state compared to the iron(II) state.

The potential diagrams for copper, silver, and gold, which are in the same column of the periodic table, are:

$$CuO^+ \xrightarrow{(1.8\ V)} Cu^{2+} \xrightarrow{0.153\ V} Cu^+ \xrightarrow{0.521\ V} Cu$$
$$Cu^{2+} \xrightarrow{0.337\ V} Cu$$

$$AgO^+ \xrightarrow{(2.1\ V)} Ag^{2+} \xrightarrow{1.98\ V} Ag^+ \xrightarrow{0.799\ V} Ag$$

$$Au^{3+} \xrightarrow{(<1.3\ V)} Au^{2+} \xrightarrow{(>1.3\ V)} Au^+ \xrightarrow{1.7\ V} Au$$
$$Au^{3+} \xrightarrow{1.42\ V} Au$$

Each of these elements forms compounds with the metal in each of the three oxidation states, +3, +2, and +1. But for each element, a different oxidation state is the principal one; these are:

Copper(II), with d^9 electronic configuration
Silver(I), with d^{10} electronic configuration
Gold(III), with d^8 electronic configuration

From the values of $\mathscr{E}^0$ given in the diagrams, it is clear that copper(III), silver(III), and silver(II) are extraordinarily strong oxidizing agents. They are produced only under extreme conditions.

Alkaline hypochlorite solution will produce copper(III) from copper(II) hydroxide:

$$2OH^- + H_2O + 2Cu(OH)_2 + OCl^- = 2Cu(OH)_4^- + Cl^-$$

But acidification of such a solution results in evolution of oxygen and production of blue hydrated copper(II) ion:

$$2H_2O + 4Cu(OH)_4^- + 12H^+ = 4Cu(OH_2)_4^{2+} + O_2$$

In these reactions, copper(III) resembles iron(VI), already discussed.

Silver(III) is produced initially if silver(I) is oxidized with ozone:

$$Ag^+ + O_3 = AgO^+ + O_2$$

but silver(III), so produced, reacts rapidly with the excess of silver(I) to give silver(II):

$$AgO^+ + Ag^+ + 2H^+ = 2Ag^{2+} + H_2O$$

The point already has been made Section 17–4 that silver(I) is a catalyst for oxidation of various reducing agents by peroxodisulfate ion. The catalytic cycles proposed to explain this action involve one or both of the higher oxidation states of silver, silver(III) and silver(II). Copper(II) also can act as a catalyst in reactions of peroxodisulfate ion; a similar cycle of steps is assumed to be responsible.

Copper(I) is unstable with respect to disproportionation:

$$2Cu^+ = Cu^{2+} + Cu$$

Using the values of $\mathscr{E}^0$ for the half-reactions, we can calculate the equilibrium constant for this reaction; the value of $\mathscr{E}^0$ for the complete reaction is

$$\mathscr{E}^0 = 0.521\text{ V} - 0.153\text{ V} = 0.368\text{ V}$$

which gives ($n = 1$):

$$K^0 = 10^{0.368\text{ V}/0.0592\text{ V}} = 10^{6.22}$$
$$K = 1.6 \times 10^6\text{ L mol}^{-1}$$

Therefore copper(I) ion cannot be prepared in aqueous solution at an appreciable concentration by treating metallic copper with a solution of copper(II) ion. However, this does not mean that metastable solutions of copper(I) ion cannot be prepared by other means. Solutions of copper(I) ion can be prepared by treatment of copper(II) with chromium(II):

$$Cu^{2+} + Cr^{2+} = Cr^{3+} + Cu^{+} \qquad \mathscr{E}^0 = +0.153\text{ V} - (-0.50\text{ V}) = +0.65\text{ V}$$

In a preparation such as this, the copper(I) ion attains a concentration higher than allowed at equilibrium in the disproportionation, and the system eventually becomes more stable by occurrence of the reaction

$$2Cu^{+} = Cu^{2+} + Cu$$

However, the disproportionation equilibrium can be displaced toward copper(I) by the presence of a complexing reagent that combines preferentially with copper(I). Copper(I) cyanide is relatively insoluble ($K_s = [Cu^{+}][CN^{-}] = 3.2 \times 10^{-20}\text{ mol}^2\text{ L}^{-2}$), and stable, soluble complex ions also form, for example, $Cu(CN)_4^{3-}$. In the presence of cyanide ion, copper(I) *does not* disproportionate. The potential diagram for the system, showing only +2, +1, and 0 states in the presence of cyanide ion, is

$$Cu^{2+} \xrightarrow{+1.78\text{ V}} Cu(CN)_4^{3-} \xrightarrow{-1.10\text{ V}} Cu$$

At cyanide ion concentrations high enough to stabilize tetracyanocuprate(I) ion, copper metal will dissolve in aqueous solution with the evolution of hydrogen:

$$2Cu + 8CN^{-} + 2H_2O = 2Cu(CN)_4^{3-} + H_2 + 2OH^{-}$$

Simple hydrated gold(III) and gold(I) ions do not exist at an appreciable concentration in acidic aqueous solution. The instability of simple gold(I) ion in solution is a consequence of its disproportionation:

$$3Au^{+} = 2Au + Au^{3+}$$

For gold(III), the concentration of Au^{3+} is limited by the solubility of the oxide:

$$Au_2O_3(s) + 6H^{+} \rightleftarrows 2Au^{3+} + 3H_2O \qquad K \cong 10^{-6}\text{ L}^4\text{ mol}^{-4}$$

Thus in a solution with $[H^{+}] = 1\text{ mol L}^{-1}$ that is saturated with gold(III) oxide, the concentration of gold(III) ion is $\sim 10^{-3}\text{ mol L}^{-1}$. This situation is changed if the acid used to dissolve gold(III) oxide is hydrochloric acid because gold(III) forms a very stable chloride complex:

$$Au^{3+} + 4Cl^- \rightleftarrows AuCl_4^- \qquad K = 2 \times 10^{21}\ L^4\ mol^{-4}$$

The net reaction that occurs when gold(III) oxide dissolves in moderately concentrated hydrochloric acid is

$$Au_2O_3(s) + 6H^+ + 8Cl^- \rightleftarrows 2AuCl_4^- + 3H_2O \quad K = 4 \times 10^{36}\ L^{12}\ mol^{-12}$$

The stability of tetrachloroaurate(III) ion is responsible also for the contrasting solubilities of metallic gold in nitric acid and in a mixture of nitric acid plus hydrochloric acid (a solution known as *aqua regia*). Metallic gold does not dissolve in nitric acid; equilibrium in the reaction

$$Au + 6H^+ + 3NO_3^- = Au^{3+} + 3NO_2 + 3H_2O$$
$$\mathscr{E}^0 = +0.80\ V - 1.42\ V = -0.62\ V$$

is very unfavorable. In the presence of chloride ion, gold(III) is converted to tetrachloroaurate(III) ion, thereby shifting the above equilibrium. The net result can be expressed in the reaction

$$Au + 6H^+ + 3NO_3^- + 4Cl^- = AuCl_4^- + 3NO_2 + 3H_2O$$

for which the equilibrium lies to the right at high concentrations of hydrogen ion, nitrate ion, and chloride ion. Gold does dissolve in aqua regia because gold is a better reducing agent in the presence of chloride ion.

Like copper(I) and silver(I), gold in both the +1 and +3 oxidation states forms very stable complex ions with cyanide ion. Their stability allows gold to be oxidized by atmospheric oxygen in the presence of cyanide ion. The reaction

$$4Au + O_2 + 8CN^- + 2H_2O = 4Au(CN)_2^- + 4OH^-$$

is used in extracting gold from its ores. Metallic gold is produced from dicyanoaurate(I) ion either by electrolytic reduction or by reaction with zinc:

$$Zn(s) + 2Au(CN)_2^- = Zn(CN)_4^{2-} + 2Au(s)$$

THE METALLURGY OF IRON

Study of the metallurgy of iron will provide us with examples of oxidation–reduction reactions in an environment very different from the aqueous solution systems that we have been considering. Iron occurs in nature in the form of the anhydrous oxides Fe_2O_3 (hematite) and Fe_3O_4 (magnetite), the hydrated oxide $2Fe_2O_3 \cdot 3H_2O$ (limonite), and the carbonate $FeCO_3$ (siderite). (Elemental iron occurs only rarely in the Earth's crust, but the core of the Earth is mainly elemental iron.) A reducing agent is required to produce metallic iron from these ores. Coke (carbon) is used. The initial stage of producing metallic iron generally is carried out in a blast furnace, pictured in Figure 19–4. A mixture of coke, limestone ($CaCO_3$), and iron oxide is added at the top of the blast furnace and preheated air (or oxygen-

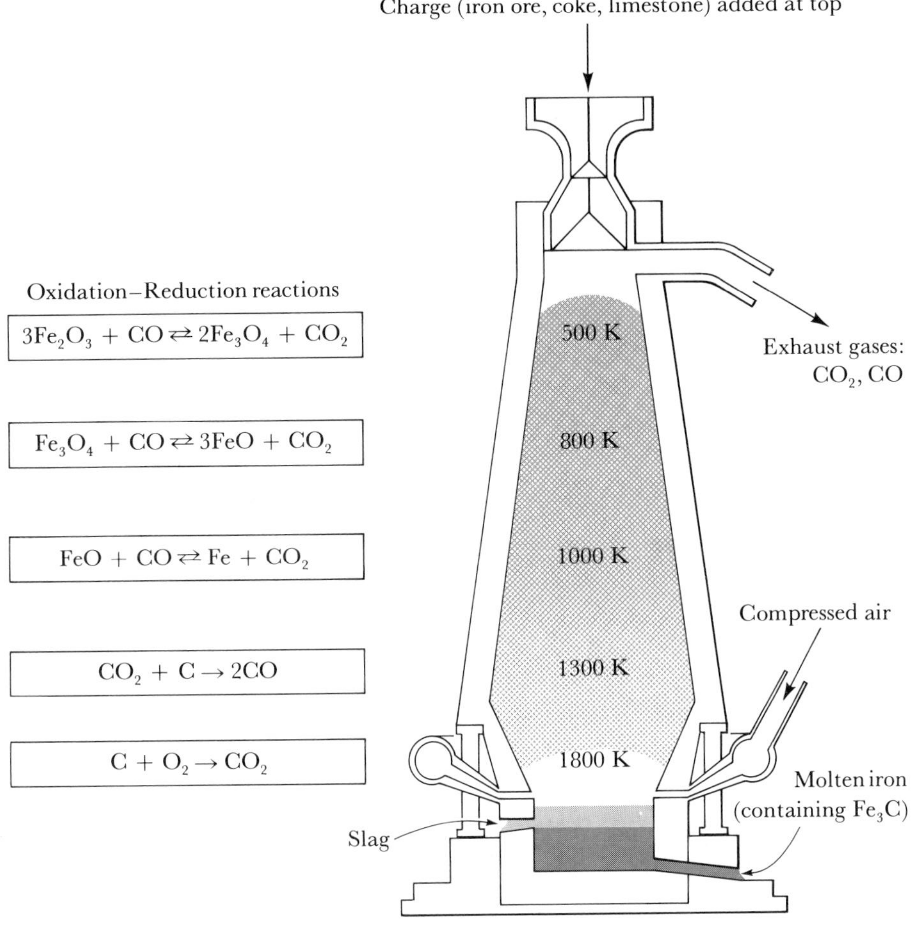

FIGURE 19–4
The blast furnace (with approximate temperatures). Near the top, the temperature is low and P_{CO_2}/P_{CO} is high. Near the bottom (after oxygen is depleted), the temperature is high and P_{CO_2}/P_{CO} is low.

enriched air) is added through the bottom. The exothermic reaction

$$C(s) + O_2(g) = CO_2(g) \qquad \Delta H^0 = -396.6 \text{ kJ mol}^{-1} \text{ (2000 K)}$$

occurs at the very bottom of the blast furnace; the evolved heat raises the temperature into the range 1800 K to 2300 K. At the top of the furnace, the temperature may be as low as ~500 K. The chemical reactions that occur in the blast furnace depend on both kinetic and equilibrium factors. Hence we cannot make the subject both simple and accurate. However, we can present a simplified version of the operation of the blast furnace by considering the equilibria involving the iron oxides and carbon oxides.

Excess oxygen exists only at the lowest part of the blast furnace; after this oxygen becomes depleted, the following endothermic reaction occurs:

$$C(s) + CO_2(g) = 2CO(g) \qquad \Delta H^0 = +167.5 \text{ kJ mol}^{-1} \text{ (1300 K)}$$

Therefore in the body of the furnace there is a mixture of gaseous carbon dioxide and carbon monoxide, which sweeps upward around the mixture of solids—iron oxides, carbon, and calcium carbonate. Carbon and carbon monoxide are reducing agents. The reactions involving reduction of iron oxides with carbon monoxide are:

1. $3Fe_2O_3 + CO(g) \rightleftarrows 2Fe_3O_4 + CO_2(g)$
2. $Fe_3O_4 + CO(g) \rightleftarrows 3FeO + CO_2(g)$
3. $FeO + CO(g) \rightleftarrows Fe + CO_2(g)$

Each of these reactions can be characterized by an equilibrium constant, which has the form

$$K = \frac{P_{CO_2}}{P_{CO}}$$

That is, the two compounds in each pair of solid phases Fe_2O_3–Fe_3O_4, Fe_3O_4–FeO, and FeO–Fe are in equilibrium at a particular value of the ratio of pressures, P_{CO_2}/P_{CO}, for each temperature. This is shown in Figure 19–5. We see in this graph that the higher oxides (Fe_2O_3 and Fe_3O_4) can exist at equilibrium only if $P_{CO_2}/P_{CO} > 10$ to 20 over the temperature range in question. The value of this ratio must be less than 0.5 to 0.1 to allow the metal to form. Low values of this ratio prevail in most regions of the blast furnace, thereby allowing the iron oxides to be reduced to the metal.

In practice, the situation is more complex than the three reduction reactions suggest. Iron melts at 1808 K and the liquid dissolves appreciable carbon. This depresses the melting point, and molten iron with dissolved carbon is the liquid that is drained from the bottom of the blast furnace. In addition, iron and carbon form a compound cementite, Fe_3C (which is

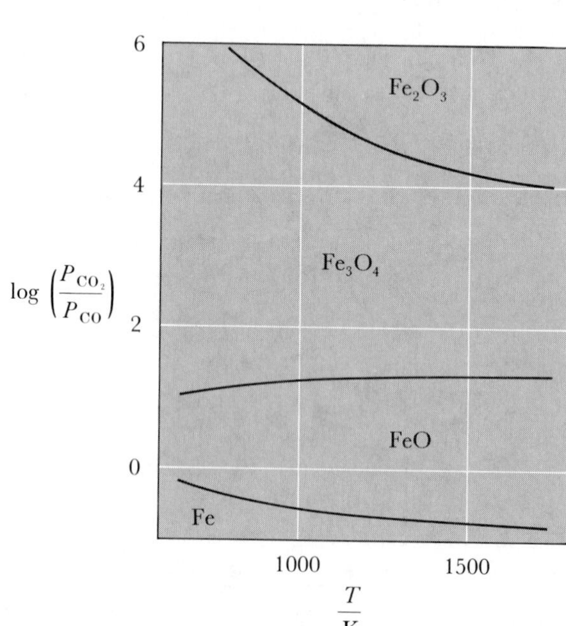

FIGURE 19–5
Ranges of P_{CO_2}/P_{CO} at which various iron oxides are stable at equilibrium. The lines separating the regions define the ratio of pressures at which two solids coexist; the lines give values of the equilibrium constants, $K = P_{CO_2}/P_{CO}$.

93.3% iron by weight). Therefore the solidified product of the blast furnace is not pure iron, but is iron contaminated with carbon and cementite; the exact nature of the solid depends on the composition of the liquid from which it formed and its rate of cooling.

The iron ore used as the raw material in the blast furnace rarely is free of silica and high-melting silicates. The addition of calcium carbonate with the iron ore and coke serves to convert these substances (represented here as SiO_2) to lower-melting calcium silicate,

$$CaCO_3 + SiO_2 = CaSiO_3 + CO_2$$

mp: 2000 K 1800 K

The calcium silicate so formed is drawn off from the bottom of the blast furnace, where it floats as a molten slag on the impure liquid iron.

Iron produced in a blast furnace contains other impurities, such as sulfur, in addition to carbon. The sulfur and carbon impurities can be removed by oxidation with a limited amount of oxygen. Pure iron has few uses; most iron is converted to varieties of steel by the addition of controlled amounts of carbon and metals such as vanadium, chromium, manganese, and nickel, each a 3*d* transition metal close to iron in the periodic table.

19–3 Coordination Chemistry of the Transition Metals

In studying equilibria in Chapter 14, we discussed the hydration of metal ions in aqueous solution and also complex-ion equilibria in which various ligands replace the water molecules coordinated to a metal ion. In that discussion, we said relatively little about structural aspects of the subject, nor did we discuss solid compounds containing coordination complexes. Now we will consider these compounds, a subject that has developed enormously since 1893, when WERNER first described the structure of coordination compounds. When Werner began his work, a compound, $CoCl_3 \cdot 6NH_3$, had been known for almost a century. The existence of this compound and related compounds presented a puzzle to chemists at that time. Both ammonia and cobalt(III) chloride were stable substances, according to existing valence rules. Therefore it was expected that these substances would not join one another to form a new compound. Additional work in the 19th century showed that by varying the preparative procedure for amminecobalt(III) chlorides, one could produce five additional different substances with varying amounts of ammonia: $CoCl_3 \cdot 5NH_3$, $CoCl_3 \cdot 4NH_3$ (two different substances with this composition), and $CoCl_3 \cdot 3NH_3$ (two different substances with this composition). These substances have different colors and different electrical conductances in aqueous solution.

Treatment of freshly prepared aqueous solutions of these compounds with silver nitrate causes the precipitation of different characteristic amounts of chloride ion per mole of compound, even though one mole of each compound contains three moles of chloride ion. Werner explained the existence of these compounds and some of their properties by proposing that cobalt was coordinated to a total of six anions (chloride ion in this example) or neutral molecules (ammonia in this example) in each of these compounds. (We now call these anions or neutral molecules *ligands.*) The total charge on anions in the compound must balance the +3 oxidation state of the cobalt(III), but the anions may or may not be part of the coordination shell of cobalt. The constituent species in each of these compounds are summarized in Table 19–3.

Werner proposed also that characteristic geometries are associated with the arrangement of atoms coordinated to the metal ion. For the compounds listed in Table 19–3, all of which involve cobalt with a coordination number of six, he proposed that the six ligands coordinated to cobalt(III) define an octahedron.

In $CoCl_3 \cdot 6NH_3$, cobalt(III) is coordinated to six ammonia molecules. This octahedral unit, shown in Figure 19–6, has a net charge of +3, and in the crystalline chloride salt there are three chloride ions (of charge −1) for each hexaamminecobalt(III) ion. The modern line formula for this substance, $[Co(NH_3)_6]Cl_3$, [named hexaamminecobalt(III) chloride] provides more information about its structure than does the older formula, $CoCl_3 \cdot 6NH_3$. In aqueous solution, this substance is a typical strong electrolyte; in dilute solution, hexaamminecobalt(III) ion and chloride ion are the predominant solute species:

$$[Co(NH_3)_6]Cl_3(s) = Co(NH_3)_6^{3+}(aq) + 3Cl^-(aq)$$

Chloride ions occupy sites in the coordination shell around cobalt(III) in the other substances listed in Table 19–3. Appropriate line formulas for the substances with five, four, or three molecules of ammonia coordinated to cobalt(III) ion are $[Co(NH_3)_5Cl]Cl_2$, $[Co(NH_3)_4Cl_2]Cl$, and $[Co(NH_3)_3Cl_3]$. The cobalt(III)-containing coordination complexes in these substances have net charge +2, +1, and 0, respectively. When these substances are dissolved in water, the chloride ion coordinated to cobalt(III)

TABLE 19–3
Coordination Compounds of Cobalt(III) Chloride and Ammonia

Composition (color)	Species present in compound
$CoCl_3 \cdot 6NH_3$ (yellow)	$Co(NH_3)_6^{3+}$ and Cl^-
$CoCl_3 \cdot 5NH_3$ (purple)	$Co(NH_3)_5Cl^{2+}$ and Cl^-
$CoCl_3 \cdot 4NH_3$ (violet)	[a]*cis*-$Co(NH_3)_4Cl_2^+$ and Cl^-
$CoCl_3 \cdot 4NH_3$ (green)	[a]*trans*-$Co(NH_3)_4Cl_2^+$ and Cl^-
$CoCl_3 \cdot 3NH_3$	[a]*fac*-$Co(NH_3)_3Cl_3$
$CoCl_3 \cdot 3NH_3$	[a]*mer*-$Co(NH_3)_3Cl_3$

[a]The meaning of these prefixes, designating isomers, is described in the legend to Figure 19–6.

$Co(NH_3)_6^{3+}$

$Co(NH_3)_5Cl^{2+}$

$Co(NH_3)_4Cl_2^+$

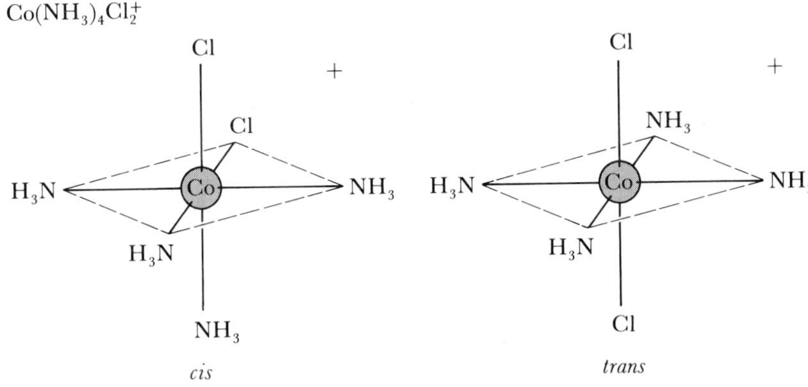

$Co(NH_3)_3Cl_3$

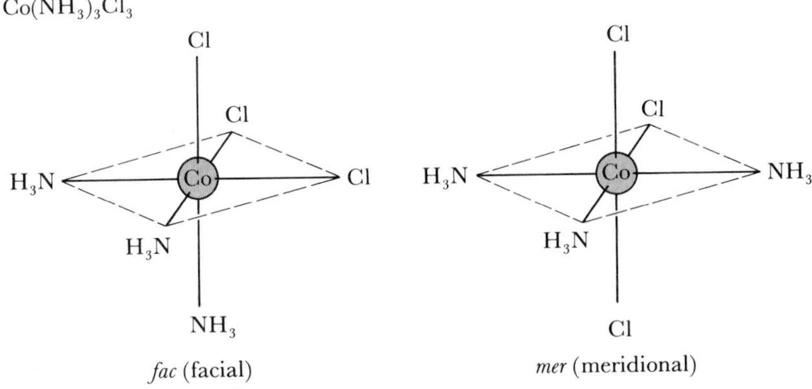

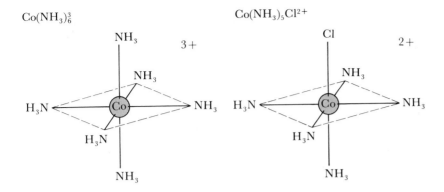

FIGURE 19–6
The structures of $Co(NH_3)_6^{3+}$, $Co(NH_3)_5Cl^{2+}$, *cis-* and *trans-*$Co(NH_3)_4Cl_2^+$, and *fac-* and *mer-*$Co(NH_3)_3Cl_3$. In each structure the six ligands define an octahedron, in the center of which is a Co atom. In cis isomers of octahedral complexes, MA_4B_2, the two B ligands are on *adjacent* corners of the octahedron. In trans isomers the two B ligands are on *opposite* corners of the octahedron. In facial isomers of octahedral complexes, MA_3B_3, the three B ligands (or the three A ligands) define a plane that is a *face* of the octahedron. In meridional isomers, the three like ligands define a *meridian* of the octahedron, that is, they define a plane that contains the central metal atom, M.

does not dissociate (except on long standing), and the observations regarding conductance and availability of chloride ion for reaction with silver(I) ion are consistent with these formulations. This is shown by chemical equations for the dissolution of these compounds:

$$[Co(NH_3)_5Cl]Cl_2(s) \rightarrow Co(NH_3)_5Cl^{2+}(aq) + 2Cl^-(aq)$$
$$[Co(NH_3)_4Cl_2]Cl(s) \rightarrow Co(NH_3)_4Cl_2^+(aq) + Cl^-(aq)$$
$$[Co(NH_3)_3Cl_3](s) \rightarrow Co(NH_3)_3Cl_3(aq)$$

The molar conductances of $[Co(NH_3)_6]Cl_3$, $[Co(NH_3)_5Cl]Cl_2$, and $[Co(NH_3)_4Cl_2]Cl$ resemble those for typical strong electrolytes of the same charge type. (Recall Figure 9–21, which showed the conductance of aqueous solutions of NaCl, $CaCl_2$, and $LaCl_3$.)

Water is a ligand that can occupy a coordination site around cobalt(III), and in dilute aqueous solution the chloride ion coordinated to cobalt(III) in pentaamminechlorocobalt(III) ion is slowly replaced by water:

$$Co(NH_3)_5Cl^{2+} + H_2O \rightleftarrows Co(NH_3)_5OH_2^{3+} + Cl^-$$

This reaction is called an *aquation* reaction. At 25°C the equilibrium constant for this reaction is 0.8 mol L^{-1}, and the half-time for the forward reaction is ~110 hours. It is this slowness of aquation that allows us to analyze the free chloride in a freshly prepared solution by precipitating silver(I) chloride without simultaneously causing the coordinated chloride ion to react. However, excess silver(I) does react directly with the complex at an appreciable rate. In experiments of this sort, which were important in the early study of the structures of these coordination compounds, it was necessary to separate the silver(I) chloride precipitate promptly from the solution, or additional precipitate would form as the coordinated chloride reacted with the excess silver(I) ion.

The slowness of ligand-exchange reactions of cobalt(III) has made their study particularly informative. If equilibrium in solution were rapidly established, a solution of ammonia, chloride ion, and cobalt(III) would contain a variety of species $Co(OH_2)_w(NH_3)_aCl_n^{3-n}$ $(w + a + n = 6)$ with a range of values of w, a, and n, depending on the concentrations of chloride ion and ammonia. The kinetic inertness of cobalt(III) allows the individual species to persist in solution for varying periods of time without being converted to other related species. This is not the case for systems in which the complex-ion equilibria are rapidly established, for example, the copper(II)–ammonia complex-ion system, depicted in Figure 14–10. Dissolving a salt containing the tetraamminecopper(II) ion, such as $[Cu(NH_3)_4](NO_3)_2$, in aqueous ammonia solution rapidly gives the equilibrium mixture of species $Cu(OH_2)_w(NH_3)_a^{2+}$ $(a + w = 4)$ for the existing concentration of ammonia.

Modern methods of x-ray diffraction allow us to determine the geometry of a coordination complex such as $Co(NH_3)_6^{3+}$, but this powerful tool was not available in Werner's time. However, he was able to establish that the characteristic geometry associated with coordination number six for cobalt(III) was octahedral, basing his theory on the number of isomers of the substances with the compositions $[Co(NH_3)_4Cl_2]Cl$ and $Co(NH_3)_3Cl_3$. Two isomers are known for each of these compositions, and this is the number of isomers for MA_4B_2 and MA_3B_3 with octahedral coordination, as is shown in Figure 19–6. Other possible geometries for coordination number six are the planar hexagon and the trigonal prism, depicted in Figure 19–7. For these coordination geometries, there are three possible isomers for the compositions MA_4B_2 and MA_3B_3. The problem of isomerism in planar hexagonal species is equivalent to the isomerism in chlorobenzenes discussed in Chapter 11. You will be asked to draw structures of the isomers for the trigonal prismatic geometry in Problem 19–23. (Examples of transition-metal complexes with coordination number six and trigonal prismatic geometry are known, but they are not common.)

Other transition-metal ions form octahedral coordination complexes resembling those of cobalt(III), a d^6 ion. Those that are kinetically inert like cobalt(III) have been amenable to the most thorough study. These are the complexes of chromium(III), a d^3 ion, and rhodium(III) and platinum(IV), d^6 ions.

Hydrated metal salts were known before the discovery of hexaamminecobalt(III) chloride, but these were not recognized as coordination compounds at that time. The series of hydrates of chromium(III) chloride—$[Cr(OH_2)_6]Cl_3$, $[Cr(OH_2)_5Cl]Cl_2$, $[Cr(OH_2)_4Cl_2]Cl$ (two different substances with this composition)—are analogous to the amminecobalt(III) chlorides already discussed. The isomers of tetraaquadichlorochromium(III) ion are

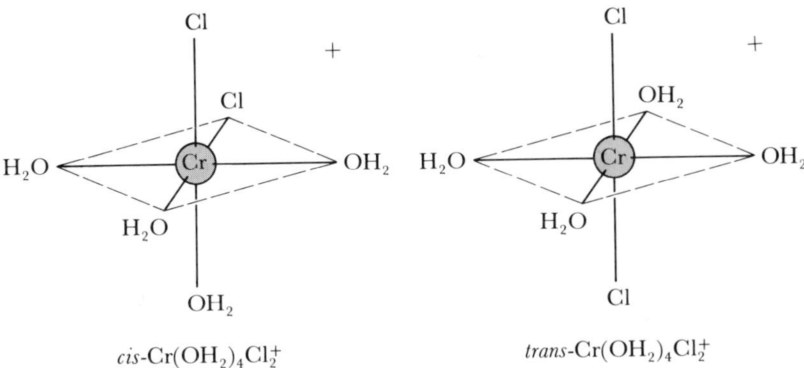

cis-$Cr(OH_2)_4Cl_2^+$ *trans*-$Cr(OH_2)_4Cl_2^+$

Both of these isomers are green. These species are not rapidly interconvertible; the slowly established equilibrium ($t_{1/2} \cong 1.4$ h at 34.8°C),

$$cis\text{-}Cr(OH_2)_4Cl_2^+ \rightleftarrows trans\text{-}Cr(OH_2)_4Cl_2^+$$

is governed by the equilibrium constant

$$K = \frac{[trans\text{-}CrCl_2^+]}{[cis\text{-}CrCl_2^+]} \cong 0.8 \qquad (34.8°C)$$

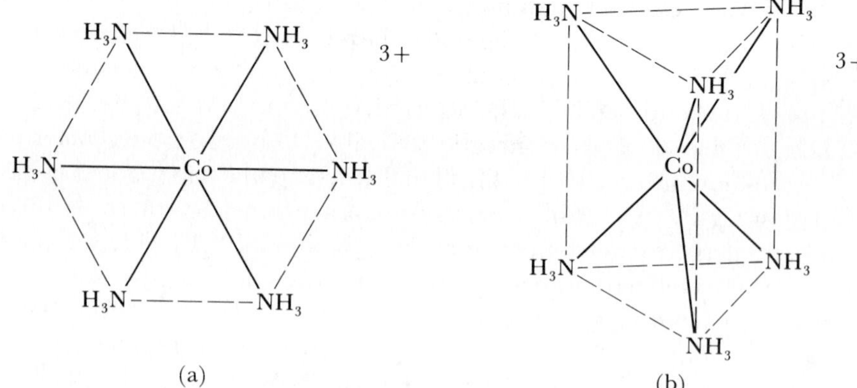

FIGURE 19–7
Other possible coordination geometries for $Co(NH_3)_6^{3+}$: (a) planar hexagon, and (b) trigonal prism.

Among the factors that determine the value of the equilibrium constant for this isomerization reaction are:

1. a statistical factor, which by itself would make $K = 0.25$. (Study the structural formulas just given and you will see that in the replacement of one water molecule in $Cr(OH_2)_5Cl^{2+}$ by one chloride ion there are four replacement sites that would give the cis isomer of $Cr(OH_2)_4Cl_2^+$ and one replacement site which would give the trans isomer.)
2. the repulsion of anionic ligands for one another, which is smaller in the trans isomer; this would make K larger than expected on the basis of the statistical factor.

The observed value, $K = 0.8$, is larger than the statistically expected value, $K = 0.25$.

Despite the fact that the cis isomer is the more stable isomer in solution, the slow evaporation of water from a solution of chromium(III) chloride in hydrochloric acid yields a crystalline salt with the composition $[Cr(OH_2)_4Cl_2]Cl \cdot 2H_2O$ containing the trans isomer of $Cr(OH_2)_4Cl_2^+$. This example illustrates two important points. One is the possibility that water in a crystalline salt may have more than one structural role. In this crystalline compound, the composition of which can be represented as $CrCl_3 \cdot 6H_2O$, two thirds of the water is coordinated to chromium(III) ion and one third of the water is hydrogen-bonded to coordinated water and to free chloride ion. The other point is the possibility that a species crystallizing from solution is not necessarily the dominant species in solution. The lower solubility of *trans*-$[Cr(OH_2)_4Cl_2]Cl \cdot 2H_2O$ is due to the greater stability of the lattice for the salt of this more symmetrical cation compared to that of the less symmetrical isomer.

Some polyatomic ligands can be bonded to a metal ion at alternative atoms. Thiocyanate ion, SCN^-, is such a ligand. The stable coordination complexes of cobalt(III) and chromium(III) with thiocyanate ion involve bonding the nitrogen end of the ligand to the metals, as in $Co(NH_3)_5NCS^{2+}$ and $Cr(NCS)_6^{3-}$. But the platinum(IV) complexes with this ligand involve bonding via the sulfur atom: $Pt(SCN)_6^{2-}$. (Complexes in which there is bonding between metal and nitrogen are called isothiocyanate complexes.)

As we learned in Chapter 14, some ligands have two or more coordination sites that are spatially arranged to allow more than one atom of the ligand to bond to the metal atom. Ethylenediamine is a ligand with two amino groups,

```
     H   H
      \ /
  H—C—C—H
     /   \
  H2N     NH2
```

each of which can be bonded to a metal ion. This bidentate ligand forms complex ions with five-atom chelate rings:

```
     H        H
      \      /
   H—C——————C—H
      |      |
    H2N      NH2
       \    /
         M
```

A transition metal having a characteristic coordination number of six can coordinate with three bidentate ligands. The structure of tris(ethylenediamine)cobalt(III) ion is given in Figure 19–8, which shows the two nonsuperimposable mirror-image isomers possible for a coordination complex with this composition. Octahedral complexes containing three bidentate chelate rings are *chiral*. Other common bidentate ligands that form chelate complexes with transition-metal ions are oxalate ion ($^-O_2CCO_2^-$), acetylacetonate ion ($CH_3COCHCOCH_3^-$), and the anions of α-amino acids, such as glycine ($H_2NCH_2CO_2^-$).

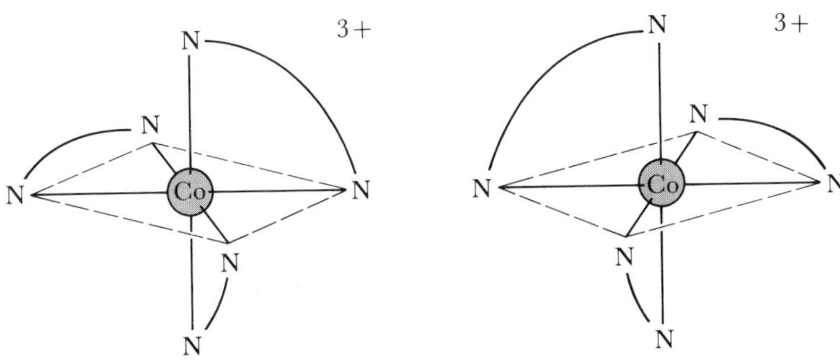

FIGURE 19–8
The chiral isomers of $Co(en)_3^{3+}$ (en represents ethylenediamine, $H_2NCH_2CH_2NH_2$, a bidentate ligand). Only the terminal nitrogen atoms of the ligand are shown in these structures. Each chelate ring contains five atoms.

Although coordination complexes of cobalt(III) and chromium(III) are kinetically inert, those of cobalt(II) and chromium(II) are not. This difference is apparent if we compare the rates at which water molecules exchange between the coordination shell of the hydrated metal ion and the bulk solvent in aqueous solution.[1] The half-times for these oxygen-exchange reactions at 25°C are:

$Cr(OH_2)_6^{2+}$	2×10^{-9} s	$Cr(OH_2)_6^{3+}$	2×10^{5} s
$Co(OH_2)_6^{2+}$	1×10^{-7} s	$Co(NH_3)_5OH_2^{3+}$	1×10^{5} s (see footnote 2)

To describe the kinetic properties of chromium(II) and cobalt(II) that contrast to the inertia of the +3 oxidation state of these metals, species of the +2 oxidation state are called *labile.* (Later, we shall learn a reason for this dramatic contrast; for the present, we simply shall use our knowledge of the contrast in a discussion of preparative chemistry for compounds of the +3 oxidation state of these metals.) Oxidation of labile cobalt(II) and chromium(II) under appropriate conditions is an efficient way to produce some coordination complexes of cobalt(III) and chromium(III).

Hexaaminecobalt(III) chloride can be prepared by the reaction of hydrogen peroxide (or oxygen) with cobalt(II) chloride in a solution containing ammonia and ammonium chloride, the cobalt(II) being present as a mixture of ammonia complexes, $Co(OH_2)_{6-a}(NH_3)_a^{2+}$; $a = 4$, 5, and 6 in solutions with moderately high concentrations of ammonia ($[NH_3] =$ 1–5 mol L^{-1}). The equation for the overall reaction [with cobalt(II) assumed to be present as $Co(OH_2)(NH_3)_5^{2+}$] is

$$2Co(OH_2)(NH_3)_5^{2+} + 6Cl^- + H_2O_2 + 2NH_4^+ = 2[Co(NH_3)_6]Cl_3(s) + 4H_2O$$

The need for ammonium chloride is indicated in the balanced equation; excess ammonium ion is needed to keep the solution from becoming too alkaline. Without ammonium ion present, the reduction of hydrogen peroxide would produce hydroxide ion, causing the formation of cobalt(III) oxide, Co_2O_3, not the desired coordination complex. (Charcoal is used as a catalyst for this reaction; without it, the principal product is $[Co(NH_3)_5(OH_2)]Cl_3$.)

It would be impossible to prepare hexaamminecobalt(III) chloride by using a salt containing hydrated cobalt(III) ion and ammonia. Most salts containing hexaaquacobalt(III) ion are not stable; hexaaquacobalt(III) ion is a very powerful oxidizing agent:

$$Co^{3+} + e^- = Co^{2+} \qquad \mathcal{E}^0 = 1.95 \text{ V}$$

which oxidizes water:

$$4Co^{3+} + 2H_2O = 4Co^{2+} + O_2 + 4H^+$$

at a measurable rate, and solutions of this species are unstable. Moreover,

[1] These exchange reactions can be studied by appropriate use of an isotopic tracer, oxygen-18.
[2] Hexaaquacobalt(III) ion, $Co(OH_2)_6^{3+}$, cannot be studied; it oxidizes water rapidly.

the addition of ammonia to a freshly prepared solution of hexaaquacobalt(III) ion causes precipitation of the oxide Co_2O_3, not formation of the ammonia complex. In the successful oxidation–reduction procedure, most of the ammonia that becomes coordinated to cobalt does so while cobalt is in the +2 oxidation state.

The coordination complexes of chromium(III) with iodide ion are very unstable with respect to aquation:

$$Cr(OH_2)_6^{3+} + I^- \rightleftarrows Cr(OH_2)_5I^{2+} + H_2O$$

$$K = \frac{[CrI^{2+}]}{[Cr^{3+}][I^-]} = 7 \times 10^{-5}\ L\ mol^{-1}\ (25°C)$$

Clearly, pentaaquaiodochromium(III) ion, $Cr(OH_2)_5I^{2+}$, cannot be prepared by mixing solutions of iodide ion and hexaaquachromium(III) ion. However, this complex forms cleanly and quantitatively in the reaction of chromium(II) ion with iodine:

$$2Cr(OH_2)_6^{2+} + I_2 = 2Cr(OH_2)_5I^{2+} + 2H_2O$$

Because iodine is an oxidizing agent of moderate strength and chromium(II) is an extremely good reducing agent, equilibrium in this reaction is very favorable ($K \cong 4 \times 10^{29}\ L\ mol^{-1}$). Presumably, the molecular iodine becomes coordinated to chromium(II):

$$Cr(OH_2)_6^{2+} + I_2 \rightleftarrows Cr(OH_2)_5I_2^{2+} + H_2O$$

and then chromium(II) is oxidized:

$$Cr(OH_2)_5I_2^{2+} \rightarrow Cr(OH_2)_5I^{2+} + I$$

The iodine atom so produced oxidizes another chromium(II) ion:

$$Cr(OH_2)_6^{2+} + I \rightarrow Cr(OH_2)_5I^{2+} + H_2O$$

The iodochromium(III) ion once formed is unstable with respect to aquation:[3]

$$Cr(OH_2)_5I^{2+} + H_2O = Cr(OH_2)_6^{3+} + I^-;$$

This reaction has a half-time of ~100 min in aqueous solution of acid ($[H^+] = 1\ mol\ L^{-1}$) at 25°C. The relative values of Gibbs free energies of the species involved in this process are shown in Figure 19–9, which illustrates both the instability of iodochromium(III) ion relative to chromium(III) ion plus iodide ion and its stability relative to chromium(II) ion plus iodine.

Many roles that transition-metal ions play in biological chemistry involve aspects of coordination chemistry. We were introduced to hemoglobin in Chapter 14, where we examined equilibrium in the successive stages of oxygen-binding. Previously in this chapter we discussed the effect

[3]It is permissible to omit the water ligands in naming a complex ion if this omission does not result in ambiguity.

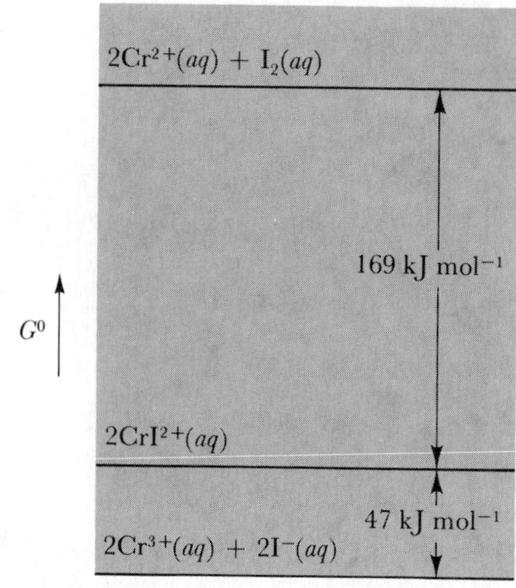

FIGURE 19–9
The relative Gibbs free energies of various combinations of 2 mol each of chromium(III) and iodide(−1) in aqueous solution, or its equivalent as chromium(II) and iodine(0). Coordinated water molecules are not shown.

coordination of hemoglobin to iron(III) and iron(II) has upon the value of $\mathscr{E}^0$ for this couple. And in Chapter 20 we shall discuss some aspects of hemoglobin as a macromolecule. Now we shall consider something of metal-ion coordination in hemoglobin. In this coordination, each of the four iron atoms is bonded to four nitrogen atoms of a planar heme group:

CH_2
CH H CH_3
H_3C— —CH=CH_2
N N
H— Fe —H
N N
H_3C— —CH_3
CH_2 CH_2
H
CH_2 CH_2
COOH COOH

A fifth coordination site is occupied by a nitrogen atom of histidine, an amino acid of the protein chain. (See Table 20–1 for the structure of histidine.) It is in this way that the iron atom and heme group in each of the four subunits of hemoglobin are bonded to the protein. In deoxyhemoglobin the iron is present as iron(II),[4] and this ion is too large to fit the opening in the planar heme group, thereby forcing the iron(II) ion to lie on one side of the heme plane; this is shown in Figure 19–10. Binding oxygen to iron raises its coordination number to six, and also allows the iron atom to become coplanar with the four nitrogen atoms of the heme.

[4]The iron(II)–heme unit is neutral, the heme unit having charge −2.

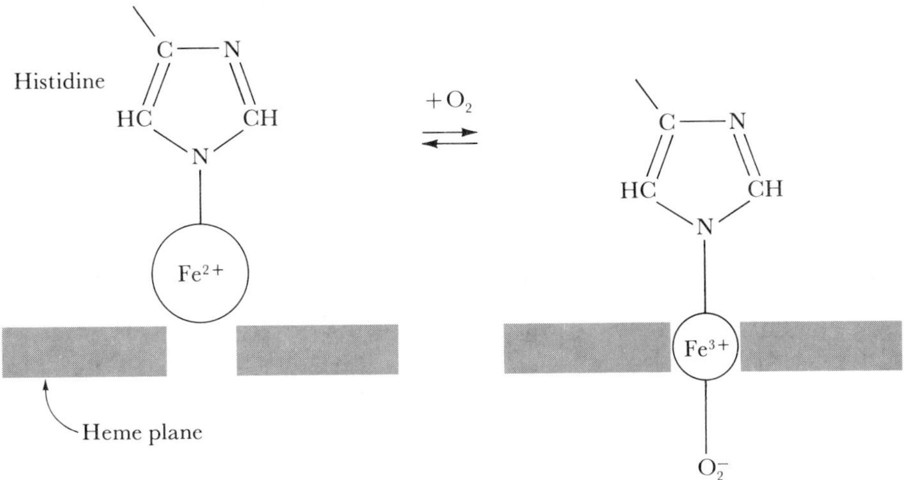

FIGURE 19–10
The binding of oxygen by iron in hemoglobin. The iron atom moves into the heme plane upon becoming oxygenated. One possible formulation of electron distribution in the iron–heme unit of oxyhemoglobin ($Fe^{3+}O_2^-$) is shown.

Details of the electron distribution in the iron–oxygen part of each subunit of oxyhemoglobin are not known with certainty. In one simple view, the unit is formulated as containing iron(III) and superoxide ion, $Fe^{3+}O_2^-$. In the framework of this model, shown in Figure 19–10, the binding and release of oxygen by hemoglobin is considered to be the operation of the reversible reaction:

$$Fe^{2+} + O_2 \rightleftarrows Fe^{3+}O_2^-$$

This model is consistent with the ability of the iron atom to fit in the plane of the heme ring in oxyhemoglobin,[5] but this fact does not settle the question because there are other data that support the other extreme formulation in which iron(II) is coordinated to oxygen, $Fe^{2+}O_2$. (We have seen other instances in which it is very difficult to establish the electron distribution in molecules or ions; this simply is an instance that has attracted much attention.) In the reaction of oxygen with hemoglobin, a particular oxygen molecule interacts with only a single iron atom, thereby precluding formation of a dimeric peroxide structure analogous to that of the peroxoiron(III) dimer discussed in Chapter 17. The cooperative nature of oxygen binding in hemoglobin (the enhanced binding of oxygen in the second, third, and fourth subunits once oxygen has been bound in the first subunit, discussed in Chapter 14) is believed to be caused by the alteration in structure of the entire protein that accompanies the binding of the first molecule of oxygen. With binding of the first molecule of oxygen to an iron(II) atom, that iron atom moves into the plane of its heme ring. This changes in small but significant ways the structures of the other three subunits, enhancing their ability to bind oxygen.

[5]The iron–oxygen distances in hydrated iron(II) and iron(III) salts are Fe^{2+}—O distance, 2.13 Å, and Fe^{3+}—O distance, 1.99 Å.

It is possible to oxidize the iron(II) of hemoglobin to iron(III) with other reagents, and one can study the coordination chemistry of this iron(III) species (called methemoglobin) with a number of ligands, but this species no longer has the capacity to bind oxygen.

For most 3*d* transition-metal ions, the predominant coordination number is six. However, we have just seen in hemoglobin an example of iron(II) with coordination number 5, and other examples of coordination to fewer than six ligands are known. A particular metal ion may have different coordination numbers, depending on the ligand to which it is coordinated. In dilute aqueous solution containing either of the weakly coordinating ligands, nitrate ion or perchlorate ion, cobalt(II) exists as octahedral hexaaquacobalt(II) ion, $Co(OH_2)_6^{2+}$. But in 10*M* HCl, cobalt(II) is present as tetrachlorocobaltate(II) ion, $CoCl_4^{2-}$, a species with tetrahedral geometry. The reaction

$$Co(OH_2)_6^{2+} + 4Cl^- = CoCl_4^{2-} + 6H_2O$$

is accompanied by a dramatic color change. The octahedral hexaaquacobalt(II) ion is pale pink, and the tetrahedral tetrachlorocobaltate(II) ion is intensely blue. In this system at intermediate concentrations of chloride ion, there are octahedral species containing a smaller number of coordinated chloride ions, such as $Co(OH_2)_5Cl^{2+}$. But in another similar system, that involving nickel(II) ion and cyanide ion, aquacyanonickel(II) species do not exist at appreciable concentrations. The reaction in which nickel(II) changes its coordination number,

$$Ni(OH_2)_6^{2+} + 4CN^- = Ni(CN)_4^{2-} + 6H_2O$$

has an equilibrium constant of 2×10^{40} L^4 mol^{-4}.

19–4 Bonding in Transition-Metal Coordination Compounds

You have learned some facts which indicate that strong bonding must be involved in coordination complexes. If it were not for the strong bonding between cobalt(III) and ammonia in $Co(NH_3)_6^{3+}$, the compound $[Co(NH_3)_6]Cl_3$ would decompose:

$$[Co(NH_3)_6]Cl_3(s) = CoCl_3(s) + 6NH_3(g)$$

because the lattice energy of $CoCl_3$ (containing small Co^{3+}) certainly is larger than that for $[Co(NH_3)_6]Cl_3$ (containing large $Co(NH_3)_6^{3+}$), and the entropy change in this gas-producing reaction certainly is positive. (Estimates of the average Co–N bond energy in hexaamminecobalt(III) ion are in the range $(3 \pm 0.5) \times 10^2$ kJ mol^{-1}.) But what is the electronic structure of these bonds between polar basic molecules and cationic metal species with unfilled orbitals of the $n = 3$ and $n = 4$ shells? According to one view, unshared electron pairs on ammonia molecules are donated to $3d$, $4s$, and $4p$ orbitals of cobalt(III):

	$3d$	$4s$	$4p$
$Co^{3+}(g)$	(⇅)(↑)(↑)(↑)(↑)	( )	( )( )( )
$Co(NH_3)_6^{3+}$	(⇅)(⇅)(⇅)(ˣₓ)(ˣₓ)	(ˣₓ)	(ˣₓ)(ˣₓ)(ˣₓ)

Electron pairs donated by six molecules of ammonia

The geometry of six bonds based upon hybridization of two $3d$, one $4s$, and three $4p$ orbitals is octahedral, so this simple theory is consistent with the coordination geometry of $Co(NH_3)_6^{3+}$. Alternately, there is electrostatic attraction between a cation and either an anion or a polar neutral molecule, and some features of coordination chemistry can be explained simply in terms of electrostatic attraction. The hydration of alkali-metal ions and alkaline-earth metal ions in aqueous solution requires no more elaborate a theory than this.

However, many aspects of the coordination chemistry of transition-metal ions are not explained by simple electrostatic interaction or by Lewis acid–Lewis base interaction. An adequate theory must explain these observations:

1. The extreme inertness (i.e., low reactivity in a kinetic sense) of octahedral coordination complexes of chromium(III) (a d^3 ion) and cobalt(III) (a d^6 ion), in contrast to the lability of many octahedral coordination complexes of iron(III), (a d^5 ion).
2. The different magnetic properties of some species containing the d^6 metal ions cobalt(III) and iron(II): $Co(NH_3)_6^{3+}$ and $Fe(CN)_6^{4-}$ are diamagnetic, but other species of the same metal ions, CoF_6^{3-} and $Fe(OH_2)_6^{2+}$, are paramagnetic.
3. The trends in the lattice energies of chlorides of $3d$ transition metals in the $+2$ oxidation state, shown in Figure 19–11. The lattice energies for the chlorides of metal ions with d^2, d^3, d^4, d^6, d^7, d^8, and d^9 electronic configurations are greater than expected on the basis of a smooth curve drawn through the values for the chlorides of calcium(II) (a d^0 ion), manganese(II) (a d^5 ion), and zinc(II) (a d^{10} ion).

These observations are explained very simply by a theory called the *crystal-field theory* (or by extensions of that theory). The crystal-field theory was developed by the physicists BETHE and VAN VLECK, in 1929 and 1932 respectively. Originally, the theory was applied by physicists to the magnetic properties of salts of transition-metal ions. It was not applied to more-chemical problems until twenty years later; now an additional thirty years have passed, and this theory and closely related theories are centrally important in explaining all aspects of transition-metal chemistry.

The crystal-field theory focuses on the energies of the five $3d$ orbitals of the transition-metal ion in the presence of the ligands that surround it. For a free transition-metal ion in the gaseous state, each of the five $3d$

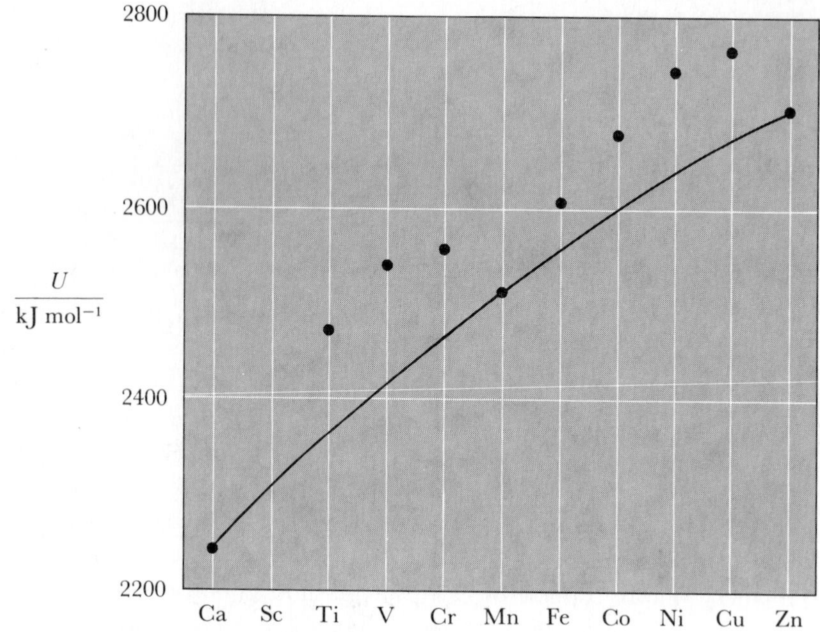

FIGURE 19–11
The lattice energy of transition-metal chlorides, MCl_2. In each of these compounds, M^{2+} is in an octahedral site in a closest-packed lattice of chloride ions. ($ScCl_2$ does not exist.)

orbitals has the same energy. Each of these orbitals also would have the same energy in the presence of a group of ligands *if the effect of the ligands were that of a spherically symmetrical field*, but this energy would be different than it is in the absence of ligands. However, the effect of some particular number of ligands (e.g., 4, 5, or 6) arranged in a particular way about the metal ion is not equivalent to that of a spherically symmetrical field. How the energies of the 3*d* orbitals of the metal ion are influenced by the ligands depends both upon their identity and arrangement. To understand this, we must consider the angular distributions of the five 3*d* orbitals, which were shown in Figure 18–4. The angular probability distribution of an electron in an orbital designated d_{z^2} or $d_{x^2-y^2}$ is highest along the orthogonal x, y, and z axes. The probability distribution of an electron in an orbital designated d_{xy}, d_{xz}, or d_{yz} is highest in regions between the orthogonal x, y, and z axes. The energies of electrons in these two sets of orbitals are influenced differently in the field created by a group of six anionic or polar ligands located on the x, y, and z axes. This is reasonable because an electron in an orbital directed toward an anionic ligand or toward the negative end of a polar ligand will be repelled by this negative charge. The energy of an electron in such an orbital is more positive (i.e., the system is less stable) than the energy of an electron in an orbital that is not directed toward a ligand. For a metal ion surrounded octahedrally by six identical ligands, the energies of the orbitals are shown in Figure 19–12; the three orbitals d_{xy}, d_{xz}, and d_{yz} have the same energy, and the two orbitals d_{z^2} and $d_{x^2-y^2}$ have a higher energy. The difference between these energies (called Δ_o) depends both upon the ligand and the metal ion.

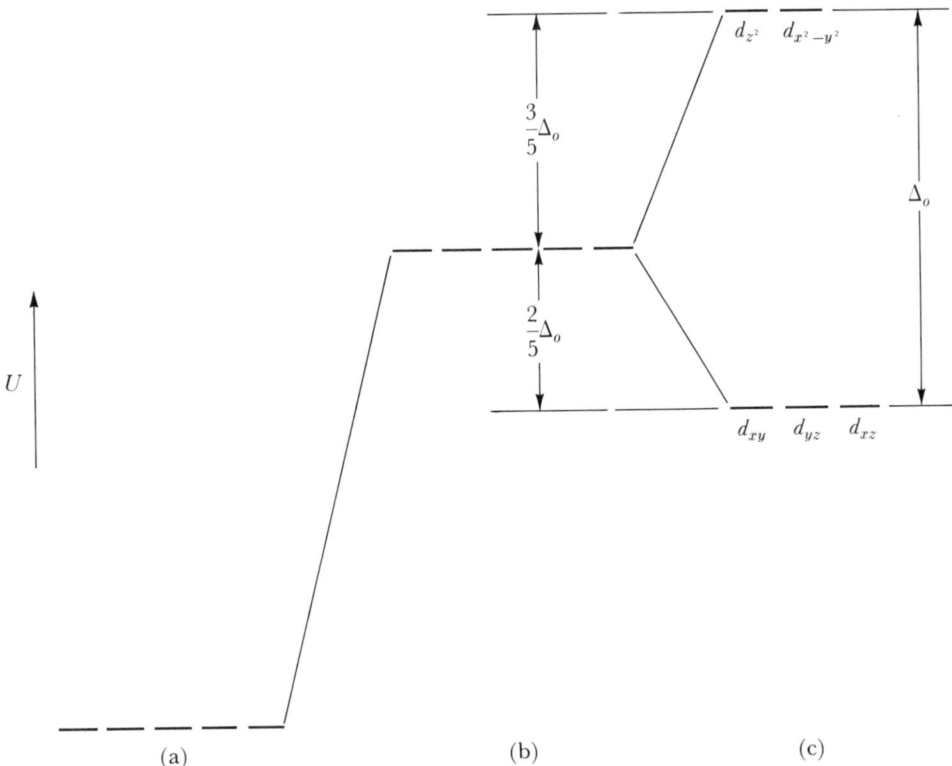

FIGURE 19–12
The relative energies of the five 3*d* orbitals of a transition-metal atom: (a) in the gaseous state; (b) in a spherically symmetrical field equivalent to that due to six identical ligands; (c) in the field of six identical ligands located along the orthogonal axes. These six ligands define an octahedron around the metal ion.

The net effect of the field of the ligands (called the crystal field in the earliest studies because crystalline compounds were being considered) depends on the occupancy of the *d* orbitals by the electrons of the metal ion and on the magnitude of Δ_o. Two tendencies must be taken into account in an explanation of the electronic structure of transition-metal complexes:

1. electrons tend to occupy the most stable orbitals available, and
2. electrons tend to have parallel spins and occupy different orbitals if possible.

These tendencies are not in opposition if the metal ion has one to three 3*d* electrons or eight or nine 3*d* electrons, but the tendencies are in opposition if the metal ion has four to seven 3*d* electrons. This generalization can be illustrated with some examples. Figure 19–13 shows the occupancy of the 3*d* orbitals by the three electrons of vanadium(II) ion or chromium(III) ion. Each electron is in a different one of the more-stable group of 3*d* orbitals. The spins of these three electrons are parallel. With each of the three electrons in a relatively stable 3*d* orbital, this configuration is particularly stable.

This stability of a d^3 ion in an octahedral field also accounts for the extreme inertness of chromium(III) ion. A reaction in which one ligand

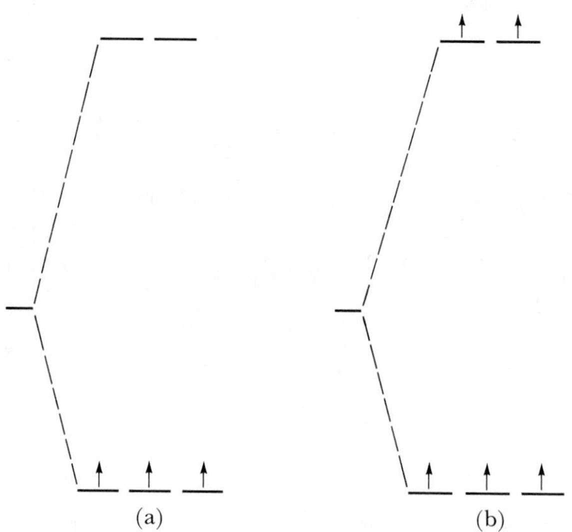

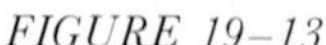

FIGURE 19–13
The occupancy of orbitals in d^3 species, V(II) and Cr(III), and high-spin d^5 species, Mn(II) and Fe(III). (a) The d^3 species; each electron occupies one of stabilized orbitals; there is a net stabilization due to the crystal field. (b) The high-spin d^5 species; each electron occupies a different orbital; there is no net stabilization due to the crystal field.

displaces another, for example,

$$Cr(OH_2)_6^{3+} + SCN^- = Cr(OH_2)_5NCS^{2+} + H_2O$$

involves an activated complex that does not have regular octahedral symmetry, and the crystal-field stabilization for chromium(III) in the activated complex is not as large as that for the regular octahedral configuration of the reactant $Cr(OH_2)_6^{3+}$ (pictured in Figure 19–13a). This high energy of the activated complex contributes to the low rate of this reaction of chromium(III), which is $\sim 10^{-8}$ times the rate of the corresponding reaction of iron(III), a d^5 ion:

$$Fe(OH_2)_6^{3+} + SCN^- = Fe(OH_2)_5NCS^{2+} + H_2O$$

The electronic configuration of iron(III) in the aquairon(III) ion and the thiocyanate complex of iron(III) are shown in Figure 19–13b. The magnitude of Δ_o for this metal ion with either water or thiocyanate ion as ligands is relatively small, and the tendency for electrons to be unpaired wins over their tendency to occupy the most stable orbitals. A species of this type is called a *high-spin* species, to indicate that as many as possible of the electrons have parallel spins. Because each of the relatively stable and each of the relatively unstable orbitals is occupied by one electron, this configuration is not given extra stability by the crystal field. This explains the relatively larger rates of complex-ion formation reactions of iron(III), compared to analogous reactions of chromium(III), and also the normal lattice energy of manganese(II) chloride (Figure 19–11).

If high-spin iron(III) ions in the reactant and product species have no net crystal-field stabilization, the distortion of the field in going to the activated complex does not raise its energy. A metal ion cannot lose crystal-field stabilization it does not have. Similarly, the lattice energy of manganese(II) chloride has no contribution from the type of crystal-field stabilization being considered here because in this compound manganese(II) is a high-spin species. Thus lattice energies of the three compounds in which there is zero crystal-field stabilization, $CaCl_2$ (involving Ca^{2+}, a d^0 species), $MnCl_2$ (involving Mn^{2+}, a high-spin d^5 species), and $ZnCl_2$ (involving Zn^{2+}, a d^{10} species) define a smooth curve in a plot of lattice energy versus atomic number. The lattice energies of other MCl_2 compounds, for instance VCl_2 which involves a d^3 cation, are larger than predicted by this smooth curve because the field of the six chloride ions stabilize these compounds.

The remaining examples to be considered involve the d^6 ions iron(II) and cobalt(III). Depending on the magnitude of Δ_o, metal ions with this number of electrons can have high-spin or low-spin configurations, which are shown in Figures 19–14a and 19–14b. Hexaaquairon(II) ion, like most iron(II) species, and hexafluorocobaltate(III) ion, unlike most cobalt(III) species, are high-spin species; hexacyanoferrate(II) ion (commonly called ferrocyanide ion), unlike most iron(II) species, and hexaamminecobalt(III) ion, like most cobalt(III) species, are low-spin species. The magnitude of Δ_o is smaller for ions of charge +2 than for ions of charge +3; this is responsible for the fact that most iron(II) species are high-spin and most cobalt(III) species are low-spin. The magnetic properties of the iron(II) and cobalt(III) mentioned previously are consistent with the orbital occupancy shown in Figure 19–14. Low-spin d^6 species have zero unpaired electrons and therefore are diamagnetic. High-spin d^6 species, with four unpaired electrons, have a net magnetic moment and therefore are attracted into a magnetic field.

The magnitude of Δ_o depends also on the identity of the ligand, and the configurations of the iron(II) species and cobalt(III) species being considered indicate that the magnitude of Δ_o is greater for CN^- than for H_2O, and is greater for NH_3 than for F^-. Detailed studies of many transition-metal-ion complexes produce a more complete ordering of the effect of various ligands on the magnitude of Δ_o; this order of values of Δ_o is:

$$CN^- > NO_2^- > en > NH_3 > NCS^- > H_2O > F^- > Cl^- > Br^- > I^-$$

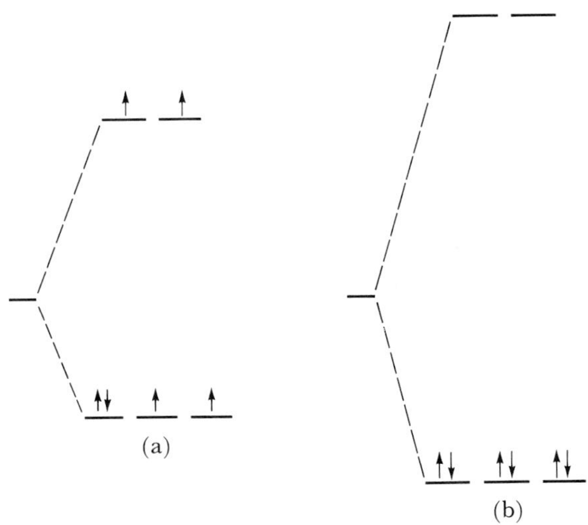

FIGURE 19–14
The occupancy of orbitals in d^6 species, Fe(II) and Co(III). (a) Small Δ_o, (ligands H_2O and F^-) with electrons occupying as many orbitals as possible; high-spin species, $Fe(OH_2)_6^{2+}$ and CoF_6^{3-}. (b) Large Δ_0, (ligands CN^- and NH_3) with electrons occupying the most stable orbitals; low-spin species with maximum stabilization due to the crystal field, $Fe(CN)_6^{4-}$ and $Co(NH_3)_6^{3+}$.

This series is commonly called the *spectrochemical series* because the value of Δ_o can be derived from observations of spectra. This relationship of spectra of transition-metal-ion species and the value of Δ_o will be discussed in Chapter 21.

With low-spin d^6 species [$Co(NH_3)_6^{3+}$ and $Fe(CN)_6^{4-}$ in the examples], all of the d electrons are paired in the relatively stable set of orbitals. This makes species with this configuration extremely inert, even more inert than the d^3 species already discussed. Although equilibrium in the reaction

$$Co(NH_3)_6^{3+} + H_3O^+ = Co(NH_3)_5OH_2^{3+} + NH_4^+$$

lies far to the right in acidic solution, hexaaminecobalt(III) ion will persist in hot hydrochloric acid solution for long periods of time (hours to days).

Although the simple crystal-field theory explains many observations in the chemistry and physics of transition-metal complexes, it does not provide for covalent character in the bonding between the central metal ion and the ligands. However, this is provided in other, more elaborate molecular orbital theories, and many of the conclusions derived from the more complete theories are the same as those that we have drawn here. The relationship between the geometrical arrangement of the ligands about the metal ion and the angular distribution of the 3d orbitals of the metal ion is of key importance in each of these theories.

This discussion of the crystal-field theory has dealt with metal ions octahedrally coordinated to six identical ligands. If the six ligands are not identical, the pattern of energies of orbitals is altered. Even more drastic alteration results if the coordination number is not six and the complex is not octahedral. If the coordination number is four and the geometry is tetrahedral, the d_{z^2} and $d_{x^2-y^2}$ orbitals are relatively more stable than the d_{xy}, d_{yz}, d_{xz} orbitals. The ideas that we have developed for octahedral coordination can be applied to other geometries, but we will not pursue this.

19–5 Metal Carbonyls

The formation of most compounds of a metal from the free element involves oxidation of the metal, for example the exothermic reaction of metallic nickel with chlorine gas to form nickel dichloride:

$$Ni(s) + Cl_2(g) = NiCl_2(s) \qquad \Delta H^0 = -316 \text{ kJ mol}^{-1}$$

The properties of this compound show it to be ionic nickel(II) chloride; its structure is a lattice of chloride ions in a cubic closest-packed arrangement, with the nickel(II) ions occupying one half of the octahedral sites. In the formation reaction just given, nickel is oxidized to the +2 oxidation state, and chlorine is reduced to the −1 oxidation state.

Some compounds of transition metals, however, do not follow this simple pattern. Metallic nickel reacts exothermically with carbon monoxide to form nickel tetracarbonyl, an extremely toxic compound:

$$Ni(s) + 4CO(g) = Ni(CO)_4(g) \qquad \Delta H^0 = -163 \text{ kJ mol}^{-1}$$

The properties of this compound (e.g., mp −25°C, bp 43°C) and the properties of carbon monoxide [$EA(CO) < 0 \text{ kJ mol}^{-1}$] make it reasonable to view nickel tetracarbonyl as a compound of nickel(0). The reaction that forms it does not involve oxidation and reduction. In the tetracarbonyl molecule nickel is bonded to carbon, with the four carbon atoms tetrahedrally arranged around the nickel atom. The angle ∠NiCO is 180°, and the bonding is described in part by assuming that each carbon monoxide molecule, **:C:::O:**, shares the lone pair of electrons on the carbon with nickel by using the unfilled 4*s* and 4*p* orbitals of nickel (0) in an excited state:

	3*d*	4*s*	4*p*
Ground state of Ni(0):			
Ni(Ar core)	(⇅)(⇅)(⇅)(↑)(↑)	(⇅)	( )( )( )
Excited state of Ni(0), a d^{10} species:			
Ni(Ar core)	(⇅)(⇅)(⇅)(⇅)(⇅)	( )	( )( )( )
Ni(0) coordinated to 4 CO molecules:			
Ni(Ar core)	(⇅)(⇅)(⇅)(⇅)(⇅)	(ˣₓ)	(ˣₓ)(ˣₓ)(ˣₓ)

In this last diagram, the electrons donated by the carbon monoxide molecule are designated by ×, and the electrons belonging originally to the nickel atom are designated by arrows showing electron spin.

This description of the bonding explains the tetrahedral shape of the molecule, but it is not completely satisfactory because it gives the nickel atom a formal charge of −4, an unusual formal charge for a metal atom. This situation can be alleviated by assuming some double-bond formation between the nickel and carbon atoms, in which the nickel atom has shared eight of its 3*d* electrons with carbon:

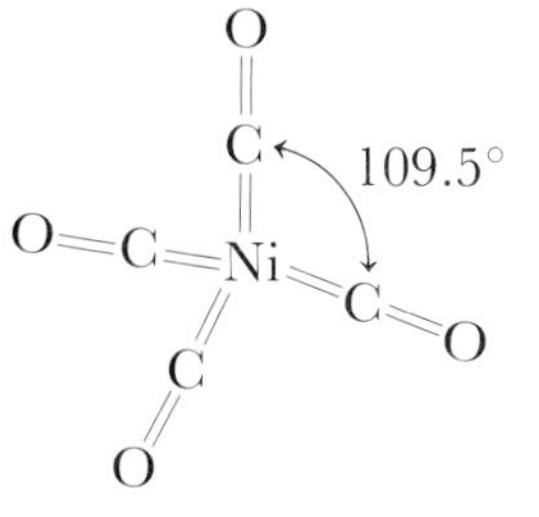

O=C=Ni(=C=O)₃ (tetrahedral; C–Ni–C angle 109.5°)

In this structure with nickel–carbon double bonds, each atom has a formal charge of zero. Various types of measurements disclose that the carbon–oxygen bonds in nickel tetracarbonyl (and other metal carbonyls) are weaker than the carbon–oxygen bonds in free carbon monoxide. This fact is consistent with the structure's having some double-bond character in the nickel–carbon bond because such structures also show the carbon–oxygen bond as a double bond (instead of a triple bond as in molecular carbon monoxide).

Nickel tetracarbonyl forms readily when carbon monoxide is passed over metallic nickel, pure or impure, at temperatures only slightly above ordinary temperatures. Since the boiling point of nickel tetracarbonyl is 43°C, the reaction carried out at 50°C is a useful method of purifying nickel. At this temperature, a stream of carbon monoxide gas will transport nickel (as nickel tetracarbonyl) away from impurities that do not form volatile carbonyls. Raising the temperature of the gaseous mixture of carbon monoxide and nickel tetracarbonyl to 230°C causes the endothermic decomposition reaction,

$$Ni(CO)_4(g) = Ni(s) + 4CO(g)$$

to occur, depositing very pure metallic nickel.

In the diagram showing orbital occupancy for nickel tetracarbonyl, the five 3*d*, the 4*s*, and the three 4*p* orbitals are fully occupied by eighteen electrons. This same feature of electronic structure is found in other 3*d* transition-metal carbonyls, for example, $Cr(CO)_6$ and $Fe(CO)_5$. The sum of the number of electrons in the 3*d* and 4*s* orbitals of the neutral metal atom plus two electrons for each coordinated carbon monoxide molecule gives 18 in each case:

Compound	Electrons in neutral metal atom		Electrons from CO		Total
$Cr(CO)_6$	6	+	(6 × 2)	=	18
$Fe(CO)_5$	8	+	(5 × 2)	=	18
$Ni(CO)_4$	10	+	(4 × 2)	=	18

Each of these metal-carbonyl species is electrically neutral, but metals with an odd atomic number cannot have the stability associated with eighteen electrons in the 3*d*, 4*s*, and 4*p* orbitals in a neutral, monomeric carbonyl. Such a metal may form a carbonyl that does not have an eighteen-electron structure, such as $V(CO)_6$. Or it may form ionic carbonyl species that have eighteen-electron structures, such as $V(CO)_6^-$, in $NaV(CO)_6$, and $Mn(CO)_6^+$, in $Mn(CO)_6Cl$ [both $V(CO)_6^-$ and $Mn(CO)_6^+$ are isoelectronic with $Cr(CO)_6$]. Or it may form dinuclear (and polynuclear) carbonyl species involving metal–metal bonds, $(OC)_5MnMn(CO)_5$. The entire field of metal-carbonyl chemistry is vast, and much more complex than these examples suggest.

Some metal carbonyls or compounds of metals coordinated with both carbon monoxide and another ligand act as catalysts in many commercially useful reactions. One is the hydroformylation reaction in which an alkene reacts with hydrogen and carbon monoxide, intially to give an aldehyde and finally to give the corresponding alcohol:

$$RCH{=}CH_2 + H_2 + CO = RCH_2CH_2CHO$$
$$RCH_2CH_2CHO + H_2 = RCH_2CH_2CH_2OH$$

This reaction, catalyzed by hydridocobalttetracarbonyl, is used on a large scale to produce saturated alcohols containing seven to nine carbon atoms. In hydridocobalttetracarbonyl, $HCo(CO)_4$, a hydrogen atom is covalently bonded to the metal. In this species, the cobalt atom has an eighteen-electron structure. In aqueous solution, this substance is a strong acid:

$$HCo(CO)_4 = H^+ + Co(CO)_4^- \qquad K_a > 1 \text{ mol L}^{-1}$$

and the conjugate base, $Co(CO)_4^-$, is isoelectronic with $Ni(CO)_4$.

A ligand isoelectronic with carbon monoxide and cyanide ion is NO^+, and there are transition-metal complexes of this ligand. If a solution of iron(II) ion is treated with nitric oxide, a brown complex ion, $FeNO^+$, is formed:

$$Fe^{2+} + NO \rightleftarrows FeNO^{2+}$$

This complex ion also can be prepared if nitrate ion is reduced with excess iron(II) in the presence of a very high concentration of acid:

$$4Fe^{2+} + NO_3^- + 4H^+ = FeNO^{2+} + 3Fe^{3+} + 2H_2O$$

(This reaction is the basis for the "brown ring" analytical test, once commonly used to detect nitrate ion.) Although this complex ion, $FeNO^{2+}$, was once considered to contain iron(II) and NO, the magnetic behavior of the complex indicates that it contains low-spin iron(I) (a d^7 species) and NO^+ (a diamagnetic ligand).

Biographical Notes

HANS A. BETHE (1906–) was born in Alsace-Lorraine and educated in Germany. After teaching at German universities for five years following receipt of his Ph.D. degree, he moved to England in 1933 and the United States in 1935. Here he joined the faculty of Cornell University. His many contributions to physics include the theory of energy production in the sun and stars, for which he received the Nobel Prize in Physics in 1967.

JOHN H. VAN VLECK (1899–), an American physicist, was educated at the University of Wisconsin and Harvard University. He served on faculties of the Universities of Minnesota and Wisconsin. He has been Professor of Physics at Harvard University since 1934. His research has dealt largely with the electronic structure of atoms and the magnetic properties of matter. His work was recognized by the Nobel Prize in Physics in 1977.

ALFRED WERNER (1866–1919) was born in Alsace before this region became part of Germany in 1871. He was educated in Zurich, where at the age of 26, as a lecturer in the university, he proposed his theory of the structure of coordination compounds. He received the Nobel Prize in Chemistry in 1913 for this work. Although Werner is best known for his study of coordination compounds, he did important work in other areas. In a publication in 1909 he recognized hydrogen bonding in the hydrogendifluoride ion, FHF^-.

Problems and Questions

19–1 Write balanced chemical equations for the reactions that occur when each of the following metal oxides is dissolved in dilute perchloric acid ($0.10\ mol\ L^{-1}$): TiO_2, FeO, V_2O_5, Cu_2O.

19–2 What oxidation state of each element, iron, cobalt, and nickel, has a d^7 electronic configuration? Which of these oxidation states is stable as a simple hydrated ion in aqueous solution?

19–3 Manganese dioxide dissolves in nitrous acid (a weak acid, $K_a = 4.5 \times 10^{-4}\ mol\ L^{-1}$) but does not dissolve in nitric acid (a strong acid, $K_a = 25\ mol\ L^{-1}$). Explain.

19–4 An acidic solution of chromium in one of its stable oxidation states is made alkaline with excess sodium hydroxide. A precipitate forms, which then redissolves. Which is the oxidation state of chromium, and what reactions account for the observations?

19–5 What must the concentration of hydrogen ion be in a solution of iron(III) to assure that 99.0% of the iron(III) is present as the simple hydrated +3 ion, $Fe(OH_2)_6^{3+}$?

19–6 In a solution with pH = 3.00, what is the total concentration of chromium(VI) if the concentration of hydrogenchromate ion, $HOCrO_3^-$, is $4.5 \times 10^{-3}\ mol\ L^{-1}$?

19–7 You wish to remove carbon dioxide from a stream of gas by passing the gas over a metal oxide. Only VO and V_2O_3 are available. Which is the better choice?

19–8 Write balanced chemical equations for the conversion of each of the polymeric vanadium(V) species, $V_2O_7^{4-}$, $V_4O_{12}^{4-}$, and $V_{10}O_{28}^{6-}$ to VO_2^+ and to VO_4^{3-}.

19–9 What transition-metal species are isoelectronic with $V_2O_7^{4-}$?

19–10 In the style of Figure 16–10, plot the value of $\mathscr{E}^{0\prime}$ for the iron(VI)–iron(III) couple as a function of the concentration of hydrogen ion from $[H^+] = 1.00\ mol\ L^{-1}$ to $[H^+] = 1.00 \times 10^{-14}\ mol\ L^{-1}$

19–11 We learned in this chapter that unstable copper(I) ion can be prepared by the reaction $Cu^{2+} + Cr^{2+} = Cu^+ + Cr^{3+}$. If you wish to achieve as high as possible a concentration of copper(I), which procedure should you use: (1) add Cu^{2+} to Cr^{2+}, or (2) add Cr^{2+} to Cu^{2+}? Why does the order of mixing reagents influence the results in this system?

19–12 Balance the following equations for oxidation–reduction reactions of chromium(VI) with organic reagents:

$$HOCrO_3^- + CH_3OH = CO_2 + Cr^{3+}$$
$$Cr_2O_7^{2-} + CH_3CH_2OH = CH_3CHO + Cr^{3+}$$

19–13 A solution of iron(VI) is prepared by dissolving K_2FeO_4 in aqueous acid. A gas evolves. What reaction is occurring?

19–14 Cerium(IV) is a one-electron oxidizing agent capable of oxidizing chromium(III) to chromium(VI) in acidic solutions; the net chemical change is

$$3Ce^{4+} + Cr^{3+} + 4H_2O = HOCrO_3^- + 3Ce^{3+} + 7H^+$$

Suggest a possible mechanism for this reaction. What type of study would shed light on the validity of your proposal?

19–15 An acidic solution of chromium(III) is treated with excess potassium permanganate. What reaction occurs, and what is the equilibrium constant for this reaction?

19–16 Draw the structural formulas of the two esters, isopropyl hydrogenchromate and isopropyl acetate. What common structural feature does each of these esters have?

19–17 What two half-reactions make up the reaction of chromium(VI) with isopropyl alcohol, discussed in this chapter?

19–18 From the relative values of $\mathscr{E}^0$ for the two Co(III)–Co(II) half-reactions,

$$Co^{3+} + e^- = Co^{2+} \qquad \mathscr{E}^0 = +1.95\ V$$

$$Co(NH_3)_6^{3+} + e^- = Co(NH_3)_6^{2+} \qquad \mathscr{E}^0 = +0.10\ V$$

answer each of the following questions:

(a) For which complex ion, $Co(NH_3)_6^{3+}$ or $Co(NH_3)_6^{2+}$, is the equilibrium constant, $K = [Co(NH_3)_6^{n+}]/([Co^{n+}][NH_3]^6)$, larger?

(b) In a solution at equilibrium containing $1M$ NH_3, $1M$ $Co(NH_3)_6^{3+}$, $1M$ $Co(NH_3)_6^{2+}$, what is the ratio of concentrations of the uncomplexed ions $[Co^{3+}]/[Co^{2+}]$?

19–19 Show how the ability of a metal oxide to oxidize carbon monoxide at high temperatures may be used to establish a scale analogous to the $\mathscr{E}^0$ scale for the oxidizing ability of oxidizing agents in aqueous solution.

19–20 If intermediate aquacyanonickel(II) species, $Ni(OH_2)_a(CN)_b^{+2-b}$ with $0 < b < 4$, are not present at appreciable concentrations, over how large a range of concentration of free cyanide ion does the ratio $[Ni(CN)_4^{2-}]/[Ni^{2+}]$ change from 0.10 to 10.0? Compare this range with that for a system of two species differing by only one molecule of ligand, for example, $Co(NH_3)_5NCS^{2+}$ and $Co(NH_3)_5OH_2^{3+}$ for which

$$K = \frac{[Co(NH_3)_5NCS^{2+}]}{[Co(NH_3)_5OH_2^{3+}][SCN^-]} = 4.7 \times 10^2\ L\ mol^{-1}$$

19–21 Draw the structures for all isomers of the following octahedral species: $Co(NH_3)_4Br_2^+$, $Co(NH_3)_3Br_3$, $Co(NH_3)_3Br_2Cl$, $Co(en)_2Cl_2^+$ (en stands for ethylenediamine), and $Co(NH_3)_2F_4^-$.

19–22 Draw the isomers of the complex of chromium(III) coordinated by the glycine anion, gly^-, in $Cr(gly)_3$. Each chelate ring involves coordination to one nitrogen atom and one oxygen atom.

19–23 Draw the structural formulas for all isomers of $Co(NH_3)_4Cl_2^+$ and $Co(NH_3)_3Cl_3$, supposing that the coordination geometry around cobalt(III) defined a trigonal prism.

19–24 Using the statistical reasoning outlined in connection with the relative stabilities of the isomers $Cr(OH_2)_4Cl_2^+$, predict the relative stability of the isomers of $Cr(OH_2)_3(NH_3)_3$.

19–25 Show the occupancy of the more and less stable sets of d orbitals for high-spin and low-spin octahedral species with d^4, d^5, and d^7 electronic configurations.

19–26 For pentaaquafluorochromium(III) ion, $Cr(OH_2)_5F^{2+}$, the rate of aquation is increased with increasing concentration of hydrogen ion in the range

0.10–2.00 mol L^{-1}, but there is no such dependence in this range of acidity for the rate of aquation of pentaaquachlorochromium(III) ion, $Cr(OH_2)_5Cl^{2+}$. Explain.

19–27 Additional metal–carbonyl anions, other than $V(CO)_6^-$ and $Co(CO)_4^-$, are $V(CO)_5^{a-}$ and $Fe(CO)_4^{b-}$. Predict the values of *a* and *b*.

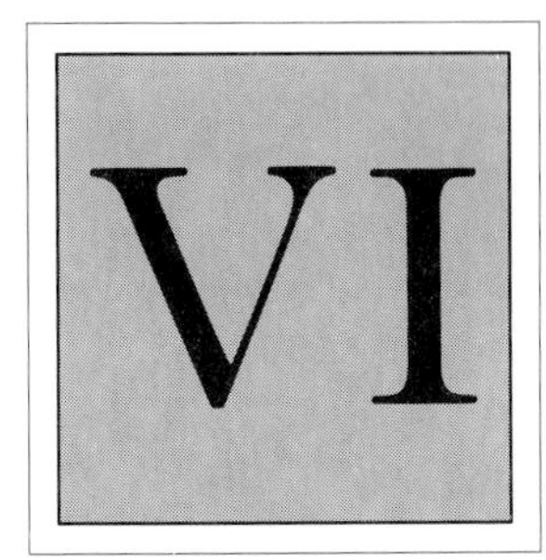

OTHER ASPECTS OF CHEMISTRY

The final three chapters of this book deal with diverse areas of chemistry that we have not encountered before. But in these chapters we will be building on a foundation laid by previous chapters.

You were introduced to organic compounds in Chapter 11; this introduction is extended in Chapter 20, which deals with more complex compounds of carbon, including those that play roles in biology. You will learn that simple principles of molecular structure, acid–base chemistry, hydrogen bonding, and oxidation–reduction to which you already have been introduced are the basis for processes in biology.

The interaction of light with matter has many aspects, and some of these will be considered in Chapter 21. Study of the spectra of molecules elucidates their structures, just as it does those of atoms, as you learned in Chapter 8. The absorption of light may produce electronically excited molecular species that are more reactive than ground-state species. As a result, photochemical reactions occur much more rapidly than analogous thermal reactions. Quantitative measurement of light absorption is a very useful method of chemical analysis.

Many aspects of chemistry can be understood without considering the atom's nucleus, even though its charge establishes the atom's identity and its mass is almost the total mass of the atom. (The ratio of nuclear mass to atomic mass ranges from 0.99946 for ^{1}H to 0.99978 for ^{254}Fm.) But there are aspects of chemistry in which the mass of the nucleus and its stability have measurable effects on chemical properties. These effects should be recognized, and they can be exploited to help solve chemical problems. The biological hazards of radiation from unstable atomic nuclei are related to the chemical reactions that this radiation causes. An important use of unstable atomic nuclei is as tracers to elucidate the mechanisms of chemical reactions, and you will be introduced to this technique.

20

Macromolecules and Biopolymers

20–1 Introduction

In this chapter we continue the study of organic chemistry started with simple molecules in Chapters 10 and 11. The emphasis here will be on more complex molecules, molecules with high molecular weights that are called *macromolecules*. We shall discuss synthetic macromolecules, prepared by the chemical industry for a variety of uses, and also naturally occurring biopolymers, the macromolecules of biology. The range of molecular weights covered by the term "macromolecule" is enormous. A typical molecular weight of polyethylene, a synthetic macromolecule, is 10^5, that of natural rubber is 10^6, and that of some varieties of deoxyribonucleic acid (DNA) is as high as 10^{10} or 10^{11}.

20–2 Polymerization Reactions

Two different types of polymerization reactions are used in producing synthetic macromolecules: *addition* polymerization and *condensation* polymerization. In the first type, small molecules add to one another,

$$nA = A_n$$

and in the second type, the polymerization is accompanied by loss of a small molecule, water in many examples,

$$nA(OH)_2 = (A{-}O)_n + nH_2O$$

In most systems these polymerization reactions do not occur rapidly, so an appropriate catalyst is needed to make the reaction rate conveniently high.

ADDITION POLYMERIZATION; HYDROCARBON POLYMERS

Ethylene and its derivatives polymerize by adding to one another; an example is the formation of 1-butene from ethylene,

$$2H_2C{=}CH_2 \rightarrow H_2C{=}CHCH_2CH_3$$

In this reaction, one carbon–carbon double bond $[E(C{=}C) = 614 \text{ kJ mol}^{-1}]$ is replaced by two carbon–carbon single bonds $[E(C{-}C) = 347 \text{ kJ mol}^{-1}]$; the reaction is exothermic because

$$E(C{-}C) > \tfrac{1}{2}E(C{=}C)$$

and this exothermicity outweighs the unfavorable entropy change accompanying conversion of two molecules to one.

Common catalysts for this type of reaction are:

1. free-radical species (species with an odd number of electrons),
2. very acidic media, which have a high proton-donating ability (i.e., media with very negative values of H_0; see Section 13–4), and
3. certain transition-metal reagents.

An example of free-radical polymerization is the high-pressure polymerization of ethylene initiated by a peroxide. Organic peroxides are not very stable; one mechanism for their decomposition involves the production of radicals as the first step:

$$R:\ddot{\underset{..}{O}}:\ddot{\underset{..}{O}}:R \rightarrow 2R:\ddot{\underset{..}{O}}\cdot$$

In the presence of an olefin (e.g., ethylene), a free radical (RO·) initiates a sequence of reactions:

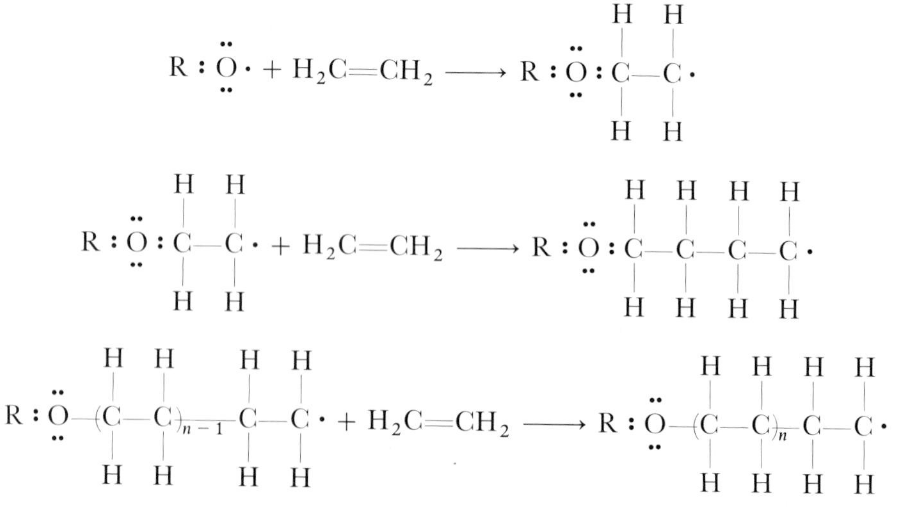

$$R:\ddot{\underset{..}{O}}\cdot + H_2C{=}CH_2 \longrightarrow R:\ddot{\underset{..}{O}}:\underset{H}{\overset{H}{C}}{-}\underset{H}{\overset{H}{C}}\cdot$$

$$R:\ddot{\underset{..}{O}}:\underset{H}{\overset{H}{C}}{-}\underset{H}{\overset{H}{C}}\cdot + H_2C{=}CH_2 \longrightarrow R:\ddot{\underset{..}{O}}:\underset{H}{\overset{H}{C}}{-}\underset{H}{\overset{H}{C}}{-}\underset{H}{\overset{H}{C}}{-}\underset{H}{\overset{H}{C}}\cdot$$

$$R:\ddot{\underset{..}{O}}{-}(\underset{H}{\overset{H}{C}}{-}\underset{H}{\overset{H}{C}})_{n-1}{-}\underset{H}{\overset{H}{C}}{-}\underset{H}{\overset{H}{C}}\cdot + H_2C{=}CH_2 \longrightarrow R:\ddot{\underset{..}{O}}{-}(\underset{H}{\overset{H}{C}}{-}\underset{H}{\overset{H}{C}})_n{-}\underset{H}{\overset{H}{C}}{-}\underset{H}{\overset{H}{C}}\cdot$$

This sequence of reactions resembles the chain reaction of hydrogen with bromine discussed in Chapter 15: each step consumes a radical and produces a radical. The chain-lengthening process continues until a reaction step occurs that *does not* give a radical. One such step is the combination of two growing chains (represented as R· and ·R′):

$$R\cdot + \cdot R' \rightarrow R:R'$$

Another chain-termination process is the reaction to give a saturated molecule plus an unsaturated molecule:

$$2R{-}CH_2{-}CH_2\cdot \rightarrow R{-}CH_2CH_3 + R{-}CH{=}CH_2$$

Although this type of reaction is said to be catalyzed by the organic peroxide, it is not strictly an example of catalysis because the peroxide is consumed. However, the chain may consist of a very large number of reaction steps, and a large amount of product results from a relatively small amount of peroxide.

A polymer-producing sequence of reactions also can be initiated by addition of a proton to an alkene. Ethylene does not polymerize by this mechanism, but details of the process can be illustrated with an alkyl-substituted ethylene; the initiating step for this substrate is

$$H_2C{=}CHR + H^+ \rightleftarrows H_3C{-}\overset{+}{C}HR$$

This type of acid–base reaction involves an alkene as a *weak* base and produces a carbonium ion (an unstable species, in which a carbon atom has only six electrons in its valence shell).[1] With a reactive substrate the carbonium ion can react with an alkene molecule to produce another carbonium ion,

$$H_3C\overset{+}{C}HR + H_2C{=}CHR \rightarrow H_3CCHRCH_2\overset{+}{C}HR$$

followed by

$$H_3CCHRCH_2\overset{+}{C}HR + H_2C{=}CHR \rightarrow H_3C(CHRCH_2)_2\overset{+}{C}HR$$

and so on. Loss of a proton by the carbon atom adjacent to the electron-deficient carbon atom gives a high-molecular-weight alkene:

$$H_3C(CHRCH_2)_nCHRCH_2\overset{+}{C}HR \rightarrow H^+ + H_3C(CHRCH_2)_nCHRCH{=}CHR$$

The proton is strictly a catalyst in this mechanism because it is regenerated.

Certain transition-metal complexes catalyze the addition polymerization of olefins. Titanium catalysts for polymerizing ethylene, developed by ZIEGLER, are assumed to act by coordinating initially to olefin, much as the proton becomes associated with the olefin. However, the steps that follow are more complex, and the mechanism of addition polymerization by

[1] In Table 12–3 you learned that ethylene has a proton affinity almost as large as that for water, 669 kJ mol^{-1} compared to 686 kJ mol^{-1}.

Ziegler-type catalysts is not completely understood. However, it is worth noting that the Ziegler process occurs at room temperature and atmospheric pressure, unlike the radical-chain polymerization of ethylene, which occurs at 1000 atm and 100°C. Polyethylene and related synthetic hydrocarbon polymers have many uses, and more than nine million tons are produced annually.

Natural and synthetic rubber are other important hydrocarbon polymers. Natural rubber dispersed in water is a fluid in a number of tropical plants. This dispersion is called a *latex*, and it contains other substances in addition to water and the rubber hydrocarbon. The empirical chemical composition of rubber purified from the dispersion is given by the formula C_5H_8. Careful heating of rubber in the absence of air gives a low-molecular-weight hydrocarbon of the same empirical composition. The compound so produced is called *isoprene* (2-methyl-1,3-butadiene), and has the structure

$$\mathrm{H_2C{=}C(CH_3){-}CH{=}CH_2}$$

Monomer units with this composition can be incorporated into a polymer chain,

$$\cdots\mathrm{{-}CH_2{-}C(CH_3){=}CH{-}CH_2{-}CH_2{-}C(CH_3){=}CH{-}CH_2{-}CH_2{-}C(CH_3){=}CH{-}CH_2{-}}\cdots$$

by redistribution of electrons in the monomer's two double bonds:

$$\mathrm{H_2C{\cdot\cdot}C(CH_3){-}CH{\cdot\cdot}CH_2}$$

Here the electrons being redistributed are shown as dots and other electron pairs are shown as solid lines. The resulting entity, with an unshared electron on each end,

$$\underset{\text{Cis}}{\mathrm{\cdot CH_2{-}C(CH_3){=}CH{-}CH_2\cdot}} \quad \text{or} \quad \underset{\text{Trans}}{\mathrm{\cdot CH_2{-}C(CH_3){=}CH{-}CH_2\cdot}}$$

can form bonds with other like entities. Geometrical isomerism is possible at each ethylene linkage, as shown, and there are naturally occurring polymers made up of monomers with each of these configurations. These are:

$$\diagup CH_2\diagdown C(CH_3){=}CH\diagup CH_2\diagdown CH_2\diagup C(CH_3){=}CH\diagdown CH_2\diagup CH_2\diagdown C(CH_3){=}CH\diagup CH_2\diagdown CH_2\diagup$$

Natural rubber (*cis*-1, 4-polyisoprene)

$$\diagup CH_2\diagdown C(CH_3){=}CH\diagdown CH_2\diagup CH_2\diagdown C(CH_3){=}CH\diagdown CH_2\diagup CH_2\diagdown C(CH_3){=}CH\diagdown CH_2\diagup$$

Gutta-percha (*trans*-1, 4-polyisoprene)

The polymer having the cis configuration is rubber, and that with the trans configuration is gutta-percha, a material that is hard and brittle at room temperature. The separate polymeric chain molecules with the trans configuration can be packed beside one another, and, as a consequence, this material is crystalline. The polymer involving the cis configuration cannot be crystalline, but is an amorphous array of randomly arranged chains. This amorphous material can be stretched, but a stretched-out array of extended chains is not its natural state, and when released, the material returns to a less extended form (which has greater entropy). Stretching the long-chain polymer does not involve lengthening its carbon–carbon bonds or appreciably distorting its CCC bond angles; rather it involves forcing the hydrocarbon chain into more extended conformations. (Figure 11–7 shows several conformations of a chain of six carbon atoms that have different end-to-end chain lengths.)

Many synthetic rubbers are built up from monomer units that resemble isoprene; in particular, 1,3-butadiene,

$$H_2C{=}CH{-}CH{=}CH_2$$

is the monomer in polybutadiene, and mixed with other unsaturated monomers, 1,3-butadiene forms polymers with rubberlike characteristics. With styrene, $C_6H_5CH{=}CH_2$, it polymerizes to give GRS rubber,[2] and with isobutylene, $(CH_3)_2C{=}CH_2$, it forms butyl rubber. (Butyl rubber has the desirable property of being especially impermeable to air.)

The polymer resulting from monomer units with two double bonds (e.g., isoprene or butadiene) still has one double bond per monomer unit after polymerization. This structural characteristic makes it possible to link

[2] A designation used in World War II meaning Government Rubber–Styrene type.

the polymer chains together. For rubber, this process is called *vulcanization*, and it occurs when rubber (natural or synthetic) is heated with sulfur. A sulfur atom (or a neutral chain S_n) lacks two electrons of having an octet per sulfur atom, and a sulfur atom (or a neutral chain S_n) can add to the double bonds of adjacent chains:

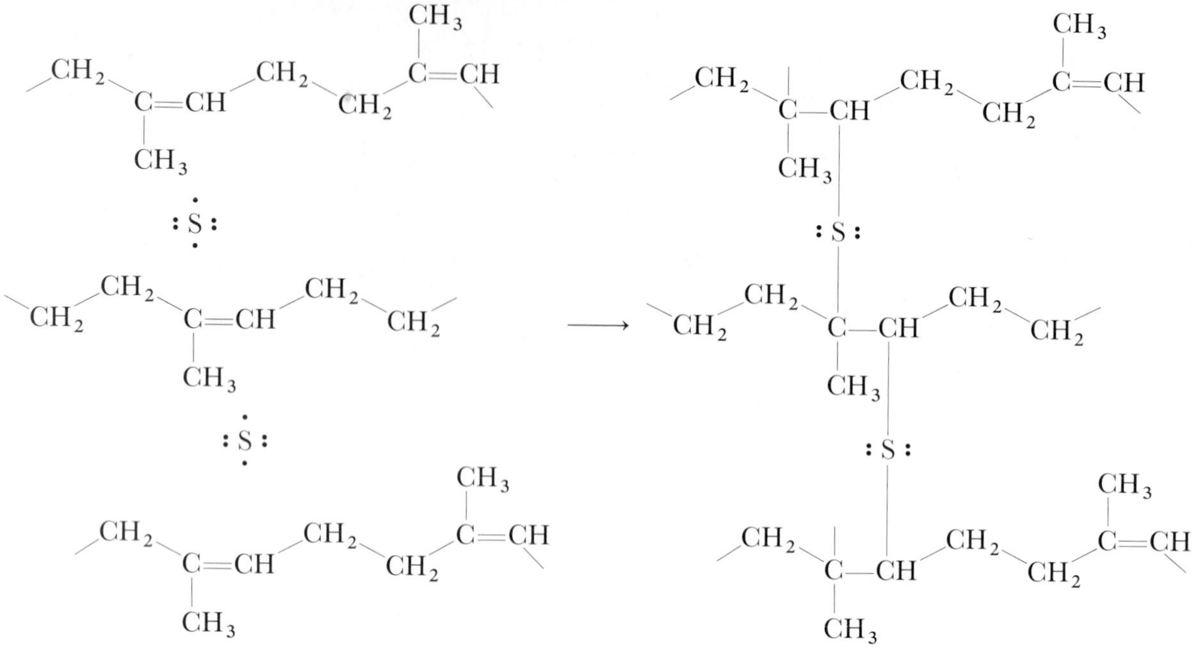

```
                                CH3                                                    CH3
                                 |                                                      |
  CH2          CH2            C=CH                     CH2   |          CH2            C=CH
     C=CH         CH2                                     C—CH             CH2
     |                                                    |   |
     CH3                                                  CH3 |
      .                                                       |
    : S :                                                   : S :
      .                                                       |
     CH2           CH2                                    CH2 |          CH2
CH2     C=CH           CH2          ——→          CH2       C—CH             CH2
        |                                                  |   |
        CH3                                                CH3 |
         .                                                     |
       : S :                                                 : S :
         .                                                     |
                                CH3                            |                       CH3
                                 |                             |                        |
     CH2           CH2         C=CH                     CH2   ||         CH2           C=CH
        C=CH           CH2                                 C—CH              CH2
        |                                                  |
        CH3                                                CH3
```

The elasticity of rubber and its strength are controlled by the amount of sulfur added. Large amounts of sulfur produce a hard rubber; this is used in making combs, cases for storage batteries, and other items.

POLYMERIZATION THROUGH ESTER AND AMIDE LINKAGES; SYNTHETIC FIBERS

In Chapter 11 you learned about the structures of esters and amides, each of which is related to the structure of carboxylic acids. The structures of acetic acid, methyl acetate (an ester), and methyl acetamide (an amide) are:

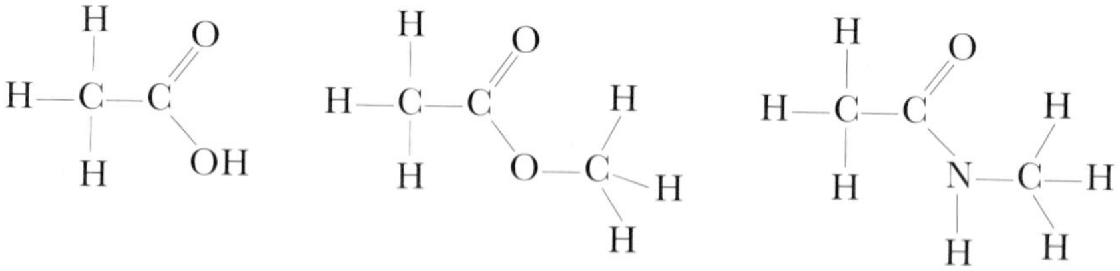

```
     H      O              H      O                       H      O
     |     //              |     //          H            |     //          H
H—C—C                 H—C—C                /         H—C—C                 /
     |     \               |     \          /              |     \          /
     H      OH             H      O—C—H                    H      N—C—H
                                     \                            |    \
                                      H                           H     H

   Acetic acid            Methyl acetate                 Methyl acetamide
```

Esters and amides can be formed from the corresponding acid by reactions in which water also is produced; equations for reactions in which methyl acetate and methyl acetamide are formed in this way are:[3]

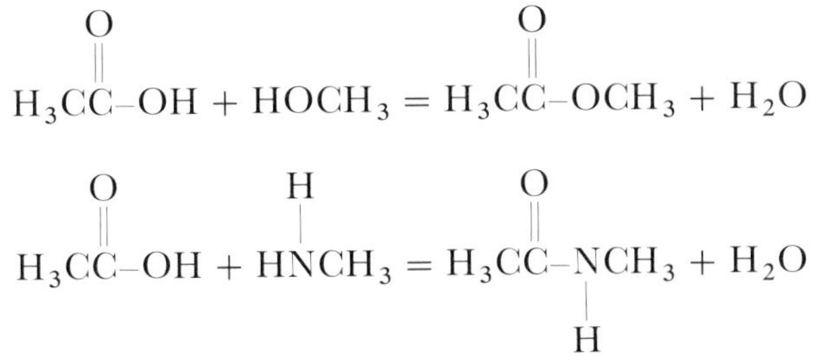

$$H_3C\overset{\overset{\displaystyle O}{\|}}{C}{-}OH + HOCH_3 = H_3C\overset{\overset{\displaystyle O}{\|}}{C}{-}OCH_3 + H_2O$$

$$H_3C\overset{\overset{\displaystyle O}{\|}}{C}{-}OH + \overset{\overset{\displaystyle H}{|}}{HN}CH_3 = H_3C\overset{\overset{\displaystyle O}{\|}}{C}{-}\underset{\underset{\displaystyle H}{|}}{N}CH_3 + H_2O$$

In dilute aqueous solution, equilibrium in these reactions does not favor the formation of the ester or amide, so practical preparative reactions use a reagent that is more reactive than the acid. Acid chlorides, such as acetyl chloride,

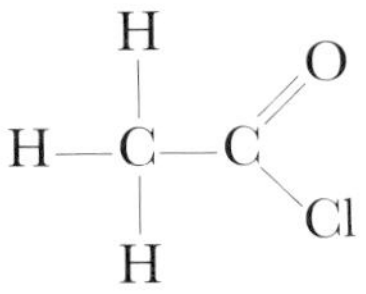

$$H{-}\underset{\underset{\displaystyle H}{|}}{\overset{\overset{\displaystyle H}{|}}{C}}{-}C\begin{matrix}{\nearrow}\!\!O \\ {\searrow}Cl\end{matrix}$$

generally are used in the production of amides:

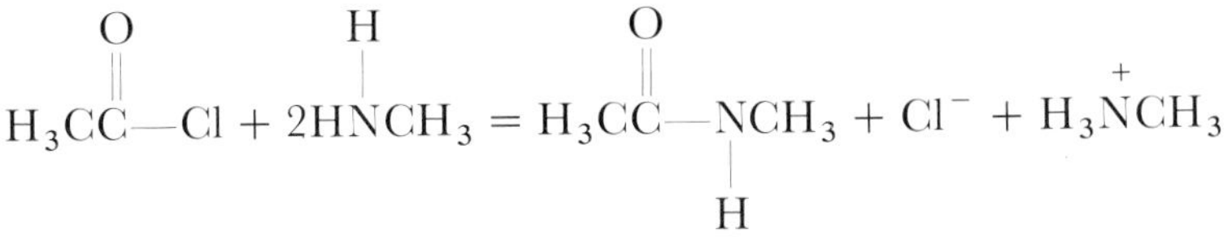

$$H_3C\overset{\overset{\displaystyle O}{\|}}{C}{-}Cl + 2\overset{\overset{\displaystyle H}{|}}{HN}CH_3 = H_3C\overset{\overset{\displaystyle O}{\|}}{C}{-}\underset{\underset{\displaystyle H}{|}}{N}CH_3 + Cl^- + H_3\overset{+}{N}CH_3$$

Reactions such as these can be the basis for producing polymeric molecules if reactant molecules with two functional groups per molecule are involved. The widely used synthetic fibers, nylon and Dacron, are polyamides and polyesters, respectively.

In the production of nylon 66, the starting materials are a six-carbon diamine and a six-carbon dicarboxylic acid:

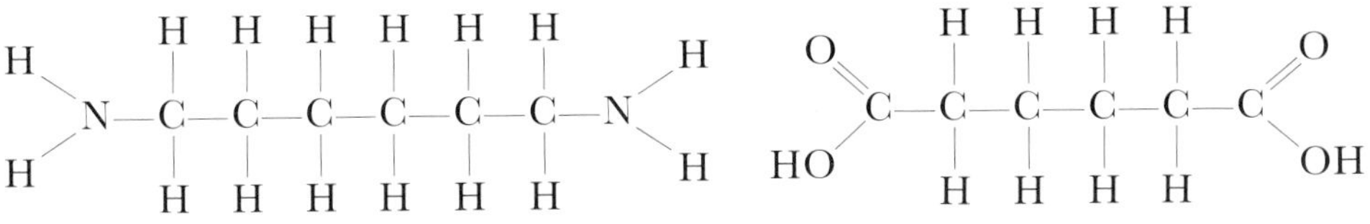

$$\begin{matrix}H \\ \searrow \\ H\nearrow\end{matrix}N{-}\underset{\underset{\displaystyle H}{|}}{\overset{\overset{\displaystyle H}{|}}{C}}{-}\underset{\underset{\displaystyle H}{|}}{\overset{\overset{\displaystyle H}{|}}{C}}{-}\underset{\underset{\displaystyle H}{|}}{\overset{\overset{\displaystyle H}{|}}{C}}{-}\underset{\underset{\displaystyle H}{|}}{\overset{\overset{\displaystyle H}{|}}{C}}{-}\underset{\underset{\displaystyle H}{|}}{\overset{\overset{\displaystyle H}{|}}{C}}{-}\underset{\underset{\displaystyle H}{|}}{\overset{\overset{\displaystyle H}{|}}{C}}{-}N\begin{matrix}\nearrow H \\ \searrow H\end{matrix} \qquad \begin{matrix}O \\ HO\end{matrix}C{-}\underset{\underset{\displaystyle H}{|}}{\overset{\overset{\displaystyle H}{|}}{C}}{-}\underset{\underset{\displaystyle H}{|}}{\overset{\overset{\displaystyle H}{|}}{C}}{-}\underset{\underset{\displaystyle H}{|}}{\overset{\overset{\displaystyle H}{|}}{C}}{-}\underset{\underset{\displaystyle H}{|}}{\overset{\overset{\displaystyle H}{|}}{C}}{-}C\begin{matrix}O \\ OH\end{matrix}$$

Hexamethylenediamine   Adipic acid

[3]The predominant species in an aqueous solution containing both acetic acid and methylamine are acetate ion and methylammonium ion. Using the principles described in Chapter 13, you can show that the acid–base reaction

$$H_3CCO_2H + RNH_2 \rightleftarrows H_3CCO_2^- + RNH_3^+$$

has an equilibrium constant that is the ratio of two acid-dissociation constants, $K = K_a(H_3CCO_2H)/K_a(RNH_3^+) \cong 10^4$.

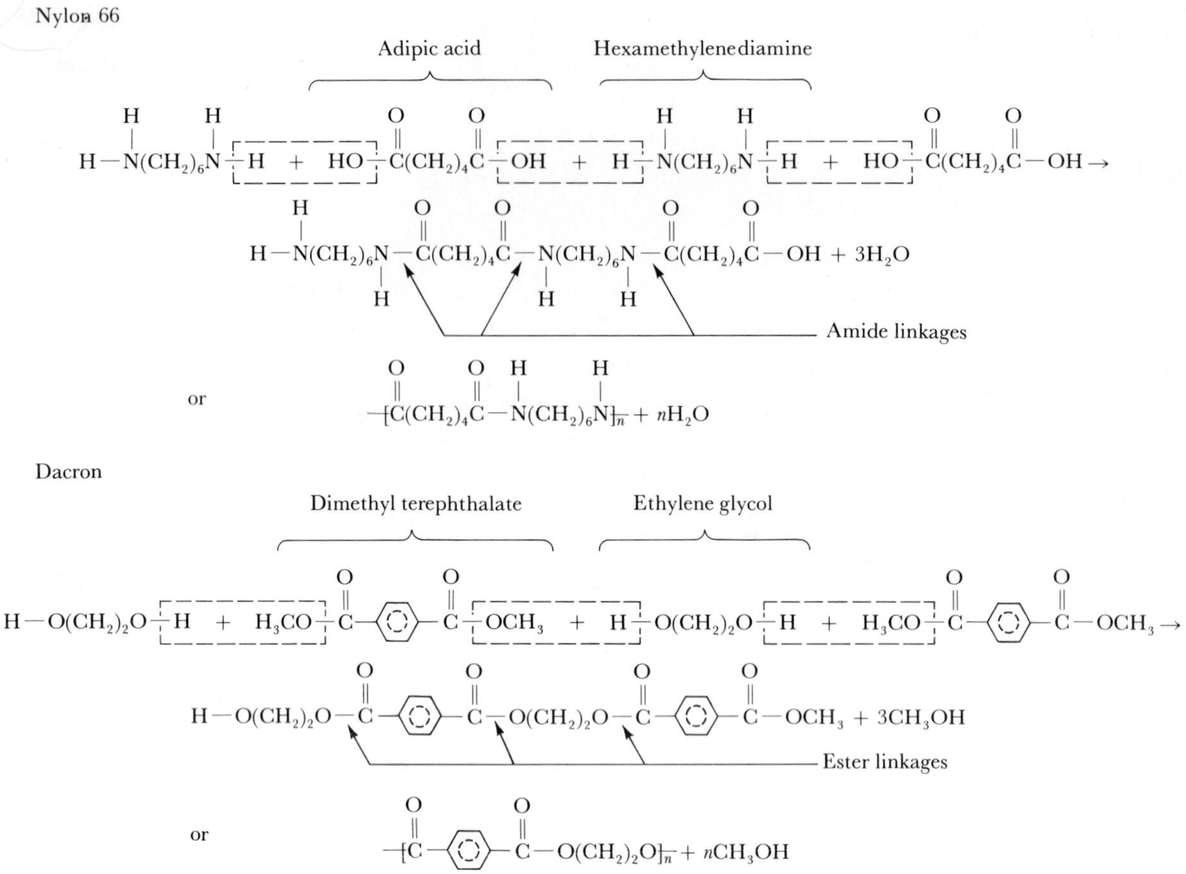

Nylon 66

Adipic acid — Hexamethylenediamine

$$\mathrm{H{-}\overset{H}{\overset{|}{N}}(CH_2)_6\overset{H}{\overset{|}{N}}{-}H + HO{-}\overset{O}{\overset{\|}{C}}(CH_2)_4\overset{O}{\overset{\|}{C}}{-}OH + H{-}\overset{H}{\overset{|}{N}}(CH_2)_6\overset{H}{\overset{|}{N}}{-}H + HO{-}\overset{O}{\overset{\|}{C}}(CH_2)_4\overset{O}{\overset{\|}{C}}{-}OH \rightarrow}$$

$$\mathrm{H{-}\overset{H}{\overset{|}{N}}(CH_2)_6\underset{H}{\underset{|}{N}}{-}\overset{O}{\overset{\|}{C}}(CH_2)_4\overset{O}{\overset{\|}{C}}{-}\underset{H}{\underset{|}{N}}(CH_2)_6\underset{H}{\underset{|}{N}}{-}\overset{O}{\overset{\|}{C}}(CH_2)_4\overset{O}{\overset{\|}{C}}{-}OH + 3H_2O}$$

Amide linkages

or

$$\mathrm{{-}\!\!\left[\overset{O}{\overset{\|}{C}}(CH_2)_4\overset{O}{\overset{\|}{C}}{-}\overset{H}{\overset{|}{N}}(CH_2)_6\overset{H}{\overset{|}{N}}\right]_n + nH_2O}$$

Dacron

Dimethyl terephthalate — Ethylene glycol

$$\mathrm{H{-}O(CH_2)_2O{-}H + H_3CO{-}\overset{O}{\overset{\|}{C}}{-}C_6H_4{-}\overset{O}{\overset{\|}{C}}{-}OCH_3 + H{-}O(CH_2)_2O{-}H + H_3CO{-}\overset{O}{\overset{\|}{C}}{-}C_6H_4{-}\overset{O}{\overset{\|}{C}}{-}OCH_3 \rightarrow}$$

$$\mathrm{H{-}O(CH_2)_2O{-}\overset{O}{\overset{\|}{C}}{-}C_6H_4{-}\overset{O}{\overset{\|}{C}}{-}O(CH_2)_2O{-}\overset{O}{\overset{\|}{C}}{-}C_6H_4{-}\overset{O}{\overset{\|}{C}}{-}OCH_3 + 3CH_3OH}$$

Ester linkages

or

$$\mathrm{{-}\!\!\left[\overset{O}{\overset{\|}{C}}{-}C_6H_4{-}\overset{O}{\overset{\|}{C}}{-}O(CH_2)_2O\right]_n + nCH_3OH}$$

FIGURE 20–1

The condensation polymerization reactions forming nylon 66 and Dacron. (The designation 66 refers to the number of carbon atoms in each of the reactant molecules.)

The reaction of these molecules to give a polymer by condensation involving the formation of water is depicted in Figure 20–1. In a laboratory or large-scale preparation of nylon, molecules of different sizes are produced. Nylon fibers are made by extruding the polymeric material as filaments at a temperature above its melting point. After cooling, the filaments can be drawn, a procedure that stretches the long-chain molecules. The strength of nylon 66 is due, in part, to hydrogen bonding between the carbonyl oxygen atoms of one chain and the amide hydrogen atoms of another chain.

In the production of Dacron, the starting materials are

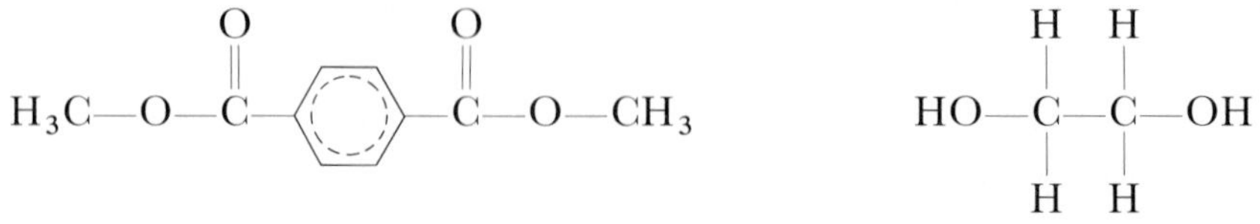

$$\mathrm{H_3C{-}O{-}\overset{O}{\overset{\|}{C}}{-}C_6H_4{-}\overset{O}{\overset{\|}{C}}{-}O{-}CH_3} \qquad \text{and} \qquad \mathrm{HO{-}\underset{H}{\underset{|}{\overset{H}{\overset{|}{C}}}}{-}\underset{H}{\underset{|}{\overset{H}{\overset{|}{C}}}}{-}OH}$$

Dimethyl terephthalate and Ethylene glycol

The reaction of these molecules to produce Dacron also is depicted in Figure 20–1. This reaction is called an *ester-exchange* reaction because the reactant is an ester, dimethyl terephthalate, rather than the acid, terephthalic acid. A factor that drives this reaction to completion is the volatility of methyl alcohol at the temperature of the process.

20–3 Proteins

You have just learned that a prerequisite for polymerization through condensation is the presence of two appropriate functional groups in each monomer molecule. You also learned that an amide linkage, as in nylon, can arise from condensation of an acid and an amine. The two functional groups are present in amino acids, and proteins are amide-linked polymers made up of the α-amino acids, the structures of which are given in Table 20–1. Thus the amide linkage, also called the *peptide linkage*, is centrally important in biochemistry.

Proteins, the most abundant organic molecules in cells, have from ~ 100 to $\sim 3 \times 10^5$ amino acid units. Some proteins, called *conjugated proteins*, contain additional organic or inorganic segments.[4] The biological functions that proteins fulfill are varied. Enzymes are an important class of proteins. They are catalysts for many reactions (see Chapter 15); over a thousand are known. Other important classes of proteins are those responsible for the structure of biological material, for example, collagen, the protein of tendons and other fibrous connective tissue. Hemoglobin, the molecule that transports oxygen in blood, is a conjugated protein that also contains an iron atom coordinated to heme (see Section 19–3). The list could be extended, but these few examples show some of the variety of proteins.

In contrast to synthetic macromolecules, proteins and other biopolymers have particular molecular weights. As already mentioned, the polyamide macromolecules making up a sample of nylon have a range of molecular weights. But all of the molecules of a particular protein have the same molecular weight. Such macromolecules are said to be *homogeneous*.[5]

The properties of a protein depend on its structure. This term "structure" applied to a protein encompasses more than it ordinarily does when applied to a small organic molecule. The aspects of protein structure are called primary, secondary, and tertiary.

Primary structure is the sequence of amino acids in the polypeptide chain and the location of covalent disulfide linkages between cysteine residues of the polypeptide chain(s).

Secondary structure is the relative spatial arrangement of amino acid residues close to one another in the linear sequence of the polypeptide chain.

---

[4]"Conjugated" used in this sense does not indicate the presence of alternate double and single bonds, as in butadiene, a "conjugated" diene.

[5]This meaning of homogeneous is not the same as that employed in Chapter 1.

TABLE 20–1
α-Amino Acids That Are Constituents of Naturally Occurring Proteins[a]

Nonpolar			
Alanine (Ala)	$CH_3-CH(NH_2)-COOH$	Proline (Pro)	$H_2C-CH_2-CH_2-NH-CH-COOH$ (ring: H_2C, H_2C, CH_2, NH, CH)
Valine (Val)	$(CH_3)_2CH-CH(NH_2)-COOH$	Methionine (Met)	$CH_3-S-CH_2-CH_2-CH(NH_2)-COOH$
Leucine (Leu)	$(CH_3)_2CH-CH_2-CH(NH_2)-COOH$	Phenylalanine (Phe)	$C_6H_5-CH_2-CH(NH_2)-COOH$
Isoleucine (Ile)	$CH_3-CH_2-CH(CH_3)-CH(NH_2)-COOH$	Tryptophan (Trp)	(indole ring: NH, =CH, C)$-CH_2-CH(NH_2)-COOH$

Tertiary structure is the relative spatial arrangement of the secondary structure in the natural protein; this determines the distances from one another of the amino acids that are not close to one another in the linear sequence of the polypeptide chain.

Most proteins can be classed either as *fibrous* proteins, which have extended conformations and generally are tough materials insoluble in water, or *globular* proteins, which have compact shapes and generally are soluble in water.

THE PRIMARY STRUCTURE OF PROTEINS

First, we will concern ourselves with the primary structure of proteins. A step in determining primary structure is the determination of gross composition. This can be done by degrading the protein into its constituent amino acids. A peptide bond is hydrolyzed in acidic solution; for a dipeptide, the chemical equation for hydrolysis is

$$H^+ + H_3\overset{+}{N}-CHR-C(=O)-NH-CHR'-C(=O)OH + H_2O = H_3\overset{+}{N}-CHR-C(=O)OH + H_3\overset{+}{N}-CHR'-C(=O)OH$$

TABLE 20–1 (Continued)
α-Amino Acids That Are Constituents of Naturally Occurring Proteins[a]

Polar			
Glycine (Gly)	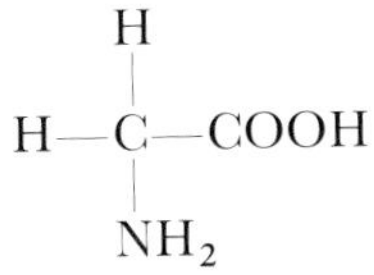 $H-\overset{H}{\underset{NH_2}{C}}-COOH$	Threonine (Thr)	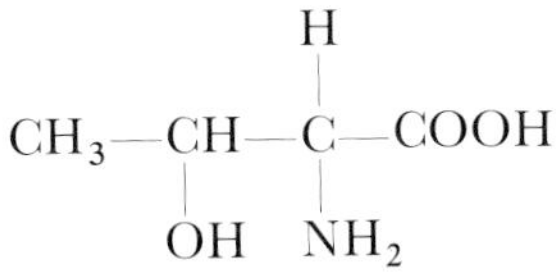 $CH_3-\underset{OH}{CH}-\overset{H}{\underset{NH_2}{C}}-COOH$
Serine (Ser)	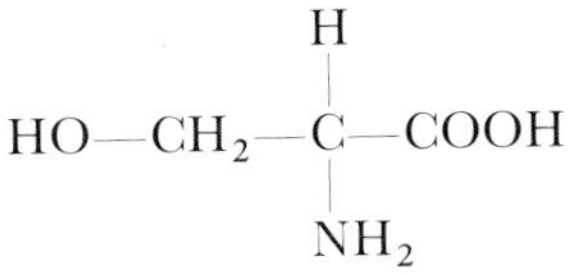 $HO-CH_2-\overset{H}{\underset{NH_2}{C}}-COOH$	Cysteine (Cys)	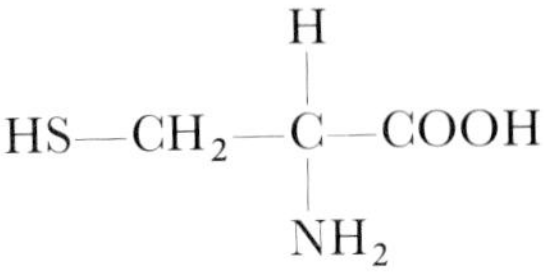 $HS-CH_2-\overset{H}{\underset{NH_2}{C}}-COOH$
Tyrosine (Tyr)	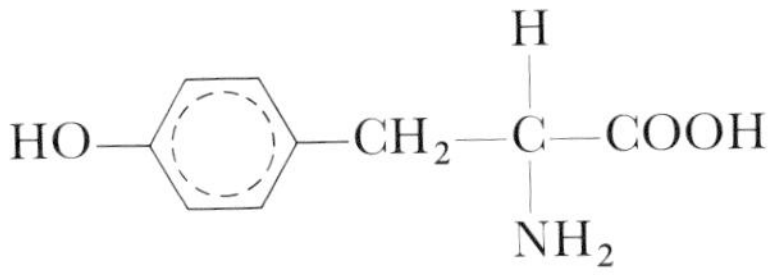 $HO-C_6H_4-CH_2-\overset{H}{\underset{NH_2}{C}}-COOH$	Lysine (Lys)	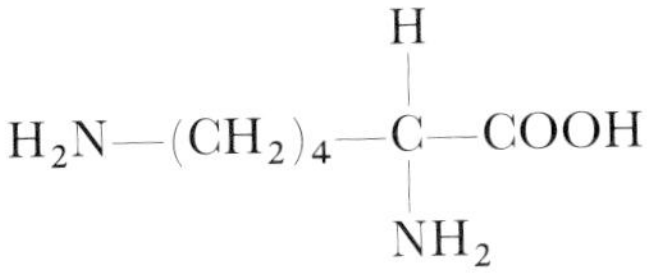 $H_2N-(CH_2)_4-\overset{H}{\underset{NH_2}{C}}-COOH$
Asparagine (Asn)	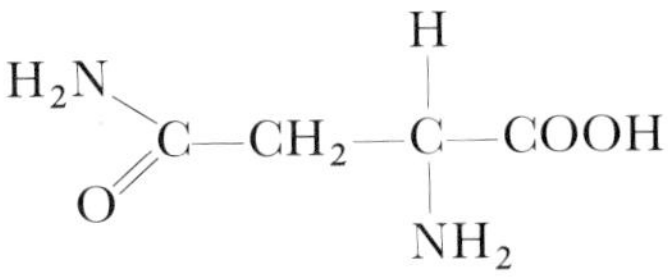 $H_2N(O{=})C-CH_2-\overset{H}{\underset{NH_2}{C}}-COOH$	Aspartic acid (Asp)	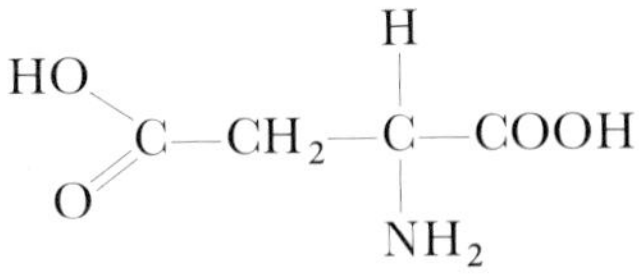 $HO(O{=})C-CH_2-\overset{H}{\underset{NH_2}{C}}-COOH$
Glutamine (Gln)	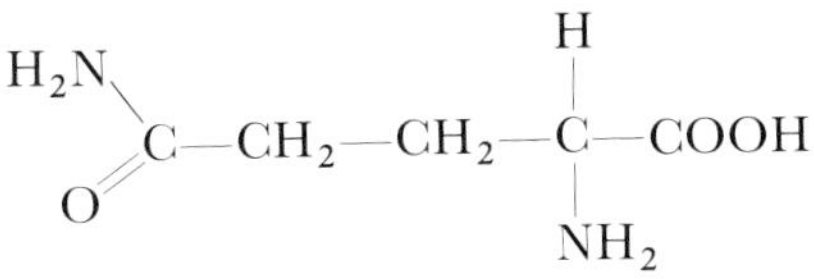 $H_2N(O{=})C-CH_2-CH_2-\overset{H}{\underset{NH_2}{C}}-COOH$	Glutamic acid (Glu)	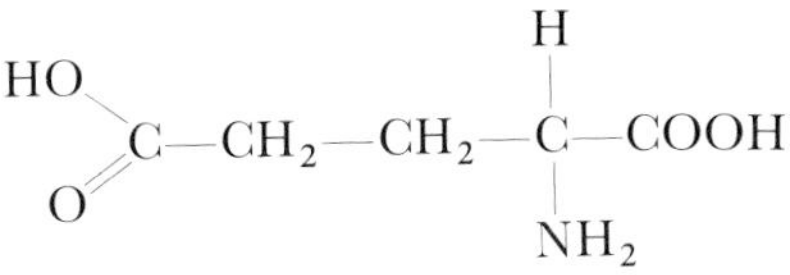 $HO(O{=})C-CH_2-CH_2-\overset{H}{\underset{NH_2}{C}}-COOH$
Arginine (Arg)	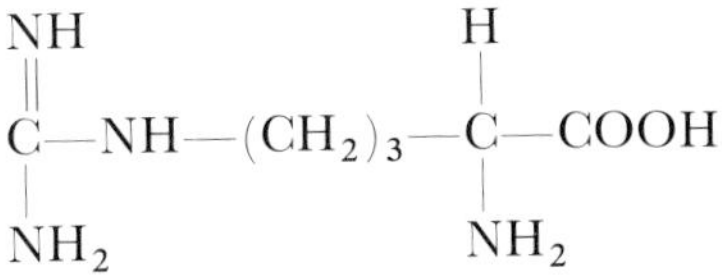 $\overset{NH}{\underset{NH_2}{C}}-NH-(CH_2)_3-\overset{H}{\underset{NH_2}{C}}-COOH$	Histidine (His)	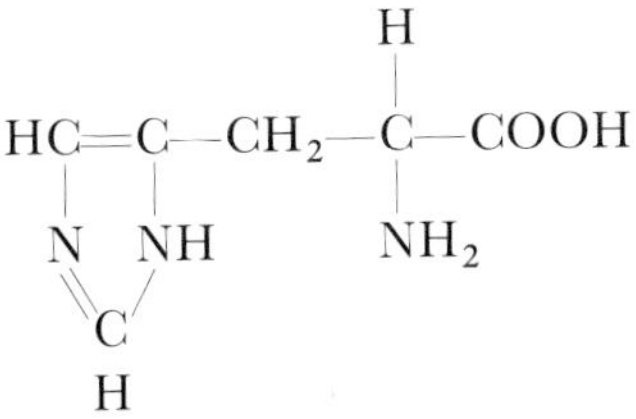 $HC{=}C-CH_2-\overset{H}{\underset{NH_2}{C}}-COOH$ (ring: HC=C–NH–CH=N–, i.e. $\underset{N}{HC}{=}\underset{NH}{C}$, N=C(H)–NH)

[a]The structures are given for the neutral forms, which are not predominant; the zwitterions are the predominant neutral forms (see Section 13–3).

Column ion-exchange procedures, like those discussed in Chapter 18, can be used to separate the different amino acids of a hydrolyzed protein from one another. Figure 20–2 shows an elution curve in which the pH of the eluting agent is changed during the course of elution. Because of their different acid-dissociation constants, each amino acid has a characteristic dependence of the average charge on the pH of the solution. The gross composition of a protein does not tell us the order of the amino acids in the protein.

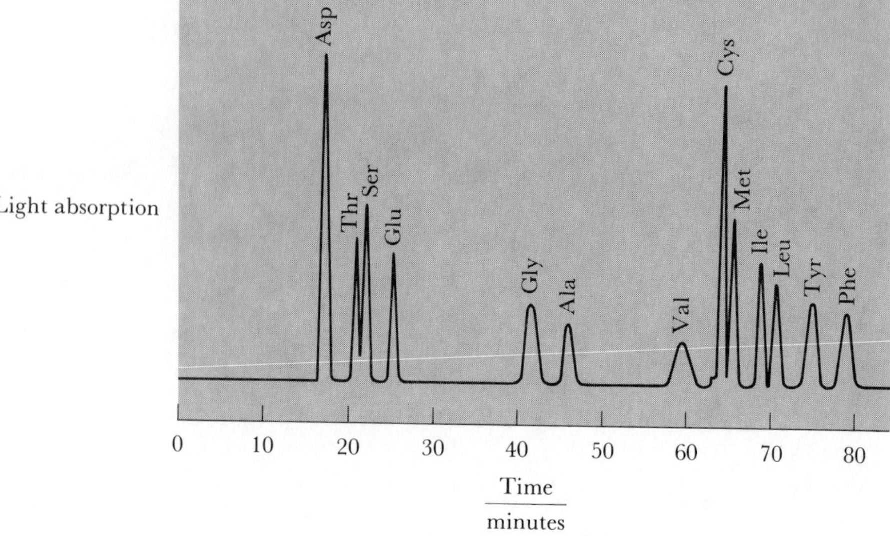

FIGURE 20–2
The elution profile for separation of the components of a mixture of amino acids. Analysis is done by passing eluent through spectrophotometer (see Chapter 21). Elution is carried out with a buffer solution.

If hydrolysis of a dipeptide gave glycine and alanine, there would be two possible primary structures of the dipeptide:

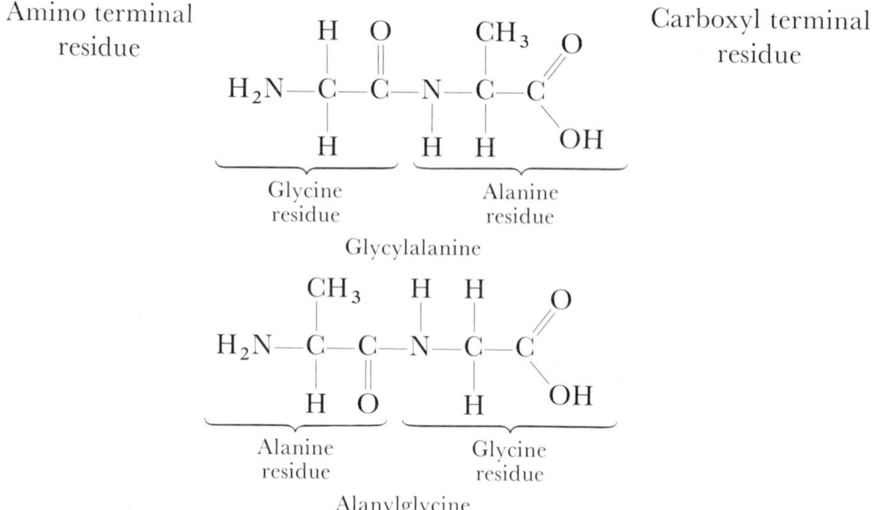

As suggested by these names, a polypeptide is named starting at the end with the free amino group. These two dipeptides are isomers, but they are different substances. Even the simplest proteins contain many different amino acids, and there are enormous numbers of isomers. However a specific biological function is associated with only that polypeptide having the correct amino acids linked in the correct order.

The first person to determine the sequence of amino acids in a protein of moderate complexity was SANGER. In 1953, he employed the reaction of

2,4-dinitrofluorobenzene with the free amino group of the N-terminal amino acid of the polypeptide to distinguish this amino acid from all of the rest. When a protein modified in this way is hydrolyzed in hydrochloric acid solution, the peptide linkages cleave, as just described, but the derivative of 2,4-dinitrobenzene and the amino acid to which it is bound *does not* decompose. These reactions, applied to glycylalanine, are:

$$O_2N\text{—}C_6H_3(NO_2)\text{—}F + H\text{—}N(H)\text{—}C(H)(H)\text{—}C(=O)\text{—}N(H)\text{—}C(H)(CH_3)\text{—}C(=O)OH \longrightarrow$$

$$HF + O_2N\text{—}C_6H_3(NO_2)\text{—}N(H)\text{—}C(H)(H)\text{—}C(=O)\text{—}N(H)\text{—}C(H)(CH_3)\text{—}C(=O)OH$$

$$O_2N\text{—}C_6H_3(NO_2)\text{—}N(H)\text{—}C(H)(H)\text{—}C(=O)\text{—}N(H)\text{—}C(H)(CH_3)\text{—}C(=O)OH + H_2O \longrightarrow$$

$$O_2N\text{—}C_6H_3(NO_2)\text{—}N(H)\text{—}C(H)(H)\text{—}C(=O)OH + H\text{—}N(H)\text{—}C(H)(CH_3)\text{—}C(=O)OH$$

Appropriate procedures allow separation and identification of the free amino acid (alanine) and the derivative of glycine. Applied to a larger polypeptide, this procedure identifies the N-terminal amino acid directly.

Incomplete hydrolysis of a protein in acid gives a mixture of smaller polypeptides. By characterizing these fragments of the parent protein molecule, including determining the N-terminal acid of each fragment, we may discover the primary structure of a protein.

EXAMPLE: A polypeptide containing eleven amino acids has an N-terminal glycine group and a COOH-terminal alanine group. After complete hydrolysis, analysis gives the composition, stated alphabetically, (Ala)(Arg)(Glu)$(Gly)_2$(Lys)$(Phe)_2$(Pro)(Thr)(Tyr), and partial hydrolysis gives the tripeptides with the primary structures Arg–Gly–Phe, Gly–Glu–Arg, Gly–Phe–Phe, Phe–Phe–Tyr, Phe–Tyr–Thr, and Pro–Lys–Ala. What is the sequence of eleven amino acids in this polypeptide?

The five tripeptides include two with an N-terminal glycine and one with a COOH-terminal alanine. Because only one alanine is present in the molecule, this tripeptide must end the molecule. But it might appear that either Gly–Glu–Arg or Gly–Phe–Phe could start the molecule. The presence of the tripeptide Arg–Gly–Phe, in which glycine is not a terminal group, rules out the latter possibility, and a start at establishing the structure is Gly–Glu–Arg–*v*–*w*–*x*–*y*–*z*–Pro–Lys–Ala. Simple matching of the other tripeptides shows the structure to be Gly–Glu–Arg–Gly–Phe–Phe–Tyr–Thr–Pro–Lys–Ala.

The hormone insulin, which regulates the metabolism of glucose, is a protein containing 51 amino acid residues. The primary structure of this molecule, which consists of two chains having twenty-one and thirty amino acid residues, is given in Figure 20–3. We see in this structure two disulfide linkages between the two chains and one disulfide linkage within one of the chains. Disulfide linkages can occur between cysteine residues. The relationship between two cysteine residues bonded together by a disulfide linkage

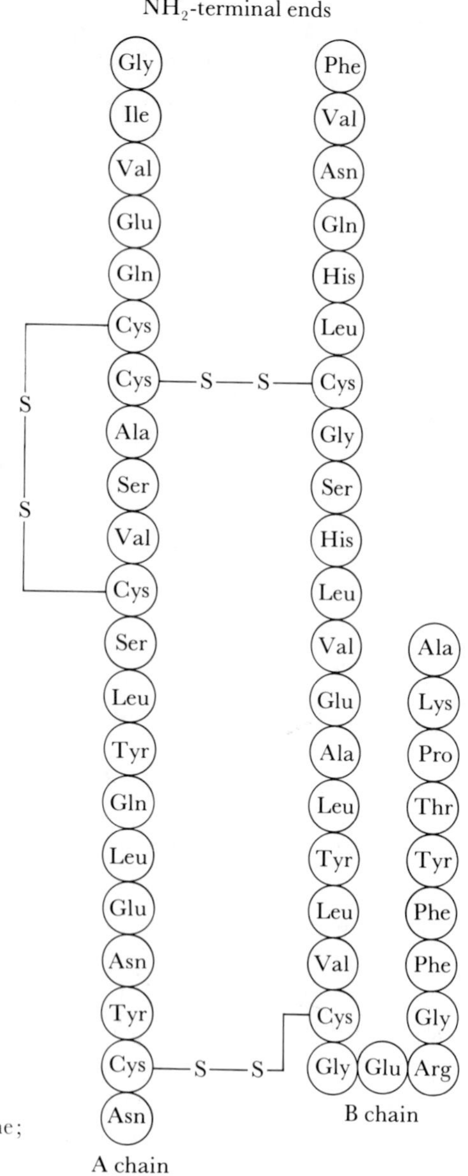

FIGURE 20–3
The primary structure of insulin. NH_2-terminal amino acids: Gly and Phe; CO_2H-terminal amino acids: Asn and Ala.

and the unlinked residues is one of oxidation–reduction; it can be shown in the half-reaction in which cystine is reduced to cysteine:

$$\begin{array}{l} H_3\overset{+}{N}CHCO_2^- \\ \quad | \\ \quad CH_2 \\ \quad | \\ \quad S \\ \quad | \\ \quad S \\ \quad | \\ \quad CH_2 \\ \quad | \\ H_3\overset{+}{N}CHCO_2^- \end{array} + 2H^+ + 2e^- = 2H_3\overset{+}{N}\underset{\underset{CH_2SH}{|}}{C}HCO_2^- \qquad \mathscr{E}^{0\prime}\ (pH = 7.0) = -0.21V$$

Cystine Cysteine

This half-reaction can be compared with the disulfide–sulfide half-reaction,

$$S_2^{2-} + 2H^+ + 2e^- = 2HS^- \qquad \mathscr{E}^{0\prime}\ (pH = 7.0) \cong -0.12\ V$$

Once we have determined the primary structure of a native protein, we may wish to synthesize it from its constituent amino acids. Polypeptides containing only a single amino acid can be prepared in a straightforward manner, but the synthesis of a polypeptide or protein with a particular primary structure of many different amino acid residues is a difficult matter. To produce a molecule with a particular sequence of amino acids, we must have a set of specific reactions, each of which must give a high yield. If each reaction gave a yield of 80%, the yield of a particular sequence of 30 amino acid residues (the length of the longer chain in insulin) would be $(0.80)^{30} \cong 0.00124$.

A recent advance in the laboratory synthesis of proteins is the solid-state technique developed by MERRIFIELD. In this procedure, the carboxyl group of the COOH-terminal amino acid first is chemically bonded to a resin particle, similar to the particles of an ion-exchange resin. Then an appropriate reaction is caused to occur between the carboxyl group of a new amino acid in solution and the amino group of the bound amino acid. At this stage, a particular dipeptide is anchored to the resin bead; then it is possible to rinse away all unreacted reagents from the reaction step that produced the dipeptide. This rinsing precedes the addition of another amino acid to give a bound tripeptide. The sequence of rinsing and adding a new reagent can be continued for as many steps as are needed. If reagents used in adding amino acids to the growing chains do not cleave the bond between the end carboxyl group and the resin bead, and if each reaction step occurs to an extent of ~100%, a yield of the desired product approaching 100% is expected. In practice, yields are not this high, but scientists have achieved yields of 85% for bradykinin, a nonapeptide, and 18% for ribonuclease, an enzyme containing 124 amino acid residues (17 different amino acids). This solid-state technique may be automated, with reagent addition, rinsing, and so on, each timed to allow the desired process to occur essentially to completion.

THE SECONDARY AND TERTIARY STRUCTURES OF PROTEINS

Keys to the secondary and tertiary structures of proteins are:

1. the geometry of the peptide linkage,
2. the hydrogen bonding between an amine hydrogen atom of one amino acid residue and a carbonyl oxygen atom of another amino acid residue of the polypeptide chain,
3. the hydrogen bonding between the solvent water and the polar and charged side groups of the protein, and
4. the tendency for nonpolar side groups to cluster together, thereby minimizing their interaction with the surrounding polar solvent (the hydrophobic effect).

Until now, we have represented the bonding in the amide (or peptide) linkage as

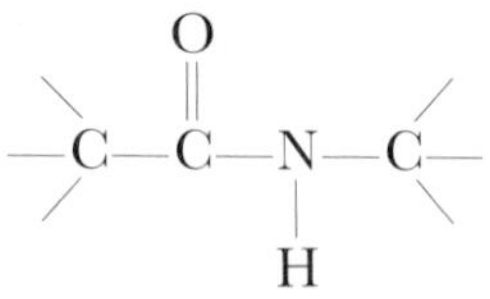

with a single bond between the nitrogen atom and the carbonyl carbon atom. X-ray crystallographic studies of crystalline polypeptides reveal that the six atoms in the amide linkage are in the same plane, with the following bond distances and bond angles:

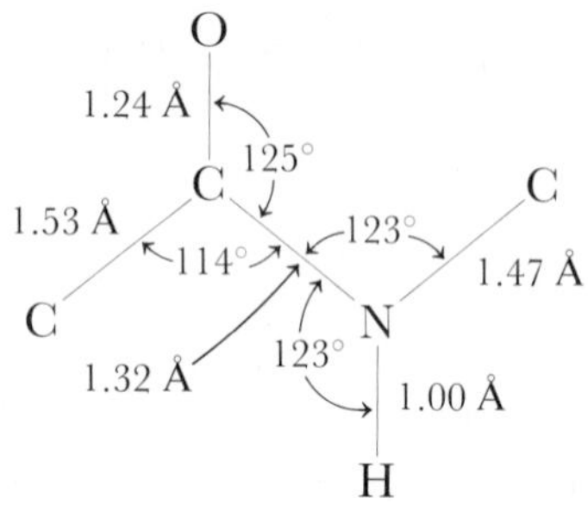

Particularly interesting is the central carbon–nitrogen bond distance (1.32 Å), which is shorter than expected for a carbon–nitrogen single bond (1.47 Å; see Table 10–7). These observations are consistent with the electronic structure of the peptide linkages being a resonance hybrid of the structures:

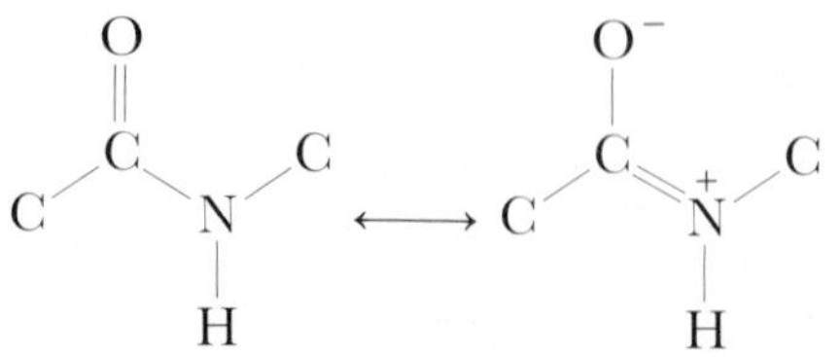

The double-bond character of the carbon–nitrogen bond prevents free rotation, and the planarity of the six atoms is analogous to the planarity of six atoms in ethylene (C_2H_4), in which the central atoms (the two carbon atoms) are joined by a double bond. In the structure contributing double-bond character to the carbon–nitrogen bond, the formal charges are -1 on oxygen and $+1$ on nitrogen. These charges enhance the hydrogen bonding mentioned in our opening summary.

However, the rigidity of the peptide linkage does not prevent free rotation about other bonds in the backbone chain of a polypeptide. Along such a backbone, every third bond is a $\text{C—N} \leftrightarrow \text{C═N}^+$ peptide linkage; the other two bonds (one C—C and one N—C) are single bonds about which free rotation is possible. This free rotation allows many conformations of a polypeptide chain including coiling. The secondary structure of many proteins involves a regular helical coil, proposed by PAULING and COREY in 1951. In this structure, called the *α helix*, the hydrogen that is bonded to nitrogen forms a hydrogen bond with the oxygen of the fourth amino acid from it in the chain. In a helical coil such as this, *all* nitrogen atoms and *all* oxygen atoms that are part of the backbone chain are held together by such hydrogen bonding (see Figure 20–4). In hemoglobin each of the four separate subunit chains has a secondary structure that is $\sim 70\%$ α helix; α keratin, the principal protein of skin and hair, is predominantly in the α-helix form.

The α-helix structure cannot exist in that part of a polypeptide chain containing certain amino acid residues. The bulkiness of the R groups of certain amino acids would cause excessive steric crowding if the protein had the α-helix secondary structure. The presence of charges on the R groups

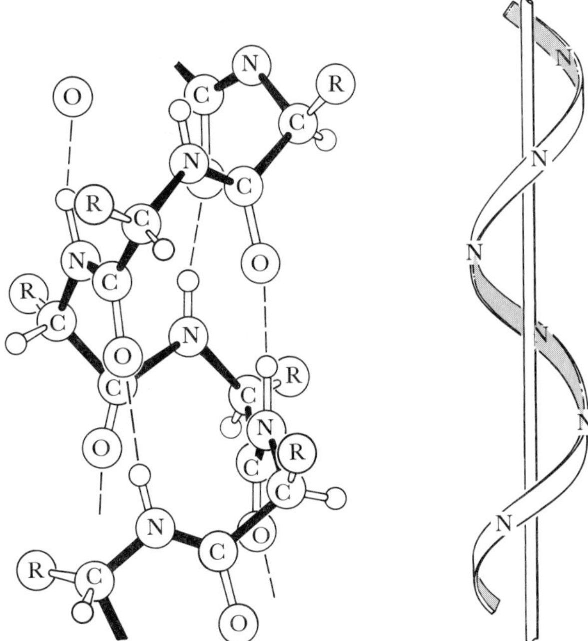

FIGURE 20–4
The right-handed α helix of a polypeptide made up of L-amino acids. There are 3.6 amino acid residues per pitch; this can be visualized by examining the nitrogen atoms shown on the helix at the right; 4 complete units encompass more than 1 pitch.

of amino acids closely spaced in the primary structure would result in excessive electrostatic repulsion in an α-helix arrangement. Thus the R groups on the side chains of the linked amino acids determines the secondary structure of the protein. For some primary structures the α helix is not a stable secondary structure.

Another type of secondary structure discovered by Pauling and Corey is called the *β-pleated sheet.* In this structure there are rows of extended polypeptide chains that are hydrogen-bonded to one another, as shown in Figure 20–5. This structure is found in the protein of silk.

In aqueous solutions of a protein there are several different types of hydrogen bonding; in addition to those of the α helix (shown in Figure 20–4) and the β-pleated sheet (shown in Figure 20–5), there are hydrogen-bonding interactions involving water:

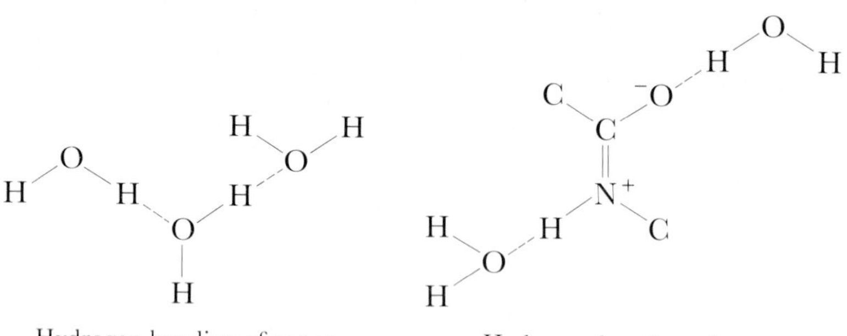

Hydrogen bonding of water molecules to one another

Hydrogen bonding of water to amino hydrogen and carbonyl oxygen

If an α-helix structure uncoils, the hydrogen bonding of the helix is lost, but it is replaced by hydrogen bonding between solvent molecules and the protein. For this to occur some hydrogen bonds between water molecules must be broken. This process of destroying the secondary structure of a protein is called *denaturation.* One means of denaturing a protein is raising its temperature because the conversion

$$\text{Protein (native form, biologically active)} = \text{Protein (denatured form, biologically inactive)}$$

is endothermic for most proteins. From the influence of temperature on the equilibrium

$$\text{Ribonuclease (native)} \rightleftarrows \text{Ribonuclease (denatured)}$$

the thermodynamic parameters for the denaturation reaction at pH = 6.5 are obtained:

$$\Delta H^0 = +213 \text{ kJ mol}^{-1} \qquad \Delta S^0 = +637 \text{ J K}^{-1} \text{ mol}^{-1}$$

The enormously positive value of ΔS^0 for the denaturation tells us that there is much order in the native protein, which is lost upon denaturation.

The equilibrium constant for the denaturation reaction is 1.00 if $\Delta G^0 = 0$,

$$K = e^{-\Delta G^0/RT}$$

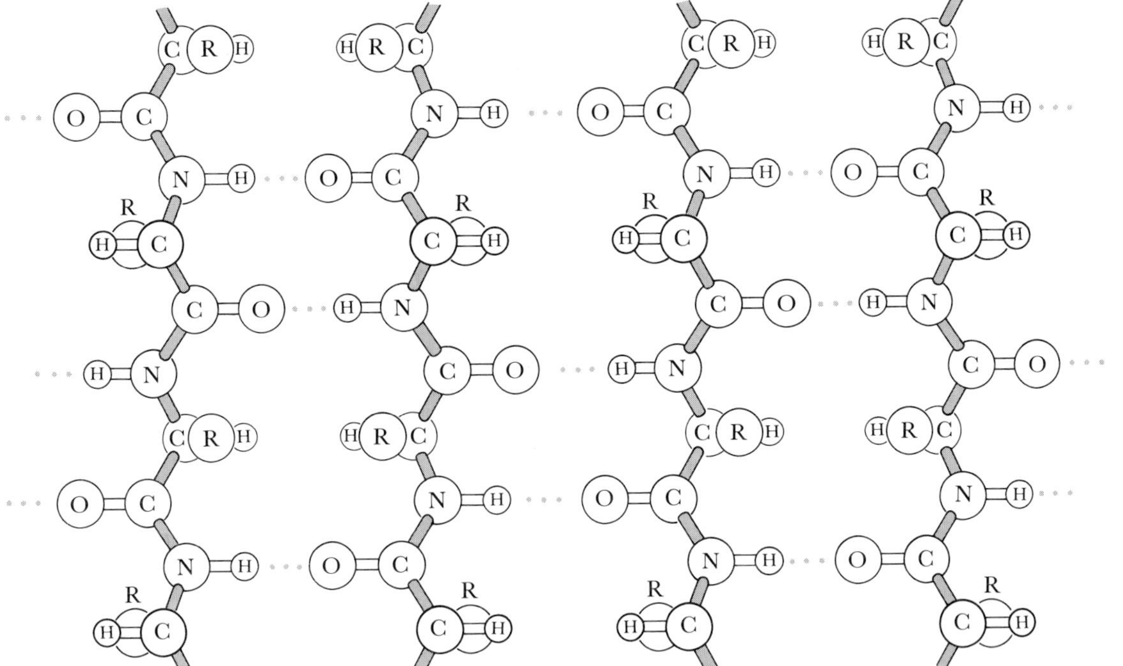

FIGURE 20–5
The β-pleated sheet. This configuration is called antiparallel because adjacent polypeptide chains have their terminal amino groups at opposite ends. This arrangement is more common than the parallel configuration, in which the chains are lined up in the same direction. [Adapted from L. Pauling, *Nature of the Chemical Bond*, Cornell University Press, Ithaca, N.Y., 3rd ed. (1960), p. 501.]

For this pH, the temperature at which $\Delta G^0 = 0$, called the *transition temperature*, can be calculated by rearranging the equation

$$\Delta G^0 = 0 = \Delta H^0 - T\Delta S^0$$

to give

$$T = \frac{\Delta H^0}{\Delta S^0} = \frac{213{,}000 \text{ J mol}^{-1}}{637 \text{ J K}^{-1} \text{ mol}^{-1}} = 334 \text{ K } (61°\text{C})$$

This denaturation, and that of some other proteins, is reversible: cooling a sample of previously heated solution restores its biological activity. Such renaturation is called *annealing*, and its slowness for some proteins results from the many, many inactive conformations through which the molecule goes in coming to the most stable conformation, which is the native form. This important point has been brought out in several types of studies. For a protein with a particular primary structure, there is under physiological conditions a particular secondary and tertiary structure that is the most stable, and this conformation is the native, biologically active form of the protein. If a protein not having the native conformation is placed in an appropriate medium, the thermal motion of the polypeptide chain eventually will bring the system into the thermodynamically stable native conformation.

A protein that contains disulfide linkages between cysteine residues can be denatured by the action of a reducing agent that breaks the sulfur–sulfur bonds. Renaturation to the native form then can be accomplished by oxidation.

Denaturation of proteins in solution also may be caused by the addition of reagents with a strong tendency to form hydrogen bonds. Urea and guanidinium ion,

$$H_2N{-}C({=}O){-}NH_2$$

Urea

$$H_2N{-}C({=}NH_2^+){-}NH_2$$

Guanidinium ion

each of which is isoelectronic with carbonate ion,

$$^-O{-}C({=}O){-}O^-$$

are such reagents.[6] As an example of denaturation by urea, we will consider hemoglobin, which contains four subunits with two different primary structures, designated α and β. These units are not bonded together by covalent bonds; rather, a combination of hydrogen-bonding interactions of the polar side chains of some amino acid residues and hydrophobic interactions of the nonpolar side chains of others gives the tertiary structure shown in Figure 20–6, in which the four subunits are combined. The addition of urea to a solution of native hemoglobin causes denaturation. The denaturation in this case involves stepwise dissociation that can be represented by the reactions

$$\alpha_2\beta_2 \rightleftarrows 2\alpha\beta$$
$$\alpha\beta \rightleftarrows \alpha + \beta$$

As we learned previously (Section 7–4), measuring the osmotic pressure of a solution is a useful way of determining the molecular weight of a macromolecule. This technique has been used on hemoglobin; at a urea concentration of 8 mol L^{-1}, the dissociation has gone past the stage of the first reaction just given. Some dissociation to simple α and β units has occurred.

The denaturation of a protein alters many of its properties, including its contribution to the viscosity of a solution.[7] For serum albumin, a globular protein with molecular weight 67,500, the intrinsic viscosity, as this property

[6]For guanidinium ion and carbonate ion only one of three equivalent resonance structures is shown. In this comparison, "isoelectronic" is used in the sense according to which hydrogen atoms are not counted. Each of these structures has 24 electrons. Guanidinium ion is a much weaker acid ($K_a \cong 10^{-14}$ mol L^{-1}) than ammonium ion ($K_a \cong 5 \times 10^{-10}$ mol L^{-1}) because guanidinium ion is stabilized by resonance.

[7]The viscosity of a liquid can be measured by determining the rate at which it flows through a tube or the rate at which a sphere falls through it.

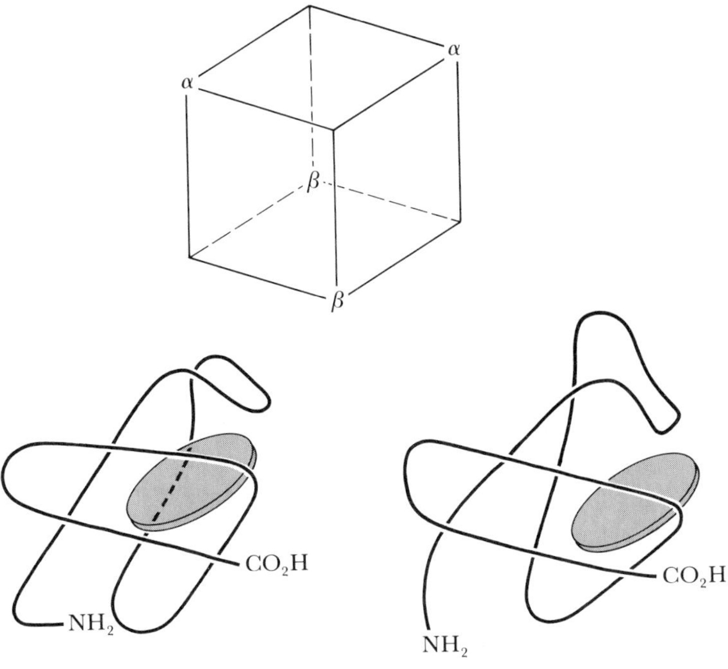

FIGURE 20–6
The structure of hemoglobin. The molecule consists of four subunits, two each of two different compositions; these are designated α and β. The four subunits in $\alpha_2\beta_2$ are in a tetrahedral arrangement (as shown at the top). The secondary structures of α and β chains are shown at the bottom.

is called, increases by a factor of ~14 when denaturation occurs. It seems reasonable that a random-coil solute molecule (the denatured protein) associated with much water or denaturing agent should retard the flow of the solution to a greater extent than does the compact native globular protein, which has very little solvent associated with it.

THE ACID–BASE PROPERTIES OF PROTEINS

In our study of equilibria in aqueous solutions of dibasic acids in Chapter 13, an amino acid, glycine, was one compound that we considered. In that discussion we also introduced the concept of isoelectric point, which you should review before you study the acid–base properties of proteins and polypeptides.

A protein has acid–base properties primarily because some of its constituent amino acids have side chains with acidic or basic groups. You will see in Table 20–1, for example, that the side chain in lysine has an amino group, $—NH_2$, and the side chain in aspartic acid has a carboxylic acid group, $—CO_2H$. Thus lysine can exist in solution with a net charge of +2, +1, 0, or −1, and aspartic acid can exist in solution with a net charge of +1, 0, −1 or −2, depending on the pH. The basic and acidic groups on the side chains of these and other amino acids are not involved in the peptide linkages of the primary protein structure, and their basicity and acidity are manifested in the protein approximately as would be expected. (We have learned, however, that the equilibrium constant for proton dissociation by a particular group, such as the carboxylic acid group, depends on the identity of neighboring groups in the molecule.) A protein also has the acidity and basicity caused by the amino or carboxyl end groups of the terminal amino acids that are not involved in peptide linkages. In large proteins these groups contribute only a small fraction of the observable basicity and acidity of the protein. The nitrogen atom of the peptide linkage has no appreciable basicity. This is consistent with the structure of the contributing resonance form in which the bond between this nitrogen atom and the carbonyl-carbon atom is a double bond, thereby giving the nitrogen atom a positive charge.

If a protein is dissolved in an acidic solution with pH $\cong$ 1 (e.g., 0.10*M* HCl), the basic side chains of the lysine, arginine, and histidine residues are protonated to cationic forms, and the acidic side chains of aspartic acid and glutamic acid residues are protonated to neutral forms. In an alkaline solution with pH 12–13, the basic side chains of the lysine, arginine, and histidine residues are not protonated; they carry no charge. In contrast, the acidic carboxyl groups are anionic, having lost their protons. Even the proton of the hydroxyl group (—OH) of tyrosine and the sulfhydryl group (—SH) of cysteine are dissociated to give anionic groups. Thus a protein has a net positive charge at pH $\cong$ 1 and a net negative charge at pH $\cong$ 13, and the average net charge depends upon the pH of the solution.

Figure 20–7 gives the titration curve for ribonuclease, a protein with 124 amino acids of which 34 have acidic or basic groups in their side chains.

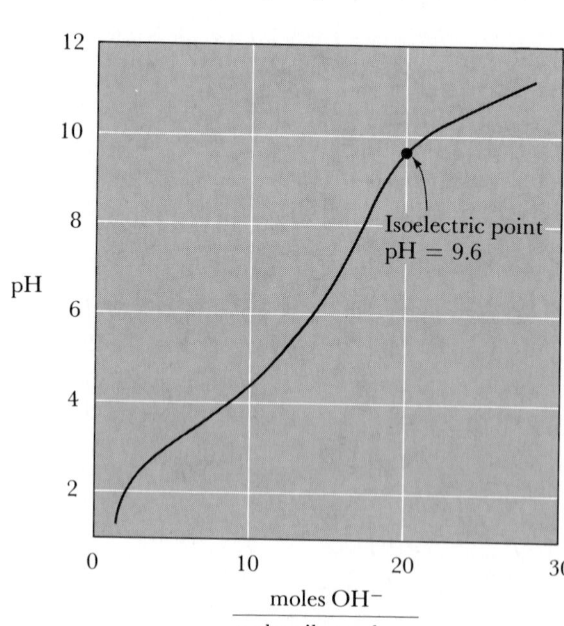

FIGURE 20–7
The titration curve for ribonuclease. Titration is started at pH $\cong$ 1, where all basic groups are protonated (i.e., in the forms $—NH_3^+$, $—CO_2H$). The base is consumed by 34 acid groups on side chains and the two terminal groups. The acid-dissociation constants of these groups range from $\sim 3 \times 10^{-5}$ mol L^{-1} to $\sim 10^{-12}$ mol L^{-1}.

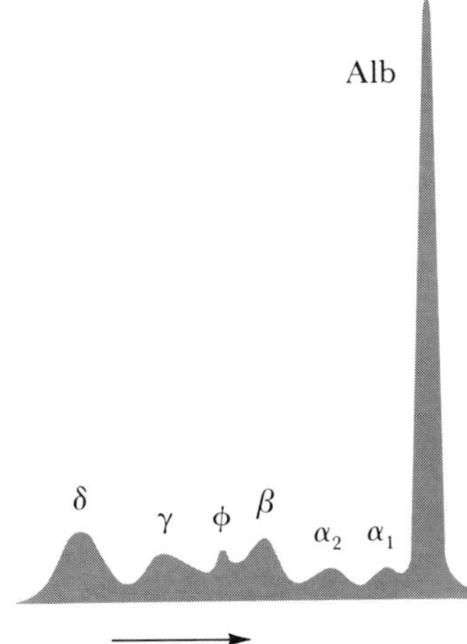

FIGURE 20–8
Protein electrophoresis. The electrophoretic pattern for normal human plasma at pH = 8.6 after 150 min at a potential gradient of 6.0 V cm^{-1}. Each peak indicates a boundary across which some protein disappears. Therefore the peak gives the velocity of a particular protein. This mixture contains six proteins, which are albumin (Alb), α_1-globulin (α_1), α_2-globulin (α_2), β-globulin (β), fibrinogen (ϕ), and γ-globulin (γ). The δ boundary does not represent an additional component. The arrow shows the direction of anion migration. [From R. A. Alberty, *J. Chem. Educ.* **25**, 619 (1948).]

Most of these side chains are basic, and the average net charge on this protein is positive until the solution is appreciably alkaline (pH = 9.6). Other proteins have isoelectric points at other values of pH; this is determined by the composition of the protein. The isoelectric point of egg albumin is 4.6 and that of hemoglobin is 6.8.

In a solution with a certain pH, each protein has a definite average charge, and it will migrate in an electric field with a characteristic average velocity for that pH. Study of this velocity of motion, called the *electrophoretic mobility*, has contributed much to our knowledge of the homogeneity of proteins. Figure 20–8 shows the electrophoretic pattern for human plasma, which contains appreciable amounts of six proteins. At the pH of this experiment (8.6), each of these proteins has a net negative charge. The most abundant protein in this material is albumin, and at this pH it has a net charge of −18. Under these conditions, the albumin molecule ion has a mobility of -5.9×10^{-9} m^2 V^{-1} s^{-1}, which can be compared with values for simple ions given in Table 9–12 (e.g., Na^+, $+5.2 \times 10^{-8}$ m^2 V^{-1} s^{-1} and SO_4^{2-}, -8.3×10^{-8} m^2 V^{-1} s^{-1}). Despite its high charge, the albumin molecule-ion migrates more slowly than simple inorganic ions in an electric field because it is strongly hydrogen-bonded to many surrounding water molecules. The unit that is migrating is very large. The principal cause of the lower migration velocity of the other proteins in human plasma is the lower net charge that they have at this pH.

The pH of the solution influences the denaturation of a protein. The hydrogen bonding partially responsible for the secondary and tertiary structures of some proteins involves the acidic and basic side chains, the state of which depends upon pH. Also, at extremes of pH where some proteins have high positive charge (at low pH) or high negative charge (at high pH), an electrostatic factor helps stabilize the denatured form. Figure 20–9 shows the influences of the pH upon the temperature range over which the reaction

$$\text{Ribonuclease (native)} \rightleftarrows \text{Ribonuclease (denatured)}$$

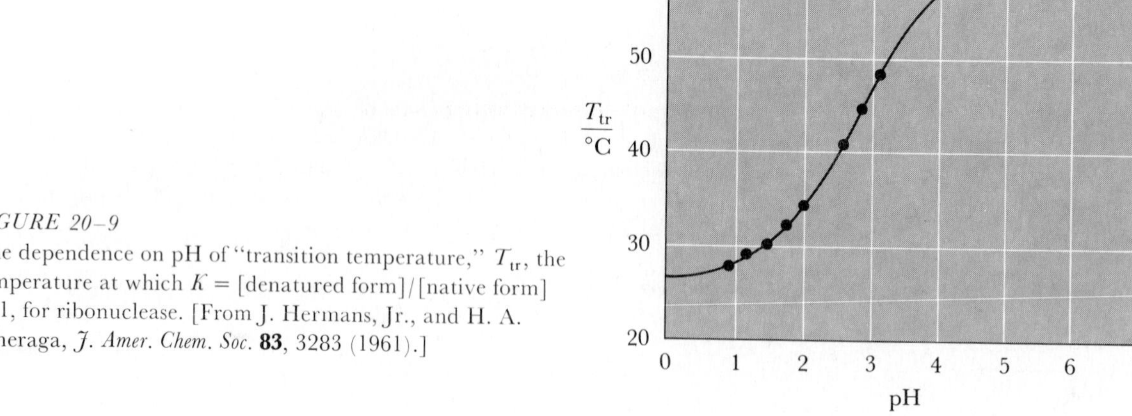

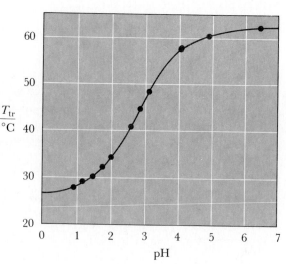

FIGURE 20–9
The dependence on pH of "transition temperature," T_{tr}, the temperature at which K = [denatured form]/[native form] = 1, for ribonuclease. [From J. Hermans, Jr., and H. A. Scheraga, *J. Amer. Chem. Soc.* **83**, 3283 (1961).]

occurs. Examination of the titration curve of ribonuclease in Figure 20–7 shows that this protein has a net positive charge at each of these values of pH. At lower pH where the net charge is more positive the denaturation occurs at a lower temperature.

20–4 Fats and Fatty-Acid Metabolism

You already have learned that esters have the structure

$$\mathrm{R{-}\overset{\overset{\displaystyle O}{\|}}{C}{-}O{-}R'}$$

and that a compound with this structure can be formed from a carboxylic acid (RCO_2H) and an alcohol ($R'OH$):

$$\mathrm{R{-}\overset{\overset{\displaystyle O}{\|}}{C}{-}\boxed{OH + H}OR' = R{-}\overset{\overset{\displaystyle O}{\|}}{C}{-}O{-}R' + H_2O}$$

Neutral fats, found widely in plants and animals, are esters of long-chain carboxylic acids and the trihydroxyl alcohol, glycerol, $C_3H_8O_3$, with the structure

$$\begin{array}{c} \mathrm{H} \\ | \\ \mathrm{H{-}C{-}OH} \\ | \\ \mathrm{H{-}C{-}OH} \\ | \\ \mathrm{H{-}C{-}OH} \\ | \\ \mathrm{H} \end{array}$$

There are fats in which one, two, or three of the hydroxyl groups of glycerol are esterified, but we will consider only those with three ester linkages:

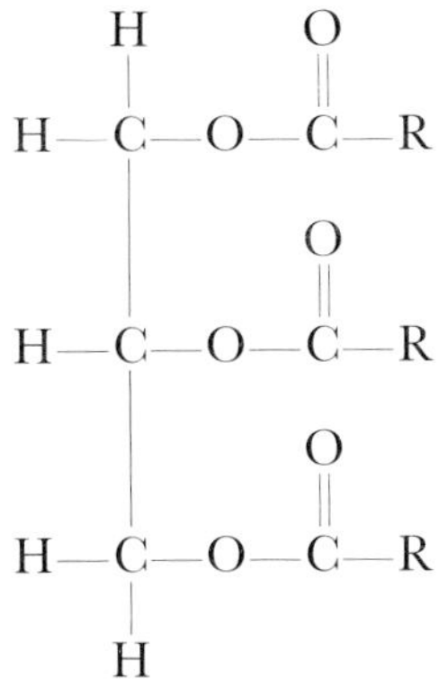

$$
\begin{array}{l}
\mathrm{H}\mathrm{O}\\
|\|\\
\mathrm{H-C-O-C-R}\\
|\mathrm{O}\\
|\|\\
\mathrm{H-C-O-C-R}\\
|\mathrm{O}\\
|\|\\
\mathrm{H-C-O-C-R}\\
|\\
\mathrm{H}
\end{array}
$$

The carboxylic acids that are combined with glycerol in some common fats are given in Table 20–2. The list includes acids with saturated alkyl groups (C_nH_{2n+1}, $C_{15}H_{31}$ and $C_{17}H_{35}$) and acids with unsaturated hydrocarbon groups. As we already have discussed, appropriately substituted ethylenes have cis and trans isomers, and this type of isomerism is possible in the unsaturated fatty acids. Oleic acid, linoleic acid, and most naturally occurring fatty acids have the cis structure at each carbon–carbon double bond. Each of these acids has an alkyl group that is unbranched (i.e., continuous or straight chain), and each acid has an even number of carbon atoms. Melting points of the unsaturated fatty acids are lower than those of the corresponding saturated acids, and the same general trend exists for the fats involving acids with these features. Thus stearin, the ester of glycerol and stearic acid (a saturated acid) is a solid at ordinary temperatures, and the ester of glycerol and oleic acid (an unsaturated acid), the major constituent of olive oil, is a liquid at ordinary temperatures. The double bonds in the glycerol esters of unsaturated acids that make up vegetable oils can be converted by hydrogenation to the single bonds present in solid fats. This is done in the manufacture of margarine from vegetable oils.

TABLE 20–2
Carboxylic Acids Derived From Neutral Fats

Name	Chemical formula	Mp °C
Palmitic acid (*n*-hexadecanoic acid)	$C_{15}H_{31}COOH$	63.1
Stearic acid (*n*-octadecanoic acid)	$C_{17}H_{35}COOH$	71.5
Oleic acid (*n*-9-octadecenoic acid)	$C_{17}H_{33}COOH$	16.3
Linoleic acid (*n*-9,12-octadecadienoic acid)	$C_{17}H_{31}COOH$	−5.0

Esterification is a reversible reaction, but it is a reaction that does not occur rapidly. The hydrolysis of esters is catalyzed by acids and enzymes. In the first step of metabolism of fats in higher animals, enzymes known as *lipases* are the catalysts that convert the fat to glycerol and fatty acid. Bases also promote the hydrolysis of fats, but it is inappropriate to call this catalysis because the base is consumed:

$$C_3H_5O_3(COR)_3 + 3OH^- = C_3H_5(OH)_3 + 3RCO_2^-$$

This reaction, employed in producing soap, which is a salt of a long-chain carboxylic acid, commonly is called *saponification*. The cleaning action of soap arises from the nature of the long-chain carboxylate ion, a hydrocarbon group attracting hydrophobic organic substances, combined with an ionic group attracting water:

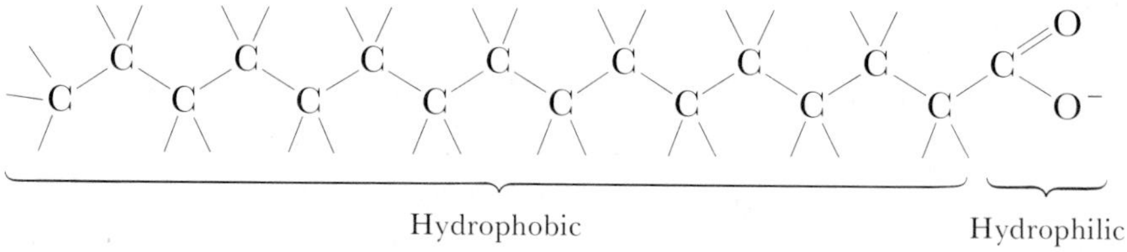

Hydrophobic Hydrophilic

Palmitate ion

The fact that most naturally occurring fatty acids contain an even number of carbon atoms suggests that the biosynthetic pathway to form them involves addition of two-carbon fragments. This has been shown to be the case, and it also has been shown that the metabolism of fatty acids involves a repeatable cycle of steps that leads to the loss of one two-carbon fragment per cycle. The mechanism for fatty-acid metabolism is complicated but each step in the cycle involves a simple transformation. These steps involve several complicated molecules:

Coenzyme A	HS**CoA**
Adenosine triphosphate	ATP
Adenosine diphosphate	ADP
Flavin adenine dinucleotide:	
Oxidized form	FAD
Reduced form	$FADH_2$
Nicotinamide adenine dinucleotide:	
Oxidized form	NAD^+
Reduced form	NADH

The structures of these molecules are given in Table 20–3. The reaction steps are:

1. *Formation* of thioester of carboxylic acid with coenzyme A (HS**CoA**):

$$R\text{—}CH_2\text{—}CH_2\text{—}CH_2\text{—}CO_2H + HS\mathbf{CoA} + ATP \rightarrow R\text{—}CH_2\text{—}CH_2\text{—}CH_2\text{—}CO(S\mathbf{CoA}) + ADP + P_i$$ (see footnote 8)

[8] P_i = inorganic phosphate (this symbol is used commonly by biochemists). This equation appears not to be balanced; the components of H_2O are combined in P_i.

2. *Oxidation* (dehydrogenation) of fatty acyl **CoA**:

$$\mathrm{R{-}CH_2{-}CH_2{-}CH_2{-}CO(S\textbf{CoA}) + FAD \rightarrow R{-}CH_2CH{=}CH{-}CO(S\textbf{CoA}) + FADH_2}$$

3. *Addition* of water to carbon–carbon double bond:

$$\mathrm{R{-}CH_2{-}CH{=}CH{-}CO(S\textbf{CoA}) + H_2O \rightarrow R{-}CH_2{-}CHOH{-}CH_2{-}CO(S\textbf{CoA})}$$

4. *Oxidation* (dehydrogenation) to convert alcohol group to ketone group:

$$\mathrm{R{-}CH_2{-}CHOH{-}CH_2{-}CO(S\textbf{CoA}) + NAD^+ \rightarrow R{-}CH_2{-}CO{-}CH_2{-}CO(S\textbf{CoA}) + NADH + H^+}$$

5. *Cleavage* of acetyl **CoA** from ketoester by another molecule of coenzyme A:

$$\mathrm{R{-}CH_2{-}CO{-}CH_2{-}CO(S\textbf{CoA}) + HS\textbf{CoA} \rightarrow R{-}CH_2{-}CO(S\textbf{CoA}) + CH_3CO(S\textbf{CoA})}$$

If stearic acid ($R = C_{14}H_{29}{-}$) participated in this series of five steps, the product would be the thioester of palmitic acid (an acid with two fewer carbon atoms) with coenzyme A plus acetyl **CoA**. Repetition of Steps 2–5 of this cycle seven more times would degrade one molecule of the acid with eighteen carbon atoms to nine molecules, each containing a two-carbon fragment from the original fatty acid. The net chemical change is

$$\mathrm{C_{17}H_{35}COOH + 9HS\textbf{CoA} + 8H_2O + 8FAD + 8NAD^+ + ATP \rightarrow 9CH_3CO(S\textbf{CoA}) + 8FADH_2 + 8NADH + ADP + P_i + 8H^+}$$

Each step in this sequence is catalyzed by an appropriate enzyme. There has been net oxidation of the fatty acid (and concommitant reduction of FAD and NAD^+). As we saw in Chapter 16, the oxidizing ability of organic reagents can be summarized in an $\mathscr{E}^{0\prime}$ value, just as for simple inorganic reagents. For these dinucleotides, the half-reactions and $\mathscr{E}^{0\prime}$ (at $[H^+] = 1.00 \times 10^{-7}$ mol L^{-1}) are

$$\mathrm{NAD^+ + H^+ + 2e^- = NADH} \qquad \mathscr{E}^{0\prime} = -0.320 \text{ V}$$
$$\mathrm{FAD + 2H^+ + 2e^- = FADH_2} \qquad \mathscr{E}^{0\prime} = -0.219 \text{ V}$$

If we compare these $\mathscr{E}^{0\prime}$ values with that for oxygen at this concentration of hydrogen ion,

$$\mathrm{O_2}(g) + \mathrm{4H^+ + 4e^- = 2H_2O} \qquad \mathscr{E}^{0\prime} = +0.815 \text{ V}$$

we see that oxygen, the ultimate oxidizing agent in biological oxidation, is capable (with a large margin to spare) of oxidizing the reduced form of either of these couples. However, the biological oxidation of NADH or $FADH_2$ is not direct; each of these highly favored oxidation–reduction reactions is coupled to the unfavorable conversion of ADP to ATP in a

TABLE 20–3

The Structures of Molecules Involved in Fatty-Acid Metabolism[a]

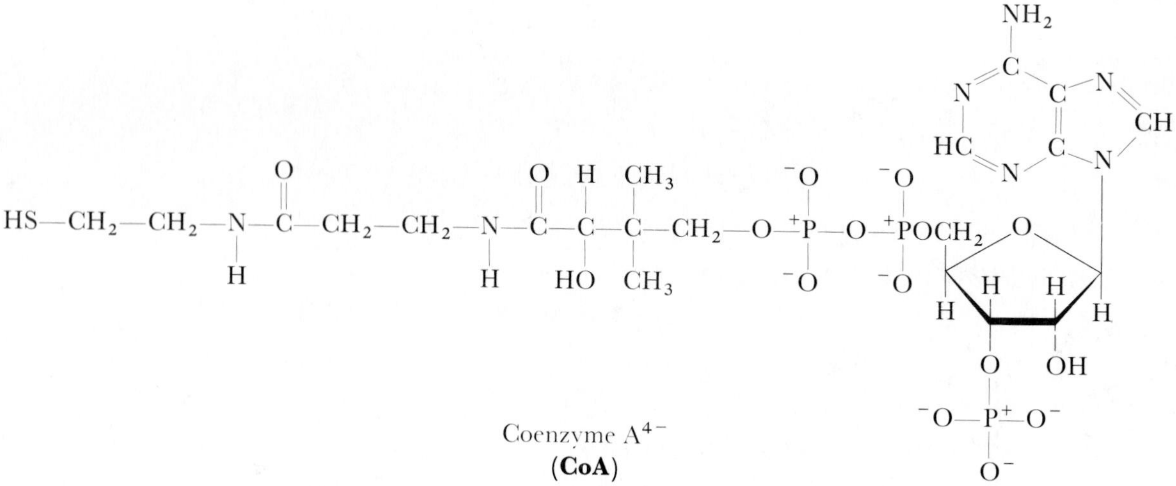

HS—CH₂—CH₂—N(H)—C(=O)—CH₂—CH₂—N(H)—C(=O)—C(H)(HO)—C(CH₃)(CH₃)—CH₂—O—⁺P(⁻O)(⁻O)—O—⁺P(⁻O)(⁻O)—OCH₂ (ribose–adenine; 3′-O—⁺P(⁻O)(⁻O)—O⁻; 2′-OH)

Coenzyme A^{4-}
(CoA)

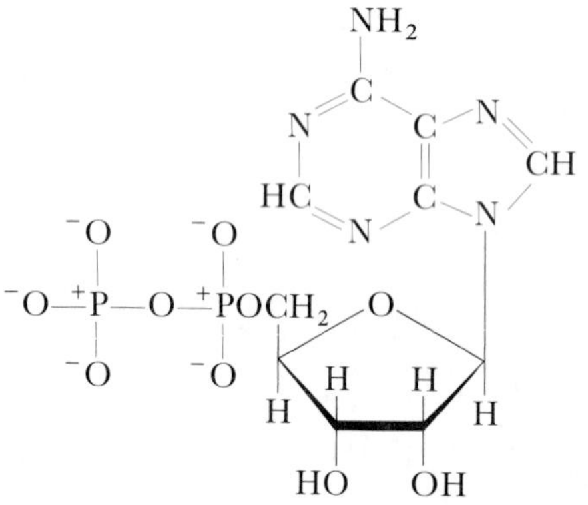

⁻O—⁺P(⁻O)(⁻O)—O—⁺P(⁻O)(⁻O)—OCH₂ (ribose–adenine; HO, OH)

ADP^{3-}

⁻O—⁺P(⁻O)(⁻O)—O—⁺P(⁻O)(⁻O)—O—⁺P(⁻O)(⁻O)—OCH₂ (ribose–adenine; HO, OH)

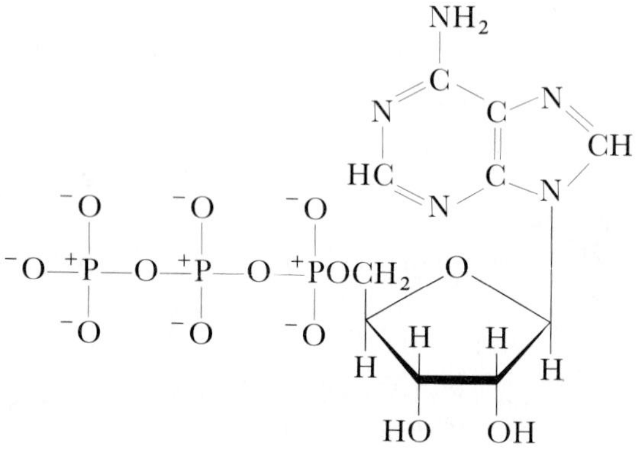

ATP^{4-}

process referred to as oxidative phosphorylation.[9] To oversimplify, we can say that the unfavorable process,

$$ADP + P_i = ATP$$

is driven by the favorable process,

$$O_2 + 2NADH + 2H^+ = 2H_2O + 2NAD^+$$

Study of these processes has shown that three molecules of ATP are formed per one molecule of NADH oxidized, and two molecules of ATP are formed per one molecule of $FADH_2$ oxidized.

[9]Recall (Section 17–4) our discussion of the coupling of the formation of HF (a favorable reaction) to the formation of XeO_3 (an unfavorable reaction).

TABLE 20–3 (Continued)
The Structures of Molecules Involved in Fatty-Acid Metabolism[a]

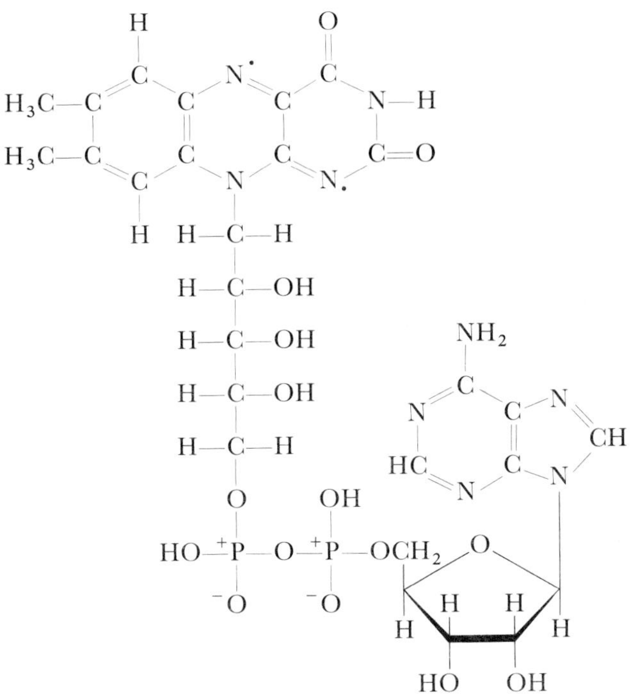

In $FADH_2$, H atoms are bonded at the two nitrogen atoms designated with dots.

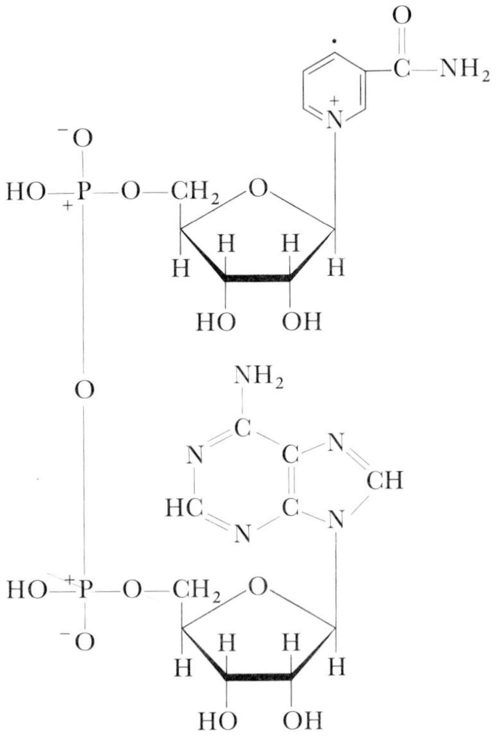

In NADH, an additional H atom is bonded at the carbon atom designated with a dot.

[a]The charges shown on N, P, and O are the formal charges associated with the single-bonded phosphate structures shown here.

20–5 Carbohydrates

Carbohydrates are a large class of organic substances, polyhydroxyl aldehydes and ketones, which includes both small molecules and macromolecules. Some small carbohydrates are important parts of our diet. Another small carbohydrate is a building block of nucleotides, which are very important biomolecules. The principal macromolecular carbohydrates are *starch* and *cellulose*. Let us first consider the nature of the simple carbohydrate molecules.

The empirical formula for simple sugars, called *monosaccharides*, is CH_2O; molecular formulas for known monosaccharides are three to eight times this simplest formula, $C_3H_6O_3$ to $C_8H_{16}O_8$. The simple sugars with which we will be concerned are those with five and six carbon atoms:

$C_5H_{10}O_5$ and $C_6H_{12}O_6$. Many isomeric monosaccharides have these compositions, but the ones we shall study are D-ribose, a pentose (a five-carbon sugar), and D-glucose, a hexose (a six-carbon sugar). The structures of these molecules are:

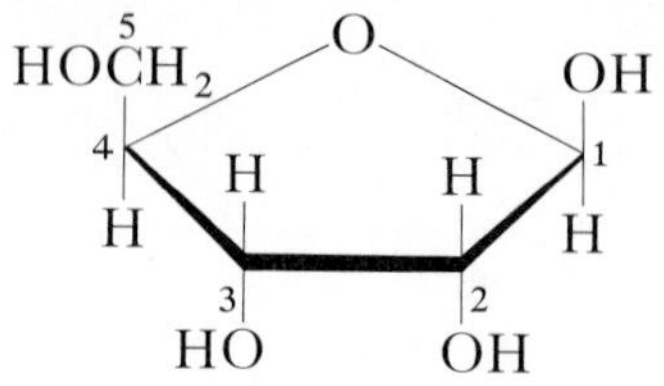

D-Ribose 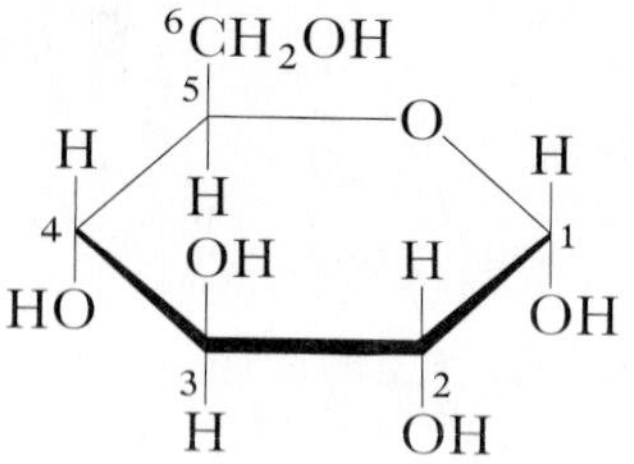 D-Glucose

The designation D for these chiral molecules refers to the configuration of the four different groups bonded to carbon atom 4 in D-ribose and to carbon atom 5 in D-glucose. (Molecules with the opposite configuration at these carbon atoms are designated as the L isomers.) Simply think of D and L as labels to distinguish optical isomers. Four of the carbon atoms in D-glucose are bonded to four different groups, therefore D-glucose is one of sixteen ($2^4 = 16$) possible optically active isomers with this structure. These molecules exist predominantly in the cyclic structures just shown, but each is related to an open-chain form with an aldehyde group:

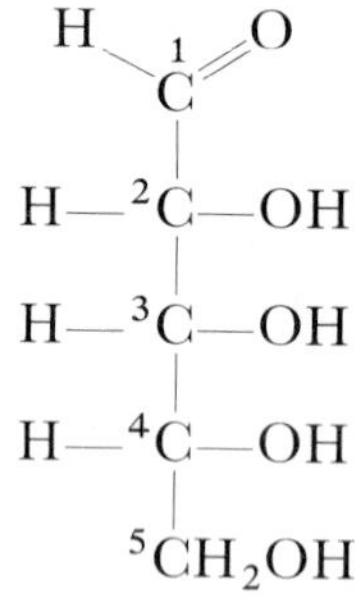

$$\begin{array}{c} \mathrm{H{-}{}^{1}C{=}O} \\ | \\ \mathrm{H{-}{}^{2}C{-}OH} \\ | \\ \mathrm{H{-}{}^{3}C{-}OH} \\ | \\ \mathrm{H{-}{}^{4}C{-}OH} \\ | \\ \mathrm{{}^{5}CH_2OH} \end{array} \qquad \begin{array}{c} \mathrm{H{-}{}^{1}C{=}O} \\ | \\ \mathrm{H{-}{}^{2}C{-}OH} \\ | \\ \mathrm{HO{-}{}^{3}C{-}H} \\ | \\ \mathrm{H{-}{}^{4}C{-}OH} \\ | \\ \mathrm{H{-}{}^{5}C{-}OH} \\ | \\ \mathrm{{}^{6}CH_2OH} \end{array}$$

D-Ribose 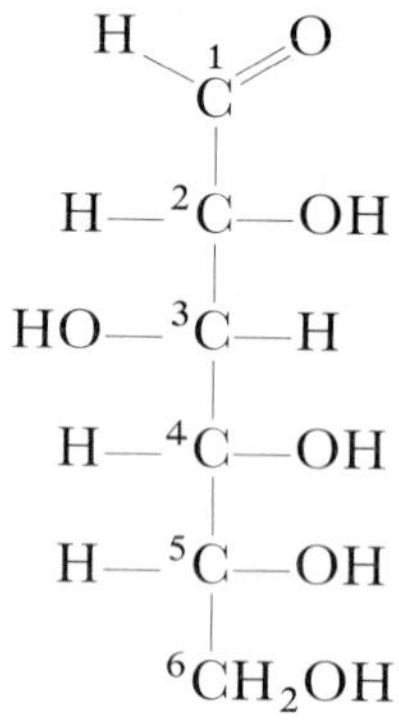 D-Glucose

Formation of the C—O—C linkage of the cyclic molecules from the open-chain molecules is analogous to the reaction of an aldehyde and an alcohol to form a hemiacetal:

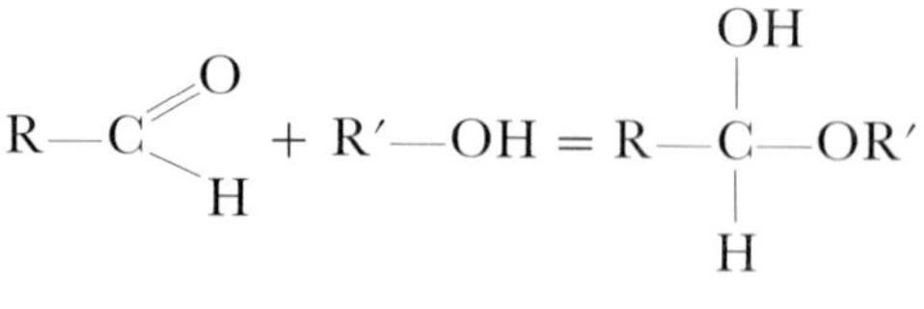

$$\mathrm{R{-}C({=}O){-}H} + \mathrm{R'{-}OH} = \mathrm{R{-}C(OH)(H){-}OR'}$$

Aldehyde Alcohol Hemiacetal

The cyclic structure of D-glucose involves a six-member ring with no carbon–carbon double bonds in the ring. Thus it resembles cyclohexane, and like cyclohexane, D-glucose exists in the chair conformation:

Although one $>CH_2$ unit of cyclohexane is replaced by an oxygen $>\ddot{O}:$ in the ring form of D-glucose, the repulsion of the unshared pairs of electrons in the cyclic sugar does not impose a special strain on the six-member ring.

Related to D-ribose is D-2-deoxyribose,

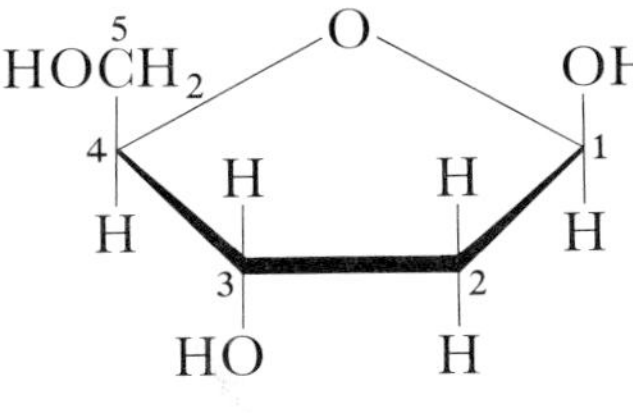

D-2-Deoxyribose

Like D-ribose, this molecule is a component of nucleotides, which will be discussed later.

Ordinary sugar, sucrose, has the chemical composition $C_{12}H_{22}O_{11}$. In acidic solution (e.g., in the stomach), this disaccharide breaks down to give glucose and another monosaccharide, fructose, a keto hexose:

$$\underset{\text{Sucrose}}{C_{12}H_{22}O_{11}} + H_2O = \underset{\text{Glucose}}{C_6H_{12}O_6} + \underset{\text{Fructose}}{C_6H_{12}O_6}$$

(Fructose has a ketone group $>C{=}O$, not an aldehyde group.) An ether linkage joins the two hexose molecules to give the disaccharide structure of sucrose:

Other disaccharides are known: for example, maltose is built from two glucose molecules.

An ether linkage of this type also joins hexose units in the naturally occurring large polymers of D-glucose, starch and cellulose. Cellulose makes up over half of the total carbon combined in organic molecules in the biosphere. (Wood is approximately half cellulose, and cotton is almost pure cellulose.) The glucose molecules in cellulose are joined at carbon atoms 1 and 4:

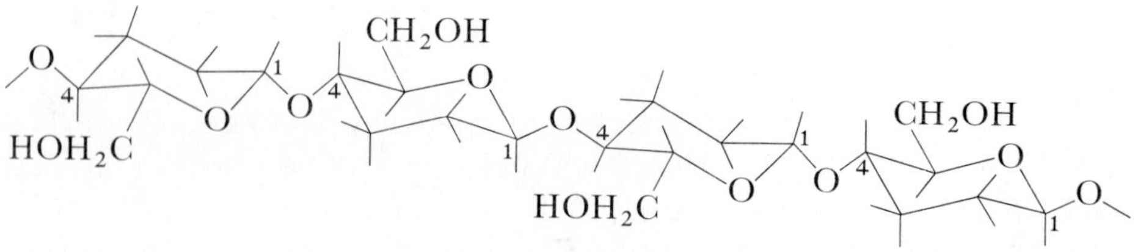

CH$_2$OH ... HOH$_2$C ... CH$_2$OH ... HOH$_2$C

The molecular weight of cellulose depends on the source; it varies from 50,000 to 500,000 or higher. The glucose residue, $C_6H_{12}O_6$ minus $H_2O = C_6H_{10}O_5$, has a molecular weight of 162; therefore the cellulose molecular weights correspond to 300 to 3000 glucose residues. The long-chain cellulose molecules are arranged side by side in bundles, called *fibrils*, which are held together by hydrogen bonding. The walls of plant cells are largely made up of cellulose.

Although cellulose fibrils are not soluble in water, the numerous hydroxyl groups of cellulose allow hydrogen bonding to water, which accounts for the absorbent nature of cotton. (This is contrasted to the non-absorbent nature of synthetic fibers, nylon and Dacron, the structures of which were given in Figure 20–1.)

The stereochemistry of linkages between monomer D-glucose units of starch differs from that in cellulose. Like cellulose, starch is a heterogeneous polymer with molecular weights ranging from as low as a few thousand to 500,000, the same approximate upper limit as in cellulose. Enzymes of our digestive tract catalyze the conversion of starch into monosaccharides. These enzymes do not catalyze the digestion of cellulose, presumably because of the different sterochemistry of the linkage connecting the glucose units in this polysaccharide.

Important for our discussion of nucleosides and nucleotides is the ability of a hydroxyl group of a sugar

1. to form an ester linkage with phosphate, and
2. to be replaced by a —NR_2 group.

The reactions for formation of methyl phosphate,

$$\mathrm{HO{-}\overset{\overset{\textstyle O^-}{|}}{\underset{\underset{\textstyle OH}{|}}{P^+}}{-}OH + HOCH_3 = HO{-}\overset{\overset{\textstyle O^-}{|}}{\underset{\underset{\textstyle OH}{|}}{P^+}}{-}OCH_3 + H_2O}$$

and ribose-5-monophosphate,

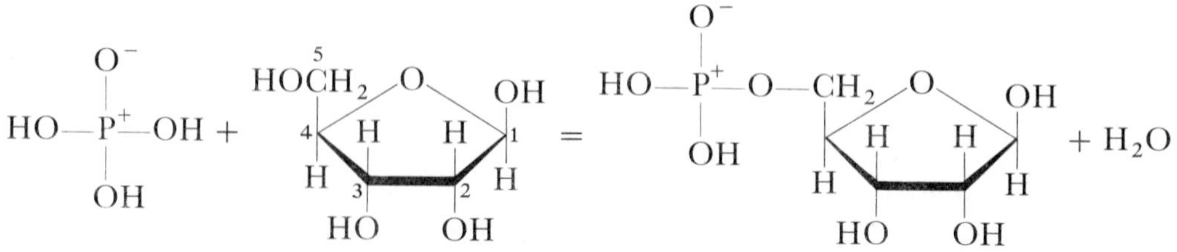

$$\mathrm{HO{-}\overset{\overset{\textstyle O^-}{|}}{\underset{\underset{\textstyle OH}{|}}{P^+}}{-}OH} + \text{ribose (}\overset{5}{\mathrm{HOCH_2}}\text{, C-1 OH, C-2 OH, C-3 OH)} = \mathrm{HO{-}\overset{\overset{\textstyle O^-}{|}}{\underset{\underset{\textstyle OH}{|}}{P^+}}{-}O{-}CH_2}\text{-ribose} + H_2O$$

are analogous to one another; the product in each case is a phosphate ester. These substances were discussed in Chapter 17 and in the preceding section of this chapter. Here we shall emphasize the ability of a ribose molecule and a phosphate group to form more than one phosphate-ester linkage. A molecule containing three phosphate groups and two ribose molecules has a small polymeric structure that is made possible by these phosphate-ester linkages:

```
           O⁻
           |
     HO—⁺P—O⁻
           |
           O
           |
          ⁵CH₂  O       OH
         4 /  H     H  \ 1
           H  3      2   H
              O     OH
              |
       HO—P⁺—O⁻
              |
              O
              |
             ⁵CH₂  O       OH
            4 /  H     H  \ 1
              H  3      2   H
                 O     OH
                 |
          HO—P⁺—O⁻
                 |
                 O⁻
```

We shall see that this type of structure is the backbone of the giant polymeric molecules DNA and RNA.

In these molecules, we also will see the hydroxyl group on carbon atom 1 replaced by the residue of a nitrogen base. The structural analogy between a hydroxyl group (or an alkoxyl group) and an amino group (or an alkyl amino group):

```
  ··           ··
 ·O : H       ·N : H
  ··           ··
               H

  ··           ··
 ·O : R       ·N : R
  ··           ··
               H
```

has been pointed out on several occastions. The structure of 1-amino-D-ribose,

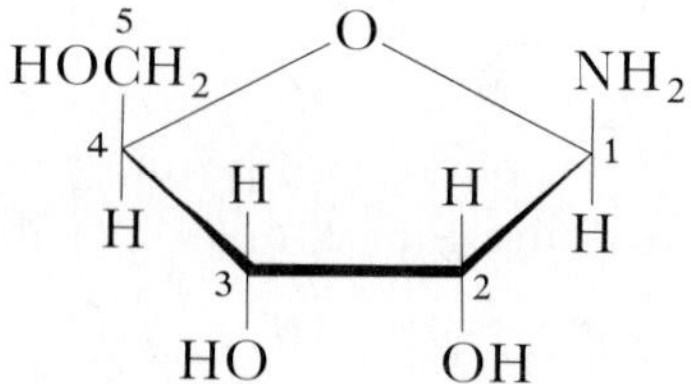

is analogous to the structure we saw in Table 20–3 and those we will encounter in our discussion of nucleotides.

20–6 Nucleotides and Molecular Biology

Nucleotides are biomolecules with an incredible variety of functions in biology. Our discussion of the types of linkages a ribose molecule can make with the phosphate group and with nitrogen bases has prepared you to understand the structure of polynuceotides, and our continuing encounter with examples of hydrogen bonding will prepare you to understand the specificity of the base pairing in double helical strands of DNA (deoxyribonucleic acid). But the topic is so vast that only an introduction can be presented here.

The nitrogen bases that are components of nucleotides are cytosine, uracil, thymine, adenine, and guanine; their structures are given in Table 20–4. The combination of one of these bases with a ribose or a deoxyribose molecule in the manner already shown is called a *nucleoside*. If the sugar is esterified with phosphate, the molecule is called a *nucleotide*. The structure of adenosine monophosphate (AMP^{2-}), an important mononucleotide, is

The structures of adenosine diphosphate (ADP) and adenosine triphosphate (ATP), molecules that we have considered in our discussion of fatty-acid metabolism, are the same as that of AMP, except that the phosphate group $—OPO_3^{2-}$ is replaced by a diphosphate group, $—OP_2O_6^{3-}$, or a triphos-

TABLE 20–4
Structures of the Nitrogen Bases

Pyrimidines		
Cytosine (C)	Uracil (U)	Thymine (T)

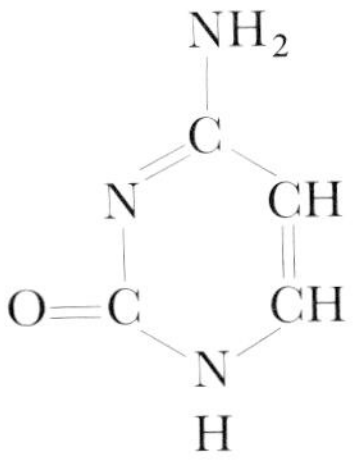

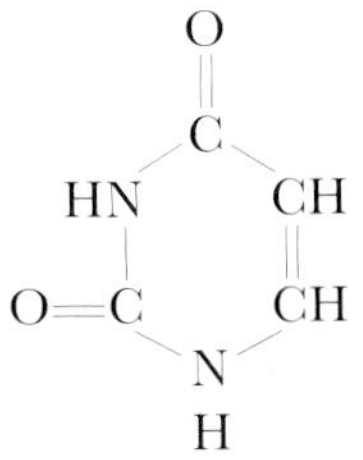

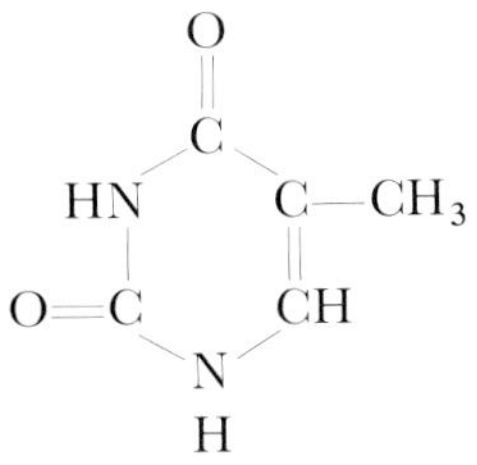

Purines	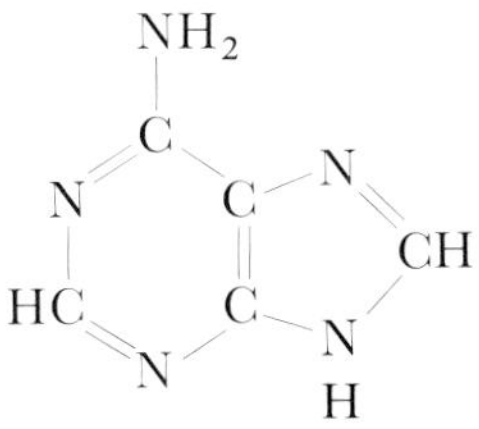
Adenine (A)	Guanine (G)

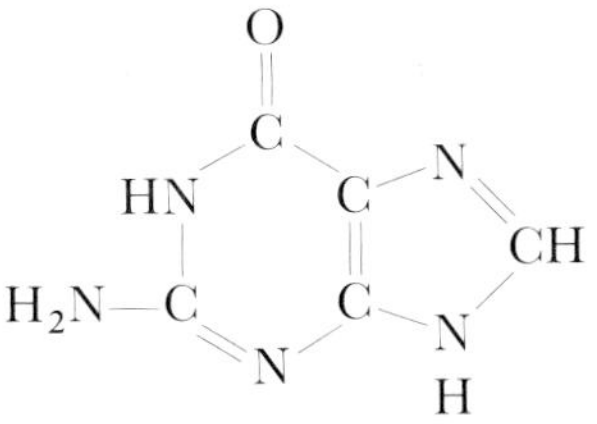

phate group, —$OP_3O_9^{4-}$, respectively. (Recall, from Chapter 17, that the extent of protonation of the various phosphate groups depends on pH.)

Polynucleotides are polymeric molecules in which nucleotide structures are bonded together by bridging phosphate groups. The basic nucleotide structures are phosphate esters of ribose or deoxyribose molecules in which a nitrogen base has replaced one hydroxyl group. The polymers that contain deoxyribose are called *deoxyribonucleic acid* (DNA) and those which contain ribose are called *ribonucleic acid* (RNA). This polymeric structure, shown in Figure 20–10, is the molecule of heredity (in DNA), in which the long polymeric strands are paired in a double helical structure, and is the molecule (in RNA) which directs the synthesis of specific proteins.

In the double-stranded DNA, the helical structure is stabilized by hydrogen bonding between the nitrogen bases on different strands of the linear polynucleotide. The hydrogen bonding is particularly favorable between the bases adenine and thymine (A–T), and between guanine and cytosine (G–C). This is shown in Figure 20–11.

Deoxyribonucleic acid was first isolated more than a century ago by F. MIESCHER, but only during the past few decades have the structure of this molecule and its role in heredity been explained. One important clue to the structure of DNA was the observation by CHARGAFF that the amounts of adenine and thymine in it are equal, as are the amounts of guanine and cytosine. These pairs are those with the favorable hydrogen-bonding inter-

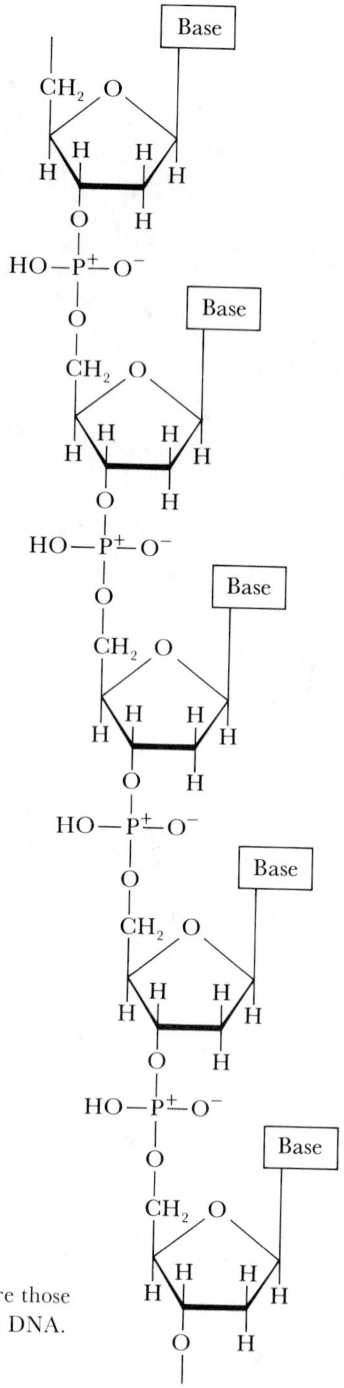

FIGURE 20–10
The polynucleotide structure of DNA. The units Base are those shown in Table 20–4. The bases C, T, A, and G occur in DNA.

actions already shown. The relative amounts of the two kinds of base pairs varies from species to species. In mammals, adenine and thymine each constitute ~30.2 mole % of the bases in DNA, and guanine and cytosine each constitute ~19.8 mole %. For *sarcina lutea*, a bacterium, the figures are ~12.9 mole % A and T and ~37.1 mole % G and C.

The landmark proposal that DNA has a double-helix structure was made by CRICK and WATSON in 1953. In arriving at this proposal, they con-

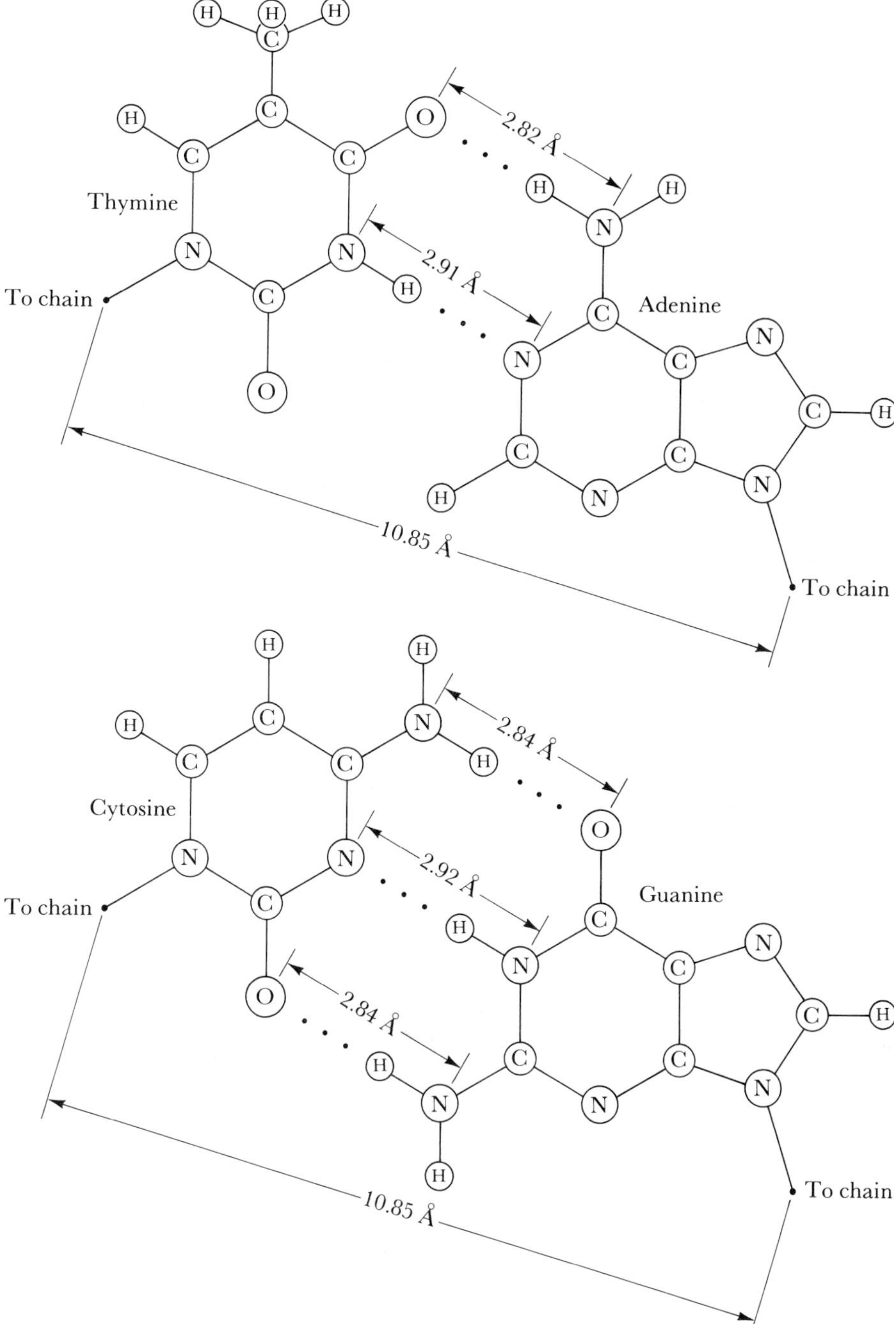

FIGURE 20–11
The hydrogen bonding in the complementary A–T and G–C base pairs. (This is the key to the Crick–Watson double helix.)

sidered known bond distances and bond angles observed in simpler molecules, plus x-ray diffraction data obtained by FRANKLIN. The genetic code is contained in the ordering of the bases on a strand of DNA. The specificity of the base pairing, A with T and G with C, dictates that information regarding the order of the bases in one strand is contained in the other complementary strand. This may be shown with a simple diagram:

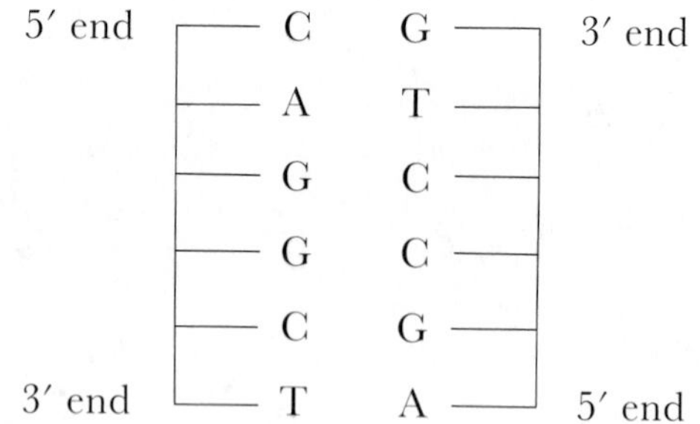

This diagram shows that the strands run in the opposite directions. The base at the end of a strand that has a terminal 5′ hydroxyl group (i.e., a 5′ hydroxyl group that is not esterfied to phosphate) is paired to a base at the end of the other strand, which has a 3′ hydroxyl group. (In writing the base sequence of a polynucleotide, the base at the 5′ end is given first.) The order CAGGCT in one strand of a hexanucleotide is stored in the order AGCCTG of the complementary strand. (Each of these orders starts with the 5′ end.) When these strands separate in replication, a new strand is built from component mononucleotides, using each parent strand as a template. The result is two double-helix DNA molecules, each with strands having the same sequence of bases as in the parent DNA molecule:

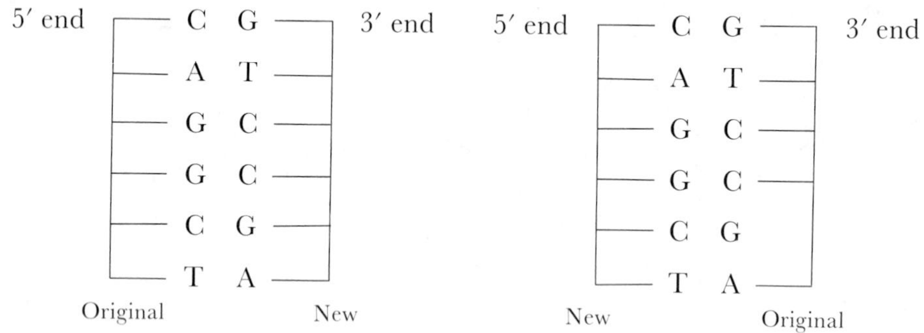

(In these diagrams, the way in which each chain spirals around the other to form the double helix is not shown.)

The chromosomes contain DNA, and the order of the bases along the DNA strand directs the construction of a particular RNA (i.e., an RNA with a particular order of bases) that in turn directs the construction of a protein with a particular sequence of amino acids. Molecules of RNA contain uracil (U) instead of thymine (T).

The sequence of processes in the biosyntesis of proteins is complex, but in the quarter of a century since the double helix was proposed, enormous, mind-boggling advances have been made. Among these is the association of particular nucleotide triplets (i.e., particular orders of three consecutive bases) in RNA as the codes for synthesis of each of the twenty essential amino acids from which proteins are built. For instance, the sequence CCG directs the synthesis of proline and AGC directs the synthesis of serine.

Therefore a segment of RNA with the sequence CCGAGC would direct the synthesis of a dipeptide segment of a protein with a proline residue followed by a serine residue. An alphabet of four letters, A, C, G, and U, generates 4^3, or sixty-four, different three-letter words. Because there are only twenty essential amino acids, there is redundancy in the genetic code (the synthesis of proline is directed by CCU, CCC, and CCA, as well as CCG), and three of the sixty-four sequences (UAA, UAG, and UGA) are words directing the termination of a polypeptide chain.

This brief discussion introduces you to some aspects of molecular biology. We have shown that simple principles of molecular structure and chemical reactions are the basis for biological specificity. You will learn more about this exciting area of science in later courses in biochemistry or biology and in reports in the popular press.

Biographical Notes

ERWIN CHARGAFF (1905–), an American chemist, was born and educated in Austria. A faculty member at Columbia University since 1935, he has pioneered in developing analytical methods for studying nucleic acids.

ROBERT B. COREY (1897–1971) was Professor of Structural Chemistry in the California Institute of Technology. He received his education at the University of Pittsburgh and Cornell University. In collaboration with Linus Pauling at Caltech, he did much to elucidate the structure of proteins.

FRANCIS H. C. CRICK (1916–), a British biologist, was educated at University College, London, and Cambridge. For his work on the structure of DNA he received, with J. D. Watson and M. H. F. Wilkins (a British biophysicist), the Nobel Prize in Medicine and Physiology in 1962.

ROSALIND FRANKLIN (1920–1958), a British crystallographer, did her work on the structure of DNA at Kings College, London, in M.H.F. Wilkins's laboratory.

R. B. MERRIFIELD (1921–), an American biochemist, was educated at the University of California, Los Angeles (B.A. 1943, Ph.D. 1949). He conducted his work on solid-phase protein synthesis at Rockefeller University, where he is Professor of Chemistry.

J. FRIEDRICH MIESCHER (1844–1895), a Swiss physiologist, is considered to be the founder of cellular chemistry. He also studied the role of carbon dioxide in blood in the regulation of breathing.

LINUS PAULING (see biographical notes in Chapter 10).

FREDERICK SANGER (1918–), an English biochemist, was educated at Cambridge, where presently he is head of the Protein Chemistry Division of the Laboratory of Molecular Biology. For his work on insulin, Frederick Sanger received the Nobel Prize for Chemistry in 1958.

JAMES D. WATSON (1928–), an American biologist, was only 25 when, as a postdoctoral student, he and F.H.C. Crick proposed their structure for DNA. His book *The Double Helix* (1968) describes this work in layman's terms. He was Professor of Biology at Harvard University, and now is Director of the Cold Springs Harbor Laboratory.

KARL ZIEGLER (1898–1973), a German organic chemist, received the 1963 Nobel Prize in Chemistry (shared with G. Natta) for his work on polymerization.

Problems and Questions

20–1 Explain the negative sign for ΔS^0 for the reaction in which gaseous ethylene at 1.00 atm gives gaseous 1-butene at 1.00 atm. If each gas were at a pressure of 1 torr, would the value of ΔS^0 be less negative, more negative, or the same?

20–2 Alkali-metal atoms are known to initiate the polymerization of olefins. Suggest a mechanism for this polymerization, using ethylene as an example.

20–3 If the carbon–carbon single bond distance is 1.54 Å and the CCC angle is 109.5°, what is the maximum extended end-to-end length of a polyethylene molecule with a molecular weight of 1.4×10^5?

20–4 Rubber tires deteriorate in an atmosphere that is rich in the components of urban smog (NO_2 and ROOH, an organic peroxide). Suggest a type of reaction that may be occurring.

20–5 Sodium hydroxide solution is accidentally spilled on a Dacron shirt. A hole appears. Explain.

20–6 Write the balanced chemical equation for each of the following reactions: (a) glycylvaline is formed from its constituent amino acids, (b) glycylvaline is hydrolyzed in acid, (c) glycylvaline is hydrolyzed in base, and (d) cystine reacts with oxygen to give cysteine.

20–7 The acid-dissociation constants for phenylalanine are $K_1 = 1.6 \times 10^{-2}$ mol L^{-1} and $K_2 = 7.9 \times 10^{-10}$ mol L^{-1}. What is the isoelectric point for this amino acid?

20–8 What fraction of phenylalanine is present in the cationic form at the isoelectric point? What are the relative concentrations of the cationic, neutral, and anionic forms at pH = 7.10?

20–9 Draw the structures of the differently protonated forms of lysine and aspartic acid.

20–10 Draw the structures and name all isomeric tripeptides containing one residue each of valine, glutamic acid, and proline. For one of these structures indicate the net charges the molecule can have as the pH of the solution is changed from ~1 to ~13.

20–11 From the molecular weight of typical amino acid residues (see Table 20–1), calculate the approximate number of amino acid residues present in a simple protein with a molecular weight of 10^4.

20–12 Complete hydrolysis of an octapeptide reveals the composition (Ala) · $(Arg)_2$(Asp)(Glu)(Gly)(His)(Pro). The molecule is shown to have arginine as the N-terminal amino acid and histidine as the COOH-terminal end. Partial hydrolysis gives the dipeptides and tripeptides Pro–Asp–His, Arg–Glu, and Glu–Gly–Arg. What is the structure of this polypeptide?

20–13 Amides (e.g., acetamide) are much less basic than are amines (e.g., ethylamine). Explain.

20–14 The synthetic polypeptides polyglutamic acid and polylysine denature very differently in relation to pH. Polyglutamic acid is denatured in alkaline solution and polylysine is denatured in acidic solution. Explain.

20–15 Draw the resonance forms for the structure of guanidinium ion.

20–16 Write the chemical equation for the saponification of glycerol tristearate.

20–17 What structural features do all of the molecules given in Table 20–3 have in common?

20–18 The equilibrium constants for the last two stages of acid dissociation of ATP are (for 25°C):

$$H_2ATP^{2-} \rightleftarrows H^+ + HATP^{3-} \qquad K = 8.7 \times 10^{-5} \text{ mol L}^{-1}$$
$$HATP^{3-} \rightleftarrows H^+ + ATP^{4-} \qquad K = 1.1 \times 10^{-7} \text{ mol L}^{-1}$$

At what pH is 96% of ATP present as ATP^{4-}?

20–19 Magnesium(II) ion forms a complex with the anion of ATP; $K = [MgATP^{2-}]/[Mg^{2+}][ATP^{4-}] = 1.0 \times 10^4$ L mol^{-1}. What fraction of ATP is complexed by magnesium(II) ion at the $[Mg^{2+}] = 5.0 \times 10^{-4}$ mol L^{-1}? (Assume the pH is high enough to make ATP^{4-} the predominant uncomplexed form.)

29–20 An acidified sample of ribonuclease (concentration unknown) is titrated, and the pH is measured as a function of the amount of added base. How can you tell from the shape of the titration curve that the substance is not pure glycine?

20–21 From the molecular weight of a typical mononucleotide, calculate the approximate number of nucleotides in a double-stranded DNA molecule with a molecular weight of 6.8×10^7.

20–22 The cells of some mammals contain $\sim 6 \times 10^{-12}$ g DNA. Using the same molecular weight for a mononucleotide as used in the preceding question, calculate the number of mononucleotides in this amount of DNA.

20–23 Write the sequence of bases in a DNA strand that is complementary to each of the strands (a) GTACCTGA and (b) CGTAAC. (Use the conventional 5′ → 3′ order.)

21

Chemistry and Light

21–1 Introduction

The interaction of electromagnetic radiation with atomic and molecular species is an important part of chemistry. Some aspects of this subject were introduced in previous chapters. Study of the diffraction of x radiation has established the arrangement of atoms in many crystalline solids (Chapter 5) and has shed some light on the structure of liquids (Chapter 6). The great theoretical work of Planck, Einstein, and Bohr, which took physics into its modern era and which was discussed in Chapter 8, was rooted in experimental measurements of the wavelength of radiation involved in various interactions with matter: the radiation given off by a heated body, the radiation that can make solids eject electrons, and the radiation absorbed or emitted by gaseous atoms. Even more energetic than x radiation is γ radiation, which is emitted in nuclear processes; this subject will be introduced in Chapter 22.

In this chapter we will be concerned primarily with the interaction of light and molecules. The internal energy of a molecule consists of rotational energy, vibrational energy, and electronic energy. The energy of each of these types in a molecule is quantized; that is, only particular values are allowed. When a molecule absorbs light its internal energy increases; a transition occurs from a state with lower energy to a state with higher energy. In the emission of light, the reverse transition occurs.

Generally, analysis of the transitions between rotational energy levels provides information about the dimensions of a molecule, and analysis of the transitions between vibrational energy levels provides information about the springiness of the chemical bonds in a molecule. Molecules in higher electronic energy levels may be more reactive than the same molecules in their ground electronic states. This greater reactivity may accelerate a thermodynamically possible reaction or may facilitate a reaction for which equilibrium is unfavorable in the absence of light. In this chapter we will touch each of these subjects. In doing so we also will continue our study of many chemical topics already introduced: bond energies, reaction mechanisms, disproportionation of intermediate oxidation states, and the electronic structure of transition-metal complexes.

THE NATURE OF LIGHT

Our previous discussion of electromagnetic radiation has omitted some points that we now will consider before discussing the interaction of light with matter. The velocity of light, a combination of oscillating electric and magnetic fields (Figure 21–1), is an important constant of nature. The value $c = 3.00 \times 10^8$ m s^{-1} is the velocity in a vacuum. But in passing through matter, the velocity, v, is lower:

$$v = \frac{c}{n}$$

in which n is the refractive index of the matter. This equation is a definition of refractive index. The value of n is determined by the effect of the light's oscillating electric field in inducing oscillating electric fields due to motion of the electrons in the medium through which the light is passing. Values of the refractive index of various materials for light with $\lambda = 589$ nm are:

Material	n
Air (at 0°C and 1.0 atm)	1.00029
Water (l at 20°C)	1.333
Benzene (l at 25°C)	1.498
Heavy flint glass (~50% PbO)	1.66
Diamond	2.42

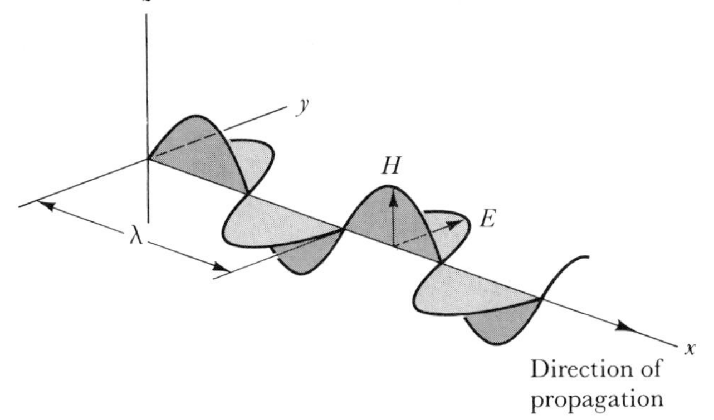

FIGURE 21–1
Light. Oscillating electric field (E) and magnetic field (H) perpendicular to one another. The direction of propagation is the x direction. The wavelength λ is defined in this figure. (See also Figures 5–6 and 8–3.)

When monochromatic light passes from a vacuum into matter, its frequency stays constant, which means that its wavelength changes:

$$\nu = \frac{c}{\lambda_0} = \frac{v}{\lambda_n}$$

in which λ_0 and λ_n are the wavelengths in a vacuum and in a medium with refractive index n. Thus

$$\lambda_n = \lambda_0 \times \frac{v}{c} = \lambda_0 \times \frac{1}{n}$$

The wavelengths for light absorption or emission by samples of matter generally are measured using instruments with a prism or grating in an air-filled chamber. In the most accurate work the fact that air has a refractive index very slightly larger than one is taken into account. The wavelength values reported in the chemical literature are values of λ_0.

For many purposes we are interested in the energy of the light photons, which we learned in Chapter 8 is related to the wavelength,

$$E = h\nu = h\frac{c}{\lambda_0}$$

We see that $E \propto \lambda_0^{-1}$, and chemists often use the reciprocal of the wavelength as a measure of the energy,

$$\lambda_0^{-1} = \tilde{\nu}$$

The dimensions of $\tilde{\nu}$ is reciprocal of length; generally $\tilde{\nu}$ is expressed in cm^{-1}, a unit called the *wave number*. Thus the energies of the limits of the visible part of the electromagnetic spectrum, 400 nm and 700 nm, expressed in wave numbers, are

$$\underline{400\text{ nm}} \qquad \tilde{\nu} = \frac{1}{400\text{ nm}} \times \frac{10^9\text{ nm}}{1\text{ m}} \times \frac{1\text{ m}}{100\text{ cm}} = 2.50 \times 10^4\text{ cm}^{-1}$$

$$\underline{700\text{ nm}} \qquad \tilde{\nu} = \frac{1}{700\text{ nm}} \times \frac{10^9\text{ nm}}{1\text{ m}} \times \frac{1\text{ m}}{100\text{ cm}} = 1.43 \times 10^4\text{ cm}^{-1}$$

In Chapter 8 we calculated that photons associated with light of these wavelengths have energies:

$$\underline{400\text{ nm}} \qquad E = 4.97 \times 10^{-19}\text{ J} \quad \text{or} \quad E = 299\text{ kJ mol}^{-1}$$

$$\underline{700\text{ nm}} \qquad E = 2.84 \times 10^{-19}\text{ J} \quad \text{or} \quad E = 171\text{ kJ mol}^{-1}$$

THE ENERGY ASSOCIATED WITH TRANSITIONS OF DIFFERENT TYPES

Magnitudes of typical separations of rotational levels are 10 J mol^{-1}, of vibrational levels are 10^2–10^3 J mol^{-1}, and of electronic levels are 10^2 kJ mol^{-1}. Table 21–1 gives the wavelengths of light (in vacuum) that correspond to these energies. By using the diagram in Figure 8–3, you can

TABLE 21–1
Relationships Between Energies of Transitions in Molecules and the Properties of Photons

Energy/J mol^{-1}	Photon properties: Energy/J photon^{-1}	λ_0/nm	$\tilde{\nu}$/cm^{-1}	Energy/k/K
1.000	1.661×10^{-24}	1.197×10^{8}	0.0835	0.120
10.00	1.661×10^{-23}	1.197×10^{7}	0.835	1.20
100.0	1.661×10^{-22}	1.197×10^{6}	8.35	12.0
1.000×10^{3}	1.661×10^{-21}	1.197×10^{5}	83.5	120
1.000×10^{4}	1.661×10^{-20}	1.197×10^{4}	8.35×10^{2}	1.20×10^{3}
1.000×10^{5}	1.661×10^{-19}	1.197×10^{3}	8.35×10^{3}	1.20×10^{4}
1.000×10^{6}	1.661×10^{-18}	1.197×10^{2}	8.35×10^{4}	1.20×10^{5}

see that transitions between rotational levels are associated with radiation in the microwave region ($\lambda_0 \cong 10^7$ nm), transitions between vibrational levels are associated with radiation in the infrared region ($\lambda_0 \cong 10^5$ nm), and transitions between electronic levels are associated with the near infrared, the visible, and the ultraviolet regions ($\lambda_0 = 100$–1000 nm), depending on the molecule. Table 21–1 also gives values of photon energy divided by the Boltzmann constant, k (1.38×10^{-23} J K^{-1}). This quotient is the temperature at which the Boltzmann factor $e^{-E/kT}$ has the value e^{-1} (0.368). As we learned in Chapter 5, this Boltzmann factor determines the occupancy of excited states by atoms or molecules in thermal equilibrium.

21–2 Energy Levels in Molecules

Now we shall develop an introduction to energy levels in molecules. In particular, we shall consider the relationship between the complexity of a molecule (the number of atoms it contains and whether it is linear) and the number of coordinates that help describe its rotation and vibration.

During the rotation and vibration of a molecule, the positions of its atoms change. These changes are described by changes of the coordinates of the atoms. To define the position of an atom in space we must specify three coordinates. These coordinates may be x, y, and z of a Cartesian coordinate system or r, θ, and ϕ of a polar coordinate system (see Figure 8–8). To define the positions of two atoms requires the specification of six such coordinates, three for each atom; to define the positions of three atoms requires the specification of nine coordinates, and so on. To define the positions of n atoms requires the specification of $3n$ coordinates. This is true whether the n atoms being considered are independent of one another or are bonded together in a molecule. When the n atoms are bonded in a molecule, the $3n$ coordinates needed to specify the positions of all of the atoms can be grouped in a certain way:

1. Three coordinates specify the position of the center of gravity of the molecule. Changes of these coordinates correspond to translation of the whole molecule.
2. Two coordinates specify the angular orientation of a linear molecule. Three coordinates specify the angular orientation of a nonlinear molecule. The angles θ and ϕ specify the orientation of a linear molecule. An additional angle must be defined to specify the tilt of a nonlinear molecule. You can visualize this by considering the angular orientations of a pencil (a linear object) and a book (a nonlinear object.) Changes of these two or three coordinates correspond to rotation of the molecule.
3. The remaining $3n - 5$ coordinates (for a linear molecule) or $3n - 6$ coordinates (for a nonlinear molecule) are internal coordinates that determine the positions of the atoms in the molecule relative to one another. Changes of these coordinates correspond to vibrations of the atoms within the molecule.

This grouping of the coordinates can be illustrated by considering molecules of different complexity:

Ar, a monatomic molecule

$3n = 3 \times 1 = 3$ coordinates, which are:
x, y, z of the atom.

H_2, a diatomic molecule

$3n = 3 \times 2 = 6$ coordinates, which are:
x, y, z of the center of gravity of the molecule, 2 angular coordinates defining the orientation of the molecular axis, and $(6 - 3 - 2) = 1$ coordinate defining the interatomic separation of the two atoms of the molecule.

CO_2, a linear triatomic molecule

$3n = 3 \times 3 = 9$ coordinates, which are:
x, y, and z of the center of gravity of the molecule, 2 angular coordinates defining the orientation of the molecular axis, and $(9 - 3 - 2) = 4$ internal coordinates defining the relative positions of the atoms in the molecule.

H_2O, a nonlinear triatomic molecule

$3n = 3 \times 3 = 9$ coordinates, which are:
x, y, and z of the center of gravity of the molecule, 3 angular coordinates defining the orientation of the molecule, and $(9 - 3 - 3) = 3$ internal coordinates defining the relative positions of the atoms in the molecule.

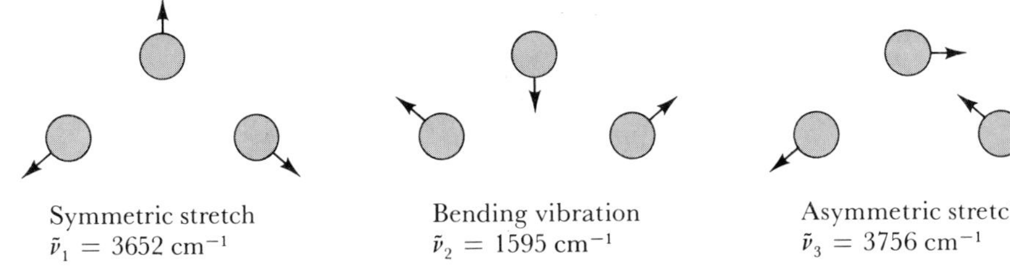

FIGURE 21–2

The normal modes of vibration of a water molecule. The frequency of the harmonic motion of the atoms in each of these vibrational modes is given in wave numbers, waves per centimeter. To convert these frequencies to vibrations per unit time, multiply by c. The frequency of the symmetric stretch is $\nu = \tilde{\nu}c = 3652\ cm^{-1} \times 3.00 \times 10^{10}\ cm\ s^{-1} = 1.096 \times 10^{14}\ s^{-1}$.

The internal coordinates used to define the relative positions of atoms in a molecule can be interatomic distances and angles defined by the lines connecting the centers of atoms. Thus in water the three coordinates could be the three distances,

$$r_{OH_a} \quad r_{OH_b} \quad \text{and} \quad r_{H_aH_b}$$

or the two distances and the angle,

$$r_{OH_a} \quad r_{OH_b} \quad \text{and} \quad \angle HOH$$

The number of internal coordinates needed to define the positions of atoms within a molecule is equal to the number of simple independent modes of vibration of the molecule. Thus for water there are three independent modes of vibration, which are called the *normal modes of vibration* of this molecule. These are pictured in Figure 21–2. All possible vibrational motions within a water molecule are made up of these three motions, each of which is quantized.

EXAMPLE: What is the number of normal modes of vibration for each of the following molecules?

CH_3I	Methyl iodide
C_3H_7OH	Propyl alcohol

For CH_3I, a 5-atom, nonlinear molecule:

$$5 \times 3 - 3 - 3 = 9 \text{ vibrational modes}$$

For C_3H_7OH, a 12-atom, nonlinear molecule:

$$12 \times 3 - 3 - 3 = 30 \text{ vibrational modes}$$

The number of normal modes of vibration of a molecule increases with an increase in the number of atoms in the molecule. This is illustrated in Figure 21–3, which shows the infrared spectra of water (dissolved in CCl_4) and of liquid methyl iodide. We see clearly three absorption bands cor-

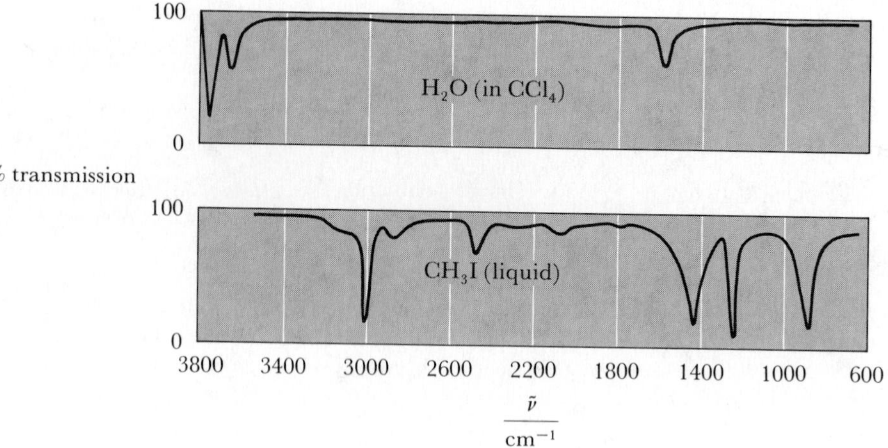

FIGURE 21–3
The infrared spectra of H_2O (dissolved in CCl_4) and liquid CH_3I.

responding to the three normal modes of vibration of the water molecule, and we see also the greater complexity of the spectrum for methyl iodide.

The interpretation of spectra has been greatly assisted by quantum mechanics, which defines the allowed energy states of a system. This is true for the electronic states of atoms, described in Chapter 8, and it is true also for the rotational, vibrational, and electronic states of molecules. In the application of quantum-mechanical calculations to rotating and vibrating molecules, simple models are used.

THE ROTATIONAL ENERGY STATES OF MOLECULES

The simplest model of a rotating linear molecule is a rigid arrangement of two masses m_1 and m_2 separated by a fixed distance R; this is a dumbbell-like object, with a moment of inertia $I = R^2 m_1 m_2/(m_1 + m_2)$.[1] Treatment by quantum mechanics of this simple model of a diatomic molecule, called the *rigid rotator*, gives the allowed rotational energies,

$$E_{\text{rotation}} = \mathcal{J}(\mathcal{J}+1)\frac{h^2}{8\pi^2 I}$$

in which the quantum number $\mathcal{J}$ can have values 0, 1, 2, · · · .

---

[1]The moment of inertia of a system made up of masses separated by particular distances is the sum of the product of each mass and the square of its distance from the center of gravity of the system. For the system of two masses, the center of gravity is a distance of $Rm_1/(m_1 + m_2)$ from m_2 and a distance $Rm_2/(m_1 + m_2)$ from m_1; therefore,

$$I = m_2\left(\frac{Rm_1}{m_1 + m_2}\right)^2 + m_1\left(\frac{Rm_2}{m_1 + m_2}\right)^2 = \frac{m_1 m_2 R^2}{m_1 + m_2} = \mu R^2$$

The function of masses $m_1 m_2/(m_1 + m_2)$ is called the reduced mass of the system; you were introduced to this function in Section 10–2, where it was represented by the symbol μ.

The spacing of rotational energy levels given by this equation is shown in Figure 21–4. Absorption of a photon can cause the rotational quantum number to change by one unit, and this figure also shows the allowed transitions for initial states with $J = 0, 1, 2, 3$, and 4. Each of these transitions has a different energy, but the differences between the energies of successive transitions are constant. This can be seen by considering the values of the product $J(J+1)$ as a function of J, and the differences between successive values of this product:

J	$J(J+1)$	Difference
0	0	
		2
1	2	
		4
2	6	
		6
3	12	
		8
4	20	

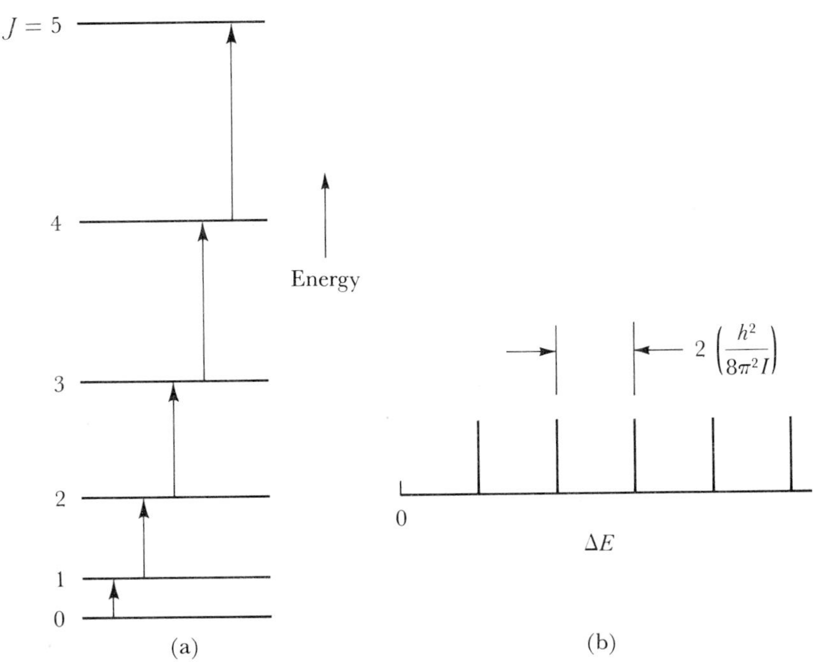

FIGURE 21–4
The rotational energy levels of a rigid diatomic molecule (a rigid rotator) and the energies of transitions between adjacent levels. (a) The energy levels:

$$E_{\text{rotation}} = J(J+1)\frac{h^2}{8\pi^2 I} \qquad J = 0, 1, 2, \cdots$$

(b) The energies of allowed transitions:

$$\Delta E(J' - 1 \rightarrow J') = 2J' \times \frac{h^2}{8\pi^2 I} \qquad (J' \text{ is some particular value of } J)$$

The successive energies of transitions between adjacent rotational states of a rigid rotator differ from one another by a constant amount,

$$\Delta(\Delta E) = 2\left(\frac{h^2}{8\pi^2 I}\right)$$

as is shown in Figure 21–4. If we know the energy difference between lines in the rotational spectrum of a diatomic molecule, we can calculate the moment of inertia of the molecule:

$$I = \frac{2h^2}{8\pi^2[\Delta(\Delta E)]}$$

Because the masses of the bonded atoms are known, an experimental value of the moment of inertia determined in this way allows us to calculate the interatomic distance in the molecule:

$$R = \sqrt{\frac{I(m_1 + m_2)}{m_1 m_2}}$$

Studies of microwave and far infrared spectra have provided some of the most accurately measured values of interatomic distances in gaseous molecules. Nonlinear molecules give more complex microwave spectra than do diatomic molecules, but the complex spectra provide information regarding the moments of inertia of the molecule rotating about different axes.

Figure 21–5 shows the spectrum of gaseous hydrogen chloride in the far infrared region; the spectrum consists of a series of approximately equally spaced absorption bands, as expected on the basis of the preceding derivation for a rigid rotator. The separation of these bands, $\Delta\tilde{\nu}$, is $\sim 20.7\ \text{cm}^{-1}$. Therefore the value of $\Delta(\Delta E)$ is

$$\begin{aligned}\Delta(\Delta E) &= hc\tilde{\nu} \\ &= 6.626 \times 10^{-27}\ \text{g cm}^2\ \text{s}^{-1} \times 3.00 \times 10^{10}\ \text{cm s}^{-1} \times 20.7\ \text{cm}^{-1} \\ &= 4.11 \times 10^{-15}\ \text{g cm}^2\ \text{s}^{-2}\end{aligned}$$

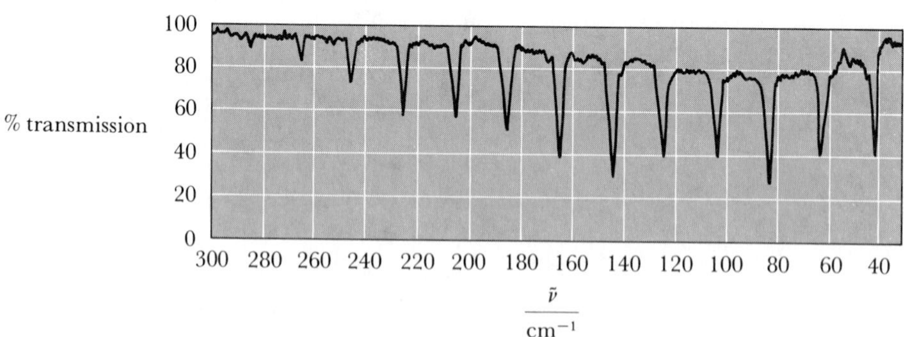

FIGURE 21–5
The far infrared region of the spectrum of hydrogen chloride. The absorption bands are due to transitions between rotational states.

From this, the moment of inertia can be calculated:

$$I = \frac{2 \times (6.626 \times 10^{-27}\ \text{g cm}^2\ \text{s}^{-1})^2}{8\pi^2 \times 4.11 \times 10^{-15}\ \text{g cm}^2\ \text{s}^{-2}} = 2.71 \times 10^{-40}\ \text{g cm}^2$$

With the moment of inertia known, the interatomic separation in hydrogen chloride can be calculated; by introducing Avogardo's constant we can use conventional atomic weights instead of the masses of individual atoms:

$$R = \sqrt{\frac{2.71 \times 10^{-40}\ \text{g cm}^2 \times (36.46\ \text{g mol}^{-1})(6.02 \times 10^{23}\ \text{mol}^{-1})}{1.008\ \text{g mol}^{-1} \times 35.45\ \text{g mol}^{-1}}} = 1.29 \times 10^{-8}\ \text{cm}$$

Because ordinary hydrogen chloride gas consists of $\sim 75\%$ $^1H^{35}Cl$ and $\sim 25\%$ $^1H^{37}Cl$, the spectrum shown in Figure 21–5 is the superposition of two very slightly different spectra. The moments of inertia of these two isotopically different molecules are almost the same because the reduced masses of isotopic molecules differ by only 0.15%. The calculation just carried out used the conventional atomic weights. If the spectra of the two isotopically different molecules were resolved, the value of $\Delta\tilde{\nu}$ from the spectrum of one of the isotopic molecules would be used with the atomic masses for that isotopic molecule to calculate the interatomic separation.

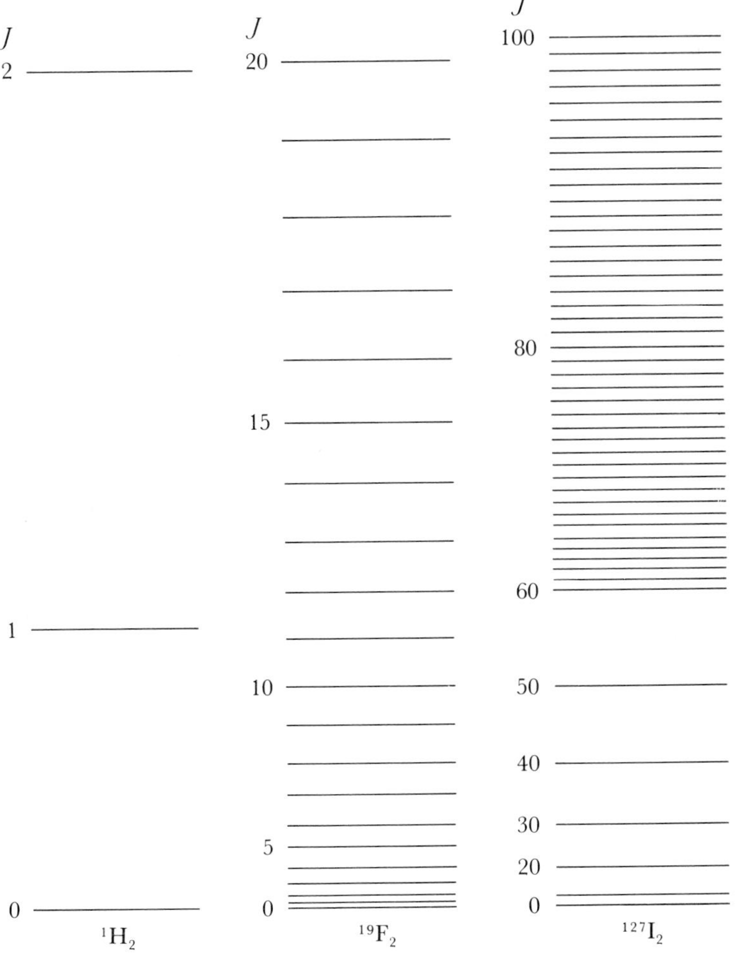

FIGURE 21–6
The spacing of rotational states for the diatomic molecules H_2, F_2, and I_2. (For I_2, only every tenth level is shown until $J = 60$.)

Because the moment of inertia of a diatomic molecule is in the denominator of the equation for the rotational energy, the spacing of the rotational levels is much closer for iodine, I_2, than for hydrogen, H_2, or fluorine, F_2. This is shown in Figure 21–6. Molecules of approximately the same mass can have rather different values of the moments of inertia, depending on the way in which the mass is distributed, as is illustrated by some diatomic molecules with molecular weight ~ 38: with $^{19}F_2$ (MW = 37.997, $r = 1.418 \times 10^{-8}$ cm, $I = 3.17 \times 10^{-39}$ g cm^2), with $^1H^{37}Cl$ (MW = 37.974, $r = 1.275 \times 10^{-8}$ cm, $I = 2.65 \times 10^{-40}$ g cm^2), and with $^3H^{35}Cl$ (MW = 37.985, $r = 1.275 \times 10^{-8}$ cm, $I = 7.49 \times 10^{-40}$ g cm^2).

In the operation of a microwave oven, radiation is absorbed by the water molecules in food to increase their rotational energy (i.e., increase the quantum numbers for rotation). The net result of this absorption of energy is an increase of temperature throughout the sample.

THE VIBRATIONAL ENERGY STATES OF MOLECULES

Just as a simplified model, the rigid rotator, is a useful first approximation in considering the rotation of diatomic molecules, so also there is a simplified model for the vibrations of the atoms within a molecule. This simplified model for a diatomic molecule is the *harmonic oscillator*, two masses, m_1 and m_2, connected by a spring, with a force constant k. We considered some aspects of this model in developing the concept of zero-point energy in Chapter 10. Now we will consider molecular vibrations more completely.

Treatment of the harmonic oscillator by quantum mechanics gives the allowed vibrational energies,

$$E_{\text{vibration}} = h\nu_0(v + \tfrac{1}{2})$$

in which the quantum number v can have values 0, 1, 2, $\cdots$, and ν_0 is the fundamental frequency of the system of spring and masses making up this harmonic oscillator. The fundamental frequency is related to the force constant for the spring by the equation

$$\nu_0 = \frac{1}{2\pi}\sqrt{\frac{k}{\mu}}$$

in which μ is the reduced mass of the system.

Example: Show that the dimensions are consistent in the equation for ν_0 just given.

The force constant k has dimensions dyn cm^{-1} (or N m^{-1} in SI):

$$k \equiv \text{dyn cm}^{-1} \equiv (\text{g cm s}^{-2}) \times \text{cm}^{-1} \equiv \text{g s}^{-2}$$

$$\frac{k}{\mu} \equiv \text{g s}^{-2}\,\text{g}^{-1} \equiv \text{s}^{-2}$$

$$\sqrt{\frac{k}{\mu}} \equiv \text{s}^{-1}$$

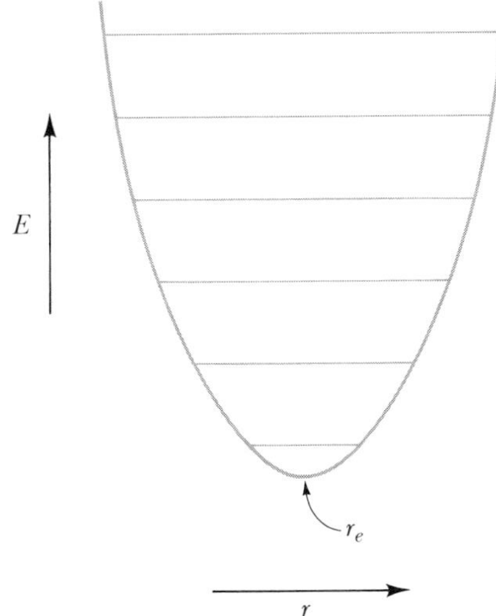

FIGURE 21–7
The potential energy of a harmonic oscillator as a function of the separation of the masses. For the harmonic oscillator, $E_p = \frac{1}{2}k(r - r_e)^2$, the equation for a parabola. The potential energy is defined as zero for $r = r_e$.

Figure 21–7 shows the parabolic dependence of the potential energy of a harmonic oscillator upon the interatomic separation of the masses. The allowed vibrational energies are shown as horizontal lines.

A real diatomic molecule is not a harmonic oscillator. The potential energy as a function of interatomic separation is not truly parabolic. This was shown in Figure 10–3. Repulsion between the bonded atoms at interatomic separations that are small compared to the equilibrium interatomic distance causes the potential energy to rise more steeply than it would in a harmonic oscillator. The weakened attraction between the bonded atoms as the bond is stretched causes the potential energy to rise less steeply for elongation of the bond. At very large distances between the two atoms, the potential energy is not a function of the separation. For a potential-energy dependence on interatomic distance of the type shown by real molecules, the separation of the allowed vibrational energies is not constant. The levels become closer together as the value of the vibrational quantum number v increases. For transitions between the vibrational levels $v = 0$ to $v = 1$, the energy is closely approximated by the harmonic-oscillator equation, and observations of this energy allows calculation of k, the force constant.

EXAMPLE: The infrared spectrum of gaseous hydrogen chloride has an absorption due to the transition $v = 0$ to $v = 1$ at 2885.6 cm^{-1}. What is the force constant for the changing of this bond length?

If $\tilde{\nu} = 2885.6\ cm^{-1}$, the value of ν is

$$\nu = \tilde{\nu} \times c$$

$$\nu = 2885.6\ cm^{-1} \times 2.998 \times 10^{10}\ cm\ s^{-1} = 8.651 \times 10^{13}\ s^{-1}$$

Rearrangement of the equation relating ν and k gives

$$k = (2\pi\nu)^2 \times \frac{m_1 m_2}{m_1 + m_2} = (2\pi\nu)^2 \mu$$

Using masses of the atoms in the principal isotopic species $^{1}H^{35}Cl$, we calculate reduced mass:

$$\mu = \frac{\dfrac{1.008\ \text{g mol}^{-1}}{6.02 \times 10^{23}\ \text{mol}^{-1}} \times \dfrac{34.97\ \text{g mol}^{-1}}{6.02 \times 10^{23}\ \text{mol}^{-1}}}{\dfrac{1.008\ \text{g mol}^{-1}}{6.02 \times 10^{23}\ \text{mol}^{-1}} + \dfrac{34.97\ \text{g mol}^{-1}}{6.02 \times 10^{23}\ \text{mol}^{-1}}} = 1.63 \times 10^{-24}\ \text{g}$$

Then the force constant is

$$\begin{aligned} k &= (2\pi \times 8.651 \times 10^{13}\ \text{s}^{-1})^2 \times 1.63 \times 10^{-24}\ \text{g} \\ &= 4.82 \times 10^5\ \text{g s}^{-2} = 4.82 \times 10^5\ \text{g cm s}^{-2} \times \text{cm}^{-1} \\ &= 4.82 \times 10^5\ \text{dyn cm}^{-1} \end{aligned}$$

The values of k for several diatomic molecules are given in Table 21–2. For the hydrogen halides, the trend is that expected: the force constant is larger the greater the bond-dissociation energy. But for the diatomic halogen molecules, this expectation is not confirmed. The force constant for fluorine is larger than that for chlorine, despite the fact that the dissociation energy of chlorine is larger. This comparison shows that there is no necessary correlation between the force constant, k, and dissociation energy, D. The value of k is related to the curvature of the potential-energy curve at the minimum, but the value of D is determined by the vertical distance from the energy of the lowest vibration level to the asymptotically approached energy of the infinitely separated atoms. The comparison of values of k for the series of diatomic molecules nitrogen, oxygen, and fluorine shows the big effect upon k as the bond order changes.

TABLE 21–2
Properties of Diatomic Molecules[a]

Molecule	$\tilde{\nu}$ / cm^{-1}	k / dyn cm^{-1}	D / kJ mol^{-1}
HF	3958.4	8.84×10^5	565
HCl	2885.6	4.82×10^5	427
HBr	2559.3	3.84×10^5	363
HI	2230.0	2.93×10^5	295
N_2	2330.7	22.6×10^5	942
O_2	1556.3	11.4×10^5	494
F_2	892	4.45×10^5	154
Cl_2	556.9	3.20×10^5	239
Br_2	321	2.43×10^5	190
I_2	213.4	1.70×10^5	149

[a] Values of $\tilde{\nu}$ and k are from G. M. Barrow, *Introduction to Molecular Spectroscopy*, McGraw-Hill Book Co., New York, 1962, p. 42.

TABLE 21–3
Characteristic Frequencies for Bond Stretching in Certain Groups[a]

Group	$\tilde{\nu}$ / cm^{-1}	Group	$\tilde{\nu}$ / cm^{-1}
—O—H	3500–3700	—C≡C—	2200–2260
>N—H	3300–3500	—C≡N	2250
≡C—H	3340	>C=O	1660–1870
=C(—)—H	3000–3120	>C=C<	1600–1680
—C(—)(—)—H	2880–3030	—C(—)(—)—O—	1000–1300

[a]From G. M. Barrow, *Physical Chemistry*, McGraw-Hill Book Co., New York, 1966, p. 352.

Each normal vibrational mode for a polyatomic molecule has some of the characteristics of the single vibrational mode for a diatomic molecule. In particular:

1. there is a zero-point energy associated with each vibrational mode,
2. the motion of atoms in each mode is, for small values of v, approximately harmonic, with a characteristic frequency ν_0, and
3. in each normal vibrational mode the amplitude of motion of the atoms increases with an increase of the vibrational quantum number.

Although motions of many or all of the atoms in a polyatomic molecule are involved in each normal vibrational mode, the stretching of a chemical bond between certain atoms is the dominant motion associated with some particular normal modes. Therefore infrared spectra of polyatomic molecules display absorption at certain energies that are characteristic of the types of bonds in the molecule. Table 21–3 summarizes the frequencies associated with transitions between vibrational states characteristic of molecules containing chemical bonds between certain atoms. The inverse dependence of vibrational frequency on the reduced mass of the bonded atoms ($\nu \propto \mu^{-1/2}$) is demonstrated clearly in this table. The stretching frequencies for bonds involving hydrogen are higher than any others. The dependence on bond order, already illustrated in connection with diatomic molecules, also is shown here in the relationship of the vibrational frequencies associated with single and double bonds between carbon and oxygen. Although infrared spectra of organic molecules are complex, the frequencies given in this table can serve as "fingerprints." Thus infrared spectra distinguish the isomers dimethyl ether (H_3COCH_3) and ethyl alcohol (CH_3CH_2OH) because the latter molecule has a hydroxyl group and the former does not.

THE ELECTRONIC ENERGY STATES OF MOLECULES

In discussing the building up of the periodic table, we focused our attention on the ground state of each atomic species. But we made the point that atoms and ions have excited electronic states, and we described excited electronic states for some monatomic species (He, K, Ca, Ca^+, Sc, Sc^{2+}, and Ti^{3+}). What is true of atomic species is also true of molecular species: there is a ground state, and there also are excited electronic states. Just as the electronic energy of an atom depends on the orbitals that are occupied

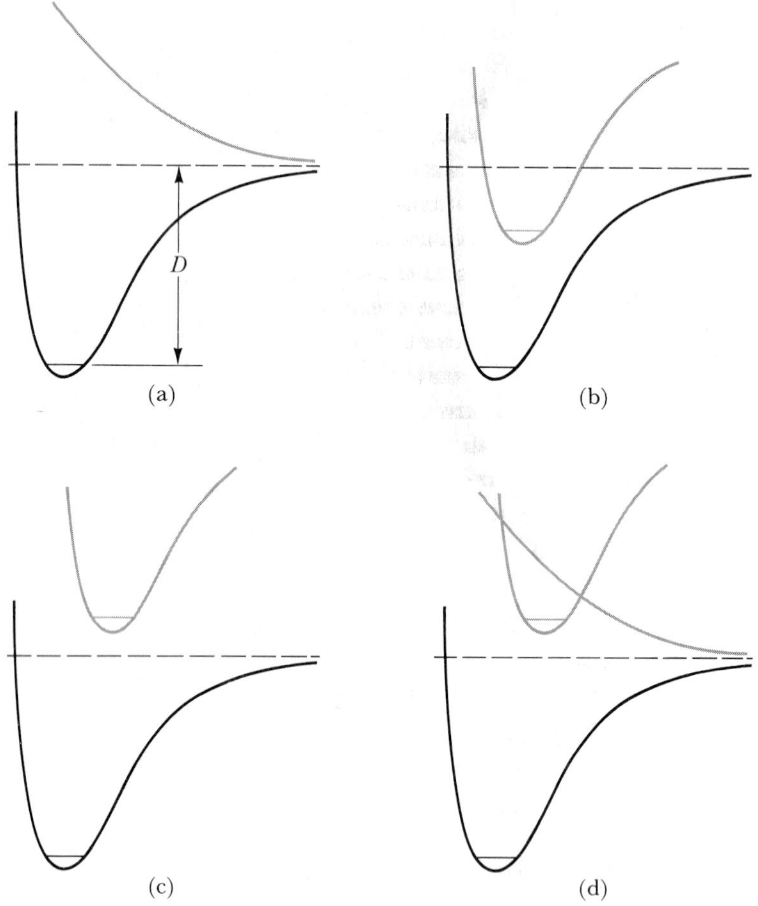

FIGURE 21–8
The potential energy versus internuclear separation of atoms in a diatomic molecule in the ground state and in excited states (the lowest vibrational states are shown). (a) The excited state is not a bound state (atoms repel one another at all separations). (b) The ground vibrational state of the excited electronic state has less energy than needed for dissociation of the molecule. (c) The ground vibrational state of the excited electronic state has more energy than is needed for dissociation of the molecule. (d) Two excited states shown.

and on the spins of the electrons, so these same factors determine the electronic energy of molecules, as Chapter 17 explained for different electronic states of diatomic oxygen.

Figure 21–8 shows examples of the potential-energy curves for two electronic states of a diatomic molecule. Four situations are depicted. In (a), the excited electronic state is one in which the atoms are not bound together. The minimum in the potential-energy curve for this excited state is approached as the atoms become separated by an infinite distance. Such a state is called an unbound state. In (b), the excited electronic state has an energy in its lower vibration states that is lower than the dissociation energy of the ground state. In (c), the energy of the excited electronic state is larger than the energy required to dissociate the molecule in the ground state. All of these situations exist in real systems, and for some systems the potential-energy curve for an unbound excited state crosses the curve for a bound excited state, as shown in (d).

21–3 The Absorption of Light by Atoms and Molecules

For light to be absorbed by an atomic or molecular species certain conditions must be met. The energy of the photon must be equal to the difference of the energy between a state of the molecule having an appreciable population and an excited state. (Light of a precisely correct energy for a possible transition is not appreciably absorbed by a sample if the sample contains too few atoms or molecules in the energy state from which that transition originates.) There are other conditions that a transition between energy states must satisfy if there is to be light absorption. Full discussion of these conditions is beyond the scope of this treatment, but it is appropriate at this level to recognize the existence of *allowed transitions* and *forbidden transitions*. Thus the transition between a singlet electronic state (one with zero unpaired electrons) and a triplet electronic state (one with two unpaired electrons having parallel spin) caused by absorption of light is forbidden. Transitions between different singlet states are allowed, as are transitions between different triplet states. The basis for this law is simple enough: the absorption of a photon does not cause a spinning electron to change its orientation in space. Transitions between vibrational states of a nonpolar diatomic molecule are not brought about by light absorption: they are forbidden. However, such transitions are allowed if the molecule is polar. The oscillating electric field of the light can transfer energy to the vibration of a polar molecule, but no such transfer of energy is possible if the molecule is nonpolar.

The existence of definite rotational, vibrational, and electronic energy states for a molecule suggest that light absorption should occur at sharply defined energies. But in a bulk sample of matter, molecules are distributed among different rotational energy states (and vibrational energy states if

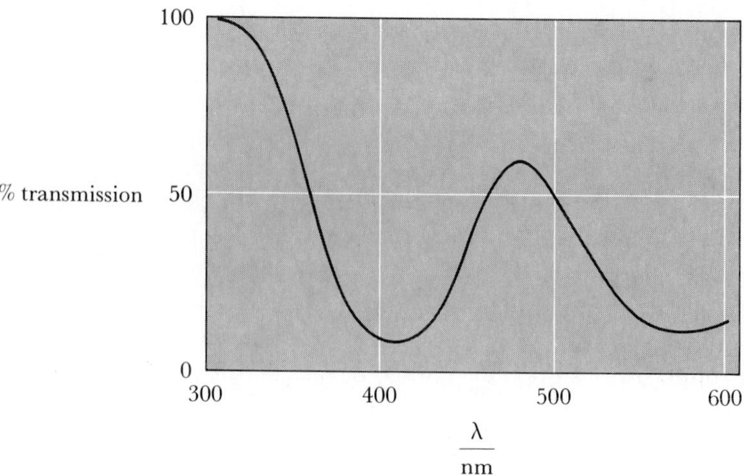

FIGURE 21–9
The spectrum of hexaaquachromiun(III) ion in aqueous solution. The ordinate gives the percentage of the light transmitted.

ν_0 is small enough or if the temperature is high enough). Conventional instruments for studying light absorption also have limits on how narrow a range of photon energies can be studied. However, these limitations are disappearing in current laser studies.

For light-absorbing materials in a condensed phase, either a pure liquid or solid or a solution, interactions of neighboring molecules blur the sharpness of the energy levels, and therefore the sharpness of the energy of transitions between levels. This results in a spectrum that is made up of absorption bands rather than sharp lines. Figure 21–9 shows the spectrum of hexaaquachromium(III) ion in acidic solution. This spectrum is plotted in the same way as the infrared spectra given in Figures 21–3 and 21–5. But visible and ultraviolet spectra generally are plotted in a way that is related to the Beer–Lambert law, which we will discuss now.

THE BEER–LAMBERT LAW

As light passes through a homogeneous medium (gas, liquid, or solid), the extent of interaction of the oscillating electric field of the radiation and electrons of the medium depends on the energy of the radiation and the nature of the matter. If the interaction is negligible, the medium is said to be *transparent*; at the other extreme, the medium is said to be *opaque* if the interaction causes the sample to absorb the light almost completely. However, we are interested in quantitative measurement of partial light absorption by the medium. This can be measured with an instrument known as a *spectrophotometer*, which allows continuous variation of the wavelength of light impinging on the sample. For each wavelength setting of the instrument, one can measure the intensity of light transmitted by a sample, I, as compared with that transmitted by a "transparent blank," I_0. In the study of liquid solutions, the blank generally used is the solvent. Attenuation of the light intensity by scattering from the walls of the container then is the same in the sample and in the blank.

The light absorption by a particular homogeneous medium depends on the number of absorbing molecules through which the beam of light passes. This is a function both of b, the length of the light path through the sample, and C, the concentration of the absorbing molecules. The fraction of the light absorbed by successive layers of equal thickness is the same, and it is independent of the intensity of the incident radiation, I_0. If 20% of the incident light of some particular wavelength is absorbed by a 1.00-cm thickness of sample, 80% of the light is transmitted through this 1.00-cm sample, and the fraction of light transmitted by a 3.00-cm thickness of sample would be

$$\left(\frac{I}{I_0}\right)_{3\text{ cm}} = \left(\frac{I}{I_0}\right)_{1\text{ cm}} \times \left(\frac{I}{I_0}\right)_{1\text{ cm}} \times \left(\frac{I}{I_0}\right)_{1\text{ cm}}$$
$$= (0.80)^3 = 0.51$$

The relationship, for a particular concentration,

$$\left(\frac{I}{I_0}\right)_b = \left(\frac{I}{I_0}\right)_{1\text{ cm}}^{b/\text{cm}}$$

is known as LAMBERT's law.

A similar relationship expresses the dependence of the intensity of the transmitted light on concentration of absorbing molecules in a solution. If 80% of the light is transmitted by a 1.00-cm thickness of solution with a concentration of 0.100 mol L^{-1}, the transmission by a 1.00-cm thickness of solution with a concentration of 0.300 mol L^{-1} would be

$$\left(\frac{I}{I_0}\right)_{0.30M} = \left(\frac{I}{I_0}\right)_{0.10M} \times \left(\frac{I}{I_0}\right)_{0.10M} \times \left(\frac{I}{I_0}\right)_{0.10M}$$
$$= (0.80)^3 = 0.51$$

Thus a 1.00-cm thickness of a solution with $C = 0.300$ mol L^{-1} absorbs the same fraction of the light as a 3.00-cm thickness of a solution with $C =$ 0.100 mol L^{-1}. The relationship relating light absorption and concentration can be made more general; for a particular length of light path,

$$\left(\frac{I}{I_0}\right)_C = \left(\frac{I}{I_0}\right)_{1\text{ mol L}^{-1}}^{C/\text{mol L}^{-1}}$$

This relationship is known as BEER's law.

Combined and expressed in logarithmic form, these two dependences of light absorption on thickness of sample and concentration give a mathematical relationship known as the *Beer–Lambert law*:

$$A = \log\left(\frac{I_0}{I}\right) = \epsilon C b$$

in which A is called the absorbance and the proportionality constant in this equation, ϵ, is known as the *molar absorptivity*, *molar absorption coefficient*, or *molar extinction coefficient*.

The molar absorptivity can be calculated from the percentage transmission of the light by a particular thickness of solution of a particular concentration. Thus 7.6% of the light of wavelength 410 nm is transmitted by a 1.00-cm thickness of a solution of chromium(III) ion at a concentration of 0.070 mol L^{-1} (a datum from Figure 21–9). If the percentage transmission is 7.6%, I_0/I is equal to $100/7.6 = 13.2$, and the calculation of ϵ is

$$\epsilon = \frac{\log\left(\frac{I_0}{I}\right)}{C \times b} = \frac{\log 13.2}{0.070 \text{ mol L}^{-1} \times 1.00 \text{ cm}} = 16.0 \text{ L mol}^{-1} \text{ cm}^{-1}$$

The Beer–Lambert law also can be derived directly from the simple premise that in passing through homogeneous matter the fractional diminution of light intensity, $-dI/I$, is proportional to the concentration of the absorbing substance, C, and the thickness of the sample, dl,

$$-\frac{dI}{I} \propto C\,dl$$

Introducing the proportionality constant ϵ', we get

$$-\frac{dI}{I} = \epsilon' C\,dl$$

Integration of this equation over the entire thickness of the cell (I_0 is incident intensity at $l = 0$, and I is the final intensity at $l = b$) gives

$$-\int_{I_0}^{I} \frac{dI}{I} = \epsilon' C \int_0^b dl$$

$$-\ln\left(\frac{I}{I_0}\right) = \epsilon' Cb$$

which, converted to decadic logarithms with substitution of $\epsilon = \epsilon'/2.303$, gives the usual form of the Beer–Lambert law:

$$\log\left(\frac{I_0}{I}\right) = \epsilon Cb$$

The direct proportionality between $\log(I_0/I)$ and concentration is easily confirmed in laboratory experiments and is very useful in analysis. Figure 21–10 shows the linear relationship between absorbance and concentration for hexaaquachromium(III) ion, $Cr(OH_2)_6^{3+}$. The figure also shows the nonlinear relationship between percentage transmission and concentration.

EXAMPLE: What concentration of hexaaquachromium(III) ion in a cell of length 5.00 cm absorbs 95.0% of light of wavelength 410 nm?

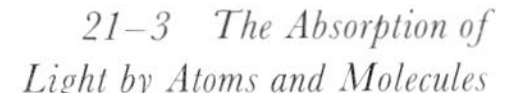

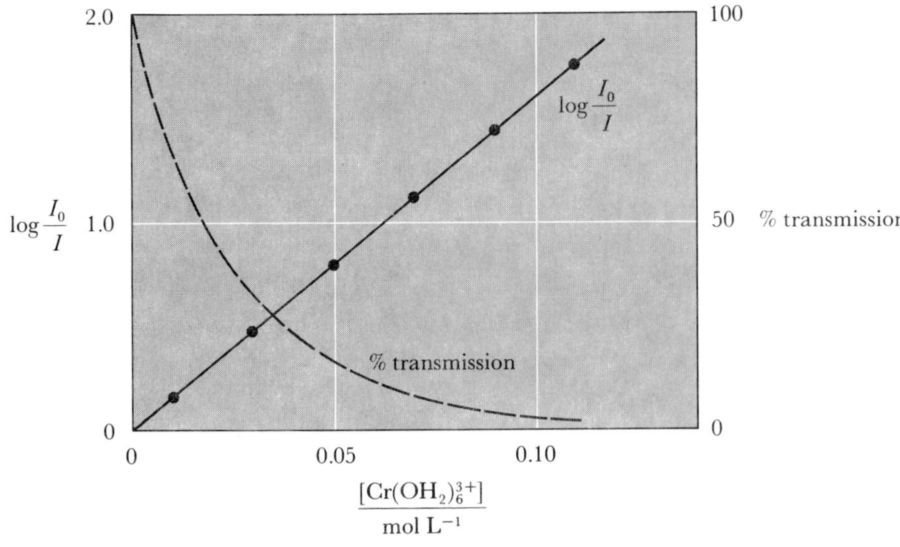

FIGURE 21–10
The relationship between absorbance, $\log(I_0/I)$, at 410 nm and concentration of chromium(III) ion in aqueous solution. The cell length is 1.00 cm; the percentage transmission is shown as a dashed line.

At this wavelength the molar absorptivity of $Cr(OH_2)_6^{3+}$ is $16.0\ L\ mol^{-1}\ cm^{-1}$.

If 95.0% of light is absorbed, $I_0/I = 1.00/(1.00 - 0.95) = 1.00/0.050 = 20$. Therefore the value of A is $\log 20 = 1.301$, and the concentration is

$$C = \frac{1.301}{5.00\ cm \times 16.0\ L\ mol^{-1}\ cm^{-1}} = 1.63 \times 10^{-2}\ mol\ L^{-1}$$

Example: A convenient analysis for chromium is based on light absorption by chromate ion (CrO_4^{2-}) in alkaline solution. At a wavelength of 372 nm, the molar absorptivity of chromate ion is $\epsilon = 4.82 \times 10^3\ L\ mol^{-1}\ cm^{-1}$. (If SI units are used, this becomes $482\ m^2\ mol^{-1}$.) What is the CrO_4^{2-} concentration in a solution in a cell of 1.00-cm length that has an absorbance of 0.807 at this wavelength?

The concentration can be calculated:

$$C = \frac{A}{b\epsilon}$$

$$= \frac{0.807}{1.00\ cm \times 4.82 \times 10^3\ L\ mol^{-1}\ cm^{-1}} = 1.67 \times 10^{-4}\ mol\ L^{-1}$$

In each of these examples we assumed that the Beer–Lambert law holds. For dilute solutions of most substances, the Beer–Lambert law does hold, but there are limits to its validity. A concentration-dependent equilibrium is one cause for failure of the Beer–Lambert law. Acidic solutions of chromium(VI) provide an example of a system that does not obey this law. In solutions with a concentration of hydrogen ion between $\sim 10^{-2}$ mol L^{-1} and 10^{-5} mol L^{-1}, two species of chromium(VI) are present, hydrogenchromate ion ($HOCrO_3^-$) and dichromate ion ($Cr_2O_7^{2-}$). The relative concentrations of these two species were discussed in Section 19–2, where it was shown that the position of equilibrium in the reaction

$$2HOCrO_3^- \rightleftarrows Cr_2O_7^{2-} + H_2O \qquad K = \frac{[Cr_2O_7^{2-}]}{[HOCrO_3^-]^2} = 35 \text{ L mol}^{-1} \; (25°C)$$

depends on the concentration of chromium(VI); thus the Beer–Lambert law is not obeyed by chromium(VI) in acidic solution. In alkaline solution, only CrO_4^{2-} is present, there is no chromium-concentration-dependent equilibrium, and the Beer–Lambert law is obeyed.

THE FATE OF ELECTRONICALLY EXCITED MOLECULES

A molecule promoted to an excited electronic state by absorption of a photon does not keep this excitation energy for an indefinite period. It may lose the excitation energy by emitting a photon. The probability that it will do so depends on the electronic structure of the excited state and its relationship to structures of the states with lower energy, including the ground state. Although this is a complex subject, we shall explain some aspects of it.

The emission of light in a transition from a singlet excited electronic state to a singlet ground electronic state, a process called *fluorescence*, typically occurs within 10^{-9} s to 10^{-6} s. This process is depicted for a diatomic molecule in Figure 21–11a. In condensed media in which excited molecules are under the influence of many neighboring molecules, a nonradiative transition may occur to a state with a different number of unpaired electrons. The occurrence of this type of transition is depicted for a diatomic molecule in Figure 21-11b. If a molecule reaches an excited state with a number of unpaired electrons different from that of the ground state, it is much less likely to return to the ground state by emitting radiation. The emission of light in this forbidden process, called *phosphorescence*, occurs typically within 10^{-3} s to 10 s. With lifetimes of this magnitude, molecules in excited triplet states have time to react chemically with other molecules before losing the energy of excitation. Thus triplet excited states of molecules with singlet ground states are important in photochemical reactions.

In the excitation transitions depicted in Figure 21–11, as the photon excites the molecule from the singlet ground state S_0 to the singlet excited state S_1, the separations between the atoms of the molecule do not change. This is so because the electronic transitions occur during a time period that is very short compared to the period of vibration of the atoms within the molecule. This generalization is known as the FRANCK–CONDON principle. Depending on the relative arrangements of the atoms in the ground elec-

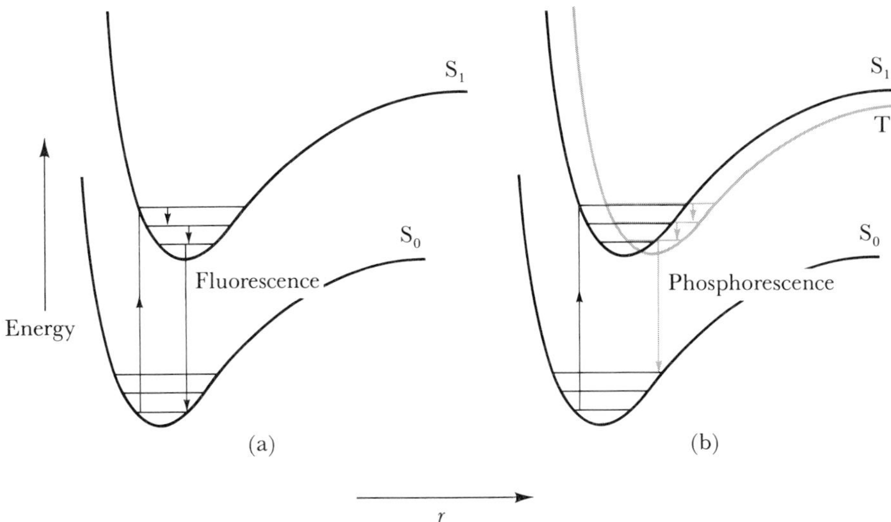

FIGURE 21–11
The processes light absorption, fluorescence, and phosphorescence. (a) Excitation to the singlet state S_1, followed rapidly (in 10^{-9} s to 10^{-6} s) by fluorescence $S_1 \rightarrow S_0 + h\nu$. (b) Excitation to the singlet state S_1, followed by process $S_1 \rightarrow T_1$, followed slowly (in 10^{-3} s to 10 s) by phosphorescence $T_1 \rightarrow S_1 + h\nu$.

tronic state and in the excited electronic state, a transition from the lowest vibrational state of the ground electronic state to an excited electronic state may produce an excited molecule with many quanta of vibrational energy.

THE DISSOCIATION OF MOLECULES BY LIGHT ABSORPTION

The absorption of light may cause a molecule to dissociate into fragments if the energy of the excited state is greater than that needed for dissociation. Whether dissociation occurs depends on the relationship among the several potential-energy curves. Figure 21–12 shows three different situations, each involving the absorption of light of the same energy, but in only two of which dissociation actually occurs. In each situation in which dissociation occurs, the energy of the absorbed photon exceeds the energy of the bond being broken. The products of bond rupture may be in the ground electronic state or in an excited electronic state. A large area of photochemistry involves study of the reaction of unstable fragment molecules, atoms, or ions formed as a result of light absorption.

A reaction brought about by absorption of light may have essentially the same mechanism as the thermal reaction with the only change being the higher concentration of reactive species resulting from light absorption, or there may be a new mechanism that was not accessible by thermal activation of the reactants. In the former category is the reaction of hydrogen with bromine,

$$H_2(g) + Br_2(g) = 2HBr(g)$$

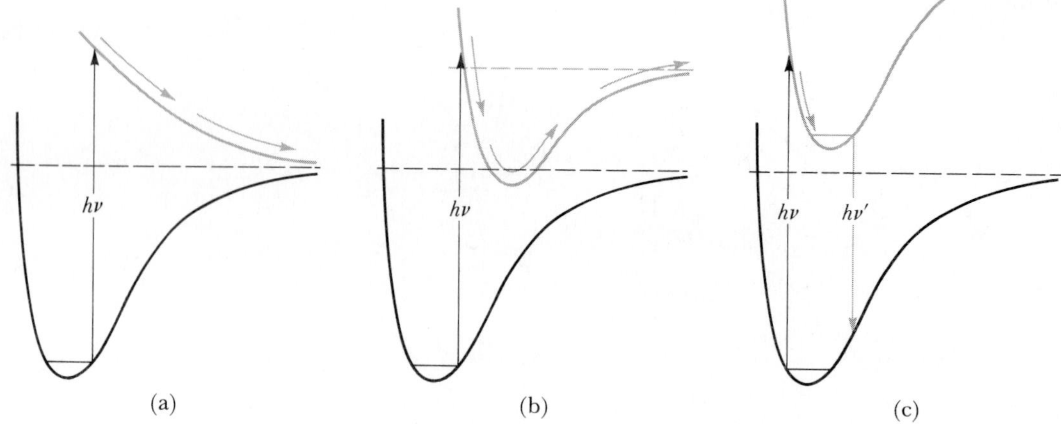

FIGURE 21–12
The consequences of light absorption by a diatomic molecule. (a) Dissociation gives atoms in their ground states. (b) Dissociation gives one atom in an excited state. (c) Dissociation does not occur; fluorescence does, with frequency of fluorescent light lower than that of absorbed light.

We discussed the mechanism for this reaction under thermal conditions in Chapter 15. With ordinary thermal excitation in the temperature range 500–600 K, the concentration of reactive bromine atoms is determined by equilibrium in the reaction

$$M + Br_2 \underset{k_{-1}}{\overset{k_1}{\rightleftarrows}} 2Br + M$$

Equating the rates of the forward and reverse reactions gives

$$[Br]^2 = \frac{k_1[M][Br_2]}{k_{-1}[M]}$$

with the result (in which the concentration of the third-body molecules, [M], does not appear):

$$[Br] = \left(\frac{k_1}{k_{-1}}[Br_2]\right)^{1/2}$$

But bromine atoms also can be produced by the absorption of light by molecular bromine, the spectrum of which in the gaseous state (given in Figure 21–13) shows absorption in the violet–blue region of the spectrum. This absorption gives bromine its reddish-orange color when viewed in white light. The maximum absorption occurs at $\lambda \cong 414$ nm; using the procedure that was outlined previously, we can calculate the energy of a mole of photons corresponding to light of this wavelength,

$$E = h\nu = h\frac{c}{\lambda}$$

$$= \frac{6.626 \times 10^{-34}\text{ J s} \times 3.00 \times 10^{8}\text{ m s}^{-1}}{4.14 \times 10^{-7}\text{ m}} = 4.80 \times 10^{-19}\text{ J}$$

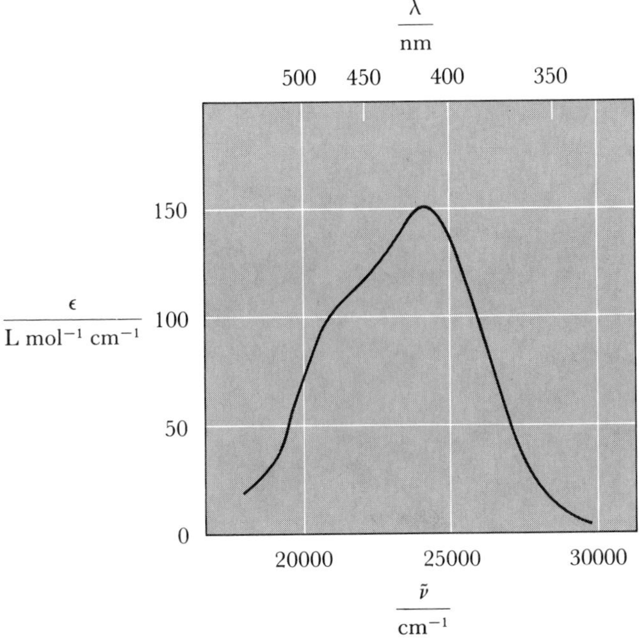

FIGURE 21–13
The absorption spectrum of $Br_2(g)$.

or (for one mole of photons):

$$E = 4.80 \times 10^{-19}\ \text{J} \times 6.02 \times 10^{23}\ \text{mol}^{-1}$$
$$= 289\ \text{kJ mol}^{-1}$$

This exceeds the bond-dissociation energy of bromine, $D(\text{Br—Br}) = 190\ \text{kJ mol}^{-1}$, so blue light dissociates bromine molecules into bromine atoms.

Under illumination, the concentration of atomic bromine is determined by the two reactions

$$Br_2 + h\nu \rightarrow 2Br \qquad \text{and} \qquad 2Br + M \xrightarrow{k_{-1}} Br_2 + M$$

Equating the rates of these two processes, we may write an equation for the concentration of bromine atoms:

$$[Br] = \left(\frac{I}{k_{-1}[M]}\right)^{1/2}$$

in which I is the rate of photon absorption (in mol $L^{-1}\ s^{-1}$). The rate of the photochemical reaction of hydrogen and bromine is proportional to the square root of the rate of photon absorption, and this reaction rate exceeds that of the thermal reaction because the concentration of bromine atoms is greater under steady illumination than in the absence of light. Because of this, the photochemically induced hydrogen–bromine reaction can be studied at ordinary temperatures, where the thermal reaction is extraordinarily slow. The same chain sequence of two reactions,

$$Br + H_2 \rightleftarrows HBr + H$$
$$H + Br_2 \rightleftarrows HBr + Br$$

occurs in the photochemical reaction as in the thermal reaction. This chain may go through many cycles before it is broken via the recombination reaction,

$$Br + Br + M \rightarrow Br_2 + M$$

A measure of the efficiency of an absorbed photon in causing chemical change is the *quantum yield*, generally designated ϕ:

$$\phi = \frac{\text{moles of chemical change}}{\text{moles of photons absorbed}}$$

As you would expect for the chain mechanism, the quantum yield in the photochemical hydrogen–bromine reaction exceeds one; quantum yields of over 100 are observed for this reaction.

A reaction that occurs photochemically but not at an appreciable rate under thermal conditions is the decomposition of acetone to ethane and carbon monoxide,

$$CH_3{-}\overset{\overset{\displaystyle O}{\|}}{C}{-}CH_3 = C_2H_6 + CO$$

The absorption of light of wavelength 313 nm causes the dissociation process,

$$CH_3{-}\overset{\overset{\displaystyle O}{\|}}{C}{-}CH_3 \xrightarrow{h\nu} CH_3{-}\overset{\overset{\displaystyle O}{\|}}{C}\cdot + \cdot CH_3$$

with the acetyl radical ($CH_3\dot{C}O$) rapidly decomposing to give carbon monoxide and another methyl radical ($\cdot CH_3$):

$$CH_3{-}\overset{\overset{\displaystyle O}{\|}}{C}\cdot \rightarrow \cdot CH_3 + CO$$

Ethane formation results from combination of methyl radicals:

$$2\,\cdot CH_3 \rightarrow C_2H_6$$

The initial photochemical process in this reaction is more complex than the chemical equation implies. The light absorption excites an electron in the carbonyl group. This excitation energy may rupture one of the carbon–carbon bonds, the weakest type of bond in the acetone molecule. However, an activated molecule of acetone can undergo other processes as well, forming other products, including methane and biacetyl ($CH_3COCOCH_3$). Quantum yields for the formation of carbon monoxide range from 0.1 to 0.5, depending on the wavelength of the light and the pressure of acetone vapor.

The photochemical reactions that we have considered so far are:

$$H_2(g) + Br_2(g) = 2HBr(g) \qquad \Delta G^0 = -110 \text{ kJ mol}^{-1} \text{ (25°C)}$$

$$CH_3COCH_3(g) = C_2H_6(g) + CO(g) \qquad \Delta G^0 = -17.1 \text{ kJ mol}^{-1} \text{ (25°C)}$$

These are reactions in which chemical equilibrium favors formation of the products. The absorption of light in these cases simply increases the rate of thermodynamically possible reactions. However, light absorption can cause thermodynamically unfavored reactions to occur. One example is the dissociation of nitrogen dioxide,

$$2NO_2(g) = 2NO(g) + O_2(g) \qquad \Delta G^0 = 69.8 \text{ kcal mol}^{-1} \text{ (25°C)}$$

which is the net result when nitrogen dioxide, an orange-brown gas, absorbs light. This overall reaction occurs via the steps

1. $NO_2(g) \xrightarrow{h\nu} NO(g) + O(g)$
2. $O(g) + NO_2(g) \rightarrow O_2(g) + NO(g)$

The resultant mixture of nitric oxide and oxygen is unstable with respect to nitrogen dioxide, and a slow thermal reaction re-forms nitrogen dioxide when the light is turned off.

The photochemical studies discussed so far typically are conducted with steady illumination of the reaction mixture. Under these conditions the balance between the rate of formation of reactive intermediates by light absorption and the rate of their disappearance in subsequent chemical reactions gives a steady low concentration of intermediates. Photochemical experiments also can be carried out under conditions in which a very intense, short burst of light dissociates an entire sample of the reactant, the reactive species so formed then undergoing their characteristic chemical changes. This technique, developed in the late 1940s by NORRISH and PORTER, is known as *flash photolysis*. The reactive molecules or atoms formed by the short flash of light characteristically react very rapidly, and studies of the rates of their subsequent reactions have proved very informative.

For instance, iodine vapor can be dissociated by light,

$$I_2(g) \xrightarrow{h\nu} 2I(g)$$

and the rate of the third-order recombination reaction,

$$I(g) + I(g) + M(g) \xrightarrow{k_3} I_2(g) + M(g)$$

can be determined directly. The efficiencies of various third-body molecules, M, differ greatly, as is shown in Table 21–4. There is a 10^3-fold range in the values of k_3 between the smallest value, that for He, and the largest value, that for I_2.

Particularly striking in these studies is the observation that these recombination reactions go more rapidly at low temperature; the third-order rate constants *decrease* with an *increase* of temperature! This unusual quality, which is observed for other atom-recombination reactions, is attributed to a two-step mechanism. The first step is formation of a loose complex between an iodine atom and M,

$$I + M \rightleftarrows IM \qquad K = \frac{[IM]}{[I][M]}$$

TABLE 21–4
Third-Order Rate Constants at 27°C for the Gas-Phase Reaction[a] $I + I + M \xrightarrow{k_3} I_2 + M$

M	$\frac{k_3}{L^2\ mol^{-2}\ s^{-1}}$	$\frac{E_a}{kJ\ mol^{-1}}$
He	1.5×10^9	-1.7
Ar	3.0×10^9	-5.4
H_2	5.7×10^9	-5.1
CO_2	13.4×10^9	-7.3
CH_3I	160×10^9	-10.7
I_2	1600×10^9	-18.4

[a]From G. Porter, *Science* **160**, 1300 (1968).

followed by reaction of this complex with an iodine atom,

$$I + IM \xrightarrow{k_2} I_2 + M$$

The rate constant, k_3, of the empirical third-order rate law is the product of the equilibrium constant, K, for the first step, which is slightly exothermic, and the rate constant, k_2, for the second step:

$$k_3 = Kk_2$$

To explain the negative activation energy associated with the empirical rate constant k_3, it is postulated that the activation energy associated with k_2 is essentially zero; this makes the apparent activation energy associated with k_3 equal to the enthalpy change associated with the exothermic equilibrium.

The photochemical production of iodine atoms has provided information about the thermal reaction of hydrogen and iodine,

$$H_2(g) + I_2(g) = 2HI(g)$$

the rate of which is governed by the rate law

$$-\frac{d[H_2]}{dt} = k_f[H_2][I_2] - k_r[HI]^2$$

The one-step bimolecular mechanism,

$$H_2(g) + I_2(g) \rightleftarrows 2HI(g)$$

is consistent with this rate law, and this mechanism was assumed correct for over a half-century. Recently, however, study of the photochemical reaction under appropriate conditions reveals a nonchain pathway involving two iodine atoms:

$$H_2(g) + 2I(g) \rightleftarrows 2HI(g)$$

Correlation of the activation energy for this reaction step with data for the thermal reaction suggests that this step also is the rate-determining step for the thermal reaction, which goes, according to the new theory, by the two steps:

1. $M(g) + I_2(g) \rightleftarrows M(g) + 2I(g)$ A rapid equilibrium
2. $2I(g) + H_2(g) \rightleftarrows 2HI(g)$ The slow step

This mechanism is consistent with the empirical second-order rate law because

$$[I]^2 = K[I_2]$$

in which K is the equilibrium constant for the first of these two steps. The rate of the rate-determining step is

$$-\frac{d[H_2]}{dt} = k'_f[I]^2[H_2] - k_r[HI]^2$$

But this rate law can be converted to the form of the empirical rate law by the substitution $[I]^2 = K[I_2]$. The rate constant k_f is equal to $k'_f K$.

In Chapter 18 the potential diagram for thallium showed stable +1 and +3 oxidation states but nothing about a +2 oxidation state. Under ordinary conditions, thallium(II) is not known, either in solution or in the form of solid compounds. However, it is possible to produce thallium(II) in detectable amounts by flash photolysis of solutions of thallium(III). In solutions of moderate acidity ($[H^+] = 0.10 \text{ mol L}^{-1}$) thallium(III) is present to an appreciable extent as hydroxothallium(III) ion,

$$Tl^{3+} + H_2O \rightleftarrows TlOH^{2+} + H^+ \qquad K \cong 0.1 \text{ mol L}^{-1} \ (25°C)$$

The photochemical process is a dissociation reaction,

$$TlOH^{2+} \xrightarrow{h\nu} Tl^{2+} + \cdot OH$$

which produces two unstable species, thallium(II) ion and hydroxyl radicals. Reactions of unstable thallium(II) can be studied in solutions of thallium(III) and various reducing agents that are subjected to flash photolysis. For instance, the reaction

$$Tl^{2+} + Fe^{2+} \rightarrow Tl^+ + Fe^{3+}$$

which is a postulated reaction step in the mechanism for the reaction of thallium(III) and iron(II), has been studied directly. Correlation of reactivity with $\mathscr{E}^0$ values allows us to estimate the value of $\mathscr{E}^0$ for the half-reaction

$$Tl^{2+} + e^- = Tl^+ \qquad \mathscr{E}^0 = 2.2 \text{ V}$$

Incorporation of the above half-reaction allows the potential diagram given in Chapter 18 to be expanded:

$$Tl^{3+} \xrightarrow{0.3\text{ V}} Tl^{2+} \xrightarrow{2.2\text{ V}} Tl^{+} \xrightarrow{-0.336\text{ V}} Tl$$

$$\underbrace{Tl^{3+} \longrightarrow Tl^{+}}_{1.27\text{ V}}$$

which in turn allows calculation of the equilibrium constant for disproportionation of thallium(II) at 25°C:

$$2Tl^{2+} = Tl^{3+} + Tl^{+} \qquad \mathscr{E}^0 = +2.2\text{ V} - 0.3\text{ V} = 1.9\text{ V}$$

$$K^0 = 10^{(1.9\text{ V}/0.0592\text{ V})} = 10^{32}$$

In a solution with the concentrations $[Tl^{3+}] = [Tl^{+}] = 0.05\text{ mol L}^{-1}$, the concentration of thallium(II) is very, very low:

$$[Tl^{2+}] = \left(\frac{1}{K} \times [Tl^{3+}][Tl^{+}]\right)^{1/2}$$

$$= \left(\frac{(0.05\text{ mol L}^{-1})^2}{10^{32}}\right)^{1/2} \cong 5 \times 10^{-18}\text{ mol L}^{-1}$$

With our knowledge of thallium chemistry expanded by this flash-photolysis study, we can see why thallium(II) is not a stable species in solutions at chemical equilibrium.

OZONE IN THE ATMOSPHERE

Ozone, the unstable allotropic form of oxygen, is produced photochemically from oxygen. The absorption of light of wavelength ~200 nm by oxygen causes dissociation,

$$O_2(g) \xrightarrow{h\nu} 2O(g)$$

which is followed by the reaction

$$O_2(g) + O(g) + M(g) \rightarrow O_3(g) + M(g)$$

If this sequence of two steps is the complete mechanism, a quantum yield

$$\phi = \frac{\text{moles of ozone produced}}{\text{moles of photons absorbed}} = 2$$

is to be expected, and this yield is observed under some experimental conditions. The photochemistry of ozone, oxygen, and nitrogen oxides in the stratosphere includes the following reactions as well as others:

$$O_3(g) \xrightarrow{h\nu} O_2(g) + O(g)$$

$$O(g) + O_3(g) \rightarrow 2O_2(g)$$

$$NO_2(g) \xrightarrow{h\nu} NO(g) + O(g)$$

$$NO(g) + O_3(g) \rightarrow O_2(g) + NO_2(g)$$

Some of these reactions produce ozone and some consume ozone, and

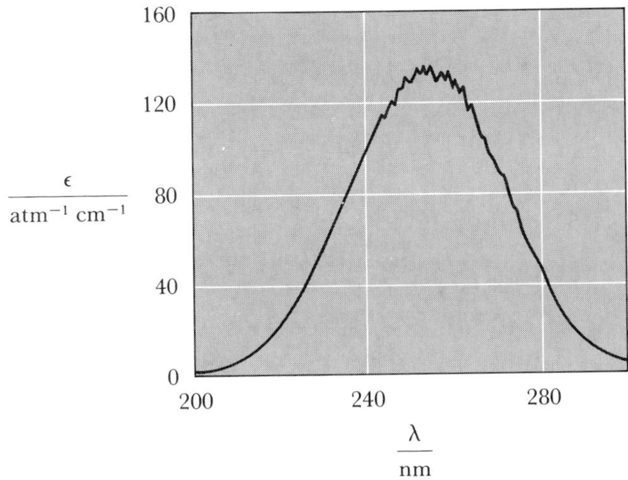

FIGURE 21–14
The spectrum of ozone in the ultraviolet region. Ozone absorbs ultraviolet solar radiation, which is not absorbed by O_2. Dimensions for ϵ, $atm^{-1}\ cm^{-1}$, result if the Beer–Lambert law for gases is written in terms of the pressure of the gas, $\log (I_0/I) = \epsilon Pb$.

under natural conditions a balance of the rates of these processes results in a steady, low concentration of ozone in the stratosphere. This natural concentration of ozone shields the Earth's surface from excessive ultraviolet radiation; Figure 21–14, giving the spectrum of ozone, shows the strong absorption of light with wavelength 220 nm to 280 nm.

People are worried that man's activities will upset the balancing reactions that determine the steady level of ozone in the stratosphere. Any substance that causes the consumption of ozone will lower this natural concentration, and will thereby increase the amount of ultraviolet radiation reaching the Earth's surface. Of particular concern are the reaction with nitric oxide, already given, and the reaction with atomic chlorine,

$$Cl(g) + O_3(g) \rightarrow ClO(g) + O_2(g)$$

which is followed by further reactions of chlorine oxides with ozone. The nitric oxide concentration level in the upper atmosphere is raised by the exhaust from supersonic aircraft. The presence of atomic chlorine in the upper atmosphere results from photochemical decomposition of chlorine compounds. A potentially large source of chlorine atoms is the photochemical decomposition of chlorofluorocarbons (e.g., Freons, $CFCl_3$ and CF_2Cl_2):

$$CFCl_3(g) \xrightarrow{h\nu} Cl(g) + CFCl_2(g)$$

The supposedly inert Freons are used as the propellant in many (but not all) aerosol sprays, and they also are used in refrigerators and air-conditioning systems. The chemical inertness of the chlorofluorocarbons assures that they will survive their slow diffusion to the upper atmosphere, where photochemical decomposition to give chlorine atoms occurs.

Ozone also is involved in the reactions at low elevations that are related to photochemical smog. It is formed as a result of reactions that already have been presented; one is the absorption of visible light by nitrogen dioxide:

$$NO_2(g) \xrightarrow{h\nu} NO(g) + O(g)$$

The atomic oxygen so produced reacts with oxygen to give ozone:

$$O_2(g) + O(g) + M(g) \rightarrow O_3(g) + M$$

The oxidation of nitric oxide by ozone,

$$NO(g) + O_3(g) \rightarrow NO_2(g) + O_2(g)$$

is exothermic enough ($\Delta H = -200$ kJ mol^{-1}) to raise the oxygen so produced to an excited singlet state. This excited form of oxygen is more effective than ordinary oxygen in oxidizing hydrocarbons to eye-irritating compounds. (The hydrocarbons also are contributed to the air by the exhaust from automobiles.)

21–4 The Spectra of Inorganic Species

The subject of spectra of inorganic species is too vast to cover briefly, but an introduction can be presented. This introduction will be concerned with two areas: (1) the spectra of halide ions, and (2) the spectra of transition metal ions.

Although transparent to visible light, halide ions in aqueous solution absorb ultraviolet light; the primary process is detachment of an electron from the halide ion and its transfer to a water molecule:

$$X^-(aq) \xrightarrow{h\nu} X(aq) + e^-(aq)$$

The spectra associated with such transitions are called *charge-transfer spectra*. The charge-transfer spectrum of iodide ion in aqueous solution, given in Figure 21–15, shows two peaks. The peak at the longer wavelength (232 nm) corresponds to production of atomic iodine in the ground state. With electron occupancy of only the $n = 5$ orbitals shown, the transition is

$$\begin{array}{lccccccc} & 5s & 5p & & & 5s & 5p & \\ I^- & (\uparrow\downarrow) & (\uparrow\downarrow)(\uparrow\downarrow)(\uparrow\downarrow) & \xrightarrow[232\text{ nm}]{h\nu} & I & (\uparrow\downarrow) & (\uparrow\downarrow)(\uparrow\downarrow)(\uparrow\ ) & + e^-(aq) \end{array}$$

The peak at the shorter wavelength (195 nm) corresponds to production of atomic iodine in an excited state,

$$\begin{array}{lccccccc} & 5s & 5p & & & 5s & 5p & \\ I^- & (\uparrow\downarrow) & (\uparrow\downarrow)(\uparrow\downarrow)(\uparrow\downarrow) & \xrightarrow[195\text{ nm}]{h\nu} & I & (\uparrow\ ) & (\uparrow\downarrow)(\uparrow\downarrow)(\uparrow\downarrow) & + e^-(aq) \end{array}$$

Because the maximum molar absorption coefficient, ϵ_{max}, for each of these transitions is very large, $\epsilon \cong 1.5 \times 10^4$ L mol^{-1} cm^{-1}, these are judged to be allowed transitions with a high probability of occurring. Support for

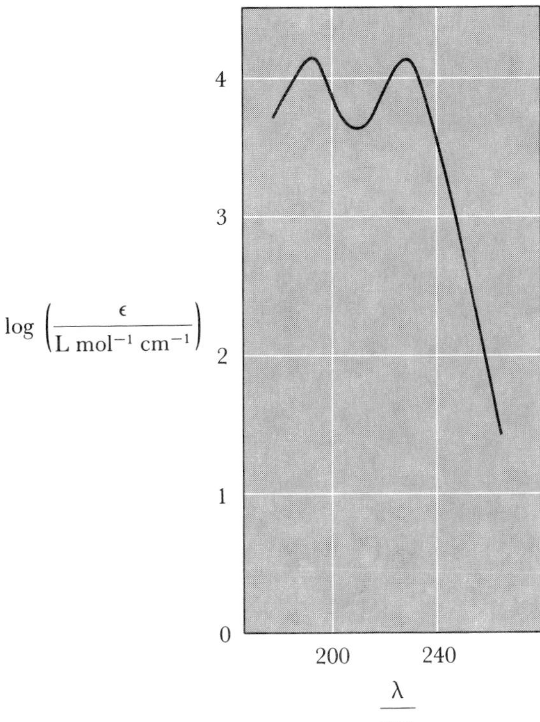

FIGURE 21–15
The ultraviolet spectrum of iodide ion in aqueous solution. The two peaks are attributed to the formation of two different states of the iodine atom.

the postulate that an electron is dissociated from a halide ion and is transferred to the solvent comes from the observation that the absorption spectrum of the hydrated electron ($\lambda_{max} = 720$ nm) is displayed by halide ion solutions that have been subjected to flash photolysis.

Following are the positions of the long-wavelength peaks in the ultraviolet spectra of chloride ion, bromide ion, and iodide ion:

$$Cl^-(aq) \qquad \lambda_{max} = 190 \text{ nm}$$
$$Br^-(aq) \qquad \lambda_{max} = 199.5 \text{ nm}$$
$$I^-(aq) \qquad \lambda_{max} = 232 \text{ nm}$$

These positions are consistent with trends in the electron affinities of the halogens,

$$EA(\text{Cl}) > EA(\text{Br}) > EA(\text{I})$$

and the hydration energies of the halide ions,

$$-\Delta H_{\text{hyd}}(\text{Cl}^-) > -\Delta H_{\text{hyd}}(\text{Br}^-) > -\Delta H_{\text{hyd}}(\text{I}^-)$$

Light absorption by a stable negative ion Y^- in the gas phase can result in an endothermic process,

$$Y^-(g) \xrightarrow{h\nu} Y(g) + e^-$$

which is the reverse of the process that defines the electron affinity of the gaseous neutral atom (or molecule). Use of adjustable lasers, which emit light of very well-defined frequencies, allows evaluation of the minimum energy needed to bring about this process. Many of the most accurately known values of electron affinities have been measured in this way, by determining the maximum wavelength of light that can detach an electron from the gaseous negative ion.

The ultraviolet spectra of many inorganic species in addition to halide ions are believed to be *charge-transfer spectra.* However the spectra of transition-metal ions include absorption bands in the visible region arising from a completely different type of transition, the excitation of an electron between different d orbitals of the metal ion. In Chapter 19, you learned a simplified crystal-field theory of transition-metal ions: In the field of six octahedrally oriented ligands, the three $3d$ orbitals with high probability between the orthogonal x, y, and z axes (d_{xy}, d_{yz}, and d_{xz}) are more stable than the two $3d$ orbitals with high probability along the x, y, and z axes ($d_{x^2-y^2}$ and d_{z^2}). The absorption of light can promote an electron from one of the more-stable d orbitals to one of the less-stable orbitals. The situation is particularly simple for d^1 ions, for example, $Ti(OH_2)_6^{3+}$, with one spectral peak, as is shown in Figure 21–16. The energy corresponding to

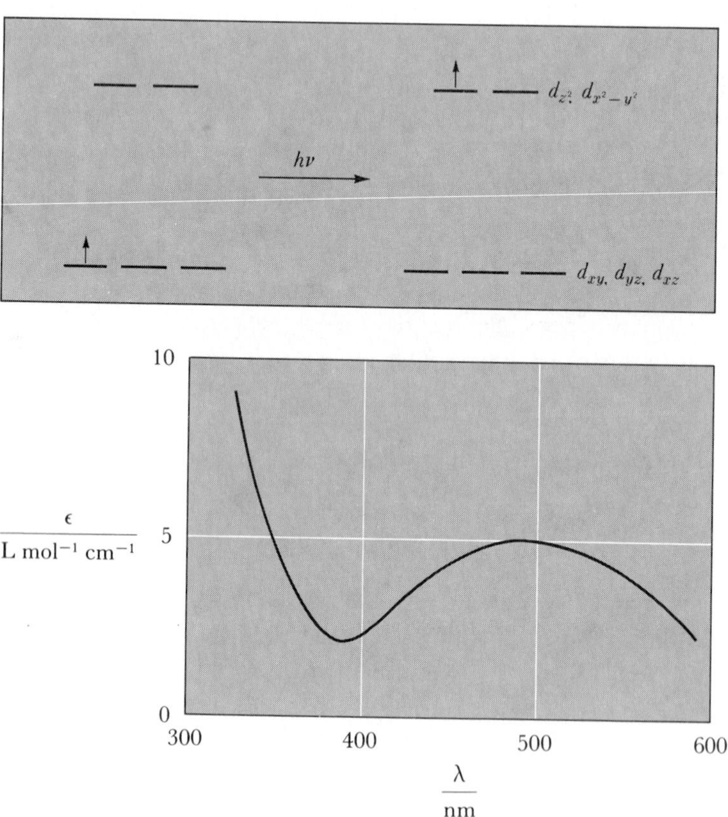

FIGURE 21–16
The d–d transition in $Ti(OH_2)_6^{3+}$, a d^1 species, and the spectrum of this species. The wavelength of the peak (493 nm) corresponds to 243 kJ mol^{-1}. (The absorption in the ultraviolet region of the spectrum is due to a charge-transfer transition.)

the wavelength of maximum absorption (493 nm) is

$$E = h\nu = h\frac{c}{\lambda}$$

$$= 6.626 \times 10^{-34}\text{ J s} \times \frac{3.00 \times 10^{8}\text{ m s}^{-1}}{4.93 \times 10^{-7}\text{ m}} = 4.03 \times 10^{-19}\text{ J}$$

or on a molar basis,

$$E = 6.02 \times 10^{23}\text{ mol}^{-1} \times 4.03 \times 10^{-19}\text{ J} = 243\text{ kJ mol}^{-1}$$

This is the value of Δ_o for this system. The molar extinction coefficient of hexaaquatitanium(III) ion at the absorption maximum is relatively low ($\epsilon \cong 5$ L mol^{-1} cm^{-1}) compared to that of iodide ion ($\epsilon \cong 1.5 \times 10^4$ L mol^{-1} cm^{-1} at 232 nm). The *d–d* transition in hexaaquatitanium(III) ion is nominally forbidden, but the charge-transfer transition in iodide ion is allowed.

The visible spectra of other octahedrally coordinated transition-metal ions also are caused by transitions of electrons between *d* orbitals. Figure 21–17 shows the spectra of nickel(II) species (each with d^8 configuration), green hexaaquanickel(II) ion, and blue tris(ethylenediamine)nickel(II) ion. Each spectrum consists of three peaks. The larger value of Δ_o associated with ethylenediamine as a ligand ($\Delta_{en} > \Delta_{H_2O}$; see Section 19–4) is responsible for the shorter wavelengths of peaks in the spectrum of $Ni(en)_3^{2+}$.

Even more dramatic changes of spectra resulting from complex-ion formation occur if complexing causes a change in the coordination number of the metal ion. Cobalt(II), a d^7 metal ion, undergoes such a change when chloride ion replaces water in its coordination shell. Hexaaquacobalt(II) ion, $Co(OH_2)_6^{2+}$, is a high-spin octahedral species. However, in concentrated aqueous solutions of chloride ion, cobalt(II) is present as tetrahedral tetrachlorocobaltate(II) ion, $CoCl_4^{2-}$. The spectra of these species are given in Figure 21–18. The transitions in hexaaquacobalt(II) ion occur at higher energy (lower wavelength) because the magnitude of Δ is larger for octahedral complexes, but extinction coefficients for tetrachlorocobaltate(II) ion are larger because these transitions are less forbidden.

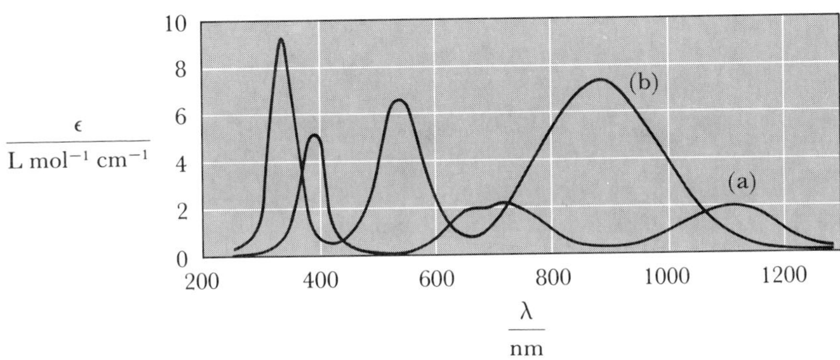

FIGURE 21–17
The spectra of nickel(II) species. (a) $Ni(OH_2)_6^{2+}$ and (b) $Ni(en)_3^{2+}$ (en = ethylenediamine).

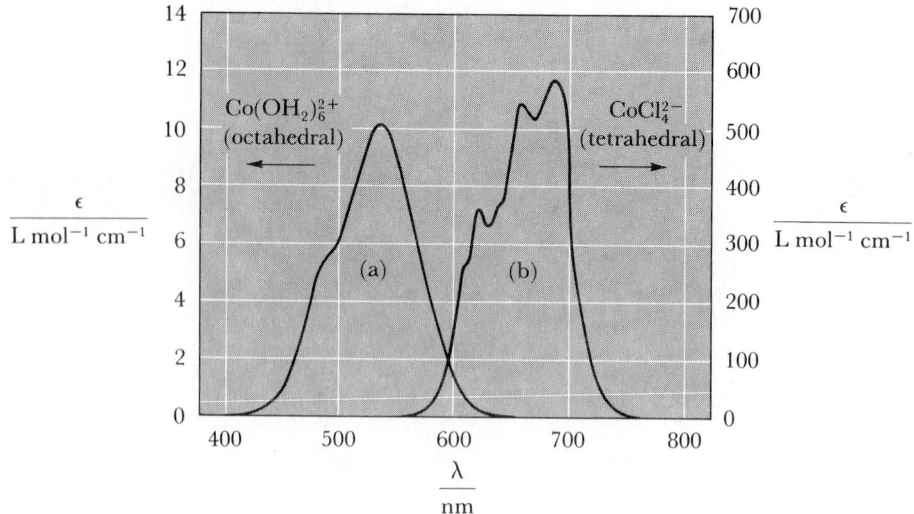

FIGURE 21–18
The spectra of (a) $Co(OH_2)_6^{2+}$ and (b) $CoCl_4^{2-}$.

The red color of the mineral ruby, which is aluminum(III) oxide (Al_2O_3) containing chromium(III) impurity, is due to the *d–d* transitions of the d^3 chromium(III) ion, which is octahedrally surrounded by six oxide ions in this material. The color of garnet, another gem mineral, is due to iron(II) and iron(III) ions in the lattice of particular alkaline-earth aluminosilicates [e.g., $Ca_3Al_2(SiO_4)_3$].

21–5 Quantitative Study of Complex-Ion Equilibria by Spectrophotometry

Because the spectrum of a transition-metal ion is influenced by the ligands that are coordinated to it, quantitative study of light absorption changes with changes of the concentrations may allow us to evaluate the equilibrium constant for the reaction in which one ligand replaces another.

Because of their usefulness in qualitative and quantitative analysis for iron or for thiocyanate ion, the blood-red thiocyanate complexes of iron(III) are of interest. With increasing concentration of thiocyanate ion, the water molecules in hexaaquairon(III) ion are replaced successively by thiocyanate ions. Attempts to study the first reaction,

$$Fe(OH_2)_6^{3+} + SCN^- \rightleftarrows Fe(OH_2)_5NCS^{2+} + H_2O$$

in solutions of low concentrations of iron(III) and high, variable concentrations of thiocyanate ion are complicated by occurrence of the second stage of complexing,

$$Fe(OH_2)_5NCS^{2+} + SCN^- \rightleftarrows Fe(OH_2)_4(NCS)_2^+ + H_2O$$

These difficulties are minimized by using solutions containing a low concentration of thiocyanate ion and a high, variable concentration of iron(III).[2] With the concentration of thiocyanate ion low (smaller than the reciprocal of the equilibrium constant for forming the second complex),

$$[SCN^-] \ll \frac{1}{K_2}$$

the relative concentration of the species containing two coordinated thiocyanate ions is low,

$$\frac{[Fe(OH_2)_4(NCS)_2^+]}{[Fe(OH_2)_5NCS^{2+}]} \ll 1$$

These inequalities are equivalent because

$$K_2[SCN^-] = \frac{[Fe(OH_2)_4(NCS)_2^+]}{[Fe(OH_2)_5NCS^{2+}]}$$

With increasing concentration of iron(III), the thiocyanate ion in solution is converted from the free ion to the complex ion. The light absorption changes from that characteristic of thiocyanate ion (transparent in the visible region) to that characteristic of pentaaquathiocyanatoiron(III) ion. This is shown in Figure 21–19, which gives the absorbance at 448 nm, a wavelength at which the only species with appreciable light absorption is the complex ion $Fe(OH_2)_5NCS^{2+}$. The equation for the line that correlates the data is[3]

$$\frac{\log\left(\frac{I_0}{I}\right)}{bC_{SCN}} = \epsilon_1\left(\frac{K_1[Fe^{3+}]}{1 + K_1[Fe^{3+}]}\right) \cong \epsilon_1\left(\frac{K_1C_{Fe}}{1 + K_1C_{Fe}}\right)$$

---

[2] Procedures for study of this reaction as a laboratory exercise are given in R. W. Ramette, *J. Chem. Educ.* **40**, 71 (1963).

[3] The derivation of this equation is:

$$\log\left(\frac{I_0}{I}\right) = \epsilon_1[FeNCS^{2+}]b$$

$$\frac{\log\left(\frac{I_0}{I}\right)}{b} = \epsilon_1 C_{SCN} \times \left(\frac{[FeNCS^{2+}]}{[SCN^-] + [FeNCS^{2+}]}\right)$$

$$= \epsilon_1 C_{SCN} \times \left(\frac{[FeNCS^{2+}]/[SCN^-]}{1 + [FeNCS^{2+}]/[SCN^-]}\right)$$

$$\frac{\log\left(\frac{I_0}{I}\right)}{bC_{SCN}} = \epsilon_1 \times \left(\frac{K_1[Fe^{3+}]}{1 + K_1[Fe^{3+}]}\right)$$

If $C_{Fe} \gg C_{SCN}$, essentially all of the iron(III) is present in the uncomplexed form, $[Fe^{3+}] = C_{Fe} - [FeNCS^{2+}] \cong C_{Fe}$, giving

$$\frac{\log\left(\frac{I_0}{I}\right)}{bC_{SCN}} = \epsilon_1 \times \left(\frac{K_1C_{Fe}}{1 + K_1C_{Fe}}\right)$$

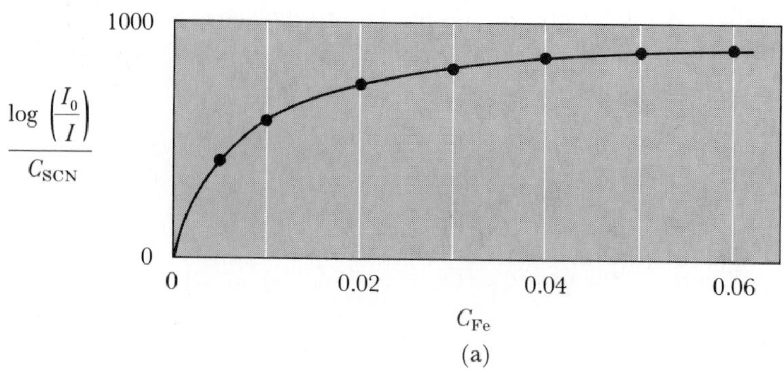

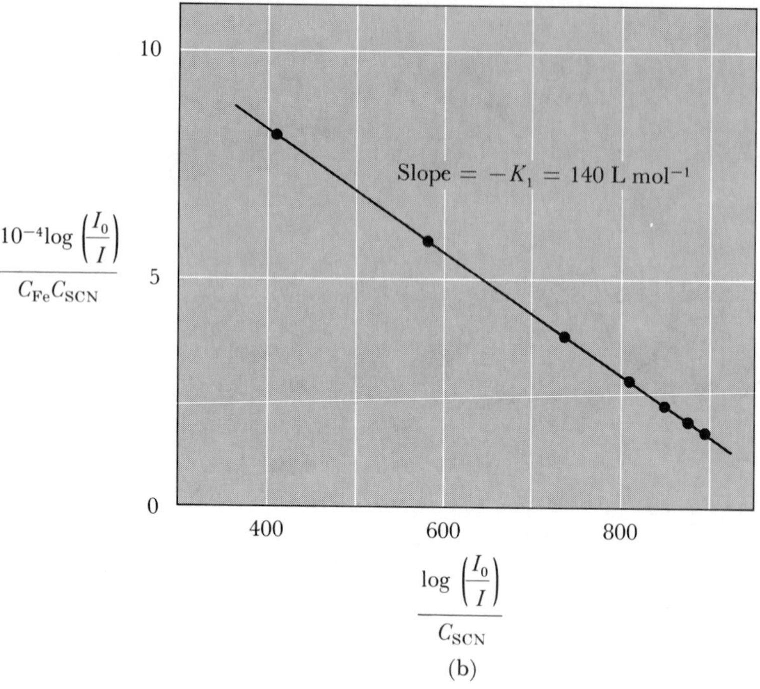

FIGURE 21–19
The light absorption of solutions of thiocyanate ion and iron(III) ion. $\lambda = 448$ nm (a maximum in the spectrum of $FeNCS^{2+}$) where $FeNCS^{2+}$ is the only light-absorbing species. $C_{Fe} \gg C_{SCN}$; $[Fe^{3+}] = C_{Fe} - [FeNCS^{2+}] \cong C_{Fe}$.

(a) $\dfrac{\log (I_0/I)}{C_{SCN}}$ versus C_{Fe}. At high C_{Fe}, the value of the quotient approaches $b \times \epsilon_1$

(b) $\dfrac{\log (I_0/I)}{C_{Fe}C_{SCN}}$ versus $\dfrac{\log (I_0/I)}{C_{SCN}}$. The slope of the line is $-K_1$; $K_1 = 140$ L mol^{-1}.

(All concentrations in mol L^{-1}.)

The factor by which ϵ_1, the molar extinction coefficient of $FeNCS^{2+}$, is multiplied in this equation is simply the fraction of the thiocyanate ion that is present as complex. Rearrangement of this equation gives

$$\frac{\log\left(\frac{I_0}{I}\right)}{C_{Fe}C_{SCN}} = -K_1 \times \frac{\log\left(\frac{I_0}{I}\right)}{C_{SCN}} + b\epsilon_1 K_1$$

This suggests a way to plot data to obtain K_1. Figure 21–19b is such a plot, $\log(I_0/I)/C_{Fe}C_{SCN}$ versus $\log(I_0/I)/C_{SCN}$. From the slope of this linear plot we obtain the value of K_1: $K_1 = -\text{slope} = 140 \text{ L mol}^{-1}$.

Procedures such as this have been used to characterize quantitatively many systems of association equilibria, both inorganic complex-ion systems and systems of organic molecules (e.g., ones in which there is association by hydrogen bonding).

We have seen in this chapter many ways in which study of the interaction of light with matter has enhanced our knowledge of chemical systems. Because the ultimate source of most of our energy is the light from the sun, increased understanding of photochemistry will help us exploit solar energy for practical purposes.

Biographical Notes

AUGUST BEER (1825–1863), a German physicist, was a faculty member at the University of Bonn. He discovered the dependence of light absorption on concentration.

EDWARD U. CONDON (1902–1974), an American physicist, received his education at the University of California, Berkeley. He worked in many areas of physics in industrial laboratories (Westinghouse Electric Co. and Corning Glass), government laboratories (National Bureau of Standards), and universities (Princeton, Washington University in St. Louis, and Colorado).

JAMES FRANCK (1882–1964), a German physicist, was educated in Germany and was associated with universities in both Europe (principally the University of Göttingen) and America (principally the University of Chicago). He received the Nobel Prize in Physics (with G. Hertz) in 1925 for his work on the interaction of electrons and atoms.

JOHANN H. LAMBERT (1728–1777), a German mathematician, astronomer, and physicist, was an extremely gifted natural philosopher. He did much in optics, astronomy, and mathematics.

RONALD G. W. NORRISH (1897–1978), an English chemist, was educated at Cambridge and later was Professor of Physical Chemistry there. He and George Porter developed the technique of flash photolysis. They shared the Nobel Prize in Chemistry in 1967 with Manfred Eigen.

GEORGE PORTER (1920–), an English chemist, was educated at Leeds and Cambridge. He was Professor of Chemistry at the University of Sheffield, and currently is Fullerian Professor and Director of the Royal Institution in London.

Problems and Questions

21–1 The refractive indexes for light from a sodium vapor lamp ($\lambda = 589.3$ nm in vacuum) of various compounds are:

NaBr(*s*) 1.64	NaI(*s*) 1.77
octane(*l*) 1.395	CS_2(*l*) 1.628

What is the velocity of light in each of these materials? What is the wavelength of this line ($\lambda = 589.3$ nm in vacuum) while passing through each of these materials?

21–2 Photons have an energy of 50.0 kJ mol^{-1}. What is the wavelength of this light, and what is its frequency in units s^{-1} and cm^{-1}?

21–3 How many normal modes of vibration are there for each of the following molecules?

$HgCl_2$ (linear)	SO_2 (nonlinear)
HCCH (linear)	NH_3 (nonlinear)
C_6H_6 (planar)	C_4H_8 (nonplanar)

21–4 Part of the microwave spectrum of a diatomic molecule is observed. The relative energies of the observed transitions are in the proportions 3:4:5. What values of J are involved in these transitions?

21–5 The moment of inertia of $^{19}F_2$ is 3.17×10^{-39} g cm^2. What is the rotational energy of 1.00 mol of $^{19}F_2$ molecules in the $J = 11$ state?

21–6 The separation of the successive absorption bands in the far infrared spectrum of HCl(*g*) is 20.7 cm^{-1}. What does this correspond to in joules per mole? At what temperature is the Boltzmann factor $e^{-E/kT}$ for this energy equal to 0.50?

21–7 Considering the general shape of the potential-energy curves for diatomic molecules, explain the fact that the average separation of the atoms increases with an increase in temperature.

21–8 Consider the bond energies of the following diatomic molecules: Cl_2, 239 kJ mol^{-1}; Br_2, 190 kJ mol^{-1}; HBr, 363 kJ mol^{-1}. What is the maximum wavelength of light that could possibly dissociate each of these molecules? Why is the wavelength of light that actually causes dissociation shorter than the values calculated?

21–9 Values of the fundamental vibrational frequency ($\tilde{\nu}_0$) for certain isotopic diatomic molecules are

$^{35}Cl_2$	560.9 cm^{-1}
$^{1}H^{35}Cl$	2886.2 cm^{-1}

What is the zero-point energy (in joules per mole) for each of these molecules? What are the reduced masses, fundamental frequencies, and zero-point energies for the other isotopic species, $^3H^{35}Cl$, $^1H^{37}Cl$, and $^{37}Cl_2$? (The atomic weight of ^3H is 3.016.)

21–10 At 414 nm, a 10.00-cm cell containing bromine gas at a concentration of 1.25×10^{-4} mol L^{-1} has an absorbance $\log(I_0/I) = 0.188$. What is the molar absorptivity of bromine at this wavelength?

21–11 The molar extinction coefficient of chromate ion at 372 nm is 4.82×10^3 L mol^{-1} cm^{-1}. What concentration of this ion is needed to absorb 50% of the light of this wavelength in a cell 1.00 cm long?

21–12 An unknown solution of chromate ion in a 5.00-cm cell has an absorbance log (I_0/I) equal to 0.845 at 372 nm. What is the concentration of chromate ion?

21–13 Values of the light absorption of two acidic solutions ($0.100M$ H^+) of chromium(VI) are compared. A solution of $1.00 \times 10^{-3}M$ $NaHCrO_4$ in a 5.00-cm cell is compared with a solution of $5.00 \times 10^{-3}M$ $NaHCrO_4$ in a 1.00-cm cell. The values of log (I_0/I) differ, even though the product $C \times b$ is the same. Why?

21–14 One wishes to have the activated solute molecules formed by light absorption distributed throughout the absorption cell as uniformly as possible. Is this more easily achieved at low, or high, concentrations of solute? Explain.

21–15 A molecule absorbs light having a wavelength of 427 nm, and fluorescence is observed ($\lambda = 507$ nm). What fraction of the excitation energy is re-emitted by a molecule that undergoes this mode of losing energy?

21–16 The photochemical reactions of hydrogen with bromine and with chlorine occur by the same type of mechanism. Explain the much greater chain length for the reaction forming hydrogen chloride.

21–17 Why does the rate of the photochemical reaction of bromine with hydrogen decrease as the concentration of an inert gas (e.g., krypton) increases?

21–18 If the third-order rate constants given in Table 21–4 are expressed in the Arrhenius form, $k_3 = Ae^{-E_a/RT}$, what are the values of A associated with the largest and smallest values of k_3 given in that table?

21–19 Using the positions of the two peaks in the absorption spectrum of iodide ion in aqueous solution, find the energy difference between the ground state and lowest excited state of the iodine atom.

21–20 Photochemical processes have been proposed as means of storing solar energy. What characteristics should a photochemical reaction have to be useful for this purpose?

22

Chemistry and the Atomic Nucleus

22–1 Introduction

In this chapter we shall concern ourselves with the atomic nucleus, and we shall need some understanding of the physics of the nucleus if we are to pursue subjects that are important to the chemist. These subjects include:

1. the abundance of the elements in nature (this subject was mentioned briefly in Chapter 1),
2. the dependence of chemical properties on atomic mass,
3. the stability of nuclei,
4. the production of unstable nuclei,
5. the modes of disintegration of unstable nuclei and the rates of such processes, and
6. the use of stable and unstable nuclides as tracers in chemical reactions.

In discussing atomic nuclei, we will use terms that we have not yet introduced and others that we have used infrequently. These terms are:

Nucleon Proton or neutron

Nuclide A nuclear species; a nucleus with a particular number of protons (Z the atomic number, also called the proton number) and a particular number of neutrons (N, called the neutron number)

Isotopes Nuclides with the same value of Z, but different values of N

Isotones Nuclides with the same value of N, but different values of Z

Isobars Nuclides with the same mass number, A ($A = Z + N$)
Positron A positively charged particle with the mass of an electron and the charge of a proton.

22–2 Relativity; the Equivalence of Mass and Energy

When we discussed various types of energy in Chapter 3 we mentioned relativistic energy, but deferred consideration of this subject. Now we must learn certain basic principles of the special theory of relativity, proposed by Einstein in 1905. The relativistic total energy, W, of a sample of matter of mass m is

$$W = mc^2$$

in which c is the speed of light in a vacuum. The mass appearing in this equation is related to the rest mass of the sample, m_0, and its speed, v, by the equation

$$m = \frac{m_0}{\left(1 - \frac{v^2}{c^2}\right)^{1/2}}$$

These equations indicate that

1. mass and energy are equivalent, the numerical conversion factor being c^2, and that
2. the mass of a sample of matter depends upon its velocity. (This velocity is measured relative to the observer who is measuring the mass.)

When the energy of a system changes, its mass changes; when the mass of a system changes, its energy changes. These equations apply to ordinary physical changes and ordinary chemical changes, as well as to the nuclear transformations that we will be discussing in this chapter. First, let us consider examples of simple physical and chemical changes.

To melt 1.00 mol of ice at 0°C requires 6.01 kJ (6.01×10^3 kg m^2 s^{-2}). The calculated change of mass of this sample of water upon melting is:

$$\Delta m = \frac{\Delta W}{c^2} = \frac{\Delta U}{c^2} = \frac{6.01 \times 10^3 \text{ kg m}^2 \text{ s}^{-2}}{(3.00 \times 10^8 \text{ m s}^{-1})^2} = 6.68 \times 10^{-14} \text{ kg}$$

This clearly is an immeasurable percentage increase of mass of this sample, which has a rest mass of 1.80×10^{-2} kg.

When 1.00 mol of liquid water is formed from its constituent elements at 25°C, the internal energy of the system decreases by 289.6 kJ (289.6×10^3 kg m^2 s^{-2}),

$$H_2(g) + \tfrac{1}{2}O_2(g) = H_2O(l) \qquad \Delta U = -289.6 \text{ kJ mol}^{-1}$$

The change of mass in this chemical change is

$$\Delta m = \frac{\Delta W}{c^2} = \frac{\Delta U}{c^2} = \frac{-289.6 \times 10^3 \text{ kg m}^2 \text{ s}^{-2}}{(3.00 \times 10^8 \text{ m s}^{-1})^2} = -3.22 \times 10^{-12} \text{ kg}$$

This decrease of mass also is immeasurably small.

FIGURE 22–1
The energy changes in some processes:
(a) $H_2O(s) = H_2O(l)$
(b) $H_2(g) + \frac{1}{2}O_2(g) = H_2O(l)$
(c) $8\,^1_1H + 8n = {}^{16}_{8}O$

But now let us consider the changes of energy and mass in the formation of 1.00 mol of oxygen nuclei from the constituent nucleons, eight protons (p) and eight neutrons (n) per oxygen-16 nucleus:

$$8\text{p} + 8\text{n} = {}^{16}_{8}\text{O nucleus}$$

The mass data that we have available (Table 8–1) pertain to neutral atoms; we can use these data directly because the total numbers of electrons in neutral atoms of the species on each side of this equation are the same. If eight electrons are added to each side of this equation, we have

$$\underbrace{8\text{p} + 8\text{e}^-}_{8{}^{1}_{1}\text{H}} + 8\text{n} = \underbrace{{}^{16}_{8}\text{O nucleus} + 8\text{e}^-}_{{}^{16}_{8}\text{O}}$$

or

$$8{}^{1}_{1}\text{H} + 8\text{n} = {}^{16}_{8}\text{O}$$

and the change in mass can be calculated:

$$\begin{aligned}
\Delta m &= m({}^{16}_{8}\text{O}) - 8m({}^{1}_{1}\text{H}) - 8m(\text{n}) \\
m({}^{16}_{8}\text{O}) &= \qquad\qquad\qquad\qquad\qquad\quad 15.99491 \text{ g mol}^{-1} \\
-8m({}^{1}_{1}\text{H}) &= -8 \times 1.00782 \text{ g mol}^{-1} = -8.06256 \text{ g mol}^{-1} \\
-8m(\text{n}) &= -8 \times 1.00866 \text{ g mol}^{-1} = -8.06928 \text{ g mol}^{-1} \\
&\qquad\qquad\qquad\qquad\quad \Delta m = -0.13693 \text{ g mol}^{-1}
\end{aligned}$$

This is an appreciable change in mass. In the process of forming nuclei of oxygen-16 from the constituent nucleons, 0.86% of the mass of the constituent nucleons disappears. The energy change in this process is

$$\begin{aligned}
\Delta W &= (3.00 \times 10^{8} \text{ m}^2 \text{ s}^{-2})^2(-1.37 \times 10^{-4} \text{ kg mol}^{-1}) \\
&= -1.23 \times 10^{13} \text{ kg m}^2 \text{ s}^{-2} \text{ mol}^{-1} = -1.23 \times 10^{13} \text{ J mol}^{-1}
\end{aligned}$$

This amount of energy is enormously greater than that associated with the physical process (the melting of ice) or the chemical reaction (the formation of water), which we have just considered. The energy changes for the three processes are shown in Figure 22–1.

The difference between the mass of the oxygen-16 nucleus and the constituent nucleons is a measure of the stability of this nuclide. We will develop this point later in connection with a survey of the stability of all stable nuclides.

22–3 Nuclides, Stable and Unstable

There are approximately 2000 known nuclides. Some are stable and occur in nature. Some are unstable and disintegrate ultimately to give stable nuclides. An unstable nuclide may occur in nature if its disintegration rate is so low that it has persisted since being formed when the Earth was formed,

$\sim 4.6 \times 10^9$ years ago, or if it is formed in the disintegration of an unstable long-lived parent nuclide, or if it is formed in some other natural process. We will consider both the combinations of protons and neutrons that are stable and also the mode of disintegration of nuclides that are unstable combinations of nucleons.

STABLE NUCLIDES

There are 266 stable nuclides; these can be classified according to whether the number of nucleons of each type is odd or even. Table 22–1 gives the number of each type of nuclide; we see that most stable nuclides are even–even nuclides. The four stable odd–odd nuclides are those with the smallest possible values of A for such nuclides:

$A = 2$	$Z = N = 1$	^2H
$A = 6$	$Z = N = 3$	^6Li
$A = 10$	$Z = N = 5$	^{10}B
$A = 14$	$Z = N = 7$	^{14}N

The data on occurrence of nuclides of the several types, as well as the relative abundances of the elements in the universe, indicate greater stability for nuclei containing an even number of each kind of nucleon (protons and neutrons). Nuclear theory, which has developed greatly in recent decades, associates energy levels and quantum numbers with each kind of nucleon in the nucleus, much as atomic theory associates energy levels and quantum numbers with electrons in an atom. Special nuclear stability is associated with nuclei having filled proton or neutron shells, just as special chemical stability is associated with atoms having filled electron shells. For nuclei, the cumulative numbers of protons or neutrons that correspond to filled shells of nucleons are 2, 8, 20, 28, 50, 82, and 126. These numbers are called "magic numbers." Elements with magic numbers for their atomic number are:

$Z = 2$	He	$Z = 28$	Ni
$Z = 8$	O	$Z = 50$	Sn
$Z = 20$	Ca	$Z = 82$	Pb

TABLE 22–1
The Nuclear Composition of Stable Nuclides[a]

	Even N	Odd N
Even Z	160	53
Odd Z	49	4

[a]You may find summaries that differ from this one as to numbers in each category. Such tabulations include as stable nuclides some very long-lived unstable nuclides (e.g., vanadium-50, an odd–odd nuclide, with a half-life of 4×10^{14} y).

The predominant naturally occurring isotope of four of these elements is a doubly magic nuclide:

^{4}He ($Z = 2$, $N = 2$)	100% of He
^{16}O ($Z = 8$, $N = 8$)	99.759% of O
^{40}Ca ($Z = 20$, $N = 20$)	96.97% of Ca
^{208}Pb ($Z = 82$, $N = 126$)	52% of Pb

It also is worth noting that the element with the largest number of stable isotopes, tin, with ten, has a proton number Z that is a magic number. The neutron number N that is represented by the largest number of stable nuclides is 82, a magic number; the seven isotones with this value of N are $^{136}_{54}Xe$, $^{138}_{56}Ba$, $^{139}_{57}La$, $^{140}_{58}Ce$, $^{141}_{59}Pr$, $^{142}_{60}Nd$, and $^{144}_{62}Sm$.

A measure of the stability of a nuclide is its mass relative to the masses of its constituent nucleons. This subject generally is handled by calculating the loss of mass when one atom is formed, the binding energy BE, dividing this by the total number of nucleons, A, and expressing the result in MeV as the energy units. The result is called the *binding energy per nucleon, BE/A*. To make these calculations, let us first calculate the energy that corresponds with one unit of atomic mass (1 amu):

$$W = mc^2$$

$$= \frac{0.001 \text{ kg mol}^{-1}}{6.0221 \times 10^{23} \text{ mol}^{-1}} \times (2.9979 \times 10^{8} \text{ m s}^{-1})^2$$

$$= 1.4924 \times 10^{-10} \text{ kg m}^2 \text{ s}^{-2} = 1.4924 \times 10^{-10} \text{ J}$$

$$= 1.4924 \times 10^{-10} \text{ V C} \times \frac{1 \text{ e}}{1.6022 \times 10^{-19} \text{ C}} = 931.5 \text{ MeV}$$

Thus for the doubly magic nuclide oxygen-16, for which the mass loss has been calculated to be 0.13693 g mol^{-1} or 0.13693 amu,[1] the binding energy per nucleon, BE/A, is

$$\frac{BE}{A} = 0.13693 \text{ amu} \times \frac{931.5 \text{ MeV}}{1 \text{ amu}} \times \frac{1}{16} = 7.97 \text{ MeV}$$

This value can be compared with the value for nitrogen-14, a nuclide for which both Z and N are odd; the mass difference is

$$\Delta m = m(^{14}\text{N}) - 7m(^{1}\text{H}) - 7m(\text{n})$$

$$\frac{\Delta m}{\text{g mol}^{-1}} = 14.00307 - 7(1.00782) - 7(1.00866)$$

$$\Delta m = -0.11229 \text{ amu}$$

[1]In our discussion of nuclear stability and nuclear transformations, it generally will be convenient to focus attention on single atoms, not moles of atoms, and to use the atomic mass unit in our calculations:

$$1 \text{ amu} \equiv \frac{1}{12} \times \text{mass of 1 atom of carbon-12}$$

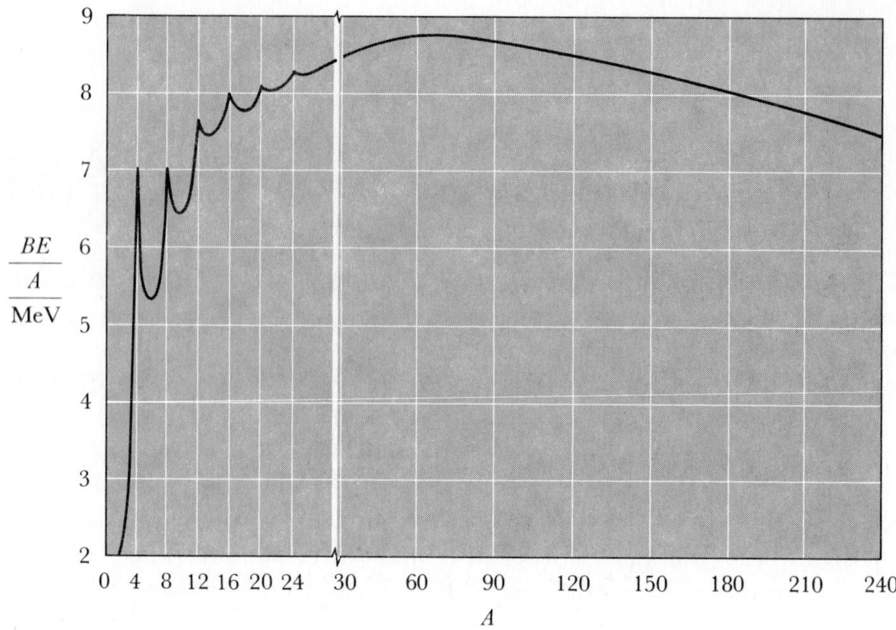

FIGURE 22–2
The binding energy per nucleon as a function of mass number; the abscissa scale changes at $A = 30$.

which gives for the binding energy per nucleon:

$$\frac{BE}{A} = 0.11229 \text{ amu} \times \frac{931.5 \text{ MeV}}{1 \text{ amu}} \times \frac{1}{14} = 7.47 \text{ MeV}$$

Thus we see that the binding energy per nucleon for doubly magic oxygen-16 is greater than for odd–odd nitrogen-14.

Figure 22–2 is a plot of the binding energy per nucleon for stable nuclides as a function of the mass number. From mass number ~16 to ~180, the binding energy per nucleon for stable nuclei is >7.7 MeV, the maximum occurring at $A = 56$ (^{56}Fe) with a binding energy per nucleon of 8.79 MeV. In light nuclei there are proportionately more nucleons on the surface of the nucleus; these nucleons do not interact with as many other nucleons as do those in the inner core of a nucleus, and this is responsible for the smaller values of BE/A for light nuclei. The coulombic repulsion of the protons in the nucleus is responsible for the decrease of binding energy per nucleon for nuclides with a large atomic number, that is, a large number of protons.

The effect of coulombic repulsion of protons in the nucleus is shown also in Figure 22–3, a plot of N versus Z with each stable nuclide represented as a point. Calcium-40 is the stable nuclide with the largest A for which $N/Z = 1$. At $A = 208$ (^{208}Pb), the value of N/Z is 1.54. Other features of this figure are revealed in the inserts. With the exceptions of $A = 5$ and 8, there is a single stable nuclide for each A in the range $A = 1$ to 20. In the range $A = 60$ to 70, there is only one stable nuclide for each odd value of A, but there are two stable nuclides for certain of the even values of A ($A =$

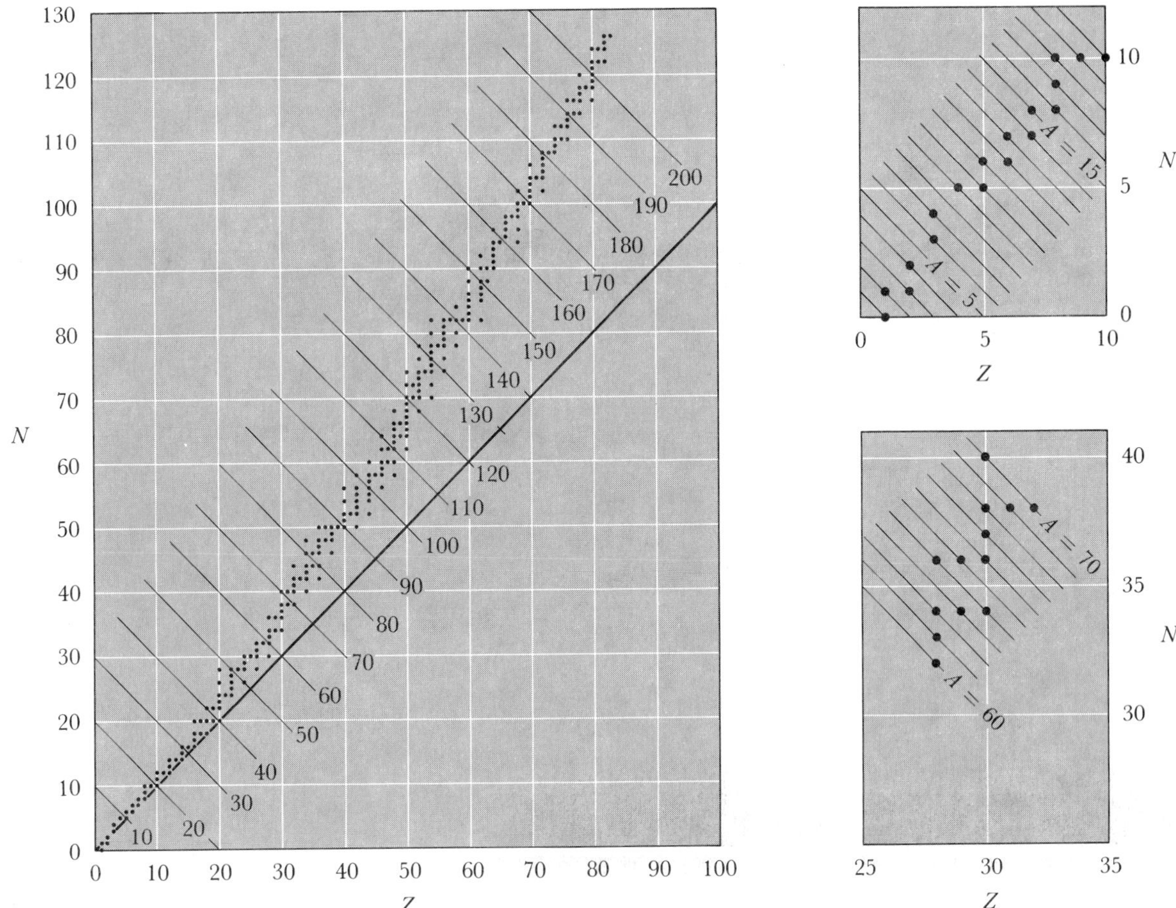

FIGURE 22–3
Stable nuclides. The solid line with slope +1 gives nuclides with $N = Z$; on this line are ^{2}H, ^{4}He, ^{6}Li, ^{10}B, ^{12}C, ^{14}N, ^{16}O, ^{20}Ne, ^{24}Mg, ^{28}Si, ^{32}S, ^{36}Ar, and ^{40}Ca. Solid lines with slope −1 for A values divisible by 10 are lines for isobars, nuclides with the same value of A. (Inserts show the $A = 1$–20 and $A = 60$–70 regions.)

64 and 70). Some points related to these are generally valid. *For each odd value of A there is no more than one stable nuclide.* (An apparent exception to this is $A = 113$, for which there are two naturally occurring nuclides, ^{113}Cd and ^{113}In. However, the heavier of these, ^{113}Cd, may be unstable with an undetectably low disintegration rate.) *For each even value of A there may be one, two, or three stable nuclides.* (The even values of A for which there are three stable nuclides are 96, 124, 130, and 136.)

UNSTABLE NUCLIDES

Over 87% of known nuclides are unstable. An unstable nuclide becomes more stable by undergoing an exothermic disintegration process, a process in which the mass of the products is lower than the mass of the parent nuclide.

The rates of observed disintegration processes vary over an enormous range, and some exothermic nuclear disintegrations are so slow that the parent nuclides are stable for all practical purposes. The observed modes of disintegration of unstable nuclides fall into two categories, that in which the mass number changes and that in which it does not. First we will consider the processes in which A changes: alpha-particle emission and spontaneous fission. Then we will consider the processes in which A does not change: emission of negative or positive beta particles and capture by the nucleus of an electron in the $n = 1$ shell.

Alpha-Particle Emission. Many naturally occurring unstable nuclides decompose by emitting an alpha (α) particle (a helium-4 nucleus). An α particle is quite stable; in Figure 22–2 you see a maximum in the binding energy per nucleon at $A = 4$ ($BE/A = 7.1$ MeV). Some of the heaviest naturally occurring nuclides gain stability by losing an α particle and transforming into a nuclide with a value of A that is 4 units smaller. Uranium-238, the predominant isotope of uranium in nature (99.27% of natural uranium), is an α emitter; the nuclear disintegration is

$$^{238}_{92}\text{U} \rightarrow {}^{4}_{2}\alpha + {}^{234}_{90}\text{Th}$$

In writing balanced equations for nuclear transformations, you should include the mass number, A, and the proton number, Z, with each symbol. This will help you balance such equations because each of these quantities is conserved. In this equation the balance of mass numbers (the superscripts) is

$$238 = 4 + 234$$

and the balance of proton numbers (the subscripts) is

$$92 = 2 + 90$$

The change of mass in this process can be calculated using the conventional atomic masses of neutral atoms. The atomic mass of uranium-238 includes the mass of 92 electrons and the atomic masses of helium-4 and thorium-234 include the mass of 2 and 90 electrons, respectively. Therefore the change of mass is:

$$\begin{aligned} -\text{AW}(^{238}\text{U}) &= -238.05081 \text{ amu [incl. } 92\, m(e^-)] \\ +\text{AW}(^{4}\text{He}) &= \quad +4.00260 \text{ amu [incl. } 2\, m(e^-)] \\ +\text{AW}(^{234}\text{Th}) &= +234.04363 \text{ amu [incl. } 90\, m(e^-)] \\ \hline \Delta m &= \quad -0.00458 \text{ amu} \end{aligned}$$

$$\Delta W = -0.00458 \text{ amu} \times \frac{931.5 \text{ MeV}}{1 \text{ amu}} = -4.27 \text{ MeV}$$

The matter that "disappears" in the disintegration of uranium-238 is transformed into the kinetic energy of the emitted α particle and the recoiling thorium-234 atom. In colliding with atoms and molecules of the surroundings, the emitted α particle and the recoiling atom give their excess energy to many atoms. This energy may simply increase the thermal energy of these atoms, but an energetic α particle also can break covalent bonds and bring about drastic chemical and biological changes in matter through which it passes. The mass of the universe is not decreased by a nuclear disintegration: the energy added to atoms of the surroundings increases their mass by the amount that "disappeared" in the nuclear disintegration.

In our discussion of stable nuclides in Chapter 1, and in the previous tabulation in this chapter, we counted bismuth-209 as a stable nuclide. Yet available mass data indicate that the α-particle emission,

$$^{209}_{83}\text{Bi} \rightarrow {}^{205}_{81}\text{Tl} + {}^{4}_{2}\alpha$$

in exothermic, with $\Delta W = -3.1$ MeV. Search for α emission by naturally occurring bismuth, which is 100% bismuth-209, has failed to reveal such a process occurring at a detectable rate.

Spontaneous Fission. Because of the decrease in binding energy with increasing A at high values of A, the disintegration of most heavy nuclides into two nuclides of similar mass number would result in a decrease of mass. This process, called *fission*, does not occur spontaneously at a detectable rate for most heavy nuclides, even though the process is possible energetically. However, for some heavy nuclides, the rate of spontaneous fission is similar to the rate of disintegration by some other mode. For uranium-235, the rate of spontaneous fission is 3×10^{-9} times the rate of disintegration by α emission. For californium-254, the predominant mode of decomposition is spontaneous fission. (We will consider induced fission at greater length when we discuss nuclear reactions.)

Emission of a Negative Beta Particle (β^- or e^-). This mode of disintegration is displayed by thorium-234,

$$^{234}_{90}\text{Th} \rightarrow {}^{0}_{-1}\beta^- + {}^{234}_{91}\text{Pa}$$

The mass number and proton number of the electron are 0 and -1, respectively. These quantities are balanced in this disintegration:

$$234 = 0 + 234 \quad \text{and} \quad 90 = -1 + 91$$

The net result of β^- emission is transformation of a neutron in the nucleus into a proton, thereby decreasing the value of N/Z. In the β^- emission by thorium-234, N/Z decreases from 1.60 to 1.57. These values are larger than the values for the heaviest stable nuclides ($N/Z = 1.52$ for ^{207}Pb and ^{209}Bi and 1.54 for ^{208}Pb). The product of this process, protactinium-234, also is unstable with respect to β^- emission.

In the transformation of thorium-234 to protactinium-234 plus a negative β particle, mass is converted to energy. To calculate the mass loss accompanying β^- emission, one need only compare the conventional atomic masses of the parent and daughter nuclides. We will derive this relationship for the present example. First, let us consider the nuclei and the emitted β^- particle:

$$\Delta m = \text{mass of protactinium-234 nucleus} + \text{mas of } 1\ e^- - \text{mass of thorium-234 nucleus}$$
$$= m(^{234}_{91}\text{Pa nucleus}) + m(e^-) - m(^{234}_{90}\text{Th nucleus})$$

This equation for the change of mass can be converted to an equation involving conventional atomic weights by *adding* and *subtracting* the mass of 90 electrons to the right side of the equation:

$$\Delta m = \underbrace{m(^{234}_{91}\text{ Pa nucleus}) + m(e^-) + 90\ m(e^-)}_{\text{AW}(^{234}_{91}\text{ Pa})} \underbrace{- m(^{234}_{90}\text{ Th nucleus}) - 90\ m(e^-)}_{-\text{AW}(^{234}_{90}\text{ Th})}$$

Thus β^- emission is energetically possible (i.e., $\Delta m < 0$) if a nuclide has a greater atomic weight than its isobar with one more proton in its nucleus. For this particular transformation:

$$\Delta m = \text{AW}(^{234}_{91}\text{ Pa}) - \text{AW}(^{234}_{90}\text{ Th})$$
$$= 234.04335 \text{ amu} - 234.04363 \text{ amu} = -0.00028 \text{ amu}$$

$$\Delta W = -0.00028 \text{ amu} \times \frac{931.5 \text{ MeV}}{1 \text{ amu}} = -0.26 \text{ MeV}$$

The free neutron itself is an unstable particle; the β^- emission to produce a proton,

$$^1_0\text{n} \rightarrow {}^1_1\text{p}^+ + {}^{0}_{-1}\beta^-$$

is accompanied by a decrease of mass:

$$\begin{aligned} m(\text{p} + \beta^-) = m(^1_1\text{H}) &= 1.00782 \text{ amu} \\ -m(\text{n}) &= -1.00866 \text{ amu} \\ \Delta m &= -0.00084 \text{ amu} \end{aligned}$$

This decrease in mass corresponds to an energy of 0.78 MeV. The disintegration reaction has been observed directly; the half-life of the free neutron is 12.8 min.

Emission of a Positive Beta Particle (β^+, a Positron). This mode of disintegration is displayed by sodium-22:

$$^{22}_{11}\text{Na} \rightarrow {}^0_1\beta^+ + {}^{22}_{10}\text{Ne}$$

In this process the ratio N/Z increases from 1.00 to 1.20. The mass-balance calculation for β^+ emission is a bit more complex than that for β^- emission.

For the example being considered,

$$\begin{aligned}\Delta m &= \text{mass of neon-22 nucleus} + \text{mass of } \beta^+ \\ &\quad - \text{mass of sodium-22 nucleus} \\ &= m(^{22}_{10}\text{Ne nucleus}) + m(\beta^+) - m(^{22}_{11}\text{Na nucleus})\end{aligned}$$

To the right side of this equation we will add and subtract the mass of 11 electrons:

$$\Delta m = \underbrace{m(^{22}_{10}\text{Ne nucleus}) + 11\, m(\text{e}^-)}_{\text{AW}(^{22}_{10}\text{Ne}) + m(\text{e}^-)} + \underbrace{m(\beta^+)}_{+ m(\beta^+)}$$

$$\underbrace{m(\text{e}^-) + m(\beta^+)}_{2\, m(\text{e}^-)}$$

$$\underbrace{- m(^{22}_{11}\text{Na nucleus}) - 11\, m(\text{e}^-)}_{- \text{AW}(^{22}_{11}\text{Na})}$$

Thus the change of mass will be negative if the atomic weight of the parent nuclide exceeds that of its isobar with one less proton in its nucleus by more than two times the mass of the electron, $2 \times 5.486 \times 10^{-4}$ amu. For this example,

$$\begin{aligned}\Delta m &= \text{AW}(^{22}_{10}\text{Ne}) + 2\, m(\text{e}^-) - \text{AW}(^{22}_{11}\text{Na}) \\ &= 21.99138 \text{ amu} + 2 \times 5.486 \times 10^{-4} \text{ amu} - 21.99444 \text{ amu} \\ &= -0.00196 \text{ amu}\end{aligned}$$

$$\Delta W = -0.00196 \text{ amu} \times \frac{931.5 \text{ MeV}}{1 \text{ amu}} = -1.83 \text{ MeV}$$

Capture by the Nucleus of an Electron in the $n = 1$ (K) Shell. This mode of decomposition, called *K-electron capture*, is displayed by beryllium-7:

$$^{7}_{4}\text{Be} + {}^{0}_{-1}\text{e}^- \rightarrow {}^{7}_{3}\text{Li}$$

In this process N/Z increases from 0.75 to 1.33. With the background of the two preceding calculations, you can do a mass-balance calculation for this process. Like the calculation for negative β-particle emission, the calculation shows that K-electron capture is exothermic if the atomic weight of the parent nuclide exceeds the atomic weight of the daughter nuclide.

Because the mass inequality requirement for positron emission, $\text{AW}(^{A}_{Z}\text{X}) > \text{AW}(_{Z-1}^{A}\text{Y}) + 2\, m(\text{e}^-)$, is more demanding than for K-electron capture, $\text{AW}(^{A}_{Z}\text{X}) > \text{AW}(_{Z-1}^{A}\text{Y})$, a nuclide that is unstable with respect to the former process also is unstable with respect to the latter process. Many such nuclides exhibit both processes for increasing their N/Z values.

The Relative Stability of Isobars. Whether a nuclide is unstable relative to its isobars is determined by the relative masses of the involved nuclides. This can be illustrated further if we consider the masses and stabilities of

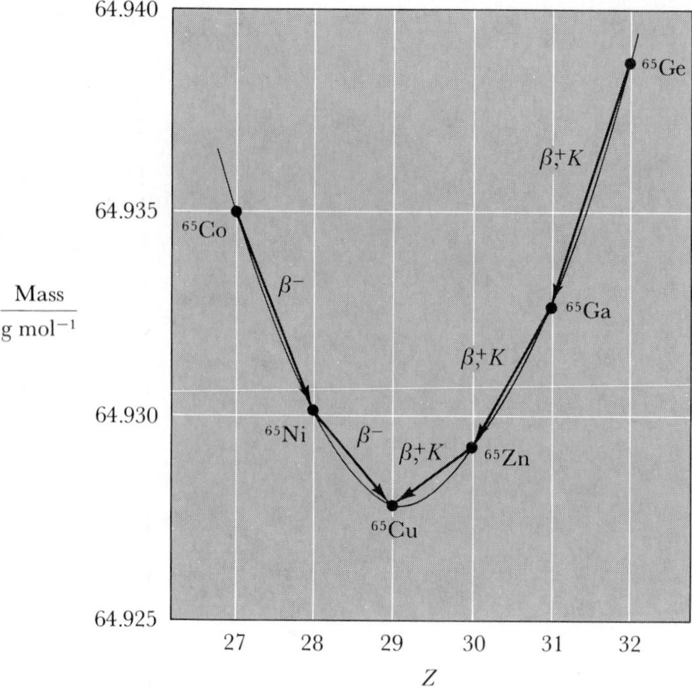

FIGURE 22–4
The masses of nuclides of mass number 65. The nuclide with the lowest mass, ^{65}Cu, is the stable representative of this mass number. Other heavier nuclides with this mass number achieve stability by β^+ or β^- emission or by *K*-electron capture.

nuclides with $A = 65$. The one stable nuclide for this mass number is copper-65 with $N/Z = 1.24$. Cobalt-65 and nickel-65 have values of N/Z equal to 1.41 and 1.32, respectively, and disintegration of these nuclides by β^- emission decreases the value of N/Z:

$$\begin{array}{lccccc} & {}^{65}_{27}\text{Co} & \xrightarrow{\beta^-} & {}^{65}_{28}\text{Ni} & \xrightarrow{\beta^-} & {}^{65}_{29}\text{Cu} \\ N/Z & 1.41 & & 1.32 & & 1.24 \end{array}$$

Germanium-65, gallium-65, and zinc-65 have N/Z values equal to 1.03, 1.10, and 1.17, respectively, and disintegration of these nuclides by β^+ emission and *K*-electron capture raises the value of N/Z:

$$\begin{array}{lccccccc} & {}^{65}_{32}\text{Ge} & \xrightarrow[K\text{ capture}]{\beta^+} & {}^{65}_{31}\text{Ga} & \xrightarrow[K\text{ capture}]{\beta^+} & {}^{65}_{30}\text{Zn} & \xrightarrow[K\text{ capture}]{\beta^+} & {}^{65}_{29}\text{Cu} \\ N/Z & 1.03 & & 1.10 & & 1.17 & & 1.24 \end{array}$$

The approximately parabolic nature of the plot of mass versus Z in Figure 22–4 is consistent with a simple theory of nuclear binding energies and with the observed occurrence of only one stable nuclide for each odd mass number; this stable nuclide has the lowest mass for that mass number.

For even mass numbers, the situation is more complex. The masses of nuclides of even mass numbers fall on two approximately parabolic curves, as is shown in Figure 22–5 for mass number 64. The masses of all odd–odd

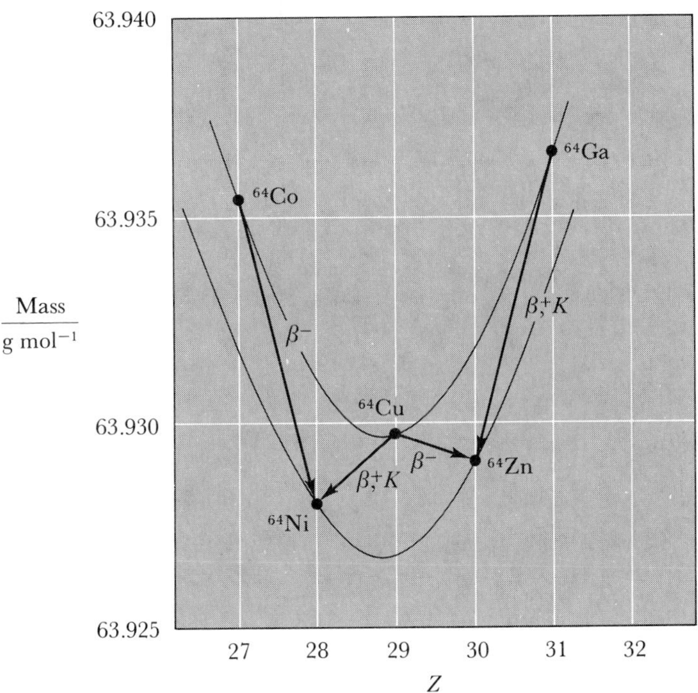

FIGURE 22–5
The masses of nuclides of mass number 64. Two even–even nuclides, ^{64}Ni and ^{64}Zn, are stable. All odd–odd nuclides with this mass number are unstable.

nuclides fall on the upper parabola; these nuclides are unstable with respect to even–even nuclides, the masses of which fall on the lower parabola. Copper-64 is unstable with respect to both nickel-64 and zinc-64; the relative importance of the modes of decay are β^- emission (39.6%), β^+ emission (19.3%), and K-electron capture (41.1%).

Gamma (γ) Radiation and Neutrinos. Because the positive electron does not exist naturally on the Earth, its fate after being emitted by an unstable nucleus is interesting. An energetic positron interacts with the matter through which it passes by ionizing bound electrons. After spending most of its kinetic energy in bringing about ionization, the positron, then relatively slow moving, is annihilated in a process,

$$\beta^+ + e^- = 2\gamma$$

in which two gamma (γ) rays are produced. For momentum to be conserved in this process in which only radiation and no material particles result, two γ rays must be emitted. If the kinetic energy of the β^+ particle is small at the time of annihilation, the energy each γ photon is that equivalent to the mass of an electron,

$$\Delta W = 5.49 \times 10^{-4} \text{ amu} \times \frac{931.5 \text{ MeV}}{1 \text{ amu}} = 0.51 \text{ MeV}$$

Emission of either a β^+ or β^- particle is accompanied by the emission of a neutrino, a neutral particle with almost zero rest mass (less than 0.0015 of the mass of the electron). The emission of such a particle in β decay was suggested by Pauli in 1931 to explain the range of energies of β particles emitted in the transformation of a particular parent nucleus into a particular daughter nucleus; FERMI named the particle *neutrino* ("little neutral one"). Because it has no charge and very low mass, the neutrino is very difficult to detect.

Gamma radiation is emitted during the disintegration of many unstable nuclides. This radiation accompanies the emission of an α particle or β particle or the capture of a K electron if the product nucleus is formed in an excited state. Previously, we calculated that the disintegration

$$^{238}_{92}\text{U} \rightarrow {}^{4}_{2}\alpha + {}^{234}_{90}\text{Th}$$

has a value of Δm that corresponds to an energy decrease of 4.27 MeV. Because of the requirement that momentum be conserved in the disintegration, the emitted α particle has most of this 4.27 MeV. In fact, α particles with two different energies, 4.195 MeV and 4.147 MeV, are observed. A thorium-234 daughter nucleus resulting from the emission of the less energetic α particle is formed in an excited nuclear state. This excitation energy is lost by emission of a γ ray, a quantum of radiant energy. This is shown in Figure 22–6. The energy of this γ radiation is the difference between the energies of the two α particles:

$$4.195 \text{ MeV} - 4.147 \text{ MeV} = 0.048 \text{ MeV}$$

Gamma radiation is very penetrating. The harmful effect of γ radiation can be understood if we convert this energy to the units in which we

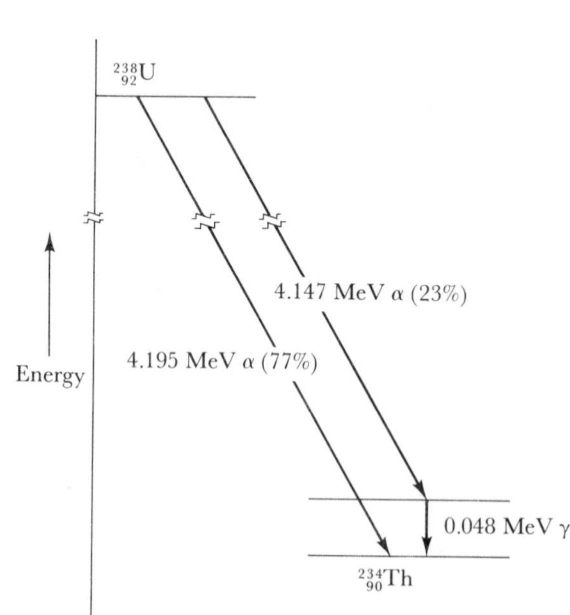

FIGURE 22–6
The energy states of parent and daughter nuclides in the disintegration

$$^{238}_{92}\text{U} \rightarrow {}^{234}_{90}\text{Th} + {}^{4}_{2}\alpha$$

Excited state of daughter $^{234}_{90}$Th is formed in 23% of the disintegrations.

express ionization energies and bond energies:

$$0.048 \text{ MeV} = 0.048 \times 10^{6} \text{ eV} \times \frac{1.602 \times 10^{-19} \text{ C}}{1 \text{ e}} \times 6.02 \times 10^{23} \text{ mol}^{-1}$$

$$= 4.6 \times 10^{9} \text{ C V mol}^{-1} = 4.6 \times 10^{6} \text{ kJ mol}^{-1}$$

which is 10^3–10^4 times typical bond energies (see Table 10–6). It also is informative to calculate from this energy the wavelength of the electromagnetic radiation. The Planck equation rearranged to allow calculation of λ is

$$\lambda = \frac{hc}{E} = \frac{6.626 \times 10^{-34} \text{ J s} \times 3.00 \times 10^{8} \text{ m s}^{-1}}{0.048 \times 10^{6} \text{ eV} \times 1.602 \times 10^{-19} \text{ C e}^{-1}}$$

$$= 2.59 \times 10^{-11} \text{ m} = 0.0259 \text{ nm}$$

The wavelength is less than 10^{-4} times the wavelength of light at the blue end of the visible spectrum ($\lambda = 400$ nm).

THE RATE OF DISINTEGRATION OF AN UNSTABLE NUCLIDE

The rate of a process by which an unstable nuclide achieves greater stability is governed by a first-order rate law,

$$-\frac{dN}{dt} = kN$$

in which N is the number of atoms, and k is the proportionality constant. As shown in Chapter 15, this equation can be integrated to give

$$\frac{N}{N_0} = e^{-kt}$$

in which N is the number of radioactive atoms at time t, and N_0 is the number at zero time. Taking the logarithm of both sides of the equation, and converting to decadic logarithms we get

$$\log N = \log N_0 - \frac{k}{2.303} t$$

Because the disintegration rate, $-dN/dt$, is proportional to the number of atoms, N, the equation just given can be transformed into an equation giving the dependence of the disintegration rate on time:

$$\log \left[-\frac{1}{k} \left(\frac{dN}{dt} \right) \right] = \log \left[-\frac{1}{k} \left(\frac{dN}{dt} \right)_0 \right] - \frac{k}{2.303} t$$

or

$$\log \left[- \left(\frac{dN}{dt} \right) \right] = \log \left[- \left(\frac{dN}{dt} \right)_0 \right] - \frac{k}{2.303} t$$

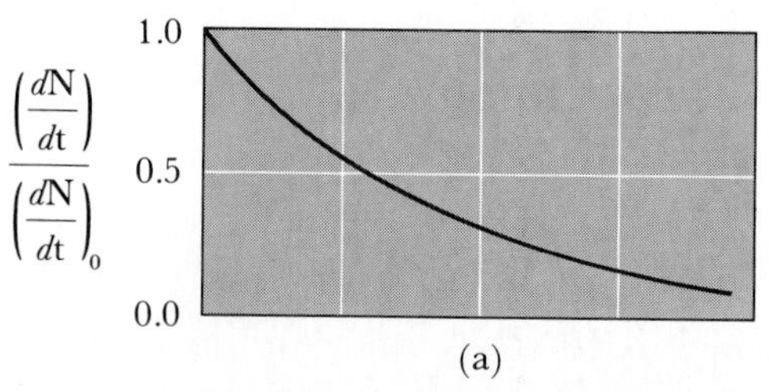

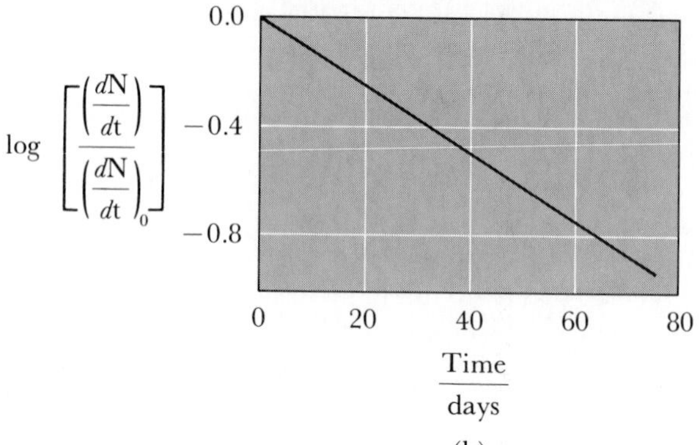

FIGURE 22–7

The disintegration rate of a sample of thorium-234 as a function of time.

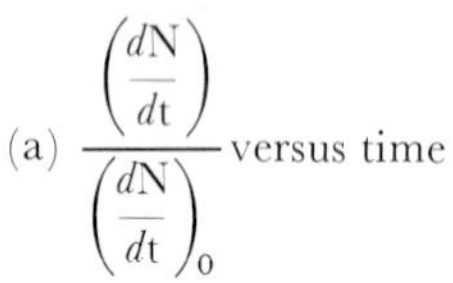
(a) $\dfrac{\left(\dfrac{dN}{dt}\right)}{\left(\dfrac{dN}{dt}\right)_0}$ versus time

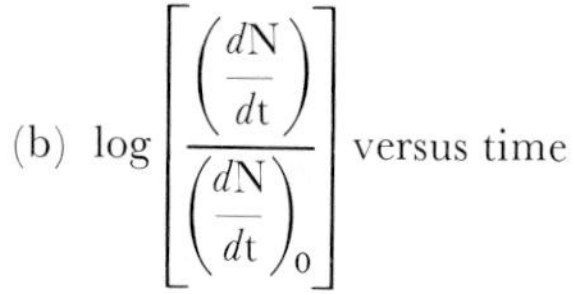
(b) $\log\left[\dfrac{\left(\dfrac{dN}{dt}\right)}{\left(\dfrac{dN}{dt}\right)_0}\right]$ versus time

(The rate is expressed relative to the value at zero time.)

It is the disintegration rate, $-dN/dt$, that is measured in most studies of unstable nuclides. Figure 22–7 shows the logarithmic decrease in the disintegration rate of a sample of thorium-234. Associated with a first-order rate law is a half-time, and, as shown in Chapter 15, the relationship of the rate constant and the half-time is

$$t_{1/2} = \frac{\ln 2}{k} = \frac{0.693}{k}$$

For the example shown in Figure 22–7, the half-time is 24.1 d, therefore the disintegration constant k has the value

$$k = \frac{0.693}{t_{1/2}} = \frac{0.693}{24.1 \text{ d}} = 0.0288 \text{ d}^{-1}$$

Generally, it is assumed that the value of k, and therefore the half-life, is characteristic of each unstable nuclide, being independent of its state of chemical combination, the presence of strong electric or magnetic fields, the temperature, and the pressure. This is approximately correct, but careful measurements have shown, for instance, that the decay constant for niobium-90 (a nuclide that is unstable with respect to K-electron capture) is slightly different in the metallic state than in the fluoride salt. Extremely high pressures also have been shown to decrease the half-life of some unstable nuclides.

One method of determining the disintegration rate constant for an unstable nuclide is that implied in Figure 22–7. Measurement of the disintegration rate over a period of time in which this rate decreases appreciably allows calculation of the half-time and of the value of k for the process. However, this method is not practical for nuclides with very long half-lives. Various methods can be used for such nuclides. For instance, plutonium-239, an α emitter, has a half-life of 2.44×10^4 y. This value has been determined by measuring the absolute disintegration rate of a weighed sample of a compound of this element. The value of k is given by

$$k = \frac{1}{\mathrm{N}}\left(-\frac{d\mathrm{N}}{dt}\right)$$

Thus a sample of PuO_2 weighing 0.0320 mg is observed to have a disintegration rate of $6.40 \times 10^4\ s^{-1}$. Therefore the calculated value of k is

$$k = \frac{1}{3.2 \times 10^{-5}\ \mathrm{g} \times \dfrac{1}{271.05\ \mathrm{g\ mol^{-1}}} \times 6.02 \times 10^{23}\ \mathrm{mol^{-1}}} \times 6.4 \times 10^4\ \mathrm{s^{-1}}$$

$$= 9.0 \times 10^{-13}\ \mathrm{s^{-1}}$$

and the calculated value of $t_{1/2}$ is

$$t_{1/2} = \frac{\ln 2}{k} = \frac{0.693}{9.0 \times 10^{-13}\ \mathrm{s^{-1}}} = 7.70 \times 10^{11}\ \mathrm{s}$$

$$= 2.44 \times 10^4\ \mathrm{y}$$

22–4 Naturally Occurring Radioactive Nuclides

The nuclide ${}^{209}_{83}Bi$ has the distinction of being the stable nuclide with the largest mass number and largest atomic number. All nuclides with $Z \geq 84$ or $N \geq 127$ are unstable. Elements with atomic numbers between 84 (polonium) and 92 (uranium) occur in nature as unstable nuclides.

Figure 22–8 shows the series of nuclides derived from uranium-238. This diagram is arranged like Figure 22–3; it is a plot of N versus Z. An α emission causes both N and Z to decrease by two units, and emission of a negative β particle causes N to decrease by one unit and Z to increase by one unit. The parent of the series, uranium-238, has a very long half-life,

$${}^{238}_{92}\mathrm{U} \rightarrow {}^{234}_{90}\mathrm{Th} + {}^{4}_{2}\alpha \qquad t_{1/2} = 4.51 \times 10^9\ \mathrm{y}$$

This half-life is close to the estimated age of the Earth. The shortest-lived isotope in this series is polonium-214, which has a half-life of 1.6×10^{-4} s; its mode of decay is

$${}^{214}_{84}\mathrm{Po} \rightarrow {}^{210}_{82}\mathrm{Pb} + {}^{4}_{2}\alpha$$

Polonium-214 is present in a uranium ore because it is produced continuously from its parent, bismuth-214,

$${}^{214}_{83}\mathrm{Bi} \rightarrow {}^{214}_{84}\mathrm{Po} + {}^{0}_{-1}\beta^-$$

which in turn is formed from nuclides derived ultimately from uranium-238.

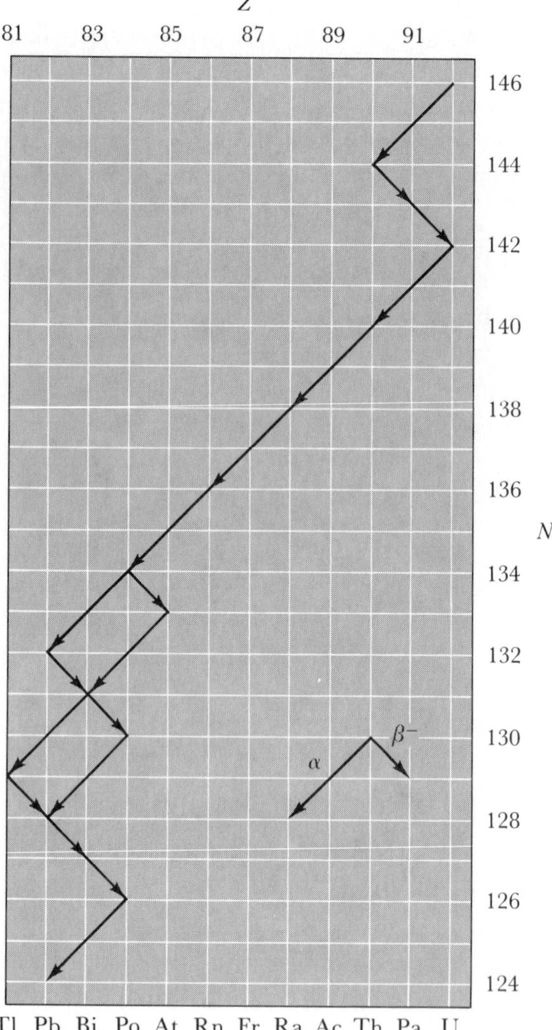

FIGURE 22–8
Members of the $2a + 2$ series of naturally occurring radioactive nuclides. The parent nuclide is ^{238}U, and the stable end product is ^{206}Pb.

In a sample of uranium ore that has not been processed to separate any intermediate daughter element, these daughter elements are present at a steady relative concentration: they are being formed at the same rate as they are disintegrating. This situation is called *radioactive equilibrium*. In the uranium-238 series at radioactive equilibrium, each of the processes

$$^{238}_{92}\mathrm{U} \rightarrow {}^{4}_{2}\alpha + {}^{234}_{90}\mathrm{Th} \qquad t_{1/2} = 4.51 \times 10^{9}\ \mathrm{y}$$
$$^{234}_{90}\mathrm{Th} \rightarrow {}^{0}_{1}\beta + {}^{234}_{91}\mathrm{Pa} \qquad t_{1/2} = 24.1\ \mathrm{d}$$
$$\vdots$$
$$^{214}_{84}\mathrm{Po} \rightarrow {}^{4}_{2}\alpha + {}^{210}_{82}\mathrm{Pb} \qquad t_{1/2} = 1.6 \times 10^{-4}\ \mathrm{s}$$

is occurring at the same rate

$$-\frac{dN_{\text{U-238}}}{dt} = -\frac{dN_{\text{Th-234}}}{dt} = -\frac{dN_{\text{Po-214}}}{dt} = \cdots$$

Each of these rates of disintegration is the product of a number of atoms and a rate constant (which is inversely proportional to a half-life):

$$-\frac{dN}{dt} = \frac{0.693}{t_{1/2}} N$$

This gives the relationship of the relative number of atoms of each kind in radioactive equilibrium:

$$\frac{N_{U\text{-}238}}{4.51 \times 10^9 \text{ y}} = \frac{N_{Th\text{-}234}}{\dfrac{24.1 \text{ d}}{365 \text{ d y}^{-1}}} = \frac{N_{Po\text{-}214}}{\dfrac{1.6 \times 10^{-4} \text{ s}}{3.15 \times 10^7 \text{ s y}^{-1}}}$$

which gives

$$\frac{N_{Th\text{-}234}}{N_{U\text{-}238}} = \frac{24.1 \text{ d}}{4.51 \times 10^9 \text{ y} \times 365 \text{ d y}^{-1}} = 1.46 \times 10^{-11}$$

$$\frac{N_{Po\text{-}214}}{N_{U\text{-}238}} = \frac{1.6 \times 10^{-4} \text{ s}}{4.51 \times 10^9 \text{ y} \times 3.15 \times 10^7 \text{ s y}^{-1}} = 1.1 \times 10^{-21}$$

An unstable nuclide may be unstable with respect to both α and β^- emission, just as copper-64 was unstable with respect to more than one mode of disintegration (see Figure 22–5). Thus bismuth-214 disintegrates by both modes:

$$^{214}_{83}\text{Bi} \begin{cases} \nearrow\ ^{210}_{81}\text{Tl} + ^{4}_{2}\alpha \\ \searrow\ ^{214}_{84}\text{Po} + ^{0}_{-1}\beta^- \end{cases}$$

In this example, branching very much favors β^- emission: 99.96% of bismuth-214 atoms decay by β^- emission.

Because the mass numbers of nuclides in a series that involves both α and β^- emitters are either the same or differ from one another by multiples of four units, four different such series are possible. These series have mass numbers that are $4a$, $4a + 1$, $4a + 2$, or $4a + 3$, in which a is an integer. Thus the uranium-238 series is the $4a + 2$ series ($238 = 4 \times 59 + 2$). The $4a$ and $4a + 3$ series also occur in nature. The long-lived parent of the $4a$ series is thorium-232, which has the mode of decay

$$^{232}_{90}\text{Th} \rightarrow ^{228}_{88}\text{Ra} + ^{4}_{2}\alpha \qquad t_{1/2} = 1.41 \times 10^{10} \text{ y}$$

The long-lived parent of the $4a + 3$ series is uranium-235, which has the mode of decay

$$^{235}_{92}\text{U} \rightarrow ^{231}_{90}\text{Th} + ^{4}_{2}\alpha \qquad t_{1/2} = 7.1 \times 10^{8} \text{ y}$$

The $4a + 1$ series does not occur in nature because the long-lived parent in this series is neptunium-237, which has a half-life short compared to the age of the Earth:

$$^{237}_{93}\text{Np} \rightarrow ^{233}_{91}\text{Pa} + ^{4}_{2}\alpha \qquad t_{1/2} = 2.14 \times 10^{6} \text{ y}$$

22–5 Man-Made Nuclear Transmutations

The alchemist's dream of transmuting base metals into gold has not been realized, but it is possible to transmute one element into another. In addition to the transmutations that occur in the disintegration of an unstable nucleus, for instance the change of uranium-238 to thorium-234, there are man-made transformations brought about by the bombardment of nuclei with projectile particles.

NUCLEAR REACTIONS

In the earliest studies of nuclear reactions, rapidly moving α particles emitted by natural radioactive substances were used as projectiles. In 1919, Rutherford bombarded nitrogen with the energetic (7.687 MeV) α particles from naturally occurring polonium-214, and observed proton production; the nuclear reaction that produced the protons was

$$^{14}_{7}\mathrm{N} + {}^{4}_{2}\mathrm{He} \rightarrow {}^{17}_{8}\mathrm{O} + {}^{1}_{1}\mathrm{H}$$

No emission of protons occurred after bombardment ceased, and the other product of this nuclear reaction, oxygen-17, is a stable nuclide (0.037% of natural oxygen). It was not until 15 years later that the JOLIOT-CURIES discovered a nuclear reaction that produced an unstable nuclide. One of the nuclear reactions that they carried out was

$$^{27}_{13}\mathrm{Al} + {}^{4}_{2}\mathrm{He} \rightarrow {}^{30}_{15}\mathrm{P} + {}^{1}_{0}\mathrm{n}$$

The product phosphorus-30 is an unstable odd–odd nuclide with $N/P =$ 1.00, a valve smaller than that for the stable nuclide with this mass number, which is $N/P = 1.14$ for silicon-30, an even–even nuclide. Positron emission by phosphorus-30 accomplishes an increase in the value of N/P and brings stability:

$$^{30}_{15}\mathrm{P} \rightarrow {}^{30}_{14}\mathrm{Si} + \beta^{+} \qquad t_{1/2} = 2.5\ \mathrm{min}$$

With the development of charged-particle accelerators, protons ($^{1}\mathrm{H}^{+}$), deuterons ($^{2}\mathrm{H}^{+}$), α particles ($^{4}\mathrm{He}^{2+}$), and nuclei of other light elements (e.g., $^{12}\mathrm{C}^{6+}$) can be used as projectiles. Before a positively charged projectile can penetrate to the nucleus to be captured, it must have sufficient kinetic energy to overcome the repulsion of the positive nuclear charge.

The neutron has no such coulombic barrier to overcome, and it is a most versatile projectile for bringing about nuclear reactions. Because it is neutral, it cannot be accelerated in any device that converts the potential energy of a charged particle in an electric field into kinetic energy of the charged particle. Neutrons are produced in some nuclear reactions, and these can be used to bring about other nuclear transformations. A convenient portable source of neutrons is a mixture of an α-emitting radioactive substance and a light element that tends to undergo a nuclear reaction in which α-particle capture leads to neutron emission, the (α,n) reaction. One such source is a mixture of compounds of radium and beryllium, in which the α particles are produced in the disintegration

$$^{226}_{88}\text{Ra} \rightarrow ^{222}_{86}\text{Rn} + ^{4}_{2}\alpha$$

and consumed in the nuclear reaction

$$^{9}_{4}\text{Be} + ^{4}_{2}\alpha \rightarrow ^{12}_{6}\text{C} + ^{1}_{0}\text{n}$$

A more widely used source of neutrons is the nuclear reactor, in which the neutron-producing reaction is fission, to be discussed.

Some nuclear reactions producing useful radioactive isotopes are:

$^{31}_{15}\text{P} + ^{1}_{0}\text{n} = ^{32}_{15}\text{P} + \gamma$ (^{32}P is a β^- emitter with a 14.3-d half-life)

$^{6}_{3}\text{Li} + ^{1}_{0}\text{n} = ^{3}_{1}\text{H} + ^{4}_{2}\text{He}$ (^{3}H is a β^- emitter with a 12.5-y half-life)

$^{14}_{7}\text{N} + ^{1}_{0}\text{n} = ^{14}_{6}\text{C} + ^{1}_{1}\text{H}$ (^{14}C is a β^- emitter with a 5730-y half-life)

$^{24}_{12}\text{Mg} + ^{2}_{1}\text{H} = ^{22}_{11}\text{Na} + ^{4}_{2}\text{He}$ (^{22}Na is a β^+ emitter with a 3.0-y half-life)

$^{11}_{5}\text{B} + ^{1}_{1}\text{H} = ^{11}_{6}\text{C} + ^{1}_{0}\text{n}$ (^{11}C is a β^+ emitter with a 20.5-min half-life)

$^{55}_{25}\text{Mn} + ^{4}_{2}\text{He} = ^{58}_{27}\text{Co} + ^{1}_{0}\text{n}$ (^{58}Co has a half-life of 72 d; its decay is branching, with both β^+ emission and K-electron capture occurring)

These particular nuclear reactions were chosen to illustrate different types of reactions, which can be designated as (n,γ), (n,α), (n,p), (d,α), (p,n), and (α,n) reactions, respectively. This notation for a nuclear reaction gives the incoming projectile and emitted light-weight particle. In this notation, the first nuclear reaction observed by Rutherford is

$$^{14}_{7}\text{N}(\alpha,\text{p})^{17}_{8}\text{O}$$

and that by which the Joliot-Curies first prepared an artificially radioactive nucleus is

$$^{27}_{13}\text{Al}(\alpha,\text{n})^{30}_{15}\text{P}$$

The Compound Nucleus. In some nuclear reactions, capture of the projectile particle and emission of a light-weight particle occur over a period of 10^{-12} s to 10^{-14} s. Short as this is, it is very long compared to the time ($\sim 10^{-21}$ s) required for the projectile particle to transverse a distance equal to a nuclear diameter ($\sim 10^{-12}$ cm). The entity that exists for this short time (the target nucleus plus the captured projectile) is called the *compound nucleus*. In the compound nucleus, the binding energy of the captured projectile (i.e., the decrease of mass) and its initial kinetic energy before capture are distributed among all of the nucleons. The compound nucleus ceases to exist when the requisite energy is concentrated in a single particle, thereby allowing its emission. If the excitation of the compound nucleus is very high, more than one particle may be emitted. If the energy is not sufficient to allow emission of a particle, the excitation energy of the compound nucleus is lost as electromagnetic radiation, that is, as γ rays.

In this picture of a nuclear reaction, proposed by Niels Bohr in 1936, there may be several possible fates for the compound nucleus. What particle or particles are emitted by the compound nucleus depends on the energy demands of the specific reactions. For instance, if an aluminum target is bombarded with neutrons, capture of a neutron by aluminum-27 (the sole stable nuclide for aluminum) gives a compound nucleus aluminum-28*. (The asterisk designates the compound nucleus.) The production of this compound nucleus and its possible fates are:

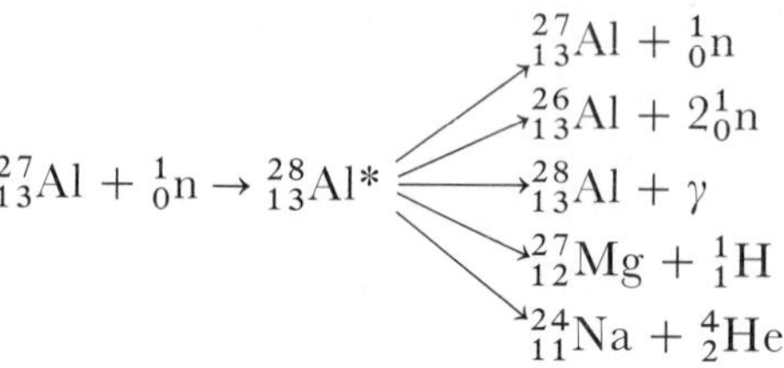

The relative yields of the various products depend on the energy of the bombarding neutron, for this determines the excitation energy of the compound nucleus.

Of the nuclei formed from the aluminum-28* compound nucleus, the first is the original stable target atom, but each of the others is an unstable nuclide:

$$^{26}_{13}Al \rightarrow ^{26}_{12}Mg + \beta^+ \qquad t_{1/2} = 7.4 \times 10^5 \text{ y}$$
$$^{28}_{13}Al \rightarrow ^{28}_{14}Si + \beta^- \qquad t_{1/2} = 2.31 \text{ min}$$
$$^{27}_{12}Mg \rightarrow ^{27}_{13}Al + \beta^- \qquad t_{1/2} = 9.5 \text{ min}$$
$$^{24}_{11}Na \rightarrow ^{24}_{12}Mg + \beta^- \qquad t_{1/2} = 15.0 \text{ h}$$

A particular compound nucleus may be formed in more than one way. For instance, in addition to being formed by the amalgamation of a neutron by aluminum-27, the compound nucleus $^{28}_{13}Al^*$ is formed in the bombardment of a magnesium target with deuterons,

$$^{26}_{12}Mg + ^{2}_{1}H \rightarrow ^{28}_{13}Al^*$$

The lifetime of the compound nucleus is long enough for its fate to depend only on the excitation energy and not on its mode of formation. For instance, magnesium-27, formed from the compound nucleus aluminum-28*, is produced in each of the nuclear reactions $^{27}_{13}Al(n,p)^{27}_{12}Mg$ and $^{26}_{12}Mg(d,p)^{27}_{12}Mg$.

There are nuclear reactions that do not involve capture of the complete projectile particle. Thus a (d,p) or a (d,n) reaction can occur if the target nucleus "strips" a neutron or a proton from a deuteron projectile. In these *stripping reactions*, as they are called, the part of the projectile particle that is not amalgamated by the target nucleus continues on its way, almost undeflected from its initial trajectory.

The Energy Balance in Nuclear Reactions. The energy balance in a nuclear reaction can be evaluated by considering the masses of the reactants and products. For instance, in the nuclear reaction first observed by Rutherford,

$$^{14}_{7}\text{N} + ^{4}_{2}\text{He} \rightarrow ^{17}_{8}\text{O} + ^{1}_{1}\text{H}$$

the atomic masses are

$$\begin{array}{ll} ^{14}_{7}\text{N (14.0031 amu)} & ^{4}_{2}\text{He (4.0026 amu)} \\ ^{17}_{8}\text{O (16.9991 amu)} & ^{1}_{1}\text{H (1.0078 amu)} \end{array}$$

The sum of the atomic masses of the products exceeds the sum of the atomic masses of the reactants:

$$\frac{\Delta m}{\text{amu}} = 16.9991 + 1.0078 - 14.0031 - 4.0026 = 0.0012$$

$$\Delta m = 0.0012 \text{ amu}$$

and the energy equivalent to this mass ($c^2\,\Delta m$) must be furnished by the kinetic energy of the projectile particle if the nuclear reaction is to occur.

A calculation such as this, appropriately modified, gives the minimum energy of a projectile required to bring about a reaction. The appropriate modification of this calculation incorporates allowance for the kinetic energy of the products of decomposition of the compound nucleus. Momentum is conserved in the nuclear reaction, and the total momentum of the emitted light-weight particle and the product nucleus must be the same as that of the projectile particle. It follows that the practical energy threshold for a nuclear reaction may be appreciably greater than the value obtained in our initial calculation. Account also must be taken of the energy needed by a positively charged projectile to overcome the coulombic repulsion of the nucleus; this it must do before the compound nucleus can be formed.

A nuclear reaction will not occur if the projectile energy is below the threshold. For energies above the threshold, the probability of occurrence varies with energy in a complex manner. Generally, a projectile with an energy slightly above the threshold is more effective than a very-high-energy projectile.

The mass of the neutron is appreciably greater than 1 amu (1.0087 amu), but nuclides differing by 1 unit in N differ in mass by close to 1 amu. Thus the (n,γ) reaction has no energy threshold and it occurs with slow neutrons for virtually all nuclides. Even "doubly magic" calcium-40 reacts with loss of mass in the (n,γ) reaction:

$$^{40}_{20}\text{Ca}(\text{n},\gamma)^{41}_{20}\text{Ca}$$

$$\Delta m = \text{AW}(^{41}_{20}\text{Ca}) - m(\text{n}) - \text{AW}(^{40}_{20}\text{Ca})$$

$$\frac{\Delta m}{\text{amu}} = 40.9623 - 1.0087 - 39.9626 = -0.0090$$

The tendency for a neutron (or another projectile) to be captured by a nucleus is described in terms of σ, a cross section for the capture process.

The larger the cross section, the greater the probability of capture. For slow-neutron capture, cross sections vary enormously:

$$^{2}_{1}\mathrm{H}(\mathrm{n},\gamma)^{3}_{1}\mathrm{H} \qquad \sigma = 5.7 \times 10^{-28}\ \mathrm{cm}^2$$

$$^{113}_{48}\mathrm{Cd}(\mathrm{n},\gamma)^{114}_{48}\mathrm{Cd} \qquad \sigma = 2.0 \times 10^{-20}\ \mathrm{cm}^2$$

(The unit of cross section commonly used in nuclear physics is the barn, 1 barn $= 10^{-24}$ cm^2. Therefore these cross sections are 5.7×10^{-4} barn and 2.0×10^{4} barns, respectively.) The large cross section for neutron capture by cadmium-113, which is 12.26% of natural cadmium, is the basis for use of cadmium rods to control nuclear reactors.

The neutrons emitted in the Be(α,n) reaction and in fission have very high energies. Therefore they must be slowed down before they can participate in a nuclear reaction as slow neutrons. These neutrons are slowed by substances called *moderators.* In each collision with an atom of a moderator, a rapidly moving neutron loses some of its kinetic energy. With enough collisions, neutrons are slowed to the thermal energy corresponding to the temperature of the moderator. Such slow neutrons are called *thermal neutrons.* Atoms with low atomic weight are more effective as moderators than are heavy atoms, and a good moderator also must have a low cross section for neutron capture. Heavy water and graphite have been very useful as moderators, and paraffin is useful in small-scale operations. The cross sections for capture of thermal neutrons by the nuclides of these moderators are:

$^{1}\mathrm{H}$	3.3×10^{-25} cm^2	$^{16}\mathrm{O}$	1.8×10^{-28} cm^2
$^{2}\mathrm{H}$	5.7×10^{-28} cm^2	$^{12}\mathrm{C}$	3.4×10^{-28} cm^2

THE TRANSURANIUM ELEMENTS

Once the neutron had been discovered, scientists strove to create elements with atomic number greater than 92, elements beyond the end of that day's periodic table. With uranium as the target element, the neutron, having no charge, is an ideal projectile for nuclear reactions. Furthermore, scientists expected that a (n,γ) reaction on uranium-238 would produce a β^- emitter, uranium-239, the daughter of which would be a nuclide of element-93. This proved to be the case, and the nuclear reaction

$$^{238}_{92}\mathrm{U}(\mathrm{n},\gamma)^{239}_{92}\mathrm{U} \qquad \sigma = 2.7 \times 10^{-24}\ \mathrm{cm}^2$$

was first observed and characterized in 1940.[2] The sequence of radioactive disintegrations,

$$^{239}_{92}\mathrm{U} \xrightarrow[23.5\ \mathrm{min}]{\beta^-} {}^{239}_{93}\mathrm{Np} \xrightarrow[2.35\ \mathrm{d}]{\beta^-} {}^{239}_{94}\mathrm{Pu} \xrightarrow[24{,}400\ \mathrm{y}]{\alpha} {}^{235}_{92}\mathrm{U}$$

[2]Before fission was characterized correctly, the products of nuclear reactions between neutrons and uranium were incorrectly interpreted as transuranium elements, when, in fact, they were fission products.

returns the system to a nuclide that occurs in nature. This nuclear reaction and the subsequent radioactive decay processes form plutonium-239, an element capable of undergoing fission with slow neutrons and therefore a nuclide that can be used in atomic weapons and in nuclear reactors.

There is much concern regarding the use of plutonium-239 in atomic reactors. This nuclide is very dangerous to health if ingested. Also, diversion of relatively small amounts of this material from a nuclear power plant could make surreptitious manufacture of nuclear weapons feasible.

The longest-lived nuclide of a transuranium element that has been produced in appreciable amounts is neptunium-237, which is produced in the (n,2n) reaction of fast neutrons on uranium-238:

$$^{238}_{92}\mathrm{U}\,(\mathrm{n,2n})\,^{237}_{92}\mathrm{U}$$

$$^{237}_{92}\mathrm{U} \xrightarrow[6.75\ \mathrm{d}]{\beta^-} {}^{237}_{93}\mathrm{Np} \xrightarrow[2.14 \times 10^{6}\ \mathrm{y}]{\alpha} {}^{233}_{91}\mathrm{Pa}$$

(Recall that neptunium-237 is the long-lived parent of the $4a + 1$ series of radioactive elements, which ultimately decay to stable thallium-205.) The disintegration rate of neptunium-237 is so low that its chemistry can be studied without the extreme precautions usually needed in working with some radioactive nuclides.

In the decades since the first transuranium elements were produced, elements with atomic numbers as high as 105 have been made. Some of these elements have been prepared by bombardment with heavy ions, for example,

$$^{238}_{92}\mathrm{U} + {}^{12}_{6}\mathrm{C} \rightarrow {}^{246}_{98}\mathrm{Cf} + 4\,^{1}_{0}\mathrm{n}$$

(Californium-246 is an α emitter with $t_{1/2} = 35.7$ h.) Successive multiple neutron capture, possible with very high neutron intensities, has been used to produce some nuclides, for example, the overall change

$$^{238}_{92}\mathrm{U} + 17\,^{1}_{0}\mathrm{n} = {}^{255}_{100}\mathrm{Fm} + 8\,{}^{0}_{-1}\beta$$

(Fermium-255 is an α emitter with $t_{1/2} = 22$ h.) Here the negative β particles shown as products are the β particles given off by short-lived intermediate nuclides. The half-lives of the transplutonium nuclides vary widely, ranging from 1.64×10^{7} y for curium-247 to a few seconds for a number of others, such as 8 s for lawrencium-257.

It is not possible to characterize thoroughly the chemistry of transuranium elements that have only very short-lived nuclides. However, chemical characterization of the man-made elements with long-lived nuclides has been as extensive as the characterization of many stable elements. For instance, the oxidation–reduction chemistry of neptunium and plutonium has been studied thoroughly, and its comparison with the corresponding chemistry of uranium is interesting. The potential diagrams for these elements in acidic solution are given in Table 22–2. Of particular interest are the strong oxidizing ability of neptunium(VII) and the oxocation form of these metals in oxidation states +5, +6, and +7. These metals in the

TABLE 22–2

Potential Diagrams for Uranium, Neptunium, and Plutonium in Acidic Solution at 25°C

$$UO_2^{2+} \xrightarrow{+0.063\ V} UO_2^{+} \xrightarrow{+0.58\ V} U^{4+} \xrightarrow{-0.631\ V} U^{3+} \xrightarrow{-1.80\ V} U$$

$$UO_2^{2+} \xrightarrow{+0.32\ V} U^{4+}$$

$$NpO_2^{3+} \xrightarrow{>2.0\ V} NpO_2^{2+} \xrightarrow{+1.137\ V} NpO_2^{+} \xrightarrow{+0.739\ V} Np^{4+} \xrightarrow{+0.155\ V} Np^{3+} \xrightarrow{-1.83\ V} Np$$

$$PuO_2^{2+} \xrightarrow{+0.913\ V} PuO_2^{+} \xrightarrow{+1.172\ V} Pu^{4+} \xrightarrow{+0.982\ V} Pu^{3+} \xrightarrow{-2.03\ V} Pu$$

$$PuO_2^{2+} \xrightarrow{1.043\ V} Pu^{4+}$$

+4 oxidation state are the simple hydrated ion of charge +4, $M(OH_2)_n^{4+}$ (shown as U^{4+}, Np^{4+}, and Pu^{4+} in Table 22–2), in contrast to the +4 transition-metal ions, with smaller ionic radius and smaller net positive charge, such as the vanadium(IV) species VO^{2+}. In the +6 oxidation state these metals are oxocations UO_2^{2+}, NpO_2^{2+}, and PuO_2^{2+}, in contrast with the anionic species for the +6 transition-metal ions, such as for chromium(VI), $HOCrO_3^-$.

NUCLEAR FISSION

The fission process ruptures a heavy nucleus to give nuclei of intermediate mass number plus neutrons. Figure 22–2 showed that the binding energy per nucleon is lower at the highest values of A than at intermediate values of A. Thus the fission of a heavy nucleus, such as uranium-235, liberates a large amount of energy:

$$^{235}_{92}U = {}^{91}_{36}Kr + {}^{143}_{56}Ba + {}^{1}_{0}n \qquad \Delta U \cong -180\ MeV$$

Although spontaneous fission in uranium-235 is very, very slow, the fission process can be caused by thermal neutrons,

$$^{235}_{92}U\,(n,\ fission) \qquad \sigma = 5.8 \times 10^{-22}\ cm^2$$

The compound nucleus uranium-236*, formed when uranium-235 absorbs a neutron, can rupture to give product nuclei with a large range of mass numbers. There is not a single fission process; rather, there are many possibilities. The relative fission yields are given in Figure 22–9, which shows also that the most probable fission processes give a heavy fragment and a light fragment. One possible fission process is

$$^{1}_{0}n + {}^{235}_{92}U \rightarrow {}^{91}_{36}Kr + {}^{143}_{56}Ba + 2\,{}^{1}_{0}n$$

The N/P ratios of the fission products in this particular fission process are 1.53 (for krypton-91) and 1.55 (for barium-143). The stable nuclides of mass numbers 91 and 143 are zirconium-91 and neodymium-143, $^{91}_{40}Zr$ and $^{143}_{60}Nd$, which have N/P ratios of 1.28 and 1.38, respectively. The N/P ratio is decreased by β^- emission, and all fission products, like the two being

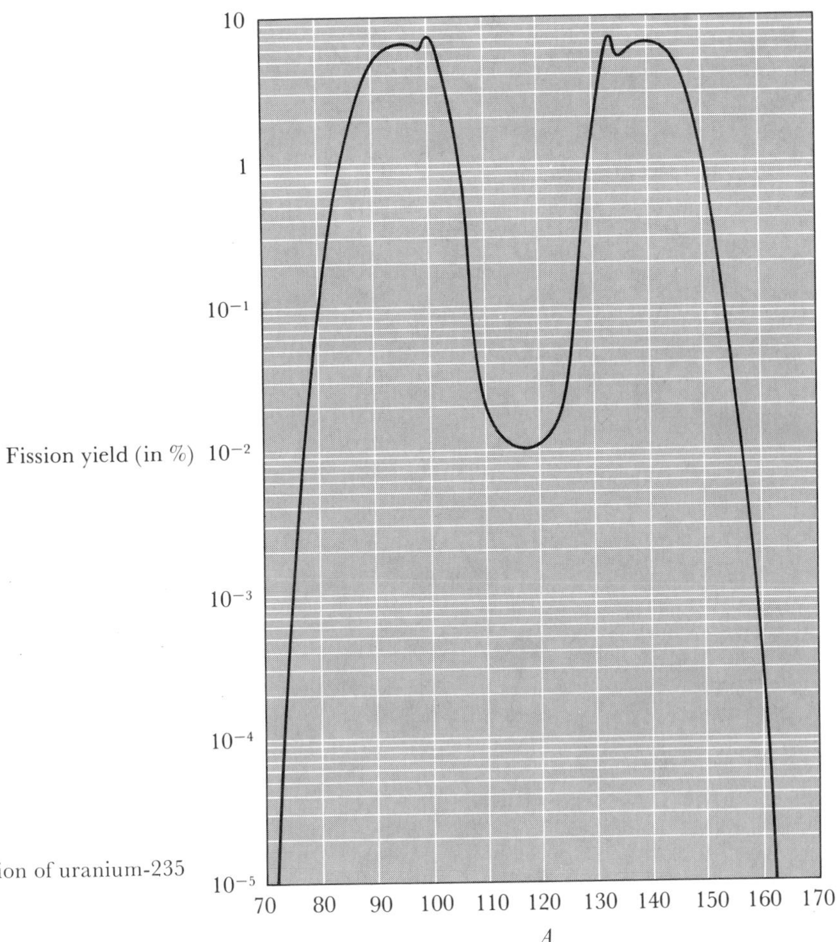

FIGURE 22–9
Yield of fission products in the fission of uranium-235 by slow neutrons.

considered, are β^- emitters. The series of disintegrations that stabilize nuclides of mass number 91, initially krypton-91, with the half-lives, is

$$^{91}_{36}\text{Kr} \xrightarrow[10\ \text{s}]{\beta^-} {}^{91}_{37}\text{Rb} \xrightarrow[1.2\ \text{min}]{\beta^-} {}^{91}_{38}\text{Sr} \xrightarrow[9.67\ \text{h}]{\beta^-} {}^{91}_{39}\text{Y} \xrightarrow[58.8\ \text{d}]{\beta^-} {}^{91}_{40}\text{Zr}$$

The series of β^- emissions for mass number 143, initially barium-143, is

$$^{143}_{56}\text{Ba} \xrightarrow[12\ \text{s}]{\beta^-} {}^{143}_{57}\text{La} \xrightarrow[14\ \text{min}]{\beta^-} {}^{143}_{58}\text{Ce} \xrightarrow[33\ \text{h}]{\beta^-} {}^{143}_{59}\text{Pr} \xrightarrow[13.7\ \text{d}]{\beta^-} {}^{143}_{60}\text{Nd}$$

Like uranium-235, artificially prepared plutonium-239 and uranium-233 have large cross sections for fission by slow neutrons. The formation of the former of these from uranium-238 has been discussed, and the latter of these is formed from thorium-232 (the abundant thorium nuclide):

$$^{232}_{90}\text{Th}(\text{n},\gamma)^{233}_{90}\text{Th}$$

followed by

$$^{233}_{90}\text{Th} \xrightarrow[22.2\ \text{min}]{\beta^-} {}^{233}_{91}\text{Pa} \xrightarrow[27.0\ \text{d}]{\beta^-} {}^{233}_{92}\text{U}$$

Neutron-induced fission is so important because of the enormous amount of energy it evolves and because extra neutrons are emitted in the process. The extra neutrons allow self-sustaining fission in a bulk sample of a fissionable nuclide. A device for controlled self-sustaining fission is called a *nuclear reactor*. Such devices contain fissionable material, and moderator to slow down the emitted neutrons. Control of the self-sustaining reaction is easier if slow neutrons propagate the reaction. Nuclear reactors may serve principally to produce energy or to release neutrons to produce new nuclides. The purpose of the reactors built during World War II was to produce plutonium-239, but reactors being built at the present time by utility companies serve to produce energy. A nuclear reactor fueled with naturally occurring uranium-235 can be designed to use some of the surplus neutrons to breed either plutonium-239 or uranium-233. Breeding of nuclear fuel in this way may help satisfy future energy needs, but serious problems of safety must be solved before it can do so.

An *atomic bomb*, a device for uncontrolled self-sustaining fission, contains no moderator, and the fission is caused predominantly by fast neutrons. Only with fast neutrons will the successive occurrence of fission, neutron production, fission, neutron production, and so on, occur fast enough to result in an explosion.

The devastating effects of nuclear weapons are produced in a number of ways. Extremely high temperatures ($>10^7$ K) are generated at the time of the explosion, and the resultant radiant energy raises the temperature of the surroundings to $>10^3$ K. (It has been estimated that over 25% of the fatalities at Hiroshima and Nagasaki were due to skin burns.) The radioactivity of the fission products, including high-energy γ radiation, persists a long time after an explosion.

NUCLEAR FUSION

Fusion is the process by which light nuclei combine to form a heavier nucleus. The overall transformation (with ${}^1_1\text{H}$ and ${}^4_2\text{He}$ representing nuclei, not neutral atoms)

$$4\,{}^1_1\text{H} + 2\text{e}^- = {}^4_2\text{He}$$

is accompanied by the disappearance of 0.02758 g mol^{-1}. This net change is responsible for production of energy in the sun and the stars. However, the transformation does not occur in a single step. Two different cycles of steps have been proposed to bring about this net change. One is the proton–proton chain:

$$\begin{aligned}
{}^1_1\text{H} + {}^1_1\text{H} &= {}^2_1\text{H} + \text{e}^+ \\
{}^2_1\text{H} + {}^1_1\text{H} &= {}^3_2\text{He} + \gamma \\
2\,{}^3_2\text{He} &= {}^4_2\text{He} + 2\,{}^1_1\text{H} \\
\text{e}^+ + \text{e}^- &= 2\,\gamma
\end{aligned}$$

The net reaction is the result of the sum of the third reaction and two times the first, the second, and the fourth reactions. To explain the dependence of energy production in stars on the mass and temperature of the stars, Bethe proposed a cycle of reactions in which carbon-12 is essentially a catalyst. The cycle involves six reactions (in addition to the reaction by which the positive electrons are annihilated):

$$^{12}_{6}C + ^{1}_{1}H \rightarrow ^{13}_{7}N + \gamma$$
$$^{13}_{7}N \rightarrow ^{13}_{6}C + \beta^{+}$$
$$^{13}_{6}C + ^{1}_{1}H \rightarrow ^{14}_{7}N + \gamma$$
$$^{14}_{7}N + ^{1}_{1}H \rightarrow ^{15}_{8}O + \gamma$$
$$^{15}_{8}O \rightarrow ^{15}_{7}N + \beta^{+}$$
$$^{15}_{7}N + ^{1}_{1}H \rightarrow ^{12}_{6}C + ^{4}_{2}He$$

The sum of these reactions (along with the annihilation reaction) is the net reaction for conversion of hydrogen to helium.

For two nuclei to react they must be moving at high velocities to overcome the coulombic repulsion of the like charges. For the nuclei to have the requisite thermal energy, a high temperature, $\sim 10^7$ K, is needed. In stars the balance between energy production and dissipation maintains such high temperatures.

Controlled fusion reactors using reactions analogous to those occurring in stars someday may produce abundant energy. Uncontrolled fusion is basis for the so-called hydrogen bomb, which contains a fission bomb to produce the temperature required for fusion of hydrogen to occur.

NEUTRON-ACTIVATION ANALYSIS

A powerful analytical method for determining the amount of a particular element in a sample of material is based upon measurement of the radioactive nuclides that the element in question forms upon neutron irradiation. The sample is irradiated with neutrons, generally from a nuclear reactor in which the flux of thermal neutrons is high (10^{12} n cm^{-2} s^{-1}). The nuclides present in the sample have different cross sections for neutron capture. The amount of each nuclear reaction depends on the neutron flux, the amount of the reactant nuclide in the sample, the cross section for the nuclear reaction, and time of irradiation. After irradiation the presence of a particular radioactive nuclide is established by the half-time for its disintegration and the energy of its emitted radiation. This discloses the presence in the irradiated sample of material of the element known to produce the detected product.

For instance, natural manganese is 100% manganese-55, and this nuclide has a relatively large cross section for capture of thermal neutrons:

$$^{55}_{25}Mn(n,\gamma)^{56}_{25}Mn \qquad \sigma = 13.3 \times 10^{-24}\ cm^2$$

The product nuclide is unstable with respect to β^- emission:

$$^{56}_{25}Mn \rightarrow ^{56}_{26}Fe + ^{0}_{-1}\beta^{-} \qquad t_{1/2} = 2.58\ h$$

Because some of the decay processes for this nuclide produce the daughter iron-56 nuclide in excited nuclear states, there are γ rays of well-defined energies (e.g., 0.84678 MeV, 1.8110 MeV, and 2.113 MeV) associated with this radioactive decay process. Methods to determine the energies of γ radiation would show the presence of the radiation characteristic of manganese-56 in the neutron-irradiated sample containing manganese. And the identification is even firmer if the γ radiation of the expected energies diminishes in intensity with the half-life characteristic of manganese-56.

This particular analysis is very sensitive: it allows detection of $\sim 2 \times 10^{-11}$ g of manganese. For quantitative analysis, chemists compare the manganese-56 radioactivity induced in the unknown sample and in known comparison samples. Thus they have found the level of manganese, an essential trace element, in particular samples of whole blood: it is 0.011×10^{-6} g cm^{-3}. Analysis of samples of a particular moon rock by this same method showed the manganese content to be $0.12 \pm 0.02\%$.

Neutron-activation analysis fails if a particular radioactive nuclide may be formed from different elements. For instance, manganese-56, the radioactive product of the (n,γ) reaction on manganese-55, also can be produced in the (n,p) reaction on iron-56 (which is 91.66% of natural iron):

$$^{56}_{26}\mathrm{Fe}(\mathrm{n,p})^{56}_{25}\mathrm{Mn}$$

Clearly, the presence of manganese-56 in an irradiated sample does not prove the presence of manganese in the sample if iron also is present and if the neutron energy exceeds the threshold for the (n,p) reaction.

The point has been made that the yield of a particular nuclear reaction depends on the duration of the irradiation. The nature of this dependence is of interest. If the rate of production of a particular nuclide is constant, the amount of this nuclide increases with time, approaching a particular level asymptotically. This is shown in Figure 22–10. The limiting amount of a nuclide is achieved when its rate of formation and rate of disintegration are equal. The amount of the nuclide achieves one half of the limiting value in a time equal to its half-life. Thus, for instance, the yield of a nuclide with a 40-minute half-life is increased greatly if the radiation period is 60 minutes instead of 10 minutes, but is not increased appreciably if the radiation period is 8 hours instead of 4 hours.

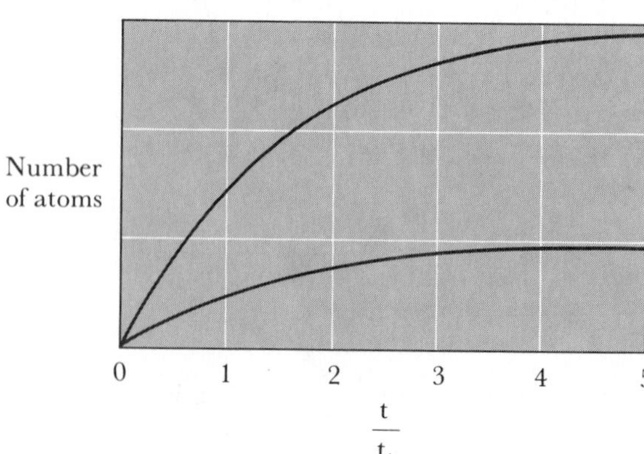

FIGURE 22–10
The yield of a nuclide being formed at a constant rate. The two curves are for different rates of formation; for each rate of formation, the number of atoms becomes one half of its limiting value at $t/t_{\frac{1}{2}} = 1$, that is, $t = t_{\frac{1}{2}}$.

22–6 The Uses of Isotopes in Chemistry and Allied Fields

Over a half century of work with stable and unstable isotopes and with the radiation associated with radioactive transformations has opened vast fields of scientific inquiry. Many problems have been solved and others are being pursued. We will discuss only some of the simplest.

TRACER STUDIES

Tracer studies are important uses of both radioactive and stable isotopes. For a stable isotope to be used as a tracer, a sample must be prepared in which a minor isotope (e.g., oxygen-18, 0.204% of natural oxygen) is enriched compared to its natural abundance. One possible enrichment procedure involves the separation of isotopes using a mass spectrometer.

In tracer studies, particular atoms in a molecular or ionic species are labeled by incorporation of the radioactive or enriched stable isotope. The labeled atoms then can be traced by appropriate analytical procedures–detection of radiation if radioactive tracers are used, or mass spectrometric methods if enriched stable isotopes are used.

Consider an aqueous solution of hydrochloric acid containing iron(II) chloride and iron(III) chloride. An electron transfer between an iron(II) ion and an iron(III) ion has the effect of exchanging the nuclei of the iron(II) and iron(III) ions, but no chemical change in the usual sense has occurred. If the iron(II) chloride initially contains some iron-59, a β^- emitter with $t_{1/2} = 45.1$ d, and the iron(III) chloride initially contains only iron atoms of natural origin, the progress of the exchange reaction,

$$Fe^{3+} + {}^*Fe^{2+} = Fe^{2+} + {}^*Fe^{3+}$$

can be followed. In time, the iron-59 atoms will be distributed randomly between the two different oxidation states. When the iron(II) and iron(III) ions in the mixture are separated from one another by an appropriate chemical technique, it is found that the exchange process at 25°C takes place over a period of seconds to minutes, depending on the concentrations. Study of the exchange as a function of the concentrations shows the rate law for exchange to be governed by a multiterm rate law:

$$\text{Rate of exchange} = k_0[Fe^{2+}][Fe^{3+}] + k_1[Fe^{2+}][FeOH^{2+}] + k_2[Fe^{2+}][FeCl^{2+}]$$

That is, the exchange of iron(II) with iron(III) in acidic solutions containing chloride ion goes by pathways that involve iron(II), Fe^{2+} (really $Fe(OH_2)_6^{2+}$), and iron(III) in each of the forms:

Fe^{3+} [really $Fe(OH_2)_6^{3+}$]
$FeOH^{2+}$ [really $Fe(OH_2)_5OH^{2+}$]
$FeCl^{2+}$ [really $Fe(OH_2)_5Cl^{2+}$]

When the iron-59 atoms have become randomly mixed between the two oxidation states, there are no further observable effects, but the electron transfer between iron(II) and iron(III) ions continues as before.

In contrast to the relatively rapid exchange between these high-spin d^5 and d^6 iron species, each with octahedral coordination shells, the exchange in acidic solution between chromium(III) [$Cr(OH_2)_6^{3+}$, a six-coordinate d^3 species], and chromium(VI) [$HOCrO_3^-$, a four-coordinate d^0 species], is very, very slow, occurring at 95°C over a period of days to weeks. The exchange has been studied using chromium-51, an unstable nuclide decaying by *K*-electron capture with $t_{1/2} = 27.8$ d. This exchange process of species with different coordination numbers must involve the making and breaking of chemical bonds between chromium and oxygen. A complex mechanism involving intermediate oxidation states of chromium is the pathway for this exchange.

In the examples of exchange reactions just discussed, no net chemical change occurred, and the tracers employed were radioactive nuclides. In the next example, the system is one in which a chemical reaction occurs and a stable nuclide is used. The reaction is the hydrolysis of an ester in an aqueous alkaline solution. For amyl acetate, the hydrolysis reaction is

$$OH^- + C_5H_{11}O\overset{\displaystyle O}{\overset{\|}{C}}CH_3 = C_5H_{11}OH + CH_3CO_2^-$$

The reaction is first order in each of the two reactants, but this information does not disclose which of the product species incorporates the oxygen of the hydroxide ion. No useful radioactive isotope of oxygen exists, but tracer studies can be made using water (therefore hydroxide ion) that contains oxygen-18 at greater than its natural abundance.[3] In such an experiment, it was found that the product amyl alcohol was not enriched with respect to oxygen-18. Thus in this reaction, and in the hydrolysis of most esters, the alkyl group–oxygen bond remains intact:

$$^*OH^- + C_5H_{11}\underset{\uparrow}{—}O\underset{\uparrow}{—}\overset{\displaystyle O}{\overset{\|}{C}}CH_3 = C_5H_{11}OH + CH_3\overset{\displaystyle O}{\overset{\|}{C}}—\overset{*}{O}^-$$

Remains intact (alkyl–O bond) — This bond breaks (O–C bond)

Tracer studies involving the long-lived carbon isotope, carbon-14, have done much to explain biochemical mechanisms. For instance, in a system of green algae in which carbon dioxide is being reduced photosynthetically to sugar, the use of carbon dioxide labeled with carbon-14

[3] The longest-lived radioactive isotope of oxygen is oxygen-15, a positron emitter with a half-life of 118 s.

allowed CALVIN to show that 3-phosphoglyceric acid,

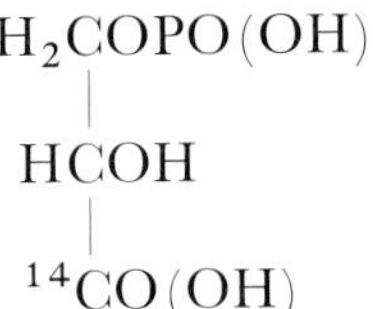

is a compound that is labeled early in the photosynthetic cycle. In a mixture as complex as that in which photosynthesis occurs, it would be practically impossible to settle the reaction mechanism without recourse to tracer studies. It was by stopping the reaction at various times after introduction of the labeled carbon dioxide that the progress of photosynthesis was monitored. The experiments involved separating the various chemical substances from one another, determining whether carbon-14 was incorporated, and then, by appropriate degradation procedures, finding the place in the molecule where the carbon-14 was incorporated.

RADIOCARBON DATING

A nuclear reaction given in a previous section of this chapter,

$$^{14}_{7}N + ^{1}_{0}n = ^{14}_{6}C + ^{1}_{1}H$$

is extremely important in establishing the age of anthropological and geological specimens, in a method proposed by LIBBY in 1946. When cosmic radiation from outer space interacts with the upper reaches of the Earth's atmosphere, neutrons are produced. (The maximum neutron intensity occurs at an elevation of approximately 40,000 feet.) These neutrons are consumed primarily by the foregoing nuclear reaction, producing carbon-14; the cross section for capture of thermal neutrons by odd–odd nitrogen-14 is 1.7×10^{-24} cm^2, much, much larger than the cross section for neutron capture by "doubly magic" oxygen-16 ($<2.4 \times 10^{-28}$ cm^2). Carbon-14 is a β^- emitter,

$$^{14}_{6}C \rightarrow ^{14}_{7}N + ^{0}_{-1}\beta^-$$

with $t_{1/2} = 5730$ y. The radioactive atoms of carbon-14 produced by cosmic radiation react with oxygen to give carbon dioxide, which in time mixes with the ordinary carbon dioxide of the atmosphere. The photosynthetic production of carbohydrates uses carbon dioxide from the atmosphere, thereby giving each growing plant its share of carbon-14 atoms. Animals eat plants, thereby giving animals their share of carbon-14 atoms. The half-life of carbon-14 is long compared to the time required for carbon-14 to spread uniformly throughout the biosphere in all living matter and in the carbonate ion of the oceans. A consequence is a constant level of carbon-14 in all living matter, the disintegration of carbon-14 being balanced by incorporation of carbon-14 from the atmosphere. This relative amount of carbon-14 in a sample of living matter is very small (~ 12 atoms of carbon-14 per 10^{13} atoms of carbon-12) but it is detectable because it is radioactive. The disintegration rate of the carbon-14 atoms in 1.00 g of living carbon is 13.6 min^{-1}. With death of a plant or animal, the carbon-14 ceases to be replenished, and the number of carbon-14 atoms in a specimen decreases with the half-life characteristic of carbon-14.

Measurement of the specific activity of a sample establishes the time elapsed since the sample was in equilibrium with the atmospheric reservoir of carbon dioxide. If this elapsed time is 5730 y, the half-life of carbon-14, the disintegration rate is

$$\tfrac{1}{2} \times 13.6\ \text{g}^{-1}\ \text{min}^{-1} = 6.8\ \text{g}^{-1}\ \text{min}^{-1}$$

In general, the specific disintegration rate, I, of a sample of age a is

$$I = 13.6\ \text{g}^{-1}\ \text{min}^{-1} \times \left(\tfrac{1}{2}\right)^{a/5730\ \text{y}}$$

This equation can be rearranged to give

$$a = 5730\ \text{y} \times \frac{\log\left(\dfrac{I}{13.6\ \text{g}^{-1}\ \text{min}^{-1}}\right)}{\log\ (0.500)}$$

The ages of samples ranging from a few hundred years to more than thirty thousand years have been established by using this method. The ages of samples younger than the lower limit are difficult to determine because the specific disintegration rate is essentially the same as that for contemporary carbon. For samples older than the upper limit, the specific disintegration rate is too small to measure. Refinement of techniques will allow the extension of this range.

The validity of the radiocarbon-dating method depends on the intensity of cosmic radiation having been approximately constant during the time span of interest, the present to 5×10^4 years ago. The data presently available suggest that this is approximately true.[4] Although the intensity of cosmic radiation varies with latitude, the mixing of the atmosphere occurs on a time scale that is short compared to 5730 years, thereby giving a constant specific activity for contemporary carbon, regardless of its geographical location.

The chemical nature of the carbon in a specimen to be studied is important. Clearly, the method is vitiated if the carbon atoms of the sample can exchange with atmospheric carbon dioxide. Large organic molecules with covalently bonded carbon atoms are particularly suitable for study, and many cellulose-containing materials or charcoal specimens have been studied by the method.

[4]However, there is evidence of errors of 700 years in radiocarbon ages of 5000 years. The errors appear to be essentially zero at ages of 10,000 years. The Earth's magnetic field has an effect on whether cosmic rays are deflected, and variations in this field are thought to be responsible for the errors in radiocarbon dating.

There are other nuclear reactions occurring in the upper atmosphere. In reacting with fast neutrons nitrogen-14 gives hydrogen-3 (tritium),

$$^{14}_{7}N + ^{1}_{0}n = ^{12}_{6}C + ^{3}_{1}H$$

an unstable nuclide with $t_{1/2} = 12.5$ y that decays by β^- emission,

$$^{3}_{1}H \rightarrow ^{3}_{2}He + ^{0}_{-1}\beta^-$$

Other processes also have introduced tritium into the atmosphere. The detonation of experimental fusion weapons during the 1950s made major changes in the tritium level.

ISOTOPE EFFECTS

The simple theory on which much of quantitative chemistry is based is that the properties of a substance do not depend on its isotopic composition; this theory is not strictly correct.

The dependence of zero-point vibrational energy on mass, discussed in Chapter 10, is the phenomenon responsible for most of the dependence of chemical and physical properties on atomic mass in isotopic molecules. In Chapter 10 we saw that the dissociation energies for $^1H_2$ and $^2H_2$ differ by 7.6 kJ mol^{-1}. Because 7.6 kJ mol^{-1} corresponds at 298.2 K to a factor of ~21 in an equilibrium constant or a rate constant [$7600 \text{ J mol}^{-1} \div (2.303 \times 8.31 \text{ J mol}^{-1} \text{ K}^{-1} \times 298.2 \text{ K}) = 1.33$; $\log^{-1}(1.33) = 21$], an isotopic substitution of deuterium for hydrogen can be expected to have detectable effects. In contrast, the difference in zero-point vibrational energies of $^{79}Br_2$ and $^{81}Br_2$ is only 24 J mol^{-1}. Effects on equilibria and rates at 298.2 K from an energy of 24 J mol^{-1} generally are trivial [$24 \text{ J mol}^{-1} \div (2.303 \times 8.31 \text{ J mol}^{-1} \text{ K}^{-1} \times 298.2 \text{ K}) = 0.0042$; $\log^{-1}(0.0042) = 1.01$]. These calculations show that isotope effects will be appreciable for light elements but generally inappreciable for heavier elements.

Many studies have been made of the effect of deuterium substitution on the rate of reaction. If a bond to hydrogen is broken in the rate-determining step, the reaction rate for the deuterium-substituted compound is expected to be less than about one fifth of that for the hydrogen compound. The rate of oxidation of isopropyl alcohol to acetone by chromium(VI) in acidic solution,

$$8H^+ + 2HCrO_4^- + 3CH_3CH(OH)CH_3 = 2Cr^{3+} + 3CH_3COCH_3 + 8H_2O$$

is reduced by a factor of ~6 if deuterium is substituted for ordinary hydrogen on the second carbon atom, $CH_3CD(OH)CH_3$. This result shows that the carbon–hydrogen bond is being broken in the rate-determining step of this reaction. Kinetic isotope effects as large as ~20 have been observed in the reaction of certain bases (B) with 2-nitropropane:

$$B + (CH_3)_2CH(NO_2) \rightarrow BH^+ + (CH_3)_2C{=}NO_2^-$$

TABLE 22–3
The Properties of Heavy Water ($^2H_2{}^{16}O$) Compared to Those of Ordinary Water (99.7% $^1H_2{}^{16}O$)

Property	$^1H_2{}^{16}O$	$^2H_2{}^{16}O$ (D_2O)
Zero-point energy/kJ mol^{-1}	55.44	75.81
Freezing point/K	273.15	276.97
ΔH_{fusion}/kJ mol^{-1}	6.008	6.280
Temperature of maximum density/°C	4	11.2
Density of liquid at 25°C/g cm^{-3}	0.993	1.104
Normal boiling point/K	373.15	374.59
Vapor pressure at 298.2 K/torr	23.756	20.643
ΔH_{vap}/kJ mol^{-1} (at 373 K)	40.656	41.535
Critical temperature/K	647.30	644.05
Critical pressure/atm	218.3	215.7

is ~20-fold faster than

$$B + (CH_3)_2CD(NO_2) \rightarrow BD^+ + (CH_3)_2C{=}NO_2^-$$

In addition to having effects upon rates of reaction, substitution of deuterium for hydrogen can have an appreciable effect on equilibrium properties of substances. This is illustrated by the properties of heavy water (^2H_2O) compared to those of ordinary water (99.7% ^1H_2O), summarized in Table 22–3.

Comparing properties of heavy water and ordinary water in Table 22–3, we see that the vapor pressure of the lighter molecule to be greater. The same is true for other isotopic molecules, and fractional distillation is used for large-scale enrichment of carbon-13, nitrogen-15, and oxygen-18. The compounds involved in these distillations are carbon monoxide for carbon-13, and nitric oxide for nitrogen-15 and oxygen-18.

22–7 Radiation Chemistry

The radiation given off in the disintegration of a radioactive nuclide (α particle, β particle, and γ radiation) causes chemical changes to occur in the matter through which it passes. The energetic radiation detaches electrons from neutral molecules to form ions, and also causes chemical bonds to break. Typical relative values of the intensity of interactions of α particles, β particles, and γ rays of the same initial energy with matter are 2,500:100:1, respectively. As a consequence, most α particles barely penetrate through a very thin sheet of paper, β particles travel many millimeters in aluminum, and γ radiation intensity is decreased appreciably only upon passing through 1–10 cm of lead.

Radiation chemistry is extremely complex, so attention here will be focused on its effect on a simple important substance, water. The primary effect of radiation on liquid water is ionization:

$$H_2O(l) \rightarrow H_2O^+ + e^-(aq)$$

The products of this process are reactive species, and they are related to other reactive species by acid–base equilibria:

$$\underset{\text{Acid}}{H_2O^+} + \underset{\text{Base}}{H_2O} \rightleftarrows \underset{\text{Base}}{OH} + \underset{\text{Acid}}{H_3O^+}$$

$$\underset{\text{Acid}}{H_2O} + \underset{\text{Base}}{e^-(aq)} \rightleftarrows \underset{\text{Base}}{OH^-} + \underset{\text{Acid}}{H}$$

Thus hydroxyl radical (OH) is the conjugate base of ionized water (H_2O^+), and atomic hydrogen is the conjugate acid of the hydrated electron. Particularly simple fates for atomic hydrogen and hydroxyl radicals can be imagined:

$$H + OH = H_2O$$
$$H + H = H_2$$
$$OH + OH = H_2O_2$$

Indeed, both hydrogen and hydrogen peroxide are found in water that has been subjected to radiation. But the situation is complicated by the reaction of these products with the intermediate species, for example,

$$OH + H_2 \rightarrow H + H_2O$$
$$H + H_2O_2 \rightarrow OH + H_2O$$
$$OH + H_2O_2 \rightarrow HO_2 + H_2O$$

Once hydrogen superoxide, HO_2, is formed, the route is open to formation of oxygen,

$$HO_2 + HO_2 \rightarrow H_2O_2 + O_2$$

although it also can react with atomic hydrogen to give hydrogen peroxide:

$$HO_2 + H \rightarrow H_2O_2$$

The primary processes occurring in dilute aqueous solutions are assumed to be the same as in pure water because the water molecules are so much more numerous than the solute species. But reactions following the primary process can involve the solute species. Thus iron(II) ion is oxidized by the radicals present in irradiated water that is saturated with oxygen:

$$OH + Fe^{2+} \rightarrow FeOH^{2+}$$
$$O_2 + H \rightarrow HO_2$$
$$H_2O + HO_2 + Fe^{2+} \rightarrow FeOH^{2+} + H_2O_2$$
$$H_2O_2 + Fe^{2+} \rightarrow FeOH^{2+} + OH$$

followed by the iron(II)–hydroxyl reaction (the first reaction above). In this mechanism, four moles of iron(III) are formed per mole of water decomposed in the step

$$H_2O = H + OH$$

The radicals formed in decomposition of water by radiation react with organic molecules including those of biological origin, and some of these reactions may be at the root of radiation damage in biological systems.

Biographical Notes

MELVIN CALVIN (1911–), an American chemist, received his Ph.D. in physical chemistry from the University of Minnesota. On the faculty of the University of California at Berkeley since 1937, he has made important contributions in many areas of physical organic chemistry and bioorganic chemistry. For his work on the pathway of carbon in photosynthesis, Professor Calvin received the Nobel Prize in Chemistry in 1961.

ENRICO FERMI (1901–1954), an Italian physicist, was a pioneer of nuclear science. After holding faculty positions in Italy (Florence and Rome), he came to the United States in 1939. He was associated with Columbia University and the University of Chicago. His work on nuclear transformations in heavy elements brought about by slow neutrons led eventually to the development of nuclear reactors. Fermi received the Nobel Prize in Physics in 1938, and the artificial element fermium ($Z = 100$, symbol Fm) was named in his honor.

FREDERIC JOLIOT-CURIE (1900–1958), a French physicist, did pioneering work in nuclear physics. For the discovery of artificial radioactivity, he and his wife Irene received the Nobel Prize in chemistry in 1935.

IRENE JOLIOT-CURIE (1897–1956), daughter of Pierre and Marie Curie, was active in research in nuclear science throughout her life.

WILLARD F. LIBBY (1908–) is Emeritus Professor of Chemistry at the University of California at Los Angeles. It was while he was on the faculty of the University of Chicago that Libby developed the radiocarbon-dating method. He was awarded the Nobel Prize in Chemistry in 1960 for this work.

Problems and Questions

22–1 Calculate the binding energy per nucleon for deuterium (AW = 2.0141). Does this calculation suggest why deuteron emission is not an observed mode of radioactive decay?

22–2 Calculate the binding energy per nucleon for the doubly magic even–even nuclide ^{208}Pb (AW = 207.9767) and an odd–odd nuclide of the same mass number, ^{208}Bi (AW = 207.9797).

22–3 Examine Figure 22–3 at the values of N and Z that are magic numbers. Compare the number of nuclides having magic number values of N and Z with the number having values of N and Z differing from magic numbers by ± 1 unit.

22–4 The stable nuclides with mass numbers between 30 and 40 are ^{30}Si, ^{31}P, ^{32}S, ^{33}S, ^{34}S, ^{35}Cl, ^{36}S, ^{36}Ar, ^{37}Cl, ^{38}Ar, ^{39}K, and ^{40}Ar. Calculate the N/P ratio for each of these nuclides. Predict the mode of decay for each of the following unstable nuclides in this same range of mass numbers: ^{31}Si; ^{30}P, ^{32}P; ^{31}S, ^{35}S, ^{37}S; ^{34}Cl, ^{36}Cl, ^{38}Cl; ^{35}Ar, ^{37}Ar, ^{39}Ar.

22–5 The atomic weights for the unstable argon nuclides and the stable isobaric nuclides are:

^{35}Ar	34.9753	^{35}Cl	34.9689
^{37}Ar	36.9668	^{37}Cl	36.9659
^{39}Ar	38.9643	^{39}K	38.9637

What are the energies associated with the disintegration of the unstable nuclides?

22–6 The energies of α particles emitted by uranium-235 are 4.597, 4.556, 4.502, 4.415, 4.396, and 4.366 MeV. What energies of γ rays do you expect to accompany this α emission?

22–7 The more energetic α particle emitted in the disintegration of uranium-238 has an energy of 4.195 MeV. What is the initial velocity of this α particle?

22–8 Technetium ($Z = 43$) and promethium ($Z = 61$) do not occur as stable nuclides. The stable nuclides of elements in these regions of the periodic table are:

$_{41}Nb$	93
$_{42}Mo$	92, 94, 95, 96, 97, 98, 100
$_{44}Ru$	96, 98, 99, 100, 101, 102, 104
$_{45}Rh$	103
$_{59}Pr$	141
$_{60}Nd$	142, 143, 144, 145, 146, 148, 150
$_{62}Sm$	144, 147, 148, 149, 150, 152, 154
$_{63}Eu$	151, 153

How do these data explain the nonexistence of any stable nuclides of elements 43 and 61?

22–9 The example given for nuclear transformation by K-electron capture was

$$^{7}_{4}Be + {}^{0}_{-1}e = {}^{7}_{3}Li$$

Show that the mass change in this transformation is simply AW($^{7}_{3}Li$) − AW($^{7}_{4}Be$). Calculate this change in mass and convert to energy units (MeV); conventional atomic weights are AW($^{7}_{3}Li$) = 7.01600 and AW($^{7}_{4}Be$) = 7.01693.

22–10 What is the total disintegration rate for 1.50×10^{-5} g ^{14}C? What percentage of a sample of carbon-14 disintegrates in 1.30×10^{4} y?

22–11 Phosphorus-32, a useful isotope in biological studies, has a half-life of 14.3 d. What fraction of a sample disintegrates in 1.00 hour, 1.00 day, 1.00 week, 1.00 month (a 30-day month), and 1.00 year (a 365-day year)?

22–12 A unit of radioactivity that has been used is the curie, an amount of radioactive material with a disintegration rate of 3.7×10^{10} s^{-1}. What number of atoms of each of the following nuclides is 1.00×10^{-6} curies: ^{32}P($t_{1/2}$ = 14.3 d), ^{226}Ra($t_{1/2}$ = 1602 y), and ^{238}U($t_{1/2}$ = 4.51×10^{9} y)?

22–13 Calculate the N/P ratio of each nuclide in the $2a + 2$ series of radioactive elements shown in Figure 22–8. Why are the daughter nuclides of α-particle emitters negative β-particle emitters, never positron emitters?

22–14 The first visible amounts of plutonium-239 were isolated in 1942. What fraction of those atoms will still exist as such in the year 2000?

22–15 The atomic weights of ^{235}U and ^{239}Pu are 235.0439 and 239.0522, respectively. What is the energy of the disintegration reaction of plutonium-239?

22–16 A sample of pure $^{239}PuO_2$ is sealed in a vessel that will trap the α particles emitted by plutonium-239. Using the energy calculated in Problem 22–15, calculate the rate of change of temperature of this sample if no heat is lost and if the heat capacity of PuO_2 is that given by the law of Dulong and Petit ($C_p \cong 3 \times 25$ J K^{-1} mol^{-1}).

22–17 An unstable nuclide is prepared in a (p,n) reaction on a stable nuclide. What type of instability is expected for this nuclide?

22–18 Although the (n,γ) reaction goes with slow neutrons, the (n,2n) reaction goes only with very fast neutrons. Explain.

22–19 Complete each of the following nuclear reactions, giving appropriate mass numbers and atomic numbers:

$$^{85}Rb(\alpha,n) \qquad ^{87}Rb(n,\gamma) \qquad ^{107}Ag(d,p) \qquad ^{107}Ag(n,2n)$$

22–20 At zero time a sample of aluminum-28 and magnesium-27 contains equal weights of the two nuclides. At what time will the disintegration rates of each of these nuclides be the same? (The half-lives are given in this chapter; assume the atomic weights are equal to the mass numbers.)

22–21 Why are the products from the fission of uranium-235 negative β-particle emitters and not positive β-particle emitters?

22–22 The atomic weights of ^{235}U and ^{238}U are 235.0439 and 238.0508. Calculate the binding energy per nucleon for each of these nuclides. Are these values consistent with the greater instability of ^{235}U toward fission?

22–23 Two nuclear reactions that occur in the upper atmosphere are $^{14}_{7}N(n,p)^{14}_{6}C$ and $^{14}_{7}N(n,^{3}_{1}H)^{12}_{6}C$. Calculate the change of mass in each reaction. Which nuclear reaction requires the more energetic neutrons? Atomic weights of these nuclides are

$$AW(^{14}_{7}N) = 14.0031 \qquad AW(^{14}_{6}C) = 14.0032$$
$$AW(^{12}_{6}C) = 12.0000 \qquad AW(^{3}_{1}H) = 3.0161$$
$$AW(^{1}_{1}H) = 1.0078 \qquad m(n) = 1.0087$$

22–24 From values of $\mathscr{E}^0$ given in this chapter, calculate the value of the equilibrium constant for the reaction

$$3Pu^{4+} + 2H_2O \rightleftarrows PuO_2^{2+} + 2Pu^{3+} + 4H^+$$

at 25°C. A small amount of pure $Pu(OH)_4$ is dissolved in a solution with $[H^+] = 0.150$ mol L^{-1}. What percentage of plutonium is present at equilibrium as Pu^{4+}?

22–25 Woven rope sandals found in a cave in Oregon were analyzed by radiocarbon-dating methods. The specific activity of the carbon was found to be 4.55 g^{-1} min^{-1}. What is the age of these sandals?

APPENDIXES

In these appendixes we summarize certain topics presented in the text and present numerical tabulations that will be useful to you in solving various problems. You also should become familiar with other, more comprehensive tabulations of mathematical and chemical data (e.g., the *Handbook of Chemistry and Physics*, published by The Chemical Rubber Co., Cleveland, Ohio). The appendixes are:

1. Physical Quantities, Units, and Conversion Factors
2. Some Aspects of Mathematics
3. Some Aspects of Chemistry
4. Answers to Selected Problems and Questions

1

Physical Quantities, Units, and Conversion Factors

Physical Quantities

In chemistry we are concerned with physical quantities, for instance, the mass and the volume of a sample of benzene. We measure these by comparing them with standard values of mass and volume, a calibrated set of weights and a calibrated pipet. In Chapters 1, 2, and 3 you were shown how to express the value of a physical quantity:

$$\text{Physical quantity} = \text{Number} \times \text{Unit}$$

For the sample of benzene, the values of the mass and volume are

$$m = 13.47\ \text{kg} \qquad V = 0.01532\ \text{m}^3$$

In expressing these quantities we have used the SI units for mass (the kilogram, kg) and length (the meter, m). Throughout your work you should write physical quantities in this way as the product of a number and a unit. Usually the unit will be derived from SI units, but this is not always the case. Thus, many chemists express energy in calories and volume in liters.

TABLE A1–1
Six Base SI Units

Physical quantity	Name of unit	Symbol for unit	Page where defined
Length	meter	m	17
Mass	kilogram	kg	16
Time	second	s	73
Electric current	ampere	A	77
Temperature	kelvin	K	20
Amount of substance	mole	mol	26

SI Units

In SI (Systeme Internationale d'Unites) there are seven physical quantities that are regarded as dimensionally independent. The six base SI units that we have used in this book are given in Table Al–1. (We have not used the seventh base unit, the candela, a unit of luminous intensity.) The definitions of these units have been given already and will not be repeated here. Table Al–2 gives SI-derived units expressed in terms of the base units by multiplication and division. There are relationships between these quantities that we have used:

$$1 \text{ Pa} = 1 \text{ N m}^{-2}$$
$$1 \text{ J} = 1 \text{ N m} = 1 \text{ Pa m}^3 = 1 \text{ V C}$$

The non-SI units that we have used most in this book are units of length (the angstrom, Å), volume (the liter, L), force (the dyne, dyn), pressure (the atmosphere, atm, and torr), and energy (the calorie, cal,

TABLE A1–2
SI-Derived Units

Physical quantity	Name of unit	Symbol for unit	Definition of unit
Force	newton	N	1 kg m s^{-2}
Pressure	pascal	Pa	$1 \text{ kg m}^{-1} \text{ s}^{-2}$
Energy	joule	J	$1 \text{ kg m}^2 \text{ s}^{-2}$
Electric charge	coulomb	C	1 A s
Electric potential difference	volt	V	$1 \text{ kg m}^2 \text{ s}^{-3} \text{ A}^{-1}$
Frequency	hertz	Hz[a]	s^{-1}

[a]Named after HEINRICH R. HERTZ (1857–1894), a German physicist. He was educated in the University of Berlin and was associated as a faculty member with universities at Kiel, Karlsruhe, and Bonn. He did important work regarding electromagnetic radiation.

TABLE A1–3
Commonly Used Prefixes

Fraction	SI prefix	Symbol	Multiple	SI prefix	Symbol
10^{-1}	deci	d	10^3	kilo	k
10^{-2}	centi	c	10^6	mega	M
10^{-3}	milli	m	10^9	giga	G
10^{-6}	micro	μ			
10^{-9}	nano	n			
10^{-12}	pico	p			

erg, and liter atmosphere, L atm). In terms of SI, these units are defined:

$$1\ \text{Å} = 10^{-10}\ \text{m}$$
$$1\ \text{L} = 10^{-3}\ \text{m}^3 = 1\ \text{dm}^3$$
$$1\ \text{dyn} = 1\ \text{g cm s}^{-2} = 10^{-5}\ \text{N}$$
$$1\ \text{atm} = 101{,}325\ \text{Pa}$$
$$1\ \text{torr} = \frac{1}{760}\ \text{atm} = \frac{101{,}325}{760}\ \text{Pa}$$
$$1\ \text{cal} = 4.184\ \text{J}$$
$$1\ \text{erg} = 1\ \text{g cm}^2\ \text{s}^{-2} = 10^{-7}\ \text{J}$$
$$1\ \text{L atm} = 101.325\ \text{J}$$

In defining the liter as the cubic decimeter, $1\ \text{L} = 1\ \text{dm}^3$, we are using the prefix "deci" meaning 10^{-1}. Commonly used prefixes are given in Table Al–3.

Expression of Physical Quantities

If you state the acceleration of gravity as 9.81 meters per second per second, you are using a traditional way of expressing this quantity. But if you write the value

$$g = 9.81\ \text{m/s/s}$$

you are creating possible confusion. A person not familiar with what you mean would be tempted to cancel s/s:

$$g = 9.81\ \text{m}/\cancel{\text{s}}/\cancel{\text{s}} = 9.81\ \text{m}$$

which is nonsense. No such confusion is possible if you write

$$g = 9.81\ \text{m s}^{-2}$$

Always express composite units in this way. Do not use slash marks.

In tabulating physical quantities, in writing certain equations, and in labeling the ordinate and abscissa of graphs, we have used a convenient device for converting physical quantities to numbers. The relationship already presented,

$$\text{Physical quantity} = \text{Number} \times \text{Unit}$$

becomes

$$\frac{\text{Physical quantity}}{\text{Unit}} = \text{Number}$$

Thus the entries in a column headed $\Delta H_f^0/\text{kJ mol}^{-1}$ are numbers; for example, if $\Delta H_f^0[H_2O(l)] = -285.9 \text{ kJ mol}^{-1}$, the quotient $\Delta H_f^0[H_2O(l)]/\text{kJ mol}^{-1}$ is a number:

$$\frac{\Delta H_f^0[H_2O(l)]}{\text{kJ mol}^{-1}} = \frac{-285.9 \text{ kJ mol}^{-1}}{\text{kJ mol}^{-1}} = -285.9$$

You should show the units for all physical quantities in doing chemical calculations. This will serve to prevent you from making careless mistakes. In formulating a calculation remember that quantities can be added and subtracted only if they have the same units. Consider the following problem.

Example: What is the concentration of chloride ion in a solution prepared by putting 13.50 g $BaCl_2(s)$ and 0.1000 L of 0.505*M* NaCl into a volumetric flask then adding water to give a volume of 500.0 cm^3?

$$[Cl^-] = \frac{n}{V} = \frac{13.50 \text{ g} \times \dfrac{2 \text{ mol}}{208.23 \text{ g}} + 0.505 \text{ mol L}^{-1} \times 0.1000 \text{ L}}{500.0 \text{ cm}^3 \times \dfrac{1.000 \text{ L}}{1000 \text{ cm}^3}}$$

$$= \frac{0.1296 \text{ mol} + 0.0505 \text{ mol}}{0.5000 \text{ L}} = 0.3602 \text{ mol L}^{-1}$$

The first term in the numerator consists of two factors, the mass of $BaCl_2$ (13.50 g) and the amount of chloride ion (2 mol) per mole of the salt (208.23 g); the product of these factors is an amount of chloride ion from this source, 0.1296 mol. The second term in the numerator also consists of two factors, the concentration of the solution of NaCl (0.505 mol L^{-1}) and its volume (0.1000 L); the product of these factors also is an amount of chloride ion, 0.0505 mol. The sum of the amount of chloride ion from the two sources is 0.1801 mol. Dividing this amount by the volume of the final diluted solution, 0.5000 L gives the result with the appropriate dimensions,

$$[Cl^-] = \frac{0.1801 \text{ mol}}{0.5000 \text{ L}} = 0.3602 \text{ mol L}^{-1}$$

Values of Some Physical Constants

Avogadro's constant	$\mathcal{N}_A = 6.0221 \times 10^{23}$ mol^{-1}
Planck's constant	$h = 6.6262 \times 10^{-34}$ J s
Velocity of light in vacuum	$c = 2.997925 \times 10^{8}$ m s^{-1}
Boltzmann's constant	$k = 1.3807 \times 10^{-23}$ J K^{-1}
Gas constant	$R = 8.3148$ J K^{-1} mol^{-1}
	$= 1.9873$ cal K^{-1} mol^{-1}
	$= 0.082061$ L atm K^{-1} mol^{-1}
Faraday's constant	$\mathcal{F} = 96{,}486$ C mol^{-1}
Charge of the electron	$e = 1.602 \times 10^{-19}$ C
Mass of the electron	$m_e = 9.1096 \times 10^{-31}$ kg
Atomic mass unit	amu $= 1.6606 \times 10^{-27}$ kg
Rydberg's constant	$R = 1.097 \times 10^{7}$ m^{-1}
Standard acceleration of gravity	$g = 9.80665$ m s^{-2}

2

Some Aspects of Mathematics

Exponential Notation and Logarithms

To express Avogadro's constant as

$$602{,}200{,}000{,}000{,}000{,}000{,}000{,}000 \text{ mol}^{-1}$$

and the mass of the carbon-12 atom as

$$0.000{,}000{,}000{,}000{,}000{,}000{,}000{,}019{,}927 \text{ g}$$

certainly would be inconvenient. Thus we express these quantities as

$$6.022 \times 10^{23} \text{ mol}^{-1} \quad \text{and} \quad 1.9927 \times 10^{-23} \text{ g}$$

To avoid cumbersome numbers and possible ambiguity in the number of significant figures associated with values of physical quantities, scientists use the exponential notation that we have just illustrated; this notation generally is called *scientific notation*. In converting

$$4931 \quad \text{to} \quad 4.931 \times 10^3$$

the decimal point was moved three places to the left. We also can consider this operation to be

$$4931 = 4.931 \times 1000 = 4.931 \times 10^3$$

In converting

$$0.0002832 \quad \text{to} \quad 2.832 \times 10^{-4}$$

the decimal point was moved four places to the right. We also can consider this operation to be

$$0.0002832 = 2.832 \times 0.0001 = 2.832 \times 10^{-4}$$

The two numbers we have been considering can be expressed as 10 raised to the appropriate power:

$$4931 = 10^{3.6929}$$
$$0.0002832 = 10^{-3.5479}$$

The exponent in each of these cases is the logarithm (to the base 10) of the number:

$$\log 4931 = 3.6929$$
$$\log 0.0002832 = -3.5479$$

The logarithm (to the base 10) of a number A is a:

$$\log A = a \quad \text{if} \quad A = 10^a$$

From your courses in mathematics you have learned that

$$10^n \times 10^m = 10^{n+m}$$

Thus if $A = 10^a$ and $B = 10^b$,

$$A \times B = 10^a \times 10^b = 10^{a+b}$$

and

$$\log (A \times B) = a + b$$

or

$$\log (A \times B) = \log A + \log B$$

You also have learned that

$$10^n \div 10^m = 10^{n-m}$$

Thus,

$$\frac{A}{B} = \frac{10^a}{10^b} = 10^{a-b}$$

and

$$\log\left(\frac{A}{B}\right) = a - b$$

or

$$\log\left(\frac{A}{B}\right) = \log A - \log B$$

Using the two numbers that we are considering, we can illustrate these relationships. For the multiplication,

$$4931 \times 0.0002832 = 1.396$$

To obtain this result using logarithms, we have

$$\begin{aligned} \log 4931 &= 3.6929 \\ \log 0.0002832 &= -3.5479 \\ \hline \log(4931 \times 0.000283) &= 0.1450 \\ 10^{0.1450} &= 1.396 \end{aligned}$$

For the division, the quotient is

$$\frac{4931}{0.0002832} = 1.741 \times 10^{7}$$

To obtain this result using logarithms, we have

$$\begin{aligned} \log 4931 &= 3.6929 \\ -(\log 0.0002832 &= -3.5479) \\ \hline \log(4931 \div 0.0002832) &= 7.2408 \\ 10^{7.2408} &= 1.741 \times 10^{7} \end{aligned}$$

Logarithms arise naturally in certain mathematical operations; in particular, the integral of dx/x is the logarithm of x:

$$\int \frac{dx}{x} = \ln x + const$$

However, this logarithm is a logarithm to the base e:

$$e = 1 + \frac{1}{1!} + \frac{1}{2!} + \frac{1}{3!} + \cdots = 2.71828$$

The relationship between logarithms to the base 10 (represented as log) and logarithms to the base e (generally represented as ln) can be obtained by considering the logarithm of a number x to each base:

$$\log x = a \quad \text{or} \quad x = 10^{a}$$

$$\ln x = \alpha \quad \text{or} \quad x = e^{\alpha}$$

Thus

$$10^{a} = e^{\alpha} \quad \text{or} \quad 10 = e^{\alpha/a}$$

But $10 = e^{2.3026}$; therefore

$$\alpha = 2.3026\, a \quad \text{or} \quad \ln x = 2.3026 \log x$$

Your pocket calculator obtains the natural logarithm of a number x by a program that takes the summation of a converging series that is a function of x. A series that converges for all finite values of x is

$$\ln x = 2\left[\left(\frac{x-1}{x+1}\right) + \frac{1}{3}\left(\frac{x-1}{x+1}\right)^{3} + \frac{1}{5}\left(\frac{x-1}{x+1}\right)^{5} + \cdots\right]$$

We can check this by taking some number, calculating values of the successive terms in this series, then seeing how many terms it takes to be within some particular limit of error of the correct value. For $x = 5.30$, the logarithm is

$$\ln 5.30 = 1.6677$$

The value of this summation after adding n terms is $n = 1$, 1.3651; $n = 2$, 1.5771; $n = 3$, 1.6363; $n = 4$, 1.6560, and so on. With four terms the summation is within 0.0117 unit of the correct value.

A particularly simple series is useful for obtaining logarithms of numbers less than 2; the series $\ln(1 + x) = x - \frac{1}{2}x^2 + \frac{1}{3}x^3 - \frac{1}{4}x^4 + \cdots$ converges for $x^2 < 1$. (This equation was used in Section 7–4 to simplify the equation for osmotic pressure of dilute solutions.) For very small values of x, the first term of the series gives a reasonably accurate value of $\ln(1 + x)$:

x	$\ln(1 + x)$
0.00100	0.0009995
0.00300	0.002996
0.0100	0.00995
0.0300	0.0296

If you have a pocket calculator that can take the natural logarithm of a number, you should use it in some problems to gain familiarity with this type of logarithm.

Accuracy, Precision, and Significant Figures

The physical quantities associated with a particular sample of matter have definite values. For some physical quantities a set of conditions must be specified (e.g., the temperature, pressure, and acceleration of gravity) before we can speak of the value for the quantity. But after the conditions are specified, there is the problem of measuring the value of the physical quantity, for instance, the weight of a particular sample of highly purified sodium chloride to be used in standardizing a solution of silver nitrate. Suppose that you weigh the sample four times with a period of drying in an oven between weighings and the successive weights are 0.1593 g, 0.1594 g, 0.1592 g, and 0.1594 g. Because there is no discernible downward trend in the values, it is reasonable to judge the salt dry from the start. The span of these values is only 0.0002 g, and this span is a measure of the *precision* of your set of weighings. We can express this precision by dividing the span by the average value:

$$\text{Precision} = \frac{0.0002 \text{ g}}{0.1593 \text{ g}} \times 100\% = 0.1\%$$

From this precision you might feel justified in taking the weight of the sample to be 0.1593 g. It would be unjustified to take the value as 0.15933 g because your weighings to the nearest 0.1 mg are not going to obtain a weight that can be given to 0.01 mg. But the *accuracy* of this average may be much less than suggested by the precision. The accuracy of the measured value is the difference between this average value and the true value of the weight of the sample. If an error were made in determining the weight of the vessel in which the sample was weighed, the weight of the sample would be inaccurate by a larger amount than suggested by the precision. To learn about the accuracy of this value, you can compare results (the concentration of the solution of silver nitrate) derived from use of this sample of sodium chloride with the results derived from other samples that involved independent weighings of the vessels used.

When expressing the result of measuring a physical quantity, the number of significant figures given *approximately* expresses its accuracy. In Chapter 14 we stated that the measured solubilities of lead sulfate and magnesium hydroxide are 0.0382 g L^{-1} and 0.00855 g L^{-1}, respectively. To express each of these quantities 3 digits are used; each value is given to 3 significant figures. However, the implied relative accuracies of the two values are different if we assume that the last digit in each value is uncertain by only 1 unit. That is, to state that the solubility of $PbSO_4$ is between 0.0381 g L^{-1} and 0.0383 g L^{-1}, implies an uncertainty of $\pm 0.3\%$, and to state that the solubility of $Mg(OH)_2$ is between 0.00854 g L^{-1} and 0.00856 g L^{-1}, implies an uncertainty of $\pm 0.1\%$. (The value 0.00855 almost has four significant figures.) When these solubilities are expressed in concentration units, mol L^{-1}, we use molecular-weight values that are more accurate than either measured solubility. The uncertainty in the molar solubilities are determined solely by the uncertainties in the solubility measurements; the values are

$$S[PbSO_4] = 1.260 \times 10^{-4} \text{ mol L}^{-1}$$
$$S[Mg(OH)_2] = 1.466 \times 10^{-4} \text{ mol L}^{-1}$$

To express the solubility of lead sulfate as 1.260×10^{-4} mol L^{-1} implies an uncertainty of $\pm 0.08\%$, but to express it as 1.26×10^{-4} mol L^{-1} implies an uncertainty of $\pm 0.8\%$. If the end result of this calculation were the molar solubility, it should have been reported as $S = 1.26 \times 10^{-4}$ mol L^{-1}. But this value of S is to be used to calculate the value of the solubility-product constant, K_s ($K_s = S^2$), and it generally is wise to keep an extra significant figure in intermediate stages of calculations and save the rounding to the final step:

$$K_s = S^2 = (1.260 \times 10^{-4} \text{ mol L}^{-1})^2$$
$$= 1.588 \times 10^{-8} \text{ mol}^2 \text{ L}^{-2} = 1.59 \times 10^{-8} \text{ mol L}^{-1}$$

This value implies an uncertainty of 0.6%, which is approximately the uncertainty that results from squaring a number uncertain to 0.3%.

To express the solubility of magnesium hydroxide as 1.466×10^{-4} mol L^{-1} implies an uncertainty of $\pm 0.07\%$, but to express it as 1.47×10^{-4} mol L^{-1} implies an uncertainty of $\pm 0.7\%$. The former implied uncertainty is much closer to the $\pm 0.1\%$ associated with the measured solubility. For this substance the calculation of K_s is

$$K_s = 4S^3 = 1.260 \times 10^{-11} \text{ mol}^3 \text{ L}^{-3}$$

Giving 4 significant figures implies an uncertainty of $\pm 0.08\%$; giving 3 significant figures, $K_s = 1.26 \times 10^{-11}$ mol^8 L^{-3}, implies an uncertainty of $\pm 0.8\%$. The uncertainty in K_s resulting from the original uncertainty of $\pm 0.1\%$ in the measured solubility is $\pm 0.3\%$. It is a close decision as to which is the more appropriate statement of the values of K_s:

$$K_s = 1.260 \times 10^{-11} \text{ mol}^3 \text{ L}^{-3} \quad \text{or} \quad K_s = 1.26 \times 10^{-11} \text{ mol}^3 \text{ L}^{-3}$$

In Chapter 14 the first value was chosen and the number of significant figure was reduced to 3 in the answers to the several problems dealing with this system.

The consequences of premature rounding of numbers in calculations can be illustrated with two examples.

EXAMPLE: What is the molecular weight of cesium sulfate, Cs_2SO_4?

The atomic weights of cesium and oxygen are known to 0.0001, but that of sulfur only to 0.01:

$$\begin{aligned} 2\ \text{AW(Cs)} &= 2 \times 132.9054 = 265.8108 \\ \text{AW(S)} &= 32.06 \\ 4\ \text{AW(O)} &= 4 \times 15.9994 = 63.9976 \\ & 361.8684 \\ & \text{or}\ \ 361.87 \end{aligned}$$

If all of the atomic weights are rounded to the nearest 0.01 before the calculation, we have:

$$\begin{aligned} 2\ \text{AW(Cs)} &= 2 \times 132.91 = 265.82 \\ \text{AW(S)} &= 32.06 \\ 4\ \text{AW(O)} &= 4 \times 16.00 = 64.00 \\ & 361.88 \end{aligned}$$

The premature rounding of values in the latter calculation results in a value that is not the best value consistent with the known atomic weights.

EXAMPLE: What is the rate of the reaction $BrO_3^- + 5Br^- + 6H^+ = 3Br_2 + 3H_2O$ in a solution with the composition $[BrO_3^-] = 0.00934$ mol L^{-1}, $[Br^-] = 0.0854$ mol L^{-1}, and $[H^+] = 0.0964$ mol L^{-1}?

The rate law for the reaction, discussed in Chapter 15, with the rate constant for 25°C is

$$-\frac{d[BrO_3^-]}{dt} = 9.0 \text{ L}^3 \text{ mol}^{-3} \text{ s}^{-1} \times [BrO_3^-][Br^-][H^+]^2$$

Using the specified concentrations, we calculate the rate to be

$$-\frac{d[\mathrm{BrO_3^-}]}{dt} = 6.67 \times 10^{-5}\ \mathrm{mol\ L^{-1}\ s^{-1}}$$

or, if this is rounded to 2 significant figures (because the rate constant is given only to this accuracy),

$$-\frac{d[\mathrm{BrO_3^-}]}{dt} = 6.7 \times 10^{-5}\ \mathrm{mol\ L^{-1}\ s^{-1}}$$

However, if each concentration factor is rounded to 2 significant figures before calculating the rate, the result is

$$-\frac{d[\mathrm{BrO_3^-}]}{dt} = 6.6 \times 10^{-5}\ \mathrm{mol\ L^{-1}\ s^{-1}}$$

As in the example involving addition, this example involving multiplication shows that premature rounding of the concentration factors leads to a calculated rate that is not correct within the uncertainty of the rate constant.

We can consider the relationship between the numbers of significant figures in a number and in its logarithm with an example involving calculated values of pH.

EXAMPLE: The pH and concentration of hydrogen ion are related logarithmically,

$$\mathrm{pH} = -\log\left(\frac{[\mathrm{H^+}]}{\mathrm{mol\ L^{-1}}}\right)$$

Because of the logarithmic relationship, an uncertainty in the concentration of hydrogen ion by a certain factor is reflected in an uncertainty in the calculated pH by a certain fractional unit of pH. This can be shown by expressing the uncertainty in the concentration of hydrogen ion by

$$\frac{[\mathrm{H^+}]}{\mathrm{mol\ L^{-1}}} = \frac{[\mathrm{H^+}]^*}{\mathrm{mol\ L^{-1}}}(1 \pm \delta)$$

in which $[\mathrm{H^+}]^*$ is the correct concentration, and δ is the fractional uncertainty. This gives

$$\mathrm{pH} = -\log\left(\frac{[\mathrm{H^+}]^*}{\mathrm{mol\ L^{-1}}}\right) - \log(1 \pm \delta)$$

$$= \mathrm{pH}^* - \log(1 \pm \delta)$$

in which pH^* is the correct value of pH. Let us consider the uncertainty in the calculated value of pH,

$$\mathrm{Uncertainty} = \mathrm{pH}^* - \mathrm{pH}$$

for selected uncertainties in $[H^+]$:

δ	$\log(1+\delta)$	$\log(1-\delta)$
0.001	0.00043	−0.00043
0.003	0.0013	−0.0013
0.01	0.0043	−0.0044
0.03	0.013	−0.013
0.10	0.041	−0.046

If the concentration of hydrogen ion is uncertain to ~1% ($\delta = 0.01$), the calculated pH must be given to 0.001 unit, and the uncertainty in the value of pH is 0.004 unit. An uncertainty in pH of 0.001 unit, 0.01 unit, and 0.1 unit corresponds to uncertainties in the concentration of hydrogen ion of 0.23%, 2.3%, and ~23%.

Graphical Presentation of Data

In many chemical studies a physical quantity, y, is measured as a function of some variable, x. (Several variables may influence the physical quantity in question, but in each series of experiments all variables except one are held constant.) Making a plot of y versus x is a reasonable first step in evaluating the data. Such a plot reveals something about the experimental scatter of the measured values, and the plot also may suggest a functional dependence, $y = f(x)$. (The functional dependence also may be suggested by a theoretical model for the system.)

Many functional relationships in chemistry can be presented in linear plots. The equation for a straight line in a plot of y versus x is

$$y = b + mx$$

as shown in Figure A2–1. The intercept on the vertical axis ($x = 0$) is b:

$$\begin{aligned} y &= b + mx \\ &= b \quad \text{if } x = 0 \end{aligned}$$

The intercept on the horizontal axis ($y = 0$) is $-b/m$:

$$\begin{aligned} x &= \frac{y - b}{m} \\ &= -\frac{b}{m} \quad \text{if } y = 0 \end{aligned}$$

We saw examples of linear plots in Chapter 4. For real gases at low pressures (see Figure 4–4),

$$PV = \alpha + \beta P$$

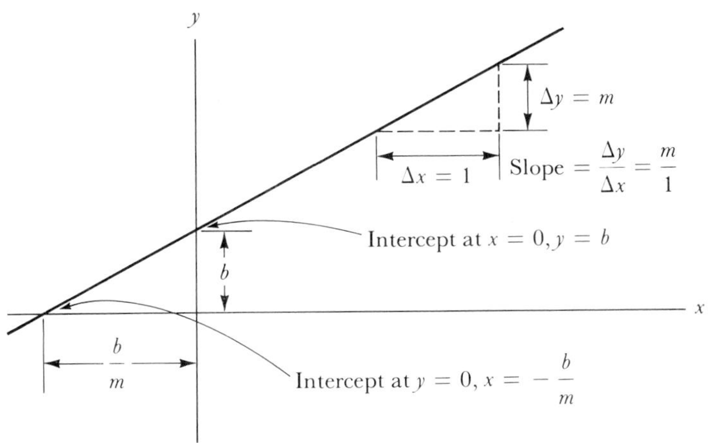

FIGURE A2–1
A linear plot, $y = b + mx$.

For carbon dioxide the extreme points in this figure are:

P	PV
0.16667 atm	22.3897 L atm
1.00000 atm	22.2643 L atm

From these data the intercept α and the slope β can be determined analytically:

$$22.3897 \text{ L atm} = \alpha + 0.16667 \text{ atm} \times \beta$$
$$22.2643 \text{ L atm} = \alpha + 1.00000 \text{ atm} \times \beta$$

Subtraction of these two equations gives

$$0.1254 \text{ atm} = -0.83333 \text{ atm} \times \beta$$

or

$$\beta = \frac{0.1254 \text{ L atm}}{-0.83333 \text{ atm}} = -0.1505 \text{ L}$$

The value of α then can be calculated:

$$\alpha = 22.2643 \text{ L atm} - 1.00000 \text{ atm} \times -0.1505 \text{ L}$$
$$= 22.4148 \text{ L atm}$$

In this calculation we have used two of the points to determine the two parameters α and β. There are analytical methods for using an entire set of data points to establish the parameters, but we will not consider these procedures. However, we will recognize that it is adequate for some purposes simply to draw with a straightedge what appears to be the best straight line; the intercept b can be read as the intersection of the straight line with the $x = 0$ axis. Then the slope can be calculated from the value of the intercept and the value of y read from the straight line at some other particular value of x, y_1 at x_1,

$$m = \frac{y_1 - b}{x_1}$$

Or the value of m can be obtained from the coordinates of two points (x_1, y_1 and x_2, y_2) on the line:

$$m = \frac{y_2 - y_1}{x_2 - x_1}$$

You should notice the dimensions of each quantity in the example we have considered:

$$PV \equiv \text{L atm}$$
$$\alpha \equiv \text{L atm}$$
$$\beta \equiv \text{L}$$
$$P \equiv \text{atm}$$

The dimensions of the two parameters are such that the dimensions of each term in the equation are the same. This must be: quantities can be equated and quantities can be added to one another only if they have the same dimensions.

Although for many relationships $y = f(x)$ only the one quadrant ($x \geq 0$, $y \geq 0$) is physically significant, this is not true for all situations you will encounter. It was not true in Figure 4–5, a linear plot of $(PV/n)_{P\to 0}$ versus t_c,

$$\left(\frac{PV}{n}\right)_{P\to 0} = 22.415 \text{ L atm mol}^{-1} + 0.08206 \text{ L atm mol}^{-1} (^\circ\text{C})^{-1} \times t_c$$

We are interested in the plot for $t_c < 0^\circ\text{C}$, but not for the two quadrants in which $(PV/n)_{P\to 0} < 0$ L atm mol^{-1}. We can calculate the intercept on the t_c axis as indicated in Figure A2–1:

$$t_c = \frac{22.415 \text{ L atm mol}^{-1}}{-0.08206 \text{ L atm mol}^{-1} (^\circ\text{C})^{-1}}$$
$$= -273.15^\circ\text{C}$$

Some Elementary Aspects of Calculus

We used simple aspects of calculus in several places in this book. If you have had calculus or are taking it currently, our usage of it will be straightforward; if you have not had calculus, a brief presentation of the meaning of *derivatives* and *integrals* will allow you to understand our applications of calculus to chemistry.

Figure 15–1 is a plot of the concentrations of bromate ion and bromine as a function of time in an experiment in which the reaction

$$BrO_3^- + 5Br^- + 6H^+ = 3Br_2 + 3H_2O$$

is occurring. The straight-line tangents drawn to these smooth curves define the instantaneous rates of change of these concentrations with respect to time; the slopes of the plots are represented by derivatives:

$$\frac{d[\mathrm{BrO_3^-}]}{dt} \quad \text{and} \quad \frac{d[\mathrm{Br_2}]}{dt}$$

Derivatives also enter discussions of chemistry in other places. In Chapter 6 we related the heat of vaporization of a liquid to the rate of change of the logarithm of the vapor pressure with respect to the reciprocal of the temperature:

$$\Delta H_{\mathrm{vap}} = -R\,\frac{\Delta \ln (P/\mathrm{torr})}{\Delta (1/T)}$$

Because the plots were very close to linear, this equation served us adequately. However, the correct equation that can be applied to nonlinear plots is

$$\Delta H_{\mathrm{vap}} = -R\,\frac{d \ln (P/\mathrm{torr})}{d(1/T)}$$

The value of ΔH_{vap} calculated in this way is the value appropriate for the temperature at which the tangent is drawn.

An equation relating variables can be converted to an equation involving derivatives by *differentiation*. The derivatives that we will use in this discussion are:

$$\frac{d}{dx}(x) = 1 \qquad \frac{d}{dx}(x^n) = nx^{n-1}$$

$$\frac{d}{dx}(e^x) = e^x \qquad \frac{d}{dx}(\ln x) = \frac{1}{x}$$

If y is a function of x and a is a constant, the derivatives analogous to those just given are

$$\frac{d}{dx}(a) = 0 \qquad \frac{d}{dx}(ay^n) = any^{n-1}\frac{dy}{dx}$$

$$\frac{d}{dx}(ay) = a\frac{dy}{dx} \qquad \frac{d}{dx}(\ln ay) = \frac{1}{y}\frac{dy}{dx}$$

$$\frac{d}{dx}(e^{ay}) = ae^{ay}\frac{dy}{dx}$$

An illustration of differentiation is provided by taking two equations, $c = f(t)$, from Chapter 15 and converting these to equations $dc/dt = f'(c)$. For the bromate ion–bromide ion reaction (with relatively high initial concentrations of bromide ion and hydrogen ion), the dependence of the concentration of bromate ion on time is

$$[\mathrm{BrO_3^-}] = [\mathrm{BrO_3^-}]_0 e^{-k't}$$

Taking the derivative of each side of this equation with respect to time, we have

$$\frac{d[BrO_3^-]}{dt} = [BrO_3^-]_0(-k'e^{-k't})$$
$$= -k'([BrO_3^-]_0e^{-k't})$$
$$= -k'[BrO_3^-]$$

which is the first-order rate law appropriate for these concentration conditions.

For the second-order Diels–Alder condensation of butadiene (Section 15–2), the dependence of concentration on time is

$$\frac{1}{[C_4H_8]} = \frac{1}{[C_4H_8]_0} + k_2t$$

Taking the derivative of each side of this equation gives

$$-\frac{1}{[C_4H_8]^2}\frac{d[C_4H_8]}{dt} = k_2$$

which can be rearranged to the second-order rate law,

$$-\frac{d[C_4H_8]}{dt} = k_2[C_4H_8]^2$$

In Chapter 15 an operation, the reverse of the differentiation that we have just illustrated, was used to convert

$$-\frac{dc}{dt} = f(c) \text{ into } c = f'(t)$$

This operation is *integration*. If

$$\frac{dy}{dx} = f(x) \quad \text{or} \quad dy = f(x)\,dx$$

the integral of $f(x)\,dx$ is y plus a constant, c,

$$\int f(x)\,dx = y + c$$

That is, the integral of $f(x)\,dx$ is that function which upon being differentiated gives $f(x)$,

$$\frac{d(y+c)}{dx} = \frac{dy}{dx} + \frac{dc}{dx} = \frac{dy}{dx} = f(x)$$

The integrals that we have used are:

$$\int a\,dx = ax + c \qquad \int x^{-1}\,dx = \ln x + c$$

$$\int x^n\,dx = \frac{x^{n+1}}{n+1} + c \qquad \int e^{ax}\,dx = \frac{1}{a}e^{ax} + c$$

(for $n \neq -1$)

You can confirm the validity of these relationships by differentiating the right side of each of these equations.

In Chapter 4 we had occasion to perform an integration and relate an integral $\int y\,dx$ between particular values of the variable x with the area under the curve of y versus x over this range. Our example pertained to the work done on the surroundings in the reversible isothermal expansion of an ideal gas; the area under the curve of P_{ext} versus V (Figure 4–21d) is this work:

$$-w = \int_{V_1}^{V_2} P_{ext}\,dV$$

For an ideal gas we can express the pressure as a function of volume, $P = nRT/V$, and for the reversible expansion, $P_{ext} = P$; this gives

$$-w = \int_{V_1}^{V_2} nRT\frac{dV}{V}$$

$$= nRT(\ln V_2 - \ln V_1) = nRT\ln\frac{V_2}{V_1}$$

3

Some Aspects of Chemistry

The Equivalent, a Unit of Amount; Normality, a Unit of Concentration

In some contexts it is convenient to describe particular amounts of different chemical substances as being equivalent to one another. This usage always pertains to a particular chemical reaction, for instance, the neutralization of acids by hydroxide ion. The chemical equations for neutralization of hydrofluoric acid (HF), chromic acid [$(HO)_2CrO_2$], and phosphoric acid [$(HO)_3PO$] by hydroxide ion (furnished by a strong base) are:

$$HF + OH^- = F^- + H_2O$$
$$(HO)_2CrO_2 + 2OH^- = CrO_4^{2-} + 2H_2O$$
$$(HO)_3PO + 3OH^- = PO_4^{3-} + 3H_2O$$

Thus

$$1 \text{ mol HF} \equiv 1 \text{ mol } OH^-$$
$$1 \text{ mol } (HO)_2CrO_2 \equiv 2 \text{ mol } OH^-$$
$$1 \text{ mol } (HO)_3PO \equiv 3 \text{ mol } OH^-$$

If we define one *equivalent* of acid as that amount which neutralizes one mole OH^-, we see that

$$1 \text{ equiv HF} = 1 \text{ mol HF}$$
$$1 \text{ equiv } (HO)_2CrO_2 = \tfrac{1}{2} \text{ mol } (HO)_2CrO_2$$
$$1 \text{ equiv } (HO)_3PO = \tfrac{1}{3} \text{ mol } (HO)_3PO$$

A general definition of an equivalent is:

> An equivalent of a chemical substance is $1/\nu$ mole, in which ν is an integer, determined by the specific reaction being considered.

Thus for the neutralization reactions being considered, $\nu = 1$ for HF, $\nu = 2$ for $(HO)_2CrO_2$, and $\nu = 3$ for $(HO)_3PO$.

In oxidation–reduction reactions, one equivalent of oxidizing agent can be defined as that amount of oxidizing agent which oxidizes one half mole of molecular hydrogen. In the reactions

$$(HO)_2CrO_2 + \tfrac{3}{2}H_2 + 3H^+ = Cr^{3+} + 4H_2O$$
$$MnO_4^- + \tfrac{5}{2}H_2 + 3H^+ = Mn^{2+} + 4H_2O$$
$$MnO_4^- + \tfrac{3}{2}H_2 = MnO_2 + OH^- + H_2O$$

the stoichiometric relationships are

$$1 \text{ mol } (HO)_2CrO_2 \equiv 3 \times \tfrac{1}{2} \text{ mol } H_2$$
$$1 \text{ mol } MnO_4^- \text{ (in being reduced to } Mn^{2+}) \equiv 5 \times \tfrac{1}{2} \text{ mol } H_2$$
$$1 \text{ mol } MnO_4^- \text{ (in being reduced to } MnO_2) \equiv 3 \times \tfrac{1}{2} \text{ mol } H_2$$

which leads to

$$1 \text{ equiv } (HO)_2CrO_2 = \tfrac{1}{3} \text{ mol } (HO)_2CrO_2$$
$$1 \text{ equiv } MnO_4^- \text{ (in being reduced to } Mn^{2+}) = \tfrac{1}{5} \text{ mol } MnO_4^-$$
$$1 \text{ equiv } MnO_4^- \text{ (in being reduced to } MnO_2) = \tfrac{1}{3} \text{ mol } MnO_4^-$$

These examples illustrate the meaning of the term "equivalent," and they also illustrate the possible ambiguity, which contrasts with the definite meaning of the term mole:

> 1 mol of chromic acid, $(HO)_2CrO_2$, has a mass of 118.01 g, but 1 equivalent of chromic acid has a mass of 59.01 g or a mass of 39.34 g, depending on whether its properties as an acid or as an oxidizing agent are being considered.
>
> 1 mol of permanganate ion, MnO_4^-, has a mass of 118.94 g, but 1 equivalent of permanganate ion has a mass of 23.79 g or a mass of 39.65 g, depending on whether it is converted to manganese(II) ion, Mn^{2+}, or manganese dioxide, MnO_2.

There is possible ambiguity in the term "equivalent," and there are no occasions when its use is mandatory. Use of balanced chemical equations and the concept of the mole will serve you well in all possible calculations of chemical stoichiometry.

With the equivalent as a unit of amount of chemical substance, there is associated *normality* as a unit of concentration:

$$\text{Normality} = \frac{\text{Amount (in equivalents)}}{\text{Volume (in liters)}}$$

The symbol N is used to stand for equiv L^{-1}. Thus a bottle of sulfuric acid with a concentration $C(H_2SO_4) = 2.00$ mol L^{-1} may be labeled 2.00M H_2SO_4 or 4.00N H_2SO_4. We have not used the normality unit in describing the concentrations of solutions, but you may encounter this unit in later study.

Nomenclature of Inorganic Chemistry

The names of many inorganic compounds have been used in this book. Some of these names describe rather clearly the composition of the compound, but some currently used names do not convey composition clearly. For instance, titanium dioxide conveys the composition TiO_2, but the mineralogical name for this compound, rutile, does not. The name sodium hydrogencarbonate is more informative than the commonly used name sodium bicarbonate for $NaHCO_3$.

SIMPLE BINARY COMPOUNDS

The name of the electropositive constituent of a binary compound often is used without modification, but in the Stock system (named for Alfred E. Stock; see Chapter 12) the oxidation state of the electropositive constituent is included with the name. The name of the electronegative constituent is modified to end in "-ide." Use of these systems are illustrated with names of two compounds of iron with chlorine:

$FeCl_2$	Iron dichloride	$FeCl_3$	Iron trichloride
	Iron(II) chloride		Iron(III) chloride
	(Ferrous chloride)		(Ferric chloride)

The names "ferrous chloride" and "ferric chloride" do not convey the proportions of iron and chlorine in these compounds, but these names are common. The "-ous" and "-ic" endings can be used without ambiguity for the electropositive constituent of a compound only if this element has but two oxidation states in addition to zero.

SIMPLE MONATOMIC IONS

Simple monatomic ions are named according to the rules just given for binary compounds. If an element exhibits only one oxidation state (in addition to zero) it is permissable to omit the oxidation state. Examples of monatomic cations are:

Na^+	Sodium ion	Tl^+	Thallium(I) ion (or thallous ion)
Mg^{2+}	Magnesium ion	Tl^{3+}	Thallium(III) ion (or thallic ion)
Al^{3+}	Aluminum ion		

(But we learned in Chapters 18 and 21 of thallium in the +2 oxidation state. Although this oxidation state is very unstable, its existence makes use of the "-ous" and "-ic" endings for the +1 and +3 states of thallium a practice to be discouraged.)

Examples of monatomic anions are:

F^-	Fluoride ion	C^{4-}	Carbide ion
O^{2-}	Oxide ion	H^-	Hydride ion
N^{3-}	Nitride ion		

POLYATOMIC IONS

Polyatomic cations generally end in "-ium"; examples are:

NH_4^+	Ammonium ion	H_2F^+	Fluoronium ion
H_3O^+	Hydronium ion (also called oxonium ion)	$HONH_3^+$	Hydroxylammonium ion
		PCl_4^+	Tetrachlorophosphonium ion

Polyatomic anions end in "-ide" or "-ate." In the -ide category are:

OH^-	Hydroxide ion	N_3^-	Azide ion
O_2^{2-}	Peroxide ion	HF_2^-	Hydrogendifluoride ion
I_3^-	Triiodide ion	CN^-	Cyanide ion

In the "-ate" category are:

CO_3^{2-}	Carbonate ion
NO_3^-	Nitrate ion
$C_2O_4^{2-}$	Oxalate ion

ACIDS

The binary inorganic acids can be named as compounds of hydrogen or, in some cases, with a "hydro-" prefix and an "-ic" ending:

HCl	Hydrogen chloride or hydrochloric acid
H_2S	Hydrogen sulfide

If the acid is derived from a polyatomic anion, it may be named as a compound of hydrogen or, in accord with custom, as an "-ous" acid (if the anion ends in "-ite") or an "-ic" acid (if the anion ends in "-ate"):

SO_3^{2-}	Sulfite ion	$(HO)_2SO$	Sulfurous acid
SO_4^{2-}	Sulfate ion	$(HO)_2SO_2$	Sulfuric acid
NO_2^-	Nitrite ion	$HONO$	Nitrous acid
NO_3^-	Nitrate ion	$HONO_2$	Nitric acid

The prefix "hypo-" is used in some cases to denote an oxidation state of the central atom lower than in the "-ous" acid, and the prefix "per-" is used in some cases to denote an oxidation state of the central atom higher than in the "-ic" acid:

ClO^-	Hypochlorite ion	$HOCl$	Hypochlorous acid
ClO_2^-	Chlorite ion	$HOClO$	Chlorous acid
ClO_3^-	Chlorate ion	$HOClO_2$	Chloric acid
ClO_4^-	Perchlorate ion	$HOClO_3$	Perchloric acid

Notice that "per-" *does not* mean a peroxide-containing species. Peroxide-containing ions or acids are denoted with a "peroxo-" prefix:

$O_2NO_2^-$ Peroxonitrate ion HO_2NO_2 Peroxonitric acid

The prefixes "ortho-" and "meta-" refer to different degrees of hydration:

$(HO)_3PO$ Orthophosphoric acid $HOPO_2$ Metaphosphoric acid

Because dehydration of a stable ortho acid often leads to polymeric species, this may be indicated:

$(HOPO_2)_n$ Metaphosphoric acid

COORDINATION COMPOUNDS

There are many polyatomic species (anionic, cationic, and neutral) with a central atom surrounded by ligands that are named as coordination complexes. The names of anionic ligands are related to the names of the anions but with an "-o" ending. Examples are:

Cl^-	chloro	CN^-	cyano
O^{2-}	oxo	NO_3^-	nitrato
OH^-	hydroxo	N_3^-	azido

Most neutral ligands keep their regular names, but the common ligands, ammonia, water, and carbon monoxide, have the names "ammine," "aqua," and "carbonyl," respectively. The names of a neutral ligand in the former group may be enclosed with parentheses.

The number of coordinated ligands of each type are designated with prefixes:

1,	mono	4,	tetra or tetrakis
2,	di or bis	5,	penta or pentakis
3,	tri or tris	6,	hexa or hexakis

The ligands are named in alphabetical order. If the entire complex is anionic its name ends in "-ate"; if neutral or cationic there is no special ending. The oxidation state of the central atom is given by a Roman numeral (or zero) in parentheses. These rules will be illustrated with some examples:

$Cr(OH_2)_6^{3+}$	Hexaaquachromium(III) ion
$Fe(CN)_6^{4-}$	Hexacyanoferrate(II) ion
$Co(NH_3)_4Cl_2^+$	Tetraamminedichlorocobalt(III) ion
$Co(en)_2Cl_2^+$	Dichlorobis(ethylenediamine)cobalt(III) ion
$Fe(CO)_4^{2-}$	Tetracarbonylferrate(−II) ion
$Mg(P_2O_7)_2^{6-}$	Bis(diphosphato)magnesium(II) ion

The prefixes bis, tris, and so on are preferred to di, tri, and so on if the name of the ligand involves a prefix, as in the ethylenediamine and diphosphato complexes.

The Sign Between Reactants and Products in a Chemical Equation

In this book chemical equations have been written involving each of the signs

$$= \qquad \rightarrow \qquad \rightleftarrows$$

There is a reason for using these different signs. Some useful information can be conveyed by the kind of sign used between the reactants and products in a chemical equation.

The equal sign is used if only the stoichiometry of reaction is given by the equation. The reaction may go in one step or by a pathway consisting of many steps; the reaction may be rapid or slow; equilibrium may favor formation of products or may not. Regardless of these features, an equal sign is appropriate in a chemical equation if only the stoichiometric relationships are of interest.

The single arrow $\rightarrow$ is used in an equation for either an elementary reaction step occurring in one direction or for the sum of a sequence of steps that is indistinguishable from an elementary reaction step. Use of a single arrow does not mean the reverse reaction cannot occur; it means only that the reverse reaction need not be considered under the conditions in question.

The double arrow $\rightleftarrows$ has a significance analogous to that for the single arrow, except that both the forward and reverse reactions are taken into account. The double arrow also is used instead of an equal sign if equilibrium aspects of a reaction are the focus of our attention.

Selected Thermodynamic Data

In Table A3–1 we bring together thermodynamic data introduced at several places in the text. The quantities tabulated (for 298.2 K) are ΔH_f^0, ΔG_f^0, and S^0. The values come from many sources; the relationships that allow derivation of these values from experimental measurements have been mentioned in the text. In particular, values of ΔH_f^0 come from:

1. calorimetric study of reactions; Hess-law calculations (Section 3–3),
2. the temperature coefficients of equilibrium constants; $d \ln K^0/d(1/T) = -\Delta H^0/R$ (Section 12–2), and
3. estimates based upon bond energies (Section 10–7).

Values of ΔG_f^0 come from:

1. evaluation of equilibrium constants; $\Delta G^0 = -RT \ln K^0$ (Section 12–2); Hess-law-type calculations (Section 12–2),
2. evaluation of emf of galvanic cells; $\Delta G^0 = -n\mathscr{F}\mathscr{E}^0$ (Section 16–4), and
3. evaluation from enthalpy and entropy values for reactants and products, $\Delta G^0 = \Delta H^0 - T\Delta S^0$.

Values of S^0 come from:

1. measurements of heat capacity from very low temperatures up to 298.2 K; third law of thermodynamics (Section 5–5),
2. calculated values using $S = k \ln \Omega$ and knowledge of spacing of energy levels in molecules (Section 5–5), and
3. calculations from values of ΔS^0 for reactions $[\Delta S^0 = (\Delta H^0 - \Delta G^0)/T]$ and knowledge of entropies of all reactants and products except one.

In Chapter 2 we pointed out that despite the simplicity of the chemical equation

$$KClO_3(s) = KCl(s) + O_3(g)$$

this reaction does not go; rather the reaction

$$2KClO_3(s) = 2KCl(s) + 3O_2(g)$$

goes. Using values from the following tabulation, we can calculate the value of ΔG^0 for each of these reactions. For the reaction that produces ozone,

$$\Delta G^0 = \Delta G_f^0[O_3(g)] + \Delta G_f^0[KCl(s)] - \Delta G_f^0[KClO_3(s)]$$

$$\frac{\Delta G^0}{\text{kJ mol}^{-1}} = 163 + (-408) - (-290) = 45$$

$$\Delta G^0 = 45 \text{ kJ mol}^{-1}$$

Under standard conditions at 298.2 K, potassium chlorate *does not* decompose to give ozone plus potassium chloride. For the reaction that produces oxygen,

$$\Delta G^0 = 3\Delta G_f^0[O_2(g)] + 2\Delta G_f^0[KCl(s)] - 2\Delta G_f^0[KClO_3(s)]$$

$$\frac{\Delta G^0}{\text{kJ mol}^{-1}} = 3 \times 0 + 2 \times (-408) - 2 \times (-290) = -236$$

$$\Delta G^0 = -236 \text{ kJ mol}^{-1}$$

Under standard conditions at 298.2 K, potassium chlorate is unstable with respect to oxygen plus potassium chloride. Of course, this calculation does not tell anything about the rate of the reaction; in fact, a catalyst and an elevated temperature are needed to make the rate appreciable.

Table A3–1 includes the values of ΔH_f^0, ΔG_f^0, and S^0 for some ions in aqueous solution. Two points must be made about these quantities:

1. The individual contributions of cation and anion to many properties of a strong electrolyte in solution cannot be evaluated. But if the value for some one ion is chosen, a consistent set of values for all ions can be derived. The convention employed for values of ions in aqueous solution is

$$\Delta H_f^0[\mathrm{H}^+(aq)] = 0\ \mathrm{J\ mol^{-1}}$$
$$\Delta G_f^0[\mathrm{H}^+(aq)] = 0\ \mathrm{J\ mol^{-1}}$$
$$S^0_{\mathrm{H}^+} = 0\ \mathrm{J\ K^{-1}\ mol^{-1}}$$

2. The entropy of a system of matter cannot be less than 0 J K^{-1} mol^{-1}; only for a perfect crystal at 0 K is the entropy equal to 0 J K^{-1} mol. But conventional ionic entropies can be less than zero (e.g., $S^0[\mathrm{Mg}^{2+}(aq)] = -137$ J K^{-1} mol^{-1}) because the system also includes the solvent. Thus the contributions to the entropy at 298.2 K by the components of a 1.00-molal aqueous solution of $MgCl_2$ containing 1.00 mol of solute are:

1.00 mol Mg^{2+}	$S^0 = -137$ J K^{-1}
2.00 mol Cl^-	$S^0 = +114$ J K^{-1}
55.5 mol $H_2O(l)$	$S^0 = +3889.6$ J K^{-1}

TABLE A3–1
Selected Thermodynamic Data

Substance and State	ΔH_f^0 kJ mol^{-1}	ΔG_f^0 kJ mol^{-1}	S^0 J K^{-1} mol^{-1}
ALUMINUM			
Al(*s*)	0	0	28
$Al^{3+}(aq)$	−531	−485	−322
$Al_2O_3(s)$	−1676	−1582	51
$Al(OH)_3(s)$	−1277		
$Al(OH)_4^-(aq)$	−1490	−1297	117
BARIUM			
Ba(*s*)	0	0	67
$Ba^{2+}(aq)$	−538	−561	10
$BaCO_3(s)$	−1219	−1139	112
BaO(*s*)	−582	−552	70
$Ba(OH)_2(s)$	−946		
$BaSO_4(s)$	−1465	−1353	132
BERYLLIUM			
Be(*s*)	0	0	10
BeO(*s*)	−599	−569	14
$Be(OH)_2(s)$	−904	−815	47
BROMINE			
$Br_2(l)$	0	0	152
$Br_2(g)$	31	3	245
$Br_2(aq)$	−3	4	130
$Br^-(aq)$	−121	−104	82
HBr(*g*)	−36	−53	199
$BrO^-(aq)$	−94	−33	42
$BrO_3^-(aq)$	−84	2	163

(*Continued*)

TABLE A3–1 (Continued)

Substance and State	ΔH_f^0 kJ mol^{-1}	ΔG_f^0 kJ mol^{-1}	S^0 J K^{-1} mol^{-1}
CADMIUM			
$Cd(s)$	0	0	52
$Cd^{2+}(aq)$	−76	−78	−73
$CdO(s)$	−258	−228	55
$Cd(OH)_2(s)$	−561	−474	96
$CdS(s)$	−162	−156	65
$CdSO_4(s)$	−935	−823	123
CALCIUM			
$Ca(s)$	0	0	41
$Ca^{2+}(aq)$	−543	−553	−48
$CaC_2(s)$	−63	−68	70
$CaCO_3(s)$	−1207	−1129	93
$CaO(s)$	−635	−604	40
$Ca(OH)_2(s)$	−987	−899	83
$Ca_3(PO_4)_2(s)$	−4126	−3890	241
$CaSO_4(s)$	−1433	−1320	107
CARBON			
$C(s$,graphite$)$	0	0	6
$C(s$,diamond$)$	2	3	2
$CO(g)$	−110.5	−137	198
$CO_2(g)$	−393.5	−394	214
$CO_3^{2-}(aq)$	−677	−528	−57
$HCO_3^-(aq)$	−692	−59	91
$CH_4(g)$	−75	−51	186
$CH_3OH(g)$	−201	−163	240
$H_2CO(g)$	−116	−110	219
$HCOOH(g)$	−363	−351	249
CHLORINE			
$Cl_2(g)$	0	0	223
$Cl_2(aq)$	−23	7	121
$Cl^-(aq)$	−167	−131	57
$ClO^-(aq)$	−107	−37	42
$ClO_3^-(aq)$	−99	−3	162
$ClO_4^-(aq)$	−129	−9	182
$HCl(g)$	−92	−95	187
CHROMIUM			
$Cr(s)$	0	0	24
$Cr_2O_3(s)$	−1128	−1047	81
$CrO_3(s)$	−579	−502	72
$CrO_4^{2-}(aq)$	−863	−706	38
$Cr_2O_7^{2-}(aq)$	−1461	−1257	214
COPPER			
$Cu(s)$	0	0	33
$Cu^+(aq)$	52	50	−37
$Cu^{2+}(g)$	3055		
$Cu^{2+}(aq)$	64	65	−111
$CuCO_3(s)$	−595	−518	88
$Cu_2O(s)$	−170	−148	93
$CuO(s)$	−156	−128	43
$Cu(OH)_2(s)$	−450	−372	108
$CuS(s)$	−49	−49	67

Substance and State	ΔH_f^0 kJ mol^{-1}	ΔG_f^0 kJ mol^{-1}	S^0 J K^{-1} mol^{-1}
FLUORINE			
$F_2(g)$	0	0	203
$F^-(aq)$	−333	−279	−14
$HF(g)$	−271	−273	174
HYDROGEN			
$H_2(g)$	0	0	131
$^2H_2(g)$	0	0	145
$H(g)$	217	203	115
$H^+(aq)$	0	0	0
$OH^-(aq)$	−230	−157	−11
$H_2O(l)$	−286	−237	70
$H_2O(g)$	−242	−229	189
IODINE			
$I_2(s)$	0	0	116
$I_2(g)$	62	19	261
$I_2(aq)$	23	16	137
$I^-(aq)$	−55	−52	106
$I_3^-(aq)$	−51	−51	239
$IO_3^-(aq)$	−221	−128	118
IRON			
$Fe(s)$	0	0	27
$Fe^{2+}(aq)$	−88	−85	−113
$Fe^{3+}(aq)$	−48	−10	−293
$Fe_3C(s)$	21	15	108
$Fe_{0.95}O$ $(s$,wustite$)$	−264	−240	59
Fe_3O_4 $(s$,magnetite$)$	−1117	−1013	146
Fe_2O_3 $(s$,hematite$)$	−826	−740	90
$FeS(s)$	−95	−97	67
$FeS_2(s)$	−178	−166	53
$FeSO_4(s)$	−929	−825	121
LEAD			
$Pb(s)$	0	0	65
$Pb^{2+}(aq)$	−2	−24	16
$PbO_2(s)$	−277	−217	69
$PbS(s)$	−100	−99	91
$PbSO_4(s)$	−920	−813	149
MAGNESIUM			
$Mg(s)$	0	0	33
$Mg^{2+}(aq)$	−462	−456	−137
$MgCO_3(s)$	−1113	−1029	66
$MgO(s)$	−602	−569	27
$Mg(OH)_2(s)$	−925	−834	64
MANGANESE			
$Mn(s)$	0	0	32
$Mn^{2+}(aq)$	−219	−223	−84
$MnO(s)$	−385	−363	60

TABLE A3–1 (Continued)

Substance and State	ΔH_f^0 kJ mol^{-1}	ΔG_f^0 kJ mol^{-1}	S^0 J K^{-1} mol^{-1}
MANGANESE *(Continued)*			
$Mn_3O_4(s)$	−1387	−1280	149
$Mn_2O_3(s)$	−971	−893	110
$MnO_2(s)$	−521	−466	53
$MnO_4^-(aq)$	−543	−449	190
MERCURY			
$Hg(l)$	0	0	76
$Hg_2^{2+}(aq)$		154	74
$Hg^{2+}(aq)$	174	165	−23
$Hg_2Cl_2(s)$	−265	−211	196
$HgCl_2(s)$	−230	−184	144
$HgO(s)$	−90	−59	70
$HgS(s)$	−58	−49	78
NICKEL			
$Ni(s)$	0	0	30
$Ni^{2+}(aq)$	−64	−46	−159
$Ni(CO)_4(g)$	−605	−587	406
$NiCl_2(s)$	−316	−272	107
$NiO(s)$	−241	−213	38
$Ni(OH)_2(s)$	−538	−453	79
$NiS(s)$	−93	−90	53
NITROGEN			
$N_2(g)$	0	0	192
$N_3^-(aq)$	275	348	108
$NH_3(g)$	−46	−16	193
$NH_3(aq)$	−80	−27	111
$NH_4^+(aq)$	−132	−79	113
$NO(g)$	90	87	211
$NO_2(g)$	34	52	240
$NO_2^-(aq)$	−105	−37	140
$NO_3^-(aq)$	−207	−111	146
$N_2O(g)$	82	104	220
$N_2O_4(g)$	10	98	304
$N_2O_5(s)$	−42	134	178
OXYGEN			
$O_2(g)$	0	0	205
$O(g)$	249	232	161
$O_3(g)$	143	163	239
PHOSPHORUS			
$P(s,\text{white})$	0	0	41
$P(s,\text{red})$	−18	−12	23
$P(s,\text{black})$	−39	−33	23
$P_4(g)$	59	24	280
$PF_5(g)$	−1578	−1509	296
$PH_3(g)$	5	13	210
$PO_4^{3-}(aq)$	−1277	−1019	−222
$H_3PO_4(s)$	−1279	−1119	110
POTASSIUM			
$K(s)$	0	0	64
$K^+(aq)$	−251	−282	103
$KCl(s)$	−436	−408	83
$KClO_3(s)$	−391	−290	143
$KClO_4(s)$	−433	−304	151
$K_2O(s)$	−361	−322	98
$K_2O_2(s)$	−496	−430	113
$KO_2(s)$	−283	−238	117
$KOH(s)$	−425	−379	79
SILVER			
$Ag(s)$	0	0	43
$Ag^+(aq)$	105	77	73
$AgBr(s)$	−100	−97	107
$AgCN(s)$	146	164	84
$AgCl(s)$	−127	−110	96
$Ag_2CrO_4(s)$	−712	−622	217
$AgI(s)$	−62	−66	115
$Ag_2O(s)$	−31	−11	122
$Ag_2S(s)$	−32	−40	146
SODIUM			
$Na(s)$	0	0	51
$Na^+(aq)$	−240	−262	59
$NaBr(s)$	−360	−347	84
$Na_2CO_3(s)$	−1131	−1048	136
$NaHCO_3(s)$	−948	−852	102
$NaCl(s)$	−411	−384	72
$NaH(s)$	−56	−33	40
$NaI(s)$	−288	−282	91
$NaNO_2(s)$	−359		
$NaNO_3(s)$	−467	−366	116
$Na_2O(s)$	−416	−377	73
$Na_2O_2(s)$	−515	−451	95
$NaOH(s)$	−427	−381	64
SULFUR			
$S(s,\text{rhombic})$	0	0	32
$S(s,\text{mono-clinic})$	0.3	0.1	33
$S^{2-}(aq)$	33	86	−15
$S_8(g)$	102	50	431
$SF_6(g)$	−1209	−1105	292
$HS^-(aq)$	−18	12	63
$H_2S(g)$	−21	−34	206
$S_2O_3^{2-}(aq)$	−652		
$SO_2(g)$	−297	−300	248
$SO_3(g)$	−396	−371	257
$SO_3^{2-}(aq)$	−635	−487	−29
$SO_4^{2-}(aq)$	−909	−745	20

(Continued)

TABLE A3–1 (Continued)

Substance and State	ΔH_f^0 kJ mol^{-1}	ΔG_f^0 kJ mol^{-1}	S^0 J K^{-1} mol^{-1}
THALLIUM			
$Tl(s)$	0	0	64
$Tl^+(aq)$	5	−32	125
$Tl^{3+}(aq)$	197	215	−192
TIN			
Sn(*s*,white)	0	0	52
Sn(*s*,gray)	−2	0.1	44
$SnO(s)$	−285	−257	56
$SnO_2(s)$	−581	−520	52
$Sn(OH)_2(s)$	−561	−492	155
URANIUM			
$U(s)$	0	0	50
$UF_6(s)$	−2137	−2008	228
$UF_6(g)$	−2113	−2029	380
$UO_2(s)$	−1084	−1029	78
$U_3O_8(s)$	−3575	−3393	282
$UO_3(s)$	−1230	−1150	99

Substance and State	ΔH_f^0 kJ mol^{-1}	ΔG_f^0 kJ mol^{-1}	S^0 J K^{-1} mol^{-1}
XENON			
$Xe(g)$	0	0	170
$XeF_2(g)$	−108	−48	254
$XeF_4(s)$	−251	−121	146
$XeF_6(g)$	−294		
$XeO_3(s)$	402		
ZINC			
$Zn(s)$	0	0	42
$Zn^{2+}(aq)$	−154	−147	−112
$ZnO(s)$	−348	−318	44
$Zn(OH)_2(s)$	−642		
ZnS (*s*,wurtzite)	−193		
ZnS(*s*,zinc blende)	−206	−201	58
$ZnSO_4(s)$	−983	−874	120

4

Answers to Selected Problems and Questions

Chapter 1

1–4 3.82×10^4 g
1–8 1.00×10^{20}, 3.76×10^{25}, 3.76×10^{25}, 1.20×10^{24}, 3.01×10^{21}, 5.42×10^{24}, 1.72×10^{21}
1–12 2.4×10^5 kg
1–14 77.73%, 72.36%, and 69.94%; w_{Fe}/w_O = 3.49:2.62:2.33 = 12.0:9.0:8.0
1–18 Fe_3C
1–23 $C_3H_6O_2$
1–26 $C_3H_4O_2Cl_2$
1–30 35.458, 74.49 g
1–32 0.308

Chapter 2

2–3 0.0703 g, 0.0684 g, 0.0722 g
2–6 178 g
2–9 $2N_2 + 5O_2 + 2H_2O = 4HNO_3$, 4.50 kg
2–12 0.52 g S
2–15 2.46 kg, 187 L, 8.35 mol
2–18 10.44 g, 0.8228 mol
2–21 0.205
2–24 2.60 mol

Chapter 3

3–2 29.4 J, 7.67 m s^{-1}, 20.0013°C
3–6 2.8×10^2 times
3–9 1.79 kJ
3–11 0.22 cal g^{-1}
3–15 −365.84 kcal mol^{-1}
3–18 17.2 kcal
3–22 9.3°C
3–25 6.4°C
3–28 10.01 kcal mol^{-1}
3–33 13.2 J K^{-1}, 470.7 J K^{-1}
3–35 −1.28 kcal mol^{-1}, 328.6 K

Chapter 4

4–1 766 torr
4–4 27.0
4–7 147.1 K

4–10 0.745, 0.295
4–12 0.0205 mol
4–15 8.9057 L atm, 0.400 mol
4–18 $1502\ m\ s^{-1}$, $1301\ m\ s^{-1}$, $139.7\ m\ s^{-1}$, $139.1\ m\ s^{-1}$, $393.6\ m\ s^{-1}$
4–21 0.9990, 0.9951, 0.9903, 0.9558
4–24 1.080, 1.080, 1.000
4–27 25.44 atm, 22.98 atm
4–31 V = constant, $218.28\ J\ K^{-1}\ mol^{-1}$, P = constant, $219.61\ J\ K\ mol^{-1}$
4–35 $-w = 3970\ J$, $q = +3970\ J$, $+19.1\ J\ K^{-1}\ mol^{-1}$, $-13.3\ J\ K^{-1}\ mol^{-1}$

Chapter 5

5–2 1.33 Å, 1.88 Å
5–4 $\frac{1}{3}, \frac{1}{6}$
5–9 4.086 Å, $6.82 \times 10^{-23}\ cm^3$, $10.51\ g\ cm^{-3}$
5–12 5.33 Å, 2.31 Å
5–15 $4.62\ r\ (6)$, $5.03\ r\ (24)$
5–18 1.61, 3.22, 4.61, 6.21
5–21 2860, 1287

Chapter 6

6–1 0.0312
6–5 2.26 g
6–8 $24.4\ kJ\ mol^{-1}$, 239.8 K
6–12 1.241 torr
6–16 Methyl alcohol
6–22 $125\ J\ K^{-1}\ mol^{-1}$, $137\ J\ K^{-1}\ mol^{-1}$
6–25 $\Delta H^0 \cong 88\ J\ K^{-1}\ mol^{-1} \times T_1 - RT \ln P_1$

Chapter 7

7–2 $333\ cm^3$, $143\ cm^3$, $35.0\ cm^3$
7–5 $0.272\ mol\ L^{-1}$, $0.289\ mol\ kg^{-1}$
7–7 0.0286, $1.633\ mol\ kg^{-1}$, $1.498\ mol\ L^{-1}$
7–9 0.767
7–11 $0.211\ mol\ L^{-1}$
7–13 89.4
7–15 $151.8\ J\ K^{-1}\ mol^{-1}$, $-2.84\ kJ\ mol^{-1}$
7–17 $0.078\ mol\ L^{-1}$
7–20 22.8 torr, 374.3 K, 269.1 K
7–22 $P_A = 476$ torr, $P_B = 152$ torr, $x_A = 0.33$, $x_A = 0.404$
7–24 85.6, 0.90 K
7–26 $2.91\ K\ kg\ mol^{-1}$
7–28 $0.094\ mol\ kg^{-1}$, 2.3 atm
7–31 3.41×10^4, 2.27×10^{-6}

Chapter 8

8–3 $1.99 \times 10^{-18}\ J$, $3.98 \times 10^{-19}\ J$, $7.95 \times 10^{-20}\ J$
8–6 341 cm, 278 cm
8–8 $4.949 \times 10^{14}\ s^{-1}$
8–11 458 nm, $2.29 \times 10^{-19}\ J$
8–14 1.875×10^{-4} cm, 1.282×10^{-4} cm, 1.094×10^{-4} cm, 1.005×10^{-4} cm
8–16 $1.181 \times 10^4\ kJ\ mol^{-1}$, $2.099 \times 10^4\ kJ\ mol^{-1}$, $3.279 \times 10^4\ kJ\ mol^{-1}$
8–19 3.17 Å, 2.65 Å, 7.14 Å, 6.61 Å, 5.55 Å
8–22 $2.083 \times 10^{18}\ s^{-1}$, 29.1
8–28 2.27×10^3 nm, 848 nm, 343.1 nm
8–33 N, O, Cl, Kr, Xe, all have IE within 10% of IE(H); therefore $Z_{eff} \cong n$

Chapter 9

9–2 $2.18\ g\ cm^{-3}$
9–5 $U = \left(\frac{4}{1} - \frac{4}{\sqrt{2}} - \frac{4}{2} + \frac{8}{\sqrt{5}} - \frac{4}{2\sqrt{2}} + \cdots\right) \times N_A \frac{e^2}{d}$
9–8 8, 4, 1
9–10 $656\ kJ\ mol^{-1}$
9–16 $344.4\ kJ\ mol^{-1}$, $15\ kJ\ mol^{-1}$
9–18 +2.2, 3.5, $Li_{0.18}Ni_{0.82}O$
9–20 $x/(1 - x)$
9–22 64.4%
9–26 −0.279°C, −0.186°C, −0.372°C
9–28 $[Na^+] = 0.104\ mol\ L^{-1}$, $[Ca^{2+}] = 0.0246\ mol\ L^{-1}$, $[La^{3+}] = 0.0216\ mol\ L^{-1}$, $[Cl^-] = 0.218\ mol\ L^{-1}$

Chapter 10

10–1 NO_2, ClO_2, SF_6^-
10–5 Planar
10–7 $\theta = 180° - \frac{360°}{n}$, $n = 5$, smaller
10–10 $51.8\ kJ\ mol^{-1}$, $36.6\ kJ\ mol^{-1}$, $29.9\ kJ\ mol^{-1}$, $439.6\ kJ\ mol^{-1}$, $443.0\ kJ\ mol^{-1}$
10–14 $r(CN^-) > r(N_2) > r(NO^+)$
10–19 $\angle$ HNC = 180°
10–22 $-37\ kJ\ mol^{-1}$, $-58\ kJ\ mol^{-1}$, $-172\ kJ\ mol^{-1}$, $-61\ kJ\ mol^{-1}$

Chapter 11

11–1 $\ddot{\underset{..}{O}}=C=C=C=\ddot{\underset{..}{O}}$

11–4 3-Methylhexane and 2,3-dimethylpentane

11–7 $4 \times r(\text{C—C}) \times \sqrt{\dfrac{2}{3}} = 5.03\ \text{Å}$

11–9 2-Butanol

11–12 Ethyl methyl ether; C_3H_7OH, bp = 97.2°C; $CH_3OC_2H_5$, bp = 10.8°C
Hydrogen bonding makes alcohol boil at a higher temperature.

11–14 *trans*-1,2-Dichloroethene

11–17 B(−1), N(+1)

11–19 $H_3C—C{\equiv}C—C{\equiv}C—CH_3$

11–24 $E(C{=}C) = 587.6$ kJ mol^{-1}, $E(C{\equiv}C) =$ 808.3 kJ mol^{-1}

11–27 *sec*-Butyl alcohol

11–30 $HN{=}C(NH_2)_2$ (guanidine)

Chapter 12

12–1 $\Delta U = D(\text{H—H}) + IE(\text{H}) - EA(\text{H}) =$ 1671 kJ mol^{-1}

12–3 $\Delta U^0 = -481.2$ kJ mol^{-1}

12–5 38.35 kJ mol^{-1}, 1.92×10^{-7} atm, 1.46×10^{-4} torr, 1.95×10^{-2} Pa, 509.0 K

12–8 114 J K^{-1} mol^{-1}, 266 J K^{-1} mol^{-1}

12–11 $P_{N_2} = 0.0098$ atm, $P_{H_2} = 0.00431$ atm, $P_{NH_3} = 0.0205$ atm

12–13 9.0×10^{-3} atm

12–16 −23 kJ mol^{-1}, 127 kJ mol^{-1}, <0

12–19 NH_3, $Fe(OH_2)_5OH^{2+}$, SO_4^{2-}, H_2O, OCl^-

12–22 84.00 g, 50.04 g, 16.99 g

12–26 $(H_2O)_2(CH_3OH)H^+$, $(H_2O)(CH_3OH)_2H^+$, $(CH_3OH)_3H^+$, $(H_2O)_2(CH_3OH)_2H^+$, $(H_2O)(CH_3OH)_3H^+$, $(CH_3OH)_4H^+$

Chapter 13

13–1 78.8 kJ mol^{-1}, 84.2 kJ mol^{-1}

13–3 1.824, 2.328, 3.143, 12.491, 10.531

13–5 1.788 ± 0.002

13–7 HCl in excess, 0.067*M* CsCl, 0.158*M* HCl

13–9 4.37 L

13–11 6.146

13–13 0.0407 mol L^{-1}, 1.8×10^{-4} mol L^{-1}, 5.6×10^{-11} mol L^{-1}

13–15 $[H^+] = [ClCH_2CO_2^-] = 4.62 \times 10^{-3}$ mol L^{-1}, $[ClCH_2CO_2H] = 0.0154$ mol L^{-1}, $[OH^-] = 2.16 \times 10^{-12}$ mol L^{-1}

13–17 2.71 g

13–19 4.91

13–21 64.49 cm^3

13–23 1.75×10^{-3} mol L^{-1}, 0.045 mol L^{-1}, 1.5×10^{-5} mol L^{-1}

13–25 Assume no NH_3 in gas phase, $[NH_4^+] =$ 0.100 mol L^{-1}, $[Cl^-] = 0.100$ mol L^{-1}, $[NH_3] = 0.108$ mol L^{-1}, $[OH^-] = 1.90 \times 10^{-5}$ mol L^{-1}

13–28 $(NH_4)_2CO_3(s) = NH_3(g) + NH_4HOCO_2(s)$

13–30 pH 2.52 to 5.07

13–32 9.3×10^{-3} mol L^{-1}

13–34 $[Cl^-] = 0.150$ mol L^{-1}, $[HCO_3^-] =$ 0.150 mol L^{-1}, $[CO_3^{2-}] = 0.250$ mol L^{-1}, $[Na^+] = 0.800$ mol L^{-1}, $[OH^-] =$ 2.93×10^{-4} mol L^{-1}, $[H^+] = 3.41 \times 10^{-11}$ mol L^{-1}, $[CO_2] = 1.15 \times 10^{-5}$ mol L^{-1}

13–37 $[H^+] = 2.01 \times 10^{-6}$ mol L^{-1}, pH = 5.697

Chapter 14

14–2 $\Delta G = 0$, $\Delta H > 0$

14–4 7.25×10^{-7} mol L^{-1}, 9.12×10^{-9} mol L^{-1}

14–6 2.15×10^{-4} mol L^{-1}, 4.44×10^{-12} mol L^{-1}

14–8 3.93×10^{-4} mol L^{-1}, 3.84×10^{-10} mol L^{-1}, −4.00

14–10 2.95×10^{-3}

14–12 2×10^{-29} mol^7 L^{-7}

14–14 0.341 g, 7.87 g

14–16 3.03 mol L^{-1}

14–18 0.018, 0.27

14–22 4.04×10^{-16} mol L^{-1}

14–24 6.0×10^{-4} mol L^{-1}

14–26 2.38×10^{-9} mol L^{-1}

14–29 70.4 cm^3

14–31 0.947

14–33 6.1×10^{-23} mol L^{-1}, pH 8 to 12

14–36 $K_s(50°\text{C}) = 1.38 \times 10^{-9}$ mol^2 L^{-2}

Chapter 15

15–2 97.6 minutes, 81.4%

15–4 $-d[\text{A}]/dt = 7.7 \times 10^{-4}\ \text{s}^{-1}\ [\text{A}] - k_r \dfrac{[\text{C}][\text{D}]}{[\text{B}]}$ or

$$-d[\text{A}]/dt = 7.7 \times 10^{-4}\ \text{s}^{-1}\ [\text{B}] - k_r \frac{[\text{C}][\text{D}]}{[\text{A}]}$$

15–6 4.1×10^{-7} mol L^{-1} s^{-1}

15–8 3.1×10^{-2} s^{-1}, 22.5 s

15–10 96 kJ mol^{-1}

15–12 1.89

15–14 0.033 mol L^{-1}

15–16 $COCl_3^{\ddagger}$, $k_r[COCl_2][Cl_2]^{1/2}$
15–18 $d[O_2]/dt = [k_1k_3/(k_2 + 2k_3)][N_2O_5]$
15–20 $N_2O_2^{2-} + 2H_2O \rightleftarrows H_2N_2O_2 + 2OH^-$
$H_2N_2O_2 \rightarrow N_2O + H_2O$
15–22 6.7×10^{-3} L mol^{-1} s^{-1}, 1.46×10^3 s

Chapter 16

16–1 +2, +2, +3, +3, +4, +6, +6, +6, +2, +5 +3, +4
16–5 Anode
16–7 1.54×10^7 A
16–10 1.115 V
16–12 2.5×10^{-9} mol^2 L^{-2}
16–14 +0.035 V, +0.36 V
16–16 −0.89 V
16–18 $V^{2+} + 2VO_2^+ + 2H^+ = 3VO^{2+} + H_2O$
$2V^{2+} + VO_2^+ + 4H^+ = 3V^{3+} + 2H_2O$
16–20 No, disproportionation must be slow
16–22 $V^{2+} + 2Ag^+ + H_2O = 2Ag + VO^{2+} + 2H^+$, 1.4×10^{25}
16–25 1.65×10^{-4} atm
16–27 5.1×10^{15}, 0.827 V, 1.8×10^{-15}, 1.24 V

Chapter 17

17–2 9.3×10^{-56} atm, no
17–4 0.028 mol L^{-1}
17–6 $O_2(g) + 4H^+ + 4X^- = 2X_2 + 2H_2O$ goes for X = Br and I
17–8 +1.08 V, 0.89 V
17–12 Pyramidal, pyramidal, octahedral, nonlinear, tetrahedral, nonlinear
17–14 Close to zero, negative, positive
17–18 3.1×10^{36} atm L mol^{-1}
17–21 553 atm, gas nonideality expected
17–24 $O_3 + 2H^+ + 2e^- = O_2 + H_2O \quad \mathscr{E}^0 = +2.07$ V
$O_2 + 4H^+ + 4e^- = 2H_2O \quad \mathscr{E}^0 = +1.23$ V
$n = 4$, $K = 5.7 \times 10^{56}$ atm
17–28 Cl_2, $S_2^{2-} + H^+ = S + HS^-$, $K_a(HS_2^-) > K_a(HS^-)$
17–32 0.230 g

Chapter 18

18–2 Eu^{2+}(Xe core) $4f^7$, Yb^{2+}(Xe core) $4f^{14}$
18–5 $K(s)$ + liquid solution ($x_{Na} = 0.17$), liquid solution ($x_{Na} = 0.47$), $Na(s)$ + liquid solution ($x_{Na} = 0.71$)
18–7 $Na(NH_3)_n^+ + Na(NH_3)_m^- \rightleftarrows 2Na(NH_3)_n^+ + 2e^- + (m-n)NH_3(l)$
2 solute particles ⇄ 4 solute particles, dilute solution favors $Na(NH_3)_n^+ + e^-$
18–10 $CO_2 + H_2O + CaCO_3 = Ca^{2+} + 2HOCO_2^-$
18–14 $AlH_4^- + 4H_2O = 2H_2 + Al(OH)_4^-$
18–17 $PbBr_4 = PbBr_2 + Br_2 \quad \Delta G^0 < 0$;
Pb(IV) can oxidize Br^-
18–20 $HgO + 2H^+ = Hg^{2+} + H_2O$ (in HNO_3)
$HgO + 2H^+ + 4Cl^- = HgCl_4^{2-} + H_2O$ (in HCl)
18–23 $ZnCl_2$(molten) + $Zn(s) \overset{?}{=} 2ZnCl$
If ZnCl is stable with respect to disproportionation it will form under these conditions.

Chapter 19

19–3 MnO_2 is reduced by HONO,
$MnO_2 + HONO + H^+ = Mn^{2+} + NO_3^- + H_2O$
19–5 0.6 mol L^{-1}
19–7 VO, the more basic oxide
19–9 $Cr_2O_7^{2-}$, Mn_2O_7
19–11 Add Cr^{2+} to Cu^{2+}
19–13 $4FeO_4^{2-} + 20\,H^+ = 3O_2 + 4Fe^{3+} + 10\,H_2O$
19–15 $MnO_4^- + Cr^{3+} + 2H_2O = HOCrO_3^- + MnO_2 + 3H^+$
$K = 6.8 \times 10^{24}$ mol^2 L^{-2}
19–18 $n = 3$, 6×10^{-32}
19–20 $[CN^-]$ changes by a factor of 3.16, $[SCN^-]$ changes by a factor of 100
19–24 $\dfrac{[fac\text{-}Cr(OH_2)_3(NH_3)_3]}{[mer\text{-}Cr(OH_2)_3(NH_3)_3]} = \dfrac{2}{3}$
19–27 $a = 3$, $b = 2$

Chapter 20

20–1 More negative
20–3 1.26×10^{-4} cm
20–7 pH = 5.45
20–12 Arg–Glu–Gly–Arg–Ala–Pro–Asp–His
20–14 Polyglutamic acid with —CO_2H side chains acquires a high negative charge in base; polylysine with —NH_2 side chains acquires a high positive charge in acid.
20–18 pH = 8.34
20–23 TCAGGTAC, GTTACG

Chapter 21

21–1 NaBr, 1.83×10^8 m s^{-1}, 359 nm; NaI, 1.69×10^8 m s^{-1}, 333 nm; octane, 2.15×10^8 m s^{-1}, 422 nm; CS_2, 1.84×10^8 m s^{-1}, 362 nm
21–3 $HgCl_2$, 4; SO_2, 3; HCCH, 7; NH_3, 6; C_6H_6, 30; C_4H_8, 30
21–5 1.39 kJ mol^{-1}

21–8 501 nm, 630 nm, 330 mm

21–10 150 L mol^{-1} cm^{-1}

21–12 3.51×10^{-5} mol L^{-1}

21–14 Low

21–17 The chain-breaking step, $2Br + Kr \rightarrow Br_2 + Kr$, occurs more rapidly the higher the concentration of Kr.

21–19 1.63×10^{-19} J, 98 kJ mol^{-1}

Chapter 22

22–2 ^{208}Pb, 7.863 MeV; ^{208}Bi, 7.845 MeV

22–4 1.143, 1.067, 1.000, 1.063, 1.125, 1.059, 1.250, 1.000, 1.176, 1.111, 1.053, 1.222; $^{31}Si\ \beta^-$, $^{30}P\ \beta^+$, $^{32}P\ \beta^-$, $^{31}S\ \beta^+$, $^{35}S\ \beta^-$, $^{37}S\ \beta^-$, $^{34}Cl\ \beta^+$, $^{36}Cl\ \beta^+$ and β^-, $^{38}Cl\ \beta^-$, $^{35}Ar\ \beta^+$, $^{37}Ar\ \beta^+$, $^{39}Ar\ \beta^-$ (all β^+ emitters also may decay by K-capture)

22–7 1.42×10^7 m s^{-1}

22–9 $\Delta W = -0.866$ MeV

22–11 0.00202, 0.0473, 0.288, 0.766, $1 - 2.07 \times 10^{-8}$

22–14 0.9984

22–15 5.31 MeV

22–17 β^+ emitter

22–20 6.07 min

22–23 −0.0008 amu, +0.0043 amu

22–25 9.05×10^3y

Index

THE TEXT OF THIS BOOK IS SET IN ELEVEN ON THIRTEEN BASKERVILLE.
THE PRINTING AND BINDING WERE EXECUTED BY
THE MAPLE-VAIL BOOK MANUFACTURING GROUP.
THE TEXT PAPER IS FIFTY POUND BOOKMAN MATTE
MANUFACTURED BY THE S. D. WARREN PAPER COMPANY.
TECHNICAL ILLUSTRATIONS WERE DRAWN BY JOHN AND JEAN FOSTER.

SOME PHYSICAL QUANTITIES

	Name of unit	Symbol for unit and its value
Force	newton[a]	$1\ \mathrm{N} = 1\ \mathrm{kg\ m\ s^{-2}}$
	dyne	$1\ \mathrm{dyn} = 1\ \mathrm{g\ cm\ s^{-2}} = 10^{-5}\ \mathrm{N}$
Pressure	pascal[a]	$1\ \mathrm{Pa} = 1\ \mathrm{N\ m^{-2}} = 1\ \mathrm{kg\ m^{-1}\ s^{-2}}$
	atmosphere	$1\ \mathrm{atm} \equiv 101{,}325\ \mathrm{Pa}$
	torr	$1\ \mathrm{torr} \equiv \frac{1}{760}\ \mathrm{atm}$
Energy	joule[a]	$1\ \mathrm{J} = 1\ \mathrm{kg\ m^2\ s^{-2}}$
	erg	$1\ \mathrm{erg} = 1\ \mathrm{g\ cm^2\ s^{-2}} = 10^{-7}\ \mathrm{J}$
	calorie	$1\ \mathrm{cal} \equiv 4.184\ \mathrm{J}$
	liter atmosphere	$1\ \mathrm{L\ atm} = 101.325\ \mathrm{J}$
Electric charge	coulomb[a]	$1\ \mathrm{C} = 1\ \mathrm{A\ s}$
Electric potential difference	volt[a]	$1\ \mathrm{V} = 1\ \mathrm{J\ C^{-1}} = 1\ \mathrm{kg\ m^2\ s^{-3}\ A^{-1}}$

[a]SI unit; the definitions of the SI units for mass (the kilogram), length (the meter), time (the second), and electric current (the ampere) are given in the text.

VALUES OF SOME PHYSICAL CONSTANTS

Avogadro's constant	$\mathcal{N}_A = 6.0221 \times 10^{23}\ \mathrm{mol^{-1}}$
Planck's constant	$h = 6.6262 \times 10^{-34}\ \mathrm{J\ s}$
Velocity of light in vacuum	$c = 2.997925 \times 10^8\ \mathrm{m\ s^{-1}}$
Boltzmann's constant	$k = 1.3807 \times 10^{-23}\ \mathrm{J\ K^{-1}}$
Gas constant	$R = 8.3148\ \mathrm{J\ K^{-1}\ mol^{-1}}$
	$= 1.9873\ \mathrm{cal\ K^{-1}\ mol^{-1}}$
	$= 0.082061\ \mathrm{L\ atm\ K^{-1}\ mol^{-1}}$
Faraday's constant	$\mathcal{F} = 96{,}486\ \mathrm{C\ mol^{-1}}$
Charge of the electron	$e = 1.602 \times 10^{-19}\ \mathrm{C}$
Mass of the electron	$m_e = 9.1096 \times 10^{-31}\ \mathrm{kg}$
Atomic mass unit	$\mathrm{amu} = 1.6606 \times 10^{-27}\ \mathrm{kg}$
Rydberg's constant	$R = 1.097 \times 10^7\ \mathrm{m^{-1}}$
Standard acceleration of gravity	$g = 9.80665\ \mathrm{m\ s^{-2}}$